Semiconductor Manufacturing Handbook

Semiconductor Manufacturing Handbook

HWAIYU GENG, CMfgE, PE

Amica Research
Palo Alto, California

Second Edition

Mc Graw Hill Education

New York Chicago San Francisco Athens London
Madrid Mexico City Milan New Delhi
Singapore Sydney Toronto

Library of Congress Control Number: 2017951483

Semiconductor Manufacturing Handbook, Second Edition

1 2 3 4 5 6 7 8 9 LPI 22 21 20 19 18 17

ISBN *9781265943424*
MHID *1265943424*

Sponsoring Editor
Michael McCabe

Editing Supervisor
Stephen M. Smith

Production Supervisor
Lynn M. Messina

Acquisitions Coordinator
Lauren Rogers

Project Manager
Kritika Kaushik, Cenveo® Publisher Services

Copy Editor
Surendra Shivam, Cenveo Publisher Services

Proofreader
Barnali Ojha, Cenveo Publisher Services

Art Director, Cover
Jeff Weeks

Composition
Cenveo Publisher Services

To our mothers who cradle the world

To our earth that gives us life

About the Editor

Hwaiyu Geng, CMfgE, PE, is a principal at Amica Research (Palo Alto, California), promoting green engineering and manufacturing programs. He has had over 40 years of diversified engineering and management experience, working with Applied Materials, Hewlett-Packard, and Intel on international high-tech projects. He has written and presented at international conferences numerous technical papers. Mr. Geng is also the editor of McGraw-Hill's *Manufacturing Engineering Handbook.* He is a patent holder.

Contents in Brief

Contents

Technical Advisory Board

Mark Camenzind *Air Liquide Electronics US, Fremont, California*

Yihlin Chan *OSHA, Salt Lake City, Utah*

Amy Geng *Institute for Education, Washington, D.C.*

Hwaiyu Geng *Amica Research, Palo Alto, California*

Michael A. Huff *MNX, Corporation for National Research Initiatives, Reston, Virginia*

Tsu-Jae King Liu *University of California, Berkeley, California*

Chris A. Mack *University of Texas at Austin*

Florian Solzbacher *University of Utah, Salt Lake City, Utah*

Manos (Emmanouil) M. Tentzeris *Georgia Institute of Technology, Atlanta, Georgia*

Michael Töpper *Fraunhofer IZM, Berlin, Germany*

Paul Wright *University of California, Berkeley, California*

Haixia Zhang *Peking University, Beijing, China*

Contributors

Mike Ameen *Axcelis Technologies, Inc., Beverly, Massachusetts* (Chap. 10)

Hal Amick *Colin Gordon Associates, Brisbane, California* (Chap. 47)

Gautam Banerjee *Versum Materials, Inc., Tempe, Arizona* (Chap. 15)

Ahmed Busnaina *NSF Nanoscale Science and Engineering Center for High-Rate Nanomanufacturing, Northeastern University, Boston, Massachusetts* (Chaps. 26, 27)

Kristin Cavicchi *Air Liquide Electronics US, Houston, Texas* (Chap. 35)

Rich Chaney *American Semiconductor, Inc., Boise, Idaho* (Chap. 23)

Yung-Hang Chang *School of Industrial & Systems Engineering and Georgia Tech Manufacturing Institute, Georgia Institute of Technology, Atlanta, Georgia* (Chap. 22)

Yi-Chia Chou *Department of Electrophysics, National Chiao Tung University, Hsinchu, Taiwan, Republic of China* (Chap. 3)

David J. Coumou *MKS Instruments, Inc., Rochester, New York* (Chap. 43)

Michael R. Czerniak *Edwards, Clevedon, United Kingdom* (Chap. 42)

Nandita DasGupta *Department of Electrical Engineering, Indian Institute of Technology Madras, Chennai, India* (Chap. 9)

Bin Dong *Sun Yat-sen University, Guangzhou, China* (Chap. 12)

M. Sky Driver *University of North Texas, Denton, Texas* (Chap. 12)

Eric T. Eisenbraun *Colleges of Nanoscale Science & Engineering, SUNY Polytechnic Institute, Albany, New York* (Chap. 13)

Mark Fessler *ASM America, Phoenix, Arizona* (Chap. 45)

Julie Fraser *Iyno Advisors, Inc., Cummaquid, Massachusetts* (Chap. 38)

Norbert Fruehauf *Institute for Large Area Microelectronics, University of Stuttgart, Germany* (Chap. 28)

Daniel Fuchs *Air Liquide Electronics US, Houston, Texas* (Chap. 33)

Michael Gendreau *Colin Gordon Associates, Brisbane, California* (Chap. 47)

Hwaiyu Geng *Amica Research, Palo Alto, California* (Chap. 1)

Barry Gotlinsky *Nanometrex Solutions LLC, Santa Fe, New Mexico* (Chap. 34)

Kenneth Grosser *Air Liquide Electronics US, Dallas, Texas* (Chap. 31)

Clint Haris *Entegris, Inc., Billerica, Massachusetts* (Chap. 37)

Miyuki Hirose *DISCO Corporation, Tokyo, Japan* (Chap. 17)

Michael A. Huff *MNX, Corporation for National Research Initiatives, Reston, Virginia* (Chaps. 2, 4)

Noriko Ito *DISCO Corporation, Tokyo, Japan* (Chap. 17)

Tracey Jacksier *Air Liquide, Delaware Research & Technology Center, Newark, Delaware* (Chap. 31)

Nesrine Kammoun *Institute for Large Area Microelectronics, University of Stuttgart, Germany* (Chap. 28)

Jeffry A. Kelber *University of North Texas, Denton, Texas* (Chap. 12)

Youngsuk Kim *DISCO Corporation, Tokyo, Japan* (Chap. 17)

John Klocke *Applied Materials, Kalispell, Montana* (Chap. 14)

Cheng-Chung Lee *Display Technology Center, Industrial Technology Research Institute, Hsinchu, Taiwan, Republic of China* (Chap. 29)

Yuh-Zheng Lee *Display Technology Center, Industrial Technology Research Institute, Hsinchu, Taiwan, Republic of China* (Chap. 29)

Larry Levit *LBL Scientific, Alamo, California* (Chap. 41)

Vyacheslay (Slava) Libman *FTD Solutions LLC, Colorado Springs, Colorado* (Chap. 32)

David N. Liu *Idearite Co., Ltd., Taichung, Taiwan, Republic of China* (Chap. 29)

Tiantao Lu *University of Maryland, College Park, Maryland* (Chap. 5)

Chris A. Mack *Lithoguru.com, Austin, Texas* (Chap. 8)

Sumio Masuchi *DISCO Corporation, Tokyo, Japan* (Chap. 17)

James McAndrew *Air Liquide, Delaware Research & Technology Center, Newark, Delaware* (Chap. 31)

Chris Muller *Purafil, Inc., Doraville, Georgia* (Chap. 40)

David L. O'Meara *Tokyo Electron Limited, Albany, New York* (Chap. 7)

Richard V. Pavlotsky *cGMP Technologist, Woodland Hills, California* (Chap. 46)

Ivy Qin *Kulicke and Soffa Industries, Inc., Fort Washington, Pennsylvania* (Chap. 19)

Jamal Ramdani *Power Integrations, Inc., San Jose, California* (Chap. 6)

Dieter Rathei *DR YIELD software & solutions GmbH, Graz, Austria* (Chap. 36)

L. P. Ren *Global Nanosystems, Inc., Los Angeles, California* (Chap. 3)

Christian Renninger *Institute for Large Area Microelectronics, University of Stuttgart, Germany* (Chap. 28)

Roey Shaviv *Applied Materials, Santa Clara, California* (Chap. 20)

Ardeshir J. Sidhwa *Quantum Global Technologies, LLC, Quakertown, Pennsylvania* (Chap. 44)

A. T. Sivaram *Advantest America, San Jose, California* (Chap. 21)

Florian Solzbacher *University of Utah, Salt Lake City, Utah* (Chap. 11)

Ankur Srivastava *University of Maryland, College Park, Maryland* (Chap. 5)

Yilin Sun *Institute of Microelectronics, Tsinghua University, Beijing, China* (Chap. 24)

Bijan K. Tehrani *Georgia Institute of Technology, Atlanta, Georgia* (Chap. 25)

Manos (Emmanouil) M. Tentzeris *Georgia Institute of Technology, Atlanta, Georgia* (Chap. 25)

Dietrich Tönnies *ROFIN-SINAR Laser GmbH, Bergkirchen, Germany* (Chap. 18)

Michael Töpper *Fraunhofer IZM, Berlin, Germany* (Chap. 18)

K. N. Tu *Department of Materials Science and Engineering, University of California at Los Angeles, Los Angeles, California* (Chap. 3)

Raymond van Roijen *GLOBALFOUNDRIES, Hopewell Junction, New York* (Chap. 39)

Bo Vanderberg *Axcelis Technologies, Inc., Beverly, Massachusetts* (Chap. 10)

Ben Wang *School of Industrial & Systems Engineering, and Georgia Tech Manufacturing Institute, Georgia Institute of Technology, Atlanta, Georgia* (Chap. 22)

Kan Wang *School of Industrial & Systems Engineering and Georgia Tech Manufacturing Institute, Georgia Institute of Technology, Atlanta, Georgia* (Chap. 22)

Dan Xie *Institute of Microelectronics, Tsinghua University, Beijing, China* (Chap. 24)

Yung-Hui Yeh *Display Technology Center, Industrial Technology Research Institute, Hsinchu, Taiwan, Republic of China* (Chap. 29)

Cihan Yilmaz *NSF Nanoscale Science and Engineering Center for High-Rate Nanomanufacturing, Northeastern University, Boston, Massachusetts* (Chaps. 26, 27)

Ardavan Zandiatashbar *Park Systems Inc., Santa Clara, California* (Chap. 16)

Chuck Zhang *School of Industrial & Systems Engineering and Georgia Tech Manufacturing Institute, Georgia Institute of Technology, Atlanta, Georgia* (Chap. 22)

Jun Zhuge *Jiangsu Seraphim Solar System Co., Ltd., Changzhou City, Jiangsu, China* (Chap. 30)

Dave Zuck *Quantum Global Technologies, LLC, Quakertown, Pennsylvania* (Chap. 44)

Foreword

Information technology (IT) has had dramatic social and economic impact in our global society, as continual advancement in semiconductor integrated circuit (IC) manufacturing technology has enabled ever-greater capacity for electronic information processing and storage at ever-lower cost per function. Key to the exponential growth of the IT industry over the past 50-plus years has been the steady miniaturization of the transistor (the basic component of an IC) to enable greater degrees of integration and hence greater IC functionality. The most advanced semiconductor chips today comprise over 10 billion transistors, with features that are finer than 10 nm. The fabrication process for such sophisticated products involves hundreds of steps, each requiring very high precision and/or ultrapure gases and chemicals. Impressive economies of scale make chips affordable: Presently the largest semiconductor "fabs" each have the capacity to process the equivalent of more than 1 million 200-mm-diameter wafers per year.

This handbook provides a broad overview of modern semiconductor manufacturing and market drivers, and detailed technical descriptions of the unit processes used today to fabricate the components and the interconnecting wires comprising an IC. Furthermore, it describes in detail manufacturing processes for photovoltaic devices, large-area electronics, and flexible electronics, which are proliferating as we enter the age of ubiquitous computing. The final sections of this comprehensive reference cover logistical aspects of semiconductor manufacturing, including cleanroom design, handling of gases and chemicals, process control, and factory automation.

This expansive volume is the fruit of Hwaiyu Geng's labor of love to rally 70 leading semiconductor manufacturing professionals from across the world, whose collective experience and expertise span the breadth of the industry, to archive their expert knowledge. I hope that you will benefit from this tremendous effort.

Tsu-Jae King Liu, Ph.D.
TSMC Distinguished Professor in Microelectronics
University of California, Berkeley
Member, U.S. National Academy of Engineering

Preface

As engineers, managers, researchers, professors, or students, we are all facing increasing challenges in a cross-functional environment. For each manufacturing problem or project, we must ask these questions: What are the unknowns, available data, conditions, and feasible alternatives to validate and find an optimum solution plan? How does one apply technical knowledge to assemble a solution plan, to support a team, and to lead and implement a project successfully?

Our challenges may include establishing microchip manufacturing processes for new products, implementing automated manufacturing processes, improving manufacturing yield, expanding capacity, or establishing quality and safety programs. A good understanding of how to manufacture microchips, as well as how it relates to other processes, will enable one to plan, design, and implement a project successfully.

The goal of *Semiconductor Manufacturing Handbook* is to provide readers with essential technical knowledge needed to work in a semiconductor fab, and the skills to establish manufacturing processes, solve problems, and improve production lines in a microchip fab. This handbook embraces both conventional and emerging microchip manufacturing technologies and includes the following parts:

Part 1 covers semiconductor fundamentals and IoT/data analytics, nanotechnology, FinFET fundamentals, MEMS fundamentals, and 3D IC design.

Part 2 gives front-end-of-line processes that include epitaxy, microlithography, etching, ion implantation, physical vapor deposition, chemical vapor deposition, atomic layer deposition, electrochemical deposition, chemical mechanical planarization, and AFM metrology.

Part 3 offers back-end-of-line processes that cover wafer thinning and singulation, packaging, bonding, interconnects reliability, and automatic test equipment.

Part 4 discusses flexible hybrid electronics, printed electronics, 3D interconnects in flexible electronics, flat-panel and flexible display technology, and photovoltaics fundamentals.

Part 5 covers gas distribution systems, ultrapure water fundamentals, process chemical handling and abatement, chemical and slurry handling, and filtration.

Part 6 reviews operations, equipment, and facilities that cover yield management, CIM and factory automation, manufacturing execution systems, advanced process control, airborne molecular contamination, ESD controls, plasma processing, vacuum systems, cleanroom design and construction, and vibration technologies.

This handbook covers topics ranging from wafer processing, final manufacturing, flexible hybrid electronics, and large-area electronics to wafer fabs and cleanrooms, contamination control, and operations management systems. The book covers the breadth and depth of semiconductor fundamentals and manufacturing technologies, and it is a useful and enlightening resource for all semiconductor professionals. This Second Edition of *Semiconductor Manufacturing Handbook* is the most comprehensive single-source guide ever published in its field.

Hwaiyu Geng, CMfgE, PE

Acknowledgments

Semiconductor Manufacturing Handbook is a collective effort by an international community of scientists and professionals comprising 70 experts from 9 countries around the world.

I am very grateful to the members of the Technical Advisory Board for their diligent reviews of this handbook, confirming technical accuracy while contributing their unique perspectives. Their guidance has been invaluable in ensuring that the handbook can meet the needs of a broad audience.

My sincere thanks go to the contributors who took many hours from their busy schedules and personal lives in order to share their wisdom and deep experiences.

Without the Technical Advisory Board members and the contributing authors, this handbook could not have been completed. This collective effort has resulted in a work that adds value to the semiconductor and Internet of Things communities.

Special thanks go to Professor Tsu-Jae King Liu at University of California, Berkeley, for her sincere and continuous encouragement to improve the handbook.

Thanks are due to the following individuals, companies, and organizations for their contributions: Dr. Paul Wright, University of California, Berkeley; Ms. Deborah Geiger, SEMI; Dr. Michael Huff, MEMS & Nanotechnology Exchange; Dr. Chris Mack, University of Texas at Austin; Dr. Haixia Zhang, Peking University; Dr. Michael Töpper, Fraunhofer IZM; Dr. Liping Ren, Global Nanosystems, Inc.; Ms. Miyuki Hirose, DISCO; Dr. Mark Camenzind, Air Liquide Electronics; Dr. Cheng-Chung Lee, ITRI; Gartner; NIST; NASA; NOAA, etc.

Thanks are also due to Senior Editor Michael McCabe and his staff at McGraw-Hill for their support and guidance. And thanks to Kritika Kaushik, Project Manager, and her team at Cenveo Publisher Services for transforming the manuscripts into a published work.

Finally, I wish to give my thanks to my wife, Limei; my daughters, Amy and Julie; and my grandchildren, Abby, Katy, Alex, Diana, and David, for their support and encouragement while I was preparing this book.

Semiconductor Manufacturing Handbook

PART 1

SEMICONDUCTOR MANUFACTURING FUNDAMENTALS

Chapter 1

Semiconductor Manufacturing, the Internet of Things, and Sustainability

BY
HWAIYU GENG *Amica Research, Palo Alto, California*

1.1 INTRODUCTION

At the end of 2016, with a world population at 7.4 billion, there were 8 billion connected devices globally.[1] In the following decades, the Internet of Things (IoT), a subset of cyber-physical systems or CPS, would connect more things around the world exponentially. With increasing processing power, reducing costs and sizes of microchips, microchips are well recognized for the advancement of electronics technology that drive the IoT with phenomenal growth and have transformed almost every facet of daily life.

"Half a century ago, a young engineer named Gordon E. Moore took a look at his fledgling industry and predicted big things to come in the decade ahead. In a four-page article in the trade magazine *Electronics*, he foresaw a future with home computers, mobile phones, and automatic control systems for cars. All these wonders, he wrote, would be driven by a steady doubling, year after year, in the number of circuit components that could be economically packed on an integrated chip,"[2] written by Dr. Chris Mack in IEEE's *The Multiple Lives of Moore's Law*.

In IEEE's Special Report titled *50 Years of Moore's Law*,[3] "Moore's law has driven technology forward at a staggering clip. We have all benefited from this miraculous development, which has forcefully shaped our modern world. In this special report, we find that the end won't be sudden and apocalyptic but rather gradual and complicated. Moore's law truly is the gift that keeps on giving—and surprising, as well."

1.2 MOORE'S LAW

Dr. Gordon Moore, a visionary engineer and innovator who cofounded Intel Corporation with Robert Noyce on July 18, 1968, forecasted a bright future of electronics that integrated electronic circuits would become less costly and more powerful over time. On April 19, 1965, 3 years before he cofounded Intel, he predicted the number of transistors on a silicon chip would double every 12 months. Moore's law is updated in 1975 that the number of transistors on a chip would double every 18 to 24 months.

In an interview with IEEE, Moore stated, "I used to give talks about how other industries might have progressed. You know, had the auto industry made progress at the same rate [as silicon microelectronics], you would have gotten a million miles per gallon of fuel, and cars that could go several hundred thousand miles an hour." Intel® 4004 processor, the first general-purpose programmable processor, was built in 1971 and contained 2300 transistors. Fast forward to 2015, the 5th Generation Intel® Core™ Processor is a 14-nanometer (nm) technology with 1.3 billion transistors inside. Comparing to the 4004 processor in 10,000 nm, Intel's 14-nm chips can deliver 3500 times the performance at 90,000 times the efficiency, and at 1/60,000th the cost.[4,5]

"No exponential ever goes on forever without some kind of disaster happening at the end. Sure it has a limit," Gordon Moore said. "Materials are made of atoms, and we're not too far from where that starts to bite us."[6] Miniaturization may be slowing down. But, through invention and innovation, there are derivative technologies such as micro-electric-mechanical systems (MEMS), flexible and printed electronics, light emitting diodes (LED), photovoltaic solar, silicon photonics, etc., bringing benefits that improve our living standards and sustainability. Those technologies play vital roles in the age of Internet of Things. Moore's law will proliferate and continue on for many years to come.

1.2.1 FinFET Extends Moore's Law

The joint work of UC-Berkeley's Professors Chenming Hu, Tsu-Jae King Liu, and Jeff Bokor created the fin field effect transistor (FinFET) technology. King Liu describes the FinFET "as a vertical transistor, kind of like a skyscraper. Conventional transistors, called MOSFETs (metal-oxide-semiconductor field-effect transistors), are fabricated along the surface of a silicon wafer, making them flat and low. FinFETs, by contrast, are fabricated along the sidewalls of a narrow, vertical fin etched into the surface of the wafer. That means they take up far less real estate, enabling a trillion transistors to be packed onto a chip."

This unusual transistor design can scale down a gate length, the distance between the source and the drain, of less than 10 nm. A smaller gate length shrinks the transistor overall and enables it to switch on and off quickly with a higher current density. Five-nanometer FinFET

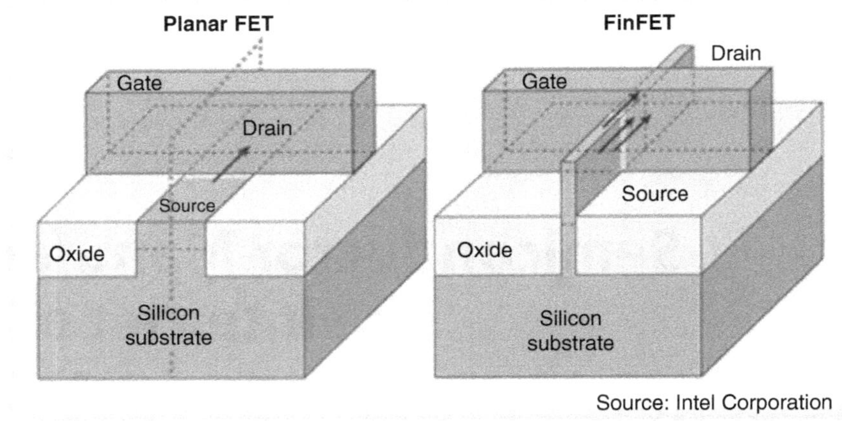

Figure 1.1 Planar FET versus FinFET.

designs are feasible and without significant modifications to the fabrication process. "Below five nanometers, you can have quantum mechanical tunneling and confinement effects," says King Liu, "That means small variations in manufacturing would result in large changes in performance." The FinFET transistor structure revitalizes Moore's law by freeing planar FET from the short-channel effects that limit device scalability (Fig. 1.1).[7,8]

1.3 INTEGRATED CIRCUITS AND DESIGN

The integrated circuit (IC), or microchip, was invented by Jack Kilby in 1958. There are many types of microchips. Microprocessors are logic chips that are used to perform computations (binary codes, logic gates, Boolean algebra, etc.). Memory chips, such as NAND (Not AND), store data. Analog chips operate over a continuous range of signal. Analog chips can be subdivided into linear IC and radio frequency IC. Digital signal processors convert between analog and digital signals. Application-specific ICs (ASICs) are customized chips for a special purpose. They are used in cars, televisions, digital cameras, or appliances and more.[9]

Based on a microchip's feature and development plan that encompasses use, power, area, cost, time to market, an IC's function and performance are determined and the logic circuit is designed. During logic circuit design, a logic circuit diagram is drawn to determine the electronic circuit required for performing requested functions. Once the logic circuit diagram is complete, simulations are performed multiple times to test the circuit's operation. If there is no issue with the operation, the actual layout pattern for the devices and interconnects is designed by computer-aided design (Fig. 1.2).

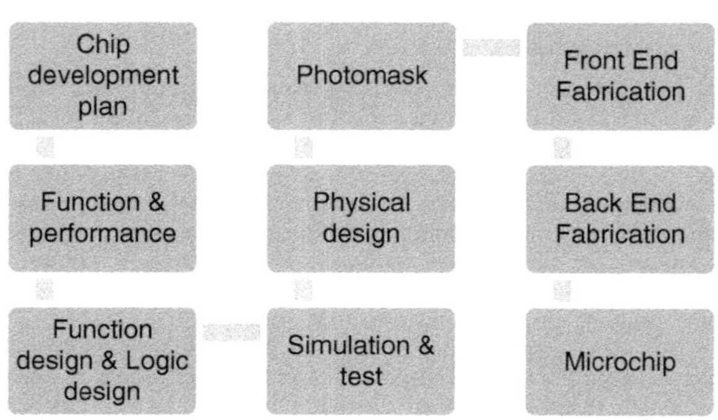

Figure 1.2 IC design and fabrication. (*Courtesy of Amica Research.*)

1.4 HOW A MICROCHIP IS MADE

A sophisticated processor can contain hundreds of millions or billions of transistors interconnected by fine metal wires. The process of making chips is extremely complex requiring hundreds of precisely controlled steps.

The base material of microchips is silicon. Silicon is called a semiconductor that relates to its molecular structure. Silicon can conduct electricity under certain conditions and nonconductive under others. This feature acts as a basic circuit of a transistor: an on/off or binary 1/0 switches either allowing or preventing electrical current to pass through a gate. The process of manufacturing microchips building millions or billions of transistors on an IC chip makes it phenomenally precise and complex.[10]

The process of manufacturing microchips includes following three main processes: (1) wafer manufacturing (2) front-end of line processing, and (3) back-end of line processing.[11–14] A microchip is formed by more than hundreds repetitive operations during front-end of line processing (Fig. 1.3). The process of manufacturing IC chip is supported by advanced manufacturing technologies that encompass material handling automation, computer-integrated-manufacturing, advanced process control, manufacturing execution system, etc.

In the fabrication of IC, wafer serves as a substrate for IC to build-on. Many steps in front-end processing, including oxidation, photolithography, etching, doping, deposition of various materials, are built over the wafer in accordance to the IC design. Inspection and testing are also included in each step of the processes. Finally, the dies are cut, packaged, tested, and ready to be used.

1.4.1 Wafer Manufacturing

A wafer, also called a slice or substrate, is a thin disc (about thickness of a credit card) made of silicon, or other semiconductor materials such as GaAs. Silicon, the principal ingredient in beach sand, is the second most abundant element in the earth after oxygen. Before it can be used in semiconductor manufacturing it is refined to nearly 100 percent pure.

SILICON INGOT

Semiconductor manufacturing begins with a silicon ingot. Silicon ingots could be made by Czochralski (CZ) process and float-zone process. Float-zone wafers are generally not larger than 150 mm due to limit of surface tension during growing process. Most crystal ingots are grown by the Czochralski method. This method begins by a piece of crystal of silicon, called *seed*, is lowered in the near 100 percent pure molten silicon in crucible. The seed and crucible are rotating in opposite direction while heating purified silicon to 1200°C in an argon

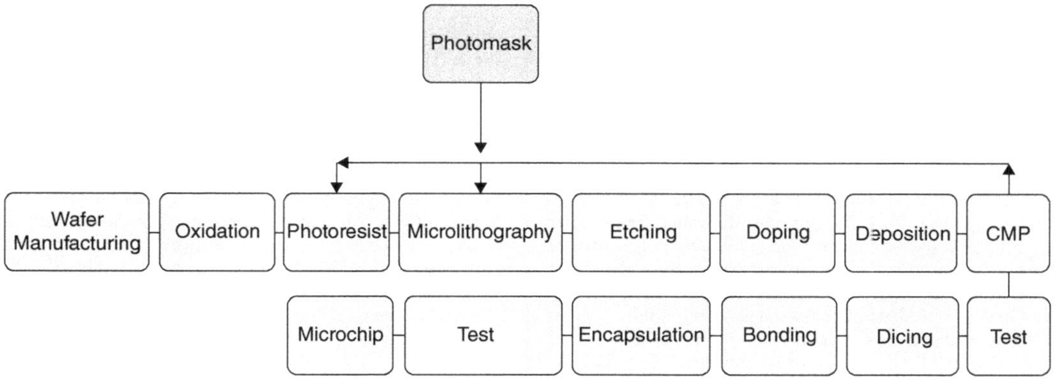

Figure 1.3 Wafer manufacturing, photomask, front-end of line process, and back-end of line process. (*Courtesy of Amica Research.*)

atmosphere. The silicon seed is slowly withdrawn from molten liquid. Silicon atoms attach to the seed, resulting in a long crystal rod or ingot, grows with a diameter of most commonly 200 or 300-millimeter (mm) across in careful control of temperature, atmosphere, and pressures. It produces electronic grade silicon that is semiconductor manufacturing quality with silicon purity to 99.9999999 percent.

Wafer Fabrication Process

A series of operations is performed to prepare silicon wafers from a silicon ingot.

Cropping, Grinding, Slicing The ends of single crystal ingot are cropped using a single-bladed diamond saw with water coolant. The ingot is then ground to a uniform diameter and the ends are chamfered to reduce the possibility of shattering the ingot. The crystal structure is determined by x-ray diffraction. A longitudinal flat section is produced by wet grinding to indicate crystal orientation of the ingot. Ingots are sliced into discs or called substrates.

Lapping After slicing, the wafers are mechanically lapped under pressure on both sides of the wafer in order to achieve flatness and parallelism. Lapping operations will remove the surface silicon which has been damaged by the slicing process and thin the wafer to desired thickness. The edge of the individual wafer is rounded by the use of wet automatic grinders.

Etching Wafer dicing and lapping degrade the silicon surface crystal structure, so wafers are etched with a solution containing nitric, acetic, or hydrofluoric acids in order to remove external damaged surface and reduce the thickness of the wafer. Etching operation could be done in manual etch tanks or automated etching machines.

Polishing After etching, the wafers are polished to a super-flat, flawless, and mirror-smooth surface with remaining roughness at atomic scale. Polishing process is usually completed in two or three polishing steps. Wafers are mounted onto a metal carrier plate that is attached to the polishing machine by vacuum. The polishing process combines with pressure mechanically and chemically using fine slurry such as Al_2O_3, SiO_2, or CeO_2.

Cleaning Finally, the wafers are cleaned to remove any particles on the surface of a polished wafer thereby making the wafer residual free. Variable ultrapure chemicals could be used for cleaning that include ammonia, hydrogen peroxide, hydrofluoric acid, hydrochloric acid, and deionized water.

1.4.2 Front-End of Line Processing

Building circuits on a wafer is an extremely complex and precise process. The process consists of hundreds of precisely controlled steps with as many as 30 layers of complex circuitry. Chips are built, taking from 10 to 30 days or more, in specific patterns and recipes to produce hundreds or thousands of die on each 300-mm wafer.

The process of making IC chips is called fabrication. The factories where chips are made are called fabrication facility or fab. In a fab, there are cleanrooms where fabrication in batch of wafers is built. Because particles of dust can ruin the complex circuitry on a chip, purified air is constantly recirculated in a cleanroom. For a class ISO 1 cleanroom, a cubic meter of air contains no more than 10 particles, measuring 0.1 micrometer or μm (millionth of a meter). Fab technicians wear bunny suits that are specially designed to keep contaminants off the wafers during fabrication.[15,16] The following are operations in the Front End of Line Processing (FEOL).

Epitaxy

Epitaxy is one of the fundamental processes used to make semiconductor devices. Epitaxy refers to deposit an overlayer of a crystal (epitaxial film) on a host crystal (substrate) in an ordered fashion. It is desired that the overlayer has a defined orientation with respect to the substrate. The epitaxial film and substrate could be same or different materials. There are three types of epitaxy: liquid phase, vapor phase, and molecular beam epitaxy.

Oxidization

Silicon oxide (SiO_2) serves as an electrical insulator that is a key component in IC circuit. Growth of oxidation can be achieved by dry or wet oxidation.

Oxidation is a diffusion process. Silicon wafers are first heated and exposed to ultrapure oxygen at elevated temperatures in a diffusion furnace. By carefully controlled conditions in the furnace, a silicon oxide (SiO_2) film, dry oxidation, with uniform thickness is growth from the SiO_2/silicon interface.

$$Si + O_2 \rightarrow SiO_2$$

Wet oxidation is preferred to dry oxidation for growing thicker oxide at a faster rate. Wet oxidation is performed by exposing the wafer to ultra-high-purity water vapor or using hydrogen and oxygen which are burned in a torch flame to form ultra-high-purity water vapor in an oxidizing furnace.

$$Si + 2H_2O \rightarrow SiO_2 + 2H_2$$

Photolithography

Photolithography is a process by imprinting a specific IC design pattern on a wafer. Before performing the photolithography process, a photomask (reticle) is prepared. A photomask has a copy of the circuit pattern drawn as an opaque pattern on a quartz plate. A quartz plate is first polished in high precision. The quartz substrate is then coated with a thin layer of chrome material by sputtering that is opaque to UV wavelengths. A photosensitive chemical called resist is spin-coated over the chrome photomask blank. Then a circuit design pattern is transcribed into the chrome film by electron beam lithography. After exposure, develop and etch, the photomask blank becomes a photomask with a copy of the circuit pattern.[17,18]

Photoresist Application A wafer is coated with a light-sensitive liquid called *photoresist*. By pouring a small amount of the photoresist to the center of the wafer while the wafer is spinning at high speed, the material will spread over the entire surface in a thin, uniform coating. The photoresist material is sensitive to "photo" (light) and it "resists" certain chemicals that are used to remove portions of the material below the photoresist.

Photomask and Exposure A photo aligner precisely aligns the photoresist-coated wafer to the photomask. The photomask, acting like a stencil, contains the pattern to be imaged onto the wafer. The aligner projects light through the photomask and optical reduction lenses to the wafer. This results in exposure to the photoresist in areas where not protected by the mask. For a positive resist, the areas exposed to light undergo a chemical reaction that will make the photoresist soluble in an alkaline solution (developer). A negative resist becomes insoluble in regions that expose to the light. The exposure tool repeats the process across the wafer to form the same image throughout the wafer. The yellow lighting is necessary in photolithography rooms to prevent unwanted exposure of photoresist to artificial light.

Resist Development and Hard Bake After exposure, a wafer is developed with aqueous solutions of either sodium hydroxide or potassium hydroxide. The developer is applied by either immersion, spraying, or atomization, causing the exposed areas of the photoresist to be dissolved and removed. This leaves a photoresist pattern similar to that depicted on the photomask.

Following resist development, an additional baking process, or "hard bake," is performed to harden the remaining photoresist. The photoresist can now protect the underlying material during etching.

Etching

The etching process removes layers of silicon, silicon oxide, polysilicon, or metals according to the patterns from mask. Etching can be done by a chemical solution (wet etching) or plasma (dry etching, reactive-ion etching).

Wet Etching Wet etching is a simple technology. It is accomplished by submerging wafers in chemical bath. Wet etching is usually isotropic with etching progressing equally in all directions that will cause undercutting of the mask layer by the same depth as desired area. Therefore, consideration of selectivity is important. Selectivity is the ratio of two etch rates: the rate for the layer to be removed and the rate for the mask layer to be protected.

Dry Etching Dry etching is one of the most used processes in semiconductor manufacturing. Dry etching may be anisotropic (direction dependent) that allows effectively etching desired surfaces or layers through the use of chemically reactive gases, or through physical bombardment of argon atoms.

Reactive ion etching (RIE) is a dry etching technology that combines chemical and physical etching. A wafer is placed in a vacuum chamber reactor. By applying radio frequency energy to a gas containing a chemically reactive element etchants, such as fluorine-based or chlorine-based gas, the plasma releases positively charged ions. The ions strike or bombard the wafer surface vertically to etch or remove materials which are then removed by the vacuum system.

After etching, the resist has served its purpose and could be removed from an etched wafer by oxygen plasma.

Doping

Intrinsic semiconductors, such as silicon, can be doped with extrinsic atoms in order to increase their conductivity. Doping can be achieved either by ion implantation or diffusion.

Ion Implantation Materials used for doping intrinsic semiconductors are typically ionized atoms or molecules for silicon with valence 4, either 3 or 5 valence electrons, such as boron or phosphorous. The dopants are accelerated in intense electrical fields to an energy ranging from sub-keV to multiple meV, and bombard the surfaces of a wafer, where not masked or exposed by etching processes. Upon striking the wafer, the dopant is embedded at various depths and precise locations within the bulk, thereby altering the dominant carrier concentration of the bulk material to p-type (hole) or n-type (electron). Ion implantation introduces damage into crystalline substrates, and a high-temperature annealing step, 800 to 1200°C for tens of seconds to 30 min, will restore the crystal structure and further incorporate the dopant atoms into the crystal lattice.

Diffusion Diffusion is a term that can be used to describe the introduction, placement, and/or activation of dopants to semiconductor in a furnace. Sources of dopants could be in gaseous, liquid, or solid form. Gaseous dopants are most widely used that include arsine (AsH_3), borane (BH_3), phosphine (PH_3), etc. Dopants move in semiconductors as a result of a concentrated gradient. Fick's laws of diffusion are equations that describe how dopant's diffusivity, concentration, temperature, and time relate in determining the dopant profile in silicon. The diffusion of dopants limits the thermal budget and processing temperatures of more advanced IC integration.

Deposition

Thin film deposition processes place thin and uniform materials such as silicon nitride, silicon dioxide, silicon, or metal onto wafers. The silicon layers are formed as an insulator. The metallic layers are formed and used in the wiring of the circuit. Some commonly used thin film depositions are physical vapor deposition (PVD), chemical vapor deposition (CVD), or electrochemical deposition (ECD).

Physical Vapor Deposition Physical vapor deposition (PVD), a method of sputtering, is performed in a high vacuum chamber. The source material (sputtering target) and the wafer substrate are connected to a high voltage power supply. When sputtering gas (an inert gas such as argon) is admitted in the chamber, argon gas is ionized and creates a plasma between sputtering target and wafer. The argon ions are accelerated and bombard the sputtering target that creates impacts knocking off target atoms or molecules. The target atoms or molecules then travel through vacuum to the wafer, where a desired thin film is formed.

Evaporation is another PVD process. First, wafers are placed in a vacuum chamber, the gases in the chamber are removed using a vacuum pump. Once the chamber is free of residual gases, the source material (metal) is heated to cause it to evaporate. The evaporated molecules are dispersed and landed on the wafer and formed a desired thin film.

Chemical Vapor Deposition Chemical vapor deposition (CVD) is a chemical process that can be accomplished in a vacuum (low pressure CVD or LPCVD) to produce high quality, high performance, large area films. Importantly, CVD is often used to provide highly conformal film coatings on trench walls and bottoms as the sizes of such features continue to shrink.

Reactant gases, or precursors, are introduced into a reaction chamber that contains wafers. In a hot-wall reactor or plasma environment, reactive species are created at the reactor walls or in the plasma, and then diffuse to the wafer surface, where film growth can take place. This is known as a homogeneous process, as reactive species are formed in the gas phase. In a cold wall reactor, the reaction energy is supplied by heating the wafer, and reactive species are created by interaction with the wafer surface—a heterogeneous process.

Atomic layer deposition (ALD) is a special type of CVD that deposits a single layer of a film each time in a tightly controlled process.

Electrochemical Deposition Electrochemical deposition (ECD) is used to deposit a thin metal layer, which will form interconnects (integrated wires) connecting various devices on an IC. ECD is an electroplating process. The substrate is submerged in electrolyte solution and serves as a cathode (negative side of power supply). A copper electrode serves as an anode (positive side of power supply) forming a circuit through electrolyte solution. The copper ions are reduced on the cathode or substrate and form a film.

Chemical Mechanical Planarization

A flat surface is required for IC processing, such as photolithograph operation. There are several ways to perform planarization. Chemical mechanical planarization (CMP), by applying abrasive chemical slurry and mechanical force (polishing pad), is one of the best planarization techniques that polishes away excess material left from previous processes and planarizes the uneven surface to create a flat surface. The depth-of-field requirement is down to angstrom level.

Repeat the Steps

There are typically over 350 process steps required to make a microchip.[19] Replicating the steps photolithography, etching, doping, deposition, CMP layer by layer until the last layer is formed. The wafer with transistors and interconnects is moved to the back-end of line processing.

1.4.3 Back-End of Line Processing

Back-end of line (BEOL) starts with testing a wafer with transistors and interconnects. An automatic, computer-driven electrical test system checks the functionality of each chip on the wafer. Chips that do not pass the test are marked with ink for rejection. The wafer is then diced into individual die. Individual die is bonded with wires, assembled, tested, and finally packaged.

Grinding (Thinning) and Dicing

The wafer is first attached by a protection tape, called back-grinding tape, to protect device layer of the wafer. The wafer is turned over and the back of the wafer is ground (thinning) to suitable thickness that would accommodate assembling and packaging.

Dicing, or singulation, is a process cutting semiconductor wafer into individual die. After thinning, the wafer is mounted on a dicing tape frame with device side upward. A disc blade with diamond grains cut the wafer into separated dies. The dicing process could be done by diamond blades, laser ablation, or stealth dicing.

Wire Bonding

Wire bonding is a process to connect IC to the substrate. Wire bonding uses very fine gold or copper wire to connect a die to the substrate or leadframe by ultrasonic or thermosonic techniques. Ball bonding is the most popular interconnect method. The formation of ball bonding starts on the chip, the wire loop, finally the bond on the substrate. Wire bonder is a fully automated machine with high-speed, high-precision servo system, an ultrasonic transducer for bonding operations, and an automatic vision system for device alignment.[20]

Packaging and Assembly

IC packaging, or encapsulation, is performed to contact and encapsulate semiconductor material in a supporting case, or package, with mostly molding resin. The package is to support IC performance by keeping IC from hostile environment, handling thermal management, and providing a reliable interconnect to the substrate which is mostly a printed circuit board or PCB, which combines a multitude of ICs and passive components to build a microelectronic system. There are packages for single chips but also for multiple dies which is referred as an MCM (multichip module) or a SiP (system in package). After packaging, printing, or engraving product information by laser is performed onto the package surface or onto the backside of the chip. Most of the packages are soldered or glued onto the substrate.

Final Test

The packaged IC chips are ready to use. However, there are many things can go wrong: die may crack during assembly, bonding wires may be poor or not connected, ICs may be affected by static discharge due to poor handling and other issues. So, the IC chip is individually tested on a tester. The chips are tested for functionality, performance, and power according to a test program prepared specifically for each type of IC in advance. In many cases these test programs are a variation of the ones used in wafer sort.

1.5 THE INTERNET OF THINGS WORLD

The first industrial revolution (IR) began in Britain in the 18th century and from there IR spread to other parts of the world. The first IR centered in textile industry and transferred farm work and manual labor in homes or small rural shops to water-powered textile machines in factories.

The second IR happened in the latter half of the 19th century until World War I. The inventions centered in steel and coal manufacturing, combustion engine, electrical power, scientific management, that led to mass production such as Ford's T-Model cars assembly lines. The first two industrial revolutions made people more productive, richer, and achieved higher living standards.

The third IR took place in 1980s going digitals. Personal computers, information technology, and material supply chains are being used in the manufacturing to automate production systems.

The IoT and data analytics are the most significant technologies to the 4th industrial revolution. They are changing the way how consumers, enterprises, and governments interact in the physical and cyber world. The IoT devices are driving a paradigm shift in the semiconductor marketplace.

1.5.1 The IoT and IIoT

In 1999, Kevin Ashton coined the phrase "Internet of Things" in his presentation to Procter & Gamble (P&G) recommending to improve its business by applying Radio Frequency Identification (RFID) information, supply chain, and the Internet.

IEEE defines IoT as "IoT refers to any systems of interconnected people, physical objects, and IT platforms, as well as any technology to better build, operate, and manage the physical world via pervasive data collection smart networking, predictive analytics, and deep optimization."[21] The Internet of Things can be expressed in a formula:

$$\text{Internet of Things} = \text{Things} + \text{Sensors} + \text{Networking} + \text{Internet} + \text{Analytics} + \text{Actions}$$

where a thing could be a computer, mobile device, people, or animal having an IP address.

The IoT and Industrial Internet of Things (IIoT) are two distinctly separate areas. The IoT connects consumer devices, and the IIoT connects high-stacks industries that include aerospace, defense, health care, automotive, semiconductor, etc. The IIoT must be built before the IoT applications can be developed.[22]

1.5.2 The IoT Anatomy

The IoT technologies consist of sensors and actuators, network, standards and protocols, and machine learning for data analytics. Sensor technologies enable IoT products to monitor, gather, and transmit information through network. The analytic work prompt people, machines, or systems to make better and orchestrated decisions. Moore's law has been driving down the size and price of sensors, bandwidth, and microchips with increasing capabilities. The price of sensors has been declining exponentially.[23] Advancements in wireless networking technology with reducing costs make it possible to connect sensors and devices ubiquitously.[24] The IoT and big data technologies have fundamentally changed how organizations and people live to make our lives better.

1.5.3 The IoT Ecosystem

In the IoT ecosystem, all physical things are cyber connected that make tremendous economic and environmental impacts to global citizens. The ecosystem could be categorized in three groups: consumer, enterprise, and government.

Consumer facing group covers products including home, entertainment, fitness, health care, shopping, vehicle, etc. Enterprise facing group covers products and services including manufacturing, transportation, automation/robots, customer services, etc. Government facing group covers smart city, public transportation, utilities, environment, energy, emergency services, defense, etc.

1.5.4 Big Data and Analytics

The characteristics of big data mainly encompass four Vs, namely volume, velocity, variety, and veracity.[25] Big data describes an enormous volume of data collecting by sensors, arriving in real time at high speed and in unprocessed format. The data could be in homogeneous (structured, e.g., mileage for car maintenance purpose) or heterogeneous (unstructured, social media, or data in text, audio, or video) forms that requires to be processed and organized. Veracity portrays many uncertainties in the data that may be noise and dirty data that must be filtered or they will cause garbage in, garbage out.

Data analytics is a comprehensive methodology that extracts value out of big data. Data analytics includes descriptive, diagnostic, predictive, and prescriptive analysis. Data analytics applies machine learning algorithm, a subarea of artificial intelligence, to scheme an optimal solution.

Computer-integrated-manufacturing, recipe management system, and manufacturing execution system[26] are essential and complicated automation elements used in semiconductor manufacturing. Applying the IIoT technologies and big data analytics can effectively orchestrate those complicated systems to better manage capital equipment's performance, improve yield, and ensure quality of products.

1.6 SEMICONDUCTOR MARKET OPPORTUNITIES IN THE IOT WORLD

The taxonomy of IoT technologies could be categorized from "things" through "analytics." Sensors are attached to "things" that have an IP address associating with a computer, mobile device, people or animal. Wireless networks and Internet transmit data to the fog and cloud. Software and hardware platforms organize big data, perform data analytics, machine learning, and provide optimum business decisions. This series of value chains demand innovated semiconductor devices ranging from MEMS, processors, memories, wifi IC, RFID, near field communication IC, power management IC, and other electronic devices.

IHS Technology reported there are 8.1 billion connected devices globally at the end of 2015. The connected devices, including smartphones, tablets, personal computers, TVs, audio devices, connected cars, equate to an astounding four devices per household. Consumers and enterprises are emerging as leading adopters of IoT solutions to lower operating costs, increase productivity, and improve life quality.

1.7 SUSTAINABILITY

Sustainable manufacturing grows from a nice to be socially responsible and becomes a major factor in business planning and decision. Sustainable manufacturing includes the following key elements:

- Life to eternity resource conservation
- Energy efficiency
- CO_2 emission
- Water conservation
- Safe working environment

The U.S. NASA's Global Climate Change reported that 2015 was the warmest earth's surface temperatures since modern record keeping began in 1880 based on two separated analyses by the National Aeronautics and Space Administration (NASA) and the National Oceanic and Atmospheric Administration (NOAA).[27] "The average global land and ocean surface temperature for January to September 2016 was 0.99°C (1.78°F) above the 20th century average of 14.1°C (57.5°F)—the highest global land and ocean temperature for January to September in the 1880 to 2016 record, surpassing the previous record set in 2015 by 0.13°C (0.23°F)," per NOAA's State of the Climate Report.[28] Substantial temperature increases and global warming are caused by CO_2 emission and greenhouse effect.[29] Most scientists agree that the global warming trend is due to greenhouse gases that humans have brought into the atmosphere. Global warming has caused unusual weather across the globe. It has damaging impacts on agriculture, fisheries, health, energy demand and air quality, and increase the risks of wildfires around the globe (Fig. 1.4).[30]

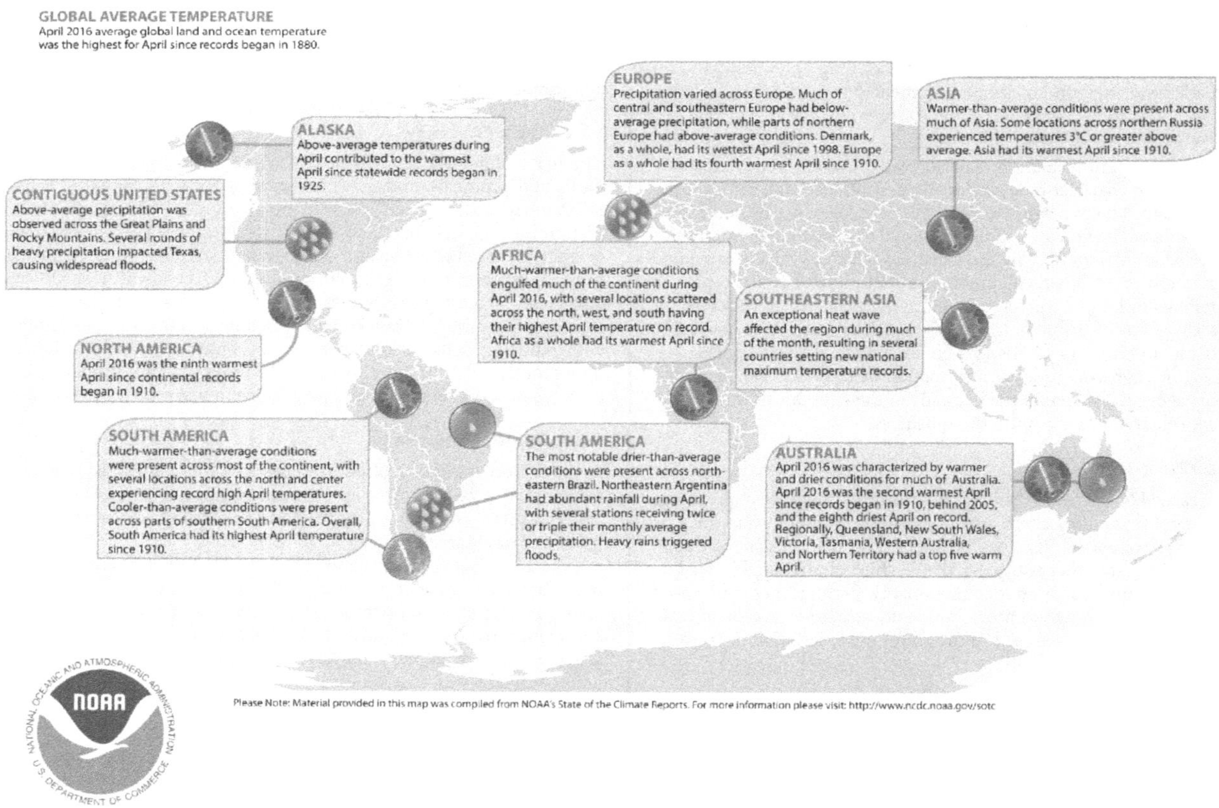

Figure 1.4 Selected significant climate anomalies and events. (*National Oceanic and Atmospheric Administration.*)

"Selected significant climate anomalies and events April 2016" by National Oceanic and Atmospheric Administration reflects natural disaster by regions due to record global temperature (http://www.ncdc.noaa.gov/sotc).

An iPhone uses about 361 kWh/year comparing to an energy efficient refrigerator at 322 kWh.[31] There are 1.7 billion mobile devices as of 2015. All data in audio, video, and text formats are processed in data centers. U.S. data centers consumed an estimated 91 billion kWh of electricity in 2013 and will be 140 billion kWh in 2020, that will emit nearly 100 million metric ton of CO_2. Data centers consume about 3 percent of the global electricity supply and account 2 percent of total greenhouse gas emission.[32] The semiconductor industry can contribute energy savings by designing chip-level and system-level power management systems, as well as using green materials.

The semiconductor industry directly contributes to power and water consumptions that are used in fabs and manufacturing processes. They can be controlled by building energy-efficient fabs and operation-efficient manufacturing processes.[33] A typical semiconductor fab with 350K ft^2 class 100 cleanroom and 225K ft^2 supporting area can consume as much as 0.9 billion kWh of electricity annually.[34,35] Applying best practices and 3D building information modeling (BIM) could build energy-efficiency facilities and cut energy consumption by 10 to 30 percent.[36]

During semiconductor manufacturing, only a very small amount of the raw materials ends up in a finished product.[37] Innovations to enhance the ability to recycle and reuse raw materials are needed with the consideration of cradle to eternity sustainability.

1.8 CONCLUSION

The IoT and data analytics technologies are leading our way to the fourth industrial revolution. The technologies have great potential to advance our economy, productivity, quality of life, and environment. Semiconductor products are the backbone and the most important component that drive the uprising of the IoT technologies.

Semiconductor manufacturing consists of extremely sophisticated capital equipment operating with complicated processes and complicated recipe management, yield management, advanced process control, and operations. Manufacturing companies have offered a business model called servitization.[38] There are two key elements in servitized business model: (1) offering advanced after-sales services beyond traditional services, and (2) proposing process optimization to achieve optimal operations. This business model ensures predictive and preventive maintenance by tracking product and process information real-time from after-sales through the end of the product life cycle. The servitization business model collects big data through embedded sensors, transmits data to the cloud, and analyzes the data for corrective actions to achieve optimal process performance. Semiconductor equipment manufacturers could apply servitization business model to increase equipment uptime, improve factory efficiency, extend product life, minimize overall cost of ownership, and reduce CO_2 emission.

The FinFET technology will further extend Moore's law. Albert Einstein once put it "Creativity is seeing what everyone else has seen and thinking what no one else has thought." Invention is to make something that has never been made before. Innovation is to make changes to something already exist by introducing new ideas, new processes, or new products.[39] The IoT technologies and products are essentially supported by semiconductor devices. The next generation of the MEMS will be driven by creative approaches and innovated new sensing technologies to volume manufacturing. Innovated wearable health-monitoring devices such as microfluidic devices (blood, sweat, tears to measure human performance) encompass monitoring, data acquisition, transmission, analysis, interpretation, and action in real time with considerations of privacy and security. Semiconductor industry must continually take a fresh look to nurture and apply creativity, invention, and innovation (Fig. 1.5) on microchips for miniaturization, higher performance, and lower power demand. The same processes should be applied to every value chain throughout semiconductor from design, manufacturing, to delivery.

Figure 1.5 Nurture creativity, invention, and innovation. (*Courtesy of Amica Research.*)

Creativity and inventions are needed in power device and power management including energy harvest from light, heat, vibration (Piezo), radio frequency, magnet, or thermoelectricity. Cradle-to-eternity sustainable manufacturing must be incorporated into IoT technologies and solutions.

In conclusion, most topics covered in this chapter have been addressed in other chapters by experts from eight countries around the world. With collective effort, we can apply best practices to accelerate the pace of innovation in semiconductor products that will improve quality of life with sustainability.

1.9 REFERENCES

1. "Cisco Visual Networking Index: Global Mobile Data Traffic Forecast Update, 2016–2021," Cisco, San Jose, February 2017.

2. Chris Mack, "The Multiple Lives of Moore's Law," IEEE, January 2017, http://spectrum.ieee.org/semiconductors/processors/the-multiple-lives-of-moores-law.

3. "Special Report: 50 Years of Moore's Law," IEEE, http://spectrum.ieee.org/static/special-report-50-years-of-moores-law.

4. D. Evans, "50 Years on, Image a World without Moore's Law," IQ, Intel, Santa Clara, April 2015, http://iq.intel.co.uk/50-years-on-imagine-a-world-without-moores-law.

5. R. Courtland, *How Much Did Early Transistors Cost?* IEEE, New York, April 2015, http://spectrum.ieee.org/tech-talk/semiconductors/devices/how-much-did-early-transistors-cost

6. Tech Visionary Gordon Moore Retires from Intel http://usatoday30.usatoday.com/tech/news/2001-05-25-moore-quits-intel-board.htm.

7. J. Kawa and A. Biddle, "FinFET: The Promises and the Challenges, Synopsys Insight Newsletter," March 2012, https://www.synopsys.com/COMPANY/PUBLICATIONS/SYNOPSYSINSIGHT/Pages/Art2-finfet-challenges-ip-IssQ3-12.aspx.

8. J. Kawa, "The Use of FinFETs in IP Design, Chip Design," January 2017, http://chipdesignmag.com/display.php?articleId=5302.

9. G. Anthes, "How to Making Chip," Computerworld, 2002, http://www.computerworld.com/article/2576786/computer-hardware/making-microchips.html.

10. "From Sand to Circuits, How Intel Makes Integrated Chips," Intel 2008, http://www.intel.com/Assets/PDF/General/308301003.pdf (accessed January 2017).

11. "How Semiconductors Are Made," Intersil, Milpitas, CA, http://rel.intersil.com/docs/lexicon/manufacture.html.

12. "Device Fabrication, Substrate Manufacture," OSHA, https://www.osha.gov/SLTC/semiconductors/substratemfg.html.

13. "From the Ingot to Finished Silicon Wafers," Microchemical, http://www.microchemicals.com/products/wafers/from_the_ingot_to_finished_silicon_wafers.html.

14. "Making of a Chip Illustrations, 14 nm, 2nd Generation 3D Trigate Transistors," Intel Corporation, Santa Clara, April 2015.

15. C. Hu, *Modern Semiconductor Devices for Integrated Circuits*, Prentice Hall, Upper Saddle River, April 2009.

16. "Substrate Manufacture," OSHA, https://www.osha.gov/SLTC/semiconductors/substratemfg.html.

17. "Photomasks for Semiconductors," Toppan, https://www.toppan.co.jp/english/products_service/pdf/photomask.pdf.

18. "The Basics of Photomask Manufacturing," Photo Sciences, 2011, http://universityphotomask.com/Photomask-101-Rev.2.pdf.

19. "How to Make a Chip," Applied Material, http://www.appliedmaterials.com/company/about/how-we-do-it.

20. I. Qin, "Wire Bonding Tutorial," Solid State Technology, http://electroiq.com/blog/2005/07/wire-bonding-tutorial/.

21. IEEE Standards Association Internet of Things Ecosystem Study, IEEE, New York, 2015.

22. Industrial Internet of Things, https://www.rti.com/industries/iot-faq.html.

23. Lee Simpson and Robert Lamb, *IoT: Looking at Sensors*, Equity Research & Strategy, Jefferies, New York, February 2014.

24. M. Chui, M. Löffler, and R. Roberts, *The Internet of Things*, McKinsey Quarterly, McKinsey & Company, New York, 2010.

25. NIST Big Data Interoperability Framework, NIST, U.S. Department of Commerce, http://nvlpubs.nist.gov/nistpubs/SpecialPublications/NIST.SP.1500-1.pdf.

26. Applied Smartfactory® MES, http://www.appliedmaterials.com/global-services/automation-software/smartfactory.

27. "Analyses Reveal Record-Shattering Global Warm Temperatures in 2015," NASA's News, January 20, 2016, http://climate.nasa.gov/news/2391/ (accessed January 2017).

28. "Global Analysis—September 2016 Year-to-Date Temperatures versus Previous Years," State of the Climate Reports, National Oceanic and Atmospheric Administration, http://www.ncdc.noaa.gov/sotc/global/2016/9/supplemental/page-2 (Accessed January 2017).

29. Spencer Weart, *The Discovery of Global Warming*, 2d ed., Harvard University Press, 2008, Timeline (Milestones), https://www.aip.org/history/climate/timeline.htm (accessed January 2017).

30. "Turn Down the Heat: Why a 4°C Warmer World Must Be Avoid," The World Bank, Washington D.C., November 2012.

31. B. Walsh, "The Surprisingly Large Energy Footprint of the Digital Economy," *Time Magazine*, August 2013, New York: http://science.time.com/2013/08/14/power-drain-the-digital-cloud-is-using-more-energy-than-you-think/

32. T. Bawden, "Global Warming: Data Centres to Consume Three Times as Much Energy in Next Decade, Experts Warn," *Independent*, London, January 2016.

33. High Performance Cleanrooms, PG&E Corporation, January 2011, http://www.pge.com/includes/docs/pdfs/mybusiness/energysavingsrebates/incentivesbyindustry/Cleanrooms_BestPractices.pdf.

34. Samsung Austin Semiconductor Fab A2 (sq ft), http://pagethink.com/v/project-detail/Samsung-Austin-Semiconductor-Fab-A2/71/.

35. "Power Consumption of Semiconductor Fabs in Taiwan," Elsevier, 2003, http://www.sciencedirect.com/science/article/pii/S0360544203000082.

36. S. Chen, A. Gautam, and F. Weig, "Bringing Energy Efficiency to the Fab," McKinsey, 2013, www.mckinsey.com.

37. Chemical Engineers in Semiconductor Manufacturing, American Institute of Chemical Engineers and Chemical Heritage Foundation, http://www.chemicalengineering.org/docs/cheme-semiconductor-manufacturing.pdf.

38. H. Geng, *The Internet of Things and Data Analytics Handbook*, John Wiley and Sons, Hoboken, 2016.

39. H. Geng, *Manufacturing Engineering Handbook*, 2d ed., McGraw-Hill Education, New York, 2016.

1.10 FURTHER READING

Castellano, R., "Cluster Tools in IC Manufacturing Has Changed the Dynamics of the Industry," *Vacuum Technology & Coating*, Mechanicsburg, Pennsylvania, March 2016.

Chip Fabrication Process, Tokyo Electron, http://www.tel.com/product/spe/making/ (accessed June 15, 2016).

Courtland, R., "The Origins of Intel's New Transistor, and Its Future," IEEE, New York, May 2011, http://spectrum.ieee.org/semiconductors/design/the-origins-of-intels-new-transistor-and-its-future (accessed June 15, 2016).

Dutta, S., T. Geiger, and B. Lavin, "The Global Information Technology Report 2015, World Economic Forum," Geneva, http://www3.weforum.org/docs/WEF_Global_IT_Report_2015.pdf (accessed June 15, 2016).

Global Society for Contamination Control (GSFCC), http://gsfcc.org/Main_Page (accessed June 15, 2016).

Hemker, D., "IoT's Divergent Needs Will Drive Different Types of Technologies," Solid State Technology, Hightech.tw (Chinese), http://hightech.tw/index.php/2012-06-06-14-12-38?start=5 (accessed June 15, 2016).

"How a Chip Is Made," Texas Instrument, http://www.ti.com/corp/docs/manufacturing/howchipmade.shtml (accessed June 15, 2016).

"How We Do It," Applied Materials, http://www.appliedmaterials.com/company/about/how-we-do-it (accessed June 15, 2016).

Hu, Chenming, Chapter 1, http://www.eecs.berkeley.edu/~hu/Chenming-Hu_ch1.pdf (accessed June 15, 2016).

Hu, Chenming, *Modern Semiconductor Devices for Integrated Circuits*, Prentice Hall, Upper Saddle River, March 2009.

Huang, X., et al., 25 nm FinFET: PMOS, DARPA AME Program, http://eecs.wsu.edu/~osman/EE597/FINFET/finfet3.pdf (accessed June 15, 2016).

IC Insights: http://www.icinsights.com/ (accessed June 15, 2016).

Kawski, J., What Is Ion Implantation? Applied Materials, Santa Clara, http://www.appliedmaterials.com/what-ion-implantation (accessed March 20, 2016).

King Liu, T., "Beyond Transistor Scaling: New Devices for Ultra-Low-Energy Information Processing," 2009, https://www.youtube.com/watch?v=U6IJoMwVbSU (accessed June 15, 2016).

King Liu, T., Sustaining the Silicon Revolution: From 3D Transistors to 3D Integration, 2015, https://www.youtube.com/watch?v=HOZqwpZALjs (accessed June 15, 2016).

Liu, S., and Y. Liu, *Modeling and Simulation for Microelectronic Packaging Assembly: Manufacturing, Reliability and Testing*, Wiley, Singapore, 2011.

Mack, C., "Fifty Years of Moore's Law," *IEEE Transaction on Semiconductor Manufacturing*, pp. 202-207, 24 (2), May 2011.

May, G., and C. Spanos, *Fundamentals of Semiconductor Manufacturing and Process Control*, John Wiley and Sons, Hoboken, New Jersey, May 22, 2006.

Microelectronic Processing and Fabrication, http://www.aplusphysics.com/courses/honors/microe/processing.html (accessed June 15, 2016).

Nishi, Y., and R. Doering, *Handbook of Semiconductor Manufacturing Technology*, 2d ed., CRC Press, Boca Raton, 2006.

Polya, G., *How to Solve It: A New Aspect of Mathematics Method*, 2d ed., Princeton University Press, Princeton, New Jersey, 1973.

Silicon Device Manufacturing, OSHA, https://www.osha.gov/SLTC/semiconductors/si_index.html (accessed June 15, 2016).

Weiss, Bettina, 50 Years of Moore's Law: A Lesson in (R)Evolution, http://www.semi.org/en/node/55471#at_pco=cfd-1.0&at_ab=-&at_pos=1&at_tot=5&at_si=553cdf4a5d3e3eb0 (accessed June 15, 2016).

Woo, S., "Dutch Company Aims to Make Chips Do More," *The Wall Street Journal*, New York, October 3, 2016.

Wu, B., A. Kumar, and S. Ramaswami, *3D IC Stacking Technology*, McGraw Hill, New York, 2011.

Xiao, H., *Introduction to Semiconductor Manufacturing Technology*, Prentice Hall, New Jersey, 2000.

Society and Education Consortium

European Semiconductor Industry Association, http://www.eusemiconductors.eu//esia/home (accessed June 15, 2016).

FABTECH, http://www.fabtech.org/ (accessed June 15, 2016).

Global Semiconductor Alliance, http://www.gsaglobal.org/ (accessed June 15, 2016).

Industrial Technology Research Institute (ITRI), https://www.itri.org.tw/eng/index.aspx (accessed June 15, 2016).

International Microelectronics Assembly and Packaging Society (iMAPS), http://www.imaps.org/ (accessed June 15, 2016).

International Solar Energy Society, www.ises.org (accessed June 15, 2016).

International Technology Roadmap for Semiconductor, http://www.itrs2.net/ (accessed June 15, 2016).

Nano-Bio Manufacturing Consortium, www.nbmc.org (accessed June 15, 2016).

SEMATECH, www.sematech.com (accessed June 15, 2016).

SEMI International Standards, http://www.semi.org/en/ (accessed June 15, 2016).

Semiconductor Industry Association, www.semiconductors.org (accessed June 15, 2016).

The Facilities 450 mm Consortium (F450C), http://www.f450c.org/ (accessed June 15, 2016).

Technical Magazines

3D InCites, http://www.3dincites.com/ (accessed June 15, 2016).

Acronyms Used at SEMATECH and Semiconductor Industry, http://public.sematech.org/Pages/acronyms.aspx (accessed June 15, 2016).

Adhesives & Sealants Industry, http://www.adhesivesmag.com/ (accessed June 15, 2016).

Air Liquide Glossary, https://industry.airliquide.us/glossary (accessed June 15, 2016).

Applied Materials Glossary, http://www.appliedmaterials.com/glossary (accessed June 15, 2016).

Association of Vacuum Equipment Manufacturers, http://www.avem.org/ (accessed June 15, 2016).

AZONetwork, www.azonetwork.com (accessed June 15, 2016).

Bodo's Power Systems, http://www.bodospower.com/ (accessed June 15, 2016).

Ceramic Industry's, www.ceramicindustry.com (accessed June 15, 2016).

Chip Design, www.chipdesignmag.com (accessed June 15, 2016).

Chip Scale Review, www.ChipScaleReview.com (accessed June 15, 2016).

Connecting the Compound Semiconductor Industry, www.compoundsemiconductor.net (accessed June 15, 2016).

Controlled Environments, www.cemag.us (accessed June 15, 2016).

EE-Evaluation Engineering, http://www.evaluationengineering.com/ (accessed June 15, 2016).

Flow Control, www.flowcontrolnetwork.com (accessed June 15, 2016).

Gases & Instrumentation, www.gasesmag.com/ (accessed June 15, 2016).

Global LEDs/OLEDs, http://globalledoled.com/ (accessed June 15, 2016).

Global SMT & Packaging, www.globalsmt.net (accessed June 15, 2016).

IEEE Spectrum, www.spectrum.ieee.org (accessed June 15, 2016).

LEDs Magazine, http://www.ledsmagazine.com/index.html (accessed June 15, 2016).

Making Silicon Chip, http://www.intel.com/content/www/us/en/history/museum-making-silicon.html (accessed January, 29, 2017)

NASA Tech Briefs, http://www.techbriefs.com/ (accessed June 15, 2016).

PC Encyclopedia, http://www.pcmag.com/encyclopedia/index/a (accessed June 15, 2016).

Photonics Spectra, www.Photonics.com (accessed June 15, 2016).

Physics Today, www.physicstoday.org (accessed June 15, 2016).

Sematech Publications, http://public.sematech.org/Pages/Publications.aspx (accessed June 15, 2016).

Semiconductor Manufacturing, IEEE Transactions, http://ieeexplore.ieee.org/xpl/RecentIssue.jsp?punumber=66 (accessed June 15, 2016).

Semiconductor Packaging News, http://www.semiconductorpackagingnews.com/ (accessed June 15, 2016).

Silicon Semiconductor, http://www.siliconsemiconductor.net/ (accessed June 15, 2016).

Solid State Technology, http://electroiq.com/ (accessed June 15, 2016).

U.S. Tech, http://www.us-tech.com/RelId/672548/ISvars/default/HOME_CURRENT_ISSUE.htm (accessed June 15, 2016).

Vacuum Technology & Coating, http://www.vtcmag.com/ (accessed June 15, 2016).

Chapter 2

Nanotechnology and Nanomanufacturing: From Silicon to New Carbon-Based Materials and Beyond

BY

MICHAEL A. HUFF *MNX, Corporation for National Research Initiatives, Reston, Virginia*

2.1 INTRODUCTION

Nanotechnology is a highly multidisciplinary field that involves the control of matter at the nanometer dimensional size scale. The ability to control atoms and molecules to form materials and products is an extremely alluring and fascinating concept because it promises the ability to make much better products and services with enormous economic potential, improved standards of living, and healthier citizens. The scientific community has recognized the importance of nanotechnology and has advocated for governments around the world to invest billions of dollars to advance and develop this technology. One area of technology with nanometer-sized dimensions already having an enormous impact on society is integrated circuits (ICs). The important dimensions of the transistor devices have steadily shrunk over the past several decades and are now approaching sizes below 10 nm, allowing many more devices to be placed on each die to enable more processing capability, storage, and faster speeds. ICs that began as die in handheld calculators expanded into personal computers, handheld wireless communication, and computational capabilities, and will continue to expand into the Internet of Things (IoT). We are at the early stages in the development of nanotechnology, and although it is very difficult to predict the implications of any technology, many experts believe that nanotechnology will be the most important technology in the history of the world in terms of its impact on the world's population and the economic growth it will provide.

2.2 WHAT IS NANOTECHNOLOGY?

Nanotechnology has been defined by the National Nanotechnology Initiative (NNI) in very general terms as the "manipulation of matter on the atomic or molecular scale where at least one dimension is sized in the range from 1 to 100 nanometers."[1] This is a very broad definition that says nothing about the things that can be implemented with this technology or their applications, but this is intentional since the purpose of this definition is to provide guidance based on dimensions. Another and perhaps more easily identifiable definition of nanotechnology is *the ability to form things from the very inexpensive and abundantly available basic building blocks of matter (i.e., atoms and molecules) that have useful purposes and applications.*

To give some perspective of the size scale of nanotechnology, a water molecule has a diameter of about 0.1 nm, a glucose molecule has a diameter of about 1 nm, a typical virus has a diameter of about 100 nm, a human hair has a diameter of about 10^5 nm, and a baseball has a diameter of about 10^8 nm. Therefore, the lower dimensional scale of nanotechnology of about 1 nm is about the same dimensional scale of basic molecules found in nature.

2.3 WHY NANOTECHNOLOGY IS IMPORTANT

Nanotechnology is important due to the enormous impact it is expected to have on society. The ability to engineer materials, and thereby products made from these materials, by accurately manipulating the position of each atom or molecule in a substance translates into extraordinary levels of deterministic control over the structure of the material that heretofore have not been available to manufacturers. This ability will allow materials to have properties that have been previously unavailable such as much higher strength, more robustness, lower weight, and exceptional material behaviors. Some believe that nanotechnology will result in the creation of the *next industrial revolution.*[2] This is partly because the breadth of nanotechnology is so enormous, but perhaps more importantly because nanotechnology will radically disrupt many or all of the existing industries, products, and processes.

2.4 HISTORY OF NANOTECHNOLOGY

The history of nanotechnology is a complicated story. Some cite a talk given by the renowned Nobel Prize winning physicist Richard Feynman in 1959 entitled, "There's Plenty of Room at the Bottom," as the first publically persuasive argument of the potential possibilities using the manipulation of matter at the atomic scale.[3] In this talk, Professor Feynman outlined a number of very interesting and far-reaching possibilities, including highly dense ICs for computers, microscopes that could observe matters at the atomic scale, and micro-miniaturized pill-formed robots that could act as doctors on the patient. Feynman also suggested the possibility of controlling the arrangement of atoms and the ability to do chemical synthesis by atomic and molecular mechanical manipulation. Looking back at this talk, we can conclude that these were incredibly insightful projections given the recent advancements of computer chips, scanning-electron microscopes, and smart pills.

The first known use of the term "nanotechnology" is now recognized as made by Professor Norio Taniguchi in 1974 who used the word to describe semiconductor fabrication processes such as physical-vapor thin-film deposition processes, whose thicknesses were measured in the tens of nanometers.[4] However, Professor Taniguchi's use of the word "nanotechnology" was virtually unknown until recently.

Later, Eric Drexler authored a book entitled, *Engines of Creation: The Coming Era of Nanotechnology*, which was published in 1986.[5] According to the author, this book was strongly motivated by Feynman's 1959 talk. Drexler's book extensively used the word *nanotechnology* and proposed a number of novel demonstrations, including: writing the entire Library of Congress on a sugar cube and a micro-miniaturized machine technology that could be used to build objects at the atomic level. Drexler also speculated on a terrifying scenario called *gray goo* wherein nanotechnology enables uncontainable self-reproducing machines that cannot be controlled. After its publication some notable scientists criticized this work as being more science fiction than science fact, but Drexler's book did get considerable attention from the scientific community as well as the general public, and consequently started people thinking more about the possibilities and consequences of nanotechnology.[6,7]

More important than talks, articles, or books to the progress of nanotechnology were the scientific advancements in patterning and imaging technology in the 1960s and 1970s that made it possible to fabricate and image features at the nanometer dimensional scales. Nevertheless, the initial scientific progress has not been fast or continuous. For example, Ernst Ruska and Max Knoll invented the electron microscope in 1931, but for decades this technology progressed very slowly and work was mostly limited to universities.[8,9] A paper published in 1961 entitled, "Newer Developments in Microminiaturization" by Mollenstedt and Speidel from the University Tubingen, reported patterning lines having widths of less than 100 nm.[10] The patterning was accomplished by patterning lines using electron-beam lithography on nitrocellulose films coated with carbon.

It is also interesting to note that during this time period several researchers were proposing using e-beam lithographic techniques to fabricate microelectronics. Two examples were a 1960 paper by Selvin and MacDonald from Sylvania Electronic Systems entitled, "The Future of Electron Beam Techniques in Microelectronic Circuitry,"[11] and a 1961 paper entitled "Electron Beams in Micro-Electronics," authored by O. Wells of Westinghouse Research.[12]

The 1980s saw some of the most significant technological advancements in nanotechnology. Specifically, the scanning tunneling microscope (STM) was invented in 1981 and enabled the imaging of individual atoms. The inventors were Binnig and Rohrer from IBM Zurich Research Labs, and they received the Nobel Prize in physics in 1986.[13] Binnig invented the atomic force microscope (AFM), and Binnig and coworkers. built the first microscope and demonstrated its operation in 1986.[14,15]

In 1985, the discovery of fullerenes was reported by Kroto and colleagues, and subsequently they were awarded the Nobel Prize in chemistry in 1996.[16] In 1991, nanotubes were reported by Sumio Iijima from NEC in the carbon soot from an arc discharge system.[17] A year later in 1992, Harrington and Maganas from Maganas Industries reported that they had synthesized carbon nanotubes using a chemical vapor deposition (CVD) process.[18]

In November 1990, the scanning tunneling microscope was used by researchers Donald Eigler and Erhard Schweizer at the IBM Almaden Research Center to manipulate individual atoms of xenon on the surface of a nickel substrate to spell out "IBM." This was the first tangible example of manipulation of individual atoms that was published.[19]

Surprisingly, it was later discovered that Radushkevich and Lukyanovich published a paper in 1952 in *Zurn. Fisic. Chim.*, showing carbon nanotubes approximately 50 nm in diameter.[20] Also in 1960, Bollmann and Spreadborough published a paper in *Letters to Nature* entitled, "Action of Graphite as a Lubricant," showing an SEM image of a multi-walled carbon nanotube.[21] And, in 1976, Oberlin et al. published a paper in the *Journal of Crystal Growth* entitled, "Filamentous Growth of Carbon through Benzene Decomposition," that reported using chemical vapor deposition to grow nanometer-sized carbon fibers.[22] Therefore, it is clear that the discovery of nanotubes predates the accepted date of 1985.

2.5 FUNDAMENTAL METHODS OF FABRICATION AT THE NANOSCALE

There are two general approaches involved in the implementation of nanotechnology. The first is the "top-down" approach and entails the use of fabrication processes and techniques to shape and form nano-dimensional scale elements from larger-sized sections of material. The top-down approach involves production techniques that are largely equivalent to those used in the IC and microsystems industries and include deposition of thin films (perhaps having thicknesses in the nanometer range), lithography to make desired patterns in a photosensitive polymer layer, and etching.[23,24]

An example of a process flow for the implementation of a nano-dimensional mechanical resonator device using these techniques is illustrated in Fig. 2.1. The fabrication process begins with a substrate that is usually made of single-crystal silicon (Fig. 2.1*a*). Semiconductor substrates have developed to unparalleled levels of flatness and smoothness and these attributes are important when attempting to make nano-dimensional-sized features. Subsequently, a thin-film material layer is deposited (Fig. 2.1*b*). This layer is termed a "sacrificial layer" since it will not be employed in the device structure, but instead it will act as a standoff between the resonator and the substrate and will be removed at the end of the fabrication process. A typical sacrificial layer would be a thin-film layer of silicon dioxide (SiO_2). Next, another thin-film layer is deposited (Fig. 2.1*c*). This layer is termed a "structural layer" because it will be the material from which the resonator is made. A typical structural layer would be polycrystalline silicon (also commonly referred to as "polysilicon").

There are a number of deposition processes that can be used to deposit the thin-film layers including CVD, physical vapor deposition that includes sputtering and evaporation, atomic layer deposition (ALD), spin coating, molecular beam epitaxy, and others. Note that some of these technologies are defined as "bottom-up" nanotechnologies.

The thin-film layer is then patterned using the methods of lithography (Fig. 2.1*d* and *e*). There are a number of different technologies that can be used in lithography including optical photolithography, electron beam (e-beam) lithography, and imprint lithography. In the present example, we assume that we are using e-beam lithography. Prior to performing e-beam lithography, a photo-sensitive polymer layer, called *photoresist*, is deposited onto the substrate surface using spin coating (Fig. 2.1*d*). The thin-film structural material layer is etched using a dry plasma reactive ion etch process. The photoresist is then removed by a plasma ashing process (Fig. 2.1*e*). Finally, the sacrificial material layer is removed using a gas-phase etchant such as vapor hydrofluoric acid. This releases the structural layer so that it can mechanically move and completes the fabrication process. The cross section and plan view of the nanoscale resonator is shown in Fig. 2.1*f* and *g*, and the width of

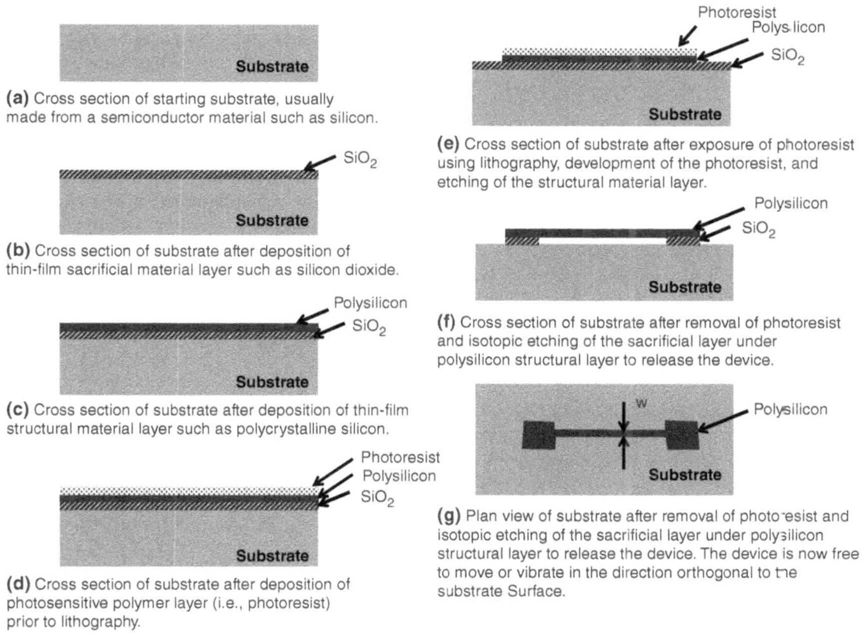

Figure 2.1 Process flow for the fabrication of a nano-mechanical resonator to illustrate a top-down approach to nanotechnology implementation. (*Copyright MEMS & Nanotechnology Exchange (MNX) at the Corporation for National Research Initiatives (CNRI); used with permission*).

the resonator is labeled as "w." It should be noted that the fabrication process illustrated in Fig. 2.1 is only one of the methods to make a nanoscale device using a top-down approach.

Perhaps the most well-known, and certainly to date, the most commercially successful example of a top-down fabrication of a nanotechnology device is the FinFET transistor technology that is used in the manufacturing of state-of-the-art ICs. A FinFET, also termed a fin field effect transistor (FET), is a nonplanar metal-oxide semiconductor (MOS) transistor technology that is extensively used for state-of-the-art microprocessors. The critical dimension of a transistor is the gate length, and the current generation of FinFET technology on the market has gate lengths of 14 nm.[25]

The notable attribute of a FinFET is the fin of single crystal silicon semiconductor material that has a relatively high aspect ratio that protrudes from the substrate surface as shown in Fig. 2.2. Figure 2.2*b* is a three-dimensional illustration of the FinFET device topology. As can be seen, like a conventional MOS transistor there is a source, drain, and gate. However, unlike a conventional MOS device, the source and drain are connected by a thin, but high aspect ratio (height is larger than the width) fin of silicon. The fin acts as the channel for the flow of charge carriers (either electrons or holes) from the source to the drain. The flow of charge carriers in the fin is modulated by the gate voltage potential. The gate overlaps the fin, but has a very thin dielectric to insulate the gate electrode from the fin as shown in Fig. 2.2*a*. The gate material is an electrically conductive material layer. Importantly, the gate overlaps the fin on three sides, the top and the two sides. Consequently, sometimes this structure is called a *tri-gate* structure. This tri-gate configuration allows the gate voltage to have improved control capability overcharge carriers in the fin. The gate length is approximately the dimension given by "w" in Fig. 2.2 and would be approximately 14 nm for the latest generation of the FinFET technology.[25]

Additionally, the FinFET structure allows multiple fins to be used in each transistor (to further improve the "on" currents), and higher packing densities. Intel claimed with the introduction of the Ivy Bridge 22 nanometer microprocessor that first used the FinFET technology,

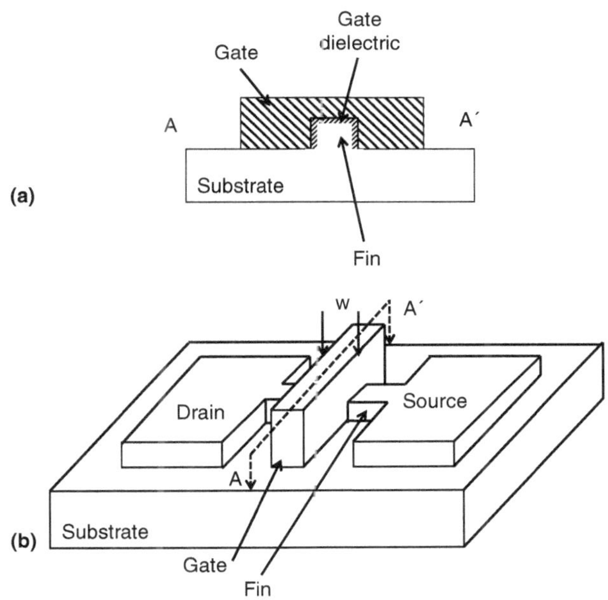

Figure 2.2 An illustration of the FinFET transistor. Shown in Fig. 2.3*b* is a 3D representation of an individual FinFET transistor with a single fin. The source, drain, and fin are all made from single-crystal silicon. The fin is a high aspect ratio channel connecting the source and drain, where the charge carriers can flow. As shown in Fig. 2.3*a*, the fin is coated in the thin dielectric layer and then with the gate material layer, which is usually an electrically conductive material. Because the voltage on the gate creates an electrical field from both sides and the top, this configuration is also called the *tri-gate* transistor. (*Copyright MEMS & Nanotechnology Exchange (MNX) at the Corporation for National Research Initiatives (CNRI); used with permission.*)

that a major redesign of the transistor structure was needed in order to continue Moore's law to the 22-nm node. The FinFET structure allowed them to extend Moore's law to 22 nm, then to the present 14-nm node, and it is expected to allow them to continue this path for several more years.[25]

It is important to note that these are only two examples of top-down approaches to the implementation of nanotechnology. Specifically, there are other tools and approaches that can be used and more technologies are being developed each year. Another important point is that top-down approaches typically do not have atomic-level of control.

The second approach to implementing nanotechnology is the bottom-up approach, which for all practical purposes means the engineered assembly of atoms or molecules to form nanoscale or larger objects. An example of the bottom-up approach would be the design of molecules having a specified shape and molecular recognition whereby individual molecules automatically *self assemble* into a particular system configuration.

One example of the bottom-up approach to nanotechnology is ALD. ALD is a thin film deposition technique that uses sequential and self-limiting gas phase chemical processes. The reactant process gases are called precursors and each reacts with the existing surface onto which the material layer is being depositing in a self-limiting fashion. Typically, two precursor gases are used in ALD and neither is introduced into the deposition chamber at the same time. One precursor is introduced and reacts in a self-limited fashion with the substrate and once completed, the next phase is started wherein the second precursor is introduced to react with the substrate in a self-limited fashion, and so on.[26]

The ALD process is illustrated in Fig. 2.3. The substrate onto which the ALD layer is to be deposited is first placed into a reactor chamber (Fig. 2.3*a*). The first precursor is introduced into the reactor chamber and reacts with the substrate surface to form a monolayer of a chemical species on the surface (Fig. 2.3*b*). After the first precursor monolayer deposition is completed, any excess precursor gas is removed from the chamber (Fig. 2.3*c*). Next, as shown in Fig. 2.3*d*, the second precursor is introduced into the process chamber and starts to deposit a monolayer on the surface of the first precursor monolayer. Once the second precursor has completed the deposition of a monolayer, the excess process gas is removed (Fig. 2.3*e*). This process can be continued any number of times to create a thin film layer of desired thickness.

The advantages of ALD over conventional CVD are that ALD layers are extremely conformal, highly uniform, and have thicknesses that can be well controlled. A number of material types can be deposited using ALD, including SiO_2, Al_2O_3, ZrO_2, HfO_2, TiN, TaN, Pt, Ti, and others. A disadvantage of ALD is the low growth rate. ALD is now one of the principal deposition technologies now used in IC manufacturing to deposit the gate dielectric material layer in modern FinFET transistors. Typically, these layers are a high-k dielectric such as HfO_2.

Perhaps the most notable examples of the bottom-up approach to nanotechnology are carbon nanotubes and other carbon-based molecules of the fullerene family. Carbon nanotubes are one of many allotropes of carbon that can be single-walled nanotubes or multi-walled nanotubes, wherein there are concentric tubes of graphene (as in the so-called Russian Doll model) or a sheet of graphene rolled around itself (as in the so-called Parchment model). Figure 2.4 illustrates a number of

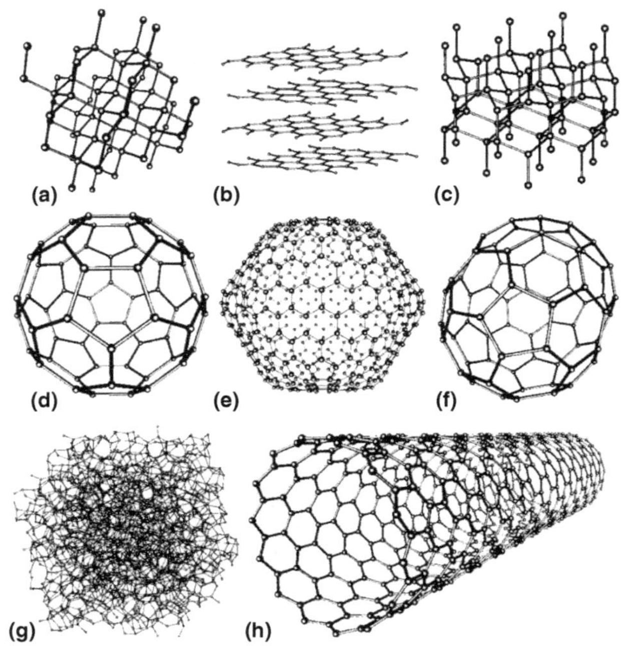

Figure 2.4 An illustration of several different types of allotropes of carbon molecules including (*a*) diamond, (*b*) graphite (single atomic layer of molecular carbon), (*c*) lonsdaleite, (*d*) C60 or Buckminsterfullerene, also known as "buckyball," (*e*) C540 fullerene, (*f*) C70 fullerene, (*g*) amorphous carbon, and (*h*) single-walled carbon nanotube. (*Released under the GNU Free Documentation License.*)

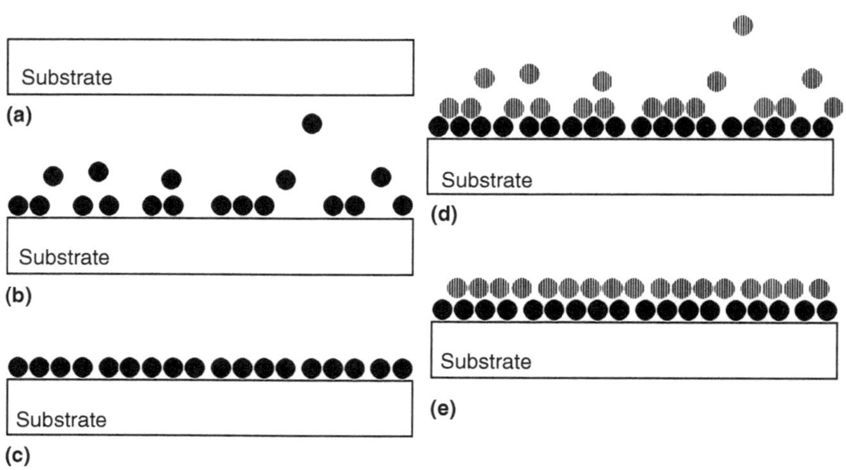

Figure 2.3 Illustration of the process in atomic layer deposition (ALD). (*Copyright MEMS & Nanotechnology Exchange (MNX) at the Corporation for National Research Initiatives (CNRI); used with permission.*)

different allotropes of carbon, including diamond; graphite; fullerenes; C60 carbon fullerene molecule, which is also called a *buckyball*; and single-walled carbon nanotube.

Single-walled carbon nanotubes have a nominal diameter of about 1 nm and have been made with truly exceptional length-to-diameter ratios of up to 132,000,000:1.[27] These materials also exhibit outstanding material properties. For example, carbon nanotubes have the highest tensile strength (e.g., between 13 and 53 GPa) and elastic modulus (e.g., between 1 and 5 TPa) of any material reported.[28–31] Moreover, carbon nanotubes have a low value of density, thereby resulting in a specific strength of 48,000 $kN{\cdot}m{\cdot}kg^{-1}$, the highest of any known material.[32] The electrical properties of carbon nanotubes depend on their structure and can be metallic or semiconducting with a theoretical maximum current density of over 10^9 $Å/cm^2$, which is about 1000 times the current density of copper.[33] It is important to note that carbon nanotubes are only conductive along the axis of the nanotube.

The thermal properties of carbon nanotubes also depend on the structure and direction of heat transport. Single-walled carbon nanotubes have been shown to have a thermal conductivity along the axis of the nanotube of about 3500 $W{\cdot}m^{-1}{\cdot}K^{-1}$ and have a thermal conductivity in the radial direction that is much lower, about 1.5 $W{\cdot}m^{-1}{\cdot}K^{-1}$.[34,35]

There are several techniques that have been demonstrated for the production of carbon nanotubes including arc discharge,[32] laser ablation,[36] and CVD,[37] with CVD demonstrating the most promise for large scale production. Using CVD growth, typically a substrate surface will be prepared by depositing a layer of a suitable catalyst material or having defined regions or sites of a catalyst on desired locations across the substrate surface. The catalysts most often used for carbon nanotubes are nickel, cobalt, iron, or a combination of these materials.[38–40] Importantly, the growth of the carbon nanotubes will be limited to only those regions on the substrate surface where the catalysts are present. Moreover, the size and type of nanotube grown is dependent on the diameter of the catalyst site on the surface of the substrate with smaller-diameter catalysts resulting in single-walled nanotubes. Therefore, the use of catalysts allows the size, types, and locations of nanotubes to be reasonably controlled by using fabrication techniques to deposit and pattern the catalyst sites to the desired diameters and locations on the substrate surface.

The CVD growth process entails heating the substrate to about 700°C and then introducing process gases into the deposition chamber, such as ammonia, nitrogen, or hydrogen, as well as a carbon-containing gas, such as acetylene, ethylene, or methane.[40] The carbon nanotubes grow from the sites of the metal catalyst on the substrate surface. If the growth process is conducted in the presence of an electrical field within the process chamber, the growth of the nanotubes will follow the direction of the electrical field and thereby make it possible to grow vertically oriented carbon nanotubes from the surface catalyst sites.[41] For example, if the process used for deposition is a plasma-enhanced CVD, then vertically oriented nanotubes can be grown. It should also be noted that under very special conditions, vertically oriented carbon nanotubes have been demonstrated without using an electric field. However, this is only possible with an extremely dense array of surface catalyst sites. Water-assisted CVD growth in addition to catalysts has demonstrated exceptionally high growth rates when compared to the use of catalysts along.[42]

Some bottom-up approaches have been termed as "biomimetic" because these methods seek to emulate biological systems used by nature. One notable example of this type of nanotechnology is "DNA nanotechnology" or "nucleic acid nanotechnology," which involves the design and implementation of nucleic acid structures.[43,44]

As mentioned earlier, some of the new carbon-based materials are expected to have a huge impact on the future development of ICs. Some of these new materials, such as graphene, are being explored to replace traditional semiconductor materials such as silicon. Graphene is one of the allotropes of carbon and has form of a two-dimensional, atomic layer thick sheet wherein the carbon atoms are arranged in a honeycomb lattice. Graphene has many extraordinary properties and behaves either like a metal or a semiconductor depending on the lattice structure. Graphene is about 100 times stronger than steel (assuming a layer of steel having the same thickness as graphene of about 3.4 Å,[45] it conducts heat and electricity extremely well, and it is nearly transparent).[46] There are two known orientations of graphene. The first is the zigzag orientation, which always demonstrates metallic behavior, and the second is the armchair orientation, which can demonstrate semiconducting or metallic behavior depending on the chirality.

Graphene exhibits remarkable electron mobility at room temperature, with reported values in excess of 15,000 $cm^2{\cdot}V^{-1}{\cdot}s^{-1}$ with[47] the potential theoretical limit of 200,000 $cm^2{\cdot}V^{-1}{\cdot}s^{-1}$,[48,49] which is about 10×10^6 times greater than copper.[50] The electron and hole mobilities are expected to be nearly the same value.[51]

Already graphene has been demonstrated by researchers to implement transistor devices. Recently, IBM researchers reported the operation of graphene FET with the highest frequencies reported to date.[52] The IBM team reported that the operation frequency increases with diminishing device dimensions and that they had measured a graphene transistor cutoff frequency of 26 GHz with a gate length of 150 nm, the highest frequency obtained for graphene so far. Furthermore, by improving the gate dielectric materials, it is expected that the performance of these graphene transistors could be further improved, possibly obtaining terahertz graphene transistors with a gate length of 50 nm.

As this graphene example demonstrates, carbon-based nanotechnology materials are enabling higher performance ICs that are possible with conventional semiconductors. However, we are at the very early stages in the development of nanotechnology, and therefore it is not possible to accurately predict every possible implication of this new technology because new materials, fabrication methods, and devices are constantly being discovered that show great promise as well.

For example, researchers from the Swiss Federal Institute of Technology have created ICs using single-atom-thick molybdenite, a substance having properties similar to that of graphene. Molybdenite is a mineral composed of molybdenum disulfide, MoS_2. Molybdenite in multiple layer form behaves as an indirect band gap semiconductor while a monolayer of this material has a direct band gap.[53,54] Mobilities of 200 $cm^2{\cdot}V^{-1}{\cdot}s^{-1}$ and room temperature current on/off ratios of 10^8 were reported. An image of the device is shown in Fig. 2.5.

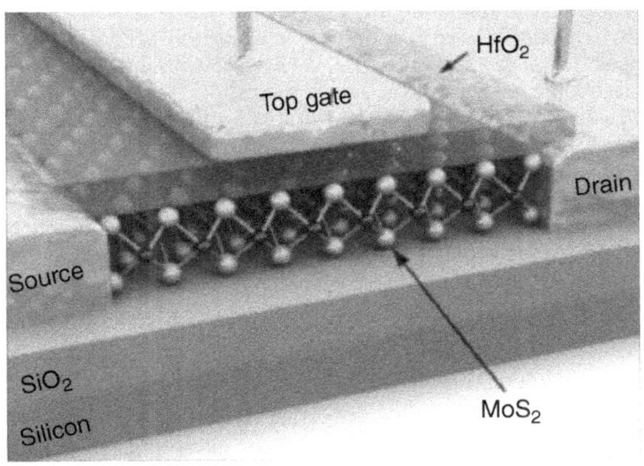

Figure 2.5 An illustration of an MoS_2 transistor reported by the Swiss Federal Institute of Technology.[63]

It is important to note that bottom-up approaches whereby larger structures or devices are made using these techniques are currently mostly limited to research endeavors and not actual products. Nevertheless, bottom-up approaches are used in the production of fullerenes and various nanoparticles (i.e., silver and gold) that are added to other materials to form compound mixtures that have enhanced material properties.

2.6 NANOTECHNOLOGY METROLOGY

Nanotechnology metrology is another important set of technologies that are used mostly for metrology purposes including the scanning tunneling microscope (STM), the atomic force microscope (AFM), the scanning electron microscope (SEM) and the transmission electron microscope (TEM), x-ray diffraction, high-resolution transmission electron microscopy, and field emission scanning electron microscopy. We discuss the STM and AFM in more detail further. The other technologies have been in existence for a far longer time and the interested reader can explore them in detail in various references.[55–63]

A scanning tunneling microscope is an extremely precise instrument that can be used to scan and image the surfaces of materials at the atomic scale.[64] This instrument consists of a very small radius needle-shaped tip (usually one atom at the apex) that is composed of a conductive material such as a metal (e.g., tungsten, gold, platinum-iridium, etc.) that is typically made using wet etching fabrication techniques (Fig. 2.6). Using piezoelectric actuation (not shown in figure), the tip is positioned in very close proximity to a material surface that is to be scanned. If a biasing voltage (labeled "V_{Bias}" in Fig. 2.6) is placed across the gap (labeled "z") between the tip and material surface and the tip is sufficiently close to the material surface (e.g., within about 5 to 10 Å), a quantum-mechanical tunneling current (labeled $I_{tunneling}$) passes across the gap between the tip and the material surface. A closed-loop feedback circuit can be used to measure the tunneling current through the tip and thereby the distance between the tip and surface as it scans across the material surface. This spatially varying tunneling tip current is used to create an image of the material surface at the atomic dimensional size scale. Alternatively, the tip current is maintained at a constant value and the actuator voltage to maintain a constant gap distance is measured. The typical tip tunneling current is on the order of a few hundred picoamps at a bias voltage of about 1 V. This instrument is extremely sensitive based on the fact that the tunneling current is exponentially dependent on the gap between the tip and material surface. Specifically, the tunneling current increases by a factor of 10 for each Angstrom the closer the tip gets to the surface. The resolution of the scanning tunneling microscope is approximately 0.1 nm in the lateral (i.e., x and y coordinates) directions and about 0.01 nm in the orthogonal (i.e., z coordinate) direction.[64]

While the scanning tunneling microscope can be used for imaging with atomic level resolution the surfaces of materials as described previously, it can also be used to perform spectroscopy as well.[64,65] That is, the dependence of the current on the voltage provides information about the electronic structure of the surface atoms.

The AFM differs from the STM in that it mostly uses a mechanical probe as opposed to an electrical probe to scan the surface of a material. The AFM uses a sharp tip at the end of a cantilever to scan across the material surface. The tip is typically made from silicon using MEMS fabrication techniques. Like the STM, the AFM tip is brought into close proximity to the material surface usually using piezoelectric actuation, although electrostatic actuation can be used as well. As the tip comes into immediate nearness to the material surface, the cantilever will deflect based on the stiffness of the cantilever, that is, Hooke's law.[66] The mechanical forces that can be measured using the AFM include the mechanical contact forces, van der Waals attraction forces, chemical bonding forces, Casimir forces, capillary forces, etc.[67] The deflection of the AFM tip is usually measured using a laser optical beam that is bounced off the top of the cantilever into a photodetector. However, the deflection can also be measured using capacitive sensors as well as other means. Using specialized probes, other quantities can be measured such as temperature and thermal conductivity.[68]

The AFM can be used in one of three modes: contact mode, tapping mode, and the noncontact mode.[69] The noncontact modes are usually preferred over the contact mode for most applications. The resolution in force of the AFM is usually on the order of a few piconewtons.[70]

2.7 NANOTECHNOLOGY MANUFACTURING

The approaches taken in nanotechnology manufacturing can also mostly be categorized as either top down or bottom up. The most notable example of products employing nanotechnology are ICs that have gate lengths of the complimentary metal-oxide semiconductor transistor devices measuring only a little over 10 nm and are composed of several billions of transistor elements on a single slab of silicon substrate.

ICs represent the largest, both in terms of volume and annual sales of top-down approach used in nanotechnology manufacturing. Including both logic and memory devices, total annual sales in 2016 expected to be about $370 billion.[71] This represents component sales and not systems sales that use ICs. Arguably the systems enabled by ICs, such as computers, cell phones, personal handhelds, etc., markets would not be possible without ICs, but even constraining the annual sales to only component sales represents an enormous market of huge economic and strategic importance.

The semiconductor industry has been reducing the size of transistors as well as increasing the number of transistors on each die steadily over several decades. This is commonly referred to as "Moore's law." The critical feature by which advanced semiconductor technologies are typically compared is the length of the gates in a complementary metal-oxide-semiconductor process. For example, the semiconductor industry reduced the lengths of gates steadily over the years, from 10 um in 1971, to 1 um in 1985, to 130 nm in 2001, 45 nm in 2008, 22 nm in 2012, and 14 nm in 2014. The next reduction in gate lengths is predicted to be 2017 with 10 nm gates.[72]

The manufacturing advancements that allow features to be made that measure only a few nanometers are quite remarkable achievements in their own right. The current methods of lithography are optical in nature whereby photons are used to expose the photosensitive polymer called photoresist. The current gate lengths of 14 nm are made using optical radiation having a wavelength of 193 nm. That is, the dimensions being printed are over 10 times smaller than the wavelength of the

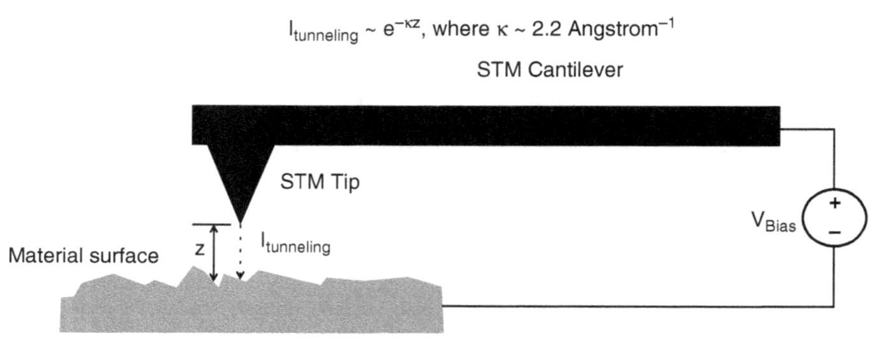

Figure 2.6 Illustration of a scanning tunneling microscope. (*Copyright MEMS & Nanotechnology Exchange (MNX) at the Corporation for National Research Initiatives (CNRI); used with permission.*)

radiation being used to make these images! The resolution of pattern that can be printed using existing photoresists and optical photolithography tools starts to distort at 45 nm. Therefore, new techniques to allow the printing of smaller features (e.g., 14 nm or less) have been developed including multiple exposures, self-aligned spacer layers, immersion lithography, etc.[73] However, it is expected that to continue to scale down the feature sizes able to be printed using lithography, reduced wavelengths such as extreme ultraviolet radiation will be required.

The bottom-up approaches in manufacturing of nanotechnology mostly include silver, gold, and titanium nanoparticles and fullerenes. Silver nanoparticles are made using wet chemistry involving the reduction of a silver salt such as silver nitrate using sodium borohydride and a colloidal stabilizer such as cellulose.[74,75] Silver nanoparticles are marketed under the claim that silver reduces external infections. However, there is no medical research demonstrating the efficacy of this material or treatment.[76,77]

Gold nanoparticles are another example of a bottom-up approach to nanotechnology manufacturing and are also used in a variety of applications. There are several methods for the production of gold nanoparticles suspended in liquids (usually water) that involve the reduction of chloroauric acid that is mixed with a reducing agent to produce gold atoms. As the number of gold atoms increase, eventually the solution becomes supersaturated with gold, which then precipitates as nanoscale clusters of gold particles. To prevent the gold from aggregating, a stabilizing agent may be added.[78,79] Gold nanoparticles are being explored for a number of medical applications including drug delivery mechanisms,[80] tumor detection,[81] gene therapy,[82] dose enhancement for radiotherapy,[83] and others.

2.8 APPLICATIONS AND MARKETS

The research and development community, the federal government, as well as the business community, predict nanotechnology to be the linchpin of high technology in the future. Collectively these technologies are expected to generate many thousands of high-paying jobs, enormous economic benefits, greatly improve our quality of life, and help us to remain secure from existing and new threats. Already nanotechnology has a large market of leading edge products including ICs (both logic and memory), health care products, coatings and surface treatments, and new materials.

The Project on Emerging Nanotechnologies (PEN) is a publically available service that tracks new products that employ nanotechnology in them.[84] As of late 2014 there are over 1800 different products listed at this service in categories ranging from appliances, automotive, electronics and computers, food and beverages, health and fitness, and home and garden. Most of the products' region of origin is the United States, with more products originating in the United States than in Europe, Asia, and rest of the world combined. It is also interesting that by far most of the products fall into the category of health and fitness products, with personal care products representing the largest segment of this product category. Additionally, the most used nanomaterial from all of these manufacturers in products is silver nanoparticles. Titanium is the second most used material type, exceeding that of carbon, which of course is the material that fullerenes are based on. Importantly, all of the products currently available are bulk material applications of nanomaterials and none of them involve the control of atoms or molecules in their production. It is also apparent that the PEN database does not count ICs in its numbers and therefore this is a notable limitation with their analysis.

The NSF recently commissioned an independent study to predict the size of the nanotechnology markets in the future. This commission concluded that the market for nanotechnology products will be in excess of $4 trillion by the year 2018.[85] Obviously this is an enormous economical potential, but is certainly possible when it is considered that every product and service could be remarkably improved using nanotechnology, not to mention that many new technologies and products that are not now possible will be realized and easily and affordably produced using nanotechnology.

Since top-down nanotechnology heavily leverages off the enormous capital investment of IC technology, its growth is expected to continue to be rapid for the foreseeable future. For example, several market research studies indicate that the sales of microminiaturized sensor devices are currently around $15 billion annually, which corresponds to approximately 13 percent annual growth rate over the past few years.[86] In addition to the sales of microminiaturized sensor components themselves, equally important is the fact that these devices enable additional $100 billion or more annual market in new or improved industrial and medical systems. The application areas considered in these studies have been typically restricted to pressure sensors, inertial sensors, fluid regulation/control, optical switching, analytical instruments, and mass data storage. Since microminiaturized sensor devices are a nascent technology, many new applications will emerge, expanding the markets beyond those which are currently identified.

2.9 IMPLICATIONS AND REGULATIONS

While there is much deserved enthusiasm and interest in the potential impact of nanotechnology, there are also concerns of the impact of this technology on society and the environment. This is based on the lack of understanding of the effects of nanoparticles on living organisms, which has not been a major consideration of the research community or of governmental funding agencies until relatively recently.

Nevertheless, recent findings have highlighted that there may be reason for concern. It was reported that nanoparticles or nanofibers that are inhaled by living organisms may lead to lung diseases, such as fibrosis.[87] In another study, mice that consumed nano-titanium oxide particles demonstrated damage to their DNA and chromosomes that could cause cancer and other long-term ill effects.[88] The results of a study of the effects of carbon nanotubes suggested that these could be as damaging as asbestos on living organisms.[89] Workers in paint factory that used nanoparticles in their manufacturing process were found to have developed serious lung diseases and subsequently these nanoparticles from the paint manufacturing process were found in the worker's lungs.[90] Moreover, there is very little understanding of the effects of nanotechnologies on the environment, or living organisms, over longer time spans.

Presently, there are very few regulations that are specifically directed at nanotechnologies. Instead, the regulation applicable to this technology is mostly covered by existing regulations and from existing overseeing bodies. That is, the regulations on nanotechnology in most countries do not differentiate between materials in their bulk forms from those same materials in nanoparticle form. Consequently, if the materials have been approved for commercial use in bulk form, nanoparticles of the same materials have no further testing or analysis of their health and safety effects and no testing or analysis of the environmental impact. Some are very critical of this approach since nanotechnology presently is essentially an unregulated area of commerce. There are increasing numbers of scientists advocating for governments to more proactively regulate nanotechnology.[91]

2.10 CONCLUSIONS

Nanotechnology is an extremely diverse, multidisciplined, and rapidly changing area of science and engineering that promises to revolutionize many products as well as enable new and better products. This chapter has provided a high-level review of nanotechnology, its somewhat complicated history, attempted to explain how things are made, given some insight into the possibilities, and provided reasons for concern of the unknown health, safety, and environment impact of this technology. Importantly, nanotechnology is enabling new developments in manufacturing to shrink ICs as well as providing new materials and capabilities. This is allowing the IC industry to continue along Moore's law thereby providing their customers with increased performance and lower costs. Nanotechnology is also promising new materials, some of which are carbon based, that may displace traditional semiconductors in the IC industry. The economic importance and extent of potential applications of nanotechnology, similar to other technologies that involve microminiaturization, promise to make it the hallmark technology of the future.

2.11 REFERENCES

1. http://nano.gov/nanotech-101/what/definition
2. Chapter 1 "From Conventional Technology to Carbon Nanotechnology: The Fourth Industrial Revolution and the Discoveries of C60, Carbon Nanotube and Nanodiamond" in *Carbon Nanotechnology*, edited by Liming Dia, Elsevier, Amsterdam, Netherlands. B.V. 2006.
3. R. P. Feynman, "There's Plenty of Room at the Bottom (data storage)," *Journal of Microelectromechanical Systems*, 1 (1): 60–66, March 1, 1992.
4. N. Taniguchi, "On the Basic Concept of 'Nano-Technology'," *Proceedings of the International Conference on Production Engineering:* Tokyo, Part II, Japan Society of Precision Engineering, 1974.
5. K. Eric Drexler, *Engines of Creation: The Coming Era of Nanotechnology*, Doubleday, New York, 1986.
6. R. E. Smalley, "Of Chemistry, Love and Nanobots," *Scientific American*, 285 (3): 76–77, September 2001.
7. R. Kurzweil, *The Singularity Is Near: When Humans Transcend Biology*. Penguin Books, New York, pp. 236–241, 2005.
8. H. G. Rudenberg and P. G. Rudenberg, "Chapter 6, Origin and Background of the Invention of the Electron Microscope: Commentary and Expanded Notes on Memoir of Reinhold Rüdenberg," *Advances in Imaging and Electron Physics,* Vol. 160. Elsevier, 2010.
9. T. E. Everhardt, "Submicron Technology—Educational Door to the Future," *Proceedings* P24, 4–13, 1980.
10. G. Mollenstedt and R. Speidel, "Newer Developments in Microminiaturization," *Proceedings* P3, 340–357, 1961.
11. G. J. Selvin and W. J. MacDonald, "The Future of Electron Beam Techniques in Microelectronic Circuitry," *Proceedings* P2, 86–93, 1960.
12. O. Wells, "Electron Beams in Micro-Electronics," *Proceedings* P3, 291–321, 1961.
13. G. Binnig and H. Rohrer, "Scanning Tunneling Microscopy," *IBM Journal of Research and Development*, 30: 4, 1986.
14. G. K. Binnig, "Atomic Force Microscope and Method for Imaging Surfaces with Atomic Resolution," United States Patent, US 4724318A.
15. G. Binnig, C. F. Quate, and C. Gerber, "Atomic Force Microscope," *Physics Review Letters*, 56: 930, March 1986.
16. H. W. Kroto, J. R. Heath, S. C. O'Brien, R. F. Curl, and R. E. Smalley, "C60: Buckminsterfullerene," *Nature*, 318 (6042): 162–163, 1985.
17. S. Ijima, "Helical Microtubules of Graphite Carbon," *Nature*, 354: 56, 1991.
18. T. C. Maganas and A. L. Harrington, "Intermittent Film Deposition Method and System," United States Patent 5,143,745.
19. M. W. Browne, "2 Researchers Spell 'I.B.M.,' Atom by Atom," *New York Times*, April 5, 1990. See: http://www.nytimes.com/1990/04/05/us/2-researchers-spell-ibm-atom-by-atom.html.
20. L. V. Radushkevich and V. M. Lukyanovich, *Zurn. Fisic. Chim.*, 111: 24, 1952.
21. W. Bollmann and J. Spreadborough, "Action of Graphite as a Lubricant," *Letters to Nature*, 186: 29–30, April 1960.
22. A. Oberlin, M. Endo, and T. Koyama, "Filamentous Growth of Carbon through Benzene Decomposition," *Journal of Crystal Growth*, 32: 335, 1976.
23. H. Geng, *Semiconductor Manufacturing Handbook*, McGraw-Hill, New York, 2005.
24. Y. Nishi and R. Doering (editors), *Handbook of Semiconductor Manufacturing Technology*, 2nd ed., CRC Press, Baco Raton, Florida, 2007.
25. "Intel Silicon Technology Innovations," Intel.com. Archived from the original March 2015.
26. Riikka L. Purunen, "Surface Chemistry of Atomic Layer Deposition: A Case Study for the Trimethylaluminium/Water Process," *Journal of Applied Physics*, 97 (12), 2005.
27. X. Wang et al., "Fabrication of Ultralong and Electrically Uniform Single-Walled Carbon Nanotubes on Clean Substrates," *Nano Letters*, 9 (9): 3137–3141, 2009.
28. S. Belluci, "Carbon Nanotubes: Physics and Applications," *Physica Status Solidi* (c), 2 (1): 34–47, 2005.
29. H. G. Chae and S. Kumar, "Rigid Rod Polymeric Fibers," *Journal of Applied Polymer Science*, 100 (1): 791–802, 2006.
30. M. Meo and M. Rossi, "Prediction of Young's Modulus of Single Wall Carbon Nanotubes by Molecular-Mechanics-Based Finite Element Modelling," *Composites Science and Technology*, 66 (11–12): 1597–1605, 2006.
31. S. B. Sinnott and R. Andrews, "Carbon Nanotubes: Synthesis, Properties, and Applications," *Critical Reviews in Solid State and Materials Sciences*, 26 (3): 145–249, 2001.
32. P. G. Collins, "Nanotubes for Electronics," *Scientific American*, 67–69, Dec. 2000.
33. S. Hong and S. Myung, "Nanotube Electronics: A Flexible Approach to Mobility," *Nature Nanotechnology*, 2 (4): 207–208, 2007.
34. E. Pop, D. Mann, Q. Wang, K. Goodson, and H. Dai, "Thermal Conductance of an Individual Single-Wall Carbon Nanotube above Room Temperature," *Nano Letters*, 6 (1): 96–100, 2005.
35. S. Sinha, S. Barjami, G. Iannacchione, A. Schwab, and G. Muench, "Off-Axis Thermal Properties of Carbon Nanotube Films," *Journal of Nanoparticle Research*, 7 (6): 651–657, 2005.
36. T. Guo, P. Nikolaev, A. Thess, D. Colbert, and R. Smalley, "Catalytic Growth of Single-Walled Nanotubes by Laser Vaporization," *Chemical Physics Letters*, 243: 49–54, 1995.
37. M. Kumar, "Chemical Vapor Deposition of Carbon Nanotubes: A Review on Growth Mechanism and Mass Production," *Journal of Nanoscience and Nanotechnology*, 10: 6, 2010.
38. N. Inami, M. Ambri Mohamed, E. Shikoh, and A. Fujiwara, "Synthesis-Condition Dependence of Carbon Nanotube Growth by Alcohol Catalytic Chemical Vapor Deposition Method," *Science and Technology of Advanced Materials*, 8 (4): 292, 2007.
39. N. Ishigami, H. Ago, K. Imamoto, M. Tsuji, K. Iakoubovskii, and N. Minami, "Crystal Plane Dependent Growth of Aligned Single-Walled Carbon Nanotubes on Sapphire," *Journal of American Chemical Society*, 130 (30): 9918-9924, 2008.
40. S. Naha and Ishwar K. Puri, "A Model for Catalytic Growth of Carbon Nanotubes," *Journal of Physics D: Applied Physics*, 41 (6): 065304, 2008.
41. Z. F. Ren, Z. P. Huang, J. W. Xu, J. H. Wang, P. Bush, M. P. Siegal, and P. N. Provencio, "Synthesis of Large Arrays of Well-Aligned Carbon Nanotubes on Glass," *Science*, 282 (5391): 1105–1107, 1998.
42. K. Hata, D. N. Futaba, K. Mizuno, T. Namai, M. Yumura, and S. Iijima, "Water-Assisted Highly Efficient Synthesis of Impurity-Free Single-Walled Carbon Nanotubes," *Science*, 306 (5700): 1362–1365, 2004.
43. J. A. Pelesko, *Self-Assembly: The Science of Things That Put Themselves Together*. Chapman & Hall/CRC, New York, 2007.
44. Nadrian C. Seeman, "Nanomaterials Based on DNA," *Annual Review of Biochemistry*, 79: 65–87, 2010.
45. "Scientific Background on the Nobel Prize in Physics 2010, Graphene," The Royal Swedish Academy of Science.
46. "Graphene Properties," www.graphene-battery.net (accessed May 29, 2014).
47. A. K. Geim and K. S. Novoselov, "The Rise of Graphene," *Nature Materials*, 6 (3): 183–91.
48. J. H. Chen, C. Jang, S. Xiao, M. Ishigami, M. S. Fuhrer, "Intrinsic and Extrinsic Performance Limits of Graphene Devices on SiO_2," *Nature Nanotechnology*, 3 (4): 206–209, 2008.
49. A. Akturk and N. Goldsman, "Electron Transport and Full-Band Electron–Phonon Interactions in Graphene," *Journal of Applied Physics*, 103 (5): 053702, 2008.
50. F. V. Kusmartsev, W. M. Wu, M. P. Pierpoint, and K. C. Yung, "Application of Graphene within Optoelectronic Devices and Transistors," *Current Topics in Applied Spectroscopy and the Science of Nanomaterials*, Springer, New York, 2014.
51. J.-C. Charlier, P. C. Eklund, J. Zhu, and A. C. Ferrari, "Electron and Phonon Properties of Graphene: Their Relationship with Carbon Nanotubes." *Carbon Nanotubes: Advanced Topics in the Synthesis,*

Structure, Properties and Applications edited by A. Jorio, G. Dresselhaus, and M. S. Dresselhaus, Springer-Verlag, Berlin/Heidelberg, 2008.

52. Y. M. Lin et al., "Operation of Graphene Transistors at Gigahertz Frequencies," *Nano Letters*, 9 (1), 422–426, 2008.

53. K. F. Mak et al., "Atomically Thin MoS2: A New Direct-Gap Semiconductor," *Physics Review Letters*, 105: 136805, September 2010.

54. B. Radisavljevic et al., "Single-Layer MoS2 Transistors," *Nature Technology Letter*, 6: 147–150, 2011.

55. D. McMullan, "Scanning Electron Microscopy 1928–1965." *Scanning*, 17 (3): 175, 2006.

56. D. McMullan, "Von Ardenne and the Scanning Electron microscope," *Proceedings of Royal Microscopical Society*, 23: 283–288, 1988.

57. Albert V. Crewe, J. Wall, and J. Langmore, "Visibility of a Single Atom," *Science*, 168 (3937): 1338–1340, 1970.

58. Jannik C. Meyer, C. O. Girit, M. F. Crommie, and A. Zettl, "Imaging and Dynamics of Light Atoms and Molecules on Grapheme," *Nature*, 454 (7202): 319–322, 2008.

59. B. Fultz and J. Howe, *Transmission Electron Microscopy and Diffractometry of Materials*, Springer, New York, 2007.

60. B. D. Cullity and S. R. Stock, "Elements of X-Ray Diffraction," 3rd ed., Prentice Hall, Upper Saddle River, New Jersey, February 15, 2001.

61. J. C. H. Spence, *High-Resolution Electron Microscopy*, Oxford University Press, Oxford, England, December 1, 2103.

62. "Intro to Field Emission." Field Emission/Ion Microscopy Laboratory, Purdue University, Dept. of Physics.

63. D. R. Stranks, M. L. Heffernan, K. C. Lee Dow, P. T. McTigue, and G. R. A. Withers, *Chemistry: A structural view*. Melbourne University Press, Melbourne Australia, p. 5, 1970.

64. C. Bai, *Scanning Tunneling Microscopy and Its Applications*, Springer Verlag, New York, 2000.

65. S. H. Pan, E. W. Hudson, K. M. Lang, H. Eisaki, S. Uchida, and J. C. Davis, "Imaging the Effects of Individual Zinc Impurity Atoms on Superconductivity in $Bi_2Sr_2CaCu_2O_8+\delta$," *Nature*, 403 (6771): 746–750, 2000.

66. B. Cappella and G. Dietler, "Force-Distance Curves by Atomic Force Microscopy," *Surface Science Reports*, 34: 1–3, 5–104, 1999.

67. P. Hinterdorfer and Y. F. Dufrêne, "Detection and Localization of Single Molecular Recognition Events Using Atomic Force Microscopy," *Nature Methods*, 3 (5): 347–355, 2006.

68. C. C. Williams and H. K. Wickramasinghe, "Scanning Thermal Profiler," *Appl. Phys. Lett.* 49 (23): 1587, 1986.

69. Q. Zhong, D. Inniss, K. Kjoller, and V. Elings, "Fractured Polymer/Silica Fiber Surface Studied by Tapping Mode Atomic Force Microscopy," *Surface Science Letters*, 290 (1–2): L688-L692, June 1993.

70. H. Butt, B. Cappella, M. Kappl, "Force Measurements with the Atomic Force Microscope: Technique, Interpretation and Applications," *Surface Science Reports* 59: 1–152, 2005.

71. "2015 IC Market: Cautious Expectations Amid a Slow-Growth Global Economy," Research Bulletins, IC Insights, February 4, 2016.

72. B. Fuller, "Fab Lite, Fewer Startups to Fuel Doubling of IC Growth Rates," *EE Times*, September 19, 2012.

73. B. Arnold, "Shrinking Possibilities—Lithography Will Need Multiple Strategies to Keep Up with the Evolution of Memory and Logic," *IEEE Spectrum*, April 1, 2009.

74. M. Sureshkumar, D. Y. Siswanto, and C. K. Lee, "Magnetic Antimicrobial Nanocomposite Based on Bacterial Cellulose and Silver Nanoparticles," *Journal of Materials Chemistry*, 20: 6948–6955, 2010.

75. M. Montazer, F. Alimohammadi, A. Shamei, M. K. Rahimi, "In Situ Synthesis of Nano Silver on Cotton Using Tollens' Reagent," *Carbohydrate Polymers*, 87: 1706–1712, January 2012.

76. M. H. Hermans, "Silver-Containing Dressings and the Need for Evidence," *The American Journal of Nursing*, 106 (12): 60–68, 2006.

77. Y. Qin, "Silver-Containing Alginate Fibres and Dressings," *International Wound Journal*, 2 (2): 172–176, June 2005.

78. J. Turkevich, P. C. Stevenson, and J. Hillier, "A Study of the Nucleation and Growth Processes in the Synthesis of Colloidal Gold," *Discussions of the Faraday Society*, 11: 55–75, 1951.

79. M. Brust, M. Walker, D. Bethell, D. J. Schiffrin, and R. Whyman, "Synthesis of Thiol-Derivatised Gold Nanoparticles in a Two-Phase Liquid-Liquid System," *Chemical Communications*, (7): 801–802, 1994.

80. G. Han, P. Ghosh, and V. M. Rotello, "Functionalized Gold Nanoparticles for Drug Delivery," *Nanomedicine (Lond)*, 2: 113–123, 2007.

81. X. Qian, "In Vivo Tumor Targeting and Spectroscopic Detection with Surface-Enhanced Raman Nanoparticle Tags," *Nature Biotechnology*, 26 (1): 2008.

82. J. Conde, A. Ambrosone, V. Sanz, Y. Hernandez, V. Marchesano, F. Tian, H. Child, et al. "Design of Multifunctional Gold Nanoparticles for In Vitro and In Vivo Gene Silencing," *ACS Nano*, 6 (9): 8316–8324, 2012.

83. S. McMahon, W. B. Hyland, M. F. Muir, J. A. Coulter, S. Jain, K. T. Butterworth, G. Schettino, et al., "Biological Consequences of Nanoscale Energy Deposition Near Irradiated Heavy Atom Nanoparticles," *Nature Scientific Reports, Scientific Reports*, 1 (18): June 20, 2011.

84. See: http://www.nanotechproject.org.

85. See: https://portal.luxresearchinc.com/research/report_excerpt/16215.

86. E. Mounier, "Future of MEMS: a Market and Technologies Perspective," Yole Development, October 2014.

87. J. D. Byrne and J. A. Baugh, "The Significance of Nano Particles in Particle-Induced Pulmonary Fibrosis," *McGill Journal of Medicine*, 11: 43–50, 2008.

88. A. Schneider, "Amid Nanotech's Dazzling Promise, Health Risks Grow," AOL On-Line News, March 24, 2010.

89. R. Weiss, "Effects of Nanotubes May Lead to Cancer," Study Says, Washington Post, May 21, 2008.

90. R. Smith, "Nanoparticles used in paint could kill, research suggests," The Telegraph, August 19, 2009.

91. D. Bowman and G. Hodge, "Nanotechnology: Mapping the Wild Regulatory Frontier," *Futures*, 38 (9), 1060–1073, 2006.

Recommended URL Links

Electronic Discussion Groups: http://www.memsnet.org/lists/

MEMS and Nanotechnology Clearinghouse: http://www.memsnet.org.

MEMS and Nanotechnology Exchange (MNX): http://www.mems-exchange.org.

Chapter 3

Fundamentals of FinFET and Recent Advances in Nanoscale Silicide Formation

BY

L. P. REN *Global Nanosystems, Inc., Los Angeles, California*
YI-CHIA CHOU *Department of Electrophysics, National Chiao Tung University, Hsinchu, Taiwan*
K. N. TU *Department of Materials Science and Engineering, University of California at Los Angeles, Los Angeles, California*

3.1 OVERVIEW

The planar MOSFETs faced various fundamental and technological challenges in the Moore's law scaling in the nanometer regime. Owing to the presence of multiple gates, fin field effect transistors (FinFETs) have emerged as their successors. FinFETs are able to tackle short-channel effects better than traditional planar MOSFETs at deep nanometer technology nodes and thus enable continuous Moore's law scaling. In this chapter, we briefly review the history, fundamentals, and future challenges for the transition from planar FET to FinFET. We then focus on one of the major FinFET challenges about source/drain parasitic resistance and survey recent advances in nanoscale silicide formation. As we know that silicide of C-54 $TiSi_2$, $CoSi_2$, and NiSi have been widely used as source, drain, and gate contacts on microelectronic Si devices. The formation of these silicides in shallow junction Si devices has been studied and reviewed. Lately, as the Si microelectronic device is advancing into 10 nm node, there is a renewed interest in nanoscale silicide formation, especially NiSi, in FinFET devices as well as in Si nanowire technology for biosensors. In this chapter, we review the grand challenge of decreasing the contact resistance in nanoscale contacts in FinFET devices. One way to do so is to develop epitaxial growth of NiSi on the vertical Si fins in the FinFET devices. In view of the fact that an epitaxial interface may have a low contact resistance, we also review the contact reactions in Si nanowires. It has been reported that nanoscale NiSi grows epitaxially in Si nanowires; there is no misfit dislocations and the misfit elastic strain can be a few percent at the device operation temperature.

3.2 FUNDAMENTALS OF FINFET

Since the advent of semiconductor transistors in which integrated circuits are designed for everything from high-speed computers to multifunction mobile devices, Moore's law has remained valid; that is, the number of transistors on a given area of silicon doubles every two years without the increase of production cost. The number of transistors on today's advanced multicore processors has reached 3 billion range from a few thousands of transistors in the mid-1970s. Such achievement is based on a complementary metal-oxide semiconductor (CMOS) technology using planar MOS field effect transistors (MOSFETs), which scales in a way that the latest technology node doubles the density of transistors by shrinking the geometry of transistors compared to the previous generation every couple of years.

As the CMOS technology approached the sub-100-nm node and beyond, static leakage became important as the amount of leakage current doubles while the density of transistors has doubled in a new technology node. Figure 3.1 shows the schematic view of a traditional planar MOSFET and its simulated subthreshold characteristics[1–2] in changing gate lengths. It can be seen that the subthreshold slope degraded as the gate length shrinks shorter. Subthreshold slope also is a function of drain voltage. In fact, when the gate length (L_g) becomes very short, the leakage current will not flow along the Si/oxide interface but nanometers below the interface. When leakage path is physically far from the gate even if the oxide were infinitely thin, the gate of transistors becomes incapable of shutting off the leakage. The leakage current I_{off} will become worse and worse as the gate length L_g continuously scales down. Scaling issues along with doping-induced variations leads to high leakage I_{off} and high-supply voltage V_{dd}, which ultimately result in excessive power consumption and design costs. With the increase in scaling challenges over time, the use of alternative transistor structures became inevitable.

The first alternative transistor structure to traditional planar MOSFETs that the semiconductor industry has adapted is FinFET. FinFET was originally invented on SOI substrates by researchers at University of California at Berkeley (Profs. Chenming Hu, Tsu-Jae King-Liu, Jeffrey Bokor, etc.).[3–6] A few years later, a FinFET built on bulk substrates was presented by Samsung, which requires heavy implant below the fin to suppress leakage in tradeoff of FinFET performance.

Figure 3.2 shows schematically FinFET structures built in (a) SOI, (b) bulk substrates, and (c) longitudinal cross-sectional view. The

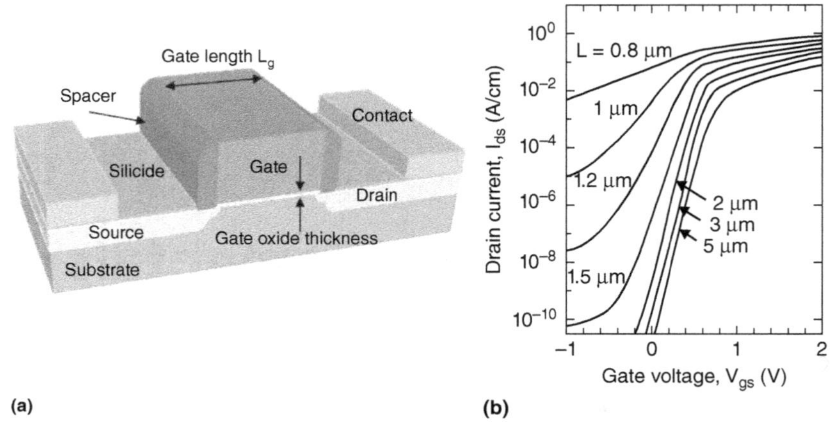

Figure 3.1 (*a*) Schematic view of a traditional planar MOSFET, (*b*) simulated leakage current increases as the gate length shrinks.[1,2]

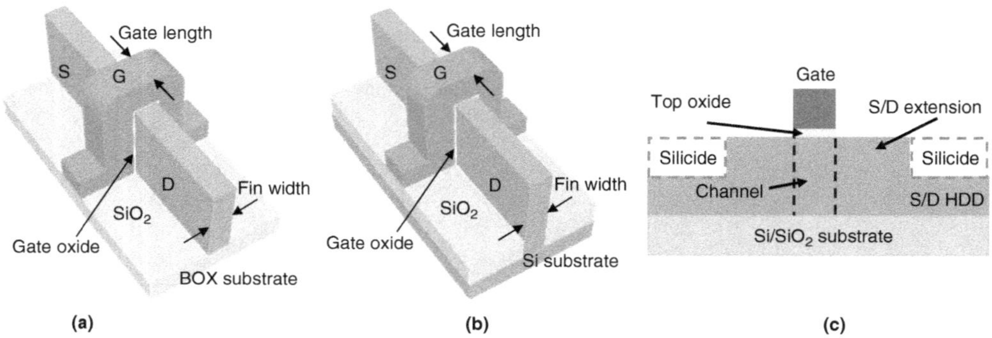

Figure 3.2 FinFET structures: (*a*) SOI substrate, (*b*) bulk substrate, and (*c*) longitudinal cross-sectional view.

distinguishing characteristic of the FinFETs is that the conducting channel is wrapped by a thin silicon fin, which forms the body of the device. The thickness of the fin measured in the direction from source to drain determines the effective channel length of the device. Compared to traditional planar MOSFETs, the body of the FinFETs is ultrathin and controlled from three sides instead of just the top from traditional planer MOSFETs. Such FinFET is easy to scale because leakage can be well suppressed if the fin thickness is equal to or less than L_g. The formation of thin fins is simple as they can be made with the same gate patterning/etching tools. In addition, because the body fin is made ultrathin there is no Si more than nanometers far from the gate(s) and the body doping becomes optional. FinFETs provide higher speed, lower leakage, lower supply voltage, V_{dd}, lower power consumption, further and better scaling capability, lower cost, better subthreshold swing, no random dopant fluctuation, less variability, better mobility, and future subthreshold design, etc.

In a multigate configuration of FinFETs, the channel is surrounded by several gates on multiple surfaces. A multigate FinFET provides a better electrical control over the channel, which not only allows more effective suppression of "off-state" leakage current but also enhances current in the "on" state known as drive current. Moreover, multigate FinFETs also have a better analog performance because of a higher intrinsic gain and lower channel length modulation. These advantages in turn translate to lower power consumption and enhanced device performance. Nonplanar devices are also more compact than conventional planar transistors, enabling higher transistor density, which translates to smaller overall microelectronics.

Figure 3.3 shows the design parameters of the thin body fin of a FinFET along with a multigate configuration. The fin width determines drain-induced barrier lowering (DIBL); the fin height is limited by etch technology and traded off by layout efficiency versus design flexibility;

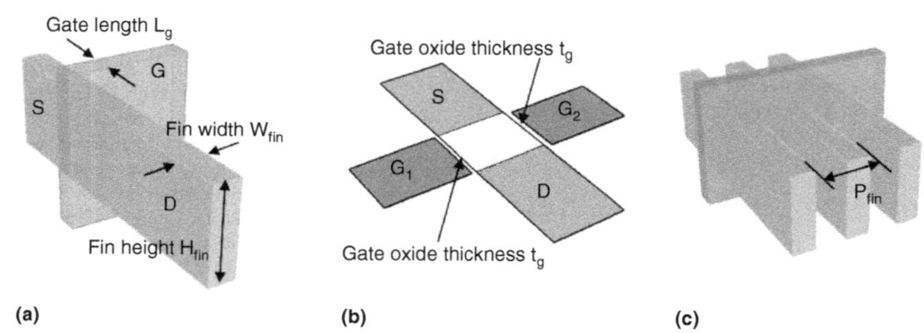

Figure 3.3 Double-gate FinFET fin parameters: (*a*) schematic view, (*b*) plan view, and (*c*) multigate configuration.

the fin pitch determines layout area, limits source/drain (S/D) implant tilt angle, and is traded off by device performance versus layout efficiency.

Similar to traditional planar MOSFETs, FinFETs also can be built using a gate-first approach. That is, the gate is first formed and then acts as a mask for the S/D implants, followed by a post-implant anneal to repair the damage introduced during implantation and achieve the desired dopant profile. The advantage of the gate-first approach is self-aligned source and drain with gate. As MOSFETS in sub-100-nm nodes and beyond, a gate stack consisting of high-κ dielectrics and metal in replacement of oxide and polysilicon must be used to form the gate because of the physical scaling limit of oxide. The gate-first approach in this case complicates the post-implant anneal step in preserving the integrity of the stack, eliminating the implant damage and achieving the desired doping profile. The solution is the S/D-first approach in which a sacrificial dummy gate is used to mask the implants, then it is removed and a new gate stack is built after the anneal step.

Figure 3.4 shows a FinFET formation in SOI substrate based on the S/D-first and body-fin-last approach.[7] The primary processes are outlined as follows: (a) The formation starts with an oxide mask layer, which is patterned and etched down to SOI. (b) The dummy gate is formed by deposition on pattered oxide layer, patterning and etching, and removal of the exposed hard mask oxide, which is followed with S/D formation and silicidation. (c) Deposition of insulator to fill contact trench followed by planarization. (d) Removal of dummy gate, which reveals the patterned hard mask through which the fin pattern is transferred to the underlying silicon layer by anisotropic etching. (e) Formation of the inner spacer on S/D sidewalls to ensure sufficient separation between the gate and the flared source and drain. (f) Finally, the gate stack is formed by deposition into the gate trench.

The semiconductor industry has been very successful with the alternative FinFET structure. Foundries have so far shipped 16-nm/14-nm

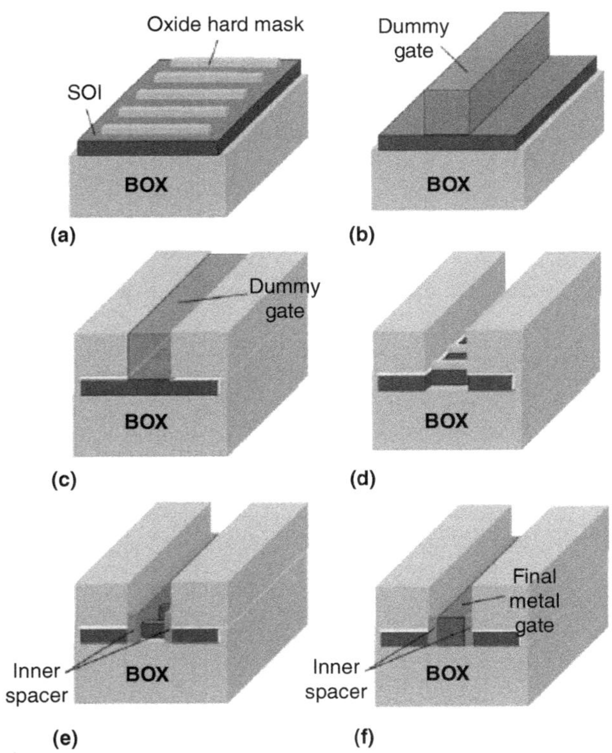

Figure 3.4 Schematic formation of FinFETs on SOI substrate with a S/D-first and body-fin-last approach.[7]

FinFETs with 10-nm FinFETs in development. As the sub-nanometer technology node continues to scale, other alternative transistors may be inevitable. Based on IMEC's roadmap, there are two transistor options for the 7-nm node, that is, the FinFET and lateral gate-all-around nanowire FET also called the lateral nanowire FET. For the 5-nm node, the industry leans toward the lateral nanowire FET. The lateral nanowire FET is basically an evolution of the FinFET from double-gate and triple-gate to all-around-gate configuration, which increases the gate area so that turns off the device more effectively. Although nanowire FET may provide better electrostatics than FinFET, it is going to be more challenge in fabrication that requires complex processes on the bottom of the device. Remaining FinFET challenges include V_{TH} adjustment, fringing capacitance between gate and top/bottom of S/D, parasitic resistance, and variability. In next section, we review the recent advances in nanoscale silicide formation for reduction of parasitic resistance of FinFETs and nanowire FETs.

3.3 RECENT ADVANCES IN NANOSCALE SILICIDE FORMATION

3.3.1 Introduction

Silicide has been an essential building block in Si-based technology owing to its outstanding physical property to serve as electrical source, drain, and gate contacts, and the precision control of the salicide process to form the contacts. The formation mechanism of silicide by solid-state interfacial reaction in forming a low-resistance silicide thin film phase such as C-54 $TiSi_2$, NiSi, or $CoSi_2$ has been reported and reviewed. In our previous review, we have emphasized the growth events of thin film silicides on Si wafers.[1,8] In this review, we change our emphasis to address the nucleation events of epitaxial silicide in Si nanowires. Why the change? This is because the trend of microelectronic Si technology in FinFET devices is approaching 10 nm node. In FinFET devices, to reduce the contact resistance in nanoscale silicide contacts is a grand challenge. To overcome the challenge, we propose to use epitaxial silicide/Si contacts. In nanoscale, we found that NiSi and $CoSi_2$ can grow epitaxially in Si nanowires.[9–11] In Sec. 3.3.2, we review briefly the structure and properties of FinFET devices. In Sec. 3.3.3, we review the nucleation of atomic layer of silicide in epitaxial growth of silicide in nanowires of Si.

Solid-state interfacial reaction is an old subject, and much of it has been covered in textbooks for bulk and thin film materials science. In interfacial reactions in nanoscale materials, it seems that the role of nanowire surface and the lack of dislocation in the nanowire have some profound effects on the kinetics of nucleation and growth in nanoscale materials.

In solid-state nucleation, the lattice mismatch between the nucleus and the matrix can impose an interfacial energy barrier to nucleation. When the nucleating phase has different unit cell volume and shape with the matrix it replaces, and when the interface between them is incoherent, elastic strain energy may not need to be considered as a part of volumetric contribution to nucleation. In other words, if the interface is nonepitaxial across a flat interface, misfit strain is insignificant. However, if the interface is coherent or epitaxial, we must consider elastic strain energy in the nucleation event. This concern applies to the nucleation of an epitaxial silicide phase in nanowire of Si. We will show that elastic energy indeed plays an important role in nucleation of the epitaxial silicide in nanowires of Si.

Homogeneous nucleation is defined as nuclei forming randomly throughout the entire matrix, and heterogeneous nucleation is defined as nuclei forming at a location of surface or internal imperfection such as microstructure defects. While the steady-state kinetic theory of fluctuation in size space is rather advanced, it is mostly for homogeneous nucleation only. However, in real events of nucleation, heterogeneous nucleation occurs predominantly because of the large difference in activation energy between the two. Thus there is a gap between theory and experiment in our understanding of the basic concepts of nucleation. To reduce the effect of heterogeneity, nucleation in small droplets was studied, for instance.[12] The crystallization of small liquid droplets in nanotubes, for example, has been found to require a much larger undercooling than that in a bulk liquid phase. It shows that homogeneous nucleation is indeed rare in real thermodynamic systems. We discuss in Sec. 3.3.3 the event of solid-state homogeneous nucleation of epitaxial transition-metal silicides,[13] especially NiSi, in Si nanowires, and the effect of elastic strain on repeating homogeneous nucleation will be presented.[14]

3.3.2 Silicide Contact Technology for Nanoscale FinFETs

FinFET and Its Resistance Components

As the channel length scaling enters sub-30-nm regime, multigate device architectures are attractive for continued transistor channel length scaling in advanced technology nodes due to their improved electrostatic control. Among the multigate device architectures, FinFET is advantageous because of its self-aligned multigate structure and better compatibility to the conventional planar CMOS process. As shown in Figs. 3.2 and 3.3, FinFET can be made in both tri-gate and double-gate configurations.

Even though multiple gate field effect transistors (MugFETs) escape some of the stringent geometric scaling requirements of their planar counterparts, they suffer large parasitic resistance owing to the extremely narrow S/D regions.[15] Application of nanoscale silicide technology to nanoscale FinFET technology is to reduce the S/D parasitic series resistance associated with the formation of nanoscale silicides in the S/D regions.

Figure 3.5*a* demonstrates schematically the resistance components at the heavily doped drain (HDD) side of a FinFET with additional structural elements added to the FinFET, that is, silicide/HDD contact, spacer, via and metal/silicide contact. The resistance components are contact resistance between via metal and silicide $R_{\text{Con}}^{\text{Via/Silicide}}$, pad resistance between pad and S/D fin $R_{\text{SP}}^{\text{Pad/fin}}$, silicide sheet resistance R_S^{Silicide}, contact resistance between silicide and HDD $R_{\text{Con}}^{\text{Silicide/HDD}}$, HDD sheet resistance R_S^{HDD}, contact resistance $R_{\text{Con}}^{\text{Silicide/Ext}}$, S/D extension resistance R_{Ext}, S/D extension overlap resistance under gate $R_{\text{Ol}}^{\text{Ext/Ch}}$, and channel resistance R_{Ch}. Among the resistance components, R_{Ext} and R_{Ch} are generally device design dependent. We have discussed the R_S^{Silicide} issues in silicide formation on Si in Refs. 1 and 8. Thus, we primarily discuss the contact issues of nanoscale silicides associated to nanoscale FinFET technology. We will start with the band diagram of metal/semiconductor contact to address the fundamental of silicide/Si contact in the section "Silicide/Silicon Contacts," then review the studies on reduction of metal/silicide and silicide/silicon contact resistances in the section "$R_{\text{S/D}}$ Reduction in Nanoscale FinFETs," discuss some design optimization of FinFETs in order to achieve low S/D parasitic series resistance in the section "Design Optimization in Nanoscale FinFETs," and finally summarize the section.

Silicide/Silicon Contacts

Silicide provides low-resistance connections between interconnection metal and active devices ever since the microelectronics technology was invented. A silicide/silicon contact is a metal/semiconductor contact because the metallic property of silicide. Depending on the Si doping level and silicide work function, the current transfer mode can in general be in rectifying (Schottky) or ohmic. Figure 3.6 shows the schematic band diagrams and I and V characteristics of silicide before and in contact with Si and NiSi is used for the discussion. In Fig. 3.6, E_0, E_c, E_v, E_i, E_{fs}, and E_{fm} are energy level in eV at vacuum, conduction band, valence band, intrinsic Si Fermi, doped Si Fermi, and NiSi Fermi, respectively. ϕ_m, ϕ_s, and χ are metal and Si work function and Si electron affinity, respectively. ϕ_B, V_{bi}, and W are Schottky barrier potential, built-in contact potential and deplete barrier region width.

When silicide (e.g., NiSi) forms, it is brought into intimate contact with Si. Charge transfer occurs until their Fermi levels align with each other. As shown in Fig. 3.6*a*, NiSi Fermi level ϕ_m is larger than *n*-type Si Fermi level ϕ_s before contact formation. To align the two Fermi levels, as shown in Fig. 3.6*b*, the electrostatic potential of the Si must be raised relative to that of NiSi. The Si bands bend upward and majority electron

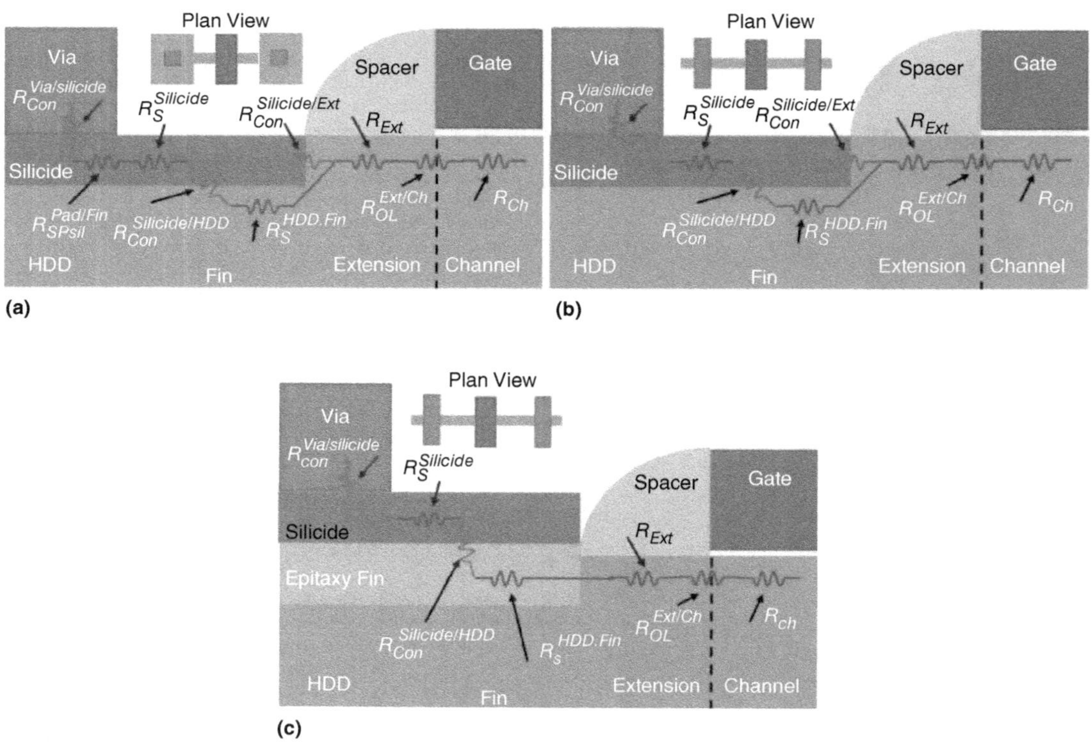

Figure 3.5 (*a*) Resistance components of a FinFET with silicide, spacer and via, (*b*) S/D resistance reduction with slot contacts, and (*c*) S/D resistance reduction with epitaxially thickened HDD and slot contacts.

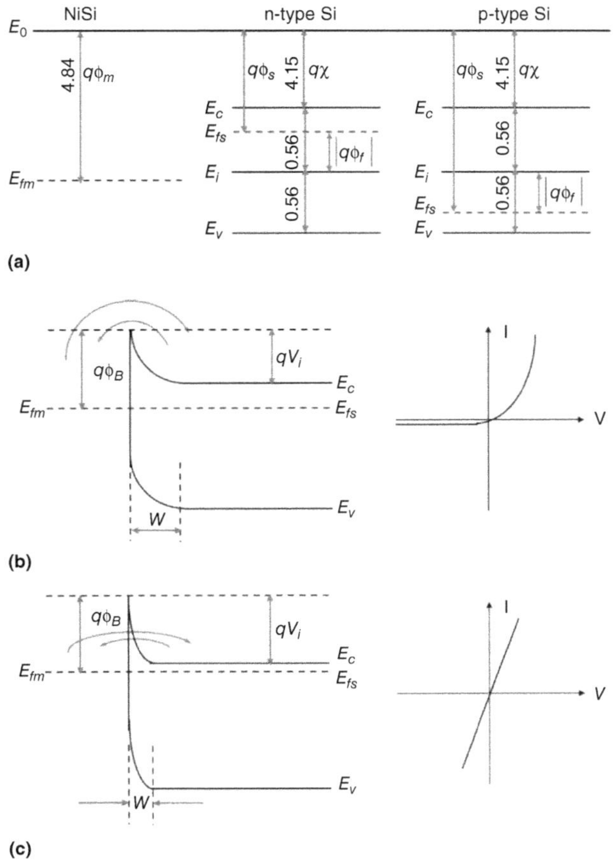

Figure 3.6 Schematic band diagrams and I-V characteristics of silicide/silicon contact by using NiSi/Si: (*a*) *n*- and *p*-type Si before contact, (*b*) Schottky contact with *n*-type Si, and (*c*) ohmic with n^+-type Si.

carriers in Si diffuse from Si to NiSi. At equilibrium, a negative charge is built up at the NiSi surface and an equal but positive charge is distributed in the deplete barrier region near the Si surface. Such a barrier blocks the majority electron carriers from moving cross the NiSi/Si contact resulting a rectifying current transfer characteristics in which there is very little current flow when applied voltage opposite to the internal built-in potential. For an ideal contact in the absence of surface states and assuming a p^+-n diode approximation, the energy barrier height, built-in potential and width of the deplete barrier region can be given by

$$\begin{aligned} q\phi_B &= q(\phi_m - \chi) \\ qV_{bi} &= q(\phi_m - \phi_s) \\ q\phi_s &= q\chi + \frac{E_g}{2} - |q\phi_f| \\ \phi_f &= \frac{kT}{q}\ln\left(\frac{N_B}{n_i}\right) \\ w &\cong \sqrt{\frac{2\varepsilon_s V_{bi}}{qN_B}} \end{aligned} \tag{3.1}$$

where N_B is Si dopant concentration, and n_i, k, q, ε_s, and T are Si intrinsic carrier concentration, Boltzmann constant, electron charge, Si permittivity, and absolute temperature, respectively.

Clearly, the primary dominant components that determine the Schottky contact properties are Si doping level and the silicide work function. With increase in Si doping concentration, the Si Fermi level moves toward conduction band, Si work function decreases, built-in potential increase and depletion barrier width decreases; with the increase in silicide work function, the Schottky barrier height increases. For a low-to-moderate doping level, the depletion barrier width is wide so that current transfer mode is in general rectifying. In a rectifying process, the current transfer mode is governed by thermionic emission in which thermal energy makes the carriers overcome the barrier and form current flow crossing the contact. It is because of high barrier that Schottky contacts have high contact resistance.

For a high doping level, as shown Fig. 3.6*c*, the Si Fermi level approximately equates to that of its conduction band ($\phi_s \approx \chi$) so that Schottky barrier height is closely equal to the built-in potential of the contact. Thus, the depletion width is reduced as

$$w \cong \sqrt{\frac{2\varepsilon_s \phi_B}{qN_B}} \tag{3.2}$$

In such a case, the width of the depletion barrier region becomes narrow enough so that the carriers can directly tunnel through the barrier by quantum mechanics. The current-voltage characteristics then show ohmic behavior in which current changes with voltage virtually proportional in both directions through a contact. Ohmic contact mode can occur when a Si doping level is greater than 6E19 cm^{-3}.[16]

The specific contact resistivity ρ_c is defined as

$$\rho_c \equiv \left[\frac{\partial V_{ms}}{\partial J}\right]_{V_{ms}=0} \tag{3.3}$$

where J is the current density across the metal/semiconductor interface and V_{ms} is the voltage across it.

Based on their current transfer characteristics, the specific contact resistivity, in $\Omega \cdot cm^2$, for both Schottky and ohmic contacts, is derived from Eq. (3.3), respectively, as follow:

$$\begin{aligned} \rho_c &= \frac{k}{qA^*T}\exp\left(\frac{q\phi_B}{kT}\right) \\ \rho_c &= \rho_{c0}\exp\left(\frac{2\phi_B\sqrt{m^*\varepsilon_s}}{\hbar\sqrt{N_B}}\right) \end{aligned} \tag{3.4}$$

where $\hbar$ is Plank's constant, m^* is the effective mass of the tunneling carrier, and ρ_{c0} is a constant depending on the metal and semiconductor in contact.

It can be seen that specific contact resistivity for both Schottky and ohmic contact modes depends exponentially on barrier height. Schottky contact mode controlled by thermionic emission is essentially independent on doping concentration while ohmic contact mode by quantum mechanical tunneling relies strongly on the doping concentration. With increase of doping level, the barrier width decreases and in turn the tunneling becomes more efficient.[16]

Contact resistance is reversely proportional to contact area. To reduce silicide/silicon contact resistance, the following general approaches can be explored: (1) choose a silicide with low barrier height to Si; (2) utilize a high doping concentration on Si; and (3) increase the area of the contact.

$R_{S/D}$ Reduction in Nanoscale FinFETs

Reduction of S/D series resistance $R_{S/D}$ for nanoscale FinFETs is to reduce the parasitic series resistance associated with the application of silicide that connects the active FinFETs as shown in Fig. 3.5. There have been numerous studies on silicide technology in planar CMOS technology and its specific application in 3D nanoscale FinFETs. Among the parasitic series resistance components, contact resistance dominates for nanoscale technology nodes.[17,18] Figure 3.7 shows the variation of S/D parasitic series resistance components with technology nodes for both NMOS and PMOS transistors.[18] It can be seen that S/D contact resistance counts for more than 60 percent of the total series resistance and increases rapidly as the technology node scales down. Thus, the best way to attack the $R_{S/D}$ issue is to reduce the contact resistance between silicide and HDD.

We have deduced three approaches in the section "Silicide/Silicon Contacts" to reduce silicide/Si contact resistance based on Schottky and ohmic contact modes. In the following, we will discuss the three approaches in general but specifically to the reduction of S/D series resistance of nanoscale FinFETs.

Electron density distribution of an *n*-channel FinFETs was simulated under linear operation, as shown in fig. 4 of Ref. 15, which highlights the need to minimize direct contact between silicide and extension fin on plane A. This can prevent current crowding and avoid silicide formation on narrow S/D extension fin of low doping level and in turn reduce spreading and contact resistances, respectively. This separation can be achieved with a controlled silicide depth, optimal spacer width, and thick epitaxial S/D regions.

As seen in Eq. (3.1), barrier height is a strong function of silicide work function. One may simply choose a silicide of low resistivity and low barrier height to Si for active device contacts. In practice, it is actually not an easy task because other requirements, for example, process compatibility with active device fabrication. In addition, surface states at silicide contact tend to pin the Fermi level deep in the Si band gap, which produces high barrier height for *n*-type Si and relatively low one for *p*-type Si.

NiSi offers numerous advantages over its precursors $TiSi_2$ and $CoSi_2$.[1,8] This includes less silicon consumption that enables shallower junction formation, low resistivity at between 10.5 and 18 μohm·cm, good narrow line sheet resistance behavior (single phase), amenable mid-gap work function for both *n*- and *p*-type silicon contacts, and low processing temperature under 700°C in which dopant diffusion in active device regions is limited so that device characteristics can be maintained.

NiSi, however, has a relative high work function of approximately 4.84 eV. This high work function not only gives a high barrier height to *n*-type Si but also higher than that of backend metals (W, Ti, TiN) that connect it before backend interconnection, which, in turn, results in high contact resistance between silicide and active silicon contact regions $R_{Con}^{Silicide/HDD}$ as well as high contact resistance between via metal and silicide $R_{Con}^{Via/Silicide}$.

Numerous studies have been carried out to reduce $R_S^{Silicide}$, $R_{Con}^{Via/Silicide}$, and $R_{Con}^{Silicide/HDD}$ under different metal/NiSi/Si contact schemes for its application in nanoscale technology nodes. This includes work function alignment between metal and silicide,[19,20] capping, surface treatment prior deposition,[21] laser spike anneal,[22,23] alloying[24,25] implantation,[21,26–28] etc.

In AlCu alloy backend metallization with W plugs, a bi-layer Ti/TiN layer is typically employed to prevent W precursor intrusion into the interlayer dielectric and hinder W plug oxidation by oxygen diffused from the interlayer dielectrics.[29] Such a backend metallization scheme forms multilayer contact interfaces with active Si device regions, that is, W/TiN/Ti/silicide/Si, or simply metal/silicide/Si. Figure 3.8*a* shows their work function alignment offsets of the metallization metals with silicide.

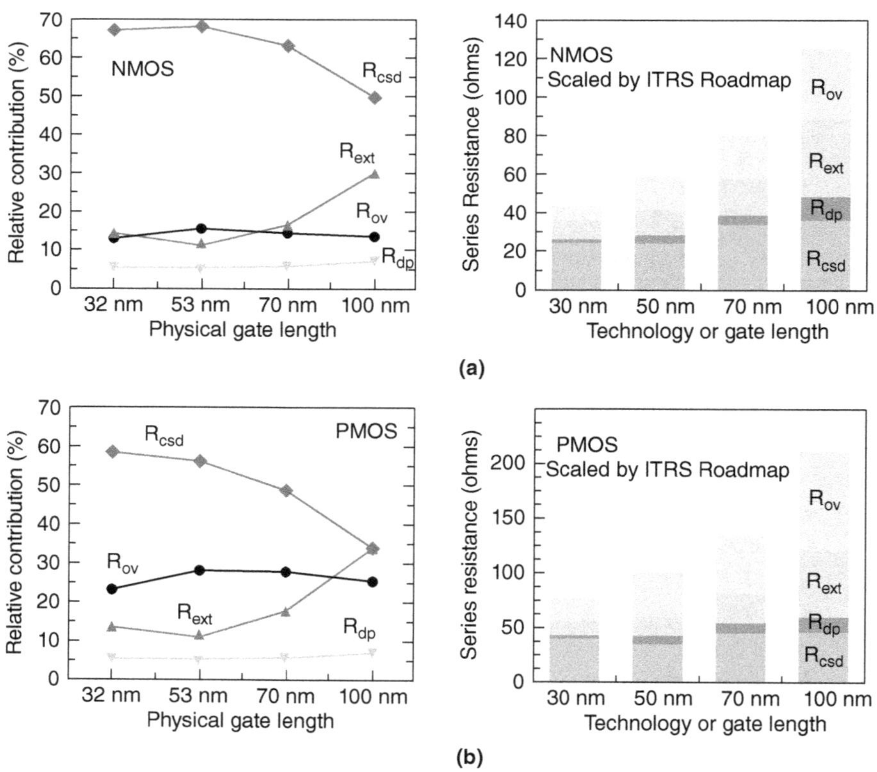

Figure 3.7 Relative contribution of S/D parasitic series resistance components with technology nodes: (*a*) NMOS, (*b*) PMOS.[18]

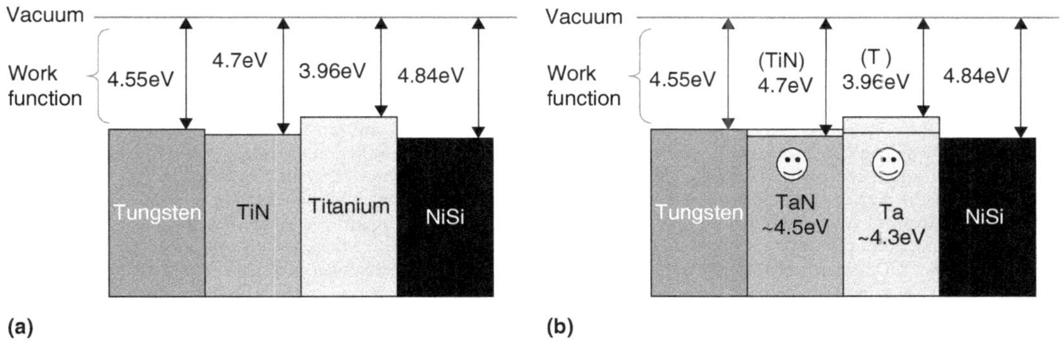

Figure 3.8 Work function alignment offsets of metallization schemes with NiSi: (*a*) W/TiN/Ti/NiSi, (*b*) W/TaN/Ta/NiSi.[19–21]

To improve the sheet resistance R_S^{Silicide} of silicide, contact resistance $R_{\text{Con}}^{\text{Via/Silicide}}$, and contact resistance $R_{\text{Con}}^{\text{Silicide/HDD}}$ of the contact scheme, an alternative TaN/Ta is used to replace TiN/Ti. Figure 3.8*b* shows the work function alignment offsets of the novel metallization scheme W/TaN/Ta with NiSi compared to the conventional one W/TiN/Ti with NiSi. Clearly, the novel metallization scheme has better work function alignment with NiSi and in turn improved $R_{\text{Con}}^{\text{Via/Silicide}}$.

The study in Ref. 21 explores the effects of capping and surface treatment prior metal deposition on NiSi sheet resistance R_S^{Silicide} and NiSi/Si contact resistance $R_{\text{Con}}^{\text{Silicide/HDD}}$ in association with the two metallization schemes. As shown in Figs. 63–64 and 67 of Ref. 21, it can be seen that (1) the use of a TaN layer as the oxidation barrier of W plug provides better NiSi sheet resistance R_S^{Silicide} and better NiSi/Si contact resistance $R_{\text{Con}}^{\text{Silicide/HDD}}$ resulting from better work function alignment; (2) degas at 250°C prior to Ni metal deposition can reduce NiSi sheet resistance as much as 30 percent, which attributes to the removal of water residue after aqueous HF-based preclean just prior to Ni deposition. The removal prevents native oxide formation and incorporation of oxide in NiSi.

The exploration of silicide alloying[24,25] and implantation[21,26–28] is intended to fundamentally lower the barrier height of silicide/silicon contact and in turn ultimately improve the contact resistance $R_{\text{Con}}^{\text{Silicide/HDD}}$. Figure 3.9 shows alloying and implant modification on silicide sheet resistance and silicide/Si barrier height. It can be seen from Fig. 3.9*a*, *b* that alloying NiSi with Er, Yb, Ti, and Al degrade the silicide sheet resistance R_s^{Silicide}. It, however, can change the silicide/n-Si barrier height significantly. For nanoscale FinFETs, the silicide/silicon contact resistance component $R_{\text{Con}}^{\text{Silicide/HDD}}$ dominates the totally S/D series resistance. Significantly lower $R_{\text{Con}}^{\text{Silicide/HDD}}$ can offset the increase of R_s^{Silicide} and results in low S/D series resistance. Generally, alloying NiSi with a low work function metal (i.e., Er or Yb) is less effective to electron barrier height of NiSi/Si contact because rapid diffusion of the modifying metal away from the Ni/Si contact. However, alloying NiSi with high work function materials (i.e., Pt, Ir, Pd) is better successful to hole barrier height of NiSi/Si contact because the alloying metal diffuses much slower than Ni; though it suffers from the increased thermal budget for silicidation and additional process complexity and contamination concerns.

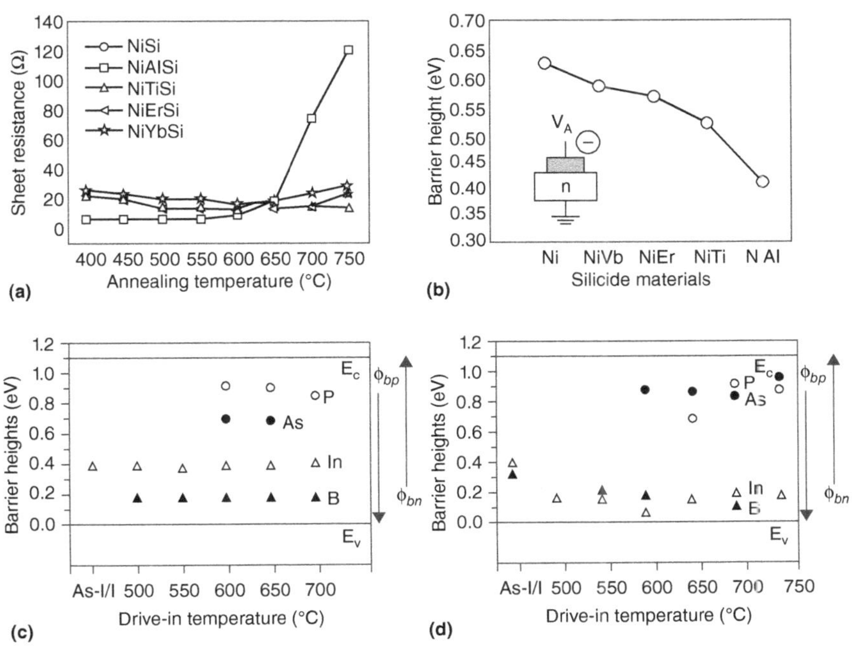

Figure 3.9 Alloying modifications on (*a*) NiSi sheet resistance and (*b*) barrier height of NiSi/n-Si[24]; implant modifications on barrier height of (*c*) NiSi/Si and (d) PtSi/Si.[27]

Implant modification of NiSi/Si barrier height in general can be done with three approaches, that is, before NiSi formation, after $NiSi_2$ formation, and after NiSi formation followed by a drive-in anneal. The primary principle behind the modification is attributed to the dopant segregation at NiSi/Si interface. As shown in Fig. 3.9*c*, *d*, the implantation of NiSi and PtSi with B, In, P, and As can significantly modify the barrier height of the silicide/silicon contacts from 0.1 to 1.0 eV so that a close-to-unity rectification rate contact is achievable. The rectification ratio of a contact is defined as the ratio of forward current to the reverse current when both forward voltage and reverse voltage are equal to 0.5 V. The process of implant modification is compatible and can be easily integrated with 3D FinFETs. Thus, it offers significant opportunities to achieve low $R_{Con}^{Silicide/HDD}$ and in turn low parasitic series resistance for nanoscale FinFETs.

Design Optimization in Nanoscale FinFETs

Design optimization is crucial to reduce $R_{S/D}$ for nanoscale FinFETs. Figure 3.5*b*, *c* shows two FinFET designs derived from the one shown in Fig. 3.5*a* with slot contacts and with epitaxially thickened HDD and slot contacts, respectively.[21] In Fig. 3.5*b*, the pad to silicide is replaced with slot contacts, which removes the spreading resistance component in Fig. 3.5*a*. In Fig 3.5*c*, in addition to slot contacts, the HDD region is thickened through epitaxial growth. The thickening of the HDD region prevents silicide formation in the narrow S/D extension fin regions of low doping level, which can significantly reduce the silicide/Si contact resistance. This can be accomplished with optimized spacer formation. The spacer width is determined by deposition of dielectrics with step coverage, which determines two important device geometry parameters, for example, S/D extension length and gate to S/D extension overlap/underlap. The S/D extension length will ultimately determine the length of silicided active material between the via/contact and S/D extension. Shorter S/D extension provides longer silicided S/D regions in which carriers are able to conduct through the metallic silicide over a longer distance and drop less voltage before breaking across the silicided S/D junction. Hence, the spacer width should be optimized as short as possible while optimal device performance of FinFETs is preserved.

Summary

The largest S/D series resistance component in nanoscale FinFETs is the contact resistance between the semimetallic silicide and the heavily doped HDD regions. Studies have showed that this component constitutes more than 60 percent of device series resistance. To reduce the contact resistance, three approaches are discussed: (1) novel design of FinFETs to maximize the contact area to silicided HDD regions, (2) increase HDD doping level and prevent silicide formation on narrow S/D extension fin regions of low doping level, and (3) more attractively, ultimately modify the silicide/Si barrier height by surface treatment prior to metal deposition, capping before silicide formation, alloying, and implantation. Modification of silicide/silicon barrier height offers significant opportunities to achieve low $R_{Con}^{Silicide/HDD}$ and in turn low parasitic series resistance for nanoscale FinFETs.

As we move to 10 and 7 nm, a new fin material will be required—probably silicon-germanium (SiGe) or perhaps just pure germanium. According to Ref. 30, SiGe will take us to 7 nm. After that, however, we're probably looking at a new transistor structure. The potential candidates can be gate-all-around (GAA) FETs and vertical tunneling FETs (TFETs) as they would allow for shorter gates and lower voltages. Both GAA FET and TFET essentially consist of nanowire source and drain surrounded by a gate. In next section, we will review the epitaxial growth of silicide in Si nanowires.

3.3.3 Epitaxial Growth of Silicide in Si Nanowires

One-dimensional semiconductor nanostructures such as Si nanowires are of interest because of their potential applications in nanoscale electronic and optoelectronic devices, especially biosensors.[31–35] Nanowires have demonstrated interesting electrical transport properties, which are different from bulk materials and thin films. This is because quantum effects play an increasingly prominent role while dimension is shrinking to nanoscale.[36] In nanowires, electrons could be quantum-confined laterally and thus could occupy discrete energy levels that are different from the energy bands found in bulk materials and thin films.[37–40] To understand the new physics demonstrated by nanowires, much effort has been dedicated to fabricating high-quality doped and undoped semiconductor nanowires. In particular, the Si nanowires with well-controlled atomic structures and contact interfaces,[41–43] and their assembly into functional devices[44,45] are the major challenges and goals for nanoscale device materials studies.

To obtain well-defined nanoscale ohmic contacts and gates on Si nanowires, it requires a systematic study of chemical reactions in the nanoscale to form these circuit components. Ni silicides are potentially useful for the electrical contacts of the Si nanowire sensors. The NiSi/Si heterostructures in the Si nanowire sensors can be synthesized by lithography on Si nanowires, where the Si region covered by Ni patterns can be transformed to NiSi to form the NiSi/Si/NiSi nanowire heterostructures. Silicon nanowire transistors have been fabricated by these heterostructures.[46,47]

Point Contact Reactions: Formation of NiSi Contacts and Heterostructures in Si Nanowires

Beside the top-down technology of fabricating small structures, self-assembly techniques allow building nanostructures beyond the limits of lithography such as VLS growth of nanowires.[48,49] Point contact reaction has been defined as reactions between nanowires or nanowire/nanoparticles via the points of contacts.[9] This opens a route of scaling down by reducing the sizes of transistor components by self-assembly and the channel length.

By dripping solutions of Si or metal nanowires or depositing metal nanoparticles on a substrate with Si nanowires, the contacts between Si nanowires and metal can be fabricated as many of them cross each other and form point contacts. The contacts are small as points; therefore, the diffusion is limited by the supply of metal into Si nanowires as supply limit reaction.[11] This mechanism of point contact reaction is different from thin film and bulk silicidations, which is either diffusion controlled or interface controlled.

After carefully controlling the reactions, a bamboo-type single-crystal NiSi grain within a straight Si nanowire was formed at 700°C, thereby the growth of silicide forming a NiSi/Si/NiSi or heterostructure.[10] Some Ni atoms may be able to diffuse through a grain of silicide; hence, additional and smaller silicide growth at the tip was observed, which is attached to the long silicide at the end.

Epitaxial Relationship between NiSi and Si and Atomically Sharp Interfaces

A set of high-resolution transmission electron microscopy (HRTEM) images of the NiSi/Si interface taken as the interface advances into the Si was shown.[15] The interface is parallel to the (111) plane of Si as well as the (311) plane of NiSi. Thus, the growth direction of NiSi is normal to the (311). The crystallographic orientation relationships between Si and NiSi, which has an orthorhombic lattice with lattice constants of $a = 0.562$ nm, $b = 0.518$ nm, and $c = 0.334$ nm, are [1-10] Si // [1-12] NiSi and (111) Si // (31-1) NiSi. Across the epitaxial interface, the misfit is around 5.6 percent so the Si is expanded under tension. However, misfit dislocations at the NiSi/Si epitaxial interface were not found and this may be due to the instability of a dislocation in a nanowire.

Reactive Epitaxial Growth of Nanoscale NiSi in Si Nanowires

The linear growth behavior of NiSi in the Si nanowire of 20 nm in diameter over the temperatures ranged from 500 to 650°C was shown.[10] The activation energy of the epitaxial growth was determined to be 1.25 eV per atom, compared to the activation energy of interstitial diffusion of Ni in Si of about 0.47 eV per atom,[50] indicating that the linear growth may be interfacial-reaction-controlled, to be discussed later.

Atomic flux by definition is the number of atoms per unit area per unit time, so that the number of Ni atoms diffusing into Si is equal to flux times area times time. If the contact area is a point, the flux or the number of Ni atoms diffusing into Si will be extremely small, and thereby the reaction can be limited by how fast the Ni atoms can diffuse into the Si nanowire. This is a unique feature of point contact reactions, and it will lead to a linear growth. Surface diffusion of Ni on the SiO_2 surface of the Si nanowire was considered whether it is a kinetic path for the supply of Ni; however, it is slower than the interstitial diffusion of Ni within Si,[50] which dominates the reactions. Nevertheless, in later sections, we show that the linear growth is due to a stepwise growth mode which is limited by the repeating nucleation events.

The atomistic mechanism of the epitaxial growth involving a moving interface is of interest. In the present case, the covalent Si-Si bonds must be broken and transformed into metallic Ni-Si bonds in order for NiSi to grow and have a moving interface of growth. Although the thermal energy at 700°C is sufficient to break the covalent Si-Si bonds, the interstitial Ni atoms are crucial in the bond-breaking process. It is known that in the interfacial reactions between metal thin films and single crystal Si wafers, near-noble metals such as Ni and Pd can react with Si to form silicide at temperatures as low as 100°C.[51,52] Near-noble metals diffuse through the silicide (not through Si as we reported here), dissolve interstitially in Si, and assist the breaking of the Si-Si covalent bonds.[53] Furthermore, native oxide on the Si surface is not an effective diffusion barrier to Ni and Pd, and therefore the metal atoms can diffuse through the native oxide and react with the Si below the oxide. Hence, in planar thin film reactions, nucleation and growth of silicide occurs right below the plane of contact. Yet in point contact reactions, owing to the lack of supersaturation below the point of contact because of the very limited number of dissolved Ni atoms and because they can diffuse away quickly, silicide in Si nanowire does not form at the location of the point contact.

Fabrication of 2 nm Nanogap of Si in NiSi/Si/NiSi Heterostructure

In Si nanowires, silicide formation occurs at both ends of the Si nanowire. If annealing is stopped before the entire Si nanowire transforms into NiSi, as shown in Fig. 3.10*a*, *b*, *d*–*g*, a nanogap is formed as shown in Fig. 3.10*g*. Figure 3.10*c* shows the whole Si nanowires has transformed to NiSi.

Based on knowing the growth rates, we can control the remaining length of the Si region or gap between the two NiSi regions. Also, we can pattern or deposit two Ni nanowires with a given spacing over a Si wire, and then utilize the reaction to bring the two NiSi grains as close as possible to form a nanogap. Furthermore, we can place a Si nanowire over two Ni pads and followed by annealing to form the nanogap.[54,55] At 500°C, the reaction rate can be controlled down to atomic scale. In Fig. 3.10*d*–*g*, a set of lattice images of the nano-heterostructure of NiSi/Si/NiSi with 11.3, 8.1, 5, and 2 nm lengths of Si nanogap is shown.

By measuring the length of the Si gap and counting the number of (111) lattice planes within the gap, we can determine the strain, and we find that the Si in the gap is highly compressed. Figure 3.10*h* shows the relationship between the compressive strain and the length of the Si gap in the nano-heterostructure NiSi/Si/NiSi at room temperature. From the difference in the interplanar spacing of Si and NiSi across the epitaxial interface, Si would be stretched radially in tension at the interface. Thus, on the basis of Poisson's ratio, Si in the nanogap will be compressed axially. Assuming a constant unit cell volume, when the x- and y-axis misfits are 5.6 percent in tension, the strain in the z-axis will be about twice in compression. On the other hand, for fixed strain energy, the average energy distributed per atomic layer in the Si would be larger in a smaller Si gap; as a consequence, the average strain increased when the length of the Si gap decreased. However, the strain can be controlled because we can control the length of the Si nanogap.

Homogeneous Nucleation of Nanoscale Silicide in Si Nanowires

In this section, we focus on the nucleation of epitaxial silicide in Si nanowires. The finding of repeating events of homogeneous nucleation of epitaxial NiSi and $CoSi_2$ in the axial direction of nanowire of (111) Si will be discussed. Using HRTEM, we observed that the growth of epitaxial silicide occurs atomic layer by atomic layer and there is a waiting period in between, indicating that it is due to the incubation time of nucleation. Hence the nucleation event is repeating for the growth

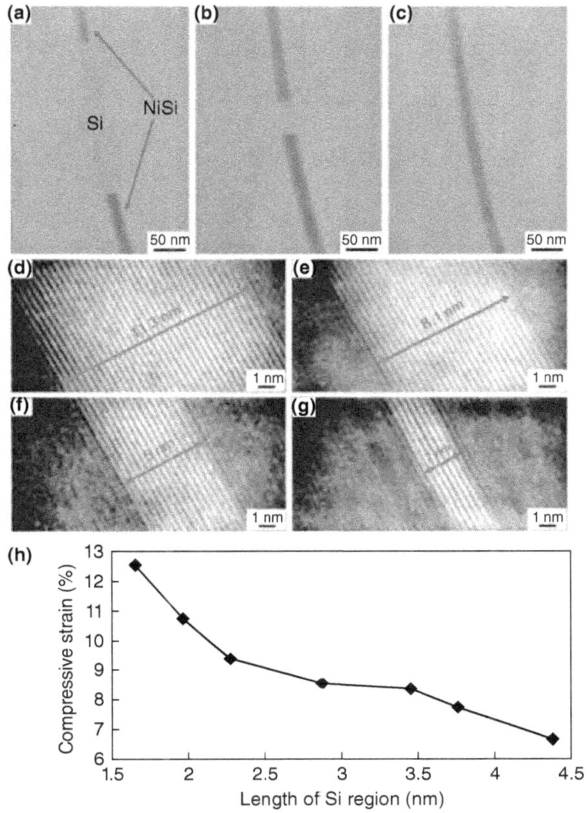

Figure 3.10 In situ TEM images showing the formation of a NiSi/Si/NiSi heterostructure within a Si nanowire, and compressive strain in the Si region. (*a*–*c*) In situ TEM images of the NiSi/Si/NiSi heterostructure. The bright area is Si and the dark area is NiSi. (*d*–*g*) High-resolution TEM images of NiSi/Si/NiSi heterostructures. The bright and dark portions of the lattice images correspond to Si and NiSi, respectively. (*h*) Plot of the compressive strain versus the length of the Si gap in the NiSi/Si/NiSi.[10]

of each atomic layer. It is homogeneous nucleation because heterogeneous nucleation has been suppressed due to the surface oxide of the Si nanowire. Knowing the incubation time, we can calculate the steady-state rate of homogeneous nucleation, that is, the number of stable critical nucleus per unit area per unit time, when the activation energy of nucleation is measured. Because it is homogeneous nucleation, we can apply Zeldovich factor to obtain the number of molecules in forming the critical nucleus in the steady-state process. In this case, we have correlated the theory and experiment of homogeneous nucleation.[13]

Stepwise Growth and Repeating Events of Nucleation

The epitaxial growth mode of NiSi, $NiSi_2$, and $CoSi_2$ in [111] oriented Si nanowires was found to be the same.[10,11,13,56] The growth in the axial direction occurs atomic layer by atomic layer with the moving of steps or kinks across the epitaxial interface, as shown in Fig. 3.11*a*, *b* for NiSi and Fig. 3.11*d*, *e* for $CoSi_2$. Figure 3.11*a*, *b* shows two consecutive HRTEM images of the motion of one NiSi atomic layer across the NiSi/Si interface, and the stepwise growth direction is the step motion in the radial direction of the wire. Similarly Fig. 3.11*d*, *e* shows two images of step motion of $CoSi_2$. During the growth of an atomic layer, the growth mode is a rapid motion of steps or kinks across the epitaxial interface. However, there is a long period of stagnation (the incubation time) before the next stepwise growth of silicide can take place. When we plot the stagnation period as well as the growth period from HRTEM videos, we obtained the stair-type curves as shown in Fig. 3.11*c*, *f*. The

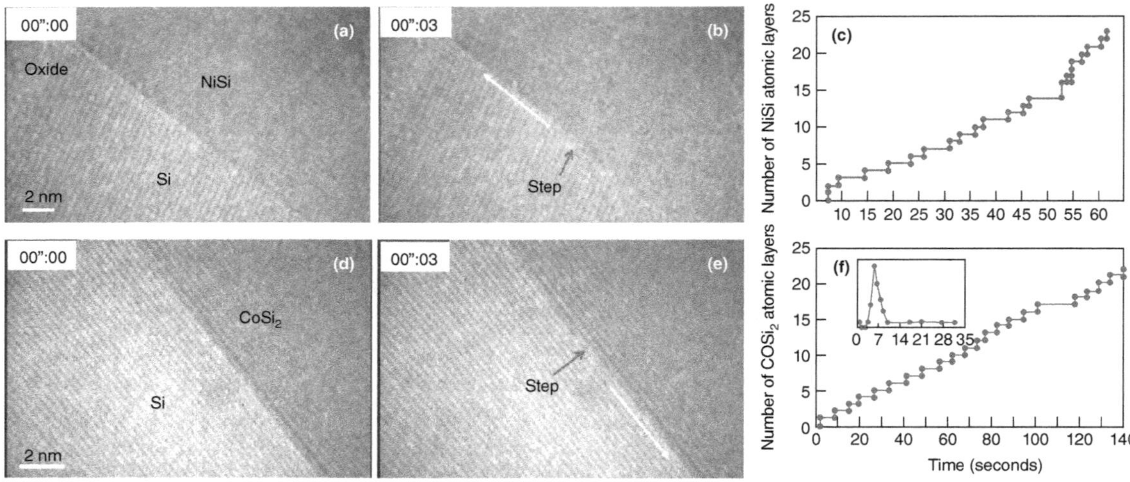

Figure 3.11 Sequence of high-resolution TEM images of motion of a step on epitaxial silicide/Si interfaces and their growth curves. (*a*, *b*) Epitaxial NiSi/Si interfaces at 600°C. (*d*, *e*) Epitaxial $CoSi_2$/Si interfaces at 800°C. The direction of the atomic layer motion is upward from the center of the nanowire to the edge in NiSi and downward in $CoSi_2$. The first two numbers are in units of seconds and the following two smaller numbers are in units of 1/100 s. (*c*, *f*) The stair-type growth curves for NiSi and $CoSi_2$, respectively. The insets are the distribution curves of incubation periods of nucleation.

vertical part of the stair-type steps shows that the growth rate or the time of growth of each silicide atomic layer is the same, which is about 0.06 s per layer for NiSi and about 0.17 s per layer for $CoSi_2$, and we note that it is just the width of the vertical line in the stair-type curves. In between the vertical lines (the horizontal part of the stair steps) is the stagnation period, which we define as the incubation time of nucleation of a new layer.

Since nucleation is a fluctuation phenomenon, the incubation periods might be different, caused by fluctuation inside the Si nanowires so that the incubation time of nucleating each atomic layer can vary slightly. Accordingly, the plots of the distribution of incubation time are shown in the inset in Fig. 3.11*c*, *f*. The average value of the incubation time of NiSi is about 3 s and that of $CoSi_2$ is about 6 s. Because the nucleation stages and growth stages can be separated, the repeating nucleation events can be isolated, and it has been used to study nanoscale silicide nucleation in Si nanowires experimentally and theoretically, as discussed further.

Homogeneous Nucleation of Epitaxial Silicide in Si Nanowires—Experimental Observation

Where is the nucleation site of the silicide? Due to surface oxide of the Si nanowire, it is likely that the energy of the interface of silicide/SiO_2 is higher than that of Si/SiO_2; therefore, the frequency of heterogeneous nucleation at the edge is low, and the nucleation of a circular disc of epitaxial silicide at the center of the epitaxial interface becomes possible. The fast radial growth of NiSi and $CoSi_2$ silicides seems to follow its nucleation from the center of the epitaxial silicide/Si interface rather than from the edge of the interface.

By the observations from in situ HRTEM recordings, we find that the fast radial growth of NiSi and $CoSi_2$ atomic layers starts from the center rather than from the edge of the silicide/Si interface. Both NiSi and $CoSi_2$ atomic layers grow toward the circular edge of the oxide wall of Si nanowires. In other words, the step moves toward the edge of the nanowire, rather than moving away from the edge.

Typically, a native oxide has low formation energy. We assume that the energy of the oxide/silicide interface is higher than that of the native oxide/Si interface. In addition, we found that when a step approaches the edge of the Si nanowire, it will slow down before it transforms the edge from the oxide/Si interface to the oxide/silicide interface because the latter has a high energy barrier. We can see a curvature of the untransformed Si near the oxide/Si edge[13] because several layers of silicide have greatly slowed down their growth within atomic distance in approaching the edge. Furthermore, we kept the electron beam at the edge region and waited for heterogeneous nucleation to take place, but we were unable to observe one.

Figure 3.12 shows that one NiSi atomic layer disc nucleates and grows from the middle region of the interface and spreads toward both ends of the oxide wall of the Si nanowire. This behavior is repeated for every atomic layer growth, which is a direct evidence of homogeneous nucleation of silicide in the epitaxial interface in Si nanowires.

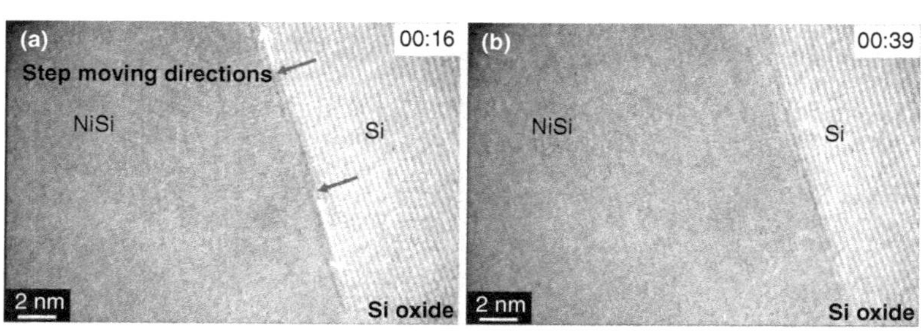

Figure 3.12 High-resolution TEM images showing motion of steps on epitaxial NiSi/Si interface at 450°C. (*a*) One NiSi atomic layer grows from the middle region of the interface and two steps are formed as indicated by black arrows. (*b*) The motion of the two steps is toward both ends of oxide as indicated by the white arrows. After two steps, reaches to the oxide ends, the layer growth ends and the interface becomes flat without steps.

The heterogeneous nucleation of a step at the edge is depicted in Fig. 3.13*a*. To form the step, the low-energy oxide/Si interface will be replaced by the high-energy oxide/silicide interface, which is energetically unfavorable. There is no microreversibility, so the heterogeneous nucleation is suppressed. In Fig. 3.13*b*, a schematic diagram of a heterogeneous nucleus is assumed with a wetting angle larger than 90°. At the triple point, we consider $\gamma_{\text{silicide/oxide}} \geq \gamma_{\text{Si/oxide}} + \gamma_{\text{Si/silicide}} \cos(180-\theta)$, where γ represents the surface energy per unit area of the interfaces. Note that the epitaxial interface between Si and silicide is a low-energy interface. When the inequality is satisfied and $\theta = 180°$, heterogeneous nucleation will not occur; instead homogeneous nucleation of a circular disk in the center of the nanowire becomes possible as depicted in Fig. 3.13*c*, *d*.

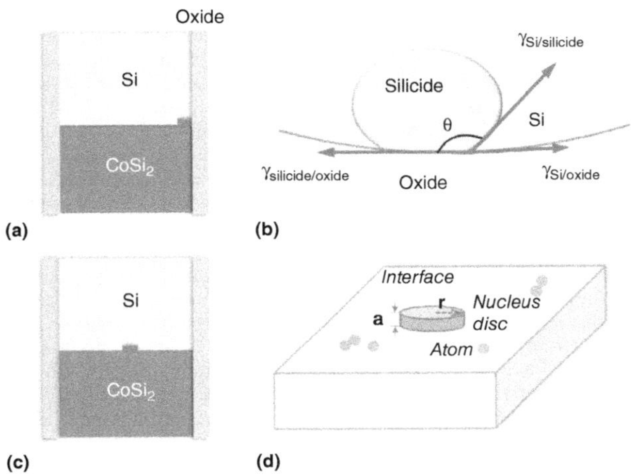

Figure 3.13 Schematic diagrams of the cross section of the epitaxial interface between silicide and Si. (*a*) It depicts a heterogeneous nucleation of a step at the right-hand side corner. (*b*) Triple point configuration of a heterogeneous nucleus. (*c*) Cross section of homogeneous nucleation of a disk in the center of the silicide/Si interface. (*d*) Schematic diagram of the nucleation of a circular disk on the interface.

In homogeneous nucleation of a circular disk as depicted in Fig. 3.13*d*, the net change in energy is $\Delta G = 2\pi r a \gamma - \pi r^2 a \Delta G_s$. The critical nucleus has a size $r_{\text{crit}} = \gamma/\Delta G_S$ and the activation energy in nucleating the critical disk is $\Delta G^* = \pi r_{\text{crit}} a \gamma$, where ΔG_s is the gain in free energy of formation of the silicide per unit volume, γ is the interfacial energy per unit area of the circumference of the disk, and a is atomic height.

Homogeneous Nucleation of Epitaxial Silicide in Si Nanowires—Correlation between Experiments and Theory

Knowing the activation energy of formation of the critical disc, the probability of nucleation of the metastable critical nucleus can be calculated, that is, the number of critical nuclei per unit area per unit time. Experimentally what we have measured in Fig. 3.11*c*, *f* is the rate of forming one stable critical nucleus on the cross section of the Si/silicide nanowire in the period of one incubation time. Thus, we have the nucleation rate of

$$I_{\text{stable-crit}} = \frac{1}{\pi R^2 \tau_i} \tag{3.5}$$

where R is the radius of the Si nanowire, and τ_i is the incubation time. Taking the diameter of Si nanowire to be 30 nm and the incubation time to be 3 s, we have

$$I_{\text{stable-crit}} = 4.7 \times 10^{10} \text{ stable nuclei/cm}^2\cdot\text{s}$$

for the nucleation of NiSi.

In the steady-state reaction, during one period of incubation, it must have dissolved a sufficient amount of Ni atoms to grow one atomic layer of the silicide. We assume that there are 10^{15} atoms per atomic layer per cm^2, and the flux of Ni needed to grow an atomic layer is

$$J_{\text{Ni}} = \frac{10^{15}}{2\times 3} = 1.67\times 10^{14} \text{atoms/cm}^2\,\text{s} \tag{3.6}$$

where the factor of 2 in the denominator is because of concentration of Ni in NiSi is half and the factor 3 is from the incubation time. While we can regard this to be the flux of Ni atoms being deposited onto the silicide/Si interface, we should consider the adatoms on the interface, and we assume that only the adatoms can take part in the nucleation process. However, the adatoms have a residence time, τ_{des}, on the interface because of desorption.

$$\tau_{\text{des}} = \frac{1}{v_s} \exp\frac{\Delta G_{\text{des}}}{KT} \tag{3.7}$$

where v_s is the vibration frequency of an adatom, ΔG_{des} is the activation energy of desorption of an adatom, and kT is thermal energy. Thus, the product of $J_{\text{Ni}} \cdot \tau_{\text{des}}$ is the effective number of adatoms per unit area involved in the nucleation process. Then, the equilibrium concentration of critical nuclei can be given as

$$C_{\text{crit}} = J_{\text{Ni}} \tau_{\text{des}} \exp\left(-\frac{\Delta G^*}{KT}\right) \tag{3.8}$$

where C_{crit} has the unit of number of nuclei per unit area. The steady-state homogeneous nucleation rate has been given as

$$I_{n^*}^s = \beta_n^* C_{\text{crit}} Z = \beta_n^* C_o e^{-\frac{\Delta G_n^*}{KT}} \left[-\frac{1}{2\pi KT} \left(\frac{\partial^2 \Delta G_n}{\partial n^2} \right)_{n^*} \right]^{1/2} \tag{3.9}$$

where β_{n^*} is the frequency of atomic jump toward an critical nucleus that converts it into a stable nucleus, and $C_{\text{crit}} = C_0\exp(-\Delta G_{n^*}/KT)$ is the equilibrium concentration of critical size nucleus, which we note is the same as Eq. (3.8). The Zeldovich factor "Z" has been included in the nucleation rate equation as a kinetic factor that stands for the percentage of critical size nuclei that become stable. The Zeldovich factor is less than 1 in all real cases.

By assuming a circular disc shape of nucleus of atomic height, the Zeldovich factor can be rewritten as

$$Z = \left[\frac{1}{4\pi KT} \cdot \frac{\Delta G^*}{(n^*)^2} \right]^{1/2} \tag{3.10}$$

where ΔG^* and n^* are respectively the activation energy in forming the critical nucleus and the number of molecules in it.

Knowing the activation energy of NiSi to be 1.25 eV per atom, we can calculate n^* at a given value of Z. Typical experimental value of Z factor is about 0.05.

Since we know the experimentally measured steady-state nucleation rate as given by Eq. (3.5), we can check it by using Eqs. (3.8) and (3.9). We have

$$\begin{aligned} I_{\text{stable-crit}} &= \beta_n^* J_{\text{Ni}} \tau_{\text{des}} \exp\left(-\frac{\Delta G^*}{kT}\right) Z \\ &= \beta_n^* J_{\text{Ni}} \frac{1}{v_s} \exp\left(-\frac{\Delta G^* - \Delta G_{\text{des}}}{kT}\right) Z \\ &= v_0 J_{\text{Ni}} \frac{1}{v_s} \exp\left(-\frac{\Delta G^* - \Delta G_{\text{des}} + \Delta G_\beta}{kT}\right) Z \end{aligned} \tag{3.11}$$

where we assume that $\beta_{n^*} = v_0\exp(-\Delta G_\beta/KT)$, and v_0 is Debye frequency of vibration and ΔG_β is the activation energy of adding an atom to the critical nucleus. It is worthwhile mentioning that the basic nature of the parameter of β_{n^*} is microreversibility. In order to maintain the equilibrium distribution of subcritical size embryos in nucleation, the frequency of adding and subtracting atoms among the embryos is high. Hence, we can assume $\Delta G^* \gg \Delta G_\beta$, so we can ignore ΔG_β.

To evaluate the products on the right-hand side of the last Eq. (3.11), we cancel v_0 against v_s owing to the fact that both are Debye frequency of atomic vibration. For ΔG_{des}, it is known from epitaxial growth of Si on Si, where $\Delta G_{des} = 1.1$ eV/atom. For the desorption of Ni, the activation energy should be lower and we assume $\Delta G_{des} = 0.7$ eV/atom. Then, we take the measured $\Delta G^* = 1.25$ eV/atom and $Z = 0.1$. Since $J_{Ni} = 1.67 \times 10^{14}$ atoms/cm$^2\cdot$s, the products on the right-hand side at $T = 700°C$ is 3×10^{10} nuclei/cm$^2\cdot$s, which is in good agreement with the measured nucleation rate of 4.7×10^{10} nuclei/cm$^2\cdot$s.

We caution that there is some uncertainty about ΔG_{des}. While we have no measured data, we note that even if we give it a high uncertainty by taking $\Delta G_{des} = 0.7 \pm 0.2$ eV/atom, it will only change the outcome by a factor about 10.

Homogeneous Nucleation of Epitaxial Silicide in Si Nanowires—Supersaturation

Nucleation requires supersaturation or undercooling. We can calculate the supersaturation in the nucleation of NiSi. The solubility of Ni in Si at 700°C is about 10^{15} to 10^{16} Ni atoms/cm^3. Since there are 2.5×10^{22} Si atoms/cm^3, the equilibrium concentration of Ni in Si is about 10^{-7} to 10^{-8}, which is extremely low. Therefore, even if we consider supersaturation of 1000, it is just 10^{-4} to 10^{-5}. When we have dissolved half of a monolayer of Ni (which has a layer thickness of 0.3 nm) into a Si nanowire of 3 μm long before homogeneous nucleation occurs, the concentration of Ni is 0.5×10^{-4}, which is about the supersaturation of 1000.

Further we show that the high supersaturation is expected for homogeneous nucleation. Considering the equilibrium solubility of Ni over the critical disc, we have from the Gibbs-Thomson equation that $n_{crit}/n_0 = \exp(\gamma\Omega / r_{crit}KT)$, where n_{crit} and n_0 are the equilibrium solubility of Ni above a flat silicide/Si interface and that above a critical nucleus, respectively.

Thus, $$r_{crit} = \frac{\gamma\Omega}{kT\ln\left(\frac{n_{crit}}{n_0}\right)} \quad \text{and}$$

$$\Delta G^* = \pi r_{crit} a\gamma = \frac{\pi\gamma^2 a\Omega}{kT\ln\left(\frac{n_{crit}}{n_0}\right)} = \frac{\pi\gamma^2 a^4}{kT\ln\left(\frac{n_{crit}}{n_0}\right)}.$$

In the last step of the above equation, we have taken $\Omega = a^3$. Since we know $\Delta G^* = 1.25$ eV/atom, we can calculate the ratio of n_{crit}/n_0, provided that we know γa^2, the interfacial energy per cross-sectional area of an atom of the epitaxial interface of silicide/Si. For reference, we know that on (111) surface of Si, each surface atom has a broken bond, so the surface energy is about 1 eV/atom$\cdot$area. Here we shall take $\gamma = 0.5$ eV/a^2 (which is about 800 erg/cm^2) and $KT = 0.084$ eV at 973 K, we have $n_{crit}/n_0 = 1800$ from the equation below. For $\gamma = 0.4$ eV/a^2, we have $n_{crit}/n_0 = 120$.

$$\Delta G^* = \frac{\pi(\gamma a^2)^2}{kT\ln\left(\frac{n_{crit}}{n_0}\right)} = \frac{3.14 \times 0.25}{0.084\ln\left(\frac{n_{crit}}{n_0}\right)}$$

Since ΔG^* is inversely proportional to $\ln(n_{crit}/n_0)$, if we take $n_{crit}/n_0 = 1000$, we obtain $\Delta G^* = 1.35$ eV/atom; $n_{crit}/n_0 = 100$ implies $\Delta G^* = 2$ eV/atom; and $n_{crit}/n_0 = 10$ implies $\Delta G^* = 4.1$ eV/atom. It shows that for a low supersaturation just over unity, the activation energy will be extremely high. This is the reason why in most real events of nucleation at a low supersaturation, it is heterogeneous rather than homogeneous nucleation.

In Situ Observation of Elastic Strain Fluctuation during Epitaxial Silicide Formation

Here, we discuss the local strain variation during silicide formation in Si nanowires at the epitaxial silicide/Si interfaces in TEM. We correlate the observations with every nucleation event for $NiSi_2$ formation and discuss the thermodynamics of the chemical reactions.

We show the high-resolution TEM image of $NiSi_2$/Si heterostructure in atomic resolution in a [110] and a [111] oriented Si nanowire, where the interface is atomically sharp with epitaxial relation of $NiSi_2[\bar{1}11]$//$Si[\bar{1}11]$ and $NiSi_2(220)$//Si(220), and $NiSi_2[011]$//Si[011] and $NiSi_2(11\bar{1})$//$Si(11\bar{1})$, respectively.[14] The misfit strain at the interface between $NiSi_2$ and Si in both cases is calculated to be 0.46 percent at room temperature, where Si is under compression due to the transformation from Si to $NiSi_2$, which has a smaller lattice constant.

We show the distribution of strain fringes with dark-field imaging mode in the $NiSi_2$/Si heterostructure nanowires from in situ videos taken during reactions.[14] Previously, we show a nanowire of $NiSi_2$/Si multiple heterostructures by point contact reactions of Ni nanoparticles and a Si nanowire at 777°C and shows another nanowire of $NiSi_2$/Si heterostructure at dark-field imaging condition, where the fringes indicate the local strain field in the nanowire. The contrast in $NiSi_2$ is much more uniform than that in Si, whereas the Si segment contains contrast fringes which indicate strain distribution. This demonstrates that the strain caused by silicidation is mostly located in the unreacted Si region and the silicide is about strain free or with low strain. A Si/$NiSi_2$/Si heterostructure in both bright-field and dark-field imaging conditions, where a short $NiSi_2$ segment is formed in a Si nanowire was shown. The bright-field image shows an overview of the heterostructure; whereas the dark-field image lights up the {110} planes and indicates the strain induced by the phase transformation from Si to $NiSi_2$. The strain is induced by the new phase, the short segment of $NiSi_2$, which acts as the center of the strain field and the strain distribution is symmetrical in the heterostructure. As different from the previous cases, the strain in this nanowire is expected to be large since the lattice mismatch at the interface of Si {220} and $NiSi_2$ {042}. The misfit at the interface was calculated to be 26 percent from RT data, which was very high so that the fringes in this case were significantly obvious in the dark-field TEM image. The smooth fringes in the dark-field image exhibit the strain is highly elastic and without plasticity such as dislocation formation. The lattice constants of $NiSi_2$ and Si are similar, with the difference about 0.46 percent, and both of the phases have the same crystal structure so they typically form epitaxial interface with little strain. The image showing large strain is an extreme case to show the symmetrical distribution of strain with high elasticity in the single crystal Si nanowire.

The strain in the Si nanowires was induced by volume change during phase transformation, from Si to $NiSi_2$, by considering the total volume of the nanowire is finite due to native oxide coverage on the nanowire surface. Due to the coverage of Si native oxide on the Si nanowire, the strain level in the Si nanogap is maintained during the new phase formation. Because the strain could possibly play a significant role in nucleating a new phase, we discuss below the implications of the effect of strain at the silicide/Si interface on nucleation and growth of the silicide.

At room temperature, the lattice constant of $NiSi_2$ is smaller than that of Si so the Si lattice near the interface is under compression. However, by considering thermal expansion of lattices during reactions at the reaction temperature,[9] say 630°C (903 K), the lattice constant of $NiSi_2$ at 630°C is larger than that of Si so that the Si lattice at the $NiSi_2$/Si interfaces, in the direction parallel to the plane of interface (say the r direction), is stretched by about 0.55 percent. Thereby, the Si lattice in the direction normal to the epitaxial interface (say the z direction) is compressed up to 1.1 percent, which is the linear sum of the strain in r direction as illustrated in Fig. 3.14*a*.

When we dissolve a high supersaturation of Ni in Si, the interstitial Ni atoms will expand the Si lattice; in other words, the Si is under compression because the Si atoms are squeezed. In the case we discussed, the Si lattice at the interfaces is under tension in the r direction by misfit, the interstitial Ni atoms will reduce the misfit strain at the interface between $NiSi_2$ and Si, as illustrated in Fig. 3.14*b*. Thus, the Si lattice in r direction near the interface acts as "sinks" for Ni atoms so that a high degree of solubility of supersaturation of Ni can take place, which is required in the homogeneous nucleation of a new layer of $NiSi_2$ on Si as described in the previous section "Homogeneous Nucleation of Epitaxial Silicide in Si Nanowires—Correlation between Experiments and Theory."

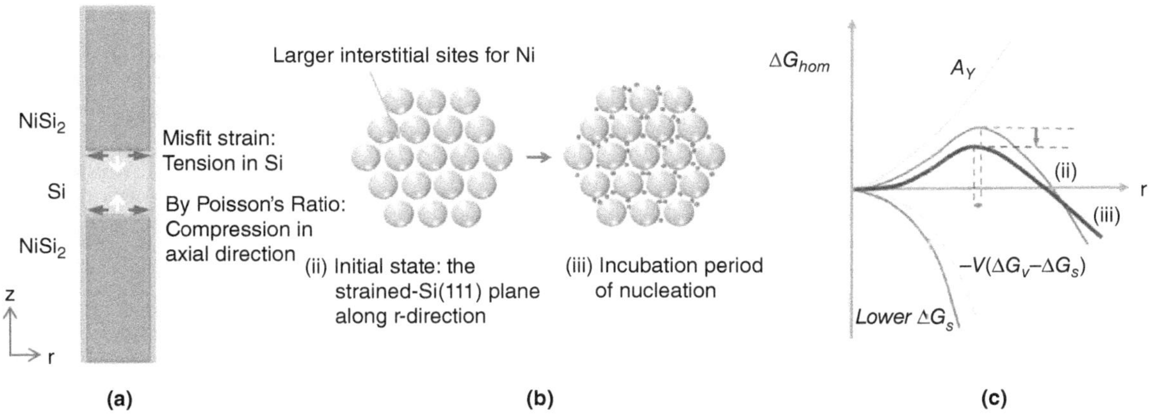

Figure 3.14 Schematic diagrams of a $NiSi_2$/Si/$NiSi_2$ heterostructure and its atomic arrangement at the cross-section interfaces. (*a*) After phase transformation of a Si nanowire to a $NiSi_2$/Si/$NiSi_2$ heterostructure. The arrows indicate the directions of strain where the black arrows show the expansion of Si lattice caused by the misfit at Si/$NiSi_2$ interfaces at the reaction temperature and the white arrows show the compression caused by the resistance of volume expansion by the native oxide confinement. (*b*) Atomic arrangement of Si(111) lattice near the interface with (right) and without (left) strain. (*c*) The variation of ΔG with radius for homogeneous nucleation by taking account of tensile strain in Si lattice at the interface as in (ii) and after the supersaturation of Ni atoms fill up the interstitial sites as in (iii).

Note that the supersaturation required for the homogeneous nucleation is about 1000.[13]

After the nucleation and growth of an epitaxial atomic layer of $NiSi_2$, it has consumed the supersaturated Ni, so the Si lattice returns to the initial state with misfit strain again and waits to nucleate a new layer during another incubation period. As shown schematically in Fig. 3.14*a*, Si lattice is under tension in *r* direction by lattice misfit which tends to expand the nanowire radius. However, the Si nanowire does not have a free surface but rather a stable native oxide, which tends to resist the radial expansion of the Si nanowire. For this reason, the misfit strain accumulates in the Si lattice. The Si nanogap therefore is under compression in *z* direction. The supersaturation of Ni solutes is supplied to the stretched Si lattices, as shown in Fig. 3.14*b*, and release the misfit strain. The thermodynamic consideration indicates that the initial state with strained Si at the interface (the deep green curve in Fig. 3.14*c*) requires high energy for reaction to take place, while the energy barrier gets lower during incubation period as shown by the blue curve in Fig. 3.14*c*.

We have shown the variation of the strain level during atomic layer reaction of $NiSi_2$ formation in a Si nanowire[14] as sequence of dark-field TEM images and the corresponding illustrations, which directly indicates the effect of strain on silicide formation. The dark-field imaging, which shows the crystallinity directly, demonstrates that the electron beam diffraction conditions, at the interface where $NiSi_2$ forms, keeps varying during the reaction. It supports the model of kinetic processes of repeating nucleation and growth at the epitaxial $NiSi_2$/Si interface. At the nucleation stage, it has tensile-strained Si at the interface and it can dissolve Ni atoms at the relatively larger interstitial sites in Si lattice until the required amount of supersaturated Ni atoms is reached for homogeneous nucleation. Note that it has compression in Si lattice along *z* direction so the Ni atoms tend to accumulate along *r* direction, which supports the fact that new layers grow along *r* direction. Once it reaches the supersaturation of Ni required for nucleation, the Ni atoms will be quickly consumed to form a critical size of nucleus (as shown in Fig. 3.13*d*) and followed by a rapid radial growth of an atomic layer of epitaxial $NiSi_2$ on Si. During the incubation time, the strain is released by dissolving Ni atoms so that the gray contrast shrinks and returns to flat interface. Once the nucleation and growth start, which consumes the supersaturation of Ni, the Si is depleted of Ni and back to the strained Si state as shown in Fig. 3.14*b*.

In homogeneous nucleation, the total free energy change is

$$\Delta G_{\text{hom}} = -V\Delta G_v + A\gamma + V\Delta G_s = -\frac{4}{3}\pi r^3(\Delta G_v - \Delta G_s) + 4\pi r^2\gamma$$

where $V\Delta G_v$ is the volume free energy reduction from forming a new phase with volume *V*, $A\gamma$ is the new interfacial energy, and $V\Delta G_s$ is the misfit strain energy. This is shown as a function of nucleus radius. The $(\Delta G_v - \Delta G_s)$ is the driving force of the phase transformation, where we note that the misfit strain energy is to reduce the effective driving force for forming a new phase. Specifically, in our case the supersaturated interstitial Ni releases the misfit strain and enhances the driving force of the chemical reaction. Taking into account of the strain energy, the critical radius of nuclei is smaller and critical activation energy of forming $NiSi_2$ is lowered, as indicated by the arrows in Fig. 3.14*c*.

Uphill Diffusion of Ni in the Strained Si

Due to the high supersaturation of Ni needed in nucleation, the diffusion of Ni to the epitaxial interface is uphill because it increases the solubility in the Si until the needed supersaturation is reached. The driving force of the diffusion is not due to the Ni concentration gradient but rather due to stress chemical potential gradient. The chemical potential of Ni atoms in the Si is affected by the strain. The interstitial Ni lowers the tensile strain in the Si induced by the epitaxial silicide/Si interface. It enables a large solubility of interstitial Ni in Si, so the uphill diffusion of Ni to go to the epitaxial interface can occur.

In a short summary, we have shown the periodic variation of misfit strain during the formation of epitaxial $NiSi_2$ in Si nanowires. The atomic layer growth occurs periodically and the period depends on the incubation time of nucleation. Each atomic layer growth of the epitaxial silicide requires an event of homogeneous nucleation. The Si lattice was under tensile strain due to the interfacial misfit whereas it was released by the interstitial solution of super-saturated Ni atoms in the strained Si lattice before nucleation. After nucleation, the strain relaxation occurred during the growth of an atomic layer of $NiSi_2$. The generation and relaxation of the misfit strain is synchronized with the nucleation and atomic layer growth. The strain effect enables a high degree of supersaturation of Ni to take place near the interface. The strain returns to the initial state once growth has occurred.

Summary

In this section, we have reviewed the growth of epitaxial silicide in nanowires of Si. We covered the repeating homogeneous nucleation during the epitaxial growth as well as the strain effect on nucleation. Because the silicide growth after each homogeneous nucleation is only atomic layer thick, it is way below the critical thickness defined in epitaxial growth of thin films on Si wafers which is controlled by misfit strain. So far, we find that the misfit strain between the epitaxial silicide and Si does not play a critical role as in thin film growth on Si wafers. Actually, we have found very larger misfit strain between the epitaxial silicide and the nanowires of Si without misfit dislocations. It seems that as long as an epitaxial silicide nucleus on Si can take place, it can grow. nanoscale epitaxial process is controlled by nucleation rather than by growth. This finding may have an important effect on the silicide contacts and gates in FinFET devices because we expect that the epitaxial interfaces may have a low contact resistance. No doubt, the study of nucleation and growth epitaxial silicide in nanowires of Si is far from complete, we need more extended and systematic studies. Clearly, in situ HRTEM is indispensible.

3.4 CONCLUSION

In the era of big data and Internet of Things, microelectronic Si devices will go less than 10 nm node, so most building blocks of the device will be in nanoscale. In this chapter, we review the challenges in S/D and gate contacts in FinFET technology as well as the epitaxial nucleation and growth of silicide in Si nanowires. The contact resistance in FinFET devices is a critical issue. We have discussed several approaches to reduce it. Especially we review the epitaxial NiSi in [111] oriented Si nanowires. This is because we expect an epitaxial interface may have a low contact resistance. However, Si nanowires can grow in [111] and [110] directions, but not in [100] direction. Yet Si wafer are all [100] oriented. Clearly there is a need to study nanoscale epitaxial growth of NiSi on [100] Si. We can do so by using lithographic technique to define the nanoscale NiSi on [100] Si. Furthermore, the contact resistance can be measured directly by using Kelvin probe. Still, how to introduce nanoscale epitaxial silicide contacts in real Si devices is very challenging.

3.5 REFERENCES

1. L. P. Ren and K. N. Tu, "Fundamentals—Silicide Formation on Si," Chapter 5 in *Semiconductor Manufacturing Handbook*, edited by Hwaiyu Geng, McGraw-Hill, New York, pp. 5.1 to 5.10, 2005.

2. N. Arora, *MOSFET Models for VLSI Circuit Simulation: Theory and Practice*, Springer Science and Business Media, Vienna, Austria, p. 288, 1993.

3. D. Hisamoto, W.-C. Lee, J. Kedzierski, E. Anderson, H. Takeuchi, K. Asano, T.-J. King, J. Bokor et al., "A Folded-Channel MOSFET for Deep-Sub-Tenth Micron Era," *IEEE Int. Elec. Dev. Meeting Tech. Dig.*, IEEE, New York, 1032–1034, 1998.

4. X. Huang, W.-C. Lee, C. Kuo, D. Hisamoto, L. Chang, J. Kedzierski, E. Anderson et al., "Sub 50-nm FinFET: PMOS," *IEEE Int. Elec. Dev. Meeting Tech. Dig.*, New York, 67–70, 1999.

5. T.-J. King Liu, "FinFET History, Fundamentals and Future," VLSI Technology Short Course, June 11, 2012.

6. C. Hu, "SOI Can Empower New Transistors to 10 nm and Beyond," Advanced substrate corners—ASN #19—Professor's Perspective, April 23, 2012.

7. D. Lu, J. Chang, M. A. Guillorn, C.-H. Lin, J. Johnson, P. Oldiges, K. Rim et al., "A Comparative Study of Fin-Last and Fin-First SOI FinFETs," 2013 International Conference on Simulation of Semiconductor Processes and Devices (SISPAD), IEEE, New York, pp. 147–150, Sep. 3–5, 2013.

8. L. P. Ren and K. N. Tu, "Silicide Technology for SOI Devices," Chapter 8 in *Silicide Technology for Integrated Circuits*, edited by L. J. Chen, Institute of Electrical Engineers, UK, pp. 201–228, 2005.

9. Y. C. Chou, K. C. Lu, and K. N. Tu, "Nucleation and Growth of Epitaxial Silicide in Silicon Nanowires," *Mater. Sci. Eng. R.*, 70: 112–125, 2010.

10. K. C. Lu, W. W. Wu, H. W. Wu, C. M. Tanner, J. P. Chang, L. J. Chen, and K. N. Tu, "In Situ Control of Atomic-Scale Si Layer with Huge Strain in the Nanoheterostructure NiSi/Si/NiSi through Point Contact Reaction," *Nano Lett.*, 7: 2389–2394, 2007.

11. Y. C. Chou, W. W. Wu, S. L. Cheng, B. Y. Yoo, N. Myung, L. J. Chen, and K. N. Tu, "In Situ Control of Atomic-Scale Si Layer with Huge Strain in the Nanoheterostructure NiSi/Si/NiSi through Point Contact Reaction," *Nano Lett.*, 8: 2194–2199, 2008.

12. D. Turnbull and R. E. Cech, "Homogeneous Nucleation of Epitaxial $CoSi_2$ and NiSi in Si Nanowires," *J. Appl. Phys.*, 21: 804–810, 1950.

13. Y. C. Chou, W. W. Wu, L. J. Chen, and K. N. Tu, *Nano Lett.*, 9: 2337–2342, 2009.

14. Y. C. Chou, W. Tang, C. J. Chiou, K. Chen, A. M. Minor, and K. N. Tu, "Effect of Elastic Strain Fluctuation on Atomic Layer Growth of Epitaxial Silicide in Si Nanowires by Point Contact Reactions," *Nano Lett.*, 15: 4121–4128, 2015.

15. A. Dixit, A. Kottantharayil, N. Collaert, M. Goodwin, M. Jurczak, K. De Meyer, *IEEE Trans. Elec. Dev.*, 52 (6): 1132–1140, 2005.

16. D. J. Plummer, M. D. Deal, and P. B. Griffin, "Silicon VLSI Technology-Fundamentals, Practice and Modeling," Prentice Hall Electronics and VLSI Series-Charles Sodini, Series Editor, 2000.

17. M. C. Ozturk and J. Liu, *Intl. Conf. on Char. and Metrology for ULSI Technology*, 2005.

18. S. D. Kim, C. M. Park, J. C. S. Woo, *IEEE Trans. Elect. Dev.*, 49: 467–472, 2002.

19. T. J. Drummond, "Work Functions of the Transition Metals and Metal Silicides," U.S. Gov. report from Sandia National Laboratories, 1999.

20. K. Choi,T, P. Lysaght, H. Alshareef, C. Huffman, H.-C. Wen, R. Harris, H. Luan et al. "Growth Mechanism of TiN Film on Dielectric Films and the Effects on the Work Function" 141–144, *IEDM Tech. Dig.*, 103, 2005.

21. C. E. Smith, "Advanced Technology for Source Drain Resistance Reduction in Nanoscale FinFETs," Dissertation of Doctor of Philosophy of University of North Texas, May 2008.

22. T. Yamamoto, T. Kubo, T. Sukegawa, E. Takii, Y. Shimamune, N. Tamura, T. Sakoda et al., *IEDM Tech. Dig.*, 143–146, Dec. 2007.

23. C. Ortolland, T. Noda, T. Chiarella, S. Kubicek, C. Kerner, W. Vandervorst, A. Opdebeeck et al., "Laser-Annealed Junctions with Advanced CMOS Gate Stacks for 32nm Node: Perspectives on Device Performance and Manufacturability," *Symp. on VLSI Tech.*, 186–187, June 17–19, 2008.

24. R. T. P. Lee, T.-Y. Liow, K.-M. Tan, A. E.-J. Lim, H.-S. Wong, P.-C. Lim, D. M. Y. Lai et al. "Novel Nickel-Alloy Silicides for Source/Drain Contact Resistance Reduction in N-Channel Multiple Gate Transistors with Sub-35 nm Gate Length," *IEDM Tech. Dig.*, 851–854, Dec. 2006.

25. K. Ohuchi, C. Lavoie, C. Murray, C. D'Emic, I. Lauer, J. O. Chu, B. Yang et al. "Extendibility of NiPt Silicide Contacts for CMOS Technology Demonstrated to the 22-nm Node," *IEDM Tech. Dig.*, 1029–1031, Dec. 2007.

26. J. Gelpey, S. McCoys, A. Kontos, L. Godet, C. Hatem, D. Camms, J. Chan, et al. "Ultra-Shallow Junction Formation Using Flash Annealing and Advanced Doping Techniques," *IWJT '08*, 82–86, 15–16, May 2008.

27. Z. Zhang, Z. Qiu, R. Liu, M. Ostling, S. L. Zhang. "Schottky-Barrier Height Tuning by Means of Ion Implantation into Preformed Silicide Films Followed by Drive-In Anneal," *IEEE Elec. Dev. Lett.*, 28 (7): 565–568, 2007.

28. G. Larrieu, E. Dubois, R. Valentin, N. Breil, F. Danneville, G. Dambrine, J.P. Raskin. "Low Temperature Implementation of Dopant-Segregated Band-edge Metallic S/D junctions in Thin-Body SOI p-MOSFETs," *IEDM Tech. Dig.*, 147–150, Dec. 2007.

29. J. G. Ryan, R. M. Geffken, N. R. Poulin, and J. R. Paraszczak, *IBM J. Res. Dev.*, 39, 1995.

30. Adam Brand, "Precision Materials to Meet Scaling Challenges Beyond 14 nm," *Semicon.*, July 2013.

31. C. M. Lieber, *MRS Bull.*, 36: 1052–1063, 2011.

32. J. Appenzeller, J. Knoch, M. T. Bjork, H. Riel, H. Schmid, and W. Riess, *IEEE Trans. on Elec. Dev.*, 55: 2827, 2008.

33. W. Lu and C. M. Lieber, *Nat. Mater.*, 6: 841–850, 2007.

34. Y. Hu, H. O. H. Churchill, D. J. Reilly, J. Xiang, C. M. Lieber, and C. M. Marcus, *Nat. Nanotech.*, 2: 622–625, 2007.

35. Y. Cui, Q. Q. Wei, H. K. Park, and C. M. Lieber, *Science*, 293: 1289–1292, 2001.

36. A. T. Tilke, F. C. Simmel, H. Lorenz, R. H. Blick, and J. P. Kotthaus, *Phys. Rev. B*, 68 (7): 075311, 2003.

37. W. Lu and C. M. Lieber, *J. Phys. D: Applied Physics*., 39: R387–R406, 2006.

38. L. Samuelson, C. Thelandera, M. T. Björka, M. Borgströma, K. Depperta, K. A. Dicka, A. E. Hansena et al, *Physica E*, 25: 313–318, 2004.

39. J. Read, R. J. Needs, K. J. Nash, L. T. Canham, P. D. Calcott, and A. Qteish, *Phys. Rev. Lett.*, 69: 1232, 1992.

40. Y. Yeh, S. B. Zhang, and A. Zunger, *Phys. Rev. B*, 50: 14405, 1994.

41. Y. C. Chou, C. Y. Wen, M. C. Reuter, D. Su, E. A. Stach, and F. M. Ross, *ACS Nano*, 6: 6407–6415, 2012.

42. C.–Y. Wen, M. C. Reuter, J. Bruley, J. Tersoff, S. Kodambaka, E. A. Stach, and F. M. Ross, *Science*, 326: 1247–1250, 2009.

43. D. E. Perea, N. Li, R. M. Dickerson, A. Misra, and S. T. Picraux, *Nano Lett*. 11: 3117–3122, 2011.

44. Z. Zhong, D. Wang, Y. Cui, M. W. Bockrath, and C. M. Lieber, *Science*, 302: 1377–1379, 2003.

45. C. M. Lieber, *MRS Bull*., 28: 486–491, 2003.

46. Y. Wu, J. Xiang, C. Yang, W. Lu, and C. M. Lieber, *Nature*, 430: 61–65, 2004.

47. W. M. Weber, L. Geelhaar, A. P. Graham, E. Unger, G. S. Duesberg, M. Liebau et al., *Nano Lett*., 6: 2660–2666, 2006.

48. R. S. Wagner, and W. C. Ellis, *Appl. Phys. Lett*., 4: 89, 1964.

49. R. S. Wagner, *Whisker Technology*, Wiley, New York, p. 47, 1970.

50. E. R. Weber, *Appl. Phys. A*, 30: 1–22, 1983.

51. L. J. Chen and K. N. Tu, *Mater. Sci. Rep*., 6: 53, 1991.

52. R. T. Tung, J. M. Gibson, and J. M. Poate, *Phys. Rev. Lett*., 50: 429, 1983.

53. K. N. Tu, *Appl. Phys. Letts*., 27: 221, 1975.

54. W. Tang, S. A. Dayeh, S. T. Picraux, J. Huang, and K. N. Tu, *Nano Lett*., 12: 3979–3985, 2012.

55. W. Tang, S. T. Picraux, J. Huang, A. Gusak, K. N. Tu, and S. A. Dayeh, *Nano Lett*., 13: 2748–2753, 2013.

56. Y. C. Chou, W. W. Wu, C. Y. Li, C. Y. Liu, L. J. Chen, and K. N. Tu, *J. Phys. Chem. C*, 115: 397–401, 2011.

Chapter 4

Foundations of Microsystems Manufacturing: An Empowering Technology for the IoT

BY

MICHAEL A. HUFF *MNX, Corporation for National Research Initiatives, Reston, Virginia*

4.1 THE DEFINITION OF MEMS AND MICROSYSTEMS TECHNOLOGIES

Microsystems, or microelectromechanical systems (MEMS), is a technology that in its most general form can be defined as miniaturized mechanical devices and structures made using the techniques of microfabrication. The critical physical dimensions of microsystems devices can vary from well below 1 μm to several millimeters. The types of devices encompassed by microsystems can vary from relatively simple structures wherein there are no mechanically moving elements, to extremely complex electromechanical systems with multiple moving elements under the control of microelectronics integrated into the same silicon substrate. Obviously, the one main criterion of microsystems is that there are at least some elements on the substrate having some sort of mechanical functionality whether or not these elements can move.

Common elements of microsystems are microminiaturized sensors and actuators, which are frequently called microsensors and microactuators. Both are properly categorized as "transducers." Transducers are defined as devices that convert energy from one form to another. In the case of microsensors, the device typically converts a measured mechanical signal into an electrical or optical signal. Over the past several decades, researchers and developers have demonstrated an extremely large number of microsensors for almost every possible sensing modality including temperature, pressure, inertial forces, chemical species, magnetic fields, radiation, etc. Many of these micromachined sensors have demonstrated performances exceeding those of their macroscale counterparts. That is, the microsystems version of, for example, a pressure transducer, usually outperforms a pressure sensor made using macroscale level manufacturing techniques. Not only is the performance of microsensor devices exceptional, but also their method of production leverages the same batch fabrication techniques used in the integrated circuit (IC) industry translating into low per-device production costs, as well as many other benefits. Consequently, it is possible to not only achieve stellar device performance, but to do so at a low cost level. Not surprisingly, silicon-based discrete microsensors have been quickly commercially exploited and the markets for these devices continue to grow at a rapid rate.

More recently, the microsystems R&D community has demonstrated a number of microactuators including microvalves for control of gas and liquid flows; optical switches and mirrors to redirect or modulate light beams; independently controlled micromirror arrays for displays; microresonators for a number of different applications; micropumps to develop positive fluid pressures; microflaps to modulate airstreams on airfoils; as well as many others. Even though these microactuators are extremely small, they frequently can cause effects at the macroscale level; that is, these tiny actuators can perform mechanical feats far larger than their size would imply. For example, researchers have placed small microactuators on the leading edge of airfoils of an aircraft and have been able to steer the aircraft using only these microminiaturized devices.[1] Furthermore, the flight capabilities of this microsystems-instrumented aircraft frequently far exceeded any expectations: the microsystems equipped aircraft at a high velocity was able to completely reverse its direction in a turning radius of nearly one wing radius!

The full potential of microsystems begins to become realized when these miniaturized sensors, actuators, and structures of any conceivable type are all merged onto a common substrate along with ICs. The sensors are able to perceive the environment, the electronics process this sensory information and make decisions about how to influence the environment, and the actuators act to influence the environment. Obviously, this is a very powerful concept and provides considerably more capability to a microchip than microelectronics alone can provide. Additionally, the integration of microsensors and microactuators with microelectronics enables microsystems to become a very important part of the Internet of Things (IoT), wherein IoT-connected microsystems are the eyes, ears, noses, etc., that are constantly gathering, storing, processing, and exchanging valuable information with other connected things to sense and control the environment.

4.2 WHY MICROSYSTEMS ARE IMPORTANT

There are many beneficial attributes of microsystems technology. First, microsystems are made using IC-like processes and this enables the ability to integrate multiple functionalities onto a single microchip. The ability to integrate miniaturized sensors, miniaturized actuators, and miniaturized structures along with microelectronics has far-reaching implications of countless products and applications, and is especially important for the implementation of the IoT. Second, microsystems borrow many production techniques of batch fabrication from the IC industry and therefore, the per-unit device or microchip cost of complex miniaturized electromechanical systems can be radically reduced—similar to the per-die cost reductions commonly seen in the IC industry. Although the cost of the production equipment and each wafer can be relatively high, the fact that this cost can be spread over many die in batch fabrication production can drastically lower the per part cost. Third, IC fabrication techniques coupled with the tremendous advantages of silicon and many other thin-film materials in mechanical applications allow the reliability of miniaturized electromechanical systems to be radically improved. The lower cost of these miniaturized electromechanical systems also allows them to be easily and massively deployed and more easily and inexpensively maintained and replaced as needed. Fourth, miniaturization of microsystems enables many benefits including increased portability, lower power consumption, and the ability to place radically more functionality in a smaller amount of space and without any increase in weight. Fifth, the ability to make the signal paths smaller and place radically more functionality in a smaller amount of space allows the overall performance of electromechanical systems to be enormously improved. In short, microsystems as a technology translates into products that have lower cost, higher functionality, improved reliability, and increased performance.

4.3 TECHNOLOGY FUNDAMENTALS OF MICROSYSTEMS

4.3.1 Microsensors Technology

Many different types of sensors have been demonstrated using micromachined devices including pressure, acoustic, temperature (including infrared focal plane arrays), inertia (including acceleration and rate rotation sensors), magnetic field (Hall, magnetoresistive, and magnetotransistors), force (including tactile), strain, optical, radiation, and chemical and biological sensors.[2-5] As discussed above, microsensors are transducers that convert one form of energy, such as mechanical force, to another form of energy, usually an electrical signal. There are several basic physical principles by which sensors function including resistive, magnetic, photoconductive, piezoresistive, piezoelectric, thermocouples and thermopiles, diodes, and capacitive. All of these sensing principles have been successfully demonstrated in microsystem devices. There are far too many variations of microsystem sensor types to review and therefore we will review only a few of the more important examples. The reader is referred to Refs. 2 to 7 for more information.

A piezoresistive material is one in which the resistance is influenced by, applied mechanical strains. This phenomenon is most prominent in semiconductors, where the strain induces changes in the electronic band structure of the material thereby making the carrier scattering rates dependent on direction of transport. This effect can be used to make a variety of sensors by placing the piezoresistive elements at a position where the strain is maximized. A useful quality factor of piezoresistive materials is the gauge factor, which is given by the normalized change in resistance divided by the strain. Silicon has a very high gauge factor approaching 200 in some configurations.[6] In comparison, metal resistors typically have gauge factors of 2 to 5. Piezoresistors are used primarily as strain measurement microsensors wherein the resistor (strain sensing element) is placed on a compliant surface or structure such as illustrated in Fig. 4.1.

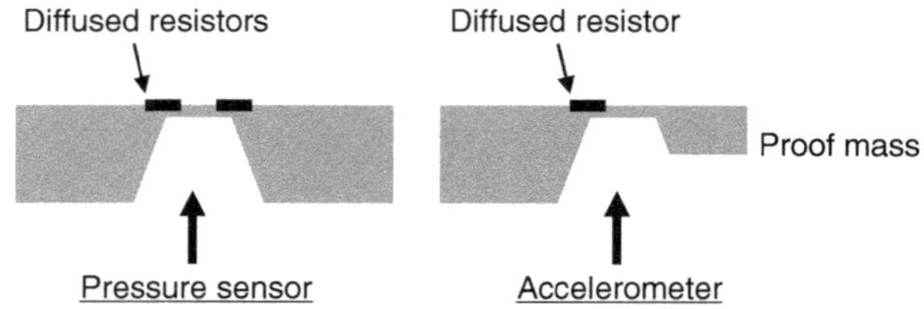

Figure 4.1 Two types of microsystems silicon microsensors made using the piezoresistive effect in semiconductors. (*Copyright © MEMS & Nanotechnology Exchange; used with permission.*)

Note that micromachining allows the substrate to be selectively removed to reduce the stiffness in the sensing region of the device. The piezoresistive effect is the most common method of converting pressure measurements into an electrical signal for microsystems pressure sensors and these devices are widely used in the automotive, medical, and industrial control markets.

Capacitive sensing is very commonly utilized in microsystem sensors due to its inherent simplicity. The capacitance, C, of a two terminal device is given by

$$C = \frac{(\varepsilon_o \varepsilon_r A)}{(d)} \text{ (farads)} \tag{4.1}$$

where ε_o is the dielectric constant of free space, ε_r is the relative dielectric constant of any material between the electrodes, A is the area of the capacitor, and d is the separation of the electrodes. In general, capacitors can be used as sensors in five different ways as shown in Fig. 4.2: (1) varying the distance between electrodes, (2) varying the position of the center electrode relative to two outer electrodes giving a differential measurement, (3) varying the overlap between electrodes, (4) varying the differential overlap between electrodes, and (5) varying the position of the dielectric in the space between electrodes.

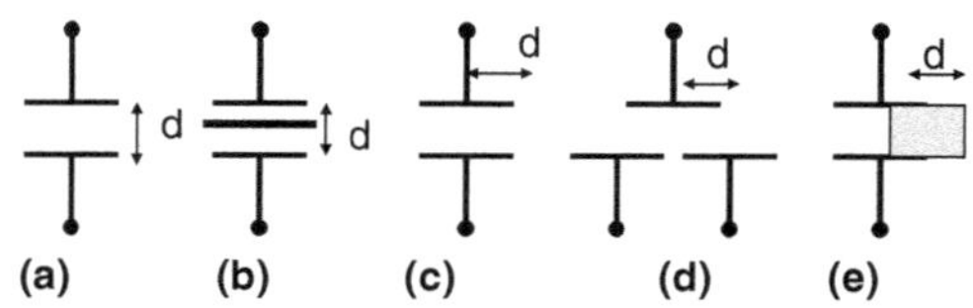

Figure 4.2 Different configurations to use a capacitor as a sensing element. (*Copyright © MEMS & Nanotechnology Exchange; used with permission.*)

Another widely used physical phenomenon for making sensors in microsystems is the piezoelectric property of some materials. Piezoelectricity is when a mechanical strain produces an electrical polarization (i.e., voltage potential) in the material. Likewise an applied electrical field will also induce a mechanical strain in the material. The first effect is primarily used for sensors while the second is usually employed for actuators. Silicon and germanium are centrosymmetric crystals and therefore display no piezoelectric effect (unless strain induced). Other materials that lack a center of symmetry such as quartz, lead zirconate titanate (PZT), or zinc oxide (ZnO) do demonstrate piezoelectricity and these materials are commonly used in microsystems.[6,7] The latter two, namely lead zirconate titanate (PZT) and zinc oxide (ZnO), can be deposited in thin-film form onto substrates.

4.3.2 Microactuator Technology

A variety of different basic principles are used to implement microsystem actuators including electrostatic, piezoelectric, magnetic, magnetostrictive, bimetallic, and shape memory alloy (SMA). Each of these has its respective advantages and disadvantages and therefore careful consideration of the specific application requirements must be part of the selection process. We shall review a few of the more popular methods for microsystem microactuators herein. The reader is referred to Refs. 2–7 for a more comprehensive review.

Electrostatic actuation is based on mutual attraction of two oppositely charged plates. The force F generated between plates under the application of a voltage potential V is given by

$$F = \frac{1}{2}(\varepsilon_o \varepsilon_r A)\left(\frac{V}{d}\right)^2 \text{ (newtons)} \tag{4.2}$$

where ε_o is the permittivity of free space, ε_r is the relative permittivity of a dielectric, A is the area of the plates, V is the applied voltage potential on the plates, and d is the separation between the plates. There are some inherent advantages of electrostatic microactuators that make them attractive for microsystems devices including relatively easy to fabricate and integrate with electronics; very low power consumption; and can exhibit very high mechanical bandwidths. Some of the disadvantages of electrostatic actuators are that the force is nonlinear with displacement and applied voltage; the resultant force is relatively small; and the operating voltages can be relatively high.

Another popular method of implementing microactuators for microsystems is based on the bimetallic effect. The bimetallic effect uses the differing thermal expansion coefficients of two different materials to realize a thermal based microactuator. When these two materials are made into a composite structure and heated, a thermally induced stress is generated that can result in the structure if it is sufficiently compliant. The thermal strain, $\varepsilon_{\text{Thermal}}$, is given by

$$\varepsilon_{\text{Thermal}} = (\alpha_{\text{filmA}} - \alpha_{\text{filmB}}) \times (T_{\text{element}} - T_{\text{ambient}}) \tag{4.3}$$

where α_{filmA} and α_{filmB} are the thermal expansion coefficients of the top and bottom films, respectively, and T_{element} and T_{ambient} are the temperatures of the bimetallic element and ambient, respectively. Some of the attributes of a bimetallic microactuator for microsystems are high power consumption for heating; low mechanical bandwidth; relatively complex design and fabrication; reasonable levels of displacements are possible; a linear deflection versus power relationship; and sensitivity to environmental conditions. A simple bimetallic microactuator can be made by depositing a thin-film of aluminum onto a thin compliant silicon cantilever and passing current through the aluminum layer. As the aluminum layer heats up by Joule heating, the cantilever will bend due to the different expansion coefficients of the two materials.

Yet another popular material for implementing microactuators is shape-memory alloys (SMAs). An SMA is a material that undergoes a martensite to austensite phase change upon heating. During this phase change, the material will return to its unstrained shape (i.e., the material exhibits a memory effect). For use as an actuator, the SMA material is in the martensite at room temperature and is unstrained. Subsequently, the SMA actuator is strained while remaining at room temperature. This can be achieved using a mechanical spring element. Then the SMA actuator is heated and the SMA material undergoes a phase change from the martensite to the austensite phase and the strain is recovered and generates a large actuator energy density.

SMAs can be sputter deposited in thin film form on silicon wafers. Heating of the film is usually achieved by Joule heating. The shape-memory effect is a reversible effect and can be repeated many times. The attributes of SMA for use in microsystems actuators include very high actuation energy densities; very large recoverable strain levels (strains of over 8 percent have been demonstrated); high power consumption and low mechanical bandwidth due to thermal activation; complicated processing; and the material can fatigue if repeatedly cycled at high strain levels.

4.3.3 Materials Used in Microsystems

There is an enormous diversity of materials used to fabricate microsystems devices including semiconductors, metals, glasses, ceramics, and polymers. Furthermore, since the functionality of microsystems devices are not only electrical, but mechanical, chemical, thermal, etc., the choice of a material or a set of materials for a given application frequently depends on its electrical as well as nonelectrical material properties. Silicon is the most popular material used in microsystems. This is partly due to the established infrastructure and extensive knowledge base of silicon as a material. Silicon is an extremely good mechanical material and has a yield strength nearly equal to that of stainless steel, and its strength to weight ratio is one of the highest of all engineered materials.[8] Nevertheless, if silicon is strained beyond its limit, it will catastrophically fail. This is unlike most metals that plastically deform if loaded beyond their yield points. Therefore, the design of a silicon-based microsystems device must take this issue into account. Another issue with silicon is that it is an anisotropic material and therefore the material properties will vary depending on the orientation of the crystal axes with respect to the load.

In addition to single crystal silicon, a variety of thin-film materials are also used in microsystems fabrication with the more popular ones being polysilicon, silicon nitride, deposited glasses, and aluminum. These thin films frequently are deposited using either a chemical-vapor deposition (CVD) processes, such as low-pressure chemical-vapor deposition (LPCVD) or plasma-enhanced chemical-vapor deposition (PECVD), or using the methods of physical-vapor deposition (PVD), such as evaporation or sputtering.

Most of the thin-film deposition technologies are very cost effective, which is why they are popular choices for microelectronics manufacturing, but they tend to have large residual (or built-in) stresses as well as stress gradients, and this often makes them difficult to work with in the implementation of microsystems devices having a mechanical functionality. The residual stresses and stress gradients in thin films are dependent on material type, temperature of deposition, method of deposition, and substrate material, and can vary from highly compressive to highly tensile.

It is extremely desirable that the mechanical properties of the materials used in a microsystems device are known before initiating design. Unfortunately, this can be difficult to achieve in practice because the material properties, especially many of the mechanical material properties, are highly dependent on the exact processing conditions and sequence used in the fabrication of the microsystems device. For example, a thin-film will have a certain residual stress level after deposition, but this residual stress can change significantly when exposed to subsequent thermal processing steps.[5]

Furthermore, as already pointed out, almost every microsystems device type has a customized process sequence, so it is very difficult to predict the values of the stresses are in these thin films at the end of the manufacturing process. Therefore, the development of a microsystem device and the manufacturing process sequence are often performed in conjunction with material properties measurements. The measured material property data can then be fed back into models in order to improve or optimize the design. However, this tends to be an iterative development process and thereby increases development time and cost.[5]

It is also important to note that the measurement of properties of thin film materials can be challenging. For example, the film cannot be removed from the substrate and have a load-deflection measurement done on it without changing the stress state of the film. Fortunately, the microsystems community has devised a number of different test structures, which can be used to measure the most important of the material properties. The reader is referred to Ref. 9 for more information about thin-film stress measurement techniques.

4.3.4 Microsystems Design Tools

Microsystems design is frequently more demanding than the design of ICs. In the ICs domain, a manufacturing process technology and the associated design rules usually already exist. The designer merely needs to build these into their computer-aided design tool, which usually only considers the electrical effects, and come up with a design. Importantly, the design tools in microelectronics have a high level of accuracy in their predictive capability of device performance.

In microsystems technology, the situation is much more complicated. As previously mentioned, a customized process sequence must frequently be developed for each device type as part of the product development effort and consequently the design rules are not known until the process sequence is finalized. Furthermore, as discussed in the preceding section, the material properties are usually not fully known and are highly dependent on the process sequence and conditions, which also are not known beforehand. Also, many microsystem devices have many physical phenomena (e.g., electrical, mechanical, thermal, chemical, etc.) occurring simultaneously which lead to many strongly coupled fields and thereby make the design effort much more complicated. Another key difference between microsystems design and IC design is that in IC design the designer usually does not need to know much about the fabrication. In microsystems design, the designer must usually be an expert in microsystems fabrication.[5]

Fortunately, there are some design tool capabilities that are now available to the microsystems community which are suitable for process, physical, device, and systems modeling. The process modeling tools are essentially the same as used by the IC industry and enable the designer to create process models and mask artwork. Numerical techniques are available to simulate the processing steps. Although these tools are quite good at predicting electrical behavior, they are less good for predicting mechanical material properties. One important feature of microsystems process design tools is the ability to create representative 3D renderings of the devices. The physical level design tools are used to model the behavior of components in the real 3D continuum using partial differential equations. These tools can be analytical or numeric and the numeric techniques include finite-element, boundary-element, and finite-difference methods. These tools are mostly modified versions of numeric modeling tools used in macroscale design. The device level models are macromodels or reduced-order models that capture the physical behavior of a component (over a limited range) and are compatible with the system level models. Care must be exercised to ensure that the dynamic range of the model is not overextended. System level models are high-level block diagrams and lumped-parameter models that describe the system as a coupled set of ordinary differential equations.[5]

4.4 HOW MICROSYSTEMS ARE MADE

4.4.1 Overview of Front-End Manufacturing Processes

Microsystems manufacturing is an extremely exciting endeavor due to the customized nature of process technologies and the diversity of processing capabilities. Microsystems manufacturing uses many of the same techniques that are used in the IC domain such as oxidation, diffusion, ion implantation, LPCVD, sputtering, etc., and combines these capabilities with highly specialized micromachining processes. Some of the most widely used micromachining processes are discussed below. Those interested in conventional microelectronics manufacturing technologies are referred to Ref. 10.

Bulk Micromachining

The oldest micromachining technology is bulk micromachining. This technique involves the selective removal of the substrate material in order to realize miniaturized mechanical components. Bulk micromachining can be accomplished using chemical, physical, or chemical-mechanical means. Wet chemical bulk micromachining has traditionally been the most widely used method of micromachining by industry.

A widely used bulk micromachining technique is chemical wet etching, which involves the immersion of a substrate into a solution of reactive chemical that will etch exposed regions of the substrate at measurable rates. Chemical wet etching is popular in microsystems because it can provide a very high etch rate and selectivity. Furthermore, the etch rates and selectivity can be modified by altering the chemical composition of the etch solution, adjusting the etch solution temperature, modifying the dopant concentration of the substrate, and even which crystallographic planes of the substrate are exposed to the etchant solution.

The basic mechanism of chemical wet etching involves the following: reactant transport to the surface of the substrate, followed by reaction at the substrate surface between the etchant solution and the substrate material, and transport of the reaction products from the substrate. If the transport of reactants to the surface of the substrate or transport of reaction products away from the substrate surface are the rate determining steps, then etching is defined as "diffusion limited" and the etch rate can be increased by stirring the solution. If the surface reaction is the rate determining step, then etching is "reaction rate limited" and the etch rate is very dependent on the etch solution temperature, etch solution composition, and substrate material. In practice, it is usually preferred that the process is reaction rate limited since this gives more repeatability and higher etch rate.

There are two general types of chemical wet etching in bulk micromachining: isotropic wet etching and anisotropic wet etching.[4,6] In isotropic wet etching, the etch rate is not dependent on the crystallographic orientation of the substrate and the etching proceeds in all directions at equal rates. The most common isotropic etchant for silicon is a solution of HNO_3, HF, and $HC_2H_3O_2$. The reaction is given by

$$HNO_2 + HNO_3 + H_2O \rightarrow 2HNO_2 + 2OH^- + 2h^+ \quad (4.4)$$

Holes and $(OH)^-$ ions are supplied by HNO_3 when it combines with H_2O and trace concentrations of HNO_2. Note that reaction is autocatalytic because of regeneration of HNO_2. Increasing the concentration moves reaction toward diffusion limited etch rate dependence and etching can be controlled by stirring. Increasing HF concentration or temperature increases surface reaction rate. In theory, lateral etching under the masking layer etches at the same rate as the etch rate in normal direction. However, in practice lateral etching is usually much slower without stirring and consequently isotropic wet etching is almost always performed with vigorous stirring of the etchant solution.

Any etching process requires a masking material to be used with preferably a high selectivity relative to the substrate material being etched. Common masking materials for isotropic wet silicon etching include silicon dioxide and silicon nitride. Silicon nitride has a much lower etch rate compared to silicon dioxide and therefore is more frequently used.

The much more widely used wet etchants for silicon micromachining are anisotropic wet etchants.[2–7] Anisotropic wet etching involves the immersion of the substrate into a chemical solution wherein the etch rate is dependent on crystallographic orientation of the substrate. The mechanism by which the etching varies according to silicon crystal planes is attributed to the different bond configurations and atomic density that the different planes exposed to the etchant solution. Wet anisotropic chemical etching is typically described in terms of etch rates according to the different normal crystallographic places, usually <100>, <110>, and <111>. In general, silicon anisotropic etching etches more slowly along the <111> planes than all the other planes in the lattice and the difference in etch rate between the different lattice directions can be as high as 1000 to 1. It is thought that the reason for the slower etch rate of the <111> planes is that these planes have the highest density of exposed silicon atoms in the etchant solution, as well as three silicon bonds below the plane, thereby leading to some amount of chemical shielding of the surface.

The ability to delineate the different crystal planes of the silicon lattice in anisotropic wet chemical etching provides a high-resolution etch capability with reasonably tight dimensional control. It also provides the ability for two-sided processing to embody self-isolated structures wherein only one side is exposed to the environment. This assists in packaging of the device and is very useful for microsystems devices exposed to harsh environments such as in pressure sensors. Anisotropic etching techniques have been around for about 30 years and are commonly used in the manufacturing of silicon pressure sensors, bulk micromachined accelerometers, and others.[5,6]

Figure 4.3 below is an illustration of some of the shapes that are possible using anisotropic wet etching of a <100> oriented silicon substrate including an inverted pyramidal and a flat bottomed trapezoidal etch pit. Note that the shape of the etch pattern is primarily determined by the slower etching <111> planes.

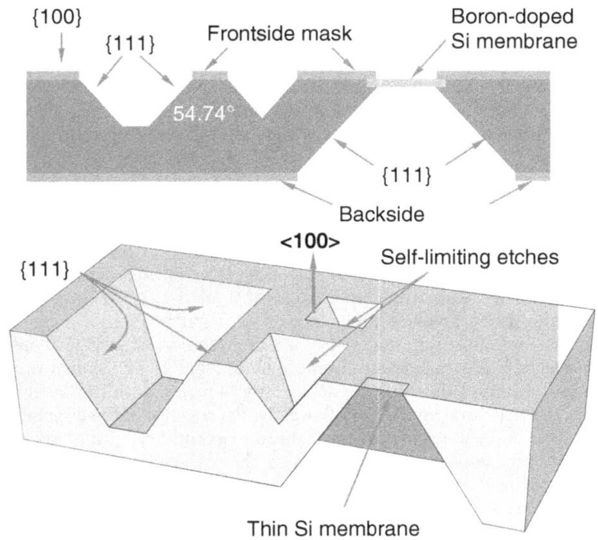

Figure 4.3 Illustration of shape of the etch profiles of a <100> oriented silicon substrate after immersion in an anisotropic wet etchant solution. (*Copyright © MEMS & Nanotechnology Exchange; used with permission.*)

Figure 4.4*a*, *b* is SEM photographs of a silicon substrate after an anisotropic wet etching. Figure 4.4*a* shows a trapezoidal etch pit that has been subsequently diced across the etch pit and in Fig. 4.4*b*, the backside of a thin membrane that could be used to make a pressure sensor. It is important to note that the etch profiles shown in these figures are only for an <100> oriented silicon wafer and substrates with other crystallographic orientations will exhibit different shapes. Occasionally, substrates with other orientations are used in microsystems fabrication, but given the cost, lead times and availability, the vast majority of substrates used in bulk micromachining have <100> orientation.

Three basic types of anisotropic wet etchants are commonly used. The first and by far the most popular anisotropic etchants are aqueous alkaline solutions such as KOH, NH_4OH, NaOH, CsOH, and TMAH. These etchants have high etch rates and a relatively high etch rate ratio between the (100) and (111) planes. Also, TMAH is sometimes preferred for use on preprocessed microelectronic wafers, since this etchant does not etch aluminum appreciably under certain conditions. The drawbacks of these etchants are that they have a relatively high etch rate of silicon dioxide, which is frequently used as a masking material. There is also the potential for alkali contamination of the wafer using these etchants, although there are cleaning procedures that can be employed to minimize these risks.

Another popular anisotropic wet etchant is ethylene-diamine and pyrocatechol or EDP. This etchant has a higher etch rate ratio of the (100) and (111) planes and has a larger variety of masking materials which can be used compared to the aqueous alkaline solutions. The drawbacks of EDP are that it is a carcinogenic material, and when using this solution it can be difficult to see the wafer etching. Also, EDP is a thick orange-yellow material which is difficult to clean up.

Useful anisotropic wet etching requires the ability to successfully mask certain areas of the substrate and consequently an important criterion for selecting an etchant is the availability of good masking materials. Silicon nitride is a commonly used masking material for anisotropic wet etchants since it has a very low etch rate in most etchant solutions. Thermally grown SiO_2 is also frequently used as a masking material, but some care must be exercised to ensure a sufficiently thick masking layer when using KOH etchants since the etch rates of oxide can be high. Photoresists are unusable in any anisotropic etchant. Several metals including Ta, Au, Cr, Ag, and Cu hold up well in EDP and Al holds up in TMAH under certain conditions.

In general, the etch rate, etch rate ratios (100)/(111), and etch selectivities of anisotropic etchants is strongly dependent on chemical composition and temperature of the etchant solution. The etch rate, *R*, obeys the Arrhenius law given by

$$R = R_o e^{-\left(\frac{E_a}{kT}\right)} \ (\mu\text{m/h}) \tag{4.5}$$

where R_o is a constant (μm/h); E_a is the activation energy (electron-volts or eV); k is Boltzman's constant (eV/Kelvin); and T is temperature in Kelvins. Both R_o and E_a will vary on the type of etchant, etchant composition, and crystallographic orientation of material being etched. Fortunately, there is a wealth of published literature characterizing many of the commonly used anisotropic etchants and the reader is referred to Refs. 2 to 7 and 11 for more information.

Frequently, when using bulk micromachining it is often desirable to make thin membranes of silicon having an accurately controlled thickness across the membrane or control the etch depths very precisely. As with any chemical process, the uniformity of the etching can vary significantly across the substrate due to loading effects, temperature variations, thickness variations in the substrate, etc., thereby making this difficult to obtain in practice. Timed etches determine the etch depth by multiplying the etch rate by the etch time. However, they are very difficult to control and the etch depth is very dependent on etchant species diffusion effects, loading effects, etchant aging, surface preparation, etc.

To allow a higher level of precision in anisotropic etching, various methods have been developed to address this problem, namely in the form of etch stops.[11,12] Etch stops are very useful to control etching processes and provide reasonably uniform etch depths across the wafer, from wafer to wafer, and from wafer lot to wafer lot. There are two basic types of etch stop methods that are used in micromachining: dopant etch stops and electrochemical etch stops.

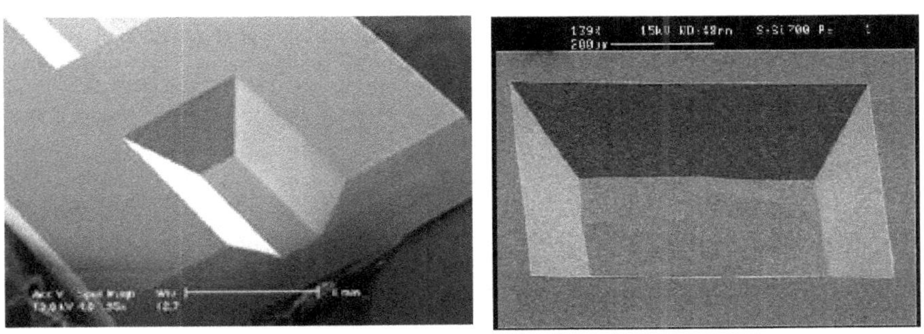

(a) (b)

Figure 4.4 (*a*) and (*b*): SEMS of a <100> oriented silicon substrate after immersion in an anisotropic wet etchant. The scale on these images has the dimension of 1 micron. (*Copyright © MEMS & Nanotechnology Exchange; used with permission.*)

Etch stops in silicon are commonly made by the introduction of dopants into the silicon material. Most popular etch stop is heavy *p*-type doping of silicon with boron (>5 × 10^{19} cm^{-3}) to create an etch stop. The lightly doped region of the wafer will etch at the normal rate and the highly doped region of the silicon will have a very slow etch rate. The dopant is introduced into the silicon using the standard techniques of diffusion or ion implantation and provides for a controlled depth and reasonable uniformity of the dopants in the substrate. Figure 4.5 is

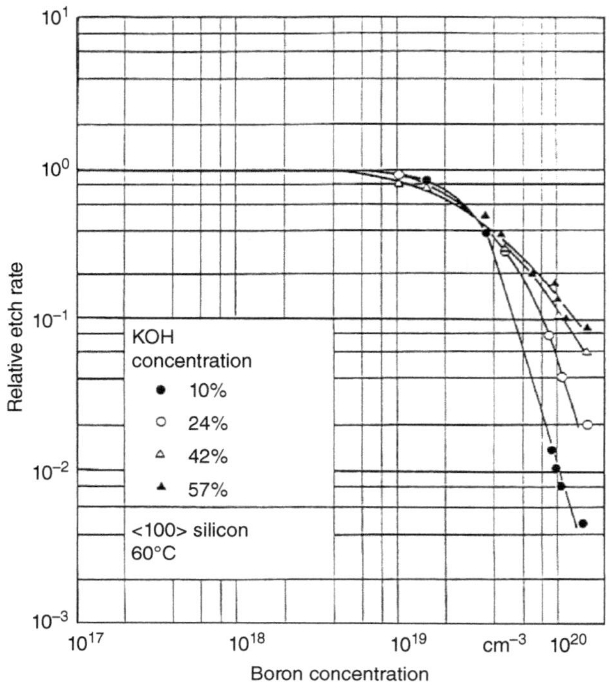

Figure 4.5 Etch rate plotted versus boron concentration for <100> oriented silicon wafer at different etchant concentrations. (*Copyright © MEMS & Nanotechnology Exchange; used with permission.*)

a graph of the normalized etch rate of <100> oriented silicon wafer in KOH at different concentrations as a function of the boron dopant concentration.[11] As can be seen, the etch rate falls off very quickly at dopant concentrations above 10^{19} boron atoms per cm^3. One of the problem with boron etch stops is that the surface layer of the silicon substrate left after etching will be so highly doped that it may not be useful for some kinds of devices. For example, the material at the concentrations required for a good etch stop would be far too high to be useful for making piezoresistive devices.

The other etch stop method used in silicon bulk micromachining is the electrochemical etch stop.[13–17] Electrochemical etching of silicon using an anisotropic etchant is useful since it provides very good dimensional control (e.g., diaphragm thickness is reproducible) and can make diaphragms with lightly doped material, which is required for high-quality piezoresistive devices. The disadvantage of electrochemical etch stops is that they require special fixturing to each wafer in order to make electrical contacts and an electronic control system is needed to control and apply the correct voltage potential to the wafer during the etch.

Surface Micromachining

Surface micromachining is another very popular technology for the fabrication of microsystem devices. There is a very large number of variations of how surface micromachining is performed depending on the materials and etchant combinations that are used.[18,19] However, the common theme involves a sequence of steps starting with the deposition of some thin-film material to act as a temporary sacrificial layer onto which the actual device layers are built, followed by the deposition and patterning of the thin-film device layer of material which is referred to as the structural layer, and followed by the removal of the temporary sacrificial layer to release the mechanical structural layer from the constraint of the underlying sacrificial layer, thereby allowing the structural layer to move. An illustration of a surface micromachining process is given in Fig. 4.6, wherein an oxide layer is deposited and patterned. This oxide layer is temporary and is commonly referred to as the sacrificial layer. Subsequently, a thin film layer of polysilicon is deposited and patterned and this layer is the structural mechanical layer. Lastly, the temporary sacrificial layer is removed and the polysilicon layer is now free to move as a cantilever.

One of the reasons surface micromachining is so popular is that it provides for reasonably precise dimensional control in the vertical direction due to the fact that the structural and sacrificial layer thicknesses are defined by deposited film thicknesses which can be fairly accurately controlled. Also, surface micromachining provides for precise dimensional control in the horizontal direction because the structural layer tolerance is defined by the fidelity of the photolithography and etch processes used. Other benefits of surface micromachining are that a large variety of structure, sacrificial, and etchant combinations can be used and some are compatible with microelectronics devices to enable integrated microsystems devices. Surface micromachining can also exploit the deposition characteristics of thin-films such as conformal coverage as seen in LPCVD deposition techniques. Finally, surface micromachining uses single-sided wafer processing and is relatively simple. This can enable higher integration density and lower resultant per die cost compared to bulk micromachining.

Nevertheless, surface micromachining does have some disadvantages, including the mechanical properties of structural thin films, particularly those deposited using LPCVD, are usually unknown and must be measured; it is common for these types of films to have a high state of residual stress frequently necessitating a high temperature anneal to reduce residual stress in the structural layer; the actual value of residual stress in a thin-film layer will change depending on the time and temperature of subsequent thermal processing steps; the reproducibility of the mechanical properties in these films can be difficult to achieve; and the release of the structural layer can be difficult due to stiction effect whereby the structural layer is pulled down and stuck to the underlying substrate due to capillary forces during release. To reduce or eliminate

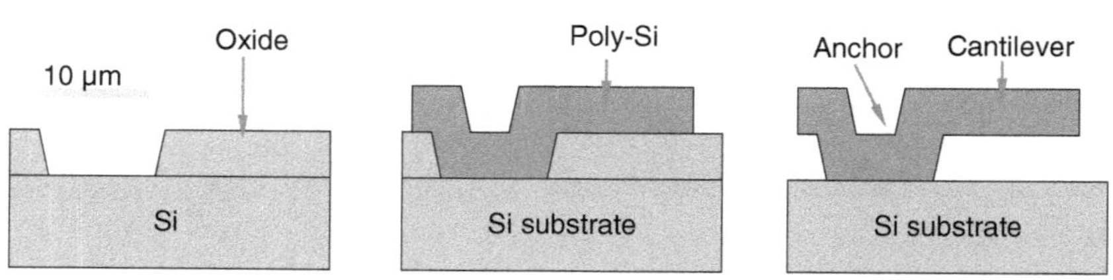

Figure 4.6 Illustration of a surface micromachining process. (*Copyright © MEMS & Nanotechnology Exchange; used with permission.*)

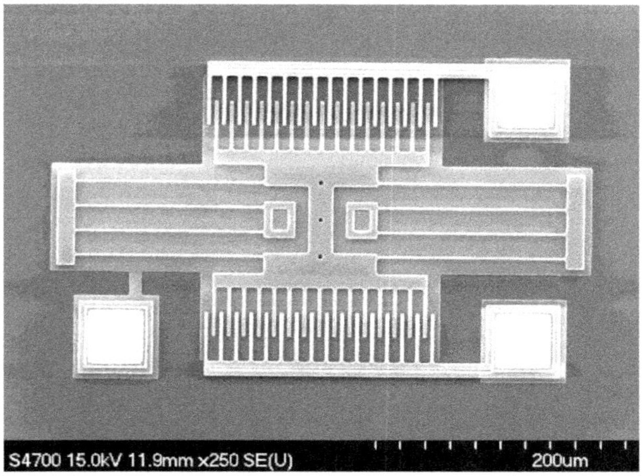

Figure 4.7 Polysilicon resonator structure fabricated using a surface micromachining process. (*Copyright © MEMS & Nanotechnology Exchange; used with permission.*)

stiction effects, special release processes and antistiction coatings may be needed. Figure 4.7 is a SEM of a surface micromachined polysilicon microsystems resonator device.

Wafer Bonding

Wafer bonding is a micromachining method that is analogous to welding in the macroscale world and involves the joining of two (or more) wafers together to create a multiwafer stack. There are three basic types of wafer bonding including direct or fusion bonding; field-assisted or anodic bonding; and bonding using an intermediate layer. In general, all bonding methods require substrates that are very flat, smooth, and clean, in order for the wafer bonding to be successful and free of voids.

Direct or fusion bonding is typically used to mate two silicon wafers together or alternatively to mate one silicon wafer to another silicon wafer that has been oxidized. Direct wafer bonding can be performed on other combinations, such as bare silicon to a silicon wafer with a thin-film of silicon nitride on the surface as well. The basic wafer bonding process has five steps[20]: (1) cleaning and wafer surface preparation—hydration and cleaning of the wafer surfaces (RCA clean, piranha immersion, etc.); (2) prebonding of the wafers—physical contacting and pressing together of wafers (must be done quickly and in a clean environment since silicon will be charged and attract particles); (3) inspection—infrared inspection of preanneal bond quality; (4) anneal to bond wafers—high-temperature anneal, typically around 1000°C; and (5) final inspection—infrared inspection of postanneal bond quality.

The wafers are initially held together after physical contacting due to the hydrogen bonds created by hydration of the surfaces. The two wafers to be bonded can be preprocessed and then aligned during the bonding procedure so as to register features on the top wafer to features on the bottom wafer. After the high-temperature anneal, the bond strength between the two wafers can be equivalent to that of single crystal silicon. There is relatively low residual stress in the bonded layers after anneal. For good bonding results, it is important that the wafers are flat, smooth, and clean prior to bonding, and the bonding process should be performed in a very clean environment. The hydration of the surfaces makes them highly charged and attractive to particles in the environment. Any particles attached to the wafer surfaces prior to bonding will result in voids between the two wafers or the inability of the wafers to bond. Although the high-temperature anneal can be performed at a lower temperature than 1000°C in direct wafer bonding, the bond strengths tend to be weaker as the anneal temperature decreases.

It is possible to treat surfaces of the wafers prior to bonding with a plasma in order to enhance the charge available for prebonding. Typically, the wafer surfaces are exposed for a very short time to an oxygen plasma. It has been reported that if wafer bonding uses plasma enhancement, the anneal temperatures can be lowered to 250 to 300°C or even lower and still achieve a high-quality bonding between the wafers.[21,22]

Another popular wafer bonding technique is anodic bonding. In anodic bonding a silicon wafer is bonded to a Pyrex 7740 wafer using an electric field and elevated temperature.[23–25] The two wafers can be preprocessed prior to bonding and can be aligned during the bonding procedure. The mechanism by which anodic bonding works is based on the fact that Pyrex 7740 has a high concentration of Na^+ ions and a positive voltage applied to the silicon wafer drives the Na^+ ions from the Pyrex glass surface, thereby creating a negative charge at glass surface. The elevated temperature during the bonding process allows the Na^+ ions to migrate in the glass with relative ease. When the Na^+ ions reach the interface, a high-field results between silicon and glass, and this combined with the elevated temperatures chemically fuses the two wafer surfaces together. As with direct wafer bonding, it is imperative that the wafers are flat, smooth, and clean and that the anodic bonding process is performed in a very clean environment. Any particles on the wafer surfaces will result in voids between the two wafers. One advantage of this process is that Pyrex 7740 has a thermal expansion coefficient nearly equal to silicon and therefore there is a low value of residual stress in the layers. Anodic bonding is a widely used technique for microsystems packaging.

In addition to direct and anodic bonding, other wafer bonding techniques are used in microsystems fabrication. One method is eutectic bonding and involves bonding a silicon substrate to another silicon substrate at an elevated temperature using an intermediate layer of gold on the surface of one of the wafers.[26] Eutectic bonding works because the diffusion of gold into silicon is extremely rapid at elevated temperatures. In fact this is a preferred method of wafer bonding at relatively low temperature.

Another wafer bonding technique used in microsystems is glass frit bonding.[27] In this process a glass is spun or screen-printed onto one substrate surface. Subsequently this wafer is physically contacted to another wafer and the composite is annealed to flow the glass intermediate layer and bond the two wafers

Lastly, various polymers can be used as intermediate layers to bond wafers including epoxy resins, photoresists, polyimides, silicones, etc.[28] This technique is commonly used during various fabrication steps in microsystems such as when the device wafer becomes too fragile to handle without mechanical support.

High-Aspect Ratio Microsystems Fabrication Technologies

Deep Reactive Ion Etching of Silicon Deep reactive ion etching (DRIE) is a highly anisotropic plasma etch process that can be used to etch very deep features into silicon with high aspect ratios. It was first introduced in the mid-1990s and has been widely adopted by the microsystems community.[29,30] The sidewalls of the etched features are nearly vertical and the depth of the etch can be tens of micrometers, hundreds of micrometers, or even completely through the entire silicon substrate. There are two basic DRIE etch processes: cryogenic DRIE and the Bosch DRIE process. The Bosch DRIE process is the most often used DRIE process in microsystems manufacturing.

The Bosch DRIE process is named after the company that developed and patented this process, namely the Robert Bosch GmbH.[29] Figure 4.8 illustrates how deep reactive ion etching is accomplished using the Bosch process. The etch is a dry plasma etch and uses a high-density plasma to repeatedly alternate between an etch cycle of the silicon and a deposit cycle wherein an etch resistant polymer layer is deposited on the sidewalls. The etching of the silicon is performed using a SF6 chemistry whereas the deposition of the etch resistant polymer layer on the sidewalls uses a C4F8 chemistry. Mass flow controllers alternate back and forth between these two chemistries during the etch. The protective polymer layer is deposited on the sidewalls as well as the bottom of the etch pit, but the anisotropy of the etch removes the polymer at the bottom of the etch pit faster than the polymer is removed from the sidewalls.

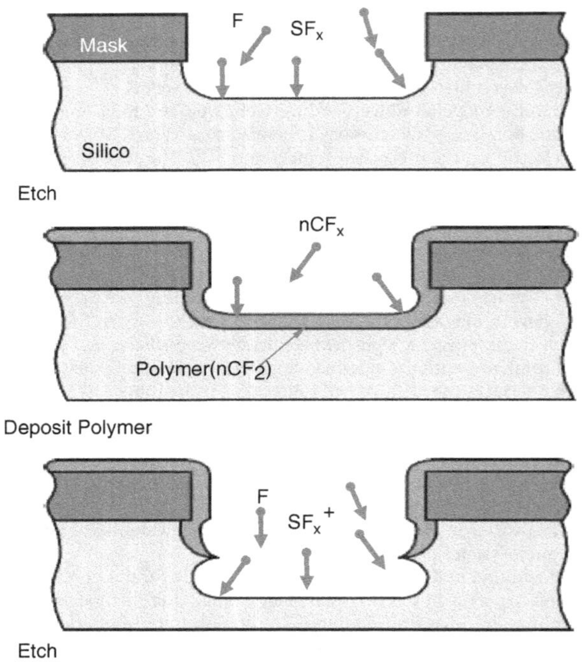

Figure 4.8 Illustration of how deep reactive ion etching works. (*Copyright © MEMS & Nanotechnology Exchange; used with permission.*)

Figure 4.9 SEM of the cross section of a silicon wafer demonstrating high aspect ratio and deep trenches that can be fabricated using DRIE technology. (*Copyright © MEMS & Nanotechnology Exchange; used with permission.*)

The sidewalls of features etched using DRIE are not optically smooth and if the sidewalls are magnified under SEM inspection, a characteristic washboard or scalloping pattern is seen. The etch rates on most commercial DRIE systems when this technology was first introduced were about 1 to 4 μm/min. However, the etch rates have been steadily improved and etch rates of more than 20 μm/min are now possible with the latest versions of DRIE tools.[31] Either photoresist or silicon dioxide is commonly used as a masking layer for DRIE etching. The mask selectivity with photoresist and oxide is about 75 to 1 and 150 to 1, respectively.

For a through wafer etch a relatively thick photoresist masking layer will be required. The aspect ratio of the etch can be as high as 30 to 1,[32] but in practice tends to be 15 to 1. The process recipe depends on the amount of exposed silicon due to loading effects in the system, with larger exposed areas etching at a much faster rate compared to smaller exposed areas. Consequently, the etch must frequently be characterized for the exact mask feature and depth to obtain desirable results. Also, in order to obtain uniform depths using DRIE etching, it is common to have the same widths for all trenches and features across the substrate. Figure 4.9 is a cross section SEM of a silicon microstructure fabricated using DRIE technology. As can be seen, the etch is very deep into the silicon substrate and the sidewalls are nearly vertical.

Deep Reactive Ion Etching of Glass Glass substrates can also be etched deep into the material with high aspect ratios and this technology has been gaining in popularity in microsystems fabrication.[33] Figure 4.10 shows a series of deep, high-aspect-ratio trenches etched into glass that are over 100 μm deep and have an aspect ratio of more than 4 to 1 using glass plasma etching technology. This etch used a nickel hard mask that was about 20 μm in thickness. The mask selectivity for nickel in this etch process is about 10 to 1. Unlike the Bosch process, this etch process is continuous and therefore there is no scalloping of the sidewalls and lateral underetching of the mask is reduced. However, this process is more susceptible to micromasking. The typical etch rates for high-aspect-ratio glass etching range between 250 and 500 nm/min. Depending on the depth of the etch desired, photoresist, metal, or a polysilicon can be used as a mask.[34]

LIGA Another high-aspect-ratio micromachining technology is called LIGA, which is a German acronym for "LIthographie Galvanoformung Abformung."[35] This is a non-silicon-based fabrication technology and requires the use of synchrotron generated x-ray radiation. The basic process starts with the cast of an x-ray radiation sensitive PMMA onto a suitable substrate. A special x-ray mask is used for the selective exposure of the PMMA layer using x-rays. The PMMA is then developed and will be defined with extremely smooth and nearly perfectly vertical sidewalls. Also, the penetration depth of the x-ray radiation into the PMMA layer is quite deep and allows exposure through very thick PMMA layers, up to and exceeding 1 mm. After the development, the patterned PMMA acts as a mold and is placed into an electroplating

Figure 4.10 SEM of high-aspect-ratio structures etched into a glass substrate performed by the MEMS and Nanotechnology Exchange. (*Copyright © MEMS & Nanotechnology Exchange; used with permission.*)

bath and nickel, or any other material that can be electroplated and is compatible with the mold material, is plated into the open areas of the PMMA. The PMMA is then removed, thereby leaving the metallic microstructure.

Because LIGA requires a special mask and a synchrotron (x-ray) radiation source for the exposure, the cost of this process is relatively expensive. A variation of the process which reduces the cost of the micromachined parts made with this process is to reuse the fabricated metal part as a tool insert to imprint the shape of the tool into a polymer layer, followed by electroplating of metal into the polymer mold and removal of the polymer mold.[36] Obviously, this sequence of steps avoids the need of a synchrotron radiation source each time a part is made and thereby significantly lowers the cost of the process. The dimensional control of this process is quite good and the tool insert can be used many times before it is worn out.

Hot Embossing Another variation of this process is to fabricate the metal tool insert using LIGA or a comparable technology and then to emboss the tool insert pattern into a polymer substrate, which is then used as the part. This process can make imprints into a polymer of over 200 μm deep with very good dimensional control. The advantage of this process is that the cost of the individual polymer parts can be very low compared to the same structures made using other technologies. Because of the overwhelming cost advantages combined with very good performance, this polymer molding process is very popular for producing microfluidic components. The mold insert can be made with LIGA, DRIE, or several other process technologies. As a general rule, the LIGA mold inserts are the most expensive, but have smoother sidewalls and are made of metal thereby providing for superior embossing results.

Other Micromachining Technologies

In addition to bulk micromachining, surface micromachining, wafer bonding, and high-aspect-ratio micromachining technologies, there are a number of other techniques used to fabricate microsystem devices. We shall review a few of the more popular methods in this section. The reader is referred to Refs. 4–7 for a more detailed overview of other microsystems processes.

XeF_2 Dry Phase Etching Xenon difluoride (XeF_2) in a vapor state is an isotropic etchant for silicon.[37] This etchant is highly selective with respect to other materials commonly used in microelectronics fabrication including LPCVD silicon nitride, thermal SiO_2, aluminum, titanium, and others. Since this etchant is a completely dry release process it does not suffer from the stiction problems of wet release processes. This etchant is popular with micromachining of microstructures in preprocessed CMOS wafers where openings in the passivation layers on the surface of the substrate are made to expose the silicon for etching.

Electro-Discharge Micromachining Electro-discharge micromachining or micro-EDM is a process used to machine a conductive material using electrical breakdown discharges to remove material.[38] A working electrode is made from a metal material onto which high voltage pulses are applied. The working electrode is brought into close proximity to the material to be machined, which has been immersed in a dielectric fluid. The minimum sized features that can be made with micro-EDM are dependent on the size of the working electrode and how it is fixtured, but holes as small as tens of micrometers have been made using this method. One issue with micro-EDM is that it is a slow serial process and therefore relatively expensive.

Laser Micromachining Lasers can generate an intense amount of energy in very short pulses of light and direct that energy onto a selected region of material for micromachining.[39] Among the many types of lasers now in use for micromachining include CO_2, YAG, excimer, etc. Each has its own unique properties and capabilities suited to particular applications. Factors that determine the type of laser to use for a particular application include laser wavelength, energy, power, and temporal and spatial modes; material type; feature sizes and tolerances; processing speed; and cost. The action of CO_2 and Nd:YAG lasers is essentially a thermal process, whereby focusing optics are used to direct a predetermined energy/power density to a well-defined location on the work piece to melt or vaporize the material. Another mechanism, which is nonthermal and referred to as photoablation, occurs when organic materials are exposed to ultraviolet radiation generated from excimer, harmonic YAG, or other UV source Similar to micro-EDM, laser micromachining can produce features on the order of tens of micrometers, but it is a serial process and therefore slow.

A recent development in laser micromachining is the use of femtosecond lasers, which uses an ultrashort pulse of laser that provides better control and results compared to conventional laser micromachining.[40] Since the pulse of the laser is shorter than the heat diffusion time and the electron-to-phonon coupling time, there is little to no heating of the irradiated material. Instead, the irradiated material transitions from a solid to vapor phase instantaneously. This reduces damage to surrounding material and provides much better control. Also, since the machining process is not linearly dependent on absorption of the laser radiation, a large number of different material types can be machined using femtosecond laser technology including semiconductors, metals, dielectrics, ceramics, and polymers.

Focused Ion Beam Micromachining Another tool for performing micromachining is the focused ion beam (FIB).[41] The accelerating voltages are adjustable from a few keV to several hundred keV. The spot sizes can be focused down to 50 nm, making it capable of making extremely small structures. The user can input a 3D CAD solid model of desired etching topology and the computer-controlled stage allows very precise registration of sample with submicrometer positional accuracy. In addition to material removal, the FIB can also be used to perform ion induced deposition, lithography, implantation doping, mask repair, device repair, and device diagnostics. Most commercial FIB tools have imaging capability and many tools can also be outfitted with a secondary column for mass analysis of particles removed from the substrate using Microscale Secondary Ion Mass Spectrometry (uSIMS).

4.5 FUTURE TRENDS

The future of microsystems includes higher levels of integration, more functionality for each device, and smaller dimensional scales. As the manufacturing capabilities continue to improve and mature, it is expected that it will be possible in the future to easily integrate a multiplicity of different sensor and actuator types on a single slab of silicon as well as the integration of microsystems with state-of-the-art electronics. This will enable enormous amounts of functionality to be squeezed into a very tiny amount of space and with a very low relative cost. Additionally, we are currently seeing the blending of microsystems and nanotechnology and this will only continue to increase into the future.

Perhaps the most important future trend is the use of microsystems in the Internet of Things (IoT), where everything and everyone will be connected together. More specifically, the IoT is a collection of physical objects that are interconnected using communication networks to allow the physical objects to collect, store, process, and exchange information. One of the most important elements that are needed in the realization of the IoT is inexpensive and unobtrusive devices that can sense and actuate to control the physical objects and/or environment in which the things are embedded. That is, microsystem devices. Therefore, microsystem devices undoubtedly have a huge role in the implementation of the IoT. No one presently knows the impact of the IoT or the resultant market sizes, but it is expected to be a major driving force for future microsystems.

4.6 CONCLUSIONS

Microsystems are revolutionizing the design of mechanical systems through miniaturization, batch fabrication, and integration with electronics. Microsystem technology is not about a specific application or device, nor a single fabrication process. Rather, this technology provides new and unique capabilities for the development of smart products for all industrial, commercial, and military applications. While microsystems devices may be a fraction of the cost, size, and weight of the products that incorporate them, they will be critical to their performance, reliability, and affordability. This chapter has reviewed microsystem technology and attempted to present this material from the perspective of the IC industry. The microsystem industry shares a common ancestry with the IC but is a much broader and more diverse technology. The diversity, economic importance, and extent of potential applications of microsystems make it the hallmark technology of the future.

4.7 REFERENCES

1. P. H. Huang, C. Folk, C. Silva, B. Christensen, Y. F. Chen, G. B. Lee, Chen, Minjdar et al., "Applications of MEMS Devices to Delta Wing Aircraft: From Concept Development to Transonic Flight Test," AIAA Paper No. 2001-0124, Reno, Nevada, Jan. 8–11, 2001.
2. G. T. A. Kovacs, *Micromachined Transducers Sourcebook*, McGraw-Hill, New York, 1998.
3. M. Elwenspoek and R. Wiegerink, *Mechanical Microsensors*, Springer, Berlin, Germany, 2001.
4. M. Madou, *Fundamentals of Microfabrication*, CRC Press, Boca Raton, Fl., 1997.
5. M. A. Huff, S. F. Bart, and P. Lin, "MEMS Process Integration," Chapter 14 of the *MEMS Materials and Processing Handbook,* editors R. Ghodssi and P. Lin, Springer Press, New York, May 2012.
6. S. Sze, *Semicondcutor Sensors*, Wiley, New York, 1995.
7. M. Gad-el-Hak, *The MEMS Handbook*, CRC Press, Boca Raton, Fl., 2002.
8. K. E. Petersen, "Silicon as a mechanical material," *Proceedings of the IEEE*, 70 (5): 420–457, May 1982.
9. R. K. Gupta et al., "Material Property Measurements of Micromechanical Polysilicon Beams," SPIE 1996 Conference (Invited Paper): *Microlithography and Metrology in Micromachining II*, Oct. 14–15, 1996.
10. R. C. Jaeger, *Introduction to Microelectronic Fabrication: Volume 5 of Modular Series on Solid-State Devices*, 2d ed., Prentice Hall Upper Saddle River, New Jersey, Oct. 2001.
11. H. Seidel, L. Csepregi, A. Heuberger, and H. Baumgartel, "Anisotropic Etching of Crystalline Silicon in Alkaline Solutions: II, Influence of Dopants," *J. Electrochem. Soc.*, 137: 3626, 1990.
12. N. F. Raley, Y. Sugiyami, T. van Duzer, "(100) Silicon Etch-Rate Dependence on Boron Concentration in Ethylenediamine-Pyrocatechol-Water Solutions," *J. Electrochem. Soc.*, 131: 161, 1984.
13. W. K. Zwicker and S. K. Kurtz, "Anisotropic Etching of Silicon Using Electrochemical Displacement Reactions," in *Semiconductor Silicon 1973* edited by H. R. Huff and R. R. Burgess, Electromechanical Society, Pennington, New Jersey, p. 315, 1973.
14. T. N. Jackson, M. A. Tischler, and K. D. Wise, "An Electrochemical p-n Junction Etch Stop for the Formation of Silicon Microstructures," *IEEE Elec. Dev. Lett.*, EDLM-2, p. 44, 1981.
15. O. J. Glembocki, R. E. Stanlbush, and M. Tomkiewicz, "Bias-Dependent Etching of Silicon in Aqueous KOH," *J. Electrochem. Soc.*, 132: 145, 1985.
16. B. Kloech, S. D. Collins, N. F. de Rooij, and R. L. Smith, "Study of Electrochemical Etch-Stop for High Precision Thickness Control of Silicon Membranes," *IEEE Trans. Electron Dev.*, ED-36, p. 663, 1989.
17. V. M. McNeil, S. S. Wang, K. Y. Ng, and M. A. Schmidt, "An Investigation of the Electrochemical Etching of (100) Silicon in CsOH and KOH," Tech. Dig. IEEE Solid-State Sensor and Actuator Workshop, Hilton Head, S.C., p. 92, 1990.
18. R. T. Howe and R. S. Muller, "Polycrystalline and Amorphous Silicon Micromechanical Beams: Annealing and Mechanical Properties," *Sensors and Actuators*, 4: 447, 1983.
19. L. S. Fan, Y. C. Tai, and R. S. Muller, "IC-Processed Electrostatic Micromotors," presented at the Int. Electron Devices Meeting (IEDM), p. 666, 1991.
20. M. A. Huff and M. A. Schmidt, "Fabrication, Packaging, and Testing of a Wafer-Bonded Microvalve," IEEE Solid-State Sensor and Actuator Meeting, Hilton Head, S.C., Jun. 22–25, 1992.
21. A. Weinhart, P. Amirfeiz, and S. Bengstsson, "Plasma Assisted Room Temperature Bonding for MST," *Sensors and Actuators A*, 92: 214–222, 2001.
22. A. Doll, F. Goldschmidtboeing, and P. Wois, "Low-Temperature Plasma-Assisted Wafer Bonding and Bond-Interface Stress Characterization," *Micro Electro Mechanical Systems*, 17th IEEE International Conference on MEMS, pp. 665–668, 2004.
23. G. Wallis and D. L. Pomerantz, "Field Assisted Glass-Metal Sealing," *J. Appl. Phys.*, 40: 3946, 1969.
24. S. Johansson, K. Gustafsson, and J. A. Schweitz, "Strength Evaluation of the Field Assisted Bond Seals between Silicon and Pyrex Glass," *Sens. Mater.*, 3: 143, 1988.
25. S. Johansson, K. Gustafsson, and J. A. Schweitz, "Influence of Bond Area Ratio on the Strength on FAB Seals between Silicon Microstructures and Glass," *Sens. Mater.*, 4: 209, 1988.
26. A. L. Tiensuu, J. A. Schweitz, and S. Johansson, "In Situ Investigation of Precise High Strength Micro Assembly Using Au-Si Eutectic Bonding," 8th International Conference on Solid-State Sensors and Actuators, Transducers 95, Stockholm, Sweden, p. 236, Jun. 1995.
27. Editorial, "Sealing Glass," Corning Technical Publication, Corning Glass Works, 1981.
28. C. den Besten, R. E. G. van Hal, J. Munoz, and P. Bergveld, "Polymer Bonding of Micromachined Silicon Structures," Proceedings of the IEEE Micro Electro Mechanical Systems, MEMS 92, Travemunde, Germany, p. 104, 1992.
29. F. Larmar and P. Schilp, "Method of Anisotropically Etching of Silicon," German Patent DE 4,241,045, 1994.
30. J. Bhardwaj and H. Ashraf, "Advanced Silicon Etching Using High Density Plasmas," *Proc. SPIE, Micromachining and Microfabrication Process Technology Symp.*, Austin, TX, vol. 2639, p. 224, Oct. 23–24, 1995.
31. A. A. Chambers, "Si DRIE for Through-Wafer via Fabrication," *Solid-State Technology*, Mar. 2006.
32. J. Yeom, Y. Wu, J. C. Selby, and M. A. Shannon, "Maximum Achievable Aspect Ratio in Deep Reactive Ion Etching of Silicon due to Aspect Ratio Dependent Transport and the Microloading Effect." *J. Vac. Sci. Technol. B*, Oct. 31, 2005.
33. ULVAC, Inc., Technical Data, 2004.
34. M. Pedersen and M. Huff, "Plasma Etching of Deep, High-Aspect Ratio Etching of Features in Fused Silica," submitted of publication to *IEEE J. Microelectromechanical Sys.*
35. W. Ehrfeld, P. Bley, F. Gotz, P. Hagmann, A. Maner, J. Mohr, H. O. Moser et al., "Fabrication of Microstructures Using the LIGA Process," *Proc. IEEE Micro Robots and Teleoperators Workshop*, Hyannis, MA, Nov. 1987.
36. W. Menz, W. Bacher, M. Harmening, and A. Michel, "The LIGA Technique—a Novel Concept for Microstructures and the Combination with Si-Technologies by Injection Molding," *IEEE Workshop on Micro Electro Mechanical Systems*, MEMS 91, p. 69, 1991.
37. P. B. Chu, J. T. Chen, R. Yeh, G. Lin, C. P. Hunag, B .A. Warneke, and K. S. J. Pister, "Controlled Pulse-Etching with Xenon Difluoride, "*Proc. Inter. Conf. Solid-State Sensors and Actuators, Transducers 97*, p. 665, Jun. 1997.
38. B. B. Pradhan, M. Masanta, B. R. Sarkar, and B. Bhattacharyya, "Investigation of Electro-Discharge Micro-Machining of Titanium Super Alloy," *Int. J. Adv. Manuf. Technol.*, 41: 1094–1106, 2009.
39. R. D. Schaefer, *Fundamentals of Laser Micromachining*, Taylor & Francis, Oxfordshire, UK, Apr. 2012.
40. R. Osellame, G. Cerullo, and R. Ramponi, "Femtosecond Laser Micromachining," *Topics in Applied Physics 123*, Springer, New York, Mar. 2012.
41. W. Driesel, "Micromachining using Focused Ion Beams," *Physica Status Solidi (a)*, 146 (1): 523–535, Nov. 16, 1994.

Useful URL Links

Electronic Discussion Groups, http://www.memsnet.org/lists/. (Accessed January 15, 2017.)

MEMS and Nanotechnology Exchange (MNX), http://www.mems-exchange.org. http://www.memsnet.org/lists/. (Accessed January 15, 2017.)

Nanotechnology Clearinghouse, http://www.memsnet.org. http://www.memsnet.org/lists/. (Accessed January 15, 2017.)

Chapter 5

Physical Design for High-Performance, Low-Power, and Reliable 3D Integrated Circuits

BY

ANKUR SRIVASTAVA *University of Maryland, College Park, Maryland*
TIANTAO LU *University of Maryland, College Park, Maryland*

5.1 INTRODUCTION

In the past few decades, the demand for high-performance and complex functionality in integrated circuits is primarily met through aggressive device scaling. Device scaling results in smaller and faster transistors and higher integration density on the same chip area. Device scaling has been fulfilling the prophecy of Moore's law, and has been proved to be a very effective solution in many aspects such as performance and cost. At the time when this chapter is written, 14-nm complementary metal-oxide semiconductor transistors have been successfully commercialized and 10-nm transistors are under development. However, the trend of transistor scaling seems to be near saturation. Further transistor scaling will inevitably encounter physical limitations and might not be cost-effective. Another huge issue of device scaling is that interconnects are becoming slower. Interconnect's RC delay has increased to such an extent that wire delay has become the primary bottleneck for further performance boost. More complicated functionalities require longer global routing thereby higher power dissipation. The wire delay and power consumption will become larger as devices are made even smaller.

Three-dimensional integrated circuit (3D IC) is one of the most promising technologies that aim to continue Moore's law as the traditional transistor scaling is approaching its physical limit. 3D ICs exploit the vertical dimension to achieve higher number of transistors per unit area. The vertical interconnect, through-silicon-via (TSV), replaces long off-chip interconnects in 2D ICs, resulting in significant improvement in bandwidth, power, and performance. While the performance advantages are pronounced, 3D ICs face several design challenges. Novel design flows and methodologies are being investigated to overcome these challenges and to reveal the true potential of 3D ICs. This chapter provides a survey of the current status of 3D IC physical design, including its mainstream design flow, the primary physical design challenges, and the state-of-the-art physical design solutions.

5.1.1 Fundamental Limitations of Transistor Scaling

Over the last several decades, the growth of computing power and IC's complexity is based on the fact that device scaling makes transistors switch faster and denser. However, in advanced technology nodes, when transistors and interconnects are made smaller and denser, several major challenges rises.

The first major challenge is that traditional transistor scaling is reaching its limit. There are two bottlenecks that hinder further transistor scaling. The first is transistor leakage, including gate leakage (due to thinner gate oxide), subthreshold leakage (lower threshold voltage), etc. These leakage currents make the transistor harder to be turned off, impacting its switching speed, and consuming more power. Novel device solutions such as high-κ and metal gate are being developed; however, it's uncertain that further transistor scaling yields any cost-effective solution for leakage control.

The second bottleneck is the significance of process variation in advanced technologies. One example is the variation in dopant concentration. Smaller devices possess fewer impurities; therefore, as device dimension shrinks, dopant concentration variation will have a more pronounced influence on transistor's threshold voltage. Another example is line edge roughness, referring to the observation that a pattern's edge deviates from a smooth, ideal shape. This effect emerges as the feather size of a device is less than the wavelength of lithography.

5.1.2 The Rising of Interconnect Delay and Power

The second major trend is that the interconnect delay and power have become a limiting factor to the performance of the planar ICs even if the transistor scaling continues. As IC's functionality becomes more complex, global wire length increases, resulting in significant increase in interconnect delay. One famous example is that the memory latency produced by long and slow buses between processor cores and dynamic random access memories (DRAMs). Reading/writing to memory takes tens to hundreds of clock cycles; therefore, memory access has become one of the major overheads in modern high-performance computing systems. Inserting more buffers is common practice to reduce the interconnect delay, but it inevitably leads to higher power consumption.

Another important interconnect-related drawback is the relatively narrow bandwidth. The number of bus channels is limited by the I/O pin counts from the external package. In the era of multicore processors and big data, huge amounts of data packets are processed and waiting to be transferred between processor cores and memories; therefore, limited bandwidth becomes a huge performance bottleneck.

5.1.3 The Concept of 3D ICs

A 3D IC stacks multiple planar wafers/dies vertically, which results in an increased transistor density and reduced wire length. The components in different layers can communicate through vertical interconnects called TSVs. A large amount of TSVs can be used, which dramatically increases chip's bandwidth and also decreases interconnect delay/power. In addition, 3D ICs provide new design options by allowing heterogeneous materials, technologies, and systems to be integrated into a single chip.

Figure 5.1 shows an illustration of a two-layer liquid-cooled 3D IC. A heat sink and a heat spreader are attached to the top of the chip to provide conventional air cooling. A DRAM layer is bonded to a processor layer, and the TSVs ensure interlayer communication. Moreover, microchannels are etched through the silicon substrate to provide liquid cooling capability. The two-layer 3D IC is connected to the package through C4 bumps.

In this chapter, we provide an overview of 3D IC design flow, discuss some major design challenges, and the state-of-the-art solutions.

5.2 3D IC DESIGN FLOW

This section gives an overview of the design flows for 3D ICs.

Figure 5.2 presents a typical 3D IC design flow. The design flow starts with 3D architectural design space exploration, where system architects decide the suitable combination of technologies and the partition of the system. After a design specification has been chosen, synthesis tools are used to convert behavioral specification into logic gate level netlist. This is followed by 3D physical design, which is composed of power/ground planning, floorplanning, placement/routing, and clock network synthesis. The significant difference between 3D physical design and its 2D counterpart comes from the placement of interlayer microbumps and power, signal, and clock TSVs. Microbumps and TSVs' reliability, signal integrity issues as well as their large dimension and manufacturing cost limit their usage, and this limitation needs to be considered during the physical design phase. The next step is to extract the electrical parameters from the physical layout and verify the extracted circuit against design rules and expected functionality. After the verification step, sign-off analysis is performed to ensure 3D IC's electrical performance (timing, power, IR-drop, etc.), thermal behavior, and reliability all meet the design specification. Finally, the 3D IC needs proper packaging as studies have shown that 3D IC's thermal distribution, power and signal integrity, and mechanical reliability are influenced by how C4 and microbumps are placed.

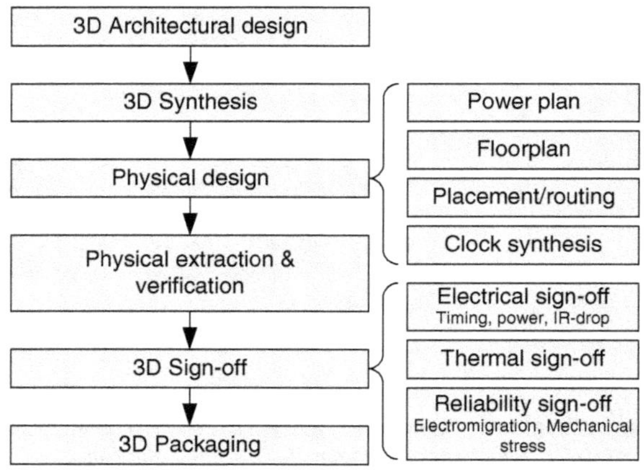

Figure 5.2 A typical design flow for 3D ICs.

In this chapter, we will focus on the physical design phase of the 3D IC design flow.

5.3 PHYSICAL CHALLENGES AND PHYSICAL DESIGN TOOLS FOR 3D IC

3D IC's physical design is a crucial step that translates a circuit abstraction (i.e., a flat circuit netlist) into an actual layout. 3D physical design tools face several unique challenges, and in this section we focus on some of the key challenges, including challenges in 3D placement, clock tree synthesis, thermal management, power delivery, and reliability concerns.

5.3.1 3D Placement

For a given standard cell netlist, the 3D placement problem aims to find the location for each logic gate in the 3D space, such that the total wire length is minimized and the usage of TSVs is under control.

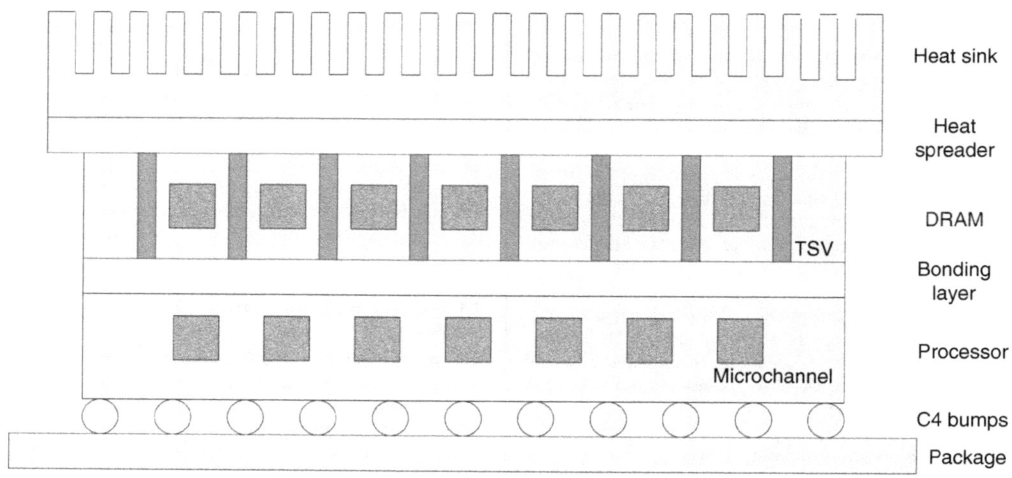

Figure 5.1 A two-layer liquid-cooled 3D IC. A DRAM layer is stacked upon a processor layer.

Several 3D IC's characteristics differentiate 3D placement with its 2D counterpart. First, each standard cell's layer index is a discrete rather than a continuous value, while traditional 2D placers deal with planar placement. 2D placers often allow certain overlaps during global placement step, which are gradually removed by subsequent detailed placement and legalization step. Rounding layer index number might result in suboptimal solution in 3D IC's placement.

Second, 3D placement needs to determine the usage of TSVs, including the count and the placement. TSVs are large placement obstacles, whose diameter ranges from 2 to 10 μm.[1] TSVs also induce thermal mechanical stress in their neighboring areas, impacting both the transistor's performance (by changing its mobility) and TSV's reliability (TSV's interfacial delamination and substrate cracking).

5.3.2 3D Clock Tree

A clock tree is a prevailing clock distribution network that delivers the clock signal from a centralized clock source to all sequential logic blocks (e.g., flip flops) called clock sinks. Some common clock tree design objectives include minimizing total wire length, total power, clock skew, etc.

Several challenges need to be addressed for 3D clock tree synthesis. First, compared to 2D clock trees in which the routing is done in a planar plane, 3D clock trees require clock TSVs to connect tree nodes in different layers. 3D clock tree synthesis algorithms need to decide the clock TSV count and their placement locations. TSV's parasitics need to be considered during delay and power calculations.

Second, it is highly recommended that "known-good-dies" are vertically stacked to ensure high yield. Therefore, sometimes each die requires separated testing infrastructure. Prebond testability requires complete clock tree on every die, and minimizing prebond 3D clock tree's overhead is one of the crucial design objectives.

5.3.3 3D Thermal Management

The vertical stacking structure of 3D IC increases the power density proportional to the number of layers. Meanwhile, both the interlayer dielectrics and the chip stacking increase the thermal resistance, leading to significant thermal challenges in 3D ICs. High temperature induces severe performance degradation and reliability loss, and also imposes the need for higher cooling capability.

In a traditional air cooled system, a heat sink is attached to the backside of a chip, offering a low resistance path to the ambient. However, air cooling might not be enough to remove the heat generated far away from the heat sink.[2]

Without better cooling schemes or better design methodologies, a large portion of the circuitries in 3D ICs has to be shut down to avoid thermal violations (known as "dark silicon"), which results in huge performance degradation.

5.3.4 3D Power Delivery

Suppressing voltage noises is the most important PDN design objective to ensure a quality and stable voltage level. Voltage noises are especially severe in 3D ICs. On one hand, the total power consumption increases as more chips are vertically stacked. On the other hand, the power I/O pin count is proportional to 3D IC's footprint area, which is relatively fixed due to manufacturing constraints. As functional units are drawing large amounts of currents from the power supply, significant power is lost during vertical transportation through IR drop. In addition, the surge of current can induce significant Ldi/dt voltage droop.

Figure 5.3 shows a two-layer 3D IC's PDN model, which consists of RLC components in the PCB board, package, and the stacking chips. The stacking chips are meshed into multiple grids, where current draws are modeled as current sources attached to each of the grids. Between grids there are RL components for planar wires and power TSVs.

Several properties of 3D power network are essential for designers to suppress the voltage noises in a 3D IC's power line. First, power consumption possesses significant spatial variations in both planar plane and along the vertical direction. Inside the planar plane different functional units dissipate various powers (i.e., an execution unit dissipates more power than a load-and-store unit). Along the vertical direction, where chips with different technologies are bonded, power consumptions might also be different. For example, a DRAM layer requires much lower power than a CPU layer.

Second, the stacking structure makes the power noise to be coupled to adjacent layers. For example, during a period when CPU cores are active, significant voltage noise are observed in adjacent DRAM layers as well.

Third, power consumptions vary dramatically in time. Heavy computational task results in higher power consumption, and run-time approaches, such as dynamic voltage and frequency scaling (DVFS), also impact the power profile dynamically

5.3.5 Reliability Concerns in 3D ICs

TSVs are crucial interconnects in 3D ICs; however, they suffer from various reliability degradations, such as electromigration (EM), stress migration, interfacial delamination, cracking, and thermal cycling, etc.

TSV's EM is driven by its high current density, forming voids inside TSVs, eventually leading to an open circuit. Due to TSV manufacturing imperfections during metal deposition stage, initial voids might already exist inside TSV, which further shortens TSV's lifetime.

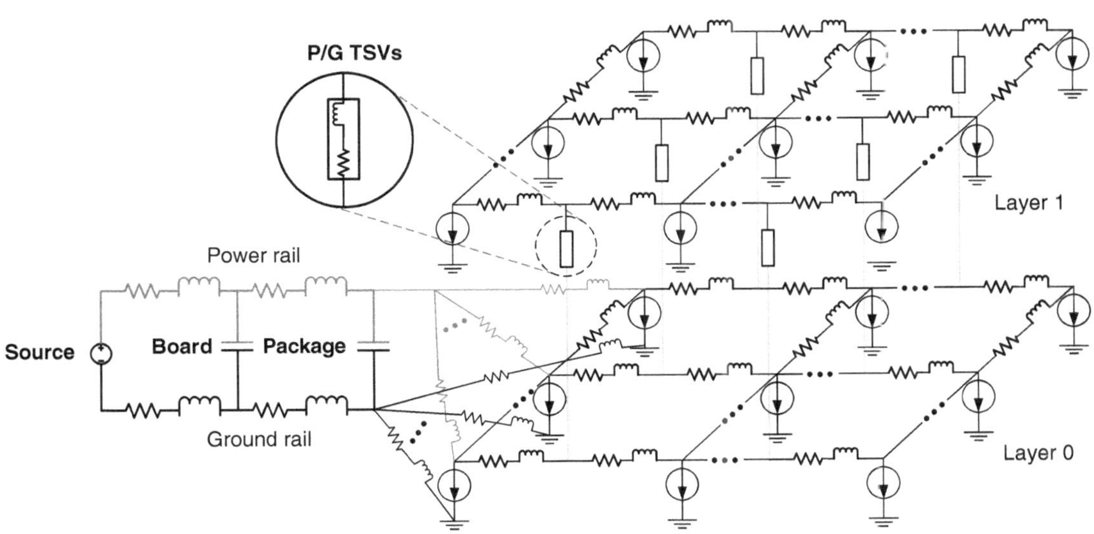

Figure 5.3 A two-layer 3D IC's PDN model.

Many other reliability loss mechanisms are related to the coefficients of thermal expansion mismatch between TSV material (e.g., copper) and silicon substrate. When 3D ICs are cooled from high-temperature manufacturing process to low-temperature operating condition, high amounts of thermal mechanical stresses are formed around TSVs, as shown in Fig. 5.4. These stresses induce stress migration, interfacial delamination between TSV and its liner, and silicon cracking once the stress exceeds the material's yield strength.

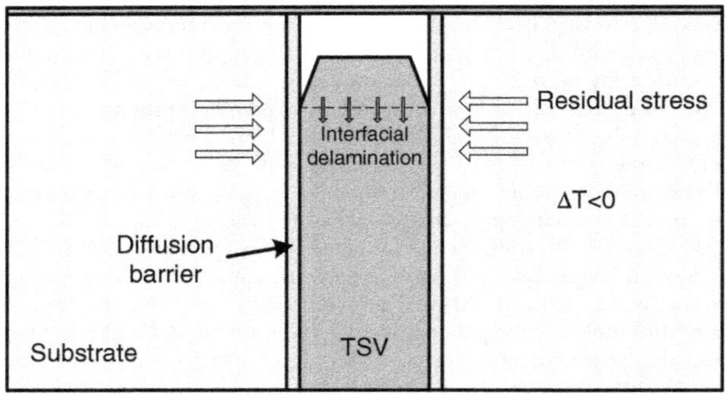

Figure 5.4 The large temperature drop between TSV manufacturing and normal operation induces significant thermal stress around TSV, resulting in stress migration, interfacial delamination, etc.

Most TSV reliability losses are heavily associated with circuit utilization and TSV temperature. For example, TSV EM is significantly accelerated when a higher amount of current flows through. TSV's stress-related reliability losses depend on the magnitude of thermal load. TSV's thermal cycling is related to how frequently temperature changes are. Therefore, TSV's reliability estimation and optimization are often coupled with thermal estimation and optimization.

5.4 PHYSICAL DESIGN SOLUTIONS FOR 3D IC

In this section, we provide the state-of-the-art solutions in the literature to address each of the design challenges discussed in Section 0.

5.4.1 3D Placement Algorithms

The most straightforward method for 3D placement follows a divide-and-conquer methodology. This method first partitions the gate-level circuit netlist into different layers while minimizing the number of interlayer cuts (e.g., using partitioning algorithms such as Fiduccia-Mattheyses[3]), and then performs regular 2D placement in each of the layers sequentially.

While the integration of partitioning and layer-by-layer 2D placement enables reusing 2D placers, the quality of the placement result heavily depends on initial partitioning. There has been a growing interest of 3D-specific placement tools, and such tools include analytical placer,[4,5] transformation-based placer,[6] partition-based placer,[7] and force-directed placer.[8,9]

Among the aforementioned placers, the analytical placers[4,5] are the most prevailing. In analytical 3D placers, Lagrange multipliers are used to minimize the total wire length and cell overlaps. The total wire length of a 3D net "e" is often estimated by half-perimeter wire length (HPWL), as shown in Eq. (5.1), where v_i is a node in e, (x_i, y_i, z_i) is v_i's coordinates and α_{TSV} represents a user-defined TSV's weight:

$$WL(e) = (\max_{v_i \in e}\{x_i\} - \min_{v_i \in e}\{x_i\}) + (\max_{v_i \in e}\{y_i\} - \min_{v_i \in e}\{y_i\}) + \alpha_{TSV}\left|z_i - z_j\right| \quad (5.1)$$

To enable differentiability, the HPWL model is often relaxed into a log-sum model, as shown in Eq. (5.2).

$$\max_{v_i \in e}\{x_i\} \approx a \cdot \log\left(\sum_{v_i \in e} e^{\frac{x_i}{a}}\right); \quad \min_{v_i \in e}\{x_i\} \approx a \cdot \log\left(\sum_{v_i \in e} e^{\frac{-x_i}{a}}\right) \quad (5.2)$$

Both TSVs and gates can be modeled as cells with given dimensions. For the cell overlap constraint, the chip is meshed into bins and each bin can have certain cell density. In each x/y dimension, the cell density is approximated by certain smoothing function, and the cell density in a 2D plane is the product of the smoothing functions in x and y dimensions.

5.4.2 3D Clock Tree Synthesis

Generally the 3D clock tree synthesis is a two-step procedure: 3D abstract clock tree generation and 3D clock tree embedding. The 3D abstract clock tree represents the connectivity from a centralized clock source to all the sequential logics. The 3D clock tree embedding determines the physical routing and buffer insertion.

The 3D abstract clock tree can be designed following either a top-down or bottom-up fashion. Top-down approaches include 3D-MMM (method of means and medians)[10] and MMM-3D,[11] and bottom-up method includes NN-3D (nearest neighbor selection).[12] The 3D-MMM[10] and MMM-3D[11] algorithms recursively partition all the clock sinks horizontally (X/Y cut) or vertically (Z cut), based on available TSV count, until there is only one clock sink in the set. The 3D-MMM algorithm is shown in Fig. 5.5 as an example. The bottom-up method (i.e., NN-3D),[12] greedily selects a pair of subtrees or clock sinks with the lowest cost and then merge that pair. The selection cost is a weighted sum of 2D Euclidean distance, TSV number, and capacitive loads.

For a given 3D abstract clock tree, the second phase of 3D clock tree synthesis, 3D clock tree embedding, determines the physical routing as well as buffer insertion. The common objectives are clock skew, clock slew, and total wire length. Kim and Kim[12] extended the DME (deferred-merge-embedding) algorithm[13] to 3D that produced a zero-skew 3D clock tree with near-optimal wire length. Further works[10,14] integrated simultaneous clock buffering and clock gating into the DME algorithm to minimize the clock slew.

Moreover, it is often desirable to provide prebond testability for 3D ICs, in which every die is tested separately before they are bonded vertically. Kim et al. used transmission gates to control whether a 3D IC is under test or normal operation. During test mode, signal transmissions

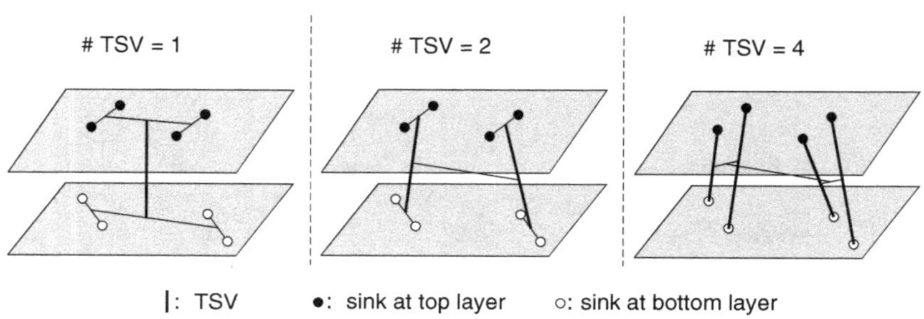

Figure 5.5 Abstract clock tree generation based on 3D-MMM.[10] Clock sinks are partitioned according to their Z indexes or in X/Y directions based on available TSV count.

via clock TSVs are forbidden; therefore, each die has its own clock delivery network. During normal operation, clock TSVs are activated and clock signal is delivered from only one centralized clock source.

5.4.3 Thermal Solutions

To address the thermal challenges in 3D ICs, microchannel liquid cooling is becoming an attractive solution. Microchannels are etched directly into the substrate of each layer thus providing a conductive heat path from the heat source to the fluid coolant (see Fig. 5.1). Coolant (e.g., deionized water) is supplied from the inlet of microchannel cavities. Heat dissipated from active layers is transferred to the coolant and pumped out of the chip. The fluid flow provides a much lower resistance path for heat to flow from junction to ambient, and provides 3D ICs with much higher heat removal capacity that scales with number of layers.[2] Despite its superior cooing capability, microchannel cooling has several negative effects: (1) The cooling capacity of the fluid flow progressively diminishes along the direction of flow as the fluid heats up, resulting in systematic thermal gradients, with degrading reliability.[15] (2) Extra power is required to pump coolant through the cavities, thus reducing the energy efficiency of the 3D IC.[16] Better microchannel designs are necessary to provide sufficient cooling while considering these two side effects.

Many other research directions have been developed orthogonal to microfluidic cooling. For example, researchers have attempted to perform thermal-aware placement[6] to minimize peak temperature and thermal gradient along with traditional placement objectives such as total wire length. Redundant thermal TSVs have been employed to further spread the heat.[17,18]

Besides thermal-aware physical design, run-time thermal management approaches in 3D ICs are also useful. Task scheduling and migration techniques have been considered to remove thermal hotspot in a 3D multicore CPU[19] and stacked DRAM circuit.[20] Furthermore, DVFS[21,22] provides fine tuning capability at processor core level, optimizing processor's thermal condition at the cost of its performance.

5.4.4 Power Delivery Network Design

To reduce the power noise in 3D PDN, several approaches have been investigated.

Power TSVs deliver power across layers but cause IR drops due to their parasitics. Researchers have increased the size and the granularity of the power TSV and achieved smaller IR drop and Ldi/dt noise.[23,24] However, such approaches inevitably impose greater power TSV usages which occupy large placement area.

Decoupling capacitance (decap) can be applied in 3D ICs to reduce dynamic voltage fluctuation such as Ldi/dt droop. However, excessive decaps can cause significant leakage power; therefore, algorithms that optimize the usage of decaps are necessary. For example, Zhou et al.[25] proposed a decap insertion algorithm for 3D IC's voltage noise reduction. In addition, 3D integration enables decap to be implemented as a separate decap layer.[26,27] Huang et al.[26] illustrated that putting a decap layer on the top of a four-layer 3D IC can achieve 36 percent reduction in worst-case peak voltage noise.

Voltage stacking is another way to tackle the power delivery challenge. With voltage stacking, the power domains are connected in series (GND-VDD-2VDD-...) such that the ground of one domain becomes the power supply for the next. If the current demand in a conventional 3D IC equals I, in an ideal voltage stacking scheme, all different power domains share the same current, which equals I/n, where n is the number of 3D layers. The reduction in current demand dramatically decreases 3D IC's power noise.

5.4.5 Reliability Solutions

TSV reliability assessment and optimization is a relatively new research field. Three categories of research have been ongoing, including TSV's reliability modeling, design-time reliability enhancement techniques, and run-time reliability management schemes.

Detailed finite-element method (FEM) was among the first TSV reliability modeling techniques. FEM involves meshing a physical structure, and using numerical method to solve a complicated physics system. Early TSV EM simulations[28] were heavily based on FEM. However, FEMs are computationally expensive; therefore, other analytical TSV reliability models have been proposed. For example, Huang et al.[29] derived a physics-based analytical model for power TSV's EM. Other works include the modeling of TSV-induced thermal mechanical stress. Ryu et al.[30] developed a semi-analytical model for TSV-induced stress. Based on this stress model, Lu and Srivastava[31] proposed an objective function to represent the intensity of TSV stress migration. In addition, Jung et al.[32] used energy release rate to model TSV interfacial cracks.

One example of a design-time approach involves insertion of redundant TSVs into the layout to enable online reconfiguration or remapping techniques once a portion of TSVs fails. For example, Jiang et al.[33] proposed a TSV repair algorithm based on TSV redundancy that was able to reconfigure the TSV usage when TSV failure occurs.

Besides redundant TSV insertion, physical design techniques such as floorplanning and placement can be implemented in a way such that TSV's reliability is optimized. For example, Zou et al.[34] proposed a TSV stress-aware floorplanning method to minimize the possibility of wafer cracking and interfacial delamination. Lu and Srivastava[35] optimized the chip's thermal profile using thermal TSVs, in order to alleviate TSV-induced stress migration. The authors also developed an EM-aware TSV placement heuristic to improve the TSV's mean-time-to-failure.[36]

There is abundant research for run-time reliability management for 2D ICs, where the essence is to find the optimal tradeoff between how fast reliability degradation happens and how much performance overhead one is willing to tolerate. Task assignment and scheduling have been used to dynamically tuning 2D IC's reliability property.[37,38] Moreover, DVFS can also tune the processor's operating points to mitigate the processor's reliability loss, at the cost of performance degradation.[38–40]

The application of run-time reliability management scheme in 3D ICs is still an ongoing effort. For example Tajik et al.[41] exploited task migration in 3D processor cores to recover 3D cache's NBTI degradation. Lu et al.[42] used dynamic frequency scaling to guarantee 3D IC's long-term reliability without losing too much of its performance benefit. Lu et al.[42] used dynamic frequency scaling to guarantee 3D IC's long-term reliability without losing too much of its performance benefit.

5.5 CONCLUSIONS AND FUTURE DEVELOPMENTS

3D ICs have provided a revolutionary solution to further boost performance and enhance energy efficiency, other than costly transistor scaling. However, compared to its 2D counterpart, the expanded design space imposes extra design complexities to the physical design domain, including placement, clock tree synthesis, etc. Furthermore, the stacking structure inevitably prolongs the power delivery paths and reduces thermal conductivity, which produces severer voltage drop and higher temperature, respectively. Moreover, TSVs introduce new reliability loss modes. In this chapter, we have provided a comprehensive survey of several challenges that 3D ICs face, and have discussed the state-of-the-art design tool development and management techniques. Further innovations are highly necessary to push this revolutionary technology into mainstream production.

5.6 REFERENCES

1. G. V. der Plas, P. Limaye, I. Loi, A. Mercha, H. Oprins, C. Torregiani et al., "Design Issues and Considerations for Low-Cost 3-D TSV IC Technology," *IEEE Journal of Solid-State Circuits,* vol. 46, pp. 293–307, Jan. 2011.

2. J.-M. Koo, S. Im, L. Jiang, and K. E. Goodson, "Integrated Microchannel Cooling for Three-Dimensional Electronic Circuit Architectures," *Journal of heat transfer,* vol. 127, pp. 49–58, 2005.

3. C. M. Fiduccia and R. M. Mattheyses, "A Linear-Time Heuristic for Improving Network Partitions," *19th Conference on Design Automation,* pp. 175–181, 1982.

4. G. Luo, Y. Shi, and J. Cong, "An Analytical Placement Framework for 3-D ICs and Its Extension on Thermal Awareness," *IEEE Transactions on Computer-Aided Design of Integrated Circuits and Systems,* pp. 510–523, Apr. 2013.

5. M.-K. Hsu, Y.-W. Chang, and V. Balabanov, "TSV-Aware Analytical Placement for 3D IC designs," *Design Automation Conference*, pp. 664–669, 2011.

6. J. Cong, G. Luo, J. Wei, and Y. Zhang, "Thermal-Aware 3D IC Placement via Transformation," *Asia and South Pacific Design Automation Conference, 2007*. pp. 780–785, 2007.

7. B. Goplen and S. Sapatnekar, "Placement of 3D ICs with Thermal and Interlayer via Considerations," in *Design Automation Conference*, pp. 626–631, 2007.

8. B. Goplen and S. Sapatnekar, "Efficient Thermal Placement of Standard Cells in 3D ICs Using a Force Directed Approach," *Proceedings of the IEEE/ACM International Conference on Computer-Aided Design*, pp. 86–89, 2003.

9. P. Spindler, U. Schlichtmann, and F. M. Johannes, "Kraftwerk2: A Fast Force-Directed Quadratic Placement Approach Using an Accurate Net Model," *IEEE Transactions on Computer-Aided Design of Integrated Circuits and Systems*, 27: 1398–1411, Aug. 2008.

10. X. Zhao, J. Minz, and S. K. Lim, "Low-Power and Reliable Clock Network Design for Through-Silicon Via (TSV) Based 3D ICs," *IEEE Transactions on Components Packaging Manufacturing Technology*, 1: 247–259, Feb 2011.

11. T.-Y. Kim and T. Kim, "Clock Tree Embedding for 3D ICs," *Proceedings of 15th Asia South Pacific Design Automation Conference (ASP-DAC)*, pp. 486–491, 2010.

12. T.-Y. Kim and T. Kim, "Clock Tree Synthesis for TSV-Based 3D IC Designs," *ACM Transactions on Design Automation of Electronic System*, vol. 16, pp. 48:1–48:21, Oct. 2011.

13. T.-H. Chao, Y.-C. Hsu, J.-M. Ho, and A. B. Kahng, "Zero Skew Clock Routing with Minimum Wirelength," *IEEE Transactions on Circuits System II, Analog and Digital Signal Processing*, vol. 39, pp. 799–814, Nov. 1992.

14. T. Lu and A. Srivastava, "Gated Low-Power Clock Tree Synthesis for 3D-ICs," *Proceedings of the 2014 International Symposium on Low Power Electronics and Design*, pp. 319–322, 2014.

15. Z. Yang and A. Srivastava, "Co-Placement for Pin-Fin Based Micro-Fluidically Cooled 3D ICs," *ASME 2015 International Technical Conference and Exhibition on Packaging and Integration of Electronic and Photonic Microsystems collocated with the ASME 2015 13th International Conference on Nanochannels, Microchannels, and Minichannels*, pp. V001T09A036–V001T09A036, 2015.

16. B. Shi, A. Srivastava, and A. Bar-Cohen, "Hybrid 3D-IC Cooling System Using Micro-Fluidic Cooling and Thermal TSVs," *2012 IEEE Computer Society Annual Symposium on VLSI (ISVLSI)*, pp. 33–38, 2012.

17. J. Cong and Y. Zhang, "Thermal via Planning for 3-D ICs," *2005 IEEE/ACM International Conference on Computer-Aided Design*, pp. 745–752, 2005.

18. B. Goplen and S. Sapatnekar, "Thermal via Placement in 3D ICs," *Proceedings of the 2005 International Symposium on Physical Design*, New York, NY, pp. 167–174, 2005.

19. A. K. Coskun, J. L. Ayala, D. Atienza, T. S. Rosing, and Y. Leblebici, "Dynamic Thermal Management in 3D Multicore Architectures," *Design, Automation Test in Europe Conference Exhibition, 2009. DATE '09.*, pp. 1410–1415, 2009.

20. D. Zhao, H. Homayoun, and A. V. Veidenbaum, "Temperature Aware Thread Migration in 3D Architecture with Stacked DRAM," *2013 14th International Symposium on Quality Electronic Design (ISQED)*, pp. 80–87, 2013.

21. R. Jayaseelan and T. Mitra, "Dynamic Thermal Management via Architectural Adaptation," *Design Automation Conference, 2009. DAC '09. 46th ACM/IEEE*, pp. 484–489, 2009.

22. J. Meng, K. Kawakami, and A. K. Coskun, "Optimizing Energy Efficiency of 3-D Multicore Systems with Stacked DRAM under Power and Thermal Constraints," *49th ACM/EDAC/IEEE Design Automation Conference (DAC)*, pp. 648–655, 2012.

23. N. H. Khan, S. M. Alam, and S. Hassoun, "Power Delivery Design for 3-D ICs Using Different Through-Silicon Via (TSV) Technologies," *IEEE Transactions on Very Large Scale Integration (VLSI) Systems*, vol. 19, pp. 647–658, Apr. 2011.

24. M. B. Healy and S. K. Lim, "Power Delivery System Architecture for Many-Tier 3D Systems," *Proceedings on 60th Electronic Components and Technology Conference (ECTC)*, pp. 565–572, 2010.

25. P. Zhou, K. Sridharan, and S. S. Sapatnekar, "Optimizing Decoupling Capacitors in 3D Circuits for Power Grid Integrity," *IEEE Design Test of Computers*, vol. 26, pp. 15–25, Sep. 2009.

26. G. Huang, M. Bakir, A. Naeemi, H. Chen, and J. D. Meindl, "Power Delivery for 3D Chip Stacks: Physical Modeling and Design Implication," *IEEE Electrical Performance of Electronic Packaging*, pp. 2037–2044, 2007.

27. E. Song, J. S. Pak, and J. Kim, "Power Delivery for 3D Chip Stacks: Physical Modeling and Design Implication," *IEEE 62nd Electronic Components and Technology Conference (ECTC)*, pp. 2037–2044, 2012.

28. J. S. Pak, M. Pathak, S.-K. Lim, and D. Z. Pan, "Modeling of Electromigration in through-silicon-via Based 3D IC," *IEEE 61st Electronic Components and Technology Conference (ECTC)*, pp. 1420–1427, 2011.

29. X. Huang, T. Yu, V. Sukharev, and S. X.-D. Tan, "Physics-Based Electromigration Assessment for Power Grid Networks," *Proceedings of the 51st Annual Design Automation Conference*, New York, NY, pp. 80:1–80:6, 2014.

30. S.-K. Ryu, K.-H. Lu, X. Zhang, J.-H. Im, P. S. Ho, and R. Huang, "Impact of Near-Surface Thermal Stresses on Interfacial Reliability of through-Silicon vias for 3-D Interconnects," *IEEE Transactions on Device and Materials Reliability*, vol. 11, pp. 35–43, Mar. 2011.

31. T. Lu and A. Srivastava, "Electromigration-Aware Clock Tree Synthesis for TSV-Based 3D-ICs," *Proceedings of the 25th Edition on Great Lakes Symposium on VLSI*, pp. 27–32, 2015.

32. M. Jung, X. Liu, S. K. Sitaraman, D. Z. Pan, and S. K. Lim, "Full-Chip through-Silicon-via Interfacial Crack Analysis and Optimization for 3D IC," *IEEE/ACM International Conference on Computer-Aided Design (ICCAD)*, pp. 563–570, 2011.

33. L. Jiang, F. Ye, Q. Xu, K. Chakrabarty, and B. Eklow, "On Effective and Efficient In-field TSV Repair for Stacked 3D ICs," *Proceedings of the 50th Annual Design Automation Conference*, New York, NY, pp. 74:1–74:6, 2013.

34. Q. Zou, T. Zhang, E. Kursun, and Y. Xie, "Thermomechanical Stress-Aware Management for 3D IC Designs," *Design, Automation Test in Europe Conference Exhibition (DATE)*, pp. 1255–1258, 2013.

35. T. Lu and A. Srivastava, "Electrical-Thermal-Reliability Co-Design for TSV-Based 3D-ICs," *ASME 2015 International Technical Conference and Exhibition on Packaging and Integration of Electronic and Photonic Microsystems*, pp. 1–10, 2015.

36. T. Lu, Yang Zhiyuan, and A. Srivastava, "Electromigration-Aware Placement for 3D-ICs," *International Symposium on Quality Electronic Design*, pp. 35–40, 2016.

37. T. Chantem, Y. Xiang, X. S. Hu, and R. P. Dick, "Enhancing Multicore Reliability through Wear Compensation in Online Assignment and Scheduling," *Design, Automation Test in Europe Conference Exhibition (DATE)*, pp. 1373–1378, 2013.

38. W. Song, S. Mukhopadhyay, and S. Yalamanchili, "Architectural Reliability: Lifetime Reliability Characterization and Management of Many-Core Processors," *Computer Architecture Letters*, pp. 103–106, 2014.

39. P. Mercati, A. Bartolini, F. Paterna, T. S. Rosing, and L. Benini, "Workload and User Experience-Aware Dynamic Reliability Management in Multicore Processors," *Proceedings of the 50th Annual Design Automation Conference*, pp. 2:1–2:6, 2013.

40. C. Zhuo, D. Sylvester, and D. Blaauw, "Process Variation and Temperature-Aware Reliability Management," *Design, Automation Test in Europe Conference Exhibition (DATE)*, pp. 580–585, 2010.

41. H. Tajik, H. Homayoun, and N. Dutt, "VAWOM: Temperature and Process Variation Aware WearOut Management in 3D Multicore Architecture," *50th ACM/EDAC/IEEE Design Automation Conference (DAC)*, pp. 1–8, 2013.

42. T. Lu, C. Serafy, Z. Yang, and A. Srivastava, "Voltage Noise Induced DRAM Soft Error Reduction Technique for 3D-CPUs", 2016 *IEEE/ACM International Symposium on Low Power Electronics and Design (ISLPED)*, pp. 82–87, 2016.

PART 2

FRONT-END-OF-LINE PROCESSES

Chapter 6 Epitaxy

BY

JAMAL RAMDANI *Power Integrations, Inc., San Jose, California*

6.1 INTRODUCTION

The last three decades have witnessed a remarkable growth in the electronic and optoelectronic industry. The growth and preparation of semiconductor materials have played a major role in this success.

The ability to prepare high-purity semiconductor materials, control interfaces and the doping profile and type, and engineer new material combinations has opened the door to new and higher-performance electronic and photonic devices. Quantum-wells-based laser diodes and photodetectors, high-electron mobility two-dimensional (2D) electron gas transistors (HEMTs), heterojunction bipolar transistors, high-brightness GaN-based light-emitting diodes (LEDs), lasers, and HEMTs are a few examples of the devices that have emerged from material growth processes.

Hand in hand with material engineering progress, equipment, and chemical manufacturers have shown similar aggressiveness in developing state-of-the-art epitaxy tools and high-purity source materials to meet the ever-challenging demands for higher-quality materials, lower costs, and safer operations.

6.1.1 Basics of Epitaxy

Royer[1] first introduced the word epitaxy in 1928 from Greek, meaning "ordered upon." Epitaxy refers to the growth of a crystal (epitaxial layer or epilayer) upon a host crystal (substrate) in an ordered fashion. In other words, the symmetry between the contacting crystal planes must be respected. The growth mode depends on lattice mismatch or the difference between the parallel lattice constant of the epilayer and that of the substrate and their chemical compatibility. Thermodynamically, the equilibrium morphology or the growth mode is governed by the relationship of the sum of epilayer surface free energy, E_{ep}, and epilayer-substrate interface free energy, E_I, with respect to substrate surface free energy, E_s, also known as wetting conditions. Three main growth regimes have been identified and are illustrated in Fig. 6.1.

1. Layer-by-layer growth

$$E_{ep} + E_I \leq E_s \qquad \text{(wetting)}$$

2. Nucleation growth

$$E_{ep} + E_I \gg E_s \qquad \text{(no wetting)}$$

3. Layer by layer followed by nucleation growth (S-K growth mode)

$$E_{ep} + E_I \geq E_s \qquad \text{(intermediate phase)}$$

The Layer-by-Layer Growth or the "Frank-Van-Der-Merwe Growth Mode"

In this regime the system is thermodynamically stable and atoms aggregate on the surface of the substrate to form monolayer islands that enlarge with deposit time in order to complete a full monolayer. Then a new monolayer island starts and the sequence continues until the desired total thickness is attained. In practice, full monolayer coverage is hardly accomplished because subsequent monolayer islands begin before the completion of the first monolayer coverage, but overall the growth process stays two dimensional. This is the case in homoepitaxy such as Si/Si and GaAs/GaAs or heteroepitaxy of lattice matched materials such as AlGaAs/GaAs, InGaP/GaAs, and InGaAsP/InP under normal growth conditions.

Extensive work has been done in various laboratories to achieve true layer-by-layer growth. Migration enhanced epitaxy (MEE) and atomic layer epitaxy or deposition (ALE, ALD) are two processes that were developed to achieve true layer-by-layer growth.[2,3] In MEE, atoms are given time to diffuse across the surface to the next available nucleation site through a cycle of source flux interruption and full surface reconstruction. In ALE or ALD, the deposition is self-limiting to one monolayer per process cycle through a combination of flux exposure and purge cycles. These two techniques are extremely useful when thickness control is required at the atomic level, but are generally limited to very thin epilayers due to the long growth rate (1 monolayer/s). ALD, for instance, is now the technique of choice for high-k gate dielectric deposition on silicon.[4]

Nucleation Growth

In nucleation growth or the Volmer-Weber growth mechanism, the system is thermodynamically unstable and growth proceeds in a three-dimensional (3D) fashion in order to lower the total surface free energy. The 3D islands increase in size as the deposition time increases until they touch and intergrow to form a continuous film. This is the case in highly mismatched materials or chemically incompatible systems. Heteroepitaxy of dissimilar materials is increasingly sought and techniques have been developed to remedy issues of lattice mismatch such GaN or sapphire, SiC or silicon substrates, also for compliant substrates and interface template engineering to overcome chemical incompatibilities.[5,6]

Layer by Layer Followed by Nucleation Growth (S-K Growth Mode)

The third growth mode or the Stranski-Krastanov (S-K) growth mode deals with intermediate cases such as low lattice mismatched materials.

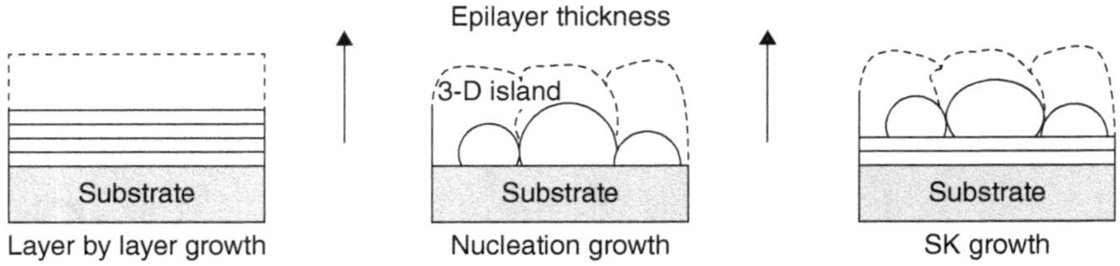

Figure 6.1 Different growth modes in epitaxy.

Initially, the system is stable for the first several monolayers and the wetting is complete. As the thickness increases, the total surface free energy increases, and a breakdown occurs followed by a 3D growth. Pseudomorphic structures such as GaInAs/GaAs and SiGe/Si are a few examples where the S-K mode growth process is applied.

6.1.2 Growth Techniques and Equipment

Varieties of methods are used to grow crystalline materials on crystalline substrates. In the semiconductor industry two major techniques are well established: chemical vapor deposition (CVD) and molecular beam epitaxy (MBE). Both techniques are of nonequilibrium character. Variants of these techniques—metal-organic chemical vapor deposition (MOCVD)—are associated with the epitaxy of compound semiconductors arsenides-, phosphides-, and nitrides-based materials, and chemical beam epitaxy (CBE) or gas source MBE (GSMBE) where the elemental sources are gaseous instead of solid sources. A combination of CVD and ultrahigh-vacuum deposition system (UHV-CVD) is generating acceptance in some applications such as the SiGe/Si materials system.[7]

6.1.3 Molecular Beam Epitaxy

In molecular beam epitaxy (MBE),[8] epitaxy is induced by the flux of atoms/molecules onto a heated substrate in a high-vacuum environment. The mean free path is large enough so that the impinging atomic or molecular fluxes are ballistic in nature and with no gas-phase interactions. The growth conditions are such that the sticking coefficient of low vapor pressure elements is near unity and high vapor pressure elements are used in excess to compensate for their lower sticking coefficient. Surface kinetics and diffusion suffice to achieve 2D growth. As illustrated in Fig. 6.2 atomic or molecular species impinging the surface undergo the following steps:

1. Migration on the surface
2. Re-evaporation

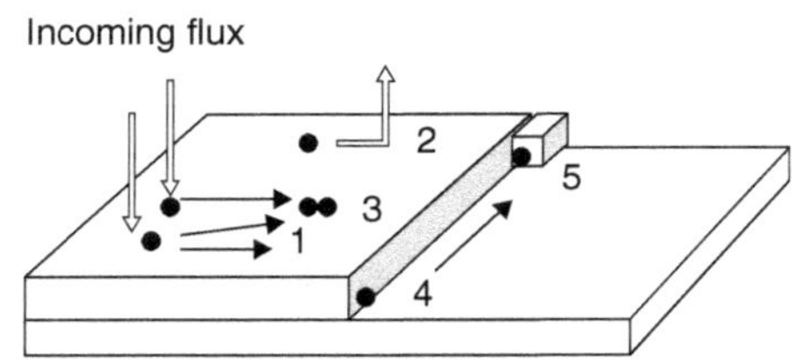

Figure 6.2 Elemental steps in the MBE process.

3. Formation of isolated 2D clusters
4. Incorporation at step edges
5. Migration along step edges and incorporation at kinks

The energetic of each step depends on the growth conditions (atomic/molecular species, fluxes, and growth temperature) and substrate surface reconstruction and orientation.

The MBE system (Fig. 6.3) is based on the evaporation/sublimation of elemental sources such as Si, Ge , Al, Ga, In, As, and P in an ultrahigh-vacuum (UHV) environment using effusion or Knudsen cells. In some instances, the e-beam evaporation method is used. These cells are the key elements of the MBE system; they must provide excellent flux stability and uniformity. This stability should be in the order of 1 percent during a working day and a daily variation of less than 5 percent. This means that the temperature control must be in the order of ±1 percent at 1000°C. Mechanical shutters are provided to switch the fluxes on and off. Since the fluxes from various cells come from different directions, the distribution across the substrate is different leading to nonuniform thickness and composition. Substrate rotation is then used to ensure high thickness and composition uniformity.

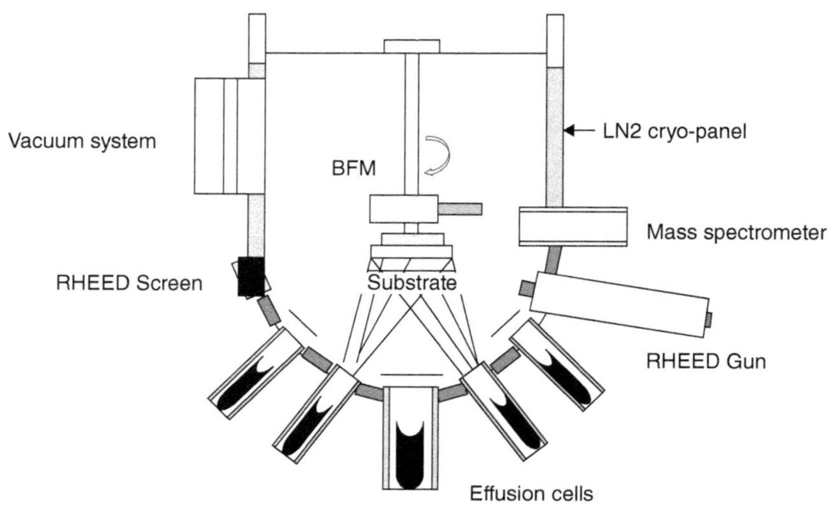

Figure 6.3 Schematic drawing on a generic MBE system.

The MBE system is made up of stainless steel and consists of a loading chamber, a transfer and analytical/treatment chamber, and a growth chamber. The MBE system is maintained in UHV using a combination of diffusion pumps, turbo pumps, cryo-pumps, or ion pumps with auxiliary titanium sublimation pumps. Liquid nitrogen (LN2) cryo-panels surround the interior of the growth chamber and effusion cell flanges to prevent re-evaporation from the internal walls and maintain thermal isolation between them. The base pressure is in the 10^{-11} torr range with hydrogen being the main residual gas.

MBE systems have evolved in the last 20 years from small production and R&D tools to large systems with multiwafer capability and wafer sizes as large as 12 in. Production tools are available from Riber (www.riber.com) and Veeco (www.veeco.com). The Riber 7000 MBE system can accommodate up to 13 sources and has a load capacity of up to 70 6-in wafers or 140 4-in wafers with a total throughput up to 24,000 6-in wafers per years.

The main advantage of this technique is the ability to use high-vacuum in situ characterization systems. The most valuable analysis tool available in MBE is reflection of high energy electron diffraction (RHEED). This technique employs a high-energy (20 keV) electron beam, directed on the substrate surface at a grazing incidence (1° to 3°). The diffraction pattern is imaged on a fluorescent screen placed opposite the e-beam gun. Due to the small incidence angle, the diffraction pattern corresponds to the first atomic layers thus giving information on the surface reconstruction during epitaxy. A detailed discussion on the principles of electron diffraction is beyond the scope of this chapter. Qualitatively, a perfectly flat surface results in a streaky diffraction pattern indicative of 2D growth, while a 3D surface will result in a spotty pattern. Figure 6.4 shows an example of a RHEED pattern of single crystal strontium titanate, $SrTiO_3$, thin epilayer film grown on silicon using MBE.[6] RHEED is also useful to monitor native oxide desorption from the substrate before deposition and to calibrate the ratio of the elements to maintain the optimum growth process. The growth rates and alloy mole fractions are monitored during the growth process using the RHEED oscillations of the specular beam. To complement the ion gauges, a quadrupole residual gas analyzer (RGA) is essential to monitor the integrity of the UHV. The RGA provides spectra of all residual gases present in the chamber, thus providing indications on possible leaks, system cleanness, insufficient bake, or inefficient pumping capability. An ion gauge placed behind the substrate manipulator is used as a beam flux monitor for daily atomic/molecular flux calibration and monitoring. This technique requires skilled technicians and operators. MBE production systems are expensive and require high maintenance, the downtime being larger than any other epitaxy system. However, the cost of ownership of MBE has been in steady decline in the last decade due to the improvement in operation, automated handling, and simplification of operation. Because of the superior quality of the AlGaAs/GaAs materials system, this technique has captured the GaAs-based electronics niche market. Moreover, because of low hazards involved in the MBE process, most GaAs epitaxy foundries have shifted their operations to MBE.

(a) (b)

Figure 6.4 RHEED patterns of epitaxial $SrTiO_3$ on a silicon substrate: (*a*) along the [110] azimuth and (*b*) along the [010] azimuth.

6.1.4 Chemical Vapor Deposition

Chemical vapor deposition occurs at or near atmospheric pressure depending on the application. The sources are gaseous and generally carried by high-purity hydrogen or nitrogen. The CVD growth process is much more complex and usually carried out at higher temperatures than MBE. It involves gas-phase and surface mechanisms.

Only recently, a full understanding of these two processes through simulation and experimental works has emerged. For example, in Si epitaxy using trichlorosilane ($SiHCl_3$ or TCS), up to eleven species of $Si_xH_yCl_z$ and nine reactions have been identified in the gas-phase reactions.[9] A first-order approximation of the TCS and H_2 reaction for silicon epitaxy is

$$SiHCl_3 + H_2 \rightarrow Si + 3HCl$$

MOCVD is very similar to CVD in most aspects.[10] The difference is in the precursors or source materials; metal-organic sources such as trimethyl gallium (TMG, $Ga(CH_3)_3$), indium (TMI, $In(CH_3)_3$), and aluminum (TMA, $Al(CH_3)_3$) are used with hydride group V elements such as AsH_3, PH_3, and NH_3. The growth conditions are slightly different from those of silicon. The growth rate can be described by the same equations (see next section) with multiple chemical species. However, for III–V compound semiconductors epitaxy an additional parameter, the V:III ratio (mole fraction ratio of group V elements over group III elements) must be taken into account. The V:III ratio needs to be high enough to compensate for the low decomposition rates of hydrides and to drive carbon and oxygen impurities from group V lattice sites. For GaN epitaxy using ammonia and TMG, the growth temperature is usually around 1050°C and requires a V:III ratio of several thousand for high-quality materials, these conditions are mainly due to low decomposition rate of NH_3.

The simplified reaction using TMG and NH_3 is given by

$$Ga(CH_3)_3 + NH_3 \rightarrow GaN + 3CH_4$$

Most likely the reaction paths of TMG and HN_3 that leads to the formation of solid GaN involves several steps in the gas phase that are difficult to quantify exactly, but it is widely accepted, at least for the GaAs compound that the decomposition of TMG follows the following path[10]:

$$Ga(CH_3)_3 \rightarrow Ga(CH_3)_2 + CH_3$$
$$Ga(CH_3)_2 \rightarrow Ga(CH_3) + CH_3$$
$$Ga(CH_3) \rightarrow Ga - CH_3$$

As for NH_3, the decomposition at high temperature should lead to atomic hydrogen and nitrogen-containing radicals. Abstraction of the first hydrogen bond is thought to be the rate-limiting factor.

$$NH_3 \rightarrow NH_{(3-x)} + xH$$

It is worth stating that only limited understanding of the exact process is available, and in general MOCVD of compound semiconductor is still purely empirical based on external outputs such as the effects of temperature, pressure, flow and substrates on the growth rate, the composition of the film, and its electrical and optical quality.

MOCVD epitaxial systems in general consist of a gas manifold, a load lock, a reaction chamber, and an exhaust system. The gas manifold is made of semiconductor-grade stainless steel tubing with pneumatic valves, mass flow controllers, and pressure regulators and transducers. The growth chamber is made of either quartz or stainless steel and consists of a susceptor and elements of heating (resistive, induction heating, and lamp heating) and finally the exhaust system with particle filters and toxic/corrosive chemicals traps.

Two reactor designs are available in the market—vertical reactors in which the gas flow is perpendicular to the substrate surface and horizontal reactors in which the gas flow is parallel to the substrate. These reactors are available in single wafer or batch (multiwafer platters) configuration with substrates sizes from 2 to 8 in diameters. Options with load-lock, a transfer chamber, and a cool-down chamber are also available.

As for the metal-organic precursors such as TMG, TMAl, they are available in varying purity and in commercial bubblers up to 5 kg. Both precursors have high vapor pressure (>1 torr) at temperatures typically less than 50°C. In the MOCVD system, precursors are set in temperature controlled baths to ensure constant vapor pressure of the source materials.

6.1.5 MOCVD of Compound Semiconductors

Compound semiconductor (CS) devices for photonic, RF, and power applications rely solely on epitaxy for design and performance, unlike Si-based devices where the process and integration drive the performance of the device.

CS devices are "epi-centric": meaning that the device active region is completely defined by the epitaxial layer, whereas Si devices are "process centric." Meaning that the critical characteristics of the device are driven by the process such as implant, etches, etc.

In particular, CS GaN material has emerged as the largest market for compound semiconductors, mostly driven by light-emitting diodes (LEDs) for displays and lighting with a forecasted total available market in excess of $15 billion by 2020. RF and power devices made of GaN are also predicted to capture a sizable market of about $1 billion in 2020 (Yole Development 2014. http://www.yole.fr/).

GaN, a wide band gap material (Eg = 3.4 eV), possess electronic and optical properties that are of great interest in photonic and electronics devices; however, epitaxial growth of high quality GaN only materialized in the late 1980s and beginning of the 1990s. The main challenges that needed to be overcome were nucleation and epitaxial growth of high quality GaN on non-lattice-matched substrate (sapphire) and *p* type doping of GaN to levels needed for high quality *p-n* junctions. I. Akasaki, H. Amano, and S. Nakamura were instrumental in the development of the epitaxial growth of high-quality GaN. In 2014, they were awarded the Nobel Prize of physics for the development of GaN blue LEDs.

In this chapter, we will review some of the characteristics of GaN epitaxial growth for RF and power transistors.

6.1.6 Epitaxy Process: An Overview

In general, the CVD process includes[11]

1. Transport of bulk gases to the growth region
2. Transport of reactants to the substrate surface through a "stagnant layer" or boundary layer. This step can be described by the following equation:

$$\frac{dh_1}{dt} = h_g(C_g - C_s)$$

where h_g is the mass transport factor of the gas in the chamber, and C_g and C_s are the concentrations of the elements in the gas phase and on the surface.

3. Adsorption of reactants on the surface
4. Surface reactions, which include chemical decomposition, surface migration to nucleation sites, and incorporation in the lattice
5. Desorption of by-products and transport to the exhaust

Steps 3 through 5 are described by the "oversimplified" equation

$$\frac{dh_2}{dt} = k_s C_s$$

with k_s, being the surface reaction rate.

In steady state

$$\frac{dh_1}{dt} = \frac{dh_2}{dt} = H \qquad \text{or} \qquad C_s = C_g\left(1+\frac{k_s}{h_g}\right)^{-1}$$

The growth rate is then given by

$$Gr = \frac{H}{N} = \frac{C_g}{N} * \frac{k_s h_g}{(k_s + h_g)} = \frac{k_s h_g}{(k_s + h_g)} * \frac{C_t}{N} * M$$

where N = number of atoms incorporated per unit volume
C_t = total number of molecules per unit volume in the gas
M = mole fraction of the reaction species in the gas

Figure 6.5 shows the boundary-layer model for GaN epitaxy using $Ga(CH_3)_3$ (TMG) and NH_3 and H_2.

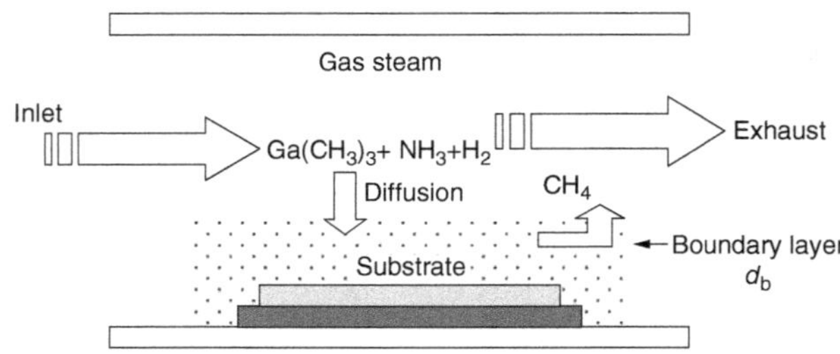

Figure 6.5 Boundary layer model schematic for GaN epitaxy.

From the growth rate equation, we see that this simple model predicts a linear relationship of growth rate and the mole fraction *M*. Experimentally, this relationship holds for low values of M ($M < 0.1$). We also distinguish two growth regimes:

1. $k_s << h_g$: The growth mode is reaction limited and since the reactions are thermally activated or of the Arrhenius type, the growth rate is given by

$$Gr = \frac{C_t}{N} * M * k_s \qquad \text{with} \qquad k_s = k_0 e^{-\left(\frac{Ea}{kT}\right)}$$

where Ea = reaction activation energy
K = Boltzmann constant
T = temperature
k_0 = a constant independent of temperature

In this regime, the growth rate is highly temperature dependent and will increase until the mass flow limits the process.

2. $h_g < k_s$: The growth process is mass transport or gas-phase diffusion controlled and the growth rate equation is reduced to

$$Gr = \frac{C_t}{N} * M * h_g \qquad \text{with} \qquad h_g = \frac{D_g}{d_b}$$

where D_g is the diffusion coefficient through the boundary layer d_b.

In practice, the mass-transport-limited regime is preferred because it is practically independent of the growth temperature and primarily driven by a square root dependence on the bulk flow, thus allowing for high growth temperature for high-quality epilayers. Figure 6.6 shows the growth regime as a function of the deposition temperature.

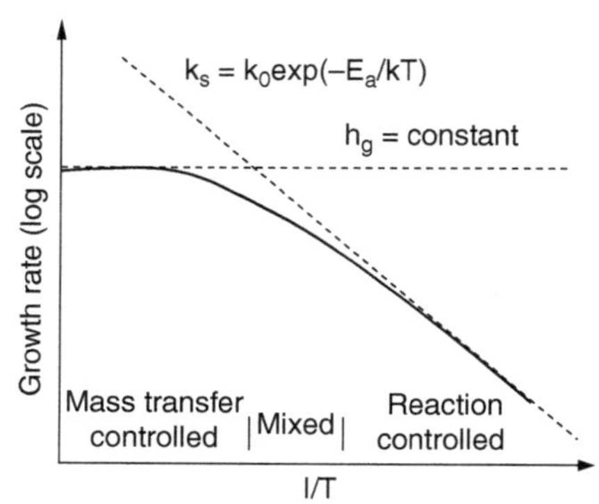

Figure 6.6 Growth regimes as a function of deposition temperature.

However, the mass transport regime depends strongly on the boundary layer thickness, which in turn depends on the reactor geometry leading to restrictions on the reactor chamber design. Horizontal-type reactors provide the ability to control the boundary layer structure by maintaining the lowest possible height-to-width aspect ratio of the reaction chamber.

According to the boundary layer theory,[11] d_b is given by

$$d_b = \left(\frac{D}{Re} * x\right)^{1/2}$$

where D is the reaction tube diameter, x is the distance along the reactor, and Re is the Reynolds numbers.

The Reynolds number depends on the gas density, the gas velocity, and gas viscosity. This configuration will retain the flow in a laminar mode, that is, the gas flows in a regular, continuous, and nonturbulent mode. Yet, the boundary layer thickness is not constant as the gas flow across the heated wafer surface leading to a nonuniform deposition rate along the flow direction and in the transverse direction. Therefore, wafer rotation in the azimuthal direction along with a special injector design for flow distribution control is added to achieve high thickness uniformity. Thorough treatment of the horizontal reactor process using computational fluid dynamics can be found in Refs. 12 and 13.

For vertical reactors type designs, in order to maintain a thin and constant boundary layer thickness Emcore (now Veeco) designed a vertical rotating disc reactor or TurboDisc®. The high speed rotation (≈ 1000 rpm) along with complex gas injectors design resulted in high velocity laminar flow and minimum gas phase reactions which in turns led to a uniform boundary layer distribution across the susceptor and thus uniform deposition rate across the wafers. Additional benefits of the TurboDisc reactor are minimum residual build up after a long campaign of epitaxy runs, stable process run to run, and low particle count when high aluminum content process is used, consequently high epitaxial yield and low cost of operation.

In practice, the growth rate is directly proportional to the mole fraction of the reaction species, that is, group III precursor that is TMG which in turns is determined by its vapor pressure of TMG in the reactor according to

$$Gr \propto F_{\text{III}} = FH_2 * \left(\frac{P_{\text{vap}}}{P_{\text{tot}}}\right) \Big/ \left(1 - \frac{P_{\text{vap}}}{P_{\text{tot}}}\right)$$

where F_{III} is the group III flow for instance TMG, FH_2 is the total H_2 flow, P_{vap} is the vapor pressure of TMG, and P_{tot} is the pressure in the reactor. If a ternary alloy such as AlGaN with an Al composition of x%, the mole fraction of Al can be easily adjusted through the group III flow according to

$$x_{\text{Al}} = \frac{F_{\text{Al}}}{(F_{\text{Al}} + F_{\text{Ga}})}$$

Epitaxy also provides the ability to control the film doping level and thus the electrical characteristics of the thin film (resistivity). GaN dopants such as magnesium for p type and silicon for n type are introduced with reaction precursors, Si in the form of diluted SiH_4 in hydrogen and Mg in a metal-organic form known as CP_2Mg (bis(cyclopentadeienyl) magnesium). Dopant incorporation is a surface kinetic process and consequently extremely sensitive to deposition temperature. Due to the complexity of the gas-phase reaction, it is practically impossible to accurately model the dopant incorporation in the epilayer from the gas-phase ratio and growth conditions. However, empirical solutions for each set of deposition parameters are established. With current epitaxy systems, the repeatability of epilayer doping concentration is excellent for typical target levels and process settings.

MOCVD production tools are available from two major tool manufacturers: Aixtron, Inc. (www.aixtron.con) and Veeco (http://www.veeco.com/technologies-and-products/mocvd-systems). The Aixtron and Veeco systems are based on three different concepts and designs, Aixtron tools use the "planetary" or "close coupled showerhead" designs and the Veeco system are based on the "turbodisc" design. All technologies lead to similar material growth performance and quite similar throughputs.

For instance the Veeco MOCVD system K465i for GaN epitaxy is a fully automated production tool capable of 12 × 4 in per platter run, 5 × 6 in and up to 3 × 8 in per platter run. This system has been designed for low maintenance (high up-time and run-time) and high throughput. The latter can be as high as 700 6-in wafers per month depending on process conditions. The system is also equipped with in situ monitoring of the epitaxial process using an integrated pyrometer-reflectometer-deflectometer (DRT-210) for wafer temperature, surface reflectance, and wafer curvature monitoring. This is a powerful tool that allows epitaxial growth calibrations, tuning, and monitoring. The wavelength used for reflectance is 930 nm. Since the refractive index of GaN is about 2.5 and that of sapphire is about 1.8, there will be strong oscillation is the GaN films as it becomes thicker (as long as the surface is smooth). The period of these oscillations is proportional to the growth rate.

6.1.7 MOCVD of GaN

As discussed earlier in the chapter, device quality GaN epitaxy needed to overcome key technological barriers due to lack of native or lattice matched substrates. GaN substrates would be impeccable for high-quality material growth; however, GaN bulk crystal growth is still in its infancy in spite of over a decade of researches, mainly due to the fact that extreme high pressures are needed to grow GaN boules. Still, GaN substrates are available in limited quantities and size (< 2 in diameter) at a very high price and are mostly used in research and development and in some applications such as for high power laser diodes and vertical power transistors. The current success seen of GaN LEDs in displays, lighting, and GaN RF/power devices stems from the ingenuity that went into of growing high-quality GaN on highly lattice and thermal mismatched substrates such as sapphire (−16 and 23 percent), SiC (−3.5 and −25 percent), and Si (17 and −115 percent).

For epitaxial growth of GaN on sapphire, the lattice and thermal mismatch between GaN and sapphire are −16 and 23 percent, respectively. A two-step growth process was invented to overcome the large lattice mismatch, forcing a Volmer-Weber growth mechanism. A low-temperature (500 to 700°C) deposition of a nucleation layer either GaN or AlN few hundred angstrom thick, this film is a highly ordered (single orientation) polycrystalline in nature with a density and size of the grains depending on the temperature, the growth pressure, and the growth rate, at these temperature the process is highly reaction limited and very sensitive to process conditions. Adequate temperature control and in situ monitoring are needed to ensure the process control of this layer. In the second step, the temperature is raised above 950°C and GaN is grown on top of the nucleation layer, during the initial stage of this step, the GaN growth proceeds in an island growth regime followed by islands coalescence through lateral growth to a thickness up to about 0.3 μm and then become 2D epitaxial growth. This model has been confirmed by in situ monitoring of the surface reflectance of the film during epitaxy. A schematic of two-step growth process is shown in Fig. 6.7.

The nucleation layer material type (GaN or AlN), the nucleation temperature, growth rate along with the temperature, V/III in the subsequent GaN buffer layer are critical to guarantee device quality materials. In Fig. 6.8, we show that, when the process window is not optimized, the growth GaN results in no recovery of the 2D growth and leads to a rough surface (oscillation damping). When the process is optimized, the GaN follows exactly the model cited above and results in a good recovery of the GaN surface with sustained oscillations corresponding to 2D growth with a specular surface.

It is worth mentioning that because of the two-stage process and the lattice and thermal mismatch, a large dislocations density is generated at the GaN thin film. At the GaN/nucleation layer interface, the dislocation density is excess of $10^{12}/cm^2$, and as the thickness of the

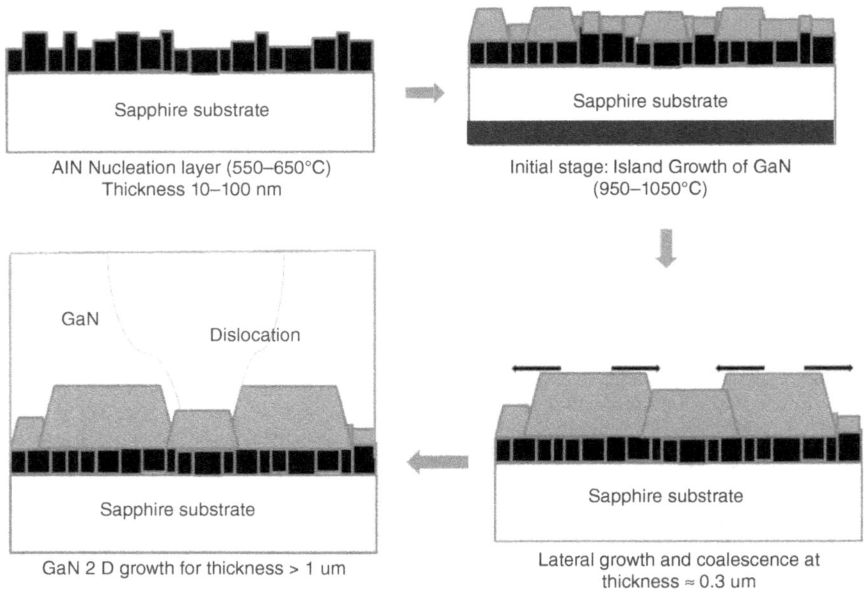

Figure 6.7 Two-step growth sequence of GaN on sapphire.

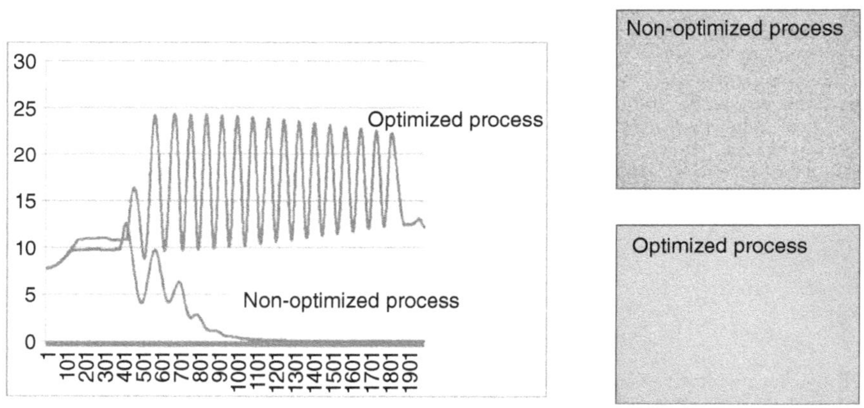

Figure 6.8 In situ surface reflectance during growth (optimized versus nonoptimized process) and microscope photographs of resulting surface morphologies.

GaN increases the dislocation, density decreases down to few $10^8/cm^2$ through a process of cancelation/annihilation. These dislocations are of three natures: screw, edge-threading, and mixed dislocations. It is generally known that, minimizing the dislocation density is needed to obtain device quality materials; however, it has been shown that with dislocation density in the $10^8/cm^2$ range only minimum impact on device performance and reliability have been reported and this is mainly due to the "inactive" nature of these dislocations.

Figure 6.9 shows a cross-section TEM of GaN/sapphire substrate. The surface morphology of GaN by atomic force microscopy (AFM) is shown in Fig. 6.10. The surface presents step-terrace formation which is indicative of step flow growth with a typical RMS roughness of less than 5 Å in a 5 × 5-μm AFM scan window for a 4-μm-thick GaN film. The dark spots at the edge of the terraces are believed to be screw and or mixed dislocations that terminate at the surface. Also it is important to note that this dislocation density would not totally relieve the strain in the GaN film, and a residual mixed type strain, tensile from in-plane GaN lattice relaxation, and compressive strain from the large thermal mismatch during cooling postepitaxial growth. This topic will not be covered in details in this chapter but for films in the 3 to 5-μm thickness range, the wafer bow can be controlled down to less than 80 μm for a 6-in-diameter substrate. The bow is monitored during the growth using the in situ deflectometer.

Figure 6.9 Cross-section TEM of GaN epilayer on sapphire substrate.

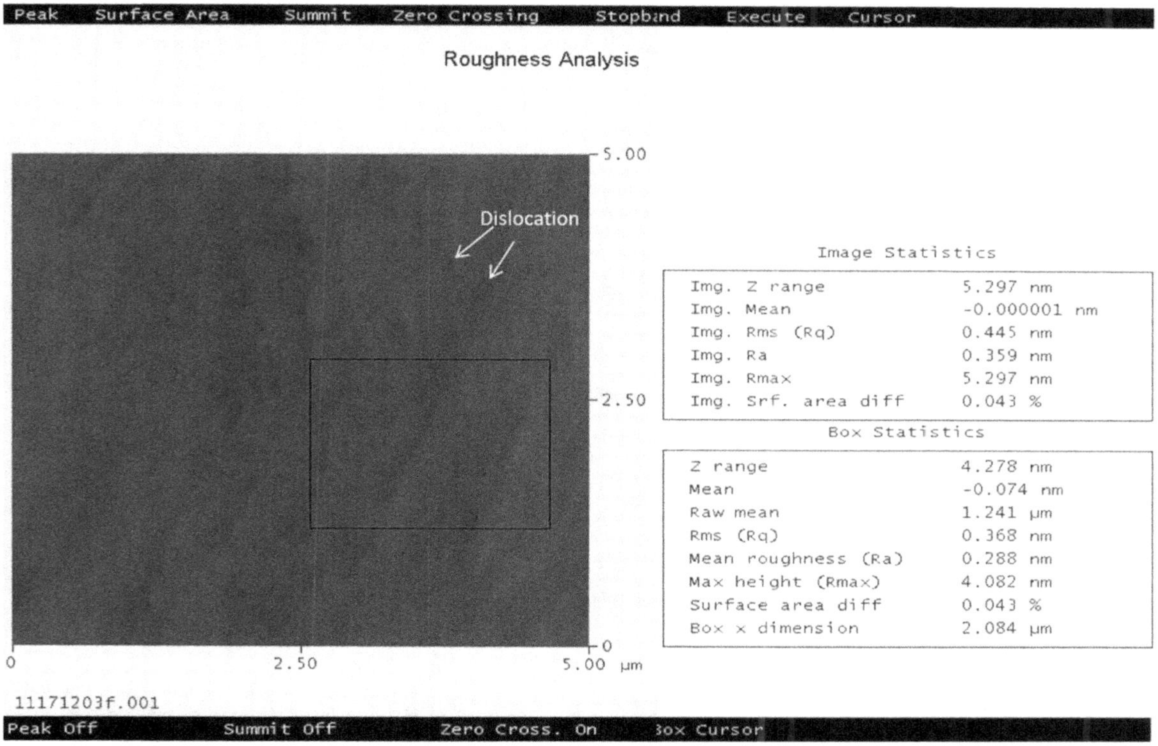

Figure 6.10 AFM image of GaN epilayer.

For a GaN HEMT device, an AlGaN (ternary) layer is grown on top of the GaN buffer layer to form a 2D electron gas that acts as the channel of the transistor, the physics of the AlGaN/GaN HFET device is out of the scope of this chapter, and readers are referred to Ref. 14 for an overview of AlGaN/GaN-based HFET device operation and applications. $Al_xGa_{(1-x)}N$ is grown in a similar fashion as with GaN; however, the process conditions such as temperature and pressure may be slightly different than that of GaN to ensure a high-quality AlGaN layer, and an abrupt interface on GaN. The interface quality dictates the quality of the Al mole fraction (x) defined only from the ratio of the TMGa and TMAl flows. The AlGaN/GaN HEMT with $x = 25$ percent results in a sheet carrier density of about $10^{13}/cm^2$ (sheet resistance of about 400 Ω/square) and a mobility in excess of 1200 $cm^2/V/s$ at room temperature, making AlGaN/GaN HFET an excellent candidate for RF and high-power devices.

6.1.8 GaN Epilayers Material Characterization and Analysis

In general, all compound semiconductors rely on large set of characterization tools to ensure the material quality needed for device performance. These tools are divided in three major sections.

1. *Physical characterization.* Such as high-resolution x-ray diffraction (XRD) to characterize the material quality such as threading dislocation density, the strain and relaxation of the epilayer film with respect to the substrate, the composition of a ternary (quaternary) alloy. AFM for surface roughness, surface pits, and other defects. TEM (transmission electron microscopy) for thickness confirmation, interface quality, dislocation density estimation. SIMS (secondary ion mass spectroscopy) for background doping, contamination monitoring, composition and thickness, and finally optical microscopy for defects, slip lines, and cracks inspection.

2. *Electrical characterization.* Van der Paw method for carrier mobility measurement. Contactless sheet resistance measurement for channel carrier density measurements, and finally C-V (capacitance-voltage) measurement for intrinsic/extrinsic carrier measurement.

3. *Optical measurement.* Photoluminescence (PL) is used to characterize the optical properties of the films. Thickness, composition, and material quality can be deduced from a PL spectra.

XRD

The crystalline quality of the film is measured by high-resolution x-ray diffraction for the (002) symmetrical and (102) asymmetrical Bragg reflections. Typical XRD spectra for the (002) and (102) are shown in Fig. 6.11. The full width at half max (FWHM) of the peak is direct measure of the crystal quality of the film. The (102) XRD peak directly correlates to the density of treading dislocations (TD) (edge and mixed type) since they distort the (102) planes. The (002) peak broadening is only affected by the screw-type dislocations. One can easily estimate the dislocation density of the GaN film by following these two simple equations.

$$D_{\text{screw}} = \frac{(\text{FWHM}_{(002)})^2}{9b_{\text{screw}}}$$

$$D_{\text{edge}} = \frac{(\text{FWHM}_{(102)})^2}{9b_{\text{edge}}}$$

where b_{screw} and b_{edge} are burger vector length of both screw and edge type dislocations.

Low values of the FWHM in the (102) and (002) reflections (266 and 170 arc/s) indicate that the TD (edge/mixed type) is low (few $10^8/cm^2$ range) and the screw-type dislocation density is even lower, meaning that the material is of high crystalline quality. In general, edge-type dislocation density needs to be less than $10^9/cm^2$ and screw-type density is less than $10^8/cm^2$ for device quality materials. These values can be achieved through epitaxy by engineering the process the nucleation layer and optimizing the growth parameters of the GaN buffer layer.

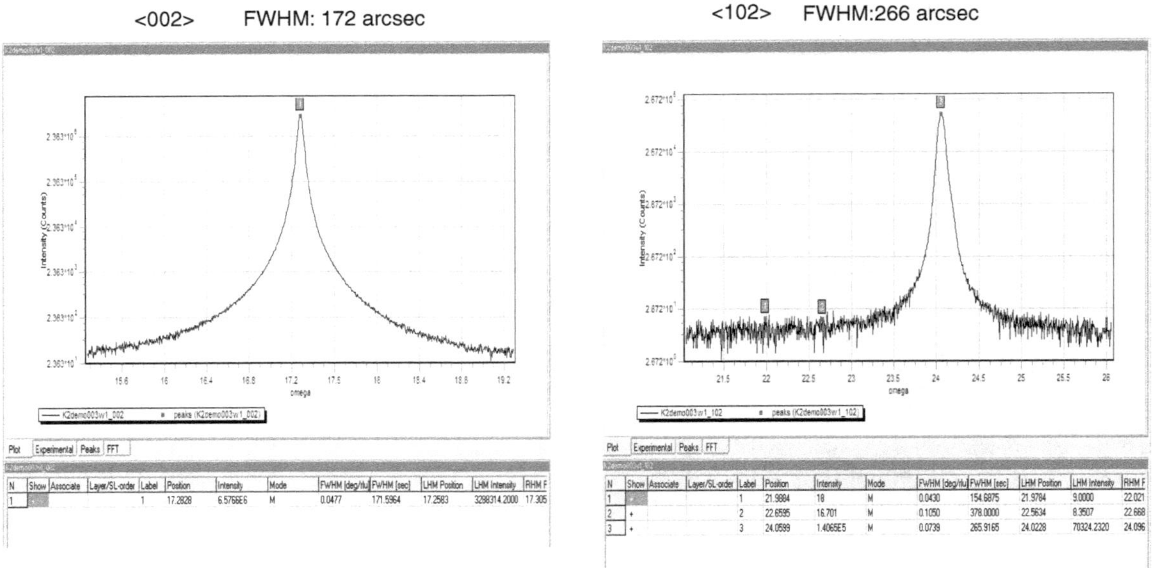

Figure 6.11 XRD rocking curve spectra of GaN/sapphire on (002) and (102) reflection planes.

PL

Figure 6.12 shows typical PL spectra at room temperature of the GaN/sapphire film. An intense emission around 365 nm is related to the band-edge (BE) emission (near band gap luminescence) can be interpreted as free exciton or a donor-bound exciton transition or donor to valence band transition. The band-edge transition is intrinsic to the optical properties of GaN (direct band gap with an Eg of 3.4 eV). In addition to BE, a broad band emission around 550 nm (2.22 eV) is usually observed and has been named the yellow band luminescence (YB) and is believed to be a radiative recombination via defects most likely Ga vacancies. The intensity oscillations of the YL band are attributed to intensity modulation by the Fabry-Perot cavity formed by the air-GaN and GaN-sapphire.

The BE position in Fig. 6.12 is at 362 nm or 3.43 eV, close to the band gap of GaN at RT, and its full width at half max is about 7.8 nm or 73.6 meV which is equal to 3 kT at RT. This FWHM is close to the theoretical limit of 1.8 kT which indicates excellent material quality (kT being the thermal energy which is equal to 25 meV at RT).

The ratio of YL:BE intensity is also important indicator of the material quality and it is used as an optimization knob of the epitaxial process. The lowest possible ratio is needed for high-quality optical and electrical devices; however, due to the different dependence of YB luminescence and BE luminescence on the PL excitation power, with BE intensity increases linearly with the incident power while the YL intensity shows a slow saturation of its intensity; therefore, it is recommended to perform PL at low intensities and at about the start of the saturation level of the YB to make sure that the YB:BE ratio is correctly defined for tuning the epitaxy process.

The thickness map of the GaN film along with the composition map of the AlGaN film can also be extracted from the PL measurements. In Fig. 6.13, we show typical thickness map of GaN and the Al composition in AlGaN thin film indicating high uniformity (2 percent) of the epitaxial process over a 6-in wafer.

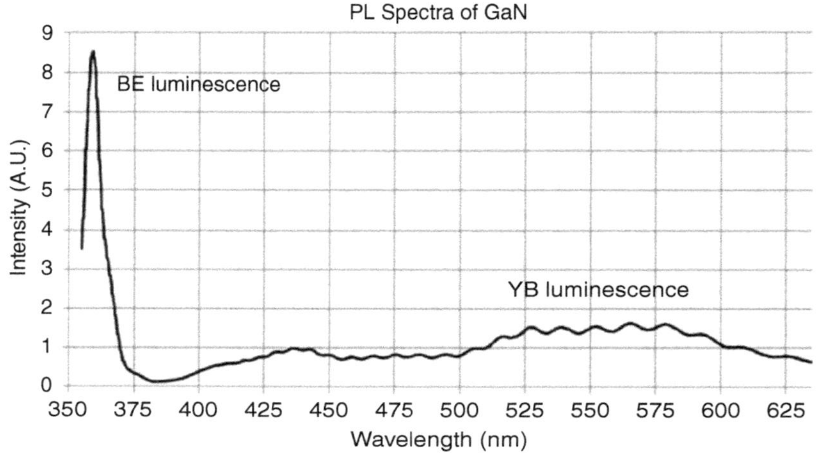

Figure 6.12 Typical PL spectrum of GaN/sapphire.

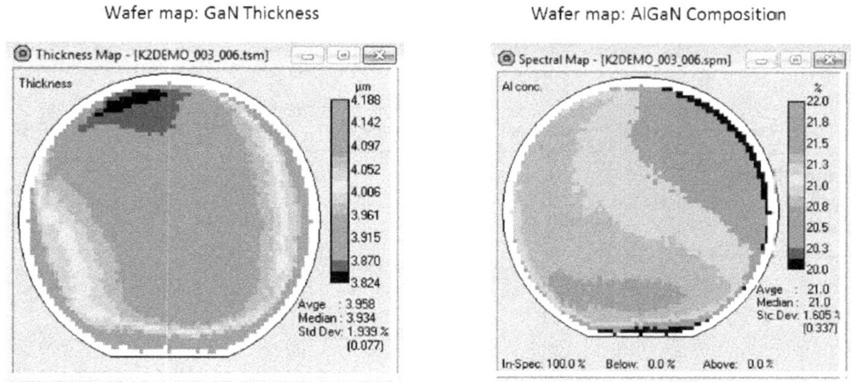

Figure 6.13 PL wafer map of a 6-in epiwafer for GaN thickness and Al composition in the AlGaN layer.

SIMS

A typical SIMS depth profile of GaN is shown in Fig. 6.14. Impurities such as oxygen and carbon have been plotted. The detected level of O is at the detection limit of SIMS. The origin of oxygen is believed to be the sapphire substrate, and its incorporation depends heavily on the growth temperature. It is important for device applications to keep the oxygen level at the detection limit in order to maintain a highly resistive buffer layer needed for GaN HEMT.

Carbon on the other hand is at $10^{16}/cm^3$ level. The source of carbon is from the decomposition of the methyl groups in TMG and its incorporation rate in the film strongly depends on the growth conditions such as temperature, V:III ratio, and the growth rate. Carbon is a deep acceptor in the GaN material system and is usually used to compensate for the deep donors, which could be either from oxygen and/or lattice defects like nitrogen vacancies to make the GaN semi-insulating.

6.1.9 Epitaxy Manufacturing

The correct balance between quality and cost is a challenge in modern epitaxial production that needs to deliver high-volume state-of-the-art epitaxial wafers with very high yield at target cost. This in turn requires tight control of the epitaxy operation and a high level of quality management. Also it is worth mentioning that unlike most processes in the silicon world, including Si epitaxy, MBE, and MOCVD of compound semiconductors are much more complex in both process and tools and are still at this time "engineering heavy" requiring highly qualified engineers with deep understanding of material growth science, material characterization, and device physics to be able to respond quickly to slight deviations in the process windows and to maintain the epitaxy operation at the highest quality level. The drawback hence is the cost associated with highly qualified personnel to run and monitor the epi operation.

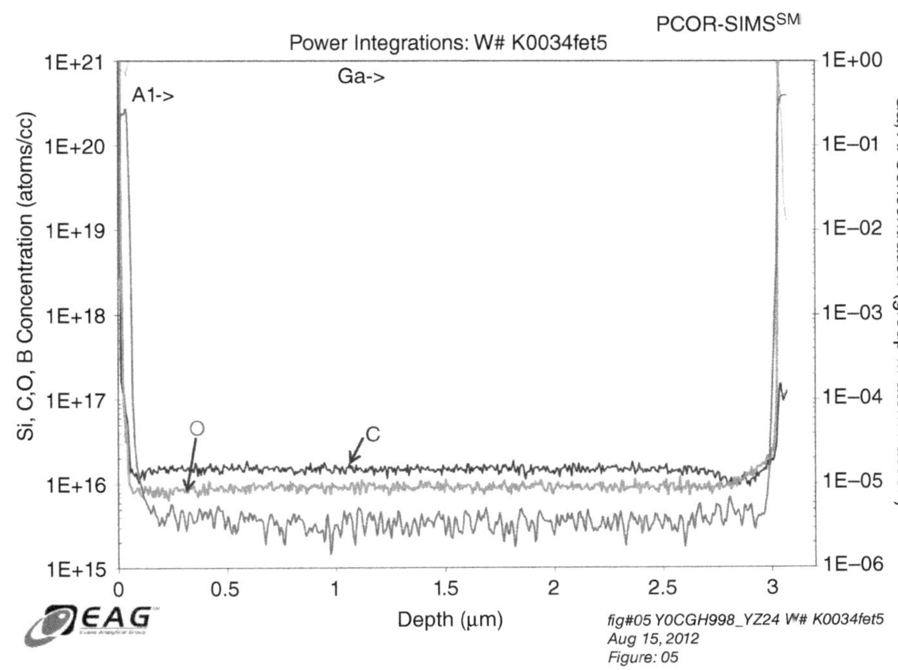

Figure 6.14 SIMS depth profile of GaN/sapphire.

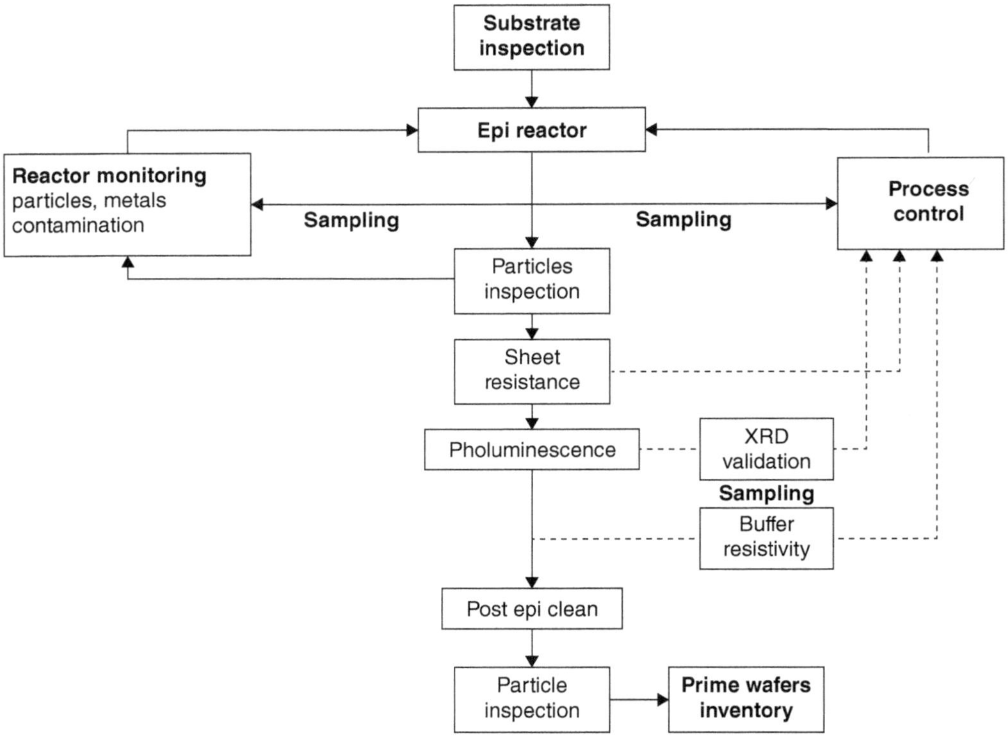

Figure 6.15 Process flow in advanced AlGaN/GaN epitaxy for HEMT applications.

6.1.10 Control Procedure

Figure 6.15 shows a typical epitaxy process flow. It is clear that quality is strongly related to the complex system of characterization and metrology. Advanced inspection systems are costly and add substantially to the overall fixed cost. The metrology allows for process development, process control, and troubleshooting and sorting.

Statistical process control (SPC) is the most common approach used in process control. The SPC chart for each parameter is recorded and used to monitor process evolution and potential upsets that occur during the process. The sampling frequency depends on the inspection tool throughput and the nature of the parameter monitored. For example, particle defects, photoluminescence properties of the film, and sheet resistance are 100 percent inspected, XRD and buffer resistivity are on a sampling inspection basis only.

Figure 6.16*a*, *b* typical example of SPC charts trends of GaN buffer layer thickness and the Al composition in the AlGaN barrier layer of the HEMT from PL measurements. As one can see, a very tight distribution of thickness, Al composition and sheet resistance are achieved over a long period of epitaxy campaign, an indication of robust process control and a wide epitaxy window design that can tolerate the slight process deviations without impacting the material specifications needed for device performances. It also shows that current MOCVD tools have progressed to a level worthy of production and manufacturing of complex semiconductor material systems for complex photonic and electronic devices.

6.2 SAFETY AND ENVIRONMENTAL HEALTH

The main concern in the operation of MOCVD epitaxy systems is safety. A substantial part of the cost associated with setting up an MOCVD operation is in the toxic/flammable gas monitoring and alarms systems along with a contingency plan. This is usually part of the facility department. In addition, since only a fraction of the inlet gases is consumed in deposition, the unreacted gases along with the gas-phase reaction by-products are run through the exhaust system. Abatement systems are installed downstream of the MOCVD tools to reduce the emission to the atmosphere. Epi-foundries must comply with Occupational Safety and Health Act (OSHA) codes, Toxic Gas Ordinance, Uniform Fire Code, Clean Air Act, and city/state regulations.

6.3 FUTURE EPITAXY TRENDS

It is obvious that epitaxy will continue evolving at the same pace as the electronic and photonic industry. Next generation devices will require tighter epilayers specifications and complex structures with abrupt transition layers and doping profiles. Heterogeneous epitaxy with low thermal budgets on patterned substrates will see more acceptances. Epitaxy systems also need to evolve in the same direction with new reactor designs to meet future device specifications, that is, long epitaxial campaigns between scheduled preventive maintenances. In addition, full automation and advanced real-time in situ monitoring and process control of material quality such as photoluminescence, thickness, and composition will be sought to improve the yield, reduce or even eliminate postepitaxy inspections in order to reduce the overall cost.

6.4 CONCLUSION

This chapter gives a bird-eye view of the fundamentals of epitaxy, the most common epitaxy processes and equipment and a condensed version of MOCVD growth of GaN for RF/power applications. We also discussed some aspects of manufacturing environment and process control. These topics are vast and require the combined knowledge of various disciplines from material science to quality control and management and to facility operation. A team of experts in all these areas makes a successful epitaxy foundry.

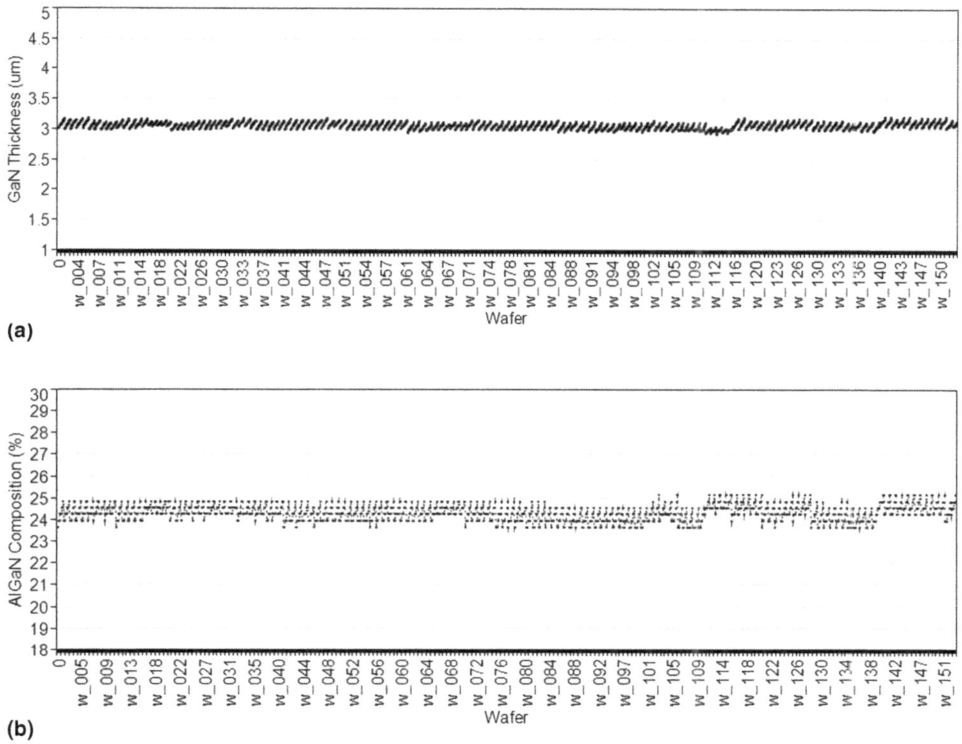

Figure 6.16 (*a*) GaN thickness from photoluminescence measurements; (*b*) Al composition in the AlGaN from photoluminescence measurements.

6.5 REFERENCES

1. L. Royer, "Some Observations on the Epitaxy of Sodium Chloride on Silver," *Bull. Soc. Franc. Miner.*, 51: 7, 1928.

2. Y. Horikoshi et al., "Growth Mechanism of GaAs during Migration Enhanced Epitaxy at Low Growth Temperatures," *Jpn. J. Appl. Phys.*, 27: 169, 1988.

3. M. A. Tishler and S. M. Bedair, "Self-Limiting Mechanism in the Atomic Layer Epitaxy of GaAs," *Appl. Phys. Lett.*, 48: 1681, 1986.

4. E. P. Gusev et al., "Ultra-Thin HfO_2 Films Grown on Silicon by Atomic Layer Deposition for Advanced Gate Dielectrics Applications," *Microelectron. Eng.*, 69: 2–4, 2003.

5. A. Bourret, "Compliant Substrates: A Review on the Concept, Techniques and Mechanism." *Appl. Surf. Sci.*, 164: 3, 2000.

6. J. Ramdani et al., "Interface Characterization of High Quality $SrTiO_3$ Thin Films on Si (100) Substrates Grown by Molecular Beam epitaxy." *Appl. Surf. Sci.*, 159/160: 127, 2000.

7. M. T. Curie et al., "Controlling Threading Dislocations Densities in Ge on Si using Graded SiGe Layers and Chemical-Mechanical Polishing." *Appl. Phys. Lett.*, 72: 14, 1998.

8. C. T. Foxon, *Principles of Molecular Beam Epitaxy, Handbook of Crystal Growth*, Vol. 3. Elsevier, Amsterdam, p. 157, 1994.

9. P. Ho et al., "Chemical Kinetics for Modeling Silicon Epitaxy from Dichlorosilanes." *Proc. 194th Mtg. Electrochem. Soc. PV*, 98-23: 117, 1999.

10. G. B. Stringfellow, *Organomettalic Vapor-Phase Epitaxy. Theory and Practice*. Academic Press, New York, 1989.

11. S. Wolf and R. N. Tauber, *Silicon Processing for VLSI Era. Process Technology*, Vol. 1, Lattice Press, Sunset Beach, CA, 1986.

12. H. Hakuba et al., "Chemical Process of Silicon Epitaxial Growth in a $SiHCl_3$-H_2 System." *J. Cryst. Growth*, 207: 77, 1999.

13. S. Kommu et al., "A Theoretical/Experimental Study of Silicon Epitaxy in Horizontal Single-Wafer Chemical Vapor Deposition Reactors." *J. Electro Chem. Soc.*, 147: 1538, 2000.

14. U. K. Mishra, "AlGaN/GaN HEMTs—an Overview of Device Operation and Applications." *Proceedings of IEEE*, Vol. 90, Issue 6, 2002.

6.6 SUGGESTED READING

Nakamura, S., Stephen Pearton, and Gerhart Fasol, *The Blue Laser diode*, Springer, New York, 2000.

Parker, E. H. C., *The Technology and Physics of Molecular Beam Epitaxy*, Kluwer Academic Publishers, Dordrecht, 1985.

Stringfellow, G. B., *Organometallic Vapor-Phase Epitaxy. Theory and Practice*. Academic Press, New York, 1989.

Chapter 7

Thermal Processing: Anneals, RTP, and Oxidation

BY
DAVID L. O'MEARA *Tokyo Electron Limited, Albany, New York*

7.1 INTRODUCTION

Semiconductor devices are created from several different materials combined in precise sequences and patterns. The materials are grown or deposited using specifically controlled ambient (temperature, pressure, and high-purity chemicals). Usually, creating the films involves elevated temperatures and subatmospheric pressures, although many oxidation and anneal processes are at atmospheric pressures. Two types of chemicals used to make the films are inert gases, for example, nitrogen (N_2), that do not react with other materials, and reactant gases that decompose into the desired film. Over time, the requirements for processes have changed with device products. Manufacturing economics in terms of cost of ownership (COO) in deference to technical need has also changed the configuration of deposition hardware. Thermal equipment, processes and controls used for annealing and oxidation in semiconductor manufacturing are described as an overview, with references provided if more detail is desired.

7.2 THERMAL PROCESSING

Advances in technology for thermal processing in semiconductor manufacturing have followed the requirements of the devices being created. In the late 1970s, silicon wafers were relatively small, less than 6 in, and device features were relatively large, with feature sizes as small as 3.0 μm. Hours at elevated temperatures were required to achieve target oxidation thickness, dopant-transfer concentrations and doped-glass flows for planarization. Because each wafer required several hours at temperature, batch furnaces were designed for effective manufacturing of many wafers at the same time. The basics of thermal effects define the thermal-anneal and oxidation processes for the batch furnaces. Heat transfer and temperature control effects on dopant diffusion, slip defects, and oxidation growth rate are applied from the early applications of thermal processing and refined to accommodate advanced semiconductor designs into the 14 nm device dimensions.

Thermal processing is based on

- Heat transfer
- Annealing, including diffusion (Fick's law)
- Oxidation (i.e., Deal and Grove)

7.2.1 Heat Transfer

Wafer temperature is essential to thermal processing for semiconductor manufacturing. Heat is transferred from heater elements to wafers through materials used to control the ambient around the wafers. Quartz (SiO_2) is used to isolate the wafers from external contamination, and permits pressure control through vacuum pump application if needed. Because the wafers are physically isolated from the heater elements, heat transfer is by radiative emission from the heater coils. Thermal mass around the heater elements and wafers prevents rapid thermal changes which is a benefit for stable temperature control, but a problem if fast processing is desired.[1]

Batch furnaces will have inert gas flowing to prevent contamination and/or to enhance the annealing effect. The gas flow can significantly change the thermal profile across the wafer and down the load, adding to thermal stability control issues for uniformity. The pressure in the furnace will also affect the heat transference between wafers and flowing gas. Figure 7.1 demonstrates three methods of cooling a batch of wafers after processing with varying effectiveness. Thermal mass restricts the rate of cooling in the case of natural cooling, which is the common method of cooling based on just reducing heater power. Large flows of N_2 can increase the cool down rate, as shown in the middle figure. The fastest cooling rate is achieved by unloading wafers directly from the furnace, but the direct unload method has risks to wafers and equipment from thermal shock, depending on the temperature of unloading.

Thermocouples (TC) control heater elements and reach the target anneal temperature by monitoring the real-time temperature. Outer TCs are placed near each section of heater elements for direct heater control, and inner TCs are positioned inside the reactor for more precise control of the temperature at the wafers, where it is most important. The TCs measure temperature by use of the seebeck effect, which translates temperature proportional to voltage potential between two dissimilar electrical conductors or semiconductors. Usually, the TCs are isolated from the wafers by SiO_2, preventing contamination from the TCs but adds some thermal delay to the temperature control system. Temperature control can be based on outer TCs, inner TCs, or a proportional combination described as ratio mix of inner and outer TCs. Typically, a calibration is required to match specific conditions of the TC output to actual temperature at the wafers.[2]

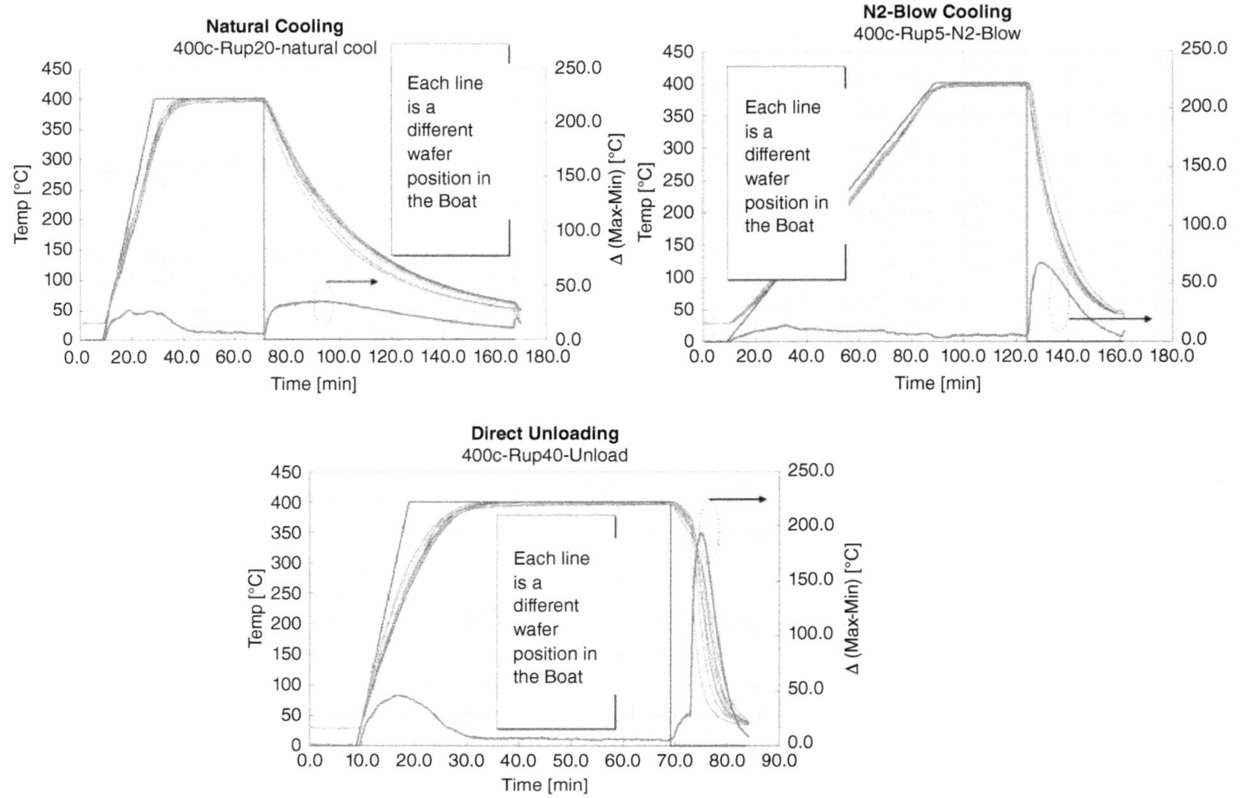

Figure 7.1 Three methods of cooling wafers in a batch furnace system, where multiple wafers are processed thermally at one time. Thermal mass from the heaters, insulation, and wafers determines the slow, natural cooling rate, which is increased with high gas flow cooling and increased further when wafers are able to be just removed from the heat.

The same semiconductor devices being manufactured are used to control the furnace heaters. Temperature control consists of a feedback loop between the TC sensors, used to compare between actual temperature and target temperature, and the heater response to the power adjustments made to meet the target temperature. The time delay between temperature measurement and the corresponding temperature change at the wafer is called the thermal response. There are many variables that affect the thermal response of a heating system, including thermal mass, thermal conductivity, heating mechanism, and control system speed. Control algorithms, such as proportional, integral, and derivative (PID), allow adjustment or compensation for specific responses of all the variables that make up a thermal system.[3]

Advances in the basic furnace temperature control systems have been made to address increasing need for temperature precision and increasingly critical requirements in thermal budget (or lower total heat-time exposure) due to smaller device-feature dimensions. Thermal mass has been reduced to improve temperature control response, and reduce wafer time at elevated temperature. New furnace configurations have been developed to include smaller, mini-batch systems to further reduce thermal mass and increase temperature change rates, and single wafer systems for the fastest temperature control at the expense of COO. In some cases, as in single wafer processing, TCs are sometimes replaced with pyrometers for temperature control, which adds a different challenge in temperature calibration based on emissivity of the material being measured. Even the performance of temperature control systems have been enhanced with the application of model-based temperature control software that anticipates the thermal response of the heater system based on results of previously determined test runs. Model-based controls can use the temperature response across a wafer to compensate reactant gradients with dynamic thermal gradients induced by actively ramping temperature during a deposition or oxidation. Figure 7.2 demonstrates the effects of various temperature control methods.

A special version of model-based temperature control, adaptive real-time temperature (ART) control, has found use especially for deposition processes. The chemistry used to deposit films can deplete as it passes across a wafer or flows down the load of wafers in a boat, which can lead to thinner films where the chemistry is most depleted. ART corrects for this problem by creating a temperature gradient to compensate for the depleting deposition chemistry. Reference 4 shows how corrective thermal gradients are made by ART. Figure 7.3, based on the paper, illustrates the precise drop in temperature during deposition to create a slightly higher temperature at the center of the wafer, where the chemistry depletes the most. The deposition is broken up into two segments to reproduce the desired thermal gradient across the wafers at the appropriate average deposition temperature. The figure also shows the predeposition steps required to set up the appropriate temperature gradients for each heater zone of the batch furnace for uniformity across the batch load as well.

7.2.2 Annealing

Both the purpose and problem with thermal processing in semiconductor manufacturing is in the effect of heat on materials. Thermal processing of materials in an inert ambient, such as N_2 or argon (Ar), and sometimes with a hydrogen or deuterium addition, is called an annealing process. There are several uses for annealing processes in semiconductor manufacturing:

- Diffusion of dopants
  - Solid source transfer
  - Fick's law

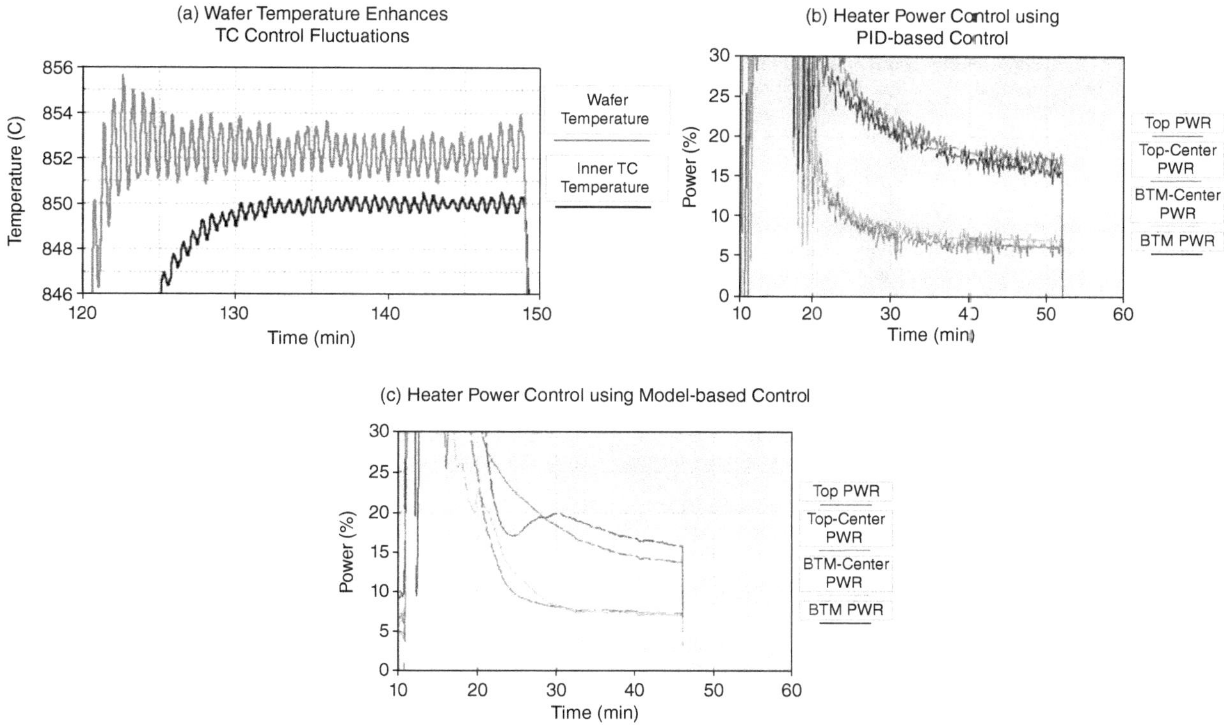

Figure 7.2 Heater temperature control effect on wafer temperature and improvements with PID and model-based temperature controls demonstrated.

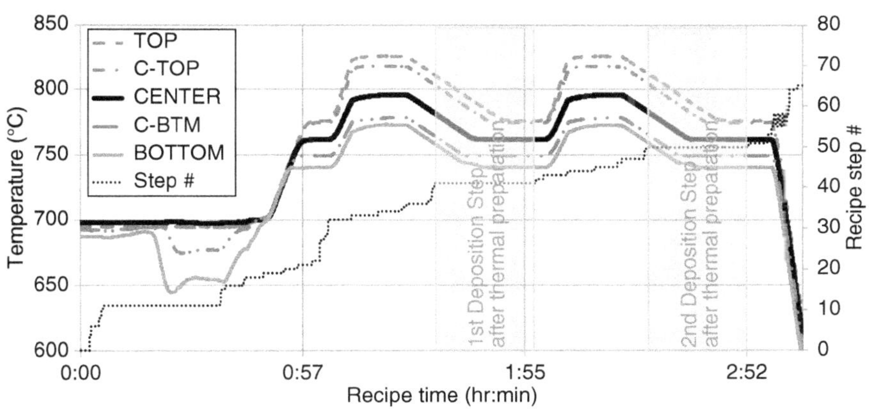

Figure 7.3 Model-based temperature control is demonstrated to improve the uniformity of a silicon nitride deposition, which is broken into two deposition cycles to set up the optimum thermal gradient across the wafers for a uniform film deposition across the wafer and boat load of wafers.[4]

- Activation of dopants (atomic bonding)
- Flow of deposited glass (BPSG flow, FCVD)
- Film densification (spin on films)
- Silicidation (TiSi, NiSi, CoSi)
- Passivation (i.e., dangling bond reduction)

7.2.3 Diffusion

Dopants are used in semiconductor materials to make them conductive. Depending on the type of dopant, either electrons or holes (electron vacancies) flow to create the conductivity. Boron is typically used with silicon to create *P*-type silicon that conducts with holes, and phosphorus or arsenic is used for *N*-type silicon that conducts with excess electrons. To dope the silicon into a conductive material, the dopant is diffused into the silicon by exposing the dopant to the silicon surface and using heat and time to determine depth and concentration of the dopant into the silicon. There are several factors affecting the diffusion of dopants:

- Dopant diffusivity (atomic size)
- Defectivity of the silicon substrate
- Barriers to diffusion, such as oxide or nitride interfaces
- Anneal temperature
- Surface dopant concentration
- Anneal time

Fick's law of diffusion is an equation that describes the effects of dopant diffusivity, concentration, temperature, and time to determine the dopant profile in silicon. Defectivity is represented in the diffusivity of the material. Other materials at the interface, such as SiO_2, can be described as a separate diffusion relationship additive to the diffusive equation into silicon.[5]

$$F = -D\partial N(x,t)/\partial x = -\partial F(x,t)/\partial x$$

where F is the rate of transfer of solute atoms per unit area of the diffusion flux density (atoms/cm^2-s). N is the concentration of solute atoms (number of atoms per units volume/cm^3), and x is the direction of solute flow. Here N is assumed to be a function of x and t only, t is the diffusion time, and D is the diffusion constant (also referred to as diffusion coefficient or diffusivity) and has units of cm^2/s.

There are several techniques for doping semiconductor materials. At one time, dopant material was introduced as a solid source shaped like a wafer inserted in front of each product wafer, from which the dopant outgassed from the solid source and diffused into the product wafer. The dopant concentration into the product wafer is determined by anneal temperature and time. Gases, such as phosphine (PH_3), arsine (AsH_3), and borane (BH_3) have also been used for larger area doping. Another common doping technique is to deposit a dopant-containing film, such as phosphorus oxychloride ($POCl_3$), followed by a thermal anneal diffusion and dopant film removal. More recent doping of semiconductor manufacturing is done by ion implant methodologies, in which the dopant species is ionized and bombard the silicon surface to force the dopants into the silicon. Ion implantation causes localized crystal structure damage that keeps the dopant profile shallow as desired. Thermal annealing is needed after Ion Implantation to reform the silicon crystallinity with dopant substituted into the crystal structure. The incorporation of dopants into the silicon lattice structure is called dopant activating, which makes the dopant electronically active.

Contemporary doping strategies may be returning to solid-source dopant transfer for dopant profile control into very small device features. Recently, the solid source is an atomic layer deposited (ALD) doped film, such as boron oxide (B_2O_3) from which the boron is diffused into silicon fin structures. The ALD is very conformal, providing excellent dopant uniformity, and the intimate proximity of the dopant allows for high dopant concentration with lower thermal budget, resulting in a shallower dopant profile.[6]

7.2.4 Densification/Flow

The construction of semiconductor devices requires materials that have specific properties at the correct time in the fabrication process. In some cases, thermal processing is used to change film properties from their state "as-deposited" to a state with better film properties for integration with other fabrication processes. For example, time at elevated temperatures, also known as annealing, has also been used to densify some deposited films, and even flow the films to planarize the final surface. Deposited SiO_2 films doped with boron and phosphorous, known as borophosphosilicate glass (BPSG), will coat the underlying structures on the silicon wafer surface. With higher temperature anneals, sometimes in an H_2O-containing ambient, the BPSG will flow like a liquid and planarize the surface, which is beneficial topography for integrating further structures on the device. The BPSG anneal process is a good example of the importance of anneal ambient and temperature ramping rates, as previous investigation[7] shows BPO_4 precipitate defects forming under inappropriate anneal ambient and cooling rate conditions. The use of BPSG-type materials for planarization has been generally eliminated due to reduced thermal budget requirements for advanced technologies.

Newer materials are being discovered that have beneficial feature-filling properties, but require annealing to make the final film properties acceptable to the device. Some of the new materials include flowable chemical vapor deposition (FCVD) and spin-on dielectric materials. Materials in this category typically deposit with an advantageous property such as conformality, planarization, selective deposition, or an inexpensive method such as spin-on materials. The deposited materials can require a thermal process after deposition to improve the film etch resistance, selectivity, as well as insulation property and/or structural property through densification, atomic bond restructuring, and/or chemical modification. For example, a typical spin-on material needs the depletion of the organic material (used to allow spin-on application) to make the film more resistant to etch for better device fabrication integration. Thermal anneals are used to convert the spin-on materials.

7.2.5 Silicidation

Thermal processing has also been used to transform the conductivity of materials by diffusion of metals into silicon and through restructuring of the alloyed material. Silicidation of contact regions in semiconductor devices lowers the overall resistance of the device, which increases drive currents and performance of the device. In early transistors, titanium (Ti) was deposited on source/drain regions and exposed to thermal anneal processing to diffuse Ti into the silicon and form titanium silicide (TiSi).[8] Thermal ramp rates and total temperature times were controlled to form the lowest resistance TiSi phase.[9] However, as transistor dimensions reduced, Ti was replaced by cobalt and nickel because of their shallower silicide formation and low contact resistivity. With the new FinFET semiconductor device architecture (~22-nm device designation), the depth dimension loss of the silicide is eliminated with more contact surface area provided by the fin structure.

7.2.6 Passivation

At the end of the semiconductor device fabrication process, there is typically a passivation step, in which the devices are exposed to approximately 400°C thermal exposure in approximately 4 percent H_2 in N_2 (forming gas). Low H_2 concentrations in N_2 are needed to prevent spontaneous combustion, and expensive hardware interlocks are required to avoid hazardous conditions. In device fabrication, the hydrogen acts like protons to passivate dangling bonds and interface traps. A balance of time at temperature is critical to prevent damage from excessive heat effects, such as excessive dopant diffusion or metal damage, while still being affective in passivating dangling bonds of damaged atomic structures.[10]

7.2.7 Defectivity

Advanced semiconductor devices are becoming more sensitive to thermal processing problems, especially due to thermal budget limitations from the critically small feature dimensions on the order of a few atomic layers. Examples of semiconductor device problems from thermal processing that are indicative of IC fabrication evolution are

- Slip dislocations
- Transient enhanced diffusion
- Thermal budget limitations

Silicon wafers are composed of atomically aligned silicon atoms that are grown to be single crystals at a specific crystallographic orientation. The atomic perfection of the silicon wafer makes the fabrication and electrical performance of the semiconductor device predictable and reliable. Early in the history of semiconductor device fabrication, when larger device features required higher temperature thermal processing, the silicon atomic structure in wafers was found to be susceptible to damage. Silicon melts at 1414°C and starts to require special consideration for thermally induced atomic structure damage above about 800 to 900°C. Above 900°C, higher temperatures make the silicon wafer more sensitive to damage from thermal gradients. Thermal gradients of a few degrees can cause the silicon atoms of the wafer to shift or slip, resulting in dislocation formations that can affect electrical performance and photolithographic alignments. Slip dislocation defects are pictorially represented in Fig. 7.4.[11] Depending on how the silicon wafer is supported, and the size of the silicon wafer, temperature ramps as low as 1°C/min may be necessary at more than 1000°C to prevent slip-related damage. In more modern IC fabrication methods, the thermal budget concern of dopant diffusion makes the thermal budget limitations well below the concern for slip dislocation formation.

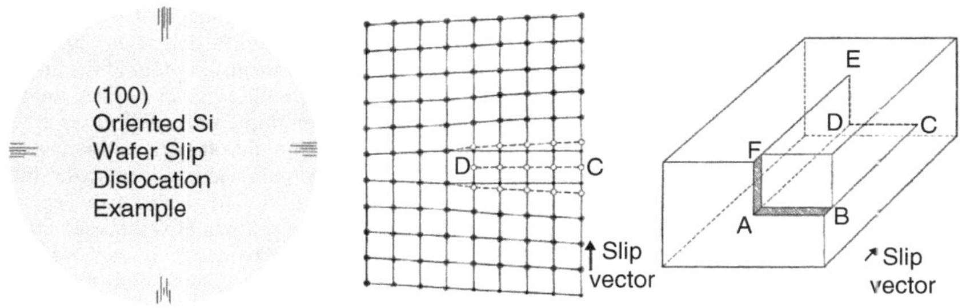

Figure 7.4 Slip dislocations are a defect in the silicon atomic structure induced by thermal gradients. Slip lines are visible on Si wafers along lines of crystallographic orientation.[11,12]

As mentioned previously, smaller features in semiconductor device design changed the requirements of thermal processing. Implant damage and diffusion during activation anneal was able to be modeled with Fick's diffusion equations until the device dimensions became small enough to consider dopant diffusion increased locally beyond the bulk equations. Transient enhanced diffusion (TED) is the dopant diffusion that occurs beyond that predicted by Fick's modeling due to diffusion paths generated by the implant crystalline damage. The atomic lattice structure of the silicon wafer sustains some damage from the ion implantation bombardment of dopant ions. Straining the atomic structure provides accelerated diffusion pathways so the dopants diffuse farther initially after implantation, making thermal annealing and activation anneals more critical for dopant diffusion control.[12]

The critical dimensions for semiconductor devices are down to 7 nm as of 2016. Several device layers are deposited one atomic layer at a time (ALD) for precise dimension control. Thermal budgets of anneal processes have been reduced to minimize, if not eliminate dopant diffusion. Dopant activation, or substitutional bonding, still requires higher temperatures (900 to 1000°C), but the time has been reduced to seconds, which makes conventional batch furnace thermal processing unusable because of slow thermal-mass response limitations. Because higher temperature anneals are required with the seconds of exposure time for smaller thermal budgets, single wafer anneals perform most, higher temperature annealing in advanced device fabrication.

7.3 RAPID THERMAL PROCESSING CONSIDERATIONS

Individual or single wafer processing has increased because of more restrictions to thermal budgets allowed for device fabrication, and the desire for fast turn-around time (TAT) in device development. TAT is the time it takes from starting the process on one wafer to starting the process of the next wafer. A typical batch furnace process will take on the order of hours to complete a process, while processing one wafer individually by RTP is typically done in a few minutes or less. Batch furnace processing allows for more thermal stability and process stability time since the throughput has the advantage of processing multiple wafers simultaneously. Disadvantages of batch processing are long thermal exposure time and long overall process time. However, the nature of single wafer processing requires fast TAT to compete with economic considerations such as COO, which describes the total cost of equipment including capital and expense costs compared to the productivity of the tool. Throughput is a typical metric of productivity for semiconductor manufacturing. A typical batch furnace has a throughput of approximately 20 wafers per hour (~5 h for a 100-wafer batch), which equates to approximately 3 min per wafer in a rapid thermal processing (RTP) system. During the 3 min of TAT, the RTP needs to transport the wafer to the processing chamber, prepare the wafer and chamber for processing, process the wafer, prepare the chamber for wafer extraction, and return the wafer to the starting point. Time is limited for RTP systems.

Semiconductor manufacturing has some applications requiring high temperatures, on the order of 1000°C for short times, typically of a few seconds or less, that are achieved through the low thermal mass configurations of RTP. The applications need atomic bonding arrangements for conductivity and/or densification, which dictate the high temperatures, while minimizing dopant diffusion limits the short process times at temperature.

7.3.1 RTP Configurations

Rapid thermal processing of wafers requires fast temperature change of the wafer. More common single wafer processing uses high doses of light or other radiation to quickly heat the wafer and surrounding susceptor, and the combination of low thermal mass and high gas flow to rapidly cool down the wafer. Some RTP schemes fix the temperature of a chamber and load the wafer at anneal temperature for the required anneal time, then withdraw the wafer to minimize thermal budget. In another configuration, a precise gradient heat zone is created, with the wafer temperature determined by the distance the wafer is inserted into the gradient zone. Regardless of the configuration, fast temperature changes create problems unique to RTP.

7.3.2 Emissivity

One challenge for RTP systems is the need for fast heat transfer, which requires fast temperature monitoring. A pyrometer is often used to measure the wafer temperature directly because of its fast response and contact-less methodology. One problem using pyrometry for temperature control is the susceptibility to errors due to changes in the light signal detected. The most direct temperature control is measurement of the wafer. Emissivity of the wafer surface observed by the pyrometer can change from the process or vary by materials on the wafer. Reference 14 describes the patterning influence on pyrometric measurements in some detail, and compares temperature control of single wafer processing with batch furnace.[13] Measuring the temperature of a susceptor holding the wafer stabilizes the pyrometer calibration, but adds thermal mass that will slow the thermal response. The optical path between pyrometer and wafer or susceptor can also degrade the temperature measurement due to changing film accumulation on the chamber windows.

7.3.3 Expediency Effects

The need for speed in RTP compromises other process conditions that can cause increased problems for semiconductor thermal processing. Considering just annealing processes, the quick time to reach process temperatures, and the short time available for time at temperature creates challenges for accuracy and defect-free processing.

- Fast thermal ramp
  - Slip dislocations
  - Wafer displacement
- Thermal uniformity (minimum stability time)
  - Surface area effects from high-density patterning

As mentioned previously, excessive thermal stress of silicon wafers can cause atomic structure dislocation defects (aka, slip dislocations).

The fast ramp rates required of RTP makes avoiding slip defects difficult. The slip-provoking stress is induced from local thermal gradients across the wafer as it is heating up. In some cases, the pattern on the wafer can enhance temperature gradients.[14] In furnace batch wafer processing, time is allowed for the heat to transfer across all the wafers for good thermal uniformity, which is not the case with RTP. The heating method has a critical requirement to actively force uniform heating across the wafer due to the lack of time for temperature stabilization.

Radial temperature gradients will cause the wafer to bow or warp in a way that can not only cause defects, but can move the wafer out of position in the RTP chamber and cause problems, including wafer breakage and difficulties for the robot to locate the wafer for removal from the chamber. RTP heating of wafers requires a uniform method of heating across a wafer, and must also compensate for emissivity differences across the wafer from patterned films to avoid local defects like slip and global problems like wafers moving out of position. The problems are made more challenging when reactive gases are added to the thermal parameters, such as in the case of oxidation.

7.4 OXIDATION

Oxidation can be considered a special case of annealing in which silicon, or silicon-containing materials, are annealed in an oxygenated ambient to grow an oxidized film. In its simplest form, silicon wafers are exposed to oxygen at elevated temperatures to achieve a silicon oxide (SiO_2) film.

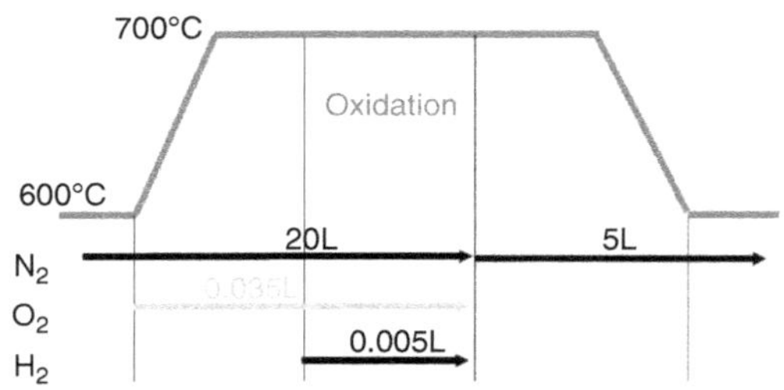

Figure 7.5 Schematic representation of a typical oxidation process, with temperature ramping and gas introduction the same as an anneal thermal process with the addition of an oxidizing ambient like oxygen.

Figure 7.5 shows the schematic representation of an oxidation process, which differs from anneal only in the oxygen flow. Heat transfer is the most important drive force for oxidation, which make thermal considerations mentioned above in anneals critical to the oxidation process. There is little difference between oxidation in batch furnace and RTP other than the thermal considerations previously discussed.

The oxidation process is described by the Deal-Grove model, which segments oxidation into three components: oxygen adsorption, diffusion, and interface reaction.[15] The attachment of oxygen gas molecules to the solid surface is considered adsorption. Oxidation is a diffusion process, because SiO_2 is grown at the silicon/SiO_2 interface as oxygen species diffuse through the growing SiO_2 film and reacting at the interface. As represented by the Fick diffusion equations, the diffusion is a function of diffusivity of the oxygen species, which affects the reaction rate at the interface along with the chemical and thermal reactivity. Since oxidation is a diffusion of oxygen and silicon, the resulting SiO_2 film is a 45 percent consumption of the original silicon and a growth of 55 percent above the original surface, unlike depositions that are 100 percent addition to the wafer surface, as in Fig. 7.6.

As the SiO_2 film thickens from oxidation, the diffusion path of oxygen from surface to interface gets longer, and the SiO_2 growth is not linear with oxidation time. The nonlinearity depends on the diffusivity of the oxidizing reactant. Since H_2O diffuses faster in SiO_2 than the larger O_2 molecules, wet oxidation is used for thicker SiO_2 targets and dry oxidation is used for thinner, higher quality dielectric property films. Figure 7.7 shows oxidation growth curves for dry and wet oxidations as parabolic trends, increasing with temperature and time.

7.4.1 Pitting Silicon

In Fig. 7.5, oxygen flow is started before the desired process temperature is reached. Usually, the reacting process chemistry is not introduced until the process temperature is stable, as is shown with hydrogen flow. Prestabilizing the temperature makes the temperature uniform across the wafers and reactor, which establishes the best uniformity for the process. In the case of oxidation, a small partial pressure of oxygen is required during the ramp up in temperature due to a potential sublimation from SiO_2 on the surface of the wafer. Without the presence of oxygen, SiO_2 on the wafer surface has a tendency to sublimate off the surface and result in pitting or surface roughening at the silicon surface. The pitting can be detected as a particle defect, or can reveal itself as degraded dielectric breakdown strength.[16] Depending on the oxidation temperature, a 0.1 to 1 percent ambient of oxygen can be enough to prevent pitting and avoid oxidation growth before thermal stability.

7.4.2 SiO_2 Film Quality

Historically, SiO_2, as the primary gate dielectric material at the heart of all transistors, was a critical material in semiconductor manufacturing. An electric field is applied across the SiO_2 gate dielectric to create an inverted channel, a depletion region, conductively connecting the source and drain of the transistor to start the electron or hole flow state. When the gate field is off, the gate dielectric and unbiased channel resistivity should keep the electronic flow to zero. Defects in SiO_2 can cause transistor degradation in current leakage, dielectric breakdown failure, reduced electron or hole mobility, band gap defect state irregularities and other performance degradations.[17,18] SiO_2 formed at higher temperatures has better dielectric properties, mainly due to more complete SiO_2 bond formation and higher film density. However, limitations in thermal budget allowance from transistor feature miniaturization compromise the oxidation temperatures permitted. Contamination is another risk to SiO_2 dielectric performance as a gate dielectric. Chlorinated chemicals like hydrochloric acid (HCl), tetrachloroethane (TCA), or dichloroethylene (DCE) are added to the oxidation ambient in some processes to tie up mobile ions in semiconductor devices and clean metal contaminants from oxidation chambers. In the case of DCE, carbon in the molecule requires a sufficient ratio of oxygen during its use to avoid carbon contamination. In an extreme case, a black carbon residue can accumulate on the SiO_2 if there is not enough oxygen for full combustion, which will be a source of particle defects. Additionally,

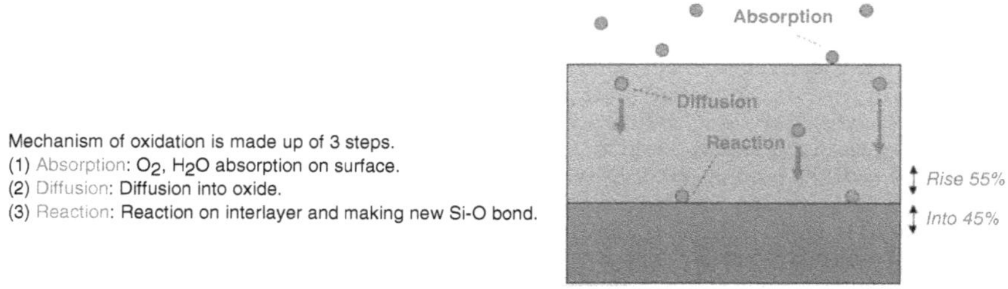

Figure 7.6 Oxidation is a combination of three actions: absorption of the oxygen atom on SiO_2 surface; diffusion of the oxygen to the silicon/SiO_2 interface; and reaction of oxygen to form more SiO_2, as described by the Deal-Grove model.

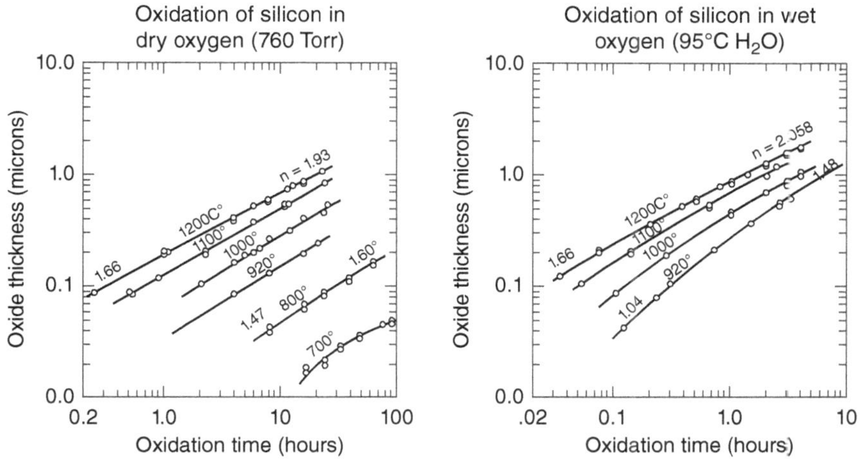

Figure 7.7 Silicon oxidation trends with temperature for dry and wet oxidation from Deal and Groves Ref. 16. (Note the order of magnitude difference in x axes.)

DCE should be kept below 5 percent of the total gas flow to avoid exceeding an explosive limit.

7.4.3 Wet Oxidation Methods

There are three methods of supplying H_2O for wet oxidation: through a torch, catalytic reaction, or vaporizer. For extreme purity, a SiO_2 torch had been commonly used for pyrogenic generation of H_2O from H_2 and O_2 flame reaction. H_2/O_2 should be close to 2 for stoichiometry (H_2O). Higher H_2 flow typically means higher H_2O density. The H_2:O_2 flow ratio is hardware limited to less than 2 to fully consume H_2 gas, which can become explosive at low (> 4 percent) concentrations. Torch hardware is designed to operate close to atmospheric pressure to prevent the torch flame from expanding to the wafers.

Generating H_2O through catalytic methods permits lower pressure wet oxidation, which can be used for growing very thin oxide films. The catalytic method reacts H_2 with O_2 on a catalytic material such as platinum. The principles of the catalytic reaction are described in Ref. 19.[19] Metal contamination can be a concern.

The most direct method of generating H_2O is to vaporize water. Purity of the water and the vaporizing equipment has been a concern historically, but is currently not a problem. High-purity water is flowed into a vaporizing element, typically a heated metal frit, to vaporize the water to gas phase. An inert carrier gas is applied to increase productivity of the water vapor. In the cases of catalytic H_2O generation and direct vaporized H_2O, care needs to be taken to keep all gas delivery lines and components between the initial vapor formation and the chamber elevated above 100°C to prevent condensation in cold spots.

7.4.4 Radical Oxidation

Another method of oxidation is to form H_2O by reacting H_2 and O_2 at low pressures to create oxygen radicals. The technique can be utilized to generate oxygen radicals of relatively small molecular size at low reactor pressure (< 1 torr). The oxidation of silicon from oxygen radicals has interesting properties such as silicon-orientation-independent oxidation and tunable oxidation selectivity between silicon and silicon nitride. As can be seen in Fig. 7.8, wet oxidations are dependent on the orientation of silicon atoms. The corner of a silicon feature can have a thinner oxidation, which can lead to an early dielectric breakdown point or leakage path. In contrast, the oxidation mechanism of radical oxygen is independent of silicon orientation. At low partial pressures of H_2, the oxidation rate of silicon and silicon nitride are similar, but at about 40 percent H_2 partial pressure the oxidation of silicon is about 50 percent faster than on SiN. The unique properties of radical oxidation enable their applications in advanced semiconductor manufacturing. Typical radical-oxidation thickness targets are 25 to 350 A.

7.4.5 Defects

Defects in silicon oxidation are film abnormalities that result in degraded dielectric performance. In defects already mentioned, points of dielectric weakness result from rough surfaces of silicon pitting from SiO sublimation as well as thinner oxidation from silicon-orientation dependence of oxidation. Silicon pitting can also reveal itself as false particle contamination. Silicon doping also impacts oxidation rates. Silicon doped with phosphorus or boron tends to oxidize faster than pure silicon, depending on the dopant concentration. Figure 7.9 summarizes some common oxidation defects.

Surface contaminants, residues, and organic materials are defects that can prevent or delay oxidation relative to other silicon surfaces. The residue can establish local regions of delayed oxidation growth, causing localized thin spots that resemble pin holes in the SiO_2 film. The thin spots are dielectric weak links that degrade the whole film capacitive performance. Some device integrations take advantage of the oxidation

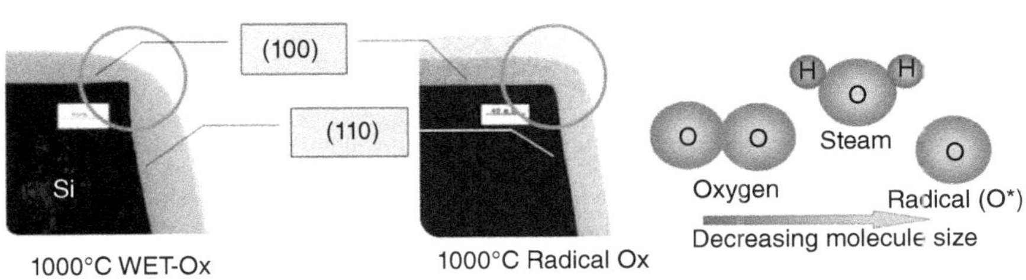

Figure 7.8 Wet oxidation rate is dependent on silicon crystallographic orientation, but the smaller reactant size of radical oxidation is silicon-orientation independent. Radical oxidation has a limited process condition window.

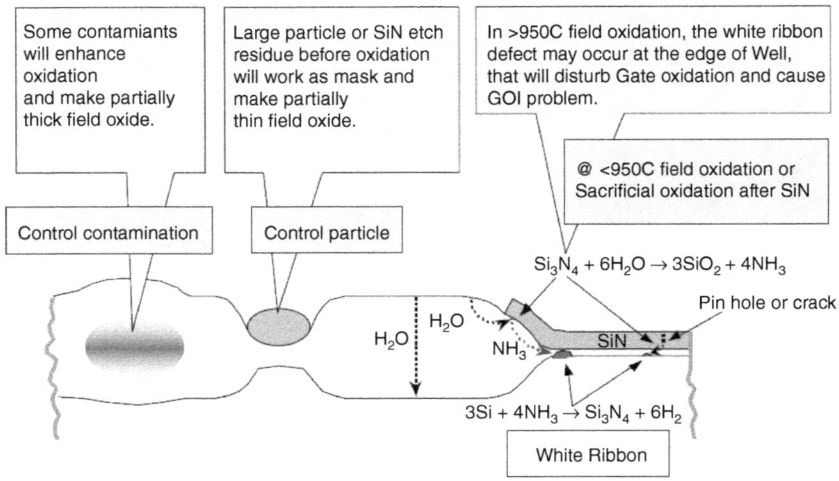

Figure 7.9 Examples of defects that can occur from an oxidation process.

suppressive properties of materials such as silicon nitride. Patterns of electrically active silicon regions are covered with silicon nitride so that oxidation grows in the areas of exposed silicon, which acts as electrical isolation. Oxidation under the edge of the silicon nitride film is called a bird beak region because of its shape, as can be seen in Fig. 7.9.

In some cases, the material being processed can outgas by-products that can leave residue in the processing chamber and/or exhaust lines. The outgassing by-product can be organic material from films that are spun-on or contain chemicals like emulsifiers to permit low temperature application of the film. The by-product accumulation may eventually cause particle or chemical contamination that could lead to defective film properties on the semiconductor product.

7.5 MANUFACTURING CONSIDERATIONS

The ultimate goal of semiconductor manufacturing is to create repeatable, reliable products that are in high demand and made at as low a cost as possible to maximize profit. High demand for semiconductor products can be observed in all the electronic products we use on a daily basis. Internet of Things (IoT) refers to a trend of making electronic devices smarter and remotely accessible by connecting them to the Internet through new semiconductor devices. Frying pans can now tell you, through your phone, when to flip your burger and when your meal is ready to eat. The semiconductor fabrication technology to make the devices has matured to a consolidating industry, where companies are merging to a few, remaining large companies because of the large development and manufacturing expenses, and to lower margins at the production level.

Market maturity has made manufacturability considerations increasingly important. The three drivers of manufacturability, or profitability in semiconductor manufacturing are

- Cost of ownership
- Statistical process control
- Yield/defect control

7.5.1 Cost of Ownership

Cost of ownership (COO) characterizes the complete cost of owning a production tool, such as an annealing or oxidation furnace. The capital costs include all hardware of the furnace that are meant to be a one-time expense, such as the furnace heaters, temperature, and gas control systems. Expense costs are consumable materials that repeat, such as the quartzware, gases, and filler or dummy wafers used to fill in batch process slots not used for product wafers. Peripheral costs, such as cleanroom space and facilities expenses, are also included in the COO analysis. The costs are balanced by the productivity of the tool which is characterized as the wafer throughput, or wafers per hour that the tool produces. The same metric is used to describe RTP tools, but the value of quick TAT has been hard to quantify in past COO models. An example of a COO analysis and a comparison between integrated circuit (IC) and photovoltaic (PV) industries is provided in Ref. 20.

In fact, it costs $271 million to design a 7-nm system-on-a-chip, which is about nine times the cost to design a 28-nm device, according to Gartner.

7.5.2 SPC

Statistical process control (SPC) is an established method to monitor the performance of semiconductor manufacturing tools for optimizing the tool productivity. The first step in SPC is to identify the critical parameter of the tool function. In the case of oxidation, the typical outputs to monitor are oxidation target thickness, uniformity, and particles added during processing. Each parameter affects the product performance. In some cases, dielectric properties are also monitored through capacitance versus voltage (C-V) measurements. The critical parameters are measured either on monitor wafers processed with product wafers or on the product wafers themselves. The process is monitored on a regular basis, for example, daily, and values for the parameter are recorded to compare against historical data from the specific tool and to compare against other tools running the same process. The data is checked for trends in performance indicating drift in process control, or other performance degradation to anticipate problems before the trends are considered out-of-control and adversely affecting the product. To that end, specification limits for the parameters are determined by what is permissible to the product, for example, a range of acceptable oxidation thickness targets. Based on the performance of a tool, control limits are established to determine when the tool is performing abnormally. The ratio of specification limit to control limit, or product requirements to tool capability for control, is called the Cp of the process. The degree that the Cp of a process achieves the target value is the CpK. CpK values are used to compare processes that make up the device product flow to focus resources on the weaker processes. Reference 21 is a link with more detail on SPC methodology.

In the case of anneals or basic thermal processing, critical parameters to monitor becomes more challenging than direct measurement of film thickness. For silicide anneals, monitor wafers need to be prepared with a specific metal deposited on silicon to represent the actual effect on the device. Factors such as interfacial contact between metal and silicon (such as native oxide thickness and uniformity) and queue-time affects between metal depositions and anneal processing must be controlled for effective process monitoring. Since temperature at time is the critical parameter, reactions sensitive to the particular temperature and time can be used as a replacement monitor, if a more robust but representative replacement can be found. Pre-implanted wafers have been used to monitor annealing processes, with the monitoring metric being the sheet resistance of the implant activation. The target sheet resistance (R_s) is calibrated to the desired product thermal budget, and sheet resistance uniformity is used to represent the thermal uniformity of the process. Caution should be taken to characterize the effects of deviation from the target thermal budget on the monitor wafer, so that an appropriate sensitivity to change can be calibrated for effective control limits.

7.5.3 Yield

Defects accumulated throughout the semiconductor production process will determine the yield or percentage of good product from each wafer. Particles are usually the main focus in the discussion of yield. If a wafer receives an additional 10 particles for each tool it uses, and there are 200 steps in a device flow to create the product, then the final particle defectivity is 2000 particles. If there are 2000 die on each wafer, it is possible that there will be no working product from that wafer. Although redundancy in circuit design can reduce some of the particulate defect yield damage, the focus on particle reduction is common in semiconductor manufacturing. Annealing and oxidation processes tend to be cleaner than other processing like depositions or etches, since there are no particles generated from the chemistry of the process. But, particles from mechanical handling or materials supplied to the chamber, such as gases and filler wafers, can still be a concern.[22]

7.6 CONCLUSION

Anneals are the most basic thermal treatment in semiconductor processing, but represent the core of thermal processing: heat and ambient control. Concerns for dopant diffusion, wafer defectivity (i.e., slip formation), and by-product outgassing are common to all thermal processes. The intricacies of precise temperature control have developed from basic, batch furnace processing to single wafer active control. The transition from batch throughput to wafer-turnaround time concerns coincides with the miniaturization of device features and the increased limitations on thermal budget. The addition of reactive gases such as oxygen and/or H_2O increases considerations for film growth, contamination and defect effects on film properties. Practical application of thermal processing for semiconductor manufacturing requires understanding of COO, SPC, and defect/yield performance for successful implementation.

7.7 ACKNOWLEDGMENTS

Thanks for contributions from Anthony Dip, Shigeki Nakatani, Gert Leusink, Teresa Low, John Gumpher, Ahmad Karim, and a wide range of Tokyo Electron Ltd. resources.

7.8 REFERENCES

1. Gregory J. Wilson, Paul R. McHugh, and Robert A. Weaver, "Numerical Heat Transfer Simulations of a Vertical Batch Furnace Under Closed-Loop Temperature Control," Electrochemical Society Proceedings, Vol 97(9), 110–117. https://books.google.com/books?id=QILlEhEGVzcC&pg=PA110&lpg=PA110&dq=Heat+Transfer+in+Semiconductor+wafer+Furnace+design&source=bl&ots=vwDnkPEkWt&sig=KzotaK25ney1lXo4F9hW7CPuH68&hl=en&sa=X&ved=0ahUKEwjikOi7wpPSAhVEwFQKHSbdAiYQ6AEIQjAH#v=onepage&q=Heat%20Transfer%20in%20Semiconductor%20wafer%20Furnace%20design&f=false, last accessed February 15, 2017.

2. Omega, Reproduced from Agilient Technologies, "Practical Temperature Measurements," Feb 2017. http://www.omega.com/temperature/z/pdf/z021-032.pdf

3. Eurotherm (Schneider Electric), "PID Control made easy," Feb 2017, Control Guru, 'fundamental Principles of Process Control, Feb 2017. http://www.eurotherm.com/pid-control-made-easy. Link to control loops: http://controlguru.com/the-components-of-a-control-loop/

4. J. Gumpher, W. A. Bather, and D. Wedel, 'LPCVD Silicon Nitride Uniformity Improvement Using Adaptive Real-Time Temperature Control,' *IEEE Trans. on Semi. Manuf.*, Vol. 16, No. 1, pp. 26–35, Feb. 2003.

5. Circuits Today, 'Diffusion of Impurities for IC Fabrication', last accessed February 15, 2017. http://www.circuitstoday.com/diffusion-of-impurities-for-ic-fabrication

6. S. Consiglio, R. D. Clark, D. O'Meara, C. S. Wajda, K. Tapily, and G. J. Leusink, 'Comparison of B_2O_3 and BN deposited by atomic layer deposition for forming ultrashallow dopant regions by solid state diffusion', JVSATA, 34(1), Jan 2016. 01A102-1 to 01A201-7.

7. M. Yoshimaru and H. Wakamatsu, *J. Electrochem. Soc.*, "Microcrystal Growth on Borophosphosilicate Glass Film during High-Temperature Annealing," 666-671, Vol. 143(2), Feb. 1996. http://jes.ecsdl.org/content/143/2/666.full.pdf

8. Arabinda Das (Solid State Technology), "How finFETs ended the service contract of silicide process Feb 2017." http://electroiq.com/blog/2016/03/how-finfets-ended-the-service-contract-of-silicide-process/

9. C.R. de Farias Azevedo and H.M. Flower, 'Microstructure and phase relationships in Ti-Al-Si system', February 2017. http://www.pmt.usp.br/LCMHC/textos/mst2.pdf

10. Joseph C. King and Chenming Hu, "Effect of Low and High Temperature Anneal on Process-Induced Damage of Gate Oxide," February 2017. http://www.eecs.berkeley.edu/~hu/PUBLICATIONS/Hu_papers/Hu_Melvyl/Hu_Melvyl_94_03.pdf

11. W. T. Read, Jr., *Dislocations on Crystals*, McGraw-Hill Book Co., Inc., New York, 1953, p. 20, figures 2.6 and 2.7. https://archive.org/details/dislocationsincr032720mbp

12. M.Y.L. Jung, R. Gunawan, R.D. Braatz, and E.G. Seebauer, "A Simplified Picture for Transient Enhanced Diffusion of Boron in Silicon," February 2017. http://web.mit.edu/braatzgroup/68_A_simplified_picture_for_transient_enhanced_diffusion_of_boron_in_silicon.pdf

13. M. Rabius, A.T. Fiory, N.M. Ravindra, P. Frisella, A. Agarwal, T. Sorsch, J. Miner, E. Ferry, F. Klemens, R. Cirelli, and W. Mansfield, "Rapid Thermal Processing of Silicon Wafers with Emissivity Patterns," Feb 2017. https://web.njit.edu/~sirenko/PapersNJIT/Ravi_JEM_2006.pdf

14. Jeffrey P. Hebb and Klavs F. Jensen, "The Effect of Patterns on Thermal Stress During Rapid Thermal Processing of Silicon Wafers," February 2017. http://web.mit.edu/jensenlab/publications/patterns_thermalstress.pdf

15. B. E. Deal and A. S. Grove, "General Relationship for the Thermal Oxidation of Silicon," *J. Appl. Phys.* Vol. 36, No. 12, pp. 3770–3778, 1965. http://w.lithoguru.com/scientist/CHE323/Deal_Grove_Model_JApplPhys_36_3770.pdf

16. J. C. Moore, J. L. Skrobiszewski, and A. A. Baski, "Sublimation Behavior of SiO_2 from Low- and High-Index Silicon Surfaces," *J. Vac. Sci. Technol. A*, Vol. 25, p. 812, 2007.

17. S. M. Sze, *Physics of Semiconductor Devices*, Wiley-Interscience, Hoboken, New Jersey, 1969.

18. Klaus F. Schuegraf and Chenming Hu, "Reliability of thin SiO_2," February 2017. http://www.eecs.berkeley.edu/~hu/PUBLICATIONS/Hu_papers/Hu_Melvyl/Hu_Melvyl_94_12.pdf

19. A. Michaelides and P. Hu, "Catalytic Water Formation on Platinum: A First-Principles Study," *J. Am. Chem.* Soc. 2001, 123, 4235-4242: https://www.ucl.ac.uk/catalytic-enviro-group/wp-content/uploads/2016/03/ja003576x.pdf, February 2017. http://www.chem.ucl.ac.uk/ice/docs/ja003576x.pdf

20. David W. Jimenez, "Cost of ownership and overall equipment efficiency: a photovoltaics perspective," February 2017. http://www.btu.com/assets/PVI6-12-Meridian-CoC-paper.pdf

21. MoreSteam.com, "Statistical Process Control (SPC)," February 2017. https://www.moresteam.com/toolbox/statistical-process-control-spc.cfm

22. Integrated Circuit Engineering Corp, "Yield and Yield Management," February 2017. http://smithsonianchips.si.edu/ice/cd/CEICM/SECTION3.pdf

7.9 FURTHER READING

P.J. Timans, "Rapid Thermal Processing," Handbook of Semiconductor Manufacturing Technology, Second Edition, CRC Press, 2008, 11-1 to 11-116. https://books.google.com/books?id=PsVVKz_hjBgC&pg=SA11-PA38&lpg=SA11-PA38&dq=Heat+Transfer+in+Semiconductor+Fabrication+Furnaces&source=bl&ots=j9sQnXtKav&sig=yLSJPtpmSqE9sKUShWhPWqQHcdE&hl=en&sa=X&ved=0ahUKEwij4YWwvpPSAhXosVQKHQxrC9s4ChDoAQhMMAk#v=onepage&q=Heat%20Transfer%20in%20Semiconductor%20Fabrication%20Furnaces&f=false, last accessed 2/15/2017

R.H. Doremus, "Oxidation of Silicon by Water and Oxygen and Diffusion in Silica," February 2017. http://home.uni-leipzig.de/energy/pdf/freume2.pdf

O_2 vs H_2O diffusivity: http://pubs.acs.org/doi/abs/10.1021/j100557a006?journalCode=jpchax

Semiconductor Device Physics: S. M. Sze, *Physics of Semiconductor Devices*, John Wiley & Sons, Hoboken, New Jersey, 1969.

Material Science: C. R. Barrett, W. D. Nix, and A. S. Tetelman, *The Principles of Engineering Materials*, Prentice-Hall, Inc., Upper Saddle River, New Jersey, 1973.

Oxidation: B. E. Deal and A. S. Grove, "General Relationship for the Thermal Oxidation of Silicon," *J. Appl. Phys.*, Vol. 36, No. 12, pp. 3770–3778, 1965.

Wolf, S. and R. N. Tauber, *Silicon Processing for the VLSI Era*, Lattice Press, Sunset Beach, California, 2000.

Chapter 8
Microlithography

BY

CHRIS A. MACK *Lithoguru.com, Austin, Texas*

The fabrication of an integrated circuit (IC) requires a variety of physical and chemical processes performed on a semiconductor (e.g., silicon) substrate. In general, the various processes used to make an IC fall into three major categories—film deposition, patterning, and semiconductor doping. Films of both conductors (such as polysilicon, tungsten, and copper) and insulators (various forms of silicon dioxide, silicon nitride, and others) are used to connect and isolate transistors and their components. Selective doping of various regions of silicon allows the conductivity of the silicon to be changed with the application of voltage (i.e., a transistor). By creating structures of these various components, billions of transistors can be built and wired together to form the complex circuitry of a modern microelectronic device. Fundamental to all of these processes is lithography, the formation of three-dimensional relief images on the substrate for subsequent transfer of the pattern into the substrate.

The word lithography comes from the Greek *lithos*, meaning stones, and *graphia*, meaning to write. It means quite literally writing on stones. In the case of semiconductor lithography our stones are silicon wafers and our patterns are written into a light sensitive polymer called a photoresist. To build the complex structures that make up a transistor and the many wires that connect the millions or billions of transistors of a circuit, lithography and etch pattern transfer steps are repeated at least 10 times, but more typically are done 40 to 60 times to make one circuit. Each pattern being printed on the wafer is aligned to the previously formed patterns and slowly the conductors, insulators, and selectively doped regions are built up to form the final device.

The importance of lithography can be appreciated in two ways. First, due to the large number of lithography steps needed in IC manufacturing, lithography typically accounts for about 30–50 percent of the cost of manufacturing. Second, lithography tends to be the technical limiter for further advances in feature size reduction and thus transistor performance and silicon area. Obviously, one must carefully understand the trade-offs between cost and capability when developing a lithography process. Although lithography is certainly not the only technically important and challenging process in the IC manufacturing flow, historically, advances in lithography have gated advances in IC cost and performance. Roadblocks in practical advancement of lithography will essentially end Moore's law, changing the very nature of the semiconductor industry, and the future of electronics.

8.1 THE LITHOGRAPHIC PROCESS

Optical microlithography is basically a photographic process by which a light sensitive polymer, called a photoresist, is exposed and developed to form three-dimensional relief images on the substrate. The general sequence of processing steps for a typical optical lithography process is as follows—substrate preparation, photoresist spin coat, post-apply bake, exposure, post-exposure bake, development, and postbake. A resist strip is the final operation in the lithographic process, after the resist pattern has been transferred into the underlying layer. This sequence is shown diagrammatically in Fig. 8.1, and most of these steps are generally performed on several tools linked together into a contiguous unit called a *lithographic cluster* or *photocell*.

8.1.1 Substrate Preparation

Substrate preparation is intended to improve the adhesion of the photoresist material to the substrate. This is accomplished by one or more of the following processes—substrate cleaning to remove contamination, dehydration bake to remove water, and addition of an adhesion promoter. One common type of contaminant—adsorbed water—is removed most readily by a high-temperature process called a *dehydration bake*. A typical dehydration bake, however, does not completely remove water from the surface of silica substrates (including silicon, polysilicon, silicon oxide, and silicon nitride). Surface silicon atoms bond strongly with a monolayer of water forming silanol groups (SiOH). Bake temperatures in excess of 600°C are required to remove this final layer of water. Since this approach is impractical, the preferred method of removing this silanol is by chemical means. Adhesion promoters are used to react chemically with surface silanol and replace the –OH group with an organic functional group that, unlike the hydroxyl group, offers good adhesion to photoresist. Silanes are often used for this purpose, the most common being hexamethyldisilazane (HMDS).

8.1.2 Photoresist Coating

A thin, uniform coating of photoresist at a specific, well-controlled thickness is accomplished by the seemingly simple process of spin coating. The photoresist, rendered into a liquid form by dissolving the solid components in a solvent, is poured onto the wafer, which is then spun on a turntable at a high speed, producing the desired film. Stringent

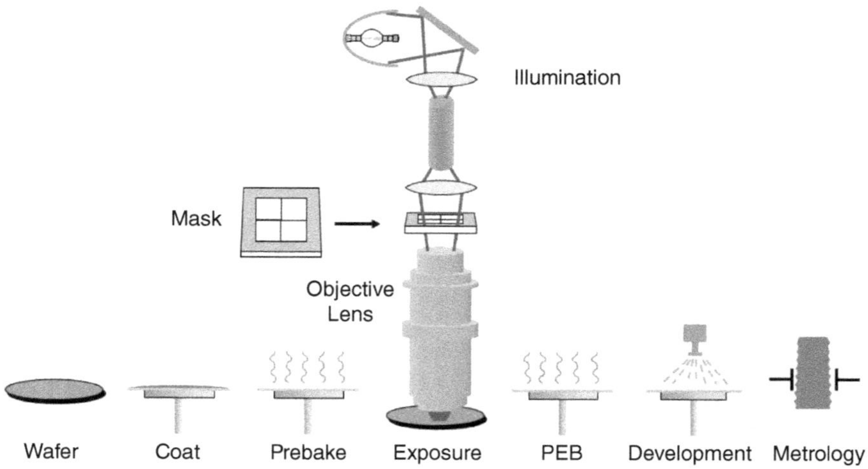

Figure 8.1 Example of a typical sequence of lithographic processing steps.

requirements for thickness control and uniformity and low-defect density call for particular attention to be paid to this process, where a large number of parameters can have significant impact on photoresist thickness uniformity and control. There is a choice between static or dynamic dispense, spin speeds and times, and accelerations to each of the spin speeds. Also, the volume of the resist dispensed and properties of the resist (such as viscosity, percent solids, and solvent type) and the substrate (substrate material and topography) play an important role in the resist thickness uniformity. At the end of this cycle a thick, solvent-rich film of photoresist covers the wafer, ready for post-apply bake. The photoresist spin speed curve (final resist thickness versus spin speed, measured experimentally) is an essential tool for setting the spin speed to obtain the desired resist thickness. The final resist thickness varies as one over the square root of the spin speed and is roughly proportional to the liquid photoresist viscosity.

8.1.3 Post-Apply Bake

After coating, the resulting resist film will contain between 20 and 40 weight percent solvent. The post-apply bake (PAB) process, also called a softbake or a prebake, involves drying the photoresist after the spin coat process by removing this excess solvent. There are four major effects of removing solvent from a photoresist film: (1) film thickness is reduced, (2) post-exposure bake and development properties are changed, (3) adhesion is improved, and (4) the film becomes less tacky and thus less susceptible to particulate contamination. Typical prebake processes leave between 3 and 8 percent residual solvent in the resist film, sufficiently small to keep the film stable during subsequent lithographic processing.

Although the use of convection ovens for the prebaking of the photoresist was once quite common, currently the most popular bake method is the hot plate. The wafer is brought either into intimate vacuum contact with or close proximity to a hot, high-mass metal plate. Due to the high thermal conductivity of silicon, the photoresist is heated to near the hot plate temperature quickly (about 5 s for hard contact, or about 20 s for proximity baking). The greatest advantage of this method is an order of magnitude decrease in the required bake time over convection ovens, to about 1 min, and the improved uniformity of the bake. In general, proximity baking is preferred to reduce the possibility of particle generation caused by contact with the backside of the wafer.

8.1.4 Alignment and Exposure

The basic principle behind the operation of a photoresist is the change in the solubility of the resist in a developer on exposure to light (or other types of exposing radiation). In the case of the standard diazonaphthoquinone positive photoresist, the sensitizer, which is not soluble in the aqueous base developer, is converted to a carboxylic acid on exposure to UV light in the range of 350 to 450 nm. The carboxylic acid product is very soluble in the basic developer. Thus, a spatial variation in the light energy incident on the photoresist will cause a spatial variation in the solubility of the resist in the developer.

Contact and proximity lithography are the simplest methods of exposing a photoresist through a master pattern called a photomask, where a shadow of the mask is used to expose the resist-coated wafer. Contact lithography offers moderately high resolution (down to about the wavelength of the radiation), but practical problems such as mask damage and resulting low yield make this process unusable in most production environments. Proximity printing reduces mask damage by keeping the mask a set distance above the wafer (e.g., 20 μm). Unfortunately, the resolution limit is increased to greater than 5 to 10 times the wavelength, making proximity printing insufficient for today's technology. By far the most common method of exposure is projection printing.

Projection lithography derives its name from the fact that an image of the mask is projected onto the wafer. Projection lithography became a viable alternative to contact and proximity printing in the mid-1970s when the advent of computer-aided lens design and improved optical materials allowed the production of lens elements of sufficient quality to meet the requirements of the semiconductor industry. In fact, these lenses have become so perfect that lens defects, called aberrations, play only a small role in determining the quality of the image. Such an optical system is said to be diffraction limited, since it is diffraction effects and not lens aberrations, which, for the most part, determine the shape of the image and the resulting resolution.

There are two major classes of projection lithography tools—scanning and step-and-repeat systems. Scanning projection printing, pioneered by the Perkin-Elmer company, employs reflective optics (i.e., mirrors rather than lenses) to project a slit of light from the mask onto the wafer as the mask and wafer are moved simultaneously by the slit. The exposure dose is determined by the intensity of the light, the slit width, and the speed at which the wafer is scanned. These early scanning systems, which use polychromatic light from a mercury arc lamp, are in the ratio 1:1, that is, the mask and image sizes are equal. Step-and-repeat cameras (called steppers for short) expose the wafer, one rectangular section (called the image field) at a time, and can be 1:1 or reduction. These systems employ refractive optics (i.e., lenses) and are usually quasi-monochromatic. Both types of systems (Fig. 8.2)

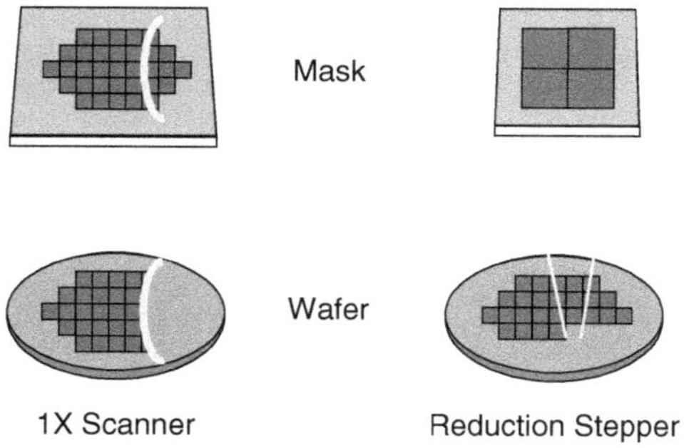

Figure 8.2 Scanners and steppers use different techniques for exposing a large wafer with a small image field.

are capable of high-resolution imaging, although reduction imaging is required for the highest resolutions due to difficulties with the manufacture of 1× masks.

Scanners replaced proximity printing by the mid-1970s for device geometries below 4 or 5 μm. By the early 1980s, steppers began to dominate as device designs pushed to 2 μm and below. Steppers continued to dominate lithographic patterning into the 1990s as minimum feature sizes reached the 250-nm levels. However, by 1990 a hybrid step-and-scan approach was introduced. The step-and-scan approach uses a fraction of a normal stepper field (e.g., 25 × 8 mm), then scans this field in one direction to expose the entire 4× reduction mask (Fig. 8.3). The wafer is then stepped to a new location and the scan is repeated. The smaller imaging field simplifies the design and manufacture of the lens, but at the expense of a more complicated reticle and wafer stage. Step-and-scan technology is the dominant technology today for below 250-nm manufacturing.

Resolution, the smallest feature that can be printed with adequate quality and control, has two basic limits—the smallest image that can be projected onto the wafer and the resolving capability of the photoresist to make use of that image. From the projection imaging side, resolution is determined by the wavelength of the imaging light (λ) and the numerical aperture (NA) of the projection lens, according to the Rayleigh criterion:

$$R \propto \frac{\lambda}{NA} \tag{8.1}$$

Lithography systems have progressed from blue wavelengths (436 nm) to UV (365 nm) to deep UV (248 nm) to today's mainstream high-resolution wavelength of 193 nm. In the meantime, projection tool numerical apertures have risen from 0.16 for the first scanners to amazingly high 1.35 NA systems today (using immersion lithography), producing features under 40 nm in size.

Before the exposure of the photoresist with an image of the mask can begin, this image must be aligned with the previously defined patterns on the wafer. This alignment, and the resulting overlay of the two or more lithographic patterns, is critical because tighter overlay control means circuit features can be packed closer together. Closer packing of devices through better alignment and overlay is nearly as critical as smaller devices through higher resolution in the drive toward more functionality per chip. Historically, overlay tolerances have been set to about 30% of the minimum feature size of the design. Today, overlay can be controlled to less than 20% of the minimum feature size, providing placement precision in the single-digit nanometers.

Another important aspect of photoresist exposure is the standing wave effect. Monochromatic light, when projected onto a wafer, strikes the photoresist surface over a range of angles, approximating plane waves. This light travels down through the photoresist, and, if the substrate is reflective, is reflected back up through the resist. The incoming and reflected light interferes to form a standing wave pattern of high- and low-light intensity at different depths in the photoresist. This pattern is replicated in the photoresist, causing ridges in the sidewalls of the resist feature. As pattern dimensions become smaller, these ridges can significantly affect the quality of the feature. The interference that causes standing waves also results in a phenomenon called *swing curves*, the sinusoidal variation in linewidth with changing resist thickness. These detrimental effects are best cured by coating the substrate with a thin absorbing layer called a bottom antireflective coating that can reduce the reflectivity seen by the photoresist to less than 1 percent.

8.1.5 Post-Exposure Bake

One method of reducing the standing wave effect for older technologies is called the post-exposure bake (PEB). The high temperatures used (100 to 130°C) cause diffusion of the exposed photosensitive material, thus smoothing out the standing wave ridges. The rate of diffusion is dependent on the post-apply bake conditions since the presence of solvent enhances diffusion during a PEB. Thus, a low temperature PAB results in greater diffusion for a given PEB temperature. For a conventional resist (like those used at 436 and 365 nm wavelengths), the main importance of the PEB is diffusion to remove standing waves. For another class of photoresists, called chemically amplified resists, the PEB is an essential part of the chemical reactions that create a solubility differential between exposed and unexposed parts of the resist. For these resists, used extensively at 248 and 193 nm wavelengths, exposure generates a small amount of a strong acid that does not itself change the solubility of the resist. During the post-exposure bake, this photo-generated acid catalyzes a reaction that changes the solubility of the polymer resin in developer. Since the photo-generated acid is not consumed in this reaction, it continues to cause more solubility changing events and thus "amplifies" the effects of exposure. Control of the PEB is extremely critical for chemically amplified resists, and like PAB the most common method is proximity hotplate baking.

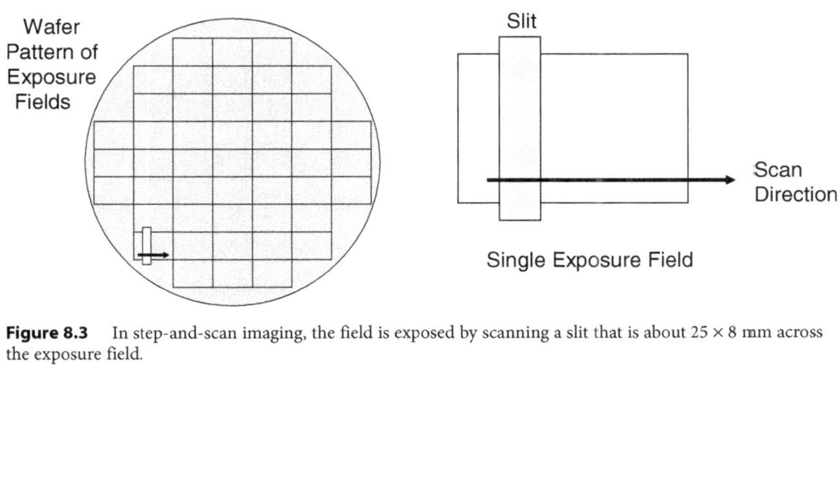

Figure 8.3 In step-and-scan imaging, the field is exposed by scanning a slit that is about 25 × 8 mm across the exposure field.

8.1.6 Development

Once exposed, the photoresist must be developed. Most common positive photoresists use aqueous bases as developers. In particular, tetramethyl ammonium hydroxide (TMAH) is used almost universally at a concentration of 0.26 N. Recently, the use of solvent developers with standard chemically amplified resists has produced negative tone development (NTD). NTD can produce better resolution and control for small hole and space features.

Development is undoubtedly one of the most critical steps in the photoresist process. The characteristics of the resist-developer interactions determine to a large extent the shape of the photoresist profile and, more importantly, the control of the sizes of the features being printed. The method of applying a developer to the photoresist is important in controlling the development uniformity (across the wafer and wafer-to-wafer) and process latitude. During spin development, wafers are spun using equipment similar to that used for spin coating, and the developer is poured onto the rotating wafer. The wafer is also rinsed and dried while spinning. Spray development has good results using developers specifically formulated for this dispense method (using proper surfactants). Using a process identical to spin development, the developer is sprayed, rather than poured, on the wafer by using a nozzle that produces a fine mist of the developer over the wafer. This technique reduces developer usage and gives more uniform developer coverage. Another inline development strategy is called puddle development. Again using developers specifically formulated for this process, the developer is poured onto a stationary wafer that is then allowed to sit motionless for the duration of the development time. The wafer is then spin rinsed and dried. Note that all three inline processes can be performed in the same piece of equipment with only minor modifications, and combinations of these techniques are frequently used.

8.2 IMAGE FORMATION IN OPTICAL LITHOGRAPHY

Projection imaging tools are sophisticated reduction cameras with stages that allow, through a combination of stepping or stepping and scanning motions, the exposure of many copies of a mask pattern onto a large wafer. The image of the mask that is projected into the photoresist defines the information content used by the photoresist to form the final resist image. Understanding the limits and capabilities of projection imaging is the first step in understanding the limits and capabilities of lithography.

Consider the generic projection system shown in Fig. 8.4. It consists of a *light source*, a *condenser lens*, the *mask*, the *objective lens*, and finally the resist-coated wafer. The combination of the light source and the condenser lens is called the *illumination system*. In optical design terms a lens is a system of (usually many) lens elements. Each lens element is an individual piece of glass (refractive element) or a mirror (reflective element). The purpose of the illumination system is to deliver light to the mask (and eventually into the objective lens) with sufficient intensity, the proper directionality and spectral characteristics, and adequate uniformity across the field. The light then passes through the clear areas of the mask and diffracts on its way to the objective lens. The purpose of the objective lens is to pick up a portion of the diffraction pattern and project an image onto the wafer, which, one hopes, will resemble the mask pattern.

8.2.1 Diffraction

The first and most basic phenomenon occurring in projection imaging is the diffraction of light. Diffraction is usually thought of as the bending of light as it passes through an aperture or past an edge, which is certainly an appropriate description for diffraction by a lithographic mask. More correctly, diffraction theory simply describes how light propagates. This propagation includes the effects of the surroundings (boundaries). Maxwell's equations describe how electromagnetic waves propagate, but result in partial differential equations of vector quantities, which for general boundary conditions are extremely difficult to solve without the aid of a powerful computer. A simpler approach is to artificially decouple the electric and magnetic field *vector* components and describe light as a *scalar* quantity, and then to use simplified (assumed) boundary conditions. Under many conditions scalar diffraction theory is surprisingly accurate, though at higher numerical apertures a full vector treatment is needed. In lithography, the distance from the mask to the objective lens is very large, so that the diffraction theory is at its simplest, called *Fraunhofer diffraction*.

In order to establish a mathematical description of diffraction by a mask, let us describe the electric field transmittance of a mask pattern as $t_m(x, y)$, where the mask is in the x-y plane and $t_m(x, y)$ has in general both magnitude and phase. The mask transmittance can be calculated using Maxwell's equations knowing the geometry and material properties of the mask, but often a simple set of assumed boundary conditions (called the Kirchhoff boundary conditions) is sufficiently accurate. For a common chrome-glass mask, the mask transmittance is assumed to be binary—$t_m(x, y)$ is 1 under the glass and 0 under the chrome. Let the x'-y' plane be the diffraction plane, that is, the entrance to the objective lens, and let z be the distance from the mask to the objective lens entrance. Finally, we will assume monochromatic light of wavelength λ and that the entire system is in air (so that its index of refraction can be dropped). Then, the electric field of our diffraction pattern $T_m(x', y')$ is given by the Fraunhofer diffraction integral

$$T_m(f_x, f_y) = \int_{-\infty}^{\infty}\int_{-\infty}^{\infty} t_m(x, y) e^{-2\pi i(f_x x + f_y y)}\, dx\, dy \tag{8.2}$$

where $f_x = x'/(z\lambda)$ and $f_y = y'/(z\lambda)$ and are called the *spatial frequencies* of the diffraction pattern.

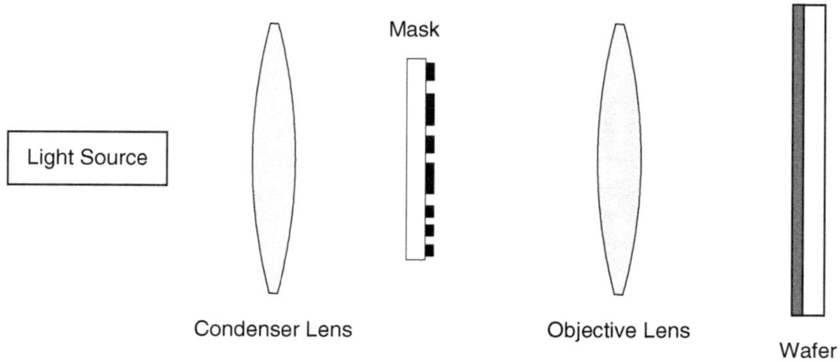

Figure 8.4 Block diagram of a generic projection imaging system.

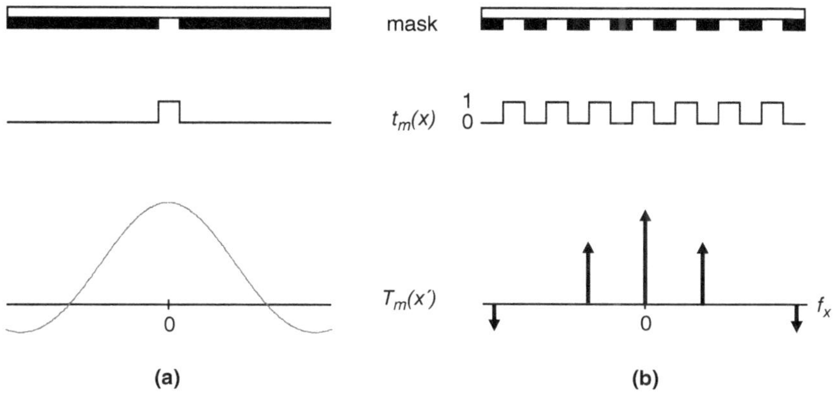

Figure 8.5 Two typical mask patterns—(*a*) an isolated space, and (*b*) an array of equal lines and spaces—and the resulting Fraunhofer diffraction patterns assuming normally incident plane wave illumination

For many scientists and engineers, this equation should be quite familiar—it is simply a *Fourier transform*. Thus, the diffraction pattern (i.e., the electric field distribution as it enters the objective lens) is just the Fourier transform of the mask pattern. This is the principle behind an extremely useful approach to imaging called Fourier optics.

Figure 8.5 shows two mask patterns—one an isolated space and the other a series of equal lines and spaces—both infinitely long in the y direction (the direction out of the page). The resulting mask transmittance functions $t_m(x)$ look like a square pulse and a square wave, respectively. The Fourier transforms are easily found in tables or directly calculated from Eq. (8.2) and are also shown in Fig. 8.5. The isolated space gives rise to a *sinc* function diffraction pattern, and the equal lines and spaces yield discrete *diffraction orders*.

$$\text{Isolated space: } T_m(f_x) = \frac{\sin(\pi w f_x)}{\pi f_x}$$

$$\text{Dense space: } T_m(f_x) = \frac{1}{p}\sum_{n=-\infty}^{\infty} \frac{\sin(\pi w f_x)}{\pi f_x}\delta\left(f_x - \frac{n}{p}\right) \tag{8.3}$$

where δ = Dirac delta function
w = space width
p = pitch (the linewidth plus the space width)

Let's take a closer look at the diffraction pattern for equal lines and spaces. Note that the graphs of the diffraction patterns in Fig. 8.5 use spatial frequency as its x axis. Since z and λ are fixed for a given optical system, the spatial frequency is simply a scaled x' coordinate. At the center of the objective lens entrance ($f_x = 0$), the diffraction pattern has a bright spot called the *zero order*. The zero order is the light that passes through the mask and is not bent. The zero order can be thought of as "DC" light, providing power but no information as to the size of the features on the mask. To either side of the zero order are two peaks called the *first diffraction orders*. These peaks occur at spatial frequencies of $\pm 1/p$, where p is the pitch of the mask pattern (linewidth plus space width). Since the position of these diffraction orders depends on the mask pitch, their position contains information about the pitch. It is this information that the objective lens will use to reproduce the image of the mask. In fact, in order for the objective lens to form a true image of the mask it must have the zero order and at least one of the first orders. In addition to the first order, there can be many higher orders, with the nth order occurring at a spatial frequency of n/p.

8.2.2 Image Formation

We are now ready to describe what happens next and follow the diffracted light as it enters the objective lens. In general, the diffraction pattern extends throughout the x'-y' plane. However, the objective lens, being only of finite size, cannot collect all of the light in the diffraction pattern. Typically, lenses used in microlithography are circularly symmetric and the entrance to the objective lens can be thought of as a circular aperture. Only those portions of the mask diffraction pattern that fall inside the aperture of the objective lens go on to form the image. Of course we can describe the size of the lens aperture by its radius, but a more common and useful description is to define the maximum angle of diffracted light that can enter the lens. Consider the geometry shown in Fig. 8.6. Light passing through the mask is diffracted at various angles. Given a lens of a certain size placed at a certain distance from the mask, there is some maximum angle of diffraction α for which the diffracted light just makes it into the lens. Light emerging from the mask at larger angles misses the lens and is not used in forming the image. The most convenient way to describe the size of the lens aperture is by its *numerical aperture*, defined as sine of the maximum half-angle of diffracted light that can enter the lens times the index of refraction of the surrounding medium. Unless we are considering immersion lithography, all lenses are in air and the numerical aperture is given by $NA = \sin\alpha$. (Note that the spatial frequency is the sine of the diffracted

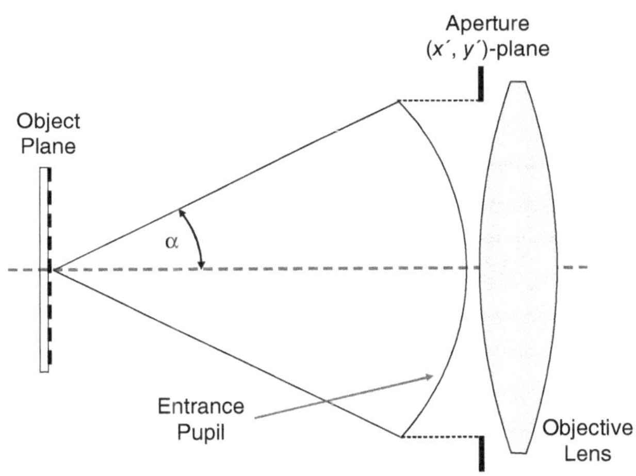

Figure 8.6 The numerical aperture is defined as $NA = n_{\text{medium}}\sin\alpha$, where α is the maximum half-angle of the diffracted light that can enter the objective lens and n_{medium} is the refractive index of the medium through which the light is traveling to the lens.

angle divided by the wavelength of light. Thus, the maximum spatial frequency that can enter the objective lens is given by NA/λ.)

Clearly, the numerical aperture is going to be quite important. A large numerical aperture means that a larger portion of the diffraction pattern is captured by the objective lens. For a small numerical aperture, much more of the diffracted light is lost. In fact, we can use this viewpoint to define *resolution*, at least from the limited perspective of image formation. Consider the simple case of a mask pattern of equal lines and spaces. As we have seen, the resulting diffraction pattern is a series of discrete diffraction orders. In order to produce an image that even remotely resembles the original mask pattern, it is necessary for the objective lens to capture the zero order and at least one higher diffraction order. If the light illuminating the mask is a normally incident plane wave, the diffraction pattern will be centered in the objective lens. Since the position of the ± 1 diffraction orders are given by $\pm 1/p$, the requirement that a lens of a finite size must capture these diffraction orders to form an image puts a lower limit on the pitch that can be imaged. Thus, the smallest pitch (p_{min}) that still produces an image will put the first diffraction order at the outer edge of the objective lens:

$$f_{max} = \frac{1}{p_{min}} = \frac{NA}{\lambda} \tag{8.4}$$

To proceed further, we must now describe how the lens affects the light entering it. Obviously, we would like the image to resemble the mask pattern. Since diffraction gives the Fourier transform of the mask, if the lens could give the *inverse* Fourier transform of the diffraction pattern, the resulting image would resemble the mask pattern. In fact, lenses are designed to behave precisely in this way. We can define an ideal imaging lens as one that produces an image at the image plane that is identically equal to the inverse Fourier transform of the light distribution entering the lens. It is the goal of lens designers and manufacturers to create lenses as close as possible to this ideal. Does an ideal lens produce a perfect image? No. Because of the finite size of the numerical aperture, only a portion of the diffraction pattern enters the lens. Thus, even an ideal lens cannot produce a perfect image unless the lens is infinitely big. Since in the case of an ideal lens the image is limited only by the diffracted light that does not make it through the lens, we call such an ideal system *diffraction limited*.

In order to write our final equation for the formation of an image, let us define the objective lens' *pupil function P* (a pupil is just another name for an aperture). The pupil function of an ideal lens simply describes what portion of light passes through the lens—it is one inside the aperture and zero outside.

$$P(f_x, f_y) = \begin{cases} 1, & \sqrt{f_x^2 + f_y^2} < NA/\lambda \\ 0, & \sqrt{f_x^2 + f_y^2} > NA/\lambda \end{cases} \tag{8.5}$$

Thus, the product of the pupil function and the diffraction pattern describes the light entering the objective lens. Combining this with our description of how a lens behaves gives us our final expression for the electric field at the image plane (i.e., at the wafer)

$$E(x,y) = \mathcal{F}^{-1}\{T_m(f_x, f_y) P(f_x, f_y)\} \tag{8.6}$$

where the symbol $\mathcal{F}^{-1}$ represents the inverse Fourier transform. The *aerial image* is defined as the intensity distribution at the wafer and is simply the square of the magnitude of the electric field.

Consider the full imaging process. First, light passing through the mask is diffracted. The diffraction pattern can be described as the Fourier transform of the mask pattern transmittance. Since the objective lens is of finite size, only a portion of the diffraction pattern actually enters the lens. The numerical aperture defines the maximum angle of diffracted light that enters the lens and the pupil function is used to mathematically describe this behavior. Finally, the effect of the lens is to take the inverse Fourier transform of the light entering the lens to give an image that resembles the mask pattern. If the lens is ideal, the quality of the resulting image is only limited by the amount of the diffraction pattern collected.

8.2.3 Partial Coherence

Although we have completely described the behavior of a simple ideal imaging system, we must add one more complication before describing the operation of a projection system for lithography. So far, we have assumed that the mask is illuminated by *spatially coherent* light. Coherent illumination means simply that the light striking the mask arrives from only one direction. We have further assumed that the coherent illumination on the mask is normally incident. The result was a diffraction pattern that was centered in the entrance to the objective lens. What would happen if we changed the direction of the illumination so that the light struck the mask at some angle θ'? The effect is simply to shift the position of the diffraction pattern with respect to the lens aperture (in terms of spatial frequency, the amount shifted is $\sin\theta'/\lambda$). Recalling that only the portion of the diffraction pattern passing through the lens aperture is used to form the image, it is quite apparent that this shift in the position of the diffraction pattern can have a profound effect on the resulting image by potentially changing which diffracted orders pass through the lens.

If the illumination of the mask is composed of light coming from a range of angles rather than just one angle, the illumination is called *partially coherent*. If one angle of illumination causes a shift in the diffraction pattern, a range of angles will cause a range of shifts, resulting in broadened diffraction orders. One can characterize the range of angles used for the illumination in several ways, but the most common is the *partial coherence factor* σ (also called the degree of partial coherence, the pupil filling function, or just the partial coherence). The partial coherence is defined as the sine of the half-angle of the illumination cone divided by the objective lens numerical aperture. It is thus a measure of the angular range of the illumination relative to the angular acceptance of the lens. If the range of angles striking the mask extends from −90 to 90° (i.e., all possible angles), the illumination is said to be *incoherent*.

The extended source method can be used to calculate partially coherent images. In this method, the full source is divided into individual point sources. Each point source is coherent and results in an aerial image calculated using the diffraction pattern appropriately shifted in the pupil for that source point. Two point sources from the extended source, however, do not interact coherently with each other. Thus, the contributions of these two sources must be added to each other incoherently (i.e., the intensities of the resulting images are added together). The full aerial image is determined by calculating the coherent aerial image from each point on the source, and then integrating (summing) the image intensity over the source.

8.2.4 Aberrations and Defocus

Aberrations can be defined as the deviation of the real behavior of an imaging system from its ideal behavior (the ideal behavior was described earlier using Fourier optics as diffraction limited imaging). Aberrations are inherent in the behavior of all lens systems and come from three basic sources—defects of construction, defects of use, and defects of design. Defects of construction include rough or inaccurate lens surfaces, inhomogeneous glass, incorrect lens thicknesses or spacings, and tilted or decentered lens elements. Defects of use include use of the wrong illumination or tilt of the lens system with respect to the optical axis of the imaging system. Also, changes in the environmental conditions during use, such as the temperature of the lens or the barometric pressure of the air, result in defects of use. Defects of design may be a bit of a misnomer, since the aberrations of a lens design are not mistakenly designed *into* the lens, but rather were not designed *out* of the lens. All lenses have aberrated behavior because the Fourier optics behavior of a single lens element is only approximately true. It is the job of a lens designer to combine elements of different shapes and properties so that the aberrations of each individual lens element tend to cancel in the sum of all of the elements, giving a lens system with only a small residual amount of aberrations. It is impossible to design a lens system with absolutely no aberrations.

Mathematically, aberrations are described as *wavefront deviations*, the difference in phase (or path length) of the actual wavefront emerging from the lens compared to the ideal wavefront as predicted from Fourier optics. This phase difference is a function of the position within the lens pupil, most conveniently described in polar coordinates. The wavefront deviation is, in general, quite complicated, so the mathematical form used to describe it is also quite complicated. The most common model for describing the phase error across the pupil is the *Zernike polynomial*, an infinite orthonormal polynomial series, usually cut off at 36 terms, with powers of the radial pupil position and trigonometric functions of the polar angle. The Zernike polynomial can be arranged in many ways, but most lens design software and lens measuring equipment employ a form called the fringe or circle Zernike polynomial. Terms of this orthogonal polynomial describe common aberrations such as coma and astigmatism. The impact of these phase errors described by the Zernike polynomial is to modify the pupil function of Eq. (8.5) to include this spatial frequency dependent phase change on transmission through the pupil.

Previous expressions for calculating the aerial image [such as Eqs. (8.2), (8.5), and (8.6)] apply only to the image at the focal plane. What happens when the wafer is out of focus? What is the image intensity distribution some small distance away from the plane of best focus? Focus errors can be described as an aberration of a sort. Consider a perfect spherical wave converging (i.e., focusing) down to a point. An ideal projection system will create such a wave coming out of the lens aperture (called the *exit pupil*), as shown in Fig. 8.7*a*. If the wafer to be printed were placed in the same plane as the focal point of this wave, we would say that the wafer was in focus. What happens if the wafer were removed from this plane by some distance δ, called the defocus distance? Figure 8.7*b* shows such a situation. The spherical wave with the solid line represents the actual wave focused to a point a distance δ away from the wafer. If, however, the wave had a different shape, as given by the dotted curve, then the wafer would be in focus. Note that the only difference between these two waves is the radius of curvature. Since the dotted curve is the wavefront we want for the given wafer position, we can say that the actual wavefront is in error because it does not focus where the wafer is located. (This is just a variation of "the customer is always right" attitude—the wafer is always right, it is the optical wavefront that is out of focus.)

By viewing the actual wavefront as having an error in curvature relative to the desired wavefront (i.e., the one that focuses on the wafer), we can quantify the effect of defocus. Looking at Fig. 8.7*b*, it is apparent that the distance from the desired to the *defocused* wavefront goes from zero at the center of the exit pupil and increases as we approach the edge of the pupil. This distance between wavefronts is called the *optical path difference* (OPD). The OPD is a function of the defocus distance and the position within the pupil and can be obtained from the geometry of the situation. Describing the position within the exit pupil by an angle θ, the optical path difference when traveling through a medium of refractive index n_w is given by

$$\mathrm{OPD} = n_w \delta(1 - \cos\theta) \tag{8.7}$$

So how does this optical path difference affect the formation of an image? The OPD acts just like an aberration, modifying the pupil function of the lens. For light, this path length traveled (the OPD) is equivalent to a change in phase. Thus, the OPD can be expressed as a phase error $\Delta\phi$ due to defocus

$$\Delta\varphi = k\,OPD = 2\pi n_w \delta(1 - \cos\theta)/\lambda \tag{8.8}$$

where k is equal to $2\pi/\lambda$, the *propagation constant* in air. We are now ready to see how defocus affects the diffraction pattern and the resulting image. Our interpretation of defocus is that it causes a phase error as a function of the radial position within the aperture. Light in the center of the aperture has no error; light at the edge of the aperture has the greatest phase error. Recall that diffraction by periodic patterns results in discrete diffraction orders—the zero order is the undiffracted light passing through the center of the lens; higher orders contain information necessary to reconstruct the image. *Thus, the effect of defocus is to add a phase error to the higher-order diffracted light relative to the zero order.* When the lens recombines these orders to form an image, this phase error will result in a degraded image.

8.2.5 Immersion Lithography

Although the scientific principles underlying immersion lithography have been known for well over 100 years (immersion microscopes were in common use by 1880), recently this technology has gained widespread use in the semiconductor industry. Immersion lithography at a wavelength of 193 nm uses ultrapure degassed water, with refractive index of 1.437, as the immersion liquid. A small puddle of water is kept between the lens and the wafer during wafer exposure. New water is constantly pumped under the lens to keep the optical properties of the fluid consistent, and to prevent the buildup of contaminants. A special wafer chuck keeps the puddle intact as the edge of the wafer is scanned under the puddle.

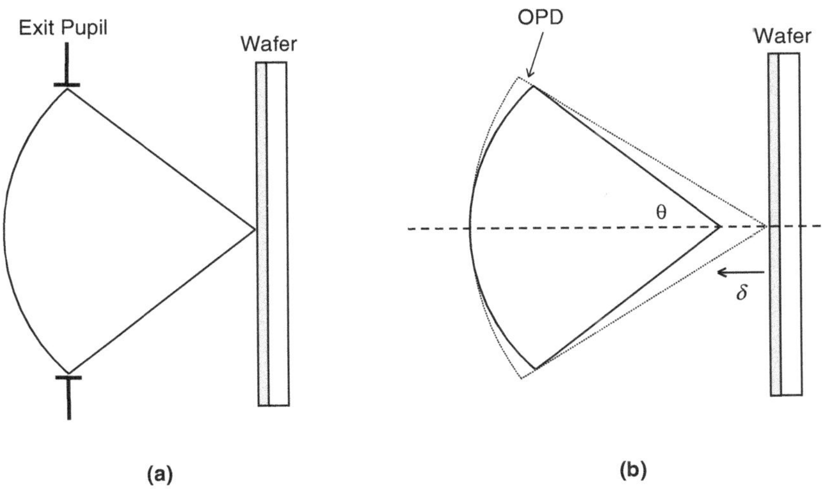

Figure 8.7 Focusing of light can be thought of as a converging spherical wave: (*a*) in focus, and (*b*) out of focus by a distance δ.

The definition of the numerical aperture includes the refractive index of the medium through which the light travels. However, adding water, with its higher refractive index, does not raise the numerical aperture by itself. Instead, it enables the design of a higher numerical aperture lens. The constraint that the sine of an angle will never exceed one cannot be violated. In practice, the largest lithographic lenses enable $\sin\alpha$ to reach 0.93, which is thus the practical limit for the numerical aperture of a dry lithography tool. For an immersion lens, this same practical constraint enables the design and construction of lenses with a numerical aperture up to 1.35. These so-called hyper-NA immersion lithography tools are now in common use in semiconductor manufacturing to print features with dimensions as low as about 40 nm.

8.3 PHOTORESIST CHEMISTRY

The formation of an aerial image is only the first step in the transfer of information from a photomask into a resist pattern. The aerial image must propagate into the resist and cause a chemical change, forming a latent image of exposed and unexposed material. This latent image, either directly or indirectly, will affect the solubility of the resist, allowing the latent image to be turned into a profile image through the process of development.

8.3.1 Exposure Kinetics

All photoresists have a light sensitive compound called a sensitizer that reacts when exposed to light of a certain wavelength. The chemistry of exposure for diazonaphthoquinone (DNQ, a popular sensitizer for near-UV resists) is given below.

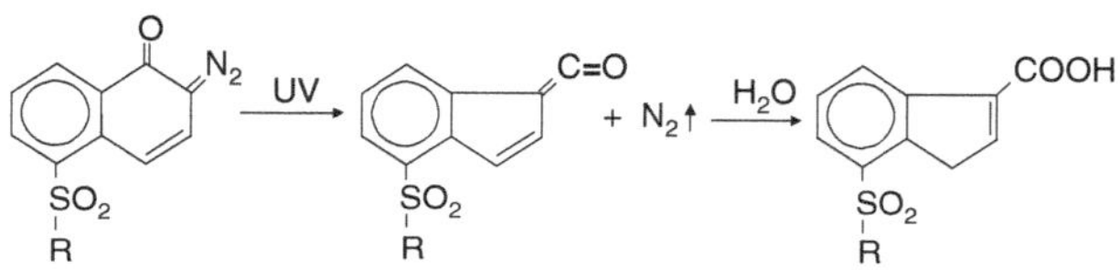

(8.9)

The diazonaphthoquinone absorbs a photon and a water molecule, releases nitrogen gas and produces a carboxylic acid (that will exhibit high solubility in a developer made of water and base).

The kinetics of the chemical reaction represented by Eq. (8.9) are first order. For light of intensity I exposing resist for a time t,

$$\frac{dm}{dt} = -CIm \tag{8.10}$$

where the relative sensitizer concentration m (the actual concentration divided by the initial preexposed concentration) has been used and C is the exposure rate constant. A solution to the exposure rate equation [Eq. (8.10)] is simple if the intensity within the resist is constant throughout the exposure.

$$m = e^{-CIt} \tag{8.11}$$

This result illustrates an important property of first-order exposure kinetics called *reciprocity*. The amount of chemical change is controlled by the product of light intensity and exposure time. Doubling the intensity and cutting the exposure time in half will result in the exact same amount of chemical change. This product of intensity and exposure time is called the exposure dose or exposure energy.

8.3.2 Chemically Amplified Resists

Unlike conventional resists, such as the diazonaphthoquinone system discussed earlier, chemically amplified resists require two separate chemical reactions to change the solubility of the resist. First, exposure turns an aerial image into a latent image of exposure reaction products. Although very similar to conventional resists, the reaction products of exposure for a chemically amplified resist do not change the solubility of the resist. Instead, a second reaction during a post-exposure bake is catalyzed by the exposure reaction products. The result of the post-exposure bake reaction is a change in the solubility of the resist. This two-step sensitization process has some interesting characteristics and challenges.

For chemically amplified photoresists, the sensitizer is called a photoacid generator (PAG). As the name implies, the photoacid generator forms a strong acid when exposed to deep UV light. The reaction of a simplified PAG is shown in Eq. (8.12).

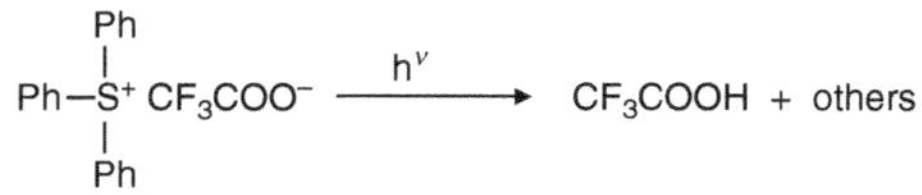

(8.12)

The acid generated in this case (trifluoroacetic acid) is a derivative of acetic acid, where the electron-drawing properties of the fluorines are used to greatly increase the acidity of the molecule. The PAG is mixed with the polymer resin at a concentration of typically 5 to 15 percent by weight for 248 nm resists. For 193 nm resists, PAG loading is kept lower at 1 to 5 percent by weight to keep the optical absorbance of the resist within desired levels. The kinetics of the exposure reaction are standard first order, as given by Eqs. (8.10) and (8.11).

Exposure of the resist with an aerial image $I(x)$ results in an acid latent image $H(x)$. A PEB is then used to thermally induce a chemical reaction. This may be the activation of a cross-linking agent for a negative resist or the deblocking of the polymer resin for a positive resist. The defining characteristic of a chemically amplified resist is that this reaction is *catalyzed* by the acid so that the acid is not consumed by the reaction and, to first order, H remains constant. A polymer such as polyhydroxystyrene (PHS) is used, which is very soluble in an aqueous base developer. It is the hydroxyl groups that give the PHS its high solubility, so by "blocking" these sites (by reacting the hydroxyl group with some longer chain molecule) the solubility can be reduced. Early chemically amplified resists employed a *t*-butoxycarbonyl group (*t*-BOC), resulting in a very slowly dissolving blocked polymer. In the presence of acid and heat, the *t*-BOC blocked polymer will undergo acidolysis to generate the soluble hydroxyl group, as shown in Eq. (8.13).

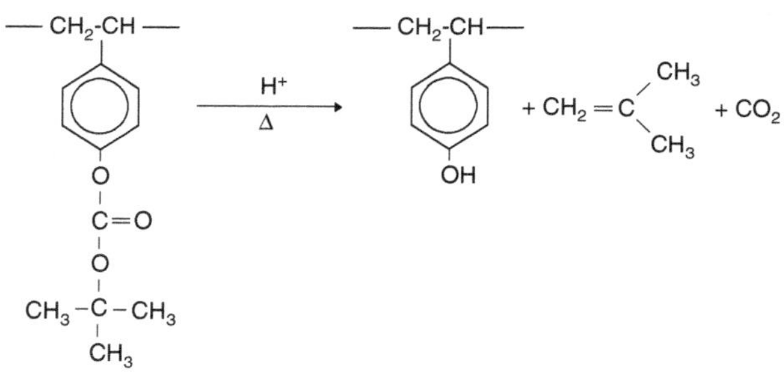

(8.13)

One drawback of this scheme is that the cleaved *t*-BOC is volatile and will evaporate, causing film shrinkage in the exposed areas. Larger molecular weight blocking groups can be used to reduce this film shrinkage to acceptable levels (below 10 percent). Also, the blocking group is such an effective inhibitor of dissolution that nearly every blocked site on the polymer must be deblocked to obtain significant dissolution. Thus, the photoresist can be made more "sensitive" by only partially blocking the PHS. Additionally, fully blocked polymers tend to have poor coating and adhesion properties. Typical photoresists use 10 to 30 percent of the hydroxyl groups blocked. Molecular weights for the PHS run in the range of 3000 to 6000 giving about 20 to 40 hydroxyl groups per polymer molecule, about 4 to 8 of which are initially blocked.

While the previously mentioned two-step dissolution enhancement mechanism is an innovative way of making the PHS polymer photosensitive using the conventional base developer, the real innovation lies in the products of the deblocking reaction. The deblocking reaction was designed to regenerate the acid as one of the products of the reaction. Thus, the acid serves as a catalyst (defined as a chemical that must be present for a reaction to occur, but which is not consumed in the reaction). This reaction has been dubbed "chemical amplification" because the impact of an absorbed photon is chemically amplified by the catalytic nature of the deblocking reaction.

While catalytic reactions are not at all uncommon in chemistry, a photo-generated catalyst opens up numerous interesting implications. What is the role of acid diffusion in determining the rate of reaction? How does this diffusion affect feature size control? What stops the reaction from continuing indefinitely? How does the average number of deblocking reactions per acid (called the catalytic chain length) influence resist performance? The answers to these questions make up the critical aspects of the design of a chemically amplified photoresist.

Through a variety of mechanisms, the acid formed by exposure of the resist film can be lost and thus not contribute to the catalyzed reaction to change the resist solubility. There are two basic types of acid loss—loss that occurs between exposure and post-exposure bake and loss that occurs during the post-exposure bake. The first type of loss leads to delay time effects—the resulting lithography is affected by the delay time between exposure and post-exposure bake. Delay time effects can be very severe and, of course, are very detrimental to the use of such a resist in a manufacturing environment. The typical mechanism for delay time acid loss is the diffusion of atmospheric base contaminants into the top surface of the resist. The result is a neutralization of the acid near the top of the resist and a corresponding reduced amplification. For a positive resist, the effects can be devastating. Sufficient base contamination can make the top of the resist insoluble, blocking dissolution into the bulk of the resist (Fig. 8.8). In extreme cases, no patterns can be observed after development. Such acid loss is mitigated in a number of ways: linking the post-exposure bake hotplate directly to the exposure tool to minimize post-exposure delay; eliminating sources of base contaminants and filtering the atmosphere around the wafer; developing photoresists with low diffusivities for common base contaminants.

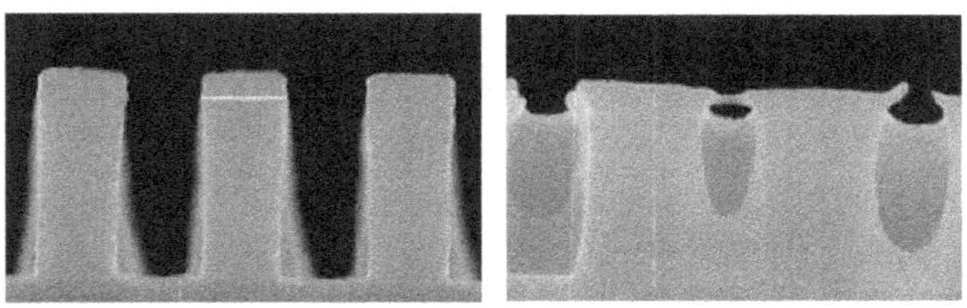

(a) (b)

Figure 8.8 Atmospheric base contamination leads to T-top formation. Shown are line/space features printed in APEX-E for (*a*) 0.275 μm features with no delay and (*b*) 0.325 μm features with 10 min delay between exposure and PEB. (*Courtesy of SEMATECH.*)

Another mechanism for acid loss is intentional rather than accidental. Most formulations of chemically amplified resists include the addition of a base quencher. Loaded at concentrations of 5 to 25 percent of the initial PAG loading, this base quencher is designed to neutralize any photo-generated acid that comes in contact with it. For low exposure doses, the small amount of photoacid generated will be neutralized by the base quencher and amplification will not take place. Only when the exposure rises above a certain threshold will the amount of acid be sufficient to completely neutralize all of the base quencher as well as cause deblocking during PEB. The main purpose of the base quencher is to neutralize the low levels of acid that might diffuse into the nominally unexposed regions of the wafer, thus making the final resist linewidth less sensitive to acid diffusion. This description of base quenching behavior is made more complicated by the fact that the quencher will, in general, diffuse during the post-exposure bake. The difference in diffusivity between the acid and the base is important in determining lithographic behavior for these types of resists.

8.3.3 Dissolution

Dissolution involves some of the most critical chemistry of the resist. The goal is to create a highly nonlinear response of the resist dissolution rate to the exposure dose, with the ideal response being a threshold "switch" between low and high development rates at some exposure level. Our discussion will focus on the development of a positive photoresist, but can be easily generalized to negative tone resists. Photoresist dissolution involves three processes—diffusion of a developer from the bulk solution to the surface of the resist, reaction of the developer with the resist, and diffusion of the product back into the solution. Generally, we can assume that the last step, diffusion of the dissolved resist into the solution, occurs very quickly so that this step may be ignored. Let us now look at the first two steps in the proposed mechanism. The diffusion of the developer to the resist surface can be described with a simple difference approximation to the diffusion rate equation.

$$r_D = k_D (D - D_S) \tag{8.14}$$

where r_D = rate of diffusion of the developer to the resist surface
D = bulk developer concentration
D_S = developer concentration at the resist surface
k_D = rate constant

We shall now propose a mechanism for the reaction of the developer with the resist. It is quite likely that this step is in fact a series of more detailed steps, including the diffusion of the developer cation into the resist to form a thin gel layer. However, we will assume a simple surface-limited reaction here. The resist is composed of large macromolecules of polymer resin along with an inhibiting species *M*, which converts to a solubility-enhancing product *P* after exposure and/or PEB. For a non-chemically amplified resist, *M* is the sensitizer and *P* is the carboxylic acid that enhances solubility. For a chemically amplified resist, *M* is the blocking group and *P* is the deblocked site. The presence of *M* acts as an inhibitor to dissolution, making the development rate very slow. The product *P*, however, is highly soluble in the developer, enhancing the dissolution rate of the resin. Let us assume that *n* molecules of product *P* react with the developer to dissolve a single resin molecule. The rate of the reaction is

$$r_R = k_R D_S P^n \tag{8.15}$$

where r_R is the rate of reaction of the developer with the resist and k_R is the rate constant. From the stoichiometry of the exposure or deblocking reaction,

$$P = M_o - M \tag{8.16}$$

where M_o is the initial inhibitor concentration (i.e., before exposure/PEB).

The two steps outlined earlier are in series, that is, one reaction follows the other. Thus, the two steps will come to a steady state such that their rates are equal. Equating the rate equations [Eqs. (8.14) and (8.15)], one can solve for D_S and eliminate it from the overall rate equation. After some algebra and letting $m = M/M_o$

$$r = r_{max} \frac{(a+1)(1-m)^n}{a+(1-m)^n} + r_{min} \tag{8.17}$$

where $r_{max} = \dfrac{k_D D}{k_D/k_R M_o^n + 1}$ and $a = k_D/k_R M_o^n = \dfrac{(n+1)}{(n-1)}(1-m_{TH})^n$.

and where m_{TH} is the value of m at the inflection point of the development rate function, called the threshold inhibitor concentration. Note that the simplifying constant a describes the rate constant of diffusion relative to the surface reaction rate constant. A large value of a will mean that diffusion is very fast, and thus less important, compared to the fastest surface reaction (for the completely exposed resist). The addition of

r_{min} to Eq. (8.17) assumes that the mechanism of development of the unexposed resist is independent of the above-proposed development mechanism. Note that the addition of the r_{min} term means that the true maximum development rate is actually $r_{max} + r_{min}$. In most cases $r_{max} \gg r_{min}$ and the difference is negligible.

Figure 8.9 shows some plots of this model for different values of n. The behavior of the dissolution rate with increasing n values is to make the rate function more "selective" between resist exposed above m_{TH} and resist exposed below m_{TH}. For this reason, n is called the *dissolution selectivity* parameter. Also from this behavior, the interpretation of m_{TH} as a "threshold" concentration becomes quite evident. Note that as the developer selectivity parameter n goes to infinity, the resist approaches the ideal step function response that is desired. Thus, the goal of resist design is to create higher values of n, which is directly related to the number of blocked polymer sites in the resist.

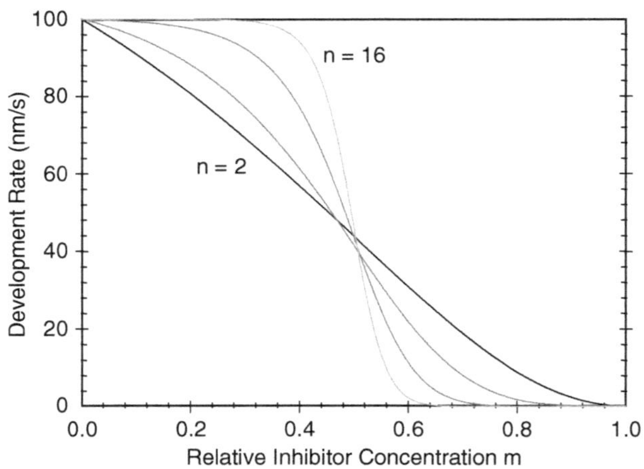

Figure 8.9 Development rate as a function of the dissolution selectivity parameter (r_{max} = 100 nm/s, r_{min} = 0.1 nm/s, m_{TH} = 0.5, and n = 2, 4, 8, and 16).

8.4 LINEWIDTH CONTROL

Historically, lithography engineering has focused on two key, complimentary aspects of lithographic quality—overlay performance and linewidth control. Linewidth (or critical dimension, CD) control generally means ensuring that the widths of certain critical features, measured at specific points on those features, fall within acceptable bounds. Overlay describes the positional errors in placing one mask layer pattern over an existing pattern on the wafer. The two types of errors are not independent in their effects on the device. Since the packing density (the closest allowed spacing between devices in a chip) is determined by the accuracy of feature edge placements, both CD control and overlay capability contribute to the design rules that determine packing density. However, by and large, the sources of errors that affect feature size and feature placement act independently so that efforts to improve overlay capability tend to have little effect on CD control, and vice versa. As a result, CD and overlay control tend to be mostly independent operations in semiconductor manufacturing.

Fundamentally, errors in the final dimension of a feature are the result of errors in the tools, processes, and materials that affect the final CD. An error in a process variable (the temperature of a hotplate, for example) propagates through to become an error in the final CD based on the various physical mechanisms by which the variable influences the lithographic result. In such a situation, a propagation of errors analysis can be used to help understand the effects. Suppose the influence of each input variable on the final CD were expressed in a mathematical form, such as

$$\text{CD} = f(v_1, v_2, v_3, \ldots) \tag{8.18}$$

where v_i are the input (process) variables.

Given an error in each process variable Δv_i, the resulting CD error can be computed from a Taylor expansion of the function in Eq. (8.18) using the total differential.

$$\Delta\text{CD} = \sum_{n=1}^{\infty} \frac{1}{n!}\left(\Delta v_1 \frac{\partial}{\partial v_1} + \Delta v_2 \frac{\partial}{\partial v_2} + \ldots\right)^n f(v_1, v_2, \ldots) \tag{8.19}$$

This imposing looking summation of powers of derivatives can be simplified if the function is reasonably well behaved (and of course we hope that our critical features will be so) and the errors in the process variables are small (we hope this is true as well). In such a case, it may be possible to ignore the higher-order terms ($n > 1$), as well as the cross terms of Eq. (8.19), to leave a simple, linear error equation.

$$\Delta\text{CD} = \Delta v_1 \frac{\partial \text{CD}}{\partial v_1} + \Delta v_2 \frac{\partial \text{CD}}{\partial v_2} + \ldots \tag{8.20}$$

Each Δv_i represents the magnitude of a *process error*. Each partial derivative $\partial CD/\partial v_i$ represents the *process response*, the response of CD to an incremental change in the variable. This process response can be expressed in many forms; for example, the inverse of the process response is called *process latitude*.

The linear error equation [Eq. (8.20)] can be modified to account for the nature of the errors at hand. In general, CD errors are specified as a percentage of the nominal CD. For such a case, the goal is usually to minimize the relative CD error ΔCD/CD. Equation (8.20) can be put in this form as

$$\frac{\Delta\text{CD}}{\text{CD}} = \Delta v_1 \frac{\partial \ln \text{CD}}{\partial v_1} + \Delta v_2 \frac{\partial \ln \text{CD}}{\partial v_2} + \ldots \tag{8.21}$$

Also, many sources of process errors result in errors that are a fraction of the nominal value of that variable (e.g., illumination nonuniformity in a stepper produces a dose error that is a fixed percentage of the nominal dose). For such error types, it is best to modify Eq. (8.21) to use a relative process error $\Delta v_i/v_i$

$$\frac{\Delta\text{CD}}{\text{CD}} = \frac{\Delta v_1}{v_1} \frac{\partial \ln \text{CD}}{\partial \ln v_1} + \frac{\Delta v_2}{v_2} \frac{\partial \ln \text{CD}}{\partial \ln v_2} + \ldots \tag{8.22}$$

Although Eqs. (8.20) to (8.22) may seem obvious, even trivial, in their form, they reveal a very important truth about error propagation and the control of CD. There are two distinct ways to reduce ΔCD—reduce the magnitude of the individual process errors (Δv_i), or reduce the response of CD to that error ($\partial CD/\partial v_i$). The separation of CD errors into these two source components identifies the two important tasks that face the photo engineer. Reducing the magnitude of process errors is generally considered a *process control* activity and involves picking the right material and equipment for the job and ensuring that all equipment is working in proper order and all materials are meeting their specifications. Reducing the process response is a *process optimization* activity and involves picking the right process settings, as well as the right equipment and materials. Often these two activities are reasonably independent of each other.

A note of caution: the derivation of Eq. (8.20) assumed that the process errors were small enough to be linear and independent in their influence on CD. This will not always be the case for some variables. One needs only to consider the two variables of focus and exposure to see that the response of CD is certainly nonlinear (it is approximately quadratic with focus) and the two variables are highly dependent on each other. Thus, the only way to judge the response of the process

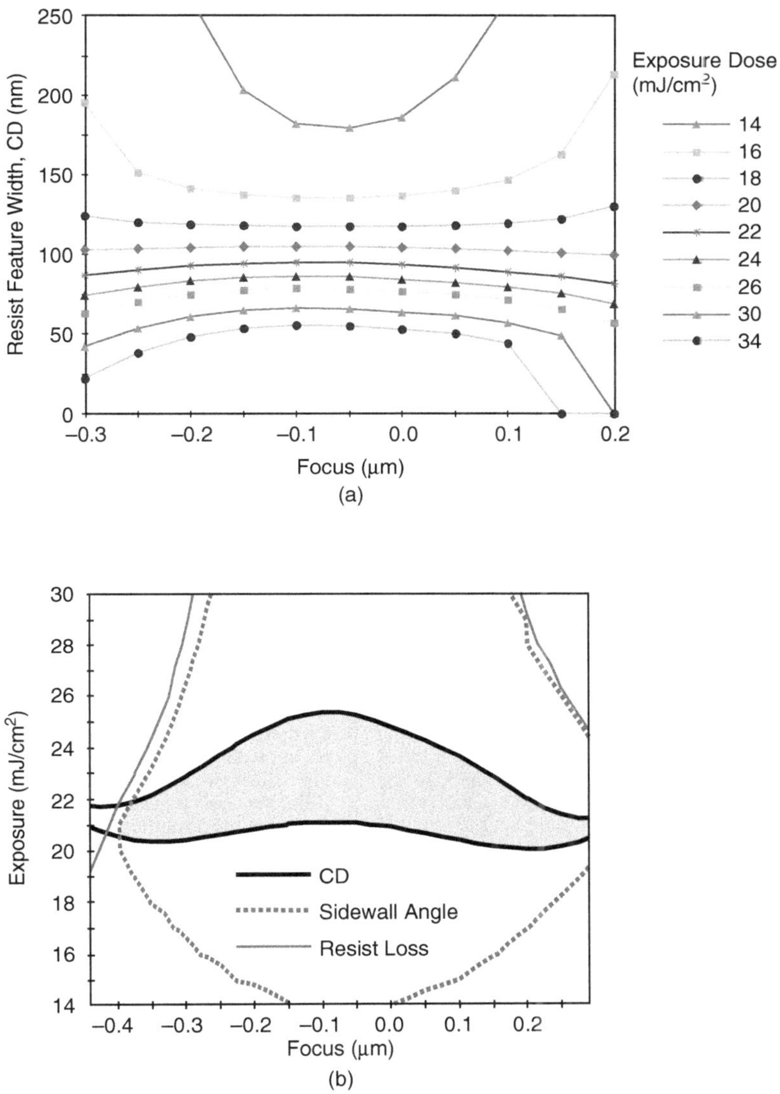

Figure 8.10 (*a*) Example of the effect of focus and exposure on the resulting resist linewidth. (*b*) Displaying the data from a focus-exposure matrix as a focus-exposure process window (shaded area) constructed from contours of the specifications for linewidth, sidewall angle, and resist loss. Here, focal position is defined as zero at the top of the resist with a negative focal position indicating that the plane of focus is inside the resist.

to focus is to simultaneously vary both focus and exposure in what is known as a focus-exposure matrix. Figure 8.10*a* shows a typical example of the output of a focus-exposure matrix using linewidth as the response in what is called a Bossung plot.

Another very useful way to plot this two-dimensional data set is a contour plot—contours of constant linewidth versus focus and exposure. The contour plot form of data visualization is especially useful for establishing the limits of exposure and focus that allow the final image to meet certain specifications. Rather than plotting all of the contours of constant CD, one could plot only the two CDs corresponding to the outer limits of acceptability—the CD specifications. Because of the nature of a contour plot, other variables can also be plotted on the same graph (such as specification limits for the sidewall angle of the resist profile and resist loss off the top of a resist line). Figure 8.10*b* shows an example of plotting contours of CD (nominal ±10 percent), 80° sidewall angle, and 10 percent resist loss all on the same graph. The result is a *process window*—inside the process window is the region of focus and exposure that keeps the final resist profile within all three specifications.

8.5 OVERLAY CONTROL

Like critical dimension control, overlay control is a vital part of lithography in semiconductor manufacturing. Errors in the overlay of different lithographic levels can directly cause a number of electrical problems, such as short or open circuits or poor transistor performance. Overlay is defined as the positional accuracy with which a new lithographic pattern has been printed on top of an existing pattern on the wafer, measured at any point on the wafer. This is opposed to the slightly different concept of *registration*, which is the positional accuracy with which a lithographic pattern has been printed as compared to an absolute coordinate grid.

To characterize overlay, measurement targets are placed at various spatial positions in the field, minimally at the four corners but often in the streets between the die (and thus within the field) as well. The spatial variation of overlay errors provides extremely important information as to the behavior of the reticle, the wafer, and the exposure tool. Overlay errors are characterized in terms of x errors and y errors separately, rather than being combined into an overlay error length. Since integrated circuit (IC) patterns are mostly laid out on a rectangular grid, the impact of an x-overlay error on device yield or performance is almost independent of any y-overlay errors that might also be present, and vice versa.

The goal of overlay data analysis is twofold: assess the magnitude of overlay errors and determine, if possible, their root causes. The overall magnitude of overlay errors is commonly summarized by the mean plus 3σ of the overlay data in both x and y directions. Root cause analysis involves explaining the data with a model that assigns a cause to the observed effect. The most common low-order model for the measured overlay errors dx and dy at wafer and field coordinates (x_{wafer}, y_{wafer}) and (x_{field}, y_{field}) is

$$\begin{aligned} dx &= \Delta x - \theta_x y_{\text{wafer}} + \Delta M_x x_{\text{wafer}} - \varphi_x y_{\text{field}} + \Delta m_x x_{\text{field}} \\ dy &= \Delta y + \theta_y x_{\text{wafer}} + \Delta M_y y_{\text{wafer}} + \varphi_y x_{\text{field}} + \Delta m_y y_{\text{field}} \end{aligned} \tag{8.23}$$

where Δx and Δy are translation errors, θ and ΔM represent the wafer (stage) rotation and magnification errors, and ϕ and Δm represent the reticle (field) rotation and magnification terms. These terms are called *correctibles*, since knowledge of their values can be fed back to the lithography tool and adjustments made before the next batch of wafers are processed.

8.6 THE LIMITS OF OPTICAL MICROLITHOGRAPHY

To date, optical lithography has been the technology of choice of IC manufacturing. What are the limits of this lithographic approach? The resolution limit as described in Eq. (8.1) outlines obvious challenges. Lower wavelengths require new light sources, lens materials, and resists. Higher numerical apertures at the existing wavelength of 193 nm require the development of higher refractive index immersion fluids and lens materials.

To make the situation worse, any improvement in resolution is always accompanied by a decrease in depth of focus (DOF). According to the Rayleigh criterion, the DOF for small features should decrease as the feature size squared. In reality, empirical results have shown the DOF to decrease as about the feature size to the first power (due to improvements in photoresists and other factors). Today's 40 nm features typically have a DOF of 100 to 200 nm. An important requirement of any improvement in the "practical" resolution of an imaging system is the ability to live within the confines of this reduced DOF. Improvements in wafer and mask flatness, autofocusing and autoleveling systems, and wafer planarization by chemical-mechanical polishing of the wafer are examples of how the industry is coping with reduced DOF.

Attempts to simultaneously improve the resolution and DOF by optical means—sometimes called optical "tricks"—include (Fig. 8.11):

- Optimization of the mask pattern shape, called optical proximity correction (OPC)
- Optimization of the angles of light illuminating the mask, called off-axis illumination (OAI)
- Adding phase information to the mask in addition to intensity information, called phase shifting masks (PSM)
- Control of the polarization of the illumination (a requirement for hyper-NA imaging tools)

Collectively, these optical approaches are known as resolution enhancement technologies (RETs) and they can reduce the resolution limit as seen in Eq. (8.4) by nearly a factor of 2.

One option to improving resolution further is to take a big leap in wavelength reduction, down to 13.5 nm and the realm of extreme ultraviolet (EUV). For more than 15 years now, conventional wisdom has put EUV lithography as the long-term answer to lithography's scaling needs. But the difficulties of that technology continue to delay its introduction, and it is still not clear if it will ever be manufacturing-ready.

In the meantime, the semiconductor industry has embraced as alternative strategy: multiple patterning. Multiple patterning takes advantage of the fact that the resolution limit of optical lithography is a pitch resolution limit: the smallest pitch that can be printed is something close to $0.5\lambda/NA$. But there is no specific limit as to how small an individual feature can be made; the feature resolution limit is based on how well process errors can be controlled. Thus, it is possible to print a pattern of lines and spaces at the minimum pitch, but with the lines having a dimension equal to about a quarter of the pitch. Such printing enables two approaches to double patterning: litho-etch-litho-etch (LELE) and self-aligned double patterning (SADP).

LELE performs lithography and etching twice, with the second pattern shifted by half of the pitch compared to the first pattern (Fig. 8.12). Conceptually simple, LELE involves significant difficulties. The first is cost: lithography and etch performed twice costs twice as much. Since Moore's Law is based on the ability to lower the cost per transistor as feature sizes decrease, a significant rise in patterning costs puts Moore's Law in jeopardy. The second problem is overlay: any positioning error of the second pattern compared to the first results in errors in the space widths between the lines. Practical use of LELE has required significant improvements in the positional accuracy of lithography tools. The advantage of LELE is its flexibility: many different shapes and pattern sizes are possible.

SADP (also called sidewall-spacer double patterning) uses a very different approach than LELE. The first lithography and etch step produces a dummy pattern, sometimes called a mandrel. A layer such as silicon dioxide is then deposited over the mandrel, often using atomic layer deposition. An etchback process removes the oxide everywhere except a thin layer on the steep sidewalls of the mandrel. The mandrel is then

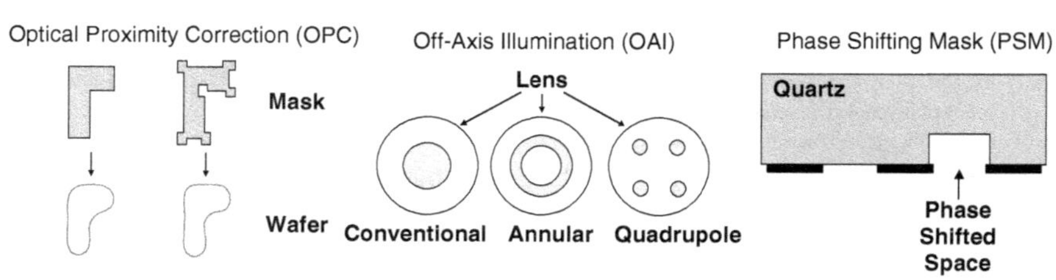

Figure 8.11 Examples of several types of resolution enhancement technologies (RETs).

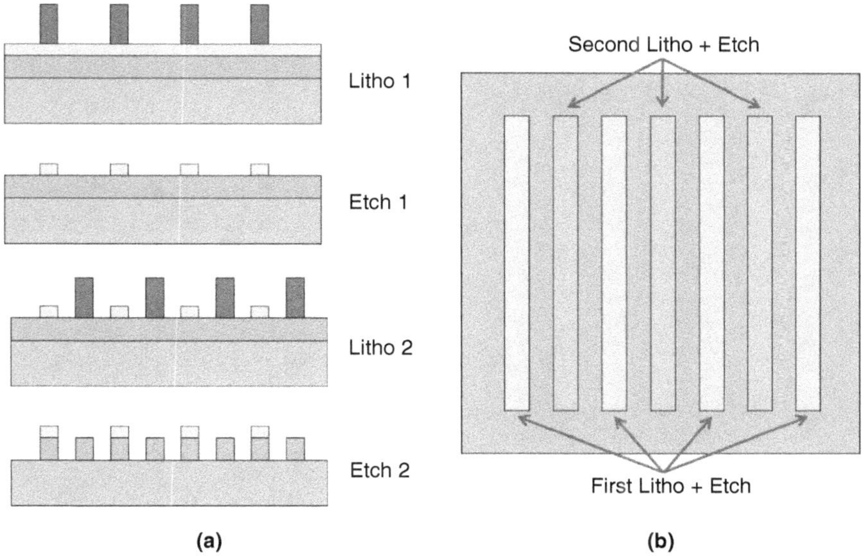

Figure 8.12 Litho-etch-litho-etch produces a final pitch equal to half the pitch of each individual lithography step: (*a*) sequence of major steps shown in cross section, and (*b*) final result shown top-down.

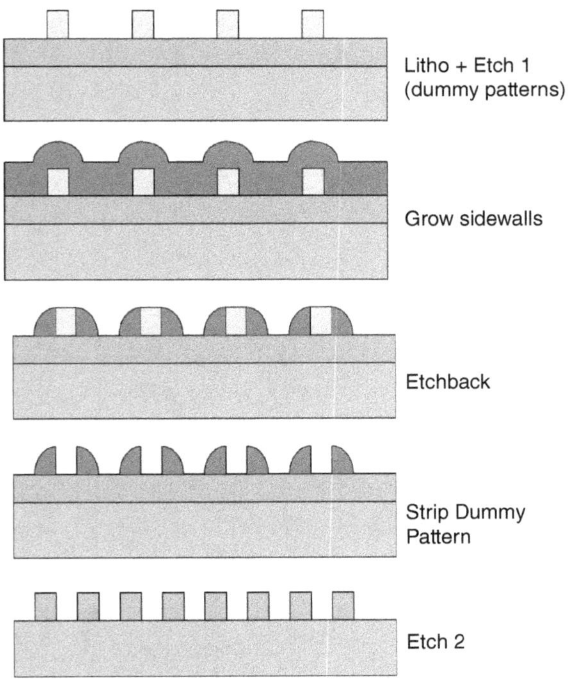

Figure 8.13 Self-aligned double patterning (SADP) sequence of steps for double patterning to produce a final pitch equal to half the pitch of the first lithography step.

chemically stripped away, leaving only the oxide sidewalls, one feature for each edge of the original pattern (Fig. 8.13). Sidewall features necessarily create connected loops, so a second lithography step (often less critical than the first) must be used to trim the ends of the loops away. SADP is cheaper and easier than LELE, and is less sensitive to pattern positioning errors. Its main disadvantage is its lack of flexibility: all features are essentially the same size, and some shapes are not possible to create using SADP.

Multiple patterning has been embraced in virtually all leading-edge semiconductor manufacturing. Memory (DRAM and Flash), with its regular, repeating patterns, is ideally suited for SADP. Some of the most advanced Flash products even use self-aligned quadruple patterning (SAQP), which essentially performs SADP twice, allowing the pitch to be reduced by a factor of 4. Logic devices built at foundries, however, require more flexibility in their designs and thus have (rather reluctantly) used LELE for dense metal patterns. There is even talk of performing more litho-etch steps (LELELE or LE^3) to reduce the pitch further.

In the end, the future of lithography will be determined by the interplay of technical capability and cost. For over 50 years optical lithography has been able to shrink feature sizes cost-effectively to keep Moore's law alive. Unfortunately, the end of this trend may be near.

8.7 FURTHER READING

Dammel, Ralph, "Diazonaphthoquinone-Based Resists," *SPIE Tutorial Texts*, Vol. TT 11, Bellingham, WA, 1993.

Mack, Chris A., "Field Guide to Optical Lithography," *SPIE Field Guide Series*, Vol. FG06, Bellingham, WA, 2006.

Mack, Chris A., *Fundamental Principles of Optical Lithography: The Science of Microfabrication*, John Wiley & Sons, London, 2007.

Mack, Chris A., http://www.lithoguru.com/

Chapter 9
Etching

BY

NANDITA DASGUPTA *Department of Electrical Engineering, Indian Institute of Technology Madras, Chennai, India*

9.1 INTRODUCTION

Etching, one of the most important process steps in semiconductor technology, essentially means selective removal of material from a given substrate. Etching can be classified in various different ways. Depending on the medium or environment of etching, it can be termed either as "wet" or "dry." In wet etching, the substrate is submerged in an etching solution. A chemical reaction between the substrate and the species in the etching solution takes place and the reaction products (usually soluble in the etching solution) are removed from the substrate. Dry etching is also often termed as plasma etching because here the substrate is exposed to a plasma (partially ionized gas) containing the etching species. The actual mechanism in a dry etching process may be quite diverse as will be discussed later.

Another way of classifying the etching process is by means of the sidewall geometry. In a typical etching process, parts of the substrate are protected (or masked) by a suitable masking material that will not be etched (Fig. 9.1*a*). When the substrate is subjected to etching, only the unprotected parts of the substrate come into contact with the etchant species. Ideally the etching should progress vertically with perfectly straight sidewalls at the edges of the mask as shown in Fig. 9.1*b*. However, in reality, the etching may progress both vertically and laterally leading to a sidewall profile shown in Fig. 9.1*c*, with a large undercut. Thus, the profile may be isotropic where the rate of vertical etching is equal (or comparable) to that of the lateral etching. The degree of anisotropy (also called directionality) in an etching process is usually defined as 1–dH/dV, where dV is the vertical etch rate and dH is the rate of lateral etch. For a perfectly anisotropic etching process, dH is very small and hence the degree of anisotropy is close to 1. By contrast, if the rates of vertical and lateral etching are nearly equal, the degree of anisotropy becomes nearly zero and the etch profile is isotropic.

While anisotropy is a desirable feature, for certain applications, a perfectly vertical sidewall may pose some problems. For example, if a film has to be deposited on an etched surface with a vertical sidewall, proper step coverage may not be achieved. A slightly sloping sidewall may be required for better step coverage. Thus, it is important that one should be able to tailor the sidewall geometry in order to achieve the desired degree of anisotropy.

Selectivity is another figure of merit associated with etching process. Selectivity (S) refers to the ratio of the etch rate of the substrate to that of the mask material. Mathematically, $S = R_1/R_2$, where R_1 and R_2 are the etch rates of the substrate and mask, respectively. A selectivity of 30 to 50 is usually considered reasonable.

The actual mechanisms involved in etching play a major role in determining the anisotropy as well as the selectivity of a particular etching process. For example, if the etching takes place purely by chemical reaction between the substrate and the etchant, then the profile is likely to be isotropic because the etchant species react with the substrate equally on all the exposed surfaces (vertical as well as lateral). However, selectivity of this process is usually very good because the etchant species react only with the substrate and not with the mask material. By contrast, if the etching process is physical (like ion bombardment), then only the surface bombarded by the energetic ions, will be etched. The process therefore has good anisotropy. However, because there is no chemical reaction involved and the material is removed by physical ablation, the etch rates of the substrate and the mask material are comparable and hence the selectivity is usually poor.

The present-day etching technology allows us to tailor the sidewall geometry as well as achieve good selectivity. A variety of etching systems are available today as will be discussed in the subsequent sections.

9.2 WET ETCHING

In the early days of semiconductor processing, wet etching was used exclusively. It is a simple and inexpensive technique. The selectivity is also very good. However, as this etching process is mostly dominated by chemical reaction, only a limited degree of anisotropy can be obtained at best. Further, the substrate is etched in corrosive acid or alkaline solution; the reaction products are also corrosive and hence proper safety protocol must be followed while carrying out wet etching. Moreover, it cannot be easily automated.

There are three subprocesses during wet etching:

1. Transfer of etchant species to the substrate surface. This process takes place by the diffusion of reactants through a stagnant layer covering the substrate.
2. Surface reaction.
3. Transport of reaction product(s) from the surface, which can be facilitated by stirring.

Any one of these three subprocesses may limit the etch rate and thus dominate the overall etch process. For example, if the etching is diffusion limited, it takes place uniformly in all directions and tends to be

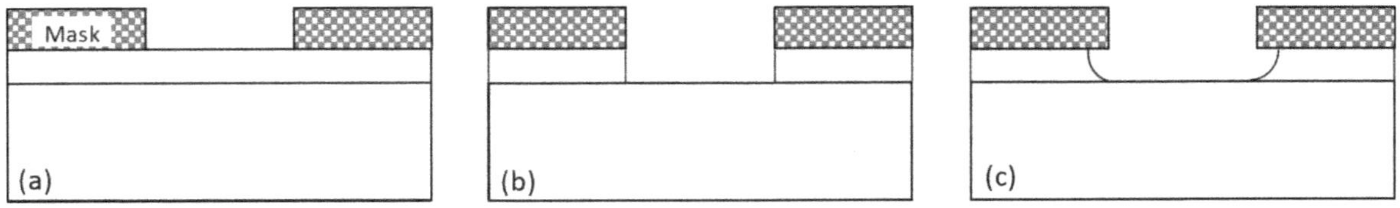

Figure 9.1 Schematic diagram of (*a*) a patterned substrate with some parts protected by mask; (*b*) ideal anisotropic etch profile showing vertical sidewall and no undercut; (*c*) isotropic etch profile with comparable lateral and vertical etch rate, showing undercut.

isotropic in nature. On the other hand, if the process is reaction-rate limited, the etching progresses differently on different crystallographic planes and hence it is anisotropic in nature.

9.2.1 Isotropic Wet Etching of Silicon

Wet chemical etching of silicon usually proceeds by oxidation followed by dissolution of the oxide. Silicon can be etched isotropically by using a solution of nitric acid (HNO_3), hydrofluoric acid (HF), and acetic acid (or water), commonly known as HNA. The three different constituents of the solution perform different actions and the etch rate can be controlled by varying the ratio of them. In this process, nitric acid oxidizes silicon, while HF etches the silicon dioxide. Acetic acid is used as a buffering agent. The oxidation reaction is essentially anodic in nature [1]. Though there is no clearly defined anode or cathode, points on the silicon surface act randomly as local anode and cathode over a period of time. At the local anode, the reaction is given by

$$Si + 2h^+ \rightarrow Si^{2+}$$

The holes required for this reaction are supplied by the reaction at the local cathode in which NO_2 plays an important role. HNO_3 contains trace impurities of HNO_2 which acts as a catalyst in this chain of reactions:

$$HNO_3 + HNO_2 \rightarrow N_2O_4 + H_2O$$

$$N_2O_4 \leftrightarrow 2NO_2$$

$$2NO_2 \leftrightarrow 2NO_2^- + 2h^+$$

HNO_2 is regenerated by reacting with the hydrogen ions created by the dissociation of water present in the etching solution. Thus the etching process is autocatalytic.

$$H_2O \leftrightarrow H^+ + OH^-$$

$$2NO_2^- + 2H^+ \leftrightarrow 2HNO_2$$

Si ions formed at the local anode reacts with the hydroxyl ions to form SiO_2 given by the chemical reaction below.

$$Si^{2+} + 2OH^- \rightarrow Si(OH)_2 \rightarrow SiO_2 + H_2$$

The SiO_2 is then dissolved in HF by the following reaction

$$SiO_2 + 6HF \rightarrow H_2SiF_6 + 2H_2O$$

The overall reaction is given by

$$Si + HNO_3 + 6HF = H_2SiF_6 + HNO_2 + H_2O + H_2$$

Etching will be uniform if each point on the surface behaves as a local anode and cathode for approximately equal duration. Doping concentration and surface defects on silicon substrate as well as the temperature of the etching solution and adsorption at the Si-etchant interface will play important roles in determining the etch rate. A large number of HNA formulations with different etch rates is available in literature and a selection is presented in Table 9.1.

Table 9.1 A Few Typical Etchant Compositions and Etch Rates for Wet Etching of Silicon

Etchant	Composition	Etch rate	Reference
HNA (CP-4)	$HF:HNO_3:CH_3COOH$::3:5:3	80 μm/min	1
HNA (planar etch)	$HF:HNO_3:CH_3COOH$::2:15:5	5 μm/min	1
KOH in isopropanol	250 g KOH with 200 g propanol and 800 g water at 80°C	1 μm/min for (100). Stops at p^{++} layer	2
EPW	Ethylenediamine 7.5 mL with 6 g of pyrazine per liter of ethylenediamine, pyrocatechol 1.2 g, water 0.5 mL at 118.5°C	0.75 μm/min for (100). Stops at p^{++} layer	3

9.2.2 Anisotropic Wet Etching of Silicon

Anisotropic wet etching of crystalline silicon can be carried out using a variety of etching solutions such as KOH, hydrazine, EPW (ethylenediamine, pyrocatechol, and water), and TMAH (tetra methyl ammonium hydroxide) [2,3,4]. The anisotropy in the etch profile is obtained by the fact that {111} planes etch at a much smaller rate compared to the other planes. The slow rate of etching of {111} planes has been attributed to two reasons: first, the density of broken (dangling) bonds is smallest in the {111} plane and hence if the etching process is reaction-rate limited, it is expected that {111} planes will etch at a very slow rate. Second, {111} planes oxidize at a much faster rate. Thus in absence of a strong dissolution agent like HF, which removes SiO_2, {111} surfaces are passivated quickly by the oxide, preventing further etching. Because {100} planes at the surface meet the {111} planes in the bulk (sidewall) at an angle of 54.74°, etching results in a V (or U) groove as shown in Fig. 9.2.

KOH-based etchants show good anisotropy and good selectivity over SiO_2 but is not compatible with CMOS process. The etch rate strongly depends on the temperature, concentration of KOH, and the doping in silicon. The etch rates of (111) and (100) silicon are typically in the ratio of 1:400. The etch rate is negligible for heavily *p*-doped silicon and

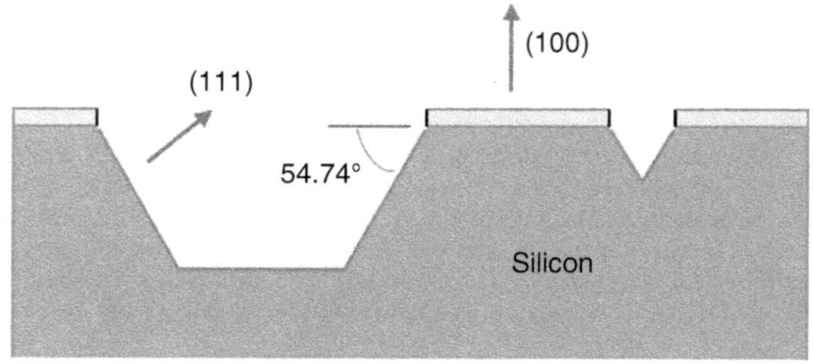

Figure 9.2 Schematic diagram of a V (U) groove etched in (100) silicon. The sidewalls are (111) while the surface plane (and the bottom in case of U groove) is (100).

hence p^{++} silicon is often used as an etch-stop layer for KOH etching. EPW (also known as EDP) with an etch selectivity of 5000:1 with SiO_2 is another popular anisotropic etchant for silicon. Addition of pyrazine is an important factor in determining the etch rate as the presence of even trace amount of pyrazine can enhance the etch rate significantly. Commercially available ethylenediamine usually contains traces of pyrazine and thus the etch rate is liable to be different for different batches of the chemical. However, the etch rate typically saturates at 6 g of pyrazine per liter of ethylenediamine. Hence the common practice is to prepare the EPW solution by deliberately adding 6 g of pyrazine per liter of ethylenediamine, so as to make it independent of batch-to-batch variation. Etching is carried out in boiling EPW solution (at 118.5°C). However, as the EPW vapors are toxic, care must be taken while using this etch solution. Typical etchant compositions and etch rates for anisotropic etching of silicon are also provided in Table 9.1.

9.2.3 Wet Etching of SiO_2 and Si_3N_4

SiO_2 is etched in HF, forming a soluble product H_2SiF_6. However, ammonium fluoride (NH_4F) is often added to the etch solution as a buffering agent. NH_4F prevents the depletion of fluoride ion and also reduces the etch rate of photoresist in the etch solution. Thus, photoresist can be used as a mask while etching SiO_2 in this etchant. This etch solution is commonly referred to as buffered oxide etch (BOE) or buffered HF (BHF). Typical etch rate of thermally grown SiO_2 in BHF is about 100 nm/min. However, the etch rate of deposited oxides is much higher and generally depends on deposition conditions.

Silicon nitride can also be etched in HF or BHF. However, the etch rate is typically much smaller than that of SiO_2 and may need elevated temperature (80 to 90°C). Consequently, photoresist or SiO_2 cannot be used as a mask against etching. To obtain selectivity over SiO_2, boiling orthophosphoric acid (H_3PO_4) can be used to etch Si_3N_4. Typically the etch rate of CVD Si_3N_4 is about 10 nm/min, while that of thermally grown SiO_2 and Si is approximately 1 nm/min and 0.3 nm/min, respectively.

9.2.4 Wet Etching of Metal Films

Metal layers such as Al, Au, Ag, and Cr are used for making Ohmic and Schottky contacts. These metal films need to be etched selectively for patterning. Al can be etched easily either in diluted HCl or in a mixture of H_3PO_4, HNO_3, and CH_3COOH. The etch rate depends on the exact composition (decided by considerations such as adhesion, line definition, etc.) and is in the range of 300 to 1000 nm/min [4]. Aqua regia ($HCl:HNO_3$::3:1) is commonly used to dissolve gold but photoresist cannot withstand this etchant. An alternative etchant for Au is 4 g of KI with 1 g of I_2 in 40 mL of water, which can be used with photoresist as mask. The etch rate is about 0.5 to 1 μm/min at room temperature. Silver can be etched in $NH_4OH:H_2O_2:CH_3OH$::1:1:4 solution. The etch rate is about 350 nm/min at room temperature and photoresist can be used as mask [1]. A standard etchant for chromium, which is often used as adherence promoter or as photomask, is a mixture of perchloric acid ($HClO_4$) and ceric ammonium nitrate ($(NH_4)_2[Ce(NO_3)_6]$. A composition of ceric ammonium nitrate:perchloric acid:H_2O = 10.9%:4.25%:84.85% provides an etch rate of approximately 60 nm/min at room temperature [5].

9.3 DRY ETCHING

Dry etching is extensively used in today's semiconductor manufacturing. The most important advantage of dry etching is the tailoring of the etch profile. Particularly as the device dimensions have been scaled down, it is necessary to be able to etch through a small window without significant undercut, to achieve smaller and more tightly packed features. Also dry etching process can be easily automated and does not use corrosive etchants like HNO_3 or KOH.

Dry etching is carried out in a plasma environment and hence is often termed as plasma etching. Plasma is a partially ionized gas with equal number of positive charges (ions) and negative charges (electrons). The degree of ionization is small (~1 in 10^6). As electrons have low mass, energy transfer from electrons to the plasma is very inefficient. Consequently, electrons can acquire very high energy (equivalent to an electron temperature of 10,000 K), while the plasma envelope remains at a low temperature (close to room temperature). This allows high temperature-type reactions to take place in a low-temperature ambient and generate free radicals. Also, the applied electric field directs the ionic species present in the plasma to bombard the substrate surface, providing directionality of etching.

Thus, the actual role of plasma in the etching process can be widely different. Depending on whether the generation of the chemically reactive species or the ion bombardment dominates during a particular etch process, a variety of etch profiles can be generated. These aspects will be discussed in the following section.

9.3.1 Basic Plasma Etching System [6, 7]

Let us first take a look at a simple plasma reactor, shown in Fig. 9.3. It consists of two parallel plate electrodes in a chamber typically held at

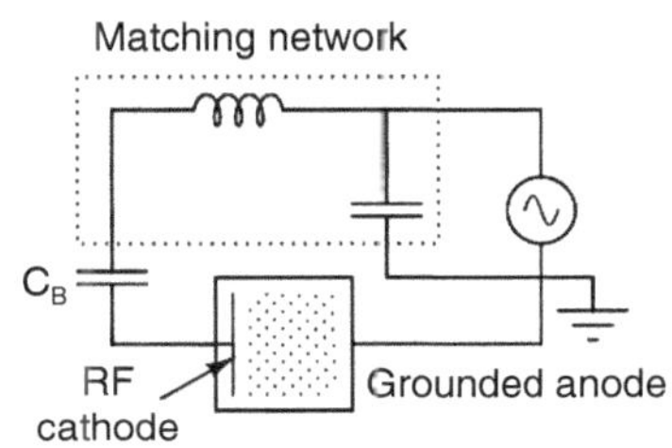

Figure 9.3 Schematic diagram of a parallel plate plasma reactor.

low pressure (1 mtorr to 1 torr) of the reactant gas. A high-frequency voltage is applied between the two electrodes, ionizing the gas. A plasma, containing both positively charged ions and electrons, is created, emitting a glow, characteristic of the gas used. The radio frequency (RF) generator used to supply the energy operates typically at 13.56 MHz. Charges diffuse toward the walls and electrodes of the chamber. As electrons diffuse faster, an excess of positive charges remain behind in the plasma. Therefore, a positive plasma potential (V_p) is created. The region adjacent to the walls (or electrodes) is depleted of charges and is called a "sheath." Since electrons move faster, insulating walls will charge negatively till the repulsive sheath field is intense enough to block further diffusion of electrons to the walls. The wall is then said to acquire a "floating potential" (V_f).

At the conducting boundaries, that is, the electrodes, the situation is a little different. Typically one of the electrodes (anode) is grounded while the other electrode is connected to the RF source through a blocking capacitor (C_B), which does not allow the electrode to discharge through the power supply. During the positive half-cycle, a large electron current will flow toward the powered electrode (traditionally called cathode). Electrons lost to the electrode will be continuously replenished from the central region of the plasma. However, during the negative half-cycle, the cathode will repel the electrons. As the ions are slow to move, a sheath (i.e., a dark region depleted of charge carriers) is developed adjacent to the electrodes, separating the electrodes from the positively charged plasma. During the first few cycles, more electrons than ions will reach the electrodes and a steady-state (DC) sheath potential, negative with respect to the plasma potential, will be developed. Although the heavy ions cannot respond to the RF signal, they do respond to the average DC sheath potential and bombard on the electrodes with high energy. As the bombardment is normal to the electrode surface, substrates placed on the electrodes will experience directional etching as discussed in Sec. 9.3.2.

As already mentioned, during the positive half cycle, electrons carry the current to the electrode with a small voltage drop. However, during the negative half cycle, continuity through the sheath is maintained by a displacement current (rather than the movement of free charge carriers) as the conductivity of the sheath is very low. Thus, the electrode-plasma equivalent circuit can be modeled as a diode connected in parallel to a sheath capacitance. When both electrodes have equal area, the sheath capacitances and the steady-state voltage distribution at both the electrodes will be equal, as shown in Fig. 9.4. However, if the powered electrode has a smaller area, the sheath capacitance will be smaller. Consequently, the impedance will be larger and the voltage developed across that sheath will be more. Therefore, ion bombardment at the smaller electrode will be more intense. Figure 9.4 shows

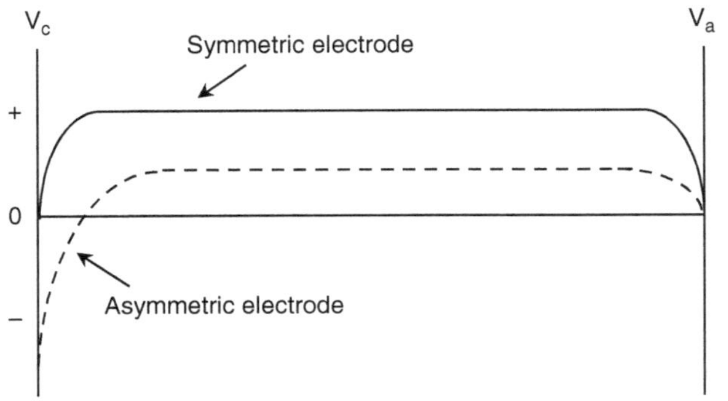

Figure 9.4 Potential profile in a plasma reactor.

the potential distribution at various points inside the plasma reactor for both symmetric and asymmetric electrode configuration. As seen from this figure, the cathode sheath potential (V_c) can acquire a very high negative value, when the area of the cathode is smaller. Most of the applied potential will be dropped across the cathode sheath and determine the maximum energy of ion bombardment. This high-energy ion-bombardment plays a significant role in etching. On the other hand, the sheath potential developed across the larger anode (grounded electrode) is comparatively less and substrates placed on this electrode will experience less intense ion bombardment.

The pressure in the reactor chamber influences the sheath potential in a complex manner. At higher pressure, as more molecules of the gas are available, the degree of ionization is more, that is, more pairs of ions + electrons are created and the plasma density increases. As the number of free electrons increases, the sheath potential also increases. At the same time, higher pressure reduces the mean free path of the electrons. Thus, the electrons are not able to acquire sufficiently high energy and the degree of ionization decreases. Two opposing mechanisms are in action here. Typically in the pressure range of 1 to 100 mtorr, the number of free electron and consequently the plasma density increases with increase in pressure. However, at higher pressure, plasma density decreases [8]. On the other hand, the sheath potential is also a function of the electron energy. Thus at lower pressure, as electrons acquire higher energy, the sheath potential usually increases. Also lower pressure means fewer ion collisions while the ions are traveling across the sheath, and hence the directionality of ion bombardment also increases.

In standard plasma etching systems, the concentration of ions and/or electrons in the plasma (also known as plasma density) and the ion bombardment energy are closely linked together, as they both depend on the RF power. As the RF power increases, degree of ionization increases, increasing the plasma density. At the same time, the sheath potential increases, increasing the ion bombardment energy. In another class of plasma etching systems, also known as high-density plasma etching, the plasma density and the sheath potential can be controlled independently by using a second excitation source, as will be discussed later.

9.3.2 Etching Mechanisms Using Plasma

As already pointed out, etching in a plasma environment can be anything from purely chemical to purely physical. The reactive chemical species created in the plasma are responsible for the chemical etching process, while the ion bombardment across the sheath provides the physical component. Let us now look at the four basic etching mechanisms that can take place in a plasma etching system [6, 9]. Figure 9.5 depicts the different etch mechanisms schematically.

Physical Etching or Sputtering

This process involves physical removal of material. A high sheath potential is created by applying a large RF power and having a cathode with smaller area. Ions bombard the substrate surface with high energy and eject surface atoms. Low pressure facilitates the process by providing high ion energy. Very good anisotropy is achieved as most of the ion bombardment is normal to the substrate surface. However, as the process involves physical removal of material, selectivity is usually poor.

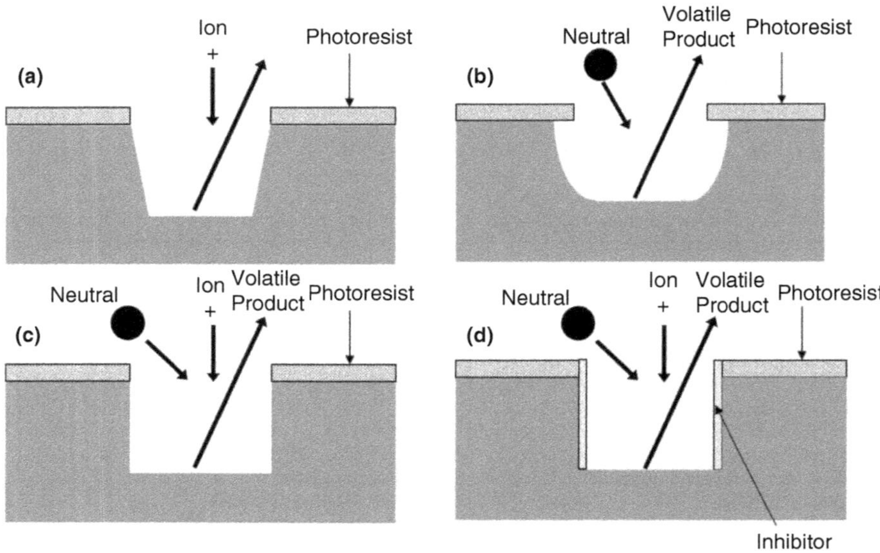

Figure 9.5 Schematic diagram showing possible etch mechanisms during dry etching: (*a*) physical etching, (*b*) chemical etching, (*c*) energetic ion-enhanced etching, and (*d*) ion-inhibited etching. (*Adapted from [9].*)

CHEMICAL ETCHING

Here the role of the plasma is limited to creating chemically reactive species. These chemically reactive species then react with the substrate forming a volatile product. A common example is etching of silicon in CF_4 plasma. Here reactive fluorine is created by the following chemical reaction:

$$CF_4 + e^- \rightarrow CF_3 + F^* + e^-$$

The F^* thus created has an incomplete bonding structure with seven electrons in its outer shell. Therefore, it reacts quickly with silicon forming a volatile by product SiF_4, which is removed from the system. The selectivity of such an etching process is excellent, but the anisotropy is rather poor because etching progresses equally on all the exposed surfaces.

ENERGETIC ION-ENHANCED ETCHING

In this process, impinging ions increase the reactivity of the substrate surface. As the ion bombardment is normal to the substrate, only the reactivity of the horizontal surface is modified. Chemical species present in the plasma can now react with this modified surface, resulting in etching. It must be emphasized that ion-enhanced etching is not merely an addition of physical and chemical etching process but a synergistic combination of the two. Here, usually the ion energy is much less than what is required for physical removal of substrate material; that is, the physical component of etching is very small. Very good selectivity is obtained as the etching is primarily chemical in nature. However, as the reactivity of the sidewalls (parallel to ion bombardment) is not affected, no etching takes place on sidewall surface and excellent directionality is achieved. Etching of undoped silicon in chlorine plasma is an example of ion-enhanced etching. In absence of plasma, undoped silicon does not react spontaneously with chlorine. However, when there is an associated ion bombardment, a rapid reaction takes place between chlorine and silicon, forming $SiCl_4$, a volatile product.

ION-INHIBITED ETCHING

In this process, two different species are required. One of them is called etchant while the other is called an inhibitor. The substrate reacts spontaneously with the etchant to form a volatile product. In absence of the inhibitor, this etching reaction would have resulted in isotropic profile. However, the inhibitor forms a thin film on the substrate surface. Ion-bombardment normal to the surface removes the film from the horizontal surfaces allowing etching to continue in the vertical direction. As there is no ion bombardment on the sidewalls, the inhibitor film is retained there and the sidewalls are protected providing excellent directionality as well as selectivity. An example of such an etching process is etching of n^+-doped polysilicon in C_2F_6/Cl_2 plasma. Here Cl_2 is the etchant while C_2F_6 is the inhibitor. n^+ polysilicon reacts spontaneously with chlorine while species from C_2F_6 form a polymer film on the sidewalls to provide directionality.

9.3.3 Dry Etching Systems

As already discussed, the sheath potential and consequently the ion bombardment will depend on the electrode area, pressure in the reactor chamber as well as whether the electrode is powered or grounded. Based on the reactor geometry and positioning of the substrates, etching systems can be broadly classified into the following four categories:

BARREL ETCHERS

This is the oldest form of plasma reactor where the reactor is a quartz chamber. RF power is supplied by an external coil wrapped around the chamber as shown in Fig. 9.6. The wafers are held in a floating cassette inside the reactor. The pressure is usually 0.1 to 2.0 torr.

Sheath potential to the wafer is small and ion bombardment plays a relatively minor role in etching. The etching is therefore, primarily isotropic. Etch uniformity is also relatively poor. The barrel etcher is used today for primarily stripping ("ashing") photoresist in oxygen plasma.

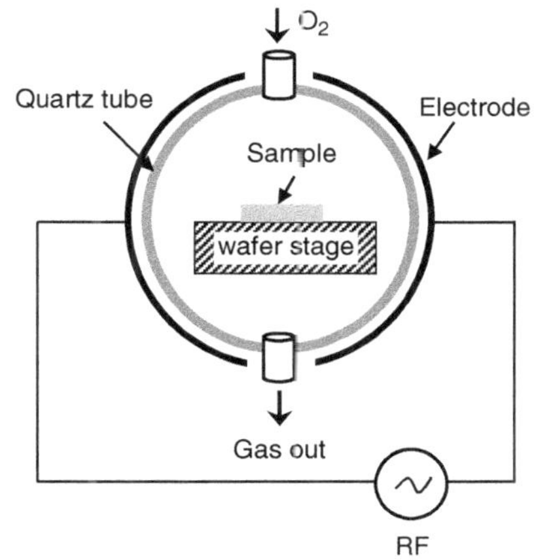

Figure 9.6 Schematic diagram of a barrel etcher.

PARALLEL PLATE ANODE LOADED ETCHER

This class of reactors is commonly known as plasma etching system, where the substrates are placed on the grounded electrode. Since the chamber walls are also usually grounded and thus electrically connected to this electrode, the effective area of the anode is much larger though the electrodes themselves may be equal in size. Thus the sheath potential developed across the anode is typically less than 20 V, and etching is essentially chemical in nature. Good selectivity is obtained but the anisotropy is relatively poor unless special inhibitor-forming species are used in the gas stream.

ASYMMETRIC PARALLEL PLATE CATHODE LOADED ETCHER

This class of reactors is also known as reactive ion etching system. However, this is a misnomer because free radicals (rather than ions) are usually the reactive species. Wafers are loaded on the smaller cathode, while the chamber itself may be the anode with a much larger area. Cathode sheath potential is usually quite high (20 to 500 V). The gas pressure is usually low (0.005 to 0.1 torr) so that ion bombardment energy is even higher. Good selectivity as well as directionality can be achieved.

HIGH-DENSITY PLASMA SYSTEM

In this class of system, the plasma density and ion bombardment energy are delinked by applying a second excitation source. The wafer is placed on a RF-powered electrode, while the plasma is created by another source usually a little above the wafer. Different types of high-density plasma etchers are available using a variety of sources for the generation of plasma, for example, inductively coupled plasma (ICP), electron cyclotron resonance plasma (ECR), surface wave plasma (SWP), and helicon [10]. Of these, ICP-RIE systems are the most popular today due to its simpler design. In these systems, the plasma generation source is inductively coupled and produces a high-density plasma (hence the name ICP or inductively coupled plasma), while an RF source is used to separately control the sheath potential. One (or more) coil(s), placed outside the chamber, is used to create a high-density plasma. As the plasma density is high, these systems can be operated at very low pressure (1 to 10 mtorr), leading to higher ion flux and larger etch rates. In a standard plasma etching system, the power has to be increased to increase plasma density and ion flux to increase the etch rate. However, at the same time, the sheath potential also increases leading to more damage to the substrate due to ion bombardment. Increasing the pressure increases the plasma density, while keeping the ion energy low but the etch profile becomes more isotropic in nature. By decoupling the sheath potential from plasma density, it is possible to achieve high plasma density (and hence high etch rate), while maintaining good directionality with low ion damage.

Deep Reactive Ion Etching (DRIE)

This class of systems employs a modified form of ion-inhibited etching. In these systems an etchant and an inhibitor are used in the gas stream and the process alternates between etching and polymer deposition. First a jet of etchant gas is introduced, which etches the substrate up to a small depth. Next, an inhibitor is introduced to coat the exposed surfaces with polymer. This cycle is repeated to achieve the desired etch depth. The ion bombardment removes the polymer from the horizontal surface while the sidewalls are protected. Excellent anisotropy as well as high selectivity can be obtained along with high etch rates. DRIE is particularly favored while etching to a large depth through a small window. In a typical DRIE process for etching silicon (schematically shown in Fig. 9.7*a* to *e*), SF_6/Ar is used for etching while CHF_3/Ar is used for polymerization. Selectivities of 100:1 and 200:1 can be obtained using photoresist and SiO_2 as masks, respectively. Etch rates of 2 to 3 μm/min are achieved with an aspect ratio up to 30:1 and sidewall angle of 90 ± 2°. However, because the etching is chemical in nature, at every cycle a small scallop is created as shown in Fig. 9.7*f*.

In another version of DRIE, known as cryogenic dry etching, no inhibitor-forming species are used. Instead, cryogenic cooling of wafer (using liquid nitrogen and liquid helium) is carried out. Reactant gases condense on the sidewalls and protect while condensing gases on bottom of the structure are removed by ion bombardment. The degree of scalloping is less and aspect ratios of 30:1 can be achieved in commercial equipment.

9.3.4 Problems Associated with Dry Etching

Loading Effect

The etch rate in a dry etching system is often nonuniform. It can be classified as (1) macroscopic loading and (2) microscopic loading. In macroscopic loading, the etch rate depends on the number of substrates loaded into the chamber. The etch rate decreases when more wafers are loaded (or more area is exposed for etching), due to depletion of the etching species. Etch rate is proportional to the concentration of the etchant species. If a significant amount of reactive species is consumed in etching reaction, the concentration of the etchant decreases with an increase in the area of the surface. Most isotropic etchants exhibit this type of loading effect. On the other hand, in microscopic loading, the etch rate varies over small distances in the same substrate. This can also be viewed as "geometry dependence"; that is, the etch rate depends on the aspect ratio of the structure to be etched. The microloading effect is seen when trenches of different aspect ratio are etched on a silicon substrate. It can be seen that the etch rate is smaller for trenches with smaller width. This can be due to various reasons such as formation of a stagnant layer leading to depletion of reactant species or geometric shadowing. Whatever be the cause, the net effect is that less etchant species reach the bottom of a narrow trench and hence the etch rate is smaller.

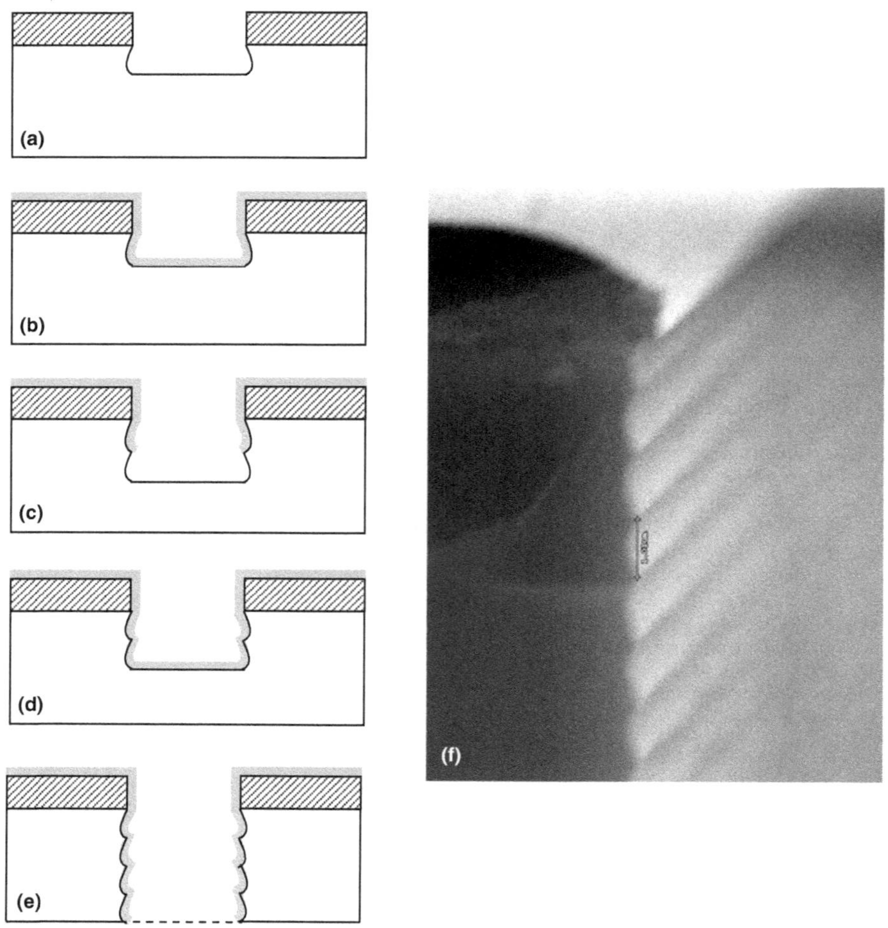

Figure 9.7 (*a*) to (*e*) Schematic diagram of the steps in DRIE. (*f*) Actual etch profile showing scallops in the silicon sample etched by DRIE.

TRENCHING

This is another manifestation of etch non-uniformity, seen mostly in dry etching dominated by high-energy ion bombardment. Some of the ions will bounce off the sidewall and hit the bottom corner, causing enhanced etching at the corner, known as microtrenching as shown schematically in Fig. 9.8.

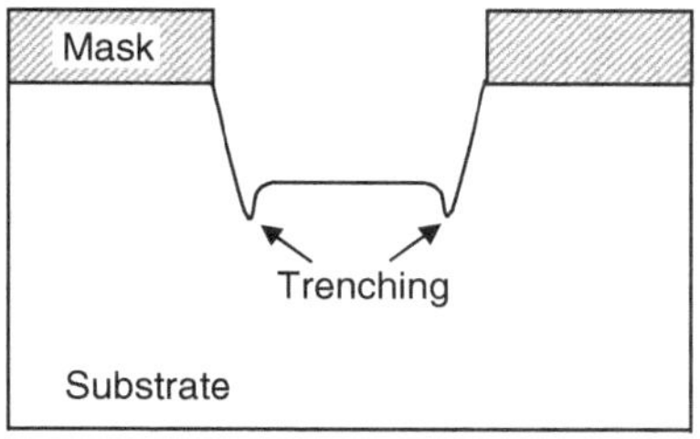

Figure 9.8 Schematic diagram showing microtrenching at the corner of the window.

ION DAMAGE

Ion bombardment, though an essential component of dry etching in many cases, can lead to various undesirable effects, such as creation of traps in the gate oxide, surface charging leading to distortion in the etch profile. This is because the path of the ions will be influenced by the charge on the surface and cause uneven etching.

UNWANTED RESIDUE

Unwanted polymeric residues are often formed in the dry etching system. Sometimes these polymers are deliberately created to obtain anisotropic etching (as in ion-inhibited etching) or they may be due to incomplete removal of the polymeric etch products. To remove these residues, a common practice is to use a selective wet chemical etching after the dry etching.

9.3.5 Dry Etching of Silicon, SiO_2, and Si_3N_4

Dry etching of silicon can be carried out using either fluorine-based or chlorine-based plasma. A variety of etch profiles can be obtained using different gas chemistries as will be discussed subsequently.

FLUORINE-BASED ETCHING

CF_4 used to be the most commonly used feed gas for fluorine etching of silicon. However, today SF_6 is the preferred feed gas though C_2F_6, C_3F_8, etc., are also used. The basic reaction is

$$CF_4 + e^- \rightarrow CF_3^+ + F^* + 2e^-$$

The free F^* radical is the reactive species. As already discussed, it reacts spontaneously with silicon and generates SiF_4, a stable and volatile reaction product.

Addition of a small amount of oxygen (1 to 10 percent by volume) enhances the etch rate as there is a rapid increase in the concentration of F^* due to the reaction between CF_3 and O_2, creating COF_2, CO, CO_2, etc. and generation of more F^*. However, too much oxygen will reduce the etch rate as oxygen will chemisorb on silicon surface making the surface more like an oxide of silicon.

Excellent selectivity over SiO_2 (typically 40:1) can be achieved while etching silicon in fluorine plasma. However, the etch profile is isotropic since the process is chemical in nature. However, using argon along with CF_4, it is possible to obtain some degree of anisotropy. This is because bombardment of argon ions enhances the etch rate in the vertical direction, while in the lateral direction the etch rate is still governed by chemical reaction. CHF_3 can also be added to SF_6 to form inhibitors on the sidewall and improve anisotropy. Figure 9.9 shows the etch profile of silicon etched in SF_6/CHF_3 plasma, where a sidewall angle of 87° has been achieved.

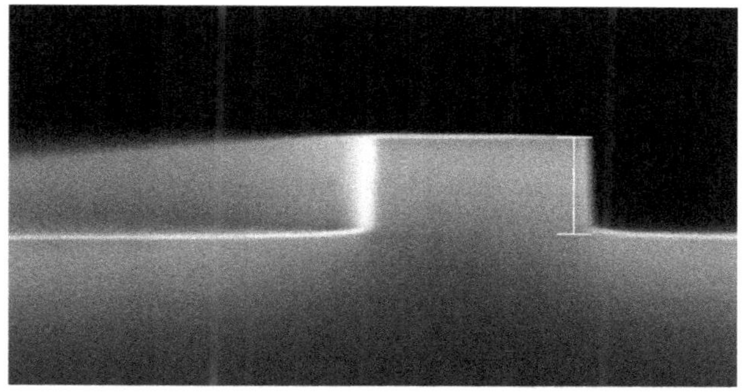

Figure 9.9 Etch profile of silicon in SF_6:CHF_3::5:18 plasma showing a sidewall angle of 87°.

The same feed gas can be used to etch SiO_2. However, CF_3^+ is the main etchant species here. F^* also etches SiO_2 but the etch rate is considerably smaller compared to that of silicon. Thus etching of windows in SiO_2 prior to diffusion/ion-implantation is difficult to control since the etch rate of Si is much larger than that of SiO_2 in the same feed gas. So, to etch SiO_2 selectively over silicon, the following techniques are adopted.

- *End-point detection.* In this technique, the intensity of F^* emission (703.7 nm) is monitored using optical spectroscopy. There is a sudden drop in the F^* intensity as the silicon layer is exposed signifying that the SiO_2 layer is etched completely.
- *Adding H_2 to the feed gas.* Hydrogen will react with F^* forming HF and suppress the etching of silicon. However, disposal of HF is a problem, so fluorine-deficient C_3F_8 is sometimes used to etch SiO_2. Highly anisotropic etch profiles are obtained.

Silicon nitride can also be etched by the free radical F^* according to the following reaction:

$$12F^* + Si_3N_4 \rightarrow 3SiF_4 + 2N_2$$

In CF_4 +4% O_2 plasma, typically the etch rates of Si, Si_3N_4, and SiO_2 are in the ratio of 17:3:2.5. To selectively etch Si_3N_4 while using a SiO_2 mask, a mixture of NF_3:Cl_2::1:2 can be used. Here, ClF or ClF_3 are the reactive species, which do not attack SiO_2. However, etching is isotropic in nature.

CHLORINE-BASED ETCHING

Etching of silicon in chlorine-based plasma is governed by three basic principles:

1. Heavily n-doped silicon (both single and poly crystalline) can be etched spontaneously in chlorine plasma leading to an isotropic profile and severe undercut.
2. However, undoped silicon cannot be etched (or etched very slowly) in chlorine plasma in absence of ion bombardment.
3. In presence of ion bombardment, undoped silicon can be etched in chlorine plasma with anisotropic profile (ion-enhanced etching).

Also since SiO_2 is not attacked by chlorine plasma, excellent selectivity can be obtained.

ETCH PROFILE TAILORING BY CHLORINE-FLUORINE COMBINATION

One of the ways the etch profile can be tailored is by the use of ion-inhibitor mechanism. For example, if Cl_2 is added to C_2F_6 in the feed gas, the etch rate of n^+ polysilicon shows a rapid, linear increase. However, the profile is anisotropic, due to polymer formation on the sidewall (inhibitor) from C_2F_6, till about 15 to 20 percent of Cl_2 is added to the feed. Also, chlorine does not etch SiO_2. Any native oxide present on silicon surface is etched by C_2F_6. Thus, C_2F_6 performs a dual role—it etches the native oxide and also forms polymer to protect the sidewalls. As more chlorine is added, there not enough polymer is formed to protect the sidewalls, and the profile becomes isotropic in nature.

Tailoring of etch profile for undoped (or moderately doped) silicon can be achieved by using a feed gas of ClF_3 and Cl_2. As already mentioned, etching of undoped silicon in Cl_2 plasma results in anisotropic profile. On the other hand, ClF_3 dissociates mostly into F^* and a much smaller percentage of Cl (or Cl_2). Thus, as the percentage of ClF_3 is increased in the feed gas, the etch profile changes from anisotropic to isotropic. Any intermediate profile can be realized by adjusting the feed gas composition.

9.3.6 Etching of III–V Semiconductors

As fluorides of group III elements (Al, Ga, In, etc.) are nonvolatile, fluorine plasma is not usually used to etch III–V semiconductors. Chlorine or bromine can be used as the feed gas to etch these materials. BCl_3 is often added to the feed gas to form an inhibitor film on the sidewalls and provide anisotropy. However, the etch rate reduces as more BCl_3 is added to the feed gas. In addition to forming the inhibitor, BCl_3 also etches Al_2O_3 and facilitates etching of aluminum-containing ternary semiconductors like AlGaAs, AlGaN, etc.

A combination of hydrogen and hydrocarbons can also be used to etch III–V materials. A feed gas of hydrogen with 5 to 25 percent of CH_4 has been used to etch GaAs and InP. High etch rate with good anisotropy has been obtained. The reaction products are organometallic compounds of group III elements [e.g., $Ga(CH_3)_x$, $In(CH_3)_x$] and hydrides of group V elements (e.g., PH_x, AsH_x). It is believed that ion-bombardment enhances the reaction between adsorbed CH_4 and the III–V substrates. However, as the percentage of CH_4 in the feed gas exceeds a certain limit (30 percent), polymer formation is observed and the etch rate reduces.

9.3.7 Etching of Other Materials

Photoresist is etched in oxygen plasma and is termed "ashing." A small amount of fluorinated gas like CF_4 or SF_6 enhances the etch rate by weakening the photoresist polymer structure and also by increasing the concentration of more reactive atomic oxygen. Atomic fluorine reacts with organic polymer and forms HF. In the process, hydrogen is removed and unsaturated sites are left in the organic polymer chain, which reacts more readily with oxygen.

As already mentioned, aluminum fluoride is nonvolatile and, hence, aluminum cannot be etched in fluorine-containing plasma. $BCl_3 + Cl_2$ feed gas is used to dry etch aluminum, with BCl_3 playing a dual role of inhibitor formation and etching of Al_2O_3, normally present on all aluminum surfaces. Other metal films such as W, Ti, and TiN can be etched in both fluorine- and chlorine-containing plasma.

9.4 SUMMARY AND CONCLUSION

Selective removal of material from substrate surface is known as etching. In this chapter, different aspects of the etching process have been discussed. The two figures of merit usually associated with the etching process are selectivity and anisotropy (directionality). If the etching is dominated by chemical reaction between the substrate and the etchant, very good selectivity is obtained. However, the anisotropy is relatively poor as the reaction takes place equally on available surfaces—vertical as well as horizontal. On the other hand, if physical ablation is the dominant etch mechanism, anisotropy is good while the selectivity is poor. Etching can be classified as "wet" and "dry." In wet etching, the substrate is immersed in a chemical solution. As chemical reaction is the dominant mechanism for wet etching, only a limited degree of anisotropy (crystal orientation dependence) can be achieved. Dry etching, where the etching medium is a partially ionized gas (plasma), is a more versatile process with a synergistic combination of physical and chemical etching. Thus, in dry etching, the etch profile can be tailored to suit a particular application.

The actual role of plasma in a dry etching process can be different, giving rise to various dry etching mechanisms. These mechanisms as well as the design of different reactors, which facilitates one or more type of etch mechanisms, have also been discussed in this chapter. Finally, the dry etching of silicon and III–V semiconductors as well as commonly used dielectrics and metals in different plasma has been discussed in detail.

9.5 ACKNOWLEDGMENTS

The author would like to acknowledge the faculty, staff, and students associated with the Centre for NEMS and Nanophotonics, IIT Madras, in general, and Dr. Shanti Bhattacharya, Dr. T. Sreenidhi, and Ms. Sumi R. in particular, for the data and figures depicting actual etch profiles. Special thanks are due to Mr. Gourab Dutta, research scholar, Department of Electrical Engineering, IIT Madras, for technical assistance, and Prof. Deleep Nair, Department of Electrical Engineering, IIT Madras, for critically reviewing the manuscript.

9.6 REFERENCES

1. S. K. Ghandhi, *VLSI Fabrication Principles*, 2d ed., Chap. 9, John Wiley & Sons, Inc., Hoboken, New Jersey, 1994.
2. R. B. Darling, EE527 Microfabrication Wet Etching https://www.ee.washington.edu/research/microtech/cam/PROCESSES/PDF%20FILES/WetEtching.pdf, February 2017.
3. K. N. Bhat, C. Yellampalle, N. DasGupta, A. DasGupta, and P. R. S. Rao, "Optimisation of EDP Solutions for Feature Size Independent Silicon Etching," *Proc. SPIE Symposium on Micromachining and Microfabrication*, vol. 4979, San Jose, USA, January 2003.
4. W. Scot Ruska, *Microoelectronic Processing*, Chap. 6, McGraw-Hill International Editions, New York, 1988.
5. *Chromium etching*, www.microchemicals.com/downloads/application_notes.html, February 2017.
6. Dennis M. Manos and Daniel L. Flamm (eds.), *Plasma Etching: An Introduction*, Chap. 1, Academic Press, Inc., London, 1989.
7. James D. Plummer, Michael D. Deal, and Peter B. Griffin, *Silicon VLSI Technology—Fundamentals, Practice and Modeling*, Chap. 10, Pearson Education Inc. publishing as Prentice Hall, Upper Saddle River, New Jersey, 2001.
8. Patrick Verdonck, "Plasma Etching," http://wcam.engr.wisc.edu/Public/Reference/PlasmaEtch/Plasma%20paper.pdf, April, 2017.
9. Marc Madou, *Fundamentals of Microfabrication*, Chap. 2, CRC Press, Boca Raton, Florida, 1997.
10. Vincent M. Donnelly and Avinoam Kornblit, "Plasma Etching: Yesterday, Today and Tomorrow," *J. Vac. Sci. Technol. A*, 31(5), doi10.1116/1.4819316, Sep./Oct. 2013.

9.7 FURTHER READING

Madou, M., *Fundamentals of Microfabrication*, Chap. 2, CRC Press, Boca Raton, Florida, 1997.

Manos, D. M. and Daniel L. Flamm (eds.), *Plasma Etching—An Introduction*, Academic Press, Inc., London, 1989.

Chapter 10
Ion Implantation

BY

BO VANDERBERG *Axcelis Technologies, Inc., Beverly, Massachusetts*
MIKE AMEEN *Axcelis Technologies, Inc., Beverly, Massachusetts*

10.1 OVERVIEW

10.1.1 Definition of Ion Implantation

Ion implantation is the placement of atoms into a bulk material using a nonthermal process, by controlling a particle beam of ionized atoms or molecules to penetrate and be deposited into the material, with the goal of changing the bulk material's properties. The particle beam is composed of ions, which permits energy control of the particles via electric fields, as well as accurate beam transport using ion optics. The process is nonthermal, in the sense that the particles distribution inside the bulk results from deceleration of the particles via collisional forces, rather than from an equilibrium of transport from thermal or random motion processes, such as diffusion; because of this, the placement of material inside the bulk can be very precise, down to the atomic level. Finally, for semiconductor manufacturing, the material modification desired is electrical doping of semiconductors, and more recently it also targets any property of the bulk material (e.g., diffusion coefficient or electron work function), as well as surface modification.

10.1.2 History

Research into doping of semiconductors was initially performed using particle accelerators developed for particle physics in the 1930s.[1] These large machines produce relatively low ion beam currents of the order of mA. During the Second World War, isotope separators were developed to produce fission fuel, and these machines had much higher beam current, up to 100 mA. In 1954, William Shockley filed a patent application describing the basic components of an ion implantation apparatus for semiconductor doping.[2] Using isotope separator technology with its high productivity, and developing it further to resolve process issues, it became possible after 1970 to perform semiconductor ion implantation on an industrial scale. Ion implantation of semiconductors has since evolved to the mature, well-established commercial application it is today.

10.1.3 Basic Components of Ion Implantation Equipment

Ion implantation equipment is designed to provide a beam of desired ions with well-controlled properties (energy, angles, beam current, etc.) to a wafer, and therefore contains the following building blocks, exemplified in Fig. 10.1:

- An ion source and extraction system. Ion sources generate desired ions from a feed gas, provided in gaseous form via a gas feed system, or from a vaporizer system.[3] Different means to ionize the source gas are used, and by far, the most common is via a Penning trap–type discharge[4] controlled by electron injection from heated cathode.[5–7]
- Because ion sources rely on a plasma chemical process, they typically produce not only the desired ion but also side products resulting, for example, from a "cracking pattern" of the feed gas.[8] To filter out any species of unsought mass or energy, the ion source is typically followed by at least one mass separation system. These "mass analyzers" rely on the difference of magnetic gyro radii[9] of different species to permit the passage of only species with the desired rigidity through a resolving aperture.
- The ion beam is transported and shaped by a train of ion-optical lenses,[10] such as quadrupoles or higher order lenses to tune beam size; electric of magnetic scanners to fan the ion beam into a wide ribbon; angle correction or parallelizing lenses; or deflection lenses for further filtering of contaminants.
- The semiconductor wafer is delivered to a process chamber via a wafer handling system (an "end station") and implanted in the process chamber. The wafer is typically positioned by a wafer holder, which will adjust the implant angle by moving the wafer around tilt and twist axes. The ion beam cross section is typically smaller than the size of the semiconductor wafer; therefore, the implant is brought about by some means of mechanically scanning the wafer through the ion beam. The most common wafer scanning architecture is hybrid scan, where the wafer is swept across a ribbon-like ion beam, whose width, but not height, exceeds the diameter of the wafer. Other architectures utilize two-dimensional mechanical scan, either by stepping the wafer horizontally and vertically,[11] or by swinging the wafer on a pendulum like mechanism[12,13] through a spot ion beam. Implanters using these types of wafer processing are referred to as serial implanters, in that wafers are being processed one at a time in a series of wafers. Older architectures, especially for high ion beam power applications, used multiwafer processing by placing many wafer on a spinning disk in so called batch implanters, but various reasons, such as particulate contamination in this type of tool[14] or the risk of increased number of damaged wafer after accidents caused a shift to serial processing in modern high current ion implantation.

In addition to these basic components, a modern ion implanter is controlled by a complex control system to ensure process control and safety. The control system monitors the ion beam with beam diagnostics; modern ion implanters include a plethora of measurements

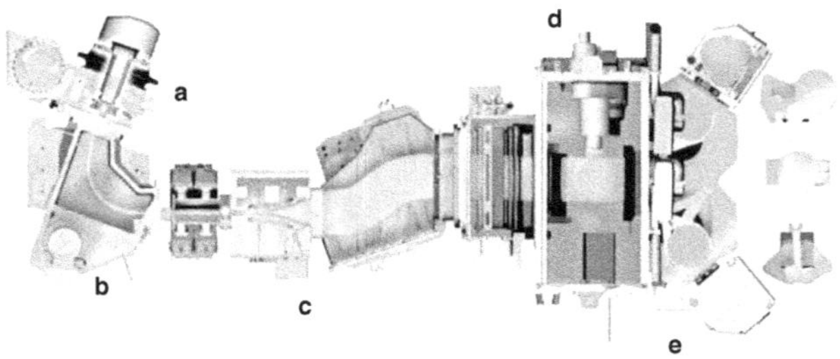

Figure 10.1 Axcelis Purion™ H, exemplifying a typical ion implanter, with serial end station, and (in this artist's impression) the ion beam in blue. (*a*) Ion source; (*b*) mass analyzer; (*c*) beam shaping optics, in this example: quadrupole doublet, scanning optics, parallelizing magnet, and beam deceleration and deflecting optics; (*d*) process chamber; (*e*) end station.

using Faraday cups, sometimes in conjunction with masks and profiling mechanisms. The wafer environment is monitored with pressure gauges, temperature probes, etc. To ensure low contamination, the wafer environment is tightly managed by judicious choice of materials of ion implanter beamline and process chamber components—the dominating engineering materials are aluminum, carbon, and silicon. Charge control systems ensure that floating structures on the semiconductor wafers do not float to uncontrolled high voltages during ion implantation, by neutralizing ion charge and current with a commensurate supply of electrons from an electron source. Modern ion implanters rely on plasma sources, commonly termed plasma electron floods (PEFs), from which electrons are extracted into the ion beam, effectively canceling both the space-charge of the ion beam, and the net electric current to the wafer.[15,16]

10.2 OVERVIEW OF MODERN ION IMPLANTATION EQUIPMENT

No single architecture can cover the five to six orders of magnitude in energy and dose of the ion implantation application space, due to power/safety concerns, and necessary tradeoffs in process quality, throughput, footprint or size, and also cost. The implant segment was therefore driven to three major equipment types: high current, medium current, and high energy.

More recently, ultrahigh-dose implanters have emerged, comprising both plasma ion immersion implantation (PIII) and derivate technologies such as plasma doping (PLAD),[17] as well as molecular ion implant using many dopant atom-molecules (e.g., $B_{18}H_{22}$ instead of atomic B)[18] to improve ion implantation productivity. PLAD can deliver very high productivity at very high doses (> 10^{16} cm^{-2}), albeit at some expense of process quality, and has therefore so far only proven suited to less demanding applications. Molecular technologies typically require the implanter to be dedicated, and can suffer from cost-of-ownership issues. For these reasons, ultrahigh-dose implanters occupy a niche market,[19] and the ion implantation market is dominated by the three types described next.

10.2.1 High Current

High dose (> 5.1014 cm^{-2}) and high wafer throughput requirements created the demand for tens of mA ion beam currents, with the original technical difficulties residing in wafer cooling. While batch tool processing and later advanced chuck cooling techniques[15] solved wafer cooling problems, the energy range of high-current implanters is limited by the maximum power of the wafer cooling system (1 to 2 kW), the beam line can be managed, and the peak energy of a high current implanter is typically limited to < 70 keV. The relatively high-current conversely also represents a limit to the lowest dose this type of implanter can deliver, because the dynamic range of the dosimetry system is finite. Modern implanters, however, have increased the range significantly, and a significant overlap with medium-current implanters exists.[20]

High-current implanters also reach the lowest implant energies dictated by production needs, today in the vicinity of 0.3 to 0.5 keV. Beam transport at these very low energies requires mitigation of beam blowup challenges,[10] and these are today overcome by several means. First, high-current beamlines tends to be the shortest with the largest cross sections; for pencil beam–type beamlines, beam heights are typically more than twice the height of medium-current or high-energy implanters. Today's most widespread high-current beamline, shown in Fig. 10.2, uses ribbon beam technology,[21] where

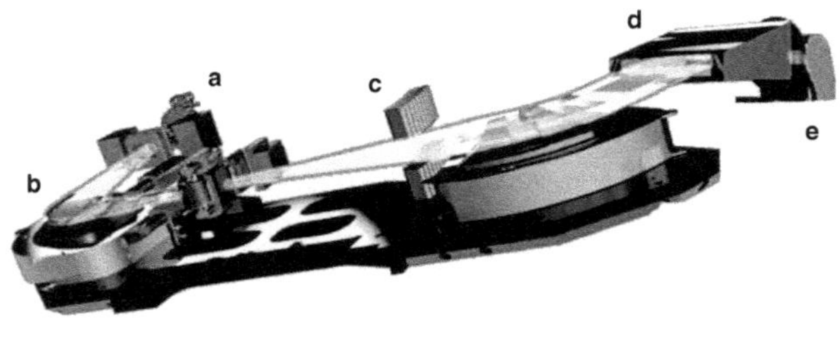

Figure 10.2 High-current beamline (Applied Materials VIISta® Trident), showing ion source (*a*); mass resolving dipole magnet (*b*); multipole lenses (*c*) for beam shaping; and deflecting deceleration stage (*d*) prior to the wafer scanning mechanism (*e*) in this serial implanter.

the ion beam is permitted to expand in one dimension to exceed the width of the wafer, thus creating relatively large beams with reduced space-charge forces. Ribbon beamlines avoid the use of beam scanning, but require more complicated beam shaping optics to ensure beam angle and spatial current uniformity. High-current beamlines can also make use of enhanced beam neutralization techniques, such as electron confinement with multicusp fields.[22] Finally, most modern high-current beamlines transport the ion beam at higher energy to a final deceleration stage, where the energy is reduced to the desired. Final deceleration can represent an energetic contamination challenge if charge-exchange processes upstream generate energetic neutrals that do not experience deceleration and thus implant the wafer at higher than final energy. To reduce this risk, modern implanters use final energy filters to deflect the decelerated beam, such that neutrals (which continue straight) do not reach the wafer.[23]

10.2.2 Medium Current

If lower beam currents, for example, a few mA, are acceptable, the maximum beam energy given by ion beam power is several times higher than high-current implanters', and the highest energy that can be attained for singly charged ions is limited by maximum field stress and steady-state voltage that can reliably be maintained to accelerate these ions; for medium-current implanters, this maximum voltage is typically of the order of several hundred keV. Medium current ion implanters make further use of both distributed acceleration and deceleration to cover a broad energy range. The lowest energies that can be achieved are limited by power supply accuracies and stabilities, typically of the order of a few keV. Multiply charged ions (with typically lower yield from the ion source) are used to further extend the energy range, and the maximum energy using triply charged ions can today reach 1 MeV. The broad energy range makes medium-current implanters the most versatile for having the capability to perform almost any implant required in typical processing, albeit at the expense of optimum productivity. A typical medium-current system is shown in Fig. 10.3.

The dosimetry system for medium-current implanters is tuned to control less than μA currents, and doses down to 10^{11} cm^{-2} can reliably be implanted.[24] Medium-current implanters have excellent angle accuracy and contamination control, and are thus suited for the most critical applications. Modern medium-current system also use charge-neutralization devices, necessary at currents greater than 20 μA.[16]

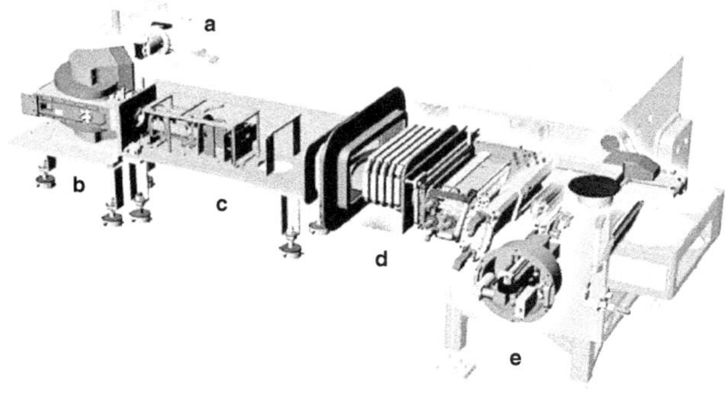

Figure 10.3 A medium-current beamline, Axcelis Purion™ M, showing: ion source (*a*); mass resolving dipole (*b*); beam shaping optics (*c*); acceleration/deceleration stage (*d*); and process chamber (*e*).

10.2.3 High Energy

At higher energies more than approximately 1 MeV, technologies other than dc acceleration are used to reach multiple MeVs. The most successful commercial high-energy system today, shown in Fig. 10.4, makes use of a radio frequency (rf) linear accelerator (linac). In an rf linac, a dc ion beam is injected into a series of 6 to 14 electrodes connected to rf resonators that first bunch, then further accelerate the beam.[25] The resonators operate at a standard industrial frequency, 13.56 MHz, and provide accelerating voltages of up 100 to 120 kV on each electrode. At optimum acceleration, a 100-keV dc-injected singly charged ion beam can thus attain more than 1.5 MeV energy. The inherent energy spread of a bunched beam is reduced by energy analyzing the accelerated beam in a final energy magnet that uses momentum dispersion (similar to mass analysis described earlier). The accelerated pencil beam is scanned and parallelized and meets the wafer in a serial process chamber.

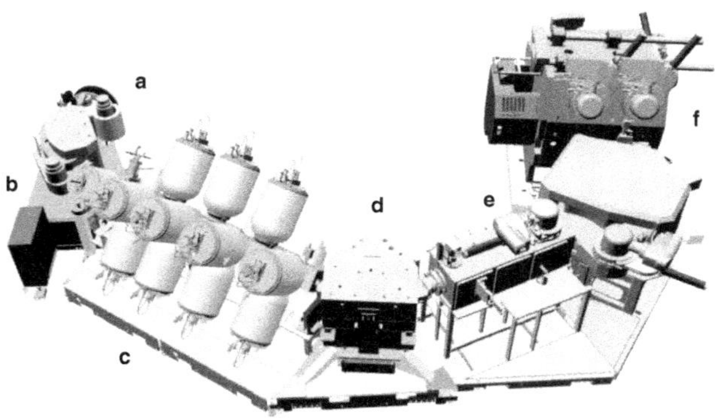

Figure 10.4 High energy system, Axcelis Purion™ Xe, using rf linac technology. The ion beam is extracted from an ion source (*a*) and mass filtered in a mass resolving dipole (*b*), then bunched and accelerated in a linac stage (*c*). A final energy filter (*d*) reduces the energy spread in the accelerated beam, then the beam is shaped, scanned, and parallelized in a beam scanning stage (*e*), before it enters the process chamber (*f*).

As an alternative to linac technology, a dc-tandem accelerator (also called tandetron) may also be used to generate beams in the range of energies required for the high-energy segment. The basic concept of a tandetron relies on a stripping cell to effectively double the accelerating capability of any given potential placed on the high-voltage terminal of the beamline.[26] Typical beam sizes and overall beamline length are comparable to the linac-based beamline,[27] although beam currents at the highest energies are significantly lower due to the relatively lower yield of highly charged ions from a stripping cell.

10.3 ION IMPLANTATION APPLICATIONS

10.3.1 Application Space

Ion implantation is one of the most established and documented processes in the semiconductor industry at this point, being fully incorporated in the 1970s and in fact, fully enabling modern microchip fabrication through the use of self-aligned processing.[28] The self-aligned process, which entails using the structure of the transistor to control the alignment of the doped and undoped areas surrounding the gate, allows reproducibility and device scaling. Arguably, the success of Moore's law[29] would not have been possible without ion implantation, which continues to be one of the highest productivity, most consistent, and cost-effective processes in use today. Figure 10.5 shows the range of implants that are used in processing of advanced devices today. A full featured device, with multiple types of devices (multiple Vt's), and input/output transistors, may require over 30 implant steps, accessing the entire dose/energy space.[30]

As such, the materials physics of implantation has books devoted to the topic, as do the variety of applications that use implantation. The reader is referred to the "Further Reading" section for detailed discussion of these topics. We will focus here on traditional and emerging applications for implantation as we enter into the sub-20 nm process regime. These applications include CMOS image sensors,[31] FINFET devices,[32] and stacked memory devices.[33]

Traditional Implant Applications

Historically, implant applications have been divided into three main categories based on energy (primary) and dose requirements of the applications. This was described above as being a main driver in designing the architecture of the implanter. The energy and dose requirements, including quality factors such as uniformity, dose precision, and angular variation dictate the beamline requirements to optimize a specific implant.

High-energy implants are required to place atoms deeper into the substrate, in order to form buried layers,[34] gettering layers for oxygen precipitation and metal reduction,[35] or isolation for prevention of latch-up in CMOS processing (the triple well process).[36] The energies required for these implants are greater than 1000 keV, and up to 4 MeV for processes requiring very deep implants. These implants are typically lower dose (E12-E13 ions/cm^2) and have stringent angle control requirements, typically less than a degree variation can be tolerated.[37]

Medium-current applications cover a similar dose range as the high-energy case, but with lower energies, less than 1000 keV to 10 keV. The applications are usually highly device sensitive implants requiring precision placement of the dopants. The energy, angle, and dose control systems in these implanters are designed with these criteria in mind. Applications include V_t adjust implants,[38] channel isolation implants (also known as field isolation, or "channel stop implants"), punch-through (or HALO) implants, and retrograde channel implants. A detailed review of these applications are found in Ref. 39.

High-current implanters are required for applications which require high concentrations of dopants or other species. These doses are typically high E14 up to E16/cm^2, with some applications requiring even higher doses. The high currents required to run these implants effectively are obtained through open beamline architectures, and often feature acceleration/decelerations to maximize beam transport, as described above.[20]

These implants are used to create degenerate silicon doping for contact purposes, including formation of low contact resistance structures.[40] These include the source and drain implants and poly-Si doping, with energies of 1 to 40 keV, and doses from 1 to 5 × 10^{15}/cm^2. Other applications include amorphizing implants, to control implant channeling,[41] co-implants to control diffusion of dopant species,[42] modification implants for etch or oxidation rates,[43] and line-edge roughness improvements.[44] The variety of implants is actually much larger than can be covered in the scope of this work, and again the reader is referred to detailed treatises available.[39]

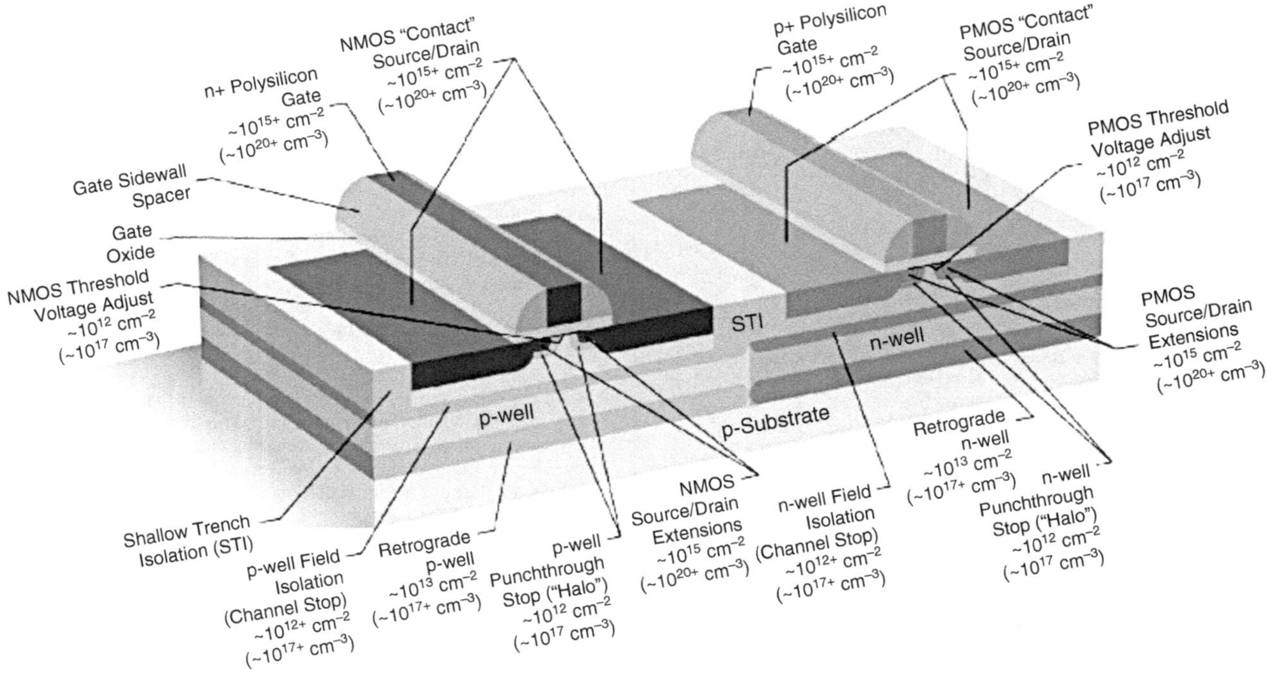

Figure 10.5 Implanted regions in advanced devices. Below the name of each region is the approximate implant dose (in cm^2) and concentration (in cm^3). (*Figure courtesy of Axcelis Technologies.*)

Emerging Applications

The impact of scaling and subsequent diversification of device architectures has placed additional challenges and constraints on implantation requirements. These devices, which include 3D devices such as FinFET,[45] require dopant that is uniform in both the horizontal and vertical direction of the fin structure. The device has raised source and drain with the channel surrounded by the gate. Implant of the source and drain (S/D) and S/D extensions (SDE) in FinFET devices is challenging, for several reasons. These include implant shadowing from adjacent devices, dopant channeling through the fin, and amorphization that leads to defects upon annealing. Performing these implants at elevated temperatures (up to 550°C) has been investigated in an attempt to maintain crystalline structure of the fin.[46]

Highly scaled devices are more sensitive to device leakage and impurities, requiring control of the damage introduced through implantation process. This had led to the introduction of additional processing capability, including cold implantation to maximize amorphous layer thickness.[47] The modifications introduced by relatively small thermal variations can be significant, as was observed by Schmeide.[48] In Fig. 10.6, the resulting damage profile after annealing was altered significantly, leading to a shift in device parameters (in this case, leakage). Other methods to control damage include use of high mass molecular implants, control of beam density, and dose rate, all of which can alter the implant damage profile.[49]

10.3.2 Process Challenges and Metrics

The landscape of process challenges for implant continues to evolve as new devices, device dimensions, and device structures place stringent requirements for improvement in the traditional areas of dose control, uniformity, charge control, and cleanliness. In order to meet these requirements, continuous improvement in both the system design parameters and corresponding metrology have been developed.

Device parametric shifts can be measured with dose, angle, or energy inaccuracy of much less than 1 percent for sensitive implants, and

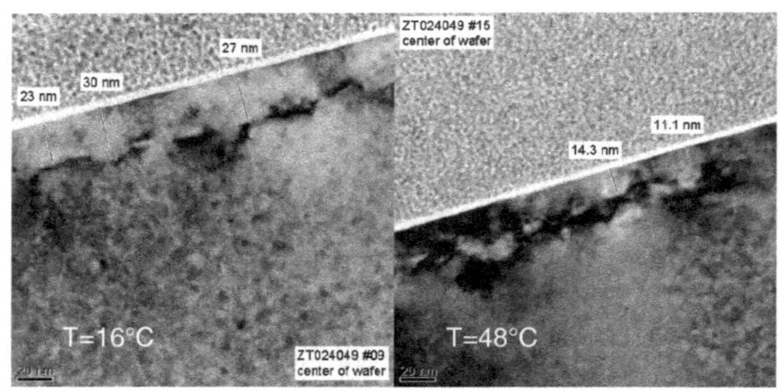

Figure 10.6 TEM profile of the residuals for 7 keV boron $5 \times 10^{15}/cm^2$ implanted at a wafer cooling temperature of 16 or 48°C.

across wafer uniformity of less than 0.2 percent variation is within the typical specification of a mid-current implanter. Choosing the proper metrology for the given energy and dose is therefore critical when monitoring the implanter performance. For mid- to low-dose implants at higher energies, Thermawave (TW) technology is used,[50] while higher dose, lower energy implants are best measured by annealing and using standard sheet resistance (R_s) measurements. A new challenge for metrology is the trend of very low energy implants, which are sensitive to surface proximity effects and depth of probing issues. Techniques developed for this regime include micro-R_s probes and optical techniques.[51,52]

Many process cleanliness issues are addressed with techniques to control ion beam transport and stability. Beams that exhibit high levels of noise, or "hash," are prone to micro-discharges, excess strike of beam guide components, and undesirable sputtering. Advances in source technology include use of optimized co-gases,[53] in addition to beam monitoring sensors for trajectory, angle, and composition.[20] Advances in graphite and other liner materials have allowed particles levels to be reduced to very low levels, less than 30 adders at 45 nm with 90 percent compliance is a typical expectation in a modern implanter.[54]

The thoroughly investigated technology of beam charging has also evolved, moving beyond the realm of simple wafer charging, which drove most of the early technologies. Now, requirements for undetectable metals levels have resulted in the elimination of filament based systems in favor of rf- or micro-wave based systems, which deliver electrons to the beam with no risk of metallic contamination from hot tungsten filaments. Wafer charging is still best monitored with device-like structures, such as SPIDER wafers,[55] though charge control requirements have largely mitigated with highly efficient charge control systems in today's implanters.[16]

10.4 OUTLOOK

Ion implantation is essential for the production of modern integrated circuits. The problems encountered in scaling devices and increasing circuit packing densities are motivating many investigations into new process alternatives that allow fabrication of deep submicron devices and ULSI circuits. The alternatives that will be adopted in future manufacturing are those that offer basic relief from scaling problems, simplicity in production, and compatibility with current processes to allow timely, cost-effective implementation in production. Furthermore, they must have flexibility for future developmental needs. Ion implantation meets these criteria, and therefore will remain essential for the production of integrated circuits for the foreseeable future.

Several trends in ion implantation applications will continue in CMOS production in the next few years.

The need for increasingly shallow SDE and S/D junctions will push the performance capabilities of ion implantation (and thermal processing) equipment. Equipment vendors will be under constant pressure to improve product performance.

Implant process control, including beam stability, beam angle repeatability, and beam profile (including density) control, along with implant temperature will become increasingly important to prevent unacceptable variations in device parameters. The need for almost undetectable levels of impurities, especially heavy metals, will drive materials and architectures that are able to meet these stringent specifications.

Implants of germanium, carbon, nitrogen, fluorine, and other non-dopant species will be increasingly employed in applications such as advanced channel engineering, preamorphization ion beam mixing for silicide S/D contact formation, diffusion control during annealing, and several other diverse applications. These high dose applications, known generically as "materials modification implants," will carry with them requirements for ultra-low particle performance, to avoid defects in the devices. Trade-offs in dosing precision, beam current (productivity), and process cleanliness will lead to differentiation of requirements for high-current and mid-current implanters.

10.5 REFERENCES

1. Freeman, J. H. (1970). Implantation Machines. *Proc. of the Europ. Conf. on Ion Implant.* (pp. 1–18). Reading: *Inst. of Physics, Physical Society and Inst. of Electrical Engineers.*
2. Shockley, W. (1957). Patent No. 2, *Forming Semiconductive Devices by Ionic Bombardment* 787, *564.* U.S.
3. Wolf, B. (1995). Production of Ions from Nongaseous Material. In B. Wolf, *Handbook of Ion Sources.* CRC Press, Boca Raton, FL.
4. Penning, F. M. (1936). Die Glimmentladung bei niedrigem Druck zwischen koaxialen Zylindern in einem axialen Magnetfeld. *Physica* 3 (9), 873–894.
5. Horsky, T. N. (1998). Indirectly Heated Cathode arc discharge Source for Ion Implantation of Semiconductors. *Rev. Sci. Instrum.*, 69, 1688.
6. Brown, I. G. (2004). *The Physics and Technology of Ion Sources.* John Wiley & Sons, Hoboken, New Jersey.
7. Renau, A. (2008). Ion Sources (Chap 6), In J. Ziegler (ed.), *Ion Implantation Science and Technology.*
8. O'Hanlon, J. F. (2003). *A User's Guide to Vacuum Technology,* 3d ed., John Wiley & Sons, Hoboken, New Jersey.
9. Chen, F. F. (1983). *Introduction to Plasma Physics and Controlled Fusion.* Plenum Press, New York.
10. Chivers, D. J. (2008). Ion Implanter Design Concepts. In J. Zieger (ed.), *Ion Implantation Technology,* pp. 5–3 to 5–28, Edgewater, MD.
11. Murrell, A., Hacker, D., Edwards, P., Mitchell, R., Banks, P., Foad, M., et al. (2004). Quantum X: Single Wafer High Current Ion Implantation Using Mechanical Scan. In L. J. Chen, J. Poate, and T. F. Lei (eds.), *Proc 15th Int. Conf. Ion Implantation Technology Part II.* Taipei, Taiwan: MRS-Taiwan, pp. 20–24.
12. Splinter, P., Graf, M., Godfrey, C., Huang, Y., Polner, D., Danis, J., et al. (2006). Optima HD: Single Wafer Mechanical Scan Implanter. *AIP Conf. Proc. 866,* Marseille, France, p. 601.
13. Kopalidis, P., Wan, Z., & Collart, E. (2010). Ribbon and Spot Beam Process Performance of the Dual Mode iPulsar High Current Ion Implanter. *Ion Implantation Technology,* 1321, p. 337. Kyoto, Japan: AIP.
14. Kawasaki, Y. E. (2004). The Collapse of Gate Electrode in High Current Implanter of Batch Type. *Proc. 4th Int. Workshop on Junction Technology.* Yorktown, NY: IEEE, pp. 39–41.
15. Mack, M. E. and Ameen, M. S. (2000). Wafer Cooling and Wafer Charging in Ion Implantation. In J. F. Ziegler (ed.), *Ion Implantation: Science and Technology,* p. 522, Yorktown, NY.
16. Vanderberg, B., Nakatsugawa, T., and Divergilio, W. (2012). Microwave ECR Plasma Electron Flood for Low Pressure Wafer Charge Neutralization. *Ion Implantation Technology.* Vallodolid, Spain: AIP, pp. 356–359.
17. Liebert, R., Walter, S., Felch, S., Fang, Z., Pedersen, B., and Hacker, D. (2000). Plasma Doping System for 200 and 300 mm Wafers. *2000 Int. Conf. on Ion Implantation Technology.* Alpbach, Austria: IEEE, pp. 472–475.
18. Glavish, H. F., Horsky, T. N., Jacobson, D. C., Sinclair, F., Hamamoto, N., Nagai, N., et al. (2006). A Beam Line System for a Commercial Borohydride Ion Implanter. *AIP Conf. Proc. 866.* Marseille, France: American Institute of Physics, pp. 167–171.
19. Dataquest. (Apr. 1016). *Wafer Fab Equipment Report,* Gartner, Inc., Stamford, CT.
20. Vanderberg, B., Heres, P., Eisner, E., Libby, B., Valinski, J., and Huff, W. (2014). Introducing Purion H, a Scanned Spot Beam High Current Implanter. *20th Int. Conf. on Ion Implantation Technology.* Portland, OR: IEEE, pp. 1–4.
21. Angel, G., Bell, E., Brown, D., Buff, J., Cummings, J., Edwards, W., et al. (1998). A Novel Beamline for Sub-keV Implants with Reduced Energy Contamination. *Proc. Int. Conf. on Ion Implantation.* Kyoto, Japan: IEEE, pp. 188–191.
22. Graf, M., Vanderberg, B., Benveniste, V., and Tieger, D. (2002). Low Energy Ion Beam Transport. *Proc. of the 14th Int. Conf. on Ion Implantation Technology.* Taos, NM: IEEE, pp. 359–364.
23. Campbell, C., Cuccetti, A., Sinclair, F., Kellerman, P., Radovanov, S., and Falk, S. (2012). VIISta Trident: New Generation High Current Implant Technology. *AIP Conf. Proc. 1496.* Valladolid, Spain: AIP, p. 296.
24. Eisner, E., David, J., Justesen, P., Kamenitsa, D., McIntyre, E., Rathmell, R., et al. (2012). Optima MDxt: A High Throughput 335 keV Mid-Dose Implanter. *AIP Conf. Proc.* 1496. Vallodolid, Spain: American Institute of Physics, pp. 340–344.
25. Glavish, H., Boisseau, P., Libby, B., Bernhardt, D., Simcox, G., and Denholm, A. S. (1987). Production High Energy ion Implanters Using Radio Frequency Acceleration. *Nuclear Instruments and Methods,* B21, 264–269.
26. Turner, N., Purser, K., and Sieradzki, M. (1987). Design Considerations of a VLSI Compatible Production MeV Ion Implantation System. *Nuclear Instruments and Methods in Physics Research,* B21, 285–295.
27. Tokoro, N., Holbrook, D., and Hacker, D. (2000). Introduction of the Varian VIISta 3000 Single Wafer High-Energy Ion Implanter. *Proc. 2000 Int. Conf. on Ion implantation Technology.* Alpbach, Austria: IEEE, pp. 368–371.

28. Fair, R. B. (1998). History of Some Early Developments in Ion-Implantation Technology Leading to Silicon Transistor Manufacturing. *Proc. of the IEEE,* 86 (01), 111–137.

29. Moore, G. (1965). Cramming More Components onto Integrated Circuits. *Electronics Magazine,* 38(8), 114–117.

30. Naito, M. T. (2011). History of Ion Implanter and Its Future Perspective. *SEI Technical Review,* V75, 22–30.

31. Bigas, M., Cabruja, E., Forest, J., and Slavi, J. (2006). Review of CMOS Image Sensors. *Microelectronics Journal,* Vol. 37, 433–451, Elsevier, London.

32. Jurczak, M., Collaert, N., Veloso, A., and Hoffmann, T. (2009). Review of FINFET Devices. *SOI Conference, IEEE International,* 1–4, Elsevier, London.

33. Nakazato, K. (2006). Future Memory Devices—from Stacked Memory, Gain Memory, Single-Electron Memory to Molecular Memory. *2006 Int. Symposium on VLSI Technology, Systems, and Applications,* 1–2.

34. Ro, J.-S. (1997). A Study of Buried Layer Formation Using MeV Ion Implantation for the Fabrication of ULSI CMOS devices, *AEPSE '97 Conference* , 349 (1,2), 130–134.

35. Agarwal, A. et al. (1996). Oxygen Gettering and Precipitation at MeV Si + Ion Implantation Induced Damage in Silicon. *Applied Physics Letters,* 69 (25), 3899–3901.

36. Tsukamoto, K., Komori, S., Kuroi, T., and Akasaka, Y. (1991). High-Energy Ion Implantation for ULSI. *Nuclear Instruments and Methods in Physics Research Section B: Beam Interactions with Materials and Atoms,* 59–60, 584–591.

37. Kapila, D., Jain, A., Nandakuma, M., and Ashburn, S. (1999). The Effect of Deterministic Spatial Variations in Retrograde Well Implants on Shallow Trench Isolation for Sub-0.18 μm CMOS Technology. *IEEE Transactions on Semiconductor Manufacturing,* 12 (4), 457–461.

38. Wolf, S. (1990). *Silicon Processing for the VLSI Era,* Vol. 2, Chap. 5, Lattice Press, Sunset Beach, CA.

39. Ziegler, J. F. (2014). *Ion Implantation, Applications, Science, and Technology,* Edition 2014, Ion Implantation Technology Co., Edgewater, MD.

40. Sze, S. M. (1981). *Physics of Semiconductor Devices,* 2d ed. John Wiley & Sons, Hoboken NJ.

41. Al-Bayati, A. (2000). Exploring the Limits of Pre-Amorphization Implants on Controlling Channeling and Diffusion of Low Energy B Implants and Ultra Shallow Junction Formation. *Conference on Ion Implantation Technology,* 54–61.

42. Pawlak, B. J. and Janssens, T. (2006). Effect of Amorphization and Carbon Co-Doping on Activation and Diffusion of Boron in Silicon. *Appl. Phys. Lett.,* 89, 062110.

43. Ya-Chin King, C. K.-J. (1998). Sub-5 nm Multiple-Thickness Gate Oxide Technology Using Oxygen. *IEDM Conference,* 585.

44. Ma, T. Y., Xie, P., Godet, L., Martin, P. M., Campbell, C. S., Xue, J., et al. (2013). Post-Litho Line Edge/Width Roughness Smoothing by Ion Implantations. *SPIE 8682, Advances in Resist Materials and Processing Technology,* San Jose, CA.

45. Fossum, J. et al. (2003). Physical Insights on Design and Modeling of Nanoscale FinFets. *2003 IEDM Tech. Digest,* 679, 29.1.1-29.1.14.

46. Onoda, H. et al. (2013). High Dose Dopant Implantation to HEated Si Substrate without Amorphous Layer Formation. *Proc. Int'l Workshop on Junction Technology,* 22.

47. Huh, T. H. et al. (2008). A Study of Implanted BF2 as a Function of Wafer Temperature during Implant. *AIP Conf. Proc.* 1066, 87.

48. Matthias Schmeide, M. S. (2011). Integration of High Dose Boron Implants—Modification of Device Parametrics through Implant Temperature Control. *AIP Conf. Proc.* 1321, 61.

49. Ameen, M. S., Harris, M. A., Huynh, C., and Reece, R. N. (2008). Dose Rate Effects: The Impact of Beam Dynamics on Materials Issues and Device Performance. *AIP Conf. Proc.* 1066, 30.

50. Curello, G. (1999). Monitoring Low Dose Implants with Advanced Therma-Wave and Capacitance-Voltage. *Proc. of Ion Implantation Technology,* 546–549.

51. Robbes, A. S., Meura, K. A., Moret, M. P., and Schuhmacher, M. (2014). Implantation and Metrology Solutions for Low Energy Boron Implant on 450 mm Wafers. *Conference on Ion Implantation Technology (IIT), 2014 20th International,* 1–4.

52. Henrichsen, H. H. (2014). Precision of Single-Engage Micro Hall Effect Measurements. *2014 International Workshop on Junction Technology (IWJT),* 1–4.

53. Hsieh, T. and Colvin, N. (2014). Improved Ion Source Stability Using H_2 Co-Gas for Fluoride Based Dopants. *2014 20th International Conference on Ion Implantation Technology (IIT),* 1–4.

54. Kirkwood, D. (2015). *Defect Control—Designing for Process Cleanliness.* Retrieved from Axcelis Technologies Knowledge Center: http://www.axcelis.com/knowledge-center/defect-control-designing-process-cleanliness, March 2017.

55. Singh, B., Elkind, A., Mack, M., and Ameen, M. S. (2000). Characterization of Charging Damage in High Power Implants Using SPIDER Wafers. *Conference on Ion Implantation Technology, 2000.* 561–564.

Chapter 11

Introduction to Physical Vapor Deposition

BY

FLORIAN SOLZBACHER *University of Utah, Salt Lake City, Utah*

Physical vapor deposition (PVD) is a high-vacuum deposition process that is typically used for metals, metal alloys, or other solid chemical compounds using thermal energy (evaporation, pulsed laser deposition) or kinetic energy of ions (sputtering, ion beam deposition) to remove material from a source (sputter target or crucible) and deposit it onto a substrate.

11.1 MOTIVATION AND KEY PROPERTIES

PVD is the most commonly used process for the deposition of metals and metal oxides in semiconductor and microelectromechanical system (MEMS) processing. The motivation for using PVD processes lies in the specific properties of the process and the deposited layers:

- A wide range of layer thicknesses ranging from nanometers to micrometers is possible.
- The layer uniformity and reproducibility is comparably high.
- Essentially, there is no limitation in the choice of source materials (such as metals, semiconductors, glass, ceramics, composites, oxides, and plastics) that can be deposited.
- Multilayer systems can easily be deposited in one vacuum process.
- The substrate temperature can be kept low (down to room temperature) and seldom exceeds 350°C during deposition. Thus damage to underlying layers is typically limited (with exception of polymers/photoresist layers, e.g., during lift-off processes).
- The layer properties (specific resistance, temperature coefficient of resistance (TCR), adhesion, structure, composition, density, refractive index) can be modified by changing the process parameters (substrate temperature, residual gas pressure, process pressure, particle energy, deposition rate, and gas atmosphere).

11.2 FUNDAMENTALS OF PVD PROCESSES

PVD processes are vacuum processes. Hence, it is necessary to briefly introduce some of the main concepts of vacuum physics such as pressure and mean free path. Pressure p can be described as gas molecules with concentration n per unit volume, mass m, and mean velocity v colliding with container walls:

$$p = \frac{nm\overline{v}^2}{3}$$

The gas molecule mean velocity v depends on the temperature and molecular mass. Assuming the particle velocity distribution follows the Boltzmann distribution, we obtain

$$\overline{v}^2 = \left[\frac{8kT}{\pi m}\right]^{1/2}$$

where k is Boltzmann constant and T is temperature, and m is the mass of gas molecule.

The mean free path of molecules between collisions is therefore

$$\overline{\lambda} = \frac{kT}{p\pi\sigma^2\sqrt{2}}$$

where k is Boltzmann constant, T is temperature, p is residual pressure, σ is particle diameter, which approximates to $\overline{\lambda} = \frac{6.3}{p}$ in air.

To obtain high-purity layers, it is imperative that the mean free path be much larger than the distance between the source and the substrate. The degree of contamination of a film depends on the purity of the source material as well as the reaction with the residual gas in the vacuum chamber. The smaller the mean free path, the higher is the possibility of vapor particles colliding on their path between the source and the substrate. These collisions can lead to additional reactions with residual gas atoms, impacting the film composition. They can also lead to randomization of travel direction for sputtered atoms and reduced net sputter rate (reflection of atoms back toward the target). In the case of reactively sputtered layers, such as, for example, metal oxides and nitrides, one makes use of this fact by initiating reactions of the vapor particles with the ambient gas. The collision probability of vapor particles drops with ratio of mean free path over distance traveled.

11.3 VACUUM EVAPORATION

Vacuum evaporation works on the principle of heating a source material in vacuum until a sufficient vapor pressure for net evaporation of the source material is reached. Equilibrium pressure is the saturation vapor pressure, that is, evaporation and condensation rates are equal. The deposition process includes three stages:

1. Evaporation of target material (evaporation phase)
2. Transport of particles through the vacuum to the substrate (transport phase)
3. Condensation on the substrate (condensation phase)

Key characteristics of this process are that the source material is heated to very high temperatures, whereas the substrate temperature can be freely chosen to influence the layer parameters. A large mean free path for vapor particles is required to prevent reaction of the source vapor particles with the residual gas or collision with other vapor particles. Thus, normally, vacuum pumps for evaporator equipment are connected directly to the main vacuum chamber with large cross-sectional tubing. At equilibrium pressure or saturation vapor pressure, evaporation and condensation rates of the source material are equal. The saturation vapor pressure is given by

$$p_s = Ae^{\left(-\frac{B}{T}\right)}$$

where A is the integration constant and B is a constant depending on heat of evaporation, that is, target material. Small temperature changes thus lead to large changes in the condensation rate (Fig. 11.1).

The degree of freedom in process parameters is limited. Low evaporation rates lead to chemical reactions with residual gas particles, and high evaporation rates (vapor pressure above source is too large) lead to collisions and subsequent partial backscattering of source vapor particles. Too high evaporation temperatures lead to building of bubbles and material will be ejected from the crucible splattering the chamber and substrate. In practice, process pressures are set between 10^{-2} and 10^{-4} Pa. The evaporation rate R can be described as

$$R = 4.43 \cdot 10^{-4} \left(\frac{M}{T}\right)^{\frac{1}{2}} p_s \ [\text{g/cm}^2]$$

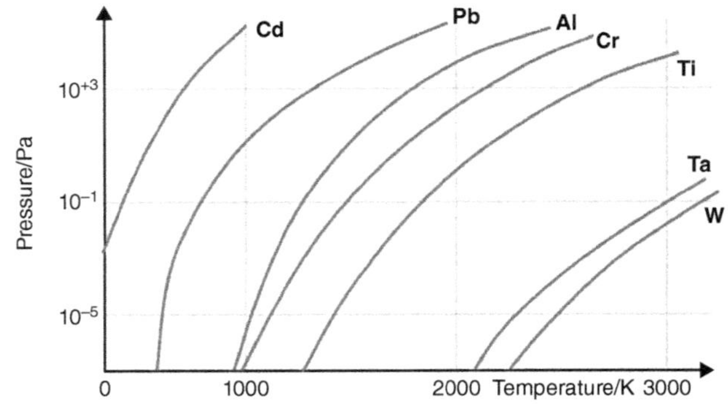

Figure 11.1 Equilibrium vapor pressures of selected metals for vacuum evaporation [1].

where M is the molecular weight and T is the absolute temperature. Examples of evaporation rates for common source materials can be found in Fig. 11.2. During the transport phase, the mean kinetic energy of the particles is

$$E_e = \frac{m}{2}v^2 = \frac{3}{2}kT_v$$

where m is the particle mass, k is the Boltzmann constant, T_v is the source temperature, and v is the particle velocity. E_e typically amounts

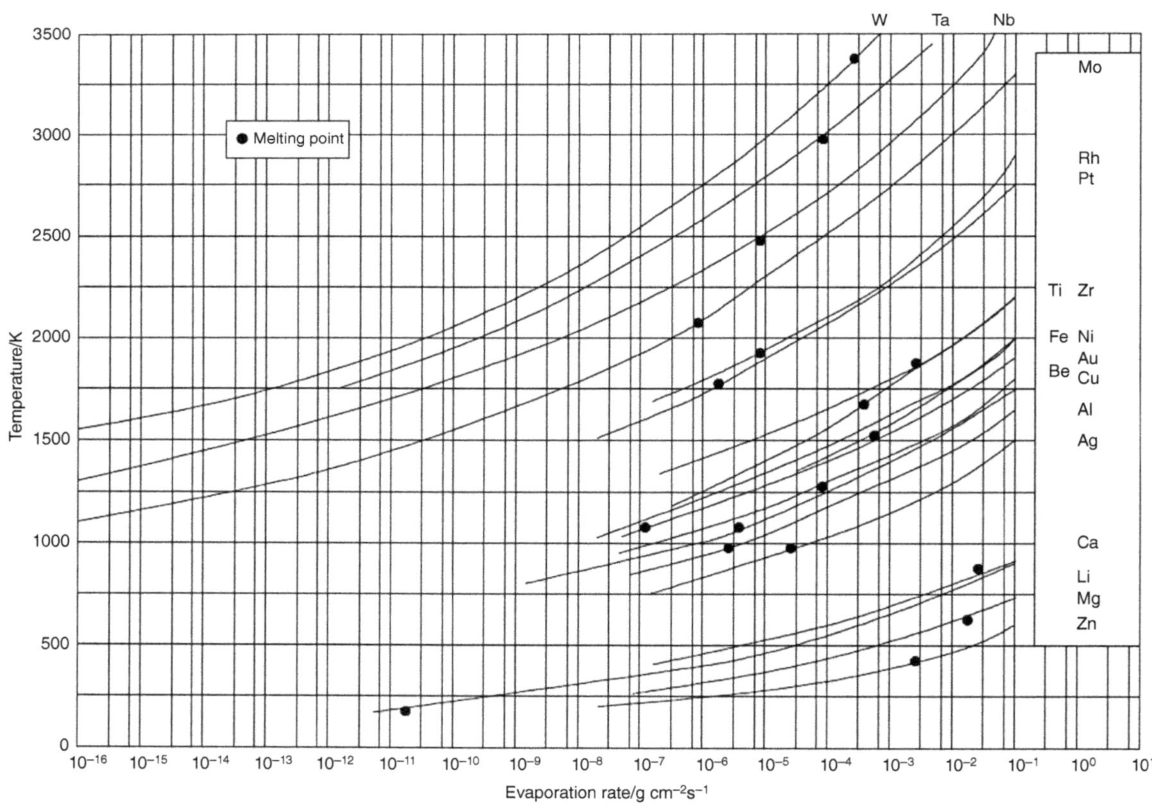

Figure 11.2 Evaporation rates of selected metals as functions of temperature [1].

to about 0.2 eV at 1500 K (0.26 eV at 2000 K); that is, the energy of the particles is small compared to sputtering, as will be described below. Also, variation in evaporation temperature allows for little variation in particle energy.

The condensation stage consists of three steps:

1. Vapor particles deposit on the substrate surface, thereby transferring kinetic energy to the substrate (adsorption).
2. Surface diffusion of the particles occurs with subsequent exchange of energy with the lattice atoms until the particle (atom) settles in a spot of low energy (nucleation/seed). Surface defects are preferred locations for nucleation; the growth of domains leads to continuous layers and the layer growth depends highly on the substrate surface and deposition conditions.
3. Volume diffusion of atoms within the lattice.

The surface structure resulting from the evaporation deposition is typically distinguished into three types or zones depending on the ratio of the substrate temperature T_s to the target melting temperature T_m (Fig. 11.3).

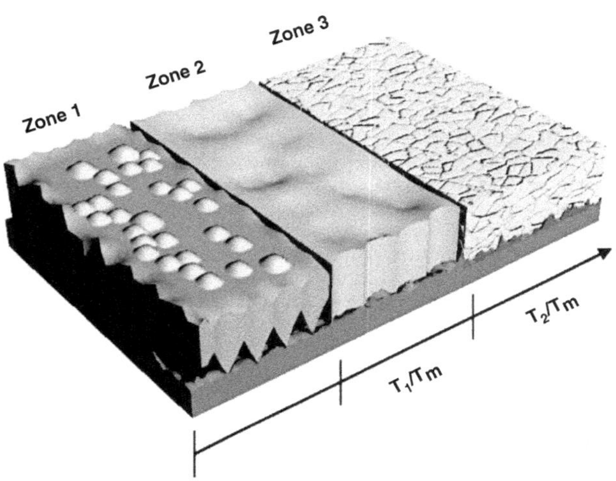

Figure 11.3 Zone model of layer surface structure for vacuum evaporation deposited layers [2].

Zone 1 (T_s/T_m < 0.25). Dendrite structure, porous, lower density than bulk material, dendrites grow with T_s

Zone 2 (0.26 < T_s/T_m < 0.45). Columnar structure, surface mobility is sufficiently large, high packing density, low surface R_a, crystal diameter grows with T

Zone 3 (T_s/T_m > 0.45). Volume diffusion in dominant layers with high density and flat surface

11.4 EVAPORATOR EQUIPMENT

For evaporation, the source material is heated using either a resistive heater or an electron beam (e-beam evaporation). The drawback of resistive heating of the vapor source is the evaporation of the heater itself leading to contamination of the layer and the limited layer thickness. Furthermore, this process does not allow the deposition of high melting temperature materials such as W, Mo, and Ta. The source material is heated either in a ceramic crucible and evaporator boat, or using an evaporator spiral with twisted tungsten wire. The heat produces an upward vapor column above which the substrate material is moved in concentric circles or planetary motion to ensure constant layer thickness across the substrate.

In e-beam evaporation, highly efficient local heating is accomplished by using an electron beam directed at the source material. The 270° deflection of the e-beam prevents contamination of the layer from the glowing e-beam cathode (Fig. 11.4). Key process parameters are the residual gas pressure, the substrate temperature, the evaporation phase, and the time. In situ layer thickness is measured using quartz micro balances or resonators. Typical deposition rates can vary between 100 nm/min

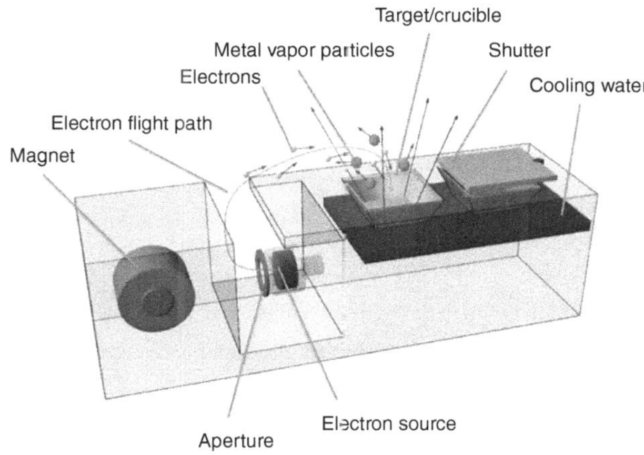

Figure 11.4 Schematic of an electron-beam evaporator.

and 5 μm/min. The core parameters for good layer quality are a sufficient thickness uniformity and good step coverage.

The residual gas pressure must not exceed approximately 10^{-2} Pa to prevent collisions of vapor molecules with residual gas molecules and contamination (adsorbed gas molecules on the substrate surface). Typical operating pressures are therefore 10^{-3} to 10^{-5} Pa. Important process parameters are residual pressure, substrate temperature, evaporation phase, and time. The evaporation rate can be controlled using the current density or the e-energy of the e-beam. Quartz thickness gauges are used for the measurement and control of the deposition rate in situ.

The core drawback of this method is the generation of x-rays at acceleration voltages >10 kV, which can damage MOS-integrated circuitry during deposition by causing trapped charges in the gate oxide. The trapped charges have to be removed using an annealing process. Turning this drawback into an advantage, the x-rays can, however, be used for in situ layer analysis (XRD, X-ray diffraction). High e-beam energies lead to bubbling of the target material and consequently the danger of splattering the substrate. For a sufficient layer quality, the following criteria have to be taken into account:

- In situ control of layer thickness
- Sufficient thickness uniformity across the wafer
- Good step coverage = $t_s/t_n \times 100$ percent; t_s—minimum layer thickness at the step edge, t_n—layer thickness in the flat part

11.4.1 Cosine Law

For good thickness uniformity across the wafer, the wafer transport follows a specific movement under the e-beam vapor. The particle flow from a small source theoretically follows the cosine law. The evaporated mass per unit area is:

$$R_D = \frac{M_e}{\pi r^2}\cos\phi\cos\theta$$

where M_e is the total mass of evaporated material and r, f, and q are according to Fig. 11.5. Since the vapor source is not a point source, the evaporation is not isotropic. In practice, a planetary motion of the substrate holder is used for most applications.

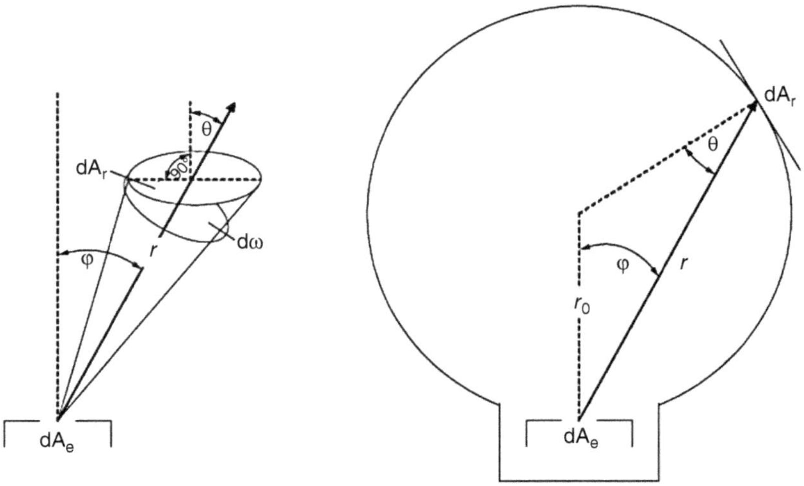

Figure 11.5 Cosine law of evaporated mass per unit area for a point source [1].

11.5 LAYERS DEPOSITED USING EVAPORATION AND THEIR PROPERTIES

Evaporation is primarily used for the deposition of metals. Principally, it allows the nonreactive and reactive deposition of metals, alloys, chemical compounds, and ceramics as single and sandwich layers.

Metals. Examples of metal layers deposited are Al, Au, Ag, Cr, Ni/Cr, Ti, Ni, Pt, and Pd. Primarily, e-beam evaporation is used.

Alloys. The evaporation of alloys is significantly more difficult than that of metals since the components of alloys seldom have the same vapor pressure at the same temperature. Therefore, sputtering is normally used for alloy deposition.

Multilayer. Multiple crucible evaporators with one or more e-beams are used to deposit multilayer structures in one deposition run.

Chemical compounds. The evaporation of chemical compounds is usually combined with dissociation. Since small changes in stoichiometry already lead to drastic changes in layer properties, chemical compounds are usually deposited using CVD or sputtering.

Reactive evaporation. The use of a reaction gas (e.g., O_2, N_2) during evaporation leads to a chemical reaction between the evaporated material and the gas and can be used, for example, to deposit oxides or nitrides of metals while using a pure metal target. CVD and sputtering allow much better process control, and are therefore primarily being used.

11.6 SPUTTERING

11.6.1 Definition

Sputtering is a plasma process, during which noble gas ions [typically argon (Ar^+)] are accelerated toward a target (cathode) from which they remove material particles. These particles build a vapor column that condenses on the substrate. Sputtering is the primary method for the deposition of metallization layers in microelectronics and MEMS. Sputtering is a four-stage process:

Stage 1. Creation of ions through collision of inert gas atoms (Ar) with electrons and acceleration of ions toward a target

Stage 2. Removing of target atoms by impact of ions with the target

Stage 3. Transport of free target atoms to the substrate

Stage 4. Condensation of target atoms on the substrate

The distance between the substrate and the target is typically 5 to 10 cm (up to 35 cm for some sputtering equipment) to allow maximum condensation rate of removed target atoms on the substrate. Seed and cluster building occurs in an identical fashion as evaporation. Atoms reaching the substrate have higher energy (3 to 10 eV) than in evaporation (0.2 eV). Collisions between the target and the gas atoms result in atoms reaching the substrate from various directions. The deposition process is therefore, to a large extent, isotropic and the step coverage of sputtered layers is better than that of evaporated layers. Sputtering exhibits a higher particle bombardment of the substrate surface than evaporation. This can on the one hand lead to crystal damage, but on the other hand also be used for in situ annealing of deposited layers. The heating of the substrate (e.g., wafer) due to the high-energy particle bombardment during the process has to be taken into account, in particular when considering deposition on polymer materials, such as during lift-off processes or for polymer MEMS devices. The layer properties can be influenced using a bias voltage.

The key differences between sputtering and evaporation are as follows:

- Sputtered atoms and molecules have a higher impact energy (3 to 10 eV, compared to 0.2 to 0.26 eV for evaporation).
- Due to collisions with gas particles (Ar), the free target atoms reach the substrate surface from various directions leading to good step coverage.
- The substrate has a higher exposure to gas than in evaporation.
- Applying a bias voltage can influence the layer parameters (e.g., planarization of surface).
- Changing the polarity of the substrate and the target, the substrate can be exposed to backsputtering. Backsputtering can be used for
  - Cleaning of the substrate surface prior to deposition
  - In situ annealing of layers in sandwich layer structures (i.e., layers are successively deposited and partially backsputtered thereby annealing the layer through physical impact and heat)
  - Sputter etching of layers that are otherwise difficult or impossible to pattern.

11.6.2 Mechanism

Ions are created through plasma discharge between two electrodes. Ionization occurs through inelastic collisions between free electrons and gas molecules (in most cases Ar). The required ionization energy (15.7 eV for Ar) stems from the electron kinetic energy. The kinetic energy of the Ar ions after the acceleration by the electric field between the cathode and anode is between 10 eV and several 1000 eV. On collision with the substrate, the Ar ions transfer a part of their kinetic energy to a confined volume (ca. 1000 atoms) of the lattice atoms through a series of quasi-elastic collisions. Some target atoms are diffracted toward the surface. If their energy is larger than the surface binding energy, they leave the lattice. The removed target atoms typically have

3 to 10 eV of energy. The mean quantity of removed atoms or sputtering yield S is given by

$$S = \frac{\text{number of removed target atoms}}{\text{number of colliding ions}}$$

The sputter yield depends strongly on the angle of impact or Ar ions hitting the target (Figs. 11.6 and 11.7). At 0°, the ions fly parallel to

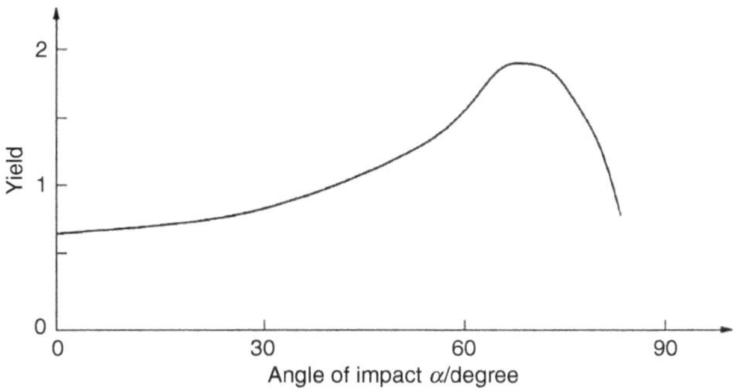

Figure 11.6 Sputter yield as a function of particle impact angle [3].

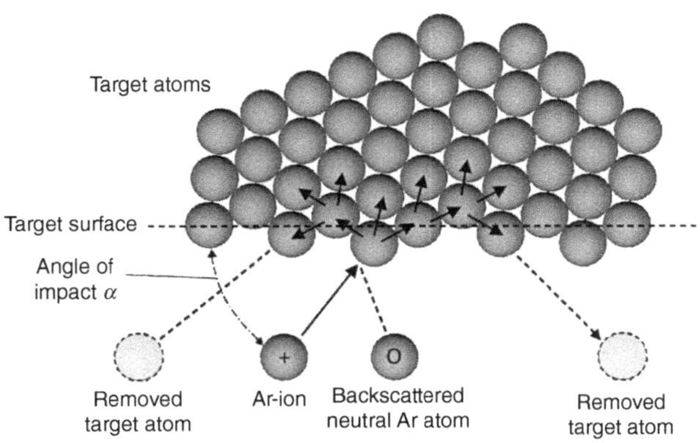

Figure 11.7 Schematic of propagation of momentum and particle movements of Ar ions hitting a sputter target surface.

the surface thus having a low probability of knocking atoms out of the target. At a 90° impact angle, the momentum translates into the bulk of the target material rather than knocking atoms out of the target. The maximum sputter yield can typically be found around 60°. S also depends on the mass relation of ion (Ar) versus the target atom as well as the binding energy between the atoms (Fig. 11.8).

11.6.3 Film Microstructure and Mechanical Properties

Depending on the deposition and annealing process parameters, PVD thin film properties can be modified to some extent. The most relevant properties are the film microstructure, crystallinity, grain size, surface texture, layer stress and hardness, and the electrical and mechanical properties (e.g., e-modulus). The layer structure, just like in evaporation, depends on the inert gas pressure and the ratio of the substrate temperature to the target melting temperature (Fig. 11.9). Typically, annealing does improve the layer crystallinity, but leaves the surface texture unaffected. Increasing sputtering power can decrease the layer crystallinity. The introduction of reactive gases can disrupt the

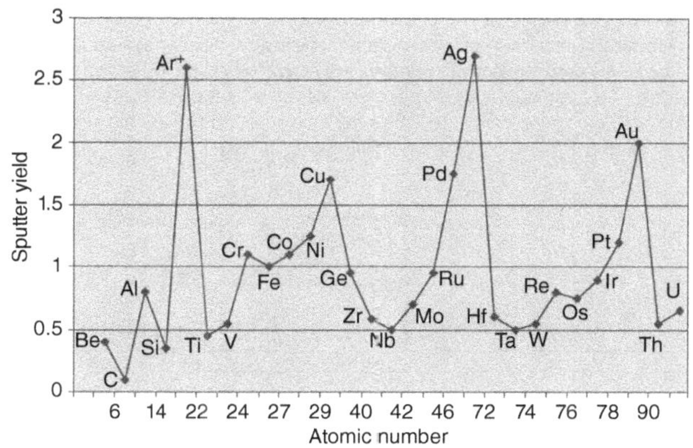

Figure 11.8 Sputter yield as a function of target atomic number [4].

crystallization process leading to smaller grain size. The most common surface analysis tools used for layer characterization are x-ray diffraction (XRD), electron dispersive x-ray spectroscopy/wavelength dispersive x-ray spectroscopy (EDX/WDX), Auger electron spectroscopy (AES), scanning electron microscope (SEM), and transmission electron microscope (TEM). XRD is used to determine the crystallinity and texture of the layers. A broader peak signifies smaller grain size. EDX/WDX is used to determine the composition of the layers, to detect contamination, stoichiometry of compound layers and reaction of donors, and dopant materials with underlying base layers.

11.7 SPUTTER EQUIPMENT

Sputtering equipment is available for a variety of different sputtering processes. Ion beam sputtering uses an ion beam to ablate material from a target, which condenses on a substrate. In this case, the ion source is separate from the vacuum chamber in which target and substrate are located. Plasma sputtering constitutes the largest and most common group of sputtering processes and systems. In this case, the ions are generated in the same vacuum chamber as the target and substrate. Plasma sputtering uses ionized gas atoms that are being accelerated

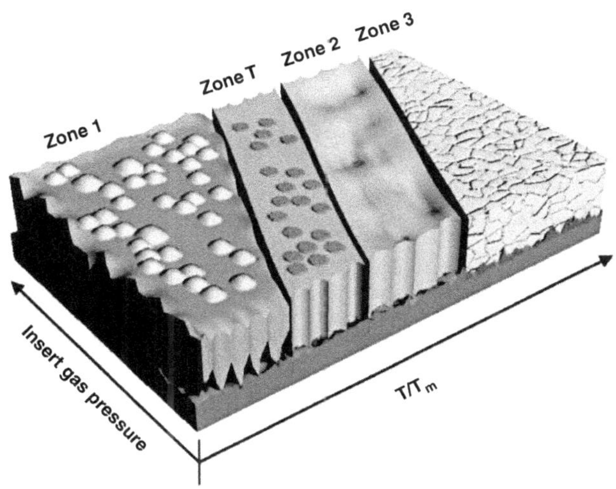

Figure 11.9 Zone model of a layer surface structure for sputtered layers [3]. Zones 1 to 3 as in evaporation, Zone T—fiber-type closely packed layer with smooth surface.

toward the target to remove material that condenses on a substrate. Depending on the type of assembly, a triode or diode system is used. The more common diode systems are operated either in DC sputtering with a DC voltage being applied across the plasma or in pulsed DC or HF/RF sputtering mode. In RF sputtering, one can choose between inert gas sputtering and reactive sputtering (Fig. 11.10).

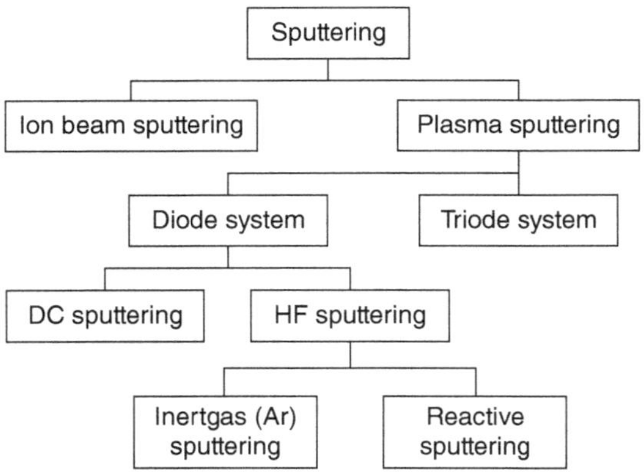

Figure 11.10 Common types of sputter processes and equipment [2].

11.7.1 DC Sputtering

DC (Direct current) sputtering is suitable only for the deposition of conducting materials. Nonconducting materials lead to charging of the target material by the Ar ions, resulting in a reduction of the negative potential. Thus, the acceleration potential for Ar ions breaks down (self-stopping process) (Fig. 11.11). DC-reactive sputtering of dielectrics

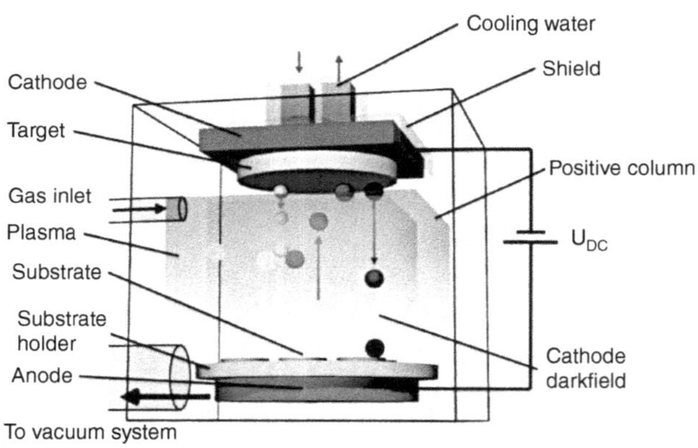

Figure 11.11 Schematic of DC-sputtering equipment.

often leads to arcing in areas of the target not exposed to intensive bombardment with ions.[1] Current understanding indicates that the positive charges accumulated on the dielectric and can cause arcing when they exceed the dielectric strength of the film. In pulsed DC sputtering, the power that maintains the plasma is cycled between on and off states, allowing the electrons from the plasma to discharge the film. If the discharge is not sufficient, then over time, charge can still build up and lead to breakdown of the film. The on/off duty cycle during pulsed DC thus has to be carefully considered.

11.7.2 HF/RF Sputtering

HF or RF sputtering prevents charging of the target material, since due to the changing field, no ions can accumulate in front of the target. Hence, the target material does not have to be conductive to allow a working sputtering process. HF sputtering allows an almost entirely free choice of target materials (such as metals and dielectric materials). The schematic layout is depicted in Fig. 11.12. The HF generator frequency is typically 13.56 MHz.

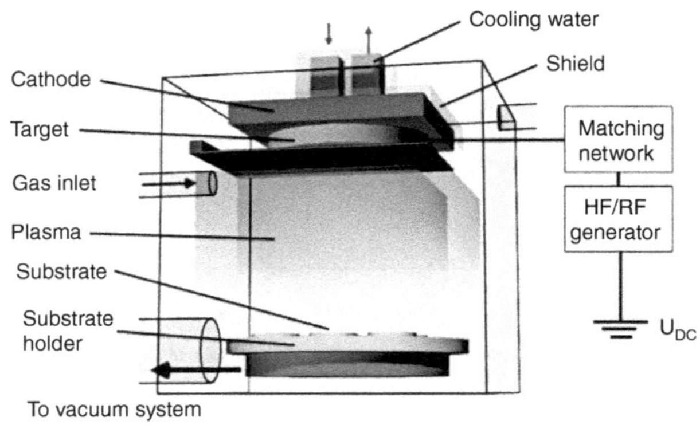

Figure 11.12 Schematic of HF/RF sputter process and equipment.

11.7.3 Self-Bias Effect

During the positive half wave of the HF voltage, more electrons than ions reach the target because of their higher mobility. Therefore, the target electrode charges negatively (self-bias) until equal numbers of ions and electrons reach the target. If the potential difference between the discharge and the self-bias is large enough, particles will be removed from the target. The significantly different sizes of the two electrodes prevent a sputtering of the substrate and susceptor (the substrate carrier is much larger than the target) during a normal sputtering operation (Fig. 11.13).

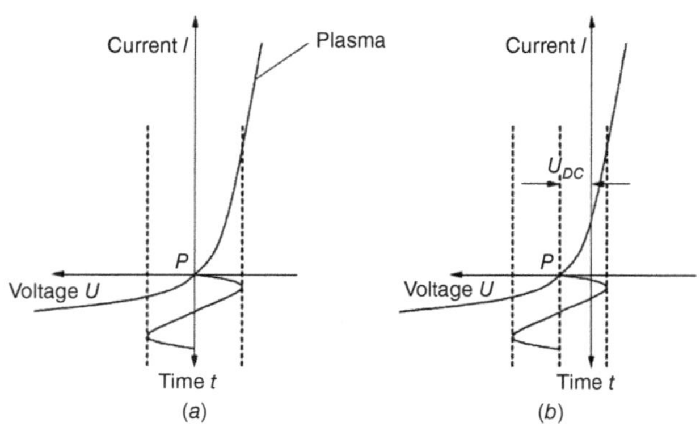

Figure 11.13 Self-bias effect in HF/RF sputtering: (*a*) initial state of current versus voltage (I–V) curve on switching on the HF voltage (excess electrons), (*b*) shift of I–V curve due to U_{DC} bias voltage (charging of the target, until the number of electrons and ions hitting the target is equal) [1].

11.7.4 Bias Sputtering

In bias sputtering (not to be confused with self-bias!), a negative bias voltage (50 to 500 V) is used for continuous bombardment of the substrate with Ar ions during sputtering. The basic mechanism is that

[1]Pulsed-DC Reactive Sputtering of Dielectrics: Pulsing Parameter Effects, A. Belkind and Z. Zhao, Stevens Institute of Technology, Hoboken, NJ; and D. Carter, L. Mahoney, G. McDonough, G. Roche, R. Scholl, and H. Walde, ©2000 Society of Vacuum Coaters 505/856-7188, 43rd Annual Technical Conference Proceedings—Denver, April 15–20, 2000 ISSN 0737-5921.

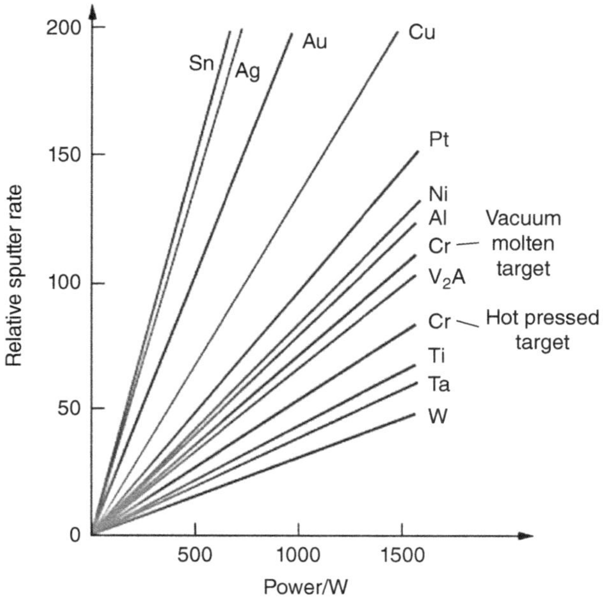

Figure 11.14 Relative sputtering rate as function of sputter power [4].

Ar ions transfer energy to the surface atoms thereby increasing their surface mobility and reaction rate. Bias sputtering (Fig. 11.14) is used in DC and HF/RF sputtering. This process is used for the following:

- Backsputtering (E_{ion} > 100 eV) is used for sputter etching to clean the substrate surface prior to metal deposition by removing native oxide over contacts or organic residue. It can also be used as a physical etching process to remove material that is otherwise difficult to pattern using wet or other dry chemical etching processes.
- In situ annealing processes to modify the layer properties during deposition due to the constant bombardment of the layer with ions leading to heating of the wafer as well as a potential layer surface damage due the impacting Ar ions.

These mechanisms influence

- Implantation of Ar ions into the layer
- Step coverage of the layer
- Layer stress
- Grain size
- Electrical properties like specific resistance and temperature coefficient of resistance (TCR)
- Surface roughness R_a
- Layer adhesion
- Layer density
- Layer hardness
- Pinhole density
- Layer composition

11.7.5 Reactive Sputtering

In reactive sputtering, a chemical reaction between target atoms and a reactive gas occurs. The resulting layer is a compound of the target and the reactive gas material (e.g., $Ti + O_2 \rightarrow TiO_2$). The reaction can take place (*a*) at the target, (*b*) in the gas-phase, or (*c*) on the substrate. Reactive sputtering is frequently used for oxides, carbides, and nitrides.

11.7.6 Magnetron Sputtering

In conventional sputtering, only a few secondary atoms contribute to further ionization of Ar atoms. Most electrons are collected at the anode leading to heating of the substrate. Magnetron sputtering increases the number of electrons contributing to ionization of Ar atoms by using an electromagnetic field. The E and B fields cause a cycloid motion of the charge carriers in the plasma. The deflection radii of the electrons are much smaller than that of the ions. Therefore, the electrons concentrate close to the target leading to a larger probability of ionization and thus, a higher sputter rate. The (**E** × **B**) drift causes electrons to flow to a special anode rather than to the substrate, therefore reducing heating of the substrate. Magnetron sputtering, naturally, has a larger impact in DC sputtering, but is used in HF/RF sputtering as well. The drawback of magnetron sputtering is the inhomogeneous use and ablation of the target (Fig. 11.15) leading to early need for replacement. New methods

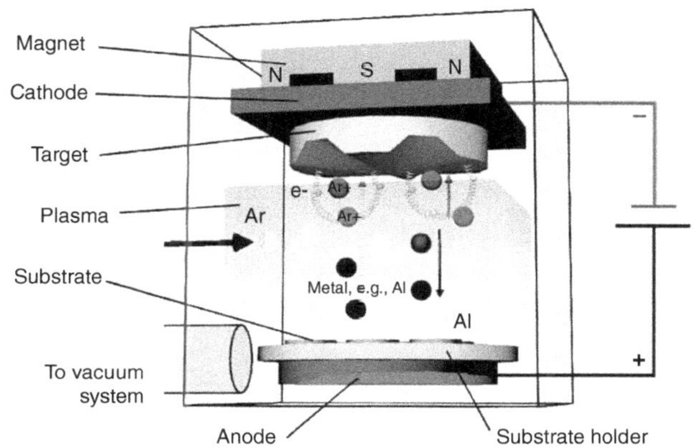

Figure 11.15 Schematic of magnetron sputter process and equipment; electrons (yellow) traveling on ellipsoidal paths, Ar (white) is ionized (red) and accelerated towards the target, releasing, e.g., sputtered Al (black) [5].

with more complex electromagnetic field geometries allow for a more efficient use of the target material. Typically, a target supplier will reclaim the remaining material on the base plate, credit the customer and charge only for the extra material as well as the sintering and bonding of the new target material.

11.7.7 Sputtering System Layout and Components

Sputtering systems have been built in a multitude of different ways. The schematic layout of the most common systems today is given in Figs. 11.16 and 11.17. The core components of such a system are the vacuum system, the RF/DC power supply, and matching network. The key components are described below.

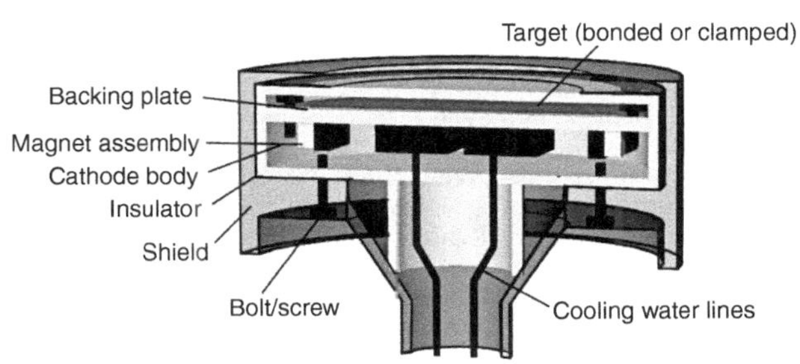

Figure 11.16 Schematic of magnetron sputter target including cooling water system.

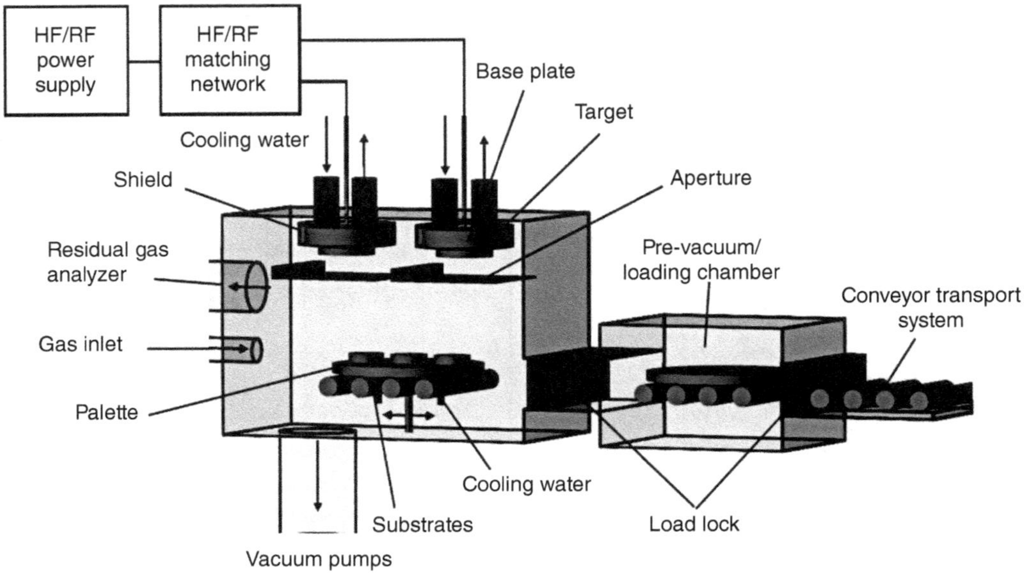

Figure 11.17 Schematic of a horizontal RF magnetron sputter machine with a loading chamber, multiple targets, and a rotating substrate pallet.

Vacuum system. The vacuum system typically consists of two chambers—the main processing chamber and an airlock (pre-vacuum chamber). The airlock is used to accelerate processing and reduce contamination of the main chamber, since venting of the main chamber is not required for substrate loading. The vacuum system is usually a two-stage system with a mechanical coarse vacuum pump and a turbo high vacuum pump. Ion current gauges are used to determine the process chamber pressure at high vacuum. A residual gas analyzer is used to determine the concentration and composition of the residual gas.

Cooling water. For chamber temperature control, a cooling water system (dual-walled chamber or welded steel tubing) and a radiation heater is used. A good chamber design eliminates the possibility of cooling water entering the chamber and getting in contact with electronics and current-carrying or high-voltage parts in case of a leakage or system breakdown.

Cathodes and targets. Most systems have multiple cathodes that allow the sputtering of two to four different materials in one run either simultaneously or sequentially. In addition to the electrical switching network, a mechanical aperture/shutter system is used to select between targets. The aperture helps define the sputter column shape and layer uniformity. The shutter will ensure that an "off" target will not only not receive a sputtering voltage, it will also be mechanically sealed to prevent exposure to accelerated Ar ions, reaction with the other sputtered material particles or simply mechanical chipping of the target that leads to deposition of macroscale contamination particles on and in the deposited layer.

Two standard configurations exist for the substrate holder and targets—horizontal and vertical sputtering (Fig. 11.20). In both cases the distance between target and substrate is typically between 15 and 35 cm, depending on the target and palette sizes and geometries as well as the sputtering parameters. Horizontal sputtering allows for a simple substrate holder (e.g., rotating disc) that passes underneath the targets. Most horizontal sputter targets have a circular shape with diameters between 4 and 12 in. Targets usually consist of a base plate with integrated water-cooling and magnets (magnetron sputtering) and the bonded target material (typical thickness 3 to 5 mm). The target base plate is reusable. In use, operators have to monitor the target use. Traces of the base plate material in the deposited layer composition indicate that the target has been ablated down to the base plate and requires replacement. Normally quality and process control will lead to replacement of a target prior to sputtering of the base plate material. The substrates are placed on the disc typically without any fixture holding them in place. The disc rotation is off center of the target to create a homogeneous layer thickness. More complex systems feature a planetary motion. An advantage of the horizontal system is its flexibility—experimental setups (e.g., fixtures holding a shadow mask aligned above a substrate) and devices with complex geometries can easily be placed on the horizontal plate without need for a special fixture. The key drawback is that bulk material that falls off the targets as well as flakes of deposited material on the chamber walls can fall onto the substrates, contaminating them.

Vertical sputtering systems eliminate this drawback. They do, however, require a more complex substrate fixture to keep wafers or samples secured to the base plate without falling off and to ensure a good thermal and electrical contact to the substrate holder. Bad electrical or thermal contact will lead to variations in layer thickness and structure across the sample. Vertical sputtering systems exist as simple linear designs where the substrate passes parallel along one or more targets and as rotary systems where substrates are mounted on a barrel-type holder. Vertical sputtering allows sequential multilayer deposition, but in most cases not simultaneous deposition, since the substrates can only face one target at any given time.

11.8 LAYERS DEPOSITED USING SPUTTERING

Sputtering allows the deposition of almost any kind of material. It allows the deposition of a free choice of metals, including those with high melting temperature. Depending on the target size and material, target prices can vary a lot (e.g., Al1Si target 10 in × 5 mm: \$1000; Pt target 10 in × 5 mm: \$18,000). All alloys can be deposited using sputtering, depending on the availability of targets for the specific sputtering system. Multilayer structures can be deposited using multiple target sputter equipment with apertures to prevent cross contamination of targets. Chemical compounds (dielectric materials, insulators, metal oxides) can be deposited either using specific compound targets (mostly sintered material) and/or using reactive sputtering. A typical configuration is using a metal or semiconductor material target and a reactive ambient

Table 11.1 Examples of Metallization Systems Used in Semiconductor Devices and Selected Physical Properties [6]

Metals: Au, Pt, Pd, Ni, Ti, Al, Cr, Mo
Alloys: NiCr, CrSi, TiW
Multilayers: Cr-Al, Ti-Au, Ti-Pd-Au, Ti-TiN-Au, Ti-TiWN-Au, NiCr-Ni-Au, SnO_2, Cr-Al
Chemical compounds: Al_2O_3, SnO_2, SiO_2, ZnO, Ga_2O_3, HfB_2, NiO, V_2O_5, Mo_2O_3, In_2O_3, IrOx glass (Pyrex)

gas atmosphere (O_2, N_2) to build oxide or nitride layers. A selection of materials and physical properties (Table 11.1) is given below:

Metals: Au, Pt, Pd, Ni, Ti, Al, Cr, Mo

Alloys: NiCr, CrSi, TiW

Multilayers: Cr-Al, Ti-Au, Ti-Pd-Au, Ti-TiN-Au, Ti-TiWN-Au, NiCr-Ni-Au, SnO_2, Cr-Al

Chemical compounds: Al_2O_3, SnO_2, SiO_2, ZnO, Ga_2O_3, HfB_2, NiO, V_2O_5, Mo_2O_3, In_2O_3, glass (Pyrex)

11.8.1 Step Coverage

An important issue in all coating processes is step coverage of textured and patterned substrate surfaces. Poor step coverage can lead to microcracks and interruption or breaking of a layer coating. Vacuum evaporation deposition follows the cosine law with the vapor particles coming straight from a point vapor source in high vacuum with little collisions and scattering. The subsequent coating process is almost anisotropic; that is, surfaces facing the vapor source are coated whereas the ones perpendicular to those remain almost uncoated, leading to poor step coverage. This characteristic, however, favors lithographic lift-off processes as patterning processes. Sputter deposition exhibits a large area vapor source and higher process pressures leading to higher frequency of collisions between vapor particles. In consequence, the vapor particles approach the substrate surface from more random directions leading to higher deposition isotropy and better step coverage. In comparison, chemical vapor deposition (CVD) layers almost always feature better step coverage than PVD layers. Process pressures in CVD are typically higher than in PVD. Furthermore, the source is a strong constant gas flow of process gases, needed as precursors for the layer (e.g., 3 sccm methylsilane in 200 cm^3/min hydrogen at 3.5 L/min total flow for LPCVD silicon carbide deposition). This leads to high particle collision frequency and partially turbulent gas flow causing a completely isotropic coating process and excellent step coverage. Only in deep high aspect ratio cavities, as caused, for example, by *deep reactive ion etching*, diffusion governs the coating characteristics in CVD (Figs. 11.18 and 11.19).

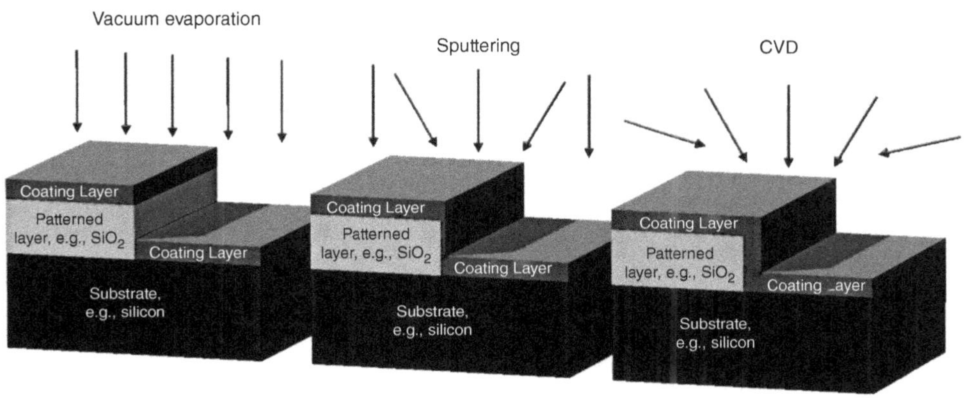

Figure 11.18 Step coverage of vacuum evaporation, sputtering, and CVD in comparison [4].

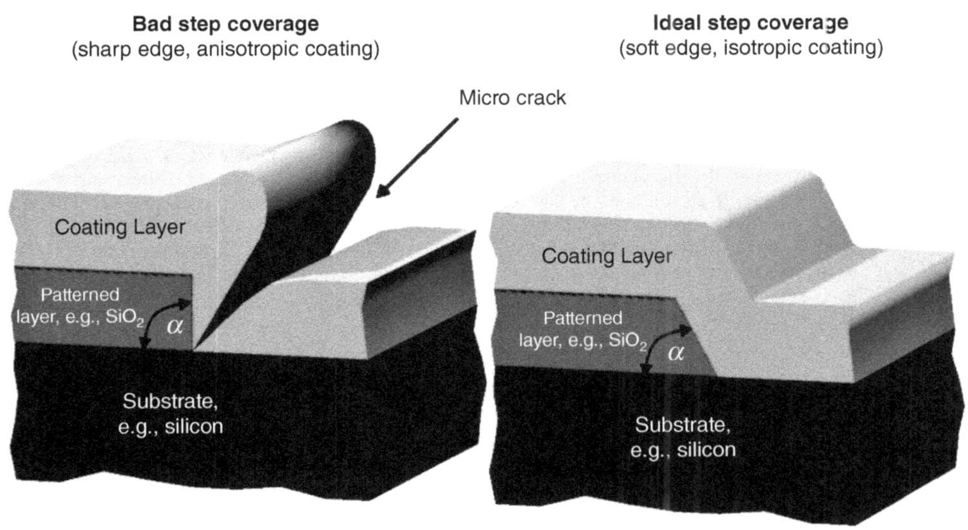

Figure 11.19 Impact of step coverage on layer coherence [4].

11.8.2 Pulsed Laser Deposition

Pulsed laser deposition is a newer technique developed in the last decade that is so far being used largely in research settings. It uses energy from a pulsed laser to ablate material in a vacuum from a target, leading to a plasma plume consisting of uncharged particles, ions, and electrons in which the material is transported at velocities of over a 1 m/s to a substrate. The deposition occurs in three phases: laser ablation/plasma creation, transport to the substrate, and condensation/film creation (Fig. 11.20).

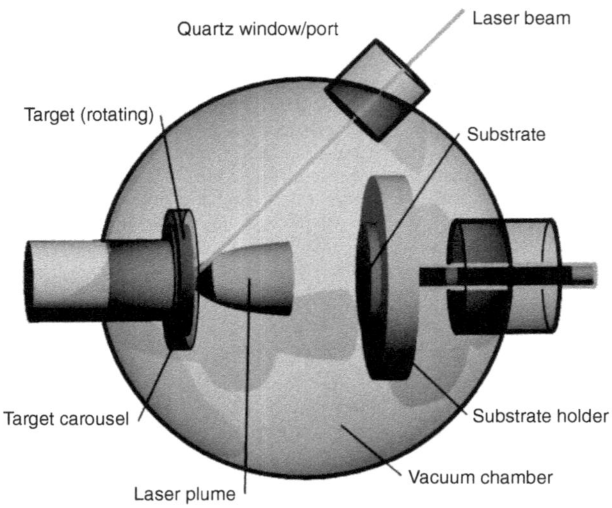

Figure 11.20 Schematic of pulsed laser deposition system with vacuum chamber, target and substrate holder, and quartz window for laser access [9].

The key driver for the development of this technique is the superior retention of stoichiometric composition of complex materials during deposition. Materials such as, for example, YBCO, Y-stabilized BaZrO, and others that are being used in electroceramic applications have significantly benefitted from this technique when transitioning these materials from their bulk, sintered form to thin films [6–8]. In addition, the high deposition rates greater than 10 nm/min, and the fact that the energy source is outside the vacuum chamber eliminating contamination from, for example, heated filaments are strengths of this technique. Laser energy and pulse rate, substrate temperature, and residual pressure impact the film properties requiring optimization which in the past was largely down empirically due to a lack of complete understanding of the processes. Challenges include the high directionality and small coverage of the plasma plume requiring rastering/scanning the laser plume or substrate movement with respect to the plume to obtain uniform deposition rates and surface coverage. In addition, the plasma often includes macroscopic molten particles that impact film growth. Mechanical velocity filters as well as off axis deposition and collision between two plasmas have been used to remove these macroscopic particles [9]. Efforts are under way to better understand and model the laser ablation and deposition process, which may constitute the basis for future industrial/larger scale use of the technology. Optical spectroscopy and laser-induced fluorescence can be used to characterize the plasma for research purposes.

11.8.3 Summary and Outlook

PVD is used for metallization layers, metal oxides, and some dielectric materials. It is essentially a "cold" process that does not require substrate heating per se. Substrate heating can, however, be used to influence the layer properties. Evaporation yields higher purity material, but the material choice is limited and the isotropic nature of the deposition process causes poor step coverage. Sputtering allows deposition of a huge choice of materials with good reproducibility and exhibits better step coverage. Most standard metallization layers are deposited using sputtering.

The dividing line between PVD and CVD processes appears to vanish. PVD and CVD processes are frequently combined in one piece of equipment. Evaporation is on a decline due to high competition from sputtering and CVD, with sputtering remaining the dominant PVD coating process in semiconductor processing. Key engineering challenges in sputtering lie in better equipment reliability (down times < 10 percent), increased throughput and automation, higher layer uniformity, and more complex multilayer deposition techniques that will require more complex sputter etching or in situ cleaning and annealing techniques. This will also lead to further consolidation of the equipment manufacturer market. From a scientific point of view, little understanding exists today of the fundamental principles underlying the impact of the process parameters on layer properties, in particular with multilayer coatings as well as the material's electrical, mechanical, and chemical properties in harsh environments (i.e., high temperatures, aggressive gas, and fluid media). This will become a focal point of future research on PVD coating processes and layers.

11.9 REFERENCES

1. W. Wehl, *Mikrotechnische Fertigung, Lecture Scriptum*, Heilbronn, Germany, 2002.

2. W. Menz, *Mikrosystemtechnik für Ingenieure*, VCH Verlag, Weinhein, Germany, 1997.

3. K. Schade, *Mikroelektroniktechnologie*, Verlag Technik, Berlin, Munich, Germany, 1991.

4. W. Pupp, *Vakuumtechnik*, Hanser Verlag, Munich, Germany, 1991.

5. R. C. West (ed.), *CRC Handbook of Chemistry and Physics*, CRC Press, Boca Raton, Florida, 1987.

6. X. Chen, L. Rieth, M. S. Miller, and F. Solzbacher, "Pulsed Laser Deposited Y-Doped $BaZrO_3$ Thin Films for High Temperature Humidity Sensors," *Sensors and Actuators B: Chemical*, 142 (1), 166–174, 2009.

7. M. Snure, D. Kumar, and A. Tiwari, "Progress in ZnO-Based Diluted Magnetic Semiconductors," *JOM*, 61 (6), 72–75, 2009.

8. G. Siegel, M. C. Prestgard, S. Teng, and A. Tiwari, "Robust Longitudinal Spin-Seebeck Effect in Bi-YIG Thin Films," *Scientific Reports* 4, article number 4429, pp. 1–6, Mar. 21, 2014.

9. *An Introduction to Pulsed Laser Deposition*, Oxford Instruments, Abingdon, Oxfordshire.

Chapter 12

Chemical Vapor Deposition

BY

BIN DONG *Sun Yat-sen University, Guangzhou, China*
M. SKY DRIVER *University of North Texas, Denton, Texas*
JEFFRY A. KELBER *University of North Texas, Denton, Texas*

12.1 INTRODUCTION

Chemical vapor deposition (CVD) and related methods, such as plasma-enhanced CVD (PECVD) and atomic layer deposition (ALD) are widely used in the microelectronics industry, and in many other areas, such as photovoltaics manufacture, ceramic protective coating, and even antireflection coating materials. The purpose of this chapter is to provide a general introductory review of this topic, with particular emphasis on applications in microelectronics fabrication. For more detailed information, readers are referred to several comprehensive reviews on this subject [1–4].

CVD is one of film deposition methods, and other choices include molecular beam epitaxy (MBE) and, particularly in the microelectronics industry, magnetron sputter deposition—often termed physical vapor deposition (PVD). CVD, however, has become more popular as the geometries of integrated circuit features have continued to shrink. While currently used in Si complementary metal-oxide-semiconductor (CMOS) microelectronics for front-end transistor fabrication, interconnect fabrication, and packaging, CVD, especially PECVD and ALD are also of interest for emerging electronic materials in nanoelectronics and spintronics.

Section 12.1 provides an introduction to the basic concepts of CVD, including homogenous and heterogeneous nucleation, and variants such as PECVD, ALD, and some other methods. Section 12.2 provides a brief history of the development of CVD and examples of its application to Si, metal, and oxide thin film deposition. Section 12.3 discusses the important issue of conformality in CVD, and associated issues of step coverage and sticking coefficient (β). Section 12.4 focuses on kinetics and thermodynamics considerations in the CVD process. Section 12.5 provides a selective description of the use and a short summary of CVD and related methods in emerging materials, including two dimensional (2D) materials.

12.1.1 Homogenous and Heterogeneous Nucleation

CVD is a type of deposition process involving the dissociation and/or chemical reaction of one or more gaseous reactants—precursors, and the formation of a solid product on the substrate. Precursor dissociation/reaction can occur in either the gas phase (homogeneous), or at the substrate surface (heterogeneous reaction and nucleation). This distinction is shown schematically in Fig. 12.1 for a generic precursor (AX_4) reacting to form a solid (A) via the generalized decomposition reaction:

$$AX_4(g) \rightarrow A(s) + 2X_2(g)$$

where (g) and (s) refer to gas and solid phase, respectively. As shown in Fig. 12.1, a hot wall reactor (as the name implies) can supply sufficient thermal energy to induce reaction at or near the reactor wall interior surfaces, resulting in the nucleation of a desired product which can then land on the substrate or the surface of the growing film. In contrast, in a cold wall reactor, where only the substrate is heated, precursor dissociation/film nucleation occurs at the surface of the substrate or the growing film. Unless reactions are carefully controlled, however, homogenous nucleation can result in the formation of particles or flakes at the surface of the film [5]. Heterogeneous nucleation is often used where formation of smooth, highly conformal, and/or epitaxial films is desired [6]. Heterogeneous nucleation, however, is not without its potential drawbacks. If the precursor reaction probability at the surface of the growing film A (Fig. 12.1) is much less than at the surface of the substrate, film growth may passivate after the formation of the first 1 to 2 monolayers. An example is the decomposition of borazine ($B_3N_3H_6$) to form hexagonal BN on various ordered transition metal substrates. In a cold wall reactor, BN film growth on a metallic substrate by borazane or borazine pyrolysis is typically limited to approximately 1 monolayer [7, 8]. Another variation of this theme is "hot-wire CVD" [9, 10] in which precursor pre-dissociation is achieved by reaction with a heated filament prior to deposition on the substrate surface. Although similar to PECVD (see further), this method can avoid potential plasma-induced surface damage [11].

12.1.2 Variants of CVD

Typically the energy supplied for precursor dissociation/reaction is thermal energy. In "hot wall" reactors, dissociation can occur at the heated reactor walls. While in "cold wall" reactors, only the substrate is heated, leading to heterogeneous reaction and nucleation mainly at the substrate surface. Alternative methods to induce precursor reaction include plasma—PECVD, and photo-assisted CVD (PACVD). The source of energy for precursor reaction can thus be used to classify CVD into thermal CVD or other methods, such as PECVD, photo-assisted CVD (PACVD), hot-wire CVD, etc.

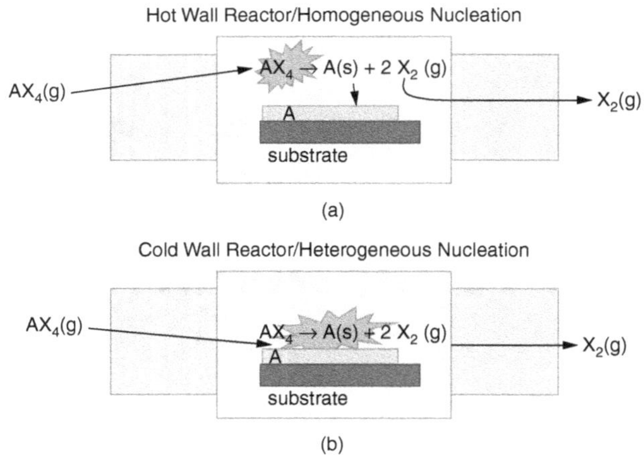

Figure 12.1 The difference between heterogeneous and homogeneous nucleation. (*a*) In a hot wall reactor, the precursor (AX_4) can react at the wall of the reactor, forming solid nuclei (A) which can then land on the surface. (*b*) In a cold wall reactor, only the substrate is heated. The precursor reaction will occur only at the surface of the substrate or the growing film.

Plasma-Enhanced CVD

Plasma-enhanced CVD (PECVD) process is shown schematically in Fig. 12.2. In PECVD, partial dissociation of the precursor is achieved by reaction in plasma prior to precursor/substrate interaction. Such a method is particularly useful when precursor thermal dissociation would require impractically high temperature, or when the deposited film or substrate becomes unstable at elevated temperatures. PECVD, for example, is commonly used in the preparation of semiconducting boron carbide films from carborane ($B_{10}C_2H_{12}$) precursors [12]. This process leads to B–H bond scission in the plasma environment, followed by precursor reaction, cross-linking, and film growth [13]. PECVD compared to other CVD methods has its benefits by being able

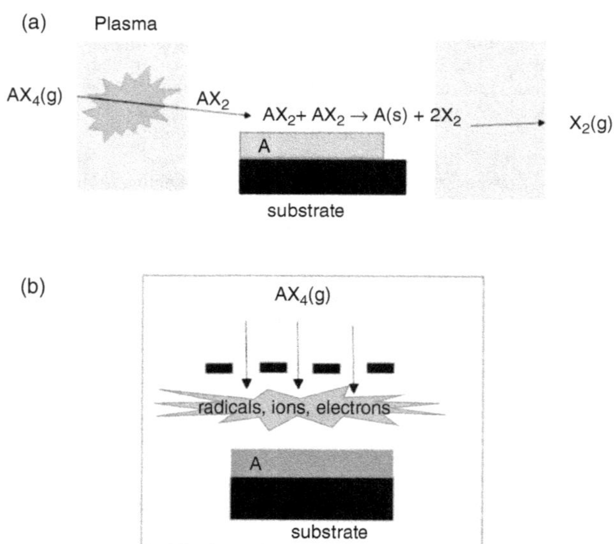

Figure 12.2 (*a*) Remote plasma-enhanced CVD. A plasma is used to achieve partial dissociation of the precursor (AX_4) prior to reaction with the substrate. Photo-enhanced CVD is similar except that the energy for dissociation is supplied by photons (typically ultraviolet). (*b*) Direct plasma-enhanced CVD. Plasma bombardment near the film surface results in direct bombardment by free radicals, ions, and electrons during deposition.

to tune the various parameters to obtain a specific and desirable chemical and electrical structure for a given film. As shown in Fig. 12.2*a*, plasma-induced dissociation can occur relatively far from the film or substrate surface (indirect PECVD), or (Fig. 12.2*b*) in the vicinity of the film or substrate surface (direct PECVD), which results in precursor dissociation, but also direct bombardment of the film/substrate by ions, energetic photons, and electrons, in addition to neutral radicals. Obviously, precursor pre-dissociation during PECVD can lead to (often) undesirable homogeneous nucleation unless careful control is exercised over plasma and deposition conditions [14]. Photo-assisted CVD is analogous to PECVD (Fig. 12.2) except that precursor pre-dissociation is induced by reaction with sufficiently energetic photons.

A variety of discharge plasmas can be applied to CVD for thin film deposition with different applications. Various sources include: dielectric barrier discharge, RF plasma, and microwave plasma, among others [15]. There are two power coupling forms by which RF discharges can be accomplished: capacitively and inductively coupled discharge. Capacitively coupled plasma (CCP) is generated between two electrodes connected with a 13.56-MHz RF power supply and ground. In a capacitively coupled RF discharge, the electron density is in the range of 10^9 to 10^{10} cm^{-3} with varied frequencies. Inductively coupled plasma (ICP) is similar to CCP but the electrode consists of a coil wrapped around the discharge volume where it inductively excites the plasma. The electron density is as high as 10^{12} cm^{-3} at lower ion energies compared with CCP [16].

Even PECVD can overcome a low rate of initial thermal CVD nucleation by introducing reactive free radicals and/or energetic ions and radicals, the problem of too extensive decomposition of precursors can be an issue. An example is the PECVD of organic films from aromatic precursors. Thiophene, pyrrole, and aniline monomers have been used to form conducting films by PECVD. Thiophene films display low conductivity because of the loss of aromatic structure in the plasma process [17]. Researchers have addressed this problem by employing pulsed PECVD to reduce the extent of precursor dissociation [18]. Studies of PECVD of aromatics like pyrimidine, pyridine, and aniline combined with boron carbide using inductively coupled plasma indicate that the precursor resonance stabilization energy is an accurate predictor of precursor dissociation during PECVD [13, 19]. Resonance stabilization energy is, roughly, the difference in calculated heat of formation of precursor's most stable valence bond structure and the experimental value [13, 19]. For aniline and pyridine, the presence of a π electron system is observed in the deposited film over a range of plasma power from 5 to 20 W. However, the low resonance stabilization energy of pyrimidine results in facile precursor aromatic ring opening during PECVD, even at plasma power of 5 W. Therefore, precautions need to be considered regarding the stability of some precursors in even the mildest of plasma environments.

Atmospheric Pressure PECVD

For efficient production in modern industrial processes, maximizing the film deposition rate while maintain good film quality is often critical. Therefore, plasma kinetics are often optimized for maximum deposition rate. Atmospheric pressure PECVD has received considerable attention for deposition of organic polymers and oxide films. A number of review articles have been published on the subject [20–24]. Atmospheric plasma discharges can be excited by a variety of methods, including RF with frequency of 13.65 MHz [25, 26] and of 150 MHz [27–29] termed very high frequency (VHF) discharge. When VHF is used for the trapping of electrons, a stable discharge is formed after gas breakdown, and deposition at higher growth rates can be often achieved than if RF discharge is used.

Photo-Assisted CVD

Photo-assisted (PACVD) can be and is often used for the UV-induced dissociation of a precursor [30]. The sources for PACVD include arc lamp, CO_2 lasers, excimer lasers and argon ion lasers, excluding the lamps as heating sources [2]. The outstanding advantages of PACVD over CVD or PECVD include the ability to focus the laser or light source on a localized area of the substrate, and to selectively dissociate certain chemical bonds but not others [2]. Low photon energies can diminish the film damage from ion bombardment which is encountered in PECVD. Lower deposition temperature in PACVD may also decrease the proportion of various defects, and also decrease the diffusion of dopants in the substrate—common disadvantages in high temperature deposition, which make the photo-assisted CVD one of the most important techniques in fabricating large-scale integrated circuits and optical coating [31, 32].

Hot Wire CVD

Hot wire CVD (HWCVD)—sometimes termed hot filament CVD (HFCVD) involves the use of a heated filament (often tungsten) to pre-dissociated precursors during the deposition process. For detailed information, including commercial adaptations, the reader is referred to a comprehensive review [9]. HWCVD is often used, for example, in the preparation of amorphous hydrogenated Si films, where hot wire-induced Si–H bond scission permits film formation and growth at lower temperatures than in conventional CVD. HWCVD is also used for formation of diamond-like C films, where filament-induced dissociation of C–H and H–H bonds forms reactive CH_3 and H radicals that promote diamond growth and inhibit graphite formation [33].

The above variants of CVD employ some methods for precursor pre-dissociation, thus avoiding high thermal temperature, and in the case of photo-excitation or hot wire pre-dissociation, avoiding sample damage that may occur in a PECVD environment. Growth rates and film conformality in general depend on other parameters of reaction conditions—precursor flow rates, carrier gases species, etc. In such cases, however, the film growth rate (at least after the initial layer), is relatively constant. Depositing a film at a pre-determined film thickness depends on accurate timing and rigorous control of reaction conditions. This can often be challenges where atomic level control of film thickness is required—as are emerging in spintronics, nanoelectronics, and electroluminescent displays [34]. In such cases, atomic layer deposition possesses certain important advantages.

Atomic Layer Deposition

An emerging technique for deposition of thin films is atomic layer deposition (ALD). ALD exploits the use of various chemical bonding environments in a process of two or more sequential gas phase chemical processes—each process being self-terminating. A typical ALD process cycle consists of four steps:

1. Exposure of the substrate surface to precursor (A), resulting in the self-terminating adsorption of A on the surface
2. Pump down or purge of the reactor
3. Exposure of the adsorbate-modified substrate to precursor (B), resulting in a self-terminating reaction with A to form the B-terminated intermediate
4. Pump down/purge of the reactor

Step 1 is then repeated.

The cyclic nature of the ALD process generally has two important advantages. First, the film thickness is in general linearly proportional to the number of A/B cycles [35–38]. This yields atomic level control over film thickness. A second important advantage is that, given sufficient pulse/purge times to allow precursor dispersion into high aspect ratio features, highly conformal films may be obtained [37]. Such considerations can limit the speed and throughput of the deposition process—important considerations in industrial applications. Thus, ALD becomes particularly important for the formation of very thin films of high quality and conformality. The technique is now widely used in the microelectronics industry for deposition of a variety of metals, oxides, and nitrides. For a detailed explanation of the various aspects of the ALD process, and the types of films produced by ALD, the reader is referred to two recent reviews [36, 37].

ALD is, as previously discussed, of increasing interest in emerging areas of nanoelectronics and spintronics. Frequently, such applications require the deposition of epitaxial, and azimuthally oriented films with precise control of film thickness [39, 40]. Multilayer epitaxial hexagonal BN(0001) films have been produced using BCl_3/NH_3 cycles, and certain transition metal dichalcogenides films have also been produced [35].

As with CVD, the ALD process is scalable to large area substrates. However, the desire to increase throughput has led to a variant sometimes known as spatial ALD [37]. This method involves precursor exposure to limited areas of the substrate, or even the exposure of different precursors over different substrate locations. This allows the near-simultaneous growth of different films over different areas of the wafer.

Given appropriate interfacial interactions and deposition conditions, ALD can be used to deposit either polycrystalline or highly ordered, epitaxial films. The latter case is often termed atomic layer epitaxy (ALE). ALE has been used to grow, for example, ZnO and Al_2O_3 films for optical applications [41]. This method has been also used to fabricate low-defect semiconductor/high-k dielectric interfaces [42]. ALD or ALE method appear to be of growing importance for growing spintronics and nanoelectronics, where proposed applications often have extraordinarily stringent requirements for film thickness and orientation [39, 40]. 2D materials, including MoS_2 [43] and h-BN(0001) [38] have been grown by ALE.

12.1.3 Summary

In summary, CVD and its variants are widely used in the semiconductor industry and related areas to deliver deposited films of high quality and conformality. Typically, CVD itself relies on thermally induced precursor dissociation, but variants such as PECVD, PACVD, and HWCVD can be used to supply energy for precursor dissociation when thermal dissociation is impractical or undesirable. ALD is a variant of CVD that is of emerging importance for depositing very thin films with rigorous control of film thickness and structure.

CVD-based methods generally have the important advantage of being able to deposit highly conformal films, even in high-aspect-ratio vias [37, 44]. This issue is explored in greater detail in Sec. 12.3, but given the increasingly high aspect ratios prevalent in modern device architectures, it is reasonable to predict that conformality will become an issue of even greater importance in the near future. Additionally, the use of such methods to deposit films of emerging electronic/spintronic materials is described in Sec. 12.5.

12.2 HISTORY

In the 1950s and 1960s, materials fabricated by CVD entered the electronics fields. A major development of this area was the replacement of Ge by Si as the basic semiconductor for device fabrication [45]. The success of Si fabrication by CVD rapidly achieved industrial importance, and Si CVD is commonly applied in the manufacture of different silicon-based semiconductor devices [46–48]. Additionally, the continued downsizing of device geometries has led to CVD of numerous metals and alloys as contacts, or as interconnects.

12.2.1 CVD in Si Microelectronics and Other Areas

Si films—crystalline, polycrystalline, or amorphous are now widely used in microelectronics fabrication [49, 50], including channels for field effect transistors, and in forming silicide electrical contacts [51], also in areas such as photovoltaics [52, 53]. The increasing complexity of "conventional" Si CMOS devices, as well as the interest in Si deposition for emerging applications, has placed increasing emphasis in deposition at low temperatures. Because of such temperature constraints, PECVD is becoming increasingly popular [54]. PECVD of Si with SiH_4 as a precursor has been accomplished at 350°C [2]. $SiF_4 + H_2$ precursors have also been used, at similar temperatures [55]. Similarly, HWCVD has been used to achieve low temperature Si epitaxy on Si (100) in this temperature range [56].

Amorphous Si (a-Si) is of interest as a junction in solar cells because strong light trapping characteristics. The use of amorphous Si for high-efficiency solar cell applications has been reviewed [57–59]. PECVD has been introduced for the deposition of amorphous silicon in large areas with film uniformity and high conversion efficiency. In a solar cell, a-Si can be applied as the intrinsic layer in a p-i-n junction. PECVD has been used for the deposition of photovoltaic layer-amorphous Si using VHF plasma [27] to achieve deposition rates of 128 nm/s, and conversion efficiencies as high as 8.25 percent have been also achieved. Other advantages of a-Si include relatively low-light-induced degradation [60].

12.2.2 Other Materials

The use of CVD to deposit transition metals in microelectronics fabrication is now extensive and well-reviewed [61], particularly with regards to the effects of surface chemistry and kinetics on the process [62, 63]. In the 1980s and early 1990s, the use of tungsten as a conductive plug to connect different levels of the device or as a diffusion barrier became of interest. WF_6 was and is the most popular precursor of choice, with reducing agents that include H_2, SiH_4, or Si_2H_6 [64, 65]. At that time, CVD of metals and metal reactions with Si to form metal silicide contacts also became of interest [66]. The introduction of Cu in the mid-1990s as an interconnect metal to replace Al, increased interest in CVD of metal fabrication for two reasons. First, CVD or variants such as ALD are now widely used to deposit thin, conformal "seed" layers of Cu for subsequent electrochemical deposition of Cu interconnect lines [67, 68]. Second, the use of Cu interconnects mandated the use of a diffusion barrier to inhibit Cu diffusion into dielectrics or Si areas of the device. This has led to the study of CVD, ALD, or PECVD of various transition metal nitrides, including TiN [69], TaN [70], and metal alloys, such as Co-W [71]. Most recently, the advent of 3D integration and the use of thru-silicon vias-TSVs has led to the investigation of oxide diffusion barriers, such as manganese oxide [72].

The current state of the art in Si-based microelectronics fabrication has been used to a broad variety of metals, nitrides, and oxides that have been deposited by CVD or other variations.

12.3 CONFORMAL CVD FILMS AND VOID-FREE FEATURE FILLING

12.3.1 General Considerations

The continued down-sizing of trenches and other features in microelectronics processing over the last four decades has resulted in very high aspect ratio (AR) features, where the aspect ratio—the ratio of trench height to width—can exceed 10:1 [37]. As shown schematically in Fig. 12.3*a*, diffusion of the precursor into the trench typically yields a higher precursor concentration at the top of the trench than at the bottom (Fig. 12.3*a*, *b*). This can result in faster film growth, closing off the top of the trench to further film growth, and the formation of a void (Fig. 12.3*c*), with degradation of electrical behavior.

Other methods of film deposition, such as magnetron sputter deposition (often labeled physical vapor deposition, or PVD), suffer the same general problem as CVD (Fig. 12.3*a*–*c*)—a higher concentration of reactants or metal atoms at the top of the trench, a faster growth rate near the top of the trench, and void formation. Indeed, PVD has difficulties in void-free filling of features with aspect ratios greater than approximately 5 [73]. A variety of methods, however, have been evolved to address this issue using CVD or variations, such as ALD. These include:

Catalyst-Based Superfill

In this method, a very thin conformal catalyst layer forms preferentially near the bottom of the trench [74]. The adsorbed catalyst helps subsequent CVD of film, leading to a high concentration of catalyst near the bottom of the trench. This contributes to faster growth rate near the trench bottom relative to the trench top, yielding void-free fill, shown in Fig. 12.3*e*, *f*. Trenches with aspect ratios approximately 5 and trench widths less than 30 nm have been filled without voids by this method [74].

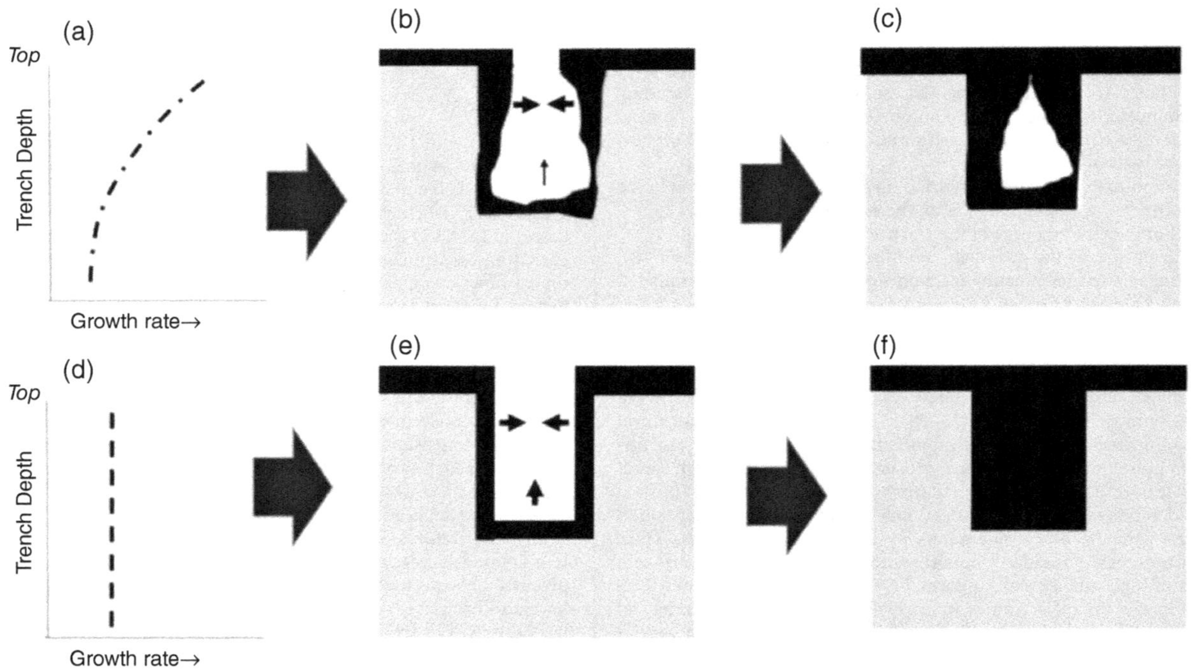

Figure 12.3 Schematics of void formation versus conformal growth in CVD. (*a*) Precursor diffusion into a trench yields a higher precursor concentration at the top of the trench than at the bottom, resulting in (*b*) faster film growth near the top of the trench and eventually (*c*) formation of a void. Conformal growth requires more uniform precursor concentration versus trench depth (*d*, *e*), resulting in void-free fill of the trench.

INHIBITOR-BASED SUPERFILL

In this approach, an inhibitor is used in conjunction with the CVD precursor to decrease the gradient in precursor concentration and growth rate (as in Fig. 12.3*c*–*e*). An example of this approach is the CVD of TiB_2 using the precursor $[Ti(BH_3)_4dme]$ (dme = 1,2 dimethoxyethane) and the inhibitor (dme), at roughly equal pressures (3 to 4×10^{-6} torr) [73]. This approach dramatically enhanced the observed conformality of films, and in principle can be applied to other systems.

ALD

Highly conformal films can in principle be formed by ALD. A key in these applications is to use low reactant concentrations and long diffusion times to achieve uniform precursor absorption during A or B cycles. Of course, this also yields very low growth rates, and ALD is often used for seed layers or liners growth, rather than for complete trench filling.

In general, the above methods seek to address the fundamental issue of a gradient in precursor concentration and growth rate difference between the bottom and the top of the trench (Fig. 12.3). This can be done by catalyst-enhanced preferential growth at the trench bottom, or by the use of an inhibitor to inhibit growth all along the trench.

12.3.2 Step Coverage

The step coverage is defined as the thickness (or growth rate) at the bottom of a trench divided by thickness (or growth rate—GR) at the top of the trench

$$\text{Step coverage} = \frac{\text{thickness(bottom)}}{\text{thickness(top)}} = \frac{\text{GR(bottom)}}{\text{GR(top)}}$$

The growth rate drops with depth of the feature (as in Fig. 12.3*a*, *b*) is caused by the precursor reaction on the walls and can be presented using a molecular diffusion coefficient (D) written as

$$D = D_0 d$$

where d is the feature size: trench width or via diameter, D_0 is diffusion constant.

For a nearly conformal CVD film, the step coverage can be calculated by considering the steady-state growth along the trench axis in the following equation [75]:

$$\text{Step coverage} = 1 - \frac{\partial(\text{GR})}{\partial p}\frac{c\rho k_B T}{2D_0}(\text{AR})^2$$

where c is a constant of 2 or 4 for trench or via, ρ is the atomic density of film, k_B is Boltzmann's constant, T is processing temperature in kelvins, P is partial pressure, AR refers to aspect ratio. This equation represents that step coverage in CVD in most cases is less than 1, which yields a situation as in Fig. 12.3*b*, *c*. Only in the cases where growth rate is poorly dependent on precursor concentration or pressure, can a conformal coating with a step coverage near 1 be achieved. An exception to this is ALD, where very low pressures and long diffusion times are used to achieve more uniform precursor adsorption at the top and bottom of the trench (Fig. 12.3*e*, *f*).

Conformal film deposition by CVD or a variant has been achieved for a large number of different materials by optimizing the reaction probability β—the ratio of reaction rate to the impingement rate of reactant. For conformal CVD of thin films deposition, there are a large range of materials have been reported. WSi_x growth into deep trenches with aspect ratio (AR) > = 50:1 has been investigated with both good step coverage and preferred stoichiometry along the trench axis [76]. $NiSi_x$ film by CVD with PF_3 injection has been formed with markedly improved step coverage quality [77].

Wenjiao B. Wang [44] and his collaborators have reported the fabrication of superconformal CVD films by controlling deposition rate along the depth of features. In their system, two molecules have been used to study the diffusion reaction.

$$\text{Step coverage} = 1 - \left(\frac{\partial(\text{GR})}{\partial p_A}\frac{p_A}{D_{0A}} + \frac{\partial(\text{GR})}{\partial p_B}\frac{p_B}{D_{0B}}\right)\frac{ck_B T}{2}(\text{AR})^2$$

where A and B present the two molecular co-reactants. For either reactant, there is maximum peak for growth rate differential versus partial pressure and smaller growth rate on both sides of maximum peak. For the sides where growth rate increases while partial pressure falls, the growth rate differential respect to partial pressure is negative, resulting a step coverage larger than 1. The reason for that is slowly diffusing reactant pressure drops faster from opening to the bottom than the quickly diffusing reactant. Whenever the step coverage is greater than 1, the step coverage is unity with a formation of superconformal coating.

12.3.3 Reaction Probability β

The ratio of reaction rate to the impingement rate is defined as the surface reaction probability β. Understanding of relationship between β and precursor pressure or temperature is another important factor for conformal film deposition. In many cases, β increases when precursor pressure is increased [73]. Because pressure is dependent on depth from the opening, the reaction probability β is similarly a function of position in trench or via. In order to get constant deposition rate everywhere inside of a recessed feature with a conformal coating, microscopic parameter-surface reaction probability β can be maintained in the range of 10^{-3} to 10^{-5} to minimize the pressure drop with depth. In the process of CVD, lower β means better diffusion of precursors into features before reaction, which gives rise to better conformality [78]. Girolami and coworkers [78] have reported conformal film deposition on vias with a high aspect ratio approximately 30. They found that thickness profile along the trench can be optimized by adding a suitable second species serving as a growth inhibitor. In the deposition at high temperature, pressure of inhibitor needs to be taken care of to maintain good conformality.

12.4 THERMODYNAMIC AND KINETIC CONSIDERATIONS

12.4.1 Thermodynamics Mechanism

Thermodynamics provide a useful guide in the selection of processing conditions by providing information on the reaction probability of. The nature and amount of solid and gaseous species present in the reaction and after deposition can be determined from the calculation of thermodynamic equilibrium.

For a given pressure and temperature, the feasibility of CVD reaction can be judged by calculating the change in Gibbs free energy for the reaction (ΔG_r^0) under standard conditions (e.g., reactant pressure of 1 atm):

$$\Delta G_r^0 = \Sigma G_f(\text{products}) - \Sigma G_f(\text{reactant})$$

For the negative Gibbs free energy ($\Delta G_r < 0$), the reaction will occur spontaneously. The equilibrium constant, K, can be calculated by

$$K = \exp(-\Delta G_r^0/RT)$$

where R is the gas constant, and T is the deposition temperature and ΔG_r^0 is the change in the Gibbs free energy for the reaction under standard conditions. Once K is known, the activity (partial pressure) of gaseous species can be evaluated by

$$K = \frac{\text{partial pressure of products}}{\text{partial pressure of gaseous reactants at equilibrium}}$$

Equilibrium constant is a key part for the selection of CVD process conditions.

The above general considerations may appear straightforward, but many CVD reactions involve more than one process. An example is the CVD of diamond-like films, where both the deposition of sp^3 carbon and etching of sp^2 carbon are important, and where reaction conditions may be far from equilibrium [79]. Such thermodynamic considerations are of particular importance for materials with a range of possible compositions, such as B_xC films formed by CVD of BCl_3, CH_4 and

H_2 [80]. The above considerations also hold for ALD, where the reactions between precursors and the substrate (initial nucleation) are of relevance [81].

12.4.2 Kinetics Mechanism

The above discussion (e.g., Fig. 12.3) highlights the critical effect of such factors as precursor concentration and reaction probability on film conformality. In practice, however, an accurate, predictive model of a deposition process—of obvious importance for conformal growth under industrial conditions—can be quite complex. Additionally, accurate modeling relies on experimentally determined growth rates and other factors, as the surface chemistry involved can be quite complex. Combinatorial approaches have been used to determine effects of reactant pressure, flow rates, and other considerations on the silane-based reduction of WF_6 to form W films even on structures with relatively low aspect ratios [82]. A more recent examination of the kinetics of W deposition on higher aspect structures relied on in-situ resistivity monitoring to obtain accurate kinetics of the growing film [83]. Detailed kinetic models, however, have been developed for such important reactions as W CVD, with reaction rates calculated from statistical thermodynamics and tabulated reaction enthalpies [84]. While such modeling affords an important degree of understanding of how reaction conditions will impact film nucleation and conformality, such models are frequently not an adequate substitute for semiempirical studies.

Important kinetic considerations in CVD—assuming that heterogeneous nucleation predominates—include the rate of transport of the precursor to the surface and the reaction rate of the precursor on the surface. The balance of these two factors determines whether a process is mass-transport limited, or reaction-rate limited. The dependence of an abstract CVD process on mass transport versus thermodynamics is illustrated [85].

The corresponding mass flux (J_{gs}) is given by

$$J_{gs} = h_g (C_g - C_s)$$

C_g and C_s are the concentration of bulk gas and at the surface interface. The concentration drops from bulk gas to the substrate. h_g is the mass transfer coefficient. The flux (J_s) of coating on the surface can be approximated as first order reaction by

$$J_s = k_s C_s$$

where the k_s is defined as reaction constant. In a steady-state conditions, $J_{gs} = J_s$, so

$$C_s = \frac{Cg}{1 + \frac{k_s}{h_g}}$$

When $h_g >> k_s$, C_s approaches C_g meaning that even for sufficient reactive species on the surface, the surface reaction rate determines the rate of CVD reaction. On the contrary, when $h_g << k_s$, only small flux of reactive species approaches the substrate. In this case, the system is mass transport controlled. The mass transfer coefficient h_g increases with increasing pressure but with decreasing temperature. Thus, with temperature decreasing, mass transfer coefficient increases and becomes higher than reaction constant k_s with reaction resistance increases. Surface kinetics control reaction is achieved at low temperature, low pressure, low concentration of reactants, but high gas flow rates.

12.5 THE FUTURE: EMERGING ELECTRONIC MATERIALS

12.5.1 2D Materials

The increasing difficulties of continued downscaling of Si-CMOS-based devices (Moore's law) has prompted the search for new devices and materials that might provide faster and lower power alternatives to the Si transistor. Considerable attention has focused on two-dimensional materials, including graphene [86], hexagonal boron nitride (h-BN(0001)), especially as an insulating barrier in conjunction with graphene [87], and transition metal dichalcogenides (TMDCs), especially in the monolayer limit [88].

These materials are of growing interest in applications such as nanoelectronics [89] and spintronics [40]. Such applications place unusual requirements for a CVD-like process. Although much has been written regarding CVD of graphene on transition metal surfaces [90], industrially practical and scalable device applications require deposition on an insulating substrate. The potential importance of h-BN (0001) as a lattice match insulator to graphene also requires the ability to grow this material beyond the monolayer limit.

Oriented films of hexagonal boron nitride (h-BN(0001)) have been formed by a number of methods, including CVD of borazine [91], magnetron sputter deposition [92], and molecular beam epitaxy [93]. Many proposed applications, however, require multilayers of BN for spin filtering [39] or spin injection applications [40], or as a tunneling barrier [89]. Borazine pyrolysis, however, is self-passivating at approximately 1 monolayer. Additionally, a number of spintronics-related applications predict that device performance could change significantly upon increase or decrease of h-BN(0001) thickness by only one atomic layer [39, 40], thus making adequate control of film thickness is difficult unless ALD is performed. An ALD process using BCl_3 and NH_3 precursors to provide continuous large area multilayers has been reported for deposition on Ru (0001) substrates [94] and more recently on Co(0001) [38].

Graphene deposition on Cu or other transition metals by CVD or various hydrocarbons has been the subject of extensive examination [95, 96]. It should be pointed out, however, that most of the studies involve deposition on metal substrates, whereas deposition on dielectric substrates would seem more applicable to device applications. To that end, monolayer graphene has been grown by CVD on h-BN under a variety of conditions using hydrocarbon precursors [97, 98]. Graphene growth has also been reported on Al_2O_3(0001), albeit at temperatures more than 1800 K [99]. The extraordinarily high growth temperatures required on this surface suggest that reconstruction of the sapphire surface plays an important role, whereas CVD growth on lattice-matched BN was achieved at much more moderate temperatures [97].

Transition metal dichalcogenides are of long standing interest in multilayer form for lubrication and related applications, and CVD has been used to achieve this [100]. Emerging electronic applications, however, require highly ordered films, often in the monolayer or few-layer limit. Highly ordered ZrS_2 films have been grown on h-BN (0001) substrates. A low-pressure CVD process involving S vapor and $ZrCl_4$ has been used to deposit ZrS_2 on h-BN(0001) [101].

The demonstrated ability to grow graphene and transition metal dichalcogenides on h-BN (0001) suggests that this material will become of increasing importance for industrially practical methods of forming devices based on heterostructures involving 2D materials.

12.5.2 Oxide and Nitride Thin Films

Oxides and nitrides also have a large appeal within the CMOS industry for uses in gate dielectrics and diffusion barriers. Typically the oxide thin films have applications [102] as electrical insulator materials, gate oxide, transparent conductors, and wide band gap semiconductors. Many of the applications that are used for oxides apply additionally to nitrides with the exception of gate oxide but replaced with diffusion barrier. The development in such applications has driven and still is driving the optimization and enhancement of deposition methods: PVD and CVD.

Thin oxide films can be fabricated by CVD in three main families: pure oxide deposition without additional oxygen supply, deposition with precursors and water (H_2O) as oxygen supply, and oxide deposition results from β-diketonate precursors requiring active oxygen supply (O_3 or O_2). The CVD of TiO_2 [103], ZrO_2 [104], $LiNbO_3$ [105], Nb_2O_5-HfO_2 [106] are easy to achieve due to the fact that the pyrolysis reaction of precursors produce oxide and volatile by-products. Alkyl precursors together with H_2O were used because the low probability of gas phase reaction, similar in the atomic layer deposition: ZnO/diethylzinc [107].

Thin nitride films can be fabricated by CVD through methods of PECVD or HWCVD by using nitrogen gas (N_2) or ammonia (NH_3) in the presence of a metal precursor or native substrates [108].

Oxide find various applications with different functions [109–111]: as electrical insulator materials, gate oxide, transparent conductors, wide band gap semiconductors. For nitride thin films, there is large lattice mismatching within the substrates. The preferred method is to grow films by diffusing the nitrogen precursor with the metal precursor and this allows the metal ligand to be removed and the nitride to be deposited conformally on the substrate.

12.5.3 Conclusions

CVD is a complex process to study due to the numerous parameters involved including temperature, pressure, precursor flow rate, the impurities of the precursor, and precursor properties relative to the substrate surface and its decomposition. With increasingly rigorous requirements for conformality and void-free filling of high aspect ratio devices structures, CVD and its variants are becoming increasingly the method of choice for conformal film deposition. Beyond Si-CMOS, however, CVD and methods such as ALD show significant promise for the deposition of novel materials for nanoelectronic or spintronic applications.

12.6 REFERENCES

1. S. Sivaram, *Chemical Vapor Deposition: Thermal and Plasma Deposition of Electronic Materials*, Springer Science & Business Media, New York, 2013.

2. K. L. Choy, *Chemical Vapour Deposition of Coatings*, Elsevier, New York, 2003.

3. D. Dobkin and M. K. Zuraw, *Principles of Chemical Vapor Deposition*, Springer Science & Business Media, New York, 2013.

4. P. M. Martin, *Handbook of Deposition Technologies for Films and Coatings: Science, Applications and Technology*, William Andrew, Norwich, 2009.

5. C. E. Morosanu, *Thin Films by Chemical Vapour Deposition*, Elsevier, New York, 1990.

6. Y. Wang, W. Chen, B. Wang, and Y. Zheng, "Ultrathin Ferroelectric Films: Growth, Characterization, Physics and Applications," *Materials*, 7(9): 6377–6485, 2014.

7. K. K. Kim, A. Hsu, X. Jia, S. M. Kim, Y. Shi, M. Hofmann, et al., "Synthesis of Monolayer Hexagonal Boron Nitride on Cu Foil Using Chemical Vapor Deposition," *Nano Lett.*, 12: 161–166, 2012.

8. G. Kim, A.-R. Jang, H. Y. Jeong, Z. Lee, D. J. Kang, and H. S. Shin, "Growth of High-Crystalline, Single-Layer Hexagonal Boron Nitride on Recyclable Platinum Foil," *Nano Lett.*, 13: 1834–1839, 2013.

9. R.E.I. Schropp, Industrialization of Hot Wire Chemical Vapor Deposition for thin film applications, *Thin Solid Films.* (2015). In press.

10. S.-T. Lee, Z. Lin, and X. Jiang, "CVD Diamond Films: Nucleation and Growth," *Mater. Sci. Eng. R Reports.*, 25: 123–154, 1999.

11. Y. Akasaka, "Expectation for Cat-CVD in ULSI Technology and Business Trend," *Thin Solid Films*, 501: 15–20, 2006.

12. R. James, F. L. Pasquale, and J. A. Kelber, "Plasma-Enhanced Chemical Vapor Deposition of Ortho-Carborane: Structural Insights and Interaction with Cu Overlayers," *J. Phys. Condens. Matter.*, 25: 355004, 2013.

13. B. Dong, R. James, and J. A. Kelber, "PECVD of Boron Carbide/Aromatic Composite Films: Precursor Stability and Resonance Stabilization Energy," *Surf. Coatings Technol.*, (n.d.). In press.

14. Y. K. Chae, H. Ohno, K. Eguchi, and T. Yoshida, "Ultrafast Deposition of Microcrystalline Si by Thermal Plasma Chemical Vapor Deposition," *J. Appl. Phys.*, 89: 8311–8315, 2001.

15. N. S. J. Braithwaite, "Introduction to Gas Discharges," *Plasma Sources Sci. Technol.*, 9: 517, 2000.

16. O. A. Popov, *High Density Plasma Sources: Design, Physics and Performance*, Elsevier, New York, 1996.

17. J. Wang, K. G. Neoh, and E. T. Kang, "Comparative Study of Chemically Synthesized and Plasma Polymerized Pyrrole and Thiophene Thin Films," *Thin Solid Films*, 446: 205–217, 2004.

18. P. A. Tamirisa, K. C. Liddell, P. D. Pedrow, and M. A. Osman, "Pulsed-Plasma-Polymerized Aniline Thin Films," *J. Appl. Polym. Sci.*, 93: 1317–1325, 2004.

19. C. Tan, R. James, B. Dong, M. S. Driver, J. A. Kelber, G. Downing, et al., "Characterization of a Boron Carbide-Based Polymer Neutron Sensor," *Nucl. Instruments Methods Phys. Res. Sect. A Accel. Spectrometers, Detect. Assoc. Equip.*, 803: 82–88, 2015.

20. S. E. Alexandrov and M. L. Hitchman, "Chemical Vapor Deposition Enhanced by Atmospheric Pressure Non-Thermal Non-Equilibrium Plasmas," *Chem. Vap. Depos.*, 11: 457–468, 2005.

21. M. Moravej and R. F. Hicks, "Atmospheric Plasma Deposition of Coatings Using a Capacitive Discharge Source," *Chem. Vap. Depos.*, 11: 469–476, 2005.

22. F. Fanelli, "Thin Film Deposition and Surface Modification with Atmospheric Pressure Dielectric Barrier Discharges," *Surf. Coatings Technol.*, 205: 1536–1543, 2010.

23. D. Pappas, "Status and Potential of Atmospheric Plasma Processing of Materials," *J. Vac. Sci. Technol. A*, 29: 20801, 2011.

24. T. Belmonte, G. Henrion, and T. Gries, "Nonequilibrium Atmospheric Plasma Deposition," *J. Therm. Spray Technol.*, 20: 744–759, 2011.

25. R. Foest, M. Schmidt, and K. Becker, "Microplasmas, an Emerging Field of Low-Temperature Plasma Science and Technology," *Int. J. Mass Spectrom.*, 248: 87–102, 2006.

26. H. Baránková and L. Bardos, "Hollow Cathode and Hybrid Plasma Processing," *Vacuum.*, 80: 688–692, 2006.

27. H. Kakiuchi, M. Matsumoto, Y. Ebata, H. Ohmi, K. Yasutake, K. Yoshii, et al., "Characterization of Intrinsic Amorphous Silicon Layers for Solar Cells Prepared at Extremely High Rates by Atmospheric Pressure Plasma Chemical Vapor Deposition," *J. Non. Cryst. Solids.*, 351: 741–747, 2005.

28. H. Kakiuchi, H. Ohmi, R. Inudzuka, K. Ouchi, and K. Yasutake, "Enhancement of Film-Forming Reactions for Microcrystalline Si Growth in Atmospheric-Pressure Plasma Using Porous Carbon Electrode," *J. Appl. Phys.*, 104: 53522, 2008.

29. K. Yasutake, H. Kakiuchi, H. Ohmi, K. Inagaki, Y. Oshikane, and M. Nakano, "An Atomically Controlled Si Film Formation Process at Low Temperatures Using Atmospheric-Pressure VHF Plasma," *J. Phys. Condens. Matter.*, 23: 394205, 2011.

30. P. J. Pernot and A. R. Barron, "Photo-Assisted Chemical Vapor Deposition of Gallium Sulfide Thin Films," *Chem. Vap. Depos.*, 1: 75–78, 1995.

31. J. Y. Tai Chen, R. C. Henderson, J. T. Hall, and J. W. Peters, "Photo-CVD for VLSI Isolation," *J. Electrochem. Soc.*, 131: 2146–2151, 1984.

32. H. Sankur, "Deposition of Optical Coatings by Laser-Assisted Evaporation and by Photo-Assisted Chemical Vapor Deposition," *Thin Solid Films*, 218: 161–169, 1992.

33. L. Schäfer, M. Höfer, and R. Kröger, "The Versatility of Hot-Filament Activated Chemical Vapor Deposition," *Thin Solid Films*, 515: 1017–1024, 2006.

34. H. Kim, H.-B.-R. Lee, and W.-J. Maeng, "Applications of Atomic Layer Deposition to Nanofabrication and Emerging Nanodevices," *Thin Solid Films*, 517: 2563–2580, 2009.

35. J. D. Ferguson, A. W. Weimer, and S. M. George, "Atomic Layer Deposition of Boron Nitride Using Sequential Exposures of BCl_3 and NH_3," *Thin Solid Films*, 413: 16–25, 2002.

36. R. L. Puurunen, "Surface Chemistry of Atomic Layer Deposition: A Case Study for the Trimethylaluminum/Water Process," *J. Appl. Phys.*, 97: 121301, 2005.

37. R. W. Johnson, A. Hultqvist, and S. F. Bent, "A Brief Review of Atomic Layer Deposition: From Fundamentals to Applications," *Mater. Today*, 17: 236–246, 2014.

38. J. B. and Y. C. and I. T. and M. S. Driver and P. A. Dowben, and J. A. Kelber, "Atomic Layer-by-Layer Deposition of h-BN(0001) on Cobalt: A Building Block for Spintronics and Graphene Electronics," *Mater. Res. Exp.*, 1: 46410, 2014.

39. S. V. Faleev, S. S. P. Parkin, and O. N. Mryasov, "Brillouin Zone Spin Filtering Mechanism of Enhanced TMR and Correlation Effects in Co (0001)/h-BN/Co (0001) Magnetic Tunnel Junction," *Cond. Matt.: Mater. Sci.*, 2015.

40. V. M. Karpan, P. A. Khomyakov, G. Giovannetti, A. A. Starikov, and P. J. Kelly, "Ni (111)| Graphene| h-BN Junctions as Ideal Spin Injectors," *Phys. Rev. B.*, 84: 153406, 2011.

41. D. Riihelä, M. Ritala, R. Matero, and M. Leskelä, "Introducing Atomic Layer Epitaxy for the Deposition of Optical Thin Films," *Thin Solid Films*, 289: 250–255, 1996.

42. R. G. Gordon, X. Lou, and S. B. Kim, "Advanced Atomic Layer Deposition and Epitaxy Processes," *VLSI Technol. Syst. Appl.* (VLSI-TSA), 2015 Int. Symp., IEEE, pp. 1–2, 2015.

43. Y. Shi, W. Zhou, A.-Y. Lu, W. Fang, Y.-H. Lee, A.L. Hsu, et al., "Van der Waals Epitaxy of MoS2 Layers Using Graphene as Growth Templates," *Nano Lett.*, 12: 2784–2791, 2012.

44. W. B. Wang, N. N. Chang, T. A. Codding, G. S. Girolami, and J. R. Abelson, "Superconformal Chemical Vapor Deposition of Thin Films in Deep Features," *J. Vac. Sci. Technol., A Vacuum, Surfaces, Film*, 32: 051512, 2014.

45. Y. Kuzminykh, A. Dabirian, M. Reinke, and P. Hoffmann, "High Vacuum Chemical Vapour Deposition of Oxides: A Review of Technique Development and Precursor Selection," *Surf. Coatings Technol.*, 230: 13–21, 2013.

46. V. S. Ban and S. L. Gilbert, "The Chemistry and Transport Phenomena of Chemical Vapor Deposition of Silicon from $SiCl_4$," *J. Cryst. Growth.*, 31: 284–289, 1975.

47. Y. Arakawa, T. Nakamura, Y. Urino, and T. Fujita, "Silicon: Photonics for Next Generation System Integration Platform," *Commun. Mag. IEEE*, 51: 72–77, 2013.

48. M. Su, D. Yu, Y. Liu, L. Wan, C. Song, F. Dai, et al., "Properties and Electric Characterizations of Tetraethyl Orthosilicate-Based Plasma Enhanced Chemical Vapor Deposition Oxide Film Deposited at 400°C for Through Silicon Via Application," *Thin Solid Films*, 550: 259–263, 2014.

49. F. Shimura, *Semiconductor Silicon Crystal Technology*, Elsevier, New York, 2012.

50. R. Baets, J. Barker, J. A. Barnard, T. M. Barton, M. E. Clarke, A. Cappy, et al., *Semiconductor Device Modelling*, Springer Science & Business Media, New York, 2012.

51. Y. Huang and K.-N. Tu, *Silicon and Silicide Nanowires*, Pan Stanford, Singapore, 2013.

52. A. M. Funde, N. A. Bakr, D. K. Kamble, R. R. Hawaldar, D. P. Amalnerkar, and S. R. Jadkar, "Influence of Hydrogen Dilution on Structural, Electrical and Optical Properties of Hydrogenated Nanocrystalline Silicon (nc-Si:H) Thin Films Prepared by Plasma Enhanced Chemical Vapour Deposition (PE-CVD)," *Sol. Energy Mater. Sol. Cells.*, 92: 1217–1223, 2008.

53. F. R. Faller and A. Hurrle, "High-Temperature CVD for Crystalline-Silicon Thin-Film Solar Cells," *Electron Devices, IEEE Trans.*, 46: 2048–2054, 1999.

54. D. He, M. Yin, J. Wang, P. Gao, and J. Li, "Deposition and Growth Mechanism of Low-Temperature Crystalline Silicon Films on Inexpensive Substrates," *J. Korean Phys. Soc.*, 55: 2671–2676, 2009.

55. T. Kaneko, M. Wakagi, K. Onisawa, and T. Minemura, "Change in Crystalline Morphologies of Polycrystalline Silicon Films Prepared by Radio-Frequency Plasma-Enhanced Chemical Vapor Deposition Using SiF_4+ H_2 Gas Mixture at 350°C," *Appl. Phys. Lett.*, 64: 1865–1867, 1994.

56. C. E. Richardson, M. S. Mason, and H. A. Atwater, "Hot-Wire CVD-Grown Epitaxial Si Films on Si (100) Substrates and a Model of Epitaxial Breakdown," *Thin Solid Films*, 501: 332–334, 2006.

57. S. De Wolf, A. Descoeudres, Z. C. Holman, and C. "Ballif, High-Efficiency Silicon Heterojunction Solar Cells: A Review," *Green*, 2: 7–24, 2012.

58. F.-M. Tseng, C.-H. Hsieh, Y.-N. Peng, and Y.-W. Chu, "Using Patent Data to Analyze Trends and the Technological Strategies of the Amorphous Silicon Thin-Film Solar Cell Industry," *Technol. Forecast. Soc. Change*, 78: 332–345, 2011.

59. L. Yu, B.O'Donnell M. Foldyna, and P. Roca i Cabarrocas, "Radial Junction Amorphous Silicon Solar Cells on PECVD-Grown Silicon Nanowires," *Nanotechnology*, 23: 194011, 2012.

60. S. Misra, L. Yu, M. Foldyna, and P. Roca i Cabarrocas, "High Efficiency and Stable Hydrogenated Amorphous Silicon Radial Junction Solar Cells Built on VLS-Grown Silicon Nanowires," *Sol. Energy Mater. Sol. Cells*, 118: 90–95, 2013.

61. B. S. Lim, A. Rahtu, and R. G. Gordon, "Atomic Layer Deposition of Transition Metals," *Nat. Mater.*, 2: 749–754, 2003.

62. T. T. Kodas and M. J. Hampden-Smith, *The Chemistry of Metal CVD*, John Wiley & Sons, Hoboken, 2008.

63. J. R. Creighton and J. E. Parmeter, "Metal CVD for Microelectronic Applications: An Examination of Surface Chemistry and Kinetics," *Crit. Rev. Solid State Mater. Sci.*, 18: 175–237, 1993.

64. J. R. Creighton, "A Mechanism for Selectivity Loss during Tungsten CVD," *J. Electrochem. Soc.*, 136: 271–276, 1989.

65. R. W. Cheek, J. A. Kelber, J. G. Fleming, R. S. Blewer, and R. D. Lujan, In Situ Monitoring of the Products from the SiH_4+ WF 6 Tungsten Chemical Vapor Deposition Process by Micro-Volume Mass Spectrometry," *J. Electrochem. Soc.*, 140: 3588–3590, 1993.

66. R. Madar and C. Bernard, "Chemical Vapour Deposition of Metal Silicides in Silicon Microelectronics," *Appl. Surf. Sci.*, 53: 1–10, 1991.

67. G. M. Nuesca and J. A. Kelber, "Cu-Metal Interfacial Interactions during Metal Organic Chemical Vapour Deposition," *Thin Solid Films*, 262: 224–233, 1995.

68. H. Kim and Y. Shimogaki, "Comparative Study of Cu–CVD Seed Layer Deposition on Ru and Ta Underlayers," *J. Electrochem. Soc.*, 154: G13–G17, 2007.

69. N. J. Ianno, A. U. Ahmed, and D. E. Englebert, "Plasma Enhanced Chemical Vapor Deposition of TiN from $TiCl_4/N_2/H_2$ Gas Mixtures," *J. Electrochem. Soc.*, 136: 276–280, 1989.

70. X. Zhao, N. P. Magtoto, and J. A. Kelber, "Chemical Vapor Deposition of Tantalum Nitride with Tert-Butylimino Tris (Diethylamino) Tantalum and Atomic Hydrogen," *Thin Solid Films*, 478: 188–195, 2005.

71. H. Shimizu, K. Sakoda, and Y. Shimogaki, "CVD of Cobalt–Tungsten Alloy Film as a Novel Copper Diffusion Barrier," *Microelectron. Eng.*, 106: 91–95, 2013.

72. K.-W. Lee, H. Wang, J.-C. Bea, M. Murugesan, Y. Sutou, T. Fukushima, et al., "Barrier Properties of CVD Mn Oxide Layer to Cu Diffusion for 3-D TSV," *Electron Device Lett. IEEE*, 35: 114–116, 2014.

73. N. Kumar, A. Yanguas-Gil, S. R. Daly, G. S. Girolami, and J. R. Abelson, "Growth Inhibition to Enhance Conformal Coverage in Thin Film Chemical Vapor Deposition," *J. Am. Chem. Soc.*, 130: 17660–17661, 2008.

74. Y. Au, Y. Lin, and R. G. Gordon, "Filling Narrow Trenches by Iodine-Catalyzed CVD of Copper and Manganese on Manganese Nitride Barrier/Adhesion Layers," *J. Electrochem. Soc.*, 158: D248–D253, 2011.

75. J. R. Abelson, "(Invited) Ultra-Conformal CVD at Low Temperatures: The Role of Site Blocking and the Use of Growth Inhibitors," *ECS Trans.*, 33: 307–319, 2010.

76. B. Sell, A. Sänger, G. Schulze-Icking, K. Pomplun, and W. Krautschneider, "Chemical Vapor Deposition of Tungsten Silicide (WSix) for High Aspect Ratio Applications," *Thin Solid Films*, 443: 97–107, 2003.

77. M. Ishikawa, I. Muramoto, H. Machida, S. Imai, A. Ogura, and Y. Ohshita, "Chemical Vapor Deposition of NiSi Using $Ni(PF_3)_4$ and Si_3H_8," *Thin Solid Films*, 515: 8246–8249, 2007.

78. Y. Yang, S. Jayaraman, D. Y. Kim, G. S. Girolami, and J. R. Abelson, "CVD Growth Kinetics of HfB_2 Thin Films from the Single-Source Precursor Hf $(BH_4)_4$," *Chem. Mater.*, 18: 5088–5096, 2006.

79. J.-T. Wang, D. Wei Zhang, and J.-Y. Shen, "Modern Thermodynamics in CVD of Hard Materials," *Int. J. Refract. Met. Hard Mater.*, 19: 461–466, 2001.

80. S. Guodong, D. Juanli, L. Hui, and Z. Qing, "Thermodynamics of Preparing BxC Condensed Phases by CVD Using BCl_3-CH_4-H_2 as Precursors," *Rare Met. Mater. Eng.*, 44: 826–829, 2015.

81. P. Violet, E. Blanquet, D. Monnier, I. Nuta, and C. Chatillon, "Experimental Thermodynamics for the Evaluation of ALD Growth Processes," *Surf. Coatings Technol.*, 204: 882–886, 2009.

82. J. C. Chang, "Chemical Vapor Deposition of Tungsten Step Coverage and Thickness Uniformity Experiments," *Thin Solid Films*, 208: 177–180, 1992.

83. S. Haimson, Y. Shacham-Diamand, D. Horvitz, and A. Rozenblat, "Resistivity Monitoring of the Early Stages of W CVD Nucleation for Sub-45 nm Process," *Microelectron. Eng.*, 92: 134–136, 2012.

84. R. Arora and R. Pollard, "A Mathematical Model for Chemical Vapor Deposition Processes Influenced by Surface Reaction Kinetics: Application to Low-Pressure Deposition of Tungsten," *J. Electrochem. Soc.*, 138: 1523–1537, 1991.

85. A. S. Grove, *Physics and Technology of Semiconductor Devices*, John Wiley & Sons, Inc., New York, London, Sydney, p. 721, 1967.

86. A. H. C. Neto, F. Guinea, N. M. R. Peres, K. S. Novoselov, and A. K. Geim, "The Electronic Properties of Graphene," *Rev. Mod. Phys.*, 81: 109, 2009.

87. C. Oshima and A. Nagashima, "Ultra-Thin Epitaxial Films of Graphite and Hexagonal Boron Nitride on Solid Surfaces," *J. Phys. Condens. Matter.*, 9: 1, 1997.

88. D. Jariwala, V. K. Sangwan, L. J. Lauhon, T. J. Marks, and M. C. Hersam, "Emerging Device Applications for Semiconducting Two-Dimensional Transition Metal Dichalcogenides," *ACS Nano.*, 8: 1102–1120, 2014.

89. L. Britnell, R. V. Gorbachev, R. Jalil, B. D. Belle, F. Schedin, A. Mishchenko, et al., "Field-Effect Tunneling Transistor Based on Vertical Graphene Heterostructures," *Science* (80), 335: 947–950, 2012.

90. S. Grandthyll, S. Gsell, M. Weinl, M. Schreck S. Hüfner, and F. Müller, "Epitaxial Growth of Graphene on Transition Metal Surfaces: Chemical Vapor Deposition versus Liquid Phase Deposition," *J. Phys. Condens. Matter.*, 24: 314204, 2012.

91. A. Goriachko, Y. He, M. Knapp, H. Over, M. Corso, T. Brugger, et al., "Self-Assembly of a Hexagonal Boron Nitride Nanomesh on Ru (0001)," *Langmuir*, 23: 2928–2931, 2007.

92. P. Sutter, J. Lahiri, P. Zahl, B. Wang, and E. Sutter, "Scalable Synthesis of Uniform Few-Layer Hexagonal Boron Nitride Dielectric Films," *Nano Lett.*, 13: 276–281, 2012.

93. S. Nakhaie, J. M. Wofford, T. Schumann, U. Jahn, M. Ramsteiner, M. Hanke, et al., "Synthesis of Atomically Thin Hexagonal Boron Nitride Films on Nickel Foils by Molecular Beam Epitaxy," *Appl. Phys. Lett.*, 106: 213108, 2015.

94. J. E. Morris and K. Iniewski, *Nanoelectronic Device Applications Handbook*, CRC Press, Boca Raton, 2013.

95. X. Li, W. Cai, J. An, S. Kim, J. Nah, D. Yang, et al., "Large-Area Synthesis of High-Quality and Uniform Graphene Films on Copper Foils," *Science*, 324 (5932): 1312–1314, 2009.

96. M. Batzill, "The Surface Science of Graphene: Metal Interfaces, CVD Synthesis, Nanoribbons, Chemical Modifications, and Defects," *Surf. Sci. Rep.*, 67: 83–115, 2012.

97. C. Bjelkevig, Z. Mi, J. Xiao, P. A. Dowben, L. Wang, W.-N. Mei, et al., "Electronic Structure of a Graphene/Hexagonal-BN Heterostructure Grown on Ru (0001) by Chemical Vapor Deposition and Atomic Layer Deposition: Extrinsically Doped Graphene," *J. Phys. Condens. Matter.*, 22: 302002, 2010.

98. M. Wang, S. K. Jang, W. Jang, M. Kim, S. Park, S. Kim, et al., "A Platform for Large-Scale Graphene Electronics–CVD Growth of Single-Layer Graphene on CVD-Grown Hexagonal Boron Nitride," *Adv. Mater.*, 25: 2746–2752, 2013.

99. M. A. Fanton, J. A. Robinson, C. Puls, Y. Liu, M. J. Hollander, B. E. Weiland, et al., "Characterization of Graphene Films and Transistors Grown on Sapphire by Metal-Free Chemical Vapor Deposition," *ACS Nano.*, 5: 8062–8069, 2011.

100. T. W. Scharf, S. V. Prasad, M. T. Dugger, T. M. Mayer, "Atomic Layer Deposition of Solid Lubricant Thin Films," *World Tribol. Congr. III*, American Society of Mechanical Engineers, 375–376, 2005.

101. M. Zhang, Y. Zhu, X. Wang, Q. Feng, S. Qiao, W. Wen, et al., "Controlled Synthesis of ZrS_2 Monolayer and Few-Layers on Hexagonal Boron Nitride," *J. Am. Chem. Soc.* 137: 7051–7054, 2015.

102. M.-G. Kim, M. G. Kanatzidis, A. Facchetti, and T. J. Marks, "Low-Temperature Fabrication of High-Performance Metal Oxide Thin-Film Electronics via Combustion Processing," *Nat. Mater.*, 10: 382–388, 2011.

103. C. J. Taylor, D. C. Gilmer, D. G. Colombo, G. D. Wilk, S. A. Campbell, J. Roberts, et al., "Does Chemistry Really Matter in the Chemical Vapor Deposition of Titanium Dioxide? Precursor and Kinetic Effects on the Microstructure of Polycrystalline Films," *J. Am. Chem. Soc.*, 121: 5220–5229, 1999.

104. Z. Song, L. M. Sullivan, and B. R. Rogers, "Study on the Initial Deposition of ZrO_2 on Hydrogen Terminated Silicon and Native Silicon Oxide Surfaces by High Vacuum Chemical Vapor Deposition," *J. Vac. Sci. Technol. A*, 23: 165–176, 2005.

105. A. Dabirian, S. Harada, Y. Kuzminykh, S. C. Sandu, E. Wagner, G. Benvenuti, et al., "Combinatorial Chemical Beam Epitaxy of Lithium Niobate Thin Films on Sapphire," *J. Electrochem. Soc.*, 158: D72–D76, 2011.

106. A. Dabirian, Y. Kuzminykh, S. C. Sandu, S. Harada, E. Wagner, P. Brodard, et al., "Combinatorial High-Vacuum Chemical Vapor Deposition of Textured Hafnium-Doped Lithium Niobate Thin Films on Sapphire," *Cryst. Growth Des.*, 11: 203–209, 2011.

107. A. B. M. A. Ashrafi, A. Ueta, A. Avramescu, H. Kumano, I. Suemune, Y.-W. Ok, et al., "Growth and Characterization of Hypothetical Zinc-Blende ZnO Films on GaAs (001) Substrates with ZnS Buffer Layers," *Appl. Phys. Lett.*, 76: 550–552, 2000.

108. A. N. Cloud, L. M. Davis, G S. Girolami, and J. R. Abelson, "Low-Temperature CVD of Iron, Cobalt, and Nickel Nitride Thin Films from bis [di (tert-butyl) amido] Metal (II) Precursors and Ammonia," *J. Vac. Sci. Technol. A*, 32: 20606, 2014.

109. J.-P. Locquet, C. Marchiori, M. Sousa, J. Fompeyrine, and J. W. Seo, "High-K Dielectrics for the Gate Stack," *J. Appl. Phys.* 100: 51610, 2006.

110. T. Minami, "Transparent Conducting Oxide Semiconductors for Transparent Electrodes," *Semicond. Sci. Technol.*, 20: S35, 2005.

111. A. Janotti and C. G. Van de Walle, "Fundamentals of Zinc Oxide as a Semiconductor," *Reports Prog. Phys.*, 72: 126501, 2009.

Chapter 13
Atomic Layer Deposition

BY
ERIC T. EISENBRAUN *Colleges of Nanoscale Science & Engineering, SUNY Polytechnic Institute, Albany, New York*

13.1 INTRODUCTION

Fabricating structures and layers for use in semiconductor and associated applications require the ability to carefully control layer thickness, composition, and structure. These properties in turn play a significant role in the performance of the layers with respect to electrical, chemical, and mechanical characteristics. As semiconductor device dimensions have trended smaller with each device generation, this has resulted in the thickness of the individual constituent layers likewise becoming thinner.[1] This evolution fundamentally dictates two outcomes: first, that the ability to control the layer needs to become more precise; and second, that the relationship between the film and the surface upon which it is grown becomes increasing important to the overall quality of the layer. Atomic layer deposition (ALD) addresses both of these criteria, and has found uses for several key layer deposition steps in leading edge IC fabrication.[2]

ALD is a film growth process that relies on iterated "half reactions" involving surface-adsorbed chemical species to form films with submonolayer precision. There are several claims to the invention of ALD, but it is widely thought to have been developed for commercial processing by Tuomo Suntola in 1974, who investigated the technique, then named as atomic layer epitaxy (ALE), for layer applications in thin film electroluminescent (TFEL) displays.[3] In this early work, metal chalcogenides such as ZnS were grown using $ZnCl_2$ and H_2S, by way of an overall parent reaction that can be separated into individual bimolecular half reactions on the surface:

Parent reaction:

$$ZnCl_2 + H_2S \rightarrow ZnS + 2HCl \tag{13.1}$$

Half reactions:

$$ZnCl^* + H_2S \rightarrow ZnSH^* + HCl \tag{13.2a}$$

$$SH^* + ZnCl_2 \rightarrow SZnCl^* + HCl \tag{13.2b}$$

where the asterisks refer to metastable "active" surface species.[4] The early use of the term "epitaxy" referred to the surface-initiated nature of the process, although a true epitaxial relationship is not necessarily involved. As such, the technique was eventually rebadged as atomic layer deposition.

Effectively, ALD is a very specialized type of chemical vapor deposition (CVD) that avoids all but specifically targeted surface chemical reactions between reactants. This is achieved by the use of cycled "pulse and purge" or "pulse and pump" steps that isolate the reactant species chemisorbed on the substrate surface. In this process, a deposition "cycle" is broken down into four individual steps:

1. Pulse of primary reactant, such as a metalorganic precursor, which chemisorbs on the surface
2. System purge with inert gas (such as argon or nitrogen) to remove gas phase and weakly bound surface species. This may be replaced with a system evacuation to achieve the same purpose
3. Pulse of the process counter-reactant, which may be a reactant gas such as H_2, H_2O, NH_3, or possibly another metalorganic
4. Another system purge with inert gas or system evacuation to remove gas phase and weakly bound surface species

This sequence is shown in Fig. 13.1.

13.1.1 The Rationale for ALD

The key concept underlying ALD is the surface-facilitated reaction. This enables superior control over the growth process, and effectively ensures that only a single layer of chemisorbed reactant participate in any ALD growth cycle. Since steric blocking effects may prevent each surface site form being occupied, this limits the potential growth rate of an ALD process to less than a single monolayer per cycle. This level of control facilitates the growth of uniform, pinhole-free films, which were critical characteristics for the original TFEL application.[5]

Perhaps as importantly, the ability to saturate all exposed substrate surfaces with a complete adsorbed layer, with the associated purge following to remove additional deposited material, enables one of the truly key features of ALD—the ability to uniformly coat very complex, high aspect ratio, and even re-entrant structures.[6] This represents a practical difference from other deposition techniques such as CVD or sputtering, where the dynamic nature of those processes results in much more difficulty in maintaining uniformity due to the need to have uniform gas species flux across the entire substrate surface. This has proven to be an important factor in the development of ALD for applications outside TFELs, as will be discussed below. The unique thickness controllability and large-area uniformity of ALD also lends itself to the use of ALD in nano-laminate applications, involving multiple layers of alternating materials. These types of structures can be used in optics applications such as Bragg reflectors.[7]

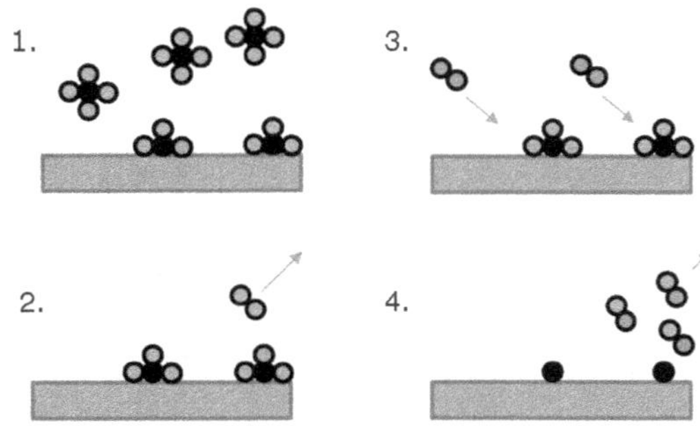

Figure 13.1 Schematic representation of the four steps in a typical ALD growth cycle. 1. Precursor adsorption on the surface. 2. Purge to remove adsorption byproducts and residual reactants. 3. Reactant pulse for surface reaction with adsorbed species. 4. Purge to remove reaction byproducts and residual reactant gases. Each cycle leads to the deposition of a fraction of a monolayer, depending in large part on the nature of the reactants employed in the process.

From a manufacturing perspective, ALD is a relatively expensive process, when compared to an approach such as sputtering and CVD, owing to its use of consumables as well as its low throughput.[8] However, it is not an ultrahigh vacuum technique like other atomic-layer growth processes such as molecular beam epitaxy (MBE), and provides unique material and growth profile capabilities. Moreover, ALD equipment can be seamlessly integrated into the manufacturing line, owing to its similarity to CVD, which can assist in lowering the barrier to introduction of ALD processing to an existing manufacturing line. In fact, ALD has been introduced into a number of manufacturing settings, as discussed further. A summary of advantages and disadvantages for ALD (when compared broadly to other deposition technologies) is provided in Table 13.1.

Table 13.1 Advantages and Disadvantages of ALD Compared Broadly with Alternative Material Processing Approaches

Advantage	Disadvantage
Thickness and compositional precision/controllability	Cost of reactants
Thickness and compositional uniformity and conformality	Low throughput/low deposition rate
Low temperature (compared with analogous CVD processes)	Deposition hardware and maintenance costs
Wide processing windows (to ensure surface saturation)	

13.2 PRIMARY COMMERCIAL APPLICATIONS FOR ALD

As noted earlier, TFEL displays were the first major commercial platform for ALD, in this case for the growth of metal sulfides beginning in the mid-1970s. The next significant commercial application was in disk drive manufacturing, specifically the Al_2O_3 dielectric layer in the magnetic head of the drive. Like the TFEL case, the key enabling ALD property for this application was the fabrication of uniform, pinhole-free films.[9]

Integrated circuit fabrication has been another major application space for ALD. This has taken advantage of various ALD characteristics for different specific layers in the chip. The application which is most associated with ALD in IC manufacturing is the deposition of the ultra-thin (< 5 nm) gate dielectric. Scaling and performance issues dictated a transition away from a conventional SiO_2-based gate dielectric to one with a higher dielectric constant, allowing the growth of a thicker layer to reduce leakage while maintaining the same effective capacitance.[10] ALD can also be applied to barrier/liner applications in interconnects and DRAM storage cell capacitor structures, which takes advantage of the superior conformality of ALD. This application is not as widely employed as the gate dielectric at this point in manufacturing.

13.3 ALD FOR FRONT END OF THE LINE SEMICONDUCTOR MANUFACTURING APPLICATIONS

As noted, the primary front end of the line (FEOL) application for ALD has been for the gate dielectric layer. As device scaling led to the need to replace SiO_2 with a higher dielectric constant (high-κ) material for the gate, this was coupled with the need to establish a manufacturable fabrication approach. Owing to the physical thickness tolerances for these layers that effectively dictate atomic-level thickness control over a full wafer surface, ALD is a clear candidate. Moreover, the need to preserve the integrity of the semiconductor channel beneath the dielectric discouraged the use of a technique such as sputtering due to potential surface damage in the presence of high-energy ions.

Intel has been reported as one of the early adopters of ALD for high-κ, employing a hybrid ALD HfO_2 layer with a nitride silicon oxide at the 45-nm node for processor chips.[9] The introduction of ALD technology by a major IC manufacturer significantly eased the introduction of ALD into the broader market. Likewise, the very high aspect ratios associated with memory chip trench capacitors has proven to be an application for ALD. For example, the metal-insulator-metal (MIM) stacks used in storage capacitors is well suited for metal and dielectric ALD.[11] For memory capacitors, oxides of Hf, Zr, and Al are among a number of materials in consideration. For the metal electrodes, Ru and RuO_2 are preferred owing to their low resistivity and refractory nature.

13.4 DEVELOPING THE ALD PROCESS

Establishing an ALD process involves developing the self-limited, surface-saturated half reactions that lead to controlled sub-monolayer material growth. This involves a number of steps to map the ALD processing regime for a particular material.

13.4.1 Establishing a Temperature Window

Process temperature is a fundamental ALD process characteristic. As the ALD process involves the controlled surface reaction of chemisorbed (or dissociatively adsorbed) reactants, the process temperature should be set such that

- It is high enough to facilitate chemisorption through precursor decomposition on the surface.
- It is low enough to avoid uncontrolled surface reactions similar to a more conventional CVD process (in this case referred to as "parasitic CVD"). In some cases, high temperature can also lead to desorption of adsorbed species, leading to a decreasing growth rate at temperatures above the ALD regime.

This temperature window is typically developed through an understanding of the nature of the reactants being used and measurements of the deposition thickness as a function of temperature, as shown in Fig. 13.2.

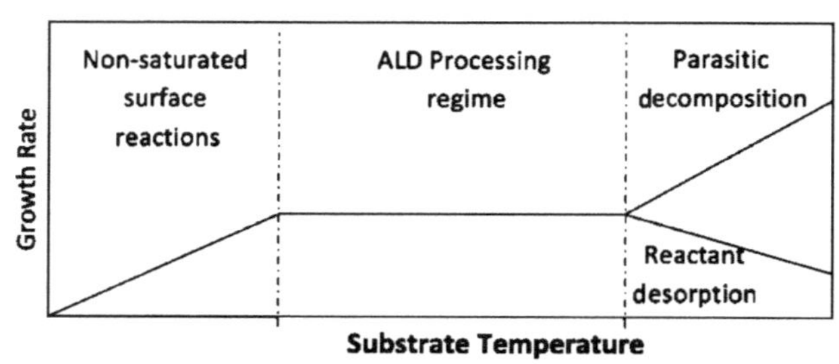

Figure 13.2 Common behavior for ALD growth rate as a function of temperature, showing the key thermal regimes.

13.4.2 Establishing Sufficient Reactant Transport to Surface

Complete surface half-reactions require that all surfaces are exposed to sufficient reactant to achieve saturation. Since a proper ALD process involves only the initial chemisorbed layer, the surface can be exposed to excess reactant without affecting growth rate provided that sufficient purging or evacuation is done to remove the excess material from the surface. This criterion applies to all reactants used in the ALD process. This is a key feature of ALD, in that this allows for extremely high aspect ratio or complex surface structures to be completely saturated, which gives rise to the superior conformality exhibited by ALD. Generally, growth rate per cycle will increase with increasing reactant pulse until surface saturation is achieved, at which point growth rate stays constant if purging is sufficient.

13.4.3 Establishing Sufficient Purge/Evacuation of the Reaction Area

In a similar way as reactant dose to the surface, the extent of the post-reactant pulse system purge or evacuation is also a key factor in establishing an ALD processing window. As noted earlier, ensuring reactant saturation to all surfaces can lead to excess physisorption (or associative adsorption) of reactants on surfaces. This requires purging or evacuation for removal, in order to leave a surface uniformly coated with a single chemisorbed layer of reactant.

In general, insufficient purging will result in "parasitic" CVD-type growth, and a locally higher growth rate, thus leading to poor thickness uniformity and conformality. In most cases excess purging/evacuation is not detrimental (besides the fact that this increases process time and thus decreases throughput), although if the chemisorbed layer is not strongly bound to the surface, it is possible for chemisorbed material to desorb with excess purging/evacuation, leading to locally low growth rate owing to loss of reactants on the surface.

As with the precursor/reactant pulse exposures, the purge/evacuation timing is important following the exposure of each of reactant to the surface.

13.5 CONSIDERATIONS IN SELECTING AN APPROPRIATE PRECURSOR AND REACTANT FOR ALD

Formally, the ALD process utilizes surface-supported half-reactions to grow a particular material. This requires that the precursor can be strongly bound to the surface by chemisorption, whereby the parent molecule dissociates, with the precursor bonding directly with the surface atoms. However, it is equally important that the adsorbed species be stable, and not undergo further decomposition spontaneously on the surface, as this would lead to uncontrolled deposition on the surface.

One well known example is the ALD growth of Al_2O_3 from trimethylaluminum and water; there the precursor methyl groups react with adsorbed hydroxyl (OH) groups on the surface ("S") to form CH_4 leaving groups and a direct Al-O surface bond[12]:

$$Al(CH_3)_3 + S\text{-}OH \rightarrow (CH_3)_2Al\text{-}O\text{-}S + CH_4 \tag{13.3a}$$

where "S" represents the substrate surface. Note that this intermediate aluminum complex is stable on the surface, and the remaining methyl groups are removed by reaction with the subsequent H_2O pulse, whereby

$$2(CH_3)_2Al\text{-}O\text{-}S + 2H_2O \rightarrow Al_2O(OH)_2\text{-}S + 2CH_4 \tag{13.3b}$$

and the remaining surface alumina molecule includes new OH groups to activate the subsequent growth cycle. This demonstrates two important precursor characteristics:

1. A clean reaction pathway leading to growth of the desired material with a volatile, thermodynamically stable by-product
2. A pathway which results in an activated surface that can allow the ALD reaction mechanism to proceed upon the newly formed surface

Precursor and reactant selection should also consider practical constraints of the application. For example, thermal constraints of the substrate material or problematic chemical interactions, which can lead to contamination or defect formation, may dictate avoiding particular types of precursors or reactants.

13.6 HARDWARE AND PROCESS INNOVATIONS TO INCREASE GROWTH RATE IN ALD

The primary "Achilles' heel" of ALD is the low growth rate, resulting in a high effective cost-of-ownership for the technology. As such, increasing growth rate is relevant for commercial users of ALD. This can be approached from the perspectives of the system hardware or the process itself.

From the hardware point of view, process efficiencies can be achieved in several ways. Minimization of system volumes and surface areas can reduce the reactant flux required to reach saturation. Reactant delivery lines should be as sort and narrow as practical to minimize wetted area without sacrificing flow characteristics. Chamber volumes should likewise be minimized. Pulsing valves should possess fast, reliable switching behavior and should be designed to minimize potential condensation of heated liquid and solid reactants.

With respect to the ALD process, there have been a number of advances developed to increase the effective growth rate and wafer throughput. When the ALD process is considered as a series of surface reactions, clearly the most direct way to increase growth rate is by increasing the rate of the surface reaction. However, the ALD growth rate is typically limited by the time required for surface saturation and the fraction of the surface covered in each adsorption cycle. Increasing precursor or co-reactant flux can achieve a faster surface adsorption rate (R), though the process is often limited by the effective adsorption rate, which is a function of the specific material systems' adsorption activation energy (E_a), partial pressure of reactant P, and the substrate temperature via

$$R = \frac{f(\theta).P}{\sqrt{2\pi mkT}} \exp(-E_a/RT) \tag{13.4}$$

where $f(\theta)$ is an empirical function of the surface coverage fraction.[13] One approach used to achieve this effect is to carefully select the reactant chemistries to favor a high interaction activity. For example, rapid ALD of SiO_2 has been achieved by using trimethylaluminum (TMA) in the fabrication of Al_2O_3/SiO_2 nanolaminates.[14]

It is also possible to operate the ALD process just outside the "formal" ALD process window, for example, by not completely removing the residual surface adsorbate during the purge step, thereby increasing the growth rate per cycle. This can lead to losses in conformality and uniformity, but depending on the application this may be an acceptable compromise.

13.7 THE USE OF PLASMAS IN ALD PROCESSES

While the majority of ALD processes are thermally activated, thermodynamics may dictate that a desired reaction pathway or a desired material structure cannot be achieved in a near-equilibrium environment. Plasmas can be used to interrupt thermodynamically favored pathways to enable growth of material phases that cannot be achieved at practical processing conditions.[15] Plasmas can also assist in the initial dissociation of the precursor and/or the process co-reactant to allow more efficient growth at relatively low temperature. This is a key advantage for applications where substrate temperature may be limited, such as polymers.

The reactivity of the plasma environment can also have unwanted effects on the process. In addition to the presence of charged particles and molecules in the plasma, the reactive nature of the plasma can have an effect of the uniformity and conformality of the coating itself. This occurs for much the same reason as for plasma-enhanced CVD, whereby the relatively high sticking coefficient of the plasma-excited species results in some loss of uniformity in adsorption across the substrate surface.[16] It should be noted that, while the conformality over very aggressive structures may not be ideal as in the case of thermal ALD processes, plasma ALD layers can often achieve relatively good step coverage, which may be an acceptable tradeoff for the enhanced material properties.

13.8 HARDWARE CONSIDERATIONS FOR ALD PROCESSING

Given the process similarities between CVD and ALD, the hardware for these processes possess some common characteristics. General chamber, gas handling and pressure control, and vacuum/abatement subsystems are similar. First generation ALD system designs employed a "hot wall" (externally heated) horizontal flow configuration, much like a CVD batch furnace. Later designs added single wafer "down-flow" systems with heated wafer chucks. The use of similar processing chemistries points to similar heating and environmental health and safety issues.

On the other hand, the fast cycling nature of the ALD process and the importance of surface-limited half reactions point to several key differences in hardware design. First is in the precursor and reactant delivery. The need to rapidly ogle delivery valves requires valves that are designed for very high cycle times between replacement. Also, as short a distance as possible between the pulse valve assembly and the deposition chamber is crucial in order to maintain best control over the reactant flux profile. In addition, since the ALD process is governed by the saturated surface reaction and not by gas-phase mixing and dynamical concerns, a minimal chamber volume is preferable in order to increase the efficiency of the process.

13.9 REVERSING THE CHEMISTRY: ATOMIC LAYER ETCHING

A recent outgrowth of ALD reaction chemistry has involved using the same principle of half reactions to essentially reverse a given deposition reaction to become an etching reaction. Most reactions that occur in ALD are reversible, and the concept of self-limiting surface reactions can be extended to layer-by-layer etching much as is the case for ALD. In atomic layer etch, a surface is exposed to a reactive species which can form a volatile product. After exposure and purging, a surface layer of the volatile etch product is left on the surface. Exposure to low energy ions in a plasma, which would not normally remove the native surface material, can facilitate desorption of the etch products, which are removed from the system leaving the substrate with an atomic layer removed.[17]

13.10 REFERENCES

1. M. Ieong, B. Doris, J. Kedzierski, K. Rim, and M. Yang, "Silicon device scaling to the sub-10-nm regime," *Science*, 306: 2057, 2004.
2. H. Kim, H. Lee, and W. Maeng, "Applications of atomic layer deposition to nanofabrication and emerging nanodevices", *Thin Solid Films*, 517: 2563, 2009.
3. T. Suntola, "Atomic Layer Epitaxy," *Tech. Digest of ICVGE-5*, San Diego, 1981.
4. J. Tanskanen, J. Bakke, S. Bent, and T. Pakkanen, "ALD growth characteristics of ZnS films deposited from organozinc and hydrogen sulfide precursors," *Langmuir*, 26 (14): 11899, 2010.
5. J. Maula, "High-quality crystallinity controlled ALD TiO_2 for waveguiding applications," *Chinese Opt. Lett.*, 8: 53, 2010.
6. S. George, "Atomic Layer Deposition: an Overview," *Chem. Rev.*, 110: 111, 2010.
7. A. Rissanen and R. Puurunen, Proc. SPIE 8249, "Use of ALD Thin Film Bragg Mirror Stacks in Tuneable Visible Light MEMS Fabry-Perot interferometers," Advanced Fabrication Technologies for Micro/Nano Optics and Photonics V, 82491A, Feb. 2012.
8. R. Johnson, A. Hultqvist, and S. Bent, "A Brief Review of Atomic Layer Deposition: From Fundamentals to Applications," *Materials Today*, 17 (5): 236, 2014.
9. A. Paranjpe, S. Gopinathb, T. Omsteadb, and R. Bubber, "Atomic Layer Deposition of AlO_x for Thin Film Head Gap Applications," *J. Electrochem. Soc.*, 148 (9): G465–G471, 2001.
10. K. Mistry, C. Allen, C. Auth, et al. "A 45 nm Logic Technology with High-k+Metal Gate Transistors, Strained Silicon, 9 Cu Interconnect Layers, 193 nm Dry Patterning, and 100% Pb-free Packaging," *Proceedings of the 2007 Electron Devices Meeting*, IEDM 2007, IEEE International, pp. 247–250, 2007.
11. Z. Karim, Y. Senzaki, S. Ramanathan, J. Lindner, H. Silva, and Martin Dauelsberg, "Advances in ALD Equipment for sub-40 nm Memory Capacitor Dielectrics: Precursor Delivery, Materials and Processes," *ECS Trans.* 16 (4): 125, 2008.
12. R. Puurunen, "Surface chemistry of atomic layer deposition: a case study for the trimethylaluminum/water process," *J. Appl. Phys.*, 97 (12): 121301–121356, 2005.
13. P. S. Kumar and K. Kirhika, "Equilibrium and Kinetic Study of Adsorption of Nickel from Aqueous Solution onto Bael Tree Leaf Powder," *J. Eng. Sci. Technol.*, 4 (4): 351, 2009.
14. G. Fong and J. Ma, "Rapid atomic layer deposition of silica nanolaminates: synergistic catalysis of Lewis/Brønsted acid sites and interfacial interactions," *Nanoscale*, 5: 11856, 2013.
15. H. B. Profijt, S. E. Potts, M. C. M. van de Sanden, and W. M. M. Kessels, "Plasma-Assisted Atomic Layer Deposition: Basics, Opportunities and Challenges," *J. Vac. Sci. Technol. A*, 29 (5): 050801, 2011.
16. H. C. M. Knoopsa, E. Langereisb, M. C. M. van de Sandenb, and W. M. M. Kessels, "Conformality of Plasma-Assisted ALD: Physical Processes and Modeling," *J. Electrochem. Soc.*, 157 (12): G241, 2010.
17. K. Kanarik, T. Lill, E. Hudson, S. Sriraman, S. Tan, J. Marks, V. Vahedi, and R. Gottscho, "Overview of atomic layer etching in the semiconductor industry," *J. Vac. Sci. Technol. A*, 33: 020802, 2015.

13.11 FURTHER READING

Hwang, C. S. (ed.), *Atomic Layer Deposition for Semiconductors*, Springer, New York, 2014.

Kääriäinen, T., D. Cameron, M. Kääriäinen, and A. Sherman, *Atomic Layer Deposition: Principles, Characteristics, and Nanotechnology Applications*, 2d ed., Wiley-Scrivener, Hoboken, New Jersey, 2013.

Puurunen, R., "A Short History of Atomic Layer Deposition: Tuomo Suntola's Atomic Layer Epitaxy," *Chemical Vapor Deposition*, 20: 332–344, 2014.

Puurunen, R. L., H. Kattelus, T. Suntola, "Atomic Layer Deposition in MEMS Technology," Chap. 26, *Handbook of Silicon Based MEMS Materials and Technologies*, edited by V. Lindroos et al., Elsevier, Amsterdam, pp. 433–446, 2010.

Suntola, T., "Atomic Layer Epitaxy," Chap. 14, *Handbook of Crystal Growth Vol. 3, Thin Films and Epitaxy, Part B: Growth Mechanisms and Dynamics*, Elsevier Science Publishers, Amsterdam, 1994.

Suntola, T., "Atomic Layer Epitaxy," *Handbook of Thin Film Process Technology*, Institute of Physics (IOP) Publishing, Inc., London, 1994.

Chapter 14
Electrochemical Deposition

BY

JOHN KLOCKE *Applied Materials, Kalispell, Montana*

14.1 INTRODUCTION

Electrochemical deposition is a critical part of modern semiconductor manufacturing and is used in many manufacturing steps. Electrochemical deposition can be tuned to the process requirements for each application including electrolytic deposition of highly conductive copper for efficient damascene interconnects. Selective deposition is achieved electrolessly to deposit Co alloy capping layers for electromigration lifetime improvement. Electrochemical deposition also provides a low-cost and high-throughput process that is very attractive for packaging applications. Thick deposits of solder alloys such as lead/tin and tin/silver are electroplated through resist for chip packaging applications.

14.2 FUNDAMENTALS OF ELECTROCHEMICAL DEPOSITION

Electrochemical deposition is the neutralization of ions or charged particles in solution to deposit material on a surface. In electrolytic deposition of metals, an external power supply provides electrons needed to reduce positively charged metal ions and form a metallic layer on the substrate (cathode) surface. In electroless deposition, the electrons come from a chemical source instead of an external power supply. In electrophoretic deposition, the deposited material (typically photoresist or paint) is neutral and contained within a charged micelle. The micelle can then be neutralized and deposited on the substrate's surface.

14.2.1 Electrolytic Deposition

The most basic electrolytic cell (Fig. 14.1) consists of an anode and a cathode immersed in an electrolyte solution and electrically connected by an external power supply. The power supply drives a potential gradient between the electrodes causing oxidation to occur at the anode and reduction at the cathode. The potential difference between the cathode and anode is referred to as the cell potential. It is the sum of surface potentials at the electrodes and any potential gradient through the system due to resistance of the electrolyte and electrodes. Typically the largest factor impacting the cell voltage is the resistance of the electrolyte filled cell between the two electrodes. A third electrode (reference electrode) can be added to the system to measure potential in proximity to the cathode or anode. The reference electrode does not carry significant current and allows more detailed examination of the potential at each electrode surface.

The reaction(s) that occurs at each electrode is determined by the availability of species in solution and their electrode potential. The standard electrode potential is typically expressed in volts, relative to standard hydrogen electrode. For a simple soluble copper anode system, solid copper is oxidized and dissolves into solution at the anode surface and copper ions are reduced and deposited as metallic copper on the cathode [Eq. (14.1)]. By applying a cathodic potential less than 0.34 V, copper ions can be reduced on the cathode surface into metallic copper. Further reductions beyond this value increase the reaction rate and current flow through the system.

$$\text{Cathode (reduction): } Cu^{2+} + 2e^- = Cu^0 \ (+0.34\text{ V}) \qquad (14.1)$$

In a typical electrochemical deposition (ECD) cell, the source of the metal being deposited is considered to be the electrolyte very near the electrode (cathode). This is in contrast to a vacuum deposition process such as sputtering, in which the mean free path is very large and the source of the deposited material is the counter-electrode, or target. The metal ions are consumed by the reduction reaction at the cathode and the concentration of these ions is reduced in a region very near the electrode surface. As the reaction proceeds, this leads to the formation of a region of reduced concentration called the diffusion boundary layer. The concentration gradient in this diffusion boundary layer is the driving force for the transport of the ionic reactant to the cathode surface. The balance of the diffusion rate to the reaction rate is a critical parameter in electrolytic systems. If the current is increased to the point at which every metal ion is consumed as soon as it reaches the cathode surface, the diffusion rate across the diffusion boundary layer is at a maximum because the concentration gradient cannot be increased. This is called the limiting current density (LCD). If the voltage is increased slightly, the current will typically not change, as a higher reaction rate cannot be supported. If the cell potential is driven higher, the current will again increase when the surface overpotential becomes sufficient to drive another reaction in the system. In acidic systems, this is often the generation of hydrogen gas. When deposition occurs at or above the hydrogen evolution potential, a portion of the applied current results in gas evolution instead of metal deposition. The percentage of current that results in metal deposition is called the efficiency of the process. In an aqueous system, decomposition of water occurs at −0.83 V. The water in the system will essentially provide an infinite source of the reactants for this reaction. Therefore, metals with reduction potentials more negative than the reduction of water (such as aluminum and titanium) cannot be electrochemically deposited from aqueous solutions. They can only be electroplated in nonaqueous systems, such as solvents or molten salts.

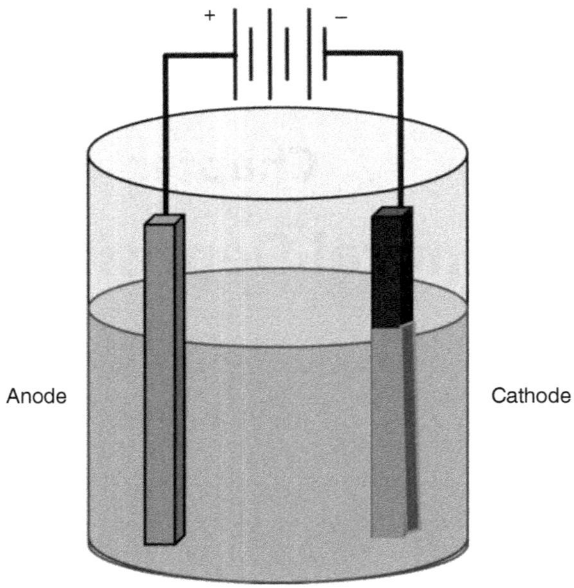

Figure 14.1 Basic electrochemical cell.

If a deposition reaction is operated near the LCD, the deposit will tend to be nodular or dendritic. Typically, industrial processes are operated at 30 to 50 percent of the LCD in order to produce smooth deposits with consistent morphology. LCD is a reasonable measure of the relative deposition speed capability of different systems. The factors that increase the diffusion rate of metal ions to the cathode surface have the most impact on increasing the LCD of an ECD system. These factors are agitation (such as flow and wafer spin speed), temperature, and metal ion concentration.[1–5]

When two different metal ions are present in solution, the deposited material will be a function of the standard electrode potential of the materials. For example, when both Cu^{2+} and Fe^{3+} are in solution, copper will readily deposit at 0.34 V. However, iron requires a more negative deposition potential (−0.04 V). Therefore, the deposit would consist entirely of copper. When the electrolyte solution contains metal ions with similar electrode potentials such as Sn^{2+} (−0.14 V) and Pb^{2+} (−0.13 V) the materials can be co-deposited as an alloy. In some cases, the reduction potential is adjusted with a complexing agent. They stabilize the metal ion in a different oxidation state effectively changing available species and the corresponding standard potential of the metal to be deposited. In other cases, chemical changes such as concentration or organic additives can be used to increase the over potential allowing simultaneous deposition of two different metals.

A rotating disk electrode is a very useful tool for analysis of electrochemical deposition systems. The spinning disc acts as a pump providing its own fluid replenishment and has a uniform LCD across its surface. The uniform deposition conditions provide a clear response to changes in deposition conditions. Variations in RPM can be used to investigate mass transport effects. Variations in applied voltage can be used to investigate current responses. Voltage current curves such as the LCD curve shown are typically generated with a rotating disk electrode. They provide a useful analytical tool for electrochemical deposition. Another variant of this called cyclic voltammetric stripping (CVS) can be used to study the amount of deposition that occurs under fixed condition in plating solutions. In CVS, a rotating disc electrode is repeatedly swept to a negative potential and then to a positive potential at a predetermined rate. This sweep is then repeated. As the potential swings negative (left on Fig. 14.2), deposition occurs on the electrode and the negative current increases. As the potential sweeps back to the right, the current becomes positive as the deposited material is stripped from the electrode. Once all the deposited material is removed, the current decreases. The peak in the upper right quadrant of the graph is called the stripping peak. The area under this curve is a quantifiable measure of the amount of deposited material for a given potential sweep. This information can then be used to analyze or control electrochemical deposition chemistry.

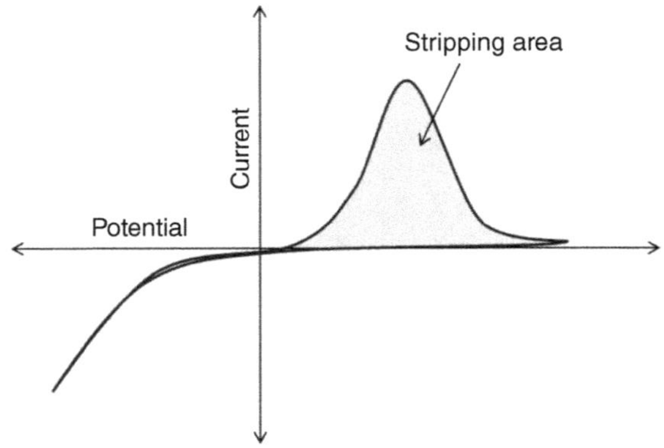

Figure 14.2 CVS curve.

Organic additives are added to the electrolyte solution to modify the deposition process. They can form complexes with ions in solution that can increase or decrease the potential at which deposition occurs. Other components block the cathode surface suppressing deposition that occurs at a given potential. Accelerator, suppressor, and levelers are classes of organic additives that are used for damascene metallization. Suppressors typically consist of polymers or copolymers of glycol ether compounds.[6,7] These organic materials adsorb on the copper surface in the presence of a small amount of chloride ion and inhibit deposition. They are called suppressors because of their effect on reducing the deposition rate (current density) at a given deposition potential. They also have the effect of increasing the overpotential required for maintaining a specific deposition rate. Accelerator increases the deposition rate or decreases the overpotential when added to a bath containing suppressors. Accelerators are typically compounds such as 3-mercapto-1-propanesulfonate (MPS) and bis (3-sulfopropyl) disulfide (SPS), or similar thiol compounds.[6,7] These materials usually provide a catalytic effect and remain adsorbed at the deposition surface as deposition proceeds.

Figure 14.3 shows the response of suppressor and accelerator during CVS analysis. The stripping peak for the "electrolyte only" solution has a large deposition peak and stripping area. When suppressor is added into

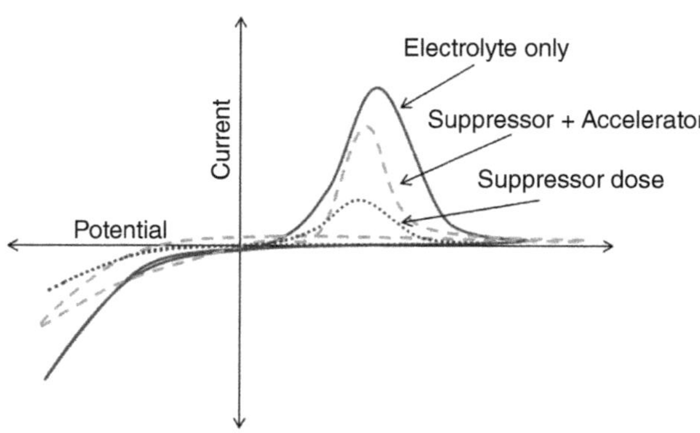

Figure 14.3 CVS response to additive dosing.

an electrolyte solution, the amount of deposition decreases for a given potential sweep. This is represented by the "suppressor dose" curve. It has a much smaller stripping area. When accelerator is added into a suppressed solution, the amount of deposition for a given potential sweep increases. This is shown in the "suppressor + accelerator" curve.

Levelers are polar (negatively charged) molecules that also perform a suppressing function. They are typically high-molecular-weight compounds with sulfonic or sulfamic acid or other nitrogen-containing functional groups.[6] They attach to the areas of the surface with higher current densities or tie up accelerator complexes (thiols) and decrease the local deposition rate. The concentrations of all these organic additives are typically very low, ranging from ppb to 0.1 percent of the bath composition. They are very susceptible to changes that impact surface concentration such as agitation (flow, liquid agitation, and spin speed) as well as temperature.

The ability of additives to suppress and accelerate deposition enables the chemistry to be tuned to deposit more rapidly at the bottom of features than at the top. Suppressors block deposition on the side walls of the features while accelerator is geometrically concentrated on the bottom surface which allows the deposition to occur more easily in the bottom of the features.[3] This creates superconformal deposition.[8,9] This is the ability of the electrochemical deposition to fill recessed features from the bottom up. Levelers are typically operated in a diffusion limited regime so that they do not interfere with fill in small features. Levelers help to limit deposition above the features once fill is complete, preventing over deposition in highly patterned areas. Extensive references are available on the theories that explain this.

The type of additive used also influences the properties of the deposited film. Properties such as impurity concentration, gran size, and film stress are modulated through the use of organic additives. Other organic components such as surfactants are sometimes added to plating baths to influence wetting.

It is also important to pay attention to the anode in electrolytic deposition systems. There are two classes of anodes available for copper deposition: consumable (copper) and inert (nonconsumable). The main differences between the types of anodes are in the anodic reactions they support. Consumable anodes promote copper dissolution at the anode. Most industrial systems utilize consumable anodes for copper deposition on semiconductor wafers. In this case, the anode reaction is the reverse of the cathode reaction, providing a low-cost source of copper and a stable copper concentration. However, the anodes must be periodically replenished as they dissolve. Inert anodes usually involve the anodic generation of oxygen gas. Oxygen gas can lead to film defects if not managed properly.[10] This reaction also requires a greater anodic potential than soluble anodes and can accelerate the oxidation and decomposition of organic additives at the anode surface. Ionic membranes can also be used to separate the anode and its electrolyte (anolyte) from the cathode and catholyte solution. Organic additives are only dosed into the electrolyte solution eliminating additive oxidation at the anode and stabilizing the plating chemistry. Any gas generated at the anode is also isolated from the wafer preventing bubble related defects.

14.2.2 Electroless Deposition

Electroless deposition involves reducing metal ions on the surface of a substrate to deposit metal. It is closely related to electrolytic deposition except that the source of reducing electrons comes from a chemical reaction that is not driven by an external power supply.

The simplest type of electroless reaction is a single displacement reaction. This reaction occurs when a reactive material is immersed in a solution containing more noble ions. Equation (14.2) demonstrates the reaction of iron in copper-containing solutions. The most classic example of this is copper cladding of iron nails. When an iron nail is immersed in a solution of copper sulfate, iron on the exposed surface loses electrons, becoming a soluble ion and copper ions receive electrons and deposit onto the surface of the nail as copper metal. This results in a copper-coated iron nail.

Copper and Iron displacement reaction: $Fe + CuSO_4 \rightarrow Cu + FeSO_4$ (14.2)

A second type of electroless deposition utilizes a reducing agent to provide electrons. Common reducing agents include formaldehyde (HCHO) and dimethylamine borane (DMAB). The deposition reaction can be either autocatalytic on an exposed metal surface or it can be catalyzed with deposition of a catalyst material such as palladium. Deposition rates are then controlled by the chemical concentrations, temperature, and mass transport to the substrate surface. Electroless deposition does not need electrical current; therefore, factors such as the terminal effect are no longer relevant. And, it is not necessary to have a seed layer. The material can be deposited electrolessly directly on to dielectric material after catalyst activation.

Electroless deposition can also provide the benefit of self-aligned and selective deposition. When the substrate has both dielectric and metallic areas exposed, a displacement reaction can be utilized to activate only the exposed metal surfaces. During the subsequent deposition step only activated metal surface receives deposition.

14.3 APPLICATION OF ELECTROCHEMICAL DEPOSITION

14.3.1 Copper Interconnect

In the early years of semiconductor processing, transistor gate delays determined chip speed. However, as devices shrank, the RC delay in the metallization (interconnect) levels became an important factor determining the chip speed. Copper replaced aluminum in interconnect metallization for many chips because of its low resistivity and improved electromigration lifetime. Copper is second only to silver in elemental resistivity at 1.67 μΩ·cm. Copper films have a much higher melting point than aluminum and are predominantly (111) in crystal orientation which is the close packed plane. Both of these factors limit mobility of the copper atoms and improves electromigration performance.[8] Copper interconnects typically offer an order of magnitude increase in electromigration lifetime over traditional aluminum copper alloys.[11]

Aluminum interconnects were formed through a subtractive metallization process by dry etching blanket physical vapor deposition (PVD) aluminum alloys. Dry etching of copper is not feasible with current technologies because the temperature required to volatilize copper compounds is not compatible with semiconductor devices. Implementing copper interconnects required switching to a damascene process flow. The damascene process involves depositing and patterning a dielectric material first and then metal is filled into the features. Finally, the excess material is removed with chemical mechanical planarization (CMP), leaving an embedded metal line. In dual damascene processing, both a lower via layer and an upper metal line layer are patterned in the dielectric material prior to metallization. Both layers can then be filled in a single metallization process, which reduces the overall number of process steps and manufacturing costs.

The standard process flow for damascene copper metallization is shown in Fig. 14.4. After the dielectric layer has been patterned, barrier and seed layers are deposited using a PVD process. Typically the barrier layer is either Ta or a TaNx/Ta bilayer and the seed layer is copper. Electrochemical deposition (ECD) is then used to fill the features with copper. The deposited copper film typically has smaller grains immediately after deposition. It is annealed at a low temperature (100 to 400°C) to increase grain size and stabilize the film. CMP is then used to remove the excess material leaving copper lines embedded in the substrate.

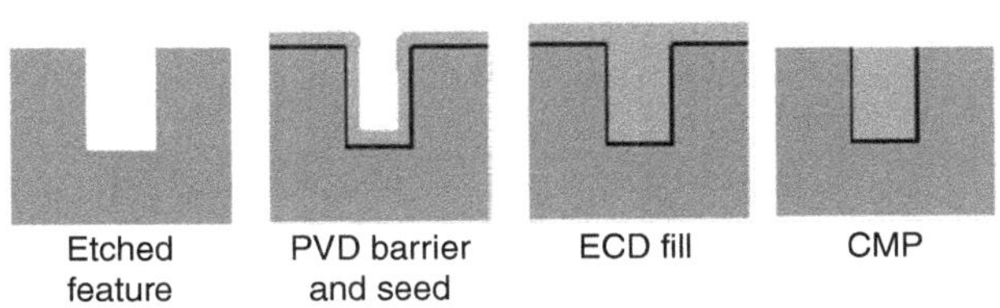

Figure 14.4 Basic damascene process flow for the formation of copper interconnects.

Electroplating is well known as a purification method for copper (electrorefining) and is capable of depositing very pure films (0 ppm of measurable impurities). However, organic additives can become incorporated into the deposit. Copper films used in interconnects typically have impurity levels that range from 10 to 500 ppm with sulfur and carbon being the main contributors. These impurities can actually increase the stability of the deposited film and improve stress migration performance. However, if the impurity content is too high, electromigration life time is negatively impacted.

Copper can be electrolytically deposited from several types of solutions—acidic copper sulfate, methane sulfonic acid, basic cyanide complexed chemistries, pyrophosphate baths, and many other specialized chemistries; some of which contain complexing agents. Most copper deposition for semiconductor processing, however, is done using relatively simple acidic copper sulfate chemistry. These baths typically consist of three inorganic components—sulfuric acid, copper sulfate, and hydrochloric acid—with organic compounds to modify the deposition and film properties. Copper sulfate provides the source of copper ions that are electrolytically deposited during the plating operation. The higher the concentration of copper ions in solution, the better the feature filling performance of the chemistry. Sulfuric acid is used to maintain an acid environment to keep copper in solution and to provide the ionic conductivity of the solution. The H^+ ions, or protons, are the preferred charge carriers in the solution and provide high solution conductivity. The pH of the solution also affects the activity of the organic additives. The concentrations of copper sulfate and sulfuric acid are not completely independent since the common ion effect limits the total solution concentration of sulfate ion.[12] This sets the upper limit on the total concentration of these two chemicals.

The primary technical challenge during electroplating damascene copper metallization is maintaining void free filling of the smallest features on the wafer. As previously discussed, bottom-up fill is achieved through organic additives. The adsorption and displacement of these additives is largely a function of the local potential or current density and the additive concentration. Therefore, it is critical to control uniformity of current across the wafer and additive concentration in the plating bath. At the beginning of the deposition process, the seed layer is very thin (200 to 400 Å of copper) to prevent pinch off of the smallest features. This results in a significant resistance across the wafer. The resistance further increases with increasing pattern density and the use of copper alloy seeds. However, as deposition proceeds, the small features are filled and the thickness of the copper layer on the wafer increases. This results in a significant reduction in the resistance across the wafer. Therefore, the potential drop across the wafer from the electrical contact at the edge of the wafer to the center of the wafer (terminal effect) decreases with time. The plating system must be able to adapt to a changing resistance in order to maintain uniform deposition and gapfill performance throughout the plating process.

The typical electroplating sequence for copper interconnects involves multiple process steps. The first and possibly most critical step (immersion) involves holding a potential on the wafer as it is immersed into plating solution. This potential must protect the copper seed layer from corrosion in the plating bath and helps to set the initial organic additive conditions on the wafer surface. The current must be controlled in such a way to ensure that the entire wafer sees similar additive adsorption. The motion profile of the wafer during immersion is also critical to allow uniform additive adsorption and prevent any bubbles from being trapped on the wafer surface.

The second step is gap fill. During this step the current applied to the wafer is modulated to maintain the right local potential for gap fill. As the smallest features fill, the surface area of the wafer can change significantly and may require significant changes in current density to maintain consistent fill. The final step of electrochemical deposition step is called either bulk deposition of cap step. During this process the critical features have already filled and current density is increased to deposit enough copper into the wide structures such as bond pads where the additives are not as effective. The majority of the copper is deposited during this step. And the primary requirement is to ensure that the final film will integrate with CMP. Factors such as uniformity, planarity, and impurities are considered during this process step. After plating is completed, the wafer is rinsed and copper is removed from the bevel area of the wafer to avoid scratching during CMP.

14.3.2 Through Silicon Vias

Through silicon vias (TSV) is a process similar to damascene copper interconnects. The feature is defined prior to metal deposition, a seed layer is deposited and then metal is electrochemically deposited into the structure using organic additives to modulate the deposition. The vias are much larger and provide a connection path through the substrate for chip-to-chip stacking. This provides shorter chip-to-chip connection lengths and a more compact package. There are many variants of TSV that range a wide scope of size, shape, and integration approaches. Most TSVs can be grouped into two general categories: backside vias and frontside vias.

Backside vias are typically etched after the device is complete and the wafer is thinned. They are etched into the backside of the wafer to make contact with the device level. This approach does not encroach on the active area of the chip and allows use of a wide tapered via that can be achieved with a low-cost etch process. These larger tapered vias are then either filled with copper or lined with copper or gold. This type of chip connection is very common in the III/V semiconductor industry.

Frontside TSVs are typically produced during device manufacturing. They consist of a high aspect ratio via that extends into the silicon of the wafer. The bottom of the vias is then revealed later in the process flow when the backside of the wafer is thinned. These narrow vias allow for a higher number of through chip connections but encroach on the device area of the wafer. These vias often have an aspect ratio that can range from 10:1 up to 20:1. Typical via depths range from 30 to 100 μm.

The larger scale of TSV drives significant changes compared to damascene vias in feature wetting time, mass transport, and process time.

Traditional copper interconnect submicrometer scale features wet very quickly.[13] The small scale results in large capillary forces that drive trapped gas into solution very quickly. Larger packaging sized structures such as TSV structures have reduced capillary forces and more gas that needs to dissolve into solution; this results in wetting times that can take several minutes compared to a fraction of a second for submicron features. In order to increase deposition tool throughput and minimize seed layer corrosion, TSV wafers are typically prewet prior to electrochemical deposition.

Mass transport is also considerably different in TSV structures than traditional copper interconnects. Traditional copper interconnects have a feature depth and scale that is substantially smaller than the diffusion boundary layer (less than 1 μm in most cases). TSV structures are large enough to allow convention in the mouth of the feature. The convective region extends approximately one aspect ratio deep into the feature. For example, a 10-μm-wide feature will have a convective region approximately 10-μm deep (Fig. 14.5).[14,15] They also have a very long

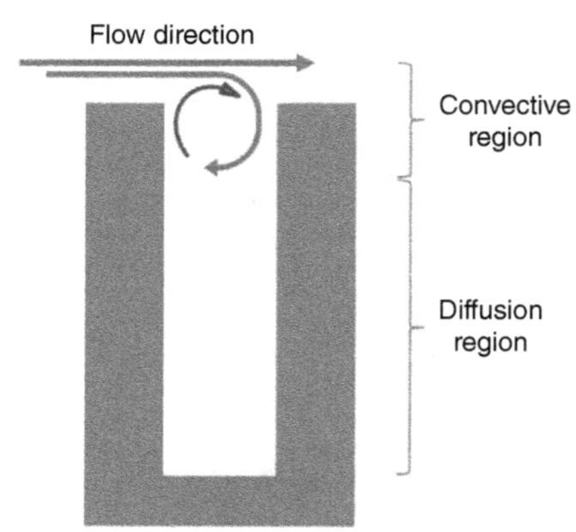

Figure 14.5 Mass transport within TSV structure.

diffusion length. Therefore, diffusion time of the plating additives and copper ions can be substantial. This is of particular importance because the features are sometimes filled with water during the prewet step. As a result, it takes a much longer time for bottom-up fill to nucleate within a TSV structure and the rate of bottom-up fill is limited by the copper transport into the feature.

Because of the slowed nucleation, copper concentration limit, and large feature size, TSV fill takes much longer to achieve than interconnect fill. This has several implications for the electrochemical process. Additives and their breakdown products have more time to accumulate on the substrate surface. Therefore, the plating bath must be carefully controlled. Because of the cost of the TSV process it is primarily used in limited applications. These tend to be high end performance devices.

14.3.3 Through Mask Plating

In wafer level packaging applications, layers of metallic features are added to the completed wafer. These features include RDL (Redistribution Layers), posts and pillars, bumps, bond pads, and passive devices. The typical scale of these features ranges from 2 μm lines to approximately 200 μm bumps. The relatively larger scale of these features allows the use of through mask plating. In this case, the seed and barrier layers can be wet etched after the photoresist is removed with minimal impact on the final feature dimensions. This eliminates the rather expensive CMP step used in damascene processes and allows the use of simple blanket PVD films. The through mask plating flow is outlined in Fig. 14.6.

Rotating disk theory shows that LCD is uniform across a rotating blanket wafer. However, in through mask deposition the mechanism changes significantly. Large regions of the substrate are masked by photoresist. These unplated areas allow more metal ions to be transported toward the edge of the wafer. Therefore, the LCD of a spinning wafer for a through mask plating application is much lower at the center of the wafer and increases toward the edge (Fig. 14.7). With smaller open areas, the difference in LCD from the center to the edge on the wafer increases. The plating rate that can be achieved is limited by the mass transport provided to the center of the wafer. In order to maintain uniform deposition at high rates across the wafer, mass transport is achieved through means other than spinning the substrate. Several methods can be used to improve mass transport at the wafer center. These include localized fluid jets to increase flow at the center of a rotating wafer and high volume cross-flows on a slowly rotated wafer. The most common and fundamentally uniform method is variations on a paddle plater concept.[16] In this case, mass transport is entirely provided by an agitator; the cycles back and forth in close proximity to the wafer surface. Wafer rotation is not relied on for mass transport and fundamentally uniform deposition can be achieved.

Because the seed layer is under the photoresist mask, bottom-up feature filling inherently occurs. Technical challenges are more related to plating uniformity. Variations in the amount of exposed surface for

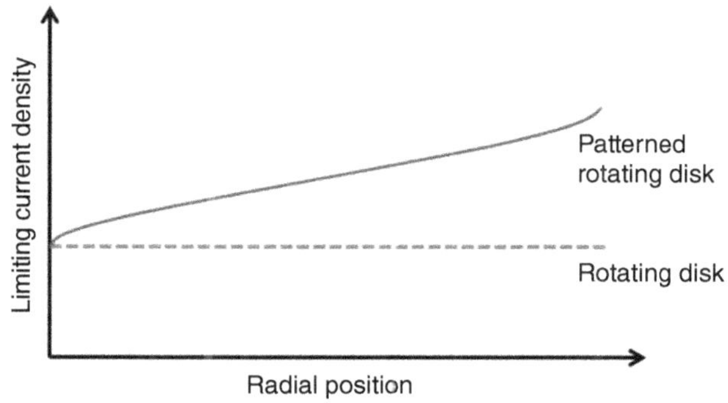

Figure 14.7 LCD of a rotating patterned substrate versus rotating disk.

plating across the die create regions of low and high current density (current crowding). This can result in significant plating thickness variation across the die (within die uniformity). Chemistries are tuned specifically to minimize within die uniformity. Typically higher acid (higher conductivity) chemistries are used because their increased throwing power better distributes current across the die. Organic additives also help to suppress deposition in high current areas further improving within die uniformity.

Maintaining uniform mass transport and current density is probably most critical when co-depositing metals to form solder alloys. For example, when tin silver deposition rate is decreased from 2 to 1 μm/min, the concentration of silver in the deposit increases from 3.5 percent eutectic concentration to 5.5 percent resulting in different material properties, including an elevated melting point.[17] This is also critical within the individual feature. When plating into the bottom of a feature mass transport is suppressed and the diffusion rate determines the allowable current density. As the feature fills, convection at the top of the feature allows for a much higher deposition right while maintaining consistent alloy composition.

Mass transport within the features and organic additives also plays a critical role on the profile of the individual plated feature. Mass transport affects the symmetry of the individual features. If flow across the feature is maintained in the same direction, the plated profile can develop a significant slope. This slope can be prevented by alternating flow direction (such as an agitator going back and forth). The concentration of organic additives combined with plating current density can result in feature profiles that range from domed to concave. Typically a flat profile is desired.

As device performance and the number of chip connections increase, the density of RDL lines is also increasing. The minimal line width is shrinking and the number of layers is increasing. This is driving the minimal line width down. In order to minimize the amount of undercut

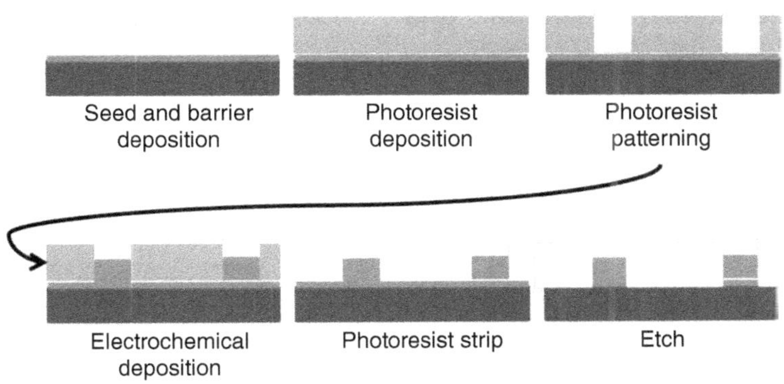

Figure 14.6 Patterned plating process flow.

and CD loss during the subsequent UBM etch, the seed and barrier layer thickness is being decreased. This results in greater terminal effect across the wafer. Because of the increasing number of lines, dense layers are seeing increasing amounts of open area. Therefore, plating uniformity control is becoming more challenging for RDL applications. More thinner resistive seed layers require more advanced plating control. Higher pattern densities require higher current densities. Greater variation in open area from layer to layer requires more flexible and uniformity control in the plating reactor.

14.3.4 Electroless Deposition

Electroless deposition is used for both packaging and interconnect processes. The most established interconnect application is selective capping layers for electromigration life time improvement. Conventional process technology utilizes either a blanket layer of silicon nitride or silicon carbide (or a combination) to cap the copper lines after CMP. This insulating layer provides sufficient protection against copper diffusion and serves as an etch stop. However, electromigration occurs primarily along this interface, reducing the device lifetime.[18] This layer also has a relatively high dielectric constant, which increases the effective dielectric constant of the dielectric stack. To achieve effective dielectric constants significantly below 3.0, the blanket-capping layer must be either changed or eliminated.[19]

Metal capping layers such as CoWP can be deposited in thin layers on top of copper interconnect lines at critical process layers. This layer effectively reduces the mobility of copper along the top interface of interconnect lines. This extends the electromigration lifetime by more than 10-fold.[20] To prevent shorts or leakage, the layer must be selectively deposited only on top of the copper. This selectivity can be achieved either through electroless deposition or CVD. Electroless deposition of CoWP or CoWB is a leading candidate for this process. The material can be selectively deposited in very thin layers to minimize the change in via resistance while increasing electromigration lifetimes (Fig. 14.8).

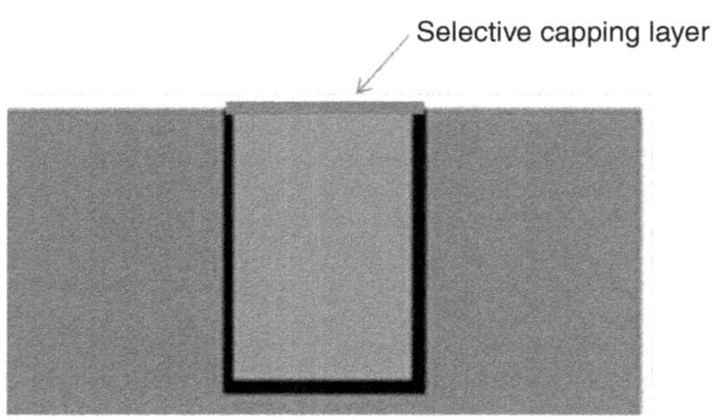

Figure 14.8 Selective capping layer.

In wafer level packaging, electroless deposition is commonly used for pad finishing applications. This can be a series of metal layers deposited through either chemical reduction process or displacement reactions. The final step in pad finishing is a displacement reaction where gold is allowed to displace an underlining layer. This provides a surface that is stable (will not oxidize or corrode) and can be wire-bonded.

Electroless processes are chemically controlled compared to electrolytic processes which are primarily controlled by an external power supply. Therefore, electroless systems have an entirely different set of critical process parameters. Temperature control is one of the most critical parameters as it drives deposition rates and determines stability of the process chemistry. If the chemistry is too cold, the deposition process will be slow and expensive. If the temperature is too hot, the chemistry can spontaneously deposit metal in locations other than on the intended substrate (plate out). If temperature is not uniform across the wafer, the deposit thickness will also be nonuniform. Mass transport must also be controlled uniformly across the substrate.

14.4 FUTURE TRENDS

With each technology node, interconnect size decreases and the current density in interconnects increases. According to the International Technology Roadmap for Semiconductors, the current density approximately doubles from the 130- to 45-nm node technology.[21] As feature sizes become smaller, the percentage of the interconnect filled by the barrier layer becomes more significant. In addition to this, the size of the interconnect feature is approaching the electron mean free path in copper. This causes the resistivity to increase due to surface scattering of electrons. To minimize line resistance and maximize electromigration lifetime, alternative materials are being investigated and integrated into semiconductor devices. These include alloy seed layers to improve copper electromigration and alternative interconnect materials that do not require a diffusion boundary layer such as cobalt. It is certain that these challenges will drive a new wave of metallization technology, including advanced packaging techniques to increase device performance.

14.5 SUMMARY

Semiconductor manufacturers continually strive to make smaller, faster chips by minimizing line resistance, feature sizes, and dielectric constants. As features shrink, the local current densities increase, creating issues with electromigration. Copper interconnects provide lower resistance and improved electromigration lifetimes relative to aluminum. To implement copper in production, manufacturers have switched to a damascene processing sequence that uses copper electroplating to fill submicrometer features. ECD prevents void formation in the deposition process by combining organic additives and electrolytic deposition parameters to produce high-yielding, reliable interconnects. To maintain consistent performance, the plating bath additives are periodically analyzed and replenished with automated chemical management systems. Semiconductor manufacturers are now researching methods to further reduce line resistance as features continue to shrink. These include switching to thinner and more conductive ALD barrier materials. Electromigration lifetimes and dielectric constants may be further improved with selectively deposited electroless capping layers. Electrochemical deposition processes are also used in wafer level packaging applications to improve finished device performance and footprint.

14.6 REFERENCES

1. H. Deligianni et al., "A Model of Superfilling in Damascene Electroplating," *Electrochemical Processing in ULSI Fabrication and Semiconductor/Metal Deposition II*, edited by P. C. Andricacos, P. C. Searson, C. Reidsema-Simpson, P. Allongue, J. L. Stickney, and G. M. Oleszek, ECS, Pennington, NJ, pp. 52–60, 1999.
2. K. Takahashi and M. Gross, "Transport Phenomena That Control Electroplated Copper Filling of Submicron Vias and Trenches," *J. ECS*, Vol. 146, No. 12, pp. 4499–4503, 1999.
3. T. P. Moffat et al., "Superconformal Electrodeposition of Copper," *ECS Electrochem. Solid-State Lett.*, Vol. 4, No. 4, pp. C26–C29, 2001. *Metallization: Materials, Processes, and Reliability*. ECS, Pennington, NJ, pp. 48–58, 1999.
4. T. Ritzdorf, D. Fulton, and L. Chen, "Pattern-Dependent Surface Profile Evolution of Electrochemically Deposited Copper," *AMC 1999*, p. 101, 2000.
5. P. C. Andricacos et al., "Damascene Copper Electroplating for Chip Interconnections," *IBM J. Res. Develop.*, Vol. 42, No. 5, pp. 567–574, 1998.
6. W. H. Safranek, *Acid Copper*, 4th ed., edited by F. A. Lowenheim, Wiley Interscience, New York, pp. 66–71, 1974.
7. A. Frank, and A. J. Bard, "The Decomposition of the Sulfonate Additive Sulfopropyl Sulfonate in Acid Copper Electroplating Chemistries," *J. ECS*, Vol. 150, No. 4, pp. C244–C250, 2003.
8. M. E. Gross et al., "The Role of Additives in Electroplating of Void-Free Cu in Sub-Micron Damascene Features," *AMC 1998*, p. 51, 1999.

9. P. Vereecken et al., "Effect of Differential Additive Concentrations in Damascene Copper Electroplating," *Meeting Abstracts of the 203rd Meeting of the Electrochemical Society*, Abstract, Vol. 606, 2003.

10. T. Ritzdorf and D. Fulton, "Electrochemical Deposition Equipment," *New Trends in Electrochemical Technology, Microelectronic Packaging*, edited by M. Datta, T. Osaka, and J. W. Schultze, Vol. 3. Francis & Taylor, London, in press. pp. 106–108, 1998.

11. M. H. Tsai et al., "Reliability of Dual Damascene Cu Metallization," *Proceedings of the International Interconnect Technology Conference*, IEEE Piscataway, NJ, p. 214, 2000.

12. R. K. Murmann, *Inorganic Complex Compounds*. Reinhold, New York, 1964.

13. M. Olim, "Liquid-Phase Processing of Hydrophilic Features on a Silicon Wafer," *J. Electrochem. Soc.*, Vol. 144, No. 12, Dec. 1997.

14. P. R. McHugh, G. J. Wilson, and T. Ritzdorf, "Simulation of Shape Evolution in Through-Mask Electrochemical Deposition," presented at 220th ECS Meeting, Abstract #E5-22350, Boston, MA, Oct. 9–14, 2011.

15. G. J. Wilson, P. R. McHugh, S. Lee, and T. Ritzdorf, "Simulation of Shape Evolution in Through-Mask Electrochemical Deposition," presented at 222th ECS Meeting, Abstract #E8-2741, Honolulu, HI, Oct. 9, 2012.

16. J. V. Powers and L. T. Romankiw, "Electroplating Cell Including Means to Agitate the Electrolyte in Laminar Flow," U.S. Patent 3652442, Mar. 28, 1972.

17. M. Bernt, P. McHugh, and G. Wilson, "Effects of Mass Transfer and Current Density on High Rate Binary Alloy Plating," presented at 210th ECS Meeting, Abstract #1637, Cancun Mexico, Oct. 29–Nov. 3, 2006.

18. E. Zschech et al., "Reliability of Copper Inlaid Structures—Geometry and Microstructure Effects," *AMC*, Vol. 18, pp. 305–311, 2002.

19. B. Kastenmeier, K. Pfeifer, and A. Knorr, "Porous Low-k Materials and Effective k," *Semicond. Int.*, Vol. 27, p. 87, 2004.

20. L. Bill, "Electroless CoWP Boosts Copper Reliability, Device Performance," *Semicond. Int.*, pp. 95–100, 2004.

21. Int'l Tech, "Roadmap for Semiconductors," September 2006, http://public.itrs.net/.

Chapter 15
Fundamentals of Chemical Mechanical Planarization

BY

GAUTAM BANERJEE *Versum Materials, Inc., Tempe, Arizona*

15.1 INTRODUCTION

Chemical mechanical planarization (CMP) is an enabling process technology for advanced sub-micrometer integration to manufacture semiconductor chips. It is the only process which can planarize on a global scale. Sub-micrometer features require the highest resolution imaging techniques. This implies short wavelength and high numerical aperture lenses resulting in small depth of focus [1]. CMP provides a flat surface at nanometer scale, which eliminates the issue of small depth of focus. Among other benefits, CMP maximizes lithographic performance, improves cleaning, reduces yield limiting defects from other processes, and permits the use of difficult-to-etch metals such as copper, which has replaced aluminum as the metal interconnect of choice by 1997, introduced by IBM. Copper wires conduct electricity with about 40 percent less resistance than aluminum wires, which results in an additional 15 percent burst in microprocessor speed. Copper wires are also significantly more durable and 100 times more reliable over time, and can be shrunk to smaller sizes than aluminum [2]. Today, several different types of CMP process applications are required to manufacture different types of devices as may be seen from Fig. 15.1 [3].

15.2 WHY IS IT IMPORTANT TO UNDERSTAND THE FUNDAMENTAL ASPECTS OF CMP?

Semiconductor device market continues to grow globally with introduction of new device types besides logic and memory [4]. Examples include various communication devices, mobiles, phablets, automobiles, and myriads of Internet-of-Things devices. For example, semiconductor content in new automobiles continues to increase for sensors, control systems, and so on. The number of sensors in automobiles currently is 60 to 100 per car and is projected to more than double in the next 8 to 10 years [3]. Along with the growth in device types, advanced nodes of manufacturing semiconductor devices continue to shrink feature sizes, which increase the number of process steps. Figure 15.2 depicts the growing number of process steps for a generic logic device. From 90 to 20 nm, the number of process steps increased due to shrinks (i.e., additional metal levels). At 14 and 16 nm, the number of process steps grew significantly, driven by transistor technology changing from conventional planar to novel 3D architectures with complex integration schemes [5].

The growth in the number of CMP process steps parallels the increase in the overall number of process steps, with the number of CMP steps doubling from 28- to 10-nm technology nodes (Fig. 15.2).

Although CMP consumables cover different materials, according to SEMI [6], the semiconductor equipment and materials industry's premier organization CMP slurry and pad revenues are expected to grow at about approximately 5 percent from 2017 to 2020, which at present is approximately \$2 billion (Fig. 15.3).

The above mentioned market value demonstrates the commercial importance of CMP as a technology enabler and, therefore, it is absolutely important to understand the fundamental aspects of CMP application processes in order to develop appropriate consumables.

15.3 BIRTH OF CMP

CMP was originally adapted from the lapping process used to polish plate glass. The original commercial practice of CMP in the semiconductor industry was to prepare raw silicon wafers. After the wafer was sawed from a single-crystal silicon rod, the mechanically damaged surface layer was removed and the surface planarized to produce a flat, scratch-free surface for VLSI devices and circuits. The wafer to be polished is mounted on a wafer carrier via back pressure or via surface tension by wetting its back surface. The wafer is pressed down against a rotating platen, which holds a compliant polishing pad. The wafer slides on the pad surface with a relative velocity generated by the rotation of the carrier and the platen [7]. At best, it could be defined as chemical mechanical polishing. Monsanto first developed this process and sold polished wafers in late 1962 and was issued the known first U.S. patent for chemical slurry formulation in February 1965 for polishing various substrates.

The origin of planarizing a patterned device wafer started with the ingenuity of Klaus D. Beyer [8] at IBM who found a new trench isolation test wafer (which was built with a trench width decreasing from 2 to 1.5 μm) on which glass trench fill produced unwanted mounds after reflow at random locations. So, after some considerations, in January 1983, he took the trench isolation test wafer with reflowed glass fill to the silicon wafer polishing area and polished the wafer together with other qualification silicon wafers. After the subsequent megasonic

1995 Qty ≤ 2 CMOS	2005 Qty~6 CMOS	2016 Qty ≥ 40 CMOS	New Apps	Substrate/E
Oxide	Oxide	Oxide	MEMS	GaAs & AlGaAs
Tungsten	Tungsten	Tungsten	Nanodevices	poly-AlN & GaN
	Cu (Ta barrier)	Cu (Ta barrier)	Direct Wafer Bond	InP & InGaP
	Shallow Trench	Shallow Trench	Noble Metals	CdTe & HgCdTe
	Polysilicon	Polysilicon	Through Si Vias	Ge & SiGe
	Low k	Low k	3D Packaging	SiC
		Capped Ultra Low k	Ultra Thin Wafers	Diamond & DLC
		Metal Gates	NiFe & NiFeCO	Si and SOI
		Gate Insulators	Al & Stainless	Lithium Niobate
		High k Dielectrics	Detector Arrays	Quartz & Glass
		Ir & Pt Electrodes	Polymers	Titanium
		Novel barrier metals	Magnetics	Sapphire
			Integrated Optics	

Figure 15.1 Growth in different process applications requiring CMP over two decades. (Courtesy of Entrepix.)

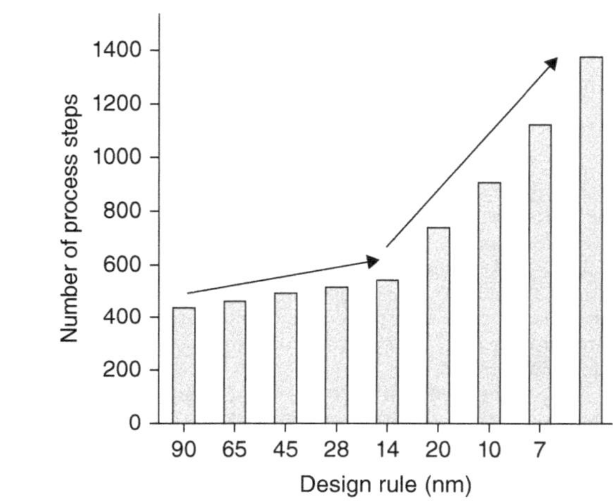

Figure 15.2 Number of process steps by design rule.

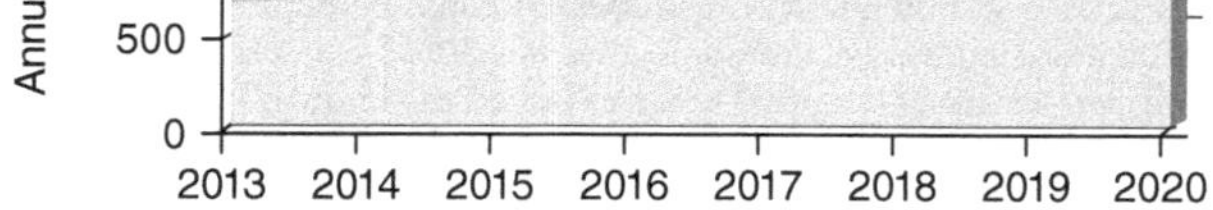

Figure 15.3 CMP slurry and pad revenue trend and forecast 2013 to 2020.

removal of the polishing slurry, the isolation test wafer exhibited an amazingly uniform green color without surface scratches like the remaining qualification bare silicon wafers. The outcome of this experiment and subsequent characterization studies showed the planarization capability of a rather crude polishing process used at that time for hundreds of ground silicon wafers each week. This unique effort led to the birth of CMP as an enabling process for oxide trench isolation for bipolar devices. One year later, an isolation trench process with a partial epitaxial silicon fill with a chemical vapor deposited SiO_2 cap formed by CMP was adopted for manufacturing.

15.4 POLISHING VERSUS PLANARIZATION

Although in various circumstances, chemical mechanical (p)olishing is still used instead of (p)lanarization, there is a significant difference between polishing a bare wafer surface of one film and planarizing a patterned wafer surface with several topographic features incorporating films of various materials. Polishing a surface means removing the top surface material at an uniform rate irrespective of the topographic features while making the surface smooth. Planarization means removing the surface of the substrate (possibly exposing different types of materials) at different rates depending on the features while achieving global planarity (flat) across the substrate surface. The process depends on both chemical and mechanical actions on the surface of the substrate being planarized. In contrast to the bare Si wafer polishing, the goal of metal CMP or STI (shallow trench isolation) CMP is to remove an overburden of one material and stop on another material, leaving a planar surface. On the other hand, the goal of inter layer dielectric (ILD) CMP is to planarize topography created by previous processing steps. This process is critical in establishing reliable multilevel interconnects in the fabrication of integrated circuits. It is now being used also for polysilicon via plugs and capacitor structures in memory devices, micro electro mechanical systems (MEMS), and through-silicon vias (TSVs) processes.

15.5 THE CMP APPLICATION PROCESS

In a CMP process, the wafer being planarized is pressed (facing down) onto a polymeric polishing pad and is abraded in the presence of a slurry, which contains active chemicals and abrasive particles. Thus, CMP can be described as a process which uses the combination of mechanical energy from the pad and abrasives and chemical energy from the slurry chemicals to polish and remove material from the wafer surface. The process flow has been described by the schematic diagram in further detail by Zantye et al. [7].

For CMP, therefore, one needs to use some consumables such as polishing pad, wafer, polishing slurry, pad conditioning disk, and post-CMP cleaning solution. There are varieties of pads, slurries, cleans, and even pad conditioning disks for different types of films to be planarized. Typical films on wafers normally planarized are Cu, Ta, W, TEOS, various low-k dielectric films, oxides, Si_3N_4, SiON, carbon-doped oxides, and nitrides of Si, Co, Mn, Ru, poly-Si, etc. However, new materials such as III-V compounds (for gate-all-around nano wire field effect transistor

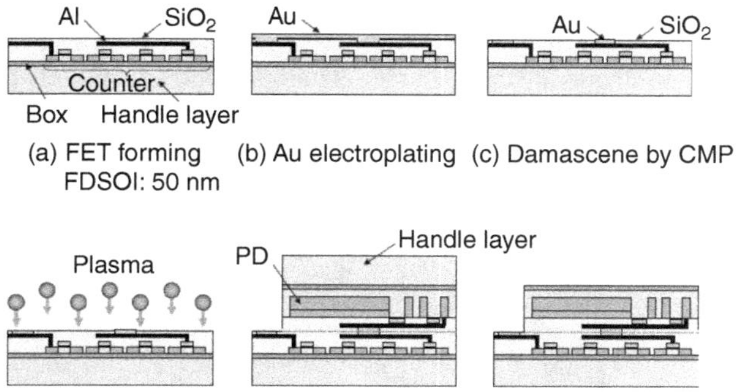

Figure 15.4 Schematic of the fabrication steps of a partial pixel CMOS sensor [9].

application) and gold (for partial pixel CMOS sensors used in super hi-vision TV camera) are also being planarized on experimental devices and, therefore, become typical materials for planarization in the future when such device architectures are commercialized (Fig. 15.4).

A complexity involved with CMP is that each material generally requires unique slurry, a properly designed pad, optimized process settings for both polish and post-CMP clean, and other factors that must be individually tailored to the application. Successful results are achieved [10] only when all the pieces of the puzzle are fit together.

One challenge is that interactions between process settings and output metrics are not simple to predict. They are seldom linear and they strongly depend on slurry, pad, film stack, process recipe settings, and design rules of devices being polished. However, certain trends generally hold true as summarized in Table 15.1 [10].

15.6 MECHANISM OF CMP PROCESS

Although CMP is now extensively used for fabricating integrated circuits, the polishing process itself was used in optical lens manufacturing for several centuries. Sir Isaac Newton in 1695 observed that scratch size decreased as abrasive size was decreased. By using very fine grain abrasives, he noted, it was possible to continually fret and wear away the glass and produce a polished surface on which the scratches become too small to be visible. Along the line of this hypothesis, Lord Rayleigh found in 1901 that the polishing process produced highly reflective, structure-less facets in a discontinuous fashion [11]. Further polishing did not improve the quality of the facets but extended their boundary. He suggested that the difference between polishing and grinding could occur by changing the character of the backing without altering the grit size. In polishing, a smaller force was applied on the abrasive by a soft, yielding backing and therefore removes material on a much finer scale,

Table 15.1 Output Metric Interactions with Process Recipe Settings

	CMP Process metrics			
Process settings	Rate	Uniformity	Defectivity	Planarization
Down force, DF	Strong	Weak	Moderate	Strong
Back pressure, BP	Weak	Moderate	Weak	Weak
Table speed, TS	Strong	Weak	Moderate	Strong
Carrier speed, CS	Weak	Moderate	Weak	Weak
Slurry flow, SF	Nonlinear	Nonlinear	Moderate	Weak
Conditioner force	Weak	Weak	Weak	Moderate
Conditioner speed	Don't care	Don't care	Weak	Weak

perhaps on the molecular level. Later research showed that polishing and grinding were not fundamentally different [12]. Very fine-scale scratches were found on the polished metal surfaces by phase-contrast illumination, even though they look perfectly smooth under ordinary light. Experiments also showed that brittle fracture occurred on ceramic surfaces under certain circumstances [13].

The material removal rate (MRR) in CMP process is empirically found to proportionally increase with the product of applied pressure and relative velocity, which may be expressed as [14]

$$\frac{dh}{dt} = kp\, pVr \tag{15.1}$$

where h is the thickness of the layer removed, t the polishing time, p the nominal pressure, Vr the relative velocity, and kp is a constant known as the Preston constant. It may be noted that the above equation can be used on both local and global scales. When it is employed to estimate the average removal rate across the wafer surface, the thickness removed should be much larger than the variation in surface roughness. It may be noted that while Preston equation has been the guiding principle in early CMP processes, it is empirical and is meant to be used for a system where material is removed by grinding. For metal and dielectric CMP, numerous models have been proposed subsequently and continued to be proposed.

For example, Zhuang et al. [15] have proposed a modified Langmuir-Hinshelwood kinetic model to simulate ILD removal rate which otherwise deviated at high $p \times V$ polishing condition and exhibited a strong correlation with the measured average pad leading edge temperature. The authors observed that when the measured average pad leading edge temperature increased from 26.4 to 38.4°C with the increase in polishing power, the ILD removal rate also increased with the polishing power, ranging from 400 to 4000 A/min. In this model, transient flash heating temperature is assumed to dominate the chemical reaction temperature. The model successfully captured the variable removal rate behavior at the highest $p \times V$ polishing condition and indicates that the polishing process was mechanically limited in the low $p \times V$ polishing region and became chemically and mechanically balanced with increasing polishing power.

In the case of planarizing, a patterned wafer, the goal is to planarize topography created by the previous process step(s). Topographical features always have different local polishing rates than global polishing rates observed on bare Si or any other blanket wafer with any type of surface film. Simply stated, pattern density is the fraction of high area to total surface area. For oxide CMP, this is driven by the underlying metal. For other layers, high area may refer to field areas between recessed features. The rate at which high features are removed is proportional to the effective contact pressure which equals the average pressure divided by the local pattern density. The region considered "local" is a direct function of planarization length which is roughly the lateral distance required for relaxation of deflections in the pad surface.

The impact of pattern density variation can be dramatic for a typical oxide CMP process. This has been described by Ouma et al. [16] in greater detail. The authors used commercially available test wafer consisting of line-space patterns in 4 × 4 mm arrays with constant pitch and varying linewidths. Initial step height of all features was 8 kA and data points were taken at the center of each array.

All of the features were planarized by the end of polish (step height near zero), but a substantial difference in oxide remaining was induced by the pattern density variation. Oxide thickness remaining over the lines in the 80 percent array was just over 11 kA while the oxide remaining over the 10 percent array was less than 8 kA—a difference of over 3 kA! This difference is nearly equivalent to the blanket film removal rate for a period of 100 s, which is how long the 10 percent array was polished after reaching local planarity [10].

In order to get a mathematical understanding of the above-mentioned differences, consider the blanket wafer polishing system, where the Preston law is a good approximation over a wide range of pressures [17]:

$$RR_0 = K_p P_0 V_0 \tag{15.2}$$

For the case where velocity V_0 and the term K_P are held constant, the local pressure determines the polish rate. This Prestonian relationship is generally valid for silica slurries polishing silica dioxide surface such as TEOS. For the situation where the entire wafer has the same pattern density, then the local pressure on top of each feature can be determined as the average pressure applied to the pad divided by the pattern density, since the force per unit area applied to the pad is applied over the area of the pattern. For the case where the pattern density is some fraction of the total area, ρ_1, then the down force is applied to this reduced area and the polishing rate of each of the features will be increased to

$$RR_1 = K_P P_1 V_0, \text{ where } P_1 = P_0/\rho_1 \quad (15.3)$$

Or for the reduced density ρ_1,

$$RR_1 = K_P(P_0/\rho_1)V_0 \quad (15.4)$$

For the sparse region where the feature density is uniform and at a density ρ_1, the features will polish at a rate determined by the local manifestation of Preston's law. For example, if $F_1 = 0.25$ or 25 percent, then the features will polish at 1/0.25 or 4 times the rate of the blanket wafer surface. The variation in local polish rate with feature density does not require a simple Preston's law relationship between pressures and polish rate. If the polish rate on a planar surface can be described as

$$RR = f_A(P) \quad (15.5)$$

for uniform features of density ρ_K, then the removal rate is described by

$$RR_K = f_A(P/\rho_K) \quad (15.6)$$

Such a pressure dependence can and does occur in metal CMP and also in nonsilica abrasive silicon dioxide polishing. More details for STI as well as metal CMP pattern dependencies can be found in the work of Bonning group at MIT [18–22]. In a subsequent paper by Ko et al. [23], experimental results were given for the removal rate variation as a function of pattern density for five dies.

15.7 CMP CONSUMABLES

15.7.1 Slurry

For CMP process, slurry is a specially formulated chemistry combining some chemicals as well as abrasive(s) in water. Typically, slurries are mixed with an oxidizer either at the time of manufacturing or at the point-of-use before loading on to the CMP tool. Today, the most common oxidizer is hydrogen peroxide and is mixed at the point-of-use. Among other ingredients in slurry, usually one finds organic complexing agents, corrosion inhibitors, and surfactants of various types.

The slurry dependence of the removal rate of a particular material that is being polished is due to slurry chemistry, that is, chemical action of the slurry chemicals of the material, the mechanical abrasion of the particles on the polished materials, interplay of the different complexing agents, oxidizers, and corrosion inhibitors. Chemicals such as oxidizers and corrosion inhibitors vastly affect the reaction rate of the slurry with similar particle nature, size, and distributions. Singh et al. [24] have described the variation of the reaction rate of the different slurry components on Cu when the reaction kinetics was studied using electrochemical chronoamperometry. According to the authors, the surface rate kinetics reaches about 60 Å/s when Cu is immersed in deionized water. The reaction rate increases to around 120 Å/s when 5 percent H_2O_2 is added. But with addition of 10 mM BTA (benzotriazole), the reaction rate came down considerably.

Depending on the end goal, different types of abrasives are used for different types of slurries such as colloidal silica, fumed silica, alpha alumina, calcined ceria, etc. For Cu-CMP process, the most common abrasive being used today is colloidal silica. Different manufacturers of colloidal

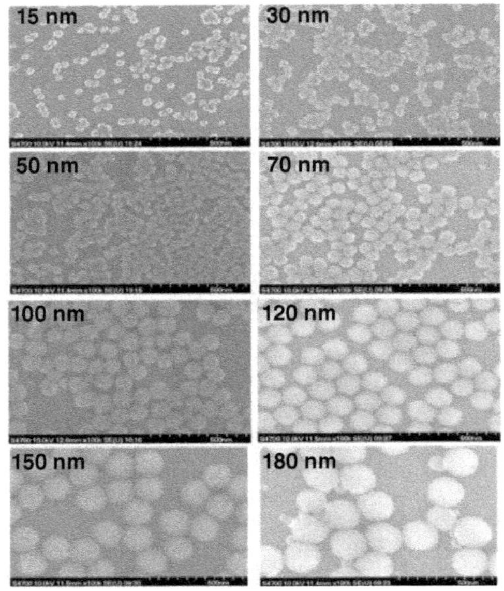

Figure 15.5 Colloidal silica with different size.

silica claim to have specific differences in physical properties (Fig. 15.5) of their end product, for example, particle size distribution, average particle diameter, purity, manufacturing process difference, etc. [25].

Typically, average particle size and concentration of abrasive particles in a slurry affect the material removal rate. Normally, the film removal rate increases with increase in concentration of abrasive particles in a slurry but at a relatively higher concentration, the rate of film removal no longer increases linearly [26].

In the commercial world of CMP slurry formulations, increase in abrasive concentration naturally increases the cost of any slurry formulation which may not be desired by the end users. Schlueter et al. [27] has demonstrated the use of chemical rate enhancer to enhance the film removal rate at a given abrasive concentration. By optimizing the abrasive concentration and the chemical rate enhancer concentration, it was possible to formulate a barrier slurry with 50 percent reduction in abrasive concentration. It also resulted in reduced post-CMP defects on wafers, which is always desired in a high-performance slurry (Fig. 15.6).

Wang et al. [28] have reported another novel way to enhance the material removal rate of GaN film. By using a hydrogen peroxide- and colloidal silica-based slurry under UV light, which was used to shine the outlet of slurry in order to catalyze the production of substance for oxidation, the material removal rate was improved significantly from 61 to approximately 103 nm/h. In summary, UVA plays a catalytic role in the polishing system to accelerate the oxidation reaction with H_2O_2. Hydroxyl radicals (OH^-) produced in the H_2O_2-SiO_2-based slurry were used to oxide GaN, and the production rate of OH^- was an important factor to influence the oxidation reaction and then influence the MRR of GaN CMP. The oxidation reaction between GaN and OH^- is listed in Eq. (15.7) [29–33].

$$2GaN + 6OH^- + 6h^+ \rightarrow Ga_2O_3 + 3H_2O + N_2 \quad (15.7)$$

When H_2O_2 is irradiated by UV light with the wavelength of 320 to approximately 380 nm, free radicals chain decomposition reaction will occur in the present system. Because of the photocatalytic reaction, the decomposition rate of H_2O_2 and the production rate of OH^* will be improved; as a result, GaN MRR will increase. The wavelength of UVA used was between 315 and 400 nm, close to the one of the photocatalytic reaction, which could accelerate the decomposition reaction of H_2O_2.

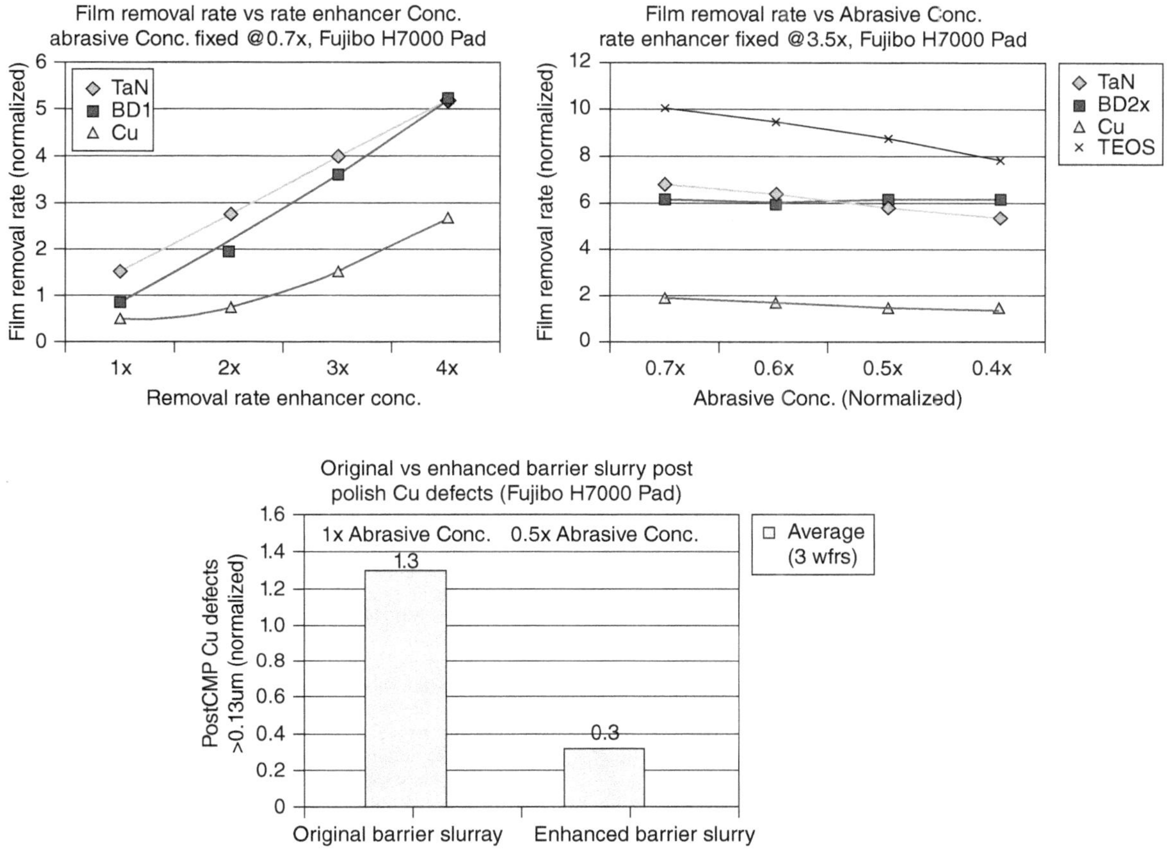

Figure 15.6 (*a*) Effect of removal rate enhancing additive; (*b*) effect of abrasive concentration reduction on film removal rate; (*c*) defectivity of original versus enhanced slurry.

While a single specific type of slurry is used for Al CMP or W CMP or oxide CMP process, for Cu CMP, typically two or even the different types of slurries are used to complete the Cu CMP process. Cu CMP process requires three distinctive steps: removal of bulk Cu, planarization of remaining Cu, or so-called "soft landing" to clear Cu and removal/planarization of barrier layer beneath the Cu layer. As the technology node shifts below 30 nm, the thickness of deposited Cu is decreasing compared to what was the norm prior to that. As a result, the first two steps may be combined into one single step in the future.

Corrosion inhibitors are used in slurries to reduce or prevent corrosion of both copper and barrier layers as appropriate. In a pattern wafer, a galvanic contact between the two films gets exposed at the end of the CMP process, which is a source of the onset of galvanic corrosion. It is essential to prevent the same, which otherwise leads to corrosion defects such as fang formation [34]. In this type of cases, copper is typically found as galvanically active in the Cu–Ta couple (where Ta is used as the barrier layer). Fangs occur when Ta is in contact with Cu as well as to the dielectric layer when exposed to slurry, and behave as galvanically active metal in the Cu-Ta-dielectric sandwich. In this type of cases, copper is typically found as galvanically active in the Cu-Ta couple. For fangs to occur, Ta which is typically used as barrier layer and in contact with Cu as well as to the dielectric layer when exposed to slurry, must behave as galvanically active metal in the Cu-Ta-dielectric sandwich. By adding suitable corrosion inhibitor additive in the slurry formulation, fang formation could be prevented (Fig. 15.7).

In the sub-1x nm technology nodes, use of Co and Ru liners are being explored in some layers of the multilevel BEOL interconnect and, therefore, similar fang issues will need to be solved using appropriate additives in the slurry formulations. After the barrier layer is planarized, dielectric layer is exposed to the slurry. So, it is also essential to formulate slurry to ensure that no damage to the dielectric layer occurs and the removal rate of this layer is in accordance with the goal set by device designers.

In general, an ideal CMP slurry should be able to achieve the required removal rates for all the films associated with the integration of a specific process, achieve excellent local and global planarization, should prevent galvanic, pitting or any other form of corrosion to any metal film in the total film stack, provide good surface finish, low defectivity, and appropriate selectivity as desired for a specific device pattern wafer. It may be noted that for all Cu and barrier slurries, colloidal silica is now used almost universally as the abrasive of choice. However, over the years, there have been modifications made to the manufacturing processes to improve the defect performances and/or adjustment of the required removal rates of various films associated with pattern wafers of specific technology nodes. A summary of this evolution of silica abrasive usage has been provided by Schlueter et al. as given in Table 15.2 [27].

Among the various slurry suppliers are Versum Material, Inc., Cabot Microelectronics Corporation, Dow Electronic Materials, Fuji Film Electronic Materials, Ferro, Fujimi Corporation, Hitachi Chemical (Japan), Anji Microelectronics (China), UWIZ (Taiwan), UB Materials (S. Korea).

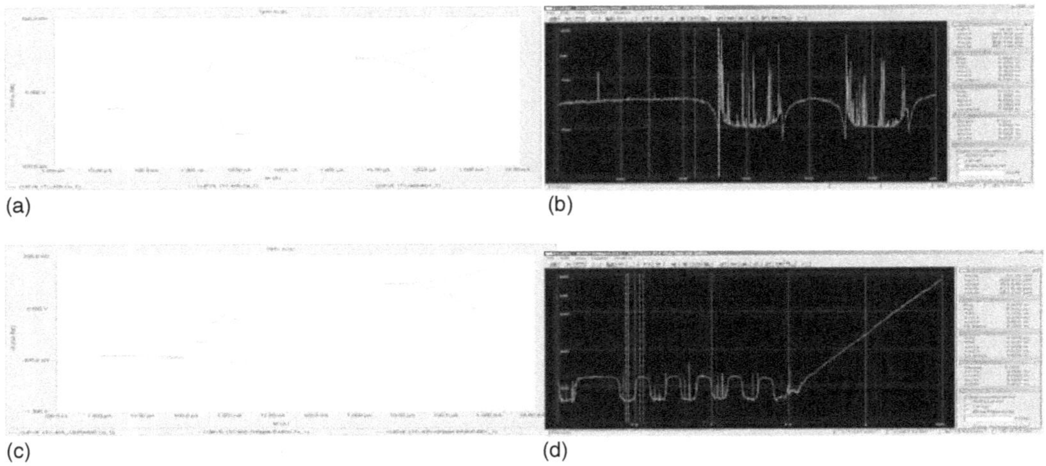

Figure 15.7 Effect of relative galvanic potentials of different films on fang formation and inhibition. (*a*) Tafel plots for BD1/Ta/Cu in barrier slurry; (*b*) AFP scan of a wafer (MIT 854) showing fangs; (*c*) Tafel plots for BD1/Ta/Cu in modified slurry; (*d*) no fang in the AFP scan with modified slurry.

15.7.2 Pad

Conventional pad technology includes a top-pad layer with or without an under-pad layer to provide a polishing surface. Polishing pads provide local planarization due to inherent material hardness and global uniformity due to the flexing action of the polishing layer. While such pads have served the industry well, inherent coupling of key output parameter—planarization and WIWNU (with-in-wafer nonuniformity)—limits the overall performance potential [35]. Physical deformation behavior of the pad surface asperities due to micro contact to the workpiece should have significant influence on the material removal process. In the past studies, Greenwood-Williamson model [36] or similar ones were utilized as a contact model to estimate the contact behavior of the pad surface asperities, although there was no strong reason that the assumptions of those models were appropriate as the contact model of the CMP pads. From the point of view of the contact model, the pad surface structures have not been evaluated and analyzed enough in the past. Assuming oval hemispherical asperities with radius distributions and considering the measured and identified geometrical features, a new contact model of pad surface asperities has been proposed by Hashimoto et al. [37]. The model is based on Hertzian elastic contact. Probability functions of the asperity height and radii are also defined based on the measured results. Analytical investigation about the asperity contact indicated that the nominal contact pressure estimated by the proposed model showed a significant disagreement with that in the conventional model.

Conventional pad systems suffer from two primary limitations. First, the continuous nature of the top polyurethane layer gives rise to an edge rebound effect as it passes under the wafer or in some cases, the (wafer) retaining ring. While retaining rings help mitigate the edge rebound effect, significant edge-related nonuniformity remains. Second, global compliance or WIWNU considerations necessitate that the total stack has limited stiffness—thus impacting within-die planarization. For example, a soft pad stack has a lower flex modulus and compression modulus, which enables good contact at low pressures. A hard pad stack, on the other hand, has a much higher flex modulus and compression modulus; hence, it offers limited contact at lower pressures. This process leads to nonuniform contact at low pressures as well as variation in contact pressure leading to potential contact "hot spots."

Table 15.2 General Cu-Barrier Slurry Abrasive Evolution

Technology nodes	Abrasive type	Conc. range	Issues
130 to 65 nm	Fumed Silica	10–15%	Scratch defects
90 to 65 nm	Standard Colloidal Silica	20–30%	More expensive (higher solids)
45 to 32 nm	Standard Colloidal Silica	10–15%	Further reduction in defects and cost needed
22 to 10 nm	High Purity Colloidal Silica	3–7%	Further reduction in defects and cost desired
10 and less	High Purity Colloidal Silica	< 5%	New materials polished

For new generation ultralow-k dielectric film integration with Cu interconnect, the down force requirement is extremely low due to the reduced mechanical strength of porous films and the conventional pad no longer provides the most optimum efficiency. New generation of pads are being explored by various CMP pad suppliers. Some of these pads overcome to a great extent the global planarity and nonuniformity issues associated with the conventional pads.

One area of concern in the conventional CMP pad is that the asperities on the pad surface contact the wafer under applied force and relative motion. With the assistance of slurry containing wet chemistry and abrasive particles, the asperities cause material removal from the wafer surface. Because polyurethane pad asperities are irreversibly deformed due to wafer contact and are also abraded by slurry particles, the pad surface must be continuously renewed with a diamond disc conditioner in order to sustain a stable process. In addition, the polyurethane pad must be relatively compliant in order to achieve uniform application of load across the wafer and across each chip, resulting in well-known die-scale layout pattern-dependent nonuniformity.

Bonning et al. [38] reported an alternate concept to overcome the above mentioned issues of the conventional pads—it is called *pad-in-a-bottle* (PIB) polishing. In PIB, polyurethane polishing pad is replaced by a stiff polycarbonate counterface and 10- to 50-μm polyurethane beads are mixed into and used in conjunction with conventional slurry and abrasive particles to achieve planarization.

The most notable CMP pad supplier in market today is Dow Electronic Materials, Cabot Microelectronics Corporation, Thomas West, Inc. (United States), Fujibo, and JSR Corporation.

15.7.3 Pad Conditioner

The CMP process is very dependent on the surface condition of the pad. During operation, the surface of the CMP pad soon gets loaded with debris from the polishing operation, which leads to glazing on the surface. This means that there are no asperities to hold the abrasive grits, which leads to inefficient polishing and possible microscratching on the surface wafer. To prevent this and extend the life of the CMP pad, it must be dressed regularly with a diamond conditioner. The diamond CMP conditioner has two functions:

1. To machine a pad initially to create a flat surface and subsequently, to remove glazing from the pad surface during operation to expose asperities. [Even a new pad will have an unacceptably rough surface—this will have to be dressed before it can be used for polishing.]
2. To restore the asperities within a pad's surface to enable the abrasive grits to perform their polishing action efficiently.

A traditional diamond pad conditioner or dresser is made by attaching diamond grits onto a metal disk. Diamond is difficult to bond with other materials due to its natural properties. The diamond industry has developed three distinct methods to hold diamond grits in place: by burying them in electroplated nickel, by retaining them in sintered metal powder, or by brazing them with a specialty alloy. The first generation of diamond pad conditioners contained chaotically distributed diamond grits, typically retained in an electroplated nickel layer. The electroplated metal only holds diamonds in place mechanically, so the displacement of diamonds from the matrix is inevitable. The loss of diamond particles during the CMP process affects wafer performance. The second-generation pad conditioners had clusters of diamond grits that were held by braze. In these clusters, the braze layer became thickened. As a consequence, not only does the distribution of diamonds become random, the diamond height also varies considerably. The third generation pad conditioners have a regular distribution of diamond grits with each diamond strongly bonded by braze. Currently it has become the world standard for most pad conditioners to be constructed [39]. Among the several traditional pad conditioning disc manufactures, popular names are 3M (United States), Morgan Technical Ceramics (United States), Entegris (United States), Kinik (Taiwan), and Asahi (Japan).

Entegris has recently brought to market a new pad conditioner based on growing diamond over the entire pad surface using chemical vapor deposition method [40]. According to Entegris, the all-diamond surface provides superior diamond retention by chemically bonding the diamond grit to the substrate. This all-diamond surface is also chemically inert to all CMP slurries, eliminating diamond pullout due to corrosion or wear of the matrix material. In addition, the high-hardness diamond prevents mechanical erosion of bond matrix by hard CMP slurry particles.

Kinik has recently reported about a new platform for next generation pad conditioning named Pyradia™ [41]. Each diamond is independently mounted on its own "stick." These sticks are positioned so as to maximize the distribution of cutting action on the pad surface. The diamond points can be precisely height controlled and the cutting behavior of individual points can be manipulated by changing the profile and orientation of the diamond face presented to the pad. Both oriented natural crystals and shaped diamonds can be utilized. Diamond configurations can be mixed to generate a range of cutting action within a single disc. Significantly, the platform is based on much larger diamond crystals with close to an order of magnitude higher strength than those used in conventional platforms, such that crystal fracture is virtually impossible.

The significant improvement in height control translates directly to performance benefits. The Pyradia disc maintains about 90 percent of its initial cut rate after 100 h, while the conventional disc has lost 40 percent of its initial cut rate in only 50 h.

Another new generation pad conditioning disk for CMP of silicon-on-insulator (SOI) has been reported by Zabasajja et al. [42]. According to the authors, the performance of a metal-free novel microreplicated pad conditioner, 3M™ Trizact™ Pad Conditioner, B5-M990-5S2, consisting of precisely engineered microreplicated abrasive structures coated with a chemical vapor deposited diamond film demonstrated pad and disk life extension by 5 to 7 times with relatively lower nonuniformity relative to a conventional pad-conditioning disk.

15.8 CMP INTERACTIONS

The targets for dishing and erosion on Cu lines continue to decrease with each successive generation of devices. Achieving these targets consistently in manufacturing will require slurries formulated to the specific materials system, pads with excellent surface texture control and processes with razor-sharp endpoint and/or high selectivity to a stop layer. Dishing and erosion, which together account for nearly 50 percent of yield losses in copper processing, are the most important effects to be minimized in the copper CMP process. Dishing and erosion are forms of local planarization where areas of the wafer polish faster than others. In dishing, the copper is dished out of the lines whereas in erosion whole sections of the die are polished faster than others [43].

The CMP process subjects the surface of the wafer to a rather hostile environment with the wafer encountering mechanical forces of both pressure and shear while immersed in aqueous chemical slurry which usually contains tiny suspended particles. As might be expected, the films deposited on the wafer must be capable of withstanding this environment without suffering undesirable irreversible harm.

In the early days of semiconductor device manufacturing, silicon dioxide itself was used as the interlayer dielectric (ILD). Interconnect structures in the subsequent advanced technology incorporate lower dielectric constant ILD materials. In the early stages of ULK exploration, it was recognized that several film types such as carbon-networked dielectrics would enable reduction in the k value but in most cases the film's mechanical properties were significantly degraded. This trend in mechanical strength is based on film class.

The industry has worked to balance the mechanical properties along with capacitance reductions due to the considerable concerns for packaging reliability failures [44]. Currently, organosilicate (OSG) glass films has been adopted both for dense OSG and now extending down to k values ranging from 2.5 (40, 28 nm nodes) to 2.2 and 2.0. The k values are scaled by addition of porosity (air has a dielectric of 1.0), enabling the industry to maintain the same class of materials but continue to reduce the capacitance impact as structures are scaled. Although, generally, the addition of porosity will reduce the film's mechanical strength, there have been several improvements in ULK films to optimize the backbone strength, whereby making the ULK film strength comparable to bulk OSG films [45].

Mechanical failure modes include delamination, cracking, and inelastic deformation. For low-k and ultra-low-k (ULK) dielectrics, the materials are characterized in terms of modulus, crack propagation velocity, and yield strength [44–46]. As the industry transitions to these materials, the CMP process must accommodate the mechanical fragility of ULK by moving toward lower pressure and lower shear process recipes. These two factors are related through the dynamic coefficient of friction, which is itself related to a large number of factors: platen rotation speed, pad composition, surface texture, pad conditioning parameters, particle size and morphology, solids concentration, slurry chemistry/composition (especially surfactant choice and concentration), among others. While the relationship among these variables is often not linear, it is generally consistent that lower pressure leads to lower shear, thus most copper and barrier polishing recipes are being developed at 2 psi and less. This maximum pressure is expected to continue dropping as ULK drives toward lower k values, which demand larger fractional pore volumes and the accompanying degradations in mechanical strength [47]. With the focus on ULK mechanical integrity, it is sometimes easy to forget that polishing slurries also contain a mixture of chemical ingredients, for example, oxidizers, complexing agents, passivation additives, selectivity modifiers, and surfactants.

Slurry developers historically benefited from having dielectric films, such as TEOS or fluorosilicate glass (FSG), which were reasonably inert and easy to clean. This is no longer the case with ULK materials, which typically have a strong tendency to absorb water (and other chemicals) from the slurry or at least trap them in any exposed pores [48]. Development of a CMP process for any of these advanced integrations must be approached as a multilevel, multidisciplinary project. The properties of the ULK film, the barrier material, the copper deposition technique, annealing cycles, and many other factors will influence the CMP process. The situation is made even more difficult by the fact that many of the ULK materials themselves are still being formulated, unlike earlier nodes where at least the dielectric, that is, FSG, was locked in at a relatively early stage. This difficulty is compounded by the complexity of developing sophisticated integrations with porous ULKs. Low-*k* materials have to be mechanically strong enough, in order to resist delamination and mechanical breakdown during the CMP process. The challenge lies not only in identifying and characterizing the candidate materials, but also in devising the best methods to integrate those materials. As material dielectric constants get closer to 1, the materials become less robust and harder to integrate.

15.9 POST-CMP CLEANING

Post-CMP cleaning is an integral part of the CMP process module to clean up the debris and other chemical residues left on wafers during CMP. There are two types of cleaning technology in use: megasonic bath and double-sided scrubber with Polyvinyl alcohol (PVA) brushes. Both have advantages and disadvantages and may be employed individually or in combination as needed to achieve effective cleaning of the wafers. The megasonic approach is sometimes useful in removing particles from narrow recesses of a polished wafer if the surface features are not completely coplanar. Brush scrubbers, on the other hand, are particularly useful at dislodging particles that may be physically attracted to the copper surface. It is relatively common to use the combination of both megasonic cleaning and brush scrubbing to give a more robust cleaning process. There are cases of fabs using only one technique and the more common single method is brush scrubbing. This practice is possible due to new generation of cleaning chemistries which enable complete removal of particles from the copper surface through chemical and electrochemical means without the need of megasonic energy to physically lift the silica particles out of copper and dielectric film surfaces.

The combinatorial application of using chemistry and electrochemistry to remove particles from a copper surface has become necessary to address the constantly shrinking size of what are considered "killer" defects. In general, this includes any particle greater than the linewidth of the smallest feature, which is now smaller than the particle sizes used in most commercial slurries. Silica particles, for example, will stick on the metal surface by electrostatic bonding if the slurry is acidic in pH. In this case, silica acquires a net positive charge while the surface charge of copper becomes negative. The electrostatic bonding is fairly strong and traditionally, megasonic energy was needed to lift the particles off the copper surface. With the acidic post-CMP cleans, the particles are assumed to be lifted off the copper surface by a very slight etching of underlying copper which is not enough to damage the copper surface topography. In contrast, if alkaline slurry is used for the final step in the CMP process, the silica particles are assumed to become negatively charged and float near the copper surface. The images in Fig. 15.8 show the ability of a new generation cleaning formulation to clean Cu surface without using megasonic cleaning. The Cu wafers were polished with a commercially available barrier CMP slurry and cleaned with (1) deionized water, (2) a previous generation clean solution that would require megasonic cleaning to aid in complete removal of particles, and (3) a new generation cleaner that does not require any megasonic cleaning step.

Post-CMP cleaning processes must address more than just particle removal. Other killer defects like unwanted metal particles or trace metal ions need to be removed as well, especially at 32 nm nodes and less. The new generations of post-CMP cleaning formulations are able to capture such metallic ions more efficiently, leaving a cleaner wafer surface than ever before. Also, all post-CMP cleaning solutions being electrochemical in nature, galvanic corrosion leading to dendrite formation is a cause to form killer defects (Fig. 15.9). However, tuning such cleaning formulations to prevent galvanic corrosion yield dendrite-free cleaned pattern wafers [49].

15.9.1 Filters

Typical other common defects on wafers are scratches (normally come from the agglomerated slurry abrasive particles), organic residues or carbon residues (from slurry components or dielectric film erosion) as can be seen from the images in Fig. 15.10.

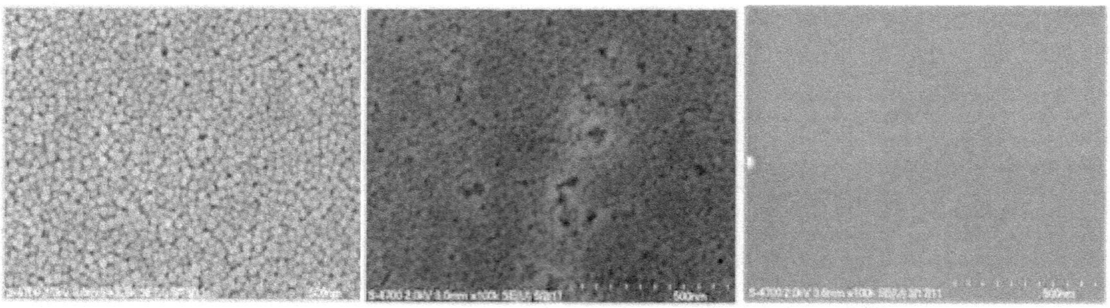

(a) (b) (c)

Figure 15.8 Comparison of two different cleans to demonstrate the ability of new formulation to clean Cu surface without the additional step of megasonic cleaning. (*a*) Polished + DI water, (*b*) polished + old clean, (*c*) polished + new clean.

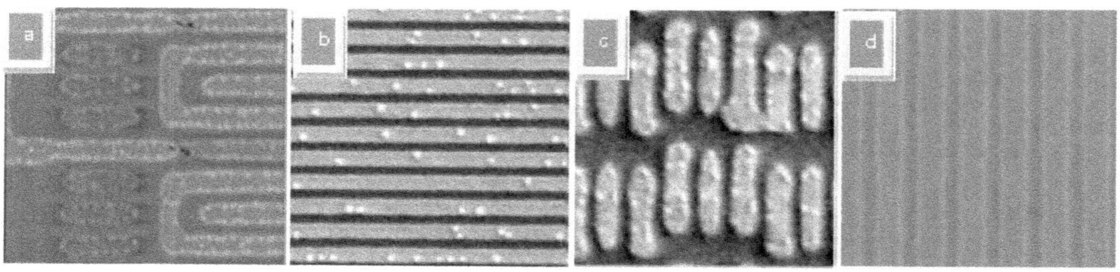

Figure 15.9 Dendrites on various wafers (*a*–*c*) and dendrite free wafer (*d*) using proprietary clean.

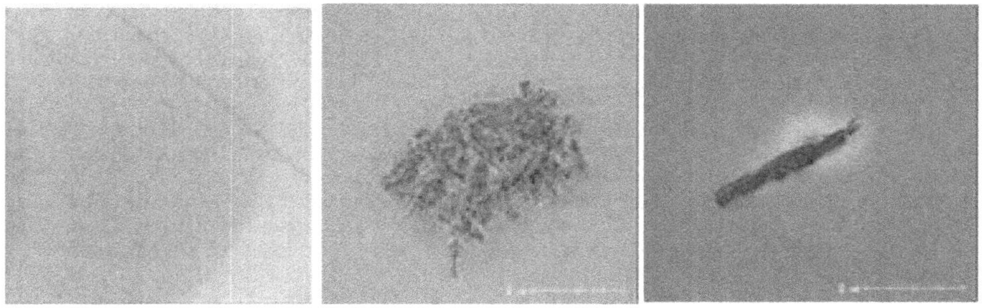

Figure 15.10 Typical other defects (scratch/organic residue/carbon residue).

Slurries are normally filtered after manufacturing and also, before use on the CMP tools to minimize the defects caused by agglomerated large particles count (LPC) without affecting the particle size distribution, solids content, or the chemistry of the slurry. Polyolefin filters produced using melt blown fibers (MBFs) with a gradient pore structure are commonly used in the industry to remove LPC from CMP slurry. Traditional MBF filters consist of wrapped webs or direct blown fiber on a core. These structures increase in fiber density from filter upstream to downstream. LPC are reduced when the CMP solution is processed through these filters. Entezarian et al. [50] have reported a composite rigid filter (CRF) which allows for LPC reduction at a significantly higher flow rate and low pressure drop compared to flow rate used with MBF filters. This CRF filter consists of a polyolefin bicomponent coarse fiber matrix and microfiber-glass web inserts that are the primary clarification and classification zones.

15.9.2 Process Tools

Last but not the least, in order to perform CMP, appropriately configured process tool is needed. Although in the early days, there were several tool makers with different tool configurations, at present only Applied Materials (United States) and Ebara Corporation (Japan) provide majority of the CMP tools for commercial semiconductor manufacturing. There are significant differences in tool configurations between the two manufacturers although the end result is always a planarized wafer. Both tool operate on dry-in/dry-out principle meaning dry preplanarized wafers go into the tool and come out dry after planarization. Ebara FREX 300 SII has 4 platens and 3 cleaners. The new Reflexion® LK Prime™ has six polishing stations and eight integrated cleaning stations with advanced process controls which enables CMP of 3D FinFET structures which have replaced planar 2D structures since 22-nm technology node.

The most commonly used CMP tools today for CMP process have 300-mm-diameter wafer holders since 65-nm pitch technology node was introduced into manufacturing. Of course, 200-mm diameter wafers and tools for the same are also being used for older technology nodes. Although not yet commercially available, 450-mm CMP tools have been operational at the tool makers' facilities and leading research consortium G450C in Albany, New York.

15.10 SUMMARY

With the continued growth of semiconductor device applications and commercial market penetration, CMP as an enabling technology to manufacture semiconductor chips has become one of the most critical process steps. With varieties of materials as thin films used in a semiconductor wafer manufacturing, the need to develop different types of consumables such as slurry, pad, conditioning disk, post-CMP clean, filter, etc., will continue to grow in demand. It is, therefore, absolutely important to understand the fundamental aspects and mechanisms of various interactions occurring during CMP processes.

15.11 ACKNOWLEDGMENTS

The author sincerely acknowledges the permission of the management of Versum Materials, Inc. to write this chapter. The author further acknowledges the permission of all the authors for allowing images, figures, and tables from their respective previous publications.

15.12 REFERENCES

1. R. Lane et al., Chemical Mechanical Planarization, March 2010, http://nanoparticles.org/pdf/Fuller-cmp.pdf (accessed January 30, 2017).

2. Copper Interconnects – The Evolution of Microprocessors, IBM Corporation to commemorate IBM 100, 2011. http://www-03.ibm.com/ibm/history/ibm100/us/en/icons/copperchip/ (accessed January 30, 2017).

3. R. Rhoades, *229th Meeting of the Electrochemical Society (CMP Symposium)*, San Diego, May 2016.

4. B. McClean, SEMI North East Forum, N. Reading, MA, March 2016.

5. S. Davis et al., Proceedings of International Conference on Planarization Technology (ICPT 2015), Chandler, AZ, Sep. 2015.

6. M.A. Fury, "It's a Material World – With Positive Forecast". http://www.semi.org/en/node/56701 (accessed January 30, 2017).

7. P. B. Zantye, A. Kumar, and A. K. Sikder, "Chemical Mechanical Planarization for Microelectronics Applications," *Materials Science and Engineering*, Vol. R45, pp. 89–220, 2004.

8. K. D. Beyer, "Plenary talk," *Proc. ICPT*, Chandler, AZ, Sep. 2015.

9. M. Goto et al., "229th Meeting of the Electrochemical Society," More-Than-Moore Symposium, San Diego, May 2016.

10. G. Banerjee et al., "Chemical Mechanical Planarization – Historical Review and Future Direction," *ECS Transactions*, 13 (4): 1–19, 2008.

11. L. Rayleigh, "Polish," *Nature*, 64: 385–388, 1901.

12. R. L. Aghan and L. E. Samuels, "Mechanisms of Abrasive Polishing," *Wear*, 16: 293–301, 1970.

13. R. Komanduri, D. A. Lucca, and Y. Tani, "Technological Advances in Fine Abrasive Processes," *Annals CIRP*, 46: 545–596, 1997.

14. F. W. Preston, "The theory and design of plate glass polishing machines," *J. Soc. Glass Technol.*, 11: 214–256, 1927.

15. Y. Zhuang et al., "Frictional, thermal and kinetic characterization of a novel ceria based abrasive slurry for silicon dioxide CMP," *Proc. CMP-MIC*, pp. 532–539, 2006.

16. D. Ouma et al., "An Integrated Characterization and Modeling Methodology for CMP Dielectric Planarization," Proc. Int. Interconnect Technol. Conf., IEEE, Piscataway, NJ. 1998.

17. M. R. Oliver, *Chemical Mechanical Planarization of Semiconductor Materials*, Springer, Berlin Heidelberg, pp. 14–15, 2004.

18. D. Ouma et al., "Wafer Scale Modeling of Pattern Effect in Oxide Chemical Mechanical Polishing," *SPIE Microelec. Mfg. Cong.*, Austin, TX, 1997.

19. D. Bonning et al., "Pattern Dependent Modeling for CMP Optimization and Control," *MRS Spring Mtg.*, San Francisco, CA, 1999.

20. T. Tugbawa et al., "CMP Symposium," *Electrochemical Society Meeting,* Honolulu, HI, Vol. PV99-37, pp. 605–615, 1999.

21. Y. Zhuang et al., "Frictional and removal rate studies of silicon dioxide and silicon nitride CMP using novel cerium dioxide abrasive slurries," *Jpn. J. Appl. Phys.*, 44 (1A): 30–33, 2005.

22. Y. Sampurno et al., "Pattern Evolution in Shallow Trench Isolation Chemical Mechanical Planarization via Real-Time Shear and Down Forces Spectral Analyses," *Microelec. Eng.*, 88: 2857–2861, 2011.

23. B. G. Ko et al., "Effects of pattern density on CMP removal rate and uniformity," *J. Korean Physical Soc.*, 39: S318–S321, 2001.

24. R. K. Singh and R. Bajaj, "Advances in Chemical-Mechanical Planarization," *MRS Bulletin*, 27 (10): 743–747, 2002.

25. F. Qin et al., "Research on the Manufacturing of Colloidal Silica Abrasive with Controlled Particle Size," *Proc. ICPT*, Chandler, AZ, Sep. 2015.

26. L. Guo et al., "Mechanical Removal in CMP of Copper Using Alumina Abrasives," *J. Electrochem. Soc.*, 151 (2): G104–G108, 2004.

27. J. Schlueter et al., "Attributes of an Advanced Node Cu-Barrier Slurry," *Proc. ICPT*, Chandler, AZ, Sep. 2015.

28. J. Wang et al., "Mechanism of GaN CMP Based on H2O2 Slurry Combined with UV Light," *ECS J. Solid State Sci. Technol.*, 4 (3): P112–P117, 2015.

29. H. Zhou et al., "Advanced Technologies in Water and Waste Water Treatment," *J. Environ. Eng. Sci.*, 1 (4): 247–264, 2002.

30. Y.-S. Shen and D.-K. Wang, "Development of Photoreactor Design Equation for the Treatment of Dye Wastewater by UV/H(2)O(2) Process," *J. Hazard. Mater.*, 89 (2–3): 267–277, 2002.

31. S. Nélieu et al., "Degradation of atrazine into ammeline by combined ozone/hydrogen peroxide treatment in water," *Environ. Sci. Technol.*, 34 (3): 430–437, 2000.

32. K. Kanamoto et al.,"Formation characteristics of calcium phosphate deposits on a metal surface by H_2O_2-electrolysis reaction under various conditions," *Col. Surf. A*, 350 (1–3): 79–86, 2009.

33. S. H. Lin and C. C. Chang, "Treatment of landfill leachate by electro-Fenton oxidation and sequencing batch reactor method," *Water Res.*, 34 (17): 4243–4249, 2000.

34. G. Banerjee, "Fang Formation During Cu CMP: A Galvanic Corrosion Problem?" *208th Electrochem. Soc. Mtg—CMP Symp.*, Los Angeles, CA, Oct. 2005.

35. R. Caprio et al., "CMP Pad Design for Ultra-Low k Compatible Cu CMP Process; p. 438," *VMIC*, Fremont, CA, Sep. 2006.

36. Q. Xu et al., "A Feature-Scale Greenwood–Williamson Model for Metal Chemical Mechanical Planarization," *J. Elec. Matl.*, 42 (8): 2630–2640, 2013.

37. Y. Hashimoto et al., "A New Contact Model of Pad Surface Asperities Utilizing Measured Geometrical Features," *Proc. ICPT*, Chandler, AZ, Sep. 2015.

38. D. Bonning et al., "Planarization with Suspended Polyurethane Beads and a Stiff Counterface: Pad-in-a-Bottle Experiments and Simulation," *Proc. ICPT*, Chandler, AZ, Sep. 2015.

39. M.-Y. Tsai, *J. Mat. Proc. Tech.*, 210: 1095–1102, 2010.

40. Planargem® CMP Pad Conditioner. http://www.entegris.com/ProductCategory_divEmerging_catCMP_Pad_Cond.aspx (accessed January 30, 2017).

41. E. Chou et al., "A Novel Platform for Next Generation Pad Conditioning," *Proc. ICPT*, Chandler, AZ, Sep. 2015.

42. J. Zabasajja et al., "Advanced CMP Conditioning for Front End Applications," *Proc. ICPT*, Chandler, AZ, Sep. 2015.

43. M. Khan et al., "Damascene Process and Chemical Mechanical Planarization," http://www.ece.umd.edu/class/enee416/GroupActivities/Damascene%20Presentation.pdf (accessed January 30, 2017).

44. E. P. Guyer et al., "Impact of Pore Size and Morphology of Porous Organosilicate Glasses on Integrated Circuit Manufacturing," *JOM*, 2006, Vol. 21, No. (4):, pp. 882–894, 2006.

45. X. H. Liu et al., *Proc. IITC*, "Fracture of Nanoporous Methyl Silsequioxane Thin-Film Glasses," San Francisco, CA, pp. 13–15, 2007.

46. D. M. Gage et al., "Chip-Package-Interaction Modeling of Ultra Low-K/Copper Back End of Line," *Proc. MRS Symp.*, San Francisco, CA, Apr. 2008.

47. M. Van Bavel et al., "The Role of Friction and Loading Parameters on Four-Point Bend Adhesion Measurements," *Future Fab Intl.*, 2004, No. 17, pp. 95–98.

48. M. R. Rao et al., "Low-K Dielectrics: Spin-On or CVD?" *MRS Spring Mtg.*, San Francisco, CA, Apr 2007.

49. G. Banerjee, *Advanced Interconnects for ULSI Technology*, edited by: M. R. Baklanov, P. Ho, and E. Zschech, Wiley & Sons Ltd., London, 2012.

50. J. Morby et al.,"Next Generation Composite, Rigid Filter for Chemical Mechanical Planarization," *Proc. ICPT*, Chandler, AZ, Sept. 2015.

Chapter 16
AFM Metrology

BY

ARDAVAN ZANDIATASHBAR *Park Systems, Inc., Santa Clara, California*

16.1 INTRODUCTION

The semiconductor manufacturing industry evolved significantly over the past half century. At its inception, the minimum feature size (i.e., node size) was at micrometer scale since the manufacturing of the first Intel 4004 chips until the announcement of the computer as the machine of the year (1983) by *Times* magazine. During that time, the high-volume manufacturing (HVM) of the chips had 20 to 40 percent yield. As part of the yield control, dimension measurements of devices were performed by optical methods. While the popularity of personal computers was rising in 1980s, the node size was shrinking to less than 1 μm. However, the optical techniques for device measurements were reaching their resolution limits. Then, scanning electron microscopes (SEM) were started being used in the integrated circuit (IC) manufacturing and was matured by the end of 1990s. At the same time, the atomic force microscope (AFM) was invented and the first experimental model was made in 1986 [1]. The miniaturization process continued as predicted according to an empirical model suggested by Gordon Moore in 1975 [2]. The node size started shrinking to below 100 nm at the beginning of 21st century and is now at less than 20 nm based on multiple reports from IC manufacturers. In fact, the semiconductor transistors were needed in order to control the AFM feedback control loop system [3]; hence, AFM's invention has been tied to the semiconductor industry. Now the time has come for AFM to help the semiconductor industry's growth by contributing to solving its metrology challenges.

The HVM yield is increased to above 80 percent and is needed to be kept at the same level or higher. A process for making a chip takes months and can have up to two thousands steps. Hence, achieving such a high yield is a great achievement for the industry since its inception. However, maintaining the high yield and improving it while the node size is shrinking is becoming more challenging. The reliance on metrology is now more than any time in the past. More than 50 percent of steps in the process for 65 nm has been dedicated to metrology and defect inspection. For 22 nm node this number goes as high as 70 percent [4]. This trend is consistent with the reported steady increase of return on metrology investment since the beginning of 21st century. Hence, the significance of metrology in device manufacturing is more than ever.

AFM was first introduced to IC fabrication plant (fab) in early 2000s (around 20 years after its invention) while the node size started going below 65 nm. Shrinking device sizes pushed the limits of the metrology techniques of that time and the need for new metrology solutions was felt more [5]. Comparing to SEM which was used in HVM around half century after its invention, AFM started to be used in manufacturing at a much younger age. Ten years later, AFM is known and accepted as a fully automated non-destructive in-line metrology tool in the fab. Automated AFM is used for critical dimension (CD) measurements [6, 7], step height/trench depth measurements, dishing and erosion monitoring [8], and defect review [9]. It also plays a critical role in hybrid metrology due to its accuracy and non-destructive nature [10–12]. In addition, AFM's multiple imaging modes are used in lab for offline characterization and failure analysis of devices at different stages of process.

In this chapter, the fundamentals of metrology in IC fabrication is briefly reviewed. It is followed by a review over the AFM technique and its latest design. The applications of AFM for fab metrology is listed and explained. In addition, some of the AFM modes for lab metrology of semiconductor devices are presented. At the end, the standard procedure for maintenance and calibration of AFM tools is reviewed briefly. Further reading materials are suggested at the end of this chapter.

16.2 METROLOGY: FUNDAMENTALS AND PRINCIPLES

Metrology, in a very simple definition, is the science of measurement. In practice, it includes both experimental and theoretical methods of measurement based on the definitions of Joint Committee for Guides in Metrology (JCGM). It can be categorized to physical and chemical metrology. By physical metrology, we refer to measuring physical properties such as length, electrical charge, and electrical current. On the other hand, chemical metrology refers to chemical composition measurement. In this chapter, we are covering physical metrology.

Metrology has a short history comparing to measurement. Measurement is referred to a single reported value and has a long history going back to the China's Great Wall or the Pyramids. On the other hand, metrology is about the uncertainty and repeatability in a series of measurements. It started to gain more attention by the development of modern assembly-lines where the parts had to be of exact sizes and shapes to fit the product on the line. This has been done by measurements of inches and feet. In case of semiconductor and IC manufacturing, the measurement scale is far smaller, hence more sophisticated measurement systems are required.

There are four main measurement systems that have been used to perform metrology in IC manufacturing: optical microscope, ellipsometer, SEM, and AFM. Optical microscope is the oldest technique which was invented more than five centuries ago. It has been around at the time of making the first semiconductor chips. On the other side of the spectrum, AFM is the youngest technique among all the four which was invented around 30 years ago and has evolved so rapidly that today it is one of available in-line metrology tools. Any of the tools used for performing a measurement is referred to as a measurement system. In a "measurement process," normally an "operator" or "appraiser" measures a "measurement target" (i.e. gauge) using a measurement system.

The measurement systems can be manual, semiautomatic, and fully automatic. The role of operator error has been already known to all metrologists, hence, many measurement systems are provided in form of an automated package. The automation has minimized the operator error; however, the operator error still exists. Most of the automated measurement systems require the operator to prepare the automation script. An automation script is generally referred to as "recipe." The preparation of recipe (i.e., recipe writing) can be slightly different from operator to operator. Hence, it has the potential for carrying the human error to the measurement. In fact, each recipe is an operator's list of commands for the measurement system. Although measurement systems can be programmed to repeat a measurement multiple times, they are not "smart" to recognize a mistake or improper measurement from an accurate one. Hence, if there is an operator error caused by a certain recipe, the measurement system continues collecting the erroneous information from the measurement targets unless the recipe is corrected by the operator.

16.2.1 Conventional Measurement System Performance Metrics

Traditionally the performance of a measurement system has been evaluated based on the following metrics:

- *Precision.* The distribution of values for a repeated measurement. It has two components of repeatability (r) and reproducibility (R) and is calculated as

$$\text{Precision} = \sqrt{r^2 + R^2}$$

- *Accuracy or bias.* The offset between the average of measured target values and the true (i.e., reference) target value
- *Stability.* Variations of a measurement system over time
- *Linearity.* Accuracy of a measurement system over its measurement range
- *Throughput.* Time needed in order to complete a measurement

Repeatability in definition is the variations in reported values for repeated measurements of same sample with same tool by the same operator. In case of automated tools, the repeatability is referred to the variations of repeated measurements on the same part using the same recipe (static precision). On the other hand, reproducibility is defined as the variations in reported values for repeated measurements of same sample using same tool by different operators. In case of automated tools, they operate without an operator. Hence, reproducibility is referred to a repeated measurement on the same sample using different recipes/conditions (also called dynamic precision).

The process of evaluating a measurement system based on the tool metrics is referred to as measurement system analysis (MSA) or gauge study. The first two metrics, precision and accuracy, has been the basic and most important parameters in the MSA. The third and fourth metrics (i.e., stability and linearity) rely on the calculated values of the first two metrics (i.e., repeatability and reproducibility). When a measurement is repeated, the reported values follow the Gaussian distribution as they have a random nature. Hence, the average and sigma (i.e., standard deviation) for the measurement are used as the representative of the measurement quality. If the sigma is small, the measurement system is precise. If the average is close to the true target value, then the measurement system is accurate. Therefore, there are four different situations for a measurement system. These four states are shown in Fig. 16.1 schematically. In an ideal situation, the measurement system (3), which is both precise and accurate, is the desirable tool for any metrologist. However, in practice a well-maintained tool is similar to (4). If a measurement system can be kept precise, the bias can be resolved by calibration and matching the tools. The true value of target is usually determined by a reference measurement system (RMS) such as AFM.

16.2.2 New Measurement System Metrics and Fleet Measurement Uncertainty

The current metrology requirements for IC fabs do need more than the conventional MSA metrics to obtain the necessary information for performing meaningful comparison between the measurements. While precision is good for a single tool evaluation, it does not provide enough information in case of a fleet of multiple tools. Hence, a new metric is defined by International Technology Roadmap for Semiconductors (ITRS) known as total measurement uncertainty (TMU) [13]:

$$\sigma^2 = \sigma_P^2 + \sigma_M^2 + \sigma_S^2 + \sigma_{\text{other}}^2$$

The uncertainty (σ) is a function of precision (σ_P), tool matching (σ_M), sample variation (σ_S), and bias and other factors (σ_{other}). It is mainly used as a value that defines the uncertainty of a measurement system. It can be used for reporting TMU for a fleet of metrology tools.

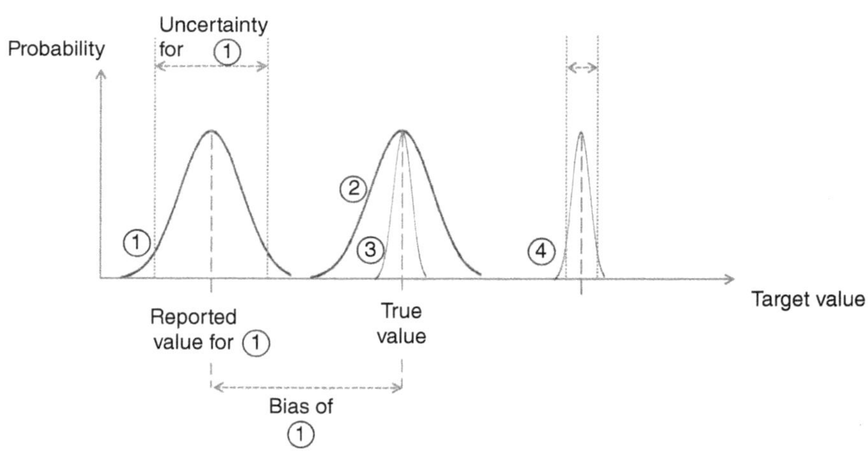

Figure 16.1 The schematic shows four different types of measurements. Each type represents a series of measurements performed on one sample using one tool: (1) inaccurate and imprecise, (2) accurate and imprecise, (3) accurate and precise, (4) inaccurate and precise.

Figure MET1 in [14] depicts schematically the difference between a single tool precision (repeatability and reproducibility) and total measurement uncertainty for a fleet of metrology tools [14]. It should be noted that TMU accounts for accuracy as well comparing to precision. Some researchers and engineers suggested to use fleet measurement precision (FMP) in measurement comparisons where the measurement accuracy is not critical [4].

16.2.3 Hybrid Metrology

There are multiple tools used for metrology applications in IC fab. Each tool has advantages and disadvantages and there is no perfect tool for all applications. As a new approach to address metrology challenges in IC fabs, hybrid metrology was introduced [10–12]. In hybrid metrology, the measurements of same parameter from multiple types of metrology tools are combined to improve the measurement of the parameter. In this methodology, a high precision tool is combined with a reference tool such as AFM to benefit from advantages of both tools while eliminating the disadvantages of each one.

16.3 AFM TECHNIQUE AND FUNDAMENTALS

AFM is one of the scanning probe microscopy (SPM) techniques in which a probe is sensing the sample surface using atomic force [15]. The first SPM technique, scanning tunneling microscopy (STM), was invented few years before invention of AFM and was used for imaging conductive samples using the electron tunneling current [16, 17]. The invention of AFM provided the ability to scan nonconducting samples. The main application of AFM was initially atomic imaging [18–20]. However, it was recognized by industry as a reference tool and used for other applications since its invention as it provides the highest resolution vertical measurement in 3D images [21, 22]. Thirty years after its invention, AFM is now being used for imaging a large variety of samples from inorganic to organic materials [23]. It is also able to collect additional data other than topography, including mechanical, electrical, magnetic, and thermal properties [23, 24].

The main components of an AFM system are probe, scanners, approach mechanism, positioning stage, laser detector, optical camera, system controller, and the computer for controlling data acquisition and image analysis. An AFM probe is a microcantilever typically made out of silicon with a relatively sharp tip located at its end. To operate the AFM system, the tip is brought to the vicinity of the sample surface using the approach mechanism. Then the probe moves over the sample surface in a raster motion while the tip-sample separation is maintained. A laser is used to monitor the cantilever response during its travel over the sample surface. The reflection of the laser from back of the cantilever is monitored using a position sensitive photo detector (PSPD).

16.3.1 AFM Scanners

In order to move the tip over the sample at nanometer resolution, piezoelectric materials are used which are called scanners. The scanners move the tip with respect to sample in X, Y, and Z directions. In the first design of AFM, three individual piezo actuators were used in form of a tripod to move the probe over the sample surface. Piezo tubes were used in commercial AFM systems soon after. In a piezo tube scanner, a set of cylindrical piezos move the probe in all three directions X, Y, and Z [25]. Although piezo tubes facilitated manufacturing of an AFM system due to its compact design comparing to the tripod design, their significant nonlinear out of plane motion especially over larger scan ranges, and the crosstalk between the scanners turned out to be their major limitations. The out of plane motion of around 50 nm over the scan area of 80×80 μm could be processed for surface roughness measurement of flat substrates using higher order flattening functions (see Fig. 6 in [26] for an example). However, it was too sophisticated to be removed from the surfaces with more complex features. To overcome this issue, flexure scanners were inverted and used [26, 27].

In this design, the Z scanner is decoupled from XY scanners to eliminate the crosstalk effect. One flexure scanner moves the probe in Z direction (Z scanner), while another scanner moves the sample in lateral directions (XY scanner). A schematic of this design is shown in Fig. 16.2*a*. In this design, the AFM probe maintains its out-of-plane orientation during the scan and regardless of the lateral scan size. Typical range of XY scanners are 100×100 μm. The typical range of Z scanners can go up to 30 μm. The basic design of an AFM head is shown in Fig. 16.2*b*. The emitted laser from its source is reflected to the cantilever using a prism. The laser beam reflects to a steering mirror, which reflects to PSPD. The vertical and lateral movements of the tip are detected using the PSPD.

16.3.2 Scanning Modes for Topography Imaging

The major scanning modes for collecting topography images with AFM are DC mode (also known as contact mode), and AC mode which is used for noncontact and tapping modes to collect topography images as well as other modes such as electrostatic force microscopy (EFM) and magnetic force microscopy (MFM) modes to detect other long-range forces such as electrostatic and magnetic [28].

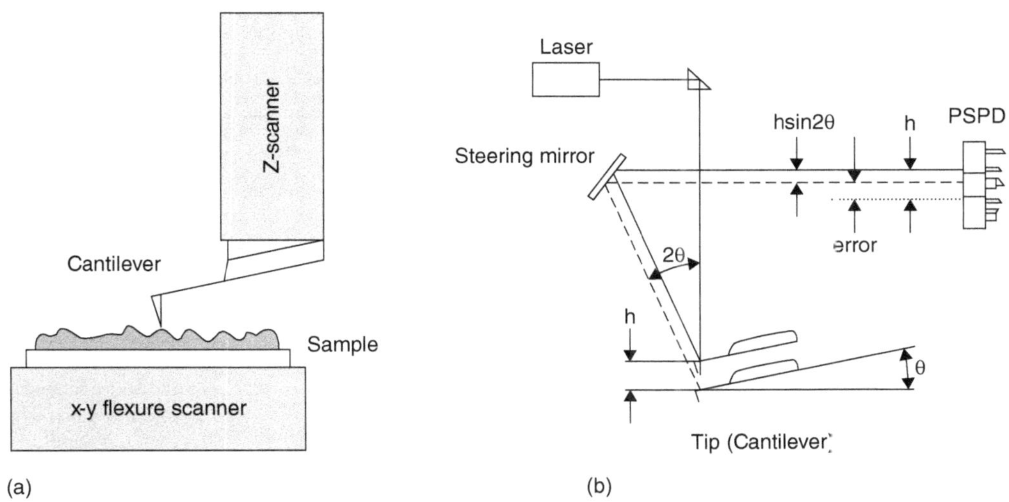

Figure 16.2 (*a*) A schematic showing the decoupled Z and XY scanner design of AFM. (*b*) The components of an AFM head. (*Reprinted with permission from Kwon et al., Review of Scientific Instruments, 74: 4378–4383 (2003).*)

In DC mode, the AFM probe is lowered until the tip is in contact with the sample and the force from its interaction with sample surface deflects the cantilever. The amount of force is linearly correlated to the deflection of the cantilever due to elastic deflection of the cantilever and is calculated as the product of cantilever deflection and its force constant. Cantilever's force constant is typically measured and reported by its manufacturer. The deflection of the cantilever is monitored by detecting movement of the laser reflection on PSPD. In DC or contact mode, tip senses the surface using short range repulsive forces. This mechanism is similar to stylus profilometer. However, in AFM the force on cantilever is maintained during the scan by adjusting the cantilever height with Z scanner. The cantilever deflection or force that is maintained during scan is referred to as setpoint. Contact mode was the initial mode used in first AFMs for collecting atomic resolution topography images. The AFM probe life is short in this mode due to strong short range and friction forces applied to the tip during the scan.

In order to maintain the force during the scan, the cantilever deflection is adjusted by moving the cantilever up and down using Z scanner. This is performed by applying a voltage to the piezo stack of Z scanner. Initially the voltage applied to Z scanner was calibrated against a standard grating to be converted to height values. Topography images were the recorded voltages applied to the Z piezo as topography values. However, piezoelectric materials have an intrinsic nonlinear response. Hence, depending on the sample height variation range, different calibration gratings were used. In addition, the creep effect of the piezo appears in the collected profile especially in case of reasonably tall features.

To address the aforementioned limitations of piezoelectric materials used in the Z scanner, sensors are added to Z scanner to measure its movement accurately. Initially the sensor's noise level prohibited its use for applications with low topography height/depth values. However, over the past decade the noise level of the sensors are decreased to less than 0.2 A. Hence, the signal collected from the sensor is used for all applications and replaced the use of topography signal based on driving voltage of Z. The sensors have linear response comparing to the driving voltage applied to Z piezo. Therefore, the need for having multiple calibration gratings for different sample heights is eliminated.

In AC mode, the tip is oscillated at a frequency close to its resonance frequency and the cantilever response is monitored using the reflected laser during the scan. In order to oscillate the cantilever, typically a piezoelectric material is used in the probe holder (i.e. bimorph). By supplying an AC signal to the bimorph, an oscillation drive provides the required energy to oscillate the cantilever close to its resonance frequency. Since an AC signal is driving the cantilever, the resulting imaging modes are generally categorized under AC mode. The resonance frequency of the cantilever is affected by the long range forces between the tip and sample surface as the oscillating cantilever is brought close to the sample surface. The resulting small change in resonance frequency leads to larger changes in oscillation amplitude (see Fig. 1 in [24]). The detected amplitude serves as a feedback (in amplitude modulated AFM systems) and is used to maintain the tip-sample separation at a constant value during the scan. The oscillation amplitude value used for scanning the sample is referred to as setpoint. Note that the setpoint in AC mode is intrinsically different from the setpoint in DC mode.

In non-contact mode, the tip-sample interaction is maintained in attractive regime during the scan. Hence, the AFM tip senses the long range van der Waals forces. In tapping mode, the tip-sample separation is decreased until the interaction enters the short range repulsive regime. Another difference between the two modes is the selected cantilever frequency. While the selected non-contact mode oscillation frequency is slightly larger than free air resonance frequency of the cantilever, it is the opposite in tapping mode [29–31].

The changes of cantilever's oscillation amplitude versus tip-sample separation is shown in Fig. 16.1 in [24]. As the oscillating tip descends to the sample surface, the amplitude decreases and tip senses longer range van der Waals interactions. If the descend continues, the amplitude decreases until it experiences a slight jump in value. At this point, the tip-sample interaction switches to repulsive regime (i.e., tapping mode). If the tip-sample separation increases again, the amplitude value increases as well. However, the slight jump will happen at a different moment. Similar behavior can be seen when looking at the phase signal. A phase signal depicts the difference in phase between the signal sent to bimorph to drive the cantilever and the signal that is detected by the PSPD. In this plot, the phase is almost zero as the oscillating tip is not in the vicinity of the sample surface. The phase possesses negative and positive values while the tip-sample interaction is in attractive and repulsive regimes respectively. (In different commercial AFM systems it could have different values and signs.) By comparison with the phase data, the area between the two jumps in amplitude value has a bistability behavior. Hence, the AFM feedback loop avoids using amplitude values in bistability region. It also can be seen that the repulsive regime has larger range over the amplitude values comparing to attractive regime. Therefore, operating in tapping mode would be less challenging for earlier AFM systems comparing to non-contact mode. Hence, conventional AFM systems used tapping mode as the main scanning mode. Eventually with more advancement of AFM systems, non-contact mode is more used and is now used as the major scanning mode in AFM imaging.

Although both non-contact and tapping modes are similar in operation and both modes provide much longer tip life comparing to DC mode, the tip life is shorter in tapping mode due to interactions between tip and sample. Short range forces are generally stronger than long range forces; hence, it becomes necessary to use non-contact mode for better repeatability and reliable metrology purposes.

16.3.3 Other Scanning Modes

In addition to topography, AFM can detect other properties such as electrical, magnetic, and mechanical using the atomic force between tip and sample [32–39]. In all of the additional modes, the AFM needs to track the sample surface properly before collecting other information. Therefore, each mode is based on either DC or AC mode imaging mechanism. A list of other major AFM modes is shown in Table 16.1. It is shown if each mode is based on AC or DC imaging mechanism. Most of these modes are currently used for physical metrology and in-lab characterization and measurements especially for failure analysis purposes.

16.4 AUTOMATED AFM FOR IN-LINE METROLOGY

Although AFM has been known widely as a manual lab tool, it was developed significantly over the past decade and automated AFM is now in use as an in-line measurement system in IC fab. An in-line automated AFM unit has typically three modules:

Table 16.1 Additional Imaging Modes That Can Be Operated Using AFM

Mode	AC	DC	Example of measurement/application
EFM, SKPM	√	√	Charge distribution on insulating layers, surface potential distribution
PFM		√	Ferroelectric domain distribution, local piezoelectric behavior
MFM	√		Magnetic domain distribution
Conductive AFM, SSRM		√	Local conductivity/resistance, failure analysis
SCM		√	Dopant profiling on semiconductor device
SThM		√	Local thermal conductivity and heat transfer
FMM		√	Local elasticity and hardness
LFM		√	Local friction/adhesion
PinPoint mechanical mapping		√	Local stiffness, adhesion, and elastic modulus

EFM, electrostatic force microscopy; FMM, force modulation microscopy; LFM, lateral force microscopy; MFM, magnetic force microscopy; PFM, piezo force microscopy; SCM, scanning capacitance microscopy; SKPM, scanning Kelvin probe microscopy; SSRM, scanning spreading resistance microscopy; SThM, scanning thermal microscopy.

1. *Equipment front end module (EFEM).* EFEM includes a robotic wafer handler, ports for loading FOUP (front open utility pod) or cassettes, and pre-aligner. The wafer handling is done by this module.
2. *Main AFM module.* The scanners are located in this module. Sample (e.g. wafer or mask) needs to be loaded to this module for imaging. It has the vibration isolation and filtered air flow for airborne particle control.
3. *Control module.* This module includes system controllers and the operating computer.

Since AFM is operating in ambient air and does not need vacuum unlike CD-SEM, the interfacing between EFEM and main measurement chamber is less challenging and does not need a pressure—controlled lock chamber.

The AFM automation capabilities can be categorized into the following groups:

- *Wafer handling.* It is mainly done in the EFEM module. Includes unloading/loading the wafer from/to its cassette or FOUP, angular pre-alignment using optical pre-aligner to orient the wafer notch, loading/unloading to/from the sample stage.
- *Automatic tip exchange.* Includes unloading/loading probes from/to AFM head, laser alignment, tip characterization.
- *Wafer navigation.* Includes global alignment and navigating to measurement locations using auto focus and optical pattern recognition.
- *Positioning, approach, and imaging.* Includes locating the location of interest with nanometer accuracy using autofocus, optical and AFM pattern recognition, fast approach, and imaging.
- *Data processing, analysis, and reporting.* Once imaging is completed, its processing and analysis and final report is completed automatically.

At the end of the measurement, the collected values for measurement target is transferred from the measurement system to its fab host (control system) using the protocols for SEMI Equipment Communication Standard/Generic Equipment Model (SECS/GEM). This enables immediate comparison with the system limits which are defined in the statistical process control (SPC). In-line communication includes the following content:

- Process information (lot ID, process ID, recipe ID)
- Remote control (process start and cancel, control status, operation status)
- Equipment information (equipment status, port status, alarms)
- Material handling information [load/unload request/complete, information on automated guided vehicle (AGV), or manually guided vehicle (MGV)]
- Process status
- Process result

Each recipe in AFM automation software includes all the information required from unloading the wafer from FOUP to perform the measurement until loading it back. Hence, the recipe includes the following information:

- System conditions during the measurement
- Alignment requirements
- Measurement locations
- Measurement method

Measurement method includes positioning, imaging parameters, processing, and analysis. The operator prepares or "writes" a recipe for a specific wafer. The recipe will be used for all the coming batches of wafers. At the end of each measurement, additional information about each step of automation is recorded and stored. The stored additional data ensures recipe portability and enables the engineers to modify and update the recipes per changes in process.

The major application of AFM include surface roughness and surface waviness measurement for monitoring CMP process, step height measurements for through silicon via (TSV) bars and holes, critical angle measurement, and CD metrology including measurement of SWA, SWR, LER, and LWR. It also attracted attention in hybrid metrology to be used as a RMS (reference measurement system). The following are some of the major applications of in-line AFM:

16.4.1 CD-AFM for CD Metrology

The trenches and structures are not perfect and a set of measurement targets are used to qualify the process. As the device size shrinks, the role of imperfections become more vital to the yield of the final product. Hence, the presence of overhangs and undercuts is inevitable. The common measurement targets are depicted in Fig. 16.3. A standard AFM has a top-down imaging method; hence, the tip may not be able to "see" the undercut of trenches or details on an almost vertical sidewall. There has been several ways suggested to address this issue. A special flared tip design was initially suggested to characterize sidewall and overhang structure and was able to get a cross-sectional profile of the structure. However, flared tips are by nature flat and wide, therefore, they have limitations in performing high resolution characterization of sidewall such as side wall roughness (SWR). In another attempt, the sample was placed on a tilting stage and sample was tilted to get high-resolution images of sidewall. Despite partial success, this method was found to be practically challenging especially for in-line AFM systems that are to handle 300-mm wafers [40].

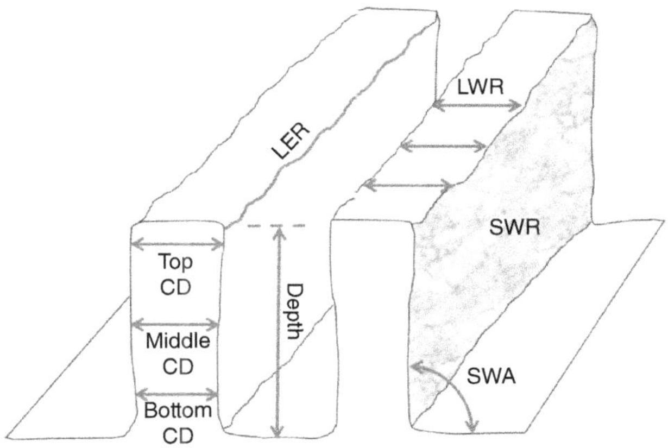

Figure 16.3 The schematic shows the common measurement targets for CD-AFM including CD, LER, LWR, SWA, SWR.

In order to address this issue, a new platform of AFM system with rotating head was designed [41]. In this platform, the Z and XY scanners are decoupled, and the AFM head can rotate sideways to scan the sample at different angles. In this method, it is possible to collect three images of the sample at three different head angles and reconstruct the final 3D image of the sample. An example of such 3D image is shown in Fig. 16.4. As it can be seen, the new CD-AFM with rotating head is capable of providing accurate cross-sectional topography as well as high resolution images for SWR, side wall angle (SWA), line edge roughness (LER), line width roughness (LWR), and CD measurements at top, middle, and bottom of the structure. The process of collecting three images, processing, and reporting final values for measurement targets is automated. The tip width is below 10 nm; hence, capable of collecting detailed resolution.

16.4.2 Atomic Force Profiler

While AFM provides local information on the surface profiles, a stylus profiler is generally used to characterize global planarity of the wafer. However, stylus profiler has limitations in the lateral and vertical resolution for characterizing sub-micrometer surface features. In addition, stylus profiler can damage the surface by dragging the tip over the surfaces with low elastic tolerance for surface tension. Atomic force profiler (AFP) addresses the major two shortcomings of stylus profiler by profiling wafer surfaces using AFM capability. In an AFP unit, a sliding stage will move the sample in the lateral direction for ranges up to 100 mm, while the AFM probe has approached the surface and maintains the interaction in non-contact mode. A XY scanner can be placed on top of the slider to combine both AFM and AFP capabilities. In AFP, the probe is controlled using the feedback loop of AFM. Hence, the force applied by probe on the sample is well controlled and minimal. AFP's vertical range is identical to AFM.

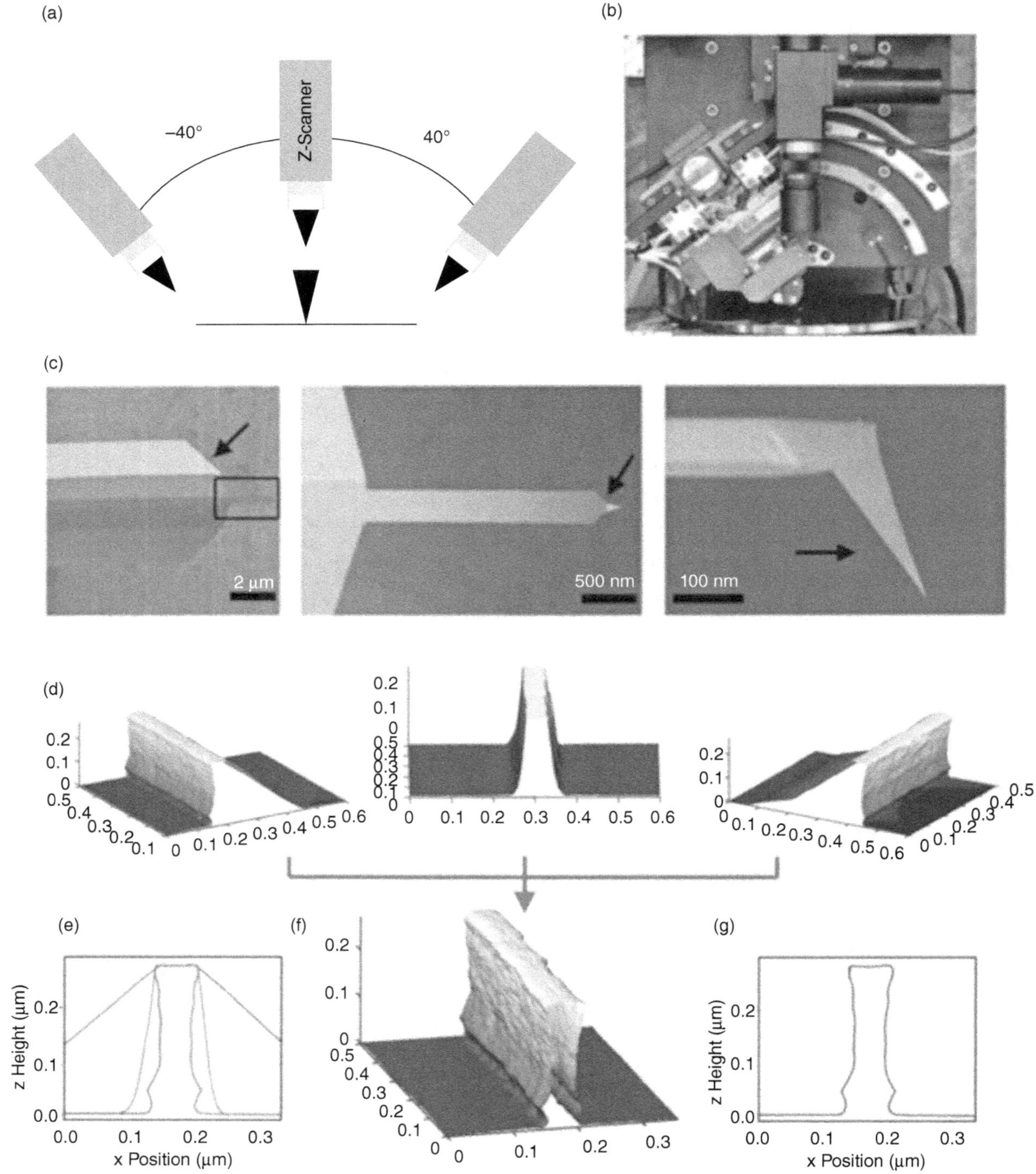

Figure 16.4 (*a*) The schematic shows the concept of CD-AFM with rotating head. (*b*) An image of the system with the AFM head at −38° position. (*c*) High-aspect-ratio tips that can be used for imaging with CD-AFM. (*d*) CD-AFM images collected at head positions of −38°, 0°, and 38°. (*e*) Height profile constructed from different scan angles. (*f*) 3D rendering of the overhang structure from the three height images. (*g*) Final cross-sectional profile. (*Reprinted with permission from Cho et al., Review of Scientific Instruments, 82: 023707, 2011.*)

The major application of AFP is in the measurements in chemical mechanical polishing (CMP) process such as dishing, erosion, surface texture, and plug recess. Another application of AFP is wafer edge profiling. Given the accuracy of motorized sample stage, it is possible to collect AFP 2D images with lateral mm scales and provide the ability to match the 3D information to optical images of the wafer edge.

16.4.3 Automatic Defect Review

Defect inspection and review is an important part of the process in chip manufacturing. As the device size shrinks, the number of defect metrology steps increases. Defect metrology steps are divided into two groups: inspection and review. In defect inspection, a map or distribution of defects on the wafer is generated and used to qualify or disqualify a lot from continuing through the process. Optical inspection tools are used at this step. These tools generally have high throughput and can operate at tens to hundreds of wafers per hour. The inspection tool has to classify the defects and need detailed information on the defects.

The required information for proper defect classification is provided by defect review tools. A review tool provides high-resolution images of the defects of the interest to either correct misclassified defects, or identify false defects. The major challenge for a review tool is successful linking between its coordinate stage and the coordinate stages of an inspection tool. This becomes more challenging for blank wafers, where no aligner or fiducial mark exists. In addition, if the review tool can provide the information in a nondestructive way, the lot can return to the process. A comparison between existing review tools is shown in Table 16.2. AFM would be an ideal solution because it can provide 3D information of the defect in a nondestructive way. However, the laborious process of finding a defect in AFM has been prohibiting it from being used for defect review effectively. By using automation, ADR-AFM addressed the common challenges and is now a tool that can be used for an automated nondestructive defect review. To operate ADR-AFM, the coordinates for defects of interest are imported. It is followed by a coarse alignment using the larger defects that can be seen using enhanced vision technique, and then the defect review process is initiated. It operates and images in noncontact mode; hence, the process is truly nondestructive.

16.4.4 In-line AFM for Mask Fabrication

In case of mask fabrication, there are two main metrology applications: CD measurement and defect metrology. Mask reticles are significantly more expensive than wafers. Hence, proper characterization, repair, and even their reuse are critical [42–44]. The in-line AFM can also be used for mask metrology as well [45]. The main changes would be in the sample handling system. Due to the high cost of mask reticles, ADR-AFM is attracting more attention in mask community.

Table 16.2 A Comparison Chart between Typical Defect Inspection and Defect Review Techniques

Methods	Available tools	Sample prep.	Destructive	Throughput	Hi-res image
Defect inspection	Optical method (bright-field and dark-field light scattering)	No	No	1–8 wph* (bright field) 30–150 wph (dark field)	No
	SEM (electron scattering)	No	Depending on the materials	N/A	2D
Defect review	DR-SEM	No	Depending on the materials	Around 300 dph†	2D
	AFM	No	No	10–16 dph (ADR) 1–2 dph (w/o ADR)	3D
	FIB	Yes	Yes	N/A	2D
	TEM	Yes	Yes	8 wpd‡	2D

*Wafer per hour
†Defect per hour
‡Wafer per day

16.5 MAINTENANCE AND CALIBRATION

Calibration is referred to a process which validates the values collected by a measurement system against a standard. There are different types of measurement standards:

- *National standards.* This type of standards is usually kept at national institutions and not available to public.
- *Primary standards.* A NIST standard that is certified against a national standard by state-of-the-art calibration process. It can be provided to any organization for any purpose.
- *Secondary standards.* A standard calibrated carefully against a primary standard by any organization.
- *Working/production standards.* The standards that are used in the fab for calibration of a measurement system.

As we go down the hierarchy of the standards, they become less accurate comparing to the national standard; however, they are cheaper and more resistant against the changes in environment. For a reliable calibration, the standard has to be traceable to a national standard or nationally accepted standard. In case of semiconductor industry, unfortunately no national standards exist for the current dimensions of nodes. In fact, the semiconductor industry has developed so rapidly that it has been very challenging to develop a national standard for the industry. However, there are companies (e.g., VLSI) that provide working standards for semiconductor industry with standard traceability to National Institute of Standards and Technology (NIST) units. Such standards should be used for calibration of measurement systems including AFM.

One common calibration issue with conventional AFMs was the nonlinearity of the Z calibration due to hysteresis behavior of Z piezo. In conventional AFM systems, the voltage applied to Z piezo is converted to length units by being calibrated against a standard. Therefore, the measurements performed over the sample with a height value of 800 nm would not be accurate for sample heights of 5 nm and less. This issue is addressed by using the sensor signal instead of Z voltage as described earlier. A sensor signal is used for topography image and it can also be calibrated against a standard. Since the sensor has a linear response, single-standard calibration can cover a large range of measurement heights. The system calibration is long lasting and may require an update once a year. However, checking the calibration is performed by metrologists more frequently.

In case of an automated in-line AFM, tip replacement on the system is performed automatically using a module called automatic tip exchanger (ATX). The tip exchange process can be performed either automatically or initiated by the operator. A reasonable number of tips are deposited in the ATX module such that the system can independently operate for few months. If different types of tips are needed, the ATX can be programmed such that the cantilever settings are updated per the tip characteristics (e.g., resonance frequency, force constant) each time a probe is replaced. Once all the tips in ATX are used, the used tips need to be removed by the operator and a cassette of new tips should be loaded.

Since the AFM has the highest vertical resolution in Z direction, it is very important that the noise level is minimized to avoid its effect on the final measurement. The noise level in AFM has two major sources: acoustic noise, and floor vibration noise. For this purpose, the automated in-line tools are provided with integrated vibration isolation units and acoustic enclosures. This becomes critical especially for in-line AFM systems which are installed in cleanroom facilities. In a cleanroom facility, the temperature, humidity, and airborne particles are well controlled. However, the floor vibration and acoustic noise could be higher than offline lab due to presence of other processes and metrology systems. In case of AFM, the noise level is measured by bringing the tip in contact with the sample surface and performing a "zero scan" test. In this test, the lateral scan dimensions are set to zero and the resulting image is the collected noise of the system. The noise level is reported as the R.M.S. (Root mean square) value of the Z values in the AFM image.

16.6 CONCLUSION

In summary, this chapter reviewed briefly the fundamentals of metrology and in-line AFM. Comparing to optical techniques and SEM, AFM is a much younger technique; however, it had a rapid growth to the level that it is recognized as an in-line RMS. With AFM's continuous growth, its role is being expanded from an RMS to a reliable metrology workhorse for profile and surface roughness measurement especially in CMP applications. It is also going to play a significant role as the in-line RMS for both CD metrology and hybrid metrology. Although AFM has been known for its tip wear and low throughput, the latest developments on non-contact mode imaging and the latest level of automation address those limitations. The literature on in-line AFM are still very limited to the knowledge of this author.

16.7 REFERENCES

1. G. Binnig, C. F. Quate, and Ch. Gerber, "Atomic Force Microscope," *Physical Review Letters*, 930–933, 1986.
2. Gordon E. Moore, "Progress in Digital Integrated Electronics," *Electron Devices Meeting*, 21: 11–13, Dec. 1975.
3. H. K. Wickramasinghe, "Scanned-Probe Microscopes," *Scientific American*, 260 (10): 98–105, 1989.
4. Bo Su, Eric Solecky, and Alok Vaid, *Introduction to Metrology Applications in IC Manufacturing*, SPIE Press, Bellingham WA, 2015.
5. Ardavan Zandiatashbar et al., "Studying Post-Etching Silicon Crystal Defects on 300 mm Wafer by Automatic Defect Review AFM," *Proc. SPIE 9778, Metrology, Inspection, and Process Control for Microlithography XXX*, p. 97782P, 2016.
6. Y. Hua et al., "New Three-Dimensional AFM for CD Measurement and Sidewall Characterization," *Proc. SPIE 7971, Metrology, Inspection, and Process Control for Microlithography XXV*, p. 797118, 2011.
7. J. Foucher, S. W. Schmidt, C. Penzkofer, and B. Irmer, "Overcoming Silicon Limitations: New 3D-AFM for CD Measurement and Sidewall Characterization," *Proc. SPIE 8423*, p. 842318, 2012.
8. Larry M. Ge and Dean J. Dawson, "Characterization of the CMP Process by Atomic Force Profilometry," *Proc. SPIE 3882, Process, Equipment, and Materials Control in Integrated Circuit Manufacturing V*, p. 112, 1999.
9. Ardavan Zandiatashbar et al., "High-Throughput Automatic Defect Review for 300-mm-Blank Wafers with Atomic Force Microscope," *Proc. SPIE 9424, Metrology, Inspection, and Process Control for Microlithography XXIX*, p. 94241X, 2015.
10. J. Foucher, P. Faurie, L. Dourthe, B. Irmer, and C. Penzkofer, "Hybrid CD Metrology Concept Compatible with High Volume Manufacturing," *Proc. SPIE 7971*, p. 79710S, 2011.
11. A. Vaid et al., "A Holistic Metrology Approach: Hybrid Metrology Utilizing Scatterometry, CD-AFM and CD-SEM," *Proc. SPIE 7971*, p. 797103, 2011.
12. N. Rana and C. Archie, "Hybrid Reference Metrology Exploiting Patterning Simulation," *Proc. SPIE 7638*, p. 76380W, 2010.
13. "International Technology Roadmap for Semiconductors, 2013 Edition, Metrology," ITRS, 2013.
14. "International Technology Roadmap for Semiconductors, 2009 Edition, Metrology," ITRS, 2009.
15. Bharat Bhushan, Harald Fuchs, and Masahiko Tomitori (eds.), *Applied Scanning Probe Methods VIII: Scanning Probe Microscopy Techniques (NanoScience and Technology) (No. 8)*, Springer, New York, 2008.
16. G. Binnig and H. Rohrer, "Scanning Tunneling Microscopy," *Helvetica Physica Acta*, 55: 726–735, 1982.
17. G. Binnig, H. Rohrer, Ch. Gerber, and E. Weibel, "Surface Studies by Scanning Tunneling Microscopy," *Physical Review Letters*, 49 (1): 57–61, 1982.
18. G. Binnig, C. Gerber, E. Stoll, T. R. Albrecht, and C. F. Quate, "Atomic Resolution with Atomic Force Microscope," *Europhysics Letters*, 3 (12): 1281–1286, 1987.
19. R. C. Barrett and C. F. Quate, "Imaging Polished Sapphire with Atomic Force Microscopy," *Journal of Vacuum Science and Technology A*, 8 (1): 400–402, 1990.
20. H. Heinzelmann, E. Meyer, D. Brodbeck, G. Overney, and H.-J. Guntherodt, "Atomic-Scale Contrast Mechanism in Atomic Force Microscopy," *Zeitschrift fur Physik B-Condensed Matter*, 88: 321–326, 1992.
21. G. T. Smith, *Industrial Metrology: Surfaces and Roundness*, Springer, New York, 2002.
22. J. Libert and L. Fei, *Method to Delineate Crystal Related Defects*, PCT Publication, WO2013055368(A1).
23. Dalia G. Yablon, *Scanning Probe Microscopy for Industrial Applications: Nanomechanical Characterization*, Wiley, Hoboken, New Jersey, 2013.
24. Ardavan Zandiatashbar, Byong Kim, Young-kook Yoo, and Keibock Lee, "Automated Non-Destructive Imaging and Characterization of the Graphene/hBN Moiré Pattern with Non-Contact Mode AFM," *Microscopy Today*, 23 (06): 26–31, 2015.
25. G. Binnig and D. P. E. Smith, "Single-Tube Three-Dimensional Scanner for Scanning Tunneling Microscopy," *Review of Scientific Instruments*, 57 (8): 1688–1689, 1986.
26. Joonhyung Kwon et al., "Atomic Force Microscope with Improved Scan Accuracy, Scan Speed, and Optical Vision," *Review of Scientific Instruments*, 74 (10): 4378–4383, 2003.
27. Fredric E. Scire and E. Clayton Teague, "Piezodriven 50-μm Range Stage with Subnanometer Resolution," *Review of Scientific Instruments*, 49 (12): 1735–1740, 1978.
28. Ricardo Garcia and Ruben Perez, "Dynamic Atomic Force Microscopy Methods," *Surface Science Reports*, 197–301, 2002.
29. Ardavan Zandiatashbar, "Sub-Angstrom Roughness Repeatability with Tip-to-Tip Correlation," *NanoScientific*, 14–16, 2014.
30. J. B. Pethica and W. C. Oliver, "Tip Surface Interactions in STM and AFM," *Physica Scripta*, T19: 61–66, 1987.
31. J. Israelachvili, "Interfacial Forces," *Journal of Vacuum Science and Technology A*, 10 (5): 2961–2971, 1992.
32. Y. Martin, David W. Abraham, P. C. D. Hobbs, and H. K. Wickramasinghe, "Magnetic Force Microscopy—a Short Review," *Magnetics Materials, Processes, and Devices*, 90 (8): 115–124, 1989.
33. C. Schonenberger and S. F. Alvarado, "Observation of Single Charge Carriers by Force Microscopy," *Physical Review Letters*, 65 (25): 3162–3164, 1990.
34. C. C. Williams, J. Slinkman, W. P. Hough, and H. K. Wickramasinghe, "Lateral Dopant Profiling on a 100 nm Scale by Scanning Capacitance Microscopy," *Journal of Vacuum Science and Technology A*, 8 (2): 895–898, 1990.
35. C. C. Williams and H. K. Wickramasinghe, "Thermal and Photothermal Imaging on a sub 100 nm Scale," *Scanning Microscopy Technologies and Applications, Proc. SPIE*, 897: 129–134, 1988.
36. A. L. Weisenhorn, P. K. Hansma, T. R. Albrecht, and C. F. Quate, "Forces in Atomic Force Microscopy in Air and Water," *Applied Physics Letters*, 54 (26): 2651–2653, 1989.
37. Yves Martin, David W. Abraham, and H. Kumar Wickramasinghe, "High-Resolution Capacitance Measurement and Potentiometry by Force Microscopy," *Applied Physics Letters*, 52 (13): 1103–1105, 1988.
38. P. Maivald et al., "Using Force Modulation to Image Surface Elasticities with the Atomic Force Microscope," *Nanotechnology*, 2: 103–106, 1991.
39. U. Durig, O. Zuger, and D. W. Pohl, "Force Sensing in Scanning Tunneling Microscopy: Observation of Adhesion Forces on Clean Metal Surfaces," *Journal of Microscopy*, 152 (1): 259–267, 1988.

40. Yves Martin and H. Kumar Wickramasinghe, "Method for Imaging Sidewalls by Atomic Force Microscopy," *Applied Physics Letters*, 64 (19): 2498–2500, 1994.

41. Sang-Joon Cho et al., "Three-Dimensional Imaging of Undercut and Sidewall Structures by Atomic Force Microscopy," *Review of Scientific Instruments*, 82: 023707, 2011.

42. James A. Folta, J. Courtney Davidson, Cindy C. Larson, Christopher C. Walton, and Patrick A. Kearney, "Advances in Low-Defect Multilayers for EUVL Mask Blanks," *Proc. of SPIE 4688, Emerging Lithographic Technologies VI, 173*, 2002.

43. Ajay Kumar and Banqiue Wu, "Extreme Ultraviolet Lithography: A Review," *Journal of Vacuum Science & Technology B*, 1743–1761, 2007.

44. Paul Morgan et al., "Computational Techniques for Determining Printability of Real Defects in EUV Mask Pilot Line," *Proc. SPIE 9050, Metrology, Inspection, and Process Control for Microlithography XXVIII*, p. 90501C, 2014.

45. Ardavan Zandiatashbar et al., "Automatic Defect Review for EUV Photomask Reticles by Atomic Force Microscope," *Proc. SPIE 9635-46, Photomask Technology*, p. 96351A, 2015.

16.8 FURTHER READING

On CD-AFM: Sang-Joon Cho et al., "Three-Dimensional Imaging of Undercut and Sidewall Structures by Atomic Force Microscopy," *Review of Scientific Instruments*, 82: 023707, 2011.

On Metrology: Bo Su, Eric Solecky, and Alok Vaid, *Introduction to Metrology Applications in IC Manufacturing*, SPIE Press, Bellingham WA, 2015.

PART 3

BACK-END-OF-LINE PROCESSES

Chapter 17 Wafer Thinning and Singulation

BY

YOUNGSUK KIM *DISCO Corporation, Tokyo, Japan*
SUMIO MASUCHI *DISCO Corporation, Tokyo, Japan*
NORIKO ITO *DISCO Corporation, Tokyo, Japan*
MIYUKI HIROSE *DISCO Corporation, Tokyo, Japan*

17.1 INTRODUCTION

This chapter covers the fundamental concepts and practical applications of thinning and singulation processes for silicon wafers, which are formed transistors and interconnects in the *front-end-of-line* (FEOL). Since the wafer thinning and singulation processes take place before assembly and packaging, it is important to provide accurate process control as well as quality for enabling the subsequent packaging processes, such as die bonding, wiring, and molding.

The semiconductor industry is now facing a major challenge, due to both economic and technical limitations, in how to realize the next generation of scaling technology. Presently, advanced package technologies such as *wafer level packages* and *three-dimensional* (3D) integration are considered promising technologies for enhancing system performance and for achieving cost reduction as an alternative to conventional *two-dimensional* (2D) scaling.[1–3] Requirements for thinner die and narrower street width will become more severe as device density increases for small form factor die and as designs aim to improve the net number of die by reducing die size, as shown in Fig. 17.1.[4] The changes to the processes, equipment, and materials used in wafer thinning and singulation in order to meet these requirements will be discussed in this chapter.

17.2 THINNING TECHNOLOGY OVERVIEW: GRINDING

Grinding, also known as *BG*, refers to backgrinding. After the wafer fabrication process, a surface protection tape, called BG tape, is attached to the device layer of the wafer prior to grinding. The wafer is turned over and the back of the device wafer is ground to thin the wafer enough for it to be assembled into a package.[5]

The overall package thickness is determined by global standards; the overall thickness has to be sufficient to accommodate the lead frame, die, molded plastic, and paste or *die attach film* (DAF).

17.2.1 Grinding Wheel

The wafer rotates on a chuck table, as shown in Fig. 17.2*a*, and is ground as the grinding wheel, as shown in Fig. 17.2*b*,[6] is lowered. The segments of grinding wheels consist of diamond abrasive grains held by a bond material. Resin and other vitrified materials are most commonly used as bond materials.

Generally, silicon grinding is composed of two steps: rough grinding and fine grinding. Rough grinding removes most of the material to be grounded at a higher rate. For this process, grinding wheels with a larger diamond abrasive (around 35 to 40 μm) are used to remove silicon and backside layers such as oxides or nitrides. Fine grinding removes the last 40 to 100 μm of silicon in order to reach the final desired thickness at a slower rate. Fine grinding typically necessitates the use of grinding wheels with smaller abrasive grains (4 μm). This is effective for reducing the damage created by rough grinding and to achieve a smoother finished ground surface.

17.2.2 Grinding Point

To begin the grinding process, a process called self-grinding is performed. A grinding wheel for initial chuck table grinding is mounted on a spindle and the chuck table, which is made of porous ceramic, is ground into a convex shape at a slight angle. This is to correct the parallelism between the inclination of the chuck table and that of the grinding wheel. The grinding wheel comes into contact with only half the diameter of the wafer during grinding, as shown in Fig. 17.2*a*, while the other half is untouched. This prevents the creation of an irregular crisscross pattern for a uniform surface finish.

17.2.3 Processing Quality

The purpose of grinding is to grind down to the desired thickness within a specified roughness. Typical evaluation criteria for grinding process quality, as shown in Table 17.1, are as follows:

- *Surface roughness* (Ra). In order to achieve the desired roughness, optimization of the grinding wheel, especially the grit size, is necessary.
- *Total thickness variation* (TTV). Insufficient adjustment of both chuck table and spindle inclination increases TTV. As the final thickness of wafers is reduced, the mass production criteria for TTV also become smaller.

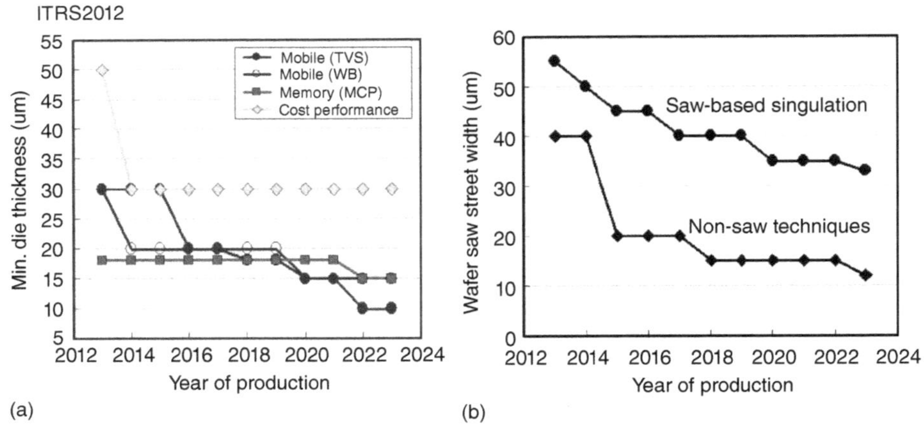

Figure 17.1 Trends in (*a*) minimum die thickness and (*b*) wafer street width from ITRS 2012 technology roadmap. Demand for thinner die and narrower street width will become more severe for advanced packaging technologies.

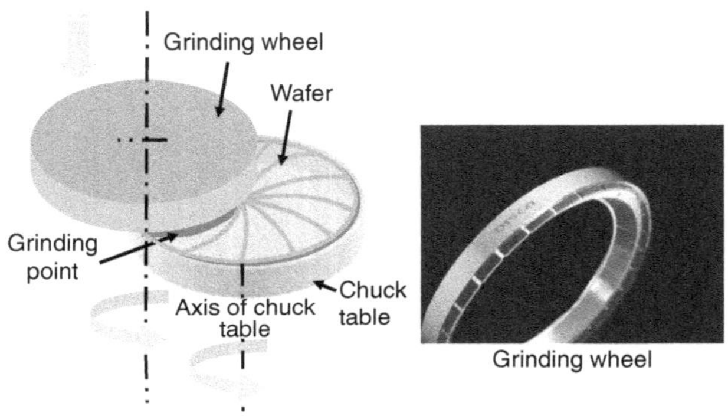

Figure 17.2 In feed: (*a*) image of in-feed grinding, (*b*) grinding wheel.

- *Edge chipping*. Grinding water or an overly aggressive feed speed can cause the wafer edge area to flutter, which can cause edge chipping. When the edge of a thin wafer contacts the transfer mechanism in a grinder, edge chipping may become the starting point for wafer breakage.

Since the original wafer edge profile is beveled,[7] as the wafer becomes thinner, the shape of wafer edge becomes increasingly sharpened like a knife edge. In order to prevent edge chipping, sometimes an additional process called "edge trimming" is used prior to grinding, which grooves the wafer edge with a dicing blade (refer to Sec. 17.5.1).

- *Die strength*.[8,9] This item is used for thin wafers and thin die to assess the strength of die assembled into a package. Factors which reduce die strength are considered to be as follows: residual grinding damage, backside chipping, thermal stress, wafer defects, and others.

TABLE 17.1 Typical Evaluation Criteria for Grinding

Evaluation criteria	Description	Typical values
Surface roughness (Ra)	The roughness of the ground surface	Around 0.01 μm for a grinding wheel with 4 μm abrasive grains
Total thickness variation (TTV)	Total thickness variation within one wafer	Less than 3 μm for a 300 mm wafer (without BG tape)
Edge chipping	Tiny cracks on the circumference of the wafer	—
Die strength	Flexural stress of die as measured by a bending test	—

17.3 THINNING PROCESS AND EQUIPMENT

A typical grinder polisher is composed of several sections performing the various functions.

17.3.1 Grinding Section

Figure 17.3 illustrates the workflow. Many fully automatic grinder polishers follow this pattern.[10] The wafer is unloaded from the wafer cassette and transferred to the processing section. There are four wafer chuck tables, which are placed on a turn table onto which the rotating spindle descends. The Z1 spindle is for rough grinding, the Z2 spindle is for fine grinding, and the Z3 spindle is for polishing (refer to Sec. 17.4.2). After cleaning, the wafer is returned to the cassette automatically.

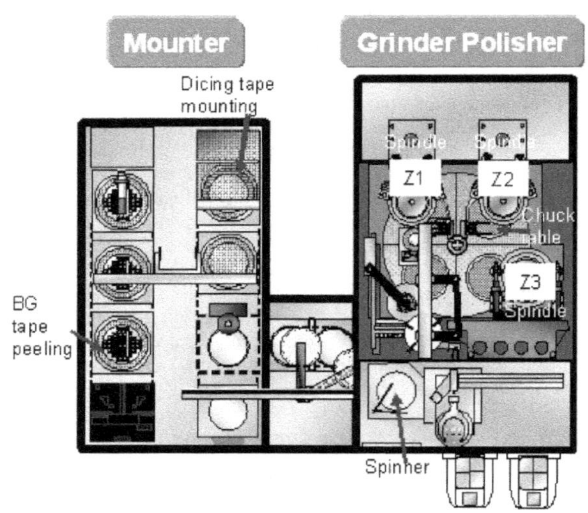

Figure 17.3 Grinding layout. (*Courtesy of DISCO Corporation.*)

17.3.2 Thickness Control with an In-Process Gauge

Wafer thickness is measured during grinding in real time using an in-process gauge and the measured thickness value is fed back by closed loop control so that the spindle stops once the thickness value reaches the desired finish thickness. An in-process gauge continuously measures the total thickness of the wafer, by measuring both the height of the upper surface of the wafer and that of the chuck table.

17.3.3 Cleaning

Since grinding produces significant particulate, cleaning is required in order to maintain the grinding quality. Once these fine contaminants are allowed to dry, they are typically difficult to remove.

- *Chuck table cleaning.* When these particles become trapped between the wafer and the chuck table they cause "dimples" (slight, localized indentations) and "crow's feet" (localized cracks) that can lead to wafer breakage. So, prior to setting the wafer on the chuck table, the chuck table is cleaned with deionized water.
- *Wafer cleaning (after grinding).* The ground wafer is cleaned with deionized water on the chuck table to remove as much particulate as possible. This prevents particles from becoming trapped between the ground wafer and the robot transport arm.
- *Spinner cleaning/drying.* In addition, the ground wafer is cleaned with deionized water again in the spinner chamber and spin dried before being returned to the wafer cassette.

17.4 THINNING TECHNOLOGY, STRESS RELIEF, AND OTHER REQUIREMENTS

17.4.1 Issues of Wafer Thinning

Since innovation in packaging technology has required thinner and smaller, wafer thinning less than 100 μm thickness is key technology to realize the demand (refer to Sec. 17.1). In general, as a wafer becomes thinner, the risk of wafer breakage increases. The main causes of wafer breakage are as follows:

Decrease of Mechanical Strength

The thinner the wafer becomes, the lower the wafer mechanical strength is. The mechanical strength of wafers thinner than 200 μm is often evaluated as die strength (refer to Sec. 17.2.3). Normally, a thick wafer has sufficient mechanical strength to resist the stress generated from the difference in tension between the device surface and the backside of the wafer. When the wafer cannot maintain its flatness due to the insufficient mechanical strength, wafer warpage[7] or wafer bowing[7] occurs. For example, warpage is often generated by grinding wafer down to 100 μm thickness in 300 mm wafer. Warped wafers are difficult to transfer, and the risk of wafer breakage increases due to unexpected contact within the equipment.

Grinding Damage

After backgrinding, damage such as fractured poly-Si, amorphous Si, dislocation, and stacking fault will remain on the ground surface. Rough grinding creates damage deeper than 5 μm, while fine grinding creates damage of around 0.2 μm depth. As wafers become very thin, even the grinding damage caused from fine grinding has a large influence on the mechanical strength of thin wafers.

Other Issues

Improved thickness accuracy is required for thinner wafers because thickness variation influences relatively larger on final thickness of thinner wafers compared with that of thick wafers. Also, secure handling methods are required to handle thin wafers during postgrinding processes.

17.4.2 Stress Relief

One of the solutions for thinning wafers to less than 100 μm thickness is "stress relief." This uses chemical and mechanical process to remove subsurface damage and warpage caused by grinding. There are three main types of stress relief (Fig. 17.4).[11]

- *Dry polishing.* Polishing without water or chemical liquid. A metal oxide that reacts with silicon in its solid state is impregnated into a bonded material as a polishing agent, held in place by resin, and used as a polishing pad. The maximum polishing rate is about 1 μm/min, and it polishes about 2 μm off of the grinding surface. After dry polishing, cleaning with only deionized water is required. This process has the lowest total cost and environmental burden.
- *Chemical-mechanical polishing* (CMP). CMP polishes about 2 μm off the ground surface using alkaline slurry and polishing pad. The maximum polishing rate is about 1 μm/min. This process requires proper disposal of the resulting waste liquid.
- *Wet etching.* A wet process which etches the silicon with a mixture of boric and nitric acids. The wafer is spun and sprayed with the acid mixture from above in order to etch it to a uniform thickness. The etching rate is about 0.5 to 2.0 μm/s, which is faster than dry polishing and CMP. Waste treatment of NO_x gas and liquid, generated during etching, is necessary. This method is well suited to ultrathin power devices which require a significant amount grinding damage to be removed after grinding. The complete removal of grinding damage via

Device	Device for mobile products (memory, logic, etc.)	Discrete device (transistor, diode, etc.)	Device for 3D-IC (BSI, memory using TSV, etc.)
Wafer thickness (μm)	150-50	200-50	200-10
Required features	Low cost High yield Thin die	Low cost High yield Thin die Improved device properties	Low cost High yield Thin die High cleanness
Recommended stress relief	Dry polishing (as well as CMP)	Wet etching	CMP

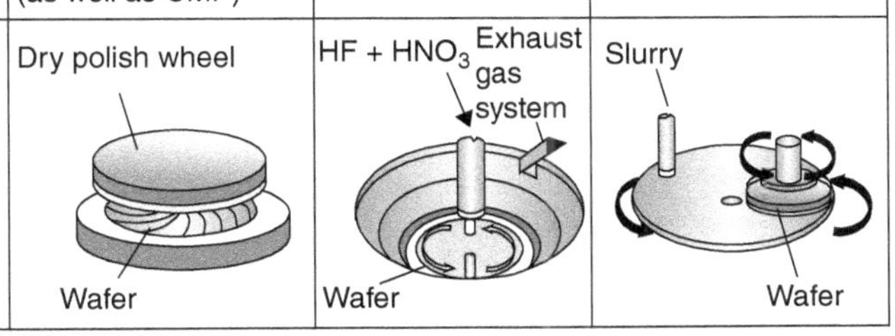

Figure 17.4 Stress relief comparison.

wet etching can contribute toward the manufacture of devices with good electrical characteristics.

- *Dry etching*. A dry process etches the silicon using fluoride gas plasma. Dry etching is often used to expose vias created in TSV devices after thinning.

17.4.3 Gettering

Management of metal impurities using gettering effect is key issue in wafer process to keep clean a device area. There are two types of gettering: intrinsic gettering and extrinsic gettering. Intrinsic gettering can be induced inside of bulk silicon during wafer fabrication. In contrast, extrinsic gettering is formed outside of the wafer. Subsurface damage caused by grinding is one of the extrinsic gettering layers.

As decreasing wafer thickness, stress relief methods should be optimized with considering both mechanical strength and gettering effect sites. Wafer thinning leads to a reduction in intrinsic gettering sites. Furthermore, when stress relief processes are used to remove grinding subsurface damage, extrinsic gettering effects are also lost. These two effects may lead to gettering issues for devices sensitive to metal contamination, such as flash memory. In order to solve this gettering issue, a method which can add extrinsic gettering effects after thinning is required. Gettering dry polishing or ultrafine grinding can create tiny damage features on the wafer backside and improve die strength. These processes have been introduced in the memory devices market.

17.4.4 Noncontact Gauge

When an in-process gauge contacts an ultrathin wafer during grinding, the contact stress may lead to wafer damage and cracking. The *noncontact gauge* (NCG) is capable of measuring the thickness of only the bulk silicon to reduce the risk of wafer breakage. NCG also contributes to improved wafer thickness accuracy. By measuring the thickness value of the bulk silicon, measurement with the NCG can eliminate other factors which may influence the thickness variation of the silicon, such as the thickness variations in the BG tape.

By swinging the NCG from the center to the outer edge of the ground wafer, the thickness profile of the ground wafer can be measured. This measurement value can be used to adjust the chuck table inclination to improve wafer TTV, which is also known as the Auto TTV function (refer to Sec. 17.9.3).

17.4.5 In-Line System

After grinding, the wafer is turned over and mounted on a tape frame using dicing tape. Then the BG tape is peeled off. When DAF is used instead of a conventional bonding adhesive, typically a 2-in-1 combined DAF and dicing tape is used. Wafers mounted onto tape frames are sent to the next dicing process where the wafer and DAF are cut into the appropriate die size.

As wafers become thinner, there is an increased risk of wafer breakage when the wafers are transferred between postgrinding equipment, such as the BG tape peeler and the dicing tape mounter. One of the proven solutions to reduce handling risks is integrating multiple units into an in-line system (refer to Fig. 17.3). The in-line system covers all of the process steps from grinding and stress relief to mounting and peeling. With an in-line system, the thin warped wafers will already be mounted securely on dicing tape when they are unloaded. This has recently seen widespread adoption of this in-line configuration for use in wafer thinning in the memory market.

17.4.6 Wafer Support System

Another solution for thin wafer handling is the *wafer support system* (WSS). WSS is a wafer handling method, which attaches a support material such as silicon or glass to the device surface of the wafer using an adhesive. This reduces both the risk of wafer breakage and wafer warpage. WSS is used in processes that require extensive handling of thin wafers, for example, in devices which have been constructed with *through silicon vias* (TSV) (refer to Sec. 17.9.2).

17.4.7 TAIKO Process

The TAIKO process, named after a Japanese drum, is a variation of the normal grinding process. The process leaves an edge of around 3 mm on the outer circumference, and only grinds the inner portion of the wafer, as shown in Fig. 17.5.[11] Because the outer support ring improves wafer strength and reduces wafer warpage, TAIKO processed wafers remain flat. It is an effective solution for handling thin wafers. Since the adhesive is not required, the TAIKO wafer can be used in high thermal process. The TAIKO process has been applied to power devices, such as insulated-gate bipolar transistors, which requires metal film deposition and thermal process after grinding.

17.5 SINGULATION TECHNOLOGY OVERVIEW AND BLADE DICING

Singulation is a process where semiconductor wafers are separated into individual die. The singulation process using diamond blades is also known as dicing. In Secs. 17.5 to 17.7, general dicing methods such as blade dicing, laser grooving, and *stealth dicing* (SD) will be discussed.

The concept of "dicing" here refers to the general blade dicing process for a mono-crystal silicon wafer after thinning via the backgrinding process. After thinning, the wafer is mounted on a dicing tape frame, with the device side of the wafer upward. The device layer on the wafer frontside is composed of dielectric and metal layers to connect internal transistors. A disc-shaped abrasive called a "dicing blade" is used to cut the wafer into individually separated die. After dicing, the wafer is proceeded to the next process called die bonding where the separated die are picked up from the dicing tape frame and placed on the lead frame or substrate.

17.5.1 Dicing Blades

Generally, a dicing blade, as shown in Fig. 17.6*b*,[12] is made of micro diamond grains (< ~5 μm) bonded together with nickel alloy plating. The size of disc-shaped blades (or hub) is approximately 50 mm, and the tip of the blade is 20 to 40 μm thick. In order to produce as many die as possible from one wafer, kerf widths must be minimized, so blades can be as thin as 10 μm.

17.5.2 Blade Dicing Process Point

Hub blades are mounted on a dicing spindle and rotated at around 20,000 to 50,000 min^{-1}. The blade cuts slightly into the dicing tape to

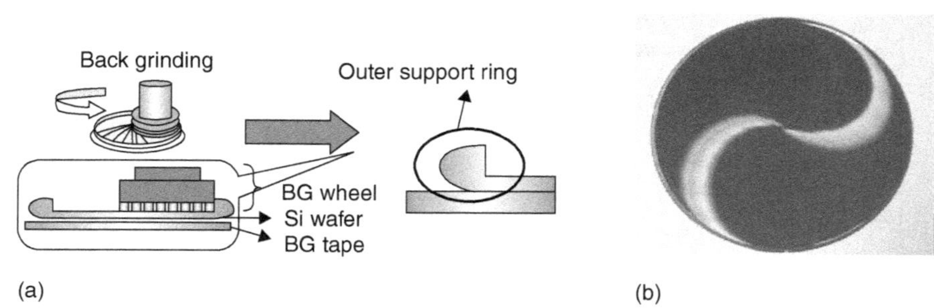

Figure 17.5 TAIKO: (*a*) TAIKO process image, (*b*) TAIKO wafer.

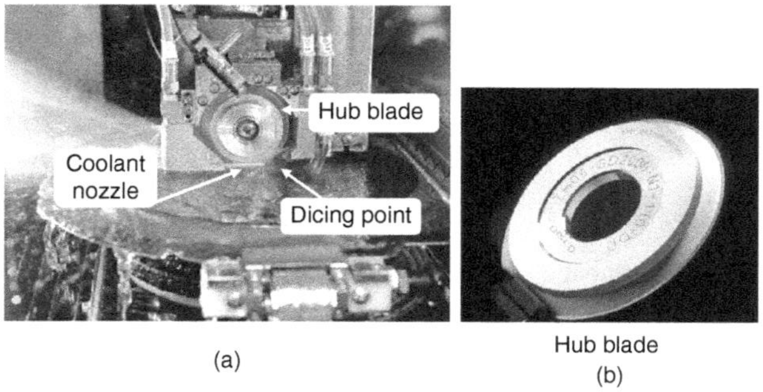

Figure 17.6 Dicing point: (*a*) image of dicing point, (*b*) hub blade.

ensure that the silicon wafer has been completely separated. This is what is known as a "full cut." The feed speed of the dicing blade varies between 50 and 75 mm/s, which is depending on the type of device wafer and criteria. During dicing, cutting water is provided to the dicing point for cooling and for removing silicon particles, as shown in Fig. 17.6*a*. The dicing quality is greatly affected by how effectively cutting water is provided to the actual dicing point.

17.5.3 Dicing Quality

The purpose of dicing is to singulate individual die from a wafer, with the least defects and maximum throughput. Typical evaluation criteria for dicing quality are as shown in Fig. 17.7.[13]

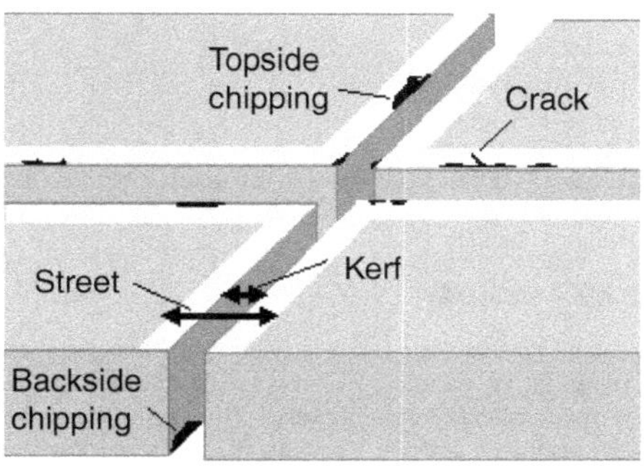

Figure 17.7 Basic terms for dicing process.

Chipping

When the rotating blade cuts through the silicon wafer, chipping is caused by the diamond grit within the blade. Generally, dicing streets are designed to be wide enough to fit both the kerf and chipping of up to a certain size. The blades have a "self-sharpening" characteristic to maintain constant dicing quality. However, if dicing conditions are not appropriate, sporadic chipping may occur due to lack of self-sharpening, and this may lead to device defects. Therefore optimization of the dicing conditions is crucial to high-volume production, where good yields are required.

Cracking

If the dicing conditions are not suitable to the wafer and the quality gets worse, cracking might occur in addition to chipping. It is caused by poor blade conditions, such as clogging or glazing. Since the cracking can be developed by heat or stress during post process and lead to the die breakage, process optimization is important to prevent yield loss.

17.6 SINGULATION PROCESS AND EQUIPMENT

17.6.1 Types of Dicing Saws

There are two types of dicing saw: fully automatic saws designed for mass production and semiautomatic saws designed for R&D and small-scale production. Dicing saws are also equipped with different numbers of spindles to suit their applications. Single spindle saws are designed for simple dicing and dual spindle saws are designed for advanced cutting methods.

Fully Automatic Dicing Saw

Fully automatic dicing saws include a handling unit for loading/unloading, auto alignment functionality, and a spinner unit in addition to the cutting function. Once the tape frame cassette is placed on the cassette stage, the fully automatic dicing saw will process all of the wafers in the cassette automatically without any operator assistance. The most commonly used model for 300 mm semiconductor wafer dicing is a dual spindle equipped fully automatic dicing saw.

Semiautomatic Dicing Saw

Semiautomatic dicing saws are basic dicing saws which only incorporate the cutting function. The tape frame–mounted wafer has to be loaded manually by the operator, and the cleaning unit is not equipped. Semiautomatic dicing saws are most commonly used for the production of the electronic components, which require less frequent load/unload (longer cutting time per wafer) as compared to silicon wafers.

17.6.2 Basic Functions of Fully Automatic Dicing Saws

The fully automatic dicing saw is the most commonly used tool in the singulation process.

Loading/Unloading

Wafers must be mounted on a tape frame using dicing tape, so that the singulated die can be transferred to the next process after dicing without separating into individual pieces. The wafers are loaded from the frame cassette automatically. The automated load/unload process is highly effective in improving the throughput in high volume manufacturing, especially for silicon wafers which require short cutting time.

Auto Alignment

In order to cut the wafer into individual die, the cutting position must be aligned. Combined with the pattern recognition system, the cutting position is detected and the blade kerf is adjusted to match the street by the dicing saw. This is called "auto alignment."

In the auto alignment, as shown in Fig. 17.8, the dicing saw recognizes a unique alignment target, which is made in wafer process, and adjusts the rotation angle of the wafer before confirming the cutting line.[14] This is to ensure that the dicing saw will cut the wafer as desired. The cut kerf is also recognized by the microscope camera, and the dicing saw applies this information to make automatic adjustments during the cutting process. This function is known as "kerf checking."

Wafer Cleaning at the Spinner Table

After blade dicing, the wafer surface will be contaminated with silicon particles and tape debris. The contamination must be removed from the wafer surface before it dries, as the contamination affects subsequent processes, including die bonding and wire bonding. Fully automatic dicing saws are equipped with a built in spinner cleaning unit to clean the wafer immediately after blade dicing. In the cleaning unit, tape frame mounted wafers are rotated and sprayed with high pressure or atomized water for cleaning, and then drying as final step.

17.6.3 Dual Spindle Applications

Recently, the most commonly used dicing tool is fully automatic dicing saw with dual spindle. The introduction of dual spindles has provided many types of blade dicing applications.

Dual Cut

The standard cutting method is called a "single cut." In single cut, the dicing saw cuts through the entire thickness of the wafer in one action using a single blade. In "dual cut" mode, identical blades are mounted to both spindles, and the dicing saw is able to cut two lines simultaneously

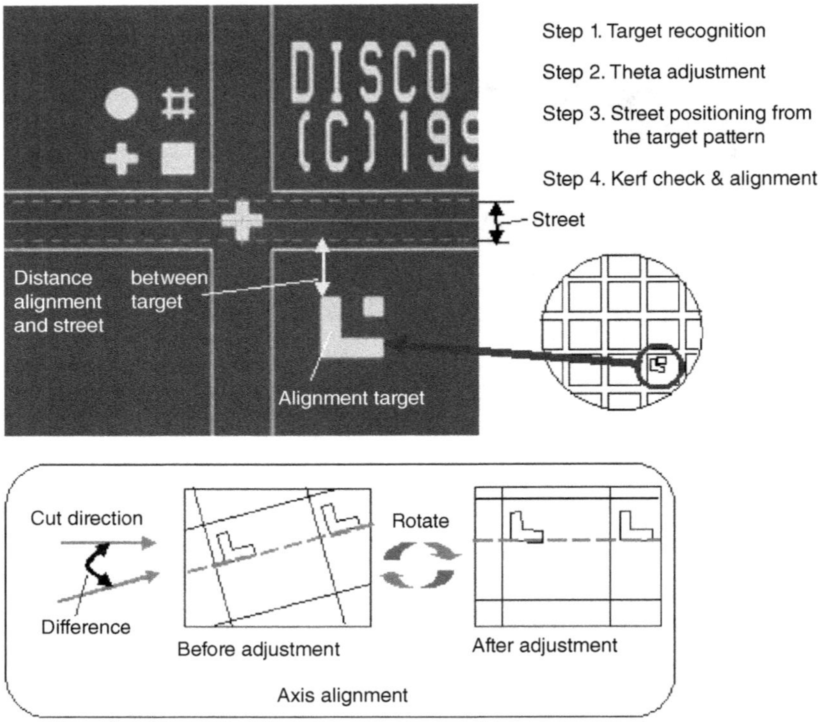

Figure 17.8 Auto alignment and kerf checking.

(Fig. 17.9*a*).[15] The cutting time for each wafer will be reduced by approximately one-half compared to single cut with a single spindle.

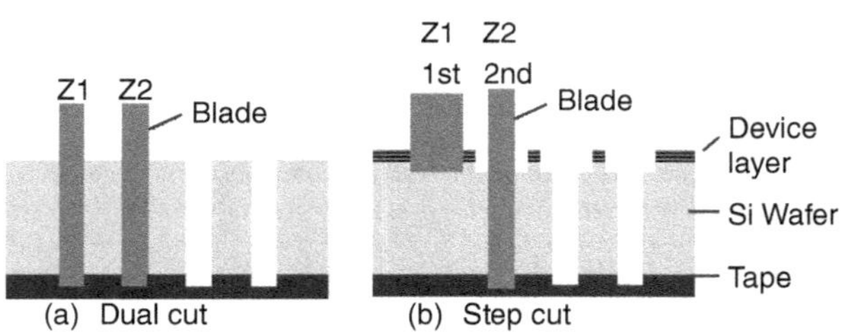

Figure 17.9 Dual spindle applications: (*a*) dual cut, (*b*) step cut.

Step Cut

Step cut is a cutting method which can improve dicing quality compared to the single- and dual-cut modes. A different type of blade is attached to each spindle and each dicing line is cut in two steps (Fig. 17.9*b*).[15] In the first step, a wide Z1 blade grooves the device layers (the TEG and surface layers), which are susceptible to chipping. In the second step, a narrow Z2 blade cuts through the silicon underneath to achieve separation. The blade types and cutting conditions can be designed for each step, such that the first step is optimized toward for cutting through the device layer and the second step is optimized for cutting through the silicon. Thus, step cut method is widely used to improve dicing quality with less frontside and backside chipping even if it requires longer process time than a single cut.

Other Dicing Methods

In addition to blade dicing, there are other methods, such as laser grooving/dicing, and advanced processes such as *dicing before grinding* (DBG). Details of these features will be explained in the following sections.

17.7 LASER TECHNOLOGY

Laser processing was introduced relatively recently for use in the semiconductor singulation process. Laser processing employs a laser beam instead of a dicing blade to achieve separation. It has been developed as an alternative way to solve the issues in blade dicing such as chipping and delamination. Currently, there are two types of laser dicing methods available in semiconductor manufacturing, one is laser ablation and another is SD. Details for both methods will be discussed in the following sections.

17.7.1 Laser Ablation

In the laser ablation process, a short-pulse laser is focused on the wafer surface, as shown in Fig. 17.10*a*.[16] The material vaporizes as it absorbs the energy at the focus point.

Ablation laser dicing has minimal thermal distortion with optimized process conditions. The process also exerts less mechanical load at the process point compared to that of conventional blade dicing.

17.7.2 Low-*k* Grooving

The most widely adopted laser ablation application is a low-*k* grooving process. Low dielectric constant (low-*k*) interlayer has been widely adopted to reduce signal delay instead of silicon oxide (SiO_2). The low-*k* insulators on the device layer suffer from poor mechanical strength due to its porous structure. The low-*k* insulator often suffers from delamination (peeling) during blade dicing even at low feed speeds. In order to solve the issue, the laser grooving has been introduced.

The device layer including the low-*k* insulator and metal interconnects is first removed by laser ablation grooving, and the remaining silicon is cut using blade dicing. This method prevents delamination,

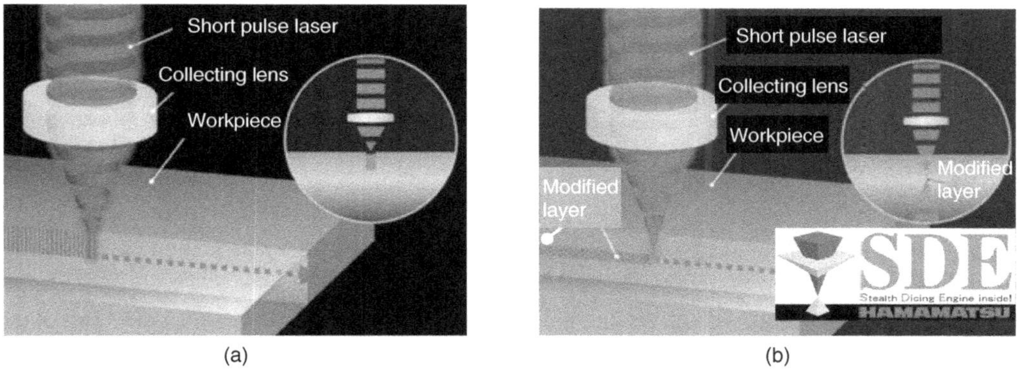

Figure 17.10 Laser process: (*a*) laser ablation, (*b*) stealth dicing.

and improves productivity by allowing a fast laser grooving feed speed and a fast blade dicing feed speed to be used.

17.7.3 Laser Full Cut

Laser dicing can be applied to cut both device layer and silicon, so-called laser full cut. The laser full cut process is effective for wafers such as brittle compound semiconductor materials (e.g., GaAs) or thin wafers with multiple layers. With laser full cut dicing, it is possible to achieve substantially higher feed speed, as well as better dicing quality with less chipping. This is a developing and promising technology for advanced dicing requirements.

17.7.4 Stealth Dicing

Stealth dicing (SD) is a process which forms a modified layer inside a wafer by focusing the laser inside the wafer, as shown in Fig. 17.10*b*.[16] Separation of the die is then achieved through tape expansion. Since the SD process only affects the material on the inside and then extends the cleaving to the surface without contact, it generates no chipping or debris. SD has several advantages that are quite effective for specific applications.

SD provides higher die strength, especially for dicing wafers thinner than 100 μm, with its chipping-less feature. It also is suitable for MEMS wafers, which has delicate surface structure vulnerable to the impact of cutting water or contaminations, because it is a perfectly dry process requiring no cutting water nor cleaning.

SD also greatly contributes to street reduction, because it requires a nearly zero kerf width. The number of die produced from a single wafer can potentially be increased, compared to blade dicing which requires a fixed street width for the kerf. It is particularly effective for wafers with small die sizes.

SD can achieve a higher feed speed than other processes in order to improve productivity while also providing greater singulation quality. As a result, SD process has been introduced into semiconductor manufacturing as a next generation dicing process.

17.7.5 Applications Potential of Laser Dicing

Laser dicing will become more flexible in the future due to its ability to cut nonlinear or noncontinuous lines unlike blade dicing. Both types of laser processes (ablation and SD) support this capability. This allows the singulation of wafers in a single process, even when multiple sizes of die are fabricated on a single wafer, as shown in Fig. 17.11.[17]

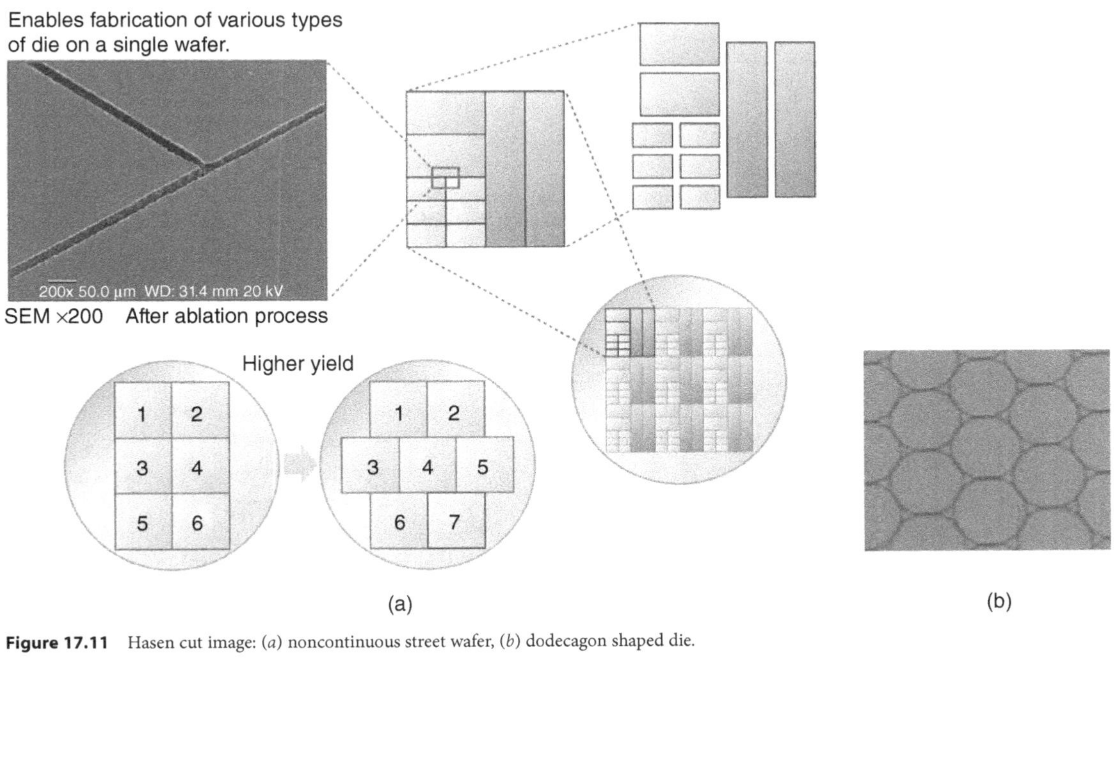

Figure 17.11 Hasen cut image: (*a*) noncontinuous street wafer, (*b*) dodecagon shaped die.

With its adaptive potential, laser dicing method has great potential and will likely be used more commonly in advanced semiconductor manufacturing in the future. However, the comparatively higher running cost (the laser oscillator and optical parts) is a factor. Once cost reductions are realized, laser dicing is projected to become a more mainstream option selected for singulation processes.

17.8 DBG AND SDBG

17.8.1 Dicing before Grinding

DBG reverses the conventional process order in which wafers are diced after grinding. In the DBG process, the device surface of the wafer is first grooved in a process known as half cut dicing. As the wafer is thinned from the backside, singulation occurs naturally once the thinning process reaches the bottom of the grooves, as shown in Fig. 17.12.

In the DBG process, because the wafer is already diced when it is thinned, wafer warpage is reduced and therefore there is a reduced risk of wafer breakage. Die singulation by backgrinding reduces the incidence of backside chipping and improves die strength compared to die singulation via normal blade dicing.

For wafers thinner than 100 μm, DBG can be combined with a stress relief process, such as dry polishing or gettering dry polishing. After grinding, the DBG wafer is mounted onto dicing tape and the BG tape is peeled away. If DAF is applied to the backside of the DBG wafer after die separation by grinding, laser scribing or an expander are used to separate the DAF. Currently, DBG is widely used in the memory device market as a process to produce thin die (30 to 100 μm).

17.8.2 Stealth Dicing before Grinding

Stealth dicing before grinding (SDBG) is part of the DBG process family. In the SDBG process, SD is performed instead of blade half cut dicing. After SD laser irradiation, the wafer is thinned and separated into die by tape expanding, as shown in Fig. 17.12. SDBG is suitable for thin wafers with narrow street widths of less than 60 μm. Since the SD process does not produce chipping, it tends to improve the die strength of thin wafers compared to that of blade dicing. Furthermore, in the SDBG process for ultrathin wafer, modified layer is formed above the final thickness so that the die is separated by the extension of the modified layer. Therefore, the modified layer can be removed by BG, leaving no mark on the die side at all. It is effective to achieve higher die strength for ultrathin die less than 30 μm thick.

17.9 3D INTEGRATION WITH THROUGH SILICON VIA

17.9.1 Advantage of 3D Integration

As mentioned in the introduction, economical, and technical limitations in current scaling trends have provided an opportunity for 3D integration with TSV to potentially become an alternative to conventional scaling technology for enhancing the performance and density of devices. Compared to 2D integration such as *system-on-chip* (SoC) technology, 3D integration has several features as shown in Fig. 17.13.[18] First, 3D integration using TSV vertical interconnects can reduce the length of interconnects from die to die, and consequently offer lower power consumption and higher bandwidth. Second, 3D integration using TSV makes vertical stacking possible. This is a promising solution for overcoming scaling limits and an alternative method to achieve high density. Third, vertical integration can reduce form factor, system volume, and footprint.

Currently there are three stacking methods: *chip-on-chip*,[19] *chip-on-wafer*,[20] and *wafer-on-wafer*,[21] as shown in Fig. 17.14. COC is based on chip-level processes and usually adopted for memory stacking in the early stages of 3D integration. For stacking memory blocks onto logic devices, COW is widely used because memory and logic devices often have different die sizes. WOW is a full wafer-level process and gives higher throughput than other stacking methods; however, there are yield considerations which need to be addressed before the method can be adopted. Choice of the appropriate stacking method is dependent upon many factors, such as cost, initial yield, and the type of device.

17.9.2 Process Integration

Major features of the *3D integration* (3DI) process compared to conventional thinning and singulation processes is the introduction of WSS, high precision TTV control, and post backside processes after thinning. There are three approaches for fabricating TSV in a 3DI process. They are known as via first, via middle, and via last.[22,23] For purposes of understanding total process integration, the via middle process, where

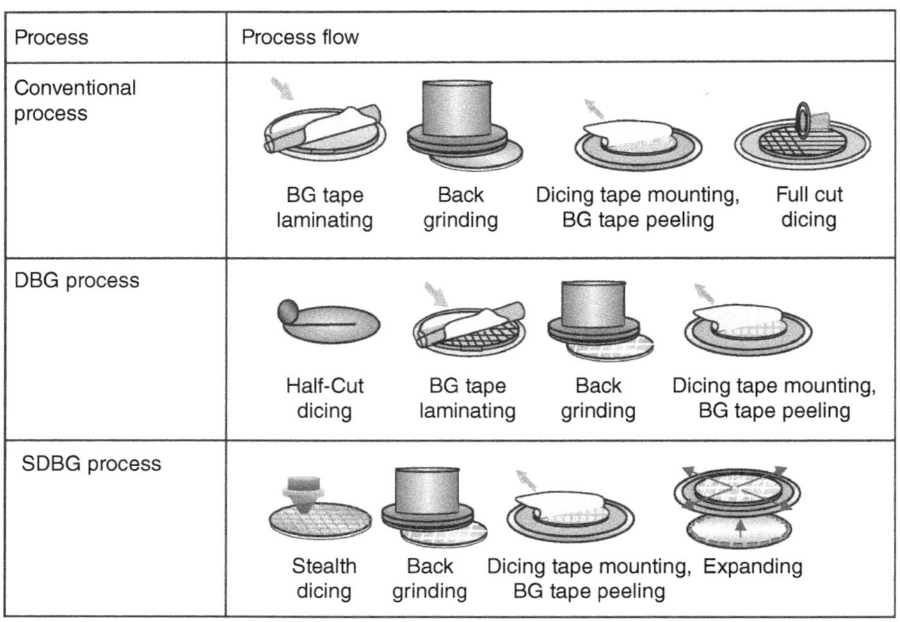

Figure 17.12 Conventional and DBG and SDBG process flow. (*Courtesy of DISCO Corporation.*)

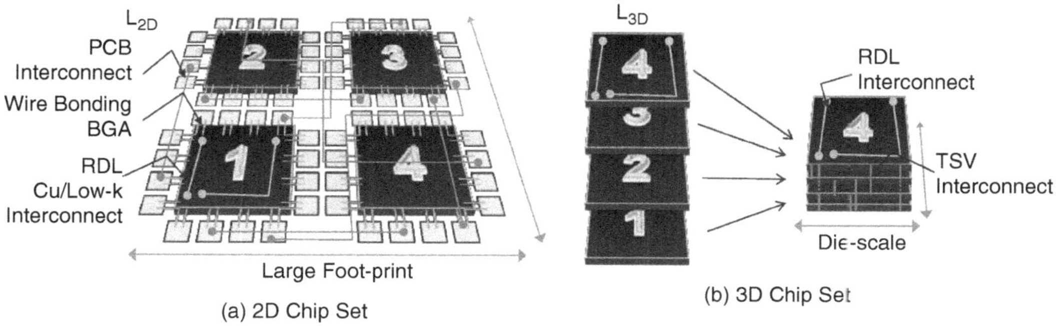

Figure 17.13 Comparison of (*a*) 2D integration (SoC) and (*b*) 3D integration using TSV.

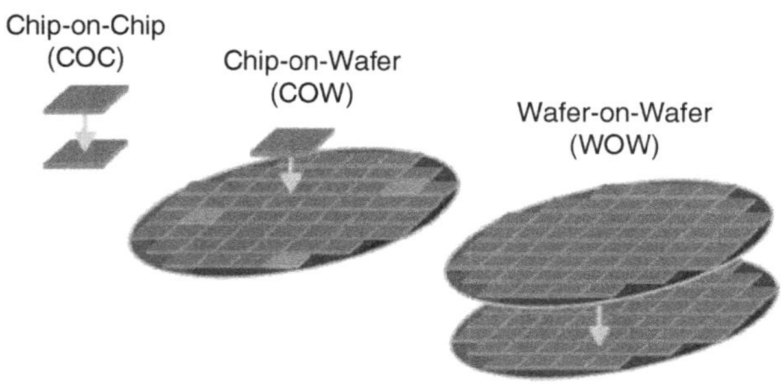

Figure 17.14 Types of stacking methods for 3D integration: COC, COW, and WOW. (*Courtesy of DISCO Corporation.*)

TSV are formed after the transistor process, will be explained as the process sees extensive use in 3DI, as shown in Fig. 17.15.

After formation of the TSV and device layers (a), edge trimming (b) is carried out to avoid edge chipping. Blade dicer is one of way to trim an edge (refer to Sec. 17.2.3). With the wafers temporarily bonded to a carrier substrate (c), that is, WSS (explained in Sec. 17.4.6), the wafer is thinned until just before reaching the TSV in order to prevent Cu contamination (d). The remaining Si is removed using a plasma or wet process (e). Next is backside process. This process is followed by the fabrication of a passivation layer and a planarization process (f–h) for this layer. After the thinned wafer is mounted onto a frame, the carrier substrate is debonded from the thinned wafer (i–j). Finally, singulation of the wafer is performed using a blade or laser (k). The next section will introduce several topics related to the thinning and singulation processes for these devices.

17.9.3 Thinning-Related Issues: Edge Trimming, Auto-TTV, and High-Level Cleaning

Edge trimming, which forms a straight wafer edge using a blade, is an effective way of preventing edge chipping, as mentioned in Sec. 17.2.3. In addition, the wafer thinning process in 3DI needs to take into account not only the edge chipping but also the potential of wafer breakage during the debonding process in which the thinned wafer and the carrier substrate are separated. Wafer breakage in this step directly affects the total yield and process cost.

More precise TTV control is very important, especially in 3D stacking devices, in order to control the amount of via protrusion. In addition, precise TTV control allows the process to be able to accommodate temporary adhesives, which often have a high TTV. High-resolution grinding processes for improving TTV, which use an automatic feedback process, have been developed. One of these processes is known as Auto-TTV using NCG[18] (refer to Sec. 17.4.4). Wafer thickness uniformity after grinding is determined by the parallelism between the wheel and the wafer surface, as shown in Fig. 17.16. After grinding away part of the Si wafer, the initial TTV is measured using an NCG. Using the initial TTV, the inclination is adjusted automatically, and the grinder proceeds to the final backgrinding. In this way, the TTV can be improved.

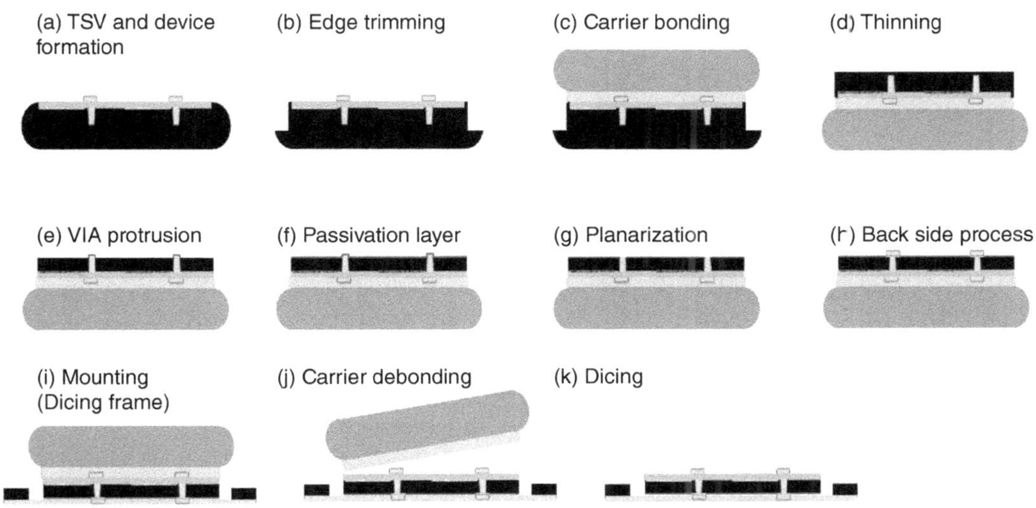

Figure 17.15 Schematic diagram of the via middle process. (*Courtesy of DISCO Corporation.*)

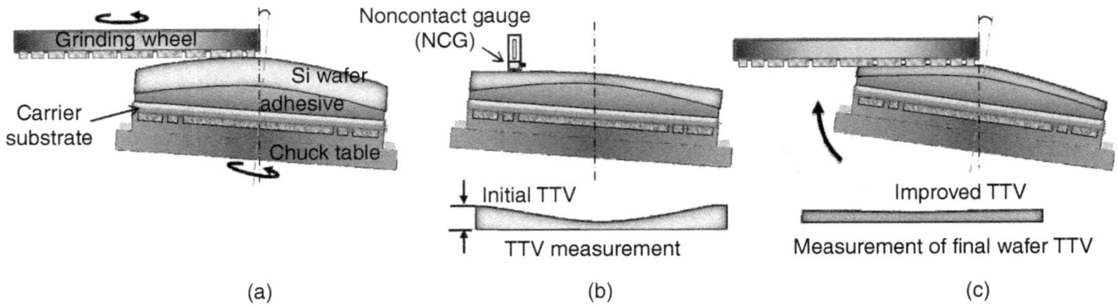

Figure 17.16 A schematic diagram of Auto-TTV process: (*a*) initial wafer grinding, (*b*) initial TTV measurement after stopped grinding, and (*c*) automatic TTV feedback and final grinding. (*Courtesy of DISCO Corporation.*)

Wafers with TSV devices also require a high level of cleanness after thinning for front-end processes such as film forming and etching. In order to remove silicon particles, organic residue, and metal impurities, additional cleaning steps such as brush cleaning, chemical cleaning, and diluted HF cleaning are usually performed after backgrinding.

17.9.4 Singulation Solutions for TSV Wafer

There are three cutting methods used for TSV wafers: cutting a single wafer into die for COC, cutting a wafer with molded die stack for COW, and dicing a wafer stack for WOW. Let us first examine the die singulation method for COC applications. The major issues when cutting a single TSV wafer into die are die strength and bumps on the wafer backside. Since many TSVs are formed vertically inside single die, residual stress is induced by Cu-TSV. As a result, die-containing TSVs have reduced die strength compared to die without TSV. Thus, more careful process optimization is required to avoid die breakage during cutting and die bonding. Wafers with TSV are typically fabricated to be very thin and have bumps on both sides. Backside bumps are a new challenge to both blade dicing and SD due to the difficulties it presents in how to mount the wafer to the dicing tape, as shown in Fig. 17.17. If the wafer with bumps is not held by the dicing tape with sufficient force, cracking may occur on the wafer during dicing. The mounted wafer must remain stable in order to maintain cutting quality in blade dicing. Dicing tapes with thicker adhesive are effective in accommodating the backside bumps, but these tapes may leave adhesive residue on the die backside, and they require a greater pick up force, which can lead to die breakage. Total optimization of both dicing conditions and tape selection is the key to solving the above issues.

SD is one of the ways to solve the above die strength and bump absorption issues. SD can improve die strength of thin wafers because it is a chipping-free process as described in Sec. 17.7.4. Since the SD process exerts a lower mechanical force on the wafer and causes less sidewall damage, it is not necessary to fully absorb the bumps into the tape during the singulation process. As an example, two types of SD process with a different debonding sequence have been illustrated in Fig. 17.18. SD before debonding is a simple process (Fig. 17.18*a*), where the stealth laser irradiates the wafer from the backside. The wafer is mounted to the dicing tape and debonded after the SD process. All of the dies are then separated via a tape expansion. Since conventional dicing tape has low thermal resistance, heat treatment during the debonding process is limited. Another method of using SD is to perform SD after the debonding process (Fig. 17.18*b*), so the results of SD will be independent of the debonding temperature. The stealth laser must be used to irradiate the backside of the wafer through the dicing tape because the laser cannot penetrate scribed lines with metal interconnects. For this process, the tape selection as well as the laser conditions should be optimized. The selected tape and laser conditions will also depend on the temporary bonding process.

A major issue for the singulation of COW and WOW is the dicing of multilayers, which can be combinations of silicon, organic, and metal. Stacked wafers or mold-encapsulated packages often contain silicon, mold compounds, and filling materials between die. Filling materials can include underfill, *nonconductive film* (NCF), and *nonconductive paste* (NCP), which is a mixture of resin and filler. Although blade dicing would be the more suitable method in these cases due to the limitations of SD laser material penetration reducing, the low process load characteristics of blade dicing are required for achieving good process quality. Another issue for the singulation of COW and WOW is the dicing of wafers with low-*k* layers especially logic devices. Each individual wafer may require laser grooving or full cut via laser ablation. These processes are still under development.

	Thicker glue dicing tape	Thinner glue dicing tape
	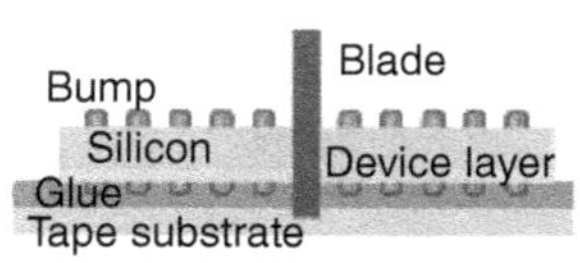	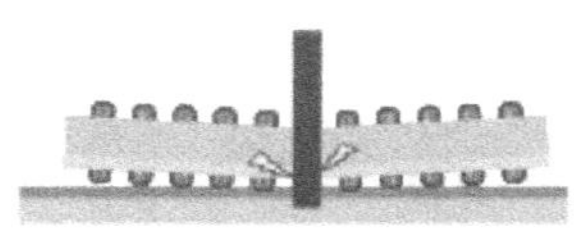
Pros	- good wafer fixing	- usage of conventional tape
Cons	- glue residue - die breaking during die pick up	- die cracks during dicing

Figure 17.17 The pros and cons of dicing tape type. (*Courtesy of DISCO Corporation.*)

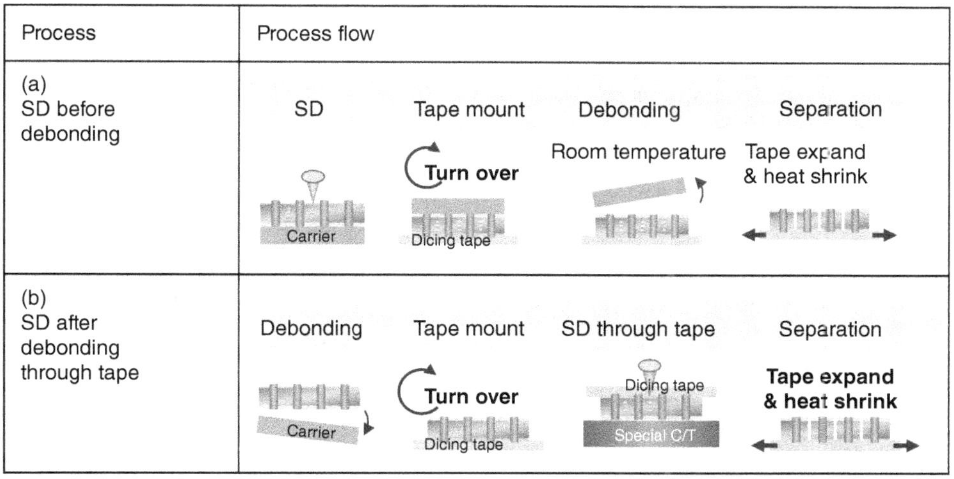

Figure 17.18 A schematic diagram of the SD process. (*Courtesy of DISCO Corporation.*)

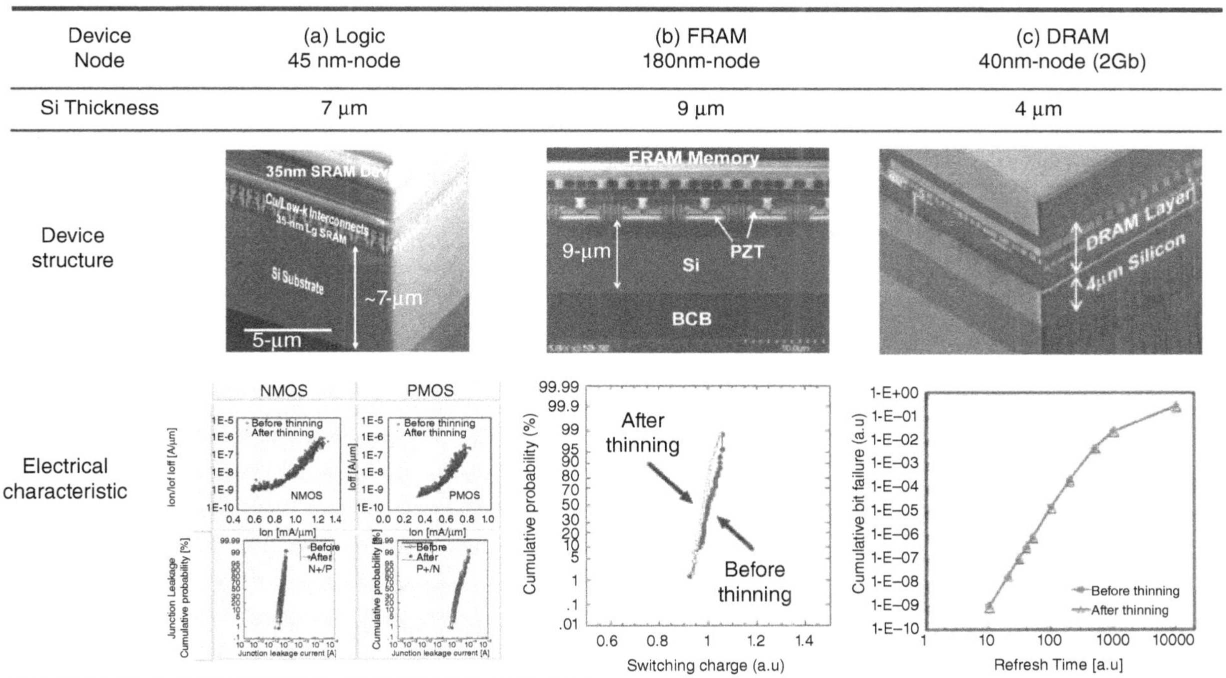

Figure 17.19 Summary of the impact of ultra-thinning on device characteristics: (*a*) logic, (*b*) FRAM, and (*c*) DRAM.

17.9.5 Impact of Wafer Thinning on Device Characteristics

This section introduces the effects of wafer thinning on device characteristics, including mobility, gettering, switching charge, and retention for CMOS logic, ferroelectric RAM (FRAM), and DRAM wafers.[24–26] A 300-mm wafer with 45-nm node high-performance logic devices containing strained transistors and Cu/low-*k* multilevel interconnects (Fig. 17.19*a*) was analyzed. A silicon wafer was bonded onto a support substrate and thinned down to 7 μm by grinding and *ultra-poligrind*. It was reported that the drive current and the carrier mobility of the ultrathin wafer were nearly the same as the wafer before thinning. In addition, there was no degradation of the interconnect characteristics, including line and chain resistances. The impact of the wafer thinning process on FRAM is shown in Fig. 17.19*b*. A FRAM wafer was thinned to 9 μm using CMP after BG. The switching charge after thinning was similar to that of initial data, and no significant failures were found within the wafer. DRAM characteristics as a function of final Si thickness from 40 to 4 μm were examined by comparing retention time and distribution before and after thinning. For all Si thicknesses, there are no significant changes before and after thinning. These results show that high performance 3D devices can be realized by using ultra-thinning technology. These devices can provide large IO and low process costs through the use of TSV with low aspect ratio or small size.

17.10 REFERENCES

1. A. Bums, B. F. Aull, C. K. Chen, C. L. Chen, C. L. Keast, J. M. Knecht, V. Suntharalingam, K. Warner, P. W. Wyatt, D. R. W. Yost, "A wafer-scale 3-D circuit integration technology," *IEEE Trans.* Electron Devices, Vol. 53, no. 10, 2507–2516, Oct. 2006.

2. M. Koyanagi, H. Kurino, K. W. Lee, K. Sakuma, N. Miyakawa, and H. Itani, *IEEE Micro.*, 18 (4), 17–22, 1998.

3. T. Ohba, N. Maeda, H. Kitada, K. Fujimoto, K. Suzuki, T. Nakamura, A. Kawai, and K. Arai, *Microelectron. Eng.*, 87 (3): 485–490, 2010.

4. ITRS Report 2012 UPDATE (see table AP2 in AssemblyPkg_2012Tables.xlsx), ITRS, 2012. http://www.itrs.net/ITRS%201999-2014%20Mtgs,%20Presentations%20&%20Links/2012ITRS/Home2012.htm (accessed Apr. 24, 2015).

5. H. Geng, *Semiconductor Manufacturing Handbook*, McGraw-Hill companies, Inc., New York, 20.2, 2005.

6. DISCO Corporation, "IF Series," http://www.disco.co.jp/eg/products/grinding_wheel/if.html (accessed Apr. 24, 2015).

7. SEMI M1-0414, "Specifications for Polished Single Crystal Silicon Wafers," *SEMI International Standards*, U.S.A., 2014.

8. SEMI G86-0303 (Reapproved 0811), Test Method for Measurement of Chip (Die) Strength by Mean of 3-Point Bending, U.S.A., 2008.

9. SEMI G96-1014, Test Method for Measurement of Chip (Die) Strength by Mean of Cantilever bending, U.S.A., 2014.

10. H. Geng, *Semiconductor Manufacturing Handbook*, McGraw-Hill Companies, Inc., 20.16, New York, 2005.

11. DISCO Corporation, "Taiko Process," http://www.disco.co.jp/eg/solution /library/taiko.html (accessed Apr. 24, 2015).

12. DISCO Corporation, "ZH05 series," http://www.disco.co.jp/eg/products/blade/zh05.html (accessed Apr. 24, 2015).

13. H. Geng, *Semiconductor Manufacturing Handbook*, McGraw-Hill Companies, Inc., 20.8, New York, 2005.

14. H. Geng, *Semiconductor Manufacturing Handbook*, McGraw-Hill Companies, Inc., 20.13, New York, 2005.

15. H. Geng, *Semiconductor Manufacturing Handbook*, McGraw-Hill Companies, Inc., 20.11, New York, 2005.

16. DISCO Corporation, "Product lineup," https://www.disco.co.jp/eg/products/catalog/pdf/product_lineup.pdf (accessed Apr. 24, 2015).

17. DISCO Corporation, "Laser Application," https://www.disco.co.jp/eg/products/catalog/pdf/laser.pdf (accessed Apr. 24, 2015).

18. Young Suk Kim, Nobuhide Maeda, Hideki Kitada, Koji Fujimoto, Shoichi Kodama, Akihito Kawai, Kazuhisa Arai, et al., *Microelectron. Eng.*, 107: 65–71, 2013.

19. G. Katti, M. Stucchi, K. de Meyer, and W. Dehaene, "Electrical modeling and characterization of through silicon via for three-dimensional ICs," *IEEE Trans. Electron Devices*, 57 (1): 256– 262, 2010.

20. L. Lin, Tung-Chin Yeh, Jyun-Lin Wu, G. Lu, Tsung-Fu Tsai, L. Chen, and An-Tai Xu, "Electronic Components and Technology Conference (ECTC)," 355–371, Las Vegas, NV, 2013.

21. T. Ohba, "3D Large Scale Integration Technology using Wafer-on-Wafer (WOW) Stacking," *Interconnect Technology Conference (IITC)*, 1–3, San Francisco, CA, 2010.

22. S. Olson and K. Hummler, "TSV reveal etch for 3D integration," 3D Systems Integration Conference (3DIC), 1–4, 2011.

23. H. S. Kamineni, S. Kannan, R. Alapati, S. Thangaraju, D. Smith, Dingyou Zhang, and Shan Gao, "Challenges to via middle TSV integration at sub-28 nm nodes," *Proc. IEEE*, Int, Interconnect Technology Conference/Advanced Metallization Conference (IITC/AMC), 199–202, Osaka, Japan, 2014.

24. Y. S. Kim, A. Tsukune, N. Maeda, H. Kitada, A. Kawai, K. Arai, K. Fujimoto, et al., "Ultra Thinning 300-mm Wafer down to 7-μm for 3D Wafer Integration on 45-nm Node CMOS using Strained Silicon and Cu/Low-k Interconnects," *IEDM Tech. Dig.*, 365–368, San Francisco, CA, 2007.

25. N. Maeda, Y. S. Kim, Y. Hikosaka, T. Eshita, H. Kitada, K. Fujimoto, Y. Mizushima, et al., "Development of Sub 10-μm Ultra-Thinning Technology Using Device Wafers for 3D Manufacturing of Terabit Memory," *IEEE Symp. VLSI Technol*, 105–106, Honolulu, 2010.

26. Y. S. Kim, S. Kodama, Y. Mizushima, N. Maeda, H. Kitada, K. Fujimoto, T. Nakamura, et al., "Ultra Thinning down to 4-μm using 300-mm Wafer proven by 40-nm Node 2Gb DRAM for 3D Multi-stack WOW Applications," IEEE, *VLSI Technol.* 26–27, Honolulu, 2014.

Chapter 18
Packaging

BY
MICHAEL TÖPPER *Fraunhofer IZM, Berlin, Germany*
DIETRICH TÖNNIES *ROFIN-SINAR Laser GmbH, Bergkirchen, Germany*

18.1 INTRODUCTION

Electronic packaging and assembly are needed to link the small dimensions of the IC to an interconnecting substrate—usually the printed circuit board (PCB). The PCB combines a multitude of ICs and passive components to build a microelectronic system. New device technologies and applications with their ever increasing performance and functionality are driving the requirements and innovation for assembly and packaging. The technology boundaries between semiconductor technology, packaging, and system design are becoming blurred. As a result chip, package, and system designers will have to work closer together than ever before in order to drive the performance for future microelectronic systems. Flip chip in package (FCIP), flip chip on board (FCOB), wafer level packaging (WLP), and system in package (SiP) are among those new packaging technologies that will enable the industry to meet future system requirements. Fan-out wafer level packaging (FO-WLP) is getting the hottest topic in the area of advanced packaging. It is currently the only process to span the limits of all existing chip packaging technologies with the potential of connecting multidies to heterogenous systems or SiP. Main advantages of FO-WLP are the substrate-less package combining lower thermal resistance with higher performance due to shorter interconnects together with direct IC connection by thin film metallization instead of wire bonds or bumps. Especially the inductance of the FO-WLP is much lower compared to flip chip ball grid array (FC-BGA) packages. In addition, cost reduction for electronic packages is a key target for the next generation of electronic products like the IoT (Internet of Things) having an extreme growth potential.

Generally, economic and technical considerations are driving packaging toward adopting wafer level processes such as wafer bumping and rerouting. The focus of this chapter is therefore on these new technologies keeping the description of the more conventional die level packaging deliberately short. Standard text books on electronic packaging are citied at the end of this chapter for more details on conventional package types.

18.1.1 Basics of Electronic Packaging and Assembly

The Role of Electronic Packaging

Packaging technology determines size, weight, ease of use, durability, reliability, performance, and cost of electronic products. The package has to protect the IC, support the IC performance (operating speed, power, signal integrity, etc.), handle the thermal management and has to compensate for stress between IC and PCB by providing a mechanically and electrically reliable interconnect. For the assembly of complex microelectronic systems, it is important that individual ICs can be tested and burned-in.

The majority of ICs are packaged in so-called single-chip packages, where the IC is electrically, mechanically, and most often thermally linked to a carrier (leadframe, interposer) which spreads the IC contacts to the outside world such that it can be assembled to the PCB. Additional plastic materials often protect the IC from mechanical damage. The increasing speed of modern IC generations calls for improved electrical performance and heat dissipation of packaging. The growing semiconductor content in electronic equipment as well as the growing demand for mobile equipment both require a reduction of the package form factor.

A further reduction in size and weight together with a performance increase can be reached by multichip packaging. An early, but commercially not very successful, adoption of this idea was the multichip module (MCM). In an MCM, bare dice are assembled on a substrate and the entire assembly is packaged. Today, the packaging industry focuses on SiP technology as an alternative multichip packaging approach. In an SiP ICs, active and passive components are combined and being packaged as a whole. Currently FO-WLP has very high potential for this kind of integration. The result is a complex microelectronic system with high functionality and better performance and smaller form factor than conventionally assembled systems. The challenge, on the other hand, is to set up a manufacturing process that provides SiP at comparable or even lower prices.

Interconnect Technologies

Bare die is assembled to a substrate or a carrier/leadframe either with its face up or its face down. Today, the majority of packages require a face-up assembly to leadframes or organic or ceramic interposers which are then assembled to the PCB (Fig. 18.1). Flip chip bonding, on the other hand, is by definition a face-down assembly technique and can be used to connect the die to a leadframe or interposer or directly to the PCB. Direct bonding of a die to a PCB (flip chip, TAB, or wire bonding) is called chip on board (COB) or direct chip attach (DCA).

The metal–metal interconnect can be made by soldering, welding, or gluing. Three major technologies have been used over the past 40 years: wire bonding, flip chip, and tape automated bonding (TAB).

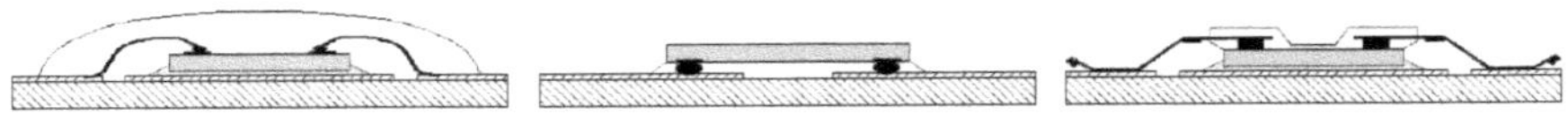

Figure 18.1 Wire bonding, flip chip, and TAB are the main interconnection technologies.

Wire bonding is the most common bonding technique. The chip is glued or soldered face-up onto a substrate or a carrier and the interconnection is made by thin gold or aluminum wires using ultrasonic or thermosonic techniques. Wire bonding is capable of fine pitch bonding and is in principle applicable with all peripheral pad pitches required by the International Technology Roadmap for Semiconductors (ITRS) roadmap (see Table 18.4). Any deformation forces are spread over the whole bottom side of the die and hardly affect the active chip area. A high degree of automation with image recognition systems have reduced the bonding time to less than 1/10 of a second per wire. Although a sequential process where each bond is made after the other wire bonding is cost competitive to flip chip and TAB, where all interconnects are made at a time. Thermosonic Au wire ball bonds between a leadframe and the chip pads is the interconnection technology for the most popular IC package: the molded plastic package.

Flip chip assembly is a face-down assembly technique originally developed by IBM and known as C4 (controlled collapse chip connection). Flip chip provides excellent performance and represents a cost effective interconnect technique for dice with high I/O counts. An overview of the different kinds of flip chip assembly is given in Fig. 18.2 [1].

A major requirement for most of the flip chip interconnections are modified pads on the IC. The so-called under bump metallization (UBM) or ball limiting metallurgy (BLM) is the basis for a low-ohmic electrical, mechanical, and thermal contact between chip and substrate. Two solutions were developed by IBM to prevent the collapse of the interconnect during soldering: Imbedding of a nonmelting Cu ball or a nonwettable surface around the pads of IC and substrate. Depending on the interconnect pitch a flip chip bonder or a highly accurate surface mount technology (SMT) assembly unit are used. A camera system or a split-field optic is used for the alignment of chip to substrate. The major advantage of flip chip assembly is the self-alignment function. Chips can be misregistered as much as 50 percent off the pad center and the surface tension of the molten solder will align the pads of the chip to the substrate metallization. The disadvantage of flip chip assembled ICs is that the bumps are the only mechanical links between chip and substrate. As a consequence, the stress caused by the CTE (coefficient of thermal expansion) mismatch of the semiconductor die and the substrate act only upon the bump interconnects. Therefore, underfillers (epoxy resins with filler particles) have to be filled in between the flip chip and the substrate which translates into extra costs in the assembly process.

Tape automatic bonding (TAB) uses polymeric tape with copper beam leads on a strip for the interconnect. These copper beams with an electroplated Sn or NiAu layer are aligned to Au-bumped dice (see the later section "Wafer Bumping Overview") and all interconnects are made in one step using thermo-compression bonding (gang bonding). TAB is a very common assembly technique for LCD drivers where a minimum pitch of 45 to 50 μm is currently in leading edge production. The minimum achievable pitch of TAB is limited by the mechanical strength of the Cu leads and is considered to be around 30 to 35 μm. Another limiting factor of TAB is the cost of the high-density beams.

ASSEMBLY

Soldering is the standard interconnect technique for assembling packaged semiconductor ICs to a PCB. Soldering allows easy repair of defective boards by desoldering defective components and replacing them with functional ones. The layout of the PCB depends heavily on the interconnection technology as described in Table 18.1 [2].

Through-hole technology (THT) was common until the late 1980s. With through-hole technology the PCB consists of a board with mechanically drilled and copper plated holes. The packaged IC has long connector pins along two sides of its periphery. The IC is assembled to the PCB by putting the pins through the holes of the PCB and soldering them at the bottom side of the board. Throughout the 1990s the higher complexity and miniaturization of electronic products has been nearly totally replacing the THT by SMT due to the higher routing capacity of

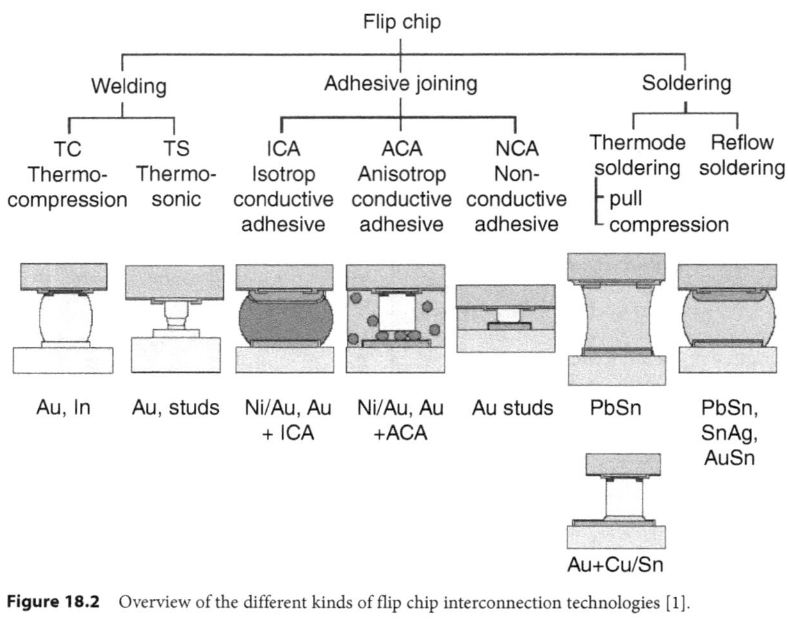

Figure 18.2 Overview of the different kinds of flip chip interconnection technologies [1].

Table 18.1 Overview of PCB Layout Rules

Technology	Interconnection pitch (mm)	Line width and spacing (μm)	Via hole diameter (μm)
Through hole	2.54	300	600
Standard SMT	1.27	150	300
Standard fine pitch SMT	0.5–0.635	60–75	130–150
High end fine pitch SMT	0.4	50	100
Extreme fine pitch SMT	0.3	35	80
Flip chip	0.2	25	50

SMT boards, higher assembly throughput and the possibility of double-side assembly. The first step for the assembly of an electronic system using SMT is to print solder onto the PCB for the surface mount devices (SMDs). With an automatic pick and place machine the fluxed SMDs are placed into position on the substrate. Reflow in a convection and IR-heating oven completes the assembly process. Unlike THT devices high pin count SMDs often have an area array pin layout. This allows designing packages with small form factor.

PCB roadmaps describe minimum line width and spacing and are driven by increasing IC pin counts and miniaturization requirements. Table 18.2 summarizes the ITRS roadmap for ball grid array (BGA, see Sec. 18.2.1) and chip scale package (CSP, see Sec. 18.2.2) boards.

Table 18.2 ITRS Roadmap for BGA and CSP Boards

Year	Ball size (μm)	Pad size (μm)	Lines (μm)	Space (μm)	Number of rows accessed
2003	400	160	48	48	3
2004	400	160	48	48	3
2005	300	120	36	36	3
2009	80	80	24	24	3
2015	60	60	18	18	3

For flip chip assembly even higher density substrates are needed. The number of I/Os per die has exponentially increased from around 500 I/Os per cm for 90-nm-node technology to over 10,000 I/Os per cm for the latest 14-nm technology. Therefore, FC assembly has been adapted mostly since 45 nm CMOS. The geometric structures for FC boards, however, can vary strongly by applications (low cost versus high-density boards for high frequency) and it is difficult to predict routing density trends. In addition, new materials will be needed for the above-GHz range. In the future low-cost PCBs for high-density applications will be a major bottleneck. Possible solutions are few-chip-BGAs/SiPs, where high-density wiring is only needed for some parts of the system. With this kind of modular system standard boards can be used as interconnection substrate of SiPs in order to save cost.

Green manufacturing has an important impact on chip packaging and assembly. As an example, the European Union has issued the "Restriction of Hazardous Substances in Electrical and Electronic Equipment (RoHS)" Directive in 2003. This directive bans the use of lead, mercury, cadmium, hexavalent chromium, polybrominated diphenyl ether from new electrical and electronic equipment, starting from July 1, 2006. Other countries have similar programs. In particular the ban of lead has important consequences for packaging and assembly because of the common use of PbSn solders. Figure 18.3 shows other areas of interest for green manufacturing in chip assembly such as the availability of halogen-free substrates and underfills. Some soldering processes, in particular high-lead flip chip for microprocessors, have reached exemptions from the ban until 2010.

For system assembly several lead-free solders are in use already. SnAgCu, SnCu, SnAgBiIn, and SnZnBi are a few examples. All these lead-free solders are rich in tin and most of them have a higher melting point than eutectic PbSn. Therefore, packages are exposed to higher temperatures during soldering which has consequences on the selection of packaging materials and the reflow process. Lead-free solders do not have as good a wetting behavior as PbSn and show different electromigration characteristics than PbSn. All this creates risks in the transition from lead to lead-free electronics.

18.1.2 Packaging Forecast

The ITRS Packaging Technology Roadmap

As the complexity of devices and systems is rapidly increasing new packaging technologies play a major role for the electronic industry. Packaging roadmaps are caught between IC roadmaps and the trends in the PCB industry. In the last few years through-hole mounting packages (DIP, PGA, etc.) were displaced by surface mount packages (TSOP, PQFP, etc.). The steady progress in IC technology increases density, speed, power, and number of I/Os. For example, gate density on a chip for high-end applications is increasing by about 75 percent per year resulting in the growth of IC chip I/O terminals by 40 percent per year. Along with the national roadmaps published by organizations such as the Semiconductor Industry Association (SIA) in the United States and the Electronic Industries Association of Japan (EIAJ), many IC companies have their own roadmaps describing the evolution of the physical parameters of ICs and electronic packaging. These roadmaps are top-down approaches anticipating changes in integrated circuit technology guiding as well the requirements for packaging. Several reasons could lead to significant changes in such roadmaps, for example completely new IC technologies, adoption of new packages or new products changing the electronic marketplace.

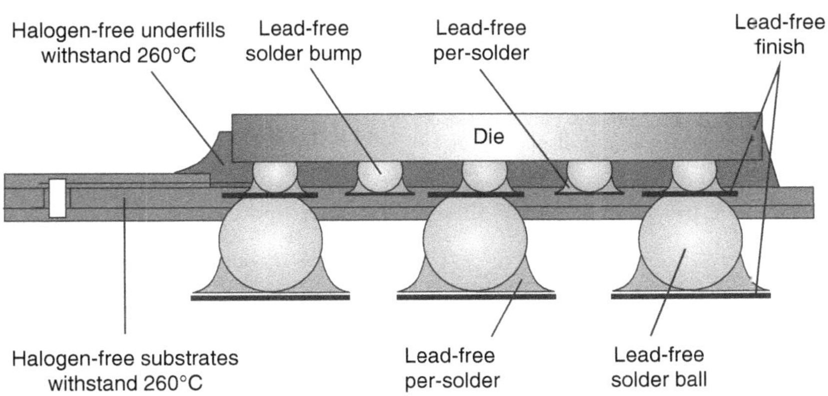

Figure 18.3 Green product legislation impact on packaging materials [3].

The following statements are based on the ITRS (2003 edition):

Electronic products are grouped together by standardized categories. These groups are:

- *Low cost/handheld.* Less than $500 consumer products, wireless products, disk drives, and displays
- *Cost performance.* Less than $3000 notebooks, desktop personal computers, telecommunications
- *High performance.* Greater than $3000 high-end workstations, servers, avionics, supercomputers, most demanding requirements
- *Harsh.* Under-the-hood and other hostile environments

The package pin count forecast for the different electronic application as proposed by ITRS is given in Table 18.3.

Table 18.3 ITRS Forecast of Pin Counts for Different Applications

Year/ application	Low-cost/ handheld	Cost-performance	High-performance	Harsh
2005	134–550	550–1760	3400	550
2010	208–777	780–2782	4009	642
2015	325–1213	1216–4339	6402	933
2018	421–1576	1581–5642	8450	1235

Since 2003, the packaging requirements have been grouped by only four product categories. These application areas cover the majority of the product development of the semiconductor industry. The technology addressed in the roadmap provides at least 80 percent of the revenue in each application area. The low-cost and handheld segments have been combined into a single category based on the realization that there will not be a significant difference in the cost or major performance requirements in these segments any longer.

The interconnection pitch is reducing in response to the increasing number of I/Os (Table 18.4).

Table 18.4 Forecast of Interconnect Pitch (on the Die) for Different Interconnect Technologies

Year/ interconnect	Wire bond ball (μm)	Wire bond wedge (μm)	TAB (μm)	Flip chip area array (μm)	Flip chip peripheral (μm)
2003	40	30	35	150	60
2004	35	25	35	150	60
2005	25	20	30	130	40
2010	20	20	20	100	20
2015	20	20	15	80	15

The wire bond pitch includes only inline configurations. Staggered pads will relax these constraints. In addition, larger pitch could be required for extremely high-current applications. Conductive adhesive flip chip is not addressed separately but may have smaller pitches for small die applications, provided high-density substrates with competitive cost are available.

Flip chip will require fan-out routing on the package substrate. Signal lines will be on the outer rows of area array to minimize package inductance. Inner balls will be used for voltage and ground to minimize voltage drop across the IC. The design will benefit from smaller bumps and reduced pitch. But higher substrate cost, substrate warpage, assembly issues, test, and electromigration have to be considered.

The ITRS identifies the following topics as the difficult future challenges:

- Improved organic substrates
- Improved underfillers for flip chip on organic substrates
- Coordinated design tools and simulators to address chip, package, and substrate co-design
- Impact of Cu/low κ on packaging
- High-current-density packages
- Package cost does not follow the die cost reduction curve
- Small die with high pad count and/or high power density
- High frequency in general
- System-level design capability to integrated chips, passives, and substrates
- New device types (organic, nanostructures, biological) that require new packaging technologies

Although assembly and packaging cost is expected to decrease over time on a cost-per-pin basis, the chip and package pin count is increasing more rapidly than cost per pin is decreasing. This explosion in pin count raises not only the absolute cost of assembly and packaging on a per-chip basis, but also the substrate and system-level packaging cost. Pin count will continue to increase in all segments while die size is expected to remain constant.

While Moore's law continues to move forward scaling in feature size to deep sub-micrometer nodes, the performance, power, bandwidth, and cost are no longer scaling at the same rate. New materials and new device types will be necessary to address this shortfall. Advances in assembly and packaging technology have helped take up the slack with new package types with improved functional density enabled by new packaging materials and processes. In 2012, it became evident that the ecosystem of the electronic industry had undergone a major change with the proliferation of the Internet, the worldwide spread usage of wireless mobile devices like smartphones and tablets and, most of all, the Internet had evolved from the IoT to the Internet of Everything. Sensors of all kinds continue to be added to the Internet and the range of remote operations is increasing by the day allowing remote operation of tools and remote medicine as an example. To fully address these changes the process of restructuring the ITRS was initiated in 2012 and completed in 2014. The 17 International Technology Working Groups (ITWGs) were replaced in 2015 by 7 focus teams in the new ITRS 2.0 (http://www.itrs2.net/). These are system integration, heterogeneous integration, heterogeneous components, outside system connectivity, more Moore, beyond CMOS, and factory integration.

Forecast by Package Type

Figure 18.4 shows the evolution of the different package types over the years with a forecast until 2020.

Until the early 1990s, through hole packages accounted for the majority of IC packages. Through the 1990s SMT packages largely replaced through hole packages. Low pin count peripheral packages such as small outline package, leadless chip carrier package, quad flat package (QFP) dominate. Especially peripheral packages with near CSP size such as the quad flat no-lead (QFN) package are rapidly gaining importance. But the share of area array packages such as BGAs and CSPs is growing. The adoption of flip chip in the form of direct chip attach, wafer level packaging or flip chip in package and 3D will continue at a rapid pace.

Forecast by Product Type

Figure 18.5 shows a breakdown of flip chip by product category. Flip chip represents high-end packages in terms of lead count and performance and is among the fastest growing packaging technologies.

Unsurprisingly, consumer and computing account for the majority of FC products in 2014. In this forecast, however, it is expected that other products such as mobile and industrial will play an important part in 2020.

18.1.3 The Electronic Packaging Industry

Semiconductor front end manufacturing is strongly driven by the need to advance to each next technology node. Processing die on wafer level and increasing the wafer diameter over the years (up to 300 mm to date) has been key to increase productivity and to meet cost targets of the industry. Packaging, on the other hand, used to be rather cost driven. In consequence, packaging is still mostly done on die level and IC manufacturers have cut cost by moving many of their packaging operations to countries with low cost labor or by outsourcing packaging to specialized subcontractors.

But with the ever-increasing semiconductor performance the electronic packaging industry is changing. Conventional (wire bond–based) packaging is starting to reach its limits when it comes to fulfill all the

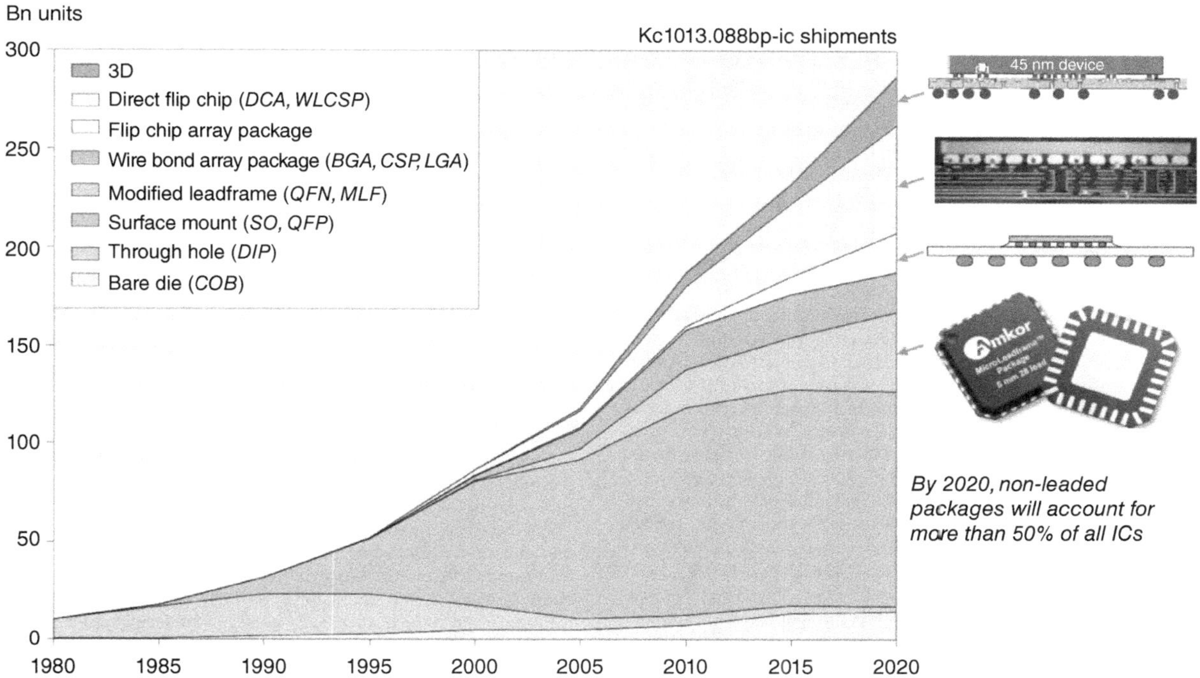

Figure 18.4 Forecast by package type. (*Courtesy of Prismark Partners LLC.*)

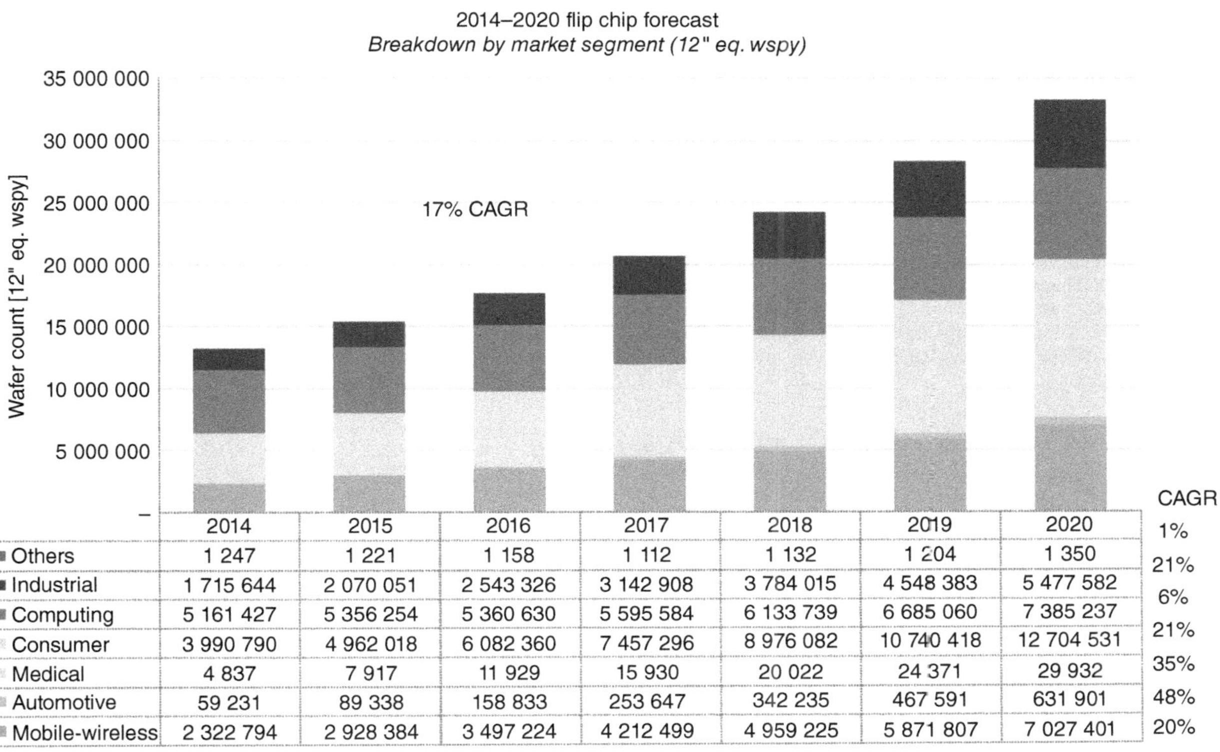

	2014	2015	2016	2017	2018	2019	2020	CAGR
Others	1 247	1 221	1 158	1 112	1 132	1 204	1 350	1%
Industrial	1 715 644	2 070 051	2 543 326	3 142 908	3 784 015	4 548 383	5 477 582	21%
Computing	5 161 427	5 356 254	5 360 630	5 595 584	6 133 739	6 685 060	7 385 237	6%
Consumer	3 990 790	4 962 018	6 082 360	7 457 296	8 976 082	10 740 418	12 704 531	21%
Medical	4 837	7 917	11 929	15 930	20 022	24 371	29 932	35%
Automotive	59 231	89 338	158 833	253 647	342 235	467 591	631 901	48%
Mobile-wireless	2 322 794	2 928 384	3 497 224	4 212 499	4 959 225	5 871 807	7 027 401	20%

Figure 18.5 Flip chip by product type for 2014 to 2020. (*Courtesy of Yole.*)

functions described under the earlier section "The Role of Electronic Packaging" for high-performance dice. Area array packaging in particular increases the complexity of package design, at the same time adding significantly more value to the chip than most other packages. The technological challenges in packaging are growing and the packaging industry has to look for means to advance in technology and to increase productivity at the same time. It is, therefore, expected that, similar to the front end, wafer level processes will be increasingly adopted in the back end to benefit from the same productivity advances.

Today, outsourcing represents an important share of the worldwide packaging, assembly and test revenue. Among the largest packaging subcontractors are Amkor Technology, Advanced Semiconductor Engineering (ASE), Siliconware Precision Industries (SPIL), and STATS-ChipPAC. Production facilities are predominantly in East Asian countries and in Taiwan in particular. Packaging foundries have been a reality long before the first silicon foundries were established (Amkor Technology was founded in 1968, TSMC was founded in 1987). Expectations in the industry are that the share of outsourcing will increase steadily to about 50 percent in the next few years. With more complex packages emerging, subcontractors see the opportunity for growth and are strengthening their endeavors to become technology leaders.

Standards are important for the packaging industry because new package types have to comply with existing standards for testing, assembly, and board technology. There are three packaging engineering standards organizations [4]: JEDEC in the United States, JEITA (Japan Electronics and Information Technology Industries Association) in Japan, and IEC (International Electrotechnical Commission) in Switzerland.

18.2 PACKAGING EVOLUTION

The evolution of single-chip packages has started from metal encapsulations and developed over the dual inline package (DIP) for through-hole assembly and SMT packages such as the plastic quad flat package (PQFP) and the BGA to the chip-size package (CSP) (see Fig. 18.6). DIP and PQFP represent packages with peripheral I/Os while BGAs and CSPs are area array packages, where a CSP is defined as a package that is only marginally larger than the chip itself. The highest level of electrical performance is achieved by direct chip attach (DCA), where a bare die is attached onto the PCB. This method has been used long before the introduction of CSPs, but the absence of any package for the die has limited the implementation of DCA into volume production.

In most area array packages the packaged die has a peripheral pad layout while the interposer/substrate provides the rerouting metallization from the peripheral pads to the area array pads. Flip chip in contrast to TAB and wire bonding has been proven as a reliable interconnection technology over active IC area. Flip chip assembly, therefore, can be used to assemble rerouted die. This is already common for microprocessors. A wafer level package provides a rerouted die with a ball grid array and combines advantages of flip chip and standard area array packaging.

18.2.1 Peripheral and Area Array Packages

The packaging industry is confronted with a wide range of pin counts and form factor requirements. Memory chips, for example, have less than 120 I/Os in contrast to logic chips like microprocessors or ASICs with often several thousand I/Os. Peripheral packages can accommodate smaller pin count devices but area array packages are necessary for large I/O numbers if the package size needs to be small.

Using area array interconnection enables a much larger interconnection pitch for the same amount of I/Os. If x and y are the die lengths and p_p and p_a are the peripheral and area array pad pitch the maximum number of I/Os is $n = 2(x/p_p - 1) + 2(y/p_p - 1)$ for a peripheral layout and $n = (x/p_a - 1)(y/p_a - 1)$ for an area array layout. A 5×5-mm large die, for example, could have a maximum of $9 \times 9 = 81$ I/Os in an area array design with a pitch of 0.5 mm. This can be assembled with standard SMT equipment. In a peripheral package the pitch would have to be 0.23 mm, which would be difficult to assemble with standard SMT and board technology and would lead to a cost increase.

The majority of all peripheral and area array packages are plastic packages. Three common packages with peripheral pins are shown in Fig. 18.6. DIP is a typical through-hole package, while the small outline package is designed for SMT. PQFP is an SMD package with pins at all four sides of the chip to maximize the possible pin count.

All these packages use leadframes. A leadframe is a sheet metal frame on which a die is attached, wire bonded and molded with plastic. The leadframe serves as a holding fixture in the assembly process and as an electrical and thermal conductor from die to PCB. In most cases, the die is mounted onto the leadframe by gluing or soldering and is electrically interconnected to the leads by wire bonding. After die attach the leadframe is overmolded with plastic to form the final package body. Most leadframes are stamped. Replacing the stamping process with an etching process makes leadframes with finer structures possible. Leadless packages, where leads are only exposed at the bottom side of the chip use such etched leadframes [5]. An important representative of leadless packages is the QFN package.

A common area array package is the BGA package. In most BGAs the rigid or flexible interposer provides the redistribution from the peripheral pads to the area array. Figure 18.7 shows the principle design of a flex-BGA (FBGA) with wire bond interconnects. The die is attached to the substrate and then encapsulated. Mechanical stiffness is provided by a metal frame and a radiation plate is added for further protection and serves as a heat spreader.

Today, in most BGAs the die is wire bonded to the interposer. The growing number of I/Os, however, is driving flip chip interconnects and the number of FC-BGAs is currently growing rapidly. The FC-BGA represents a typical flip chip in package technology.

18.2.2 Chip Scale Packages

A chip scale (or chip size) package (CSP) is defined as a package with an outline not more than 1.2 times the size of the die. CSPs are typically BGAs with a grid pitch between 1 and 0.5-mm (with a trend down to 0.3 mm). Therefore, full compatibility to SMT is given and no additional

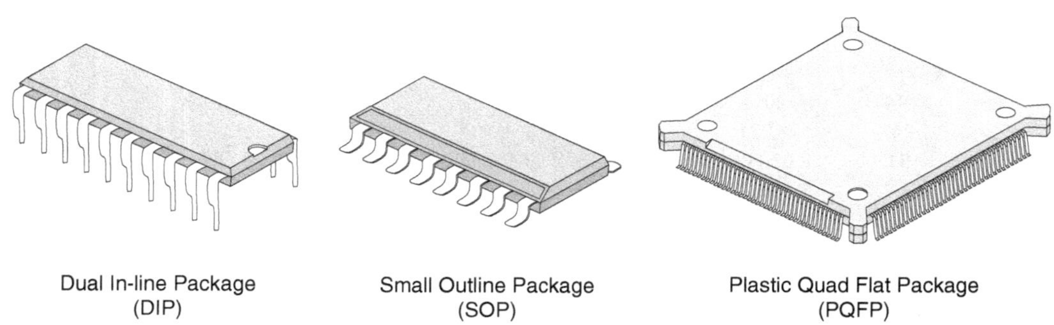

Figure 18.6 Standard molded plastic packages.

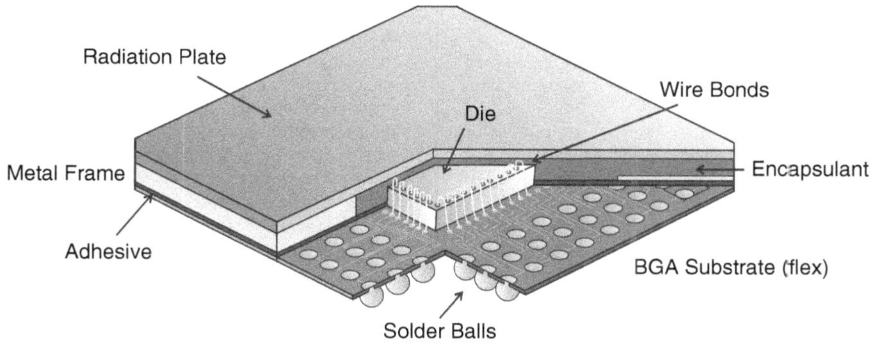

Figure 18.7 Principle design of a BGA with flexible interposer.

processes or extra equipment is necessary for the assembly of CSPs. CSP is suited for low and medium pin count devices. Products that are packaged in a CSP include flash and other memory chips, DSPs, and analog devices. It is believed that CSP will become the standard package solution in the future because they can be tested like other packaged devices.

Today, a large variety of CSP designs exists. An important requirement on all CSP designs is that the CTE mismatch between PCB and die needs to be compensated with either interposers, leadframes, or polymer layers. Most CSPs are currently packaged using standard die level packaging technology based on leadframes, flexible or rigid interposers. All common interconnect techniques (wire bonding, TAB, and flip chip) are used in CSPs. The rigid interposer concept, for example, uses wire bonding or flip chip bonding on an FR-4 board. Wafer level packaging represents an entirely new concept and allows true chip size packages. The adoption of CSP technology was initially difficult because of the unavailability of PCBs for these small packages with their narrow ball grid arrays. High-density substrates such as micro-via boards are now available. The packaging industry owes the commercial success of CSP to the efforts of mostly Japanese companies that demonstrated the first commercialized products using CSPs [6].

18.2.3 Wafer Level Packaging

Economic considerations are driving the packaging industry toward adopting wafer level packaging (WLP). The idea of WLP is to finalize as much of the packaging sequence on wafer level as possible. WLPs are by definition true chip size packages. Wafer level processes are independent of the number of dice and the number of bond pads per die and wafer. With WLP the back end will benefit from productivity advances in the front end such as larger wafer diameters and die shrink. This is not the case with all the other types of CSPs, where each die has to be individually mounted on a carrier or an interposer [7–9].

Like other CSPs wafer level packages come with area array pads. If the dice on the wafer have a peripheral pad layout a redistribution process becomes necessary to reroute the peripheral pads to the area array pads (Fig. 18.8). These area array pads need to be solderable pads and are usually built up using thin film technology. Normally, a redistribution layer (RDL) is a combination of polymer and metal layers. Polyimide (PI) or benzocyclobutene (BCB) are used for dielectric isolation. If the process is done at the back end of a front end line also inorganic interlayers like silicon nitride are used. Aluminum or copper are the metals of choice for the rerouting metallization.

The first redistribution technology was published in 1994 by Sandia. In 1995, Fraunhofer IZM and Technical University of Berlin (TUB) initiated several German and European projects to explore this technology for different applications. By the years more and more companies started to offer WLP services in large volume. Flip Chip International (formerly Flip Chip Technology) in Phoenix, Arizona, and Unitive (starting out of MCNC) at Research Triangle Park, North Carolina (and since 1999 also in Taiwan), created standards in this technology under

Figure 18.8 Wafer level package: The peripheral pads are redistributed into an area array. Solder balls are attached to the pads of the redistribution. In this example, the dielectric polymer is benzocyclobutene.

the trade names UltraCSP (FCI) and Xtreme (Unitive) and are now shipping WLPs in million pieces per week.

18.2.4 Embedding Technologies

There are two main approaches for embedded die technologies: Fan-out wafer level packaging (FOWLP), where dies are embedded into polymer encapsulants and 3D vertical integration, where dies are embedded into PCBs. A lot of activities are running worldwide, dealing with fan-out wafer level integration. Main drivers are here: the embedded wafer level ball grid array (eWLB) by Infineon and the redistributed chip package by Freescale.

FOWLP has been started in volume production for mobile and wireless applications (mainly wireless baseband) and is now moving to automotive and medical applications. Infineon started with a 77-GHz radar IC chip set being the first 77-GHz solution based on FOWLP (eWLB) instead of a bare die.

FOWLP has a high potential in significant package miniaturization concerning package volume but also in thickness. Main advantages of FOWLP are the substrate-less package, lower thermal resistance, higher performance due to shorter interconnects together with direct IC connection by thin film metallization instead of wire bonds or flip chip bumps and lower parasitic effects. Especially, the inductance of the FOWLP is much lower compared to FC-BGA packages. In addition, the RDL can also provide embedded passives (R, L, C) as well as antenna structures using a multilayer structure. It can be used for multichip packages for SiP and heterogeneous integration. Hence, technology is well suited for RF applications.

For FOWLP, two basic process flows are encountered the "mold first" or the "RDL first" approach. Process flows of both options are depicted in Fig. 18.9.

"Mold first" starts with die assembly on an intermediate carrier followed by overmolding and debonding of the molded wafer/panel from the carrier. The RDL based on thin film technology is finally applied on the reconfigured molded wafer/panel. For the "RDL first" approach the RDL is applied first of all on an intermediate carrier and the bumped dies are assembled by chip to wafer bonding on the RDL. Afterward, the assembly is overmolded and underfilled and the molded wafer/panel, including RDL, has to be released from the carrier.

Both options have advantages and challenges. Where "mold first" approach has to deal with die shift, warpage and related fine line and spaces issues. "RDL first" option needs to handle an RDL layer on a temporary carrier as well as die assembly by micro-bump soldering and underfilling.

18.2.5 Packaging with Higher Integration

MULTICHIP PACKAGING AND 3D INTEGRATION

Multichip modules (MCMs) combine multiple bare dice on a substrate, which are packaged as a whole. MCMs are similar to hybrid microelectronic systems but achieve higher performance. First work was done middle of the 1970s for mainframe computers at IBM. Where the main drivers have been electrical performance and circuit density. The different MCM technologies are based on thick film ceramics, thin film and PCB techniques. MCMs are classified accordingly as MCM-C (ceramic), MCM-L (laminated), MCM-D (deposited = thin film–based), and MCM-E (embedding types using chip first technologies originally developed by General Electric) [10, 11]. Mainly flip chip bonding and wire bonding are used to interconnect the bare dice to the substrate. In an MCM-E the chip is directly linked by thin film metallization to the substrate circuitry which guarantees impedance controlled interconnection.

The major challenge which hindered MCM for mass production was cost and chip yield. Test and burn-in of bare dice is much more complicated than full functional testing of packaged dice using test sockets. Known good die (KGD) became the synonym of the barrier for the commercialization of MCMs. The main applications of MCMs have been military, aerospace, and mainframe computer applications. The situation changes, however, if tested WLPs or even CSPs are used instead of bare dice. Such an assembly is not an MCMs according to the original classification but a promising packaging concepts included in the different SiP approaches which will be discussed later in more detail.

Stacking of chips can be done using wiring bonding, flip chip or vertical integration by wafer stacking or chip-to-wafer stacking. The main advantage of wiring bonding is the existing infrastructure. In addition, there is no need of modification of the chip pads. The final aluminum metallization of the IC can be used directly for the bonding process. For example, Amkor, ASE, Hitachi, and Sharp use this approach mainly for memory dice (Fig. 18.10).

The disadvantages are the limited electrical performance and the lack of addition integration of passives components.

Infineon Technologies has developed a stacking technology for face-to-face interconnection using very thin metallizations. The technology is called SOLID, which is an acronym for diffusion soldering: solid-liquid interdiffusion [12]. In contrast to the preceding generation, where chips were attached with adhesive in a cost-intensive and technically unsatisfactory process, two chips are simply placed on top of each other face-to-face, as in a "sandwich." The next step is the soldering. Both contact surfaces are coated with thin copper (3 μm) before being permanently bonded at temperatures of around 270°C.

To avoid the FC bonding process the thin chip integration (TCI) concept can be used. Key element of this approach is the use of extremely thin ICs (down to 20 μm thickness), which are incorporated into the redistribution. This technology offers excellent electrical properties of the whole microelectronic system [13].

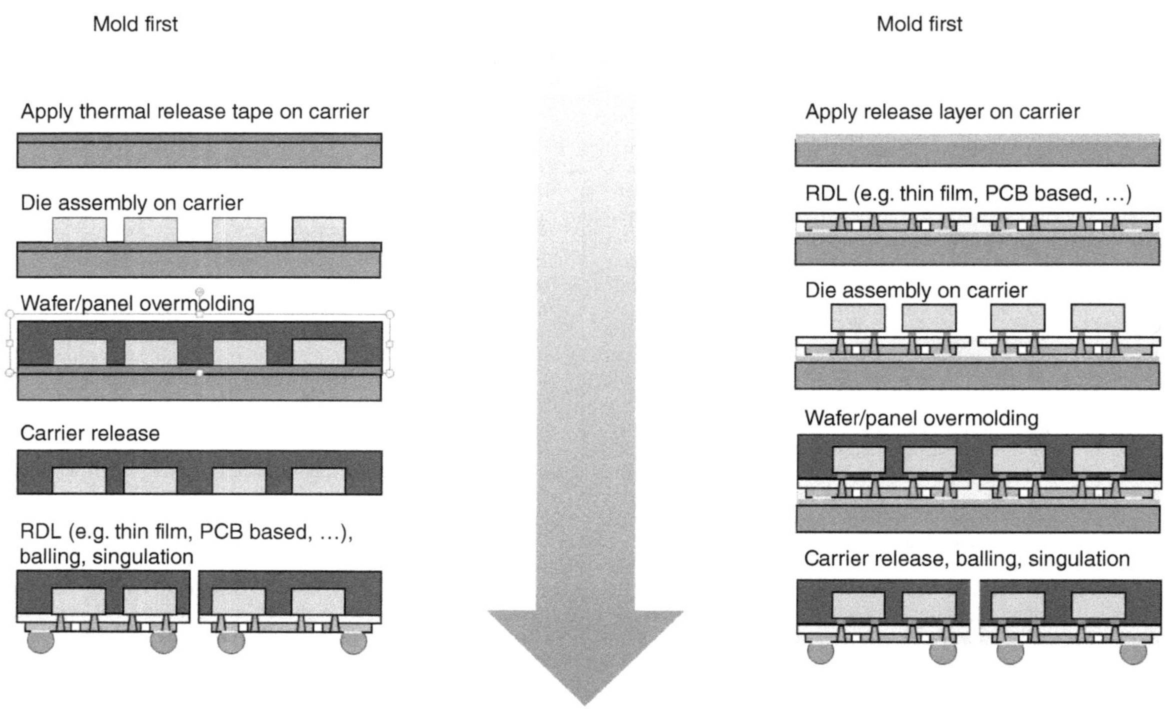

Figure 18.9 Process alternatives for FOWLP/FOPLP.

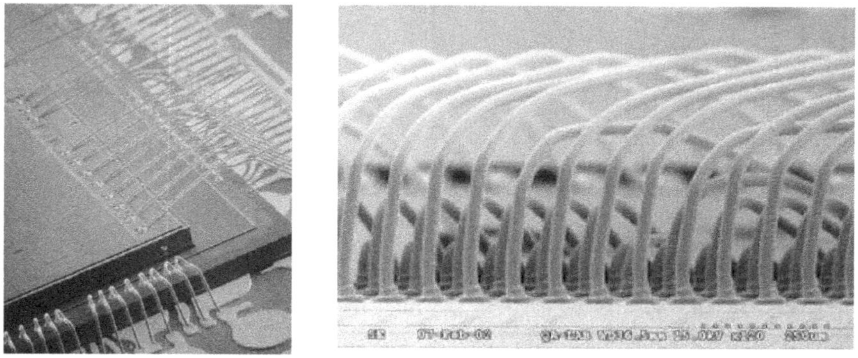

Figure 18.10 3D packaging using wire bonding. (*Courtesy of Sharp, left; ASE, right*)

Integration of Passives

Resistors, capacitors, and inductors are generally referred to as passive components. Compared to the developments in IC technology passive components at the circuit board level have made only marginal advances in decreasing size [14]. Therefore, they are a major hurdle for further miniaturization of electronic products. For example, decoupling capacitors play a major role in reducing the switching noise for high-speed digital electronic systems. As switching speed is well above 1 GHz and power supply voltage is decreasing at the same time, the power supply noise is not negligible. Large capacitance density and small leakage current are required. They should have very thin capacitor dielectric thickness and should be mounted as close to the chips as possible to minimize parasitic inductance.

The smallest passive SMT component today is 0.5 × 0.25 mm (generally referred as 0201). Over a trillion passives were used in 2000, which costs about 0.5 cent per SMT piece. But 1.3 cents have to be paid for assembly, inspection, test, and rework. Especially, analog and mixed signal systems require lots of passives. The ratio of passives to ICs is in the range of 12:1 to 25:1 in cellular phones. This can increase to over 30:1 for camcorders or pagers. The number of passives has increased from 369 for early Pentium™ designs to 2195 for the Pentium™ III.

Ulrich and Schaper [14] are classifying the different integration level as follows:

- *Discrete passive component* as a single passive element which is packaged for THT or SMT
- *Integrated passive component* which consists of multiple passives sharing a package or a substrate
- *Embedded passive component* which is inside substrate for active components
- *Passive array* which consists of passives of like function in one package or one chip
- *Integrated passive network* consists of multiple passive components of more than one function formed on the surface of a separated substrate and packaged as a single SMD. Simple functions like filters or terminators are formed.
- *Integrated passives subsystems* are similar to passive networks but are more complex and could include active devices packaged together in a single SMD. These subsystems are close to SiP concepts.

The potential of integrated passives is obvious if one compares it with the revolutionary change in electronics going from single transistors to integrated circuits. Moore's law was the result of the constant development in on-wafer technologies. The main difference is that passives cannot be scaled down to sub-micrometers due to physical limits. In addition, the possible reduction in footprint is limited for integrated passives. For example, discrete ceramic SMT capacitors are using multilayer structures (10 to 20 layers) which are unacceptable for integrated passives. Therefore, SMT capacitors can achieve higher capacitance values at a given board space.

System in Package

The definition of SiP has not been well established across the literature. Basically, SiP requires to break up the boundaries between system design, semiconductor front end and back end in order to have more flexibility for overall product design in order to enhance the product performance and reduce cost. A paradigm shift in product design and manufacturing is necessary because Moore's law is now more and more restricted by packaging and substrate technology than by nanotechnology. A close cooperation between semiconductor, interconnection, integrated passives, substrate, and system designers will be necessary [15]. Stacked memory dice should not fall into the class of SiP, a minimum combination of logic and memory is required [16]. Even though the SMT and area array technology were major developments in the electronic industry, the gap between semiconductor technology and packaging has been increasing. System in package (SiP) will be the packaging wave for the next 10 years to keep Moore's law alive. SiP bares many similarities with MCMs. The possibility to use tested CSPs or WLPs instead of bare dice without compromising form factor and performance considerations compared to bare die assembly is a new aspect of SiP. Chip stacking and integration of passives are also included in the SiP concept. Tummala from PRC (Packaging Research Center, Atlanta) is postulating a system on package (SoP) as an extension of SiP, including optical interconnects to the substrates [17]. An alternative approach to SiP that has been discussed in the last few years was system on chip (SoC) technology, where the system functionality is fully integrated into a single chip. The industry, however, now commonly agrees that SiP has many advantages over SoC, which include

- Combination of different semiconductor technologies
- Integration of microelectromechanical systems (MEMS), microoptoelectromechanical systems (MOEMS), biosensors, etc.
- Faster design
- Faster prototype cycles
- Higher flexibility
- Reduced mask costs
- Higher yield due to KGD concepts

The ever decreasing size of electronic packages has brought the technology from simple plastic housing to material and equipment challenges touching soon the nano world. Simulation of electrical, thermal, and thermomechanical nature will gain higher importance to identify the interaction of packaging material. The concept of SiP will not only be the key to further miniaturization but also to higher reliability due to fewer interfaces and interconnects.

Silicon interposers with through silicon vias (TSVs) are becoming a very important element for the realization of 3D SiPs. Main advantages of silicon interposers are the decoupling of front end and back end processing for the implementation of TSVs, RDLs, and the integration of active and passive devices. Besides independent manufacturing, interposers can be realized in a short time-to-market time frame and

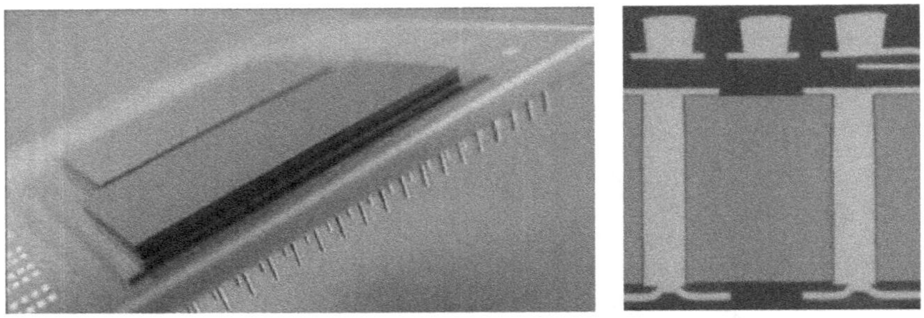

Figure 18.11 Left: Organic package with TSV interposer with flip chip assembled devices (ca. 43k Cu-TSVs and 4 metal layer wiring, Cu-pillar). Right: Cross section of an interposer with Cu TSV (20/100 µm).

in a cost effective way. TSV interposers are designed and manufactured for the different application areas, which also results in different technical specifications ranging from high-density TSV integration and high-density RDL for digital applications to interposers for RF application as well as MEMS integration and optical interconnects. The TSV interposer serves as a high-density wiring carrier for high IO count devices and as an interface to a standard organic package. Interposers have multiple functions including fanning-out interconnect pitches of high-pin count devices to match organic substrate technology as well as providing high-speed interconnects between processors and memory or other logic chips for network and computing applications. They also provide mechanical stability and thus improve the reliability of stacked devices. Concerning technical features, interposers deal with high-density wiring RDLs, TSVs, and top/bottom side interconnect formation (bumping). This allows the electrical signal routing across the top side and to the interposer back side.

Figure 18.11 shows an organic package (PGA) with TSV interposer and flip chip assembled devices with high IO counts. The cross section of the 100-µm-thick silicon interposer with copper through silicon vias (20 µm diameter) and a multilayer redistribution line as well as copper pillar interconnect structures on top side is shown on the right side of Fig. 18.11.

18.3 WAFER BUMPING AND REDISTRIBUTION TECHNOLOGY

18.3.1 Process Technology

WAFER BUMPING OVERVIEW

A bump is defined as a usually conducting three-dimensional interconnect element between die and substrate. In most cases the bump is formed on the wafer prior to wafer dicing and die attach. This process is called wafer bumping. Interconnect methods are soldering, thermo-compression bonding, and adhesive bonding (Fig. 18.2). Depending on the application, many different bump metallurgies are used ranging from pure Au, Cu, Sn, or In to alloys such as PbSn, AuSn, AgSn, and AgSnCu. Various parameters need to be looked at when choosing the best bump metallurgy. These parameters include the melting point, bump hardness and shear strength, the conductivity, the resistance to electromigration, the tendency to form oxides, the bond pad metallization of substrate and wafer, and finally the cost to deposit the metal. Common bumping technologies include gold bumping, solder bumping, and copper posts.

Gold Bumping Gold interconnects do not oxidize or corrode and have excellent electrical and thermal conductivity. Because of their high melting point gold bumps are not reflowed. Gold bumped dice are typically attached to flexible substrates or tapes (TCP: tape carrier packaging) by thermo-compression bonding (TAB: tape automatic bonding) or adhesively bonded to glass substrates (COG: chip on glass) using conductive films. Gold bumping allows very tight bump pitches. Today, LCD driver ICs with their high I/O count and peripheral bump layout account for the majority of gold bumped dice.

Figure 18.12 shows a typical gold bumping process. A TiW-Au under bump metallization (UBM) is sputtered onto the wafer and a liquid photoresist film (~20 µm thick) is spin coated, exposed, and developed. The gold is deposited by electroplating followed by resist strip and UBM etch. Finally, the gold bumps are hardened by an annealing step. Devices are tested after die attach to the tape. Rework of TAB, COF, or COG interconnects is difficult. Today, minimum gold bump pitch for LCD drivers in mass production is around 35 µm with a 10-µm gap between bumps (COG applications allow finer pitch than TAB). These are the tightest pitches of any bumping process in mass production.

Solder Bumping Soldering is the most common SMT assembly technology and allows to form reliable interconnects between die and rigid

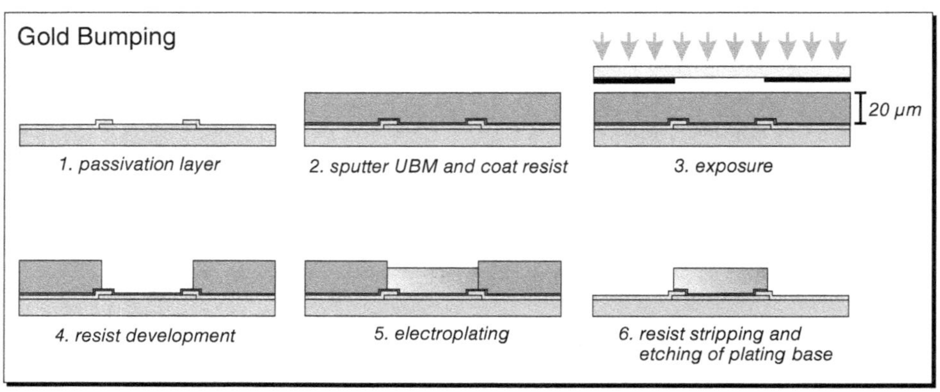

Figure 18.12 Gold bumping process flow.

substrates that often are uneven by a few 10 μm. Soldering, therefore, largely avoids any stress build up between die and substrate during assembly. In addition, solder joints allow for easy repair of defective devices. On the other hand, electrical and thermal conductivity of solder is not as good as that of gold or copper interconnects. Soldering a bare die to a substrate is known in the industry as C4 (controlled collapse chip connection) process, an acronym and technology introduced by IBM in the early 1960s. Solder bumps are the common interconnect element of high-end flip chip assembly. Today, flip chip technology is used in mass production for high-end dice, such as microprocessors and graphic chips.

Solder bumping requires a wettable surface on the IC pad and on the substrate. Often nonwettable solder stop layers are added to the wafer (polyimide or BCB) and to the board (epoxy). To reduce thermally induced stress on the solder interconnects the gaps between the interconnects are underfilled by a filled epoxy. The maximum solder bump height is limited by ball pitch while the minimum solder bump height must not compromise interconnect reliability and has to allow reliable flux cleaning and underfilling. Flip chip die have typical solder bump heights of approximately 100 μm (Table 18.5).

Solder interconnects are formed during a reflow process with a reflow temperature of about 220°C if eutectic PbSn (Pb: 37%, Sn: 63%, melting point: 183°C) is used. This temperature range is acceptable for all common substrate types. High-lead PbSn bumps, on the other hand, are more stable against electromigration and are commonly used for applications requiring high-current densities like is the case for microprocessors. Flip chip bonding with high-lead PbSn (Pb: ~95%) solder joints with a reflow temperature of about 350°C is possible on ceramic substrates or high temperature polymeric board materials but not on many others. High-lead PbSn bumps can be soldered to substrates by providing eutectic PbSn on the substrate and reflowing the eutectic solder only. This allows to remove the flip chip by melting the eutectic solder joint maintaining the integrity of the high lead PbSn bumps on the die. Another advantage is that such a flip chip interconnect has a very defined stand-off height because the high-lead bump does not melt.

Table 18.5 Selection of Solders for Flip Chip Interconnects

Solder	Melting point (°C)	Remark
63Pb37Sn	183	Eutectic PbSn, low melting point, compatible with organic PCBs, commonly used for most SMDs.
95Pb5Sn (or similar)	315	High-lead, good electromigration behavior, highly reliable thermomechanical interconnect, flip chip on ceramic substrate; no reflow of high-lead bumps during chip attach on PCB (eutectic PbSn on PCB side), flux-free reflow in H_2 atmosphere.
96.5Sn3.5Ag (or similar)	221	Currently most common binary lead-free solder for flip chip. Typically used in conjunction with electroplating.
97Sn/3Cu	227	Difficult to electroplate. Short bath lifetime.
95.5Sn3.9Ag0.6Cu	218	Common lead-free solder paste, Cu content reduces Cu consumption from UBM.
80Au20Sn	280	Common for flux-free optoelectronic assembly on gold finishes, controlled stand-off height.
In	157	Allows very low reflow temperatures.
Sn	232	Tin whisker formation during plating.

Legislations by the European Union and other countries demand the ban of lead from electronic products by 2006 with the exception of high-lead solders for microprocessor applications. However, the industry faces certain technological barriers replacing PbSn flip chip interconnects with lead-free solders. This is explained in more detail under the later section "Solder Deposition."

Various techniques to deposit solder exist. Their basic process steps are shown in Fig. 18.13. It is important to control the bump volume across the wafer as well as the solder composition in order to achieve

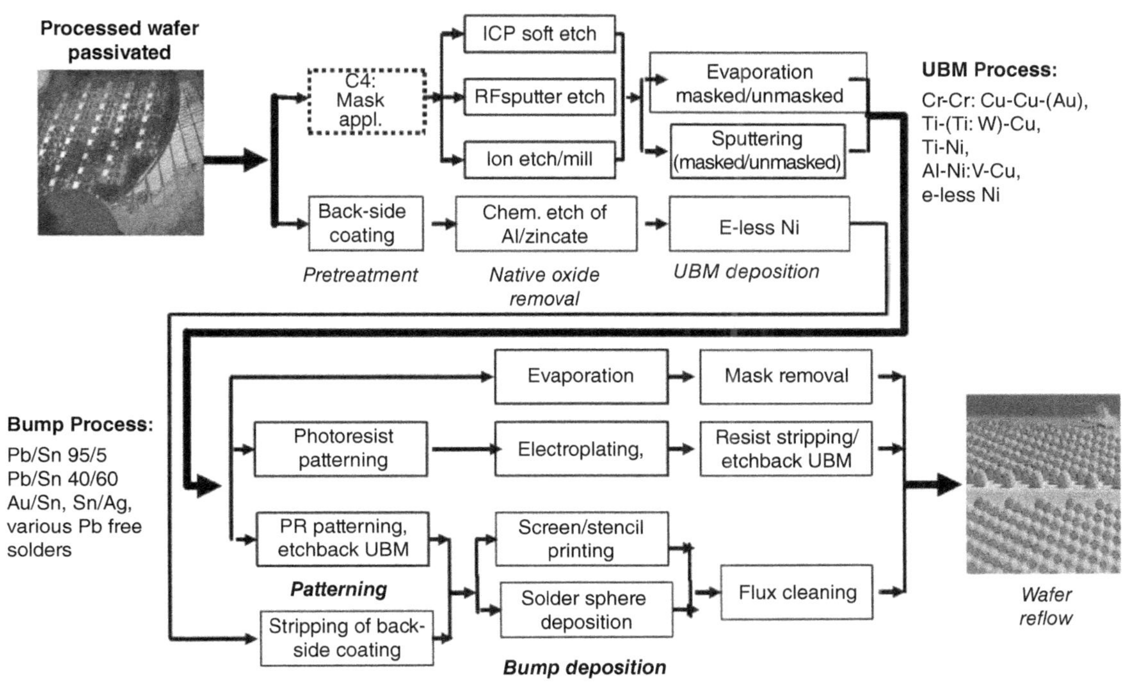

Figure 18.13 Basic process flow of solder bumping technologies. (*Courtesy of Unaxis and Fraunhofer IZM.*)

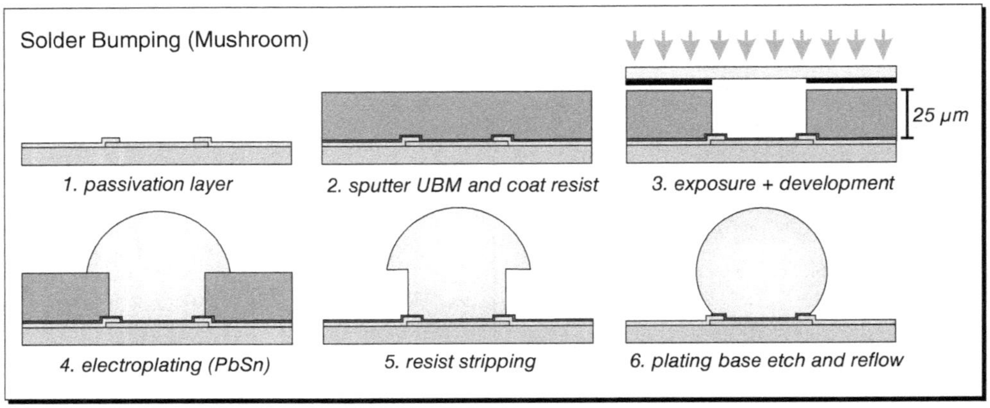

Figure 18.14 Solder bumping process flow (electroplating).

a uniform bump height distribution and to avoid an incomplete reflow process.

The most important deposition techniques for solder bumping are evaporation, electroplating, and solder paste printing. Figures 18.14 and 18.15 depict the typical wafer bumping process flows for electroplated and solder paste bumps. The process flow shown in Fig. 18.14 uses electroplating to deposit the solder. In this case, the UBM that provides a wettable surface is sputtered and a thick liquid or dry film resist is coated onto the wafer. The resist is then exposed and developed. Electroplating allows to plate the solder over the top of the resist layer to form mushroom like structures. Then the resist is stripped and the UBM is etched between the bumps. The reflow process transforms the solder into nearly ball shape and leads to the formation of intermetallic compounds at the UBM/solder interface, which is important for a reliable adhesion of the bump to the UBM (see the later section "Under Bump Metallization"). The advantage of mushroom plating is that the photoresist layer can be significantly thinner than the final solder ball height and that solder deposition is fast because the solder surface is growing when being plated over the resist edge. A disadvantage is that with mushroom bumping plating control becomes more demanding. Photoresist thickness in mushroom plating typically is between 25 and 60 μm. On the other hand, mushroom plating becomes a problem for finer bump pitches. Thicker resist (~100 μm) is used for fine pitch bumping, where the solder is completely plated in the bump mold.

The process shown in Fig. 18.15 uses solder printing. Either metal or resist stencils are used, where resist stencils are required for finer pitches and are the more common solution for flip chip bumping. With stencil printing the UBM is patterned and etched before solder deposition. Next, a thick photoresist (~70 μm and larger) is coated onto the wafer. After exposure and development solder paste is screened into the molds. As the UBM pad defines the final bump base the molds for the stencil process can have a larger footprint than the final bump in order to have more solder paste being screened into the mold to achieve a larger bump height. Before resist stripping the solder paste has to be heated to be transformed into solid solder. After resist strip the solder is reflowed to take on ball shape.

Copper Pillar Copper has approximately a 10 times better electrical and thermal conductivity than PbSn solders. Copper posts are therefore attractive whenever high currents pass through the interconnect like is the case with power devices. This is particularly important for fine pitch bumping which reduces the size of the bump base and increases the resistance of the bump. Copper posts are not reflowed and therefore can be used to provide high-aspect-ratio structures for fine pitch bumping without reducing the bump stand-off height. For the interconnect to the

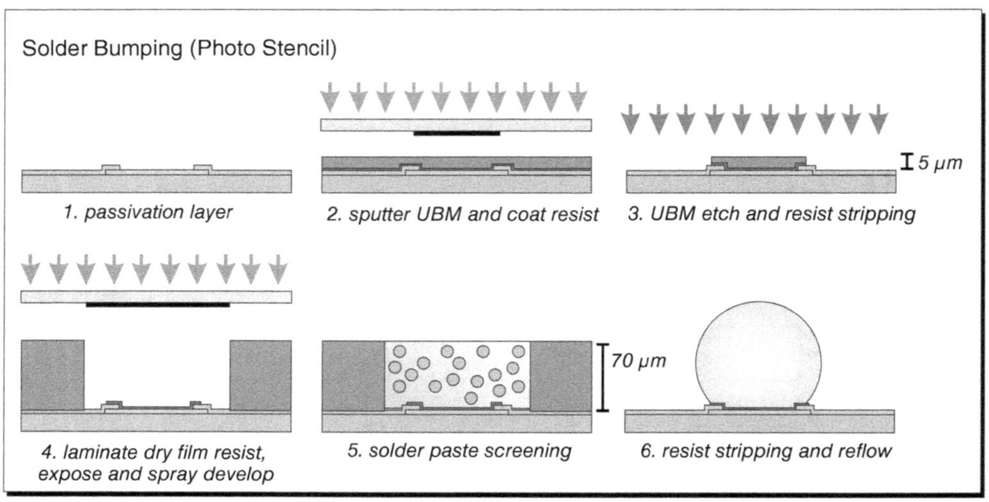

Figure 18.15 Solder bumping process flow (photo stencil).

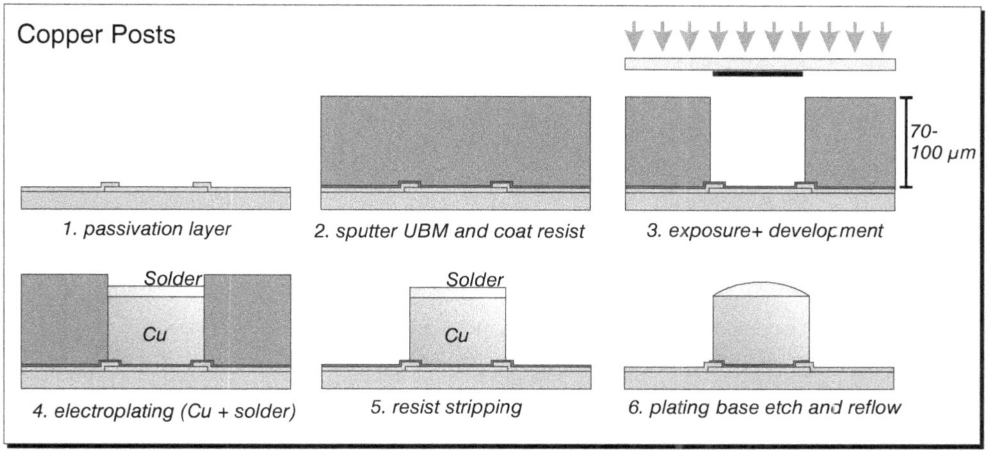

Figure 18.16 Copper pillar process.

substrate a solder tip is added on top of the post. Copper posts are plated and a typical process flow is shown in Fig. 18.16. Unlike solder plating copper post require in-via plating with very thick resist layers of greater than 70 μm. Copper posts are discussed as future bump interconnects for microprocessors.

Other Bumping Technologies All bumping techniques described above use either solder printing or electroplating. Alternative bump technologies include stud bumping, solder jetting, and evaporation. Electroless bumping is actually an under bump metallization technique and is described under the later sections "Under Bump Metallization" and "Electroless Plating Tools."

Stud bumps are made with wire bonders by cutting of the wire right after bonding it to the IC pad. The bump can be either left with a spike or can be coined creating a flat surface or can be sheared off across the top directly after bonding. This technique is fairly flexible in regards to the desired bump metallurgy and can be used for example for gold bumping and even for solder bumping. Because stud bumping is a serial process, it has little importance for mass production but is an important technique for flip chip prototyping and low volume manufacturing. Stud bumping is the major technology applicable to singulated dice.

Solder jetting is a serial and maskless solder deposition technique, where solder droplets are ejected from a print head onto the wafer. High ejection frequencies are possible. However, overall process control is difficult and solder jetting has not yet been adoption by the industry.

Evaporation is the deposition method of the classic IBM C4 process. In terms of quality, evaporated solder bumps are excellent but the overall cost of the evaporation process is very high. The industrial importance of evaporation, therefore, currently decreases quickly especially with the transition from 150 to 200 mm and from 200 to 300 mm wafers.

Under Bump Metallization

The under bump metallization (UBM) is the direct interface between the interconnecting solder and the final chip metallization. The UBM has to provide a low contact resistivity to the chip pad and the solder, good adhesion to the chip metallization and the chip passivation and a hermetic seal between UBM and IC pad. It has to be a reliable diffusion barrier between IC pad and bump with low film stress and it needs to be sufficiently resistant to stress caused by thermal mismatch or during die assembly. In case of PbSn bumping, common UBM stacks are Cr-Cr:Cu-Cu-Au (original C4 from IBM); Ti-Cu; Ti:W-Cu; Ti-Ni:V; Cr-Cr:Cu-Cu; Al-Ni:V-Cu; Ti:W(N)-Au. Usually, these UBM stacks are subsequently deposited by sputtering. A schematic draw of Ti:W-Cu for PbSn is given in Fig. 18.17.

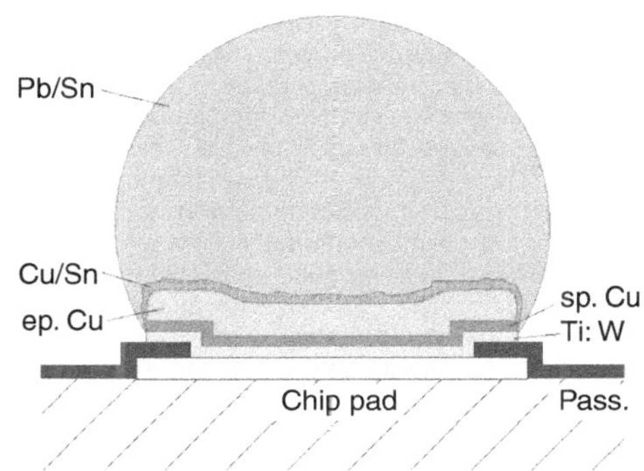

Figure 18.17 Example of a UBM stack (Ti:W-Cu) for PbSn bumps [18].

In the case of PbSn bumps deposited onto a copper-based UBM, intermetallics compounds (IMCs) between Sn and Cu are formed by the reflow process, providing the required adhesion of the bump to the chip pad. IMCs are brittle in nature due to the ordered crystal structure, which is in contrast to the solid solutions like the PbSn. The metals which are mostly used in packaging—Cu, Ni, Au, and Pd—are forming binary intermetallics with Sn-based solders of the Hume Rothery type [19]. These compounds are based on electron valence bonding. The crystal structure is controlled by the number of electrons in the bonding. The composition of each phase can be calculated based on the concentration of the valence electrons. For example, the Cu_3Sn and the Cu_6Sn_5 phases are found for intermetallics of Cu and Sn and Ni_3Sn_4 and Ni_3Sn phases are formed between Ni and Sn. The growth rate depends on temperature, the different activation energies of compound forming and diffusion processes. In general, the intermetallic growth rate is much higher for Cu compared to Ni. This is becoming more important for the lead-free solder due to their higher Sn content.

Several methods have been developed for the UBM deposition. As described early the UBM deposition can be part of the sputtering/plating process (see the later sections "Solder Deposition" and "Vacuum Metallization Systems"). All bumping processes shown in Figs. 18.12 to 18.16 include a UBM etch process. Some devices have sensitive surface areas and do not allow such an etching process. A vacuum-based technology for the depositions of UBM avoiding the metal etch process is the so-called lift-off process. This technology is not restricted to UBM

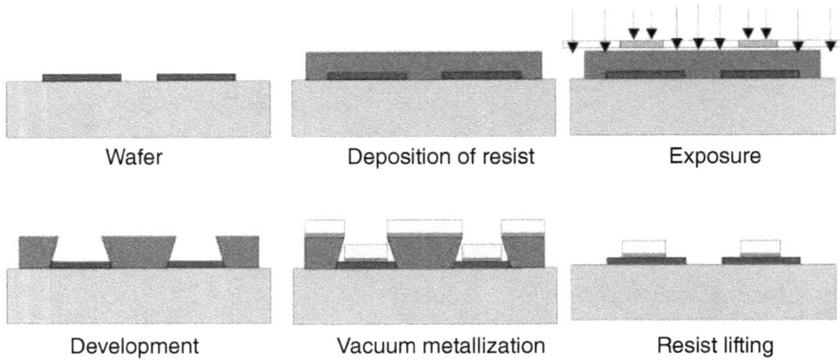

Figure 18.18 Process flow of lift-off.

deposition but can also be used for other thin film metallization. A general process flow is given in Fig. 18.18.

The lift-off photoresist is deposited by spin coating on the wafer. Only negative type resists can be used because an under-cut in the opening has to be created which acts as an aperture for the vacuum metallization. The key step is the separated metal deposition on the resist and inside the opening. By stripping the resist the metal is lifted leaving a well-defined metallization. Mostly evaporation is used for this metallization because the sputtering processes usually have a temperature budget over the glass transition temperature of the resist.

Electroless Ni is a complete wet deposition process avoiding lithography and vacuum-based processes. The process is based on autocatalytic deposition, which is necessary to plate more than monolayers without any current. The energy is supplied internally inside the plating bath by the oxidization of adsorbed hypophosphite. The released electrons are able to reduce the Ni^{2+}. The phosphorus converted from hypophosphite is built into the Ni layer. This can change the mechanical and electrical property of the Ni. To start the electroless Ni deposition, the Al surface of the bond pad has to be zincated. The Al is, therefore, cleaned and microetched to remove the oxide and improve the adhesion to the Ni. Due to the active electrical nature of circuits the bottom side of the wafers has to be protected by resist, which needs a cure to withstand the wet chemicals. The deposited Ni has a standard height for solder bumping of typically 5 μm. Plating uniformity is better than 5 percent on wafer and batch to batch. The phosphorus content should be around 10 percent. A flash of Au (< 100 nm) is deposited onto the Ni to prevent the formation of oxides, which would decrease the wettability for solder deposition. The Au content should be as low as possible to avoid brittle Au intermetallics at the solder interface.

The Ni UBM has been extensively tested. No failures were detected even after 10000 h thermal storage at 300°C, 10000 cycles AATC (−55/+125°C), and 10000 h humidity storage (85°C/85 percent r.h.) [20].

Solder Deposition

The three most important solder deposition techniques in wafer bumping are electroplating, solder paste printing, and preformed solder ball attach. Table 18.6 compares advantages and disadvantages of all three technologies. Please note that the industry tries to push the limits of each technology and that this table does not necessarily represent a common consensus in the industry.

Plating An electroplating process as a low-cost alternative to evaporation was introduced by Hitachi in 1981 [21] and is now widely used for fine-pitch bumping. Electroplating is a relative slow deposition technique with typical plating speeds ranging from 0.2 μm to only a few μm per minute depending on the deposited material. The setup of a plating system is described under the later section "Electroplating Systems." An important challenge in the plating process is the design and maintenance of composition and purity of the electrolyte. In the case of PbSn plating, tin and lead salts (see Table 18.10) are dissolved in the electrolyte and dissociate into their anions and cations. Due to the applied voltage the positively charged Sn^{2+} and Pb^{2+} cations migrate to the cathode (wafer) and are deposited by a discharge reaction on its surface. For this to occur, the cations need to be reduced to the metallic state by accepting electrons from the cathode.

Table 18.6 Comparison of Different Solder Deposition Techniques

	Flexibility of use with different solder types	Cost of solder deposition	Fine pitch capability	Typical application
Electroplating	Limited	Medium	Very good	Fine pitch flip chip, expensive die
Solder pasting	Good	Low	Medium	Low to medium pin count flip chip
Ball attach	Very good	Low	Poor	BGAs of wafer level packages

Plating time can be described by Faraday's law:

$$\tau = \frac{\rho}{JE\alpha} \cdot T$$

where T is the plating thickness, ρ the density of the metal deposit, E the electrochemical equivalent, α the current efficiency ratio of actual/theoretical weight deposited, and J the current density. In fact, the electroplating process is a much more complex process. For example, the mechanism of the metal deposition consists of several steps: The hydrated metal ions have to diffuse to the wafer surface covered by a Helmholtz double layer. In addition, the metal ions can be chemically attached to complexing molecules. Additives are controlling the metal growth to achieve fine-grained solders. Obviously, the plating time can be shortened by increasing the plating current density. However, the maximum current density is limited because high plating speeds translate into higher challenges in plating bath maintenance.

For the deposition of solders the plating speed of the individual constituents must be balanced according to the desired final solder composition. The electrochemical equivalent is a material dependent constant and a function of atomic mass and the number of valence electrons of the individual metal. Therefore, the only parameter that allows balancing the partial plating speeds is the plating current density J or, in other words, the resistivity of the electrolyte. With a given voltage, the current density of each individual metal is obviously a function of the density of the corresponding cations in the plating bath. Another important factor that strongly influences the current density is the electrochemical potential of the cations. This electrochemical potential describes the energy required to reduce a cation to its metallic state. It is standardized to the normal hydrogen electrode which has 0.00 V by definition. In case of PbSn these standard electrochemical potentials are

$$Pb^{2+} + 2e^- \rightarrow Pb^0 \ (-0.13 \text{ eV})$$

$$Sn^{2+} + 2e^- \rightarrow Pb^0 \ (-0.14 \text{ eV})$$

The probability for each of these reaction is a function of the energy it takes to reduce the cation. In the case of PbSn, the natural tendency to reduce Pb^{2+} is slightly higher than the natural tendency to reduce Sn^{2+}. For other solders, the electrochemical potential of each constituent can be significantly different, which can lead to difficulties to design the electrolyte.

Furthermore, the equilibrium between the metal and its ions is controlled by the activity (or approximated by concentration) of the products and reactants, which is described by the Nernst equation. The actual cell potential departs from the equilibrium value by the over-potential or cell polarization, which includes all concentration, diffusion, and other effects. These can be minimized by bath agitation, high ion content, lower current density, etc. In addition, the seed layer must be thick enough to avoid voltage drops below a few mV. A deposition rate of over 4 μm/min with a uniformity of less 3 percent (1 sigma) over 300 mm can be achieved with optimized conditions using automatic plating systems.

The electrolyte composition has to take all these factors into account. This can be done by optimizing the cation concentration ratio or by organic additives that modify the electrochemical potential of the individual cation types. Designing proper electrolytes for alloys, however, is difficult and becomes more difficult and expensive the more elements are involved. Another difficulty is that constituents of the electrolyte can react with each other thereby modifying the electrolyte performance. Therefore, binary alloy compositions like PbSn, SnAg, or SnCu are common, while ternary alloys are difficult and expensive to plate. This is a certain limitation of plating technology for lead-free manufacturing because ternary or even quaternary alloys might become the preferred lead-free solution in the industry.

Plating techniques can use constant voltage (potentiostatic), constant current (galvanostatic), or pulse plating. Pulse plating is the method of choice for fine-pitch application providing a more uniform, smooth deposit with less porosity.

Solder Printing Solder printing is a standard process technology in the manufacturing of PCBs. The typical setup of a stencil printer. The solder pastes that are screened into the apertures of the stencil consist of solder particles of 15 to 45 μm diameter in binders and flux. Small grain size is important for fine pitch bumping. The grain size, however, should not be too small in order to minimize the active solder surface to avoid excessive oxidation and to prevent solder being taken away with the flux. The reflow process forms the solder ball from the solder particles. But at the same time it significantly reduces the volume of the originally deposited solder bump. Especially for fine pitch area array designs this limits the total solder volume that is deposited by solder printing. Another difficulty of stencil printing for fine pitch bumping is the formation of voids during the reflow processes. On the other hand, solder printing is a fairly simple and inexpensive process step when compared to electroplating. An important advantage of stencil printing is the large variety of available solder pastes. This offers flexibility and is of particular importance for selecting lead-free solders.

Ball Attach Direct ball attach is only used for large solder balls of typical diameters of 300 μm or more. In some cases a vacuum head serves as a template for the ball layout on the wafer and picks up preformed solder spheres from a reservoir. The spheres are dipped into flux and are placed on the wafer and are reflowed. The advantage of ball attach is that it is an inexpensive way to provide large solder volumes and that it is easily applicable to any solder type. Ball attach is often used to provide solder balls on redistributed dice were the pitch of the area array pad layout is larger or equal to 500 μm. Equipment makers are trying to push the limits to finer geometries.

Process Technology for Pad RDLs

Several different redistribution processes have been developed but main process steps are similar to each other. Differences exist mainly in the material selection. As an example, the redistribution technology of Fraunhofer-IZM/TU Berlin will be described in more detail (Fig. 18.19).

First a dielectric layer is deposited on the wafer to enhance the passivation layer of the die. Pinholes in an inorganic passivation would give shorts in the rewiring metallization. The polymer layer under the rewiring metallization acts also as a stress buffer layer for the bumping and assembly processes. Using photosensitive polymers requires fewer processing steps for thin film wiring than non-photosensitive materials that have to be dry etched. Fraunhofer IZM/TU Berlin uses photopolymers. The rewiring metallization consists of electroplated copper traces to achieve a low electrical resistivity. A sputtered layer of Ti:W-Cu (200/300 nm) serves as a diffusion barrier to Al and as a plating base. A positive acting photoresist is used to create the plating mask. After metal deposition the plating base is removed by a combination of wet and dry etching. A second photopolymer layer is deposited to protect the copper and to serve as a solder mask. Most of the polymers can be deposited directly over the copper metallization without any additional diffusion barriers. Electroplated Ni/Au is used for the final metallization. Solder balls (high melting or eutectic PbSn) are deposited by solder printing directly on the redistributed wafers. Then the solder paste is reflowed in a convection oven under nitrogen atmosphere and the flux residues are removed in a solvent adapted to the used solder paste. Mean value of the solder ball diameter can be adapted to the assembly and board requirements between approximately 100 and 250 μm depending on the

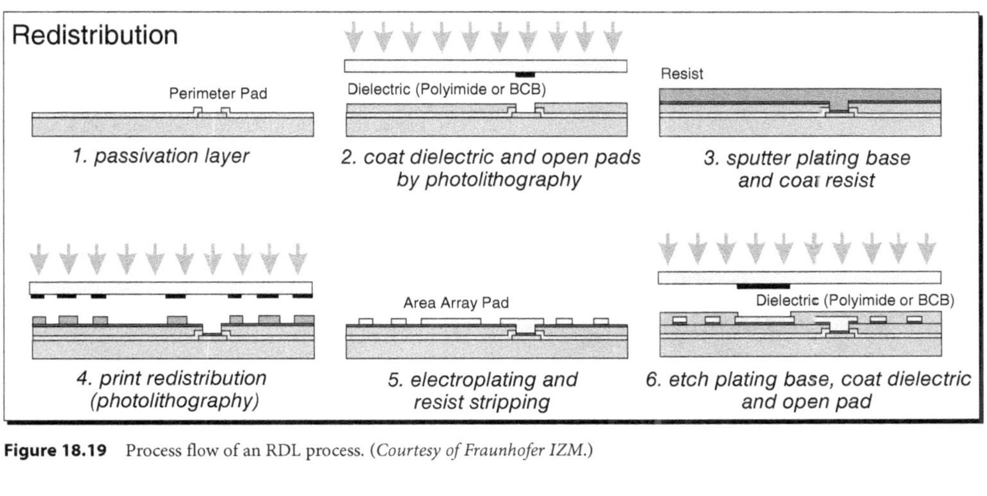

Figure 18.19 Process flow of an RDL process. (*Courtesy of Fraunhofer IZM.*)

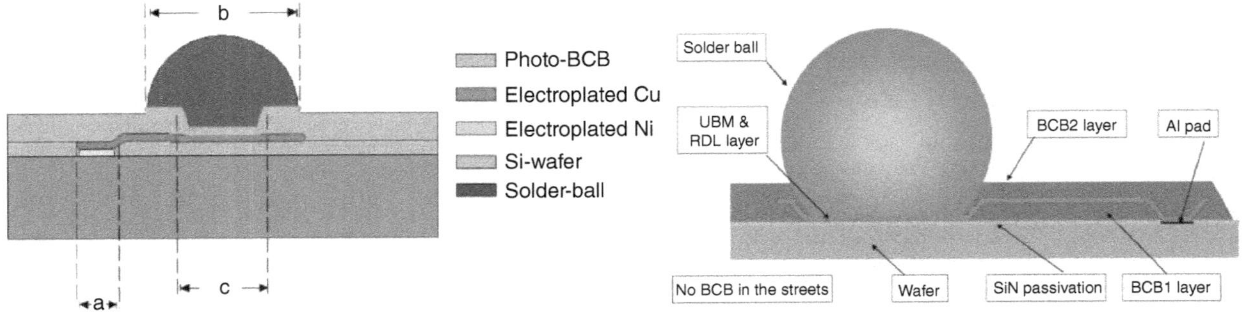

Figure 18.20 Two different RDL layouts: Polymer between chip and bump (left, *courtesy of Fraunhofer IZM*) and UBM directly on top of the inorganic passivation (right, *courtesy of FCI*).

ball pitch. Shear testing is the method of choice for a first quality check. Values for these ball diameters should be higher than 130 cN per bump. Dicing the wafer with a standard wafer saw completes the CSP-WL buildup. The reliability of the RDL was evaluated for consumer, medical, automotive, and space applications.

RDL Design

There are two major different designs for redistribution: The bump and the UBM are separated to the chip surface by an additional polymer layer or the UBM is deposited directly on top of the inorganic chip passivation (Fig. 18.20).

The company FCI (formerly FCT) came up with the classifications of bump on polymer (Fig. 18.20, left) and bump on nitride (Fig. 18.20, right) for the different built-up structures. It is still under discussion whether the type of polymer under the bump plays a critical role for the reliability. The bump on polymer type is preferred if underfiller is used. Different geometric values are possible for the design. The opening in the first polymeric passivation (a in Fig. 18.20, left) should be similar but not the same size than the peripheral pad passivation. The UBM and the ball size (b and c in Fig. 18.20, left) can be adjusted to the pitch of the area array. The maximization of the ball size is limited by the pitch. Today minimum ball pitches of 500 µm and even 400 µm are in production.

Layer Adhesion

The reliability of a multilayer thin film structure strongly depends on the adhesion between the different layers [22]. In Fig. 18.21, the different interfaces in a photo-BCB (as a photopolymer) /TiW/Cu/PbSn technology-based WLP are shown.

1. Photo-BCB to the inorganic chip passivation (silicon-oxide, nitride, etc.) and to metal [chip pad (Al) or redistribution metallization (Cu)]

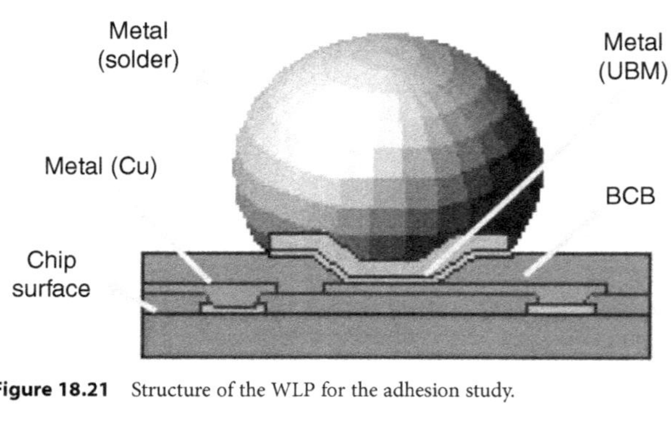

Figure 18.21 Structure of the WLP for the adhesion study.

2. Metal (i.e., redistribution metallization or UBM) to photo-BCB
3. Photo-BCB to photo-BCB
4. Solder to UBM

The adhesion in these thin film structures can be described with three mechanisms: Roughness, chemical bonding using adhesion promoters and chemical interlocking/diffusion. Important for a reliable package is the integrity of the interfaces during the lifetime of the microelectronic product. For interface (1) organosilane-based adhesion promoters are used to create essential layers to couple the organic dielectric to the inorganic surfaces. The theoretically ideal structure would be a monomolecular layer, coupling on one side to the inorganic surface and the other to the polymer. Different adhesion promoters for photo-BCB have been evaluated. High adhesion strengths on several inorganic surfaces were obtained using a vinylsilane, which is spun directly on the wafer before BCB deposition. There is a strong indication that chemical bonding through Si-O bonds is responsible for the adhesion. The thickness of the adhesion promoter layer is in a range of 0.5 to 5 nm. This adhesion promoter layer improves the adhesion to values over 60 MPa on Al, Cu, and different inorganic chip passivations.

The metal to BCB interface (2) depends strongly on the metallization technique [23]. Electroless deposited metals have no adhesion on untreated BCB films with its very smooth surfaces (roughness in the Å range). Sputtering is used as a reliable metallization process for the redistribution of the WLP because the metal atoms have a penetration depth of around 100 nm. This guarantees the strong adhesion of over 80 MPa between the sputtered metal on the photo-BCB. Paik et al. [24] described the stable interface between Cu and BCB. In addition, there is nearly no influence of the descum process (RIE) using a fluorine gas/oxygen mixture, which is necessary to achieve a 100 percent electrical yield in via holes. Only pure O_2 plasma reduces the adhesion due to an oxidization of the BCB surface [25, 26].

For the photo-BCB to photo-BCB interface (3) high adhesion is obtained by performing a partial cure of the underlying BCB layers followed by a final full cure of the whole stack. A correlation between the adhesion strength and the degree of cure was found. The roughness and the surface chemistry of the photo-BCB layer modified by RIE had no significant effect on adhesion. The mechanisms of chemical interlocking and interdiffusion are the driving forces for the adhesion between BCB layers. Therefore, the degree of cure is the key to high adhesion.

The solder to UBM interface is based on the formation of intermetallics (see the earlier section "Under Bump Metallization"). Due to the brittle nature of those intermetallics, there should be minimum growth of these layers during the operation lifetime. Ni is preferred over Cu in the case of Sn-based solders because of the slower growth rates of NiSn-phase compared to CuSn intermetallics.

In conclusion the surface chemistry and physics should be carefully analyzed for the reliable built-up of thin film structure. Special emphasize has to be made to all modification processes like sputtering, plasma, etc.

Integrated Passives

As described under the earlier section "Integration of Passives," the integration of passive components on board or wafer level will be an important step toward further performance enhancement and system miniaturization. Three basic material classes are needed for the realization of integrated passives elements: conductors, resistors, and dielectrics [15]. These can be made out of metals, polymers, or ceramics. The main difference between polymer and ceramic technologies is the maximum process temperature, which can be up to 300°C for polymers but can reach 700°C and above for the firing of ceramics. Metals like Cu, Au, or Al or metal-filled polymer thick films with a resistivity of less than 0.1 Ω/square are used for the conductors to avoid high parasitic resistance. Alloys like NiCr, CrSi, or TaN, cermets (ceramic–metal composites) or carbon-filled polymers are the materials of choice for resistors having values of 100 to 10,000 Ω/square. Polymers with a dielectric constant κ of 2 to 5, amorphous metal oxides (κ = 9-50) or crystallographic ordered mixed oxides with $\kappa > 1000$ are the central building blocks for capacitors. A wide variety of deposition technologies are used for integrated passives depending on the material types and the structuring process (additive versus subtractive): sputtering, evaporation, spin-on, lamination, sol-gel, chemical conversion like oxidization, etc. The different processes can be classified according to the MCM types: thin film (MCM-D), polymer thick film (MCM-L) and ceramic thick film (MCM-C). Highest performance is achieved using thin film technology. Although this technology is still struggling with cost the use of 300 mm substrates (glass, metal sheets or Silicon), will allow to reduce the cost in the future. Polymer thick film is closely linked to the PCB infrastructure. Large substrate size or even roll-to-roll processing enables lowest cost processing. In addition, there is a well-established PCB infrastructure, which has no principal barrier for the integration of passive elements. The main hurdles, on the other hand, are the less controllable tolerances and the limited material spectrum for very high frequency applications. A trade-off has to be made whether the additional cost and complexity to the manufacturing process that add extra internal layers for passives are worth to free board space. Ceramic thick film is cost wise in between. Further cost reductions are achieved by photosensitive materials and improvements in printing but the substrate size is generally limited to 6 in^2. The high-temperature compatibility is an interesting feature for automotive and other applications under extreme conditions.

An integrated passive filter based on thin film technology is given as an example. The built-up process is similar to the redistribution for WLP; therefore, the same production line can be used. Copper/BCB technology in conjunction with NiCr sputtering is the core process steps. In Fig. 18.22, the built-up process for inductors, resistors, and capacitors are shown.

The electrical characterization of the different elements is given in Fig. 18.23.

The low capacitance values of 0.2 to 2.5 pF are due to the BCB usage with its low dielectric constant of 2.6. Sputtered high κ materials have to be used for the nF range. A combination of these passive elements can be used for the realization of filter components. An example for Bluetooth band (2.4 GHz) is given in Fig. 18.24.

One filter consists of three inductors of 3.9 nH and 2 capacitors of 1.8 nF. Microstrip lines with 50 Ω impedance are used for the interconnect of the single elements resulting in two low pass filters of second order and one single inductor within an area of 1.3 × 2.6 mm. PbSn or lead-free solder balls are used for the board assembly.

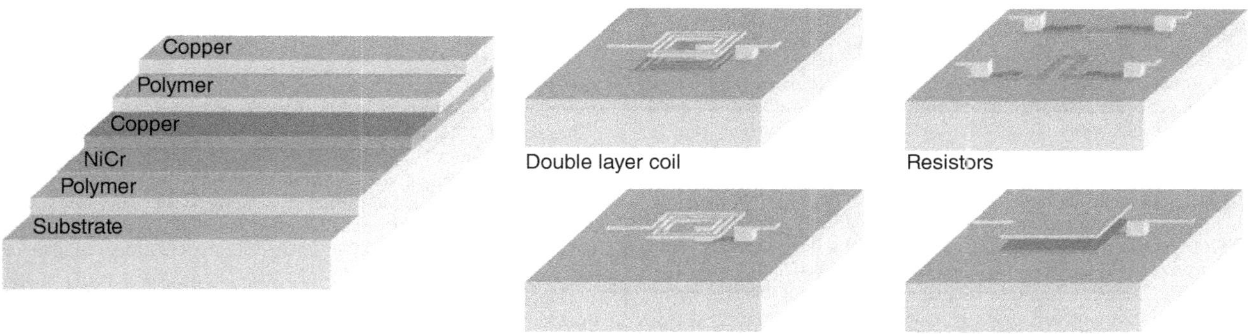

Figure 18.22 Thin film built-up for integrated passive components [27]. (*Courtesy of Fraunhofer IZM.*)

	Inductors	Capacitors	Resistors
Structure	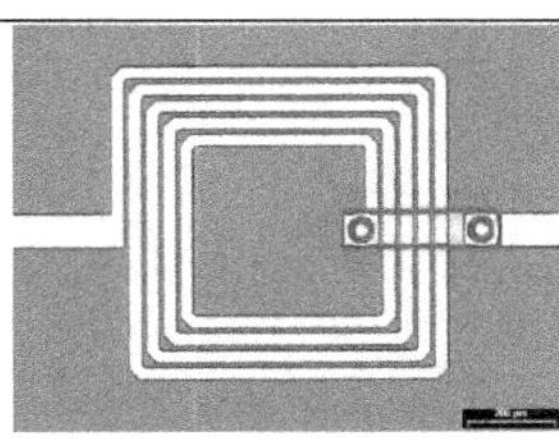 BCB/Cu/BCB/Cu/BCB	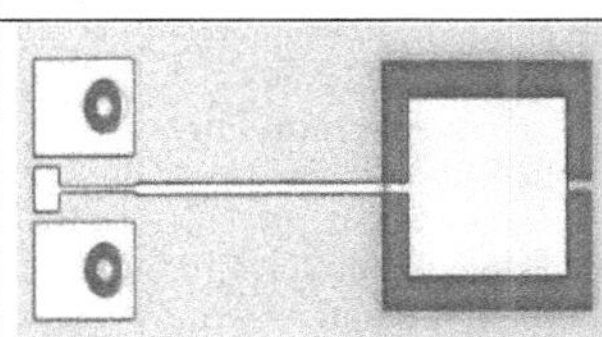 BCB/Cu/BCB/Cu	 BCB/NiCr/BCB/Cu
Values	0.7 – 80 nH Q-Factor up to 35	0.2 – 2.5 pF	100Ω – 150kΩ

Figure 18.23 Thin film passive components. (*Courtesy of Fraunhofer IZM.*)

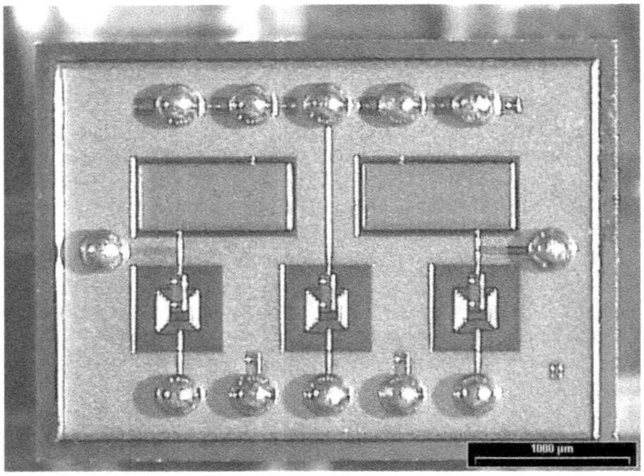

Figure 18.24 An FC mountable integrated filter for Bluetooth band.

A similar technology using BCB/Cu as the core process has been developed by IMEC [28, 29]. The joint venture Intarsia (Dow Chemical and Flextronics) had installed a manufacturing line for integrated passives using BCB/Cu/sputtered TaN on 350 × 400-mm glass panels. It was closed in 2001. At that time the SiP concept was not yet at the horizon and Intarsia was therefore far ahead of the needs in electronics industry. SyChip (United States) and Telephus (Korea) are recently founded companies for the commercialization of these promising technologies.

18.3.2 Equipment for Wafer Bumping

Many of the wafer level processes described under Sec. 18.3.1 appear to be similar to front-end processes. But process requirements are very different. Therefore, standard front-end equipment is often not a good choice for wafer bumping or wafer level packaging because it leads to over-investment and over-engineering. This has to be kept in mind when setting up a wafer level packaging line. Today, merchant bump services offer wafer bumping at a price depending on production volume, wafer size, and technology ranging from approximately US$50 to US$200 per wafer. In average, a typical bumping line processes about 3000 wafers a week. If one assumes US$100 per wafer to be the cost target for a wafer bumping line total annual revenue of that line would be US$15.6 million. If 1/3 of this revenue is used for equipment depreciation and if the equipment has to be depreciated within 5 years, the total equipment cost should not exceed US$26 million. Obviously, this budget is small compared to a front end line. Therefore, dedicated wafer bumping equipment is necessary.

Vacuum Metallization Systems

Since the introduction of wafer bumping and WLP thin film metallization has gained a strong momentum for packaging. Sputtering and evaporation are the key thin film metallization technologies for the fabrication of integrated circuits. The main difference between sputtering and evaporation is the kinetic energy of the deposited atoms which is in the range of 0.1 to 0.5 eV for evaporation and 1 to 100 eV for sputtering, which guarantees a much higher adhesion. In addition, the uniformity of the deposited metals is much higher. For 200- and 300-mm wafers, the evaporation distance has increased to a nearly unacceptable level which decreases further the deposition efficiency being proportional to the square of distance. Therefore, only sputtering will be discussed in detail.

Sputter systems are used to deposit UBM stacks and wiring metallization (typically Al and Cu) for rerouting layers. Sputter systems are vacuum tools (10^{-3}–10^{-5} torr) that generate a plasma of argon ions. Accelerated by an electrical field these ions bombard a target leading to the sputtering of target atoms. These atoms are deposited on the wafer. Sputter systems used for UBM deposition often use magnetrons to efficiently maintain the plasma.

Figure 18.25 shows a dedicated sputter system for wafer bumping applications. It consists of a deposition chamber with five sputter targets,

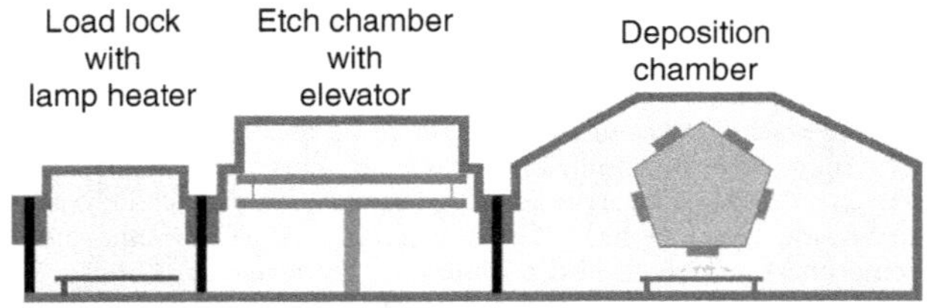

Figure 18.25 Design of a dedicated sputter system for wafer bumping. (*Courtesy of NEXX Systems.*)

an etch chamber to remove oxides or other contaminants from the wafer surface and a load lock chamber. The five different sputter targets are of interest for bumping applications as different technologies (gold bumping, PbSn bumping, lead-free bumping, re-routing) require different UBMs. Batch processing usually leads to sufficient results in packaging applications. A combination of batch processing with single wafer handling (not shown) is therefore an interesting option to reduce cost of the sputtering process. An important aspect for the cost of sputtering is the efficient utilization of sputter targets. In addition, a process has to run in each chamber at the same time.

Important parameters that have influence on the properties of the deposited film and need to be controlled by the equipment are the sputter pressure and the substrate bias.

Resist Processing Equipment

Any photoresist process consists of the following process steps: resist coat (including resist bake), UV exposure, and develop.

Resist Coat Laminators are used to coat dry film resist. The lamination process typically applies heated rolls. Control of film adhesion is particularly important when dry film resist is used in conjunction with electroplating. For reliable further processing of the wafer it is important to precisely cut the laminated dry film at the wafer edge and to avoid any delamination. Fully automated lamination processes often use laser cutters.

The coating process with liquid resist is more complex. Figure 18.26 shows the schematics of a spin-coating process. First, photoresist is

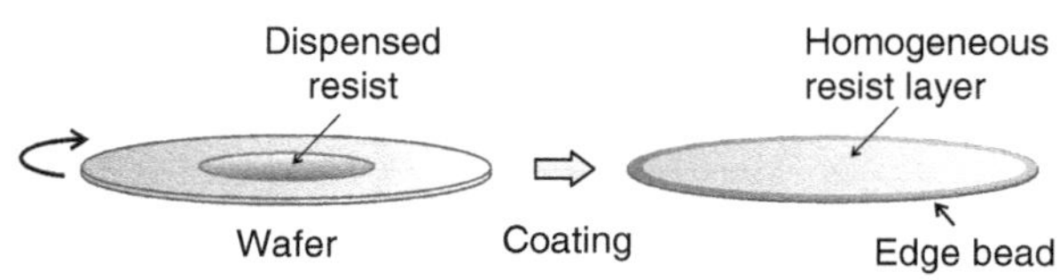

Figure 18.26 Spin coating process.

dispensed onto the wafer. The dispense pattern can be important for the final film uniformity and for the material consumption. In most cases, a spiral dispense from the edge to the center is the method of choice. Then the spin coater rotates the wafer at a speed of several hundred to thousand rotations per minute to uniformly spread the resist film over the entire wafer surface. During this step most of the solvent in the resist evaporates. The thickness of the final resist film is controlled by resist viscosity, spin speed, and surface tension. The resist thickness (50 μm) can vary less than 1 percent over 300 mm wafers by optimized processing conditions. For a given resist formulation the centrifugal force adjustable by the spin speed is the main parameter for adjusting the resist height. A complete mathematical model for describing the spinning process is still not available. For practical use the so-called spin curves (i.e., resist thickness versus rotation speed) are given by the material suppliers for adjusting a spin coater. The polymer thickness h is a function of the angular velocity ω and two parameters K and m:

$$h = K \cdot \omega^{-m}$$

K describes the coater design and the solid content in the resist, while the interaction of the polymer and the solvent is given by m [30]. Solvents which evaporate during the spinning process, which is, for example, typical for positive tone AZ4620 resists from AZ Electronic Materials, lead to $m = 0.5$. Higher values are typical for polyimides, which are dissolved in solvents with lower vapor pressure. The evaporation of solvents can be suppressed by a rotating lid over the spinning chuck (Gyrset™ coating, trademark of Suss MicroTec). This leads to a wider thickness range for a given resist formulation.

Spin coating always leads to the formation of a resist edge bead due to surface tension. Depending on the subsequent process steps, this edge bead remains on the wafer or has to be removed (full or partial edge bead removal is, for example, necessary for electroplating). The edge bead is often removed chemically on the spin coater by solvent dispense or by UV exposure and development, where the latter is the more accurate method. The coating process has to be optimized for film thickness, film uniformity, minimum edge bead, best exposure results, and minimum resist consumption.

After spin coating, the resist is baked in order to remove the remaining solvent from the resist. Baking temperature depends on the resist type but is typically between 80 and 150°C. The baking process has a big influence on the exposure result observed after resist development.

Even if the production volume is low, it is advantageous to use full automatic coating system for good process control and to obtain repeatable results. An automatic system has at least to consist of a handling module, a spin coater with a dispense unit and a hotplate/coolplate. Often cluster tools are used that combine several of these process modules in order to increase throughput and in order to be able to run different chemistries in one system. In particular, multiple hotplates are important since baking is often the most time-consuming step when processing thick resist.

A special challenge is the resist coating of 3D structures such as trenches, via holes, or resilient bump elements. Electrophoretic resist plating is a common technology to coat severe topography whenever a metal surface is available. An emulsion of photoresist is used for electrophoretic resists. These are commercially available from Rohm and Haas (formerly Shipley) as Eagle and PEPR. The organic phase consists of small micelles (50 to 150 μm). Each charged micelle contains the resist. The negative-tone Eagle has 1 to 15 percent solvent and is cathodic. The anodic PEPR is positive tone and has 1 percent solvent. The deposition can be done in conventional electroplaters. The deposition is very fast. Within less than a minute 5 to 10 μm resist is deposited uniformly over a plating base. The process is self-limiting. The disadvantages are the high voltage (100 to 300 V), which has to be applied and the leveling bake after resist plating which limits the application for 3D structure. In addition, thin copper plating bases will be strongly attacked by the chemicals that are corrosive.

Another deposition technology for severe topography uses spray technology and can be used for metal and nonmetal surfaces. The resist or polymer is modified by dilution and sprayed by a special nozzle over the wafer. Even nearly vertical sidewalls can be coated. An example is given in Fig. 18.27.

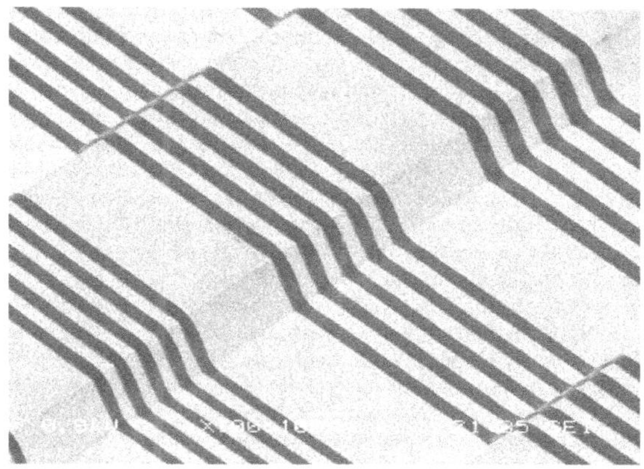

Figure 18.27 A 30-, 35-, and 40-μm line and space patterns across a 100-μm-deep trench. The dark lines are photoresist. (*Courtesy Suss MicroTec.*)

A further advantage of spray coating is the reduction of material consumption. In spin coating, most of the resist is spun over the edge of the wafer. Using spray coating only the wafer surface is covered with resist. For thicker layers multiple spraying steps have to be performed. The film quality (uniformity, flatness) of sprayed resist is not as good as with spin coating because there are no centrifugal forces leveling the film. But for most of the packaging applications, resists are used for etching or as plating molds and polymers for electrical isolation were only a given thickness has to be reached.

UV Exposure The most common exposure tools for wafer bumping/wafer level packaging are mask aligners and 1× steppers (as opposed to reduction steppers that are used in the front end). Figure 18.28 shows

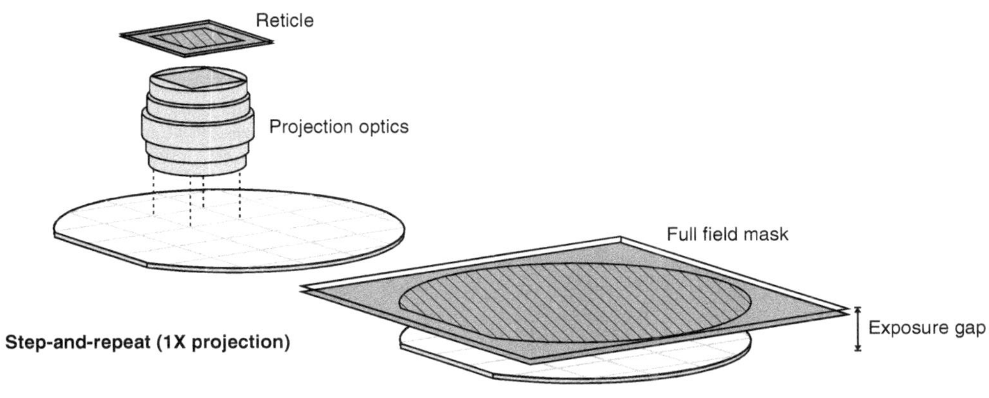

Figure 18.28 Proximity printers (mask aligners) and 1× steppers are the most common exposure tools in wafer bumping.

the characteristic difference between the two exposure techniques. Mask aligner are proximity printing tools, where mask and wafer are separated by an exposure gap of approximately 50 μm. Proximity printing allows even 300-mm wafers to be exposed in one shot. A mask for mask aligners is larger than the wafer and contains the full wafer layout. Typical exposure intensity ranges between 20 and 100 mW/cm^2 (350 to 450 nm wavelength). Proximity aligners offer a low-cost lithography solution for high-volume production. Care has to be taken that the exposure process is optimized in a way that minimizes mask contamination such that mask cleaning is required only every few wafer lots. The full field mask layout requires a careful mask design in order to avoid run-out on large wafers.

1× steppers are projection systems that image a 1:1 projection of the reticle onto the wafer. Steppers expose the wafer with multiple shots. Maximum field size is about 20 × 40 mm. Therefore, more than 30 shots may be necessary to expose a full 200-mm wafer depending on the wafer layout. This translates into a throughput disadvantage compared to the mask aligner especially for large wafer sizes. Because of the small field size the typical exposure intensity per shot is about 1500 to 3000 mW/cm^2.

Mask aligners and 1× steppers use mercury short arc lamps. In order to achieve short exposure times for thick resist applications, the use of the full spectrum between approximately 350 and 450 nm is desirable. This spectral range includes the g-line (436 nm), h-line (405 nm), and i-line (365 nm) of the mercury spectrum.

Depending on the process overlay accuracy in advanced packaging has to be hardly better than ±1 μm. Mask aligners and steppers are capable of achieving this overlay accuracy.

Mask design has to be made carefully to provide the required accuracy. Most often only the layout of one chip of the wafer is available. The step-and-repeat data from the front-end manufacturing have to be taken into account. A small variation can result in run-out. GDSII is the preferred data format in packaging with the benefit of small files, hierarchal data format. Circles are reproduced by segments. The DXF is used in lots of other applications. True circles are possible but there are no standard design rules. Gerber, which is common in the PCB industry, should not be used for mask design due to the low data integrity and the resulting extremely large files.

Resist Develop Most thick photoresist today use aqueous developers, which translates into lower material cost than solvent-based developers (see the later section "Photoresists"). The developer has to be applied at elevated temperatures to show best efficiency and to allow short process times. Photoresist can be either developed by puddle or spray development, where spray development allows better process control especially when very thick resist layers have to be developed. With spraying the mechanical pressure of the developer allows to develop structures with higher aspect ratio. Puddle develop, on the other hand, is faster and lower cost and is therefore used for thinner resists below approximately 20 μm.

Plasma Processing

Plasma treatment is used for a wide range of process steps: Surface preparation to enhance the adhesion, preplate resist wettability enhancement, descumming (removal of resists or polymer residues in vias or openings), resist strip, metal etch, rework, etc. Only ceramics and copper cannot be etched in practice. In general, plasma processes modify the surface by etching or depositing materials. Therefore, this process has to be used carefully to avoid unwanted reactions. But the surface activation results are often a must for some of the process steps.

Mostly RF-RIE (reactive ion etching) and microwave downstream technology is in use for this dry etching process. The inert gas sputtering etching is used in sputtering tools for pretreatment of surfaces prior to metal deposition. The advantage of dry etching compared to wet etching is the clean nature of the process. No residual ions are left on the surface. Process control is excellent because the etching process is started and stopped by switching power. In general, the material consumption is low compared to wet etching processes.

RIE is based on parallel plates, which are capacitively coupled. This gives a highly directional etching process with a medium power of up to 600 W. No charged ions are at the wafer surface in case of downstream plasma. Much higher power can be achieved for isotropic etch profiles with a high etching rate. Single wafer processing is preferred for both technologies especially for a wafer size larger than 150 mm due to uniformity issues in batch processing. In addition, wafer temperature control cannot be given in batch processing tools. Highly reactive radicals and ions are created in the plasma. As the electron energy of plasma used in the microelectronic industry is in the range of 2 to 10 eV mostly dissociation reactions occur because ionization typically only starts above 10 eV. The total etching/cleaning process is controlled by the absorption of the etchant on the surface, the reaction rate at the surface and the desorption of the products. A short plasma descum is necessary for the photo-BCB after cure (final polymerization) to guarantee a yield close to 100 percent. Small residues on the bottom of the vias can be easily removed by a 30-s descum. A fluorine gas (CF_4, SF_6, or NF_3) has to be added to the oxygen because a pure oxygen plasma oxidizes the BCB to a SiO_2-like surface due to the silicon inside the BCB backbone. This etching characteristic is also responsible for the different etching rate and the resulting surface roughness of BCB as shown in Fig. 18.29.

BCB has a very smooth surface with a roughness of less than 0.8 nm. If the surface is etched by RIE with only 10 vol% SF6 the roughness is increased to over 20 nm. Increasing the SF6 content to 25 vol% reduces the roughness to less than 3 nm. Therefore, the plasma process has not only to be optimized regarding the etching rate. Surface roughness and resulting surface chemistry have to be monitored for a reliable built-up of thin film structures.

Electroplating Systems

The basic electroplating cell consists of an anode, a cathode, a power supply, and the electrolyte. The wafer serves as the cathode and the seed layer (as part of the UBM) acts as the plating base. Two common types of plating cells are rack platers and fountain platers (Fig. 18.30). In a rack plater the wafer is vertically mounted in the electrolyte, wherein a fountain plater the wafer is horizontally mounted with the seed layer facing down.

The plating speed and plating control critically depends on several factors including electrolyte composition, impurity levels and plating current uniformity. Equipment and electrolyte maintenance are, therefore, important and add to the cost of plating. It is very important to constantly refresh the electrolyte near the wafer surface in order to prevent local variations of electrolyte efficiency. In a rack plater bath

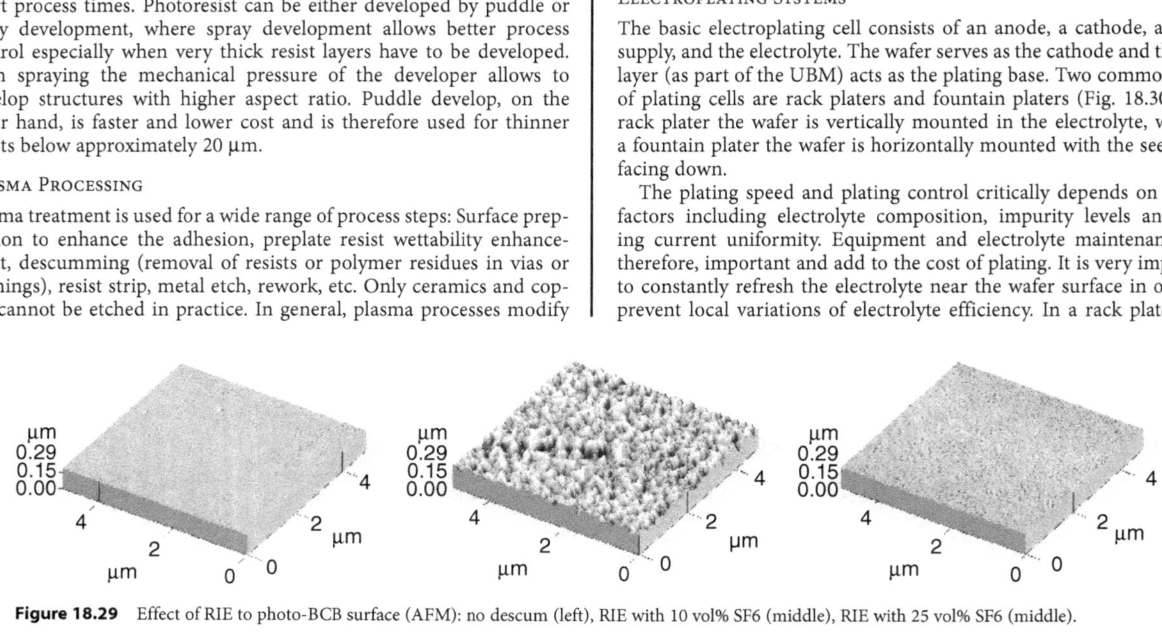

Figure 18.29 Effect of RIE to photo-BCB surface (AFM): no descum (left), RIE with 10 vol% SF6 (middle), RIE with 25 vol% SF6 (middle).

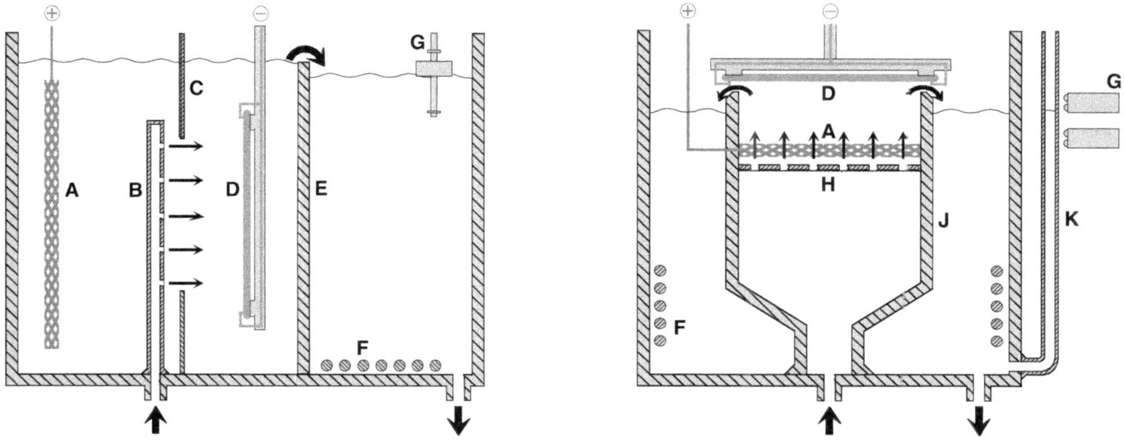

Figure 18.30 Construction scheme of the electrolytic cell of a rack plater (left) and a fountain plater (right) [31]. A: anode, B: spray tubes, C: shielding, D: wafer, E: overflow, F: immersion heater, G: level switch, H: diffusion plate, J: cylinder, K: riser.

agitation by spray tubes is necessary. In a fountain plater it is important to constantly provide the same amount of electrolyte on each location of the wafer. This is controlled by a diffusion plate.

Among the most important parameters that influence the uniformity of plating height and solder composition as well as bump morphology is the electrical field distribution across the wafer as it defines the plating current. Simple platers often only inject the current through few points on the wafer. This, however, does not allow good plating control especially for a large wafer size. The alternative is to apply the voltage over many points along the wafer parameter (Fig. 18.31). In this case,

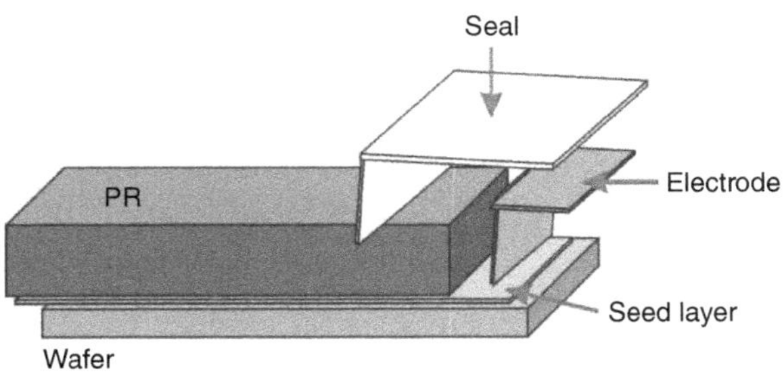

Figure 18.31 In a cup plating system the resist at the wafer edge needs to be removed and the electrode makes contact to the seed layer along the entire wafer perimeter. A sealing ring is attached onto the resist surface to prevent electrolyte from touching the electrode.

the photoresist is completely removed along the perimeter of the wafer (edge bead removal) and the electrode in form of a ring is attached to the wafer. A sealing ring is put on top of the resist surface to prevent the electrode from being contaminated by the electrolyte.

The current distribution has approximately rotational symmetry but can show a radial variation. In this case, the anode design in fountain platers offers another means to control the plating current uniformity by compensating for radial field variations. In addition, the ratio of open area (this is the total plated area) versus full wafer area influences the plating current uniformity. It is important to have a uniform distribution of bumps across the wafer surface, which may require to put bumps in some areas on the wafer without any die underneath (dummy bumps). Likewise, the precision of edge bead removal has an important influence on the plating result.

Stencil Printers

Figure 18.32 shows the principal setup of a stencil or screen printer. These printers are commonly used to deposit solders onto PCBs. The technology was further developed for wafer bumping. A rigid support pallet should be used to protect the wafer. The screens are typically made by laser drilling of metal sheets or by electroforming. The printer aligns the stencil to the substrate or wafer. The printer's alignment system should be able to recognize small fiducials on the wafer which can be pad corners, emblems, or other metallized features that have a good contrast to the rest of the wafer [32]. Then solder paste is dispensed and a squeegee screens the paste across the stencil and into the

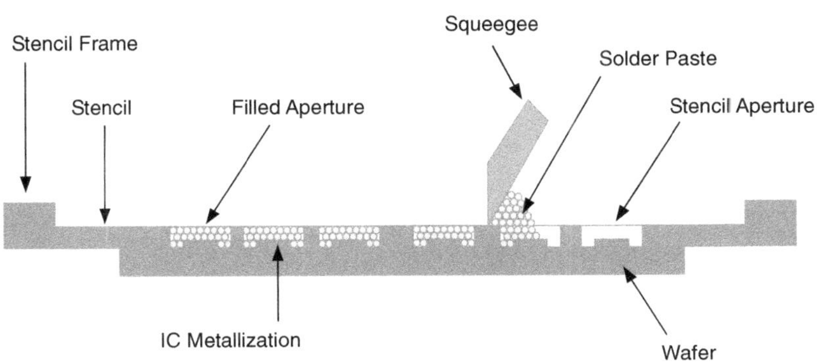

Figure 18.32 Principal setup of stencil printing apparatus. (*Courtesy of EKRA GmbH.*)

apertures. Within seconds the paste is distributed through the stencil and thousands of pads are bumped at a time. One of the key issues is the requirement that all the paste has to be transferred from the stencil to the wafer. Any solder paste residues in the screen will reduce the final bump height uniformity. For wafer bumping often photoresist stencils are used instead of metal stencils (Fig. 18.15) because of the fine pitch flip chip interconnects. Photoresist stencils take advantage of the high-resolution capabilities and high overlay accuracy of thin film technology.

Then the stencil is removed followed by a reflow to form the final solder ball. The metal stencils have to be cleaned by solvents. Mostly water-soluble solder pastes are used in production. If photoresist is used for the stencil process, the paste is partly reflowed before stripping the resist. A second reflow has to be done afterward, which requires an additional fluxing step.

The stencil or screen bumping process is limited by the layout of the chip pads, which are designed mostly for wire bonding. Therefore, the pads are rowed at the periphery of the die. Rectangular openings have to be designed for the stencil to maximize the amount of solder paste transferred to the solder. To calculate the size of the reflowed bump as a function of pad size and geometry, the following equation is used for a reflowed bump as a truncated sphere: $V = (1/2)A \cdot H + (\pi/6) \cdot H^3$ with V as the solder volume, A as the pad area and H for the bump height. The advantage of stencil printing is the wide variety of solder paste materials. It is possible to adjust the solder composition for the requirements of lead-free.

Electroless Plating Tools

A process line for electroless Ni/Au consists of a set of wet benches. Typically, 14 tanks are necessary for full production (25 wafers at a time for a wafer size up to 200 mm and 12 wafers for 300 mm technology). The total size is roughly 6.0 × 1.5 m. Extra space has to be reserved at the back of the line for services like refilling or cleaning. Only DI-water supply, compressed air, nitrogen, electricity, and exhaust have to be added. A robot cassette handler can be added for full automation. Monitoring of the chemical composition of the bathes is a critical task for high quality. It has to be done by manual chemical analysis or by an automatic analyzing unit. The Ni content and additives have to be added by the time of operation. The tanks have to be cleaned by strong acids according to the utilization of the equipment. Specification of the Fraunhofer IZM process is given in Table 18.7.

Table 18.7 Specification of the e-Less Ni/Au Process

Property	Specification
Wafer material	Si
Bond pad material	AlSi1%, AlSi1%Cu0.5%, AlCu2%
Pad metal thickness	≥1 μm
Passivation	Defect-free Nitride, Oxide, Oxi-Nitride, Polyimide, BCB
Residues on bond pads (nonorganic/organic)	<5 nm not acceptable
Wafer size	100–300 mm
Wafer thickness	>200 μm (>150 μm)
Bond pad geometry	Any (square, rectangular, round, octagonal)
Passivation opening	>40 μm
Bond pad spacing	>20 μm
Passivation overlap	5 μm
Wafer fabrication process	CMOS, BiCMOS, Bipolar
Ink dots	Acceptable Stability depend on ink
Probe marks	Acceptable
Scribe lines	Must be passivated (thermal oxide) Test structures acceptable
Laser fuses AL fuses	Not acceptable
poly si fuses	Acceptable (with limitations)

Source: Courtesy of Fraunhofer IZM.

Control of the UBM quality is monitored by shear strength, which has to be around 150 MPa (min 100 MPa). The Al etching process has to be restricted to less than 0.5 μm to avoid damage to the silicon devices.

UBM Etch Equipment

The UBM etch process removes the UBM metallization between the bumps. For cost and technology reasons, wet chemical etching is common. For a UBM stack consisting of different metal layers different etch chemistries are required for each layer. Among the requirements for the etching step are a uniform etching result, a minimum bump undercut and to monitor the remaining metallization thickness in order to stop the etching process or to switch the etch chemistry in case a layer of a UBM stack is fully removed. For best process control spray etchers are used either in a single wafer or batch configuration. It is important to design the etching process in such a way that the bump surface is not oxidized or modified in any other way. In addition, the design of the UBM stack has to take the UBM etch process into account in order to achieve reliable and good process results.

Reflow Ovens

In the reflow process the deposited solder is melted and a mechanically stable interface between the solder and the terminal metal surface on the wafer (reflow after wafer bumping) and the substrate (reflow during flip chip assembly) is generated. Reflow after wafer bumping transforms the solder into a sphere-like shape. Like in any other soldering process oxidation of the solder leads to an insufficient flow behavior of the liquid solder and to the formation of voids and to a mechanically and electrically poor solder joint quality. The solder is, therefore, cleaned from oxides by either reflowing in nitrogen atmosphere and providing flux or by reflowing in a reducing atmosphere (e.g., H_2 or formic acid) without flux. A new technology uses plasma-assisted reflow. The use of flux can have a detrimental effect on the formation of the solder joint or, in case of MEMS or optoelectronic devices, can even contaminate the product. Flux-free reflow is therefore a desired option. In the case of PbSn, flux-free reflow with hydrogen is only applicable to high-lead PbSn with its high reflow peak temperatures around 350°C because the reaction with hydrogen requires a certain activation energy. For eutectic PbSn solder or lead-free solders with lower melting points, flux-free reflow in reducing hydrogen atmosphere is not possible and application of flux is required.

Various furnace designs and reflow techniques exist: Convection ovens with heated process gas, hotplates, or IR heaters. The oven design and heating process influence the reflow process and flux activation. A certain disadvantage of hotplates is that a controlled and reliable heating and melting of the solder can become difficult if a wafer is contaminated on its bottom side (e.g., by flux) or if wafers are warped. In convection furnaces, on the other hand, the solder is heated directly by a well-controlled heated gas stream. With the introduction of lead-free solders with their usually higher melting temperatures compared to eutectic PbSn, it can become desirable to have means to only expose the solder to these higher temperatures while wafer or board are exposed to lower temperatures. A combination of convection oven and IR heating might become the best solution for this requirement. The maximum reflow temperature is typically a few 10 K above the melting point of the solder and it is important that an optimized temperature profile is maintained during the reflow process. Different sections of the furnace have different temperatures and the wafer—depending on the speed of the belt—is exposed to a certain temperature profile over time.

For fine pitch bumping flux management and flux cleaning is important to avoid particle generation and to maintain a good yield. A flux management system allows the processing atmosphere to be evacuated from the process chamber with the flux still in the gaseous state. This allows the process chamber to stay clean. Remaining flux on the wafer has to be cleaned after reflow.

18.3.3 Materials

Photoresists

Photoresists are photosensitive materials that are applied temporarily on the wafer. In most cases, they serve as etch masks or plating molds.

Table 18.8 Classification of Photoresist and Their Characteristics

Class	Resin/dev. chem.	Pre-bake	Rehydration	Exposure spectrum	Dose	Sidewall	Resolution	Stripping	Application
Pos. liquid	Novolak/ aqueous	Yes	Yes	Broadband (chemically amplified)	High	45°–85° (~90°)	+	Easy	Gold bump, mushroom solder bump, UBM
Neg. liquid	Acrylate, epoxy/aqueous, organic	Yes	No	Broadband	Medium	~90°	+	More difficult	Gold bump, in-via plated solder bump
Neg. dry film	Acrylate, epoxy/aqueous	No	No	Broadband	Low	~90°	–	More difficult	In-via plated solder bump, photo stencil

Table 18.8 summarizes different types of photoresist materials used in wafer bumping and wafer level packaging. The base resin of positive-tone photoresists is typically Novolak, whereas negative-tone resists are based on acrylate or an epoxy resin. Positive-tone resists have photoactive compounds that make exposed areas soluble in a diluted alkali (base) solution such as sodium hydroxide (NaOH) or metal ion–free TMAH. Negative dry film resist is developed with aqueous carbonate developer. These are about the lowest cost and most environmentally friendly developer chemistries possible since most of the developer consists of water. For negative liquid resist, aqueous-based developers as well as more expensive organic developers are used. The base chemistry of resist, solvent, and developer is important for the configuration of coat/develop clusters as different chemistries should not be used in the same process module.

Liquid photoresists—positive and negative tone—require a pre-bake step to drive the solvent out of the resist layer. The baking step can be very time consuming especially for very thick layers. Insufficient baking leads to trapped solvent in the middle of the resist layer. Especially with positive-tone resist, this is detrimental to sidewall control as belly-shaped resist profiles appear because the developer becomes more efficient when solvent is still present. In addition, trapped solvent can contaminate a plating bath over time, which will deteriorate plating control. With positive-tone resists in particular, it is important to carefully ramp up baking temperature in order to avoid stress induced into the long cross-linked polymer molecules. Proximity hot plates are usually required to process thick positive resist.

An important difference between positive- and negative-tone resists is that positive-tone resist requires a rehydration step after baking to allow water to penetrate into the resist layer again. This is necessary because the resist will not work without sufficient water present in the resist layer. This rehydration step can take hours for very thick layers.

Most common thick photoresists are exposed by broadband (combined g, h, and i line of the mercury spectrum). The strong absorption of the i line in positive resist makes it difficult to achieve vertical sidewalls. The sidewall angle, therefore, strongly depends on the resist thickness and typically ranges from 45° to 85°, where steep angles are easier to achieve in thicker layers than in thinner ones. However, new chemically amplified positive-tone resists show near 90° sidewalls with broadband exposure as well.

An important parameter to look at in regards to tool productivity is the exposure dose as it determines the exposure time and thus the throughput of the exposure tool. Here the different resist classes show very different characteristics: Dry film resist requires very low doses even for very thick layers, while the exposure dose for a very thick positive resist is often very high, resulting in very long exposure times making it difficult to use this resist class for layers thicker than 50 μm. Negative liquid resists have acceptable dose requirement even for layers thicker than 50 μm and are a good alternative to dry film resist whenever the film thickness exceeds approximately 50 μm.

An advantage of positive photoresist is in the stripping of this material. Positive resist can be easily stripped using standard solvents such as NMP. Negative resists have a stronger resistance to organic solvents and require more aggressive stripping chemicals, which increases cost and requires a more careful treatment to guarantee, for instance, a good wetting behavior of the UBM during plating.

From an application point of view the following trends can be observed: Positive-tone resist is still very popular because of its ease of use. It is often selected whenever resist thickness is typically less than 50 μm and near 90° sidewall angle is not a requirement. Usually positive resist are used for mushroom solder bump plating, gold bumping, rerouting, and UBM patterning (in case solder is applied by stencil printing). However, since straight sidewalls are important in gold bumping, negative-tone resists are increasingly used in this application. Resist layers thicker than 50 μm are necessary for in-via solder plating and photo stencil printing. In most cases, negative resists are used. Thick dry film resists dominate as they are commercially available since many years. However, their limited resolution capabilities are driving the interest toward negative liquid resist for fine pitch solder bumping.

Spin-On Dielectrics (Photosensitive)

Thin film polymers have proven to be an integral material basis for many different types of advanced electronic applications. First used as IC stress buffer layers, then established for MCMs, they are now used in various new packages, especially in the field of WLP, SiP, and MEMS. The requirements for the selection of a given polymer are quite broad: High decomposition or glass temperature for the high-temperature processes in packaging like solder reflow, high adhesion, high mechanical and chemical strength, excellent electrical properties, low water up-take, photosensitivity, and high-yield manufacturability. Only thermosets are, therefore, the polymer class for packaging applications. An important process difficulty is due to the fact that these high-end thermosets are nearly insoluble in organic solvents Therefore, prepolymers are manufactured, which have a molecular weight in the 100 thousands and are dissolved in organic solvents. These solutions are commonly called precursors and are ready for the spin-on process. The final polymerization is done on the wafer by thermal curing after photo processing.

In addition, the selection of the optimal polymer for a given application depends not only on its physical and chemical properties and processability, but also on its intrinsic interfacial characteristics.

Table 18.9 gives a selection of common types of photosensitive spin-on dielectric materials. Unlike photoresists, these materials remain permanently on the wafer and act as repassivation (buffer coat) or as insulation layer for redistribution traces. Repassivation is used as additional protection of the active chip area wherever required (mechanical damages, α-particles, and flux media) in order to increase chip reliability. In the case of flip chip, it also reduces the influence of underfill stress on the die surface. PI (polyimide), BCB (benzocyclobutene), and PBO (polybenzoxazole) are common repassivation materials and are used as insulation for RDLs or integrated passives (copper coils) as well. However, BCB has gained a certain dominance in the market for these applications. BCB and polyimides (with only a few exceptions) are negative acting materials requiring organic developer while PBO is positive acting. All materials require a bake to remove solvents but no rehydration step. In some cases a post-exposure bake is necessary to enhance the photo-initiated polymerization. All materials are exposed by the broadband spectrum and achieve a sidewall angle of approximately 40° to 60°. Resolution is usually not an important requirement because only vias of 20 μm or larger in diameter have to be opened over larger I/O pads.

Table 18.9 Classification of Dielectrics and Their Characteristics

	Trade name	Photo-sensitivity	Developer	Base chemistry	Curing T	Diel. const	Loss factor	Dielectric strength	Tg / Decompo-sition T	CTE	Tensile strength	Elongation to break	Residual stress	Youngs modulus	Water uptake
					[°C] for ~1–2 hrs	[1 kHz–1 MHz]	[1 kHz–1 MHz]	[V/μm]	[°C]	[ppm/K]	[MPa]	[%]	[MPa]	[GPa]	[%]
Asahi Kasei	Pimel G7621	negative	organic	Polyimide	>350	3.3	0.003		355	40–50	150	30	40–50		0.8
	MA 1000			Phenolic	200–250	3.9				40	120	15	16		
Asahi Glass	ALX 211	negative	organic	Fluoro-Polymer	190	2.6	0.001	410	>350	60	90	20	32	1.3	0.4
Dow Chemical	Cyclotene 4000	negative	organic	Benzocyclobutene	210–250	2.65	0.0008	n.a.	>350	45	87	8	28	2.9	<0.2
	Cyclotene 6000	positive	aqueous	modified Benzocyclobutene	210–250	3.2	0.015	500	290	80	95	13	28	2.9	1.1
	Cyclotene Dry Film 14-P005	negative	organic	Benzocyclobutene	200–250	2.57	0.0032	510	250/>300	63	80	13	28	2.5	0.1
Dow Corning	Photoneece PWDC 1000	positive	aqueous	Polyimide	320	2.9	n.a.	n.a.	290	36	130	40	28	3	n.a.
	WL 5150	negative	organic	Silicone	250	3.2	0.07	39	<250	236	6	37	26	0.16	0.24
Dupont	WPR	negative	aqueous	permanent dry film, Acrylate	140	3.8	0.042	n.a.	75	69/88 over Tg	48	2	n.a.	2.1–3.3	1.4
	PerMX	negative	organic	dry film, Epoxy	150–250	2.9	0.006	46	220/300	50	75	5	n.a.	3.2	<1
Fuji Film	Durimide 7000 Series	negative	organic	Polyimide	>350	3.3	0.007	340	>350/510	27	170	73	30	2.9	1.3
	Durimide 7500/7400/7300 Series	negative	organic	Polyimide	350	3.2	0.003	345	285/525	55	215	85	n.a.	2.5	n.a.
	AP2210A	negative	aqueous	n.a.	350	3.1	0.003	272	325/518	45	147	77	n.a.	2.4	0.5
	AN 3310	negative	aqueous	n.a.	350	3.0	n.a.	n.a.	305/475	53	125	45	n.a.	2.2	0.5
	FB 5000	positive	aqueous	PBO	350	2.98	n.a.	280	305	48	161	91	n.a.	2.8	0.5
	LTC 9000	positive	aqueous	PBO	200	3.0–3.6	0.01	n.a.	n.a.	n.a.	n.a.	n.a.	n.a.	n.a.	n.a.

HDM (Hitachi-Dupont)	PI 2730	negative	organic	Polyimide	>350	2.9	0.003	n.a.	>350	16	170	n.a.	18	4.7	>1.0
	PIX 3400	negative	organic	Polyimide		3.4		n.a.	450	43	112	100	33	2.6	1.6
	HD 4000	negative	organic	Polyimide	375	3.3	0.001	250	410	35	200	45	34	3.5	1.3
	HD 7000	negative	organic	Polyimide	375	3.2	0.002	250	410	50	175	70	30	2.6	1.7
	HD 8820	positive	aqueous	Polybenzoxazole	320	3.0	0.009	470	320	60	170	100	37	2.3	0.5
	HD 8820 low T	positive	aqueous	Polybenzoxazole	250	3				55	114	25	35	n.a.	n.a.
	HD 8921	positive	aqueous	Polybenzoxazole	250	3.0	n.a.	n.a.	260	70	130	110	33	2.1	n.a.
	HD 8930	positive	aqueous	Polybenzoxazole	200	3.1	n.a.	n.a.	240	80	170	80	25	1.8	n.a.
Hitachi	AH-1000	positive	aqueous		<200	n.a.	n.a.	n.a.	207	57	100	43	20	2.0	n.a.
JSR	WPR 1201	negative	aqueous	nanofilled Phenol resin, CA, contains	190	3.6	0.03	n.a.	210	56	90	6.4	n.a.	2.2	1.5
	WPR 5100	positive	aqueous	nanofilled Phenol resin, NQD based	190	3.5	0.02	n.a.	210	54	80	6.5	n.a.	2.5	1.5
MicroChem Corp.	SU 8	negative	organic	highly branched Epoxy	100	3	0.08	n.a.	>200	50	n.a.	n.a.	16–34	4.1	n.a.
Nippon Steel	Cardo VPA	negative	aqueous	modified acrylic	200	3.4	0.03	n.a.	180	80	7.5	11.5	14–30	2.5	1.6
Rohm and Haas	Intervia 8023	negative	aqueous	Epoxy	200	3.2	0.033	>100	160	59	90.5	9	17	4.3	0.28
Shin Etsu	SINR	negative	organic	Siloxane + others	<250	2.9–3.0	0.003	280–300	200	80–150	20–30	15–40	0.2–2.6	0.05–0.6	<0.2
Silecs	SAP 200	negative	aqueous	Siloxane	175–190	3.7	n.a.	400	>200	18	30–40	6	15	3.5	<1
Sumitomo Bakelite	Excel CRC 8000	positive	aqueous	Polybenzoxazole	320	2.9	0.01	260	550	55	118	65	n.a.	2.8	0.3
	PNBn	negative	n.a.	Polynorbornene	180–250	2.4	n.a.	4.11	280	180	18	32	n.a.	0.8	0.1
Toray	Photoneece BG 2400	negative	organic	Polyimide	>350	3.2	n.a.	n.a.	255	25	180	40	n.a.	3.9	n.a.
	Photoneece UR 5480	negative	organic	Polyimide	>350	3.2	0.002	n.a.	>350	16	150	40	n.a.	4.2	n.a.
	Photoneece PW 1000	positive	aqueous	Polyimide	>350	2.9	n.a.	n.a.	290	36	130	40	n.a.	3	n.a.

Table 18.10 Overview of selected electrolytes [31]

	Cu electrolyte	Ni electrolyte	PbSn electrolyte	Au electrolyte	Sn electrolyte
Contents	$CuSO_4$, sulfonic acid, chloric acid, grain refiner and leveler, wetting agent	$Ni(NH_2SO_3)_2$, boric acid, grain refiner and wetting agent (if necessary)	$Sn(CH_3SO_3)_2$, $Pb(CH_3SO_3)_2$, methane sulfonic acid, grain refiner, wetting agent, oxidation inhibitor	$(NH_4)_3[Au(SO_3)]$, ammonium sulfite, ammonia, organic grain refiner and leveler, complexing agents, and stabilizers	$Sn(CH_3SO_3)_2$, methane sulfonic acid, grain refiner, wetting agent, oxidation inhibitor
Metal concentration	20 g/L Cu	45 g/L Ni	Total of 28 g/L	12 g/L Au	20 g/L Sn
Temperature	25°C	50°C	25°C	55°C	25°C
pH value	<1	4.0	<1	7.0	<1
Current density	10 … 30 mA/cm²	10 … 30 mA/cm²	20 mA/cm²	5 … 10 mA/cm²	7 … 15 mA/cm²
Current efficiency	Nearly 100%	>95%	Nearly 100%	>95%	Nearly 100%
Anode material	Phosphorus alloyed copper	S-activated nickel pellets	Appropriate Pb/Sn alloys	Platinum-covered titanium	pure tin

Metals

Sputter equipment uses sputter targets of high-purity metals. Their geometrical shape of the targets has to optimized by the equipment manufacturer and is adjusted to the electrical field of the sputter equipment. For cost saving reasons, it is important to achieve maximum target utilization. Magnet targets such as Ni need special tool designs.

Electrolytes are aqueous solutions with dissolved metal salts and additives as shown in Table 18.10. The electrolytes are shipped as ready to use or as a set of concentrates that will be mixed and diluted on site. Depending on the electrolyte different kinds of analysis have to be made during the operation time to monitor the composition of the electrolyte. The chemical suppliers offer application specific solutions to maintain a long lifetime of the electrolyte.

Solder pastes are a mixture of small solder spheres (<15 µm), flux and binders. They are classified by JEDEC according to the particle size of the solder (Table 18.11).

Table 18.11 Solder Paste Classes [32]

Class	1	2	3	4	5	6	7	8
Solder sphere size [µm]	75–150	45–75	20–45	20–38	15–25	5–15	2–11	2–8

An additional function of the paste is the prefixing of the placed components on the substrate before reflow. The viscosity of the pastes should be between 250 and 550 Pa·s for screen printing and 400 and 800 Pa·s for stencil printing.

Underfiller

The reliability of flip chips assembled on carriers is limited mostly by the thermal fatigue lifetime of the solder joints. The CTE mismatch of the different materials generates stress at the interfaces and the solder joints. This has been getting worse by the introduction of FC assembly on organic substrates due to the higher CTE of the PCB compared to ceramic substrates. In 1987, Nakano et al. [33] published the highly positive influence of epoxy resin filled with silica dispensed between flip chip and board. This use of underfillers opened the introduction of the flip chip on board concept to consumer products without the need of first level package. The challenge for these filled polymeric materials is the combination of a fast process (i.e., enhanced capillary flow) and the final thermomechanical property mechanically interlocking chip and substrate. The CTE of underfillers should be close to the CTE of the solder (i.e., 23 ppm/K for eutectic PbSn) and an elastic modulus of about 10 GPa to provide an optimum solder joint creep strain relief [34]. The decreasing area array pitch will reduce the gap which has to be filled. For a 100-µm pitch the gap will be less than 50 µm. The addition of volatile additives is less favorable because this would lead to voiding and shrinkage. Most of the underfillers are based on epoxy, hardeners, and 60 to 70 percent inorganic fillers which adjust the modulus and the CTE to the required performance. Adhesion to the chip and board surface is essential to maintain the interlocking mechanism during life time. The underfiller fillet height should be at least 50% of the die thickness to minimize stresses. The underfiller is dispensed along the chip in L- or U-shape (Fig. 18.33).

One side of the die has to be open during the dispensing process otherwise air would be trapped under the chip. This can be monitored

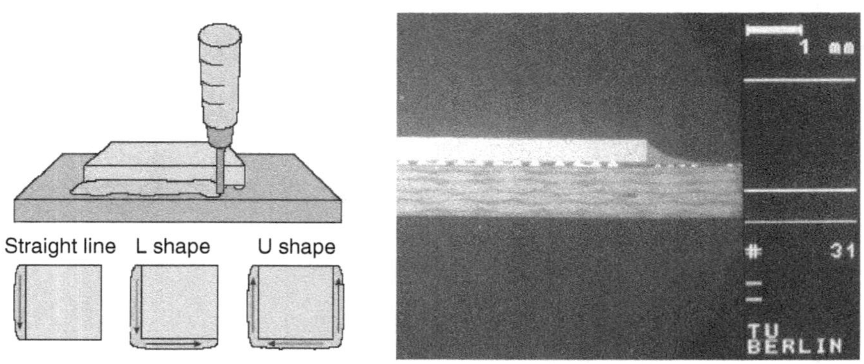

Figure 18.33 Underfill dispense and cross-cut of underfilled flip chip. (*Courtesy of Shandong Engineering Machinery Co. Ltd.*)

by acoustic microscopy. New technologies like deposition of underfiller before FC assembly are under development. In this case, the chip has to be held for around 100 ms in place to avoid the rebounding from the board. Unfortunately, the self-alignment feature is suppressed by these no-flow underfillers. In general, the main issue for the use of underfiller is the extra process step and time after reflow, which is not needed for the assembly of other SMT components. In addition, rework of thermoset-based underfiller is still not a practical manufacturing method.

FLUXES

Fluxes are needed to remove oxides during the assembly and soldering process. In addition, the tackiness allows to hold the placed components in place before the reflow and the final joining process is completed. Fluxes can be inorganic acids, organic acids, rosin, and no-clean resins. The J-STD classification describes both flux activity and flux residue activity as follows: L = low or no flux/flux residue activity; M = moderate flux/flux residue activity, and H = high flux/flux residue activity. These classes are further labeled for activity or corrosiveness.

18.4 CASE STUDIES

18.4.1 Pure Redistribution for WLP (BioTronik, Micro System Engineering, Fraunhofer IZM)

In a joint project Biotronik company, known as one of the world's leading companies of medical implantable devices, together with Fraunhofer IZM and TU-Berlin have proofed the application of WLP for pacemakers. Especially for medical implantable devices size reduction of electronic packaging is coupled with a high functionality and high reliability. Biotronik and Fraunhofer IZM jointly developed a new packaging technology for an implantable microelectronic system, where all single chip packages are WL-CSPs mounted on a 3D rigid-flex substrate [35, 36]. The WLPs were assembled on a two-layer rigid-flex board with an automatic SMT assembly line. The full assembled board for the pacemaker is shown in Fig. 18.34.

Figure 18.34 Fully assembled board for a pacemaker (WLP are marked with a dot).

Implantable microelectronic devices have to pass reliability tests which are comparable to automotive and space applications. In Table 18.12, the test conditions and the results are summarized.

Table 18.12 Test Conditions and Results for Implantable Medical Electronic Devices

1000 h HTOL	@125°C, 100 mA	Passed
1000 h high temp	@150°C	Passed
Autoclave	121°C, 2 atm, 168 h	Passed
Temp. humidity bias	(1000 h 85/85 RH)	Passed
AATC 1000 cycles	(−55°C/+125°C)	Passed
Mechanical shock	(1500 G, 0.5 ms 6 axis)	Passed
Centrifuge	(10,000 G)	Passed

The challenge of adapting WLP for a next generation pacemaker was successfully demonstrated.

18.4.2 High-Density Multichip Module for a Pixel Detector System (ATLAS Consortium)

The multichip module which is described here is a prototype for a pixel detector system for the Large Hadron Collider LHC at CERN, Geneva. The project is part of the ATLAS experiment [37]. It will study proton-proton interactions. For the pixel detector a modular system is needed which can be put together to build the large detector system. Each module is an excellent example showing highest wiring capacity together with FC assembly with a 50-μm pitch.

A module consists of a sensor tile with an active area of 16.4 × 60.4 mm, 16 read out chips, each serving 24 × 160 pixel unit cells, a module controller chip, an optical transceiver and the local signal interconnection and power distribution busses. The sensor wafer is 4 in with a thickness of 250 μm. 288 electronic chips are on one 200-mm wafer.

This extremely high wiring density, which is necessary to interconnect the read-out chips, was achieved using a thin film copper/photo-BCB process above the pixel array [38]. The silicon diode array is used as the substrate, that is, the basic building block for the detector system. The bumping of the read out chips was done using electroplating PbSn. The bump pitch is 50 μm. All dice are then attached by flip-chip assembly to the sensor diodes and the local busses (Fig. 18.35).

Two-pixel detector substrates were on each 4-in wafer. A four-layer thin film metallization is necessary for routing the power, ground and signal layers. The signal lines were designed in microstrip line configuration. The signal layers are bus lines for the interconnection of the read-out chips. The link between the silicon pixel cells and the flip chip contact pads were realized using vias of staggered and staircase type. This was only achievable by a high-density multilayer metallization. A maximum of via diameter of 25 μm was allowed by the design rules.

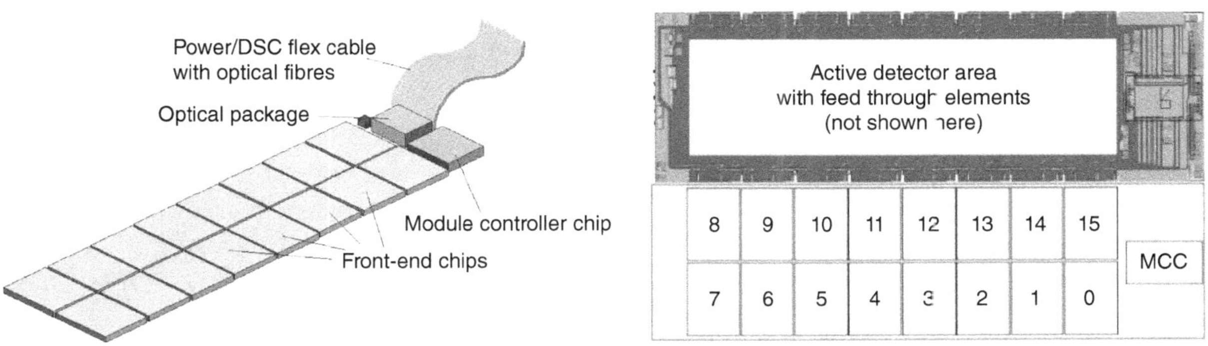

Figure 18.35 MCM-D concept for the particle detector.

A minimum isolation layer of 3 μm thickness between the metal layers was demanded. Photo-BCB was chosen as dielectric because of matching physical parameters and because of its process properties. Three micrometer copper was electroplated for the metallization. For copper lines of width $w = 20$ μm, thickness $t = 2.2$ μm with a line spacing $s = 30$ μm, a BCB dielectric thickness $h = 8$ μm, one computes for a microstrip line configuration a typical line capacitance of 1.2 pF/cm with a time of flight of about 55 ps/cm. The characteristic impedance is around 50 Ω and the voltage coupling to the neighbored line is estimated to be −20 dB. For a 7-cm long line, the signal attenuation is about 30 percent.

The process flow of the bumping technology for the electronic wafers consists of incoming inspection, sputtering/plating process including lithography and etching, bump inspection, thinning (180 μm final thickness), dicing, cleaning, FC bonding, x-ray inspection, module test and rework (in case of not working dice, a rework of chips is possible). The FC assembly can only be done using a bonder with a placing accuracy of close to 1 μm. The results prove that it is possible to build high-density MCM-Ds with more than 6000 I/Os per cm^2 as a multilayer BCB/Cu silicon substrate. Four via layers are needed for the feed through connections from the sensor pad to a pad in the uppermost Cu layer (to be used for the bump connection to the read out chips). As there are more than 61,000 interconnection structures with nearly 250,000 vias in a single module, a test program has been set up to determine experimentally the via yield of the thin film multilayer and to study the procedure of flip-chip assembly onto the MCM layers. Here four 2-μm-thick Cu layers separated by 5-μm-thick photo-BCB layers have been deposited onto monitor sensor substrates. The vias in the uppermost BCB layer (component layer) are opened to allow the solder joining of the read-out chips. Full scale modules with 16 read-out chips bump bonded to the substrate have been built. From more than 1.1 million monitored vias a defect rate of less than 10^{-5} has been monitored. An important factor of such complex module is the importance of KGD and the possibility of repair. In Fig. 18.36, the yield of modules is calculated versus the chip yield:

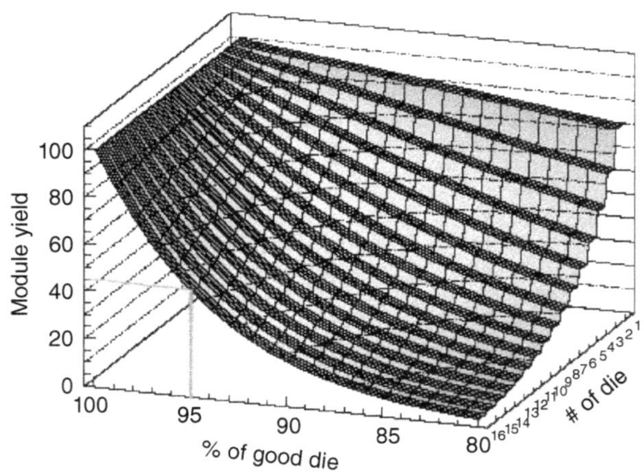

Figure 18.36 Module yield as a function of KGD and number of dice [4].

A total module yield of only 44 percent can be achieved if the chip yield is 95 percent for this 16 chip module.

18.4.3 Die Size SAW Package

Surface acoustic wave (SAW) filters are RF devices that are used in mobile communication systems. SAW devices are ideal candidates for wafer level packaging: On the one hand the increasing functionality of smart phones demands a continuous miniaturization of SAW devices. On the other hand, the bare SAW filter die is very small, often having a footprint of less than 1 mm^2, and wafer-level packaging thus is economically attractive. Also from a technical perspective, SAW devices are well suited for wafer-level packaging because, due to their small size, stress between SAW device and PCB is low and no underfill is needed.

A SAW filter consists of a piezoelectric substrate (e.g., quartz, lithium niobate, lithium tantalate) with fine electrode structures. With one electrode electronic signals are converted into acoustic (mechanical) waves that propagate along the surface of the substrate. These acoustic waves are then converted back into an electronic signal by a second electrode. The performance of the filter is defined by the electrode structures and by the mechanical properties of the piezoelectric substrate.

A wafer-level package of a SAW device can, therefore, not simply consist of an RDL added directly on top of the functional wafer. Such an RDL would alter the mechanical properties of the surface and, in consequence, the characteristics of the SAW filter. Figure 18.37 depicts

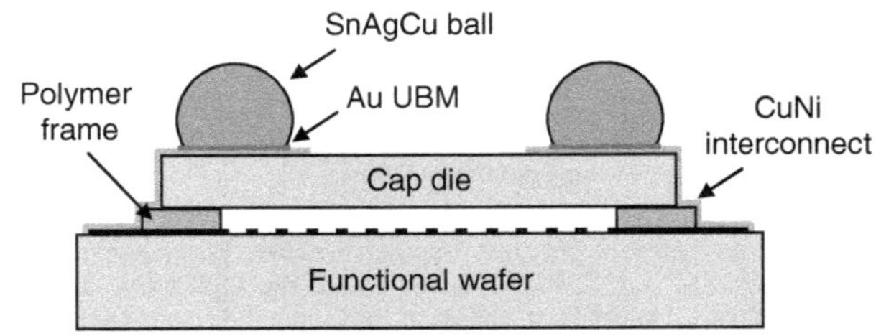

Figure 18.37 Schematic cross section of a DSSP package. (*Courtesy of EPCOS.*)

the cross-section of the die size SAW package (DSSP) developed by EPCOS. The electrode structures are protected by a cap die that is made from of same material as the functional wafer. First, a polymer frame is added to the functional wafer by a photolithography process and then the cap die is bonded onto this frame without touching the fingers of the electrodes. The cap dies produce a severe wafer topography of several hundred micrometers. A special process technology is required to realize the RDL with traces that have to interconnect the I/O pads of the original SAW device with the ball grid array on top of the cap die.

The solution is a spray coating process in combination with an angular exposure system [39]. Spray coating (s. Fig. 18.38) allows depositing a more conformal, uniform, and defect-free resist film compared to spin coating. When exposed to UV irradiation under a 90° angle, the resist film deposed along the sidewalls would require a very high exposure dosage because of its large effective film thickness. An angular exposure system exposes resist films that are coated across topographies with vertical sidewalls more efficiently and ensures better process control.

Figure 18.38 SEM image of DSSP before solder ball attach. (*Source: EPCOS.*)

Figure 18.38 shows an SEM image of a DSSP wafer after die bonding and redistribution. Solder balls are not yet attached. The metal traces that route the I/O pads of the SAW filter to the 4 × 2 ball grid array on top of the cap dies are conformal with the wafer topography. Pitch of the ball grid array is only 220 μm. Footprint of an individual die is thus only about 880 × 440 μm with a total thickness of only 250 μm. DSSP technology, therefore, offers SAW devices with an extremely small form factor.

18.4.4 Image Sensor Wafer-Level Packaging

Most wafer-level packages are flip-chip devices. Redistribution layer and solder balls are put directly onto the active area of the wafer so that the chip is assembled with its active area facing toward the PCB. For purely electronic devices flip-chip WLPs are usually suitable. A non-flip-chip WLP has been developed by ShellCase [40]. It has been adopted primarily for the wafer-level packaging of image sensors. Image sensors—unless they are backside illuminated—must be assembled to the PCB with their sensitive area facing up. A flip-chip design, therefore, is not suitable. In general, many optical and MEMS devices are not designed for flip-chip assembly because the chip surface has to interact with the environment without any restrictions caused by packaging.

Several designs have been developed by ShellCase. As an example, Fig. 18.39 depicts the cross section of the ShellOC wafer-level package [41]. The silicon wafer is sandwiched between two thin glass wafers. Between the silicon wafer and glass wafer 1 there is a structured epoxy layer that provides frames around the individual image sensors. Especially glass wafer 1 has to offer excellent optical properties and the wafer bonding process has to be accurate enough to avoid any optical interference with the performance of the image sensor. An RDL routes the I/O pads from the active device area to the bottom side of the wafer. Solder balls are then dropped onto the grid array defined by the RDL at the backside of the wafer. One advantage of this WLP design is that the silicon wafer—after being permanently bonded to glass wafer 1—can be ground extremely thin without having to deal with the common issues involved with the handling of very thin wafers.

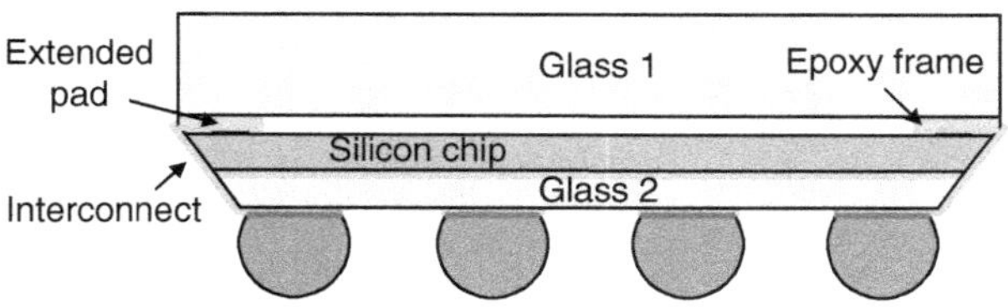

Figure 18.39 Schematic cross section of the ShellOC wafer-level package. (*Courtesy of ShellCase.*)

In the case of the ShellOC package, top and bottom side of the chip are interconnected by routing metal traces along the lateral surfaces of the individual dies. This is realized by first extending the original pads toward the edge of the die followed by notching the wafer stack (through glass wafer 2 and silicon wafer) in order to expose the extended pads. Next, an RDL is defined by photolithography that routes metal traces from within the notches onto the bottom side of the wafer stack. This is done in such a way that the redistribution traces are electrically connected to the extended pads forming so-called T-contacts.

In the last few years, new WLP technologies were developed for image sensors aiming at replacing the T-contacts with through-silicon-vias (TSVs). One example is the OPTO-WLP [42] developed by SCHOTT. A schematic cross section of this WLP is shown in Fig. 18.40. After bonding the silicon wafer to a glass wafer, the silicon wafer is thinned to a final thickness of about 50 to 100 μm. Tapered TSVs are then etched into the silicon at the location of the original bonding pads. The ball grid array on the bottom side of the wafer is electrically connected to the bonding pads though the TSVs. One advantage of this WLP design is that the dicing streets can be narrower offering more space on the wafer for the image sensor devices.

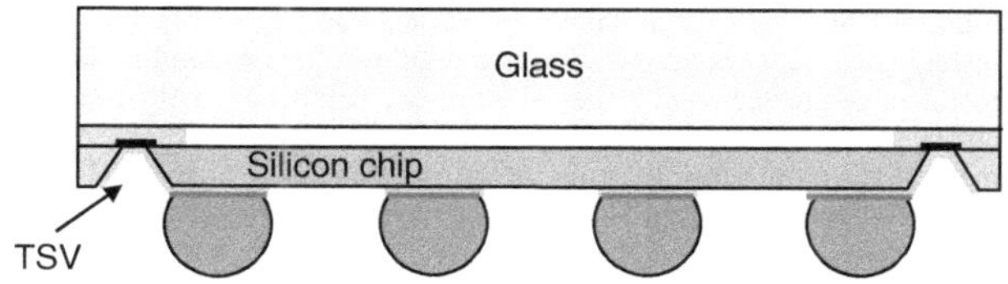

Figure 18.40 Schematic cross section of the OPTO-WLP wafer-level package. (*Courtesy of Schott.*)

The redistribution traces have to be isolated from the silicon. Therefore, usually a combination of inorganic and organic passivation layer is deposited on the backside of the wafer and within the TSVs, where the organic layer also serves as a stress buffer. The inorganic layer will often be silicon dioxide, while the organic passivation may, for example, consist of an epoxy resin. Similar to the DSSP technology discussed under Sec. 18.4.3, organic layers, photoresist, and inner dielectric layer have to be deposited conformally across the TSVs. This is usually accomplished by spray coating. A further improvement of this packaging concept will be achieved by the usage of microstructured glass. This new deposition technology will add hermetic sealant to MEMS WLP [43].

18.5 OPTOELECTRONICS AND MEMS PACKAGING

Optoelectronics and MEMS impose new challenges on the packaging, test, and assembly of micro devices. Typical MEMS devices include inkjet printer heads, magnetic read/write heads, micro mirror arrays, pressure, and acceleration sensors and gyroscopes. Optoelectronic systems range from discrete components such as LEDs and semiconductor laser diodes to complex modules combining active and passive optical and electronic components on one board (Fig. 18.41). Optical and mechanical

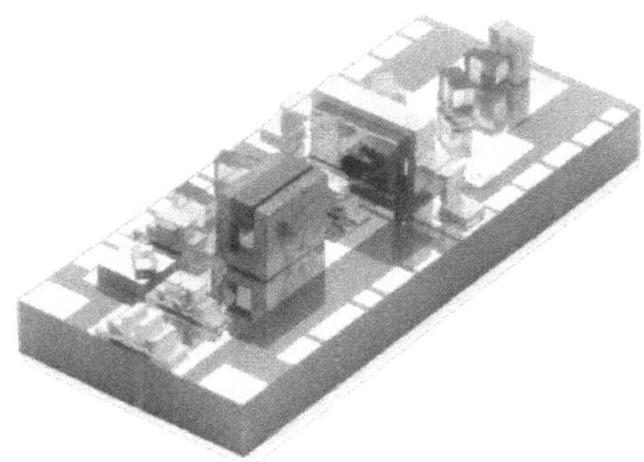

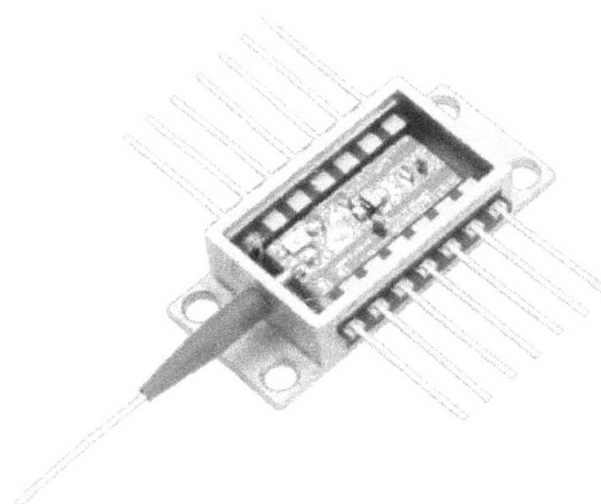

Figure 18.41 Optoelectronic device with active and passive optical components assembled on a silicon substrate. (*Courtesy of Axun Technologies.*)

functions have to be maintained by package and assembly. Chemical or biological sensors need packages that allow chemical and biological reactions of the device with the environment. All this has a big influence on the package design, material selection and—in consequence—on the manufacturing process. Existing infrastructure is used to package these devices wherever possible. However, packaging often represents a barrier that prevents new devices from being commercialized.

Placement accuracy is often more critical than for pure electronic devices. Semiconductor lasers, for example, often have to be aligned to a fiber with a precision better than 1 μm in order to get sufficient optical coupling between both components. This can be accomplished by either passive alignment or high accuracy active alignment, which requires different tool sets than standards SMT assembly. High precision assembly of a laser diode, for example, can be accomplished by measuring the light coupling into the fiber and feeding this information back to the alignment system. Another important aspect of optoelectronic and MEMS packaging is that standard pick and place equipment can damage delicate devices with functioning parts at its surface.

For the assembly of complex modules the possibility of bare die assembly and the self-alignment capability of flip chip bonding make this interconnect technique attractive for optoelectronics and MEMS. Many components, however, do not allow flux reflow as flux residues can deteriorate the device performance. In addition, the metal coating of the board is often different from pure electronic products. Optoelectronics, for example, often use boards with gold pads in order to avoid corrosion and to have good electrical and thermal conductivity. Gold coatings are ideal for flux-free soldering because no oxides form on the gold surface. The different board metallization and the requirement for flux free reflow lead the use of different bump metallurgies than in standard flip chip. When reflowing high-Sn solders on Au films the brittle $AuSn_4$ phase is formed during reflow as soon as sufficient Au is dissolved by the solder. An alternative are solders with low Sn content such as $Au_{80}Sn_{20}$. AuSn bumps are extensively used in the assembly of optoelectronics since AuSn bumps are relatively hard and do not tend to deform allowing to maintain a high alignment accuracy over time. Flux-free assembly of AuSn bumps is accomplished by reflowing the solder in a reducing atmosphere.

Hermetic packaging is a key function for most of the MEMS packages. Therefore, the cost of packages based on expensive sealing techniques sometimes exceeds the cost of the bare MEMS device. Besides hermeticity, the controllability of the cavity ambient can be an important issue for sensors [44]. For example the time response of a microrelay depends on the ambient pressure. Wafer bonding of structured glass wafers offers the opportunity of combining low cost WLP with hermetic seal. Different techniques have been developed over the past few years.

18.6 REFERENCES

1. H. Oppermann, *Vorlesung: Aufbau- und Verbindungstechnik für Mikrosysteme*, TU Cottbus, 2003.

2. Lit W. Jillek and G. Keller, *Handbuch der Leiterplattentechnik*, Eugen Leuze Verlag, Germany, 2003.

3. J. Lau and K. Liu, "Lead-free Soldering," *Advanced Packaging Magazine*, p. 28, Feb. 2004.

4. *Advanced IC Packaging Markets and Trends*, 7th ed., Electronic Trend Publication, 2003.

5. E. G. Combs, "Leadless Plastic Packages, Such as the DFN and QFN, Have Inspired a Renaissance in a Mature Technology," *Chip Scale Review*, 75–77, Mar. 2003. https://scholar.google.com/scholar?q=Chip+Scale+Review%2C+75%E2%80%9377%2C+Mar.+2003.&btnG=&hl=en&as_sdt=0%2C5&as_vis=1.

6. K. Iwabuchi, "CSP Mounting Technology," *Proc. SEMI Technology Symposium* 1996, Chiba/Japan and K. Kosuga, "CSP Technology for Mobile Apparatuses," *Proc. International Symposium on Microelectronics*, Philadelphia, Oct. 1997.

7. P. Garrou, "Wafer Level Packaging Has Arrived," *Semiconductor International*, 119, Oct. 2000. https://www.highbeam.com/doc/1G1-67000661.html.

8. P. Garrou, "Wafer Level Chip Scale Packaging (WL-CSP): An Overview," *IEEE Trans. Adv. Packaging*, 23(2), 2000. http://ieeexplore.ieee.org/document/846634/.

9. M. Töpper, J. Simon, and H. Reichl, "Redistribution Technology for CSP using Photo-BCB," *Future Fab International*, 363, 1996.

10. R. A. Fillion, R. J. Wojnarowski, and W. Daum, "Bare Chip Test Techniques for MCM," *Proc. EIA/IEEE Electron. Comp. Technology Conf.*, Las Vegas, p. 554, 1989.

11. M. Töpper, K. Buschick, J. Wolf, V. Glaw, R. Hahn, A. Dabek, O. Ehrmann, et al., "Embedding Technology—A Chip First Approach using BCB," *Advancing Microelectronics Vol. 24*, No. 4, Jul./Aug. 1997.

12. Infineon Presents SOLID, A World First 3D Chip Integration Technology—"Chip-Sandwich" Offers Way Out of the Wiring Crisis, Aug 12, 2002. http://www.infineon.com/cms/en/about-infineon/press/market-news/2002/128975.html.

13. M. Töpper, K. Scherpinski, H.-P. Spörle, C. Landesberger, O. Ehrmann, and H. Reichl, "Thin Chip Integration (TCI-Modules)—A Novel Technique for Manufacturing Three Dimensional IC-Packages," *Proceedings IMAPS 2000*, Boston, US, Sep. 2000.

14. R. Ulrich and L. Schaper. "Integrated Passive Component Technology" Wiley Interscience/IEEE Press 2003.

15. van Roosmalen, "There Is More Than Moore," *5th Int. Con. on Mech. Sim. and Exp. in Microelectronics and MST*, EuroSim, 2004.

16. B. Levine, "System-in-Package: Growing Markets, Ongoing Uncertainty," *Semiconductor International*, 47–61, Mar. 2004.

17. R. Tummala, "System-on-Package Integrates Multiple Tasks," *Chip Scale Review*, 53–56, Jan./Feb. 2004.

18. L. Dietrich, J. Wolf, O. Ehrmann, and H. Reichl, "Wafer Bumping Technologies Using Electroplating for High-Dense Chip Packaging," *Proc. Third Intern. Symposium on Electronic Packaging Technology* (ISPT'98), Beijing (China), Aug. 17–20, 1998.

19. U. Müller, "Anorganische Strukturchemie," Teubner Verlag, 3. Auflage, 1996.

20. S. Anhöck, A. Ostmann, H. Oppermann, R. Aschenbrenner, and H. Reichl, "Reliability of Electroless Nickel for High Temperature Applications," *Intern. Symposium of Advanced Packaging Mat. Conf.*, Braselton, US, Mar. 1999.

21. T. Kawanobe, K. Miyamoto, and Y. Inaba, "Solder Bump Fabrication by Electrochemical Method for FC Interconnection," *Proc. of IEEE Electronics Components Conf., S.* 149, May 1981.

22. M. Töpper, A. Achen, and H. Reichl, "Interfacial Adhesion Analysis of BCB/TiW/Cu/PbSn Technology in Wafer Level Packaging," *Electronic Components and Technology Conference (ECTC)*, New Orleans, Louisiana, pp.1843–1846, May 2003.

23. M. Töpper, Th. Stolle, and H. Reichl, "Low Cost Electroless Copper Metallization of BCB for High-Density Wiring Systems," *Proc. of the 5th Internat. Sympos. on Advanced Packaging Materials*, Braselton, US, Mar. 1999.

24. K. W. Paik, R. J. Saia, and J. J. Chera, "Studies on the Surface Modification of BCB Film," *Proc. MRS*, Boston, Nov. 1990.

25. F. Krause, K. Halser, K. Scherpinski, and M. Töpper. "Surface Modification due to Technological Treatment Evaluated by SPM and XPS Techniques," *Proc. MicroMat*, Berlin, Apr. 2000.

26. P. Chinoy, "Reactive Ion Etching of Benzocyclobutene Polymer Films," *IEEE Trans. Comp. Packaging and Manufact. Tech., Part C*, 20 (3): 199–206, 1997.

27. K. Zoschke, J. Wolf, M. Töpper, O. Ehrmann, Th. Fritzsch, K. Scherpinski, F.-J. Schmückle, et al. "Thin Film Integration of Passives—Single Components, Filters, Integrated Passive Devices," *Proc. of the 2004th ECTC Conf.* 54(1), July 2004.

28. G. Carchon, S. Brebels, K. Vaesen, P. Pieters, D. Schreurs, S. Vandenberghe. W. DeRaedt, et al. "Accurate Measurement and Characterization of MCM-D Integrated Passives up to 50 GHz," *Proc. Int. Con. and Exh. on HDI and System Packaging*, Denver, CO, pp. 307–312, 2000.

29. G. Carchon, P. Pieters, K. Vaesen, W. DeRaedt, B. Nauwelaers, and E. Beyne, "Multilayer Thin Film MCM-D for the Integration of High-Performance Wireless Front-End Systems," *Microwave Journal*, 44: 96–110, 2001.

30. W. Daughton, "An Investigation of the Thickness Variation of Spun-on Thin Films Commonly Associated with the Semiconductor Industry," *J. of Electrochemical Society*, 129 (1): 173–179, 1982.

31. L. Dietrich, J. Wolf, O. Ehrmann, and H. Reichl, "Wafer Bumping Technologies Using Electroplating for High-Dense Chip Packaging," *Proc. Third International Symposium on Electronic Packaging Technology (ISPT '98)*, Beijing, China, Aug. 17–20, 1998.

32. J. Schake, "Stencil Printing for Wafer Bumping," *Semiconductor International*, pp. 133–134, Oct. 2000.

33. F. Nakano, T. Soga, and S. Amagi, "Resin Insertion Effect on Thermal Cycle Resistivity of FC Mounted LSI Devices" *Proc. of ISHM Conf.*, pp. 536–541, Sep. 1987.

34. A. Schubert, R. Dudek, D. Vogel, V. Michel, and H. Reichl, "Thermo-Mechanical Reliability of FC Structures Used in DCA and CSP," *Proc. of the International Symposium on Advanced Packaging Materials*, pp. 153–160, 1998.

35. M. Töpper, M. Schaldach, S. Fehlberg, C. Karduck, C. Meinherz, K. Heinricht, V. Bader, et al. "Chip Size Package—The Option of Choice for Miniaturized Medical Devices," *Proc. IMAPS Conf.*, San Diego, US, 1998.

36. M. Schaldach, M. Töpper, J. Müller, M. Starke, T. Tessier, O. Ehrmann, and H. Reichl, "State-of-the-Art Technology Development for Medical Implantable Systems," *Proc. IMAPS 98*, San Diego, US, Oct. 1998.

37. K.-H. Becks, E. H. M. Heijne, P. Middelkamp, L. Scharfetter, W. Snoeys., "A Multi-Chip Module, the Basic Building Block for Large Area Pixel Detectors," *Proc. of IEEE MCM Conf.*, 1996.

38. M. Töpper, P. Gerlach, L. Dietrich, S. Fehlberg, C. Karduck, C. Meinherz, J. Wolf, et al., "Fabrication of a High-Density MCM-D for a Pixel Detector System using a BCB/Cu/PbSn Technology," *Intern. J. of Microcircuits & Electronic Packaging*, 22 (4): 305–311, 1999.

39. B. L'huillier, M. Hornung, D. Tönnies, M. Jacobs, C. Bauer, F. Hammer, T. Heuser, et al., "Application of an Angular Exposure System to Fabricate True-Chip-Size Packages for SAW Devices," *18th European Microelectronics and Packaging Conf.*, 94–99, Brighton, Sep. 2011.

40. A. Badihi, "Ultrathin Wafer Level Chip Size Package," *IEEE Trans. on Advanced Packaging*, 23 (2): 212–214, May 2000.

41. M. Bartek, S. M. Sinaga, G. Zilber, D. Teomin, A. Polyakov, and J. N. Burghartz, "Shellcase-Type Wafer-Level Packaging Solutions: RF Characterization and Modeling," *Proc. of the 1st IWLPC Conf.*, San Jose, Oct. 2004.

42. J. Leib and M. Töpper, "New Wafer-Level-Packaging Technology Using Silicon-Via-Contacts for Optical and Other Sensor Applications," *Proc. 54th Electronic Components and Technology Conf.*, Las Vegas, pp. 843–847, Jun. 2004.

43. D. Mund and J. Leib, "Novel Microstructuring Technology for Glass on Silicon and Glass-Substrates," *Proc. 54th Electronic Components and Technology Conf.*, Las Vegas, pp. 939–942, Jun. 2004.

44. E. Parton and H. Tilmans, "Wafer-Level MEMS Packaging," *Advanced Packaging Magazine*, pp. 21–23, Apr. 2004.

18.7 FURTHER READING

Elshabini-Riad, F. D. Barlow, eds., *Thin Film Technology Handbook*, McGraw Hill, New York, 1997.

Garrou, P., and I. Turlik, *Multichip Module Technology Handbook*, McGraw-Hill, 1998.

Hwang, J., *Modern Solder Technology for Competitive Electronic Manufacturing*, McGraw-Hill, New York, 1996.

Lau, J., *Ball Grid Array Technology*, McGraw-Hill, New York, 1995.

Lau, J., and S. W. Ricky Lee, *Chip Scale Package*, McGraw-Hill, New York, 1999.

Messner, G., I. Turlik, J. W. Balde, and P. Garrou, *Thin Film Multichip Modules*, ISHM Publication, Sandborn, 1992.

Puttlitz, K., and P. Totta, *Area Array Interconnection Handbook*, Kluwer Academic Publishers, Cambridge 2001.

Qu, Shi, and Yong Liu, *Wafer-Level Chip-Scale Packaging*, Springer, New York, 2015.

Reichl H., *Hybridintegration*, Verlag Hüthig, Heidelberg, 1988.

Reichl, H., Direktmontage, Springer Verlag, 1998.

Seraphim, D. P., R. Lasky, and Che-Yu Li, *Principles of Electronic Packaging*, McGraw-Hill, New York, 1989.

Tummala, R., E. Rymaszewski, and A. Klopfenstein, *Microelectronic Packaging Handbook*, Parts 1–3, Chapman & Hall, London, 1997.

Chapter 19
Bonding Fundamentals

BY
IVY QIN *Kulicke and Soffa Industries, Inc., Fort Washington, Pennsylvania*

19.1 INTRODUCTION

19.1.1 History and Current Capability

One of the most important steps in semiconductor packaging is to connect the IC to the substrate. This step is called *interconnect.* There are several interconnect technology including wire bonding, flip chip, and TAB bonding. Among the different interconnect technologies, wire bonding is the most popular interconnect method due to its low cost, flexibility, and large installed base. Wire bonding uses conductive wires such as Au and Cu to connect the die to the substrate.

The first wire bond ever made can be traced back to the first transistor born in Bell Labs in 1947. In the early years, there are only a few wires in each package. Wire bonding is largely a manual operation. The operator will move the XY stages and Z axis manually to perform various operations. Figure 19.1 shows an early device bonded with wire bonder (around 1962) [1].

Wire bonding has evolved over the years. In 1972, K&S introduced the first automated wire bonder. Today's wire bonding is much more sophisticated. Figure 19.2 shows advanced high pin count wire bonded device with over one thousand wires [2]. Wire bonder has evolved into fully automatic machines with high speed, high precision servo system, ultrasonic transducer for the bonding operations, and automatic vision system for device alignment. The throughput and pitch capability have improved dramatically over the years.

19.1.2 Different Type of Wire Bonding

There are two different wire bonding technologies: ball bonding and wedge bonding. The main difference between a ball bonding and a wedge bonding process is that the ball bonding process starts with a ball that is formed from the wire, and the first bond is a ball bond; while the wedge bonding don't form a ball and bond the wire on the first bond location. Figures 19.3 and 19.4 show the first bond and wire loop for a ball bonding and a wedge bonding process. Most of this chapter focuses on ball bonding process flow and equipment details because ball bonding is the most popular wire bonding method.

For ball bonding process, the main steps are illustrated in Fig. 19.5. Bonding cycle starts with a ball formation. The ball is then seated inside the bonding tool that is called a capillary. The initial unbonded ball is sometimes referred to as a free air ball (FAB). Then bonding tool descends to the first bond location, for example, the bond pad and makes the first bond. After the first bond, the bonder performs a set of looping motions to form the wire bond loop. The bonding tool descends to the second bond, normally the substrate lead finger. After the second bond, the bonding tool rises to tail length and then breaks the wire tail that is used to form the free air ball for the next wire.

The basic steps for a wedge bonding process are similar to ball bonding that including bonding the first and second bond, and performing the looping motions. Wedge bonding does not form a free air ball. Wedge bonder often has additional mechanism to help tear the wire and feed the tail after the second bond.

The bonding tools used in ball bonding are called capillaries, which are axial-symmetric ceramic tools with vertical feed holes. The bonding tools used in wedge bonding are called wedges, which are generally made of tungsten carbide or titanium carbide with an angled feed hole at the bottom of the wedge. Figure 19.6 shows wire bonding tools used in fine pitch applications. The tool tips are shaped to give the clearance needed in fine pitch bonding.

Since the ball bonder can loop in different angles from the first bond, no rotational (theta) axis is required. The table and bond head move in X, Y, and Z directions. A rotary bondhead is needed on the wedge bonders to bond wire at different angles. The rotational motion in the rotary bond head requires a theta motor and additional moving parts. The moving mass on a wedge bonder is much heavier than the bond head on a typical ball bonder, resulting in slower speed. Currently, the state-of-the-art ball bonder speed is more than double that of the fastest wedge bonder. Due to its higher speed, lower cost, and more flexible looping capability, ball bonding is the most commonly used interconnect method today.

On the other hand, wedge bonding has several advantages over ball bonding. Large wire sizes up to 500 μm in diameter enable high current capability in power semiconductor devices. Ribbon wires can further extend the current-carrying capabilities, especially in high-frequency applications because of the skin effect. Wedge bonding process enable bonds on Al, Au, and Cu, where both wire and bond pad are the same materials. Interconnects between the same materials are known to be highly reliable. Because of the high-reliability bonds, the wedge bonding technology is used in a number of space, medical, military, and automotive applications. Deep access bonding tools enable high pitch bonds and very low loops are possible with small wire feed angles.

Figure 19.1 K&S 422 model wire bonder bonds a 5-mil aluminum wire to a silicon transistor chip.

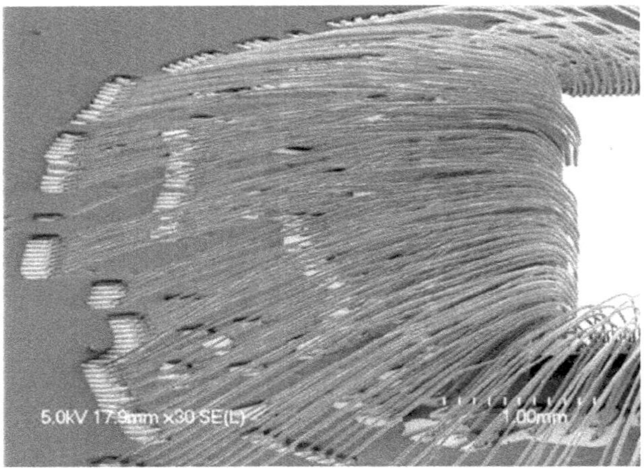

Figure 19.2 Advanced high pin wire bonded device (2015). Wire is 0.7-mil Pd-coated Cu wire.

19.2 WIRE BONDING EQUIPMENT

19.2.1 Wire Bonder Subsystems

To be able to accomplish wire bonding process at high speed, very fine scale and with highest reliability, the design of the wire bonders have gone through many years of innovation. An automatic ball bonder is a precision machine with many advanced hardware, software, servo control algorithms, vision algorithm, and advanced process controls. A picture of a K&S fully automatic bonder is shown in Fig. 19.7. Some of the key subsystems are labeled. The main subsystems and their functionality are explained below.

Wire Handling System

The main purpose of the wire feed system is to feed the wire from the wire spool to the tip of the capillary. A second function of the wire feed system is to provide adequate tension to the wire to prevent excessive wire vibration (whipping) during the high speed XYZ motions. Another function of the wire-handling system is to provide a ground path to the wire. The key components of the wire-handling system include spool mount; air guide that provides the control of wire feeding and provide tension to the wire; wire tensioner that provides additional wire tension and also constrains the wire above the wire clamp; and wire clamp that can close on the wire when needed.

For the wire-handling system to perform well, all surfaces the wire touches should be smooth to avoid wire damage and wire feeding issues. To prevent wire damage, wire clamp parallelism, proper tension, and alignment are needed.

XY Table, Bondhead, and Optics System

The motions during a bonding cycle is carried out by the XYZ servo systems. The XY table provides the motions in X and Y axis. The high-powered XY table is capable of moving at well over 10 G of acceleration. The position of the system is controlled by high-precision encoders with sub-micrometer resolution. The most advanced wire bonders have rigid roller bearing table design to maintain accuracy at speed.

XY table carries the bondhead and optics systems. The bondhead is the mechanism that performs the bonding motion and it travels in Z direction in high speed. The bondhead travels in z direction and has its own encoder. Limiting the weight of the moving mass is important to achieve high speed. Special lightweight, high stiffness, low CTE

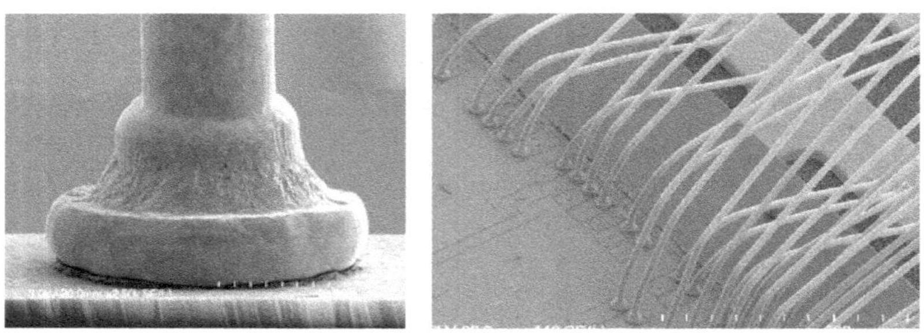

Figure 19.3 First bond and loop shape from a ball bonding process.

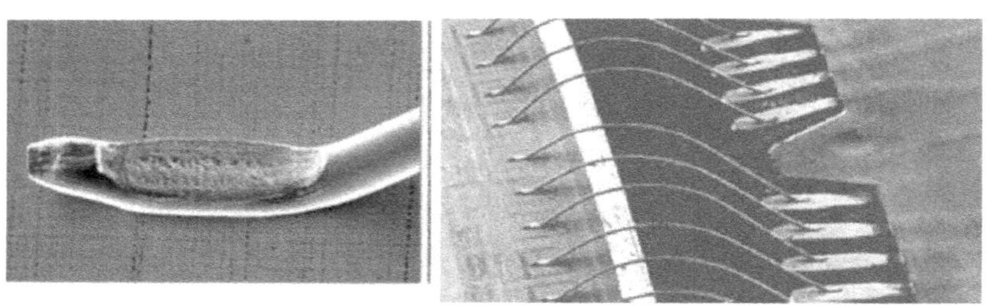

Figure 19.4 First bond and loop shape from a wedge bonding process.

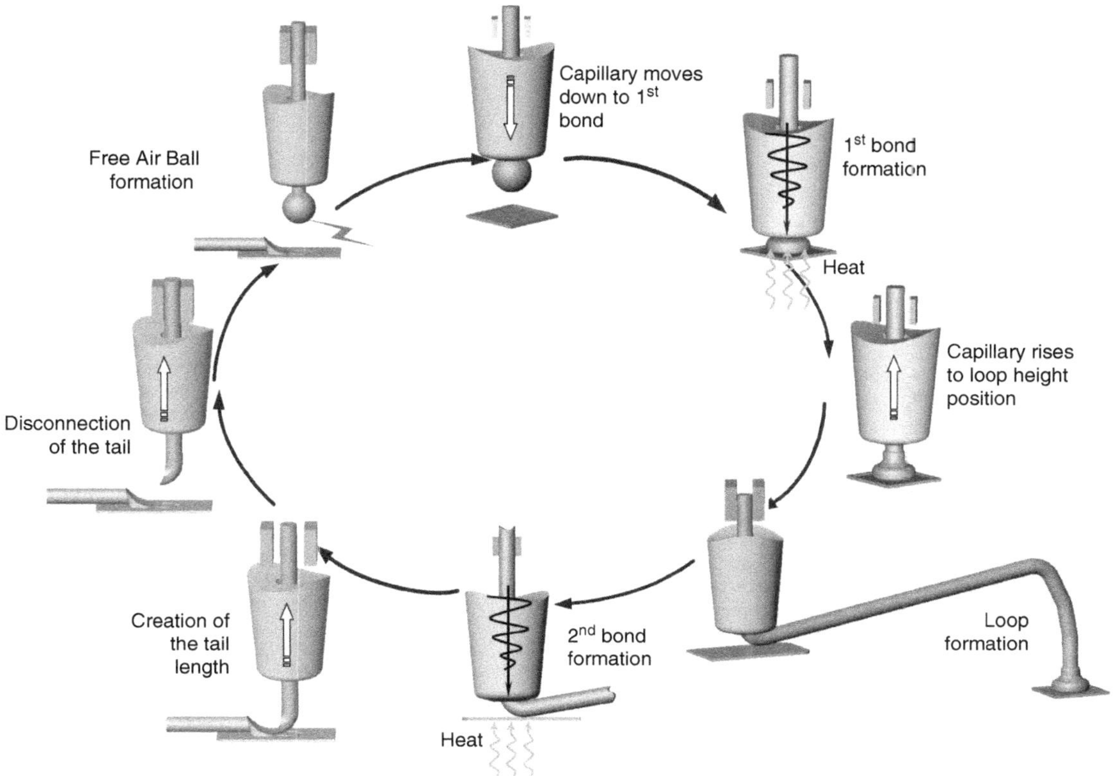

Figure 19.5 A ball bonding wire cycle.

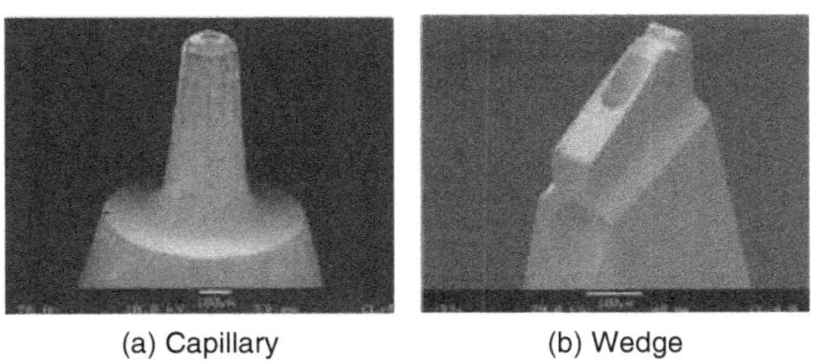

Figure 19.6 Ball bonding and wedge bonding tools.

materials are chosen for the moving components. Another important function of the bondhead is to apply accurate and repeatable bond force needed for the bonding process. The most advanced wire bonder has the ability to control the Z axis bonding force down to levels of 1 g or less.

The optics system provides the vision function to the bonder. Optics system is used to detect the location of the die and substrate and align the device so that the bonding locations are known. Sophisticated pattern recognition software (PRS) can learn the features on the die and substrate, these features are called the "eye points." After a program is taught, the vision system can automatically find these eye points and align the device accurately. The bonder then knows where each bond should be placed.

Material Handling System

Material handling system transports the unbonded devices from the input magazine to the bond site, after bonding is completed, it transports the bonded device to the output magazines. Wire-bonded devices are normally come in strip format such as leadframe and BGA substrates. They are transported in magazines.

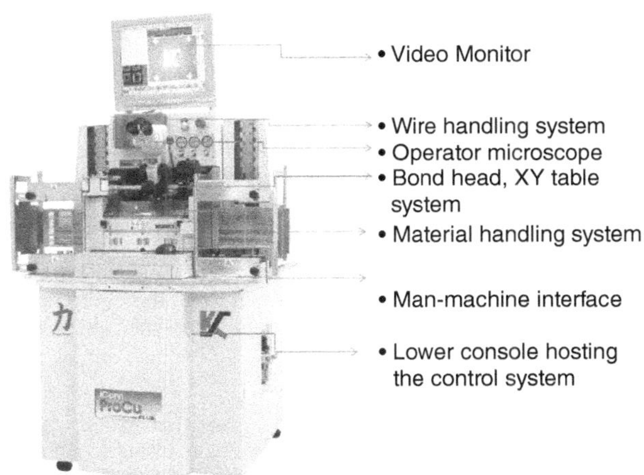

Figure 19.7 Picture of K&S IConn Bonder with the key subsystems labeled.

One of the key specifications for the material system is the capability of handling different size leadframes. Large format leadframes with hundreds or even thousands of devices per strip are more and more common. Figure 19.8 shows wide QFN (quad-flat no-leads) and leadframes that contain many devices on a single strip. The wire bonders in production today can handle up to a 100-mm-width leadframe.

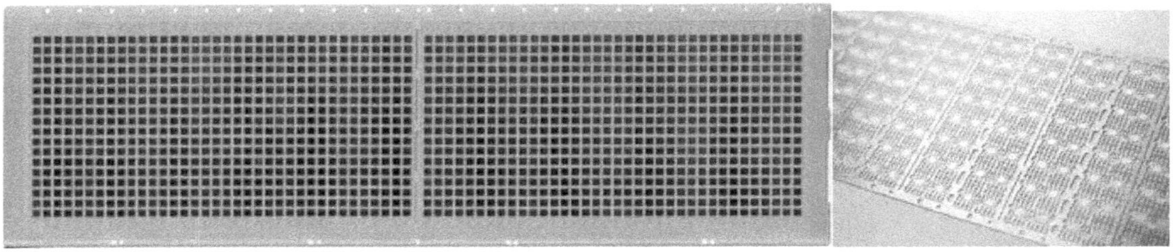

Figure 19.8 Large format leadframe and substrate.

Software and Control System

Wire bonder software performs many functions such as teaching the device dimensions, teaching wiring diagrams, programming bonding parameters, performing various calibrations, setting up machine configurations, aligning the device, performing various XYZ motions, controlling bond force and ultrasonic outputs, and collecting performance data such as productivity and assist statistics. There are a vast collection of different features on the machines for basic and advanced functions for very diverse applications. These applications range from single-wire LED device with thousands of devices on each strip, to mobile processors with over 1000 wire and a few hundred parameter sets per device. There are many electronic boards to support various software and hardware functions of the machine. The electronic boards include main CPU board, vision CPU board, servo amplifier boards, servo controller boards, ultrasonic generator board, wire clamp driver board, temperature controller boards, and material handling system boards. High-speed communication lines connect these boards to ensure timely communications between them for proper function of the bonder.

19.2.2 Wire Bonder System Consideration

The most important wire bonder system requirements are throughput, accuracy, fine pitch bonding capability, alternative wire bonding capability (such as Cu and Ag wire) and wire size capability. Through the years, wire bonders went through many technology advances to improve the speed of wire bonding and fine pitch capability. Twenty years ago, the state-of-the-art wire bonders had 80-μm bond pitch capability (reference bonder K&S 1488 Plus), today the most advanced wire bonders can support 35-μm bond pitch in production. The finest pitch device in mass production is about 40 to 45 μm pitch with both Au and Cu wire. The speed of the wire bonder also more than doubled during the same period of time. The cost of wire bonder also comes down over the years. Coupled with the throughput increase, the cost for wire bonding interconnect is getting cheaper. All these improvements keep the wire bonder as the most popular interconnect method. Next we will go into more detail about these system performances.

Throughput

Wire bonder throughput is the key to support the volume requirement of our industry. The total device cycle time includes into three components including the bonding time, material indexing time, and vision operation time. Bonding time is the time the bonder spends bonding wires while going through the wire cycle as shown in Fig. 19.5. Indexing time and vision time are considered the nonbonding time. Indexing time is the time the machine spends on transporting the material. Vision time is the time the bonder spends for eye point finds and alignment of the device. The percentage of each time component is highly depended on the application. For very low pin count devices, index and vision time can take up significant amount of time. For high pin count devices, bonding time takes up majority of the time. For wire count over 500, bonding time percentage is normally over 98 percent of the total time. To improve throughput of high pin count device, reducing the bonding time is the key.

Wire cycle time can be further broken down into the following components.

- *High-speed motion time.* This is the time traveling from bond to bond. This time mainly depends on the raw servo speed of the bonder. Faster wire bonders have more powerful motors, lighter pay load that enables faster speed.
- *Bond time.* This is the time to spend bonding first bond and second bond. This time depends on the surface condition of the bonding surfaces. For example, in the case of Cu wire bonding, if the ball is even slightly oxidized, it is hard to form intermetallic bond, it may require longer bond time. Transducer performance also affects time. Higher-frequency bonding may reduce bonding time compared to lower-frequency bonding because more scrubbing can be achieved in shorter amount of time.
- *Looping time.* This is the time the bonder performs the looping motion to form the desired loop shape while traveling to second bond location. Raw machine speed, wire feed system as well as loop shape and performance requirements affect this time.

On the fastest automatic wire bonder, easy applications with simple loop profile and easy to bond surfaces has wire cycle less than 100 ms, which means that the wire bonder can bond more than 10 wires per second. More complex applications may have a wire cycle time more than 100 ms. For some special configuration such as die-to-die bonding with extra bump underneath the second bond, wire cycle time is longer. The low pin count applications tend to have shorter wire cycles because the loops tends to be simple and bond time tends to be shorter for simple first and second bond requirements. On the other hand, the high pin count applications require more complex loop shapes that have longer looping time.

Accuracy

The accuracy requirement of the wire bonders is closely linked to the fine pitch requirement. To be able to place the bond inside small bond pads, the machine has to be accurate. Like the speed of the wire bonder, accuracy was also improved over the years. Accuracy is often defined as 3σ XY placement repeatability. Around 20 years ago, the best wire bonder has 10 μm accuracy (e.g., K&S 8020 model ball bonder). Today, the most accurate ball bonder models have 2 μm accuracy and with 35 μm pitch capability.

Accuracy measurements are normally done with an optical microscope. The distance from the ball center to pad center are measured in both X and Y direction. Large sample size is requirement to have a good assessment of the accuracy. Accuracy measurement can be performed manually with operators manually measure the pad location and the ball location. There are also automatic measurement equipment that can automatically find the bond pad and ball. Automatic measurement equipment is often used in large-scale manufacturing facilities while the manual or semiautomatic measurements are often used in R&D and small production environment. A quick assessment of the placement accuracy is often done on the wire bonder by examining the live video image of the bonded ball inside the bond pad opening.

Some of the key accuracy contributors include XY table accuracy, vision system teach and find accuracy, and machine thermal stability over time and temperature.

FINE PITCH CAPABILITY

The pitch capability of the wire bonder is one of the key specifications. The most advanced wire bonders have 35 µm pitch capability while finer pitch such as 30 µm and even 25 µm have been demonstrated in a small lab scale. Thirty-five-micrometer bond pad pitch capability had been demonstrated on the K&S ProCu Plus bonder using 13-µm (0.5-mil) Pd-coated Cu wire [3]. In mass production, the finest pitch process is about 40 µm pitch for Au and Cu wire. The majority of the fine pitch devices running in mass production today are 50 µm bond pad pitch or above. In fact, 50-µm bond pad pitch with 18 µm (0.7 mil) PdCu wire are one of the most common configurations in high pin count application for mobile devices.

MACHINE PORTABILITY, ASSIST RATE, AND YIELD

Each year, hundreds of billions of devices are made with wire bonders. In the large assembly houses, thousands of wire bonders are being used at any given time to produce these parts. In order to ensure smooth production, bonder to bonder portability is required. Bonder portability means all the bonders can use the same process recipe (process programs) to produce parts with acceptable outcome. To ensure bonder portability, the design of the hardware needs to be robust, the manufacture of the hardware needs to be repeatable. There are also many calibrations on the bonder to ensure repeatable behaviors. These calibrations include optical illumination level, bond force, ultrasonic output, servo system calibration, temperature calibration of the heatblock, XY table map calibration, etc.

Assist rate is another important consideration. Some failure types are vision alignment issue, material indexing issues, and bonding issues. Proper bonder setup and calibration are need to lower assist rate. The cleanness of the bonding material is a major factor. Some of the assists can be recovered by operators without yield loss, some will result in yield loss.

Automatic assist recovery schemes have been developed recently to recover from bonding errors such as no stick on pad, no stick on lead and short tails. These routines can help reduce the number of times operator has to assist the machine. It has shown to improve net throughput and yield. We will discuss these automatic recovery features in the first bond and second bond optimization sessions.

19.3 WIRE BONDING PROCESS

There are several important aspect of wire bonding process. They will be discussed next.

19.3.1 Free Air Ball Process

Free air balls used in ball bonding are formed by the electronic flame off (EFO) discharge. The EFO system consists of an EFO box that supplies the high voltage needed for the arc over, the EFO cable, EFO wand, which serves as the first electrode, and bonding wire which serves as the second electrode. Negative voltage is applied to the EFO wand at the beginning of the EFO fire. Since the wire is grounded, the EFO wand is the negative electrode in this case. The air gap between EFO wand and the grounded wire is ionized by the high voltage. Heat generated by the electrical discharge melts the wire from the wire tip upward, surface tension causes the molten metal to roll up into a ball. Numerical and experimental studies of the electronic discharge and subsequent heat transfer can be found in [4–6].

The common specifications for the EFO box are high voltage capability of 4 to 5 kV, current capability to 20 to 250 mA for regular application and up to 500 mA for heavy wire application.

The wire "tail" is formed from the second bond process from the previous wire. Figure 19.9 illustrates how the capillary forms the second bond, which includes a stitch bond and a tail bond. The stitch bond is what is left behind as the second bond for the previous wire. The tail bond is separated from the substrate after the bond head lifts off for the next wire's flame off position.

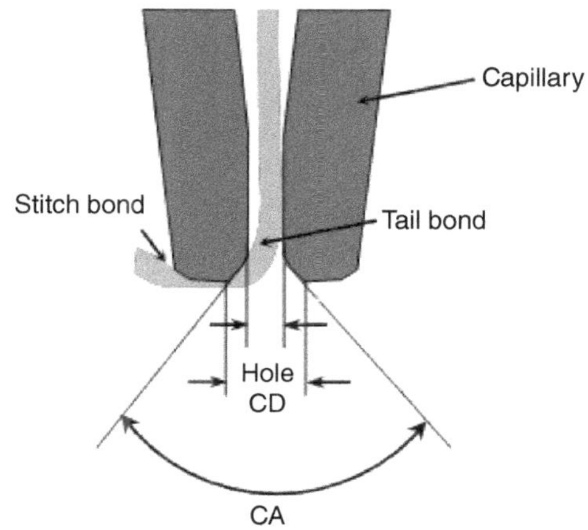

Figure 19.9 Illustration of a tail bond formation with a capillary. The capillary dimensions affect the tail shape, including hole, chamfer diameter (CD), and chamfer angle (CA).

Tail shape can affect free air ball formation. The ball diameter required to completely consume the bend tail is 1.5 to 1.6 times the wire diameter. We refer to the ball diameter to wire diameter ratio as ball size ratio (BSR). For a more repeatable free air ball and first bond, we recommend a minimum BSR 1.5 to 1.6 for fine pitch applications.

The free air ball can be examined and measured by bonding the wire with the free air ball attached and flipped upward. Some automatic wire bonders have the functionality to bond the free air ball for the engineers to study the free air ball formation. Figure 19.10 shows free air ball formed from PdCu wire using the "form FAB" process. In this case, the resulting free air balls were checked and the results show they are consistent, round, and oxidation free.

The key metrics for measuring free air ball are the following:

- *Ball size repeatability*. This is measured by the ball size standard deviation, minimum and maximum Variation in ball size will affect the first bond repeatability such as bonded ball diameter and height. Too small of free air ball could result in capillary hit on the pad (considered yield loss for some applications). Too large of free air ball could result in low bond strength and ball lift.
- *Ball symmetry (concentricity)*. This can be measured by the free air ball offset, which is the offset of the ball center to the wire center [7]. The direction of the offset is typically biased toward the EFO wand. Ball concentricity is important because poor concentricity can lead to poor centering of the ball on the pad.
- *Ball oxidation for Cu and Ag wire*. Au wire ball formation is done in atmosphere. For Cu and Ag wire, inert gas environment is needed to

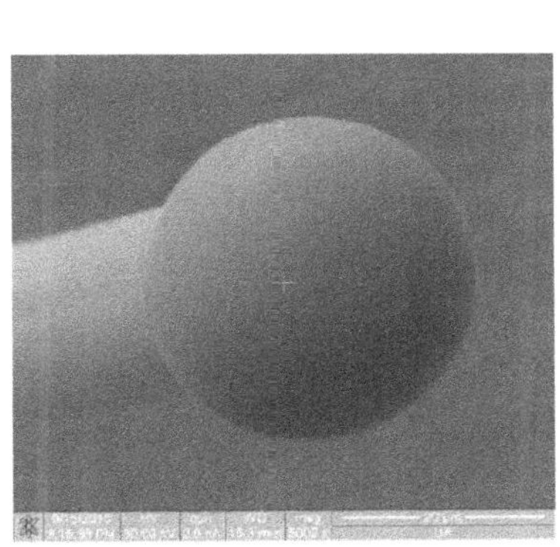

Figure 19.10 28-µm free air balls formed by the "form free air ball" process on the wire bonder. Wire is 18 µm PdCu wire. Cover gas is nitrogen.

form oxidation free balls. The better protected the ball is from oxygen, the easier the bond forms during the first bond process [7]. Ball oxidation can be assessed from the optical inspection by looking for discoloration of the ball. Oxidized ball is often less round and sometimes has pointed tip. The design of the inert gas delivery system plays a key role in the oxidation protection.

The main EFO parameters affecting ball formation are the EFO firing time and EFO firing current. The EFO firing time is on the order of 1 ms. For small wire diameter and small ball diameter, the EFO firing time could be around 100 µs. For larger wire and ball, it can take several milliseconds. In modern EFO systems, the duration of EFO fires can be tightly controlled and is not normally a factor for ball size variation. EFO current variation has a significant influence on free air ball size. EFO current should be optimized for different wire diameters and wire types. Lower current is normally used for smaller wire diameter.

Other parameters that affect the ball formation including EFO gap (the distance between the wand and the wire tip), the tail extension (the wire tail length outside the capillary), the gas flow rate and gas type if inert gas is used as in Cu and Ag wire bonding. EFO gap optimization is important because it affects ball size repeatability and concentricity.

Besides the equipment and process factors to the free air ball process, other material variations have impact on the ball formation as well. Wire variation is an important factor that can affect free air ball diameter. For 20 µm PdCu wire, study has shown actual FAB diameter varies by ±2 µm for a ±1 µm difference in wire diameter. To control ball size in production, wire diameter needs to be tightly controlled.

Contamination is another source for ball size variation. Wire contamination produces erratic ball size variation. There are other sources of contamination. The wire is pressed down upon the substrate during the tail formation. Contamination on the lead is transferred to the wire, end up causing the kind of wire contamination failures discussed above.

19.3.2 First Bond Process

After the free air ball is formed, the capillary travels to the first bond location to make the ball bond. Thermosonic bonding process is the most common method to make the electrical connections between the ball and the bond pad [8–10]. In this process, the free air ball is bonded to a metalized pad of an IC using heat, ultrasonic energy and normal force. Figure 19.11 shows a bonded ball in relationship with the capillary. The capillary inner chamfer (IC) plays an important role in the shape and size of the bonded ball. The key dimensional measurement for first bond (ball bond) is the ball diameter and ball height of squashed ball shown as "V1" in Fig. 19.11.

The heat for the bonding process is applied from the heat block that supports the device. The normal range of wire bonding temperature is around 150 to 250°C. For substrates such as BGA, the bonding temperature needs to be lower than the glass transition temperature (T_g) of the material; BGA device typically uses temperature around 150°C. For leadframe devices, the temperature is around 200°C. The temperature uniformity is important to form a consistent bond. For wide leadframes, the heat block design needs to be optimized for good temperature uniformity across the large bonding area.

Ultrasonic bonding has been used in wire bonding for over 50 years and the ultrasonic system has been improved over the years to provide faster, more repeatable and stronger wire bonds. The ultrasonic system on a wire bonder consists of the ultrasonic generator and the transducer. The ultrasonic generator provides electrical power to the transducer at a given frequency. On state-of-the-art wire bonders, the ultrasonic generator employs phase lock loop and amplitude control circuitry to provide automatic adjustment to track the changes during wire bonding. The transducer is a piezoelectric resonator that translates this electrical energy into mechanical vibrations. The clamping mechanism of the transducer provides the mounting of the bonding tool, which is typically a ceramic capillary in ball bonding.

The main effect of ultrasonic application is the sweeping of the surface oxidation by the ultrasonic scrubbing motion of the capillary. Al bond pad has a thin layer of oxidation. The presence of the oxidation layer inhibits the intermetallic formation between the ball and pad. The

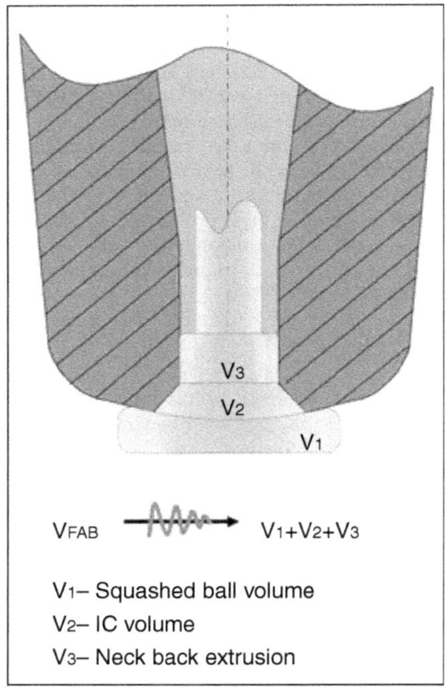

Figure 19.11 First bond formed by the capillary. The bonded ball consists of three parts: V1 is the squashed ball volume, V2 is the volume of the ball inside the inner chamfer (IC), V3 is the volume of the ball that extruded inside the capillary hole.

relative motion between the bottom of the ball and the bond pad helps removes the thin oxidation layer on the Al pad as previous chapters have described. This allows metal to metal contact and the intermetallic compound formation.

For a very long time, the most popular operating frequency for wire bonding was around 60 kHz. In recent years, higher frequency became more popular for possible advantages of shorter bond times and lower temperature bonding [11].

Au wire bonding is well understood with a long history in electronics packaging. The metrology for quantifying Au wire bond quality is mature and standardized. In Au wire first bond process, only ball shear, ball diameter, height, and intermetallic coverage (IMC) are routinely checked to qualify a wire bonding process. Starting around 2009, fine pitch Cu wire bonding has taken off mainly driven by the sky rocketing Au price. Although more expensive than bare Cu wire, Pd-coated Cu (PdCu) wire gained popularity mainly as a result of a more robust second bond process. PdCu wire has also shown better reliability under humid conditions than bare Cu wire [12, 13]. In addition to the traditional Au first bond metrology, Cu wire first bond process often requires other quality checks such as wire pull (at first bond), cross sectioning, Al splash, and pad crack check. Only devices known not to crack are not required to run the dielectric crack test. The metrology for first bond is summarized in Table 19.1 for different bonding wire types. For new Ag wire applications, the more extensive Cu wire metrology is recommended.

To optimize the process, a multiparameter DOE (Design of Experiments) can be used to optimize impact force, bond force, and Ultrasonic current. A single parameter sensitivity test can be done to understand the process window for that parameter. For the Au process, as the USG current goes up, the shear/areas increases until it levels off, the ball size also increases with the ultrasonic current [14]. After USG current reaches certain level, the shear/area peaks. Further USG current increase actually reduces shear/area; this is because the ball diameter increases without further shear strength increase when the USG current is too high.

Table 19.1 Wire Bond Process First Bond Metrology Comparison

	Au wire	Cu, PdCu, and Ag wire
Ball diameter and ball height	Primary dimensional/size measurement	Key dimensional/size measurement
Aluminum splash	Rarely measured because it is not usually an issue	Key dimensional/size measurement Often defines upper limit of process for fine pitch device
Ball shear shear/area	Primary bond strength measurement	Secondary bond strength measurement
Wire pull (at first bond)	Secondary process response Mostly breaks at ball neck	Key bond strength measurement Ball lifts to pad peeling window is an indicator of process robustness
Intermetallic coverage (IMC)	Bond strength measurement Checked on the bottom of the ball Typically target 70% or more	Primary bond strength measurement Checked on the bond pad Typically target 80% or more
Cross sectioning	Not typically done Mainly used to check for Kirkendall voiding in bake tests	Frequently used Uniform thickness of Al remain is desirable
Dielectric crack	Rarely a constraint on the bonding process	Significant constraint on bonding process for many devices

The optimal USG current setting can be chosen for the highest shear/area while still meeting the bonded ball diameter upper limit.

Au ball bonding to Al pads is relatively straightforward—the optimal USG is set to maximize ball shear while still meeting the ball size specification. On the other hand, Cu ball bonding to Al pads is challenging due to pad damage associated with high shear. The conventional Au wire optimization for maximal shear does not work with Cu wire bonding. A typical Cu process window is limited by low IMC and ball lift on the low end and pad peeling, Al splash, and crack at the high end. A Cu ball bond process development typically requires careful process design and optimization of a far more number of process parameter settings, which can often be time consuming and challenging. The latest Cu specific process, such as the ProCu process, simplifies Cu ball bonding process development and provides higher IMC, lower splash, and effectively eliminates crack. This type of process is response driven, where the user specified desired response such as target ball diameter and the process model automatically calculates the optimal parameter settings. Fine tuning adjustment parameters are available to the processes engineers so that they still have full control to optimize the bonding parameters for the application [15–17].

The complexity of Ag wire process development is somewhere between Au and Cu. Because Ag alloy wire is harder than the Al pad, most of the process issues encountered with Cu wire, for example, Al splash, pad peeling, and dielectric crack are also encountered with Ag wire. To simplify Ag wire process optimization and improve process capability, new Ag specific process is developed (such as the ProAg process) [17].

19.3.3 Second Bond Process

After the first bond is completed, the capillary travels to the second bond location to make the second bond. A second bond consists of a stitch bond and a tail bond. Figure 19.12 shows a second bond with the tail bond still attached to the stitch bond [1]. The stitch bond strength can be measured by pull test. The tail bond was broken off right after the second bond was formed. Thus, it is hard to assess tail bond strength after the device was already bonded. Tail bond strength is critical for a stable second bond process. When tail bond strength is too low, a short tail is likely to occur. On more advanced bonders, a tail pull test SW is available. It uses the Z axis servo to measure the break strength

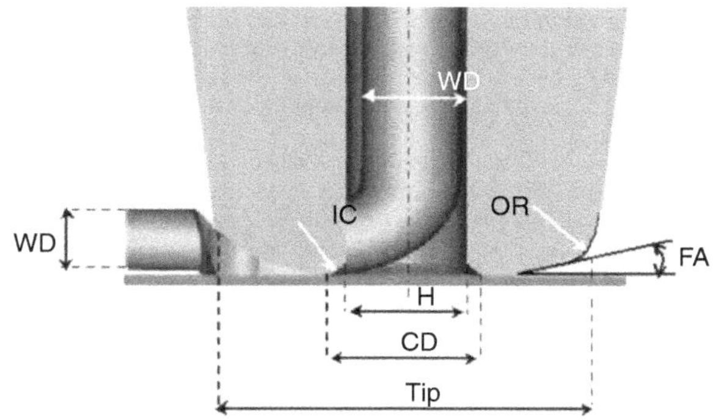

Figure 19.12 Picture of second bond in relationship to the capillary dimensions. H is capillary hole diameter, CD is chamfer diameter, OR is outside radius of the capillary, and FA is angle of the capillary bottom face. IC stands for inner chamfer. WD stands for wire diameter.

of the tail bond. It is a valuable tool to evaluate tail bond integrity [18]. Figure 19.12 also shows the stitch bond and tail bond in relationship with the capillary geometry. The stitch bond is mostly affected by the capillary outside dimensions such as capillary tip, face angle, and outer radius. The tail bond is mostly affected by the capillary inside features such as capillary hole, chamfer diameter, and inner chamfer angle.

Thermosonic bonding process is the most common method for first bond. Thermosonic process is also widely used for the second bond process. Unlike the first bond process, the nonultrasonic process is more common for second bond especially for leadframe and QFN devices. The nonultrasonic processes include thermocompression process and thermo-scrub process. With thermo-compression process, only heat and normal force is applied. This process can be used for easy to bond material. In thermo-scrub process, heat, normal force, and XY table scrubbing motions are used to make the second bond. Compared to ultrasonic scrubbing, XY table scrubbing is much lower in frequency but higher in amplitude. Ultrasonic scrubbing amplitude is on the order of tenth of a micron, XY scrubbing amplitude is one order of magnitude higher at several microns. XY table scrubbing frequency is much lower, for example, 100 to 200 Hz instead of 60 to 120 kHz frequency for ultrasonic scrubbing. The reason for using nonultrasonic processes is because the stitch bond location on many leadframe and QFN devices are sensitive to ultrasonic vibration. Severe resonant oscillation has been observed upon applying ultrasonic energy which could result in unstable stitch bond processes and faulty bonds. To avoid such problem, processes without ultrasonic vibration are used [19]. Thermo-scrub process is often slower than the thermosonic process because of the lower frequency.

For second bond metrology, stitch bond strength, tail bond strength and stitch remain after pull test are the main quality measure. Stitch pull strength is the primary strength measurement for second bond. Pull test can be done at different locations along the wire. Pull test at ball neck region is normally used to evaluate first bond strength. Pull test near stitch bond (second bond) is the best pull test method to evaluate second bond strength. Sometimes it is difficult to position the pull test hook close to second bond. Mid Span Pull can be used to evaluate second bond as well.

Acceptable pull strength depends on the wire type and wire diameter. It also depends on the geometry of the wire such as wire length, loop height and die height. For example, low loop, very long wire span may give low pull strength even though the bonds are very strong. Breaking mode for the pull test is also important. For breaking mode, breaking at the hook position, at the ball neck and near the second bond stitch with high break load are indications of good first and second bond strength. Ball lift and pad peeling during pull test are indications of first bond problem. Second bond lift is indication of weak second bond. Ball neck break with low strength could be an indication of neck damage that may be results of poor looping formation.

Second bond optimization mainly includes optimizing the bonding force and scrub amplitude. For ultrasonic process, the scrub amplitude is controlled by the USG current parameter. For a thermo-scrub process with XY table scrubbing, the scrub amplitude is the XY table motion amplitude.

The important factors influencing the bonding are diffusion and cold welding [19]. Cold welding has been extensively investigated and has been shown that for a successful cold weld, a certain amount of plastic deformation at the interface is required [20, 21]. Both ultrasonic scrubbing and XY table scrubbing provide the plastic deformation at the wire and lead interface. For a second bond process, choosing the process type is the first step. For bonding on rigid substrate such as BGA strips, thermosonic process is still very common. For many applications where the lead finger moves due to device compliance, ultrasonic process is prone to resonant problem. Table scrub is often used to avoid resonant issue as well as providing large plastic deformation for the bonding surface [19].

19.3.4 Looping Process

Wire bond loops connect the first bond and the second bond. Wire bond loops are formed through a series of looping motions of the capillary. The motions are controlled by wire bonder software and the servo controller. Each motion is executed precisely to form repeatable loops.

A couple of examples of formed wire loops are shown in Figs. 19.13 and 19.14. Figure 19.13 shows a simple loop profile with a single bend near the ball bond, this bend is referred to as the kink bend, with the height at this point as the kink height. The single bend loop is also called a "standard" loop; it can be used for shorter wire lengths. More complicated loop profiles can have more than one bend. Figure 19.14 illustrates a loop profile with two bends, a kink bend near the ball bond and a second bend at around 50 percent of the wire span. The second bend is referred to as the span bend, with the height at this point as the span height. The two-bend loop is also called a "worked" loop; it can be used for longer wire spans and taller loops.

Figure 19.13 Examples of formed loop profile with one kink bend (standard loop).

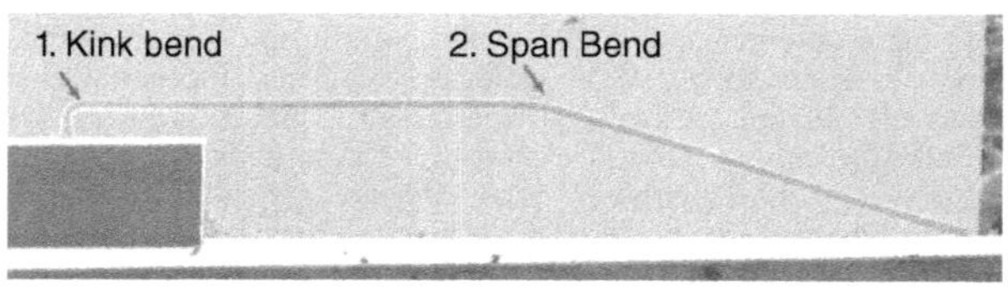

Figure 19.14 An example of two bend loop with kink bend and span bend (worked loop).

Much more sophisticated loops can be programmed on the wire bonder to meet special design requirements such as ultralow loop, die-to-die loop, and loops over SMT components. Lateral bends can also be programmed when the design calls for side bend profiles to achieve necessary clearances. The most advanced loop trajectory can form up to 10 bends in the loop with fine controls of each motion. More detailed discussion on complex loop shapes can be found in [2, 22]. A few examples of different loop profiles are shown in Fig. 19.15.

The key metrics of wire bond looping performance are loop height and wire straightness measurement. Loop height is measured as the maximal height of the loop which is often at the kink bend position (see Fig. 19.13). Loop height is the kink height in most cases. Span height is measured if the span bend location needs to be controlled for clearance reasons, for example, when loop needs to clear other components such as SMT or die edge. Wire straightness can be measured at different locations along the wire. Wire leaning amount is measured as the off axis distance at kink bend, and wire sway is measured as the maximal off axis distance along the wire, usually at span bend. For packages with plenty space between wires, rigorous looping measurement are not normally required. A cursory measure of the loop height and a visual check of loop shape, loop interference, and loop damage are enough for many wire bonded packages. For fine pitch device, the wire straightness requirement is stringent to prevent wire short. Wire leaning and wire sway should be measured for this type of devices. The reason to have two different wire straightness measurements is due to the smaller wire spacing near the first bond. Normally, the first bonds on the die are much closer to each other than the stitch bonds on the substrate. Therefore, the leaning specification is usually tighter than the wire sway specification. For packages with stringent loop height requirement such as ultralow loop and high-density multiple tier applications, the loop height and the loop profile need to be tightly controlled. In these applications, loop heights including kink height and span height should be measured (Table 19.2).

19.4 CONCLUSIONS AND FUTURE DEVELOPMENTS

Wire bonding is a mature technology that makes more than 10 trillion bonds each year. New technologies continue to extend the life of wire bonding by further reducing the cost and providing more capability. For cost reduction, alternative wires such as Pd-coated Cu wire, Cu alloy wire, Ag alloy wire, and lower-cost substrate technology, such as PPF (Pre-plated finish) QFN, should be considered. Advanced equipment and process technologies can help retain the material cost savings by ensuring robust and fast bonding processes. For today's advanced node devices with complex looping requirements, the 3D looping tool can help shorten the design cycle, reduce costly redesign, and improve wire bond yield, which in turn, enables wire bonding more complicated advanced node devices at a lower cost.

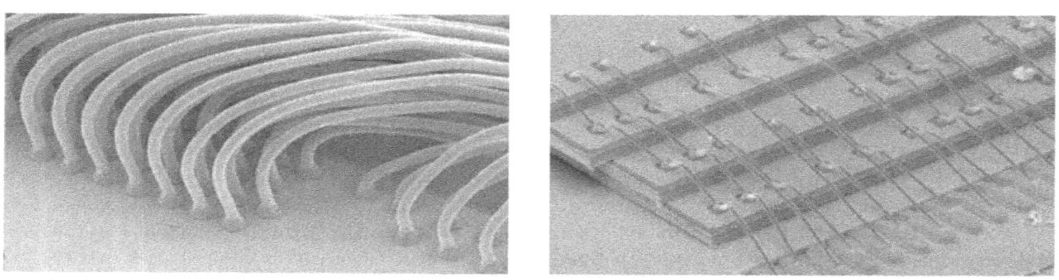

Multi-tier looping for high pin count fine pitch devices

Die to die bonding for stack die package

Figure 19.15 Various loop profiles for different applications.

Table 19.2 Looping Metrology

Measurement	What to measure	How to measure
Loop height	Primary height measurement. Measured at the highest point of the loop or at kink bend.	Optical scope or on bonder measurement
Span height	Secondary height measurement. Measured at the span position.	Optical scope or on bonder measurement
Loop profile	Overall profile of the loop	On bonder measurement, optical scope or SEM
Leaning	Wire straightness measurement. Measure the off axis distance at the kink position.	Optical scope or on bonder measurement
Sway	Wire straightness measurement. Measure the maximum off axis distance.	Optical scope or on bonder measurement
Neck damage	Inspect any visual damage at the ball neck region.	SEM or pull test result

19.5 REFERENCES

1. Jeffrey Rodengen, "50 years of Innovation: Kulicke and Soffa," Write Stuff Syndicate, 2012.
2. I. Qin, B. Milton, G. Schulze, C. Huynh, B. Chylak, and N. Wong, "Wire Bonding Looping Solutions for Advanced High Pin Count Devices," 66th Electronic Components and Technology Conference, pp. 614–621, IEEE, 2016.
3. I. Qin, "Advances in Wire Bonding to Lower the Cost of Interconnect," *Chip Scale Reviews*, Nov./Dec. issue, pp. 26–29, 2015.
4. Wei (Ivy) Qin, I. M. Cohen, and P. S. Ayyaswamy, "Fixed Wand Electronic Flame Off for Ball Formation in the Wire Bonding Process," *Journal of Electronic Packaging*, 116 (3): 212–219, 1994.
5. Wei (Ivy) Qin, P. S. Ayyaswamy, and I. M. Cohen, "Ball Formation from Deformed Wire End in Wire Bonding Process," *Thermal Processing of Materials: Thermo-Mechanics Control, and Composites*, HTD Vol. 289, Ed.: V. Prasad, et. al., Published by the ASME, New York, 107–115, 1994.
6. Wei (Ivy) Qin, I. M. Cohen, and P. S. Ayyaswamy, "Ball size and HAZ as Functions of EFO Parameters for Gold Bonding Wire," *Proceedings of the Pacific Rim/ASME International Intersociety Electronic and Photonic Packaging Conference*. American Society of Mechanical Engineers, New York, pp. 391–398, 1997.
7. A. Shah et al., "Advanced Wire Bonding Technology for Ag Wire," Electronic Packaging Technology Conference, IEEE, Singapore, Dec. 2015.
8. B. Langenecker, "Effects of Ultrasound on Deformation Characteristics of Metals," *IEEE Transactions on Sonics and Ultrasonics*, Vol. SU-13,1, Mar. 1966.
9. L. Levine, "The Ultrasonic Wedge Bonding Mechanism: Two Theories Converge," ISHM, pp. 242–246, 1995.
10. G. Harman, *Wire Bonding in Microelectronics*, 2d ed., McGraw-Hill, New York, pp. 23–26, 1997.
11. Bob Chylak, Ivy Wei Qin, and Jim Eder, "Achieve Optimal Wire Bonding Performance through Ultrasonic System Improvement," Semiconductor Singapore Technical Conference, SEMI Singapore, 2004.
12. T. Uno, S. Terashima, and T. Yamada, "Surface-Enhanced Copper Bonding Wire for LSI," Electronic Components and Technology Conference, pp. 1486–1495, 2009.
13. H. Abe, "Reliability Evaluation of Copper Wire Package," SEMICON Japan, Tokyo, Dec. 2010.
14. I. Qin et al., "Ball Bond Process Optimization with Cu and Pd-Coated Cu Wire," ECS Transactions–CSTIC, Vol. 44, Packaging and Assembly, 2012.
15. I. Qin et al., "Process Optimization and Reliability Study for Cu Wire Bonding Advanced Nodes," 64th Electronic Components and Technology Conference,IEEE, 2014.
16. I. Singh et al., "Pd-Coated Cu Wire Bonding Reliability Requirement for Device Design, Process Optimization and Testing," 45th International Symposium on Microelectronics IMAPS, San Diego, Sep. 2012.
17. I. Qin, A. Shah, H. Xu, B. Chylak, and N. Wong, "Advances in Wire Bonding Technology for Different Bonding Wire Material," 48th International Symposium on Microelectronics, IMAPS Conference, Orlando, Oct. 2015.
18. I. Qin et al., "Wire Bonding of Cu and Pd-Coated Cu Wire: Bondability, Reliability, and IMC Formation," 61st Electronic Components and Technology Conference, IEEE, 2011.
19. A. Rezvani et al., "Stitch Bond Process of Pd-Coated Cu Wire: Experimental and Numerical Studies of Process Parameters and Materials," 46th International Symposium on Microelectronics IMAPS, Orlando, Oct. 2013.
20. Bay, Niels. "Mechanisms Producing Metallic Bonds in Cold Welding," *Welding J.*, 62 (5): 137, 1983.
21. Li, Long, Kotobu Nagai, and Fuxing Yin, "Progress in Cold Roll Bonding of Metals," *Science and Technology of Advanced Materials*, 9 (2): 023001, 2008.
22. Stephen Babinetz and James Loftin, "Looping Challenges in Next Generation Packaging," Advanced Packaging Technology Symposium, SEMICON Singapore, Singapore, 2001.

Chapter 20
Interconnects Reliability

BY
ROEY SHAVIV *Applied Materials, Santa Clara, California*

20.1 INTRODUCTION

20.1.1 Fundamentals

Reliability is the ability to consistently perform a designated task over a specified period of time. Therefore, a reliability test would measure the extent at which a given operation would yield the same result over time. The failure to produce the specified result would constitute a *reliability failure.*

For interconnects the designated task is the transfer of an electrical signal from point A to point B. The laws of physics would dictate the expected signal at point B relative to the input signal at point A. Thus, if a given current is passing between these two points a given voltage drop is expected to be measured, and that voltage drop is directly proportional to the resistance of the conductor. Similarly, if a given voltage drop is placed between points A and B, the current flowing would be dictated by the resistance of the medium.

Simply described, interconnects are made by conductors and the insulators that separate them. Thus, one would require from the conductors a given conductivity that is stable over time, and from the insulators high resistivity, that is also stable over time. Hence, the reliability of the interconnect system is determined by the stability over time of the conductivity of the conductors and the resistivity of the insulators. A *reliability failure* can, therefore, be defined as any change in conductor's conductivity or insulator's resistivity that is larger than a predetermined allowable deviation from the initial measurement.

It is, therefore, simple to devise a test for reliability assessment. For a conductor, one needs to define a representative interconnect test structure and measure its conductivity over time. This would be done by passing a known current through a given line and measuring the voltage drop across that line. Following the relation $R \equiv V/I$, where R is the resistance, V is the voltage drop, and I is the current, one can monitor resistance as a function of time and define failure when resistance increases or decreases by more than a specified value, typically 10 or 20 percent. Similarly, for insulator reliability one needs to define a representative test structure, typically a capacitor; set a known voltage drop across that capacitor, and monitor current over time. Variations of these two test methods exist and serve particular needs but in general the principle of operation is similar in most cases.

20.1.2 Requirements and Standards

Typically, the lifetime requirement is 10 years of operation. Of course, specific applications, like aviation, safety components, military components, space operations, and the like would have stricter requirements but in general, the rule of thumb is 10 years of operation. However, that criterion is a vague one because it is not clear what "10 years in operation" really means. A more elaborate definition is in order and a clear standard is required. These would allow for a quantitative assessment of reliability performance and an understanding of the statistics involved.

Reliability standard for the microelectronic industry are published by JEDEC and are universally accepted.[1] Accepted lifetimes and operation conditions are specified therein. For example, lifetime requirement of 100,000 h is specified for electromigration.[2] Moreover, the acceptable failure rate at that lifetime and the operation conditions are also specified.

Obviously, reliability testing cannot go on for 100,000 h, or about 11 years. A clear reliability assessment is needed in a timely manner in order to ascertain the suitability of a given product to the market at the time of product release. Thus, accelerated reliability testing is required. This is accomplished by first establishing a physical model for the failure mode. Once a physical parameter that influences the failure mode is identified the reliability experimenter can change that parameter to cause failures at a faster rate. Second, a statistical model for the failure mode is required. This would allow establishing a test sample that would allow for an assessment of the failure statistics at the accelerated conditions. Once physical and statistical models are in place, a calculation of expected lifetime prediction is at hand.

Interconnect reliability exhibits three main failure modes: electromigration, stress migration, and dielectric breakdown. In the following sections we will review each mode and the physics and statistics that govern them.

20.2 ELECTROMIGRATION

Electromigration is a physical phenomenon that results from the transfer of momentum between the charge carriers and the conductor medium. In a way it is similar to erosion, where the flow of water over a hard rock is capable of forming the Grand Canyon. Electromigration was identified as a failure mechanism for interconnects by I. A. Blech and H. Sello in 1966.[3] In interconnects the conductor medium is a metal, typically Cu or Al, and the charge carriers are electrons. As electrons flow through the medium they collide with the stationary metal atoms. The accumulated momentum transfer creates a drift of atoms in the direction of the electron wind, opposite to the direction of the current.

Over time, the drift would cause conductor atoms to accumulate at the anode and voids would form at the cathode. Eventually, the voids would create a high-resistance path or even a discontinuity at or near the cathode, resulting in a failure. The drift velocity would, therefore, govern the kinetics of void growth which, in turn, determine the expected lifetime.[4]

The physics governing electromigration is well understood. The time to fail is predicted by the Black Equation.[5]

$$t_{50} = Aj^{-2}e^{\frac{E_a}{kT}} \tag{20.1}$$

where t_{50} is the mean time to fail, j is the current density, E_a is the activation energy, k is the Boltzmann constant and T is the temperature. A is a constant that is dependent on the material properties, bonding strength, atomic mass, interface and grain boundary properties, ionic scattering, resistivity, electron mean free path, and mean time between collisions. Typically A is determined experimentally because it is structure and material dependent. Electromigration time to fail follows log normal statistics, which allows for a straight forward determination of the mean time to fail, (t_{50}), and the distribution, σ. The activation energy can be determined from the slop of the Arrhenius plot of $\ln(t_{50})$ versus $(kT)^{-1}$, as illustrated in Fig. 20.1*a*.[6] The Black equation postulates a $1/j^2$ relationship between t_{50} and the current density. Recent studies confirm the $1/j^1$ dependence for void nucleation failure modes whereas a $1/j^1$ relationship is found for failure mechanisms associated with void growth.[7] Because in practice both mechanisms operate simultaneously, Eq. (20.1) is rewritten as

$$t_{50} = Aj^{-n}e^{\frac{E_a}{kT}} \tag{20.2}$$

where n denotes the "current acceleration coefficient" and is between 1 and 2. A plot of $\ln(t_{50})$ versus $\ln(j)$ at a given temperature would result in a straight line whose slope is n, as illustrated in Fig. 20.1*b*.[6]

The drift velocity of the metal atoms in the metal line can be expressed as[8]

$$v_d = \frac{D_{\text{eff}}}{kT} Z^*_{\text{eff}} \, e\rho j \tag{20.2}$$

where D_{eff} is the effective diffusivity, Z^*_{eff} is the effective charge number, k is the Boltzmann constant, T is the temperature, ρ is the resistivity, and j is the current density. For a typical copper interconnect four different diffusion paths can be identified resulting in four different drift velocities. These are (1) drift along the interface between the copper metal and the metal barrier, typically Ta, TaN, or a Co liner, (2) drift along the interface between the copper metal and the dielectric diffusion barrier, typically SiC or SiCN, (3) drift along grain boundaries, and (4) drift in the bulk of the copper metal. Consequently, one may identify four different activation energies for Cu electromigration, associated with each of the four drift velocities along the four paths mentioned above. The activation energy for bulk Cu was studied by Sullivan and was reported as 2.1 eV.[9] For Cu conductor the activation energy was measured at 0.79 eV by Park and Vook.[10] Activation energies between 0.7 and 1.2 eV were reported by C.-K. Hu et al.,[11] whereas this author reported activation energies between 0.91 and 0.96 eV.[6] The wide range of reported activation energies is associated with the different drift paths highlighted above as characterized by Fischer et al.[12] who confirmed the activation energy of bulk Cu as reported by Sullivan[8] and identified the interface between the dielectric diffusion barrier, or cap layer, and the Cu conductor as the weak interface for electromigration, with activation energy of 0.8 to 0.9 eV. A comprehensive study of electromigration-related failure mechanisms was published by Li et al.[13] A correlation between the adhesion of the cap layer to Cu, void growth rate, and activation energy was established by Lane et al.[14] Thus it is established that the critical factor for electromigration in copper interconnects is the integrity of the conducting metal and the barrier layers encapsulating it. Strategies for improvement of the drift velocities were reviewed by Oates.[4]

Recent studies demonstrate that electromigration lifetime is also strongly dependent on the dimensions of the conductors.[15] As line dimension scale with the evolution of technology in accordance with Moore's law, smaller voids become critical and cause electromigration failures. Several remedies to this problem were explored by Christiansen et al.[16] who found that a selective Co cap or Mn doping of the Cu conductor significantly increase activation energies and consequently electromigration lifetime, as illustrated in Fig. 20.2. Modern manufacturing technology employ either Mn doping of Cu or a Co cap over the Cu conductor as remedies to electromigration failures. Other methods, such as Al doping,[17] and plasma self-aligned barriers (PSAB),[6,18] are often utilized toward the same goal as well.

Much attention is being given to short-line electromigration[4,19] and electromigration under AC or pulse-DC[20,21] conditions for application beyond the 22-nm technology generation. It is concluded that the mechanisms governing electromigration under AC or pulse-DC conditions are the same mechanisms discussed above for pure-DC conditions. Hence, only the net-"on" time for pulse-DC operation affects lifetime. Since most devices do not operate only under DC conditions, the "off" time may prolong lifetime. The short-line effect emerges as a concern as technology scaling continues beyond the 14-nm technology generation. The mechanism is shown to be related to defect density and

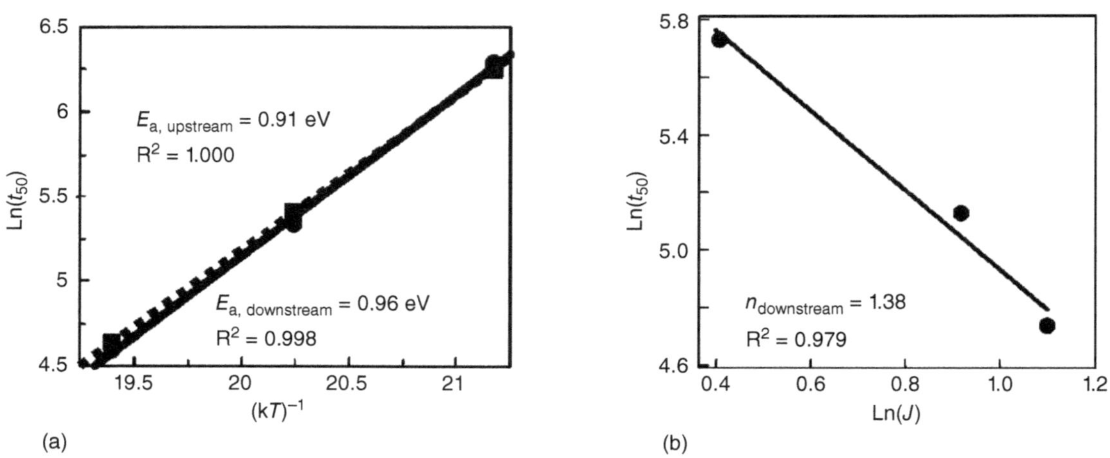

Figure 20.1 (*a*) Determination of the activation energy: A $\ln(t_{50})$ versus $(kT)^{-1}$ plot. The activation energy is the slop of the line. (*b*) Determination of the current acceleration coefficient: A $\ln(t_{50})$ versus $\ln(j)$ plot. The current acceleration coefficient is the slope of the line.

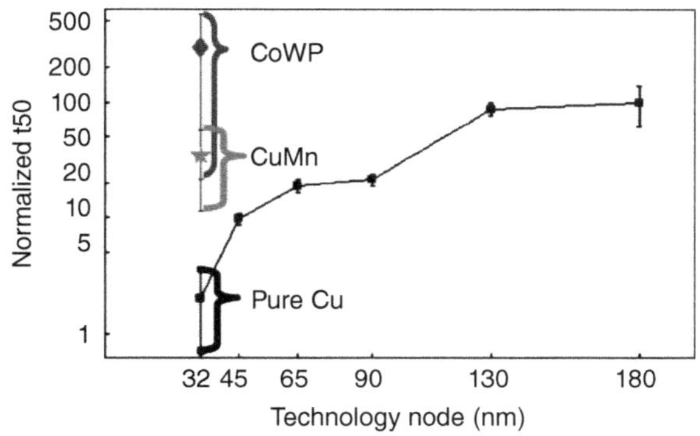

Figure 20.2 Normalized median failure times versus technology node for six technology generations. The symbols are as follows: black square for pure Cu, blue diamond for Cu with CoWP capping layer, and red star for Cu with Mn doping. (*Reproduced with permission.*)[15]

copper encapsulation. Controlling these failure modes is critical for the continuation of technology scaling along Moore's law.

20.3 STRESS MIGRATION

Stress migration is a physical phenomenon that results from the tendency of voids to merge as means to minimize the surface tension and surface-to-volume ratio in order to achieve a state of lower free energy. The phenomenon seen in copper metallization is not fundamentally different from the merger of bubbles in soap or beer,[22] or even on the cosmic scale.[23] In copper metallization the manifestation of stress migration is typically with voids that congregate near or under a via. These voids are often referred to as "stress-induced voiding" (SIV) or "via stress migration" (VSM) failures.

The physics of SIV was described in details by Ogawa et al.[24] It was observed that vias connecting wide metal lines are more prone to SIV. This leads to defining an "active diffusion volume," where vacancies diffuse and congregate. The creep rate follows the model proposed by McPherson and Dunn[25] and the maximum void formation rate is found at about 190°C.[24] A strong correlation between SIV and the thermal history of the copper metallization between copper plating and CMP is reported, as illustrated in Fig. 20.3. The stress hysteresis of Cu was studied by Alers et al.[26] as well as by Gan et al.[27] A typical set of stress hysteresis curves of unpassivated copper is shown in Fig. 20.4.[27] We note that the stress decreases with temperature. That is because the thermal expansion coefficient of copper is larger than that of silicon.[28] However, at about 190°C the stress hysteresis becomes flat and stress is no longer a function of temperature. That is because plastic deformation of the copper metal compensates for the differences in the thermal expansion,

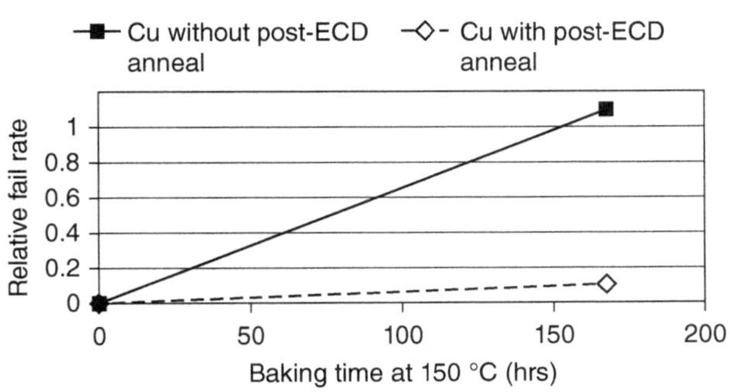

Figure 20.3 The effect of thermal history of copper metallization on the failure rate due to stress migration. A 150°C anneal between copper plating (ECD) and CMP is shown to significantly reduce the failure rate. (*Reproduced with permission.*)[24]

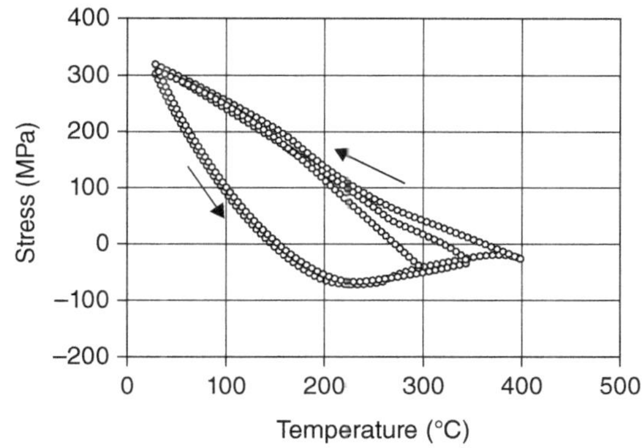

Figure 20.4 Stress hysteresis plot for unpassivated Cu. Plastic deformation is evident by the flat stress between 200 and 300°C. Tensile stress increases upon cooling. The film exhibits about 300 MPa following the anneal process. (*Reproduced with permission.*)[27]

keeping the stress almost constant. It is interesting to note that the onset of plastic deformation coincides with the maximum rate of creep and void formation found by Ogawa et al.[24]

Thus, we see that SIV is prevalent where a large reservoir of metal is available, where vacancies preexist, and where the temperature is high enough for vacancies diffusion but not too high where plastic deformation mitigate voids formation. Voids would also tend to congregate near or under a via due to micro voids formation there during the via etch and subsequent processes, creating a sufficient vacancy density at the vicinity of the via to allow for vacancies diffusion and congregation during anneal.[24] Minimizing the formation of these microvoids during processing or can be an effective way to mitigate stress migration failures. One way to do this is to better the interface between the PVD metal barrier surrounding the via and the copper metal in the preceding layer as demonstrated by Alers et al.[25] and Shaviv et al.[6]

A second manifestation of stress migration is reported by Lee and Oates[29], who found voids formation in very narrow interconnect Cu wires close to large Cu plates on the same layer, with no nearby vias. The mechanism for this failure mode is found to be related to stress gradients forming in the large Cu reservoir and close to the intersection between the narrow and the wide lines. This creates a force that pulls the metal out of the narrow line and into the wider reservoir and in effect breaks the narrow line. The phenomenon is found to be related to the properties of the metal barrier and adjustments to that barrier mitigate and resolve it.

As all stress migration phenomena require a large reservoir that provides the so-called "effective diffusion volume"; thinner copper is less susceptible to stress migration.[26] Moreover, the probability of failure can be diminished by redundancy. Typically, redundancy is not desired because of the extra space needed for it. However, because stress migration is found only at or near large reservoirs, adding redundancy there should be achievable. As technology progresses metal thickness is reduced at each technology generation. Thus, stress migration may be mitigated by design in advanced nodes by using thinner copper and including redundancy.

20.4 DIELECTRIC BREAKDOWN

As discussed in the introduction to this monograph, a typical interconnect consists of conductors and the insulators that separate them. While most of the "activity" takes place inside the conductors, one can never overlook the insulators since they provide the mechanical infrastructure as well as the necessary separation between conductors, without whom short circuits would prevail.

Everyone who had witnessed a thunderstorm must be familiar with the dielectric breakdown phenomenon. The accumulation of charge in cumulonimbus clouds creates strong electric fields. The electrostatic discharge that neutralizes these fields is the familiar lightning and thunder phenomenon, which is caused by a dielectric breakdown of the insulating air. In principle, there is no fundamental difference between the dielectric breakdown during lightning and thunder and the dielectric breakdown in interconnects. In both cases an electric field across an insulator leads to a catastrophic failure of dielectric properties and to a formation of a conduction channel in which charge is conducted to neutralize the electric field. One significant difference between solid state insulators and liquid or gas insulators is that in the solid state the breakdown it typically irreversible.

The physics of dielectric breakdown had been studied at length for almost a century. However, fundamental understanding of the phenomenon is still incomplete. A theory of cold and hot electron discharge was proposed by Schottky in 1923.[30] In 1928, Fowler and Nordheim proposed a model for electron emission in intense electric fields.[31] That phenomenon is often referred to as Fowler-Nordheim conduction and leads to a current driven mechanism for dielectric breakdown, which leads to a linear dependence between $1/E$ and $\ln(t_{BD})$, where E and t_{BD} are the electric field and the time to breakdown, respectively.[32] This model became to be known as the "$1/E$ model." An alternative thermo-chemical model to explain dielectric breakdown in integrated circuits had been advocated by McPherson and Mogul.[33] In this model, typically referred to as the "E model," the time to breakdown (t_{BD}) exponentially decreases with the electric field E. Another model, commonly referred to as the "$\sqrt{E}$" model, had been proposed by Chen et al.[34], who found that t_{BD} exponentially decrease with $E^{1/2}$. A power-law dependence, $t_{BD} \propto A^{-1/\beta}$, where A is the area of the capacitor and β is the Weibull shape parameter, had been observed and proposed by Ernest et al.[35] Yet another model for the dependence of t_{BD} on E was proposed by Haase et al.[36] One may, therefore, ask why is it that in spite of the vast body of work on dielectric breakdown no consensus about the physics emerged to date?

The key difficulty is that although the t_{BD} of interest is at operation conditions, that is in low electric fields, testing must be conducted at high electric fields. Testing at low electric fields is impractical for it would last longer than the lifetime of the tester. Thus, a method for a reliable extrapolation from testing at high fields to a projected lifetime at low field is a prerequisite to any assessment of dielectric reliability. An added difficulty is that at high fields the models above converge. All provide a good fit to the experimental results and all stand on a sound physical ground. At low fields the models diverge and extrapolated lifetimes at operating conditions can vary by orders of magnitude from model to model. This author and others have found that those extrapolations often result is lifetime estimates that exceed the actual lifetime of the universe. However, a close look at the models described above would show that upon extrapolation to low fields both the E model and the $\sqrt{E}$ model would predict a finite lifetime when $E=0$. This prediction is not physical because at $E=0$ lifetime *must be* infinite, as dictated by the definition of the dielectric breakdown phenomenon. Simply stated, the experiment measures the effect of electric field on t_{BD}. No field means, *by definition*, no breakdown, and an infinite t_{BD}. Therefore, at low field t_{BD} *must be* greater than that calculated by either the E model or the $\sqrt{E}$ model. Consequently, these two models provide a convenient method to calculate a lower limit to t_{BD} and are commonly used.

In recent years an attempt to decipher the actual dependence of t_{BD} on E at low fields had been conducted by Croes et al.[37], who used very long test times to generate actual results at low electric fields. Experimental results from an extended run lasting several years are shown in Fig. 20.5. Results show that, indeed, at high electric fields all models converge. However, at low electric fields, as experimental t_{BD} reaches months or even years, the best agreement between theory and experiment is found with the power-law model and with the impact damage model, proposed by Lloyd et al., which is a combination of the $\sqrt{E}$ and the $1/E$ models.[38] That model is particularly interesting because it postulates that the breakdown is related to threshold electron energy, associated with material properties, and to the statistics associated with

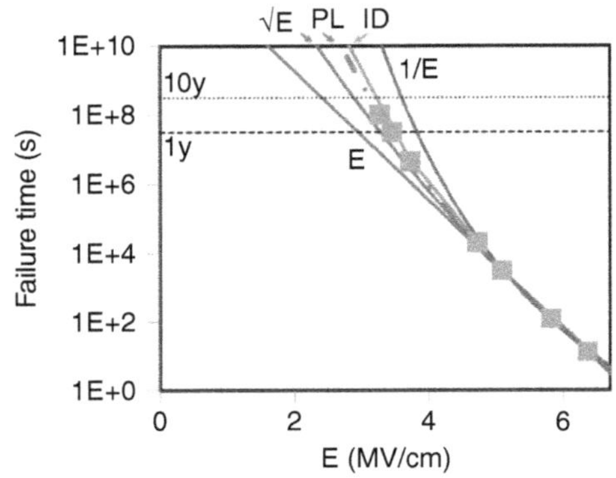

Figure 20.5 Experimental results from low field testing over a prolonged period of time and the differences between the various models for the field dependence of t_{BD}.[37]

electron behavior. Both the Power-law and the impact damage model would predict t_{BD} at operation conditions that is longer than prediction of the, commonly used, E model or $\sqrt{E}$ models.

What influences t_{BD} and how can we prolong it? All models show a strong dependence between the electric field and t_{BD}. Thus, controlling the electric field should be the first line of defense. According to Haase,[36] at a given operation potential the variations in the electric field are strongly dependent on patterning control, and in particular, variations in critical dimensions (CDs) and line edge roughness. Thus, a good control of CDs and minimizing line edge roughness is the first line of defense for preventing dielectric breakdown failures.

Three mechanisms for dielectric breakdown were identified by Alers.[39] These are breaching of the metal barrier, adhesion failures along the dielectric barrier and the low-k dielectric, and breaching of the dielectric barrier. Thus, a second line of defense would be the integrity of the barrier encapsulating the copper metallization.[40] A weak barrier would be easier to breach, leading to a lower breakdown field or a shorter t_{BD} at any given field. Similarly, good adhesion between the various dielectric films, and in particular the dielectric diffusion barrier and the low-k dielectric between the lines, is critical to prolonging t_{BD}.[41] In addition, the integrity of the dielectric diffusion barrier must be sufficient to sustain the electric field and prevent dielectric breakdown.

20.5 CONCLUSION

No one likes to step out of a store with a new purchase just to have it break a short while afterward. We all expect a product to perform its designated duties for a reasonable length of time. Thus, reliability is an integral part of product specifications. Unreliable products are never successful in the market place. In order to achieve a reliable product, all elements that make up the product must be reliable. The saying that a chain breaks at the weakest link is indeed correct in real life.

Here we took a close look at the reliability of interconnects in integrated circuits. We looked at requirements and standard and then at the causes for failures. Three failure modes were considered: electromigration, stress migration, and dielectric breakdown. We looked at the physics of the phenomena and the models that describe them. We examined ways to mitigate them so as to prolong lifetime.

Electromigration is caused by the momentum transfer from the charge carriers, typically electrons, to the metal atoms during the conduction of electrical current. The time to fail is affected by material properties, the current density and the temperature as described in Eq. (20.2),

$$t_{50} = Aj^{-n} e^{\frac{E_a}{kT}}.$$

Thus, for a given current, density and material set the higher the activation energy the better the electromigration performance is. Methods to achieve high activation energy were discussed.

Stress migration is a result of creep and consolidation of vacancies. It is found that stress migration in copper metallization is fastest at about 200°C. It is also found that stress migration is dependent on the design and the thickness and width of the metal conductors. Thus, the best method to avert stress migration is to establish design rules that forbid stress migration sensitive structure and to add redundancy where needed. In so doing stress migration may be designed out.

Dielectric breakdown is affected by the material properties and the electric field. The physics governing dielectric breakdown is not yet fully established. However, since all the leading physical models agree well with the experimental results at high fields, it is reasonable to establish a methodology for accelerated testing and for extrapolation to operation conditions. It is found that mitigating dielectric breakdown is achievable by maintaining good CD control and preventing line edge roughness, with a high integrity dielectric and metal barriers encapsulating the copper conductor, and by maintaining good adhesion at all interfaces.

A long list of references is provided herein. The reader is encouraged to explore this subject by further reading.

20.6 REFERENCES

1. See http://www.jedec.org/. Reliability standards are in http://www.jedec.org/standards-documents/results/reliability.
2. JEDEC/FSA joint publication. Foundry Process Qualification Guidelines (wafer fabrication manufacturing sites), jp001.01, section 8.1. May 2004.
3. Ilan A. Blech and Harry Sello, "The Failure of Thin Aluminum Current-Carrying Strips on Oxidized Silicon," IEEE, *Fifth Annual Symposium on the Physics of Failure in Electronics*, Piscataway, NJ, pp. 496–505, 1966.
4. A. S. Oates, "Strategies to ensure electromigration reliability of Cu/Low-k interconnects at 10 nm." *ECS Journal of Solid State Science and Technology*, 4 (1): N3168–N3176, 2015.
5. James R. Black, "Electromigration—A Brief Survey and Some Recent Results," *IEEE Transactions on Electron Devices*, 16 (4): 338–347, 1969. James R. Black, "Electromigration Failure Modes in Aluminum Metallization for Semiconductor Devices," *Proceedings of the IEEE*, 57 (9): 1587–1594, 1969.
6. Roey Shaviv, Sanjay Gopinath, Marcelle Marshall, Tom Mountsier, and Girish Dixit, "A Comprehensive Look at PVD Scaling to Meet the Reliability Requirements of Advanced Technology," *IEEE International Reliability Physics Symposium*, 2009.
7. R. G. Filippi, P.-C. Wang, A. Brendler, and J. R. Lloyd, "Correlation between a Threshold Failure Time and Void Nucleation for Describing the Bimodal Electromigration Behavior of Copper Interconnects," *Applied Physics Letters*, 95 (7): 2009 and references 10–19 therein. http://dx.doi.org/10.1063/1.3200233.
8. E. Liniger et al., "In Situ Study of Void Growth Kinetics in Electroplated Cu Lines," *Journal of Applied Physics*, 92 (4): 1803–1810, 2002.
9. George A. Sullivan, "Search for Reversal in Copper Electromigration," *Journal of Physics and Chemistry of Solids*, 28 (2): 347–350, 1967.
10. C. W. Park and R. W. Vook, "Activation Energy for Electromigration in Cu Films," *Applied Physics Letters*, 59 (2): 175–177, 1991.
11. C.-K. Hu and B. Luther, "Electromigration in Two-Level Interconnects of Cu and Al Alloys," *Materials Chemistry and Physics*, 41 (1): 1–7, 1995.
12. A. H. Fischer, A. Von Glasow, S. Penka, and F. Ungar, "Electromigration Failure Mechanism Studies on Copper Interconnects," *Interconnect Technology Conference, 2002. Proceedings of the IEEE 2002 International*, pp. 139–141, 2002.
13. Baozhen Li, Timothy D. Sullivan, Tom C. Lee, and Dinesh Badami, "Reliability Challenges for Copper Interconnects," *Microelectronics Reliability*, 44 (3): 365–380, 2004.
14. M. W. Lane, E. G. Liniger, and J. R. Lloyd, "Relationship between Interfacial Adhesion and Electromigration in Cu Metallization," *Journal of Applied Physics*, 93 (3): 1417–1421, 2003.
15. C.-K. Hu, D. Canaperi, S. T. Chen, L. M. Gignac, B. Herbst, S. Kaldor, M. Krishnan et al., "Effects of Overlayers on Electromigration Reliability Improvement for Cu/Low k Interconnects," *Reliability Physics Symposium Proceedings, 2004. 42nd Annual. 2004 IEEE International*, pp. 222–228, 2004.
16. C. Christiansen, B. Li, M. Angyal, T. Kane, V. McGahay, Y. Y. Wang, and S. Yao, "Electromigration-Resistance Enhancement with CoWP or CuMn for Advanced Cu Interconnects," *Reliability Physics Symposium (IRPS), 2011 IEEE International*, pp. 3E-3, 2001.
17. S. Yokogawa and H. Tsuchiya, "Effects of Al Doping on the Electromigration Performance of Damascene Cu Interconnects," *Journal of Applied Physics*, 101 (1): 013513, 2007.
18. Hui-Jung Wu, Wen Wu, Roey Shaviv, Mandy Sriram, Anshu Pradhan, Kie Jin Park, Jennifer O'Loughlin et al., "Electromigration Improvement for Advanced Technology Nodes," *ECS Transactions*, 18 (1): 269–274, 2009.
19. A. S. Oates and M. H. Lin, "The Impact of Trench Width and Barrier Thickness on Scaling of the Electromigration Short-Length Effect in Cu/Low-k Interconnects," *IEEE International Reliability Physics Symposium (IRPS)*, 2013.
20. Roey Shaviv, Gregory J. Harm, Sangita Kumari, Robert R. Keller, and David T. Read, "Electromigration of Cu Interconnects under AC and DC Test Conditions," *Microelectronic Engineering*, 92: 111–114, 2012.
21. M. H. Lin and A. S. Oates, "AC and Pulsed-DC Stress Electromigration Failure Mechanisms in Cu Interconnects," *IEEE International Interconnect Technology Conference (IITC)*, 2013.
22. Neil E. Shafer and Richard N. Zare, "Through a Beer Glass Darkly," *Physics Today*, 44 (10): 48–52, 1991.
23. R. K. Sheth and R. Van De Weygaert, "A Hierarchy of Voids: Much Ado about Nothing," *Monthly Notices of the Royal Astronomical Society*, 350 (2): 517–538, 2004.
24. E. T. Ogawa, J. W. McPherson, J. A. Rosal, K. J. Dickerson, T.-C. Chiu, L. Y. Tsung, M. K. Jain et al., "Stress-Induced Voiding under Vias Connected to Wide Cu Metal Leads," *IEEE 40th Annual Reliability Physics Symposium Proceedings*, pp. 312–321, 2002.
25. J. W. McPherson and C. F. Dunn, "A Model for Stress-Induced Metal Notching and Voiding in Very Large-Scale-Integrated Al- Si(1%) Metallization," *Journal of Science and Technology*, B5 (5): 1321–1325, 1987.
26. Glenn B. Alers, J. Sukamto, P. Woytowitz, X. Lu, S. Kailasam, and J. Reid, "Stress Migration and the Mechanical Properties of Copper," *Reliability Physics Symposium, 2005. Proceedings. 43rd Annual. 2005 IEEE International*, pp. 36–40, 2005.
27. Dongwen Gan, Paul S. Ho, Yaoyu Pang, Rui Huang, Jihperng Leu, Jose Maiz, and Tracey Scherban, "Effect of Passivation on Stress Relaxation in Electroplated Copper Films," *Journal of Materials Research*, 21 (6): 1512–1518, 2006.
28. G. K. White, "Thermal Expansion of Reference Materials: Copper, Silica and Silicon," *Journal of Physics D: Applied Physics*, 6 (17): 2070, 1973.
29. Chang-Chun Lee and Anthony S. Oates, "A New Stress Migration Failure Mode in Highly Scaled Cu/Low-Interconnects," *IEEE Transactions on Device and Materials Reliability*, 12 (2): 529–531, 2012.
30. W. Schottky, "Cold and Hot Electron Discharge," *Z. Phys*, 14: 63, 1923.
31. Ralph Howard Fowler and L. Nordheim, "Electron Emission in Intense Electric Fields," *Proceedings of the Royal Society of London A: Mathematical, Physical and Engineering Sciences*, 119 (781): 173–181, 1928.
32. Ih-Chin Chen, Stephen E. Holland, and Chenming Hu, "Electrical Breakdown in Thin Gate and Tunneling Oxides," *IEEE Journal of Solid-State Circuits*, 20 (1): 333–342, 1985. N. Klein and P. Solomon, "Current Runaway in Insulators Affected by Impact Ionization and Recombination," *Journal of Applied Physics*, 47 (10): 4364, 1976.
33. J. W. McPherson and H. C. Mogul, "Underlying Physics of the Thermochemical E Model in Describing Low-Field Time-Dependent Dielectric Breakdown in SiO_2 Thin Films," *Journal of Applied Physics*, 84 (3): 1513–1523, 1998.
34. F. Chen, O. Bravo, K. Chanda, P. McLaughlin, T. Sullivan, J. Gill, J. Lloyd et al., "A Comprehensive Study of Low-k SiCOH TDDB

Phenomena and Its Reliability Lifetime Model Development," *44th Annual Reliability Physics Symposium Proceedings, IEEE International*, pp. 46–53, 2006.

35. Ernest Y. Wu, A. Vayshenker, E. Nowak, J. Sune, R.-P. Vollertsen, W. Lai, and D. Harmon, "Experimental Evidence of TBD Power-Law for Voltage Dependence of Oxide Breakdown in Ultrathin Gate Oxides," *IEEE Transactions on Electron Devices*, 49 (12): 2244–2253, 2002.

36. Gaddi S. Haase and Joe W. McPherson, "Modeling of Interconnect Dielectric Lifetime under Stress Conditions and New Extrapolation Methodologies for Time-Dependent Dielectric Breakdown," *Proceedings of the 45th annual. IEEE International Reliability Physics Symposium*, pp. 390–398, 2007.

37. K. Croes, C. Wu, D. Kocaay, Y. Li, Ph Roussel, J. Bömmels, and Zs Tőkei, "Current Understanding of BEOL TDDB Lifetime Models," *ECS Journal of Solid State Science and Technology*, 4 (1): N3094–N3097, 2015.

38 J. R. Lloyd, E. Liniger, and T. M. Shaw, "Simple Model for Time-Dependent Dielectric Breakdown in Inter- and Intralevel Low-k Dielectrics," *Journal of Applied Physics*, 98: 084109, 2005.

39. Glenn B. Alers, *45th Annual Reliability Physics Symposium Tutorial*, 2007.

40. S. B. Law, S. L. Liew, C. S. Seet, E. C. Chua, Y. K. Kim, B. C. Zhang, B. J. Tan et al., "Impact of Barrier Film Characteristics on Vramp & TDDB Lifetime of 65 nm Node Cu/Low k Interconnects," *AMC Symposium*, pp. 133–134, 2007.

41. K. Chattopadhyay, B. van Schravendijk, T. W. Mountsier, G. B. Alers, M. Hornbeck, H.-J. Wu, R. Shaviv et al., "In-Situ Formation of a Copper Silicide Cap for TDDB and Electromigration Improvement," *Reliability Physics Symposium Proceedings, 2006. 44th Annual, IEEE International*, pp. 128–130, 2006.

20.7 FURTHER READING

A. H. Fischer, A. von Glasow, S. Penka, and F. Ungar, "Process Optimization–The Key to Obtain Highly Reliable Cu Interconnects," *Interconnect Technology Conference, 2003. Proceedings of the IEEE 2003 International*, pp. 253–255, 2003.

C.-K. Hu, B. Luther, F. B. Kaufman, J. Hummel, C. Uzoh, and D. J. Pearson, "Copper Interconnection Integration and Reliability," *Thin Solid Films*, 262 (1): 84–92, 1995.

I. A. Blech, Physics of Failure in Electronics, Electromigration in Thin Aluminum Films on Titanium Nitride," *Journal of Applied Physics*, 47 (4): 1203–1208, 1976.

Chapter 21

Automatic Test Equipment

BY

A. T. SIVARAM *Advantest America, San Jose, California*

21.1 AUTOMATIC TEST EQUIPMENT INTRODUCTION

Semiconductors are key components of all electronic products. As the prevalence and variety of electronic products increases, so do the requirements and complexity of semiconductor manufacturing. The worldwide sales for semiconductor ICs were about $338 billion in 2015. For 2016, the semiconductor market is expected to grow by 0.3 percent (Source: Semiconductor Industry Association). Semiconductor manufacturers produce many types of devices, such as power regulators, DACs, ADCs, RF transceivers, and microprocessors, by etching millions or even billions of transistors and other electronic components on several layers of the semiconductor wafers. Although the steps in manufacturing these devices vary for different device types, the general process involves wafer manufacturing, wafer fabrication, packaging, and final test (Figs. 21.1 and 21.2). The manufacturing process is summarized as follows:

- Wafer manufacturing
  - Silicon polycrystalline growth
  - Wafer polishing
  - Epitaxy
- Wafer fabrication
  - Photolithography
  - Etch and strip process
  - Diffusion, implant, and metallization
  - Chemical mechanical planarization (CMP)
- Packaging and final test
  - Wafer probe test
  - Packaging
  - Final test

Historically, every semiconductor device manufactured has been tested. In general, a device will be tested at several different points during the manufacturing process. IC manufacturers perform testing for the following main reasons:

1. Increase confidence, the device will meet its published specifications.
2. Identify and reduce manufacturing defects to improve the yield of the process.
3. Detect assembly errors after packaging to reduce customer returns.
4. Improve overall quality and reliability of semiconductor manufacturing.

Testing is done using automatic test equipment (ATE), which is a computer that is designed to perform tests on different devices. Techopedia explains automatic test equipment: "Automated test equipment is a computer-operated machine used to test devices for performance and capabilities." The unit that is being tested is known as device under test (DUT). ATE can include testing for electronics, hardware, software, semiconductors, or avionics. There are uncomplicated ATEs such as volt-ohm meters that measure resistance and voltage in PCs. There are also complex ATE systems that have several test mechanisms that automatically run high-level electronic diagnostics such as wafer testing for semiconductor device fabrication or for integrated circuits. Most high-tech ATE systems use automation to perform the test quickly. The objective of ATE is to quickly confirm whether a DUT works and to find defects. This testing method saves on manufacturing costs and helps prevent a faulty device from entering the market. Because ATE is used in a wide array of DUTs, each testing has a different procedure. One actuality in all testing is that when the first out-of-tolerance value is detected, the testing stops and the DUT fails the evaluation. In the semiconductor manufacturing area ATE systems are integrated with material handling systems to move wafers and packaged parts through various test steps. The combination of ATE and material handling system is called a test cell. The SEMI standards (http://www.semi.org/) define the framework for the automated test cell used in semiconductor manufacturing.

There are various types of electrical and reliability tests conducted at different stages of assembly and for different purposes once die are created. Tests include die sort test on wafers, electrical and environmental testing, and burn-in on packaged devices.

Wafer sort identifies bad die prior to assembly by electrically testing each chip on the whole wafer after fabrication. The test is for electrical performance and circuit functioning so as to prevent the assembly of bad die. During wafer sort, a probe card, which consists of many tiny probes, connects to each bonding pad on the die to provide the electrical connection between the die and the IC tester. Power is supplied through the probes to test the circuit and record the results. Die with circuits that fail the test are marked with an ink dot, and the wafer sort results are then analyzed to improve yield management of the chip fabrication process.

After assembly and packaging, the completed package must pass more extensive electrical tests as well as environmental tests prior to being sent to a customer. Environmental tests are performed to weed

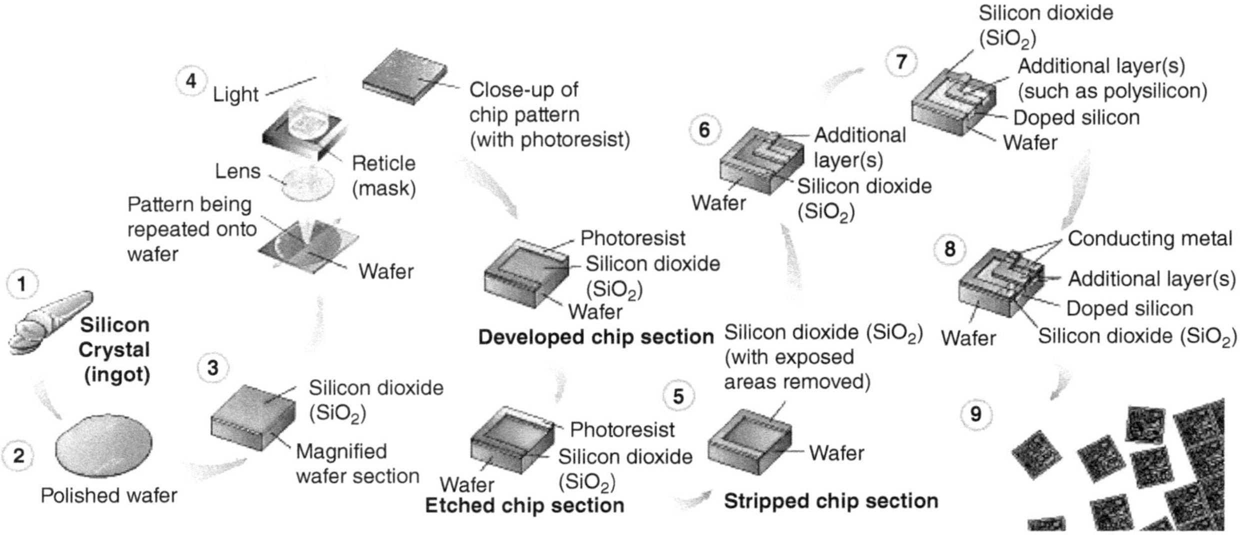

Figure 21.1 Wafer fabrication. (*Adapted from www.sematech.org.*)

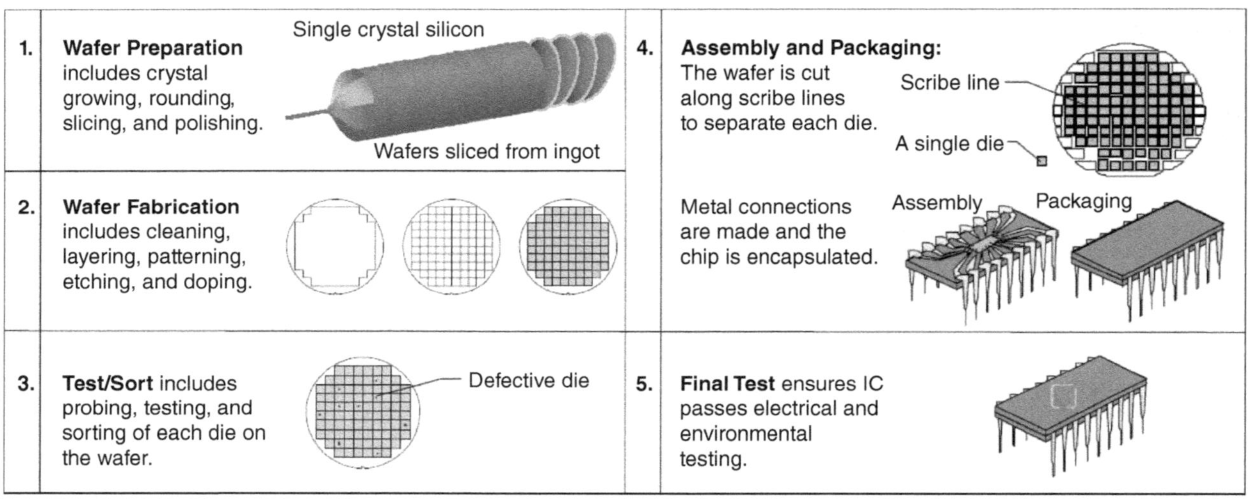

Figure 21.2 Packaging and final test (Semiconductor Manufacturing Technology by Michael Quirk and Julian Serda, Prentice Hall).

out packages with defects (e.g., due to possible contamination or existence of low-quality die attach or wire bonds) that may cause reliability problems over the lifetime of the package. Typical environmental tests include temperature cycling, thermal shock, mechanical shock or vibration, temperature humidity bias testing (THBT), autoclave (ACLV), or moisture sensitivity testing.

Electrical tests verify that chips perform to specification. There are two steps of this final test. Parametric tests check the general performance of the device or circuit and ensure that it meets certain input and output voltage, capacitance, and current specifications. Functional tests verify the specified function of the chip; logic chips are put through logic test, while memory chips go through tests for data storage and retrieval capabilities. The test equipment at this point consists of computer-controlled probers and handlers for automatic operation of the test process.

For burn-in tests, packages are inserted in sockets under an electrical bias to stress the electrical interconnection of the chip. The test is conducted in temperature cycling chambers. The purpose of burn-in is to sort out marginal chips and circuits that will fail during the "infant mortality" phase of a product life. Devices that pass burn-in are statistically more reliable.

21.2 ATE HISTORY

For the first two decades of the industry, semiconductor manufacturers custom-built most of their own tools and equipment in-house (Fig. 21.3). In the early 1960s, Fairchild Semiconductor, Signetics, Texas Instruments, and others began to sell their specialized semiconductor test equipment to their customers and competitors.

TI engineers developed a centralized automatic tester (CAT) transistor-testing machine in 1958. TI's Transistor and Component Tester (TACT), introduced in 1962, served the company and its customers for many years. One of the first IC testers offered commercially was the Signetics Model 1420.

Vic Grinich (1924-2000) **TI TACT (1962)** **Nicholas DeWolf (1928-2006)** **Teradyne J259**

Figure 21.3 ATE history. (*Credit: http://computerhistory.org/semiconductor/timeline/161-test.html.*)

Under cofounder Vic Grinich, Fairchild produced a series of in-house transistor testers, beginning with the type 1A in 1959. The type 4 tester was introduced for public sale at the March 1961 IRE Show. This group grew into an important instrumentation division that produced several generations of machines, including the Model 4000M Automatic Test System, a workhorse of the industry in the 1960s, and the Sentry 400, one of the first high pin-count testers for large-scale integrated (LSI) circuits.

Nicholas DeWolf, chief engineer at semiconductor manufacturer Transitron, Wakefield, MA, teamed up with Alex d'Arbeloff in 1961 to found Teradyne in Boston to build the D133 diode tester, one of the first commercial test systems to use semiconductors in place of vacuum tubes. In 1966, DeWolf and Milt Collins designed the model J259 based on the Digital Equipment Corporation PDP-8 minicomputer as the industry's first computer-controlled test system. This machine laid the foundation for today's automatic test equipment (ATE) industry and Teradyne as major player. Born in Philadelphia on July 12, 1928, he went on to graduate from the Massachusetts Institute of Technology at the tender age of 19. He started in the industry as an engineer for General Electric in 1948 when the world was trying to master the solid state diode. The challenge was not so much in the making of them, but in the testing of them. DeWolf mastered the technology and quickly became an expert. Because GE could test them better than anyone, they became a leader and the company also marketed a tester designed by DeWolf. In 1952, he wrote the following canons of design:

The design of factory production test equipment:

1. It must be reliable.
2. It must be safe to use, and almost impossible of causing injury.
3. It must be reliable and use good components and circuits.
4. It should be simple, to ease maintenance.
5. It must be easily calibrated with simple gear.
6. It must operate in a wide range of supply voltages.
7. It must operate in a wide range of temperatures.
8. It must be reproducible. If several sets are built, they *must* read the same thing, even if it involves a loss in overall absolute accuracy.
9. It must be simple to maintain and accessible.
10. The operating portion of the gear at the operator's station must be small to prevent overcrowding.
11. It must test only the desired characteristics and not be responsive to other parameters, or varying quality may result as product changes are made.
12. It must be accurate.
13. It must be stable.
14. It must be capable of a complete test in a short period of time, or testing expense will increase.
15. It must require a minimum of thought on the part of the operator.
16. It must not fatigue the operator, or mistakes will occur.
17. It should be reasonably neat in appearance. This instills confidence in the equipment.
18. It should be flexible enough to allow for specification changes that inevitably occur.
19. It must favor the rejection of bad units over their acceptance, in respect to predictable drifts and failure. This applies particularly to automatic test equipment fail safe!
20. It should be reasonable quiet and cool and not damage tested units.
21. Remember the operator, who is 2/3 of the equipment.
22. It must give repeatable results.

There is much to learn from these, as most still apply more than a century later. Amazingly, he was only 24 when he wrote them and it was at a time when solid state electronics were in their infancy. Most people live a lifetime without such brilliance and clarity of thought. Another gem of DeWolf's was his belief that what customers want is seldom what they need and you should design to the latter. In those days, few knew this basic principle of product marketing.

Since the introduction of the first tester in the 1960s, semiconductor test systems were developed to keep pace with the semiconductor advances following Moore's law, which 60 years ago predicted cost per transistor would halve every two years. During the late 1990s, there were as many as seven major semiconductor ATE manufacturers serving the needs of the then \$200 billion semiconductor chip manufacturing industry. Table 21.1 shows the various test equipment makers and the models from the 1960s till now. Just like the semiconductor industry it was serving the test equipment industry went through consolidation over the years as the business changed and today the \$3.5 billion semiconductor ATE market is shared by two major companies Teradyne and Advantest. Teradyne generated \$1.68 billion in the financial year 2014, compared to \$1.43 billion during the previous four quarters ending in 2013. Heavily focused on the cyclical semiconductor test market, the upturn witnessed in the system on chip (SOC) industry is expected to benefit the company in the near future. "The fourth quarter wraps up a very strong year for Teradyne from both a market share and cash flow perspective," said CEO and President Mark Jagiela. "Revenues grew 15% driven by strong SoC demand, our operating profit grew 26%, and we generated over \$300 million in free cash flow. Despite an expected softening of SOC test demand in 2015, our cycle hardened operating model, combined with our long term outlook for the markets we serve, provides us the confidence to substantially increase our capital return plans for 2015." From a product perspective, Teradyne's UltraFLEX system has been in high demand, as it addresses complex SOC chips used in new mobile devices. The company also foresees growth coming from its storage test product line, specifically its 2.5-in drive test product, as well as wireless test, which is a market the company serves through Litepoint.

Similar to Teradyne, Advantest depends heavily on the semiconductor test business, which represents close to 75 percent of the company's total revenues. Advantest's semiconductor test division reached \$1.63 billion in fiscal year 2014, which ended in March 2015. This was an increase of 46 percent over the previous fiscal year. Increased shipments of smartphones and tablets as well as the use of various types of semiconductors in these mobile devices will be a key growth driver for the company in the future. Five markets account for over 85 percent of worldwide sales of semiconductor manufacturing equipment creating a

Table 21.1 Test Equipment Manufacturers

Teradyne	Digital	J259	J283	J384	J387	J389	J900	J941	J953	J971	J750	Magnum from Nextest ETS800 from Eagle Test System
	Memory	J386	J937									
	Linear	A300	A360	A370	Magnum 5		Magnum CIS					
	MXSL	ETS800										
	SoC	Catalyst	Flex	Ultraflex								
Advantest (Takeda Riken)	Digital	T310	T320	T3380	T3340	T3320	T3310	T66XX				
	Memory	T318	T3370	T3331	T3332	T3340	T3324	T55XX	T53xx	T57xx	T58xx	
	SoC	T2000	T2000IPS									
Verigy (Agilent, Hewlett Packard)	Digital	HP82000	HP83000	HP330	HP660							Merged with Advantest
	Memory	V5000	V6000									
	Linear	HP94000										
	SoC	V93000										
Xcerra (LTX, Trillium)	Linear	MTS77	LTX77	LTX90	HI T	Synchro						Changed to Xecerra
	Digital	Array master	Micro master	Delat master	Valid master							
	SoC	Fusion										
Schlumberger (NPTest, Sentry, Fairchild, Schlumberger)	Digital	FST4000	FST5000C	Sentry 400/500	Sentry II/V	Sentry VII/VIII	Sentry VII/VIII	Sentry Series 20/21	Series 50 S1040	Series 15 S1650	S9000 FX/ GX/IXIX/ KXST2000	Changed to NPTest then merged with Credence
	Memory	Xicom 558	Xincom 55	S90	S92							
	Linear	S80										
Credence	Digital	STS256	IMS Series	SX Series	D10 Series							Merged with LTX
	MXSL	Duo	Quatro									
Tektronix	Digital	S3260	S3270	S3280	S3295							Merged with Credence
	MXSL	Vista Series										
Genrad	Digital	GR16	GR18	GR14	GR125	GR160	GR180					Merged with Teradyne
Megatest	Digital	Q8000	QII	Mega 1	Impact	Polaris						Merged with Teradyne
	Memory	Q2/50	Genesis									

very concentrated market: Taiwan, South Korea, China, Japan, and the United States. Europe and the Middle East represent another 6 percent, and other markets (primarily in Southeast Asia) account for 6 percent of the world market.

21.2.1 ATE Classification

In the early to mid-1990s, ATE systems used to be broadly classified into three types based on cost of the tester: "big-iron" ATE, typically costing above $1,000,000; low-cost production tester, costs ranging between $399,000 and $1,000,000; and low-cost DFT (design for test)–focused tester, under $100,000. The difference in cost of these three types of ATE comes from the difference between structural, functional, and system level testing. Structural testing involves developing a set of test vectors to detect specific faults that may have been introduced in the design by errors such as DSM (deep sub-micrometer) effects and processing defects. By applying specific test vectors to a circuit, and then capturing and comparing the actual responses against the expected responses is what is known as functional testing. Big-iron testers were designed to test the device functionally; at speed functional testing is performed at every pin. An example of a big-iron ATE is the Schlumberger's ITS9000. Low-cost production testers combine structural and functional testing. Some tradeoffs in speed, performance, or flexibility are made with these testers in comparison to the big-iron testers such that overall ATE cost is reduced. An example of a low-cost production tester is the Credence SC212. Low-cost DFT-focused testers have limited or no functional test capabilities, they are designed to support test methodologies such as boundary scan or BIST. This type or architecture has a lower cost due to lower pin count and accuracy. The DFT-focused tester can play a key role in design verification and this class of tester has survived over the years mainly due to its ease of use. An example of the DFT tester in the mid-1990s is the Invoy's Personal Ocelot while the Advantest CTS (cloud testing service) tester is a contemporary DFT tester.

Semiconductor devices are classified based on the type of circuits they contain such as digital, linear, mixed-signal, memory, flash, RF, microwave, and finally SoC, which contain all types of circuits. For each of these device types ATE systems exist, which have the optimized hardware architecture for efficiently testing the particular device type. ATEs are also classified in terms of the number of channels and the density of the gates in the device they tested. The testers were called high pin count (HPC), reduced pin count (RPC), LSI, and VLSI testers. This type of classification was superseded by the application segment for devices such as microprocessor, graphic processor, application processor, microcontroller, ASIC, FPGA, DSP, and telecommunication. Each one of these segments drove the manufacturing of a very large volume of devices of the order of millions of units per month justifying testers with unique capabilities for testing in each of these device segments. Finally, in today's market ATE is broadly classified in functional areas as logic, memory, mobile, power, automotive, image sensor, and SoC. These days the SoC tester by far dominates the ATE landscape in high-volume manufacturing of semiconductor devices.

21.3 DIGITAL TESTER

21.3.1 Digital Testing

Historically, logic devices followed the discrete device design in the early 1960s. Early testers designed for testing digital devices used a stored program memory which essentially verifies the truth table of combinational logic circuits by applying patterns on the input pins of the DUT and comparing the sampled logic levels of the output pins to the expected pattern stored in the program memory. This simple digital tester gave rise to the development of most of the rest of the ATE concepts over the past five decades. Digital data has two states—on/off, high/low, or logic "1" or "0." A stimulus on one or more of the inputs to the device causes a result on one or more of the outputs of the device. It is made of transistors, or more specifically *field effect transistors* (FETs), configured together to build gates. A gate is the simplest of all digital building blocks. Gates include AND, OR, XOR, and NOT among others and can be combined to build logic for specific functions or blocks. A device may have several blocks that, together, form a digital device. These individual bits of data can be combined to form a byte or a word of digital data, and 1 bit has two possible states (0, 1). Binary words on a digital device are also called a bus. They can be used to address memory, as the data that go into some memory location, or as control lines. Many companies sell digital building blocks as their *intellectual property* (IP). This gives the designer access to the highest performance design and allows a designer to build a device using proprietary digital blocks along with a microprocessor block from one company and an interface block from another company to create a new device that fills a specific need. Digital signals are received or transmitted to a device based on the device clock. The rising or falling edge or both edges of a clock signal are used to latch data into flip-flops in the device. The faster the clock can run, the faster the device can use data, and the faster the device can complete a given task. Processors and ASICs may have several clocks. The device may be able to do computations internally with a clock of one frequency, yet transmit and receive data at much different speeds. In this case, the device may generate its own clock, using a phase-lock loop, and transmit data to other devices according to that clock. The highest clock frequencies are generally used inside a device to do computations; other clock speeds are used to reduce the risk of bit errors in communicating to other components.

21.3.2 Functional, Structural, and Defect-Based Tests

A common way to test a digital device is by using *functional tests* that exercise the digital inputs the same way as they would be used in the end application. For instance, with an "AND" gate, the following sequence of inputs—00, 01, 10, 11—would produce these outputs—0, 0, 0, 1. IC designers simulate device operation to make sure their design is correct. Test vectors are then generated from this simulation data. As the number of gates on a digital device grows, the time it takes to simulate the device and the number of vectors required to test it grow exponentially. Devices with millions of transistors cannot feasibly be simulated completely because of time to market (TTM) and time to volume (TTV) pressures and the costs of added simulation and test equipment (such as larger memory and more processing power required). Therefore, engineers can design test circuitry into the device and use a combination of functional and structural tests to verify the device's operation. Simple functional tests are written to test interconnects between different blocks within a device and structural tests are written to test the individual gates. A design-for-testability review can determine which of these methods are needed for a particular device. *Structural tests* seek to verify that all the pieces of the product circuit are present and working. Alternatively, a structural test verifies that the DUT is free of faults but does not attempt to verify that it performs to all its specifications. The most common digital structural test is scan. With scan tests, the designer must use scan flip-flops and when the chip is placed in scan mode, special input pins on the device can shift data serially through chains of these registers and into particular flip-flops. Data that are required to test for a specific fault can be easily applied to the gate inputs. After the correct data are placed in the flip-flops, the device is clocked once and data are captured by scan flip-flops at the output of these gates. That data are shifted out serially and compared against the data predicted by simulation. Multiple "chains" of scan registers can be tested in parallel for added throughput. Any data can be placed into any gate so the fault coverage is very high. The down side is that the data for all these different serial scan chains require lots of vector memory and, since the scan lines have typically lower speed than the rest of the device, the test time can be long. The IEEE 1450 Standard Test Interface Language (STIL) standard addresses "design" related aspects of digital test data. Since its introduction in 1999 STIL has evolved over the years to support advanced applications such as (1) embedded cores, (2) families of test patterns, (3) mapping to ATE systems, (4) mapping to simulation, and (5) devices with advanced DFT features. A third test strategy is known as *defect-based testing* (DBT). DBT attempts to detect defects in a circuit by measuring any of its outputs while looking for unusual behavior, even if it is within specification. The most widely used class of DBT methods is based on measuring the circuit's quiescent power supply current. Called IDDQ, these methods are most useful in *complementary metal oxide semiconductor* (CMOS) circuits. Ideally, in a CMOS device no current should flow from the supply when the device is powered on but not running. Any current that flows is the result of leaky FETs or defects, usually bridges between interconnect lines that bias gates of the series connected *n*-channel MOS (NMOS) and *p*-channel MOS (PMOS) FETs away from the supply rails such that both FETs are weakly turned on and current can flow from *VDD* to Gnd. Once a device is characterized to have a certain *I*DDQ value, any part with *I*DDQ above can be considered a bad part. The device is usually put into its quiescent state(s) by scanning in patterns so *I*DDQ tests are built on top of structural scan tests. As FETs get smaller, their natural leakage increases rapidly, which makes it harder to distinguish the defect current from the defect-free current. More advanced *I*DDQ tests have been developed for small geometry devices including delta *I*DDQ and *I*DDQ current signatures. These tests make multiple *I*DDQ measurements with the device programmed to be in various "quiet" states. Extreme changes indicate a failure. These more advanced methods require considerable characterization. A common alternative to the accessibility, resources, and time required to read patterns into a device from an external tester is *built-in self-test* (BIST) that can execute either functional or structural tests. A very high-speed or very dense digital block may not be able to be tested with scan. BIST is implemented as a small state-machine circuit built into the device itself. This circuit generates inputs and compares outputs of a particular block in a device. Programming the device to drive random data can test a high-speed circuit using the device's own input pin to receive that data and compare it to an expected result. BIST can be implemented as part of the device or off the device employing so-called *built-out self-test* (BOST) as a small circuit on the wafer that is physically cut off when the device is packaged.

21.3.3 High-Speed Digital Testing

Advances in silicon, such as smaller gate widths and superior materials, and fabrication techniques have enabled extremely fast on-chip communication. In many cases, the bottleneck of the ICs has become transporting data between chips and other system building blocks. The need to create faster chip-to-chip interfacing has resulted in many different standards for data transportation. Unlike sub-gigabit-per-second (Gbit/s) testing, high-speed testing involves no tristating—as mentioned in an earlier section—because tristating causes additional capacitance. Termination is always used to avoid reflections, and the data bus effectively becomes a high-speed transmission line. Understanding the particular high-speed device and its challenges is an important first step to outlining a test plan and determining the fit to a particular piece of ATE. For example, some standards utilize embedded clocks, whereas others, such as HyperTransport, provide a clock synchronized with the data and the internal *phase locked loop* (PLL). The latter may require an interface with source synchronous capabilities, whereas the former may require a clock-recovery circuit. Many cases exist where the digital tester is asked to synchronize to a bit stream that is not deterministic or contains seemingly random data packets. Many high-speed bus protocols are differential based. Differential swings are typically extremely small and require a digital tester with the ability to provide a linear signal with a well-defined, controlled slew rate at low swings. Some applications require a dual-transmission line implementation to enable fast write-to-read transitions or to enable dedicated drive-only or receive-only ATE to interface with a bidirectional device interface. An example is *double data rate* (DDR) devices that must use a driver and a receiver to interface bidirectional DDR data pins (single ended) with ATE. Setup and hold and skew measurements are typically important in the world of high-speed tests and the ATE hardware must be able to accurately measure skews and setup and hold times. One of the most important high-speed parameters is jitter. The impact jitter has, as it is introduced to a device, is to cause a connected receiver to detect an incorrect bit transition, causing a fault. Generally, the tester needs to be able to inject and measure jitter. A jitter tolerance measurement requires that the tester inject a known amount of jitter while the device's outputs are monitored for an effect on the bit error rate as a functional test. Jitter transfer, on the other hand, is a measure of how a device amplifies or transfers jittery input to its outputs. The ability of the device's PLL to filter out the jitter is of key importance. For this test, the tester provides a jittery source and the device's transfer of the jitter is characterized. Eye diagrams are another commonly used, graphical way to view the total jitter present on a device, as measured by the test system. An eye diagram is a composite view of all the bit periods of a captured waveform superimposed on each other. To create an eye diagram with a digital tester, the pattern is typically executed multiple times; each time the edge position and/or strobe levels are modified. The horizontal size and overall shape of the eye indicates the jitter present on the device. The shapes of the eye crossings indicate whether or not deterministic jitter is present in addition to random jitter. Devices with source-synchronous outputs can require a different approach to tests than traditional techniques. The main difference in testing these outputs is the nondeterministic placements of the transitions of output signals. Traditional devices have specifications that define the timing relationship of the inputs to the outputs. Source synchronous devices do not have such a specification; instead, the data output is also accompanied with a clock signal that is coupled to the data. For ATE, this means that the output clock of the device has to be used to sample the data. Today, many of the source synchronous parts can tolerate huge shifts in output placement, two to four data bit times in the system, and are able to be tested without the use of source synchronous capability. In this case, searching for the best place to capture the data is one way of using testers without source-synchronous capability.

21.3.4 Deterministic versus Nondeterministic Behavior

Traditional functional testing has been done using deterministic patterns that can be defined prior to testing. These patterns typically came from simulations of the device and contain stimulus and response data. The stimulus signals are intended to exercise the DUT and the response signals are the results of the simulations based on the stimulus. The part being tested is exercised to validate that it matches the simulated responses; for a given input there is only one correct output. Nondeterministic behavior has multiple forms, but the basic definition is that an algorithm or protocol has to be used to determine the correctness of the outputs and cannot be predicted prior to execution, thus there can be no previously generated "expect data." This can take the form of either the data or the execution timing being the variable. For data, this means with a given input, there is a range of valid outputs for a good device. For timing, this means that timing of the comparator strobe is set by finding the timing placement of some pin and measuring additional pins relative to this timing location.

21.3.5 Basic Setup for Digital Test

The basics of setting up and running a digital test include defining the pin configuration, or DUT pin to tester resource mapping, the voltage levels, the timing of the pin edges, and bringing the definitions of the correct drive and receive states, or vectors into a framework to develop a program. Most of this data can be leveraged from EDA. The dc tests are explained in the common test section.

Additional ac tests include functional tests, where vector inputs are given and outputs compared to, timing tests to find out where the edges produce a passing result and where they do not, level tests to dynamically find input and output sensitivity, propagation delay that determines the amount of time taken to get through a circuit, and setup and hold measurements that determine how long an input needs to be set up before a clock and how long it needs to be held after a clock cycle occurs. Other tests include frequency measurements, jitter measurements, and sweep tests. A sweep test moves either a voltage, timing edge, or frequency through a range of values, and then determines where the device passed and where it failed. Usually this is a characterization test because it is time consuming. Often in production a single test functional pattern will be run with critical timing and/or levels and a pass or fail status is determined.

21.3.6 Protocol-Based Tests

The traditional ATE functional test method uses a stored stimulus response pattern and this does not work well due to the following characteristics of industry standard interfaces, which are getting integrated into digital devices:

1. Elaborate training sequence devices go through during which the device state transitions can take variable number of device cycles due to the nondeterministic nature of the protocol.
2. Devices use packet formats which include CRC fields and this makes pattern generation for the testers, which typically use the lowest level 01LHX formatted data, very complicated.

Examples of low- and high-speed interfaces are SPI, USB, PCIe, and SATA to name a few. "Protocol aware ATE" have hardware packet sequencers for generating stimulus in the native packet format to the DUT. The ATE simultaneously captures and assembles the response from the DUT into packets for testing in real time at the interface speed.

21.4 LINEAR TESTER

21.4.1 Linear Device Testing

Linear devices are those class of devices that produce a continuous time varying output for a corresponding input. Unlike digital devices, they have an infinite number of states. Examples of linear devices are voltage regulators, operational amplifiers, and analog switches. Testing of linear devices originated from logic testing. The linear device test methods were developed from the application of digital testers to operational amplifiers (OPAMPs). The digital tester used a linear test adapter to test the voltage and currents involved in OPAMP testing.

21.4.2 Linear Device Tester

The architecture of a linear device test system is very dependent on the type and variety of linear device that will be tested. Often the first choice to be made in selecting a tester for linear devices is between *automated test equipment* (ATE) and "rack and stack," that is, a combination of individual instruments. Since the architecture of a test system for linear devices is highly dependent on device functionality and these devices require a very low cost of test (COT), many times the architecture is custom tailored to a specific family of devices and is not a general purpose one. However, some of the typical tester building blocks that are needed include a switching matrix, which provides a method of connecting the source and measurement devices to the different pins of the DUT. The number of pins, their function, and parameter specifications must be considered when specifying (or designing) the switching matrix requirements. Power supplies will be required to provide power to the device for testing. Power supply considerations include the number of supplies and their voltage and current specifications. All traces, wires, and fixtures must withstand the maximum expected voltage and current. Some tests can be dangerous to humans or even the test equipment. For example, testing the breakdown voltage of the DUT exposes it to a very high voltage (hundreds of volts) to ensure compliance to specification. Often a pulsed high-current source will be used to keep the test equipment to a reasonable size and to prevent destroying the DUT, particularly in wafer form when no heat sinks can be attached to the device. Even at package test, any kind of heat sinking capability would be minimal. Digital logic input and output pins may be used as digital control pins needed to access the DUT or configure it for certain test modes. Control, signal lines, or identification pins may be used to control relays or switches on the DUT fixture. Different devices require different fixtures and often the test program will electronically check the identification information of the fixture to ensure it is the correct one. The prober/handler interface will allow indexing to the next die or package as well as sending category or binning information that allows the sorting of devices into different grades. Often the interface will also integrate equipment for temperature testing. The integration of a prober or handler into a tester is often more time consuming than expected due to the mixing of mechanical and electrical issues. There is a standard set of signals expected for this interface that can be provided with simple TTL logic or via a standard interface like the *general purpose instrument bus* (GPIB). System controllers are required to run and coordinate all the operations for the test cell. Considerations for the system controller include the operating system used, networking capabilities, processing power or speed, upgradeability, the testing language(s) supported, and whether the operating system language is an industry standard or a proprietary. A final factor to be considered is that often test systems are installed in clean areas or areas with limited access. Because it is inconvenient to carry out test development in these limited-access areas, it is important to find a way to do remote development.

21.4.3 Basic Test Setup for Linear Devices

The test setup for a linear device is dependent on its function and the level of testing required. For instance, testing a simple diode will require a programmable voltage and a current source, in addition to a voltage and current measuring device (typically called a parametric measurement unit or PMU). An oscilloscope or programmable threshold counter will also be required to measure its ac switching characteristics. The architectural decisions on the tester will depend upon how much automation is needed and what level of accuracy is desired. For example, to test a small batch of diodes, a human operator can manually test with a power supply, a voltmeter, a pulse generator, an analog oscilloscope, a pencil, and notepad. For a large quantity of diodes, with the requirements of data logging and temperature testing, one would need to add automation—comprising a system controller, computer controllable PMU, signal generator, scope/counter, a temperature controllable device handler, and some form of device-marking system to allow the data log to track the device.

21.5 MIXED-SIGNAL TESTER

Mixed-signal (MXSL) devices are the class of devices that have a good amount of digital logic built into them along with analog functions. Best examples are DAC/ADC converters, phase lock loop (PLL) circuits, and digital filters. MXSL devices can be quite complex and the MXSL testers are classified by application because each application is a unique test problem.

21.5.1 The Basic Building Blocks of a Mixed-Signal Device

The digital section functionality and organization are similar to that described in Sec. 21.3.1 of this chapter. A more specialized digital function includes DSP, which is a mathematical technique used to process a signal according to a specific need. A typical use of DSP is to convert the time domain data into data more suited to CPU math operations; for example, converting time domain data into the frequency domain (*fast Fourier transform*) and then performing the math to arrive at an actual test parameter number such as *total harmonic distortion*. Other uses include complex signal processing such as signal modulation or demodulation or digital filtering. Other digital blocks include CPUs, digital logic, and embedded memory. The analog is the section of the device that deals with signals that vary in time as well as amplitude. These signals are usually considered real-world type signals, because all signals of this nature usually are not related to a fixed timed event but can occur randomly and are nonperiodic. The most common analog building blocks are amplifiers and filters. The converters section of the device can convert an analog signal, or one that is varying in time and amplitude into a digital representation of the applied signal (A to D converter) or can take a digital signal and convert it into an analog or time and amplitude-varying signal (D to A converter). The converters provide the interface between the analog domain and the digital domain. The digital domain data can be modified or transferred without any deterioration and are more suited to digital logic and computation while the analog representations are more suited for human use, such as audio and video signals and are vulnerable to noise and distortion. The filtering section uses techniques to remove or limit unwanted signals from the signal of interest. This filtering can be in the form of analog filtering, usually occurring in the time domain, and using active components such as op amps or passive components such as resistors, inductors, and capacitors. Digital filtering is a math process using either a convolution operation in the time domain or FFT-based operation in the frequency domain and will either subtract unwanted or enhance desired components. Most *system-on-chip* (SOC) devices include PLL sections to generate appropriate clock signals to be distributed to various functional blocks. An external oscillator may supply a basic clock signal to the device and it may be multiplied or divided into suitable frequencies for each functional block. In a serial data transmission SOC, an incoming receiver signal may contain its own clock that is extracted by a PLL.

The power section is one of the most critical sections of the device that is frequently overlooked. Optimum power supply, distribution, and bypass filtering are key to keeping noise and ripple to a minimum to allow the device to function properly.

21.6 MEMORY TESTER

21.6.1 Memory Testing

Just like digital testing the testing of memory components started in the sixties with J-K flip-flops and shift registers. Memory densities and speeds are continually driven by market demands. Memory is prevalent in almost all applications that use semiconductors. In 2016, the sales of DRAM of all densities is expected to be around 150 billion units. Semiconductor memories can be categorized into two general types: volatile and nonvolatile. A volatile memory loses its data contents when its power is switched off while a nonvolatile memory retains its data without power. The most common nonvolatile memory is also known as flash memory and is discussed in Sec. 21.7. Volatile memory cells can be further classified as SRAM or *dynamic random access memory* (DRAM). To retain data, a DRAM cell requires a data refresh operation at regular intervals and if power is removed, all data are lost. An SRAM cell will retain data as long as a power source is applied. The reason for this behavior, in both cases, can be found in the design. In order to maximize memory density, high-density DRAM memory arrays are designed with a single transistor with a capacitor load to hold the data. The charge stored in a DRAM cell would eventually leak off if not for the refresh operation. A refresh reads the current data and writes it back to the DRAM, which renews the stored charge in the DRAM cell hence, the term dynamic memory.

An SRAM cell is implemented with multiple transistors to latch the data and eliminates the need for a refresh operation. Semiconductor memories are typically organized as an array of cells, each one containing a logic bit "1" or "0," of information. These arrays are typically very regular. The dimensions of the array are denoted by X and Y, or rows and columns. Through row and column select lines, each cell may be individually enabled for reading and writing. The X dimension address corresponds to the row enables and the Y dimension address to the column enables, although for some manufacturers these X and Y definitions are reversed. An address is placed on the address pins of the device, which is then internally decoded to enable the required cells.

More than one cell may be enabled by an address so that multiple bits of data may be manipulated at the same time. For instance, a memory device with a 1 Mbit capacity has 1048576 cells in its array. These cells may be organized as 131072 addresses that enable eight cells at a time. This would commonly be known as a 128K × 8 memory device. Likewise, the same memory might instead be organized as a 64K × 16 device.

21.6.2 Basic Memory Test Setups

Similar to other types of digital semiconductor devices, the test setup includes pin configurations, voltage levels, timing conditions, and test patterns determined by the design of the device. Different logic designs require different timing and level setups. For example, a device with HSTL outputs would require one set of voltage and timing conditions while a device using SSTL 1.8 compliant off chip drivers would require a completely different set of conditions. In general, there are three steps involved in the testing process—determining the fault coverage requirement, planning and implementing the test setup, and finally analyzing the test results. A good understanding of the fabrication process, such as historical data and a good understanding of the design, determines the fault coverage requirements. This was discussed in Sec. 21.3.1. Early in the design process, decisions must be made about the level of BIST. In one extreme, designers could implement BIST engines to accomplish most of the functional memory tests, leaving the standard dc parametric test to ATE. Or, the design might not provide enough extra space for the BIST circuitry on the silicon die, in which case all the testing, including functional dc and ac, would be performed with ATE. The functional testing of a memory device can be viewed as three parts—the memory core consisting of memory cells, the address decoders consisting of logic used to select a particular section of the memory array, and finally the logic used to control device operations such as read, write, and refresh. Each section of the memory requires different patterns to test for proper functionality. Some of the fault models are the same as any other logic device such as stuck-at faults or stuck-at logic "0" or "1." Because of the way a memory design is laid out, other faults also apply. These are called neighborhood sensitive faults that have to do with the proximity of one element to another. Some examples of this are coupling faults defined as cross talk between data lines or electrical shorts between data lines, and address decoder faults defined as inaccessible address lines or multiple cell access faults.

21.7 FLASH MEMORY TESTER

21.7.1 Flash Memory Testing

Flash memory technology is a mix of EPROM (erasable programmable read-only memory) and EEPROM (electrical EPROM) technologies. The term flash was chosen because a large chunk of memory could be erased at one time. The name, therefore, distinguishes flash devices from EEPROMs, where each byte is erased individually. Now-a-days, flash memory technology is a mature technology. It is a strong competitor to other nonvolatile memories such as EPROMs and EEPROMs, and to DRAM applications. Flash memory allows users to electrically program and erase information and is a nonvolatile memory; that is, power can be removed from the circuit, and it will retain the information. Flash memory is similar to EPROM, but with a much thinner oxide layer between the gate and the source/drain. Each individual flash memory cell is a FET with a modified gate structure. The gate structure has a floating gate where electrons are trapped or stored. By trapping or storing these electrons, the threshold voltage of the FET is raised and it is considered programmed. When the electrons are removed from the floating gate of the FET, it is considered erased. The process of storing and removing charge is complicated. It requires that certain voltages be placed on the contacts of the FET for controlled periods of time. It is possible to over-program or over-erase the FETs causing the memory device to cease functioning. Users typically want to deal with bytes or words instead of individual bits. Because of this, the flash memory device contains a state machine that is a microcontroller that acts as a user interface. The user issues predetermined commands to the flash memory device, such as "program this word," or erases some portion of the flash memory, and the microcontroller does the low-level work of moving the charge onto or off the floating gate of each affected FET. Different flash memory manufacturers have various features and gate structures that enhance this basic behavior. One approach is to store different levels of charge to allow multiple bits of information per cell. Another is to modify the FET structure by allowing the floating gate to store different charges at each end of the gate thus allowing multiple bits per FET. In any case, the trend is to store more bits in less space thus reducing the cost per bit. There are two different types of flash memory technologies: NOR and NAND. *NOR*: The data words can be accessed in any order (random access). This type of flash memory is typically used for code or code and data storage. Codes can be executed directly from the flash memory because the random access allows looping and branching to different locations in the device. Typical end applications are for cell phones, PDAs (internal storage—not the flash memory storage cards), and PC BIOS. *NAND*: The data words are accessed a page at a time, with the data clocked out of a buffer. This type of flash memory is typically used for data storage. Because the access is somewhat serial, the data are usually buffered to some other memory (SRAM/DRAM) if the code is to be accessed. Digital cameras and MP3 players typically use this type of memory because the data access is usually in order. This is the memory in all the flash memory storage cards—Compact Flash (CF), SmartMedia, Secure Digital (SD), eXtreme Digital (XD), USB memory sticks, SSD storage, and mobile devices.

21.7.2 Basic Test Setup for a Flash Memory Device (Typical Tests)

The basic test setup for a flash (memory) device includes common test parameters, for example, opens, shorts, leakage, static/dynamic *Icc* current, and voltage reference. Functional tests include *program/erase tests* and can be done in two ways, either customer or test mode. Customer mode uses the built-in microcontroller to do the programming or erasing. This is not usually the most efficient method, but is one that is guaranteed to function in the end application. Test mode is the manual method where the test program controls all the DUT voltages directly. This method of programming and erasing can be optimized for the current temperature and other conditions. Test mode is usually used for sort testing because it can save time and allows the test program to determine the appropriate settings for the DUT's resources that are then saved into the DUT memory itself for use by the customer mode algorithms. In this way, the test defines the parameters for its end application functionality. The *margin test* will allow the manufacturer to check all the FETs by reading the whole DUT to ensure that they can properly detect a logic "0" and a "1." Conceptually, this is a pass or fail (go/nogo) voltage threshold (Vt) test. Since there are millions of FETs, to do an individual Vt test on each FET is not cost-effective. The margin test will allow the manufacturer to determine the distribution of the Vts across the DUT without actually measuring each Vt individually. There are many other tests that can be done on flash memory devices. They include the *program/erase suspend* to see if the program or erase steps can be suspended and resumed, testing block locking that checks the ability to prevent overwriting certain data when the part is locked, and other data security feature functions. These tests are usually specific to the feature set of the flash memory device.

21.7.3 Typical Production Test Flows

Sort 1 is the first time a flash (memory) device is checked on a wafer after the IC fabrication steps. This is where the DUT's voltage reference cells are trimmed. This allows the DUT to determine the difference between a logic "0" and a "1" state. After setting up these thresholds, the DUT is checked for basic program and erase functionality as well as checking for slow or fast programming bits. Any FETs that do not stay within predetermined limits for Vt and other characteristics are marked for potential repair with redundant elements. Other tests may or may not be done depending on the feature set. High fallout tests are done at Sort 1 to weed out the bad DUTs early. Devices that pass Sort 1 are programmed with a pattern and sent to the burn-in ovens; this step is called the retention bake. Because the oxide layers are so thin, any anomalies that prevent the proper storage of the charge on the floating gates are judged defective.

The retention bake step is a form of accelerated aging to determine if the DUT will be able to retain data for its specified lifespan. Sort 2 is a wafer level test that quantifies the amount of charge loss or gain to see if it falls within the allowed limits. Retrimming of the reference cells is done and the Vt distributions are again checked with margin tests. Any bits that do not pass this step are repaired with redundant elements. A custom pattern is then programmed into the part, and it is sent to packaging. Sort 3 or *package test* runs a simple functional test to make sure that the packaging step did not damage the DUT. For KGD sales, the package step is not done, but the equivalent is done on the wafer prior to die separation. The last step is to program any of the custom features of the DUT and program any data desired by the end customer.

21.7.4 What Makes Testing Flash Memory Difficult?

Flash memory testing is difficult because the program and erase execution times are long. In order to get the electrons onto or off of the floating gate, the three terminals of the FET need to be held at certain voltage levels. This allows electrons to migrate across the barrier oxide and get trapped or released. Since program/erase time and data retention are both affected by the thickness of the barrier oxides, a tradeoff between the two characteristics must be made. To achieve the goal of 20 years of data storage, the program and erase times end up being much longer than equivalent times for SRAM or DRAM, for example, μS or mS instead of nS. This means that total test time for flash memory is much longer than that of other volatile memory technologies. Testing all the characteristics of a flash memory device can take on the order of minutes. Because of this, flash algorithms are designed to make the best use of several test patterns that overlay to minimize unnecessary erases. For this reason, patterns common with SRAM and DRAM testing such as *walking 1s and 0s* are not used in flash testing.

With flash densities ever increasing and the number of individual FETs in a device growing, the number of memory cells is large. Densities of several gigabytes are available today, meaning that several billion FETs per device must be checked. Combining long program and erase times with billions of FETs results in even longer test times. This is why most manufacturers resort to as much parallelism as the latest tester and probe card technologies will allow. Higher parallelism increases complexity because more resources must be coordinated and more data must be managed. There is a need for redundant memory cells because with billions of FETs, it is virtually impossible to guarantee that each one will be fully functional after all the complex fabrication steps. One speck of dust can damage an entire device. For this reason, each flash memory device is made with redundant rows and/or columns to allow the device to be repaired during a test. This means the redundant elements must also be checked for proper functionality and the test program must manage the swapping and tracking of errors in the main array as well as the redundant elements. The redundancy analysis and repair routines are arguably the most difficult portion of the test process. Some devices even allow repairing of the repair elements, which further increases this complexity.

21.7.5 Future Trends in Flash Memory

As mentioned previously, flash memory is an offshoot of EPROM. It has virtually replaced EPROM and EEPROM because of its better capability and lower cost. Inevitably, this will also happen with flash memory. Several technologies are being investigated that could replace flash memory, such as *ferroelectric RAM* (FRAM), *magnetic RAM* (MRAM), *ovonic unified memory* (OUM), *polymer memory,* and *spin memory.* There is also a desire to combine the capabilities of SRAM, DRAM, and flash memory into one unified memory type that does what all three can do, but is cheaper to manufacture.

21.8 RF TESTER

21.8.1 RF Device Testing

The types of devices that will be included in this consumer market segment are at a high integration level and include RF reception or transmission, or both. This includes devices that go into smartphones, TV tuners, WLAN Internet, Bluetooth, and ultra-wideband applications. The term *RF* refers to an analog signal, called a carrier, typically well above 100 MHz that is modulated to contain information. This signal is propagated through the airways and then the carrier signal is removed and the modulated data are decoded. There are many devices manufactured today that can be considered RF devices. RF techniques are used in both the wireless and wired, or wireline industries. The wireless RF industry comprises cellular phones, wireless LAN or WLAN, Bluetooth, ultra-wideband, and other standards of sending information from point a to point b without wires. In contrast, the wireline RF industry comprises cable TV tuners and other appliances used to send information through cables.

21.8.2 Building Blocks of RF

There are all kinds of devices that use RF signals, but a "pure" RF device is one that has only RF in, RF out, and a power supply. This kind of device requires a traditional RF test approach and, therefore, a specific type of test system that has sufficient RF source and measure capabilities to meet the device specification. As devices evolve, more and more functionality is being placed on a single substrate, thus the emergence of RF SOC devices. RF SOC devices have RF in, RF out along with digital and analog controls. These kinds of devices will have functions such as registers, digital interfaces, memories, mixed-signal blocks, as well as RF inputs and outputs. This kind of device requires a much more flexible test system that can handle the variety of signal types mentioned. All RF tests start with understanding the basic building blocks of RF SOCs. These basic building blocks are amplifiers that take an input signal and increase its power, mixers that take an input signal and translate it from one frequency into another, filters that take an input spectrum and modify certain frequency components, and *voltage controlled oscillators* (VCOs) that convert a voltage into a frequency.

21.8.3 System-Level Tests

System-level tests, for example, testing the IC as a radio rather than a series of blocks, provide true performance verification because the device is tested under conditions similar to its end-use environment. Examples of these types of measurements include *adjacent channel power* (ACPR), *error vector magnitude* (EVM), and *bit error rate* (BER). *Bit error rate* is the number of error bits divided by the number of transmitted bits. The bit sequence to be sent to the DUT is coded into an RF modulated signal. The DUT then demodulates the signal to the *in-phase and quadrature* (I and Q) signals. The I and Q signals are decoded to get the digital data (bit sequence) that were originally sent to the DUT. The digital data from the DUT are then compared to the original bit sequence to determine the BER. *Error vector magnitude* is a measure of modulation or demodulation accuracy. The ideal modulation vector is compared to the resulting vector coming from the DUT. Error vector magnitude is the root mean square value of the error vector over time at the instant of the symbol clock transitions.

$$\text{EVM} = \frac{\text{rms error vector}}{\text{outermost most magnitude}} \times 100\%$$

Or expressed in decibels:

$$\text{EVM} = 20 \times \log \frac{\text{EVM\%}}{100\%} \text{dB}$$

Using *orthogonal frequency division multiplexing* (OFDM) as an example, these calculations are made for each burst and data vector, resulting in an individual EVM (the error vector for a particular instant in time relative to the symbol clock). *Adjacent channel power*—when two or more signals (f1 and f2) at different frequencies are applied to an amplifier, nonlinearities in the amplifier cause the output to contain additional mixing frequencies called *intermodulation* (IM) products. These occur at frequencies ($mf1 \pm nf2$) and at ($mf2 \pm nf1$), where $m + n$ is the order of the IM product. For example, the second-order IM products would be at f1 + f2, f1 − f2, and f2 − f1. Even-order products are located near the baseband, while odd-order products are located on either side of the input signals. As an example, if two input tones are applied to the input of the amplifier, the output consists of amplified versions of the two input tones, plus odd-order intermodulation products caused by amplifier nonlinearity. If the frequency spectrum is divided into channels of equal bandwidth located next to each other, and the two input tones are applied, distortion in the amplifier causes signals to appear in the output adjacent channels. In the digital-modulation world, the two tones at the input are really replaced by an infinite number of tones (or noise) within the defined channel bandwidth. The energy generated by amplifier nonlinearity in adjacent channels is called spectral regrowth and is measured as adjacent channel leakage. The actual specifications take the measured spectral regrowth within a specified bandwidth and ratio it as a relative quantity to the desired power in the main channel. This is done for each of the two adjacent channels above and below the desired transmit channel, expressed in dB and is called *adjacent channel leakage ratio* (ACLR).

21.9 SoC TESTER

21.9.1 SoC Device

System-on-a-chip (SoC) devices contain digital, memory (SRAM, DRAM, and FLASH), analog, low/high-speed interfaces (USB, PCIe, SATA, SSI, ...), and RF blocks. The primary market drivers for SoC devices are consumer electronics with complex products, shrinking product life cycles, low cost, high reliability, and miniaturization. Examples of the products are smartphones, tablets, and mobile devices. SoC devices are made possible by advances in semiconductor technology making it possible to have extremely large counts of transistors on a single chip, mixed technologies on the same chip such as logic, memory, analog, and RF. Creation of intellectual property (IP) and core-based design has also contributed to the emergence of SoC devices.

The Marvell® Avastar™ 88W8787 (Fig. 21.4) is an example of a highly integrated SoC. The Avastar family of wireless devices delivers single-function and multifunction radios for the entire spectrum of always-on consumer electronics devices. This SoC is designed for both simultaneous and independent operation of the following:

- IEEE 802.11a/g/b and 802.11n for wireless local area network (WLAN)
- Bluetooth 3.0 + high speed (HS) (also compliant with Bluetooth 2.1 + EDR)
- FM transmit and receive (digital encoder/decoder FM radio with RDS/RBDS)

The 88W8787 has generic interfaces SDIO, G-SPI, high-speed UART, and PCM for connecting WLAN, Bluetooth, and FM to the host processor. For FM Tx/Rx, the device supports inter-IC sound (I2S)/analog stereo audio interfaces. An I2C-compatible interface is available to connect FM Tx/Rx to the host processor, as well. FM Tx/Rx can also share the host interface with Bluetooth. The device is packaged in a TFBGA or CSP flip chip.

21.9.2 SoC Testing

Functional testing represented the early methodology for IC testing in the late 1970s to the early 1980s. The shift from functional test to scan test represented the next generation in testing. SoC testing introduced two main challenges the size of the test suites and the access to the embedded cores. This led to the development of test compression techniques and test access mechanisms within the SoC. SoC testing is a composite test composed of individual tests for each core, user defined logic (UDL) tests, and interconnect tests. Testing SoC is best done by testing block by block to verify each block and then performing a system level test to verify the operation of the chip as a whole. Today designers can embed test blocks in SoC devices, which can be turned on by the test system to test and debug each block. To facilitate SoC testing the IEEE 1500 standard was approved in late March 2005 after the IEEE P1500 Standard for Embedded Core Test (SECT) was started 10 years earlier in 1995. The scope of the standard defines a mechanism for the test of core designs within a SoC. This mechanism constitutes a hardware architecture and leverages the IEEE 1450.6 Core Test Language (CTL) to facilitate communication between core designers and core integrators. The IEEE 1500 has two levels of compliance: a wrapped and unwrapped compliance. The wrapped compliance caters for a core that comes with an IEEE 1500 wrapper function and a CTL program. The unwrapped compliance refers to a core that does not have an (complete)

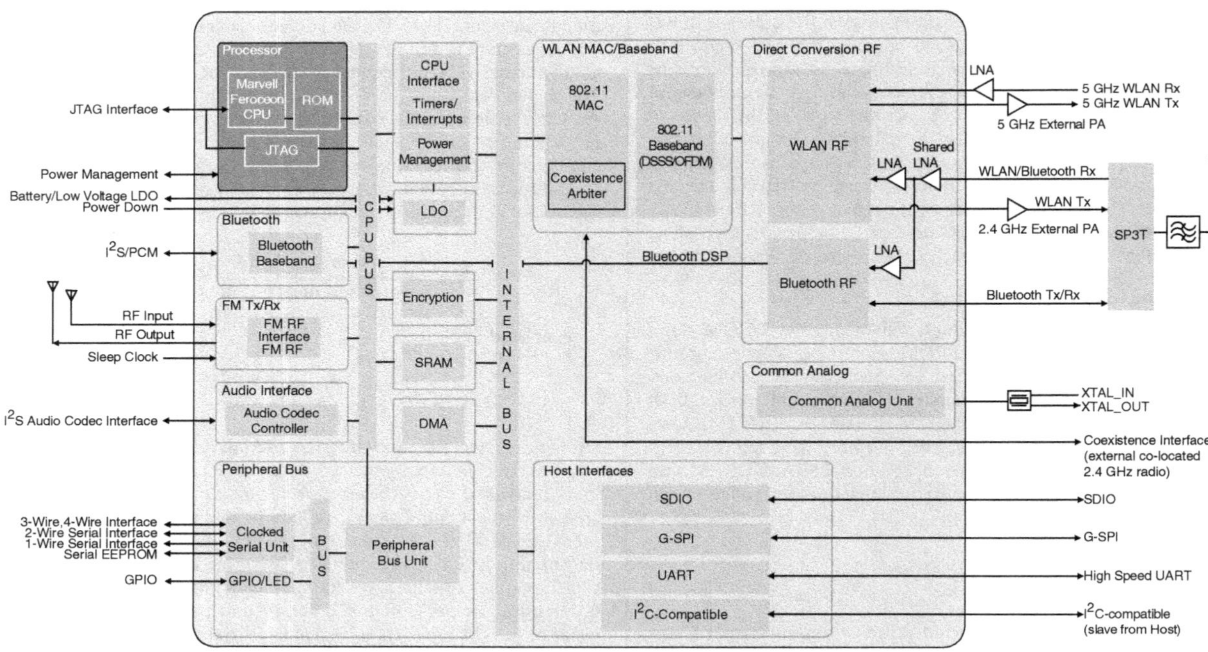

Figure 21.4 Marvell Avastar 88W8787 SoC. (*Credit: www.marvell.com/wireless/avastar/88W8787.*)

IEEE 1500 wrapper but does have a CTL description. The IEEE 1500 is independent of the functionality of the IC or the embedded cores. SoC test architectures have been used by the test industry prior to release of the IEEE 1500 standard. A good description of TESTRAIL, AMBA, TESTSHELL, and TESTCOLLAR can be found in SoC Test: Trends and Recent Standards (Michael Higgins and Ciaran MacNamee, Circuits and Systems Research Centre (CSRC), University of Limerick, ITB Journal). It also shows ETM10 (Embedded Trace Macro) by ARM (Advanced Risc Machines) has the IEEE 1500 wrapper built around its core. Software tools and test vector compression techniques that have been used are also presented in section four, which include schemes such as IBMs STUMPS, Philips TR-Architect, SmartBIST, and IEEE 1450.6.

21.9.3 SoC Test Architecture

Zorian et al. showed a conceptual architecture for testing SoC devices, as illustrated in Fig. 21.5. Test source and sink provide test stimuli and compare test responses. The ATE can act as the source and sink. The test access mechanisms forward test data from the source to the core under test (CUT) and from the CUT to the sink. Core test wrapper provides the connection to the core terminals to the rest of the chip and to the TAM, isolating the CUT from its environment during test.

Because the SoC will have a multitude of IPs, the ATE will have to provide capabilities found in digital, linear, memory, and RF testers. However, by designing in test capabilities within the SoC and using the processing capabilities available in the cores the ATE architecture can be simplified lowering the overall cost of test.

21.10 BURN-IN TESTER

Burn-in is an electrical stress test that employs voltage and temperature to accelerate the electrical failure of a device. Burn-in essentially simulates the operating life of the device, since the electrical excitation applied during burn-in may mirror the worst-case bias that the device will be subjected to in the course of its useable life. Depending on the burn-in duration used, the reliability information obtained may pertain to the device's early life or its wear-out. Burn-in may be used as a reliability monitor or as a production screen to weed out potential infant mortalities from the lot. Burn-in is usually done at 125°C, with electrical excitation applied to the samples. The burn-in process is facilitated by using burn-in boards where the samples are loaded. These burn-in boards are then inserted into the burn-in oven (tester), which supplies the necessary voltages to the samples while maintaining the oven temperature at 125°C. The electrical bias applied may either be static or dynamic, depending on the failure mechanism being accelerated (Cagatay Bozturk, Kent State University).

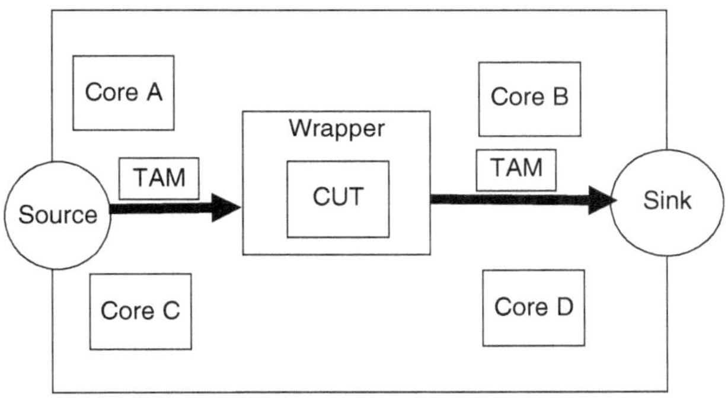

Figure 21.5 SoC test architecture.

21.11 DESIGN DIAGNOSTIC EQUIPMENT

These are classified into circuit probers and modification tools. Circuit probers are beam (E-beam or laser)–based systems used in R&D for failure analysis during new silicon bring-up or root causing field returns due to failure in the system. Modification tools are focused ion beam (FIB) systems used for circuit edits while debugging new silicon, saving the designer time and money.

21.12 ATE MARKET SIZE

The total ATE market size is driven by the volume and type of devices manufactured, the total test time for each device, and the number of steps in the manufacturing process where testing takes place. Figure 21.6 shows the various manufacturing steps.

Test equipment is first used in wafer sort to identify and package all good die and discard the bad die. The wafer sort can save manufacturing cost if it can screen all the bad die. In case of very advanced technology and high-frequency device designs it is not always possible to test the device completely on the wafer. After packaging the good die the assembled IC package is tested on the ATE once before burn-in and a second time after burn-in. The package test would include the tests that were not done during wafer probe on the die due to device complexity and speed.

The ATE semiconductor test segment is composed of several distinct types of testers:

- Digital/logic tester
- Analog/linear tester
- Mixed-signal tester
- Memory tester
- System-on-chip tester
- Burn-in tester
- Design diagnostics

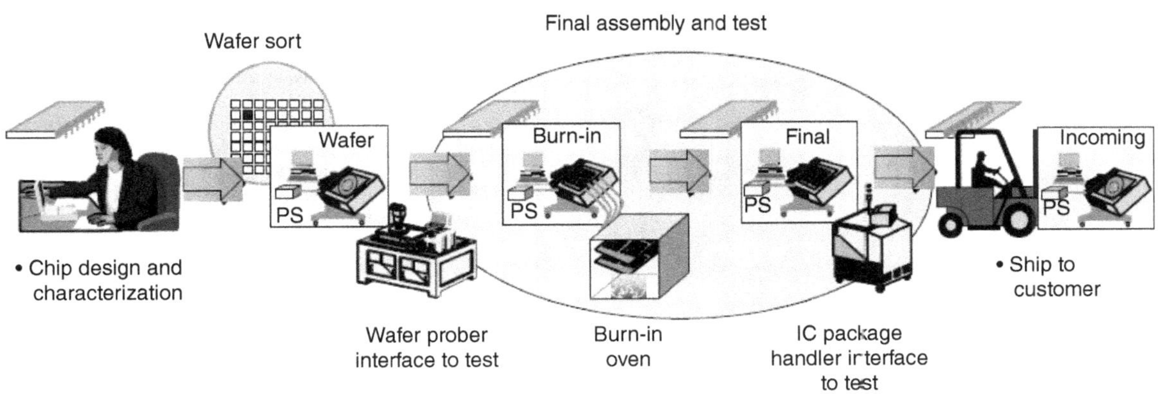

Figure 21.6 Semiconductor manufacturing testing steps.

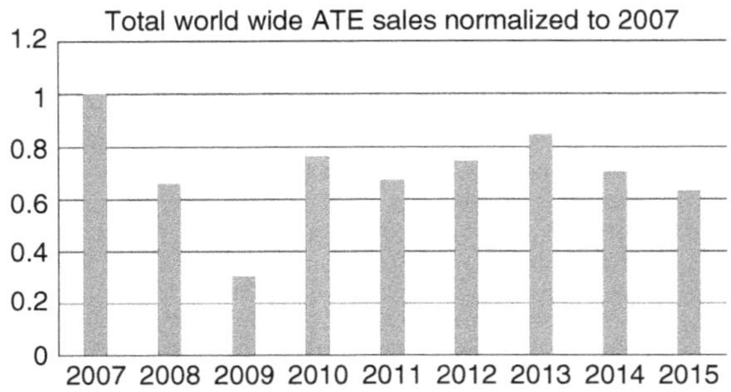

Figure 21.7 Worldwide ATE sales normalized.

The ATE/semiconductor test space is cyclical with upturns and downturns similar to the semiconductor industry. Figure 21.7 shows the world wide ATE sales from 2007 to 2015 in a normalized form.

The following factors determine market trends and dynamics in the ATE space:

- ATE vendors are tied to the boom and bust cycle of semiconductor market.
- Test equipment orders are driven by CapEx spending.
  - Capital utilization rates drive capital equipment orders.
    - Customers buy on technology, low-failure rates (i.e., equipment reliability), and ability to deliver systems on time.
  - Device manufacturers' sentiment on the future of chip volumes.
- Low test time equates to more devices tested which leads to higher revenues.
- Customer testing methodology.
- Customers want:
  - To extend return on investment (ROI) of existing systems—need for in-field upgrades
  - To test different technologies from day to day—need for flexibility in the system
- ATE vendors have a limited customer base (only 25 major independent device manufacturers and 10 independent test houses worldwide).
- ATE vendors attempt to tie development of system(s) to one "teaching" customer and ultimately leverage sales to many "production" customers.

21.13 ATE ARCHITECTURE

For several decades since the first digital tester appeared in the market the architecture of the production ATE used in the manufacturing test floor was made up of a mainframe unit and a test head. The test head was mounted on a manipulator for docking to wafer prober and device handler for production testing of wafers and packaged parts. Figure 21.8 shows a typical production ATE.

Figure 21.8 Production ATE.

The mainframe housed the tester computer, timing generation and distribution hardware, pattern memory, and power supplies for the mainframe elecronics. The test head housed the pin electronics cards, which interfaced to the device under test mounted on a fixture called the performance board, load board, or a device interfcae board. The test head had slots arranged in a circular or rectangular fashion to house the pin electronics (PE) cards. The early mid-1970s PE card had just enough electronics to support two channels of I/O for digital device testing and also a power supply with very modest curent capability for powering up the device under test. By the early 1990s such test equipment supported upto eight I/O channels per PE. The early testers belonged to the shared resource architecture, where timing and level were assigned to groups of pins to opitimize the size, cost, and complexity of the tester electronics. This worked well for testing simple logic devices with low pin count. But as semiconductor integration fueled by Moore's law increased the pin count and complexity of ICs, the tester architecture quickly evolved to per pin architecture. By the mid-1990s there were several tester companies each with a different architecture competing for the manufacturing dollars: Advantest with shared resource, Trillium with Timing per pin, Schlumberger with sequencer per pin, Hewlett Packard with processor per pin and Megatest with tester pin architecture. Each architecture had its beneifits and performed well with the customers that adopted it. In the long run the shared resource architecture was seen to be less flexible and with large pin count devices ran out of resources requiring multipass testing. Due to shrinking feature size in semiconductors moderns testers have all reources per pin. In addition, testers of this millenium have all the tester related hardware from the mainframe to the test head something discussed in detail under the SoC ATE architecture.

21.13.1 Architecture of a Digital Tester

Digital tester architectures can vary widely, depending on the target market of devices to cover. Some digital testers are aimed at specific test market segments, such as DFT or low-end production; other digital testers have a scalable platform, allowing changes in speed and memory depth to enable the end user to expand the system as needs arise or to reconfigure systems for a particular job without having to invest in different platforms, DUT boards or test program translations. A high-end digital tester per-pin architecture is shown in Fig. 21.9 and will be described in this section. The heart of the digital tester is where the device *input and output* (I/O) pins of the PE connect to the tester I/O pins of the PE. The PE contains many functions that will be described here. The driver sets the logic "1" and logic "0" voltage levels that will be driven into the device and sets the pin to a high impedance or "tristate" function in case the DUT needs to drive the pin. The comparator sets the logic "1" and logic "0" voltage levels that will be measured from the device. If the level is between these two logic levels it is called a "mid-band" level. DC sourcing and measuring are also accomplished for level setting and biasing statically with a *parametric measurement unit* (PMU). Since the DUT will expect ac dynamic signals, the timing edges have to be generated by a *timing generator* to meet the DUT needs. There are common formats for both the driver and the comparator that are a combination of levels and timing within a device period or cycle. Usually the device cycle is 1/clock rate at which the device is running and represents a period of time to take 1 bit of data, process it and send out 1 bit of data. The drive formats are shown in Fig. 21.10. The pattern data is combined with the timing generator edges in a formatter circuit in the tester to generate the different drive waveforms.

Each type of driver waveforms is needed for setting up the different types of the DUT pins. The NRZ and SBC formats are used for driving data bus pins of the DUT. The RZ and RTO formats are intended for positive and negative clock input pins of the DUT, respectively. The expect modes include expect 1, expect 0, expect tri state, or do not care. These modes sample DUT response over a time interval (window strobe) or a very short duration (edge strobe). The sampling is done against logic "0" and logic "1" preset levels in the PE and a hardware

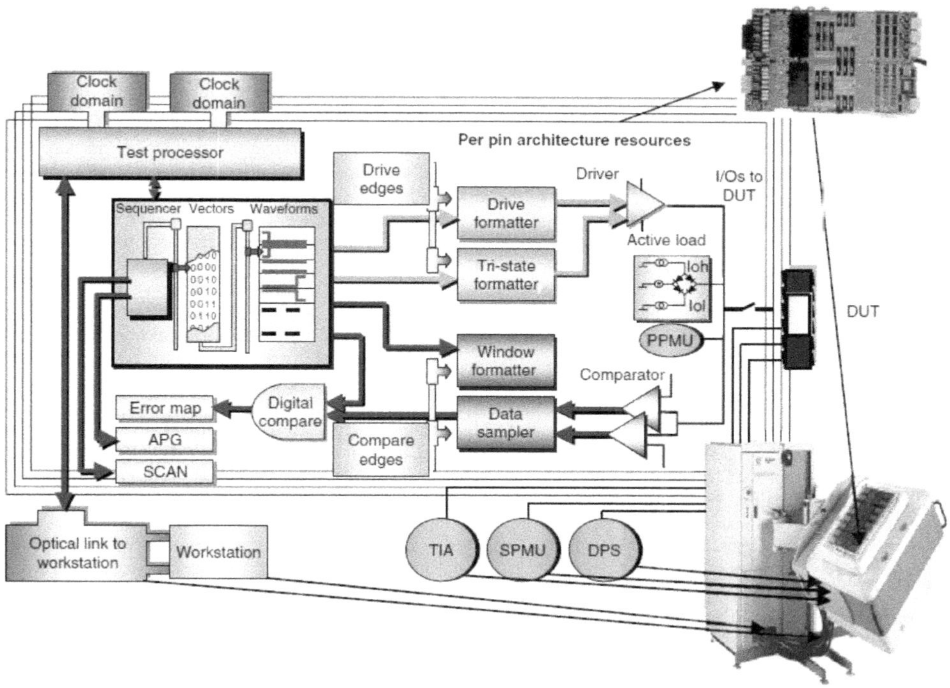

Figure 21.9 Digital tester architecture.

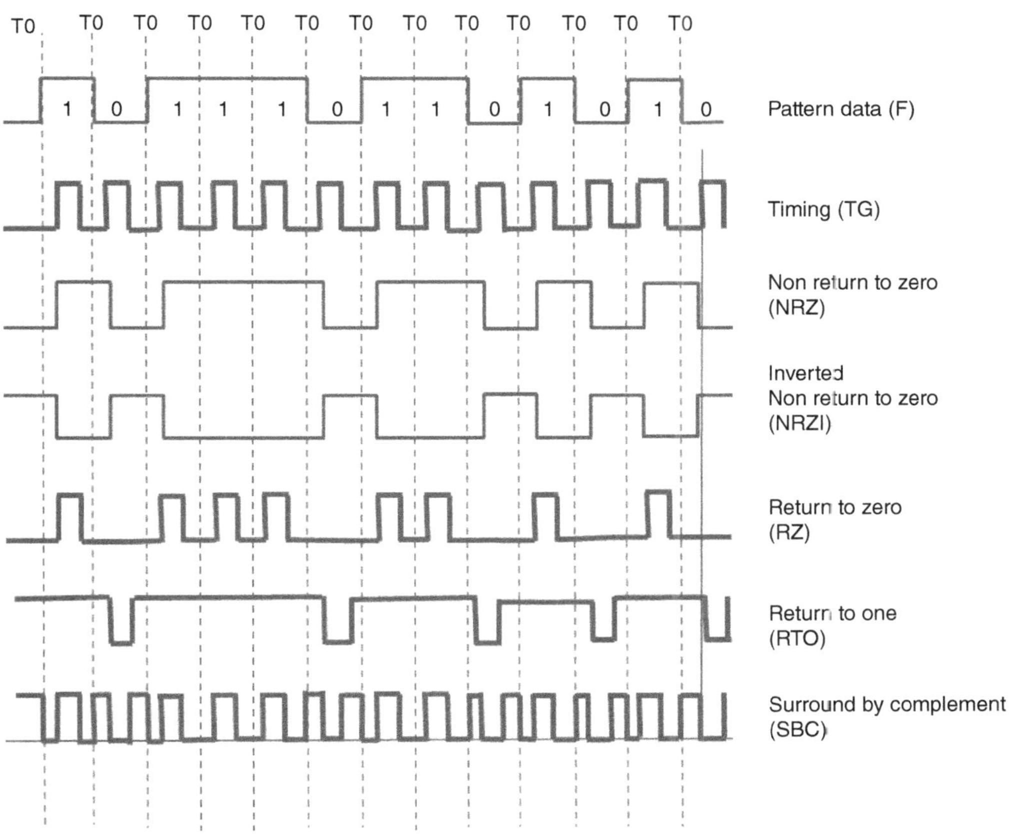

Figure 21.10 Digital drive data formats.

fail signal is generated if the DUT response does not meet expected data. The more edges available per device cycle, the more flexible and potentially faster the signal can be. The vector memory contains the logical "1" or "0" data for each device cycle input from the tester. The compare memory is a table of what the tester expects to see from the DUT for each cycle and the error memory logs the pins and locations that fail. These all interact so that a logic "1" drive coming from the vector memory has the correct timing and level so that the DUT recognizes it as a valid input, and the DUT can send the appropriate response that the comparator decodes to a logic "1" or "0," which is then compared against the compare memory to judge if it is the correct response. The test may or may not run through the memory addresses in a linear fashion. A sequencer is a vector address generator that will allow the memories to be accessed in structured ways like subroutines or randomly move with jump operations. Also, the flexibility of being able to change from one format or setting of edges to another in the next device cycle is desirable; this is called changing *on-the-fly* (OTF). All these operations need to be orchestrated to put the DUT into its end application environment, and a test processor accomplishes this task on an individual pin-by-pin basis. All the functionality described under pin electronics is available on every pin, or per-pin to be able to meet the flexibility and complexity demands of many different digital devices.

Scan testing has become a key methodology for structural testing with digital devices. Very large lengths, for example, over 100 Mbytes, of scan memory need to be available in the configuration to accommodate this need. Other specialty functions such as *capture memory* and *digital source memory* will be covered in the mixed-signal test section, and *algorithmic pattern generator* will be covered in the memory section. Some other critical parts of the architecture include clocks and timing. Clocks are what drive the DUT and it may need more than one domain. A different domain may use a frequency unrelated to others used by the DUT, or be nonphase related. *Overall timing accuracy* (OTA) and *edge-placement accuracy* (EPA) are widely considered to be the most important and fundamental specifications of a digital tester. These specifications indicate how accurately an edge can be launched and measured from the test system. Edge timing resolution is also critical to DUT performance as well as the jitter, or the undesirable movement of edges from one device cycle to the next.

Other functions in the architecture include power supplies, a higher precision dc parametric measurement unit to make more exacting measurements than the per-pin PMU can make, a system controller or CPU to run the program, and a *time interval analyzer* that is used to making timing measurements on edges. The software environment is interactive and *graphical user interface* (GUI) based. Because the increasing speeds and number of pins per PE generate more heat, the testers of today are liquid cooled. The coolant runs through cold plates mounted on both sides of a PE card and pressing on components smeared with a conductive thermal paste.

21.13.2 Architecture of a Linear Device Tester

The architecture of a linear device test system is very dependent on the type and variety of linear device that will be tested. Because the architecture of a test system for linear devices is highly dependent on device functionality and these devices require a very low cost of test (COT), many times the architecture is custom tailored to a specific family of devices and is not a general purpose one. However, some of the typical tester building blocks are needed. A switching matrix provides a method of connecting the source and measurement devices to the different pins of the DUT. The number of pins, their function, and parameter specifications must be considered when specifying (or designing) the switching matrix requirements. Power supplies will be required to provide power to the device for testing. Power supply considerations include the number of supplies and their voltage and current specifications. All traces, wires, and fixtures must withstand the maximum expected voltage and current. Some tests can be dangerous to humans or even the test equipment. For example, testing the breakdown voltage of the DUT exposes it to a very high voltage (hundreds of volts) to ensure compliance to specification and safety considerations must be adhered to. Often a pulsed high-current source will be used to keep the test equipment to a reasonable size and to prevent destroying the DUT, particularly in wafer form when no heat sinks can be attached to the device. Even at a package test, any kind of heat sinking capability would be minimal. Digital logic input and output pins may be used as digital control pins needed to access the DUT or configure it for certain test modes. Control, signal lines, or identification pins may be used to control relays or switches on the DUT fixture. Different devices require different fixtures and often the test program will electronically check the identification information of the fixture to ensure it is the correct one. The prober/handler interface will allow indexing to the next die or package as well as sending category or binning information that allows the sorting of devices into different grades. Often the interface will also integrate equipment for temperature testing. The integration of a prober or handler into a tester is often more time consuming than expected due to the mixing of mechanical and electrical issues. There is a standard set of signals expected for this interface that can be provided with simple TTL logic or via a standard interface like the *general purpose instrument bus* (GPIB). System controllers are required to run and coordinate all the operations for the test cell. Considerations for the system controller include the operating system used, networking capabilities, processing power or speed, upgradeability, the testing language(s) supported, and whether the operating system language is an industry standard or a proprietary. A final factor to be considered is that often test systems are installed in clean areas or areas with limited access. Because it is inconvenient to carry out test development in these limited-access areas, it is important to find a way to do remote development.

21.13.3 Architecture of a Mixed-Signal Tester

The architecture of a tester that can test all the diverse functionality of the basic building blocks has to match the stimulus expected and has to be capable of measuring the expected outcome. Fundamental digital tester requirements for mixed-signal tests are similar to that of digital tester as shown in Fig. 21.9. However, there are two added digital capabilities that are necessary to provide the digital stimulus waveform data and to capture the results of the analog-to-digital conversion. These two capabilities are digital source memory and digital capture memory. The added digital functionality and added analog subsection are shown in Fig. 21.11. Digital capture memory is the ability of the ATE system to store the results of the vector compare action into DSP accessible memory as data rather than a pass/fail condition. *Analog-to-digital converters* (ADCs) require that the sample data be stored into tester memory because their outputs are nondeterministic in nature; further DSP is necessary to determine a pass or fail condition. This can require a large capture memory for high-resolution devices. Additionally, in order to minimize the tester resources, the tester requires the capability to selectively store sample data on a per vector address basis. This capability is referred to as selective capture. This requirement is demonstrated by audio *coder/decoders* (CODECs) that are used frequently in the telephony and multimedia audio industries. These devices require a 256-bit frame, while only 20 bits of this frame represent the results of one channel. Digital capture functionality requires the ability to capture only the relevant sample bits within this frame to reduce the unnecessary overhead of capturing and postprocessing the unnecessary 236 bits per sample. For adequate dynamic test coverage, the digital capture rate must be able to operate at the maximum frequency at which the device under test samples. As in digital and memory devices, there is a requirement of digital vector drive/compare, digital capture, and PMU capability on the same DUT pin, depending on which mode it is in for a specific test in the flow. Therefore, the trend in the ATE industry has been toward a solution that provides this capability on a per-pin basis to reduce the necessity of multiplexed tester resources. By providing advanced per-pin architectures with these capabilities native to each pin, cost and performance are optimized. *Digital source memory* (DSM) is the ability to segment a portion of digital vector memory as a contiguous block of data that can be called from standard test vectors. In most cases, this memory is utilized in frame-based, serial device interfaces. This capability simplifies the effective management of source waveforms for DAC testing. Similar to capture memory, the necessary size is relative to the resolution and

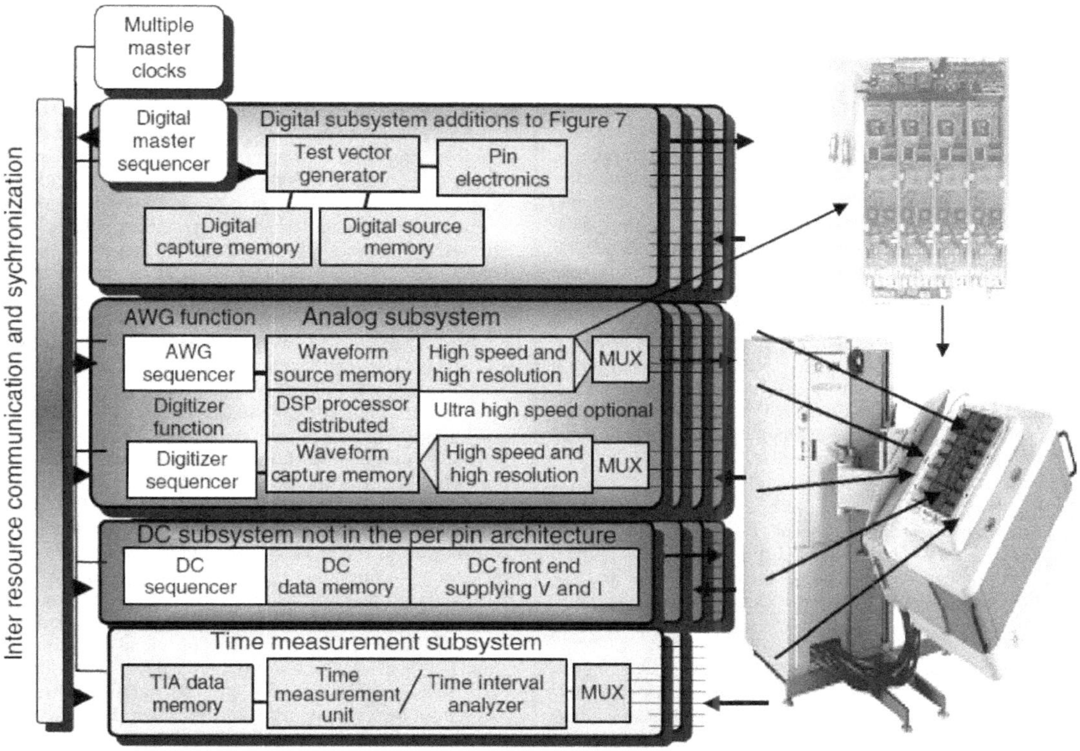

Figure 21.11 Mixed-signal tester architecture.

number of waveforms to be sourced. As this capability is used primarily in frame-based interfaces, the speed requirements are not stringent. In the current market, devices that require speeds of over 50 MHz do not have frame-based interfaces and as such do not require DSM capability. *Analog subsection*—there are two analog module requirements for mixed-signal tests: waveform generators and digitizers. Both are usually specified by their resolution and sample rate. An *arbitrary waveform generator* (AWG) provides a sample memory that the user programs to allow the generation of any type of waveform that can be defined mathematically. Output frequency changes can be accomplished by either modifying the sampling clock or waveform data. The sampling clock of the AWG can be derived from a common master clock with digital vectors to ensure coherency, or from another master clock. Usually in mixed-signal testers at least two master clocks are provided for supporting multiple frequency combinations in tests. Waveform digitizers sample the device-generated analog output and store the discrete voltage values in the local memory. A local DSP within the digitizer will then calculate the performance relative to the specifications and return the results to the user's computer. For cases of sampling, a waveform whose frequency is over the maximum available Nyquist frequency of the digitizer, under sampling can be utilized. This methodology takes advantage of a high-input bandwidth of the measure unit to alias the signal of interest into the spectrum. A digitizer that utilizes this sampling methodology is referred to as a sampler and is usually specified by bit resolution and bandwidth at the −3 dB cutoff point. Many mixed-signal implementations have multiple functional blocks. As such, ATE test resources must provide an autonomous test capability within each tester module to enable parallel testing with minimal overhead. This capability within each module, as shown in Fig. 21.11, includes:

- *Analog sequencer.* Similar to the digital pin sequencer, the analog sequencer steps through a series of instructions to control the operation of the module. It is in effect an address generator for access to the AWG or digitizer memory locations. This allows an AWG to repeat a block of waveform or switch waveforms from one cycle to the next (OTF), and a digitizer/sampler to do multiple measurements sequentially without any interruption from the system.
- *Memory.* These comprise AWG source or digitizer capture waveform memory within the module. For complex mixed-signal devices such as multimedia, cellular phones, and hard disk drives, test signal waveforms can be long and require many different stimuli and measurement segments requiring deep memory.
- *Digital signal processor.* A local DSP provides the ability to calculate results without requiring the transfer of the capture results to the tester controller workstation for processing. FFTs and filtering are the most common operations.
- *Parallelism.* There is an architectural advantage to having multiple autonomous units to provide multisite and parallel IP block testing capability.

Other necessary pieces to the architecture include highly accurate dc source and measure capability to be used for voltage and current setup and measures. A per-pin architecture is the most flexible and contributes to parallelism that is necessary to satisfy COT economics. Clock distribution in a system is necessary because different resources running concurrently will often run at different frequencies, so they must be capable of dividing a master clock rate down as needed. In fact, sometimes frequencies are not multiples of each other and it may be necessary to support multiple unrelated (frequency) clocks within the system. This is also desirable when implementing the under sampling technique mentioned previously. Power supplies deliver the necessary voltage and current to power up the device and keep it running. Usually, some relays are used on the DUT board to facilitate signal multiplexing and other uses; these require drivers to turn on the relay coils.

21.13.4 Architecture of a Memory Tester

Stand-alone memory devices are typically tested with *algorithmic pattern generators* (APGs). An APG generates tester cycles that contain a pattern to execute a series of write and read operations of data to a predetermined sequence of addresses. The tester will capture results from the read operations and compare them to the expected data. There are two memory tester architectures to consider—per-pin and centralized. The centralized architecture implements the APG using hardware architected as a collection of programmable hardware registers. The contents of the registers are gated to dedicated tester buses for address pins, data pins, and control pins. In contrast, a tester with a per-pin architecture provides identical functionality using a software APG that programs each tester channel to execute a memory test. The centralized APG architecture is a traditional tester implementation that does a good job of testing standard memory devices. The per-pin architecture provides a flexible memory test solution that does not use dedicated tester buses and can generate APG patterns as well as logic patterns. This allows the per-pin architecture to be used to test standard memory devices as well as new memory architectures that also contain logic. This capability is also very useful in testing embedded memories.

Test Result Analysis

A requirement when characterizing memories is to represent the pass/fail results in a graphical format. The graphical display, known as a bitmap display, shows a two-dimensional depiction of the memory array and the passing and failing bit represented by different characters or colors.

Memory Built-in Self-Test

Embedded memories are a big part of SoC devices occupying a major portion of the total silicon area. The yield of embedded memories are very critical to the overall production ramp of the product. Memory built-in self-test (MBIST) uses dedicated test hardware implemented in silicon to test the embedded memory. The MBIST architecture can use the processing power in the SoC to completely test the memory and repair the memory as well.

21.14 ARCHITECTURE OF A FLASH MEMORY TESTER

The typical flash memory tester architecture is shown in Fig. 21.12 and will be described in this section. Because flash memory consists of millions of similar FETs that require a similar algorithm to test, the APG allows the manufacturer to create an algorithm for one FET and then simply set up ranges for the amount of rows, columns, and blocks (X, Y, and Z address generators) to test. *Crossover or scrambling* circuitry is needed to map the address generators to the correct address pins because flash memory devices have different numbers of rows, columns and blocks, circuitry. *Parametric measurement units* (PMUs) are used to measure voltage or current parameters on each FET. This architecture also allows "trimming" the device, where the reference cells and comparators are programmed to differentiate between a logic "1" and a "0" state. Because there are so many FETs and either 8-, 16-, or 32-bit wide data buses, multiple PMUs, or even *per-pin measurement units* (PPMUs) are configured into the architecture to speed the testing. *Parallelism*—testing a flash memory device takes a large amount of time because of the relatively low speed and large size. To reduce cost, flash memory manufacturers typically test multiple devices at the same time. This requires that the resources to test each device be duplicated and that the system must support multiple APGs and have enough pins for all the address, data, and control pins of the DUTs. The individual *test site controller* is where the test program runs. It is responsible for quickly controlling the APG and the other test resources. It saves the data and is also responsible for displaying the test results. The overall *system controller* is where the tester control software runs and coordinates all the test sites. This is where the test program is developed and downloaded to each individual test site controller for parallel execution. Often the distinction between the test site controller and the system controller is blurred with different test systems having these resources responsible for different roles. Typically the system controller has access to the manufacturer's network and production floor control systems so

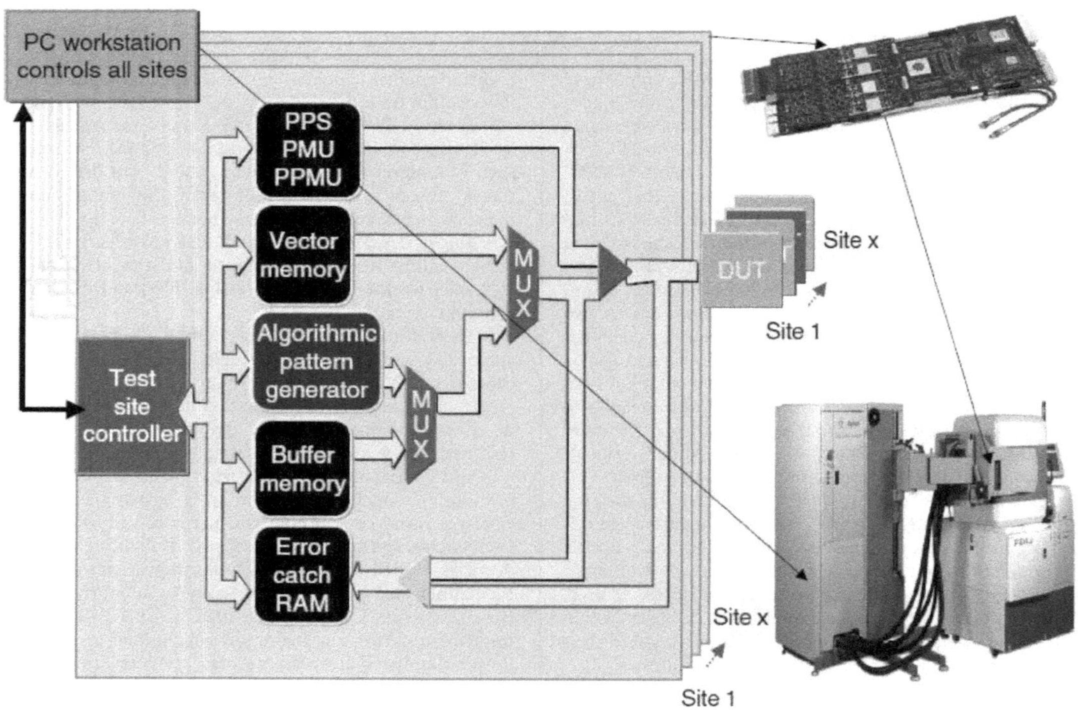

Figure 21.12 Flash memory tester architecture.

test jobs can be automatically dispatched and the test results can be saved. The *prober/handler* interface will be used in the equipment to handle either a wafer or a package. A prober/handler interface will allow control to index to the next die or IC package, as well as send category or binning information to signify a good DUT or a bad DUT. This allows sorting devices into different grades. If temperature testing is required, there is a standard set of signals expected for this interface, which can be provided with simple TTL logic or via a standard interface like GPIB.

21.15 ARCHITECTURE OF AN RF SYSTEM

The RF SoC device will need the same class of analog and digital test capability as outlined in these sections, respectively. Additionally, the RF part of the test system will have to be able to source and measure RF signals. The capability required will need to match the needs of the device under test in terms of specifications, for example, frequency, bandwidth, power, purity, and dynamic range. The number of resources and how many are active at the same time will also need to match the device requirements. A suitable architecture is shown in Fig. 21.13. These can be simple, single-frequency sources or more complicated ones such as multifrequency sources that modulate data onto a carrier or even frequency-hopping sources that instantaneously change frequency a certain number of times per second. The standards for the modulation schemes, which are numerous, need to match the device requirements as well as the number of frequency hops per second. The amount of noise present on the system is always a consideration. RF, because of the frequency involved, has special fixture considerations, not only for the DUT board, but also for the construction and location of system resources. Also with today's COT pressures, the architecture needs to support multisite configurations both in performance and resource availability as mentioned earlier.

21.16 ARCHITECTURE OF A SOC TESTER

Because the SoC device contains microprocessor, memory, analog blocks, high- and low-speed external interfaces, and wireless the tester for these devices consist of test hardware from each of the tester architectures described so far. To efficiently test the high volume of SoC devices in production, the ATE architecture has evolved to a tester in the test head consisting of different type test modules or instruments. The mainframe of the tester just houses the system computer, site controllers cooling, and power needed for the rest head electronics. The site controllers offload tester-related activities from the system computer and is used in production ATE to improve test throughput. Figure 21.14 shows a SoC tester form factor with the test head containing all the tester instruments. The test head is mechanically designed to have slots for housing the instruments. Each slot provides power for the instrument operation and cooling to take away the heat generated by the instrument. Depending on the power and cooling requirements, an instrument can occupy more than one slot. A sync bus and clock distribution connect the instruments for proper operation. A high-speed bus connects all the slots for fast data transfer between the instruments and the system (or site) controller. The sync bus allows communication between test modules whereby a digital instrument can trigger a power supply measurement in the power module or start an A/D conversion in the analog module. The test modules are designed in such a way that they can easily be shared among multiple sites during production testing. In this way, there are no waste (unused) resources in any module by going from single to multisite testing. The clock generation and distribution achieves seamless synchronized multitime domain execution, where each interface of the DUT can be tested at its rated frequency. Per-pin/port architecture of the instruments is a key enabler for this.

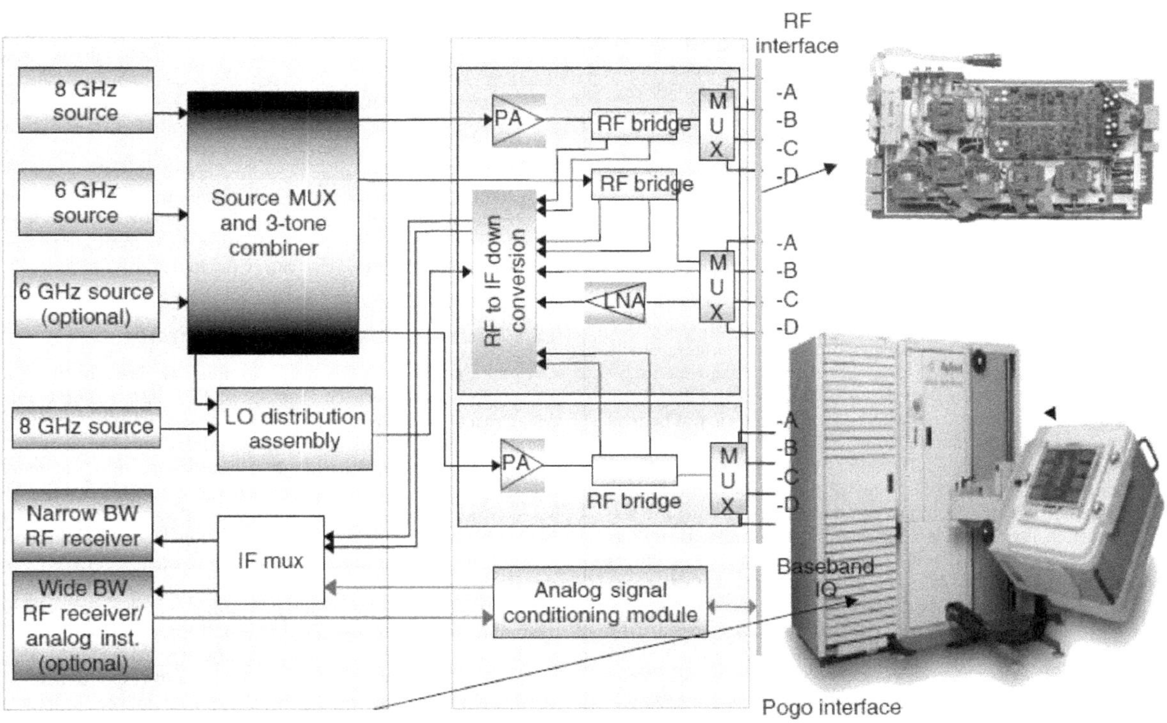

Figure 21.13 RF tester architecture.

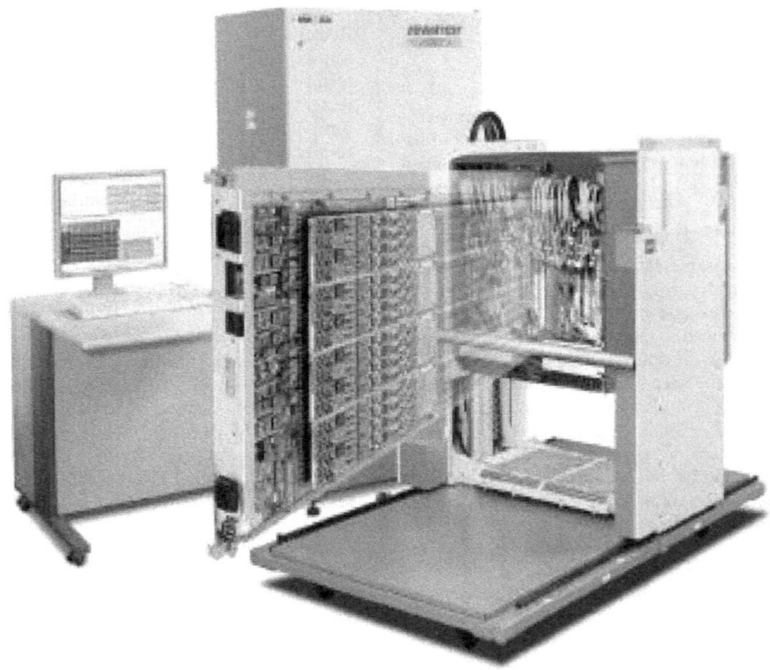

Figure 21.14 SoC tester.

21.17 DFT TEST TECHNIQUES

With the advances made in DFT a lot of test capability is built into the SoC device itself. A very important key to this is the processing power in the device. And with logic gate densities doubling every 24 months following Moore's, chip designers have been adding test-related circuitry to the device and in some cases redundant copies of a circuit, cell, or core on the same die. This is a technique that has been in use in memory design for many years. Also, with this extra space, built-in self-test and design configuration circuitry can be employed on silicon for self-diagnosis. If the BIST and design circuitry can do all the testing, and the circuitry is working as designed, there is less need for ATE in terms of the accuracy and fidelity needed for testing the SoC device. This is especially useful in large die because making the die tolerate a defect under certain conditions can materially increase yields as well as lower the test time by allowing only the BIST and configuration circuitry to be tested by ATE. Another test strategy used for SoC devices with flash is to download a control program into the flash through an interface, usually I2C or SPI, on the device using the ATE as the master and then have the microprocessor execute the program which is actually a diagnostic for the system inside the chip. The results of the diagnostic can be read by the ATE to make the pass/fail decision. This way the tests run at the native of the system internally while the ATE needs to run only at the slow interface speed. Designing analog circuits for testing is intrinsically more difficult than for digital circuits because analog circuits are less regular and analog signal paths are more difficult to isolate than their digital counterparts. Most analog DFT is done to isolate analog blocks so that they can be characterized individually and to enable the application of the test stimulus and the measurement of the test response at the internal nodes of these blocks. This test access is provided by inserting analog multiplexors or configuring the analog switches that are already part of the design. Such switches can also be used in some cases to reconfigure the circuit blocks to test each other. For example, an ADC can be connected to test a DAC. Particularly useful in testing radio transceivers and digital *serializer/deserializers* (SerDes), which are really analog circuits, is the concept of loop-back, where an on-chip receiver tests an on-chip transmitter and vice versa. The loop-back connection can be made by switches internal to the DUT or through the test fixture. Sometimes, it is even possible to add feedback around an analog circuit in test mode to cause it to oscillate. In this "oscillation BIST," the health of the DUT is indicated first by the presence of an oscillation and, second, by this oscillation being at the correct frequency. Finally, there are often opportunities to alleviate the difficulties of transmitting high-frequency and high-quality signals between the DUT and the tester's pin electronics by generating the stimulus and measuring the response with additional test circuits on the chip or on the DUT interface, that is, the probe card or load board. Often these test circuits serve to convert signals, rather than to generate or measure them, by mixing frequencies up or down, amplification, or converting to dc.

21.18 EMERGENCE OF CLOUD-BASED DFT TESTER

Over the past several years tester vendors have brought to the market test systems that have been classified as DFT testers. These testers are distinguished by having a capital cost below $500,000, digital frequencies of 100 MHz or less, and large amounts of vector memory. These can either be designed as low-cost testers with this functionality or a cost-optimized scaled down version of a functional performance-based tester. Figure 21.15 shows the generally accepted view of the difference between a DFT tester and a functional tester. A DFT tester will generally not have the speed and accuracy to do functional tests and will also not have analog instruments that allow it to do SoC level testing. DFT testers require a sophisticated IC design environment and a design methodology that supports DFT. One of the goals of using a DFT tester is reducing the capital cost, but test times are generally longer due to the need to load and unload test stimulus and response data through serial scan ports. These ports generally run at a maximum frequency of 40 MHz. A particular DFT tester from Advantest lowers capital cost even further by providing the tester for free and charging the user

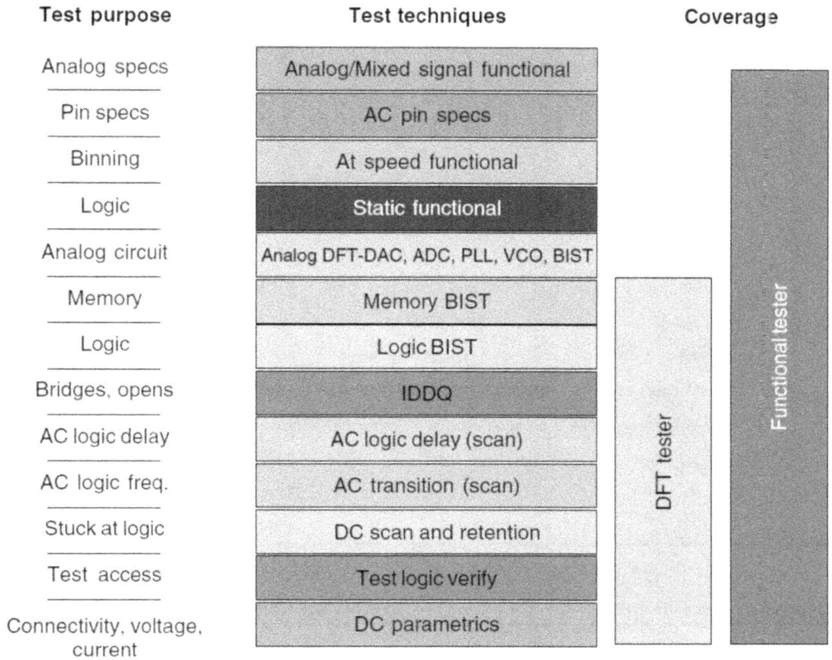

Figure 21.15 DFT tester test techniques.

for the test IPs downloaded for testing the DUT. Instead of buying and maintaining or renting expensive test equipment this innovative product allows the user to sign up for a test service. The user can then download test IPs resident in the cloud and exercise them on the new silicon for testing and characterization. Each month the user is billed for all the IPs used. The test equipment for executing the test algorithms, a compact desktop size unit, is provided to the user for the entire duration of the service. The users will only bear the cost of interfacing their new device to the test equipment which is done by mounting the device on a user designed performance board with connectors and cables purchased from an e-commerce website. The main benefits of using this service are immediately apparent:

- No capital costs.
- Each user can afford to have own personal tester.
- User pays only for the algorithms used on a monthly basis.
- Switching to later generation HW is at no extra cost to the user.

21.19 ATE SPECIFICATIONS

The *International Technology Roadmap for Semiconductors* (ITRS) is an organization dedicated to tracking and reporting future technology requirements in the semiconductor segment and test. The ITRS enjoys wide sponsorship and contacts including the Semiconductor Industry Association (SIA), the European Electronic Component Association (EECA), the Japan Electronics & Information Technology Industries Association (JEITA), the Korean Semiconductor Industry Association (KSIA), Taiwan Semiconductor Industry Association (TSIA), and the International SEMATECH. The ITRS forecasts test requirements for different device technologies both near term and long term. These forecasts in conjunction with market requirement directly drive target specifications for ATE introduced to meet the manufacturing test challenges in the future years. As device pin count, I/O frequency, analog requirements, and power demand increase, the interface between the tester and the DUT becomes increasingly complex. The tester manufacturer will guarantee the specifications of the tester to the connection point, the pogo pins in the test head. The ATE also has calibration and verification programs, which are used to calibrate the tester hardware at regular intervals and verify the accuracy of the tester. To compensate for the fixture delay, focused calibration software using time domain reflectometry (TDR) is made available by the ATE manufacturer. Complex simulation capability is needed to allow optimization of interface layout routing and geometries versus instrument location and path performance. Simulations require detailed models of the test equipment instrument, electrical delivery path, probe card or load board and contactor, and the DUT. Such simulations are needed to guarantee signal and power performance at the die. Tables 21.2 to 21.5 show tester specifications published by ATE manufacturers for different types of testers.

21.20 ATE DATA FORMATS

In a manufacturing test program flow, the device under test is set up with levels, timing, pattern, analog waveform, and RF signals. The tester then measures the response of the DUT. Depending on the test the measured result is a pass/fail flag, digitized current/voltage, or a measurement of a frequency. The result is logged in a file and at the end of production the lot summary is saved with the test data for later analysis by the semiconductor manufacturer. The data formats used for test input and output are mostly custom and defined by the tester manufacturer. However, there are several common formats including a standard for pattern data that allow test engineers to port test vectors and timing. For test results, there is test data format (STDF), which is used by device manufacturers to analyze test result data from different ATE.

IEEE Std. 1149.1 STIL was released in 1999 and provided a way for test engineers to create scan and other DFT patterns for the test flow from STIL files generated from design simulation. After fixing device fails during silicon debug, the latest and corrected STIL files can be used to create the production test program on different production ATE. There are two other popular formats for test vector data, WGL (waveform generation language) and VCD (value change DUMP). While WGL is a cyclized, format like STIL VCD is an event-based format.

Table 21.2 Typical Tester Specifications of High-End Microprocessor and SerDes

Item	1990–2000	2000–2010	2010+	Comment
Digital pin count	1024	2048	3200+	
Max input clock frequency	800 MHz	1.6 GHz	2 GHz	
Max data rate	1.6 Gbps	3.2 Gbps	6.5 Gbps	Drive/strobe
Overall timing accuracy	100 ps	75 ps	50 ps	
Vector memory per pin	64M	256M	256M	
Scan data memory/scan chain	1G	8G	16G	
Scan max data rate	100 Mbps	200 Mbps	200 Mbps	
APG max frequency	200 MHz	200 MHz	200 MHz	Cache testing
High-current device power	1.3–3.3 V	0.7–1.5 V	0.5–1.1 V	20–100 A
DC precision measure unit	Voltage: −1.5 to 6 V accuracy: 0.15% Current: ±150 mA accuracy: 0.2%			Per module
DC precision measure per pin	Voltage: −1.5 to 6 V accuracy: 0.15% Current: ±20 mA accuracy: 0.2%			Per pin
Serial max data rate	3.2 Gbps	8 Gbps	16 Gbps	
Lanes per instrument	8	16	32	
Max jitter injection	±16 UI	±64 UI	±128 UI	

Table 21.3 Typical Mixed-Signal Tester Specifications

Item	Baseband	Audio	Comment
Channel per instrument	4 × 2 ports	4 × 2 ports	
AWG sampling rate	400 Msps	190 Ksps	
Resolution	16 bits	24 bits	
Signal amplitude pk-pk	7.8125 mV–2 V	2 mV–8 V	Differential
Signal amplitude accuracy	< ±0.6 dB	< ±0.5 dB	
Sine wave distortion	−78 dBc	−103 dBc	
Digitizer sampling rate	2–256 Ms/s	32–820 ks/s	
Resolution	16 bits	18 bits	
Input voltage range	±2.8 to ±0.25 V	±4 to ±0.7 V	
Signal amplitude accuracy	< ±0.6 dB	< ±0.23 dB	
Sine wave distortion	< −60 dBc	−105 dBc	

Table 21.4 Typical Memory/Flash Memory Tester Specifications

Item	1990–2000	2000–2010	2010+	Comment
Max. number of MUT	64/station	64/station	128/station	Two stations maximum
Overall timing accuracy	800 ps	400 ps	400 ps	
ALPG maximum frequency	70 MHz	140 MHz	266 MHz	
X, Y address, Instrumental memory	16X, 16Y, 1K	18X, 18Y, 2K	24X, 24Y, 2K	

The WGL format syntax was originally developed by TSSI and has become a de facto industry standard for generating test vectors. The VCD format was defined along with the Verilog hardware description language (VHDL) by the IEEE Standard 1364–1995 in 1995. An extended VCD (eVCD) format defined 6 years later in the IEEE Standard 1364–2001 supports the logging of signal strength and directionality.

Table 21.5 Typical RF Tester Specifications (2010+)

Item	Vector signal generator	Vector signal analyzer	Comment
Frequency range	100 MHz to 6 GHz	100 MHz–12 GHz	
Output level	−120 to +5 dBm	−120 to +17 dBm −120 to −20 dBm	
Input Level		−120 to +17 dBm −120 to −20 dBm	Preamp on Preamp off
Sampling clock	60 KHz to 204.8 MHz	60 KHz to 204.8 MHz	
Waveform resolution	14 bits	16 bits	
Waveform memory	2.75 MW/I, Q		
Data		64–2M	
No. of channels	4/module	4/module	

STDF was originally developed by Teradyne and is now used by the semiconductor test industry for collecting and analyzing device test data. During wafer sort in high-volume manufacturing, data is needed on failing die to detect systematic failures by statistically processing the commonality of failures that occur during production. Test results data is also needed for reliability improvement, process monitoring for yield improvement, adaptive control of test low to improve test efficiency, product characterization, test floor statistical process control and monitoring tester calibration, and test repeatability. STDF is flexible enough to meet the needs of the different testers that generate raw test data, the databases that store the data, and the data analysis programs that use the data. The fact that it is a single, coherent standard also facilitates the sharing and communicating of the data among these various components of the complete ATE system. In STDF data is stored in the form of records. Each record has a header consisting of three sub fields described below:

REC_LEN. The number of bytes of data following the record header. REC_LEN does not include the 4 bytes of the record header.

REC_TYP. An integer identifying a group of related STDF record types.

REC_SUB. An integer identifying a specific STDF record type within each REC_TYP group.

The header of each STDF record contains a pair of fields called REC_TYP and REC_SUB. Each REC_TYP value identifies a group of related STDF record types. Each REC_SUB value identifies a single STDF record type within a REC_TYP group. The combination of REC_TYP and REC_SUB values uniquely identifies each record type. This design allows groups of related records to be easily identified by data analysis programs, while providing unique identification for each type of record in the file. All REC_TYP and REC_SUB codes less than 200 are reserved for future use by Teradyne. All codes greater than 200 are available for custom applications use. The codes are all in decimal values. The official list of codes and documentation for their use is maintained by Teradyne's Semiconductor CIM Division (SCD). In STDF files test data is stored in binary but STDF files can be converted to ASCII format called ATDF. Software tools are also available to perform statistical analysis and other operations on test data.

RITdb (rich interactive test database) is the next generation STDF being proposed by the Collaborative Alliance for Semiconductor Test (CAST) special interest group of SEMI. The main purpose behind this activity is to provide standards-driven data environment for semiconductor test. This will include simple standards-based data capture, transport, and relationship model for E-test, probe, final test data. Equipment configuration and performance data will also be stored in the RITdb. This new format is intended to support real-time monitoring of status and performance, data analysis that requires accurate, consistent data from multiple-related sources. The Adaptive Test Roadmap requirements of the ITRS Adaptive Test Roadmap will also be supported. Because it is designed to succeed the STDF format, RITdb will have better field definitions, faster access, and easier extensibility.

21.21 MANUFACTURERS AND ATE MODELS

There are around 50 ATE companies worldwide engaged in the business of making and selling hardware, software, and services. It is not the intent to document them all here in this section. The list below summarizes the top 10 ATE companies based on their sales volume, longevity in the business, and number of employees.

21.21.1 Teradyne, Inc.

Teradyne was the leader in 2015 selling testers for logic, RF, analog, power, mixed-signal, and memory devices. Founded in the 1960s at the start of the semiconductor device manufacturing revolution, they have been in business for more than 50 years. In addition to making semiconductor device testers they make production board test systems for electronics manufacturers. Teradyne is shipping the following semiconductor ATE models currently:

- ETS-800 for analog testing. Teradyne acquired Eagle Test Systems in 2008.
- Magnum 5, Magnum CIS for flash memory, and CMOS image sensor (CIS) testing. Teradyne purchased Nextest Test Systems in 2007.
- J750, J750HD-Litepoint, and IP J750HD for Micro Controller Units, RF-SOC and CIS testing. In 2011 Teradyne acquired Litepoint a provider of test solutions for the development and manufacturing of wireless devices.
- UltraFLEX Test System Family for SoC, RF connectivity, APU, high-end MCUs, and mobile power management ICs (PMIC) testing.

21.21.2 Advantest Corporation

Ikuo Takeda founded Advantest, then known as Takeda Riken Industries, in Tokyo's Itabashi Ward in 1959. In 1971, the company developed the first IC tester capable of testing at 10 MHz. In 2015, Advantest products include semiconductor ATE and mechatronic equipment for IC manufacturing. The following ATE models are available from Advantest in 2015:

- V93000 SoC Smartscale family for digital, analog, precision DC, and RF-SoC testing. A variety of test instruments can be configured in four different test head sizes for optimized COT. V93000 HSM for high-speed memory testing. In 2011, Advantest purchased Verigy, a spinoff of Agilent Technologies, which came out of Hewlett Packard, the original designer of the 93,000 tester architectures in the late 1990.
- T2000 and T2000IPS family for SoC, automotive, and analog device testing.
- T53xx, T55xx, T57xx, and T58xx series for testing DRAM, SRAM, and FLASH.
- MPT3000 for SSD protocol testing.
- EVA100 for low pin count analog device testing.
- Cloud Testing Service CX1000 family for design verification, new silicon debug, and diagnosis.

21.21.3 Xcerra

Xcerra was created in 2014 following the LTX-Credence acquisition of Everett Charles Technologies (ECT) and Multitest from Dover Corporation in December of 2013. LTX-Credence itself was created in 2008 following the merger of LTX Corporation and Credence Systems Inc. both ATE manufacturers. In 2015 Xcerra ranked third behind Teradyne and Advantest. Their sales composed of semiconductor ATE, semiconductor device handlers, PCB test equipment, and test fixtures. Xcerra's current tester offerings are

- Diamond series for testing 3G/4G baseband, applications processors, integrated RF transceivers, RF-SOC ASSP, and SiP
- X-series for testing RF power amplifiers (PA), PMICs, and data converters
- ASL series for lower pin count PMICs, voltage regulators, and op-amps

21.21.4 UniTest

UniTest makes low cost testers for DDR/DDR2/SDRAM memory devices and modules. The component tester supports testing 256 MUTs in parallel and the modules can be tested by eight in parallel with the module tester. The company also provides application level production tester for memories and modules by providing a system level test environment for the memory device or module in a motherboard. UniTest was established in 2000 in the Republic of Korea. The ATE models provided by them are

- UNI500 series of DDR/DDR2 memory testers
- CSTA4000 DDR/DDR2 memory system level tester in PC board environment
- UNI840 GDDR2/GDDR3/GDDR4 graphic memory system level tester
- UNI820 HDD SDRAM buffer memory system level tester
- UNI400 series of DDR/DDR2 and FB-DIMM memory module testers

21.21.5 FEI

FEI completed acquisition of DCG systems in 2015 to become a leader of semiconductor device electrical fault characterization, localization, and editing equipment. FEI also sells transmission electron microscope for process monitoring in semiconductor wafer manufacturing. The equipment offered by FEI for the semiconductor device diagnostics are

- *Meridian*-IV electrical failure analysis system provides for fault localization and analysis in integrated circuits.
- *Meridian* WS-DP system for full-wafer electrical fault analysis during production.
- *Meridian*-V laser voltage probing (LVP) and laser voltage imaging (LVI) for timing and functional analysis for integrated circuits.
- *Hyperion* system for transistor nanoprobing for electrical characterization.

21.21.6 Chroma ATE, Inc.

Headquartered in Taiwan, Chroma was founded in 1984. They supply semiconductor ATE and turnkey test and automation solutions serving the LED, photovoltaic, Li-battery, electric vehicle (EV/EVSE), semiconductor/IC, laser diode, flat panel display, video and color, power electronics, and passive component market. The following ATE models are sold by Chroma:

- *Model 3650EX/CX* for SoC and analog testing
- *Model 33xxD/P* for general purpose VLSI testing

21.21.7 National Instruments

Founded in 1976, NI's core business was in providing instruments for the bench for engineers and scientists. In 2014, NI entered the semiconductor ATE market with the announcement of their Semiconductor Test System (STS) series. These PXI-based automated test systems are designed for testing RF and mixed-signal devices by deploying PXI-based modules in semiconductor production test environments.

21.21.8 SPEA

With its headquarters in Italy, SPEA has been building testers for the component and board test industry since 1976. SPEA's testers are used in the manufacturing test of MEMS, discrete devices, stand-alone integrated circuits, systems-on-a-chip (SoC), and systems-in-package (SiP) chips.

- *SPEA* offers CTS1000-10/20/30 tester models for semiconductor digital, analog, and RF device testing.

21.21.9 Aehr Test Systems

Headquartered in Fremont, California, Aehr Test Systems is a worldwide supplier of systems for burning-in and testing memory and logic integrated circuits. Aehr Test's newest system is for packaged part test during burn-in for both low-power and high-power logic as well as all common types of memory devices. They also have a full wafer contact test and burn-in system. Their systems can effectively burn-in and functionally test complex devices, such as digital signal processors, microprocessors, microcontrollers, and systems-on-a-chip. They provide a reusable carrier, temporary package that enables IC manufacturers to perform cost-effective final test and burn-in of bare die. The following models are provided by Aehr Test Systems:

- ABTS-L and ABTS-LI for logic and high power logic device burn-in.
- ABTS-M, ABTS-P, and ABTS-PI for memory and high-power memory device burn-in.
- FOX-1P and FOX-15 wafer burn-in system.

21.21.10 Micro Control Company

Micro Control Company was founded in 1972 as a manufacturer of test equipment for the electronics industry. The microprocessor-controlled test system produced in 1973 was the first in the industry. Micro Control is selling the high-power burn-in and test systems to customers worldwide. The current product line includes high-power burn-in systems for memory and logic applications, automatic test:

- HPB-4A/5C—high-power burn-in with test system
- LC-2—logic and memory production burn-in system

21.22 FUTURE ATE DIRECTIONS

Throughout the 50-year history of the semiconductor industry, its innovation and growth have been fueled by rapid technical evolution based on Moore's law, which in 1965 predicted that the number of transistors that could be incorporated in an integrated circuit would double every year. The model 500 transistor tester was one of the early programmable automatic testers, which used vacuum tubes to test transistors. Over the years, testers kept pace with the development of the ICs, but the tester technology was always older than the device technology it was supposed to test. Once built-in self-test (BIST) concept was introduced, test hardware was designed into the IC itself making the tester technology comparable to the device. The semiconductor industry is now well established in nanometer manufacturing. This has led to the proliferation of intellectual property cores (IP cores). IP cores, also known as IP blocks and "virtual components," are designs that can be used to build integrated semiconductor devices and "systems-on-chip." It is typically impossible to create new IC designs without predesigned IP blocks as a starting point. These design components are called *intellectual property* blocks because they are traded as rights to use and copy the design. Firms that focus on this business model are often called *chipless* semiconductor firms. They are widely marketed by European, American, and Asian firms, and they are critically important building blocks in current and future digital products. Firms can reuse internally developed IP cores in their own products or they can gain revenues through licensing, royalties, and customization of these predesigned components. At present, the globally leading vendor is the ARM Holdings plc, based in the United Kingdom, whose IP cores were used in about every fourth programmable electronic device manufactured in 2007. With the reduction in feature size of nanometer technology, there is also quite a bit of area available in silicon to implement test hardware. This brings the tester very close to the device under test something the external ATE has been striving to achieve over the years but has been limited by the physical size of the tester, the test fixture and the device package.

In the next several years, the IC industry could approach fundamental technological disruption in its history. The rapid incremental innovation that has led to exponential growth in the number of transistors on a chip and expanded the applications of IC technology to all areas of human life can slow down due to limits of physics. The ITRS has formulated a new roadmap, the ITS2.0, to focus more on the end applications of semiconductors rather than the semiconductor devices. The main reason for this is the semiconductor industry is following a new trend "More than Moore," where added value to devices is provided by incorporating functionalities that do not necessarily scale according to Moore's law. More than Moore (MtM) refers to a set of technologies that enable nondigital micro/nanoelectronic functions. These are regarded as a complement of digital signal and data processing provided by the "More Moore" domain the current advanced CMOS technologies scaling according to Moore's law. This includes the interaction with the outside world through an appropriate transduction (sensors and actuators) and the subsystem for powering the product. These functions may imply analog and mixed-signal processing, the incorporation of passive components, high-voltage components, micromechanical devices, sensors and actuators, and microfluidic devices enabling biological functionalities. ITRS emphasizes that "More-than-Moore" technologies do not constitute an alternative or even competitor to the digital trend as described by Moore's law. In fact, it is the heterogeneous integration of digital *and* nondigital functionalities into compact systems that will be the key driver for a wide variety of test application fields, such as communication, automotive, environmental control, health care, security, and entertainment. Whereas "More Moore" may be viewed as the brain of an intelligent compact system, "More-than-Moore" refers to its capabilities to interact with the outside world and the users. Under the new ITRS 2.0 roadmap the following areas will set new directions for ATE:

21.22.1 Cost of Test

As cost per transistor scales down with more integration ATE equipment maker is always finding new ways of reducing test cost. In the SoC tester architecture, configuration of the tester with the right complement of test instruments for testing the device can reduce the tester cost. ATE manufacturer will offer a variety of scalable instruments with different test rates for digital and sample rates for analog to choose for testing the device. Another major factor is multisite efficiency. Testers with CPU per site architecture typically have a higher test throughput. Number of sites and the ability to share test instruments across sites is also key in reducing the cost of test.

The next step in COT is in implementing parallel test capability in an individual device. *Concurrent test* (CCT) is the parallel test methodology that allows the simultaneous testing of multiple IP cores within a device. Concurrent test requires a good understanding of the test equipment capabilities and also the way the test program generation is done. Successful CCT is based on careful planning and implementation discipline. It calls for a cross-functional team of design and test engineers to work up front, planning the solution, then it requires more test program integration on completion of the individual core test programs. The trend is toward test program reuse, which requires that the software related to test programs be capable of merging test programs of previously tested cores with new ones. Combining this parallel test solution with the multisite model described previously allows even more efficiency. The benefits include reduction of test time, greater usage of test resources, and a more effective utilization of them, improving the return on invested capital. Other benefits include improvement on the quality of parts by running the device closer to its end application environment and the reduction of the hidden cost of returns, as well as improved customer satisfaction with the delivery of products with a reduced statistical factor of defects measured in parts per million. Because CCT applies to single device execution, the individual cores within the device need to be properly designed for parallelism and isolated so that the testing of one IP core minimally impacts others. Core isolation for any form of parallel test application is a new test methodology linked to DFT tools offered by EDA companies. This expands the linkage between design and testing. EDA companies that traditionally provide design tools are currently expanding their product offering to include design methodologies that assist design for testability. In addition, there is movement in the *design for manufacturing* (DFM) area where new design methodologies interact with ATE for fast diagnosis of failures. This shows the strength of the new supply chain's links with feedback loops between EDA and ATE for faster TTM and TTV.

21.22.2 Alternative Test Methods (Dfx/BIST/SLT)

Dfx stands for DFT (design for test), DFM (design for manufacturing), and design for yield. These are test techniques using structures implemented in the IC to improve time to volume in high-volume manufacturing. They have been in use in the industry for the last 20 years. The amount of test vectors used and test data generated by these methods is on the rise as the number of IP cores increase in every new generation of the SoC. In addition, test results for each device is not just limited to pass fail information but a lot of process monitor data. Adaptive test methods will be used by ATE to collect test data generated during HVM for and use it for the following applications:

1. Statistical analysis to look for trends for yield improvement
2. Real-time monitoring to change flow for better test throughput
3. Feed results forward to the next step in HVM for test optimization
4. Offline data analysis for improving test program efficiency
5. Monitor ATE stability and schedule ATE calibration

BIST hardware for memory (MBIST) and logic (LBIST) are necessary for testing embedded cores in SoC devices. With the Internet of Things (IoT) phenomenon just around the corner ramping to 50 billon smart objects around the world in no time BIST and self-repair (BISR) will be designed into these devices. Multisite testing with large site counts will be implemented on future ATE for these devices.

System level test (SLT) will be coming back to manufacturing for SoC testing. In the early 1960s and 1970s system boards were tested functionally in a systems level tester. The functional test was also combined with an in-circuit test. In the in-circuit test, PCBs with mounted components or electronic components are contacted and tested for faults in printed conductor routing, soldering and component faults test. Today's SoC devices are miniature versions of system boards but implemented in a chip.

ATE can test the system components with scan test similar to the in-circuit test of the boards. ATE can also test each IP core of the SoC with test access mechanisms designed into the SoC. P1000SECT discussed earlier makes this possible. Future ATE will need to support SLT for SoC to detect subsystems fault. The support for SLT will include test generation, system boot up, and debug environment.

21.22.3 Power and Thermal Management

Current ATE handles power requirements for MPU, GPU, and APU devices very well in terms of high current requirement at low core voltages with good accuracy. However, future devices for mobile and cell phone are focused on precise temperature control and efficient power operating modes. They require ATE to monitor temperature and measure low currents under different application modes. Thermal response of the device to a stimulus is also a test for cell phone devices. For automotive devices, temperature testing from −40°C to more than 100°C during production is a requirement.

21.22.4 MEMS and Sensor Testing

COT challenges for these extremely low-cost devices continue to increase from one year to the next. Very high volumes are required with a very high site count to lower manufacturing cost.

21.23 ACKNOWLEDGMENTS

Donald Blair, Robert J. Smith, Jeff Brenner, George Redford, Neils Poulsen, Edwin Lowery, Hans Verleur, Asad Aziz, Ariana Salagianis, Shawn Klabunde, Michael Kozma, Gary G. Raines, Peter O'Neil, Gina Bonini, and William T. O'Grady.

21.24 FURTHER READING

Agrawal, V., ATT Bell Labs, general reference papers website: http://www.informatik.unitrier.de/~ley/db/indices/atree/a/Agrawal:Vishwani_=.html.

Bozturk, Cagatay, "Burn-in, Reliability Testing, and Manufacturing of Semiconductors," research paper, Kent State University, 2006.

Burns, M. and G. W. Roberts, *An Introduction to Mixed Signal IC Test and Measurement*, Oxford University Press, New York, 2001.

Bushnell, M. L., D. Vishwani, and V. Agrawal, *Essentials of Electronic Testing for Digital, Memory & Mixed-Signal VLSI Circuits*, Kluwer Academic Publishers, New York, 2000.

Chang, L. L., "Systematic Methodology with DFT Rules Reduces Fault-Coverage Analysis," *EE Design*, Aug. 2001, available at http://www.eedesign.com/isd/features/OEG20010803S0032.

Geng, Hwaiyu, *Semiconductor Manufacturing Handbook*, McGraw-Hill, New York, 2006.

Goor, A., *Testing Semiconductor Memories, Theory and Practice*, Wiley Online Library, 1991.

Kao, W. and K. Hasebe, "Simulation of Tester Environment Improves Design to Test Link for Mixed Signal ICs," *Proceedings of the ATE Instrument Conference*, Anaheim, California, Jan. 14–17, 1991.

Laeson, Eric, Protocol Aware ATE 2008 Beijing Advanced Semiconductor Technology Symposium.

Mahoney, M., "DSP Based Testing of Analog and Mixed Signal Circuits," *The Computer Society of IEEE*, 1987.

Maxfiled, C., *Boolean Boogie*, HighText Publications, Solana Beach, CA, 1995, http:/www.Harrington-institute/downloads/whitepapers/UnderstandingSix-Sigma.pdf.

Parker, K., *The Boundary-Scan Handbook, Analog and Digital*, 2d ed., Kluwer Academic Publishers, Boston, 1998.

SoC Test: Trends and Recent Standards (Michael Higgins and Ciaran MacNamee. Circuits and Systems Research Centre (CSRC), University of Limerick. ITB Journal).

Tocci, R. J., *Fundamentals of Electronic Devices*, Charles E. Merrill Publishing Company, Columbus, Ohio, 1975.

2015 Top Markets Report Semiconductors and Semiconductor Manufacturing Equipment. Department of Commerce USA.

Vision ATE 2020 at ITC 2007, Andy Evans.

VLSI Test Symposium 2007, Innovative Practices, Eric Larson.

Zorian, Y., E. J. Marinissen, and S. Dey, "Testing Embedded-Core-Based System Chips," *IEEE Computer*, 32 (6): 52–60, Jun. 1999.

21.24.1 Information Resources

http://www.computerhistory.org/siliconengine/dedicated-semiconductor-test-equipment-enters-commercial-market/

http://www.evaluationengineering.com/

http://www.fabtech.org/industry.news/9907/05.04.shtmlw

http://www.ieee.org/portal/index.jsp

http://www.itctestweek.org/error404.shtml

http://www.itrs2.net/itrs-reports.html

http://www.marvell.com/wireless/avastar/88W8787/

http://www.reedelectronics.com/tmworld/index.asp?layout=Community&industry=Semiconductor+Test&industryid=19033&rid=0&rme=0&cfd=1

http://www.semi.org/wps/portal

http://www.semitest.org/site/Membership/index_html

http://www.testandmeasurement com/content/homepage/default.asp?VNETCOOKIE=NO

https://www.techopedia.com/definition/2148/automatic-test-equipment-ate

PART 4

TECHNOLOGIES FOR FLEXIBLE HYBRID ELECTRONICS AND LARGE-AREA ELECTRONICS

Chapter 22

Printed Electronics: Principles, Materials, Processes, and Applications

BY

KAN WANG, YUNG-HANG CHANG, BEN WANG, AND CHUCK ZHANG *School of Industrial & Systems Engineering and Georgia Tech Manufacturing Institute, Georgia Institute of Technology, Atlanta, Georgia*

22.1 INTRODUCTION TO PRINTED ELECTRONICS

Flexible hybrid electronics (FHE) has experienced rapid development during the past decade, leading to the development of precise, cost-effective, and scalable processing techniques [1–3]. This allows for the production of flexible, stretchable, or wearable electronic devices that can be used in everyday life. The printed electronics (PE) technology has attracted significant attention in recent years due to its potential to simplify process steps and reduce costs when compared to conventional photolithography [4, 5]. Conventional photolithography consists of multiple steps, including thin film vapor deposition, photoresist deposition, masking, and wet etching. PE technology, on the other hand, is a one-step process. Furthermore, PE technology can be readily scaled to large-area production with high throughput. These features can be applied in a way that lowers cost and offers flexibility for work in large areas. This allows for new ways to develop novel electronics and accelerate their use in other areas of research. Examples include organic light emitting diodes, organic solar cells, thin-film transistors, logic circuits, radio frequency identification (RFID) tags, and sensors. However, the fact that the majority of electronic devices are still fabricated using traditional silicon technology indicates the practical limitations of PE in reliability and cost-effectiveness. In addition, the complexity, the difficulty in patterning large areas, and the limited range of materials that can be directly patterned all represent significant disadvantages of PE. This chapter discusses the fundamentals of the rapidly evolving PE technology and both the advantages and limitations of the major processes in this category.

PE technology is currently available on the market. Some commercialized products are enabled or heavily rely on PE technology. These market products can be categorized into the following groups:

- Organic Light-Emitting Diodes (OLEDs)
- Organic/inorganic photovoltaics
- Integrated smart systems (RFID, sports fitness/health care devices, smart cards, sensors, and smart textiles)
- Electronics and components

Each device category has a potentially huge market. For instance, the worldwide new lighting market can grow by up to \$150 billion by 2020 [6]. The features of some typical PE products and their current status are summarized in Sec. 22.6.

22.2 PRINTED ELECTRONICS: PRINCIPLES AND FUNDAMENTALS

Many printing methods have been applied to conventional electronics fabrication for a long time, such as screen printing, gravure printing, flexography printing, and offset printing. They have also been applied to the manufacturing processes of many advanced PE products. Depending on the nature of the PE product, a suitable choice of manufacturing method has to be made regarding the ink, substrate, designed device structure, pattern geometry, manufacturing speed, yield, quality, and production cost. The fundamental printing parameters are as follows:

- *Resolution*. Display application for smart phone/tablets, among the most popular applications today, requires fine patterning above 300 pixels per inch (ppi). A resolution of a few micrometers with ±5 μm position accuracy will be required. Multilayer printing accuracy is also a key factor.
- *Wettability*. Flatness within a few nanometers to several tens of nanometers is required for many OLED applications such as TV and lighting since the typical OLED layer thickness is less than 100 nm. Sharpness at pattern edges and bonding with substrates are strongly dependent on the underlayer (acceptance layer) material and its design.
- *Uniformity*. Variations from a few centimeters to more than 1 m in size area is required in combination with the designing ink composition and the drying process.
- *Compatibility*. The compatibility of inks with printing components such as rollers, masks, doctor blades, and inkjet heads has a significant effect on yield and quality in mass production.
- *Scalability*. One of the great benefits of PE technology is its capability for mass production at a reasonable cost. The high speed and high quality of printed patterns should be maintained for up to hundreds of printings.

Roll-to-roll printing is one of the active research areas in PE technology because it enables large-scale production by high-speed web handling. Roll-to-roll printing allows for large-scale production of such items as RFID antennas or keyboard membranes. However, the roll-to-roll process is not mature enough to be applied to many areas where PE technology is used because adjustments have not yet been established among materials, printing methods with suitable web handling, accurate positioning, and inspection methods with definitions of defect criteria. Sheet-fed production is still a major printing method for most PE products. To shift from sheet-fed printing to roll-to-roll printing will require time to develop the printing technologies with suitable parameters, including materials development.

PE technologies that are not currently suitable for roll-to-roll manufacturing scenario are sometimes called direct-write technology. Those relatively new technologies include inkjet printing, aerosol jet printing (AJP), syringe dispensing, laser-assisted chemical vapor deposition (LCVD), laser particle guidance (LPG), matrix assisted pulsed laser evaporation (MAPLE), focused ion beam (FIB), etc.

Direct-write processes are fast, flexible, and with high tolerance for errors. It grants circuit designers much freedom by directly converting patterns in CAD/CAM files into prototypes. It also has a much wider range of inks and substrates. Most importantly, it largely simplifies the manufacturing processes. Masking and screen-printing processes take several steps to fabricate a circuit. Making a mask or a new screen can take days, even weeks. A direct-write process has the possibility of turning weeks of prototyping into hours. With this kind of improvement, it seems inevitable that direct-write processes will revolutionize the electronics industry.

While this is true, competing with masking production rates has proven to be an onerous task. Consumers of direct-write technologies appreciate its superior speed and demonstrated superior results. The main complaint, however, is repeatability on an industrial level. While many direct-write techniques have proven to be very effective for demonstrating a concept or device, the prototyping process was not repeatable in a mass-production manner. This disadvantage would completely remove the benefits obtained from the fast fabrication and characterization results. The large variation in products has heavily damaged the throughput of direct-write technologies. While throughput may not be the only issue in the electronics industry, it is the only one that matters if a new process cannot reach specified standards. To replace masking techniques such as screen printing or photolithographic patterning, direct-write methods will have to complete these tasks in seconds, not hours. If the scalability issue cannot be solved soon, the only field that direct-write technologies will be competitive in would be rapid prototyping (RP). Figure 22.1 summarizes the throughput and resolution of various PE technologies [4, 7].

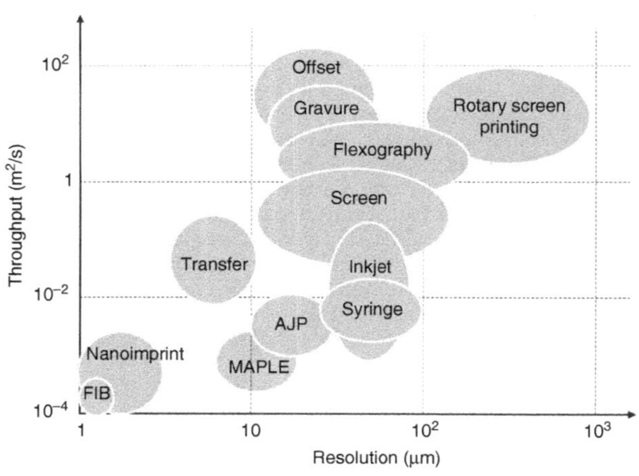

Figure 22.1 Throughput versus resolution for various PE technologies. (*© Chuck Zhang. Used with permission.*)

22.3 MATERIALS FOR PRINTED ELECTRONICS

The choice of printing methods is the major decision that has to be made before launching research projects or before building up production lines. It is a complex decision that involves deciding which inks, substrates, and printers are suitable. The choice of substrate plays an important role in this decision. The viscosity/surface tension of the ink, device structures, and whether the device line/layer is thin or narrow will affect the pattern quality obtained. The cross-section profile of a printed circuit or a device has a distinctive shape. A square cross section is desirable in order to obtain most electronics properties. Unfortunately, this does not happen with PE technology except with the use of high-viscosity inks such as those used in screen printing. In printed lines created with inkjet printing or aerosol jet printing, low-viscosity ink droplets land on a substrate. Its cross section sometimes has a coffee-ring effect, depending on the viscosity of the ink, its wettability on a substrate, and the vaporization uniformity of the solvent [8]. When the ink droplet is drying, the segregation of solute content toward the outside edge of a droplet sometimes occurs, which causes the coffee-ring effect. This shape is not desired in most cases because many defects are likely to form in the concave central area. Thus, the droplet shape must somehow be made flat. At the very least, a half-cylinder shape is desired.

The desired shapes of printed lines are obtained by wetting control on substrate faces, which can promote or prevent the spread of printed inks according to the objective [9, 10]. The wettability of a specific type of ink on a specific type of substrate can be measured by depositing a single ink drop on the substrate, as shown in Fig. 22.3. The wetting angle is a good indicator for the wetting condition. The underlying governing physics are the surface energies of the liquid and the substrate, and the Young's equation. Table 22.1 summarizes the surface energy ranges for typical polymeric substrates. The wettability of the substrates can be adjusted by either chemical or physical treatment of the substrates.

Table 22.1 Surface Energies of Typical Polymeric Substrate Materials

Surface energy (dyn/cm^2)	Polymeric substrate material	Wetting property (water)
10–20	Silicone Fluorocarbon polymers	Water repellent
20–35	Polyethylene Polypropylene	Hydrophobic
35–50	Polyester Nylon Epoxy Acrylic resin Polyethylene terephthalate Polyimide	Polar
50–60	Polyvinyl alcohol Cellulose Polyvinylpyrrolidone	Hydrophilic

In addition to controlling wettability, the drying condition is also important to the cross section. Some PE technologies such as inkjet, offset-gravure, flexography, aerosol jet, etc., use inks with low concentrations of metallic nanoparticles. The viscosity of those inks is very low. Because of the amount of solvents, the evaporation has to be controlled to achieve suitable solid sediments. Failure to do that may result in unexpected high electrical resistance. An effective method to overcome this problem is to put an absorption layer on top of the pristine polymer surface to reduce the flow of solute.

Aerosol jet printing is a special case as the ink deposited on the substrate consists of many small aerosol droplets instead of a large droplet. The coffee-ring effect is minimized unless multiple layers are printed before the first layer is dried. Because of this property, aerosol jet printing can be used on a wide range of substrates including polymers, metal, composite prepregs, etc.

In most cases, PE technologies are used to create conductive patterns. Metallic nanoparticle pastes or dispersions are used as inks in those cases, including three most common: silver, gold, and copper nanoparticles. Carbon-based inks are also a popular family that has been recently adopted for use with many direct-write technologies. Those include carbon nanotubes, graphite, graphene, decorated carbon nanotubes, and the mixture of two or more contents. Some researchers have reported that mixing these inks with carbon-based nanomaterials and metallic nanoparticles have potential in stretchable electronics printing.

With more and more complex functions and designs of printed electronics, there are demands for specialized inks other than conductive ones. For example, boron nitride nanotubes (BNNTs) can be dispersed into certain solvents to create a piezoelectric ink. There are many applications for a thin layer of patterned dielectric materials. Both inorganic and polymeric dielectric inks have been developed. There are also semiconductor nanoparticle inks and polymer semiconductor inks in the market. Recently, researchers have developed biological inks that can be printed by aerosol jet printing. The most common inks used in various PE technologies include metal (such as silver) particles, carbon nanotubes, graphene, and polymer.

22.4 MANUFACTURING PROCESSES FOR PRINTED ELECTRONICS

Current PE technologies can be categorized into two major groups based on their ink deposition techniques: contact printing and noncontact printing. Figure 22.2 shows the category tree of the printing

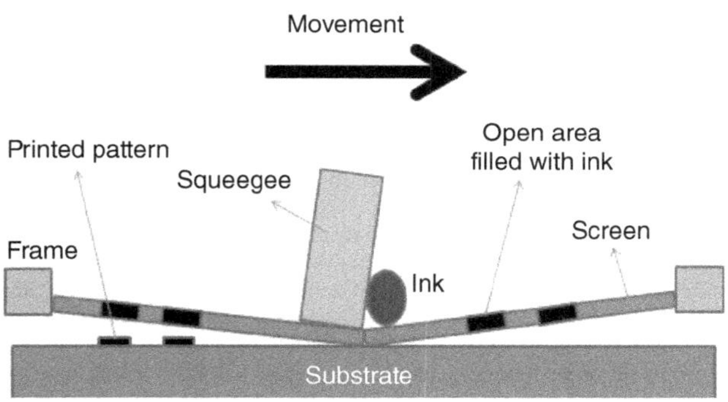

Figure 22.2 An illustration of the screen printing process. (© *Chuck Zhang. Used with permission.*)

electronic process. Each technology has its own advantages and disadvantages. Contact printing is best suited for applications that require low resolution and high throughput. Noncontact printing is flexible in design and uses less material. It is more suitable for rapid prototyping and novel sensor development. The throughput of noncontact printing is typically slower than the contact printing process. However, the feasibility of scaling a noncontact printing process has been proven with the parallel nozzle array. The representative contact printing technologies are screen printing and gravure printing. The representative noncontact printing technologies are inkjet printing and aerosol jet printing. Brief descriptions of those processes are listed below.

22.4.1 Screen Printing

Screen printing is a mature and commonly used technology for printed electronics and has been adopted for use in making printed circuit boards (PCB), interconnect fabrications, and printing current collectors for solar cells [11]. Compared with other printing methods, the most distinct feature of screen printing is its high aspect ratio of printed objects [12]. The usual thickness of a screen-printed image is in the range of several tens of micrometers, but the thickness can exceed 100 μm with a single pass of printing, especially when a thick screen mesh is used. This cannot be done by any other printing method. For other methods such as inkjet or flexography printing, the typical thickness is less than 5 μm. The process of screen printing is illustrated in Fig. 22.2. A doctor blade forces ink through a stencil to create patterns on a target substrate. After transfer, the ink remains in the voids and the stencil is lifted. Screen printing has been used for fabricating numerous printed electronic devices including organic thin film transistors, photovoltaics, and sensors.

Large-scale screen printing that is beyond widths of 2 m has also been achieved in the industry, especially for plasma display panels. A variant method of screen printing, rotary screen printing, is used to achieve large-scale mass production. This process is very fast with a limited printing resolution.

22.4.2 Gravure Printing

Gravure printing is a promising technology for printing high-resolution devices [13]. The process of gravure printing is shown in Fig. 22.3.

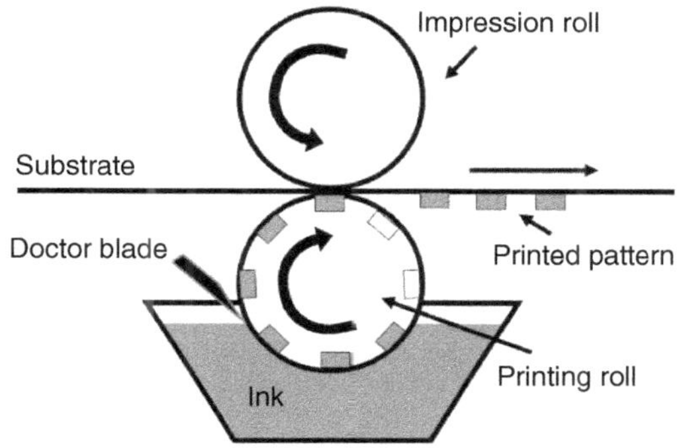

Figure 22.3 An illustration of the gravure printing process. (© *Chuck Zhang. Used with permission.*)

Gravure printing is used in industries that require high-quality and high-speed printing such as the magazine industry and the printing of currency [14]. The working mechanism for gravure printing begins with a cylinder patterned with cavities that goes through an ink reservoir to coat the cylinder uniformly. A doctor blade removes the excess ink outside of the cavities. The gravure cylinder then presses onto a target substrate to transfer the desired pattern. The gravure process has high throughput and high resolution. It is suited for large area and low-cost device fabrication. Several researchers have demonstrated using gravure printing for fabricating printed electronic devices.

22.4.3 Inkjet Printing

Inkjet printing is a drop-on-demand additive printing technology, or direct-write technology [15]. Unlike screen and gravure printing, inkjet printing does not require a patterned stencil or cylinder. Instead, digital layout files are used to define the printing patterns [16]. In inkjet printing, the printing head moves along the stage as drops are jetted onto the substrate. Three principal energy sources are used for inkjet droplet creation: heat, piezoelectric deflection, and an electric field (Fig. 22.4). In thermal-bubble inkjet printing, a heating element is rapidly warmed and any ink that is near it vaporizes and creates a pressure wave that forces a droplet from a small nozzle. A disadvantage to this jetting mechanism is that the inks used must be compatible with the temperature pulses. An alternative inkjet printing mechanism that does not thermally stress the ink is piezoelectric jetting. With this method, a piezoelectric plate is deformed by an applied electric field. This deformation creates an acoustic wave in the ink reservoir, which travels to the nozzle and forces out a droplet This jetting mechanism works with a broad range of inks of intermediate viscosity and surface tension and scales to large industrial arrays composed of thousands of nozzles.

Inkjet printing has been widely adopted in many printing electronics research groups because of the low cost of inkjet printers and the

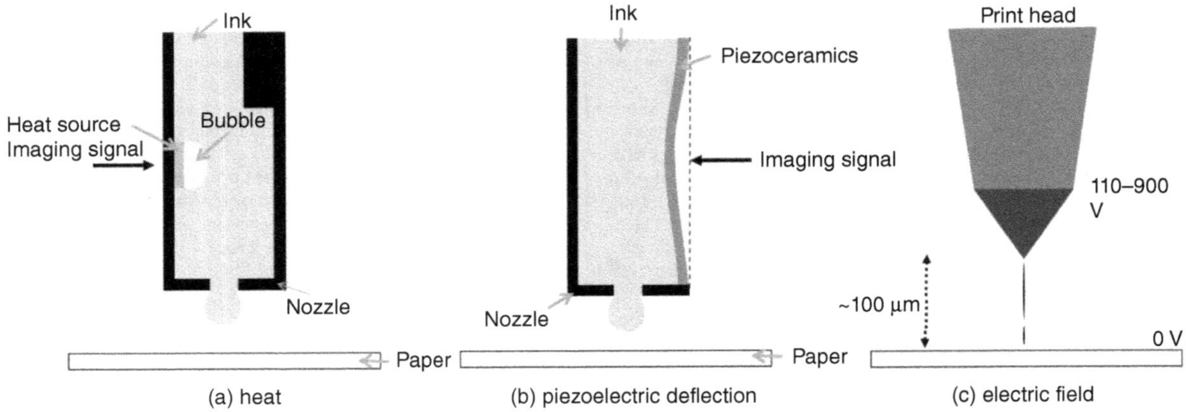

Figure 22.4 Three major working mechanisms of inkjet printing. (© *Chuck Zhang. Used with permission.*)

flexibility to work on different materials. Inkjet printing has been used to fabricate thin film transistors consisting of zinc oxide and polyvinylpyrrolidone as the active device region and gate dielectric, respectively. An inkjet printing and vacuum drying process can be used to develop a high-performance n-channel transistor that has a uniform amorphous C_{60} Fullerene. Resistors, capacitors, and inductors are developed using inkjet printing on polyimide substrate with various functional inks. Recently, carbon nanotubes and graphene materials were successfully printed using inkjet technology for flexible transistor and advance sensors.

22.4.4 Aerosol Jet Printing

Aerosol Jet® printing is a relatively new noncontact printing method in printed electronics [17]. In the aerosol jet printing process, functional inks are aerosolized with an atomizer. A gas stream is used as carrier gas to direct aerosol particles toward the deposition head. A sheath gas flow is used in the deposition head to aerodynamically focus the aerosol flow. Figure 22.5 illustrates the overall process of aerosol jet printing using an ultrasonic atomizer.

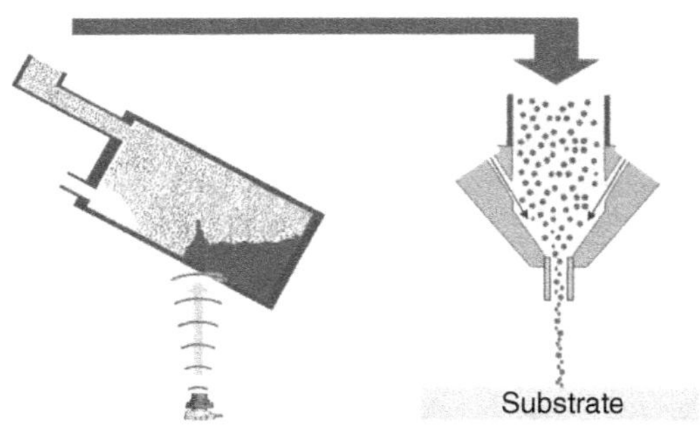

Figure 22.5 An illustration of the aerosol jet printing process using ultrasonic atomizer [18]. (© *Chuck Zhang. Used with permission.*)

One important benefit for AJP is the long standoff distance between the printing head and the substrates [19]. Therefore, the substrate does not have to be planar in order to print the desired pattern on it. The only limitation for the substrate being used in AJP is mostly governed by the ink properties. The limitations on the ink for AJP are the physical and chemical properties that enable the ink to be atomized. A wide range of materials have been developed and used for printing conductors, semiconductors, and dielectrics.

22.5 MAJOR CHALLENGES AND POTENTIAL SOLUTIONS

The field of PE is fast evolving with the newest material technologies and mechanical engineering technologies. One reflection of this is the rapid expansion of the range of materials the PE technologies can process. Piezoelectric inks, semiconductor inks, biological inks, and other functional inks have been added to this set in recent years. The viscosity of the ink is also less of a limitation with the newly developed processes. For example, aerosol jet printing, which is one of the newest members of PE technologies, can theoretically process inks with any viscosity within a range of 1 to 3000 cP, whereas inkjet printing typically requires the ink viscosity to be between 8 and 12 cP. However, there are still a number of considerations regarding printability when choosing an ink for a PE process [20, 21].

One serious consideration for choosing an ink material is its stability [22, 23]. The majority of inks for PE processes are in a form of suspension, such as metallic nanoparticle inks, ceramic inks, and carbon-based nanomaterial inks. To disperse the nanoparticles homogeneously, an appropriate solvent must be identified. Usually, surfactants are also needed to keep the nanoparticles in suspension from reaggregating. However, surfactants can adversely affect the properties of printed devices, especially the electrical conductivity. The dilemma is that surfactants are required to separate nanoparticles in the ink, but are not desired in the final product where the nanoparticles are supposed to be in touch with each other without barriers. Even with surfactants, agglomeration is destined to happen in most suspension inks. This leads to the most common problem with PE machines: nozzle clogging. Because of the size of nanoparticles and possible agglomerates, the reduction in diameter of the printing nozzle is limited; therefore, the print resolution is also limited. A possible solution to this issue is to develop solution inks to replace suspension inks, like the particle-free reactive silver ink developed by Lewis et al. This ink can print through 100-nm nozzles, which is one order of magnitude smaller than the suspension inks.

Another challenge with the inks is the material lifetime [24]. Some inks, or the solvents of the inks, are reactive when exposed to light, moisture, oxygen, or heat sources. After synthesis, the contents of those inks are continuously changing. This makes the performance of printed devices unpredictable. It is even a larger problem when considering the supply chain of the entire PE industry. There are some good formulas for conventional metallic inks that have been tested in the market for many years. However, it is a very common issue among newly developed inks, especially biological inks.

Other than the issues with the materials, there are other challenges in the printing processes like the registration of printed patterns for multilayer devices [25]. For precisely designed electronics, the layout of different layers has to overlap exactly in order for the device to work properly. However, the running part needs to be taken out from the printing machine for sintering/drying/curing after each layer is printed. Positioning errors are inevitable even with the best part handling system because of the precision limits of the printing stage and the dimension change of the printed pattern after sintering/drying/curing. There are some image-based systems for registration, but the improvements are not satisfactory. The true solution to this problem is yet to be found.

The inks and substrates used in PE devices cause many issues at the process level. For example, the sintering/drying/curing of certain ink has to be compatible with the thermal tolerances of the corresponding substrate. This is an important issue in 3D printing/printed electronics integration as many 3D printable materials are plastics with a low working temperature. Conductive inks usually require high-temperature postprocessing. To solve this problem, research has been done on transferring printed patterns from a temporary, high-temperature substrate to the destiny surfaces, using the differences in surface energies [26]. Another possible solution is to develop low-temperature sintering conductive inks.

The uniformity of the printed pattern is another issue with the processes, especially the direct-write processes. The lines printed by PE processes often have ill-defined edges, discontinuity, overspray, etc., as shown in Fig. 22.6. The areas filled by those lines therefore have voids, uneven thickness, rough surface, etc. Many of these are not effectively

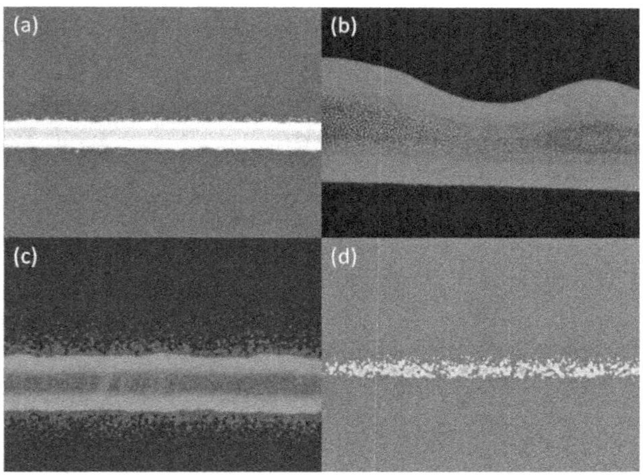

Figure 22.6 Uniformity issues in printed lines. (*a*) A typical printed line by aerosol jet printing; (*b*) voids and ill-defined edges; (*c*) overspray; (*d*) discontinuity. (© *Chuck Zhang. Used with permission.*)

controlled in current practice, which leads to another challenge—quality control and repeatability.

Compared to conventional electronics manufacturing, PE technologies are far below the current standards of quality control in the industry. The poor repeatability of printed electronics devices is a combined effect of many factors, such as unstable inks, substrate wettability, defects in printed patterns, ink/substrate adhesion, dynamic machine condition, uncontrolled printing atmosphere, etc. With the development of in situ monitoring technology and automatic process control methodologies, the repeatability of PE technologies is expected to gradually improve with the amount of ongoing research being done.

On the manufacturing level, PE technologies face another set of challenges: scalability, yield, environmental impact, health and safety, supply chain design, etc. Different PE processes are in different stages of scale-up, from the fully matured screen printing to the recently demonstrated large-scale aerosol jet printing. The overall outlook for the PE field suggests that attention to scaling-up novel PE technologies has dramatically increased in the last decade. It is reasonable to believe more and more industry scale PE manufacturing processes will be demonstrated in the near future.

22.6 APPLICATION CASES

22.6.1 Interconnections

In electronics manufacturing, an electric interconnection in a system integration step is always required before releasing products. All PE products require certain kinds of system integration. For instance, displays fabricated using PE technology always require several logic circuits, an energy supply, wireless/wired connection circuits, and other features. Photovoltaic cells require series interconnections and connectors to cables to a battery or a power conditioner. Thus, an interconnection method is one of the essential technologies for PE products [27, 28]. Several important requirements for PE interconnection technology follow:

- Low temperature: less than or equal to 80 to 130°C
- Task time: less than a few seconds to a few minutes
- Resistance: approximately 10^{-5} to 10^{-4} Ω cm
- Thermal conductivity: as better as possible

The first point is a process temperature. For PE's mass production market, most products will have low-cost substrates such as polyethylene terephthalate (PET), polyethylene naphthalate (PEN), papers, or even acrylonitrile-butadiene-styrene (ABS) resin. The heat-resistance temperatures of PET, papers, and ABS are in the 100 to 130°C, 150 to 200°C, and 80 to 100°C ranges, respectively. Task time must be short if mass production is intended with fast roll-to-roll printing. As for resistivity, even though a lower value is better for electronics, it is expected that the level of current standard interconnection materials such as solders and conductive adhesives will be kept at around $1.2^{-10} \times 10^{-5}$ Ω cm. Thermal conductivity will sometimes be required to prevent organic semiconductor devices from heating up because many of them are susceptible to heat and moisture.

For many years, interconnection methods have emerged as a result of the revolution in electronics products. Particularly in the last few decades, the most commonly used interconnection method, soldering has undergone a major transition from being lead to lead-free. Following this transition, lead-free alloys must be sought out even for PE technology. Isotropic and anisotropic conductive adhesive technologies, which can have low process temperatures, are attractive for PE products. Nevertheless, the process temperature range required for PE products is far below those of existing low-temperature interconnection materials and processes.

22.6.2 Organic Light-Emitting Diode

OLEDs have great potential for many applications such as multicolor displays, indicator lights, and signage. The major advantages of OLEDs include excellent color-tuning capability with various emission wavelengths, which are easily obtained by changing the chemical structure of the organic compounds, and good processability of the conjugated organic material.

OLEDs are conventionally fabricated by thermally evaporating organic materials in an ultrahigh-vacuum environment, assisted with common polymer processing techniques such as spin-coating [29, 30]. There are many disadvantages associated with these simple technologies, including high solution wastage and lack of patterning capability. Selective distribution of the materials over the substrate is usually not allowed with conventional manufacturing. All of these issues limit their commercial applications because essentially only one material can be used to coat the entire surface, hence the surface can only display one color [31]. The integration of multiple organic layers (to fabricate red, green, and blue emitters for color displays) would require the patterning of the individual organic layers. However, the patterning of organic materials by conventional photoresist is difficult because of their solubility in and sensitivity to aqueous solutions and many solvents.

To fully realize the potential of OLED, individual patterning is required during the deposition of the different polymers. The limitation of the conventional deposition technologies can be overcome by using the PE technology. Particularly, inkjet printing has emerged as an attractive deposition technique for organic semiconductors in optoelectronic applications. Inkjet printing does not require any chemical processes that employ wet etching of photoresist or other related material that might induce defects in the functional organic layers.

The major advantages of inkjet include

- Only a small amount of functional materials is deposited on substrates, which minimizes waste and eliminates expensive mask-making and photolithography.
- It is a noncontact technology, so it is insensitive to substrate defects.
- It can process almost any low-viscosity organic solutions.
- It is possible to build very thin films.

Because of its potential for printing on a range of nonflexible and flexible substrate types, inkjet technology has been considered to have a key role in the field of the optoelectronic applications where the manufacturing of flexible electronic devices is a dynamic, continually evolving industry that is changing the way the world interacts with electronics. In research on the development of flexible electronic technology, there is an increasing trend to replace conventional glass substrates with a polymer substrate.

The inkjet technology has some intrinsic limits, such as the nozzle clogging, the surface roughness due to overlapping drops, and bad uniformity in thicknesses due to drying processes [32, 33]. The accuracy of organic layer thicknesses is an important OLED quality index because it directly affects brightness and color consistency. Moreover, it is crucial to fabricate devices with uniform thicknesses in each layer because deformities may lead to electric currents peaks, which induce localized overheating that may gradually damage the device over time.

It is still too early to say all-organic flexible electronic devices will become dominant in the OLED industry. However, a hybrid of organic and inorganic layers is certainly achievable in the foreseeable future.

22.6.3 Ultra-High Frequency Radio Frequency Identification

RFID is a rapidly developing wireless technology used for identifying objects. RFID plays an increasingly important role in modern inventory management, asset control, and toll collection [34]. Areas of application are expected to increase in industries including health care, Internet of Things, and security. Recently, research interest regarding improvements of the performance and reliability of RFID systems has increased. In an RFID system, objects to be identified are equipped with tags consisting of an antenna and an application specific integrated circuit (ASIC). The information is stored in the memory of the ASIC. An RFID reader transmits an encoded radio signal to interrogate the tag. The RFID tag receives the message and then responds with its identification and the stored information.

UHF RFID tags use electromagnetic coupling in the far field communication, and data transfer between a tag and a reader is typically established by modulation of backscattered radiation from the tag antenna using load modulation. In an active UHF RFID system, the tag is equipped with an internal energy source, for example, a battery, whereas a passive UHF RFID tag is totally dependent on the power-carrying incident wave from the reader to the supply power to its IC chip. The read range of a passive RFID system is, therefore, much more limited compared to an active RFID system, which is considered to be one of the obstacles to wide adoption of passive RFID systems [35].

RFID provides similar functions as a bar code system, but with additional advantages:

- RFID tags can be read through some obstacles.
- RFID systems are more resistant in challenging environments such as dirt, etc.
- RFID systems allow multiple tags to be read simultaneously.
- RFID tags convey more information than a typical bar code [36].

RFID tag performance is affected by many factors. The important RFID tag read range characteristic has been widely studied [36–39]. Another important factor is the influence of electromagnetic properties of objects near, or in contact with, the RFID tag. For example, it has been shown that the tag read range and radiation patterns are strongly affected by nearby metal or water [36, 40].

Etching is typically used for manufacturing the conductive antenna trace for the RFID tag. Even though etching is generally considered an efficient process, it has many drawbacks:

- Etching is a complex manufacturing process that includes many phases.
- The chemicals used in the process are not environmental friendly.
- The substrate selection is limited by the chemical process.
- Etching creates waste and material loss when the unwanted metal is removed from the substrate.

All of those drawbacks can be addressed by PE technologies such as inkjet printing or aerosol jet printing, in which the expensive material is deposited only on desired locations. PE makes it possible to integrate electronics on new types of flexible and stretchable printed circuit board materials. Furthermore, PE technologies have the potential for providing a fast and simple one-stop method. PE technologies also allow the RFID tag antenna to be printed directly on the product package and enable tunable antenna conductor thickness. This increases the flexibility of the design of a UHF-RFID as it is not possible with the conventional etching process [41–43].

22.7 CONCLUSION

As a summary, the major benefits of PE technology are listed as follows:

- The products can be thin, lightweight and have a large functional area. PE technologies can make large products up to several tens of meters wide, which is significantly larger than those made with conventional silicon (Si)-based technology. This is an important feature for certain applications such as displays, solar panels, and lighting equipment.
- It reduces preproduction time and cost. A huge investment is required for the establishment of the production foundry of the conventional silicon-based technology. There are considerable risks associated with manufacturing short-lifetime products like cell phones, tablets, and PCs. The most advanced semiconductor foundry cannot be maintained by a single enterprise. PE-based production requires less than approximately one-tenth, sometimes even hundredths, of the investment, and the cycle time is also reduced considerably.
- It enables the ability to create wearable devices. There is an increasing demand for wearable devices. Current Si-based "wearable" devices are large, heavy, stiff, and consume power quickly. Truly wearable devices must be lightweight, thin, comfortable, and must power themselves.
- It promotes the idea of Internet of Things. In the future, all products, even pencils, will be equipped with some sort of intelligent device capable of communicating with the outside world wirelessly. Si dyes cannot be used in most of them because of silicon's high cost. PE products can be easily integrated into those products because of the flexibility in design and fabrication.
- PE technology is eco-friendly. Eventually, all electronic devices must be environmentally friendly. Eco-friendly means free of toxins and rare earth elements and requiring low energy, both in the manufacturing process and in operation. There is also a strong demand for eco-friendly manufacturing processes. The reduction of solid and liquid waste in manufacturing is a key feature of PE technology, as is the fact that it uses less energy.

More exciting developments in the field of PE are under way, which is reflected in the rapidly growing academic reports and commercialized products regarding PE. It may take a little more time to develop, on a large scale, the expected huge markets such as displays, photovoltaics, and lighting with printable OLED technology. There are many interesting issues that remain to be experimentally and theoretically addressed in this field. Clearly, the possibility of investigating the intrinsic limits of organic and oxide semiconductors will continue to interest researchers investigating new materials and further expanding our understanding of the fundamental electronic properties of printed semiconductors.

Wiring technology is also of interest. Room-temperature wiring is made possible using either the new ink or postprocessing. There is no doubt that new transparent conductive films (TCFs), which will replace the conventional ITO (indium tin oxide), will play a key role in opening the huge PE market.

Standardization is another task that needs to be addressed before PE technology can take the lead in the electronics industry. Standardization usually happens when a technology has reached a certain point in its development, resulting in a certain level of mass production in the market. PE technology has not yet reached this level. The primary aim in the standardization of PE is to help open new markets by first establishing a common platform for PE technology.

22.8 REFERENCES

1. A. Dodabalapur, "Welcome to Flexible and Printed Electronics," *Flexible and Printed Electronics*, 1 (1): 010201, 2016.
2. J. U. Meyer et al., "High Density Interconnects and Flexible Hybrid Assemblies for Active Biomedical Implants," *IEEE Transactions on Advanced Packaging*, 24 (3): 366–374, 2001.
3. E. Fortunato et al., "High-Performance Flexible Hybrid Field-Effect Transistors Based on Cellulose Fiber Paper," *IEEE Electron Device Letters*, 29 (9): 988–990, 2008.
4. E. Cantatore, *Applications of Organic and Printed Electronics*, Springer, New York, 2013.
5. A. C. Marques, J.-M. Cabrera, and C. de Fraga Malfatti, "Printed Circuit Boards: A Review on the Perspective of Sustainability." *Journal of Environmental Management*, 131: 298–306, 2013.
6. T. Baumgartner et al., *Lighting the Way: Perspectives on the Global Lighting Market*, McKinsey & Company, 2012.
7. D. Lupo et al., "OE-A Roadmap for Organic and Printed Electronics," in *Applications of Organic and Printed Electronics*, Springer, U.S., pp. 1–26, 2013.
8. S. Jung, S. D. Hoath, and I. M. Hutchings, "The Role of Viscoelasticity in Drop Impact and Spreading for Inkjet Printing of Polymer Solution on a Wettable Surface," *Microfluidics and Nanofluidics*, 14 (1–2): 163–169, 2013.
9. M. Liu et al., "Inkjet Printing Controllable Footprint Lines by Regulating the Dynamic Wettability of Coalescing Ink Droplets," *ACS Applied Materials & Interfaces*, 6 (16): 13344–13348, 2014.
10. P. Q. Nguyen et al., "Patterned Surface with Controllable Wettability for Inkjet Printing of Flexible Printed Electronics," *ACS Applied Materials & Interfaces*, 6 (6): 4011–4016, 2014.
11. X. Cao et al., "Screen Printing as a Scalable and Low-Cost Approach for Rigid and Flexible Thin-Film Transistors Using Separated Carbon Nanotubes," *ACS Nano*, 8 (12): 12769–12776, 2014.
12. M. Li et al., "Recent Developments and Applications of Screen-Printed Electrodes in Environmental Assays—a Review." *Analytica Chimica Acta*, 734: 31–44, 2012.
13. S. Khan, L. Lorenzelli, and R. S. Dahiya, "Technologies for Printing Sensors and Electronics over Large Flexible Substrates: a Review," *IEEE Sensors Journal*, 15 (6): 3164–3185, 2015.
14. J. Cen, R. Kitsomboonloha, and V. Subramanian, "Cell Filling in Gravure Printing for Printed Electronics," *Langmuir*, 30 (45): 13716–13726, 2014.
15. G. Cummins and M. P. Desmulliez, "Inkjet Printing of Conductive Materials: A Review," *Circuit World*, 38 (4): 193–213, 2012.
16. C. Ru et al., "A Review of Non-Contact Micro- and Nano-Printing Technologies," *Journal of Micromechanics and Microengineering*, 24 (5): 053001, 2014.
17. J. M. Hoey et al., "A Review on Aerosol-Based Direct-Write and Its Applications for Microelectronics," *Journal of Nanotechnology*, vol. 2012, Article ID 324380, 1–22, 2012.
18. T. Seifert et al., "Additive Manufacturing Technologies Compared: Morphology of Deposits of Silver Ink Using Inkjet and Aerosol Jet Printing," *Industrial & Engineering Chemistry Research*, 54 (2): 769–779, 2015.
19. http://www.optomec.com.
20. J. Chang, T. Ge, and E. Sanchez-Sinencio, "Challenges of Printed Electronics on Flexible Substrates," *IEEE 55th International Midwest Symposium on Circuits and Systems (MWSCAS)*, New York, 2012.
21. J. E. Anthony, "Organic Electronics: Addressing Challenges," *Nature Materials*, 13 (8): 773–775, 2014.
22. Y. Aleeva and B. Pignataro, "Recent Advances in Upscalable Wet Methods and Ink Formulations for Printed Electronics," *Journal of Materials Chemistry C*, 2 (32): 6436–6453, 2014.
23. A. Kamyshny and S. Magdassi, "Conductive Nanomaterials for Printed Electronics," *Small*, 10 (17): 3515–3535, 2014.
24. C. Sekine et al., "Recent Progress of High Performance Polymer OLED and OPV Materials for Organic Printed Electronics," *Science and Technology of Advanced Materials*, vol. 15, 2016.
25. J. Lee et al., "Register Control Algorithm for High Resolution Multilayer Printing in the Roll-to-Roll Process," *Mechanical Systems and Signal Processing*, 60: 706–714, 2015.
26. Y.-H. Chang et al., "A Facile Method for Integrating Direct-Write Devices into Three-Dimensional Printed Parts," *Smart Materials and Structures*, 24 (6): 065008, 2015.
27. D. M. Dickirson and R. E. Digiacomo, Jr, *Printed Circuit Board Interconnection*, Google Patents, 1993.
28. V. Kantola et al., "1.3 Printed Electronics, Now and Future," *Bit Bang*, 63–102, 2009.
29. J. Burroughes et al., "Light-Emitting Diodes Based on Conjugated Polymers," *Nature*, 347(6293): 539–541, 1990.
30. D. Braun and A. J. Heeger, "Visible Light Emission from Semiconducting Polymer Diodes," *Applied Physics Letters*, 58 (18): 1982–1984, 1991.
31. T. Hebner et al., "Ink-Jet Printing of Doped Polymers for Organic Light Emitting Devices," *Applied Physics Letters*, 72 (5): 519–521, 1998.
32. R. D. Deegan et al., "Capillary Flow as the Cause of Ring Stains from Dried Liquid Drops," *Nature*, 389 (6653): 827–829, 1997.
33. R. D. Deegan, "Pattern Formation in Drying Drops," *Physical Review E*, 61 (1): 475, 2000.
34. J. D. Griffin et al., "RF Tag Antenna Performance on Various Materials Using Radio Link Budgets," *IEEE Antennas and Wireless Propagation Letters*, 5 (1): 247–250, 2006.
35. Z. Tang et al., "The Effects of Antenna Properties on Read Distance in Passive Backscatter RFID Systems," *Networks Security, Wireless Communications and Trusted Computing, 2009. NSWCTC'09. International Conference on, IEEE*, New York, 2009.
36. D. M. Dobkin and S. M. Weigand, "Environmental Effects on RFID Tag Antennas," *Microwave Symposium Digest, 2005 IEEE MTT-S International*, IEEE, New York, 2005.
37. H. Rajagopalan and Y. Rahmat-Samii, "Platform Tolerant and Conformal RFID Tag Antenna: Design, Construction and Measurements," *Applied Computational Electromagnetics Society Journal*, 25 (6): 486–497, 2010.
38. Y. Tanaka et al., "Change of Read Range for UHF Passive RFID Tags in Close Proximity," *RFID, 2009 IEEE International Conference on*, IEEE, New York, 2009.
39. P. V. Nikitin and K. Rao, "Antennas and Propagation in UHF RFID Systems," *Challenge*, 22: 23, 2008.
40. P. Raumonen et al., "Folded Dipole Antenna Near Metal Plate," *Antennas and Propagation Society International Symposium, 2003. IEEE*, New York, 2003.
41. J. Virtanen et al., "The Effect of Conductor Thickness in Passive Inkjet Printed RFID Tags," *IEEE Antennas and Propagation Society International Symposium (APS)*, 2010.
42. E. Koski et al., "Performance of Inkjet-Printed Narrow-Line Passive UHF RFID Tags on Different Objects," *Antennas and Propagation (APSURSI), 2011 IEEE International Symposium on*, IEEE, New York, 2011.
43. K. Koski et al., "Inkjet-Printed Passive UHF RFID Tags: Review and Performance Evaluation," *The International Journal of Advanced Manufacturing Technology*, 62 (1–4): 167–182, 2012.

Chapter 23
Flexible Hybrid Electronics

BY
RICH CHANEY *American Semiconductor, Inc., Boise, Idaho*

23.1 INTRODUCTION

Ultrathin, conformal, and even physically flexible electronic products have been under development for years but have recently seen increased focus. Flexible hybrid electronics (FHE) are a new area of development that addresses the needs for ultra thin, conformal, and physically flexible electronics. FHE exists at the intersection of two industries: the electronics industry and the high-performance printing industry.

Printed electronics have been around for a long time with concepts shown as early as 1903 and various forms of production since then.[1] The current concepts of printed electronics are focused on using methods such as screen, flexo, gravure, offset, and inkjet, alone or in combination to print both conductive and dielectric materials on a variety of substrates. Printing is done on both sheet and roll-to-roll platforms. These printing concepts are focused on additive manufacturing techniques, where only the material desired is deposited. This is different than the subtractive techniques generally used for semiconductors and traditional PCBs, where a blanket layer of material is deposited, masked, and etched to form the desired patterns.

One of the attractions for printed electronics is that the methods mentioned are additive manufacturing processes. Eliminating the masking and etching processes holds the promise of faster, cheaper manufacturing, with some of the printing methods holding the promise of on-demand fabrication.

In order to be viable for most applications, printed systems require transistor-based functionality such as sensor readout, logic, memory, and communications that are generally provided by silicon-based integrated circuits (ICs). Traditionally packaged ICs are thick, rigid, and require attachment and connection methods that are not directly compatible with flexible printed circuits. Printed transistors have very large feature sizes are significantly larger than silicon-based technologies. Printed transistor line widths are measured in tens of micrometers while silicon transistor lines are measured in tens of nanometers. Additionally, flexible printed transistors have very low electron mobility and are orders of magnitude slower than silicon-based ICs.[2] Without the ability to integrate ICs or print IC-type functionality, flexible systems lack the features and capabilities users demand. This gap has limited the usefulness and market adoption of flexible electronics.

Flexible hybrid electronics combine the best characteristics of printed electronics and silicon ICs to create high-performance, ultrathin, and physically flexible systems.

This chapter represents a rapidly evolving space with regard to both the products produced and the manufacturing techniques used to fabricate them.

23.2 WHAT ARE FLEXIBLE HYBRID ELECTRONICS?

Flexible hybrid electronics are created by taking the best of two industries: the fast, high-density ICs from the semiconductors and the large form factor additive manufacturing printing industry. Flexible product concepts have been discussed for many years with limited product realization. Recent advances and investments are rapidly accelerating FHE. One example is the recent investments in creation of NextFlex, the U.S. Government's Manufacturing Innovation Institute supported by industry and academia. FHE will enable novel products with form factors as shown in Fig. 23.1 and will support applications that include

- Wearable health monitoring for lifestyle and fitness
- Medical health monitoring for improved health care
- Soft robotics for the elderly or wounded soldiers
- Sensor monitoring for structures, aircraft, or automotive
- Light-weight rugged sensors for harsh environments

Flexible printed electronics and silicon IC technologies each have properties and characteristics that are desirable for conformal and physically flexible products. However, none can meet the full set of requirements alone. The flexible printed electronics deliver large form factor elements such as sensors, interconnects, and antenna elements. Silicon-based technologies are needed to meet the high-performance logic, memory, and mixed-signal requirements. Conventionally packaged ICs and even thinned die are generally too thick and rigid for aggressive bending. Figure 23.2 illustrates the advantages of FleX ICs to traditionally thinned die with a FleX IC of approximate 25 μm thickness on the left and a traditional die with 280 μm backgrind on the right. Both die in the image are 2.5 × 2.5 mm on a 5-mm radius mandrel.

These limitations have been addressed by the FleX™ Silicon-on-Polymer™ process, where standard full thickness silicon-on-insulator (SOI) wafers are transformed into flexible wafers and subsequently ultrathin, physically flexible singulated die. The IC becomes flexible and ultrathin, enabling new attachment and connection methods that are compatible with flexible circuit manufacturing.

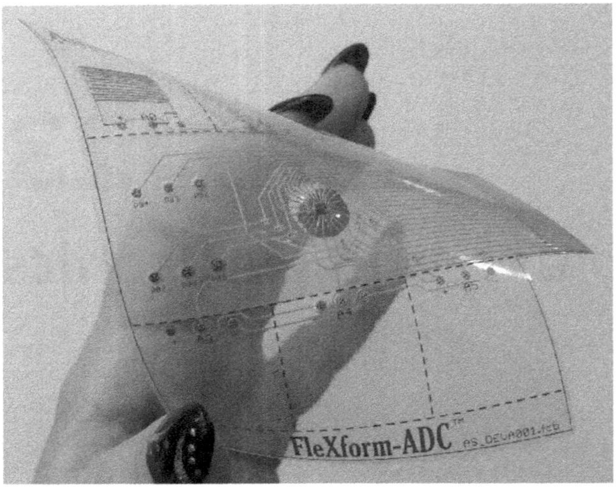

Figure 23.1 Flexible hybrid electronics.

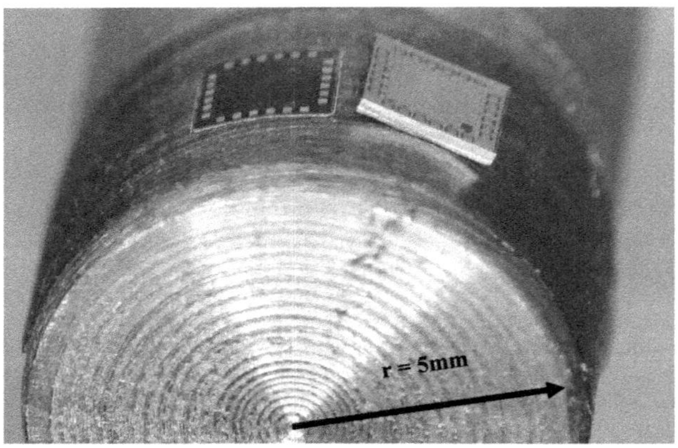

Figure 23.2 FleX IC and thinned die on 5-mm radius mandrel.

23.3 WHY ARE FLEXIBLE HYBRID ELECTRONICS NEEDED?

Flexible hybrid electronics have been proposed to revolutionize multiple markets: health care, environmental monitoring, displays and human-machine interactivity, energy, communication, and wireless networks. Their concept appears to provide a likely path to truly all-embracing electronics. The potential benefits of printed and flexible electronics include thinness, lighter weight, greater durability, and the ability for conformal integration. However, printed and flexible electronics have thus far failed to achieve widespread adoption; broad commercialization continues to be perennially "just around the corner."

What are the applications for flexible hybrid electronics? This is a common question for emerging technologies.

There are some well-defined answers. NextFlex describes FHE as the intersection of additive circuitry, passive devices, and sensor systems that may be manufactured using printing methods (sometimes referred to as printed electronics) and thin flexible silicon chips or multichip interposer structures into devices. These hybrid systems can take advantage of the power of silicon and the economies and unique capabilities of printed circuitry to form a new class of devices for IoT, medical, robotics, and communication markets. While we primarily use the term flexible, markets are interested in products that fall into the categories of flexible, stretchable, and conformable.

There are some lesser-defined answers. The author believes FHE will become pervasive in ways we do not fully appreciate yet. This is similar to the phenomena of touch displays. Touch-sensitive displays had been around for many years before the iPhone launched in 2007, but it was the iPhone that is generally credited with the explosion of touch based input devices. The Palm Pilot devices first launched in 1996 achieved great success with resistive touch screens. Before that were the Apple Newton in 1993 and the Simon in 1994, which was the first product to combine a touch screen input with a phone. Prior to that were the HP-150 and Atari 520ST computer systems launched in 1983 and 1986, respectively. Touch input goes all the way back to 1966.[3] Even though the technology had existed for a long time, it took the first "killer app" to bring widespread market adoption. Because of the iPhone, a great number of products have adopted touch displays: smartphones, computers, automobiles, toys, appliances, and more. FHE will follow a similar path. The technologies exist today but are just being integrated into early products. Once a "killer app" emerges, many products designed to be desired by consumers will be physically flexible and are likely to be hybrid between printed and silicon content.

The benefits of flexible hybrid electronics are clear. Combining printed electronics with FleX ICs creates very thin, very flexible circuits and systems in a form factor that is compatible with conformal and bendable requirements. An example can be envisioned with sensor systems. Printed sensors are inherently flexible, low cost, and can have very large footprints, up to wing or fuselage size in aircraft structural health monitoring applications. Printed sensors can be made with additive manufacturing techniques allowing customized production, potentially even in the field.

Flexible hybrid development is occurring in multiple areas of interest. Body worn medical devices have garnered interest from a variety of developers including the Nano-Bio Manufacturing Consortium (NBMC), a partnership formed by the FlexTech Alliance for the Air Force Research Laboratory (AFRL), in collaboration with a nationwide group of industry partners. NextFleX is a public-private consortium of industry, academia, nonprofits, and government agencies with a mission to advance the US manufacturing of FHE. NextFlex recently submitted a call for proposals focused on flexible hybrid array antennas and asset monitoring for time critical inventory as two specific areas of interest.

Consumer wearable devices are a natural market for FHE. Current devices such as smart watches and fitness trackers primarily consist of a "slab-on-a-strap" architecture with a box of rigid display and electronics attached to a flexible strap. Inherently flexible displays such as OLEDs or e-Ink type displays married with printed sensors and FleX ICs would yield new form factors and capabilities for an entirely new generation of wearable devices.

Similar to the applications described above, FHE is well suited to address sensor needs for the Internet of Things/Internet of Everything (IoT/IoE). These very small sensor nodes will have one overarching application requirement, which is to be extremely low power. There are several system design considerations to achieve low power. One of these is to locally process the sensor data to reduce the amount of data transmitted by the sensor system.

Without local data processing, all sensor data must be transmitted to the processing unit. Without any IC capability at the sensor node, all data must be transmitted in real time. This level of communication is relatively high powered and is detrimental to battery life of remote systems. With a minimal IC capability at the sensor node, it is possible to convert the sensor signal into digital data and queue it for burst transmission. This uses less power than constant transmission but comes at the expense of real time data.

FleX ICs provide the ability to both locally process the sensor data as well as provide a level of autonomous control. In this way, only sensor data that triggers an exception event needs to be transmitted, significantly reducing system power. In this way, remote battery powered sensor systems can greatly extend their operational life without increasing the battery capacity or energy harvesting requirements.

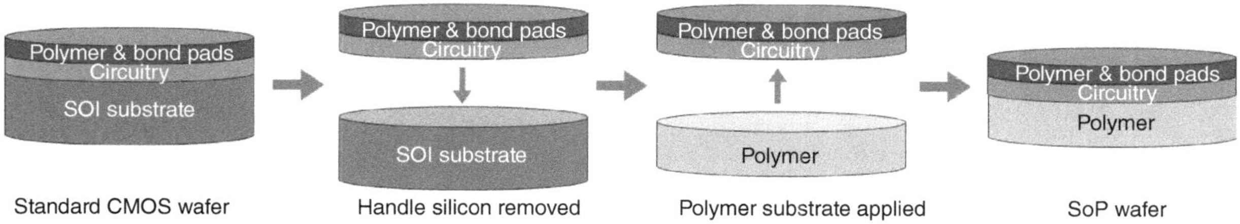

Figure 23.3 Simplified silicon-on-polymer process.

23.4 HOW ARE FLEXIBLE HYBRID ELECTRONICS MADE?

As described earlier in this chapter, flexible hybrid electronics are a combination of printed electronics and silicon-based ICs. While the FHE approach has many appealing properties, it also presents significant challenges and limitations. The absence of well-matched adhesive and conductive ink combinations makes attaching rigid packages to flexible substrates very difficult. In addition, maintaining that attachment while flex and band stresses are applied elevates the mechanical integrity challenge to levels not usually encountered in electronics. Finally, package design, especially for complicated ICs is seldom compatible with an overall flexible assembly. These challenges have to date limited the success of hybrid systems and lead innovators to seek a more fully flexible system construction.[4]

Major challenges have been overcome to enable FHE such as making silicon ICs thin enough to have an adequate level of flexibility and developing methods to fabricate reliable interconnects between the disparate technologies. Likewise, the printing side of FHE has been developing new capabilities for fine line widths and pitches that better integrate with silicon ICs.

A variety of silicon-based thin ICs are available from the semiconductor industry. Which method is used may be determined by availability or by the degree of flexibility required. The semiconductor industry has thinned die for many years. Silicon wafers are commonly 200 or 300 mm in diameter and 725 μm thick, although 150-mm-diameter wafers are still in production for many products. Wafers are commonly thinned to varying thickness as little as 100 μm. It is becoming more common for wafer grinding as thin as 50 to 70 μm, and wafers as thin as 5 μm have been demonstrated.[5] However, as wafers get thinner a variety of challenges occur.

Edge chipping is an issue with aggressive wafer thinning. The edge of a silicon wafer is rounded or beveled. As the wafer is ground thin, the mechanical strength of the wafer edge is significantly weakened.[5] During grind these weak wafer edges can chip, which causes further weakening of the wafer and leads to wafer breakage. As wafers become thinner, issues around edge chipping become more severe. Different methods exist to mitigate edge chipping such as edge trim, where a groove is cut into the outer edge of the wafer prior to grind to prevent the wafer from becoming sharp.

Wafer stress is another issue for thin wafers. The grind process can weaken the wafer, which also translated into weaker die after singulation. A variety of methods have been developed to provide stress relief for thin wafers such as chemical mechanical polishing (CMP), wet etching, dry etching, and dry polishing. Each of these methods removes the damage layer that remains after fine grinding, leaving the backside of the wafer with a mirror finish. This increases the wafer and die strength and reduces wafer and die warpage.[6]

Once the wafer is thinned, the individual die must be singulated or diced. Dicing is straightforward on thicker wafers but presents a variety of challenges for thin wafers. Similar to the edge chipping at the wafer edge, individual die can chip during the dicing process and the issue becomes worse as the wafer gets thinner. For some thin wafer applications, simply moving to a finer blade may be sufficient although this may require using step cuts instead of single pass dicing to prevent blade clogging.[7] As the wafer becomes thinner, it may be desirable to utilize dicing before grinding (DBG) methods, where the wafer is partially cut through for dicing prior to backside grinding. DBG, generally, results in improved die strength while reducing the risk of wafer breakage.[8]

As backgrind target thicknesses reduce, the wafers get more difficult to handle during processing and after the processing is complete. As wafers get thinner, it is often desirable to have a support carrier that provides mechanical strength. This may be in the form of carrier tapes, tapes on frame, or carrier substrates temporarily bonded to the thin wafer.

Novel thin wafer methodologies have been developed to eliminate the grind and singulation issues. The FleX™ Silicon-on-Polymer™ process developed by American Semiconductor converts standard silicon wafers into fully flexible wafers.[9,10] The FleX process has produced physically flexible, ultrathin, single-crystalline CMOS with multilayer metal interconnects.[9,11] With this process, CMOS is fabricated using a standard SOI process on 200 mm wafers. Then the FleX process, as shown in Fig. 23.3, completely removes the handle silicon and adds a polymer substrate to create ultrathin flexible wafers as shown in Fig. 23.4.[12] The resulting FleX wafers are singulated into individual FleX ICs. The FleX process eliminates the issues of wafer and die edge chipping and post-grind stress on the wafer and die.

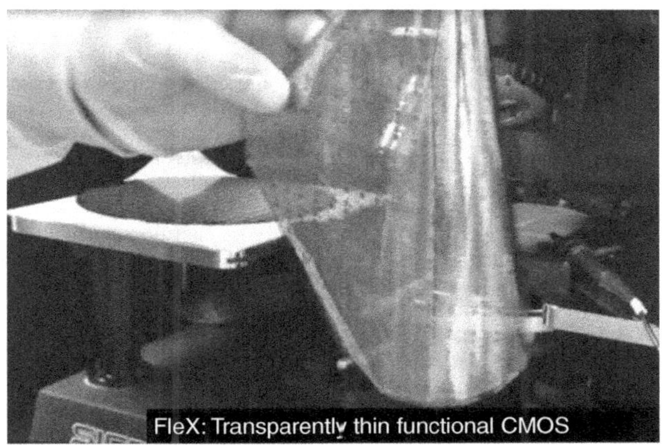

Figure 23.4 130-nm CMOS FleX silicon-on-polymer wafer.

The other side of FHE is the contribution of printed electronics. Printing has developed over centuries with inks on paper and the printing process is clearly additive through the selective deposition of materials on thin, flexible substrates.

By virtue of the available form factors, printed techniques enable electronic functionality to be delivered in locations and formats that cannot be achieved with conventional electronics packaging and assembly approaches. Printed electronics are created using standard printing techniques to apply materials to define circuit interconnects and devices. Conductive inks include printable materials that can be composed of metal nanoparticles. Substrate and dielectric materials include materials such as polyethylene terephthalate (PET). Active devices such as thin-film transistors (TFT) and sensors can also be printed using a variety of materials. Circuits, substrates, and devices can be printed using a variety of methods that include screen printing, inkjet, and a wide variety of emerging roll-to-roll (R2R) methods. To the extent that these techniques can be executed on existing printing and curing (drying) systems, the approach provides the opportunity for electronics to be manufactured at low manufacturing costs by utilizing a large installed capital equipment base.[4]

Printing electronics is not without challenge. Materials must be functionalized for conductors, insulators, and other printed content such as sensors. These materials must be compatible with the selected flexible substrate as well as properties desirable to the application such as flexibility, stability over temperature, and application compatibility. The application compatibility may vary widely such as sweat resistant and nonirritating for human worn sensors or solvent resistance for certain structural health monitoring systems.

Another issue facing printed electronics is the capability to produce continually smaller features on continually smaller pitches with increasing precision for registration between layers. Printed line widths of 160 μm on a 260-μm pitch have been achieved and are repeatable on recent demonstrations. However, to integrate higher capability FleX IC content, these dimensions need to shrink by half. High-end FHE integration will also require highly precise and repeatable registration between printed layers for multilayer circuit boards. The good news is that these areas of development are making rapid progress and will evolve quickly.

The final components of FHE manufacturing and integration are placing the thin ICs onto the flexible substrate and making electrical interconnects between the silicon and printed content.

Placing thin ICs onto flexible circuit boards (FCB) is a multistep process. First, the ICs must be picked off of a suitable carrier substrate, then placed on the FCB with a high degree of accuracy with respect to the X, Y, and theta of the printed features. This is not significantly different from what is done with standard packaged ICs and thick IC die but does pose unique challenges for extremely thin ICs. Standard die pick involves deforming the adhesive carrier material such that the die may be removed without damaging it. A variety of methods exist such as using small ejector pins to lift the die thereby deforming the tape and make release possible. These methods start to fail with thin die that crack and fracture under the stress generated. These methods are not viable with FleX ICs as the die is thinner and more flexible than the tape substrate and simply deforms around the pin(s) without releasing from the adhesive.

Fabricating the electrical interconnects between the printed and silicon content is a critical function in FHE manufacturing. The interconnects must form high-quality electrical interfaces to different types of materials while remaining physically flexible. The electrical interconnects for FHE differ from those for standard packaged or bare die ICs. Wire bonding is not desirable for FHE as wire bonds are not inherently flexible and have a relatively large z-height that must be encapsulated or otherwise protected, diminishing from the thinness and flexibility FHE target. Solder interconnects are generally not desirable as many FHE substrates are not compatible with solder processes.

FHE interconnect is an area of continued focus for manufacturing maturity and is expected to improve rapidly in a short time. Some methods have already been demonstrated such as directly printing interconnects between the FleX IC and printed substrate as shown in Fig. 23.5 on a FleX-ADC with 200 μm bond pads on a 200-μm pitch.

The direct printed interconnect method has demonstrated reliability through initial testing that is suitable for many FHE applications. Flexibility testing to a 15-mm radius of curvature demonstrates over 20,000 convex/concave bends before failure. A complex curvature axial torsion test demonstrates over 100,000 deformation cycles without failure. The failure modes of these tests show the electrical interconnects to be the issue, with some failure points visually identifiable as shown in Fig. 23.6.[13]

This initial reliability testing is interesting and promising for FHE interconnects but more must be done. Standards are lacking in this space and must be defined for what level of reliability FHE should achieve and how testing should be done.

Figure 23.5 FleX IC with direct printed interconnects.

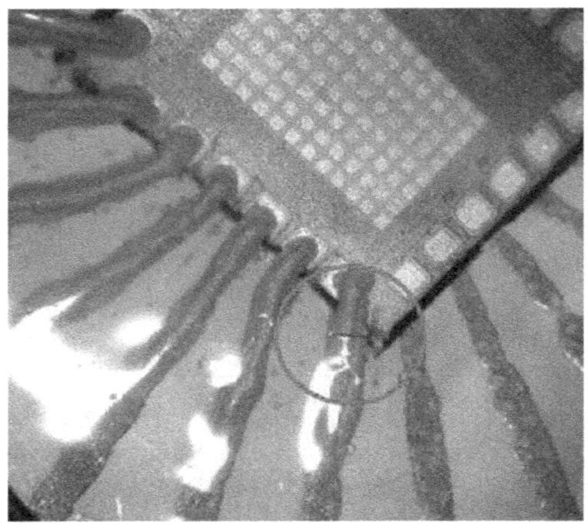

Figure 23.6 Crack in direct printed FHE interconnect after approximately 26,900 15-mm-radius concave/convex bend cycles.

Other methods for die attach and interconnect are currently in development. One area of promise is to use anisotropic conductive materials, commonly known as z-axis adhesives, that serve as both the die attach and electrical interconnect materials. There are a variety of z-axis materials that come in ink, adhesive, and film formats have a variety of cure methods such as temperature or UV exposure. The main challenge of existing z-axis materials is that they generally require some level of compression in the area to be electrically conductive. This poses an issue because IC bond pads are typically recessed. There are a variety of methods to create a positive height IC bond pads, which can be as simple as plating the pads. Once the necessary topographies are formed, the die can be attached to the FHE substrate with interconnects formed as the z-axis material is cured.

The overall goal to flexible hybrid electronics is to maintain the cost, additive nature, and relatively large form factor of printed electronics with effectiveness and efficiency. Then the hybrid approach is used for the maximum manufacturing efficiency strategy of printing what has value in printed form, and attaching integrated devices that cannot be printed.[4]

23.5 CONCLUSIONS AND FUTURE TRENDS

Flexible hybrid electronics hold the promise to revolutionize a variety of existing electronics devices and new markets by enabling physically flexible products largely produced with additive manufacturing methods. FHE is a relatively new space that will rapidly evolve and mature such that flexible products become ubiquitous.

23.6 REFERENCES

1. Circuits_100Years.pdf, www.et-trends.com/files/Circuits_100Years.pdf, *accessed 5/24/16.*
2. R. Chaney et al., "Physically Flexible High Performance Single Crystal CMOS Integrated with Printed Electronics," 2014 IEEE Workshop on Microelectronics and Electron Devices. pp. 28–31, Apr. 2014.
3. E. Johnson, "Touch Displays," U.S. Patent 3482241, Dec. 2, 1969.
4. D. Hackler et al., "Enabling Electronics with Physically Flexible ICs and Hybrid Manufacturing," *Proceedings of the IEEE*, vol. 103, No. 4, pp. 633–643, Apr. 2015.
5. Ultra-Thin Grinding, Disco: http://www.discousa.com/eg/solution/library/thin.html.
6. Stress Relief, Disco: http://www.discousa.com/eg/solution/library/strelief.html.
7. Dicing Thin Wafer, Disco: http://www.discousa.com/eg/solution/library/dicing_thin.html.
8. DGB DAF Laser Cut, Disco: http://www.discousa.com/eg/solution/library/dbg_daf.html.
9. R. Chaney and D. Hackler, "Semiconductor on Polymer Substrate," U.S. Patent 9 082 881, Jul. 14, 2015.
10. R. Chaney, D. Hackler, D. Wilson, and B. Meek, "FleX Silicon-on-Polymer: Flexible (Pliable) ICs from Commercial Foundry Processes," GOMAC, Mar. 2013, www.americansemi.com/GOMAC2013_31.2.pdf.
11. R. Chaney and D. Hackler, "Performance Electronics Integration in Flexible Technology," GOMAC, Mar. 2010, www.americansemi.com/GOMAC2010_28-18.pdf.
12. R. Chaney and D. Hackler, "High Performance Single Crystal CMOS on Flexible Polymer Substrate," GOMAC, Mar. 2011, www.americansemi.com/GOMAC2011_30-16.pdf.
13. D. Leber et al., "Electromechanical Reliability Testing of Flexible Hybrid Electronics Incorporating FleX Silicon-on-Polymer ICs," *2016 IEEE Workshop on Microelectronics and Electron Devices*, pp. 38–41, Apr. 2016.

Chapter 24
Flexible Electronics

BY
DAN XIE AND YILIN SUN *Institute of Microelectronics, Tsinghua University, Beijing, China*

Flexible electronics or circuits is a novel technology characterized by assembling electronic devices on flexible plastic substrates, such as polyethylene terephthalate (PET), polyimide (PI), or regular paper. A remarkable trend for flexible electronics to replace rigid boards has appeared due to its higher productivity and lower costs. Flexible circuits can be manufactured using identical flexible materials, allowing the board to adjust to be a desired shape or to be flexible during its use. Considering its unique flexible properties, flexible circuits exhibit great potential in various applications, including organic light-emitting diodes (OLEDs), solar cells and CMOS circuits, and so on.

24.1 FLEXIBLE APPLICATIONS

Flexible circuits are often needed in various applications where flexibility, space savings, or production constraints limit the serviceability of rigid circuit boards or hand wiring.

24.1.1 Flexible Display

OLEDs are widely used for flexible displays or making a flexible OLED display instead of a backlight. S. Ummartyotin et al. have fabricated a nanocomposite film composed of bacterial cellulose (10 to 50 wt%) and polyurethane (PU) based resin, which is utilized as a substrate for flexible OLED display.[1] The OLED-based flexible displays could achieve light emission when it is bended and stand out for many unique advantages, such as low energy cost, lightweight, surface emission, and self-luminous.

24.1.2 Flexible Solar Cells

Flexible photovoltaics are paving the way to low-cost electricity and developed as powering devices utilized in many professional engineering, such as satellite systems. To meet the demand of renewable energy, organic, inorganic, and organic-inorganic solar cells are fabricated over flexible substrates by high-throughput (often roll-to-roll printing) technologies to afford lightweight, economic solar modules that can be integrated into, not installed on, various surfaces.[2] Compared with conventional rigid solar cells, it can be intentionally rolled up and easily deployable, which makes them adapt to a variety of applications and could be embedded into wearable devices.

24.1.3 Flexible Sensors

Faced with environmental issues, smart and wireless sensor networks for environmental, health, safety and security purposes, such as the detection of hazardous gases, environmental gas monitoring, air quality control and chemical process control, become an aroused general interest.[3] As shown in Fig. 24.1*a*, interdigitated electrode structures have been fabricated on a flexible PET substrate as gas sensors. The sensors over flexible substrate contribute to the development of wearable, handheld, and portable consumer electronics.

24.1.4 Flexible Transistors and CMOS Circuits

Flexible field effect transistors, constituting the basic cell of large scale integrated circuits, executive different function as diverse logic gate elements, sensing units or memory cells. An example of flexible graphene or carbon nanotubes (CNTs)-based field effect transistors has been shown in Fig. 24.1*b* and *c*. Based on the transistor structure, significant processes in the COMS integrated circuits over flexible substrate have been made such as inverters,[4] which are the basic building block for integrated circuits. Meanwhile, flexible circuits contribute to the prevailing smart and wearable consumer electronics in cameras, personal panel computers, mobile phones, calculators or exercise monitors, and so on.

Besides, flexible circuits are also used to improve the connection problems in various applications, for example, in computer keyboards, where the flexible circuits are used as the switch matrix. Above all, flexible circuits, especially smart, wearable, and portable consumer electronics, are entering into our daily life.

24.2 KEY MATERIALS FOR FLEXIBLE CIRCUITS

The elements of the flexible circuits should meet certain requirements on the life of the product, the complexity of the process and the reliability under bending conditions. A typical flexible circuit is constituted of a number of basic material elements: substrate films (base materials), functional layers such as the conducting channel or gate dielectrics in TFT (thin-film transistors) structures, electrodes, and so on. Here, some key materials are introduced, which could be appropriately compatible with the manufacturing and operation of flexible circuits.

24.2.1 Base Materials

The base materials provide the fundamental mechanical support for the flexible circuits above, ensuring the flexibility in working condition. Under normal circumstances, the base materials maintain the basic forms of the flexible circuit and determine most primary physical and electrical properties of the flexible circuit, especially in tack-free circuit constructions, where the characteristics depend on the quality of the adopted base materials.

(a)

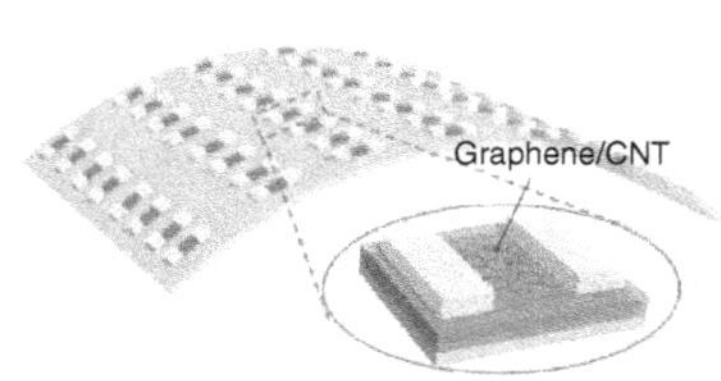

(b) (c)

Figure 24.1 (*a*) Flexible gas sensors in interdigitated electrode structures on the PET substrate; the schematic of flexible field effect transistor arrays based on graphene/CNTs (*b*) and the photograph of the fabricated devices (*c*). (© *Xie Dan. Used with permission.*)

When choosing base materials, there are several conventional considerations: the thickness of the substrate films, the mechanical strength of base materials, the temperature, and chemical stability as well as light transmittance when used in transparent flexible circuits. Although a wide range of thickness is permitted, the thickness of most substrate films are to be in a thinner size and selected in a narrow range from several tens to hundreds micrometers. Generally, thinner materials mean better flexibility. Another crucial factor is temperature and chemical stability. During the device fabrication, some temperature-dependent processes are need, which require the base materials to possess relatively high temperature–bearing characteristics. Above all, the major criteria and properties required of a suitable substrate are high-dimensional stability, good thermal resistance, mechanical bearing capacity, good electrical properties, flexibility, low moisture absorption, chemical resistance, low cost, consistency from batch to batch, wide availability.[5]

Generally, organic polymers are used as the base materials such as polyester (PET), polyimide (PI), polyethylene naphthalate (PEN), aramid papers, and various composites. PI films are widely used as the flexible substrate due to their advantageous electrical, mechanical, chemical, and thermal properties under bending conditions. Here, we conclude the related properties of these base materials as listed in Table 24.1.

24.2.2 Organic Semiconductors and Dielectrics

In 1964, the field effect in a small-molecule organic material was demonstrated.[6] Since then, the discovery of new organic semiconductors and dielectrics contributed to the development of the organic thin film transistors (OTFTs) since then.[7] The organic materials for the flexible electronics could be prepared by different deposition methods, which are compatible with the low-temperature flexible substrates. In general, organic materials could be deposited on the plastic substrates by thermal vapor methods such as pentacene or solution processing like P3HT. Several typical organic materials and their deposition methods are shown in Table 24.2.

Table 24.1 Characteristics of Typical Base Films

Name	Flexibility	Operation temperature	Tensile strength	Chemical resistance	Cost
Polyester	Excellent	~105–120°C	Excellent	Good	Low
Polyimide	Excellent	~-200–300°C	Excellent	Good	High
Polyethylene naphthalate	Excellent	~150–160°C	Excellent	Good	Low
Aramid papers	Good	~220°C	High	Fair	Moderate
Composites	Good	~150–180°C	Good	Very good	Moderate

In flexible field effect transistors (FETs), two device structures are designed and widely utilized. As shown in Fig. 24.2, one is the bottom gated and another is the top gated. To be compatible with organic semiconductor channels over flexible substrates, the gate dielectrics should also be prepared at low temperature. Usually, the organic dielectrics are made by spin-coating method such as PMMA, PVDF, etc., and inorganic dielectric oxides by ALD, sputtering, or PECVD, such as HfO_2, Al_2O_3, etc.

The roughness of gate dielectrics directly affects the performance such as mobility of organic semiconductor materials, especially for bottom-gated FETs. For example, the pentacene film structures have been found to significantly affect the performances of pentacene FETs. It has been reported that the activation energies for pentacene island nucleation and molecule re-evaporation onto the substrate surface were dramatically reduced with the dielectric surface roughness increasing.[16]

24.2.3 Two-Dimensional Materials

Since two-dimensional (2D) materials are reported, they have attracted much attention due to the unique properties and have been developed into various applications. Among these excellent 2D materials, graphene with single layered carbon atoms in a honeycomb structure stands out for ultrahigh carrier mobility up to 2×10^5 cm^2 $V^{-1}s^{-1}$ [17,18] superior mechanical strength,[19] and excellent thermal conductivity.[20] Based on such advantages, graphene has been widely used in various electronic device applications, such as semiconducting channels for high-speed transistor and logic circuits, transparent electrodes for flexible energy-harvesting devices, and displays, and sensing materials for tactile sensors, and so on.[21–23]

In recent years, transition-metal-dichalcogenides (TMDs) have been concerned after graphene.[24–27] Being a typical layered TMDs, MoS_2 has the sizeable band gap compared to the zero band gap of graphene and has been widely investigated due to their unique properties, which would be an excellent candidate in the flexible electronics applications.

Besides, some other carbon-based materials such as carbon nanotube (CNT) networks, also contribute to the development of flexible electronics due to their outstanding electrical properties and flexible behaviors. Here, some typical 2D-based electronic devices over flexible substrate have been listed as shown in Table 24.3, where they have been used as channel layer in FET structures or transparent electrodes.

24.2.4 Flexible Conductive Materials

The electrodes work as an important medium for conveying electrical signals, which have a direct impact on the reliability of the circuit. The contact failure or fracture of electrodes will cause an abnormal operation of the electronic devices, especially for flexible electronics. In order to ensure the good electrical signal transmission, the electrodes with good mechanical properties are needed. Here, typical flexible electrodes will be introduced.

Indium Tin Oxide

Indium tin oxide (ITO) is a ternary composition of indium, tin, and oxygen in varying proportions, which is transparent and colorless in thin layers and yellowish to grey in bulk form. ITO is one of the most widely utilized transparent electrodes due to its electrical conductivity and

Table 24.2 Typical Organic Semiconductors

Semiconductor	Molecular structure	Field-effect mobility (cm^2/V·s)	Deposition method	Applications
Pentacene	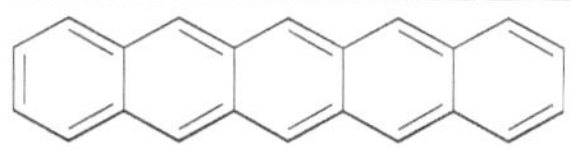	~1[8]	Evaporation	Flexible disposable DNA hybridization sensor[9]
P3HT	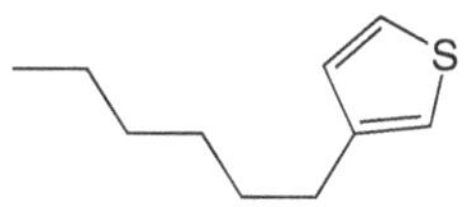	~0.1–0.3 [10]	Solution	Flexible solar cells[11]
α-6T	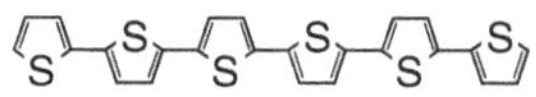	0.015 [12]	Evaporation	Flexible solar cells[13]
TIPS-pentacene	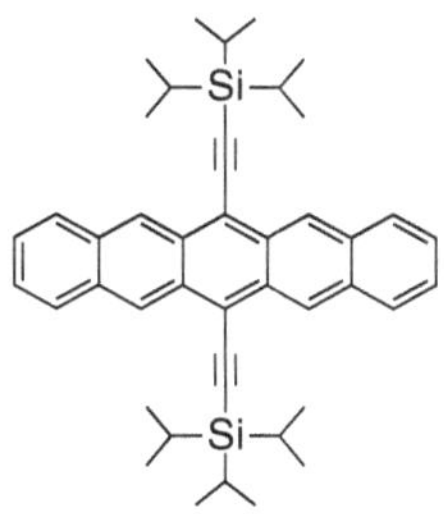	~1 [14]	Solution	Flexible gas sensors[15]

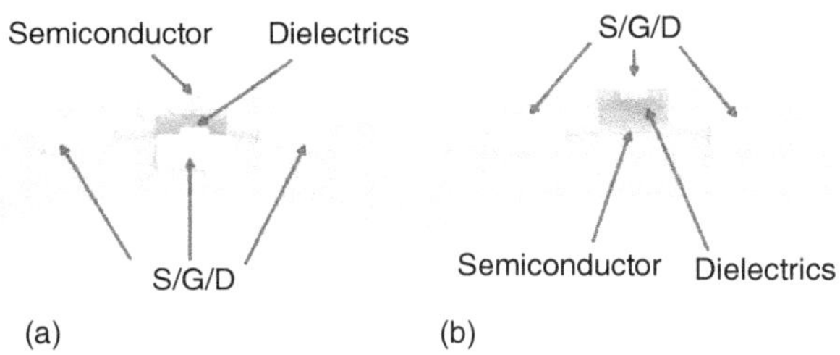

Figure 24.2 The schematic of flexible field effect transistors. (*a*) Bottom-gated FETs; (*b*) top-gated FETs. (*© Xie Dan. Used with permission.*)

optical transparency. In common, there is a tradeoff between conductivity and transparency, because increasing the thickness and increasing the concentration of charge carriers will increase the material's conductivity, but conversely decrease its transparency. Generally, a 10 Ω/□ in ITO film could be achieved with thickness of 100 to 200 nm while a 100 to 300 Ω/□ could be measured with thickness of 20 to 30 nm. Another important reason why ITO is chosen as transparent electrodes is that it

Table 24.3 2D Materials in Flexible Electronic Devices

2D materials	Functional layer	Applications	References
CVD graphene	Sensing materials	NO_2 sensors	28
CVD graphene	Channel layer	Radio-frequency (RF) FETs	29
CVD graphene and graphene oxide (GO)	Channel layer, electrodes (graphene), and gate dielectrics (GD)	All graphene-based thin film transistors	30
Exfoliated MoS_2 and CVD graphene	Channel layer (MoS_2) and electrodes (graphene)	Flexible transistors	31
CNTs	Electrodes	Biosensor	32

can be deposited as a thin film. Thin films of ITO are usually deposited on substrates by physical vapor deposition, electron beam evaporation, or a range of sputter deposition techniques.

Metal Foils

A metal foil is a conductive material, which is widely used as the conductive element of a flexible circuit. Among various metal foils with a wide range of thickness available for flexible circuits, copper foils account for the vast majority of all flexible circuit applications. The reason why copper foils are chosen is due to their excellent balance between cost and electrical properties. It is also the case that there are many different kinds of copper foil, which can be matched to specific applications. In general, copper foils suitable for use in flexible circuits are divided into two typical categories: electrodeposited (ED) and wrought (W). On the one hand, electrodeposited copper is usually applied in the applications where the minimum dynamic curvature is needed for the flexible circuits. The flexibility of such ED cooper is weaker due to its vertical grain boundary structure. These grain boundaries extend through the whole material and contribute to the fast spread of cracks or other mechanical damages, resulting in reduced electrical and mechanical performance. On the other hand, wrought copper is produced by heating and mechanically rolling ingots of pure copper to the desired thickness through rollers. Therefore, W copper is also named as rolled and annealed (RA) copper. Although the copper thickness has been strictly limited by the processes, an overlapping plate-like grain structure could be achieved. Such plate-like structure allows a longer crack propagation path compared with ED copper, which makes W copper possess a higher mechanical strength and a better stability after repeated bending operations.[5]

In the actual production of circuits, a specified alternative metal foil, such as a special copper alloy or other metal foil will be used in the construction.

24.3 MANUFACTURING TECHNOLOGY FOR FLEXIBLE CIRCUITS

The manufacture of flexible circuits is related with the initial design flexibility and in combinations of manufacturing stages for the basic types of circuit constructions. In the manufacturing of flexible circuits, some processing steps are in a high degree of similarity with the traditional manufacture of rigid boards. However, considering the large-scale production of the flexible circuits, the amenability of flexible circuits to roll-to-roll (R2R) processing makes it significantly different with that of rigid circuits.

Roll-to-roll processing, also referred as web processing, is the process of creating electronic devices on a roll of flexible plastic or metal foil. At present, this R2R processing has been widely applied in the fabrications of large-area electronics, such as solar cells, flexible screen or display, etc., which is in urgent need of cost reduction. In other active electronic circuit manufacturing, a large number of patterning steps is required by the additive processes of directly printing the active materials, shadow masking, or subtractive patterning by photolithography. All of these techniques can be adapted to web processing.[33] However, facing the demand for the miniaturization of feature size in electronic devices, the roll-to-roll photolithography and etching tools are not capable of 2 μm resolution and overlay registration. In that case, the innovations in equipment and process design, and system integration are the next important research goals of R2R processing for the flexible circuit manufacturing.

Besides, there are several special fabrication methods of flexible devices in the laboratory. Screen printing is a high-throughput process, which could be compatible with R2R processing. In fact, many crucial materials used in the fabrication of flexible circuits could be printed, such as metallic paste, organic conductors and insulators, and so on.

In order to avoid the mechanical damage during the fabrication process, a transfer technology has been developed by separating the flexible circuits from the silicon-based substrate. As shown in Fig. 24.3, the flexible substrate such as PI can be deposited on the Si-based substrate followed by a standard curing process. Then the traditional manufacturing processes of rigid boards can be applied on this composite substrate to form the electronic circuits. Finally, the flexible circuits could be separated from the Si-based substrate by mechanical stripping or chemical corrosion method. The whole fabrication process is conducted on the Si-based substrate, which helps to be compatible with conventional CMOS processes.

24.4 CONCLUSIONS AND FUTURE TRENDS

Flexible circuits offer a significant way to overcome the limit of rigid boards, which are honored with weight reduction, excellent mechanical performance and thermal compliance in diverse working environment. Particularly with the rapid development of smart and portable electronics, flexible circuits are becoming innovative solutions to a wide range of problems limited by rigid boards. However, further improvements are needed to deal with the substrate degradation during the manufacturing processes, thermomechanical stability, and compatibility among the different functional units. Commercially available R2R fabrication process

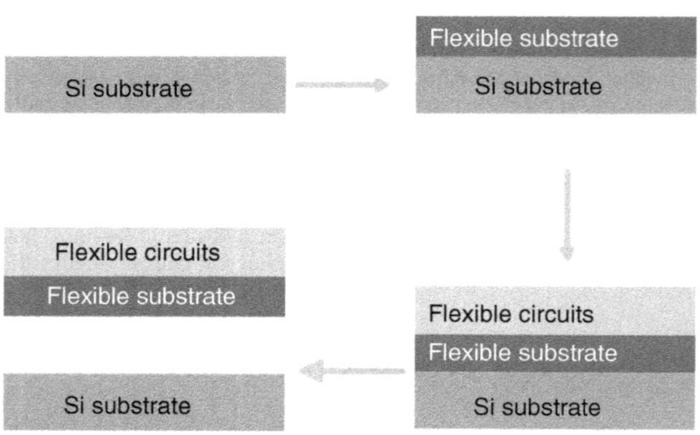

Figure 24.3 The transfer method for flexible circuit manufacturing. (© *Xie Dan. Used with permission.*)

with a high precision is the significant research goal. In conclusion, it is important and significant for the development and achievement in the flexible electronics technology, which may bring a completely new and smart life in the future.

24.5 REFERENCES

1. S. Ummartyotin, J. Juntaro, M. Sain, et al., "Development of Transparent Bacterial Cellulose Nanocomposite Film as Substrate for Flexible Organic Light Emitting Diode (OLED) Display," *Industrial Crops and Products*, 35 (1): 92–97, 2012.

2. M. Pagliaro, R. Ciriminna, and G. Palmisano, "Flexible Solar Cells," *ChemSusChem*, 1 (11): 880–891, 2008.

3. S. Claramunt, O. Monereo, M. Boix, et al., "Flexible Gas Sensor Array with an Embedded Heater Based on Metal Decorated Carbon Nanofibers," *Sensors and Actuators B: Chemical*, 187: 401–406, 2013.

4. Y. Chen, Y. Xu, K. Zhao, et al., "Towards Flexible All-Carbon Electronics: Flexible Organic Field-Effect Transistors and Inverter Circuits Using Solution-Processed All-Graphene Source/Drain/Gate Electrodes," *Nano Research*, 3 (10): 714–721, 2010.

5. P. Macleod, "A Review of Flexible Circuit Technology and Its Applications," PRIME Faraday Partnership, 2002.

6. G. H. Heilmeier and L. A. Zanoni, "Surface Studies of α-Copper Phthalocyanine Films," *Journal of Physics and Chemistry of Solids*, 25 (6): 603–611, 1964.

7. C. D. Dimitrakopoulos and P. R. L. Malenfant, "Organic Thin Film Transistors for Large Area Electronics," *Advanced Materials*, 14 (2): 99–117, 2002.

8. D. J. Gundlach, Y. Y. Lin, T. N. Jackson, et al., "Pentacene Organic Thin-Film Transistors-Molecular Ordering and Mobility," *Electron Device Letters, IEEE*, 18 (3): 87–89, 1997.

9. J. M. Kim, S. K. Jha, D. H. Lee, et al., "A Flexible Pentacene Thin Film Transistors as Disposable DNA Hybridization Sensor," *Journal of Industrial and Engineering Chemistry*, 18 (5): 1642–1646, 2012.

10. H. Sirringhaus, "Device Physics of Solution-Processed Organic Field-Effect Transistors," *Advanced Materials*, 17 (20): 2411–2425, 2005.

11. S. Y. Chuang, C. C. Yu, H. L. Chen, et al., "Exploiting Optical Anisotropy to Increase the External Quantum Efficiency of Flexible P3HT: PCBM Blend Solar Cells at Large Incident Angles," *Solar Energy Materials and Solar Cells*, 95 (8): 2141–2150, 2011.

12. W. A. Schoonveld, J. B. Oostinga, J. Vrijmoeth, et al., "Charge Trapping Instabilities of Sexithiophene Thin Film Transistors," *Synthetic Metals*, 101 (1): 608–609, 1999.

13. L. Bormann, F. Nehm, N. Weiß, et al., "Degradation of Sexithiophene Cascade Organic Solar Cells," *Advanced Energy Materials*, 6 (9): 1502432, 2016.

14. M. M. Payne, S. R. Parkin, J. E. Anthony, et al., "Organic Field-Effect Transistors from Solution-Deposited Functionalized Acenes with Mobilities as High as 1 cm^2/Vs," *Journal of the American Chemical Society*, 127 (14): 4986–4987, 2005.

15. X. Yu, N. Zhou, S. Han, et al., "Flexible Spray-Coated TIPS-Pentacene Organic Thin-Film Transistors as Ammonia Gas Sensors," *Journal of Materials Chemistry C*, 1 (40): 6532–6535, 2013.

16. H. G. Min, E. Seo, J. Lee, et al., "Behavior of Pentacene Molecules Deposited onto Roughness-Controlled Polymer Dielectrics Films and Its Effect on FET Performance," *Synthetic Metals*, 163: 7–12, 2013.

17. X. Du, I. Skachko, A. Barker, et al., "Approaching Ballistic Transport in Suspended Graphene," *Nature Nanotechnology*, 3: 491–495, 2008.

18. K. I. Bolotin, K. J. Sikes, Z. Jiang, et al., "Ultrahigh Electron Mobility in Suspended Graphene," *Solid State Communications*, 146 (9): 351–355, 2008.

19. C. Lee, X. Wei, J. W. Kysar, et al., "Measurement of the Elastic Properties and Intrinsic Strength of Monolayer Graphene," *Science*, 321: 385–388, 2008.

20. A. A. Balandin, S. Ghosh, W. Bao, et al., "Superior Thermal Conductivity of Single-Layer Graphene," *Nano Letters*, 8: 902–907, 2008.

21. N. Petrone, I. Meric, J. Hone, et al., "Graphene Field-Effect Transistors with Gigahertz-Frequency Power Gain on Flexible Substrates," *Nano Letters*, 13 (1): 121–125, 2012.

22. J. Liang, L. Li, K. Tong, et al., "Silver Nanowire Percolation Network Soldered with Graphene Oxide at Room Temperature and Its Application for Fully Stretchable Polymer Light-Emitting Diodes," *ACS Nano*, 8 (2): 1590–1600, 2014.

23. B. Zhu, Z. Niu, H. Wang, et al., "Microstructured Graphene Arrays for Highly Sensitive Flexible Tactile Sensors," *Small*, 10 (18): 3625–3631, 2014.

24. K. S. Novoselov, A. K. Geim, S. V. Morozov, et al., "Electric Field Effect in Atomically Thin Carbon Films," *Science*, 306 (5696): 666–669, 2004.

25. Q. H. Wang, K. Kalantar-Zadeh, A. Kis, and J. N. Coleman, "Electronics and Optoelectronics of Two-dimensional Transition Metal Dichalcogenides," *Nature Nanotechnology*, 7, 699–712, 2012.

26. K. S. Novoselov, D. Jiang, F. Schedin, T. J. Booth, V. V. Khotkevich, S. V. Morozov, and A. K. Geim, "Two-dimensional Atomic Crystals. Proceedings of the National Academy of Sciences of the United States of America," *PANS*, 102(30), 10451–10453, 2005.

27. K. I. Bolotin, K. J. Sikes, Z. Jiang, et al., "Ultrahigh Electron Mobility in Suspended Graphene, Solid State Communications," *Solid State Communications*, 146(9), 351–355, 2008.

28. G. Yang, C. Lee, J. Kim, et al., "Flexible Graphene-Based Chemical Sensors on Paper Substrates," *Physical Chemistry Chemical Physics*, 15 (6): 1798–1801, 2013.

29. N. Petrone, I. Meric, T. Chari, et al., "Graphene Field-Effect Transistors for Radio-Frequency Flexible Electronics," *Electron Devices Society, IEEE Journal of the*, 3 (1): 44–48, 2015.

30. S. K. Lee, H. Y. Jang, S. Jang, et al., "All Graphene-Based Thin Film Transistors on Flexible Plastic Substrates," *Nano Letters*, 12 (7): 3472–3476, 2012.

31. J. Yoon, W. Park, G. Y. Bae, et al., "Highly Flexible and Transparent Multilayer MoS_2 Transistors with Graphene Electrodes," *Small*, 9 (19): 3295–3300, 2013.

32. Y. T. Chang, J. H. Huang, M. C. Tu, et al., "Flexible Direct-Growth CNT Biosensors," *Biosensors and Bioelectronics*, 41: 898–902, 2013.

33. William S. Wong and Alberto Salleo (eds.), *Flexible Electronics: Materials and Applications*, Springer Science & Business Media, New York, 2009.

24.6 FURTHER READING

Rim, Y. S., S. H. Bae, H. Chen, et al. "Recent Progress in Materials and Devices toward Printable and Flexible Sensors," *Advanced Materials*, 28 (22): 4415–4440, 2016.

Chapter 25

RF Printed Electronics: Communication, Sensing, and Energy Harvesting for the Internet of Things and Smart Skin Applications

BY

BIJAN K. TEHRANI AND MANOS (EMMANOUIL) M. TENTZERIS *Georgia Institute of Technology, Atlanta, Georgia*

25.1 INTRODUCTION

The current trend of emerging wireless technologies is pushing toward the miniaturization of wireless devices along with the simultaneous enhancement of on-demand application-specific functionality. Throughout the past few decades, the driving force behind this advancement has been the maturity of active semiconductor-based devices. Operational frequencies are pushing into new untapped spectra while noise levels are dropping, bandwidth is increasing, and modulations schemes are multiplying in intelligence.

As the production of wireless consumer devices continues to follow ever-growing market demands and emerging technological needs, attention must be drawn to the means by which this massive volume of devices is realized, specifically the industrial processes used to manufacture these systems. Up to now, most industrial standards for the fabrication of wireless systems have been based around subtractive manufacturing processes, fundamentally the consecutive addition, patterning, and removal of material from a host. Materials are deposited in bulk and patterned through photolithography, where sacrificial material is then removed, followed by additional tasks of etching, stripping, and polishing to prepare for the next processing step. With each iteration of the lithographic process, material waste is created typically in the form of harmful etchants and environmentally detrimental by-product solutions. For this reason, any reduction in the use of subtractive processes is considered a major step in the direction of reduced waste, cost, and processing complexity.

With the current and predicted scale of the production of electronic systems, additive manufacturing methods are gaining an increasing interest as an alternative to subtractive methods for the reduction of material waste and production cost in the manufacturing of the next generation of robust, flexible, easy-to-scale, reconfigurable wireless systems. In this chapter, the processes and materials of additive printing fabrication technologies will be discussed, outlining the current achievements, limitations, and future direction of the technology as a whole. The integration of printing technology with wireless radio frequency (RF) systems will be presented, highlighting integral components such as printed antennas, sensors, and energy harvesting sources. Practical real-world applications will be presented for each of these components, including real-time environmental gas monitoring and self-powered wearable electronics, demonstrating the many enabling features of additive printing technologies for wireless applications.

25.2 PRINTING PROCESSES AND MATERIALS

The processing schemes and material considerations behind additive printing methods are directly responsible for the efficient and scalable fabrication of low-cost, highly robust electronic devices. When compared to standard subtractive processes, additive methods offer a considerable reduction in discrete processing actions; however, careful attention must still be paid in order to adhere to existing standards for practical production-level efficiencies. Parameters such as curing temperature and ultraviolet (UV) cross-linking energy are to be shared with that of standard microelectronic fabrication schemes. Being a purely additive process, electronic printing methods additionally highlight several material characteristics, such as ink viscosity, surface energy, composite particle volume, and glass transition temperature. The following section will outline several of the key processing and material considerations used with inkjet and 3D printing methods for electronic fabrication.

25.2.1 Inkjet Printing

The most widely used method of inkjet printing for electronic fabrication is the piezoelectric drop-on-demand method. This technology involves the expulsion of ink droplets from a cartridge through the actuation of a piezoelectric membrane in a selectively patterning fashion. Ink materials must adhere to certain ranges of viscosity, surface tension, and composite particle diameter to achieve jettability in a piezoelectric-driven system. Parameters such as ink temperature, jetting voltage, and cartridge back pressure have to be controlled in order to allow for the efficient and reliable expulsion of ink droplets onto a host substrate. Once a droplet is expelled, its interaction with a host substrate and the surrounding ink droplets must be considered. The total free surface energy of the host substrate must be higher than the surface tension of the ink droplet to ensure good wetting between the two phases to discourage the ink from "balling-up"; however, too great of a differential can cause potentially undesired ink migration or spreading. Typically a difference of approximately 10 mN/m between the surface tension of ink and the surface energy of the substrate is desired to achieve good wettability [1].

Standard materials for the most common electronic inkjet printing processes can be separated into three general categories: conductive, dielectric, and sensing. Conductive materials are typically composed of suspensions of metallic nanoparticles in a low-viscosity solvent solution. Noble metals, such as silver and gold, are preferred for nanoparticle-based inks because of their high resistance to oxidation in the ambient processing conditions that are typically used in inkjet printing environments. Furthermore, metallic nanoparticle inks are advantageous for low-temperature processing with thermal sintering at temperatures less than 200°C, as well as options for sintering at ambient temperature with laser-based and photonic solutions. Dielectric inks are commonly realized with organic polymer solutions with thermal or photoinitiated cross-linking agents. Polymeric dielectric inks are typically classified by their per-layer printed thickness, where a PVP polymer-based ink can realize films with thicknesses in the range of hundreds of nanometers and an SU-8 polymer-based ink can achieve films with thicknesses in excess of one hundred micrometers [2]. In fact, dielectric materials that are well established and characterized within standard microelectronic fabrication processes, such as the two aforementioned polymers, are popular targets for adoption with inkjet printing systems in order to facilitate integration with existing production-level technologies. Sensing inks are composed of materials that intentionally respond to certain environmental stimuli, such as temperature, strain, humidity, and the presence of a gas. Many popular solutions for sensing inks consist of dispersions of carbon nanomaterials, such as graphene oxide flakes and carbon nanotubes (CNTs). With these inks, nanomaterials are suspended within an aqueous solution of solvents and surfactants to allow for low-temperature processing and good wetting with host substrates. The nanomaterials within these inks have the potential to be easily functionalized to allow for selective sensitivity to a desired sensing target.

25.2.2 3D Printing

In general, a printing method is considered to be "3D" when the printable scales of all three dimensions are within the same magnitude. The most active areas of development with both consumer and industrial-grade 3D printing methodologies can be categorized into the following technologies: fused deposition modeling (FDM), aerosol jet, and stereolithography (SLA). FDM printing methods involve the vector-based deposition of thermoplastic materials through the heating of a filament beyond its glass transition temperature and then extrusion through a small nozzle in a layer-by-layer fashion. Aerosol jet technology involves the atomization of an ink material, which is then focused into a narrow beam with a sheath gas and deposited onto a host substrate, offering a wide range of possible ink material characteristics, such as viscosity and particle size, for high resolution 3D printing. SLA printing utilizes the selective photonic exposure and cross-linking of a photopolymer solution to create a solid 3D object, where resolution is determined by the minimum increment of the platen stepper motor and the resolution of the projection light or laser source.

As with the methods of the 3D printing, the materials utilized throughout cover much larger characteristic ranges than those used with inkjet printing. FDM printing typically focuses on the thermomechanical extrusion of solid plastic materials, which includes popular consumer-grade thermoplastics like polylactic acid (PLA) and acrylonitrile butadiene styrene (ABS); however, ceramic composites and other high-viscosity materials can be printed through syringed-extrusion methods with the same FDM motor control system. Aerosol jet systems have the advantage to print a wide variety of organic and nanoparticle-based materials due to the atomization process of the ink materials; however, surface energy considerations similar to those of inkjet printing must be addressed in order to allow for good wetting and adhesion of the aerosolized ink to the host substrate. SLA printing materials consist of photopolymer resins that react to exposures of UV and near-UV radiation (300 to 400 nm), typically composed of a solution of photoinitiators, monomers, and oligomers.

25.3 PRINTED RF APPLICATIONS

Wireless technologies have become a staple of modern society, where connectivity is often considered a necessity in any given environment. Four integral components of a practical wireless system are antennas, interconnects, sensors, and power sources. Antennas are integral components that are responsible for capturing and radiating electromagnetic waves in an efficient fashion necessary for wireless operation. Interconnects are responsible for interfacing the components of a wireless system with a focus on preserving the integrity of the RF signals throughout. Sensors are used to interface a system with an ambient environment, whether it be the strain on a support of a bridge, the humidity of a moisture sensitive environment, or the quality of the water in a public utility distribution network. Finally, the peripheral components of a wireless system must be powered through an efficient means that will not greatly degrade the autonomous nature of a wireless solution. Additive printing technologies enable each of these wireless components in unique ways, where fully printed vertically integrated multilayer structures are combined with low-temperature processing techniques for novel materials and topologies unachievable with conventional lithographic fabrication processes [3, 4].

In the following sections, the integration of additive printing fabrication technology with these fundamental wireless components will be discussed in detail with the goal of realizing printed wireless systems with a reduction in fabrication cost and an increase in design robustness and reconfigurability for a broad range of emerging wireless applications, such as the Internet-of-Things (IoT), vehicle-to-X (V2X) connectivity, smart skins, and smart cities.

25.3.1 Antennas

An essential component of any wireless system, the antenna is responsible for converting the radio signal from the wireless transceiver to a radiating electromagnetic signal, and vice versa. In conventional wireless technologies, an antenna is fabricated on a low-loss laminate material, such as a printed circuit board (PCB), where it is limited to the standard milling or lithography techniques typically employed for the fabrication of the remainder of the circuit [5]. Through the utilization of additive printing techniques, antenna structures can be integrated with wireless transceivers in ways that are highly reconfigurable, offer a reduced spatial footprint, and drastically reduce lossy interconnections' lengths.

One proof-of-concept demonstration of this technology is the system-on-package (SoP) design scheme. In an SoP design, antenna structures and other peripheral components are integrated with a wireless transceiver directly within the same external package, reducing interconnect lengths and overall device footprints. An example of one of the first reported inkjet-printed on-package antennas is shown in Fig. 25.1 [6]. In this effort, a 30-GHz patch antenna structure is fabricated directly on top of the dual inline package (DIP) molding of an

Figure 25.1 Fully inkjet-printed on-package 30-GHz patch antenna (left) perspective view and (right) top view [6].

integrated circuit (IC). Utilizing silver nanoparticle and SU-8 polymer inks, a ground plane, a thick (> 100 μm) RF substrate, and one patch antenna topology are all additively manufactured in a low-temperature postprocessing fashion. The inclusion of a printed ground plane allows for on-demand electrical isolation from the molding compound and from the RF circuitry below the antenna if it is a desired design feature. Though this demonstration does not directly interconnect the antenna to the packaged IC, integration can be made possible through the use of through-package vias (TPVs) and wireless coupling.

Through the selective patterning of thick dielectric inks and thin metallic traces, inkjet printing is used to realize multilayer structures reaching into the vertical (third) dimension, often referred to as 2.5D technology. This technology is useful in the fabrication of high-frequency printed interconnects on a chip-level scale. A recent work demonstrates for the first time the fabrication of 3D first-level RF interconnects interfacing an IC die to a packaging substrate as an alternative to traditional wire bond interconnects [7]. In this design, shown in Fig. 25.2, an SU-8 polymer ink is inkjet printed to pattern two features: a die attach adhering the IC die to glass the packaging substrate and dielectric ramps bridging the surface of the packaging substrate up to the surface of the die. Finally, inkjet-printed silver nanoparticle inks are used to pattern coplanar waveguide (CPW) transmission line interconnects yielding low parasitic inductance and a small insertion loss of 0.6 to 0.8 dB/mm at 40 GHz. These interconnects can be used to integrate various ICs, sensors, passives, and antennas into a single package as a fully printed wireless SoP solution.

While inkjet printing can be used to realize planar and multilayer antenna structures, there is a great interest in the realization of truly 3D electronic structures through additive manufacturing. In contrast to conventional planar and 2.5D laminate-based antennas designs, 3D antenna designs offer an additional degree of freedom to be taken advantage of for the implementation of advanced wireless solutions including enhanced omnidirectional radiation and 3D beam steering capabilities. Furthermore, an additional degree of freedom can be added to an antenna structure: physical reconfigurability. With the ability to physically change the shape of an antenna structure, a reconfigurable or "origami" (4D printed) wireless system can furthermore change in real time its effective electrical shape for application-specific operation and deployment.

This physically and electrically reconfigurable technology has been recently utilized in the development of 4D (3D foldable) wireless systems. In order to highlight the advantages of these advanced systems, a 2.3 GHz origami wireless system package is realized using a combination of inkjet and 3D printing technologies, shown in Fig. 25.3 [8]. The premise of this work begins with the design of a 3D wireless system with multidirectional radiation capabilities for diversity-enhanced wireless energy harvesting and communication, realized through a 3D cube-shaped package. In order to reduce the system cost, material volume, and fabrication time of the system, a planar cross-shaped shell is printed with the goal of folding into a complete 3D cube package. Foldable hinges are realized through the use of a 3D-printed thermoset shape memory polymer (SMP). When the printed SMP hinges are

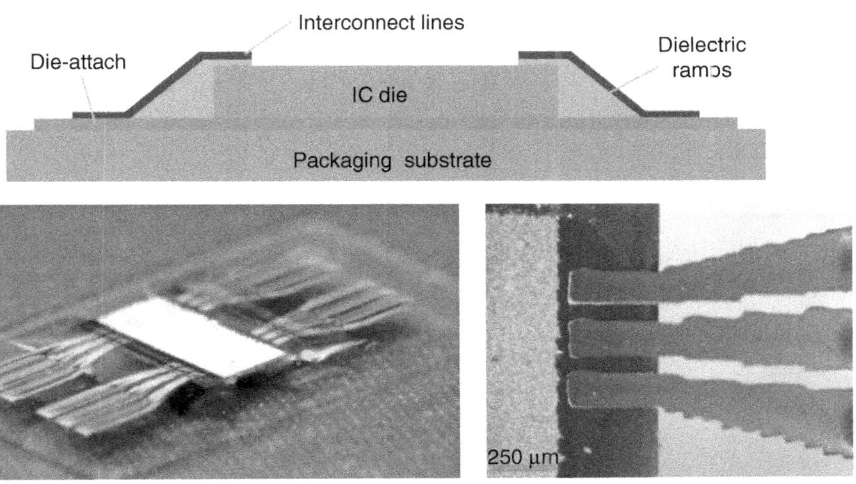

Figure 25.2 Fully inkjet-printed millimeter-wave CPW interconnects for SoP solutions: (top) diagram of packaging material stack-up and (bottom) inkjet-printed 3D interconnect samples demonstrating low insertion low beyond 40 GHz [7].

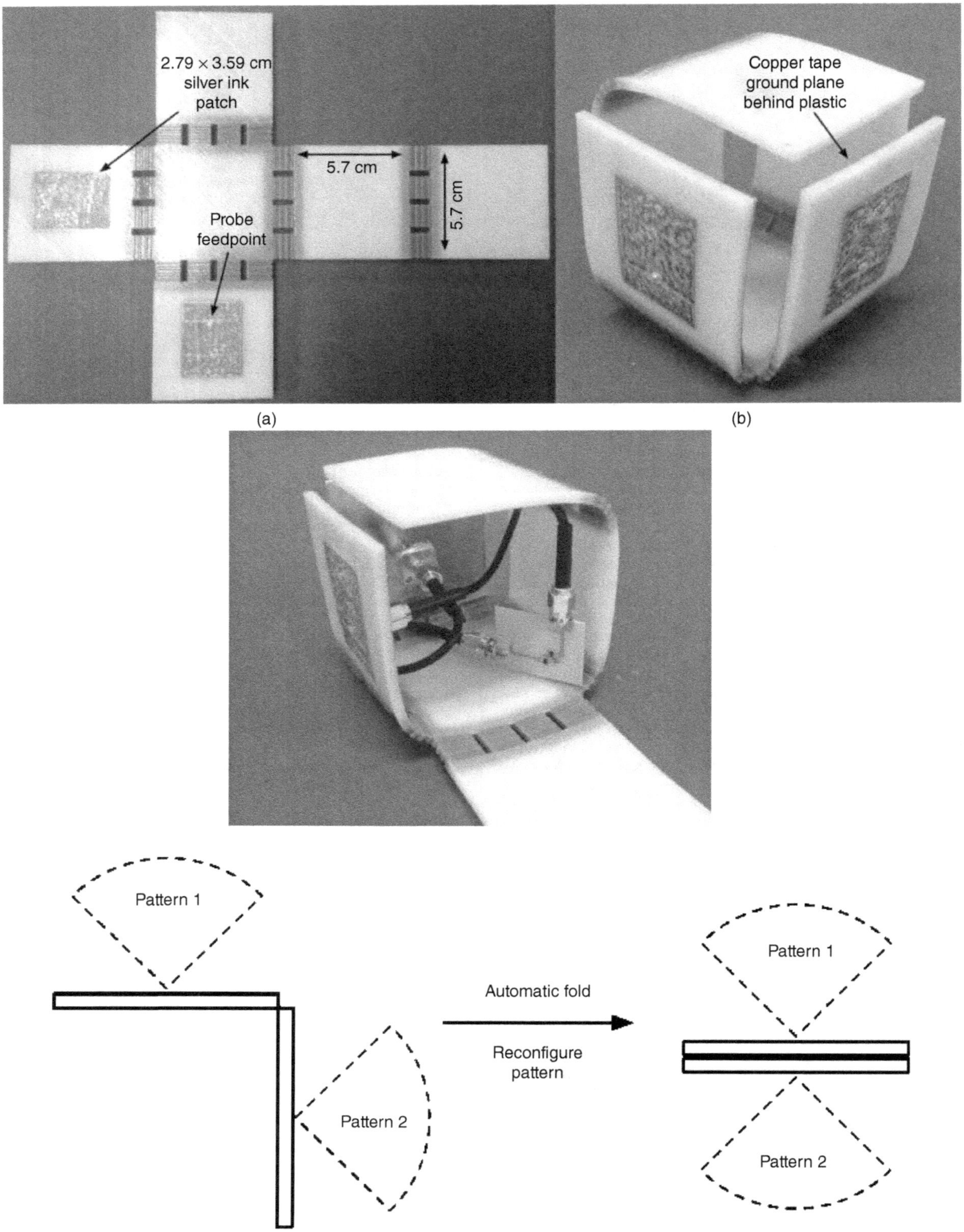

Figure 25.3 Fully printed reconfigurable "origami" wireless system package featuring a reconfigurable multiantenna diversity realized through 3D and inkjet printing fabrication technologies [8].

heated past their glass transition temperature of 55°C, they become bendable, allowing the planar shell to be folded into a cubic shape. Once heat is removed from the hinges, the rigid integrity of the hinges returns, realizing a fully printed thermally controlled origami wireless system package.

In addition to the advantages of fabrication time and cost, this novel enclosable 4D system allows for the easy integration and shielding of internal electronic systems, shown in Fig. 25.3. In this specific demonstration reported in this chapter, internal wireless energy harvesting circuitry is directly integrated with the patch antennas on the outer surface of the origami cube, providing diverse self-powering solutions for zero-power sensing and communication technologies.

25.3.2 Sensing Platforms

Environmental sensors are already widely utilized for a variety of both consumer and industrial applications, including personal biomedical health monitoring, structural mechanical integrity monitoring, climate condition monitoring, and hazardous multiphase material detection. Ideally, these sensors would be distributed in a ubiquitous fashion throughout a targeted environment to allow for optimum broad-range monitoring and would be critical for the development of the first real-world IoT implementations. However, conventional sensing technology either requires the use of wired electrical connections interfaced with a central sensing platform or bulky wireless systems requiring an external power supply, such as a battery. As a solution, low-cost wireless sensing platforms can be realized through the integration of passive sensing mechanisms and novel ink materials taking advantage of the additive nature of inkjet and 3D printing fabrication technologies.

The growing field of microfluidic systems is a branch of traditional microelectromechanical systems (MEMS) that involves the channeling of microliter-range volumes of fluid for numerous applications such as tunable electronics, semiconductor chip-level heat passivation, and sensing. From a sensing standpoint, microfluidics has highlighted advantages for biomedical and environmental fluidic sensing, where small samples of a target environment can be sampled and analyzed through a miniaturized lab-on-a-chip solution to reduce target contamination and waste. Whereas these systems typically rely on standard lithography-based MEMS fabrication processes, recent efforts have demonstrated the feasibility of realizing low-cost flexible microfluidic systems through the use of inkjet printing technology [9]. One highlighted work, shown in Fig. 25.4, utilizes inkjet printing for two features of a microfluidic system: the patterning and molding of the microfluidic channel and the patterning of the metallic conductor for the sensing resonators [10]. When various liquids with varying constants of relative permittivity are loaded into the microfluidic channels interfacing the resonators, the resonators detune and the frequency of the RF structure experiences a shift, which can be quantified and used a sensing parameter.

Inkjet printing is used to fabricate a soft lithography mold on a glass substrate which is then used to pattern a flexible polydimethylsiloxane-based (PDMS) microfluidic channel. The physical flexibility of the channel allows the system to be bent and conformed to a variety of hosts, including the human body. Metallic nanoparticle-based ink is then printed onto a flexible polyethylene terephthalate (PET) substrate to realize the conductive pattern of the spiral-shaped slot resonators of the system. The PDMS channel mold is then joined with the resonator substrate in a replaceable fashion for "peel-and-replace" capabilities, an advantageous feature for sensitivity switching and reusability. This microfluidic sensing platform is used to sense such liquids as hexanol, ethanol, glycerol, and water, achieving a resonance shift of 43.8 percent when an empty channel is filled with water while maintaining its operational efficiency through flexing tests with a range of various radii of curvature down to 7 mm, shown in Fig. 25.5.

Another area of interest for practical sensing applications lies in the development and fabrication of gas sensing platforms for real-time environmental monitoring and quality control. Recent attention has been directed toward the use of carbon nanomaterials, such as graphene and carbon nanotubes (CNTs), as a highly responsive gas sensing material. One of the most pronounced benefits of using CNT sensing films is the ability to chemically functionalize the nanotubes, allowing for target selectivity in a typical multi-sensing-agent environment. The most common process currently used to fabricate carbon nanomaterials is chemical vapor deposition (CVD), where the materials are grown on a metal catalyst. However, due to the fact that this process requires a thermal profile reaching up to 1000°C, the catalog of potential host substrates is greatly restricted, excluding popular low-cost flexible organic substrates. In order to realize these carbon nanomaterial films on flexible substrates, inkjet printing is an attractive technology.

Carbon nanomaterials, such as graphene and CNTs, can be formulated into inks and printed for the low-temperature fabrication and processing of efficient nanomaterial sensing films. A recent effort demonstrates the development of two fully inkjet-printed gas sensing platforms with these novel materials: reduced graphene oxide-based (rGO) ammonia sensors and CNT-based dimethyl methylphosphonate (DMMP) sensors [11]. The sensing component consists of a nanomaterial film printed onto a flexible polyimide host substrate interfaced with an array of interdigitated conductive electrodes, where the printed sensing film forms a resistive path between the electrodes, shown in Fig. 25.6. When a gas is introduced to the sensor, the resistivity of the sensing film changes with a certain slope dependent on the concentration and duration of exposure. The rGO-based ammonia sensor was able to achieve a sensitivity of 2.8 percent/10 ppm, demonstrating a new state-of-the-art performance within the field of inkjet-printed rGO-based ammonia sensors. This work also introduced for the first time an inkjet-printed solution for DMMP sensing with a functionalized CNT film. For both the rGO- and the CNT-based sensors, high sensitivities

Figure 25.4 Additively manufactured microfluidics-based "peel-and-replace" RF fluid sensor for wearable applications, utilizing inkjet printing patterning and soft lithography with an inkjet-printed mold [10].

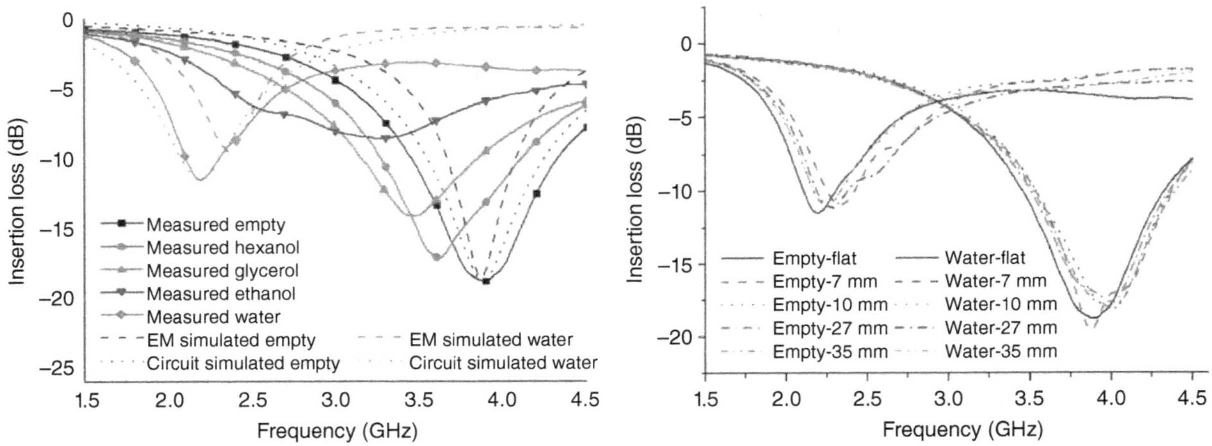

Figure 25.5 (Left) Measured and simulated insertion-loss parameters for the additively manufactured microfluidic sensor with different fluid channel fills and (right) measured insertion-loss parameters with different radii of curvature [10].

are reported with a very short response time (< 5 min), which is a valuable metric for real-world implementation in practical scenarios of real-time monitoring. Further investigation is taking place into the characterization of the RF behavior of these printed nanomaterial sensing films for utilization in RF sensing platforms [12]. The potential to integrate these fully printed sensing films into wireless systems allows for the realization of flexible wireless gas sensing platforms with drastically reduced cost and increased functionality as well as for the implementation of real-world multi-sensing smart skins.

25.3.3 Energy Harvesting Systems

The fundamental goal of wireless energy harvesting is the efficient conversion of wireless electromagnetic radiation into a stable, useable source of power. In contrast to wireless power transfer techniques,

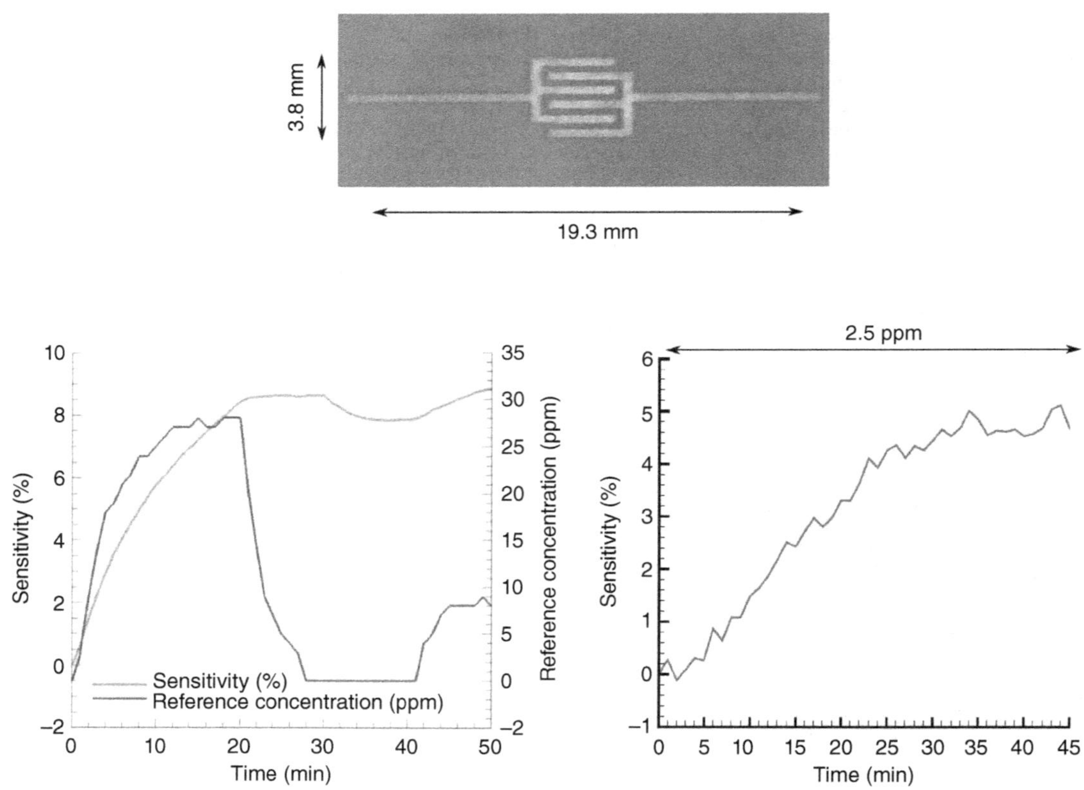

Figure 25.6 (Top) Fully inkjet-printed carbon nanomaterial-based sensor on flexible polyimide substrate and experimental sensitivity measurements for (bottom left) rGO-based ammonia sensors and (bottom right) CNT-based DMMP sensors [11].

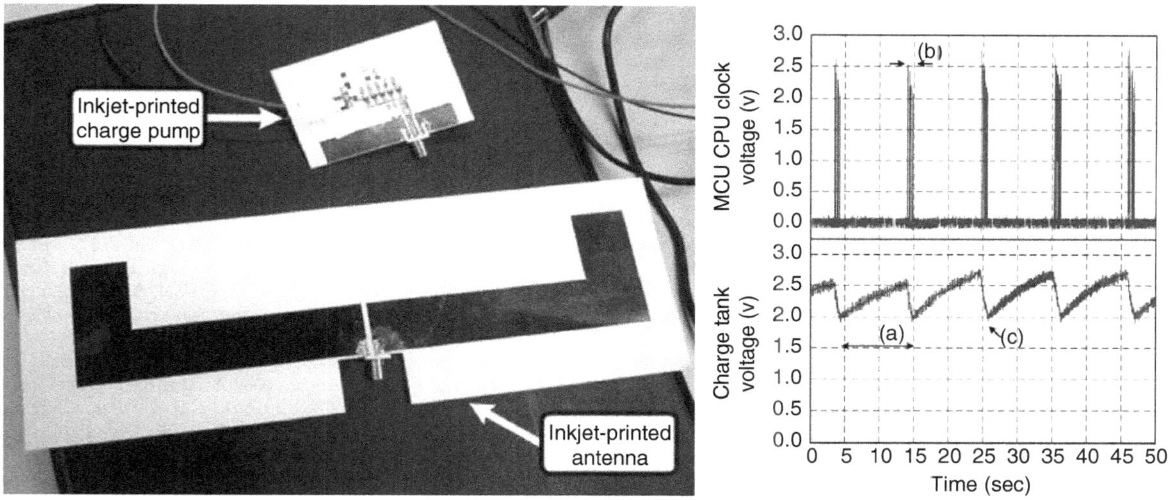

Figure 25.7 (Left) Inkjet-printed charge pump and antenna for ambient far-field wireless (TV) energy harvesting and (right) measurements from an experimentally implemented harvesting system in a real urban environment (Atlanta, GA) used to autonomously power a microcontroller system [13].

wireless energy harvesting systems target unintentional, or "ambient," electromagnetic radiation, for example, from communication and broadcast sources like radio, over-the-air television, and Wi-Fi. Due to the fact that the energy received from these wireless sources is typically very small (in the order of microwatts), each component of the energy harvesting system must be carefully tuned for maximum efficiency in order to allow for practical implementation as a wireless system power source. Energy harvesting technology has the ability to enable the coming generation of fully autonomous ("zero-power") wireless systems for IoT, where wireless sensor and communication nodes are distributed ubiquitously throughout an environment to realize universal connectivity. Inkjet printing technology allows for energy harvesting systems to be developed in a low-cost, highly efficient fashion with a focus on reducing the environmental footprint of typical lithographic processes.

The methodology of wireless energy harvesting systems can be divided into two electromagnetic classifications: near field and far field. Near-field harvesting systems rely on the proximity coupling of coil antenna structures to harvest wireless radiation from a nearby source, such as a handheld gadget (cell phone, walkie-talkie, tablet, etc.). Far-field harvesting systems are designed to operate at variably large distances from a radiating source (several kilometers), where the power of wireless signals and the magnitude of voltage collected at an antenna are relatively low. Because of the resulting low power and voltages of far-field energy harvesting, these systems rely on networks of capacitors and diodes, known as charge pumps, to convert low-voltage AC signals to a DC voltage of a useful magnitude. These integral charge pump circuits can be realized through the integration of lumped surface mount components, inkjet-printed circuitry, and recyclable organic substrates outlined in a recent effort [13]. This work focuses on the development of ambient energy harvesting systems for self-sustaining wireless sensor platforms, highlighting the use of inkjet printing to realize low-cost yet robust systems. The target frequency range is one used for digital TV broadcast (512 to 566 MHz), a collection of bands that is commonly populated in most urban areas although it can be easily extended to higher bands, such as Wi-Fi. In the demonstrated proof-of-concept prototype shown in Fig. 25.7, both the charge pump and the harvesting antenna are realized with inkjet printing on paper substrates. With the system fully implemented, the harvested ambient power is used to fully power a microcontroller programmed to temporarily wake up and report.

In contrast to far-field energy harvesting systems, near-field energy harvesting systems operate much closer to a radiating source (within meters) and as a result are capable of gathering power in the range of milliwatts. One emerging application of these relatively close-proximity systems is their use in wearable electronics, where a conformal wireless system can be powered through stray radiation from another personal wireless device. A recent work highlights this novel methodology, where near-field energy harvesting systems are designed and optimized to gather energy from a handheld two-way talk radio operating at 464.55 MHz [14]. This harvesting system is implemented using an inkjet-printed near-field coil antenna and circuit with integrated surface mount components to reduce fabrication cost and allow for application-specific prototyping, shown in Fig. 25.8. The scavenged power from

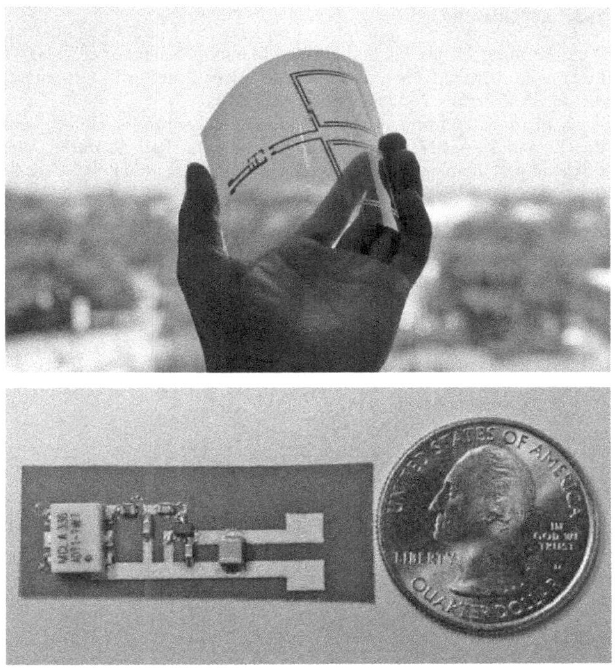

Figure 25.8 Inkjet-printed near-field energy harvesting system on a flexible polyimide substrate for wearable harvesting applications [14].

the communicating radio could be used to power peripheral electronic devices worn by the holder of the radio, for example, powering head-up display (HUD) goggles for a soldier in the field, activating biological monitoring sensors, or reporting GPS coordinates to a central base.

25.4 CONCLUSION

Throughout the various emerging fields of wireless electronics, printing technology has established itself as an efficient means of fabricating advanced wireless systems in a low-cost, environmentally advantageous, and highly robust fashion. Beginning with the most integral components of any wireless system, fully printed antenna structures are fabricated through purely additive multilayer processes to realize novel 3D/4D reconfigurable system-on-package solutions for the next generations of wireless technology to come. Novel selective wireless sensing platforms utilizing microfluidic MEMS and carbon nanomaterial films are fabricated with inkjet printing processes outside of a cleanroom, allowing for the feasible implementation of ubiquitous wireless sensing networks over a broad range of environmental and biomedical monitoring applications. And finally, the last aspect of the autonomous wireless system is demonstrated through inkjet-printed solutions for far- and near-field wireless energy harvesting systems to realize self-powered wireless sensing and communication platforms in a physically and conceptually flexible fashion.

Through the selective patterning of polymer inks, nanomaterial dispersions, thermoplastic filaments, and photopolymer resins, additive printing technology enables the fabrication of vertically integrated 3D wireless systems previously unachievable through the utilization of standard subtractive lithographic processes alone. This low-cost and robust fabrication technology is essential for the realization of emerging wireless technologies such as ubiquitous IoT networks, V2X connectivity, and smart skin systems. However, the future of advanced electronic manufacturing is not a singular technology that must be chosen over another, it is the integration of these fabrication technologies in a smart, reliable, and conscious approach to truly realize the coming generation of wireless systems.

25.5 REFERENCES

1. Jiantong Li, Fei Ye, Sam Vaziri, Mamoun Muhammed, Max C. Lemme, and Mikael Östling, "Efficient Inkjet Printing of Graphene," *Advanced Materials*, 25 (29): 3985–3992, 2013.
2. Bijan K. Tehrani, Chiara Mariotti, Benjamin S. Cook, Luca Roselli, and Manos M. Tentzeris, "Development, Characterization, and Processing of Thin and Thick Inkjet-Printed Dielectric Films," *Organic Electronics*, 29: 135–141, 2016.
3. J. G. Hester, S. Kim, J. Bito, T. Le, J. Kimionis, D. Revier, et al., "Additively Manufactured Nanotechnology and Origami-Enabled Flexible Microwave Electronics," *Proceedings of the IEEE*, Vol. 103, pp. 583–606, 2015.
4. S. A. Nauroze et al., "Additively Manufactured RF Components and Modules: Toward Empowering the Birth of Cost-Efficient Dense and Ubiquitous IoT Implementations," in *Proceedings of the IEEE*, vol. 105, no. 4, pp. 702–722, 2017.
5. B. K. Tehrani, J. Bito, J. G. Hester, W. Su, R. A. Bahr, B. S. Cook, et al., "Advanced Antenna Fabrication Processes (MEMS/LTCC/LCP/Printing)," in *Handbook of Antenna Technologies, N. Z. Chen (ed.), Springer, Singapore*, pp. 1–24, 2014.
6. B. K. Tehrani, B. S. Cook, and M. M. Tentzeris, "Post-Process Fabrication of Multilayer mm-Wave on-Package Antennas with Inkjet Printing," in *Antennas and Propagation & USNC/URSI National Radio Science Meeting, 2015 IEEE International Symposium on Antennas and Propagation*, pp. 607–608, 2015.
7. B. K. Tehrani, B. S. Cook, and M. M. Tentzeris, "Inkjet-Printed 3D Interconnects for Millimeter-Wave System-on-Package Solutions," *in Microwave Symposium (IMS), 2016 IEEE MTT-S International, 2016.*
8. J. Kimionis, M. Isakov, B. S. Koh, A. Georgiadis, and M. M. Tentzeris, "3D-Printed Origami Packaging with Inkjet-Printed Antennas for RF Harvesting Sensors," *IEEE Transactions on Microwave Theory and Techniques*, Vol. 63, pp. 4521–4532, 2015.
9. C. Mariotti, W. Su, B. S. Cook, L. Roselli, and M. M. Tentzeris, "Development of Low Cost, Wireless, Inkjet Printed Microfluidic RF Systems and Devices for Sensing or Tunable Electronics," *IEEE Sensors Journal*, Vol. 15, pp. 3156–3163, 2015.
10. W. Su, B. S. Cook, and M. M. Tentzeris, "Additively Manufactured Microfluidics-Based 'Peel-and-Replace' RF Sensors for Wearable Applications," *IEEE Transactions on Microwave Theory and Techniques*, Vol. 64, pp. 1928–1936, 2016.
11. J. G. D. Hester, M. M. Tentzeris, and Y. Fang, "Inkjet-Printed, Flexible, High Performance, Carbon Nanomaterial Based Sensors for Ammonia and DMMP Gas Detection," in *Microwave Conference (EuMC), 2015 European*, pp. 857–860, 2015.
12. J. G. D. Hester, M. M. Tentzeris, and Y. Fang, "UHF Lumped Element Model of a Fully-Inkjet-Printed Single-Wall-Carbon-Nanotube-Based Inter-Digitated Electrodes Breath Sensor," in *Antennas and Propagation & USNC/URSI National Radio Science Meeting, 2016 IEEE International Symposium on Antennas and Propagation*, 2016.
13. S. Kim, R. Vyas, J. Bito, K. Niotaki, A. Collado, A. Georgiadis, et al., "Ambient RF Energy-Harvesting Technologies for Self-Sustainable Standalone Wireless Sensor Platforms," *Proceedings of the IEEE*, Vol. 102, pp. 1649–1666, 2014.
14. J. Bito, J. G. Hester, and M. M. Tentzeris, "Ambient RF Energy Harvesting from a Two-Way Talk Radio for Flexible Wearable Wireless Sensor Devices Utilizing Inkjet Printing Technologies," *IEEE Transactions on Microwave Theory and Techniques*, Vol. 63, pp. 4533–4543, 2015.

Chapter 26

Printing of Nanoscale Electronics and Power Electronics

BY

CIHAN YILMAZ AND AHMED BUSNAINA *NSF Nanoscale Science and Engineering Center for High-Rate Nanomanufacturing, Northeastern University, Boston, Massachusetts*

26.1 INTRODUCTION

Commercial high-performance electronic device manufacturing is still mostly an expensive vacuum-based top down conventional CMOS process conducted in a planar fashion on rigid substrates [1]. Over the last few years, there has been a paradigm shift in manufacturing of low-end electronic devices and other nanomaterial-based devices away from conventional fabrication processes and toward printing technologies. Inkjet, screen printing, and gravure are printing technologies that are commercially used for printing electronics, flexible displays, and RFID tags. Current printed electronics (currently a 50 billion dollar industry [2]) offer significant savings over similar conventional silicon-based electronics. For example, the cost of a printed integrated sensor-plus-digital-readout device is 1/10th to 1/100th the cost of current silicon-based systems [3]. However, the smallest features that can be printed using inkjets today is about 20 μm, although this is still sufficient for many applications, its scale leaves it far behind today's silicon electronics. In fact, the last time silicon devices had a 20-μm linewidth was in 1975 [4]. The next generation of printed devices requires a technology leap from 1975 to 2015 by printing features at today's silicon electronics linewidth, which is 1000 times smaller (about 20 nm). Many researchers have shown techniques that can print down to 100 nm using jets but using aerosol jets has severe rate limitations even if thousands of jets are used. There is a need for nanoscale printing technology that is a high rate and one that can utilize many nanomaterials and is capable of printing heterogeneous structures. However, for the nanoscale printing process to offer a viable and economical alternative approach for making nanoscale devices, it has to print nanoscale features in order of magnitudes faster than current electronic printing but at a comparable cost. In addition, these nanoscale devices should have a superior performance at high yield as compared to electronics currently made using nanomaterials.

Organic and inorganic nanomaterials such as nanoparticles, nanowires, and nanotubes could be ideal building blocks for nanoelectronics due to their superior electrical transport properties and ability to tune them through doping [5, 6]. Many fundamental and applied studies (including studies by the proposing team) focused on exploiting the novel properties of 1D and 2D nanomaterials for creating novel devices. These studies resulted in advancing the fundamental understanding of these materials and their applications. However, much more research work remains in order to commercialize these devices because of several issues such as inconsistencies in the 1D and 2D raw material properties, scalability, and yield of the devices fabricated, etc. The use of silicon nanoparticles for printing flexible nanoscale electronics, on the other hand, introduce minimal variations and will have high reliability and consistent supply of raw nanomaterials. It can also be doped with various elements to the desired extrinsic semiconductor properties, which have been studied for many decades throughout the history of CMOS industry. One of the many structures studied is nanoscale silicon particles and wires, has exhibited better consistent properties than 1D and 2D nanomaterials currently in use in research and development laboratories.

26.2 NANOSCALE DIRECTED ASSEMBLY AND TRANSFER

We have developed directed assembly and transfer processes were specifically developed to be scalable and to enable precise and repeatable control of assembly and orientation of many nanoelements at high rate. By doing so, we have successfully assembled and printed ordered arrays and networks of nanoparticles [7, 8], conducting polymers [9], polymer blends [10], and SWNTs [11–16], molybdenum disulfide [17, 18] into various nonuniform structures [10, 19], including three dimensional architectures [11, 20, 21] at multiple length scales [22] that are patterned on rigid or compliant substrates [11, 23] and transferring them [20, 24]. Figure 25.1 shows that the multiscale offset printing system consists of two steps: (1) the electrophoretic directed assembly of nanomaterials on a template [25, 26] ("inking") followed by (2) transfer of the assembled nanoscale structure intact onto a recipient substrate. Each step takes about two minutes or less regardless of the density or the scale of the pattern. The steps take place at

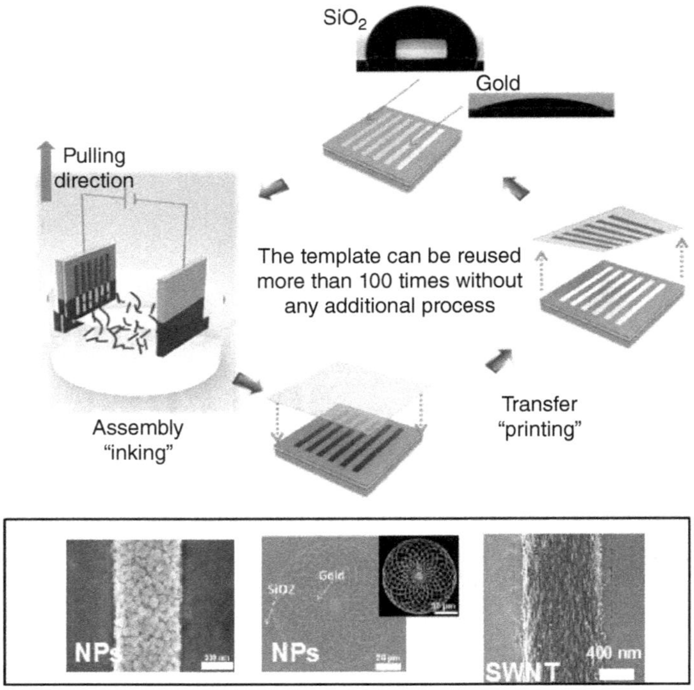

Figure 26.1 Schematic of the nanoscale offset printing of nanotubes and nanoparticles.

room temperature and pressure and has been shown to print using 0D, 1D, or 2D nanomaterials (such as nanoparticles, nanotubes, MoS_2, or polymers) suspended mostly in water, thus offering a greener manufacturing approach for making highly ordered nanostructures as shown in Fig. 26.2. Furthermore, through the directed assembly-based nanoscale printing approach, the team has shown that individual nanoparticles fuse and form highly homogenous polycrystalline structures exhibiting excellent electrical transport properties [7].

Several researchers have achieved assembly of arrays of zero dimensional nanoparticles to lengths of micrometer [27–32]. Ligands, polymers, and surface functionalizations that are employed on the nanoparticle for achieving these highly ordered nanoparticle arrays

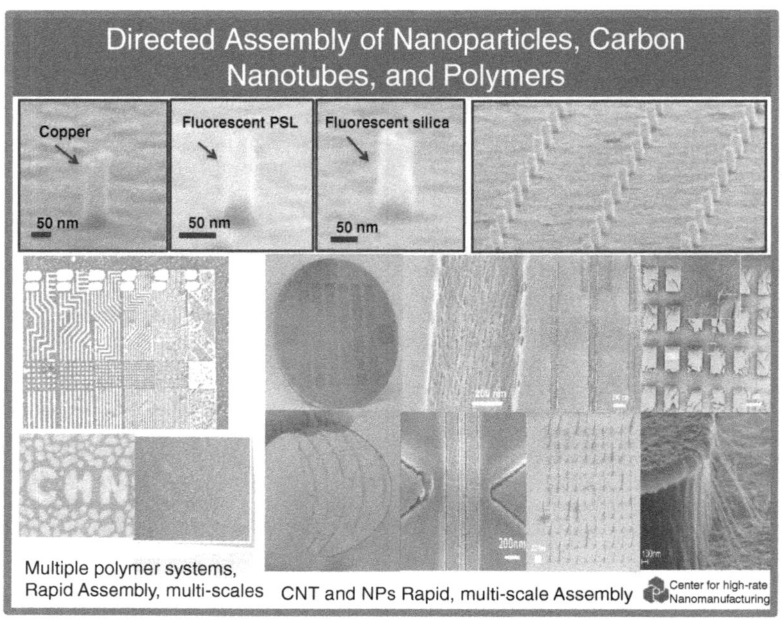

Figure 26.2 Examples of directed assembly of NPs, CNTs, and polymers.

inhibit them from exhibiting high conductivity [33]. Even in the absence of these surface functionalizations, the nanoparticles in these linear arrays are only in point contact resulting in huge potential drop (influence by hoping conductivity) during charge transport [34, 35]. In addition to this drawback, when used for flexible electronic application, they exhibited a huge variation in electrical properties from that of the virgin state during stretching and bending cycles [36]. Thus such linear chains of nanoparticles would not be conducive to applications mentioned above due to lack of electrical and mechanical integrity.

The developed directed assembly processes were utilized to manufacture various nanomaterial-based device components such as transistors [17, 18], chemical [37, 38], and biosensors [39, 40] and interconnects [7, 8]. The process utilizes the embedded nanoscale patterns in the template that utilize electrophoresis to attract the nanomaterial to the template. The processes take place at room temperature and pressure. The key to the success of the process is the damascene template (fabricated using conventional semiconductor damascene techniques [41–44]), which could be rigid or flexible for printing on soft or hard substrates, respectively [25]. These damascene templates have two-dimensional (2D) conductive nano and microscale features separated by an insulating film as shown in Fig. 26.3. The templates are topographically flat overall and a conductive film underneath connects all the patterned conductive features to ensure a uniform charge on all patterns or wires to yield a uniform assembly on the substrate patterns with uniform thickness on nano or microscale patterns. The template is functionalized such that only conductive patterns will be wetted. The functionalization lasts for at least a hundred printing cycles before the need for reapplication. Figure 26.3*a* shows the steps of making a template for a one-layer assembly, Fig. 26.3*b* shows an image of a rigid template, and Fig. 26.3*c* shows an image of a polymer-based flexible template.

26.3 APPLICATIONS IN POWER ELECTRONICS

26.3.1 Printed High-Performance Logic Circuits

The complementary metal-oxide semiconductor (CMOS)-based inverter is one of the basic units of logic elements in modern integrated circuits (ICs), comprising p-type and n-type transistors, with exceptional properties such as extremely low power consumption and a small noise margin. Due to its low power consumption, a complementary inverter is superior to other inverter configurations (e.g., resistive-load inverters). However, transistor miniaturization down to 22 nm and below by the current top-down approach imposes many challenges, such as short channel effects and gate leakage, etc. [45]. To overcome many of these barriers toward miniaturization of ICs, novel materials in conjunction with bottom-up fabrication methods can serve as a viable alternative. In addition, room temperature and ambient processing capabilities of this bottom-up directed assembly approach could also contribute to lowering the fabrication cost. To build ICs with specific functionalities via the bottom-up approach, both p-type and n-type nanoscale semiconductor materials are required. While CNTs with specific chirality exhibit a p-type behavior at room temperature under ambient conditions with excellent characteristics, a stable complementary n-type material has not been successfully demonstrated until recently. Efforts toward converting the p-type CNTs to n-type via potassium doping [46], using functional groups existing in polymers and molecules [47], and by using metals of different work function have resulted in limited success. All these methods suffer from either stability or fabrication compatibility issues, for example, potassium-doped CNT field-effect transistors (FETs) show a degraded performance with time [48]. A highly stable n-type material with excellent FET characteristics compatible with the bottom-up approach would drastically alter the manufacturing protocols for future nanoelectronic devices beyond current scaling limits. MoS_2, an inherent n-type semiconductor, with high thermal stability, resistance to oxidation, absence of dangling bonds, and the presence of a band gap (1.2 to 1.8 eV) [49], promises to be an excellent n-type counterpart for the p-type carbon nanotube for developing logic circuits. Belonging to the family of layered transition metal dichalcogenides, MoS_2 forms a 2D structure via strong covalent bonding among S–Mo–S atoms whereas the weak van der Waals force holds the layers together [49]. Single or few-layer MoS_2 nanostructures have been obtained using mechanical exfoliation [49], liquid exfoliation [50], and lithium intercalation [51], and hence it is compatible with bottom-up approaches. So far, however, a complementary inverter,

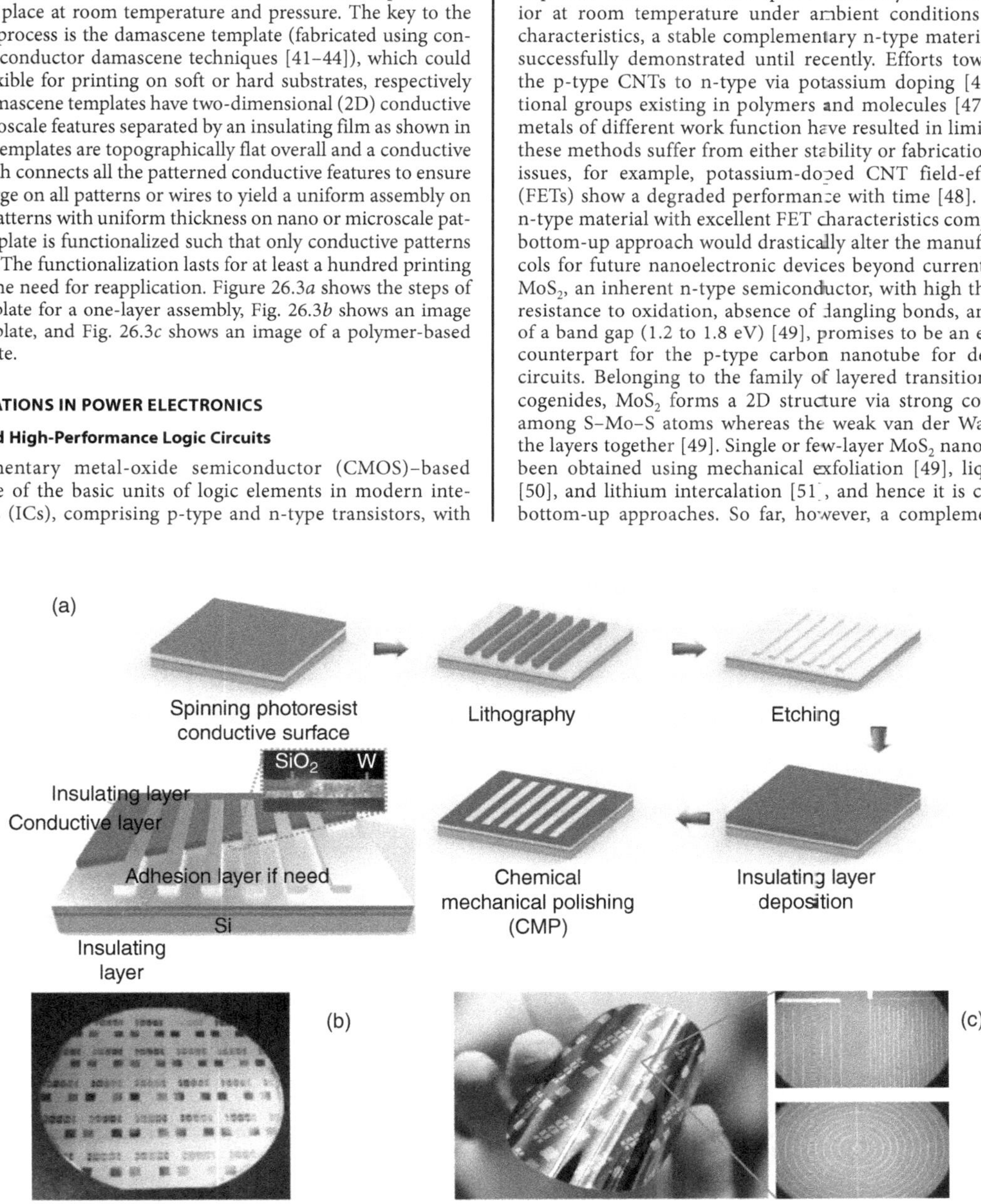

Figure 26.3 (*a*) Making of a damascene template, (*b*) 4-in rigid template, (*c*) polymer-based template.

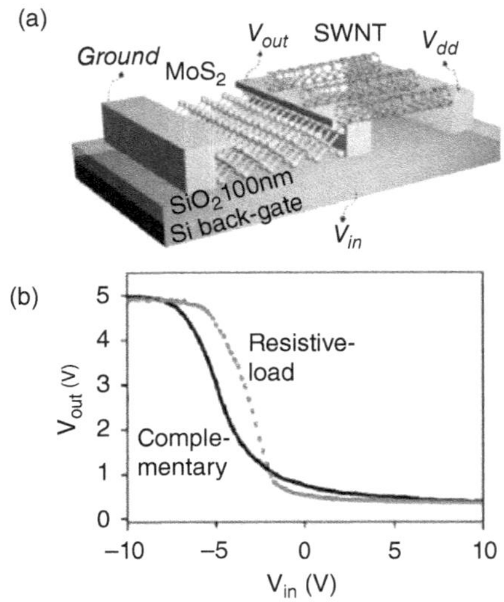

Figure 26.4 (*a*) A schematic illustrating the design of complementary inverter that consists of a p-type CNT FET and an n-type MoS_2 transistor; (*b*) voltage transfer characteristics (V_{in} versus V_{out}) of an inverter under a driving voltages (V_{dd}) of 5 V.

which is desirable for its low power consumption, has not been shown yet. In the current study, we have successfully fabricated high-performance CNT and MoS_2 thin film transistors via a bottom-up approach and have used them to demonstrate a complementary inverter device [18]. The fabricated p-type CNT thin film transistor showed an ON/OFF ratio greater than 103 with mobility 0.4 $cm^2\ V^{-1}\ s^{-1}$, whereas the n-type MoS_2 transistor had an ON/OFF ratio greater than 105 with mobility approximately 3 $cm^2\ V^{-1}\ s^{-1}$. The gain of MoS_2-based resistive load inverter depends on both load and driving voltage (V_{dd}) having a maximum value of approximately 1.7 at 5 V. The MoS_2/CNT transistor–based complementary inverter exhibited a gain approximately 1.3 at a driving voltage of 5 V (Fig. 26.4).

An inverter, the fundamental logic element in an IC, converts low input (logic 0) to high output (logic 1), or vice versa. Sophisticated logic gates can be realized by proper design using inverters as a basic element, for example, two inverters in parallel generate a NOR gate. A typical MoS_2 transistor has an ON-state resistance in the range of hundreds kilohm and an OFF-state resistance in the range of a gigaohm. Hence, for the resistive-load configuration, an external 5.1 MoS_2 resistor was connected to the drain of a MoS_2 transistor as shown in Fig. 26.3. To achieve low power consumption logic gate components, we subsequently studied a complementary inverter comprising MoS_2 and CNT transistors, which is schematically depicted in Fig. 26.3*a*. The input (V_{in}) is connected to the gate terminal of both p-type CNT and n-type MoS_2 transistors. The output (V_{out}) is connected to the drain terminal of both transistors. The source of the p-type CNT FET is connected to a positive V_{dd} while the source of the n-type MoS_2 transistor is connected to ground. When V_{in} is negative, the p-type CNT transistor is in the ON state while the n-type MoS_2 transistor is in the OFF state. Hence, V_{out} reaches the V_{dd} value, which indicates an operation from logic 0 to logic 1, that is, inverter operation. The resultant inverter voltage transfer characteristic is shown in Fig. 26.3*b*, which clearly demonstrates an inverter characteristic. The inverter gain is approximately 1.3. The inverter gain is moderate due to the thick dielectric layer used, imperfection between the conduction channel and the gate oxide of the MoS_2 transistor, and the performance mismatch of the MoS_2 and CNT transistors. A single layer MoS_2 transistor integrated with thin high-k dielectric (30 nm HfO_2) has a mobility of over 200 $cm^2\ V^{-1}\ s^{-1}$ [8]. The current study has employed a back-gate geometry and relatively thick dielectric layer (100 nm SiO_2), which inevitably result in low mobility and compromise the subthreshold swing.

26.3.2 Printed 3D Interconnects for Power Electronics

The interconnects in electronics are currently fabricated by conventional techniques such as electrodeposition and vacuum deposition. These methods typically require high temperature and pressure processing with excessive use of chemical additives, which could make them expensive and inefficient for fabricating. For example, the current trend in semiconductor technology toward 3D stacking of devices provides the advantages of short interconnections, miniaturization, and compact packaging [52]. Through-silicon-via (TSV) is an emerging technology that would allow the interconnection of the circuits in

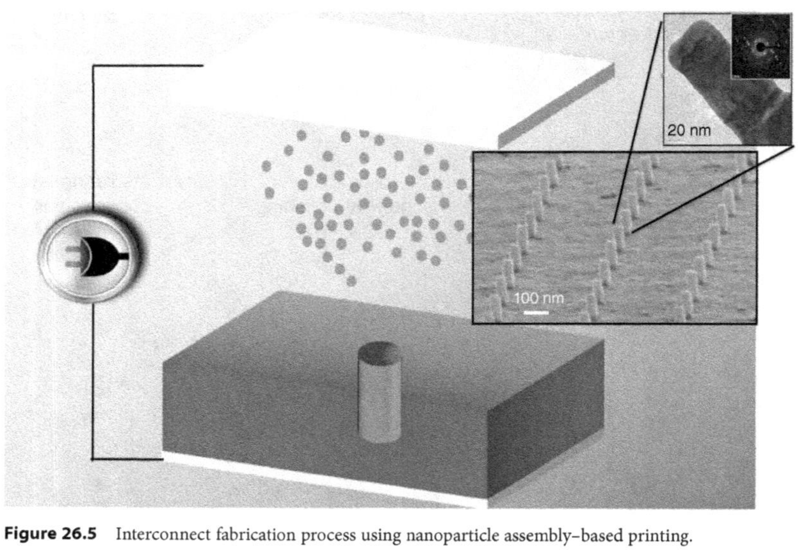

Figure 26.5 Interconnect fabrication process using nanoparticle assembly–based printing.

3D stacked chips [53]. By using TSV, it is possible to save the space that would be necessary for bonding wires and to make wiring lengths shorter. The interconnection using TSV requires filling very large vias, which are tens of micrometers in diameter and hundreds of micrometers in depth. However, fabrication of such large and high aspect ratio TSV interconnects by utilizing conventional electroplating or thin film deposition is costly and technologically challenging. For example, producing void-free TSV interconnects using electroplating requires bottom-up fill, which involves the use of seed layers and many chemical additives. This increases the complexity and the cost involved in the process. To overcome this issue, International Technology Roadmap for Semiconductors (ITRS) has outlined the necessity for the development of manufacturable and cost-effective alternative filling techniques [52].

We introduced an entirely new, material independent, room temperature and pressure printing process for fabricating 3D interconnects in electronics [7, 8]. In this process, colloidal nanoparticles (NPs) are precisely assembled and fused (printed) into 3D nanostructures in a single step using an externally applied electric field (Fig. 26.5). Compared to conventional fabrication methods, this method is carried out at room temperature and pressure without the need for an intermediate seed layer and chemical additives. This significantly reduces the cost and complexity associated with conventional manufacturing. In contrast to electroplating or thin film deposition, this method can fabricate solid nanostructures from inorganic or organic NPs made of conducting, semiconducting or insulating materials with feature sizes from several micrometers down to 25 nm, and do it in less than a minute over a large area (wafer scale). Using this technique, we have fabricated interconnects made from Cu, W, Au, Si [7]. TEM (transmission electron microscopy) and electrical characterizations reveal that printed Au interconnects have polycrystalline nature without any voids and very low resistivity (1.96×10^{-7} Ω·m) [7]. These results indicate that this novel approach could be considered as an excellent candidate for fabrication of TSV interconnects using colloidal assembly-based printing technique.

26.4 REFERENCES

1. P. S. Peercy, "The Drive to Miniaturization," *Nature*, 406 (6799): 1023–1026, 2000.
2. F. Sullivan, "Printed Electronics—Technologies and Applications (Technical Insights)," http://www.frost.com/prod/servlet/market-insight-print.pag?docid=108885683, 2007.
3. J. Ernst, "Printed Electronics Memory: Challenges of Logic and Integration," EDN Network, 2012.
4. A. I. Kingon, J.-P. Maria, and S. Streiffer, "Alternative Dielectrics to Silicon Dioxide for Memory and Logic Devices," *Nature*, 406 (6799): 1032–1038, 2000.
5. J. Hu, T. W. Odom, and C. M. Lieber, "Chemistry and Physics in One Dimension: Synthesis and Properties of Nanowires and Nanotubes," *Accounts of Chemical Research*, 32 (5): 435–445, 1999.
6. C. Dekker, "Carbon Nanotubes as Molecular Quantum Wires," *Physics Today*, 52: 22–30, 1999.
7. C. Yilmaz, A. E. Cetin, G. Goutzamanidis, J. Huang, S. Somu, H. Altug, D. Wei, et al., "Three-Dimensional Crystalline and Homogeneous Metallic Nanostructures Using Directed-Assembly of Nanoparticles," *ACS Nano*, 8(5), 4547–4558, 2014.
8. C. Yilmaz, T.-H. Kim, S. Somu, and A. A. Busnaina, "Large-Scale Nanorods Nanomanufacturing by Electric-Field-Directed Assembly for Nanoscale Device Applications," *Nanotechnology, IEEE Transactions on*, 9 (5): 653–658, 2010.
9. M. Wei, Z. Tao, X. Xiong, M. Kim, J. Lee, S. Somu, S. Sengupta, et al., "Fabrication of Patterned Conducting Polymers on Insulating Polymeric Substrates by Electric-Field-Assisted Assembly and Pattern Transfer," *Macromolecular Rapid Communications*, 27 (21): 1826–1832, 2006.
10. M. Wei, L. Fang, J. Lee, S. Somu, X. Xiong, C. Barry, A. Busnaina, et al., "Directed Assembly of Polymer Blends Using Nanopatterned Templates," *Advanced Materials*, 21 (7): 794–798, 2009.
11. P. Makaram, S. Selvarasah, X. Xiong, C.-L. Chen, A. Busnaina, N. Khanduja, and M. R. Dokmeci, "Three-Dimensional Assembly of Single-Walled Carbon Nanotube Interconnects Using Dielectrophoresis," *Nanotechnology*, 18 (39): 395–204, 2007.
12. S. Somu, H. Wang, Y. Kim, L. Jaberansari, M.G. Hahm, B. Li, T. Kim, et al., "Topological Transitions in Carbon Nanotube Networks via Nanoscale Confinement," *ACS Nano*, 4 (7): 4142–4148, 2010.
13. X. Xiong, C.-L. Chen, P. Ryan, A. A. Busnaina, Y. J. Jung, and M. R. Dokmeci, "Directed Assembly of High Density Single-Walled Carbon Nanotube Patterns on Flexible Polymer Substrates," *Nanotechnology*, 20 (29): 295–302, 2009.
14. L. Jaber-Ansari, M. G. Hahm, T. H. Kim, S. Somu, A. Busnaina, and Y. J. Jung, "Large Scale Highly Organized Single-Walled Carbon Nanotube Networks for Electrical Devices," *Applied Physics A*, 96 (2): 373–377, 2009.
15. L. Jaber-Ansari, M. G. Hahm, S. Somu, Y. E. Sanz, A. Busnaina, and Y. J. Jung, "Mechanism of Very Large Scale Assembly of SWNTs in Template Guided Fluidic Assembly Process," *Journal of the American Chemical Society*, 131 (2): 804–808, 2008.
16. S. Selvarasah, X. Li, A. Busnaina, and M. R. Dokmeci, "Parylene-C Passivated Carbon Nanotube Flexible Transistors," *Applied Physics Letters*, 97 (15): 153120, 2010.
17. J. Huang, A. Datar, S. Somu, and A. Busnaina, "Modulating the Performance of Carbon Nanotube Field-Effect Transistors via Rose Bengal Molecular Doping," *Nanotechnology*, 22 (45): 455202, 2011.
18. J. Huang, S. Somu, and A. Busnaina, "A Molybdenum Disulfide/Carbon Nanotube Heterogeneous Complementary Inverter," *Nanotechnology*, 23 (33): 335203, 2012.
19. L. Fang, M. Wei, C. Barry, and J. Mead, "Effect of Spin Speed and Solution Concentration on the Directed Assembly of Polymer Blends," *Macromolecules*, 43 (23): 9747–9753, 2010.
20. B. Li, M. G. Hahm, Y. L. Kim, H. Y. Jung, S. Kar, and Y. J. Jung, "Highly Organized Two- and Three-Dimensional Single-Walled Carbon Nanotube–Polymer Hybrid Architectures," *ACS Nano*, 5 (6): 4826–4834, 2011.
21. S. Selvarasah, A. Busnaina, and M. Dokmeci, "Design, Fabrication, and Characterization of Three-Dimensional Single-Walled Carbon Nanotube Assembly and Applications as Thermal Sensors," *IEEE Transactions on Nanotechnology*, 10 (1): 13–20, 2011.
22. J. Chiota, J. Shearer, M. Wei, C. Barry, and J. Mead, "Multiscale Directed Assembly of Polymer Blends Using Chemically Functionalized Nanoscale-Patterned Templates," *Small*, 5 (24): 2788–2791, 2009.
23. X. Xiong, P. Makaram, A. Busnaina, K. Bakhtari, S. Somu, N. McGruer, and J. Park, "Large Scale Directed Assembly of Nanoparticles Using Nanotrench Templates," *Applied Physics Letters*, 89 (19): 193108, 2006.
24. B. Li, H. Y. Jung, H. Wang, Y. L. Kim, T. Kim, M. G. Hahm, A. Busnaina, et al., "Ultrathin SWNT Films with Tunable, Anisotropic Transport Properties," *Advanced Functional Materials*, 21 (10): 1810–1815, 2011.
25. H. Cho, S. Somu, J.-Y. Lee, H. Jeong, and A. Busnaina, "High-Rate Nanoscale Offset Printing Process Using Directed Assembly and Transfer of Nanomaterials," *Advanced Materials*, 2015. Published : E pub date Feb. 3, 2015; doi:10.1002/adma.201404769.
26. A. Busnaina, H. Cho, S. Somu, and J. Huang, "Damascene Template for Directed Assembly and Transfer of Nanoelements," U.S. Patent 20,140,318,967, 2014.
27. M. G. Warner and J. E. Hutchison, "Linear Assemblies of Nanoparticles Electrostatically Organized on DNA Scaffolds," *Nature Materials*, 2 (4): 272–277, 2003.
28. A. N. Shipway, E. Katz, and I. Willner, "Nanoparticle Arrays on Surfaces for Electronic, Optical, and Sensor Applications," *ChemPhysChem*, 1 (1): 18–52, 2000.
29. R. Shenhar, T. B. Norsten, and V. M. Rotello, "Polymer-Mediated Nanoparticle Assembly: Structural Control and Applications," *Advanced Materials*, 17 (6): 657–669, 2005.
30. J. H. Fendler, "Chemical Self-Assembly for Electronic Applications," *Chemistry of Materials*, 13 (10): 3196–3210, 2001.

31. I. Matsui, "Nanoparticles for Electronic Device Applications: A Brief Review," *Journal of Chemical Engineering of Japan*, 38 (8): 535–546, 2005.

32. S. Maenosono, T. Okubo, and Y. Yamaguchi, "Overview of Nanoparticle Array Formation by Wet Coating," *Journal of Nanoparticle Research*, 5 (1–2): 5–15, 2003.

33. F. P. Zamborini, M. C. Leopold, J. F. Hicks, P. J. Kulesza, M. A. Malik, and R. W. Murray, "Electron Hopping Conductivity and Vapor Sensing Properties of Flexible Network Polymer Films of Metal Nanoparticles," *Journal of the American Chemical Society*, 124 (30): 8958–8964, 2002.

34. B. D. Gates, "Flexible Electronics," *Science*, 323 (5921): 1566–1567, 2009.

35. J. H. Jun, B. Park, K. Cho, and S. Kim, "Flexible TFTs Based on Solution-Processed ZnO Nanoparticles," *Nanotechnology*, 20 (50): 505201, 2009.

36. D. Huang, F. Liao, S. Molesa, D. Redinger, and V. Subramanian, "Plastic-Compatible Low Resistance Printable Gold Nanoparticle Conductors for Flexible Electronics," *Journal of the Electrochemical Society*, 150 (7): G412–G417, 2003.

37. C.-L. Chen, C.-F. Yang, V. Agarwal, T. Kim, S. Sonkusale, A. Busnaina, M. Chen, et al., "DNA-Decorated Carbon-Nanotube-Based Chemical Sensors on Complementary Metal Oxide Semiconductor Circuitry," *Nanotechnology*, 21 (9): 095504, 2010.

38. H. Y. Jung, Y. L. Kim, S. Park, A. Datar, H.-J. Lee, J. Huang, S. Somu, et al., "High-Performance H2S Detection by Redox Reactions in Semiconducting Carbon Nanotube-Based Devices," *The Analyst*, 138: 7206, 2013.

39. A. Malima, S. Siavoshi, T. Musacchio, J. Upponi, C. Yilmaz, S. Somu, W. Hartner, et al., "Highly Sensitive Microscale in Vivo Sensor Enabled by Electrophoretic Assembly of Nanoparticles for Multiple Biomarker Detection," *Lab on a Chip*, 12 (22): 4748–4754, 2012.

40. S. Siavoshi, C. Yilmaz, S. Somu, T. Musacchio, J. R. Upponi, V. P. Torchilin, and A. Busnaina, "Size-Selective Template-Assisted Electrophoretic Assembly of Nanoparticles for Biosensing Applications," *Langmuir*, 27 (11): 7301–7306, 2011.

41. A. Busnaina and N. E. McGruer, "Directed Assembly of Carbon Nanotubes and Nanoparticles Using Nanotemplates with Nanotrenches," Google Patents, 2006.

42. C. M. Barry, A. Busnaina, J. L. Mead, T. Zhenghong, and M. Wei, "Directed Assembly of a Conducting Polymer," Google Patents, 2006.

43. A. Busnaina, C. Yilmaz, T. Kim, and S. Somu, "Nanoscale Interconnects Fabricated by Electrical Field Directed Assembly of Nanoelements," Google Patents, 2010.

44. A. Sirman, A. Busnaina, C. Yilmaz, J. Huang, and S. Somu, "High Rate Electric Field Driven Nanoelement Assembly on an Insulated Surface," Google Patents, 2011.

45. I. Ferain, C. A. Colinge, and J.-P. Colinge, "Multigate Transistors as the Future of Classical Metal-Oxide-Semiconductor Field-Effect Transistors," *Nature*, 479 (7373): 310–316, 2011.

46. A. Javey, R. Tu, D. B. Farmer, J. Guo, R. G. Gordon, and H. Dai, "High Performance n-Type Carbon Nanotube Field-Effect Transistors with Chemically Doped Contacts," *Nano Letters*, 5 (2): 345–348, 2005.

47. C. Klinke, J. Chen, A. Afzali, and P. Avouris, "Charge Transfer Induced Polarity Switching in Carbon Nanotube Transistors," *Nano Letters*, 5 (3): 555–558, 2005.

48. H. Ryu, D. Kälblein, R. T. Weitz, F. Ante, U. Zschieschang, K. Kern, O. G. Schmidt, et al., "Logic Circuits Based on Individual Semiconducting and Metallic Carbon-Nanotube Devices," *Nanotechnology*, 21 (47): 475207, 2010.

49. B. Radisavljevic, A. Radenovic, J. Brivio, V. Giacometti, and A. Kis, "Single-Layer MoS_2 Transistors," *Nature Nanotechnology*, 6 (3): 147–150, 2011.

50. K. Lee, H. Y. Kim, M. Lotya, J. N. Coleman, G. T. Kim, and G. S. Duesberg, "Electrical Characteristics of Molybdenum Disulfide Flakes Produced by Liquid Exfoliation," *Advanced Materials*, 23 (36): 4178–4182, 2011.

51. D. Yang, S. J. Sandoval, W. Divigalpitiya, J. Irwin, and R. Frindt, "Structure of Single-Molecular-Layer MoS_2," *Physical Review B*, 43 (14): 12053, 1991.

52. I. R. Committee, *International Technology Roadmap for Semiconductors, 2011 Edition*. Semiconductor Industry Association, http://www.itrs. net/Links/2011ITRS/2011Chapters/2011ExecSum.pdf, 2011.

53. C. Li, M.-Q. Zou, Y. Shang, and M. Zhang, "Study on the Thermal Transient Response of TSV Considering the Effect of Electronic-Thermal Coupling," *Journal of Semiconductor Technology and Science*, 15 (3): 356–364, 2015.

Chapter 27

3D Interconnects in Flexible Electronics

BY

CIHAN YILMAZ AND AHMED BUSNAINA *NSF Nanoscale Science and Engineering Center for High-Rate Nanomanufacturing, Northeastern University, Boston, Massachusetts*

27.1 INTRODUCTION

Flexible electronics are lightweight, rugged, bendable, rollable, portable, and potentially foldable. Much work has been done to address the issues such as patterning and materials combinations to provide optimum performance of thin-film transistors (TFT) and p-i-n photodiodes for flat panel display and image sensor backplanes. However, fabrication of multilayered interconnects in these devices is still challenging. Conventional interconnect fabrication techniques such as electrodeposition and vacuum deposition require high temperature and pressure processing with excessive use of chemical additives, which makes them incompatible with plastics used in flexible electronics. There is a need for a room-temperature and pressure-manufacturing method to create multilayered interconnects for flexible electronics. Herein, we introduce a new, material independent, room-temperature and pressure-manufacturing process for printing three-dimensional (3D) interconnects. In this process, colloidal nanoparticles (NPs) are precisely assembled and fused into 3D nanostructures in a single step, using an externally applied electric field. Compared to conventional fabrication methods, this method is carried out at room temperature and pressure without the need for an intermediate seed layer and chemical additives. This significantly reduces the cost and complexity associated with conventional manufacturing. In contrast to electroplating or thin film deposition, this method can fabricate solid nanostructures from inorganic or organic NPs made of conducting, semiconducting, or insulating materials with feature sizes from several micrometers down to 25 nm, and do it in less than 1 min over a large area (wafer scale). Using this technique, we have fabricated nanopillars made from Cu, W, Au, Si, SiO_2, and their hybrid derivatives. TEM (transmission electron microscopy) and electrical characterizations reveal that manufactured metallic nanostructures have polycrystalline nature without any voids and very low resistivity ($\sim10^{-7}$ Ω·m). This novel approach can find application in main control panels (MCPs), multichip modules (MCMs), chip-on-board (COB), wafer-level packaging (WLP), flexible electronics, hybrid silicon to printed electronics attachment.

27.2 DIRECTED ASSEMBLY OF NANOPARTICLES

Nanostructures with complex geometries and 3D architectures have recently received considerable interests in many fields, including electronics [1], optics [2], energy [3], and biotechnology [4] because they enable superior device performance and miniaturization. Most approaches for fabricating these nanostructures rely on vacuum-based thin film deposition or electroplating, which requires a seed layer and many chemical additives. Directed assembly of nanoparticles (NPs) has been shown to be a promising alternative approach for building functional nanomaterials and nanostructures in aqueous solutions at room temperature and pressure [4–6]. Much progress has been made in assembling NPs into one-, two-, and three-dimensional nanostructures by utilizing electric [7, 8], magnetic [9], and fluidic forces [10]. However, there is a dearth of studies on fabricating crystalline, solid, and homogenous nanostructures. This is mainly due to the difficulties in controlling the assembly and fusion of NPs into desired geometries. Depending on the particle composition, functionalization, and size, the forces and energies governing the assembly and fusion of NPs can be different. For example, based on the suspension medium, NPs could possess different surface properties such as surface energy and charge, which can affect the assembly process and the NPs interaction with the substrate. Similarly, larger size NPs have higher melting temperature compared to smaller ones making them difficult to fuse into solid structures. To successfully fabricate nanostructures with the desired material and geometry, it is essential to identify the governing parameters that control the forces involved in the assembly process.

In this work, we developed a directed assembly technique that enables assembly and fusion of various metallic NPs to fabricate highly organized 3D crystalline, solid nanostructures on rigid or flexible surfaces. The process is as fast as electroplating, shown to be scalable to the millimeter scale and conducted at room temperature and pressure without requiring chemical additives making it highly suitable for cost-effective, green nanomanufacturing.

27.3 3D INTERCONNECT FABRICATION PROCESS

The fabrication process involves directing the assembly of colloidal NPs, using dielectrophoresis (DEP) [11], onto a substrate. The substrate consists of a conductive film (on a rigid or flexible substrate) coated by any patterned insulator [e.g., poly(methyl methacrylate), PMMA or SiO_2] that features nanoscale or larger patterns such as vias, trenches, or other geometries as shown in Fig. 27.1*a*. An ac electric field is applied between the substrate and a counter electrode positioned millimeters away from the substrate in the NP suspension. The electric field creates a dielectrophoretic force on the NPs, moving them toward the vias, where the magnitude of electric field is highest. NPs under the effect of DEP also experience a pearl-chaining force [12], which arises from the interaction of induced dipoles between the NPs. The particle-particle interaction becomes very important when the NPs assemble on the bottom surface of the vias allowing the NPs to attach to already-assembled NPs. A dc offset can also be coupled with an ac voltage to increase the force on the particles [13]. As the NPs assemble, the applied electric field induces their fusion, which arises from localized Joule heating, induced by the applied ac voltage at the NP junctions [14]. Following the fusion, solid interconnects are formed in the vias (Fig. 27.1*b*). Figure 27.1*c* shows the scanning electron microscopy (SEM) images of the fabricated gold interconnects after the patterned insulator is removed. The inset image in Fig. 27.1*c* demonstrates the successful fabrication of interconnects down to 25 nm with controlled dimensions.

To estimate the fusion mechanism and the Joule heating at nanoparticle junctions (Fig. 27.2*a*), we first determined the contact radius between the gold particle and the gold surface due to the adhesion-induced deformation. The average van der Waals pressure p_0 due to the van der Walls force, F_a, was calculated by [15, 16]

$$p_0 = F_a/\pi a^2 \approx 2w/z_0 \tag{27.1}$$

where w is the work of adhesion and z_0 is the separation distance between the particle and the surface. Work of adhesion between the particles is

$$W_A = 2\sqrt{\gamma_1\gamma_2} \tag{27.2}$$

where γ_1 and γ_2 are surface free energy of two contact materials.

According to Maugis and Pollock (MP) model in plastic deformation, the contact radius can be calculated by the following equation [17]:

$$a = \sqrt{\frac{F_{\text{external}} + 2\pi W_A R}{3Y}} \tag{27.3}$$

where Y is the yield strength of the particle and F_{external} is any externally applied force other than van der Walls force.

Once the contact radius is known, the contact resistance (R_c) can be calculated by the following equation:

$$R_e = \frac{\square_c \times z_0}{A_c} \tag{27.4}$$

where $\square_c$ is the resistivity of gold at the contact, and A_c is the contact area. It has been reported that the resistivity of metals at nanoscale dimensions is higher than their bulk resistivity [18–20].

The RMS value of ac current was measured using an ammeter at an applied ac voltage of 12 V_{pp} at 50 kHz with a 2-V dc offset. After determining the electrical resistance and the current passing through the contact, Joule heating can be calculated by

$$Q = I_p^2 \cdot R_c \cdot t \tag{27.5}$$

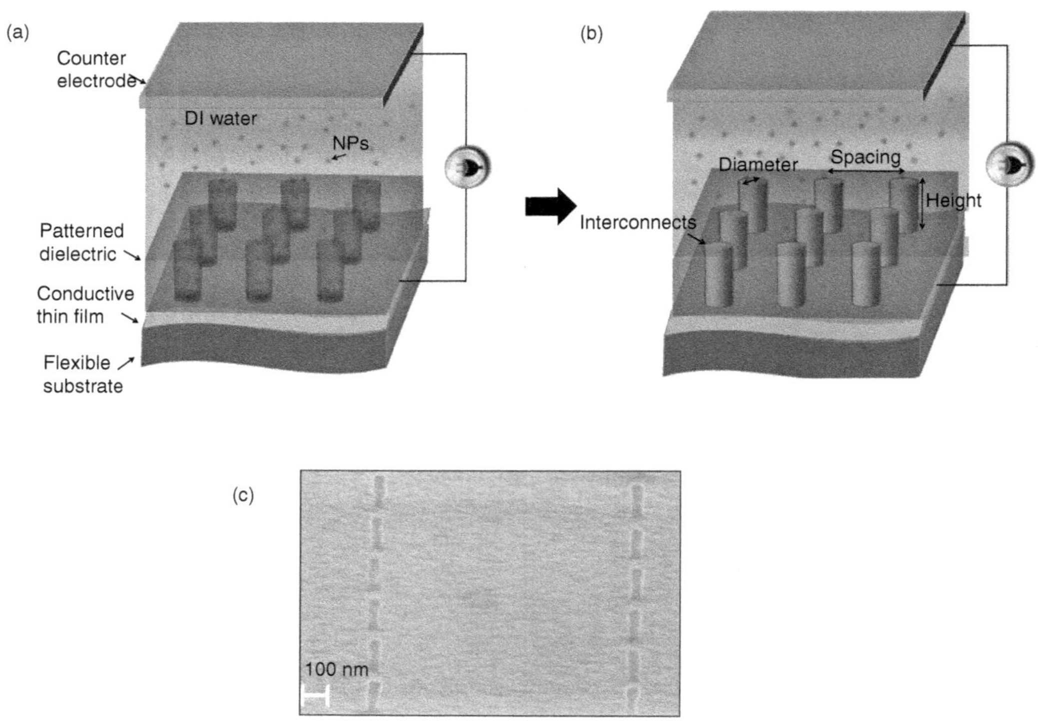

Figure 27.1 Fabrication of 3D interconnects through electric field-directed assembly of NPs.

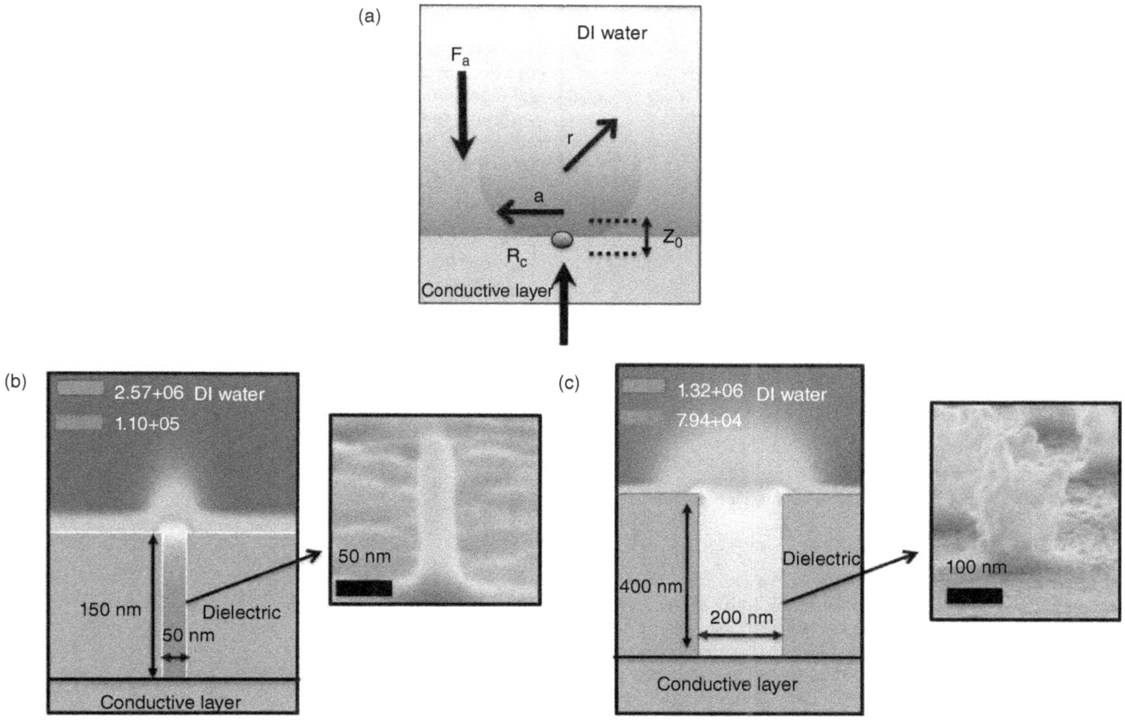

Figure 27.2 (*a*) Inset shows the current passing through a single particle. An electric field simulation in (*b*) a 50-nm diameter and 150 nm deep, (*c*) 200-nm diameter and 400-nm deep via. SEM images of the nanostructures made in these vias are shown next to the simulation results.

where I_p is the magnitude of the applied current and t is the time period of the current passing through the particle-surface contact. The time required to fuse the monolayer of particles was determined by assembling the particles at different time steps.

The rise in nanoparticle temperature due to Joule heating was calculated by

$$\Delta T = \frac{Q}{v \cdot c_{v,\text{gold}}} \tag{27.6}$$

where v is the particle volume, $c_{v,\text{gold}}$ is the specific heat capacity of bulk gold, and ΔT is the temperature increase. Combining Eqs. (27.4) to (27.6), we obtain

$$\Delta T = \frac{3I_p^2 \rho_c z_0 t}{4\pi^2 c_{v,\text{gold}} \alpha^2 r^3} \tag{27.7}$$

The temperature increase at the junctions of 5-nm particles was calculated to be 593°C. Considering the fact that small size particles have much lower melting temperature compared to their bulk melting points [21], complete fusion of 5 nm particles at 593°C is possible. Similarly, we found that if 50-nm particles are used, the temperature of 50-nm particles only increase by 3.23°C (due to low electric field and large contact area, Fig. 27.2*c*), which may lead to incomplete particle fusion.

To experimentally verify these calculations, we assembled 50-nm gold particles in 200-nm-diameter vias. We applied an ac voltage of 12 V_{pp} at 50 kHz with a 2-V dc offset during 1990s for all experiments. It is evident from the SEM images in Fig. 27.2*c* that in contrast to complete fusion of 5-nm gold particles, 50-nm gold particles were nonfused as it was predicated in the calculations However, we also found that the partially fused or nonfused nanoparticles can be completely fused into solid nanopillars by passing a high current through them [22].

27.4 MATERIAL CHARACTERIZATION

We placed the fabricated nanopillars on a copper TEM grid using a low-energy lamella lift-out process prior to TEM characterization. The bright-field images indicate that the gold NPs were completely fused during the assembly process, transforming them into homogenous interconnects lacking air voids or gaps. The small-area electron diffraction (SAED) pattern obtained from the entire interconnect presented in the inset to Fig. 27.3 reveals the polycrystalline nature of the interconnects. Only two grains, each having its lattice orientation in one direction, existed over the 30 × 30-nm area of the nanopillar. Recalling that the nominal diameter of an individual NP was 5 nm, we infer that large number of NPs fused into a single grain. The formation of a single-crystalline material might have resulted from the recrystallization of multiple NPs during the fusion process [23].

27.5 ELECTRICAL CHARACTERIZATION

An SEM-based in situ nanomanipulator (Zyvex S-100) was used to determine the electrical characteristics of the interconnects. We performed these measurements using two tungsten probes having a tip diameter of 20 nm; one of the probes contacted the interconnect while the other contacted the thin gold layer under the PMMA. The quality of the contact between the probe and the nanopillar was an important

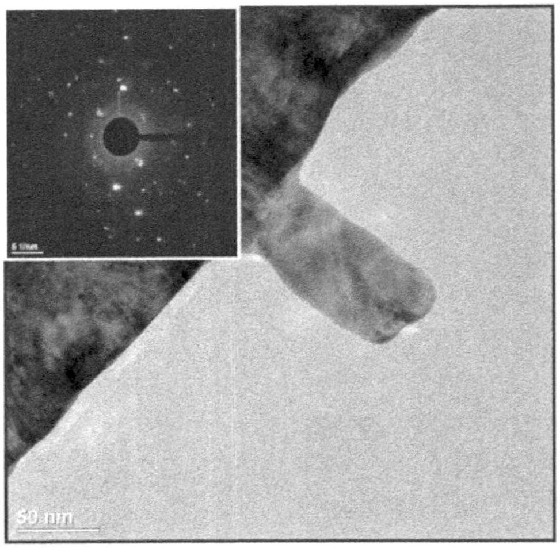

Figure 27.3 Bright-field TEM image and corresponding electron diffraction pattern of the gold interconnect.

parameter affecting the reliability of the measurements. In our experiments, achieving good contact (slight penetration into pillars) between the probe and a 50-nm-diameter nanopillar was difficult because of the small pillar diameter. Indeed, we obtained large variations in resistance (from tens of ohms to hundreds of kilohms) for small size nanopillars, depending on the quality of the contact. The over-deposited nanostructures, however, consistently produced lower values of resistance (hundreds of ohms or lower), indicating that a larger contact area, due to probe penetration, improved the reliability of the measurements.

The measurement was carried out for two chips each having hundreds of nanopillars. From each chip ten nanopillars were randomly chosen and tested (Fig. 27.4). The resistance of bottom gold surface was measured as 10 Ω using two probes. We obtained similar resistance value when we changed the distance between the probes. Hence, we assumed that obtained resistance of 10 Ω is due to the contact between the probe and the metal. Therefore, we subtracted 10 Ω from the measured resistance while calculating the resistivity of nanopillars. Based on our measurements, the lowest resistivity (calculated) for 20 different NP-based interconnects is 1.96×10^{-7} Ω·m. This value is even lower compared to some of the previously obtained values of resistivity for electroplated or vacuum-deposited gold nanowires having similar dimensions [19, 24].

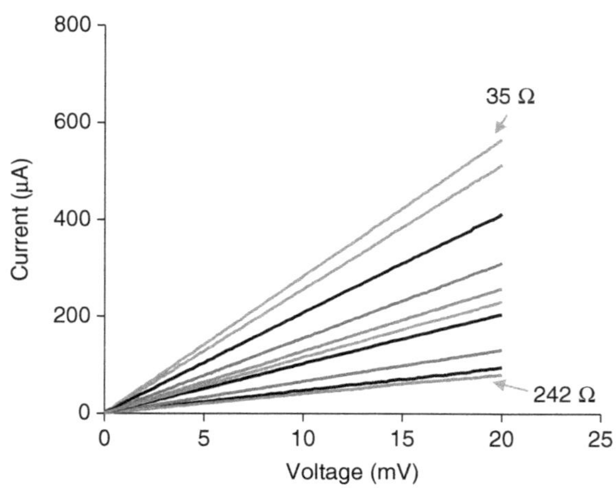

Figure 27.4 *I–V* measurements of 10 different gold interconnects.

27.6 CAPABILITIES OF THE FABRICATION PROCESS

Using the presented directed assembly approach, any other metallic NPs can be fabricated on any conductive surface. Figure 27.5 shows successful fabrication of interconnects from aluminum, copper, tungsten NPs. Although the melting temperature of tungsten is much higher than the copper and aluminum, it was possible to fuse 10-nm tungsten particles into homogenous interconnect structures by passing an additional amount of current following the assembly process [22].

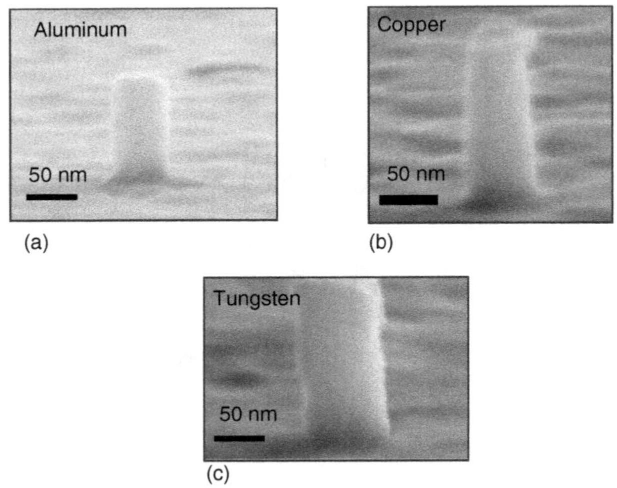

Figure 27.5 SEM images of various types metallic and nonmetallic interconnect structures manufactured by directed assembly of nanoparticles.

For each type of particles, control of the governing parameters (voltage, frequency, time, and nanoparticle concentration) enabled successful fabrication of nanopillars with controlled dimensions over large areas. The fabricated nanopillars arrays showed good uniformity over a large area. For example, the uniformity of 35 nanopillars measured at different locations of nanopillar arrays was 90.3 percent. The nanopillars in the corner of the arrays were taller than the ones in the center due to the electric field edge effect. When the nanopillars in the corners were excluded, the uniformity was over 95 percent.

27.7 COMPARISON WITH OTHER APPROACHES

The developed fabrication process enables water-based and room-temperature manufacturing of 3D interconnects without the need for high-vacuum or chemical-based processing. Therefore, this method not only significantly reduces the cost and complexity associated with conventional manufacturing, but it also provides various advantages over the conventional interconnect fabrication techniques (Table 27.1). For example, although electroplating has the advantages of being simple and cost effective in general, not every metal can be electroplated on any conductive surface [25, 26]. This makes electroplating limited in terms of the number of structures that can be fabricated on desired surfaces [25]. In addition, in some cases, electroplating process involves the use of seed layers and many chemical additives [27], which could increase the complexity and the cost involved in the process. Because the formation of the interconnects in our process is governed by the physical assembly followed by fusion of NPs on the surface and not by the chemical nucleation as in electroplating, any conducting materials can be directly fabricated on rigid or flexible plastic surfaces without requiring an intermediate seed layer or chemical additives. We also showed that interconnect formation rate of our directed assembly process is as fast as electroplating for the same interconnect geometry, density, and area [22]. This together with the scalability of the presented process makes it an excellent candidate for fabrication of interconnects in flexible electronics.

Table 27.1 Advantages of Directed Nanoparticle Assembly Method Over Conventional 3D Interconnect Fabrication Processes

Interconnect fabrication method	Room temperature	Atmospheric pressure	Chemical additives	Fabricate any conductive material	Fabricate on any plastic substrates
Electroplating	Yes/No	Yes	Yes	No	No
Thin film deposition	No	No	Yes/No	No	No
Directed nanoparticle assembly	**Yes**	**Yes**	**No**	**Yes**	**Yes**

27.8 REFERENCES

1. J. M. Blackburn, D. P. Long, A. Cabañas, and J. J. Watkins, "Deposition of Conformal Copper and Nickel Films from Supercritical Carbon Dioxide," *Science*, 294 (5540): 141–145, 2001.
2. G. A. Wurtz, R. Pollard, W. Hendren, G. Wiederrecht, D. Gosztola, V. Podolskiy, and A. V. Zayats, "Designed Ultrafast Optical Nonlinearity in a Plasmonic Nanorod Metamaterial Enhanced by Nonlocality," *Nature Nanotechnology*, 6 (2): 107–111, 2011.
3. Z. Fan, H. Razavi, J.-W. Do, A. Moriwaki, O. Ergen, Y.-L. Chueh, P. W. Leu, et al., "Three-Dimensional Nanopillar-Array Photovoltaics on Low-Cost and Flexible Substrates," *Nature Materials*, 8 (8): 648–653, 2009.
4. A. Kabashin, P. Evans, S. Pastkovsky, W. Hendren, G. Wurtz, R. Atkinson, R. Pollard, et al., "Plasmonic Nanorod Metamaterials for Biosensing," *Nature Materials*, 8 (11): 867–871, 2009.
5. O. D. Velev and S. Gupta, "Materials Fabricated by Micro- and Nanoparticle Assembly—the Challenging Path from Science to Engineering. *Advanced Materials*, 21 (19): 1897–1905, 2009.
6. V. Liberman, C. Yilmaz, T. M. Bloomstein, S. Somu, Y. Echegoyen, A. Busnaina, S. G. Cann, et al., "A Nanoparticle Convective Directed Assembly Process for the Fabrication of Periodic Surface Enhanced Raman Spectroscopy Substrates," *Advanced Materials*, 22 (38): 4298–4302, 2010.
7. K. D. Hermanson, S. O. Lumsdon, J. P. Williams, E. W. Kaler, and O. D. Velev, "Dielectrophoretic Assembly of Electrically Functional Microwires from Nanoparticle Suspensions," *Science*, 294 (5544): 1082–1086, 2001.
8. G. Li, C. Yilmaz, X. An, S. Somu, S. Kar, Y. Joon Jung, A. Busnaina, et al., "Adhesion of Graphene Sheet on Nano-Patterned Substrates with Nano-Pillar Array," *Journal of Applied Physics*, 113 (24): 244-303, 2013.
9. R. M. Erb, H. S. Son, B. Samanta, V. M. Rotello, and B. B. Yellen, "Magnetic Assembly of Colloidal Superstructures with Multipole Symmetry," *Nature*, 457 (7232): 999–1002, 2009.
10. T. Kraus, L. Malaquin, H. Schmid, W. Riess, N. D. Spencer, and H. Wolf, "Nanoparticle Printing with Single-Particle Resolution," *Nature Nanotechnology*, 2 (9): 570–576, 2007.
11. H. A. Pohl, *Dielectrophoresis*, Cambridge University Press, Cambridge Massachusetts, 1978.
12. X. Xiong, A. Busnaina, S. Selvarasah, S. Somu, M. Wei, J. Mead, C.-L. Chen, et al., "Directed Assembly of Gold Nanoparticle Nanowires and Networks for Nanodevices," *Applied Physics Letters*, 91: 063101, 2007.
13. C. Yilmaz, T.-H. Kim, S. Somu, and A. A. Busnaina, "Large-Scale Nanorods Nanomanufacturing by Electric-Field-Directed Assembly for Nanoscale Device Applications," *IEEE Transactions on Nanotechnology*, 9 (5): 653–658, 2010.
14. R. J. Barsotti, M. D. Vahey, R. Wartena, Y.-M. Chiang, J. Voldman, and F. Stellacci, "Assembly of Metal Nanoparticles into Nanogaps," *Small*, 3: 488–499, 2007.
15. S. Krishnan, A. A. Busnaina, D. S. Rimai, and L. P. Demejo, "The Adhesion-Induced Deformation and the Removal of Submicrometer Particles," *Journal of Adhesion Science and Technology*, 8: 1357–1370, 1994.
16. D. S. Rimai, D. J. Quesnel, and A. A. Busnaina, "The Adhesion of Dry Particles in the Nanometer to Micrometer-Size Range," *Colloids and Surfaces A*, 165: 3–10, 2000.
17. D. Maugis and H. M. Pollock, "Surface Forces, Deformation and Adherence at Metal Microcontacts," *Acta Metallurgica*, 32: 1323–1334, 1984.
18. U. Ramsperger, T. Uchihashi, and H. Nejoh, "Fabrication and Lateral Electronic Transport Measurements of Gold Nanowires," *Applied Physics Letters*, 78: 85–87, 2001.
19. M. Calleja, M. Tello Ruiz, J. V. Anguita, F. Garcla, and R. Garcla Garcla, "Fabrication of Gold Nanowires on Insulating Substrates by Field-Induced Mass Transport," *Applied Physics Letters*, 79: 2471, 2001.
20. Z. Zou, J. Kai, and C. H. Ahn, "Electrical Characterization of Suspended Gold Nanowire Bridges with Functionalized Self-Assembled Monolayers Using a Top-Down Fabrication Method," *Journal of Micromechanics and Microengineering*, 19 055002, 2009.
21. S. H. Ko, I. Park, H. Pan, C. P. Grigoropoulos, A. P. Pisano, C. K. Luscombe, and J. M. Frechet, "Direct Nanoimprinting of Metal Nanoparticles for Nanoscale Electronics Fabrication," *Nano Letters*, 7: 1869–1877, 2007.
22. C. Yilmaz, A. E. Cetin, G. Goutzamanidis, J. Huang, S. Somu, H. Altug, D. Wei, et al., "Three-Dimensional Crystalline and Homogeneous Metallic Nanostructures Using Directed Assembly of Nanoparticles," *ACS Nano*, 8 (5): 4547–4558, 2014.
23. Z. Tang, N. A. Kotov, and M. Giersig, "Spontaneous Organization of Single CdTe Nanoparticles into Luminescent Nanowires," *Science*, 297 (5579): 237–240, 2002.
24. Y.-J. Chen, J.-H. Hsu, and H.-N. Lin, "Fabrication of Metal Nanowires by Atomic Force Microscopy Nanoscratching and Lift-Off Process," *Nanotechnology*, 16: 1112, 2005.
25. M. W. Lane, C. E. Murray, F. R. McFeely, P. M. Vereecken, and R. Rosenberg, "Liner Materials for Direct Electrodeposition of Cu," *Applied Physics Letters*, 83 (12): 2330–2332, 2003.
26. A. Radisic, J. G. Long, P. M. Hoffmann, and P. C. Searson, "Nucleation and Growth of Copper on TiN from Pyrophosphate Solution," *Journal of The Electrochemical Society*, 148 (1): C41–C46, 2001.
27. A. Radisic, Y. Cao, P. Taephaisitphongse, A. C. West, and P. C. Searson, "Direct Copper Electrodeposition on TaN Barrier Layers," *Journal of The Electrochemical Society*, 150 (5): C362–C367, 2003.

Chapter 28

Materials for the Manufacturing of an Inkjet Printed Touch Sensor

BY

NESRINE KAMMOUN, CHRISTIAN RENNINGER, AND NORBERT FRUEHAUF *Institute for Large Area Microelectronics, University of Stuttgart, Germany*

28.1 INTRODUCTION

Printed electronics is an additive process technology, which implies that the different functional layers of an electronic system are deposited in their structured form. It aims at replacing the subtractive conventional structuring process, which is based on the idea of depositing a whole layer and then in a second time removing parts of it in order to obtain the functional structures. The subtractive process is mostly achieved through multiple photolithographic steps, using vacuum equipment and needing the manufacturing of masks, which are necessary for the etching of the layers. Therefore, printed electronics represent a promising alternative, allowing lower costs and less waste of material. Besides, the process is faster and more flexible because no mask is needed: layout adjustment or changes can simply be executed before printing. This is why the printed electronics are becoming an attractive technology to be explored and established.

Touch screens or touch sensors are generally among the most suitable candidates for the switching to printing methods. Compared with other large-scale microelectronic circuits and especially unlike integrated circuits, the touch sensors' development and progression is not ruled by Moore's law. Touch sensors don't follow the integration density rules here, because they are correlated to the properties of the human body, particularly the size of our finger and the electrical and physical properties of the human touch. This specificity of touch sensors minimalizes the functional disadvantage of the additive processes (compared to conventional processes), which basically consists in the fact that only larger structures are achievable because of the relatively low resolution of the currently available printer.

Although in the foreseeable future classical lithography and etching technique will continue to represent the most important process technology to manufacture electronics devices, printed touch sensors could significantly gain in importance in the near future. The demand for touch screens, especially in mobile devices such as smartphones and tablets, continues to rise over the world, which also creates the need to alternative manufacturing technologies such as the printing methods, in order to achieve competitive low prices and easy attractive processes.

There are different ways of printing electronics, in this study an inkjet printer: the Dimatix DMP-2800 Materials Printer is used. Inkjet printers have existed since the 1960s and are often encountered in the household as affordable color printer. They operate on the principle of selectively applying small droplets of a hydraulic fluid (in the classical case ink) to a certain surface. Thus, the printed image is created from a grid of points, similarly to the image created from the individual pixels on a screen. The distance between two different points is the resolution of the image and is measured in dpi (dots per inch). This printing method has an additional advantage: it is a contactless structuring process.

There are two main technologies in use in contemporary inkjet printers: continuous inkjet printer inkjet (CIJ) and drop-on-demand (DOD). A CIJ is based on creating a continuous ink stream by pressing it through the fine nozzles. Continuous inkjet is the oldest inkjet technology and it is an established one, it is a fast efficient printing way but the resolution remains limited compared to the DOD technology, which is actually used by the Dimatix printer. Whereas many home color printers use the thermal DOD, the Dimatix printer is based on the piezoelectric DOD principle. By applying a voltage deforming a piezoelectric membrane a drop from the ink is shaped and pressed away through the nozzle. The piezoelectric technology allows a wider range of usable inks because there are no thermal requirements on the ink as in the case of the thermal DOD.

Inkjet printers are very flexible. So they are well suited for many different materials, which can be applied in several layers on different substrates. And because for printing there is no mask required, every change in the layout of the structures can be made quickly and easily, which is a huge flexibility advantage compared to the lithographic masks, which can't be changed after being manufactured with high costs and time-consuming processes. Thus, inkjet printers are also ideal for use in rapid prototyping and for applications with high processing flexibility requirements.

To meet the expectations on printing electronic systems, material and process parameters are strongly correlated. It means that the ink has to fulfill different requirements like controlled drying, appropriate viscosity, suitable surface energy, suitable surface roughness and stability. The process parameters have to be accordingly adjusted to the particular properties of the used ink. In this study, following the necessary optimization of the different materials especially indium tin oxide (ITO), the corresponding printing process was developed and fully printed touch panels were manufactured [1].

28.2 MATERIAL AND PROCESS OPTIMIZATIONS

ITO is a transparent conductive oxide and typically consists of 90 percent indium and 10 percent tin oxide. Because of its good optical and electrical properties, it mainly occurs as an electrode material in different screen manufacturing applications, as in liquid crystal displays and touch screens. The ITO thin layers are then usually deposited by sputtering and then etched. For printing purpose an ITO ink is prepared as a mixture of nanoparticles present in a solvent. During annealing, the solvent evaporates, leaving only the nanoparticles, which form a solid ITO layer.

For these experiments, an ITO ink is used, based on ITO nanoparticles, which represent 40 percent of the ink. The ITO ink layers have to be annealed after being deposited or printed, in order to attain the necessary electrical and optical properties. The ITO ink contains a binder, which has to be UV-activated.

The standard postannealing process for the ITO ink layers is

1. Drying: for 30 min at 120°C
2. UV curing: 2 h in the UV cabinet (UV A-B-C, 200–400 nm, 33 mW/cm^2)
3. Thermal annealing: for 90 min at 300°C

The ITO ink is first filled in the corresponding syringe with the appropriate filter (1 μm). The visual aspect of the so-coated and annealed substrates presents no detectable coloration, and has a high transparence level.

The main measurements that were performed on the layers are four-point resistivity measurement and transmission measurement. The main results are summarized in Table 28.1, which shows the electrical and optical properties of the 800-nm ink layer.

It is important to note that the ITO layers can reach better resistivity values when using higher temperatures: at 650°C, the measured resistivity is only of 0.0378 Ω·cm. But taking into account the fact that such high temperatures are not compatible with the display manufacturing processes, mainly because the commonly used glass cannot withstand temperatures more than 300°C, it seems clear that the goal of the optimizations is to reach acceptable electrical and optical properties, without using a high-temperature annealing step.

However, there are still problems concerning the ITO layer properties and the process:

- Postannealing process duration is too long (~4 h).
- Sheet resistance is relatively high: 0.72 kΩ/□ for 800-nm ITO layer (resistivity: 0.057 Ω·cm).
- Aging problem: The stability of the ITO ink–based layers is a crucial issue for the further processing.

Indeed the electrical properties undergo significant changes, especially under the normal storing conditions: air. Under a temperature of 80°C or under nitrogen atmosphere, the so-stored substrates show a better aging behavior, the changes of the measured values are significantly lower, as shown in Fig. 28.1. Still, the achieved electrical properties of the layers undergo a considerable negative aging effect, which is due to the relative instability of the inks used, which react with humidity, light, and oxygen. The organic bonds, mainly those achieved through the UV curing in the ink layer seem to be the cause of this aging problem.

Table 28.1 Electrical and Optical Properties of the Ink-Based Layers

Sheet resistance (kΩ/□)	0.72
Resistivity (Ω·cm)	0.057
Transmission (%)	92.96

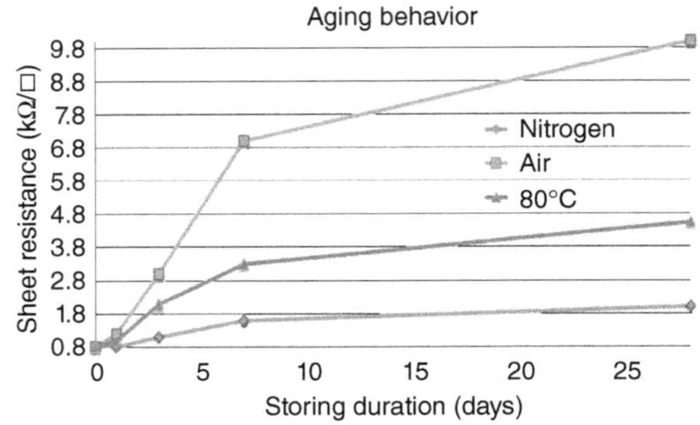

Figure 28.1 Aging behavior of the ITO ink layers.

To investigate the ITO ink layers, roughness measurements were performed, using an atomic force microscope (AFM). This equipment allows the measurement of surface properties on a nanometer scale, with a maximum range of 40 × 40 μm. The achievable resolution is highly dependent on the needle geometry and that is why many experiments with different needles and with different measurement speeds are needed to obtain meaningful results. The AFM pictures in Fig. 28.2 are roughness measurements of a typical ITO ink layer, on 10 × 10 μm patterns.

The general aspect of the ITO ink layer is shown on the left AFM image and presents the "grainy" profile of the layer. On the left side, a focused image on a 2 × 2-μm squared pattern shows the typical sponge-like aspect of the ITO layers. The ITO ink layer has considerable roughness values of around 200 nm (maximum height), which is also coherent with the high resistivity of the layers compared to sputtered ITO.

This visual aspect of the ITO layers was the basis of the optimization idea; to improve the resistivity of the layers, more chemical bonds have to be created in order to *smoothen* the layer and thus facilitate the electron transport. This is basically the doping principle. ITO is already doped with tin, so experiments were performed to investigate the effect of more tin doping on the properties of the ITO layers. The investigated tin proportions are 5, 7.5, 10, 15, and 20 percent (relatively to the ITO powder mass in the ink). Over 20 percent tin proportion, a brown coloration on the coated substrates is visible. This is why the investigated proportion of tin does not go over 20 percent.

Figure 28.3 shows the sheet resistance of a 800-nm ITO layer with different additional tin shares. As expected, the electrical properties of the ITO ink layers improve with increasing tin content. The sheet resistance decreases from 0.7 to 0.18 kΩ/□.

This shows clearly that tin proportions in the range of 20 percent are efficient to achieve better sheet resistance values without using a high-temperature process. However, as expected, adding more tin to the ink is not necessarily advantageous for the optical properties of the layers.

Although visually, there is no detectable brown coloration on the substrates with higher tin proportions, preliminary spectral measurements proof that the additional tin has mostly a negative effect on the transmission.

Based on these first results, the inks with 15 and 20 percent tin content were further investigated and the transmission of the corresponding layers was measured over a large spectrum and plotted in Fig. 28.4, which shows that the transmission decreases drastically with the increasing tin content, which leads to low transmission values for both inks in the visible wavelength interval. For 20 percent additional tin, the transmission is clearly less than 90 percent, which is a too low value for display applications, such as touch panel.

Because of the correlation between material and process optimization, the aim at this point was to adjust the process parameters to the ink with additional tin, so that the properties of the layers can be improved, and an acceptable compromise between optical and

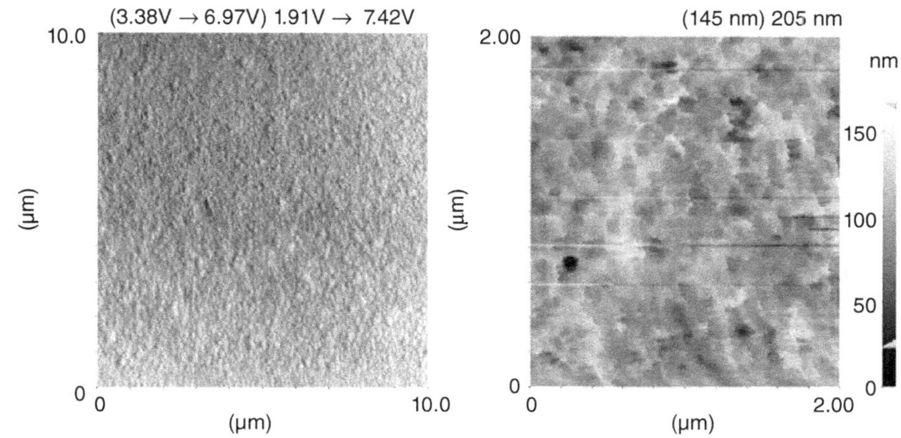

Figure 28.2 AFM pictures of the ITO layers.

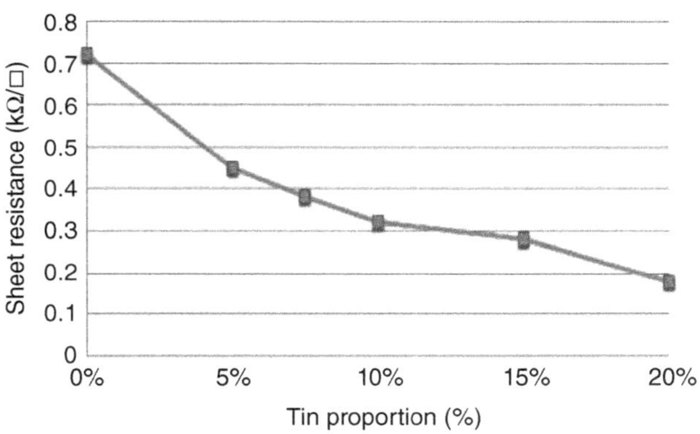

Figure 28.3 Sheet resistance depending on tin content of an 800-nm layer.

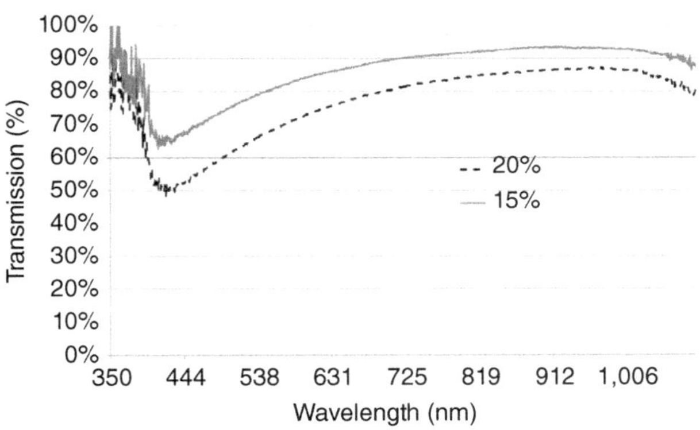

Figure 28.4 Transmission depending on tin content.

electrical properties can be achieved. Concerning the aging problem of the ITO ink layers, it can be assumed that the organic bonds formed through the UV activation of the binder system of the ITO ink are unstable, which may lead to the rapid aging of the layers. So the first process optimization was to omit the UV-activation step, and to compare the results concerning the layer characteristics and the aging behavior. This experiment was carried out on the standard ITO ink and the results are shown in Fig. 28.5.

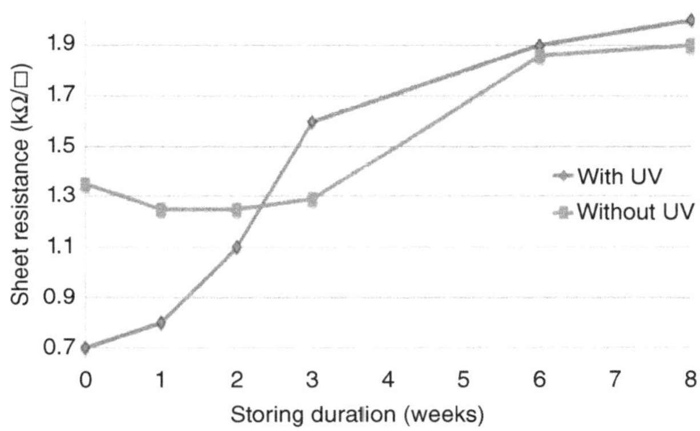

Figure 28.5 UV-curing effect on the aging behavior.

For the standard ink, the optimum sheet resistance is not achieved without UV activation. The immediate sheet-resistance measurement gives a value of 1.35 kΩ/□ instead of 0.72 kΩ/□. But observing the aging behavior curves of the layers with and without UV confirms the hypothesis, that the UV activation leads only to temporary, fast-aging organic bonds, which makes the electrical performance better at first, but already after the first 2 weeks this "positive" UV effect progressively disappears. Besides, increasing the tin content of the inks compensates this positive effect of the UV activation on the sheet resistance, in a more stable way. Indeed, it seems that the tin-doping mechanism is more durable than the through UV-curing created bonds. The aging behavior of the layers obtained without UV activation, depending on the tin content of the ink, is shown in Fig. 28.6: for 15 and 20 percent tin shares the optimal sheet resistance is immediately achievable without UV step and the ITO layers with 20 percent tin content have a significantly improved aging behavior.

Concerning the low transmission problem, the goal was to find a tradeoff between good electrical and optical properties, which means improving transmission without loosing much conductivity. Based on this idea three different changes of the postannealing process were successively experimented and the corresponding transmission values of the ITO layer with 20 percent tin are then shown comparatively on Fig. 28.8.

1. *Omitting the drying step*. The transition from drying at 120°C to annealing at 300°C without UV step probably leads to the formation of additional haze making the transmission drop further. This is why a process without drying was tested to possibly improve the transmission.

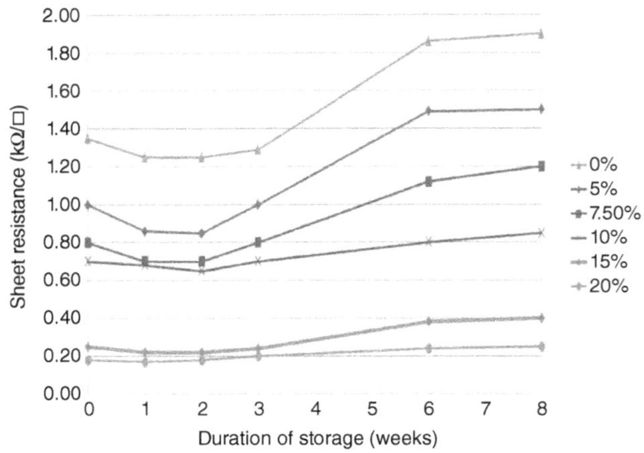

Figure 28.6 Aging behavior depending on sheet resistance.

The thick continuous curve in Fig. 28.8 shows that the transmission is improved in comparison to the standard process values shown by the thin continuous curve. The whitish appearance of the coated substrates is less visible without the drying step which indicates less haze.
This explains the positive effect on the measurable transmission values.

2. *Shortening the annealing duration to only 30 min.* The duration of the annealing step is also affecting the optical characteristics, because it is the origin of additional haze and particle agglomerations inside the layer, which leads to lower transmission. The initial duration of 90 min annealing was the optimal duration to reach the good sheet-resistance values for the standard ink, but the higher tin content makes the layers more conductive, which allows the reducing of the duration to only 30 min. The dots curve in Fig. 28.8 shows the corresponding further improved transmission (compared to both of the continuous curves).

A comparison of the necessary total process duration to obtain the optimal sheet resistance for the different inks depending on tin contents is shown in Fig. 28.7.

This figure shows that reaching the optimal sheet resistance after 30 min is only possible for the highest tin contents. For 15 and 20 percent tin content, the final postannealing process duration is 30 min, only a thermal annealing step, at 300°C is necessary. Drying and UV activation are both omitted.

3. *Thin layer.* As a compromise between good optical and electrical properties, choosing a thinner layer provides the advantage of having a better transmission, with still acceptable sheet resistance. The interrupted line curve in Fig. 28.8 shows the transmission of a 400-nm layer

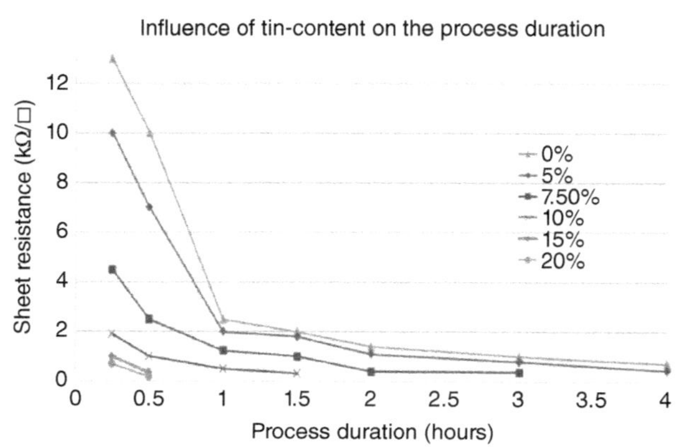

Figure 28.7 Total process duration depending on tin content.

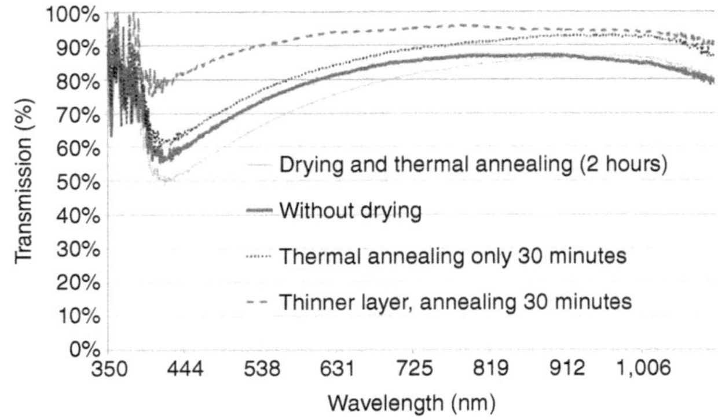

Figure 28.8 Comparison of the postannealing process alternatives.

after the annealing step of 30 min at 300°C, and it is clearly improved over the whole spectrum.

Figure 28.9 is, like Fig. 28.4, a comparison of the transmission spectra of the ITO inks with 15 and 20 percent additional tin, after the different process optimizations listed above were achieved. It shows that the 20 percent tin ink has the best transmission performance with the corresponding adapted process.

Besides, the 20 percent ITO ink leads to a resistivity of 0.014 Ω·cm, which is even a better value than the 0.0378 Ω·cm resistivity achieved at 650°C. The comparison results between the standard ITO ink and the ITO ink with 20 percent additional tin are summarized in Table 28.2.

The performed material optimizations and process improvements allowed the achievement of a process time reduction from 4 h to 30 min, and the layers obtained are thinner, with better conductivity and acceptable transmission (> 90 percent) and an improved aging behavior.

Resulting layer characteristics are good enough for the intended applications, that is why the optimized process, and the 20 percent ink will be used in the next steps.

28.3 ADDITIVE PROCESS PARAMETERS

For all the following experiments, the Dimatix DMP-2800 Materials Printer is used.

For each to-be-printed-ink a separate print head, the cartridge, is required. It has 16 nozzles in a row with 254 μm spacing. The nozzle diameter is about 21.5 μm. To make the nozzles jet the ink, the Dimatix printer makes use of the piezoelectric effect: applying a voltage deforms a membrane, so that a drop is jetted out of the nozzle.

This voltage can be changed accordingly to the jetting aspect and also to the printing quality. For the printing experiments performed during

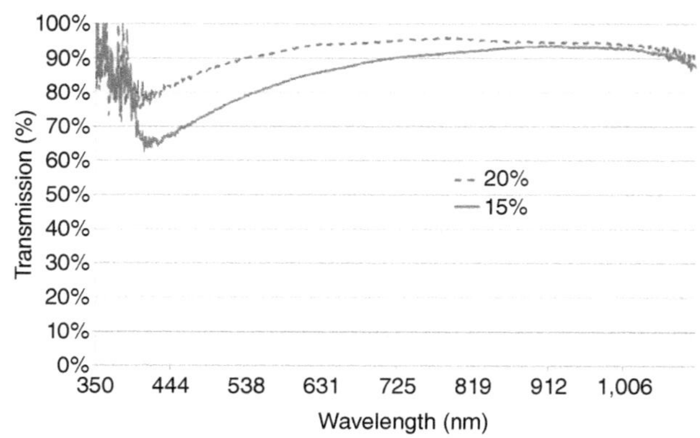

Figure 28.9 Transmission depending on tin content after the process optimizations.

Table 28.2 Electrical and Optical Properties of the Ink-Based Layers with and without Additional Tin after the Process Optimizations

Tin content	0%	20%
Layer thickness	Sheet resistance (kΩ/□)	
800 nm	0.72	0.18
400 nm	1.42	0.35
Resistivity (Ω·cm)	0.057	0.014
Transmission (%) at 550 nm	92.96	9054

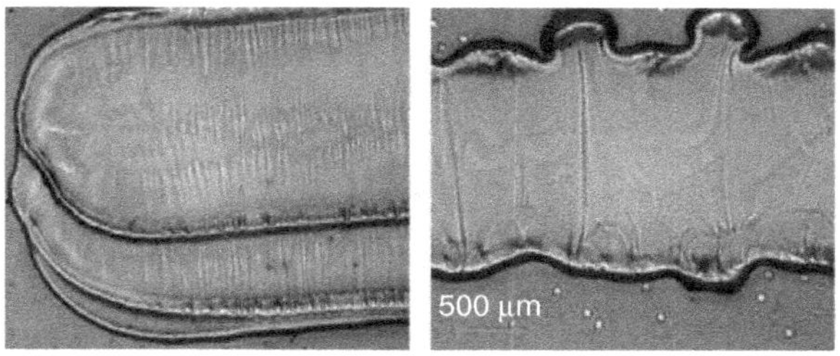

Figure 28.10 Printed structures: 60 versus 30°C.

this study, the main parameters adjusted to achieve the desired printing quality were the temperature, the drop spacing, the pretreatments, and the ink viscosity.

28.3.1 Temperature

The temperature of the substrate is variable from 28 to 60°C. Figure 28.10 shows the printed structures depending on the temperature. Printing under high vacuum plate/substrate temperature (60°C) leads mostly to relatively well-defined structures. The edges are regular enough and there is no overwhelming of the fine patterns. The problem is that the temperature is so high that the drying of the single drops or structure parts happens too fast, and the resulting printed structure is not homogeneous enough. Printing under lower vacuum plate/substrate temperature makes the drying of the single parts of the structure slower, so that the different drops flow together into one homogeneous structure before the drying occurs. This eliminates the overlapping. However, the lines are no longer straight, the edges of the structures are rippled, and the smallest structures are totally overwhelmed in this case, which strongly affects the printing resolution.

To summarize, high temperatures lead to better defined edges of the structures, but an overlapping shapes pattern is not avoidable (Fig. 28.10, on the left). Low temperatures give a better flowing of the printed material into one homogeneous layer, but the edges are fuzzy, it deteriorates the printing resolution (Fig. 28.10, on the right). The optimum is a compromise depending on the dilution of the ink (viscosity) and the drop spacing used. The printing experiments showed that a substrate temperature between 40 and 50°C is suitable for an efficient printing of the ITO ink.

28.3.2 Drop Spacing

Drop spacing (DS) is the center-to-center distance of the drops deposited by the printer. It can be changed in 5 μm increments, and it determines the resolution on the X-direction and the necessary angle of the cartridge to get the same resolution. Changing the drop spacing changes the amount of jetted ink per area, which affects the resolution and the transmission of the printed structures

Experiments are performed by changing the drop spacing from 5 to 40 with 5-μm steps. In the following only the extreme cases are presented to understand the effect of this parameter. Figure 28.11 shows, on the left, the printing results by 60°C with a drop spacing of 40 μm, this aspect of the structures is highly representative of the printed structures in case of drop-spacing values equal or above 30 μm. These high values do not allow the structure to be printed as a continuous layer, and the individual ink drops can be seen on the substrate. This shows that a high drop spacing like 40 μm implies an insufficient amount of printed ink so that the single points of the printed lines are visibly separated, there is no continuous printed layer development in this case, which makes it unusable.

On the right side of Fig. 28.11, the other extreme case: drop spacing 5 μm is presented. The figure shows that besides the problem of overwhelming the fine structures, the problem with the low drop-spacing values (especially 5 and 10 μm) is that the printed structures are not "transparent" anymore. Due to such low drop spacing, the amount of ink dropped is relatively high, and the layer thickness is increased so that the aspect of the structures is dark and not transparent, making the use of such low drop-spacing values also impossible. The printed structure shown in the figure is flooded and the layer is too thick to get a good transmission.

In conclusion, it appears that moderate drop-spacing values going from 20 to 30 μm are the most suitable for the printing of ITO ink, the value 25 μm is the optimum, which guarantee a compromise between the good printing quality, a good transmission of the layer and the attainable printed structures resolution.

28.3.3 Pretreatments

There are two major pretreatments, which were tested:

1. *The UV-O_3-cleaning.* At a wavelength of 254 nm, and with a power of 18 mW/cm^2, the substrates are exposed for 10 min just before they are printed. This is a simple and economical way to clean the surface of the substrate, which increases its surface energy and thus improves the printing resolution.

2. *The nonreactive plasma pretreatment.* This is a sputtering form where the substrates are bombarded with energetic nonreactive gas particles (argon), which leads to the etching of the organic impurities

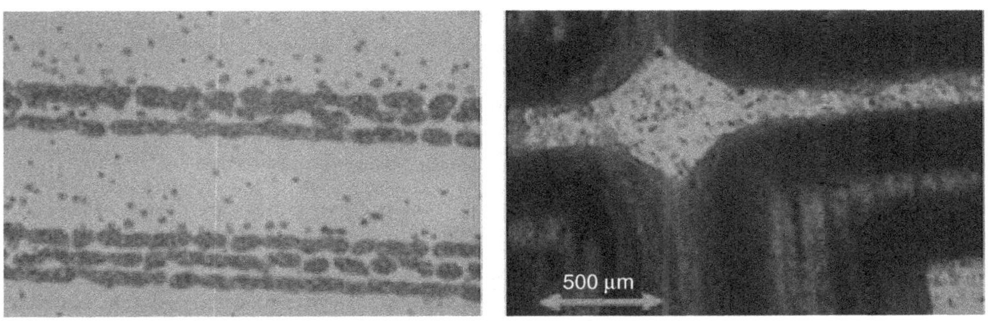

Figure 28.11 Printed structures: drop spacing 40 μm versus 5 μm.

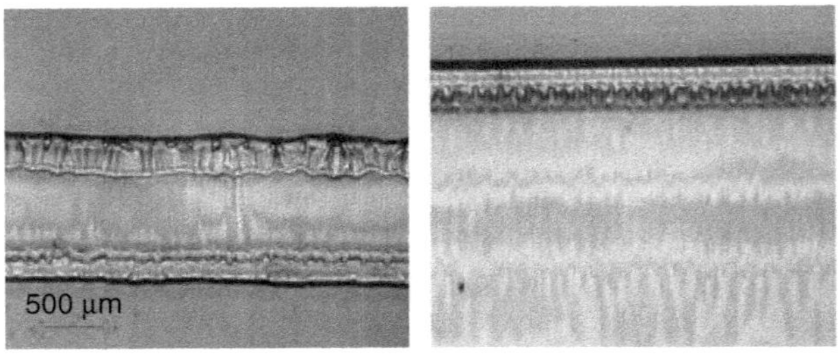

Figure 28.12 Printed structures: UV versus plasma.

on the substrate surface. This mechanical cleaning activates the surface of the substrate. Hence, the printing quality is improved.

Both of these processes lead to increasing the surface energy of the substrates, which yields to better wetting characteristics of the printed materials. Figure 28.12 shows the printed structures depending on the pretreatment used. The UV-based one (on the left) leads to a more controlled drying behavior of the ink drops on the substrate compared to plasma, where the ink typically flows over the edges of the wanted structures. Hence, the UV cleaning leads to better defined structures and better printing resolution. Besides it is a simple, economical process to activate the substrates on a vacuum-free way.

28.3.4 Viscosity and Waveform Editing

Figure 28.13 shows some fine conductors (200 µm), which were printed as interrupted lines, though using optimal temperature and drop-spacing values and appropriate pretreatment.

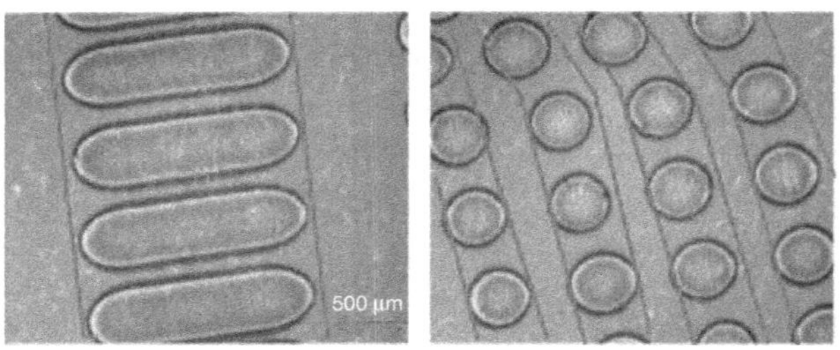

Figure 28.13 Fine conductors printed as interrupted lines, red lines indicate the boundaries of the intended printed structures.

Switching to smaller drop spacing leads to a total flooding of these fine structures. Changing the temperature and pretreating of the substrates is also obviously not the solution in order to optimize the printing quality. The other crucial parameter which can be changed is the viscosity. The previous printing experiments are performed with a 25 percent diluted ink. The ink was diluted with an appropriate solution and the diluted ink makes the appropriate jetting of the nozzles easy to achieve. The low viscosity helped the jetting, but was not appropriate for the printing quality: the wetting of the droplets was not enough to achieve continuous layers.

Hence, at this point the challenge was to increase the viscosity by omitting the dilution of the ink, and then still make the nozzles jet. This is possible by using the waveform editing application, which allows the control of the shape of the piezoelectric pulse to the nozzles whose jetting behavior is seen and adjusted by the drop watcher of the Dimatix printer. Experimenting with several possible waveforms led to the determination of an optimized piezoelectric pulse adapted to the drop jetting behavior of the undiluted ITO ink. This optimization allowed for the increased material viscosity a reliable and reproducible jetting of the nozzles.

Figure 28.14, on the left, shows the structures obtained with the diluted ink, printed by 60°C, without any pretreatment. It is obvious for all the reasons explained above that under these conditions the

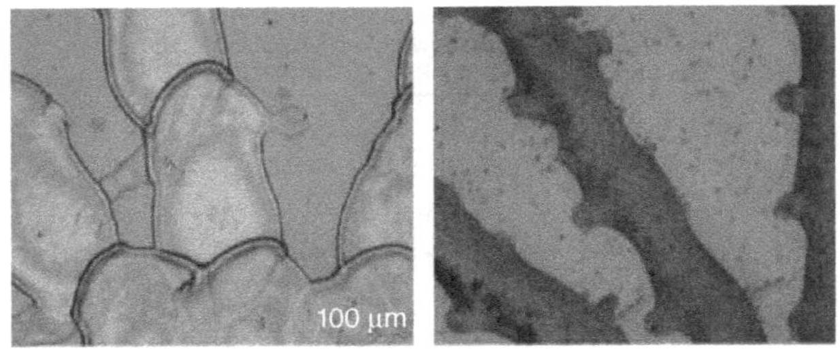

Figure 28.14 Printing with diluted versus undiluted ink.

ink flow too much on the substrate and the resolution is not good enough: the fine conductors cannot be printed. Under the same conditions, just switching to the undiluted ink, the printed structures are changed. The definition of the structures' edges is improved (Fig. 28.14, on the right). But, with this viscosity and under high temperature (60°C), the flowing of the ink drops is not enough.

The right combination of the above individually investigated parameters led to an established additive process for the ITO ink: the final parameters were: temperature 45°C, drop spacing 25 µm, UV-ozone pretreatment, and undiluted ink (viscosity about 20 cP).

The so-printed structures are shown in Fig. 28.15. The resolution of 100 µm to 200 µm was achieved, the printing accuracy was optimized and the frequency of drops and leaks was minimized, so that the ITO-Ink was reliably inkjet-printed with a significantly improved reproducibility.

28.4 TOUCH-PANEL DEMONSTRATOR

A PCAP (**p**rojected **cap**acitive) touch sensor is based on the principle of projecting the touch of the finger on a matrix of capacities. The touch is essentially sensed as a change of the corresponding capacitance at a definite point of this matrix. The touch-sensor consists hence of several layers, two of them are each a conductive, transparent material patterned as an array of parallel electrodes, which is rotated in the second plane to 90°, forming the rows and the columns of the sensor matrix. Between these rows and columns there is a sandwiched electrically insulating layer, ensuring the capacitive functioning of the sensor [2].

In the most cases ITO is used as a transparent conductor. Because of the rarity and the high costs of ITO, there are different alternatives which were tested: different new materials such as metal nanowires, new transparent conductive oxides, carbon nanotubes, and graphene, are proposed for next generation PCAP touch screen [3–5]. In the case of this study, the printing of the ITO serves the purpose of material and cost saving through the additive form of the process. Therefore, several experiments were performed to use the optimized ITO ink and the corresponding additive process parameters in order to manufacture a fully printed touch sensor.

A Microchip Projected Capacitive Development Kit (Type MPMXV3) equipped with test controller MTCH6301 and a "conventional" reference demonstrator were used to characterize the printed touch sensor. The reference demonstrator was replaced by the printed touch sensors whose functionalities were measured using the corresponding software utility.

The microchip test kit uses two main characterization mechanisms: the self-capacitance measurement which measures the capacitance of a conductor in regard to its surroundings and the mutual-capacitance measurement which determines the capacitance between two conductors at each intersection of a row and a column in the sensor matrix [6].

The self-capacitance is only suitable for one finger per line touch. When two fingers are simultaneously on one line, it gives the same result as one finger, potentially causing ambiguity of touch point, the so-called *ghost*-positions problem. To make multitouch possible, the mutual-capacitance measurement is necessary because it addresses each node of the sensor individually so that all touch points are unambiguously sensed.

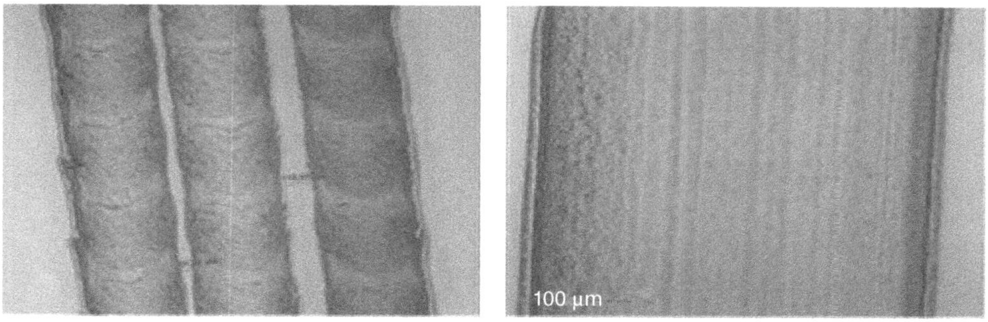

Figure 28.15 Optimized printing of the ITO structures.

28.4.1 Layout of the Touch Sensor

The purpose of the study was to manufacture and characterize a functional, fully printed demonstrator, which consists of a 9 × 12 electrode sensor (approximately 7 × 9 cm). The layout, shown in Fig. 28.16, is composed of multiple diamond-shaped elements, each connected to the next with a conductive neck. Necks connect diamond elements at crossings, diamond elements are used to maximize the exposure of sensor electrodes to the touch.

The capacitor layer structure at each intersection of the ITO rows and ITO columns is composed of the polyimide (PI) sandwiched between the ITO row and column necks. The changes of this capacitance occurring due to the finger touch are on the basis of the functioning principle of the PCAP sensor.

28.4.2 The Printed Materials of the Touch Sensor

To print the touch sensor different materials were used: obviously the for-this-purpose-optimized ITO ink was used to print the electrode matrix. As for the conducting lines connecting the touch sensor to the microchip, silver in the form of a fluid ink was used. Small silver nanoparticles are mixed in a solution and then combined during annealing (by laser or in the oven) to a continuous silver conductive layer. The silver ink (Type Metalon JS-B40G) was printed to obtain silver conductors with a 2.54-mm raster allowing the sensor to be connected to the controlling and characterizing unity. An annealing temperature of 180°C is already enough to get a conductive silver layer.

However, using the silver ink comes with a relatively increased resistivity compared with the obtained values in case of conventional silver. This is also generally the case of the actually existing inks and printable solutions, which cannot ensure exactly the same good properties of the original material. The relative quality decreasing of the used materials has to be compatible with the chosen printed application as in this case for the touch sensor.

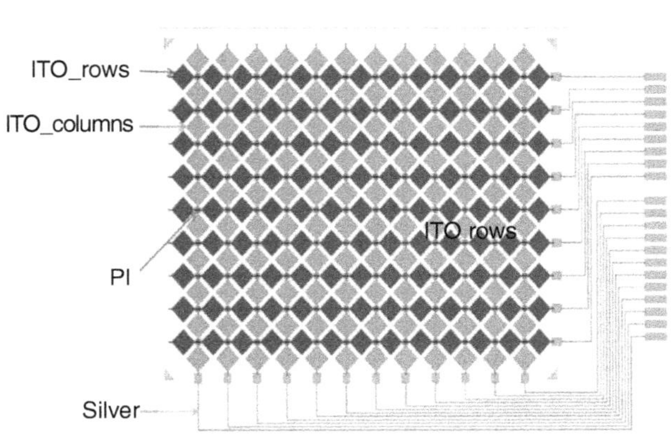

Figure 28.16 Touch-sensor layout.

As for the insulator layer between the ITO electrodes, there were different possibilities investigated, such as photoresists, which did not fulfill the thermic requirements for the necessary insulator for this touch sensor. Indeed, the insulating material has to be not only a good dielectric in a printable solution (suitable viscosity required) but must also be almost transparent and thermally resistant, because it has to withstand the postannealing temperature of ITO, 300°C, without being degraded. A suitable material category was the polyimides. A solution based on the polyimide SE-4811 (Type 0526, Nissan Chemical Industries), diluted with the solvent: ethylene glycol monobutyl was used in the printing experiments of this study. To reach a good printability, different formulations of the polyimide mixture were tested and the optimal solution was the one with only 5 percent polyimide. Besides improving the viscosity and the printing quality, this low content allows actually to get thinner printed layers. After printing, the layer is annealed at 220°C to achieve the imidization of the polyimide.

28.4.3 Printing Process of the Touch Sensor

The printing of the silver connecting conductors was uncritical: after adjusting the printing parameters, the printed structures were well-defined and functional.

The printing of the ITO rows and columns as well as the insulating polyimide squares was successfully performed after optimizing the corresponding parameters. The individual printed components presented a good printing quality, the optical control and the microscope visualizing of the printed layers show that the desired resolution is achieved, and the layers are continuous without any breaks or interruptions. The resistivity measurements of the ITO diamond structure were corresponding to the expected values as described above.

Figure 28.17 shows the dimensions of the diamond-shaped elements as they were designed in the layout (Fig. 28.16) and as they are actually printed, after optimizing the printing parameters to obtain a maximal resolution. The printed structures dimensions are about 20 percent bigger than the design dimensions, leading to smaller gaps. This dimension change is, as proofed through many printing experiments, a reliable mechanism which has to be considered when designing the printing layout. As in this case, the difference between printed structures dimensions and layout dimensions was taken into account in the designing phase, so that the actual layout was 20 percent bigger than the desired one.

28.4.4 Capacitive Performance of the Touch Sensor

Because the structures had acceptable printing quality, printing experiments were then performed to characterize the capacitance structures in order to evaluate the insulating quality of the printed polyimide.

The capacitances of the conventional touch sensor demonstrator (which is included in the microchip development kit) were measured by an LCR meter, to have a reference value for a functional touch sensor. The measured values are found in the range from 10 to 15 pF. When approaching a finger to the touch sensor the measured capacity

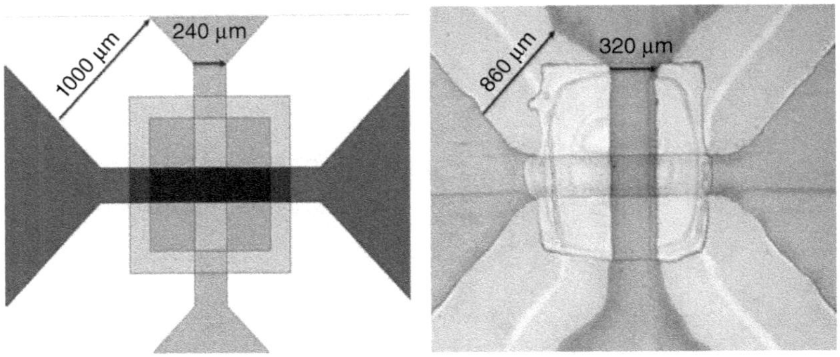

Figure 28.17 Dimensions of the designed and the printed diamond-shaped elements.

decreases by approximately 0.5 pF, when the touch happens on the cover glass which is a protection layer for the sensor. Without the cover glass the capacitance decrease is about 6 pF. For the printed demonstrators detecting the touch on the frontside is equivalent to the touch measurement without protection glass, whereas the touch on the backside of the printed sensor is comparable detecting the touch on the conventional sensor with cover glass, because the glass substrate itself is here used as a protection, when the touch does not happen directly on the electrodes.

On the basis of these measurements the goal was to achieve printed capacities, which are as matched as possible with the reference values of the microchip sensor. Capacitance calculations based on the geometric and electric properties of the capacitor (based on the well-known capacitor formula: $d = \varepsilon_r \varepsilon_0 \cdot \frac{A}{c}$) show that the polyimide thickness has to be about 300 nm to achieve the capacitance reference values of the conventional demonstrator.

However, the actual measurements of the printed capacitors showed that they mostly present leakage currents and short circuits and are thus unusable. To better understand the origin of these short circuits, measurements were performed with a profilometer, to investigate the thickness profile of the printed layers.

Figure 28.18 shows the so-obtained profiles of the ITO neck structure and the overlaying polyimide square. And the first observation is that the profile of the ITO conductor is mountain shaped, with 3 µm maximal height whereas the polyimide is only 200 nm layer, which corresponds to the ITO roughness (see Fig. 28.2). This means that the printed polyimide likely flows down from the ITO electrodes leading to several pinholes and leakage currents. This also shows that printed thin polyimide layers in the desired range of 300 nm have insufficient insulating quality, which means that thicker layers are necessary for a functional capacitor.

The second observation made on the basis of these measurements was that the printed polyimide drops dry too fast on the edges of the squared structure leading to a massive coffee stain effect. This phenomenon occurs during the evaporation of liquids containing solid particles resulting in a typical annular-shaped pattern because almost the entire material is transported to the edge of the structure [7].

On the basis of the established explanations for the insulation problem, different approaches were tested to help the polyimide stay on the ITO neck, and to obtain a sufficient insulator layer thickness on the electrode (> 500 nm). Some of the experiments aimed to improve the wetting of the polyimide on the ITO, through UV activation or a nonreactive plasma treatment, but to no avail. Polyimide printing experiments with increased substrate temperature were also performed in order to achieve a fast drying of the polyimide, preventing its flowing of the ITO electrode, but the coffee stain effect was just more pronounced, confirming the fact that the fastest drying occurs indeed at the edges of the structure. Similarly, the experiments with drop spacing less than 25 µm, which is the optimum value for the polyimide printing, resulted in thicker edges instead of a homogeneous, sufficient thickness on the ITO electrode.

Because the coffee stain effect was unavoidable when printing the polyimide, the idea was to use it as a part of the solution. For this purpose multiple layers printing was tested, based on the principle of using the first printed layer coffee stain as a wall confining the further printed layers which land inside of this wall structure and thus stay on the top of the electrode. Different variations of this approach were tested, like printing the layers on each other or in each other, with or without UV activation. Figure 28.19 shows that even printing four polyimide layers is not a success guaranty, because, as in the case of the figure, it can induce successive coffee stain edges where the supplementary polyimide preferentially flows, without wetting in the middle of the structure.

To help the wetting of polyimide on polyimide, an intermediate annealing step for 30 min at 220°C was performed. This step worked as an initiation of the polyimide internal bonds inside of the first layer, which helped the second one adhere to the already thermally cured crystal structure. In Fig. 28.20, on the left, the layer thickness profile of the resulting polyimide layer with an annealing step between the

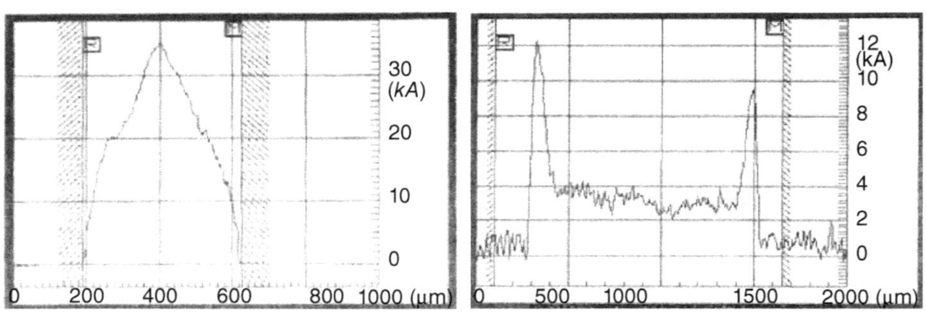

Figure 28.18 Profile of the ITO layer and the polyimide layer.

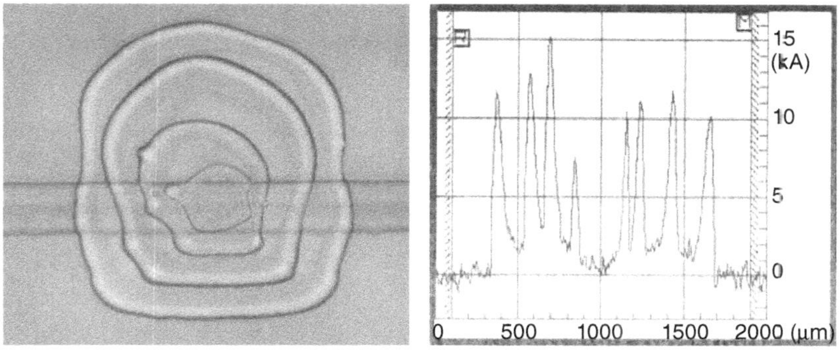

Figure 28.19 Profile of four in each other printed polyimide layers.

two printing processes, is presented proofing the success in achieving thicker insulator layers (here 2 µm) using this method.

Some test runs allowed to achieve a polyimide total thickness of up to 3 µm. However, together with the lower ITO layer the so-obtained total thickness of the stack is about 5 µm or more. The upper ITO film is thus printed on an extremely uneven surface with a considerable thickness variation of 5 µm between glass and layer stacks. Hence, this situation represented a mechanical stress issue for the upper ITO layer, which was visible at the intersection where several cracks of the ITO column were detected (Fig. 28.20, on the right).

Hence, thick polyimide layers, about 2 µm and more, induced problems with the upper ITO neck. This latter cannot withstand the differences of the thicknesses between glass and printed layers without showing damages. As long as the cracks occur parallel to the electrode direction, only the conductivity deteriorates, but the electrode is still functional. However, the perpendicular cracks can cut into the electrode, which becomes idle and is not functional anymore.

For this reason, the film thickness of the polyimide must be carefully set. Too thin layers lead to leakage currents and short circuits whereas too thick layers lead to breakthroughs of the upper ITO electrode and to idle connections. Many experiments showed that the optimal value is in the range between 600 nm and 1 µm.

The final polyimide printing parameters were based on printing two squares, respectively, of 1.08 mm and 0.9 mm with an intermediate annealing step. The first square is printed at 60°C, so that the edges, which are the wall structure, are well defined, and the second one is printed at 40°C to allow the polyimide to flow and form a homogeneous layer inside of the surrounding square. Figure 28.21 shows the so-printed polyimide insulating structure, and its corresponding profile, which presents homogeneous layer thickness around 1 µm between the electrodes.

Following the development of this successfully adjusted polyimide process, the capacitance measurements confirmed the functionality of the printed capacitors. Two functional touch sensor demonstrators were printed and characterized. The first one was printed using the classic design (as in Fig. 28.17) and the second one uses a hybrid design based on printing a silver neck instead of ITO in the upper electrode bridge part. This modification improves the capacitance value range and their homogeneity on the whole sensor because of the better conductivity and stability of the printed silver compared to the printed ITO.

Figure 28.22 shows how the printed touch sensor actually looks like.

Depending on their touch sensibility, the touch panels were operated on the backplane (as in the case of a protection glass) or in the frontplane. Table 28.3 summarizes the performances of the fully printed demonstrators in comparison to the reference "conventional" one.

Figure 28.23 shows an example of the multitouch-sensing performance of demonstrator 2, which is almost as good as the one of the reference sensor and includes its different touch functionalities.

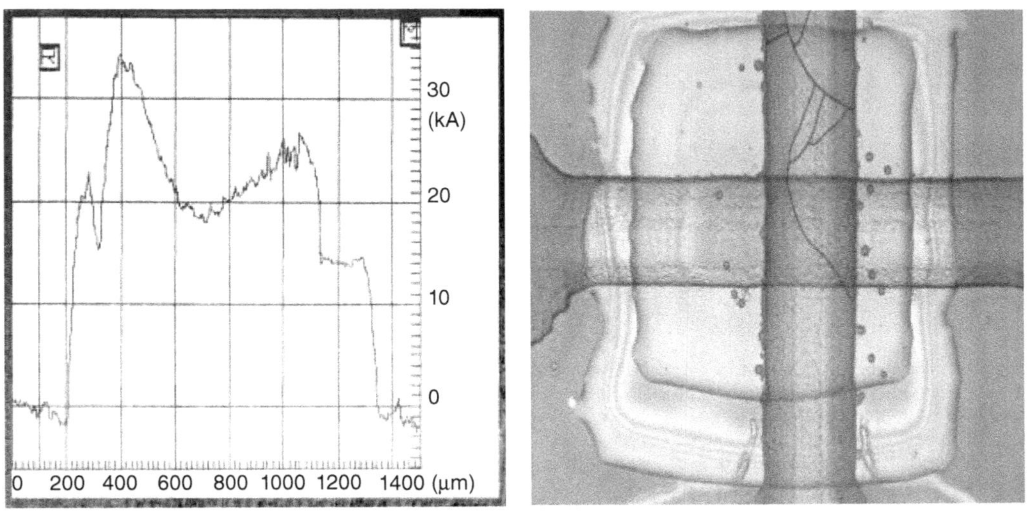

Figure 28.20 Profile of two in each other printed polyimide layers with an intermediate annealing step (left) and the resulting upper ITO layer, which is clearly breaking because of the high thickness of the polyimide (right).

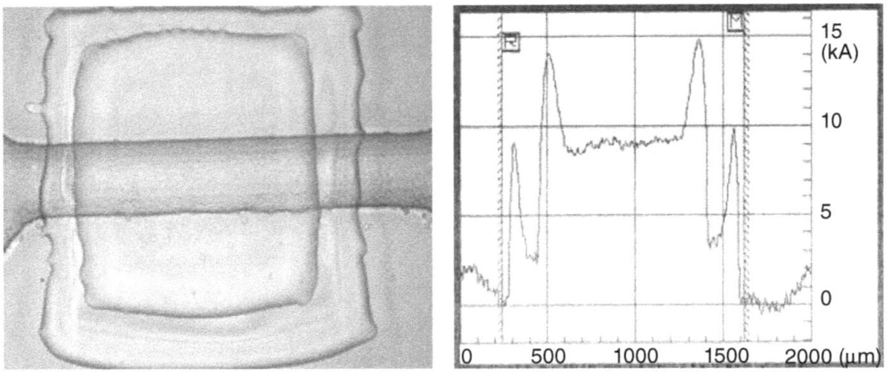

Figure 28.21 Optimized printed polyimide layer profile.

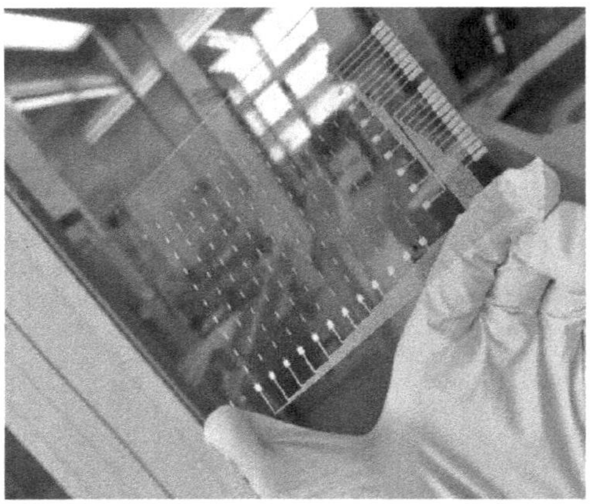

Figure 28.22 A manufactured printed touch sensor demonstrator.

Table 28.3 Performances of the Printed Demonstrators: D1 with Classic Design and D2 with the Hybrid One

	D1	D2	Microchip-D
Capacitance	70 to 160 pF	32 to 41 pF	10 to 15 pF
Contacting	Frontplane	Backplane	Backplane
Performance	acceptable	very good	perfect
	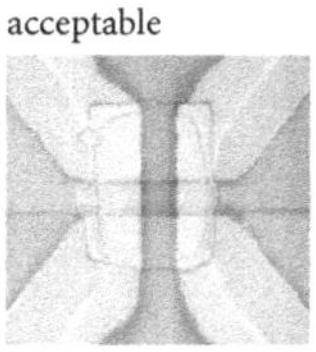	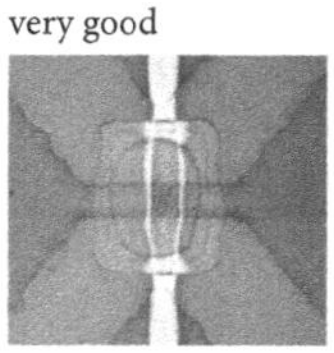	

28.5 CONCLUSION

In this study, an ITO ink has been optimized in correlation with the parameters of the corresponding postannealing process. These material optimizations and process improvements lead to

- Thinner layer with better conductivity (compared to the reference conductivity obtained with high temperature process), and with an acceptable transmission value
- Process time reduction from 4 h to only 30 min
- Significant improvement of aging: under appropriate storage conditions (almost) no more aging

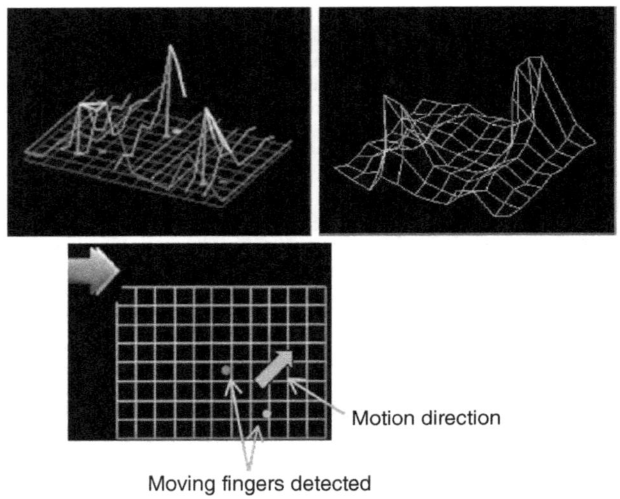

Figure 28.23 Multitouch sensing with three fingers and gesture recognition with two fingers.

It was demonstrated that through the addition of 20 percent tin to the material, the properties of the ITO-layers, mainly the conductivity and the aging of the electrodes, are considerably improved. The electrical resistivity of the ITO layers is decreased from the initial value of 0.0378 Ω·cm attained after an annealing step under 650°C to the value of 0.014 Ω·cm achieved under an annealing temperature of 300°C which is a more appropriate temperature for the further industrial processing.

The ITO-material was then optimized to be inkjet printed, using a Fujifilm-Dimatix printer. The parameters of the additive process have been investigated and the printing of functional ITO ink structures with a good resolution was reliably performed. An acceptable and functional printing quality was achieved: the optimized resolution is about 200 μm.

These additively structured ITO electrodes are further used for the manufacturing of a fully printed touch sensor on glass substrate. For this application, optimizing the printing process of the polyimide insulating layer to ensure functional capacitances on the PCAP sensor was necessary. The printing of polyimide as an insulator between the diamond-mustered ITO electrodes was optimized through using the coffee stain effect as a wall structure, which leads after the filling to a homogeneous isolation layer. The 7 × 9-cm printed demonstrators were characterized and their performances were confirmed: multitouch functionality as well as reliable detecting of gestures and patterns is ensured.

In conclusion printing touch panels is an attractive possibility, provided the optimized materials and processes are ensured. The so-manufactured printed touch sensors can also be further optimized: less variation of the capacitance values will certainly improve the touch-sensing quality. This homogeneity of the capacitance range can be achieved, when the printing process is improved to achieve thinner printed layers with comparable or better resolution and quality.

Besides the next step is to transfer and to adapt this developed process to plastic foils, in order to achieve printed flexible touch sensors. The challenge will be to achieve acceptable properties of the electrodes with only using plastic foils compatible temperatures.

Hence, the research experiments in this field have to be further carried on, because there is still a huge improvement potential which can be used to create the printed display applications for the future.

28.6 ACKNOWLEDGMENT

Special thanks to Solvay Fluor GmbH in Hannover who provided the ITO-containing inks and supported the project financially.

28.7 REFERENCES

1. N. Kammoun, C. Renninger, and N. Frühauf, "Investigation of a Transparent Electrode Ink for the Manufacturing of an Inkjet-Printed Touch Sensor on Glass Substrate, Flexible and Printed Electronics Conference, Electronic Proceedings," 2016.

2. Gary L. Barrett, "Projected Capacitive Touch Screens/Touch International," 2009 (Doc. Number 6500468).

3. T.-K. Ho, C.-Y. Lee, M.-C. Tseng, and H.-S. Kwok, "Simple Single-Layer Multi-Touch Projected Capacitive Touch Panel," *SID Symposium Digest of Technical Papers*, 40: 447–450, 2009.

4. K. A. Sierros, D. R. Cairns, D. S. Hecht, C. Ladous, R. Lee, and C. Niu, "Highly Durable Transparent Carbon Nanotube Films for Flexible Displays and Touch-Screens," *SID Symposium Digest of Technical Papers*, 41: 1942–1945, 2010.

5. Y. S. Yun, D. H. Kim, B. Kim, H. H. Park, and H.-J. Jin, "Transparent Conducting Films Based on Graphene Oxide/Silver Nanowire Hybrids with High Flexibility," *Synthetic Metals*, 162: 1364–1368, 2012.

6. Todd O'Connor, "mTouchTM Projected Capacitive Touch Screen Sensing Theory of Operation/Microchip," Technology Inc. 2010 (Doc. Number DS93064A).

7. Robert D. Deegan, Olgica Bakajin, Todd F. Dupont, Greb Huber, Sidney R. Nagel, and Thomas A. Witten, "Capillary flow as the cause of ring stains from dried liquid drops," *Nature* 389, 827–829, doi:10.1038/39827, London, England, 1997.

28.8 FURTHER READING

Das, R. and Peter Harrop, "Printed, Organic & Flexible Electronics: Forecasts, Players & Opportunities 2013–2023," Research Paper, 2014.

Derby, B., "Inkjet Printing Ceramics: From Drops to Solid," *Journal of the European Ceramic Society*, 31(14), 2543–2550, Elsevier, London, 2011.

Friederich, A., J. R. Binder, and W. Bauer, "Rheological Control of the Coffee Stain Effect for Inkjet Printing of Ceramics," *Journal of the American Ceramic Society*, 96(7), 2093–2099, 2013.

Lee, J., M. T. Cole, J. C. S. Lai, A. Nathan, "An Analysis of Electrode Patterns in Capacitive Touch Screen Panels," *Journal of Display Technology*, 10(5), 2014.

Lupo, Donald, Wolfgang Clemens, Sven Breitung, and Klaus Hecker, *Applications of Organic and Printed Electronics: A Technology-Enabled Revolution*, Cantatore Eugenio, ed., Springer Science+Business Medsia, New York, 2013.

Suganuma, Katsuaki, *Introduction to Printed Electronics*, Springer Science+Business Media, New York, 2014.

Chapter 29

Flat-Panel and Flexible Display Technology

BY

CHENG-CHUNG LEE, YUNG-HUI YEH, AND YUH-ZHENG LEE *Display Technology Center, Industrial Technology Research Institute, Hsinchu, Taiwan, Republic of China*

DAVID N. LIU *Idearite Co., Ltd., Taichung, Taiwan, Republic of China*

29.1 INTRODUCTION

Display has been used in various applications such as computer, communication, and consumer electronic vehicles. It is an interface between human eyes and information. Without display, information is not visible. In addition, display is preferred thin and flat because display space can be limited. Therefore, flat-panel display has become a very important and popular electronic device in the world. The various types of flat-panel displays are classified in Fig. 29.1. Besides rigid substrate–based display, flexible substrate–based displays are emerging as a key component to electronic products with innovative form factors in the near future. Because of their special feature, flexible displays can make future devices more attractive and to provide rich user experience.

Display technologies are classified into two major groups: direct view and projection. The direct view group is typically further classified into emissive and non-emissive subgroups. Organic light-emitting diode (OLED) display, plasma display panel (PDP), flat cathode ray tube (CRT), and electroluminescent display (ELD) are common in the emissive subgroup while liquid crystal display (LCD) and electrophoretic display (EPD) are common in the non-emissive subgroup. The projection group consists of liquid crystal on silicon (LCoS), digital mirror display (DMD), LCD, and CRT.

29.2 DEFINITIONS

This section defines some of the terminology that is commonly used in various types of flat-panel display. The terminology, structure, and characteristics of various displays are discussed in this chapter.

Alternating current (ac) mode. AC mode is an approach to operating the display panel. When this approach is used, a dielectric layer is usually needed as part of the panel.

Brightness. Brightness is the intensity of the light or image. The unit is nit (cd/m^2) or foot-lambert (English unit).

Candela (Cd). This is the luminous intensity which is the luminous power per unit solid angle. The unit is lm/sr.

Cd/m^2. This is a unit of brightness.

Color filter. A color filter is typically composed of red, green, and blue sections that are polymer resin films with dispersed pigments and able to pass red, green and blue light, respectively. The color filter is popularly used in color LCD.

Color temperature. Color temperature is used to indicate the position of the curve of a black body for a white color. High color temperature indicates a blue color tendency in the Commission International d'Eclairage (CIE) chromaticity diagram while low temperature indicates a red color tendency in the CIE chromaticity diagram.

Contrast ratio. Contrast ratio indicates the ratio of the strongest and the weakest light intensity of a display. The weakest light intensity of a display is very small in a dark room environment.

Direct current (dc) mode. The dc mode is an approach for operating a display panel.

Direct addressing. This is a technique in which each pixel is directly connected to a driver. Therefore, each pixel needs independent wiring. It is not a practical technique when the pixel numbers are large.

Direct view. This mode indicates that a display image is viewed directly without changing the display size. This mode is the opposite of the projection mode.

Flexible display. Flexible display is fabricated on a flexible substrate instead of a rigid glass substrate. It can have curved, conformal, bendable, foldable, or rollable form factors.

Foot-lambert (fL). A foot-lambert is an English unit of brightness and is defined as lumen per square foot.

Full color. Full color is defined as a display with 8 bits (or 256 gray levels) for each red, green, and blue color.

Gray level. Gray level is a mechanism of leveling the image intensity between the strongest and the weakest emission for each pixel.

Lumen (lm). The lumen is a unit of luminous flux.

Lux. Lux is defined as one lumen per square meter.

Matrix addressing. Matrix addressing is a technique that uses a row and a column lead to select a pixel. The lead is shared with other pixels. Therefore, this technique is the most commonly applied method for displays using large pixel numbers.

Monochrome color. Monochrome color is defined as only one color, which can be green, or red, and so on

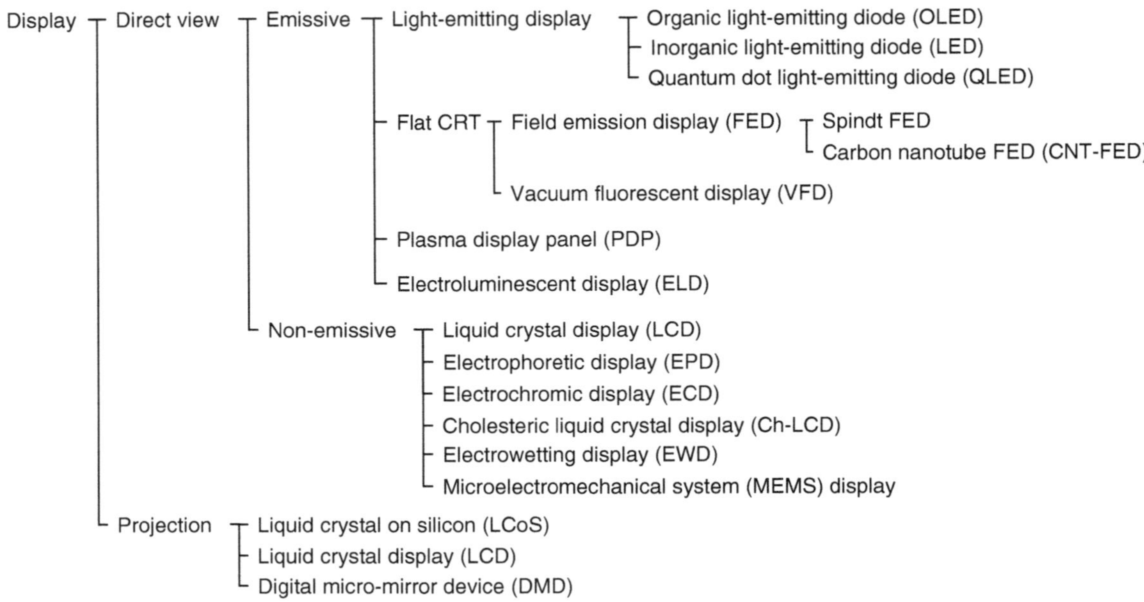

Figure 29.1 Classification of flat-panel display.

Multiple color. Multiple color is more than two colors in a display.

Optical efficiency. Optical efficiency indicates the efficiency of the conversion of electric power to the light output. The typical unit is lm/W.

Pixel. A pixel is a picture element that is typically composed of three primary color subpixels—red, green, and blue.

Projection. This mode indicates that the image is not directly viewed by the display itself but by a screen on which an enlarged image is projected. This mode is the opposite of the direct view mode.

Resolution. This term indicates how display is resolved. The typical unit is ppi (pixel per inch). A high value of ppi indicates that the display has a higher resolution.

Thin film transistor (TFT). This is a transistor made on a substrate using thin-film technology. It is used in TFT LCD or TFT OLED as a switching component or a driving component of a pixel so that the display performance can be increased and a higher video content display is possible. A TFT array can also be made on a flexible substrate for flexible display application. The material of flexible substrate can be plastic, thin glass, or thin metal foil.

Touch panel. A touch panel is an input device positioned on top of a display. A user can interact with the information processing system through simple or multitouch gestures. It can be resistive, capacitive, optical, or acoustic type.

Video format. This is used to indicate how many pixels are in a display. Larger pixel numbers for a fixed size of display show a higher capacity of video content. The typical format is VGA (640 × 480), SVGA (800 × 600), Full HD (1920 × 1080), and so on.

29.3 WHAT ARE THE FUNDAMENTALS AND PRINCIPLES OF DISPLAY?

The major factors for defining the qualities of a display, such as pixel size/resolution, brightness/optical efficiency, contrast ratio, color/gray level, and response time, will be dealt with in this section. The basic structures, characteristics, and benchmarks of various displays will be also discussed later in this section.

29.3.1 What Are Display Qualities?

The purpose of this subsection is to discuss the important factors of display qualities.

Pixel Size/Resolution

In flat-panel display, a pixel is a picture element that is typically composed of three primary color subpixels—red, green, and blue. The subpixel is typically called a dot. In other words, each pixel usually has three dots. The pixel size is important because it is the smallest picture element. A display composed of many smaller pixels indicates that it is able to generate a relatively fine image. There is a relevant term of display quality related to pixels—*display panel resolution.* A display panel resolution indicates the number of pixels per inch. For example, a typical display panel resolution of 100 ppi indicates 100 pixels per inch. The pixel size is around 250 μm. Another term associated with pixels is *display video format.* Display video format indicates the amount of pixels in the display that can be used. For a 55-in display with Full HD resolution, the display has 1920 × 1080 pixels. For the same size of display, more pixel numbers indicate that a more detailed image can be displayed.

Brightness/Optical Efficiency

Brightness is the intensity of the image. The unit of brightness generally used is nit (or cd/m^2) or foot-lambert. The typical brightness value of flat-panel display is 500 cd/m^2 for TV, 250 cd/m^2 for monitor applications and 300 cd/m^2 for mobile phone applications.

Optical efficiency (OE) is defined as

$$\text{OE} = \frac{\text{brightness or luminance}}{\text{power}}$$

A higher value of OE is needed because it indicates a higher brightness with less power.

Contrast Ratio

There are two major contrast ratios: one is the dark-room contrast ratio (display contrast ratio) and the other is the bright-room contrast ratio (ambient contrast ratio). The *dark-room contrast ratio* (DCR) is defined as

$$\text{DCR} = \frac{\text{on-pixel brightness}}{\text{off-pixel brightness}}$$

On-pixel and off-pixel brightness are defined as the brightness due to pixel turning on and off, respectively. The off-pixel brightness must be (and usually is) very small so that the DCR can be large. A large value of DCR indicates a good contrast ratio of a display. The typical DCR value is 1000:1.

However, ambient brightness is an additional important factor that affects the contrast ratio in practical use. This contrast ratio is called the *bright-room contrast ratio* (BCR). BCR is defined as

$$\text{BCR} = \frac{\text{on-pixel brightness} + \text{ambient brightness}}{\text{off-pixel brightness} + \text{ambient brightness}}$$

The value of DCR is usually higher than BCR because the ambient brightness in BCR is usually higher than the off-pixel brightness. The typical BCR value is 30:1. The BCR for paper is around 10:1.

Color/Gray Level

All flat-panel displays are able to display either monochrome or multiple colors. Chroma indicates the saturation degree of a color either from monochrome or multiple colors. A typical color display is composed of three basic colors (red, green, and blue). In other words, the display is able to display various colors using these three basic colors. It is also important to know the chroma values of these red, green, and blue colors because the maximum of color capacities is determined by these values. This indicates that a higher saturation of basic colors has a higher capacity of image colors to display.

A white color is composed of three colors—red, green, and blue. A different color temperature occurs due to the different color intensities of the red, green, and blue colors. The color temperature is higher when the color blue is rich, and it is lower when the color red is rich. However, the white color should be used to follow the curve of the white balance (black-body curve) in the chromaticity diagram. The color temperature is typically designed in the range of 6500 to 11,000°C.

Gray level is a mechanism of leveling the image intensity between the strongest and the weakest emission for each pixel. More gray levels generate more levels of image intensity.

Speed (Response Time)

A typical display image is designed with at least 60 frames per second (~16 ms per frame) so that most human eyes can observe the image properly. Therefore, each frame must be completed within 16 ms. For a display with a large video content, the time sharing of each pixel will be very limited. Therefore, the response time of each pixel to turn on or off is very important for a display. If the pixel needs a long response time, it will not be possible to display a large video content. The response time is particularly important for the application of motion picture display. The response time of LCD is relatively slow—around 1 to 10 ms. Response time of other displays, such as PDP, organic LED (OLED), and field emission display (FED), are around the microsecond scale or less.

Addressing

Direct and matrix are two common types of addressing techniques in flat-panel displays. Direct addressing is a technique in which each pixel is directly connected to a driver. A pixel is selected by an independent lead that is not shared with other pixels. Therefore, this technique is limited for displays with small pixel numbers, such as a display of a few alphanumeric characters. Matrix addressing is a technique that uses a row and a column lead to select a pixel. The lead is shared with other pixels. As a certain row is selected via electronic scanning, a bench of display data is sent to all columns. The pixels are therefore displayed. Compared with direct addressing, the number of leads is reduced. Therefore, the matrix addressing technique is the most commonly used method for displays with large numbers of pixels, such as graphic display.

Passive matrix and active matrix are two major types of matrix addressing techniques. The passive matrix type applies passive components while the active matrix type applies active components for pixel switching. By using the active component, the active matrix technique is able to increase the pixel performance. Higher quality or higher video contents of display are used in this addressing technique.

29.3.2 Basic Structures and Characteristics of Various Displays

Many different flat-panel displays are developed and commercialized, and we will discuss as many displays as possible in this subsection. The displays that will be discussed are LCD, PDP, OLED, FED, LCoS, DMD, and EPD. The main direction of the discussion will be the basic structure and characteristics of these displays.

LCD

LCD is the most commonly used flat-panel display currently. The operational principle of LCD is the use of an electric field to control the *liquid crystal* (LC) molecular arrangement. The pass percentage of light to the viewer is determined by different LC molecular arrangements. Some LCs need polarizers to control the light output properly. Since LCD is a non-emissive display a backlight source is needed for the transmissive type. For color LCD, a color filter is typically needed for most types of LCs. The most popular LC in flat-panel display is the *nematic*-type LC. When a higher image quality or larger video content is applied, transistors in pixels become necessary. An LCD with a transistor in each pixel is called a TFT LCD and is popularly used in most commercial personal computers, mobile devices, and televisions. The cross section of a typical transmissive type of TFT LCD is shown in Fig. 29.2.

In the figure, a spacer is needed to keep the distance between the upper plate (color-filter substrate) and lower plate (TFT-array substrate) so that there is space for LC molecules to be arranged. The space is typically 2 to 4 μm. Alignment layers are used to align the LC molecules in a certain orientation. A storage capacitor helps to store voltage so that the LC has enough voltage to be arranged properly. A black matrix is used to cut off some of the ambient light so that the contrast ratio is

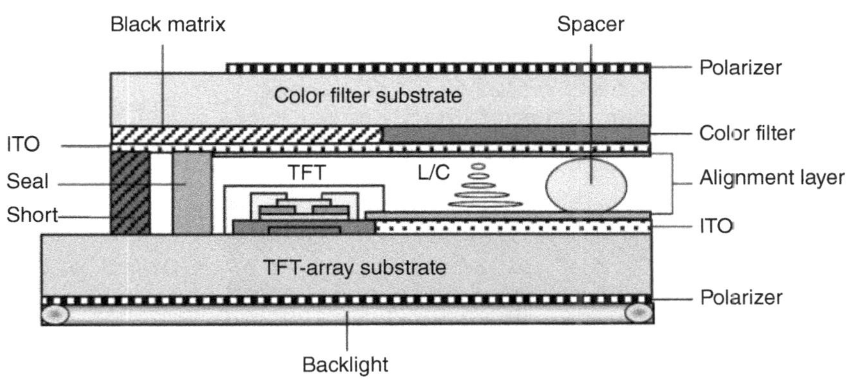

Figure 29.2 A typical structure of transmissive type of LCD panel.

increased. The pixel electrode connects the TFT and LC parts and the common electrode connects the LC part to ground via a "short."

The reflective type of LCD is widely used in portable devices ever because it consumes less power than the transmissive type of LCD. For this type of LCD, a backlight is not needed and ambient light is the light source for display. A reflective layer is usually additionally needed and implemented in the lower plate (TFT-array substrate).

For LCD, there are three major kinds of alignment type for liquid crystal in the cell structure. One is twisted nematic (TN) which is the most common and has the shortest response time. The weak point of this type is the color shifts at wider viewing angles. The second is vertical alignment type, which derived to multidomain VA (MVA) and patterned VA (PVA) after further developments. MVA and PVA mode panels offer good viewing angles and better contrast. However, color reproduction is its weakness. The other one is in-plane switching (IPS) technology. The switching electrodes for liquid crystal arrangement are positioned on the same plane in the liquid crystal cell. Further developments have led to super IPS (S-IPS) and fringe field switching (FFS). This type offers better color reproduction and better viewing angle almost to 180°.

In addition, there are three common types of transistors that are used in each pixel of TFT LCD. One is *amorphous silicon* (a-Si) TFT which is widely used in active matrix LCD (AMLCD). The second one is *low temperature polysilicon* (LTPS) TFT. The third is amorphous metal oxide (MOx) TFT. In recent advancement of amorphous compound oxide semiconductor for TFTs, primarily indium gallium zinc oxide (IGZO) can achieve LTPS-like performance at cost levels better than LTPS. Since LTPS TFT has naturally higher mobility than other TFTs, the size for LTPS device is smaller. Therefore, LTPS TFT–based LCD has a higher aperture ratio and brightness. In addition, LTPS is able to integrate discrete integrated circuits (ICs) such as *digital analog converter* (DAC), memory, timing circuit, and so on into the panel. The stability of LTPS TFT is also superior to a-Si TFT, especially for current driven application, such as TFT OLED. However, LTPS TFT has poor uniformity compared with a-Si and metal oxide TFTs, particularly in large displays.

PDP

The PDP operational principle is similar to that of a fluorescent lamp. They both use a gas discharge mechanism. The gas discharge generates ultraviolet (UV) light. The gases most commonly used in fluorescent lamps are mercury and argon while those most commonly used in PDPs are neon and xenon. The UV wavelength is typically 254 nm (main peak) for fluorescent lamps and 147 nm/173 nm (two main peaks) for PDPs. The wavelengths of 147 nm/173 nm are called vacuum UV (VUV) and do not exist in air. RGB phosphors in PDP are excited by UV light. There are two types of PDPs. One is the dc-type PDP (DC PDP) and the other is the ac-type PDP (AC PDP). Since AC PDP has a longer life and higher performance, it is the most popular PDP used in the market. The typical structure of AC PDP is shown in Fig. 29.3.

In the figure, the rib is very important because it serves not only as the spacer between the upper and lower plates but also as crosstalk isolation among R, G, and B.

The rib height is typically hundreds of micrometers. MgO is most commonly used for the protection layer because it is a good material for ion bombardment and has a higher secondary electron emission. These characteristics of MgO help the PDP to have a longer life and lower the operation voltage. The dielectric layers used in both the upper and lower plates serve as capacitance as ac power is applied. A surface discharge mechanism in PDP is commonly used and successfully keeps the plasma on the surface of the upper plate so that the plasma will not damage the phosphor that is located in the lower plate. Therefore, the display life can be extended. However, for the surface discharge type of PDP, the arrangement of electrodes is critical. The electrode in the upper plate should preferably be as transparent as possible because most of the area of the upper plate is occupied by electrodes. Therefore, *indium tin oxide* (ITO), being a transparent conductive material, is commonly used. However, the conductivity of ITO is not as high as that of a typical metal. A conductive metal with an electrode of relatively small linewidth is applied to the ITO electrode so that the electrode conductance is increased. The metal electrode on the ITO electrode is called the auxiliary electrode or bus electrode.

PDP has a high quality of picture and simple manufacturing process steps. The need for high-voltage driving (typically 200 V) is the main drawback affecting the cost and the application.

OLED

The operational principle of OLED is the use of electron and hole injections into the luminance layer to combine together and then to generate light. Since the luminance layer is an organic material, the display is called OLED. OLED is a solid-state display and does not comprise vacuum, liquid, or gas. Therefore, the structure is relatively simple compared with other displays. There are two types of OLED—small molecule OLED and polymer OLED (PLED). Small molecule OLED uses small organic semiconductor molecules as luminescent materials, which are typically deposited by vacuum thermal evaporation. PLED uses polymer semiconductors as luminescent materials which are typically deposited by spin coating or inkjet printing. The most commonly used approach to form a pattern of color OLED is the shadow mask approach. Another approach is mainly inkjet printing. Those luminescent materials can be fluorescent or phosphorescent emissive species. Typically, OLED with phosphorescent emissive materials has better device efficiency.

In order to have a higher performance of OLED, TFT is additionally used. The typical structure of TFT OLED is shown in Fig. 29.4.

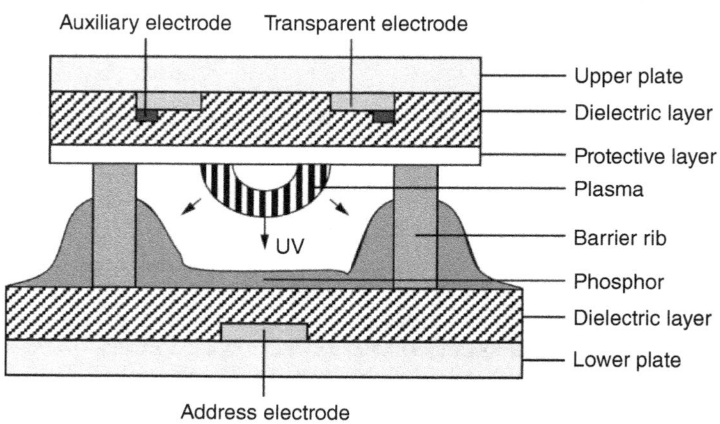

Figure 29.3 A typical structure of AC PDP display panel.

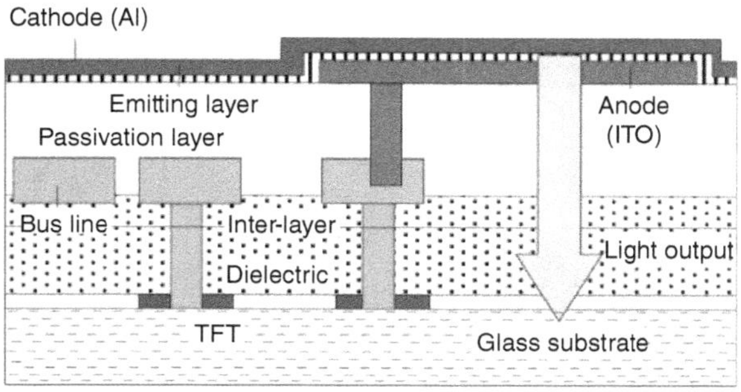

Figure 29.4 A typical structure of TFT OLED display panel.

In the figure, aluminum (Al) is used as the cathode and ITO as the anode. The device structure is bottom emission (downward emission) type since Al topmost is reflective and opaque.

Top emission is another approach in which the anode serves as a reflective layer and the cathode should be semi-transparent. The material selection of electrodes in the top emission approach is critical. MgAg alloy is the common cathode for top emission OLED. However, the aperture ratio for top emission is higher than that for bottom emission. Regarding the TFT backplane design, the pixel arrangement for the top emission approach is more flexible than for bottom emission.

OLED is a solid-state thin film device. It can be fabricated on a thin flexible substrate, such as plastic or thin metal foil. With good water vapor and gas permeation resistant thin film encapsulation on top of OLEDs and gas barrier on substrates, the resulting OLED can be flexible and foldable with acceptable reliability. This capability is OLED's unique feature to other emissive display technology.

Although OLED has all the good characteristics of display technology, such as high optical efficiency, simple and thin device structure, and low driving voltage, the major drawback of OLED is the nonuniform lifetime for red, green, and blue colors. Especially, blue-color devices always have the shortest lifetime (LT50%) around tens of thousands hours at the initial brightness of 1000 nits.

FED

FED is a display using electrons of field emission to excite phosphors and generate luminance. Field emission uses a high electric field rather than the thermionic approach to extract electrons in vacuum. There are many types of FEDs. Spindt-type (sharp-cone-type) FED was intensively studied a decade ago. As the evaporation needed to form the sharp-cone emitter is difficult in displays with large areas and the uniformity of the sharp-cone emitter is also poor, most researchers of Spindt-type FED have shifted to research of *carbon nanotube* (CNT) FED and other approaches of FED.

In the device structure, electrons are emitted from the CNT material and excite RGB phosphors to generate light. The space between the upper and lower plates is typically from hundreds to thousands micron. CNT as an emitter is the most promising approach to FED display in the last few years.

Although FED has CRT-like picture quality and high optical efficiency, one of the major drawbacks of FED is the poor emission uniformity.

LCoS

LCoS is a reflective microdisplay technology in which a projection display combining LC and silicon semiconductor technologies. The lower plate including the TFT is made on silicon wafer. Therefore, the TFT can be made as small as typical semiconductor devices, and the resolution of the display can be as high as possible. In addition, the TFT mobility is the same as a single crystal since the TFT is made on a silicon wafer.

The upper part is a typical LCD structure that is made on a glass substrate. The lower part is a typical TFT structure but is made on silicon wafer. In the LCD cell, the reflective electrode is needed to reflect light as the pixel is switched on. The pixel size can be made as small as several μm. The die size of the lower plate is designed as 1 in. Therefore, an 8K × 4K format is easy to implement. LCoS displays can be of the single panel or three panel type. The panel size is typically 1 in. The cost of single-panel type LCoS is lower. However, the technology barrier is higher. The major issue is the yield of mass production.

DMD

DMD is another popular type of projection display called digital light processing (DLP) projector. A digital mirror device is fabricated by microelectromechanical systems (MEMS) technology. In combination with an optical system, this kind of display uses many digital mirrors to reflect light in response to input signal to generate a video or image pattern. By using proper voltage, every single mirror can be tilted so that the light hitting each mirror can be deflected individually to create a dark or bright pixel on the projected screen.

The DMD panel is made of silicon wafer. Using a proper silicon machining approach, many mirrors and certain electronics are formed inside the silicon wafer. After these components are formed, a silicon wafer cover is used as a seal so that a DMD panel is completed. The major challenge of the DMD device is to manufacture it with high resolution and high yield since the device includes electronics complex with lots of mirrors.

EPD

EPD is a reflective type display with a display media having switchable bi-stable states. The display media comprises microcapsules in which charged white particles of titanium oxide and black dye particles are dispersed in oil. These microcapsules mixed with a polymer host are coated as a film sandwiched between two parallel plates with patterned conductive electrodes. By using proper voltage, the electric field can drive the charged particles to the electrodes with the opposite charge from that on the particles. As the white particles are positioned at the front side of the display, they can reflect the incident ambient light and appear white. As the black particles are located at the front side, the display appears black. The typical structure of EPD is shown in Fig. 29.5.

The EPD panel can be simple segment or active matrix type. The latter one is that an EPD film is sandwiched between a TFT backplane and a top conductive plate. The most important feature of EPD is the capability of being bi-stable switching. It means that the panel can stay at an addressed image after a proper driving scheme. No power is needed to hold that image just like ink on paper It is a power saving display unlike the conventional display technology. Instead of the rigid glass substrate with flexible substrates, EPD display can also have flexible form factor as the OLED. However, the slow switching speed is the major drawback of EPD. The response time is more than 500 ms, which is suitable for displaying static image. The other one issue is the color reproduction not adequate for color image application.

29.3.3 Benchmarks of Various Displays

In this subsection, various displays will be discussed. Although there are many different advantages of different displays, certain disadvantages are common. Table 29.1 shows the benchmarks of various displays.

Although LCD needs many components, such as backlight, color filter, polarizer, and so on compared with other displays, the manufacturing cost has been significantly reduced due to mature technology and large-scale mass production. In addition, other traditional disadvantages, such as low view angle and long response time, are improving dramatically. PDP has two fundamental drawbacks: large pixel size and high driving voltage. As PDP is applied in large-sized displays, the drawback of large pixel size can be reduced. However, as the TV sets go into 4K version, it is hard to produce cost-effective PDP compared to LED-backlit LCD. OLED is a kind of solid-state display that has no need of vacuum, liquid, or gas inside the device. It also has the advantage of high optical luminance and self-emission. At present, OLED is competitive to small-sized LCD and becomes a key component for mobile devices. FED is generally believed to have a CRT-like quality with flat characteristics. However, the emission uniformity has not significantly

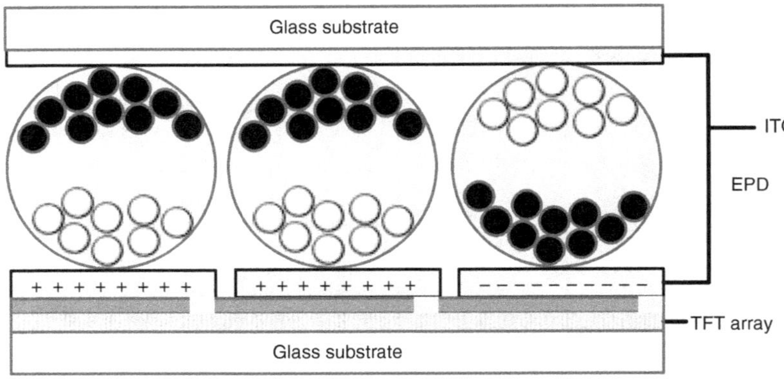

Figure 29.5 A typical structure of EPD display.

improved during the last few years. Besides, its color lifetime is not long enough. Single panel LCoS will be very competitive in terms of cost. Single panel DMD has been cost and space effective. For a projection application using either LCoS or DMD, the projection size can be conveniently adjustable based on the user needs.

For flexible display, various types of displays such as cholesteric LCD, EPD, and OLED are developed. The first two technologies are reflective displays for electronic paper (ePaper) application. Cholesteric LCD uses cholesteric liquid crystal as the display medium, which has good characteristics in bi-stability and color but it is poor in contrast ratio. The stacked color ChLCD fabricated on plastic substrates showed the best still color image among ePaper technology. However, the complex pixel structure and its related driving components and system make it hard to be cost-effectively produced. For EPD, the response time and colorization of panel are two major challenges for the future extended application. For emissive flexible display technology, OLED is the most potential candidate due to its simple solid-state structure and good device performance. The one of major challenges is to have a cost-effective and robust flexible encapsulation approach.

29.4 WHAT IS THE MANUFACTURING PROCESS?

FPD manufacturing needs two plates (the front and bottom plates) at the beginning of the process. These two plates are aligned and sealed followed by electronic board/film bonding and mechanical assembly. Before shipping, aging and a series of tests are also performed in order to make sure that the display is reliable. Among the various displays, TFT LCD is the most popular information display and covers from small- to large-sized electronic products.

The manufacturing process starts with TFT array and color filter glasses. The structures of the TFT array and color filter glasses are shown in Fig. 29.2. The TFT array glass is processed not only by the transistor process but also by the LC alignment layer process. The color filter glass is not only processed by the color filter process but also the LC alignment layer process and the photo spacer process. The two plates are then aligned and sealed. LC is vacuum-injected and the end seal is performed. Instead of vacuum injection method, one drop filling (ODF) is the major approach for liquid crystal cell process for larger-sized mother glass production line (larger than 1800 mm × 1500 mm of sixth generation).

Table 29.1 Benchmarks of Various Displays

	Advantage	Disadvantage	Application size	Technology status
LCD	• Technology mature and low cost • Low voltage drive • Low power consumption • Long lifetime	• Needs many components (backlight, color filter polarizer, and so on) • Limited operation temperature • Long response time	• ≦ 110 in	• Large-scale mass production
PDP	• Easy to scale-up size • High picture quality • Fewer process steps (thick-film process)	• High voltage driving • Limited pixel size	• ≦ 152 in	• Mass production stopped in 2016
OLED	• High optical efficiency • Relatively simple process • Low driving voltage • Flexible	• Short lifetime in blue color • Not easy to scale-up size	• ≦ 77 in	• Mass production
FED	• CRT-like picture quality • High optical efficiency	• Poor emission uniformity • Short color FED lifetime	• < 60 in	• R/D
LCoS	• High resolution • Low cost of one panel type • Projection size adjustable	• High cost of three-panel type • Low yield	• > 100 in	• Mass production
DMD	• High response time • High resolution • Projection size adjustable	• High cost • Complex device process	• > 100 in	• Mass production
EPD	• Energy saving • High contrast ratio • Flexible	• Slow response time • Poor color saturation	• ≦ 32 in	• Mass production

By use of ODF, the liquid crystal is discharged on the bottom plate and then the top and the bottom plates are assembled. ODF can dramatically shorten the processing time from several days to several minutes for a 30-in panel. After that, a polarizer is attached and the tape carrier package (TCP) is bonded. As the resolution of display continues growing and the border edge of display becomes narrower, the package of LCD driving system changes from TCP to chip on film (COF) and chip on glass (COG). The next steps performed are electronics, backlight, and chassis assembly. Aging and various tests are performed in between so that display panel performance and reliability are confirmed.

Since different displays have different processes, we are not able to discuss the detailed process for each display in this chapter. However, most processes of displays use thin-film, thick-film, and cell-formation processes, aging, and tests. Therefore, in the following sections we will discuss these processes separately.

29.4.1 Thin-Film Processes

Thin-film processes normally consist of deposition, photolithography, and etching techniques. Sputter, evaporation, and chemical vapor deposition (CVD) are the common techniques for deposition. The photolithography system includes a photoresist coater, exposure system, developer, and stripper while the etching system includes dry and wet etchers. TFT processes of TFT LCD and TFT OLED use thin-film processes. The backplanes (bottom plates) processes of LCoS and DMD use thin-film processes too.

29.4.2 Thick-Film Processes

Deposit and cure systems are two common types used in thick-film processes. Screen printing and inkjet printing to deposit a paste or slurry are very common techniques for the deposit system, while a furnace is the equipment typically used to cure the paste or slurry. Most processes of PDP and FED use thick-film processes.

29.4.3 Cell Formation Processes

At the beginning of cell formation processes, sealing material is dispensed on the plates. Alignment between the two plates follows. With proper curing of the sealing material, the cell is formed. Most of various displays use these processes for cell formation.

29.4.4 Aging and Testing Processes

Aging is very important. Using this process, the device can be stabilized and its reliability can be ensured. Testing is performed to sort out a not good (NG) panel so that it will not leak NG panel to next steps. Most of the various displays need these processes to control their quality and reliability.

29.5 FUTURE TRENDS AND CONCLUSIONS

In this section, we will discuss future trends in two aspects of flat-panel display (FPD), technology and applications, and make concluding remarks.

29.5.1 Technology Trends

Although there are many technologies emerging and developing, we will list the most common trends in this subsection. Table 29.2 shows the technology trends of flat-panel and flexible display.

There are two approaches for reaching large display sizes. One is by direct view and the other is by projection. The direct view approach typically provides a higher contrast ratio than the projection approach. However, in the projection approach, it is easy to scale up the display size compared with the direct view approach.

Undoubtedly, high image quality is a continuous goal for the development of display technology. The factors affecting image quality are pixel size/resolution, brightness/optical efficiency, contrast ratio, color/gray level, and switching speed, which we have discussed in a previous section.

Driver number reduction and production scale up are two major ways to reduce cost. LTPS is one of the approaches for reducing the driver cost. This is because the high mobility of LTPS enables the integration of drivers and most discrete integrated circuits (ICs) into the display panel. The cost is therefore reduced.

One of the most exciting future trends is the need for light and flexible displays. Flexible displays will be needed because this kind of display can be applied on nonflat backgrounds. In addition, flexible display can be manufactured by use of the present invested sheet to sheet production line with modification. This can transform the small-sized mother glass line to produce new and innovative flexible display without huge capital investment. The manufacturing cost and entrance barrier can be therefore reduced.

Flexible display is typically fabricated on the flexible substrate instead of rigid glass substrate. The flexible substrate includes plastic substrate, thin-glass substrate, and thin metal foil substrate. A flexible active-matrix organic light-emitting diode (AMOLED) display has features of thin, light, unbreakable, and flexible. Among various flexible displays, flexible AMOLED is the most promising information display in the future.

Comparing to the rigid substrate AMOLED display, the flexible AMOLED display typically consists of TFT array fabricated on flexible substrate, OLED deposition, thin-film encapsulation and flexible encapsulation, integrated circuit (IC)/flexible printed circuit (FPC) bonding, and a de-bonded process from the glass carrier substrate.

Table 29.2 Technology Trends of Flat-Panel Display

	Item	Specification		Relevant tech.
Display size	Projection	Front	40–200 in	LCoS, DMD
	Direct view	40–100 in		TFT LCD/TFT OLED/copper process/oxide TFT
		4–50 in		a-Si TFT/LTPS TFT/LCD/OLED
Display quality	Resolution	Full HD → Ultra HD		LTPS, oxide TFT, copper process
	Brightness	150 cd/m^2 → > 500 cd/m^2		LTPS, Super-IPS (in Plane Switch), VA (vertical Alignment), OLED
	Contrast ratio	500:1 → > 5000:1 >100,000		Super-IPS, VA OLED
	Color/gray level Color gamut	8-bit → 12-bit gray level 90% NTSC → 150% NTSC (CIE1976)		Driving New phosphor
	Response time	40 ms → 25 ms → < 8 ms		LC material, driving
Cost-effective	Driver reduction			LTPS
Light/flexible	Reflective, flexible substrate, COG, COP, SOP			OLED/EPD
Ecology	Power consumption, production waste reduction			Design/process improvement

Table 29.3 Application Trends of Flat-Panel Display

Size	Major technologies	Application
Large size (30 ~ 300 in)	• a-Si TFT LCD • Oxide TFT LCD/OLED • Micro Display (LCoS, DMD)	• Wall-mounted TV • Desktop TV • Front projection
Medium size (8 ~ 30 in)	• a-Si TFT LCD • LTPS TFT OLED/LCD • a-Si TFT EPD	• Monitor • Laptop PC • E-book • Flexible display
Small size (below 8 in)	• a-Si TFT LCD • LTPS TFT OLED/LCD • OLED • EPD	• Mobile phone • Wearable device • View finder • Virtual reality • Flexible display

A de-bonded process typically use mechanical or laser lift-off method to separate a flexible display from the carrier substrate. In addition, in order to increase the flexibility of flexible AMOLED display, the total thickness of flexible thin-film encapsulation should be as thin as possible.

29.5.2 Application Trends

It is obvious that FPD has various applications since each kind of FPD has features that are suitable for different applications. Table 29.3 shows the application trend of flat-panel display, which is classified according to the display size.

The display size is an important factor in deciding the applications of the display technology. LCD is the most popular flat-panel display in the existing market. The display size of LCD ranges from less than 1 in to more than 100 in. Therefore, the application of LCD is wide. After more than 20 years of development in material, process, and related production equipment, OLED gradually penetrated into the application market dominated by LCD for a long time. The majority of OLED ranging from 1 to 10 in is used in mobile devices, such as wearable devices, mobile phones, and tablets. It also keeps moving forward to large-size display for TV application. Projection display is generally used in screen sizes of more than 100 in. Its applications include large meeting room display and home entertainment display.

29.5.3 Conclusions

There are many new display technologies extensively explored in the past decades. Among the various display technologies, LCD apparently has become the main stream of flat-panel display technology since around 2000. Although its architecture is sophisticated, LCD attracts much more attention and effort than any other display technology. In the past few years, many efforts have been dedicated to improve its weak performance issues, such as viewing angle limitation, high cost, power consumption, color gamut, and low response time. Besides LCD, OLED has great potential to become more popular in the near future due to its better performance in many aspects compared to other displays. The lifetime and reliability of OLED have been improved dramatically in the past few years. Plastic substrate–based OLED can be flexible, foldable, rollable, and conformable. This flexible feature makes OLED more competitive to be designed into future products. We believe that attractive electronic devices with innovative form factors coming from flexible display will be unveiled in the future.

29.6 FURTHER READING

Armitage, D., Underwood, I., and Wu, S.-T., *Introduction to Microdisplays*, Wiley-SID, Hoboken, 2006.

Bhowmik, A. K., Li, Z., and Bos, P. J., *Mobile Displays: Technology and Applications*, Wiley-SID, Hoboken, 2008.

Castellano, J. A., *Handbook of Display Technology*, Academic Press, Cambridge, 1992.

Chen, J., Cranton, W., and Fihn, M., *Handbook of Visual Display Technology*, Springer, New York, 2012.

Crawford, Gregory P., *Flexible Flat Panel Displays*, Wiley-SID, Hoboken, 2005.

den Boer, W., *Active Matrix Liquid Crystal Displays*, Elsevier B.V., New York, 2005.

Hatalis, M. K., et al., "Flat Panel Display Materials II," *Material Research Society*, PA, 1997.

Jensen, K. L., "Electron-Emissive Materials, Vacuum Microelectronics and Flat-Panel Displays," *Material Research Society*, PA, 2000.

Keller, P. A., *Electronic Display Measurement: Concepts, Techniques, and Instrumentation,* Wiley, Hoboken, 1997.

Lee, J.-H., Liu, D. N., and Wu, S.-T., *Introduction to Flat Panel Displays*, Wiley-SID, Hoboken, 2008.

MacDonald, L. W. and A. C. Lowe, *Display Systems: Design and Applications*, Wiley-SID, Hoboken, 1997.

Matsumoto, S., *Electronic Display Devices*, Wiley, Hoboken, 1991.

Nelson, T. J. and J. R. Wullert, "Electronic Information Display Technologies," *World Scientific*, 1997.

O'Mara, W., *Liquid Crystal Flat Panel Displays: Manufacturing Science and Technology*, Van Nostrand Reinhold, New York, 1993.

Refioglu, H. I., *Electronic Displays*, IEEE Press, New York, 1983.

Sasaki, A. and C. J. Gerritsma, *Optoelectronics*, Vol. 7 (2), Mita, Tokyo, 1992.

Sherr, S., *Electronic Displays*, 2nd ed., Wiley, Hoboken, 1993.

Stokes, A., "Display Technology: Human Factors Concepts," Society of Automotive Engineers, 1998.

Tannas, L. E., *Flat-Panel Displays and CRTs*, Van Nostrand Reinhold, New York, 1985.

Tsujimura, T., *OLED Displays-Fundamentals and Applications*, Wiley-SID, Hoboken, 2012.

Weston, G. F. and R. Bittleston, *Alphanumeric Displays*, McGraw-Hill, New York, 1982.

Whitaker, J. C., *Electronic Display: Technology, Design and Applications*, McGraw-Hill, New York, 1994.

Wu, S. T. and Yang, D. K., *Reflective Liquid Crystal Displays*, Wiley-SID, Hoboken, 2001.

Yang, D. K. and Wu, S. T., *Fundamentals of Liquid Crystal Devices*, Wiley-SID, Hoboken, 2006.

Chapter 30

Photovoltaics Fundamentals, Manufacturing, Installation, and Operations

BY

JUN ZHUGE *Jiangsu Seraphim Solar System Co., Ltd., Changzhou City, Jiangsu, China*

30.1 INTRODUCTION

30.1.1 What Is Solar Energy?

Solar energy is using technologies to convert sunlight into other forms of energy, such as heat and electricity, directly or indirectly. Solar energy is clean, free, and safe. It is an endless energy source without geographical boundary.

In today's world, a variety of technologies are available to utilize of solar energy. It is used to power airplane flying around the world without a drop of fuel. It is used to heat up water and raise temperatures inside buildings without burning fossil fuels. It is used to produce electricity instead of burning coals or gas. Solar becomes a part of our daily lives at every corner of the world.

30.1.2 Why Solar Energy

Fossil fuels are one of the world's most important energy resources that include coal, petroleum, and natural gas. However, fossil fuels are non-regenerative natural resources and are generally buried undergrounds, because they are formed by natural processes such as anaerobic decomposition of buried dead organisms. Thus, they are limited in supply and their exploitation and burning cause serious environment problems.

Demand for energy has raised since the expansion of global industrialization. According to the U.S. Energy Information Administration (EIA) data [1], at least 4000 power plants are required to establish by 2035 to meet the increasing global electricity demand even if the annual growth rate is only 1.7 percent.

The exploitation process of these conventional fuels, especially in coal mining can cause land and village subsidence and water pollution. Burning of coals and other fossil sources can emit environmental hazardous gases into atmosphere, such as sulfur dioxide, nitrogen oxides, and carbon dioxide. Carbon dioxide is the main ingredient of greenhouse gases and about 90 percent of the anthropogenic CO_2 emissions are generated by fossil energy consumption [2]. The largely increased concentration of greenhouse gas is a main leading cause to global warming. Since 1860, the global average temperature has increased 0.8°C [3]. In addition, emitted sulfur dioxide and nitrogen oxide could react with the water molecules in atmosphere to produce acid rain that destroys plants, harms aquatic animals, and erodes infrastructure.

Energy crisis and environment pollution have becoming this century's two main global issues. Therefore, an alternative green energy is greatly anticipated to solve these problems.

30.1.3 Global Solar Market Overview

Because its clean and inexhaustible characteristics, solar energy has attracted widespread attention since it was invented. Its application history can be categorized into five stages [4].

The first period (1954 to 1973): Bell Labs made a single solar cell with 6 percent photoelectric conversion efficiency. Solar cell then started its slow development.

The second period (1973 to 1980): In 1973, the first worldwide oil crisis broke out caused by Middle East wars. After that, many countries, especially the developed industrial countries, have tried to strengthen government support in utilizing solar and other renewable energy technologies. In the same year, the United States increased solar energy research funding. Japan government announced the "Sunshine Plan" in 1974.

The third period (1980 to 1992): In the 1980s of the 20th century, international oil prices fell sharply; however, the solar energy products still sustained higher prices and had photoelectric low conversion efficiency. Hence, many countries with the United States being the most prominent one have cut the solar energy research funding significantly.

The fourth period (1992 to 2000): Burning mineral energy has resulted global environmental pollution and ecological damage and posed serious threats to human survival and development. In 1992, the United Nations Conference on Environment and Development (UNCED) conference passed a series of important documents, such as "The Rio Declaration," "Agenda 21," and "The Framework Convention on Climate Change" to protect global environment. Governments around the world announced supporting policies in clean energy technology development after this meeting. This has stimulated the international cooperation in the field of solar energy and brought solar energy technology into a new level. The landmark events in this stage include

Japan re-enactment "Sunshine Project" in 1993, the United States proposed "Million Solar Roof Program" in 1997, and in 1998, University of New South Wales created a new world record for monocrystalline silicon solar cell with 25 percent efficiency.

The fifth period (2000 to present): The crude oil prices skyrocketed from less than $30 per barrel in 2000 to $150 per barrel in 2008. The alarming global environment issues have raised people's desire to solar energy. A series of globalized renewable energy subsidies were initiated or enhanced, especially after the Chernobyl nuclear accident and the Fukushima Daiichi nuclear disaster. The solar industry entered a new era with vigorous development and rapid growth in capacity.

According to the International Energy Agency (IEA) statistics and forecasts [5], annual worldwide solar cell productions increased more than 6 times in the past 10 years and the average annual growth is 50 percent. Annual worldwide productions were 2500 MW in 2006, reached 15200 MW in 2010, and were at 50 GW in 2014. The world annual installed capacity of solar PV production raised from 6629 MW in 2008 to 45.6 GW in 2014. Solar energy is expected to contribute more than 10 percent global energy demand by 2030, and supply 20 percent of worldwide electricity by 2040. By the end of the 21st century, solar power will account for more than 60 percent of overall power supply. These figures prove the prosperous prospect in the development of solar photovoltaic industry and its strategic position in further energy supply.

30.2 FUNDAMENTALS OF PHOTOVOLTAICS

30.2.1 What Is Photovoltaics?

Photovoltaics is the process of converting harvestable sunlight into usable consumer electricity. The basic device that allows for this conversion to take place is called a *solar cell*. Currently, there are several different types of cells being used successfully: crystalline silicon, amorphous silicon, thin film, CIGS, CdTe, gallium arsenide, and others. Among these cells, crystalline silicon is the most widely employed with the highest energy conversion efficiency.

Crystalline Silicon Molecular Structure and Electrical Properties

Silicon is one of the most abundant elements on earth. The crystalline molecular structure can be altered to control the electrical property. Natural occurring silicon is a nonconductor of electricity.

Here is why: In general, matters consist of individual atoms. Each atom has a nucleus where positively charged particles called protons reside. The nucleus is surrounded by multiple orbits of electrons, which are negatively charged particles. The charge of the protons and electrons cancel one another such that the atom is electrically neutral. A silicon atom can be visualized to have three layers of electrons revolving around the nucleus, two in the innermost orbit, eight in the middle orbit, and four in the outermost orbit. The electrons in the two inner orbits are bound tightly to the atoms, but those in the outer orbit can be moved around by outside sources of energy. In crystalline silicon, each atom is surrounded by four neighboring atoms. These neighboring atoms each share one electron to the outer orbit of the center atom, bringing the total number of electrons in the orbit to eight. This is known as covalent bonding.

The electrons are stable and require photons with energy greater than 1.1 electron-volts to break free. Typically, this is the sunlight near the infrared region of the electromagnetic spectrum. Pure silicon does not have enough free electrons that can be directed to move together as an electric current under the influence of an electrical potential. This explains why pure silicon cannot conduct electricity.

During the manufacturing of crystalline silicon, certain impurities can be intentionally added in very precise amounts to alter the electrical conduction property. The intentional addition of impurities to the silicon is called *doping*. During the crystalline formation process, the impurity atoms enter the crystal lattice structure through a substitution process. An example of a dopant is phosphorous, which has five electrons in the outer orbit of the atom. That is one more electron than the silicon atom. The phosphorus covalent bond has nine electrons. As a result, the bond becomes unstable and very little energy is needed to break one of the electrons free. Environmental thermal energy present at room temperature is sufficient to break the bond. Silicon with phosphorous doping is known as *N-type silicon*.

Boron is another common impurity used to dope silicon. The boron atom has three electrons in its outer orbit, one less than the silicon atom. This leaves the covalent bond with only seven electrons, or, a "hole" is created that allows an outside electron to move in and settle. Silicon with boron doping is known as *P-type silicon*.

Silicon (Si) P-N Junction and Barrier Voltage

The formation of a P-N junction is when you connect P-type Si to N-type Si material (Fig. 30.4). After the junction is formed, the thermal energy in the surrounding atmosphere is strong enough to break free an electron around each phosphorus atom on the N-type side of the junction. While on the P-type side, there is a hole in the orbit around each boron atom waiting for a free electron to occupy and settle. The N-type side has a high concentration of free electrons. This is based on the theory of diffusion. The electrons will have a tendency to move across the junction to the P-type side in order to equalize the concentration. Most of the electrons that first move across the junction to the P-type side will have a high probability of settling into a "hole" in the orbit around a boron atom as described above. It is natural to assume that this movement of electrons will continue until the concentration of free electrons becomes uniform across the entire silicon substrate.

However, there is a very important side effect that occurs and is caused by the movement of free electrons from N-type side to P-type side. Before any movement of the electrons, the silicon, phosphorus, and boron atoms most all have the correct amounts of protons and electrons to be electrically neutral. As the electrons begin to leave the N-type side and move into the P-type side, a positive potential begins to form on the N while a negative potential begins to form on the P. As the movement continues, the potentials increase and move further away from the junction (Fig. 30.1). Electrons are attracted to and driven toward positive potentials.

So now there are two opposite mechanisms driving these free electrons: diffusion from high concentration on the N side to P side, and electrical potential driving them from P side (negative) to N side (positive). At some point, equilibrium occurs. A steady state is reached where the diffusion movement is exactly balanced by the electrical potential drive. The end result is that a steady electrical field occurs across the junction. This is called the *barrier voltage*. This barrier voltage is a basic property of a silicon crystalline P-N junction device. This is a key element in the operating theory of the silicon crystalline solar cell.

Sunlight-Induced Current and Potential through Silicon P-N Junction

As discussed at the beginning of this chapter, in order to break the covalent bond of a Si atom and release an electron, it requires a photon with energy of greater than or equal to 1.1 eV or sunlight near the

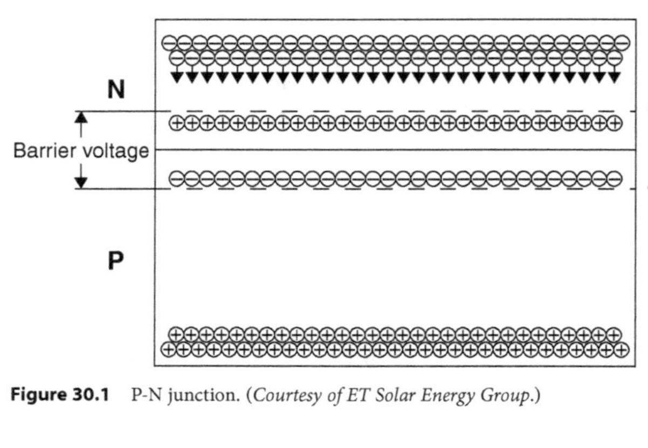

Figure 30.1 P-N junction. (*Courtesy of ET Solar Energy Group.*)

infrared region of the spectrum. When the Si P-N junction device is placed under sunlight, some electrons will be randomly released in various parts of the device. "Holes" will be created in the absence of the released electrons in addition to those created by the boron dopants on the P-type side. Some of the released electrons will fall back into some of the "holes" in a process called *recombination*. Other electrons will move randomly throughout the device *until* they reach the area near the junction where the barrier voltage exists. The barrier voltage provides these randomly moving electrons with a common direction, toward the N-type side! If the N-type side is connected to the P-type side through an external wire, then the electrons can exit through the wire. The electrons will become an electrical current arriving at the P-type side and will begin their random movements once again. A complete circuit is possible as more electrons are continuously freed by the sunlight, which move through the junction and directed by the barrier voltage. This is known as the *short circuit current* of the solar cell (Fig. 30.2).

If the P-N junction device is left open and unconnected under the sunlight, then the free electrons that wander into the junction area without recombination will still be influenced by the barrier voltage. These electrons will move to the N-type side further away from the junction and will create "holes" at P-type side. This will occur further away from the junction. Eventually, a steady-state electrical potential will be generated across the entire device, positive on the P end and negative on the N end. This is known as the *open circuit voltage* of the solar cell. This is not to be confused with the barrier voltage across the junction (Fig. 30.3).

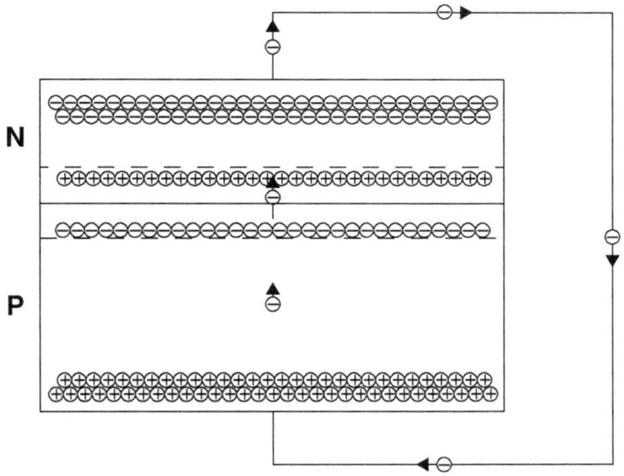

Figure 30.2 Solar cell short circuit current. (*Courtesy of ET Solar Energy Group.*)

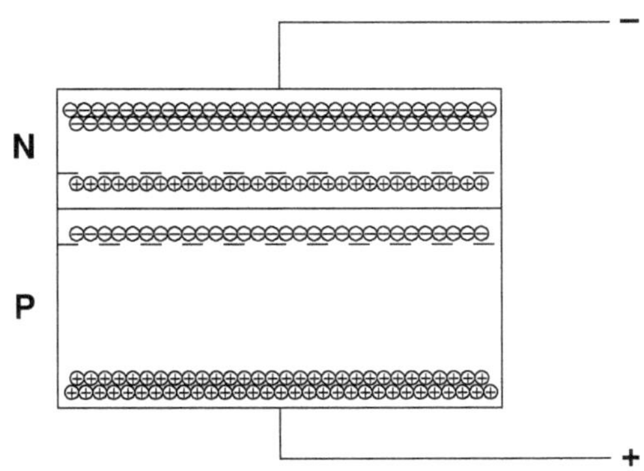

Figure 30.3 Solar cell open circuit voltage. (*Courtesy of ET Solar Energy Group.*)

CRYSTALLINE SILICON SOLAR CELL OPERATION OVERVIEW

A crystalline silicon solar cell is actually a P-N junction device fabricated from a silicon wafer. Efficiency enhancement coatings and metal contacts are added to the device so that the cells can be linked and connected together to form a solar module. Multiple solar modules are linked together, which form an array to generate a planned amount of current and voltage. Often, the output array is converted to alternating current (AC) electricity and added to the local utility supply circuit for that service area or region.

30.2.2 How Solar Cells Are Built

CELL PROCESS

A solar cell is made from a slice of silicon wafer. There are two types of silicon wafer: monocrystalline (mono) and polycrystalline (poly). It is very easy to distinguish its difference from physical dimension. The mono cell made from a 6-in-diameter wafer. The wafer edges are cut off to form a square. The poly cell is formed as a 6×6 in square from the beginning.

A solar cell can be regarded as a large-area diode with a relatively thin emitter (n-type, thickness $d \approx 0.2$–2 μm) and a thick base (p-type, $d \approx 50$–500 μm). Solar cell is made of five layers: front contact, antireflection, n type, p type, and back contact layers. A model of solar cell structure is shown in Fig. 30.4.

The emitter is on the front surface where the incident light is coupled into the cell to create electron-hole pairs by absorbing photons of sufficient energy. The carriers diffuse to the space charge region of the p-n junction, where they are separated and are converted from minority carriers to majority carriers. Those can be extracted at the metal contacts on the front and rear of the cell to deliver electrical power.

Our starting material is p-type polycrystalline "as cut" silicon wafer with a thickness of 200 μm and a resistivity of around 1 to 3 Ω·cm.

Figure 30.5 is an example of the manufacturing process flow for a solar cell.

Cleaning and Texturing The sawing operation leaves the surfaces as cut wafers with a high degree of damage. This presents two problems: the surface region is of a very bad quality and the defects can lead to wafer bulk material fracturing during processing. About 10 μm are etched off from the surface to remove the damage.

Bare silicon has a high surface reflection rate more than 30 percent. The reflection is reduced by texturing and applying antireflection coatings (ARC) to the surface. "Roughening" of the surface reduces the reflectivity by increasing the chances of the light bouncing back onto the surface, rather than out to the surrounding air.

Wet acidic texturing is a process that removes the surface damage, fractures, and improves the overall cell's light trapping capability. Wet chemistry performed with solutions containing hydrofluoric acid, hydrogen nitrate $HF/HNO_3/H_2O$ tends to etch isotopically, and can result in features with rounded surfaces. Approximately 5 to 10 μm of silicon will be removed from each surface. The resultant structures are small rounded etch pits with diameters of up to 10 pm are uniformly.

Diffusion Phosphorus is universally used as the n-type dopant for silicon in solar cells. Since solid-state diffusion demands high temperature, it is very important that the surfaces are contamination free before processing. After texturing, the wafers are subjected to acid etch to neutralize alkaline that remains and eliminates adsorbed metallic impurities.

The cells are loaded in quartz boats, which are heated and held at temperature during processing. The boats enter and exit the furnace through one end. Processing gases are dispersed through the opposite end. Phosphorus itself can be supplied by bubbling nitrogen through liquid phosphorus oxychloride $POCl_3$ before injection into the furnace. All surfaces of the wafer will be diffused. The primary benefit of using quartz material in the furnace is cleanliness. The quartz tube acts like a micro environment that protects from possible contamination during processing. Though this is a batch step, high throughput can be achieved because many wafers can simultaneously diffuse in each tube, commercial furnaces consisting of stacks of four tubes.

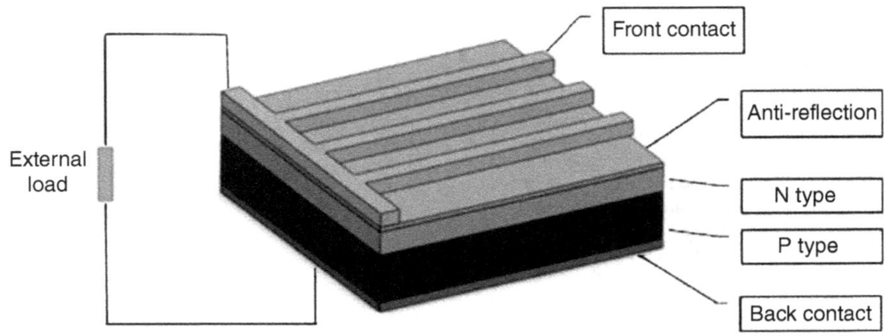

Figure 30.4 Solar cell structure. (*Courtesy of ET Solar Energy Group.*)

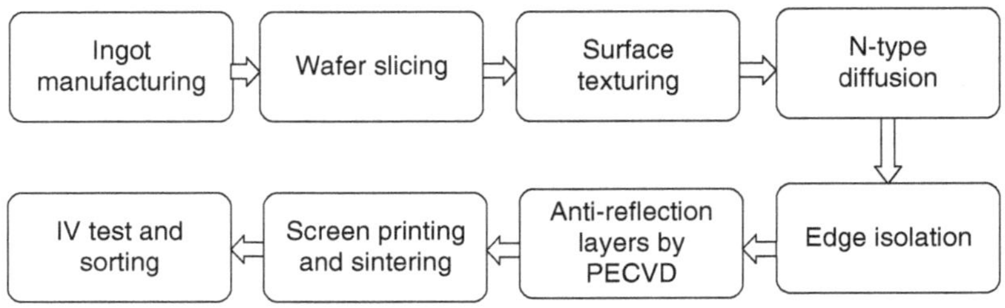

Figure 30.5 Solar cell manufacturing process flow. (*Courtesy of ET Solar Energy Group.*)

Edge Junction Isolation and PSG Removal After diffusion, an amorphous phosphosilicate glass (PSG) remains on the surface of the wafer. This is usually removed by etching it off in diluted HF because it can hinder subsequent processing steps. The n-type region at the wafer edges would interconnect the front and back contacts: the junction would be shunted by this path translating into a very low shunt resistance. To remove this region, acid solution is used. The wafers flow on the surface of the acid solution, only the backside and the edge contact with the solution.

PSG removal and junction isolation process steps are completed in one single tool. This improves the process integration and workflow. Junction isolation is performed by etching the rear side of the cells while leaving the front surface dry. Automation systems are available for loading and unloading the wafers. The wafer floats on the surface of the acid solution. The acid concentration must be chosen carefully to control the density of the mixture. The other part of process is similar to the texturing process.

Antireflection Layers Antireflection coatings on solar cells are similar to those used on other optical equipment such as camera lenses. This consists of a thin layer of dielectric material, with a specifically chosen thickness. The desired top surface coating thickness causes the wave to be out of phase with the wave reflected from the semiconductor surfaces. These out-of-phase reflected waves destructively interfere with one another, resulting in zero net reflected energy. In addition to antireflection coatings, interference effects are also commonly encountered when a thin layer of oil on water produces rainbow-like bands of color.

An antireflection of silicon nitride is typically deposited using chemical vapor deposition process (CVD). Precursor gases of silane (SiH_4) and ammonia (NH_3) are fed into a chamber and break down due to temperature (LPCVD) or due to a plasma enhancement (PECVD). Other systems use microwaves to cause the silane/ammonia reaction to take place.

Surface recombination can have a major impact both on the short-circuit current and on the open-circuit voltage. High recombination rates at the top surface have a particularly detrimental impact on the short-circuit current because top surface also corresponds to the highest generation region of carriers in the solar cell. Lowering the high top surface recombination is typically accomplished by reducing the number of dangling silicon bonds at the top surface by using a passivation layer on the top surface. The electronics industry relies on the use of a thermally grown silicon dioxide layer to passivate the surface due to the low defect states at the interface. For commercial solar cells, dielectric layers such as silicon nitride are commonly used.

One of the primary benefits of processing monocrystalline material with PECVD is that it produces hydrogenation. Atomic H interacts with impurities and defects in the bulk of Si, neutralizing their recombination properties to a certain extent, a phenomenon that is usually expressed as bulk passivation. Optimizations of processing parameters (temperature, plasma excitation by power and frequency, gas flow rate) are necessary and a compromise should be reached. The processing gasses are excited by means of an electromagnetic field, and the wafers are located within the plasma, the graphite plate and the wafer as electrode.

Screen Printing and Sintering The requirements for the front metallization are low contact resistance to silicon, low bulk resistivity, low line width with high aspect ratio, good mechanical adhesion, and solderability. Resistivity, price, and availability considerations make silver the ideal choice as the contact metal. Copper offers similar advantages, but does not qualify for screen printing because subsequent heat treatments are needed. Copper is high diffusible and will produce contamination on the silicon wafer.

During the actual printing, the screen and the wafer are not in contact, but a distance apart called the snap off. A paste is dispensed and pressure is applied to rubber squeegee as it is moved from one side of the screen to the opposite one. Dragging and pressing the paste in front of it. When an opening is reached, the paste fills it and sticks to the wafer, remaining there after the squeegee has passed and the screen has elastically retired. Squeegee moves the paste across the screen. This action causes a decrease in the viscosity of the paste, which in turn allows the paste to pass through the patterned areas onto the substrate. As the squeegee passes, the screen peels off and the paste viscosity returns to normal.

It should be noted that silver and aluminum pattern is different.

Aluminum is required because silver does not form ohmic contacts to P-Si, thus it cannot be soldered. Though in principle a continuous contact will give better electrical performance, most wafers feature a back contact with a mesh structure: apart from paste-saving considerations, this is preferred to a continuous layer because the different expansion coefficients would cause the cell warp during the subsequent thermal step.

A high temperature–sintering step is required to burn off unwanted organic components and allow for the metallic grains to form a good conductor. This will ensure good conductance with the underlying silicon. The front paste is deposited on an insulating layer and the back contact on the substrate. Upon firing, the active component of the front paste must penetrate the ARC coating to contact the n-emitter without shorting it. In order to comply with these requirements, the composition of the pastes and the thermal profile of this critical step must be adjusted very carefully.

I-V Test and Sorting The illuminated I-V curve of finished cells is measured under an artificial light source with a spectral content similar to sunlight at a controlled temperature of 25°C (Fig. 30.6). Defective devices are then rejected, and the good cells are classified according to their output. Modules will subsequently be built with cells rated in the same class to guarantee minimal mismatch losses.

30.2.3 How Is a Solar Panel Manufactured?

A conventional solar panel is made of five layers: glass, EVA (ethylene vinyl acetate), cells, and EVA and backsheet (photovoltaic polymeric material). Each layer in each of its uniqueness serves the purpose of protecting the solar cell through its 25 years of lifespan under different harsh working environment from the 120°F temperature of Africa desert to the icy-cold Alaska region, or under the humid Florida sky. So, the materials used to construct a solar panel require high quality of EVA and backsheet. The solar cells are sandwiched by EVA layers, covered by glass on the front and by backsheet on the back. Please see below illustration of a solar panel construction (Fig. 30.7).

Panel manufacturing process is shown as below (Fig. 30.8).

Cell Interconnection through Layout

Cells are first connected in series in units called *strings*. To assemble the strings, cells are individually placed on a processing unit. Tabbing strips are positioned on cell surfaces, which have designed contact locations. The contact locations are the n+ busbar on the front of the cell, and strips of silver on the back. Strings of up to 10 or 12 cells are

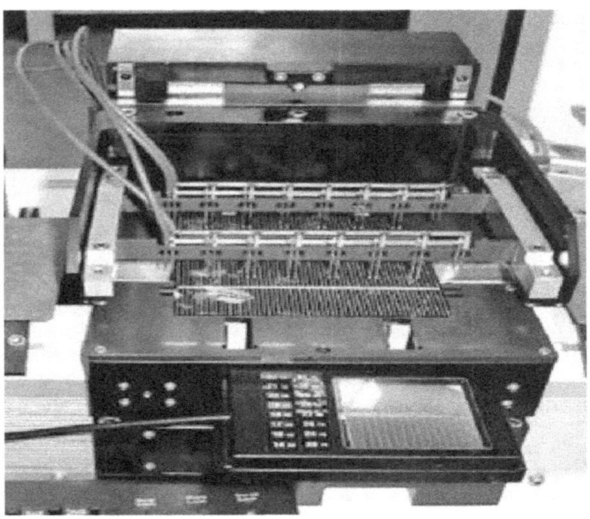

Figure 30.6 I-V test. (*Courtesy of ET Solar Energy Group.*)

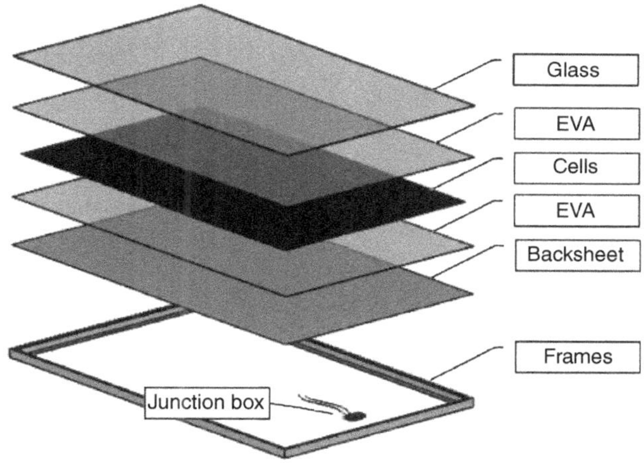

Figure 30.7 Solar panel construction diagram. (*Courtesy of ET Solar Energy Group.*)

typically incorporated into a single laminated panel. Individual strings are combined in series by tabbing to form an array of up to 60 or 72 cells in a sequential process. For example, a module configuration of 72 cells in series includes six individual strings, each of 12 cells, and connected by tabbing strips across the ends of adjacent strings alternating from end to end.

The Layers of Panel

Once a string of solar cells have been interconnected, the string is moved to a "layup" station. At the layup station, a mechanical pick holding an entire string is used to integrate the strings into the desired electrical grid with materials needed to complete the laminated panel.

In the front cover assembly step, tempered glass is placed on a table. Then, an encapsulant (EVA) (Fig. 30.9) is placed on the front cover. Strings of solar cells are assembled, including the tabbing strips that connect adjacent solar cells. Next, a further layer of encapsulant (EVA) and a back sheet/cover are placed on the solar cell strings.

Lamination and Curing

Lamination is a process using high pressure and temperature sufficient to melt the encapsulant (a curing process solidifying the liquid encapsulant) to form a panel. Then a trimming action is done to remove the extra EVA and backsheet.

Panel Assembling and I-V Curve Test

Before lamination, wiring from the interconnected cells is positioned outside of the laminate. This is required, so that the panel can be complete by attaching a junction box for electrical connections and a frame to support and protect the edges of the laminate.

Once the silica gel has been cured, panels washed, the panel will tested for the I-V curve. The I-V curve test measures and calculates critical parameters such as short circuit current, current density, open circuit voltage, fill factor, and maximum output power (Fig. 30.10).

Quality Control

Before modules are packaged, manufacturers will perform a final check to ensure their quality.

The visual check is a necessary and common inspection. Through this process, the appearance defects can be detected. These defects include scratches or flaws on the glass or frame, dent and wrinkle on the back-sheet, inclusions and defects on the module inner.

In addition to the visual inspection, an EL (electroluminescence) test will be performed in searching for visually undetectable microcracks and contact breaks.

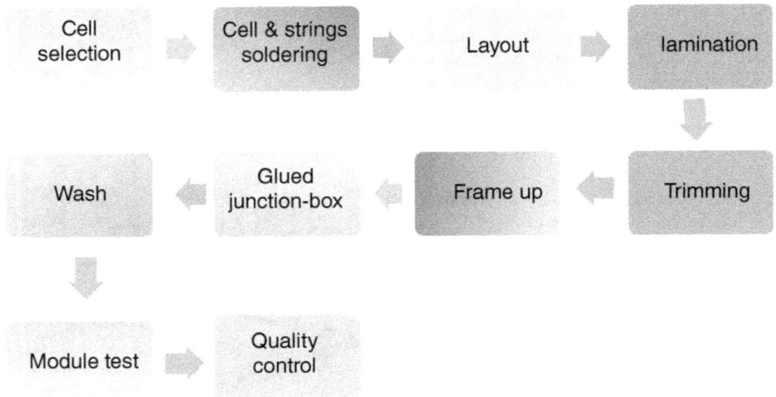

Figure 30.8 Panel manufacturing process flow. (*Courtesy of ET Solar Energy Group.*)

Figure 30.9 Layer of encapsulate. (*Courtesy of ET Solar Energy Group.*)

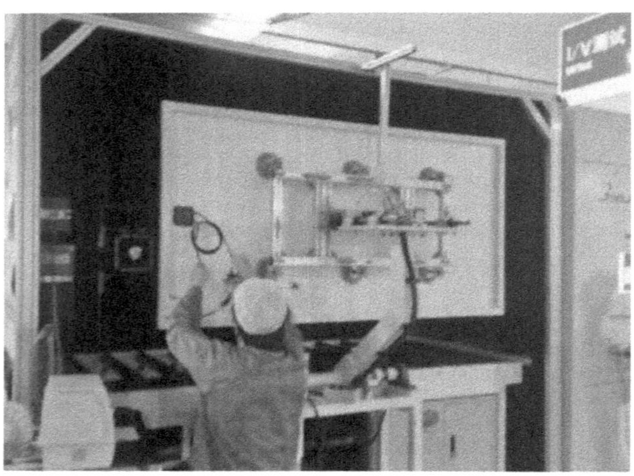

Figure 30.10 I-V curve measurement. (*Courtesy of ET Solar Energy Group.*)

30.2.4 Types of Solar Panels

All PVs are packaged and connected assembly of individual solar cells. Solar modules types are based on solar cell substrate material composition and assembly method.

There are currently three types of high-volume manufacturing panels in production: polycrystalline solar panel, monocrystalline solar panel, and thin-film solar panel.

Polycrystalline Solar Panel

Polycrystalline solar panel is made of polycrystalline silicon. Polycrystalline materials are solids that are composed of many crystallites of varying size and orientation.

Polycrystalline silicon or polycrystalline silicon (multi-Si) cells are made from cast square ingots, which are large blocks of molten silicon, then go through cooling and solidifying process.

Because of lower cost, polysilicon modules are the most common type used in photovoltaics, yet follow less efficient than monocrystalline modules.

Monocrystalline

Monocrystalline solar panel in bulk material is manufactured as a single crystal silicon ingot. The crystal lattice structure of the entire sample is continuous and unbroken through the ingot with no grain boundaries. So monocrystalline can show unique properties, particularly mechanical, optical and electrical, and is anisotropic.

Usually monocrystalline or single crystal silicon is used in the manufacturing of high-performance solar cells. Solar panels manufactured from this type of bulk material are more efficient and expensive than most other types of solar panels.

Thin Film

Over time, the photovoltaic solar cells have evolved. The crystalline silicon cell is the first generation. Thin-film solar cell is the second generation. Thin-film cells are made of depositing one or more layers of photovoltaic material on a substrate, such as glass, plastic, or metal. Most designs sandwich active material between two panes of glass, so those modules are approximately twice as heavy as crystalline silicon panels. Thin-film cells are flexible, lower in weight, and have less drag. These are used in building integrated photovoltaics and as semitransparent, photovoltaic glazing material that can be laminated onto windows.

Thin film has always been cheaper but less efficient than conventional crystalline-Si technology. But they have a smaller ecological impact determined from life cycle analysis.

New Innovations in Solar Panel Industry

Smart Module with Maximum Power Point (MPPT), Monitoring, and Shutdown Over the past several years, dc optimizers have become an important technology ingredient in many residential and commercial system designs. By adding these devices to each module, the system designer can reduce the power loss from shade obstructions, and safeguard the system against long-term module mismatch from uneven soiling or debris.

Such as "ET cell optimizer module" takes dc optimization to the next level by including a highly integrated power regulator on each cell-string within the solar module. Also, there are many different kinds of product, which can optimize the power output of PV module on panel level or cell-string level, monitor the performance of your PV sites, detect the electrical parameters, auto-shut down on emergency situations, and tell you the energy production of PV system. What industry calls these advanced modules as "smart module" (Fig. 30.11).

AC Module The newly developed solar inverter has become more popular. Its solar microinverter, which is a device used in photovoltaics that converts direct current (dc) generated by a single solar module to alternating current (ac). The primary advantage of microinverter system is that shading, soiling, snow, or catastrophic failure of any given panel will not disproportionately reduce the output of the entire array.

In the early days, installers would have to install the PV module and microinverter separately; microinverters are always installed on the racking frame of the solar system. Today, the solar panel can ship as a fully integrated assembly with the microinverter as an ac module. The shipping ac module must be certified in compliance with UL solar standard (UL1741). The ac module has many benefits such as saving labor costs, Potential Induced Degradation (PID) risk avoidance, mismatch reduction, and optimized single power output. It also functions as an integrated monitoring system with communication capability, monitoring hardware, software, so that real-time pale performance can be viewed remotely.

Dual-Glass Module The structure of the traditional solar panel is designed with glass/EVA/cells/EVA/backsheet, and aluminum frame. The difference between the traditional and dual glass module is that the backsheet is replaced with glass and frameless design. For long-term reliability, there are two layers of tempered glass. The dual-glass module has a wide variety of applications. The BIPV configuration uses transparent dual-glass module to build the wall or rooftop. This maintains optimal light transmission and power generation.

Figure 30.11 Smart module. (*Courtesy of ET Solar Energy Group.*)

30.3 PHOTOVOLTAIC POWER PLANT

The PV power plant is designed to generate electricity. It consists of several components, including solar panel, solar inverter to change the electrical current from dc to ac, racking, wiring, and other accessories that complete a working system.

PV systems range from small residential rooftop projects to large-scale commercial building applications. Power generation can run from a few kilowatts to large utility-scale power plant of hundreds of megawatts. Today, most PV systems are grid-connected, while off-grid, stand-alone, or hybrid systems only account for a small portion of the market.

30.3.1 How to Design a Solar System

A solar system consists of three main parts: PV array, balance of system equipment (BOS), and inverter and metering.

The PV array is made up of solar panels, which convert sunlight to electricity. BOS includes mounting systems, wiring systems, disconnects, ground-fault protection, over-current protection, solar inverter, combiner box, and so on. The inverter is a device that takes dc power from the PV arrays and converts it into standard ac power. Finally, the metering component is an electronic device that measures and reports the solar system performance.

Customer site inspection and preparation: investigate local building and electrical codes. Verify the customer requirements. The solar system designer can use few of PV design software such as PVsyst and PVSOL simulation tools to construct array system that meets the customer requirements. Additionally, these tools can be used to calculate the total energy production, performance ratio, etc.

30.3.2 Design Software and Tools

PVsyst

PVsyst is the most common design tool used by PV architects, engineers, and researchers. PVsyst can be viewed in wide variety of formats: full report, specific graphs, tables, and data can be exported for use in other software.

To start a PV system simulation, you must define the site and meteo data. For a given project (a defined site and meteo), users can construct several variations for calculation. On the system design board, users can define the orientation of modules, desired power or available area, choose PV module and inverter, loss factor, then PVsyst will propose the array configuration which allows users to conduct a simulation.

The simulation calculates the distribution of energies throughout the year: total energy production [MWh/y], performance ratio (PR [%]), specific energy [kWh/kWp], main energies, and gains/losses involved in the simulation.

PVSOL

PVSOL is another common design tool in the industry used for system design and simulation.

PVSOL divides the system into three different types: off-grid system, grid system, and hybrid system. It has the functionality to construct and simulate a battery storage system with load. The abundant load types and detailed parameters are one of the important features. Additionally, it can calculate the self-consumption.

Users can choose and define the orientation, module types, inverter, and loss factor, then reach a simulation results. Except of energy production data, PVSOL can carry out an economic efficiency calculation, which is important and valuable for PV investors.

30.3.3 How to Build a Solar System

There are three phases in building a solar system: engineering, procurement, and construction. Engineering designs a system that meets customer needs and complies local building code. Procurement purchases all necessary materials and delivers them to project site on time to meet schedule. Construction builds the project based on the engineering design and layout that has been approved by the local authorities.

Designing and installing a PV system is very complex process and involving many variables such as geographical location, weather profile, construction environment, etc. The process flow to build a solar system as below:

- Conduct a site inspection
- Review local authority building and electrical codes
- Design and layout PV array system
- Simulate annual energy output and project ROI
- Construction and installation
- Commissioning and interconnection

30.3.4 How to Install a Solar System

Installation Safety

Solar modules will produce electrical energy when exposed to sunlight, and for each module, dc voltages may exceed 30 V and the current can reach up to 8 A. If modules are connected in series, the total voltage is equal to the sum of the individual module voltages. If modules are connected in parallel, the total current is equal to the sum of individual module or series combination currents. To maintain site safety during installation, all modules are covered with an opaque material to prevent electricity from being generated.

Each module manufacturer publishes installation instructions and safety precautions via their website. Before installation occurs, you should read and familiarize yourself with their installation and safety protocols. Modules should be connected with the same rated output current in series or series combinations of modules with the same voltage in parallel.

For other components used in PV system, suitable equipment must be chosen to meet ratings and local codes: connectors, wiring, and mounting hardware, etc. Installation instructions and safety precautions for all other components must be adhered too. When determining component voltage ratings, modules short-circuit current and open-circuit voltages should be multiplied by a factor of 1.25, conductor ampacity, fuse sizes, and size of controls connected to the module or system output. An additional multiplying factor of 125 percent (80 percent de-rating) may be applicable.

Installation Orientation and Location

When installing the modules, the face of the units should be placed where they are highly exposed to the sun. It is recommended that the modules usually face the equator; thus, in the Northern Hemisphere the surface should be oriented toward the south, and in the Southern Hemisphere, toward the north. Usually the angle between modules and the ground should be local latitude ± (5°–10°).

Modules should not be installed near equipment or in locations where flammable gases can be generated or collected. Modules should not be installed in a location where they will be immersed in or continually exposed to water.

Solar Panel Mounting Methods

There are two preferred methods for mounting solar panels: bolts or clamps.

Bolt Method The module must be attached and supported by at least four bolts through the indicated mounting holes. Most installations will use the four inner mounting holes on the module frame.

Depending on the local wind and snow loads, additional mounting points may be required (e.g., if the snow load is as high as 5400 pa, you may use eight bolts).

Clamping Method If module clamps are used to secure the module, the torque on the clamp bolt should be around 8 to 10 Nm.

A minimum of four module clamps should be used, two on each long/short frame side, in the general clamping areas denoted by the wide arrows on the drawing.

Depending on the local wind and snow loads, additional module clamps may be required (e.g., if the snow load is as high as 5400 pa, you should use eight clamps on the long frame).

Wiring and Connection

There are two cables used in the junction box, which are marked for polarity. At the opposite end of the cables, the positive (+) cable connects male (+) connector and the negative (–) cable connects female (–) connector. This allows for easy connection of adjacent modules by firmly inserting the male connector of a module into the female connector of an adjacent module until the connector is fully seated. The cables and connectors have been designed to speed installation so that the modules can easily be interconnected in series configuration.

Precaution should be taken when cabling during the installation. The connectors should be kept dry and clean, and ensure that connector caps are handed tight. Do not attempt making an electrical connection with wet, soiled, or otherwise faulty connectors. Avoid direct sunlight exposure and water immersion. Faulty connections can result in arcs and electrical shock. Check all electrical connections and make that fully engaged and securely lock.

When connecting module strings in parallel to the distribution box, use suitable third-party PV system connectors with suitable cable (PFV) length. Field wiring should be done with suitable cross-sectional areas that are approved for use at the maximum short-circuit current of the PV module. It is recommended that installers use only sunlight resistant cables (PVF1 type) for dc wiring in PV systems. The recommending minimum wire size should be 4 mm^2 and must be subject to the local national codes and regulations.

Grounding

All module frames must be properly grounded complying with all local electric codes and regulations. A bonding or toothed washer is required to make proper and reliable electrical grounding connection with the anodized aluminum frame. Devices listed and identified for grounding metallic frames of PV modules are permitted to ground the exposed metallic frames of the module to grounded mounting structures. For equipment grounding, it is recommended that the negative pole of PV module array is connected to earth during all PV system installations. That will keep optimal performance of PV power plants, which are located in a hot, high humidity climate and high maximum system voltage.

30.4 MAINTENANCE AND OPERATION

30.4.1 Plant Maintenance

Schedule visual checks of key equipment (PV panels, rails, grounding terminals, inverter, dc combine boxes) to ensure the equipment is under good condition for operation. Soiling and snow cleanings are required for PV panels and vegetation abatement is required to avoid shading PV panels.

Soil disbursed on a PV panel surface causes energy loss. Cleaning maintenance is a balance of the cost and energy loss. In practice, more frequent cleaning is needed during dry periods when dirt and dust may accumulate. Rainfall provides natural cleaning.

Specific environmental conditions such as industrial soot or greasy residue around PV plant may require more frequent cleaning and a special detergent.

Use clean water and a soft sponge or cloth for cleaning. A mild, non-abrasive cleaning agent can be used if necessary. Do not use dishwasher detergent. Do not leave water standing on the glass surfaces of PV modules for a long time.

In order to keep maximum energy generation, PV module cleaning is normally arranged in the morning or evening or on non-sunshine days.

Snow Clearing

Snow clearing is rare due to the risk of damaging the module or rack plus the high cost. In practice, the snow clearing is considered in rooftop PV system when the weight of snow threatens the roof and may cause the collapse.

Vegetation Abatement

Grass and trees are growing fast in specific climate and causes energy lose due to shading. Local regulation has to be considered if chemical treatment is allowed. Alternative approach has to be considered even before EPC when the cost of cutting or trimming is high.

Test and Measurement

Portable PV I-V measurement equipment and insulation tester are readily available. Infrared imager can be used to look for hot spots when required.

Safety and Security

There is high dc or ac voltage in PV system, which may cause high risk of personal injury. Each aluminum frame PV panels is grounded and all electrical equipment (combine boxes, inverters, transformers, switch gear cabinets, etc.) have grounding cables. A regular visual check and measurement is required to ensure all grounding points are in good condition.

Video surveillance is popular in large scale PV plant to keep the expensive assets safe. Intrusion detection can be applied at the fence line and entrance gates. The requirements or design changes depend on plant owners or local regulation. Maintenance of the security surveillance systems is necessary as well.

30.4.2 Advanced Maintenance and Operation for Large Power Plants

From small- to large-scale PV plants, it is extremely important to visually check every single solar panel and key equipment—mission critical for optimum performance. For example, reference the ET Solar 30MW of solar plant built in North Carolina. A total of 35,000 pieces of solar panel and 1000 of string inverters were installed on 200 acre site. This large-scale PV plant operates with excellent performance of maximum power generation with minimum down time, which is very crucial to the investor.

A remote monitoring and diagnostic operation control center is needed for monitoring every solar panel daily, monthly, and yearly performance and diagnosing problem remotely to reduce the operating cost and in timely manner.

To achieve above objectives, below are the relevant activities:

- *Operation.* Defined as monitoring of key parameters of PV power plant, fault diagnostic and identification, set up maintenance schedule, review maintenance results and create monthly, quarterly, yearly reports.
- *Maintenance.* Defined to fix faults and execute preventive regular maintenance of PV module cleaning, snow clearing, and vegetation abatement. Also include test and measurements, PV module performance, grounding and insulation of inverter, combine box, transformers, etc., according to maintenance manual of manufacture. Security check is necessary of daily maintenance.
- *Advanced operation and maintenance.* Defined as operation and maintenance, which requires high-skilled engineering capabilities of retrofit, revamping, remote monitoring and diagnostic and predicative maintenance based on data mining. Asset, warranty, insurance, and financing management are also critical for the power plant.

Plant operation covers general on-site PV plant operation by monitoring parameters of key equipment (such as power meter, transformers, inverter, dc combine box, PV string, and communication device) according to operation instruction.

These parameters can be viewed by trend plots and data tables provided by the monitoring software system.

Understanding the operation principle of equipment and normal range of data is critical to identify if they are in good condition. Monitoring system may provide warning or alarm to the user by SMS or email.

30.5 FUTURE PROSPECTS OF PHOTOVOLTAICS

Global warming is a major threat to our daily life. Around the world, people are facing more frequent and severe drought, floods, fires, and storms. Temperature is rising globally. Animals are forced to migrate to cooler area. Risen sea levels become a new threat to human life.

Solar energy is currently the most effective alternative solution in today's global warming crisis. Governments and private enterprises are jointly developing technologies to improve photovoltaic efficiency and make solar system smarter and cost-effective. The newly developed PERC solar cell technology can improve solar cell efficiency to 20 percent from today's 17 percent. It is projected that by the year of 2025, cell efficiency will reach as high as 40 percent.

From the system perspective, companies like ET Solar have developed smart modules such as solar cell optimizer. A smart integrated circuit (IC) chip assembled between solar cell strings can enhance the power output from solar panels. Some solar panels also contain built-in communication device to communicate with power plant control center for improving power output efficiency.

Because of increased solar cell efficiency coupled with lower prices, photovoltaic technologies have been widely adopted by consumer products such as building cooling and heating system, power electrical cars, street lights, etc. The future of the photovoltaic industry is promising.

30.6 REFERENCES

1. U.S. Energy Information Administration, Annual Energy Outlook 2014 with projections to 2040, Washington D.C., http://www.eia.gov/forecasts/aeo/pdf/0383(2014).pdf.
2. U.S. Energy Information Administration, Natural Gas 1998: Issues and Trends, Washington D.C., http://www.eia.gov/pub/oil_gas/natural_gas/analysis_publications/natural_gas_1998_issues_trends/pdf/chapter2.pdf.
3. NASA Earth Observatory, Michael Carlowicz, Global temperatures 1885–2014, http://earthobservatory.nasa.gov/Features/WorldOfChange/decadaltemp.php.
4. Javier Campillo and Stephen Foster, "Global Solar Photovoltaic Industry Analysis with Focus on the Chinese Market," Mälardålen University, Sweden, May 2012, http://www.diva-portal.org/smash/get/diva2:127961/FULLTEXT01.pdf.
5. International Energy Agency, "Technology Roadmap of Solar Photovoltaic Energy," 2014 edition, Paris, France, https://www.iea.org/media/freepublications/technologyroadmaps/solar/TechnologyRoadmapSolarPhotovoltaicEnergy_2014edition.pdf.

PART 5

PROCESS GASES AND CHEMICALS

Chapter 31
Gas Distribution Systems

BY

KENNETH GROSSER *Air Liquide Electronics US, Dallas, Texas*
JAMES MCANDREW AND TRACEY JACKSIER *Air Liquide, Delaware Research & Technology Center, Newark, Delaware*

31.1 INTRODUCTION

Gas distribution systems have become increasingly more critical to providing high-purity gases to the semiconductor tools. The primary function and goal of a distribution system is to convey the gas from the source of the product to the process tool without adding contaminants. Customers are now expecting some gases to have very low ppb (part per billion) levels of each atmospheric impurity as delivered to the process tool.

In a gas supply system the major subsystems are

- Gas production, either production of the gas off site and delivered to the semiconductor manufacturer's facility, or on-site production of the gas.
- Point-of-delivery (POD) is typically after the source product has been pressure regulated and filtered and then "delivered" to the customer's facility.
- Distribution system to a network of main and submain lines with tool-hookup valves below the manufacturing floor, often referred to as the point-of-use (POU).
- Tool hookup, or the connection between the submain valve and the process tool.

Gas distribution systems apply to both "atmospheric" bulk gases (i.e., Ar, N_2, He, H_2, O_2) and to specialty gases (typically from high-pressure cylinders), refer to Sec. 31.10; however, bulk specialty gases are growing in use. Typical bulk specialty gases include anhydrous NH_3, anhydrous HCl, silane, NF_3, CO_2, and more.

Bulk atmospheric gases are provided by numerous methods depending largely on the volumes and purities required by the semiconductor customer, including

- Cryogenic (refrigerated liquid) nitrogen, hydrogen, argon, and oxygen brought to the customer's site in insulated tanker trucks and stored on site in cryogenic vessels.
- More recently cryogenic helium is being supplied to very high volume consumers.
- High-pressure compressed gas tube trailers (150 to 200 bar) are commonly used for hydrogen and helium.
- On-site nitrogen generating plants for gaseous nitrogen only.
- On-site air separation plants, in which air is liquefied and separated into nitrogen, oxygen, and (optionally) argon. Such plants make both cryogenic liquid and gaseous nitrogen, and cryogenic liquid for the other molecules. The cryogenic liquid nitrogen can then be stored on site for backup, by vaporization of the liquid.

Bulk specialty gas supply systems have several advantages over the more conventional method of high-pressure cylinders. Advantages of bulk specialty gas systems include lower unit cost of the product, a more comprehensive analysis of the product prior to delivery, and fewer changes of the supply of gas.

One of the benefits of larger volumes includes greater consistency of product for many tools over a longer period of time. This facilitates better control and greater stability of the process at each tool. Manufacturing processes are very susceptible changes in gas quality and can result in reduced product quality and consistency, therefore reducing changes in the materials supporting the processes enhances yield.

Another benefit of fewer changes is related to reducing or eliminating the introduction of moisture and air contamination into the system. The inherent reactivity of corrosive gases makes them useful in the semiconductor manufacturing process. However, this reactivity can contribute to challenges in handling and distribution which can compromise the purity of the gases and the systems. Reactive gases generally react with oxygen and moisture to form particles or other undesirable contaminants. For example, SiH_4 reacts rapidly with O_2 to form SiO_2 particles. Corrosion of distribution systems will also occur when oxygen and moisture are present with reactive gases such as halide gases (HCl, HBr, WF_6, etc.), releasing both particles and volatile metal impurities. In corrosive gas systems a key factor in mitigating internal corrosion of the gas panel and gas distribution system is low moisture content in the system. The major factors for limiting moisture in a system are

- Low moisture content in the source of the product.
- Moisture removal purifiers (i.e., molecular sieves, if compatible with the product) located between the gas source and the gas supply panel.
- Effective, automated purging systems in the gas supply panel, including purified purge gas and engineered vacuum/pressure purge cycling of the connecting line between the product source and the gas supply panel (often referred to as the "pigtail").
- For liquefied gases, sufficient heating of the cylinder or bulk source. If sufficient heat is not provided, liquid droplets are easily entrained in the gas system. These droplets typically carry higher levels of moisture and other contaminants that are present in the homogenous stream of vapor.

Because moisture adsorbs on surfaces, it can enter a distribution system when the system is exposed to air, and remain there even after the air is purged out. This moisture remains adsorbed on surfaces of piping and components and can react with gases that are introduced later. Thus, it is very important to have rigorous purging procedures and design characteristics that minimize adsorbed moisture as well as air infiltration.

Semiconductor customers often have more than one distribution system for each gas species in order to segregate systems to mitigate or eliminate the possibility of cross contamination of systems. Typical systems include

- Process N_2 or ultrapure N_2 (gas that comes into contact with the product during processing).
- Nonprocess N_2 (usually with a less rigorous purity specification).
- Utility N_2 (with the least rigorous purity specification).
- Process Ar.
- Weld Ar (WAr). This is a separate distribution system specifically for orbital welding of tubing, typically for the addition of process tools.
- Process O_2.
- Utility or nonprocess O_2.
- Process He.
- Helium for leak testing.
- Process H_2.
- H_2 for abatement (burn box) systems.
- CDA (clean dry compressed air) for nonprocess applications.
- Stepper-CDA specifically for lithography tools with special high-purity requirements, and which cannot be backed up by N_2 because the "air" constituents (especially refractive index) are critical to the lithography process.

It should also be noted that the terms "ultrahigh purity" and "high purity" are used copiously throughout the semiconductor and electronics industries and neither have a clear definition. Each semiconductor manufacturer has their own specifications defining their expectations for leak integrity, gas purities, pressures, temperatures, flow rates, etc. These specifications vary from user to customer, and even vary from site to site with the same user.

Bulk gas purity specifications from customers have also become more rigorous over the years. In the 1960s specifications were in "ppm" (or even "nines") whereas today they are in the low "ppb's." Cleanroom cleanliness has also improved over the years. To put this into perspective, one of the authors' father worked in a cleanroom in the 1960s (2 in or 51 mm wafers, ≥ 10 μm geometry), and he came home one day to tell the family that the new cleanroom protocol requirements prohibited *smoking* in the cleanroom. In the 1990s, there was arguably a tendency to over specify, by automatically advancing purity requirements with each node of the semiconductor manufacturing roadmap. This led, for example, to the expectation that part-per-trillion purities would be required "soon." While these requirements have indeed come to pass in some cases, there has been an overall maturing to have a scientific basis for purity specifications.

Although the specification of "nines" (i.e., 6 Nines, 6N means 99.9999 percent pure, or a total "budget" of 1 ppm of impurities) is still used for cylinder gases, the shortcoming of this specification method is that the *impurities of interest* still must be defined at specific ppm or ppb levels. Where it becomes especially uncertain is when some impurities are not listed at all. For example, some customers may or may not list N_2 as an impurity in H_2; therefore, the "nines" specification lends no comparison between the two.

In summary, the use of the terms "high purity" and "ultrahigh purity" are *relative* or *qualitative* terms and should only be used in conjunction with specifically defined criteria for a given customer or product.

31.2 DESIGN PRINCIPLES

The effective design of a gas distribution system is critical to meeting user expectations and should consider purity of the gas, ease of maintenance, flexibility for adding or changing tools and tool layouts throughout the life of the facility, and reliability for the lifetime of the facility.

The basic principles for purity, reliability, and longevity are as follows (each discussed in more detail later in the text). Note that for the purposes of discussion, the term "free" is used to mean *approaching the minimum or best attainable metric for each principle.*

- *Leak-free* systems to eliminate atmospheric infiltration or aspiration into the system.
- *Outgas- and permeation-free* systems to mitigate extraneous impurities from entering the system.
- *Dead zone–free* systems to minimize the contributions of archived impurities in stagnant regions of the system.
- *Particle-free* systems including the elimination of particle generating sources and systems that readily clean up and eliminate any particles that do enter the system (mostly after the initial installation).
- Moisture-free systems that avoid moisture intrusion by allowing for good purging practices, and use materials that dry relatively rapidly (e.g., stainless steel tubing with low roughness to minimize surface area).
- *Passivated (nonreactive) surfaces* to reduce adsorption of moisture, enhance moisture removal, and as a means to reduce corrosion in reactive gas systems.
- *Control of flow dynamics* to prevent static regions in the system that could archive and release impurities.
- For reactive gases, *maintaining vapor phase* by ensuring that adequate heat is provided at the gas source and avoiding cold spots in the distribution system. Liquid droplets of reactive gases can accelerate corrosion significantly and entrain contaminants from the bulk source.

The leading principles to address these attributes of a gas distribution system are

- Distribution system layout configuration (e.g., loop, radial, fishbone).
- Valve types and layout.
- Pressure drop maintained within customer specifications.
- A design that minimizes startup and commissioning time (i.e., meeting customer specifications without delay).
- Improving long-term operation and maintenance of systems by the involving operations personnel well before turnover of systems.
- Long-term management of systems to preclude contamination and to maintain purity over the life of the facility.

Bulk gas distribution system designs generally fall into one of three categories with the following monikers (shown in Fig. 31.1):

- Loop or ladder distribution system
- Radial distribution system
- Fishbone distribution system

The loop or ladder distribution system has a "main" that either encircles the entire distribution system or has two mains that connect all of the submains or laterals. This is often utilized with the expectation to improve the balance of pressures throughout the systems.

This type of system is often perceived as improving the gas quality throughout the system but this is not always the case. Experience as well as computer simulations [1] have shown that static regions can exist in a loop or ladder type distribution system, resulting in difficulties purging the system both at startup and during ongoing operation of the system.

This can be understood intuitively by considering that it requires a differential pressure to cause gas to flow in a system, and that the equalization of pressure throughout the system reduces differential pressures in some regions, and therefore reduces the flow in those regions, hence static areas.

The benefits and liabilities of a loop or ladder type of distribution system should be carefully considered, evaluated, and understood before implementing this type of system.

A radial distribution (Fig. 31.1) consists of one main line, and numerous submains or laterals to supply gas to tools often with a sample or vent valve at the end of each submain (also known as "end of line" or EOL).

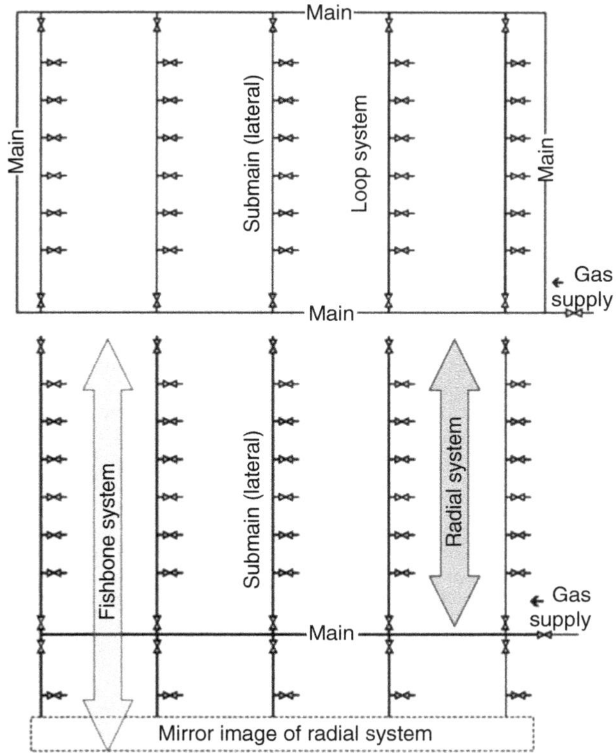

Figure 31.1 Loop/ladder, radial, and fishbone distribution systems. (© *Air Liquide. Used with permission.*)

Although this type of system is also susceptible to static regions if tools are not consuming product, or no tools are connected near the end of each submain, it lends itself to efficient purging during startup and any time that contamination may be introduced into the system during operation. This is accomplished by safely purging each submain through the EOL vent valve, starting with the submain closest to the source of the product gas and progressing down the main to the furthest submain last.

Pressure drop concerns can be mitigated by properly sizing the submains and using a point-of-use (POU) regulator at each tool connection.

The "fishbone" type of gas distribution system is similar to the radial system but it has submains extending from both sides of the main (Fig. 31.1). The main then resembles the spine of a fish with "bones" (laterals) extending from both sides of the spine.

This type of distribution system has essentially the same characteristics as a radial system for purging and system cleanup by purging each corresponding submain simultaneously and then using the same sequencing as described for the radial system.

31.2.1 Leak-Free Systems

It is important that product not leak out of a system, but just as important, it is essential to prevent atmosphere from aspirating into a system (adding mostly H_2O, O_2, and N_2, and to a lesser degree CH_4, CO, CO_2, and H_2).

It is a common misconception that if a line is under pressure that the movement of molecules will only be from inside the system to outside the system. The primary driver for the movement of gas is *partial pressure*. For example, the partial pressure of N_2 in atmosphere is about 0.78 bar. The partial pressure of N_2 in a purified Ar system (e.g., < 1 ppb N_2) is nearly zero so there is a driving pressure of 0.78 bar N_2 partial

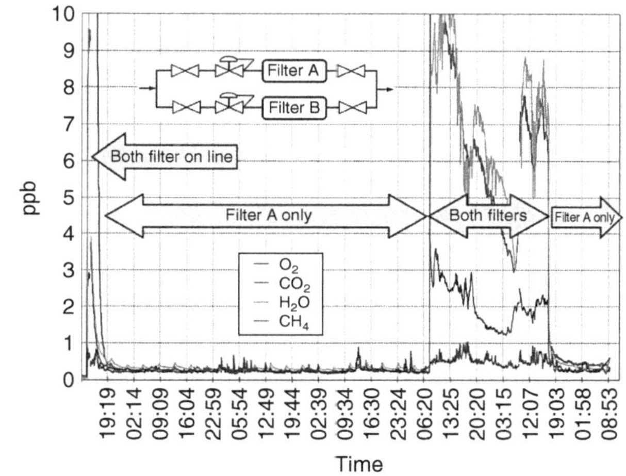

Figure 31.2 Impurities aspirated into system through a leak. (© *Air Liquide. Used with permission.*)

pressure into the system. Depending on the nature and shape of the leak path atmospheric aspiration can be significant.

Figure 31.2 shows the purities at the outlet of two filters in parallel (analyzed using an APIMS [2], atmospheric pressure ionization mass spectrometer). With high purity N_2 flowing through only Filter-A all impurities were < 1 ppb. With gas flowing through both filters, some of the atmospheric impurities were over 10 ppb. A leak (not identified by a pressure decay test) in the filter housing gasket was identified as the root cause.

A common leak rate specification for semiconductor manufacturers for process systems is $\leq 1 \times 10^{-9}$ atm·cc/s *inboard helium leak test*. The test is conducted by using a helium leak mass spectrometer device that puts the system under test under vacuum. Subsequently, helium is *sprayed* (applied by a portable source of helium released from a hand nozzle) at each potential leak point (valves, welds, fittings, gauges, etc.), and if a leak is present, helium aspirates into the system, diffuses, and is detected by the helium mass spectrometer and converted to atmospheric cubic centimeters per second.

To put the magnitude of leak rates into perspective:

- At a leak rate of 1 atm·cc/s, it would take 1 s to leak 1 cc (cm^3) of gas at standard temperature and pressure.
- A leak rate of 1/10 atm·cc/s (or 1×10^{-1}) would take 10 s to leak 1 cc of gas.
- A leak rate of 1/100 atm·cc/s (or 1×10^{-2}) would take 100 s to leak 1 cc of gas.
- A leak rate of 1×10^{-9} atm·cc/s would take:
  - 10^9 s to leak 1 cc of gas
  - 31.7 years to leak 1 cc of gas
  - 897,921 years to leak 1 ft^3 of gas
  - 31,709,799 years to leak 1 m^3 of gas

Inboard helium leak detectors typically have a detection limit of 1×10^{-11} atm·cc/s; therefore, they are well suited for testing to a criteria of 1×10^{-9} atm·cc/s.

Although outboard helium tests are also sometimes conducted (where the system under test is pressurized with helium or percent helium and a portable helium "sniffer" is used to detect leaks), there are limitations to this type of test. The maximum sensitivity of the test is no more than the ambient level of helium in the area (typically converted to an equivalent leak rate of around 1×10^{-6} atm·cc/s). Once the system is filled with helium, then helium permeates into seat materials in the system (valves, regulators, etc.) and is subsequently desorbed into the system. It will then appear as a virtual leak in a subsequent inboard leak test, rendering the inboard test less sensitive and less valuable.

In summary, the objective is to provide systems that are leak "free" not only at the time of installation but by the proper selection of components, installation methods, and maintenance procedures remain without leaks for the lifetime of the facility.

31.2.2 Dead Zone Free Systems

Dead zones are regions of a distribution system where gas flow can be static. Examples of system dead zones, and examples of dead legs from "Tee-Offs" to valves are shown in Fig. 31.3.

Dead zones or dead legs in gas distribution systems are not always immediately obvious and include lengths of small tubing used to connect to pressure gauges or pressure transmitters (PT), branches of a distribution system where few or no tools are demanding gas from the submain, future connection points, etc. Examples of dead legs are illustrated in Fig. 31.3.

Static regions in gas distribution systems can become areas to archive impurities, especially by desorption (or outgassing) of moisture from the tube walls and/or seat materials in valves, regulators, check valves, etc. When turbulence or pressure transients occur in the system, the static or dead regions are upset and the archived impurities are released into the active regions of the system, resulting in elevated levels of impurities delivered to the tools.

Dead zones and dead legs can make it difficult to clean a system following the installation (commissioning) or if the system becomes contaminated with impurities during production.

Designing systems with minimum dead legs, dead zones, and static regions expedites the initial cleanup, as well as maintain purity levels at the tools, and facilitates system cleanup from an unexpected contamination event.

If some parts of the distribution cannot be purged with a direct flow of gas, the most effective strategy is to use pressure cycling (pressurizing the system, then releasing the pressure, but not to atmospheric pressure). During the low-pressure part of the cycle, contaminants in the dead volume will be removed as gas flows out of it. This approach requires manual intervention but is substantially faster than passive purging of dead volumes.

31.2.3 Particle-Free Systems

Particles are known to severely impact device performance; therefore, clean manufacturing processes must be maintained to eliminate particle inclusions in gas streams. A typical gas distribution system has filters at the POD (where the gas enters distribution system). More information on filter types and performance is given in Sec. 31.3. To facilitate the maintenance of bulk gas systems, it is a good practice to have two filters and regulators in parallel.

With two filters and regulators in parallel, each is sized to accommodate the full design flow. With this configuration, one filter/regulator can be isolated and serviced or replaced if necessary without interrupting the flow of gas to the facility.

A common complexity of having two filters in parallel is that flow will tend to favor one leg of the system owing to unequal pressure drops in the two legs and because of the difficulty in setting both regulators at exactly the same pressure. When uneven flow occurs there can be both a near-dead-leg and also uneven loading of particles trapped by the filters. The latter can then result in an additional pressure drop difference between the two filters. To balance this, the regulators can be readjusted

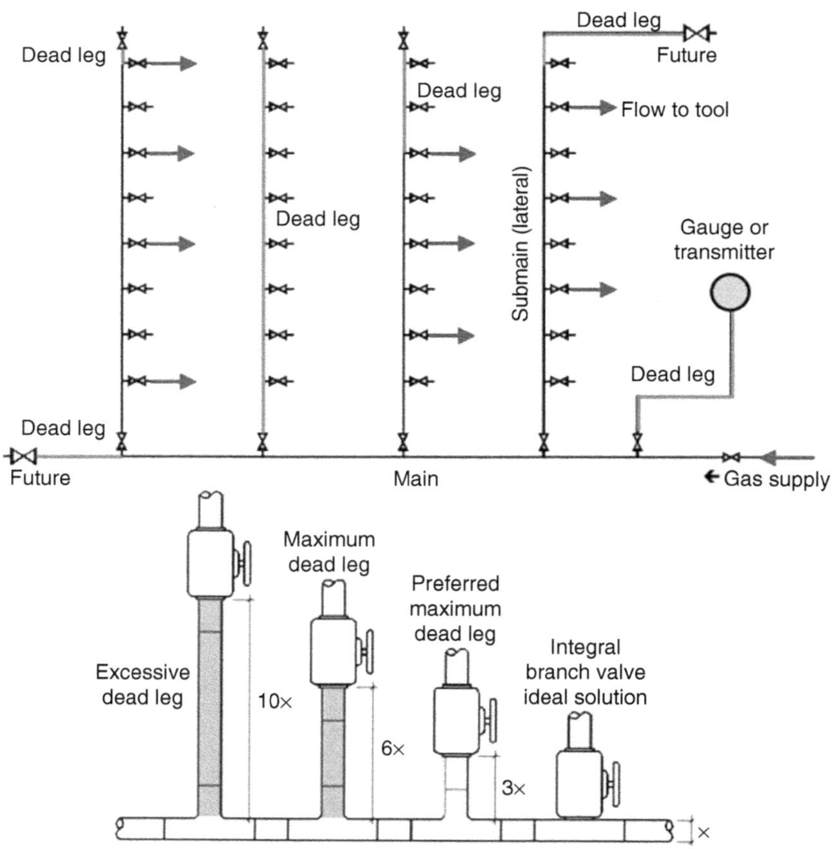

Figure 31.3 Examples of dead legs and dead zones. (© *Air Liquide. Used with permission.*)

on a regular frequency to alternate the flow path between the two filters (i.e., the lead-lag concept).

The selection of materials and components (Sec. 31.3) and installation practices (Sec. 31.5) play an important role in controlling particles in distribution systems. Three primary considerations include

1. Tubing should have smooth, passivated surfaces (i.e., electropolished) that do not produce, archive, or release particles. This principle supports both the initial cleanup of the systems as well as maintaining low levels of particles throughout the life of the system.
2. Components should have the same features as the tubing as well as having flow dynamics that flush out particles, and seat material that does not archive particles.
3. Utilization of point-of-use (POU) filters at the inlet of each tool is important. These ensure that virtually no particles are sent to the tools. Given the low challenge of particles from a properly designed gas distribution system, the POU filter should maintain their expected performance for the lifetime of the tool without the need for replacement.

Another source of particles comes from the process tool gas processor itself, as well as the process. Even if particle-free gas is delivered to the inlet of the process tool, it is important that the process tool manufacturer design the internal gas processor or gas handling system using the same considerations applied to gas distribution systems. If not then the particles can be introduced into the process regardless of the quality of the gas distribution system supplying the tool. This applies not only to particles, but all impurities that could damage or affect the process, and hence the product yield.

Particles can come from many sources including those left in the components from the manufacturing process; exposure to atmosphere during the initial installation; inadequate cleaning after cutting tubing for welding; the welding process; any time two surfaces chafe (i.e., movement of valve, regulator, or flow controller internals); breakthrough of a filter; the addition of valves or modifications to distribution systems during operation, etc.

31.2.4 The Unique Nature of Moisture as an Impurity

The interaction of moisture with gas distribution systems and all of the related components and materials is different from all other impurities, thus deserving of a dedicated discussion.

Moisture is an *adhering* molecule because of its polar molecular characteristics that tends to bond with surfaces. Water molecules also have a strong attraction to one another, which is the phenomenon behind surface tension (seen as water "beading" on a surface) and also contributes to its adhesion properties. As a result, moisture is difficult to remove from gas distribution systems. The most common methods of moisture removal are turbulent flow purging with a dry inert gas (i.e., N_2 or Ar at < 1 ppb H_2O) and the application of heat. Heat can be applied to a gas distribution system either by heating the purge gas or applying heat directly to the system (i.e., heat guns or heat trace tape), although it should be noted that heating the purge gas is effective only over a very short distance and provides only moderate heating. Direct heating is far preferable but caution must be exercised not to exceed the thermal ratings of components, such as seat material, gaskets, gauges, pressure transmitters, filters, etc.

The amount of moisture adhering to *wetted* surfaces (areas of the gas distribution systems in contact with the product gas) is related to both the micro-surface area and the type of material. More will be discussed in Sec. 31.3.1, but in general, the greater the surface area the more moisture is adsorbed.

As a result of moisture adsorbing onto surfaces and then desorbing (also known as outgassing or off-gassing), the level of moisture detected at the end or outlet of a tube or distribution system is not necessarily representative of the amount of moisture entering the system. Figure 31.4 (upper half) shows a "transfer function" of a pulse of moisture introduced into a tube and the moisture analyzed at the end of the tube [3].

The tube used in this evaluation was a 6.1-m length of 316L electropolished (EP) stainless steel ¼-in tubing. The pulse of moisture

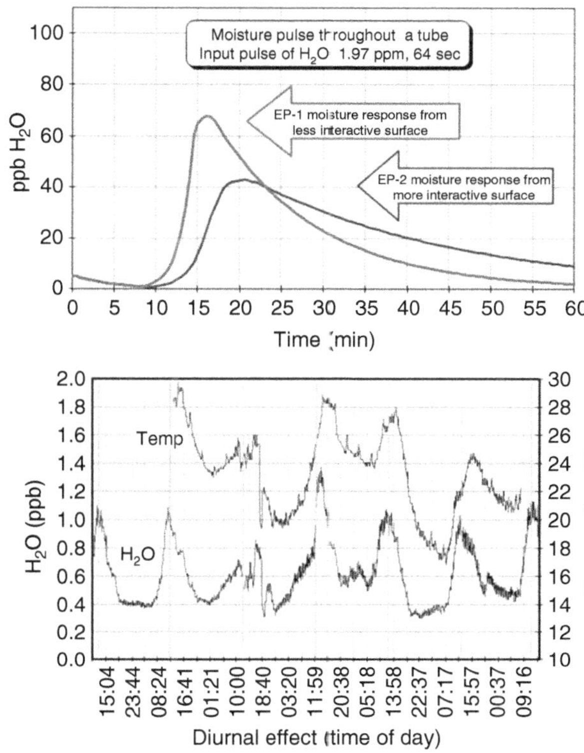

Figure 31.4 Moisture pulse through a tube and moisture diurnal effect. (© *Air Liquide. Used with permission.*)

was 1.97 ppm for 64 s, flow was 800 sccm. The EP-1 and EP-2 notations represent two different EP processes from two different suppliers.

The evaluation shows that both types of tubing interact with moisture (adsorption/desorption) such that the gas exiting the tube does not represent the profile of the gas pulse entering the tube. The EP-1 passivated tubing interacts less with moisture, hence more faithfully transmitting the pulse, whereas the EP-2 material interacts more with moisture, hence attenuating the input pulse. In both cases, the areas under the tube outlet curves closely approximate the area under the pulse curve (minus the amount of moisture retained on the tube surface).

Distribution systems, especially small tubing used as sample lines display the "diurnal" effect as a result of heating and cooling of the lines (day versus night temperatures and direct sunlight). As the tube is heated, moisture adsorbed on the surfaces of the system tend to desorb (or off-gas) and the moisture is added to the product gas. As the system cools, some moisture in the line is adsorbed on the surfaces of the system and removed from the product gas. This principle is well understood for molecular sieve material commonly used to dry gases. The molecular sieve media adsorbs moisture because of its large surface area, and then when saturated, it is heated with a dry gas purge and the moisture is driven out of the media and the process repeats. The diurnal effect is shown in the lower half of Fig. 31.4.

The data for Fig. 31.4 was obtained using APIMS with an LDL for H_2O of 50 ppt. The temperature was taken from the outdoor POD filter housing of an N_2 supply system. The tubing was 304L stainless steel tubing with unknown passivation. The vertical yellow lines represent noon of each day. The graph shows how the measured moisture in the system varies in relationship to the temperature of the system.

The diurnal effect is especially pronounced in lengthy sample lines (typically ¼ in) from a sample point on a system to remote instrumentation. The apparent changes in moisture can be misinterpreted as changes in the product gas. To mitigate this effect either insulation and/or heat trace tape can be applied to the sample line. Insulation is especially helpful where the sample line is subject to direct sunlight. Excessive diurnal variations may indicate the need to replace or regenerate a gas dryer.

In summary, moisture interacts with surfaces and permeable materials (i.e., valve seats, filter media, etc.) unlike most other impurities and, as such, deserves special considerations in the design, installation, commissioning, and operation of high-purity gas distribution systems.

31.2.5 Passivated Surfaces

There are numerous types of surface passivation, but most commonly used process for tubing and components used in the semiconductor industry is referred to as "electropolishing" (EP). Passivation alters the surface of a material to make it less reactive with the product gas. This can include making it less interactive with moisture, reactive gasses (cylinder or specialty gases), and particles.

The EP process is essentially the opposite of the electroplating process. When plating a piece of material with chromium, a piece of chromium (the anode) and the item to be plated (the cathode) are placed in an electrolyte bath. A direct current is applied, and chromium migrates from the piece of chromium to the cathode, resulting in the familiar mirror like chrome surface. This is a simplified explanation and the process requires considerable control over the chemistry and pH of the electrolyte, temperature of the bath, voltage, current, spacing between the cathode and anode, etc., in order to maintain the consistency and quality of the finished product.

Although electropolishing can be performed on a variety of materials, for those used in semiconductor gas distribution systems, the material is typically stainless steel (i.e., 316L for tubing and components, "L" is for low carbon, 304 may be used for CDA and nonprocess system components). The constituents of stainless steel (commonly abbreviated "SS") vary depending on the designation, but generally have 16 to 18 percent Cr, 10 to 14 percent Ni, 2 to 3 percent dopants (C, Mn, Si, P, S), and balance Fe.

The EP process is typically applied after surfaces have already been mechanically polished to low roughness. To EP a component (tubing, valves, regulators, filter housings, etc.), the component becomes the anode (contributor of surface material) and another piece of metal (Cu or other good conductor) becomes the cathode (recipient of material). Maximum current flows at sharp features, because they concentrate charge (higher current density), with the result that they are selectively removed from the surface. With the careful selection of process parameters Fe can be selectively removed from the surface, leaving a Cr enriched layer typically 25 to 40 Å. The Cr layer then becomes Cr-Ox (chromium oxide), which has desirable corrosion resistant characteristics. It also has less affinity for moisture. Both corrosion resistance and reduced interaction with moisture are related to the thickness and quality of the Cr-Ox layer, as well as the purity of the base metal (low in inclusions).

Another desirable quality of the wetted surfaces is smoothness (discussed further in Sec. 31.3). Rough, especially micro-rough, surfaces not only provide more area for moisture interaction but also provide features that will tend to archive and release particles. The removal of sharp features by the EP process is particularly important in reducing particle trapping.

It should be noted that the smoothness of and EP surface is not necessarily attained by the EP process, but by mechanical polishing of the surfaces prior to EP. If the EP process is too aggressive then excessive material will be removed from the surface, resulting in increased roughness.

Other forms of passivation include post-installation conditioning of systems. Examples include fluorine or tungsten hexafluoride (WF_6) systems, where controlled levels of fluorine or WF_6 mixed with an inert gas are purged through the system to intentionally complete any reactions with the product gas and *passivate* the surface such that further reactions are mitigated. The process may include purging with a mix of the reactive gas, followed by purging with inert gas, and repeating the process until tests conclude that reactions are essentially abated. Such specialty gases are extremely hazardous, therefore this type of passivation should only be performed by properly trained and qualified personnel using safe and reliable procedures.

In summary, for high-purity systems passivated surfaces on tubing and components have become the standard of the industry. This treatment reduces moisture interaction, improves corrosion resistance, and reduces particles in gas distribution systems.

31.2.6 Flow Dynamics

Similar to maintaining flow in high-purity DI (deionized) water systems (DIW), if a gas distribution system is expected to deliver low impurity levels (e.g., < 5 ppb each impurity at the process tool branch valve or POC, Point of Connection), it is important to ensure that gas is moving throughout the system at all times to mitigate static or dead zone regions (discussed in Sec. 31.2.2). The ideal gas distribution design for system criteria << 5 ppb would be as shown in Fig. 31.5.

Figure 31.5 shows a "radial" type of gas distribution system with a flow controlling type of device at the end of each submain (or lateral)

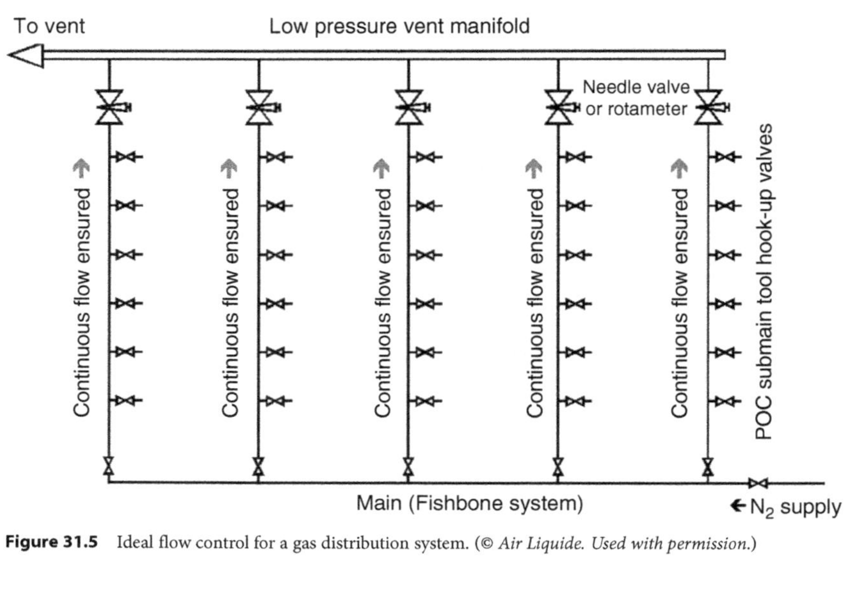

Figure 31.5 Ideal flow control for a gas distribution system. (© *Air Liquide. Used with permission.*)

that ensures a continuous purge through the gas submains at all times, regardless of the gas flow demands of the process tools, thus ensuring no static regions in the submain. By doing this with judicious control of flows, the gas purity at the POC will approach the purity of the supply of gas at the point of delivery (POD).

In summary, for low ppb systems it is essential that the flow dynamics of a gas distribution system are considered to be part of the design and operation of the system.

31.2.7 Tool Branch Valve Layout

As with any business intended to be profitable, it is well understood that cost control of a new facility is a significant factor. However, cost-of-ownership (COO) is often just as, or even more, important.

When considering the design of bulk gas distribution branch tool POC valves, one method is to install valves at locations that will support the initial tool layout; however, this has a distinct limitation. Semiconductor facilities are constantly undergoing modification, addition of tools, removal of tools, etc. If a submain shut down every time a valve has to be added or moved, it can be costly and will also interrupt production. Additionally each invasion into an operating system is an opportunity to allow contamination to enter the system or to produce a leak point.

When considering cost of ownership, it is prudent to install branch valves at a regular spacing without respect to the initial tool layout. This also facilitates the prefabrication of 20 in (6.1 m) subsections of the submain at an off-site facility, where cleanliness and welding can be better controlled. Leak testing of subsections can also be performed while "on the bench" of the prefabrication shop, reducing the time for leak testing in the field. Subsections can then be moved to the semiconductor facility in protective bagging and assembled on the utility racks with improved efficiency and reduced chances of contamination. Of course, the general functional layout of the facility (i.e., lithography, implant, deposition, etc.) will dictate where certain bulk gas submains will be located (i.e., helium, hydrogen, argon, etc.), indicative of which gas species are required to support specific processes.

31.3 MATERIALS

Material selection is very important for the design of a high-purity gas distribution system and can impact leak integrity, particle counts, purity, and long-term quality and reliability of the system for the life of the facility. There are numerous materials used in the construction of gas distribution systems, but this discussion will be limited to those most commonly used.

31.3.1 Tubing

There is a difference between tube and pipe. The size of a tube is the exact outside diameter (OD) of the tube, whereas the size of a pipe is somewhere in between the OD and ID (inside diameter) and varies with the "schedule" or thickness of the pipe. Tube is most commonly used in the gas distribution systems inside the semiconductor facility. Pipe is sometimes used for larger sizes (often > 6 in or 152 mm) from the production plant for nitrogen to the facility. Although a misnomer, systems are often referred to a "piping system" even though they may be constructed with tubing.

The most common material for process gas distribution systems and nonreactive specialty gas systems is 316L, EP (refer to Sec. 31.2). Another attribute of the EP surface is the surface roughness (or smoothness), given as "Ra" or average height of the micro-features of the surface test area expressed in micro-inches. A less commonly used measure is "Rmax" or the maximum height of the micro-features either in μ-inches or μm.

A common device used to measure surface roughness is known as a *contact stylus profilometer*. This device uses a small stylus that is displaced over the test piece and a plot is traced showing vertical displacement (with an exaggerated scale for easier visualization) versus the horizontal displacement. Noncontact profilometers are also available.

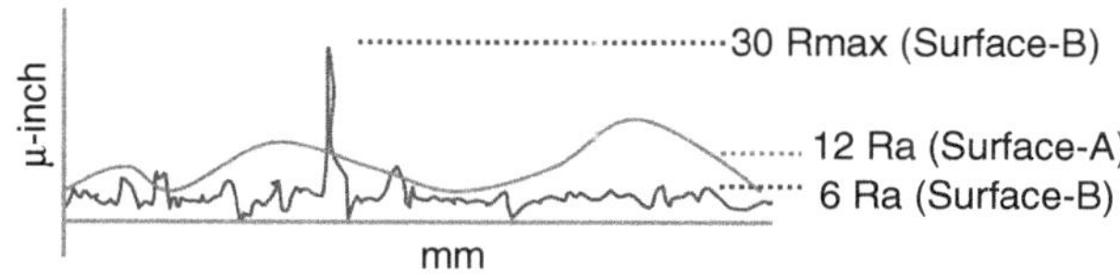

Figure 31.6 Representation of surface profiles from a profilometer. (© *Air Liquide. Used with permission.*)

Figure 31.6 is a depiction of two typical surface profiles and also shows the difference between Ra and Rmax.

A limitation of the Ra method of determining surface roughness originates from when surface smoothness first became of interest to the semiconductor industry. At that time the most easily adopted surface roughness measurement metric came from the machining industry in the form of a surface roughness comparator reference plate that contained a series of small areas, each with a different roughness (typically ranging from 50 to 200 Ra). These were used by customers to specify their expectation for the surface finish on a machined part. The machinist would process the workpiece to visually and/or tactilely match the specified area of the reference plate.

Since this method was first introduced, surface roughness requirements by the semiconductor industry (typically 5 to 15 Ra) have exceeded the effectiveness of the exemplar plate. The Ra limitation is that it does not consider either the true surface area or sharpness of the micro topography. Figure 31.6 depicts examples of two surfaces (A and B) using the Ra method (the representation has been exaggerated to illustrate the point). In this example, Surface-B (6 Ra) has sharper micro-features and more surface area than Surface-A (12 Ra). Surface-A (12 Ra) would actually be preferable due to less surface area and fewer micropeaks to archive and release particles.

Fractal dimensional analysis of surfaces (Rf) [4] provides a better measurement of the surface morphology because it considers both amplitude and frequency of the features. In this approach, the "fractal dimension" of a surface profile is defined in terms of how the measured length of the profile varies with the scale used. If the surface has a lot of small features, the length measured increases as a finer and finer scale is used. The classic example of a profile with a high fractal dimension is the coastline of a country. A study [4] took profilometer readings of four surfaces (as Rf, Ra, and Rmax). The surfaces were then contaminated with 0.6 μm polystyrene beads and then cleaned with DIW and swabbed. A SEM (scanning electron microscope) was used to inspect the surfaces for effective removal of particles.

The conclusion of the study was that the ease of removing particles is more reliably represented by Rf (fractal) values, than by Ra values.

For reactive (specialty) gases, various metallurgical processes are available from manufacturers (VAR, vacuum arc remelt, VIM vacuum induction melt, EBR e-beam remelt, etc.), all of which provide a more pure base metal, with fewer nonmetallic inclusions which then, when electropolished, yields a more robust and consistent Cr-Ox layer that is more resistant to corrosion.

For highly corrosive gases, other types of alloys are used, including nickel-only; nickel-chromium-molybdenum; nickel-copper; and nickel-chromium in varying degrees of content and other dopants (for strength, malleability, and other properties). Low Mn 316L SS has also been shown to reduce corrosion near the welds on tubes [5]; however, this material can be cost prohibitive and requires special weld techniques.

Specialty gas systems (reactive, toxic, etc.), depending on code and site-specific requirements, often use "double containment" tubing systems. This involves installation of the product tubing with a second larger tube surrounding the product tube (the region in between the two tubes is referred to as the annular space). This is a safety consideration to ensure that if the product gas leaks, that it is contained and disposed of safely. To provide an alarm method in the event of leak, three methods are commonly used:

- The annular space is maintained at vacuum and a detected loss of vacuum (due to product gas leaking into the annular space) is alarmed.
- The annular space is maintained at a pressure above the product line and a detected drop in pressure (due to the annular space gas leaking into the product line) is alarmed.
- The annular space is left open in the specialty gas control panel and the hazardous gas alarm system that monitors the panel will alarm if a leak is detected.

For CDA (clean dry compressed air) systems, copper tubing was used originally for many years, but because of the hazards associated with brazing copper, especially in an operating facility, and the increased purity of CDA systems, 304 or 304L SS BA (hydrogen bright annealed) used more frequently today, and it is installed using the same equipment and techniques employed for 316L high-purity systems (e.g., orbital welding, discussed further in Sec. 31.5).

Oxygen systems require that all piping and components be cleaned to remove hydrocarbons that may present a fire hazard when exposed to pure oxygen (cleaned for oxygen service, CFOS). In general, piping systems that are cleaned for semiconductor applications will exceed oxygen cleanliness requirements, but fire or other regulations may impose specific requirements on oxygen systems that must be respected.

Regarding "poly" tubing (e.g., any nonmetallic material), this is not an acceptable material for high-purity systems because atmosphere will permeate through the tubing as a significant rate. This is a well-known phenomenon from the permeation tube industry. These are small segments of polymeric tube containing a specific molecule (H_2O, O_2, N_2, He, etc.) that permeates through the tube at a consistent rate and a purified gas is passed over the tube (or by the tube in a "T" configured device) to produce a known level of impurity in the gas stream for calibration purposes (Fig. 31.7).

31.3.2 Valves

For high-purity gas systems, the key factors for selecting valves are surface finish (Ra), leak integrity (from atmosphere), leak integrity (across the seat), not archiving or contributing particles (especially when operated), and Cv. Cv is the *flow coefficient* of a component and is determined by flowing water through the component with a prescribed head pressure and then measuring the amount of water passing through the component for a given time frame. In compressible fluids (gases) Cv is also an indicator of pressure drop across the component, where the higher the Cv, the less pressure drop for a given flow rate.

For high-purity gas distribution systems the two most commonly used valves are stainless steel *diaphragm* and *bellows* valves.

The diaphragm valve uses a curved metal disc that impinges on a seat to isolate the flow of gas. In general, diaphragm valves have a lower Cv, lower dead volume, and less seat material than bellows valves and are used for lower flow applications such as tool hookup, sample lines, gauges, pressure transmitters, and the like.

Bellows valves (Fig. 31.8) have a convoluted sleeve (that isolates the gas stream from atmosphere) that facilitates the extended travel of the seating mechanism and, as a result, can be made into designs that completely remove the seat from the flow path and are often referred to as a "full throat" bore. As such, bellows valves have a very high Cv and are typically used in larger systems (i.e., high-volume nitrogen supply lines). The valve shown in Fig. 31.8 is not a full-throat design.

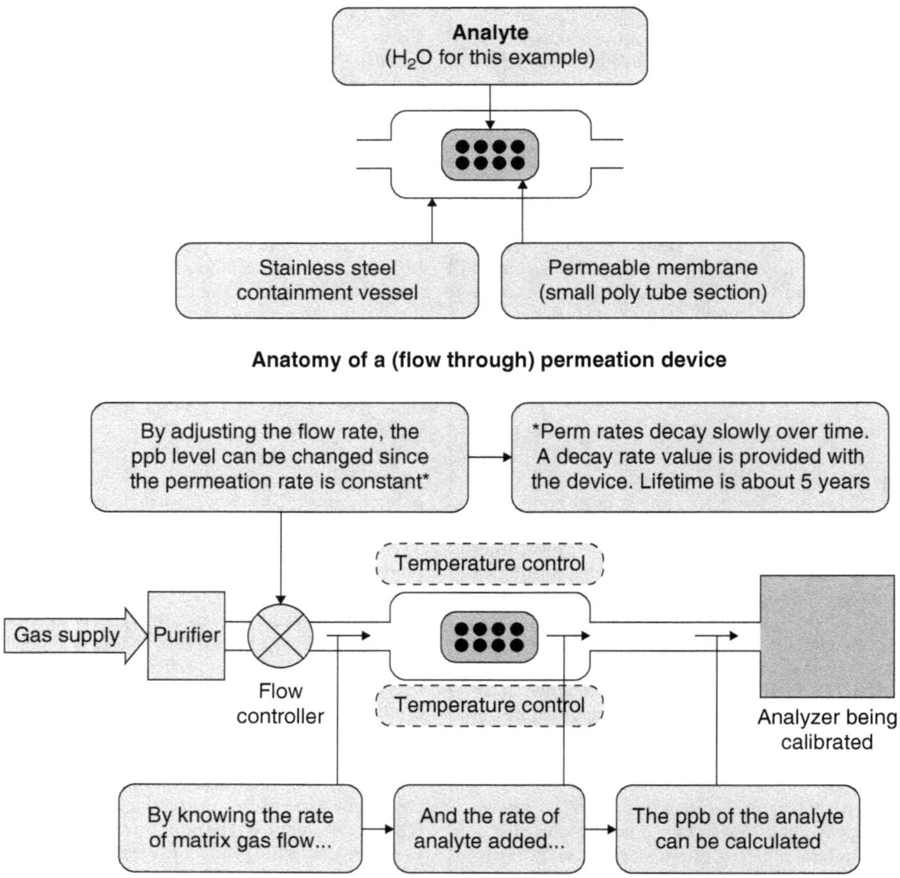

Figure 31.7 Permeation device for generating a calibration standard. (© *Air Liquide. Used with permission.*)

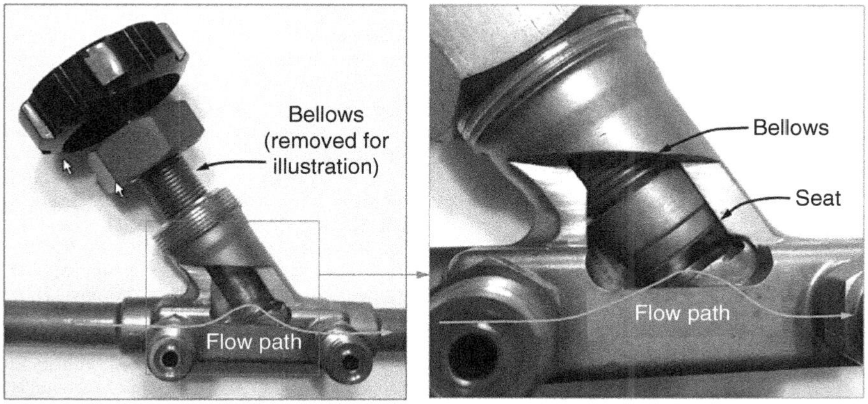

Figure 31.8 Example of [brass] bellows valve. (© *Air Liquide. Used with permission.*)

Ball valves are also used for their exceptionally large Cv; however, these are typically for large diameter systems found in nitrogen plants and they are generally located upstream of purification systems. They are also used throughout nonprocess gas distribution systems and CDA systems, where purity and particle specifications are not as rigorous.

31.3.3 Mechanical Connections

Mechanical connections are most commonly used where removal and replacement of components is necessary (i.e., gauges, flow meters, pressure transmitters, etc.). The three most commonly used components are flat face o-ring seal, compression seal, and flat face (metal) seal. Nonprocess and CDA systems often use the o-ring and compression fittings.

A compression fitting uses a tapered ferrule fitted on the tube that is compressed and wedges tightly against the tube and the device throat resulting in a reliable seal. They are available in brass and stainless steel. The advantages to these are simple installation (not requiring welding), cost-effectiveness, and reliability. The disadvantages are that they are not "zero clearance" (e.g., they require a physical separation in order to remove a component or disconnect lines), and leak integrity can diminish after numerous remakes (disconnect and reconnect).

A flat face o-ring seal is most commonly used for nonprocess and CDA systems where a zero clearance assembly and disassembly is desired. This device uses a flat metal face that impinges on a captive nonmetallic o-ring. Since o-rings are permeable, they are not commonly used in high-purity systems because atmosphere can permeate through the o-ring and enter the gas system.

For high-purity systems, the most common mechanical connection is the flat face seal fitting. This device comes in two primary variations. One type uses a flat metal gasket (of varying materials), that is compressed between two toroid (donut-shaped) surfaces that impinge on the gasket and form a seal with a leak rate integrity in the range of the detection limit of leak detectors (1×10^{-11} atm·cc/s). The other type is essentially the opposite design, where two flat faces impinge on a toroid gasket, with similar leak integrity.

The advantages to flat face seal fittings are zero clearance to facilitate removal and replacement of components bordered by rigidly fixed components (i.e., in specialty gas control panels), and if properly operated and maintained, can reproduce the initial leak integrity over many remakes. There are also no permeable materials in the fitting.

31.3.4 Filters

Filters are used to remove particles and are generally located at the POD and at the tool connections (downstream of the POC). The common media are membrane, metal, and ceramic.

Membrane filters are a fibrous structure of a nonmetallic material such as hydrophobic PTFE. Although the media resists adsorbing and absorbing moisture, the housing does. There is a dry down time associated with membrane filters, but they are offered with a preconditioning treatment that removes particles from the outlet side of the face of the media and dries them. Since the membrane is flexible, it can be susceptible to transient pressure events, resulting in rupture and break through. Membrane filters are known for low-pressure drops.

Metal filters come in a variety of alloys (nickel, stainless steel, etc.) and are either fibrous or sintered metal. Sintering is a process that uses metal powder, granules, or pellets and applies heat and compression to fuse the metal pieces together to form a porous metal structure.

Metal filters, with their considerable (unpassivated) surface area will readily adsorb moisture, but they can be baked at elevated temperatures for drying. Due to their rigidity, these filters are less susceptible to damage from pressure transient events.

Ceramic filters use a porous ceramic media that can have benefits for compatibility with corrosive gases. These filters also adsorb moisture more readily than membrane filters [6], and can be baked.

Filters are rated by their efficiency at removing particles at the size most likely to penetrate the specific media (most penetrating size, or MPS). A common rating is 9-log efficient at removing 0.003 μm MPS. This means that if challenged with 10^9 particles that one, at 0.003 μm, would likely penetrate the media.

There is a significant difference between particle removal in liquids and gases. There are more removal mechanisms in gases than in liquids. This means that the same filter is more efficient at removing particles in gas service than in liquid service. In gas applications filters remove particles both larger and smaller than the MPS size rating.

31.3.5 Check Valves

Check valves are designed to prevent the gross flow of gas in a direction opposite to the components designated flow direction. It is noteworthy to understand that with a given minimum back pressure (reverse flow pressure stated by the manufacturer), that check valves are only considered to be "bubble tight." The term is subjective but generally results in a leak rate of between 1×10^{-1} to 1×10^{-3} atm·cc/s across the seat. On smaller sizes they also contain a polymeric seat material with all of the aforementioned concerns.

It is important to realize that when gas is flowing through a check valve, that the device is essentially an open valve and therefore offers no impedance to the *back diffusion* of molecules. This condition increases in significance as the flow rate decreases.

Check valves are frequently oversized. This results in the check valve "chattering" noisily, that is, rapidly cycling open and closed instead of remaining constantly open. If a steady gas flow is required on a gas line with a check valve, the check valve should be sized carefully, with a small enough Cv that the pressure drop across the check valve is sufficient to keep it open. A chattering check valve will also release particles.

31.4 INSTALLATION SPECIFICATIONS

To ensure a high-quality gas distribution system, a clear and unambiguous specification should be developed that defines the methods and requirements for all aspects of the installation process. These are often in the form of the CSI (Construction Specifications Institute) format and numbering system. Some of the more important topics follow with a brief discussion on each.

31.4.1 Safety

Safety is paramount, always first and deserves references to related industrial codes, municipal codes, and site specific safety requirements.

31.4.2 Quality Assurance

Quality assurance usually includes a quality assurance representative (QAR) and, depending on the size of the project, a team of technicians to perform field observations and report deficiencies. The following should be considered:

- Auditing installing contractors
- Source inspections for critical equipment and systems (filter/regulator skids, purifiers, cryogenic vessels, etc.)
- Qualification of weld technicians, weld programs, and weld criteria
- Incoming materials inspections for compliance with materials specifications, inspection for damage, dimensional verification, surface quality, quarantine of unacceptable materials, etc.
- Weld coupon inspections and field weld inspections for compliance with weld acceptance criteria
- Compliance with installation specifications for workmanship
- Cleanliness installation techniques (to minimize moisture and particle contamination)
- A reporting, tracking, and corrective actions system for quality deficiencies

31.4.3 Materials Storage and Handling

Materials stored or staged for installation on site should be protected from contamination, damage, weather, and the like. It is common practice to require racks to support subassemblies to keep them off the floor and stage them for proper sequencing of the installation.

31.4.4 Installation of Fittings

Mechanical connections represent the most frequent occurrences of leaks in a newly installed system. For both compression and flat face seal fittings it is recommended that the specification include a requirement for all installers to have completed a formal training session on the proper makeup of these fittings. Training is commonly offered by the manufacturers of the fittings.

For compression fittings, it is also prudent to equip all installers and QA personnel with a gap-gage offered by the fitting manufacturer. This device is a go/no-go gage that is inserted between the compression gland and nut and it will identify if the proper amount of compression displacement has been applied.

Sometimes nylon ferrules are used for temporary connections (to avoid scratching or compressing the tube), but this can be a dangerous practice and it is not recommended without substantial supervision, controls, and tracking. Nylon ferrules used on a stainless steel tube can and will slip under pressure, turning the fitting into a dangerous projectile.

For flat face seal fittings, there are numerous mechanisms that can result in leaks. This is ironic because this device, when installed properly, provides the best possible leak integrity for a mechanical fitting. The following have all been observed (resulting in leaks):

- Inadequate tightening.
- Excessive tightening.
- Lack of a gasket.
- Damaged toroid bead. This can result from making up the connection without a gasket or allowing the surfaces to come into contact with anything other than the gasket. A scratch on bead surface is a path to leak.
- Multiple use of a single use gasket. The first use results in the indentation of the toroid bead into the gasket and the second use compresses in a slightly displaced area resulting in cross tracking between the two grooves and a leak.
- Use of two gaskets. This can occur if distracted during the connection process or if a retainer type of gasket is used and not removed when the fitting is disconnected.

A variety of gasket materials are available, and material compatibility with the product gas (especially reactive gases) must be considered. Nickel gaskets are frequently used because they more easily conform to the toroid bead and help mitigate leaks from a scratched bead. However, as an example of material incompatibility, nickel gaskets exposed to carbon monoxide will form nickel carbonyl [7], resulting in contamination of the product gas and degradation of the process using this gas.

Specifications should require that any fitting using a special gasket material be conspicuously labeled with appropriate warnings on the proper material to be used if the connection is serviced.

31.4.5 Purge and Weld Gasses

Purge gases are typically nitrogen or argon, and weld gas is nearly always argon (sometimes with a small percentage of helium or hydrogen to enhance the weld bead and minimize surface damage at the heat affected zone, HAZ). Purge gases and weld gases should have specifications to ensure quality.

31.4.6 Tubing Preparation (for Welding)

When automated orbital welding is used, it is an autogenous process (e.g., fusing the two work pieces together without the use of filler material to make the weld), and it is critically important to prepare the tube ends properly before welding.

Specifications should include criteria for how the tube is

- Cut (lathe, saw, etc.) and that the cutting surfaces be kept clean and free from lubricants that can be embedded into the material during cutting and welding.
- Faced to be flat, smooth, and precisely perpendicular.

Cleaned of particles from cutting, and any ink marks near the weld region.

31.4.7 Orbital Welding

The most commonly used welding technique on stainless steel tubing for high-purity gas systems is known as automated, orbital, autogenous, TIG (tungsten inert gas), butt fusion welding. The process involves an inert gas purging through the inside of the tube (back-seal gas), a purged envelope around the weld arc (arc-seal gas), an electrode from which an electric arc strikes the butt interface of the two tubes being welded, controlled rotation of the electrode around the circumference of the tubes, and fusion of the two tubes.

31.4.8 Weld Evaluation and Criteria

These expectations need to be clearly defined in the project specification because there is a wide range of criteria, and the evaluation of welds as compared to the criteria can be subjective.

31.4.9 Certification Criteria

Project specification should have clear, quantitative criteria for the performance of the completed gas distribution system. The paradigm for this is "to meet or exceed customer expectation." In order to achieve this without ambiguity, the expectations must be identified. This benefits the end user (semiconductor manufacturer), the designing authority, and the installing contractor.

31.5 QUALITY ASSURANCE

The assurance of quality is a process that includes the system design, detailed specifications, proper selection of materials, good materials handling, clean installation methods and techniques, exceptional welding, and diligent, ongoing inspections by qualified quality assurance representatives (QAR).

QARs are often employed by the end user, the design authority, and the installing contractor. Typical activities of a QAR include

- Conformance to the contractor's published internal procedures
- Conformance to design authorities published specifications and criteria
- Conformance to industry accepted practices for high-purity systems
- Conformance to local and national codes and ordinances
- Conformance to overall design drawings, layout, component quantities, and placement
- Incoming materials inspection procedures and thoroughness
- Cleanliness handling and storage techniques
- Evaluation and qualification of weld technicians
- Welding program development and conformance to weld evaluation criteria
- Maintaining lines under continuous purge except for leak testing
- Prefabrication inspections and conformance to specifications
- Proper tube preparation (cutting, facing, cleaning)
- Conformance to project specific welding practices and procedures
    - Coupon in/out compliance to weld evaluation criteria
    - Cataloging and retention of coupons
    - Field welds conformance to weld evaluation criteria
    - Logging and marking welds
    - Purge gases, filtration, purification, and supply lines conformance to specifications
- Lines are installed plumb, level, and square/perpendicular to building features
- Purge ports, valve/regulator handles, etc., are readily accessible and functionally usable
- Gauge faces are readily visible
- Corrective actions for a mechanical damage to lines and components incurred during installation
- Flat face seal makeup (installer certification, correct rotation, rotation offset marking)
- Compression connections proper make-up (correct rotation, gap gauge check)
- Conformance to leak testing procedures, data recording method, and interpretation of results
- Distribution lines conditioning (purging for dry down and elimination of impurities to specification criteria)
- Correct analytical test protocols
    - Instrument calibration is current and instruments are operated by trained, qualified personnel
    - Ensure the accuracy and integrity of collected and reported data
    - Pressure integrity and pressure decay test procedures and accuracy of data
    - Overall presentation of data and conformance with project documentation requirements
- Daily documentation and reporting of observed deficiencies in quality or deviations from specifications, criteria, and/or expectations

Everyone involved in the quality assurance process must be familiar with all aspects of the governing specifications and customer expectations. Performing field inspections with complete objectivity is essential. Construction deadlines are always a predominant element during installation, so special precautions must always be taken to ensure that safety and quality are not compromised.

31.6 CERTIFICATION

The *quality* of a system is *built* into it. This includes the design, selection of materials, materials handling, installation methods and techniques, and welding, and the quality assurances or scrutiny of the installation process. *Certification* (Sec. 31.7) is a quantitative process that objectively verifies that the quality specifications and expectations for the project have been fulfilled.

It is essential that all instrumentation used in the certification process be operating properly, up to date on preventative maintenance, in current calibration, and are operated by trained and qualified personnel.

31.6.1 Helium Leak Testing

For inboard helium leak testing, documentation can be enhanced by recording the leak rate on a time based plot. Figure 31.9A shows leak testing that revealed a leak at a weld.

When leak testing, it is important to identify an onset of a potential leak point. Once it has been identified as a rise in the natural decay of the leak detector background then the leak should be repaired. It is not always necessary to determine the absolute value of the leak by spraying helium on the leak until steady state is achieved, refer to Fig. 31.9B. This takes additional time, allows more helium to enter the system and hence desensitizing subsequent tests, and does not add value to the process.

Figure 31.9 shows [C] the curve of the expected background decay of an inboard helium leak detector. Curve [B] shows the onset of a brief application of helium at a potential leak point. Curve [A] represents the results of continuous application of helium until steady state is achieved. Although knowing the steady-state leak rate may be of interest, it is only necessary to identify the onset of a leak [B] to determine that a leak is present and requires corrective actions.

By using a timebased plot of the leak rate, it is not necessary to have an individual attend the leak detector while another person applies (sprays) helium along a lengthy line. This not only reduces personnel, but provides a permanent, objective record of the leak test of a specific line or branch of the distribution system.

Helium leak detectors are calibrated using a calibrated leak standard. This is an apparatus similar to the permeation device described in Sec. 31.3.1, although for this device a stainless vessel is filled with helium and a thin glass envelop (frit) allows helium to permeate to an outlet fitting that is connected to the leak detector inlet. Leak detectors are also available with an internal leak standard that can be activated manually from the control panel or automatically upon each startup. Leak standards decay with time and the decay rate is customarily shown on the device (i.e., 10 percent decay per year), and should be taken into consideration when establishing calibration procedures.

It is important to address all potential leaks (e.g., any rise above the natural background decay curve) with the goal of a leak-free system. This is essential to the long-term reliability of a system because it should be assumed that leaks will only worsen with time.

31.6.2 Moisture and Oxygen Analyses

When analyzing for moisture and oxygen at the end of a main or submain to verify compliance with project purity criteria, data can be difficult to interpret if the levels are high and/or are not on a downward trend. One way to determine if the data represents the actual condition of the line under test, or if there is an anomaly with either the instrumentation or the sample line (tube from the sample point to the instrument) is to increase the flow of gas in either the line under test or the sample line.

Leaks or outgassing are relatively constant over a few hours or days. If the purge gas for testing is purified or very stable then an increase in flow would likely not show a change of purity by the instrument; however, if there is a leak or outgassing then the increase in flow would dilute the impurity in the gas and result in a decrease in impurities analyzed. A close approximation is that a doubling of flow rate will halve the impurity level, *if the level of impurity is from a leak or outgassing*. Conversely, if the flow rate is doubled and the impurity remains constant then the impurity level is either from the product source or an instrument anomaly.

To diagnose moisture and oxygen instrumentation if there is a concern about high readings, a POU ambient purifier can easily be inserted into the line at the inlet to the instrument. If the instrument tends toward zero then the high reading of the line under test is likely correct. If the instrument reading remains high then it is likely an instrument anomaly.

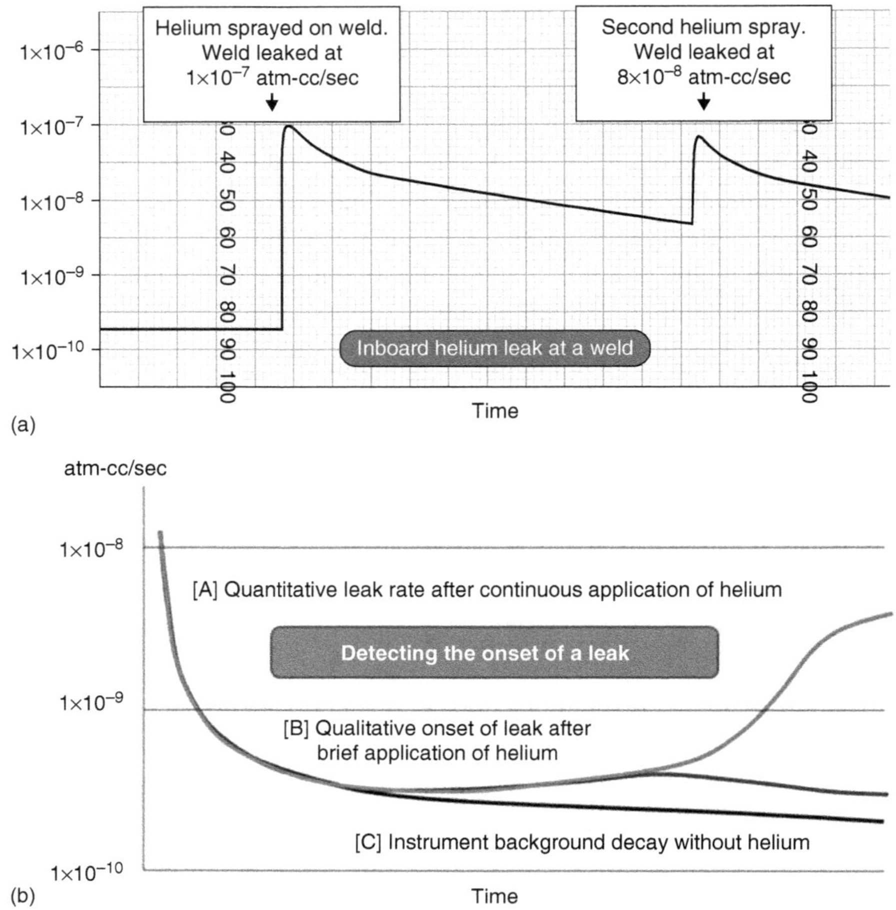

Figure 31.9 Inboard helium leak test of a weld and detecting the onset of a leak. (© *Air Liquide. Used with permission.*)

When testing lines for corrosive gas systems it is important to understand that the moisture measured at the effluent of the line under test is not necessarily an indicator of moisture adsorbed on the interior of the system. Corrosive gases will interact with moisture adhering to the surfaces of the distribution system and become sites for corrosion. For this reason, it is important to consider using heat to remove moisture from surfaces and mitigate corrosion before charging the system with the product gas.

31.6.3 Particle Testing

Particles are usually analyzed at either greater than or equal to 0.01 μm by a condensation nuclei counter (CNC) or at greater than or equal to 0.1 μm using an aerosol laser particle counter (LPC).

The CNC requires a condensing fluid (i.e., butanol, DIW, glycerin, etc.) and commercial models do not normally have size discrimination of the particle counts. This instrument condenses vapors on to the particles and make them larger and more easily identifiable by a laser detection system.

LPCs have size discrimination (i.e., 0.1, 0.2, 0.3, 0.5, 1.0, and ≥ 5 μm). If high particle counts are encountered when testing a line, this can be a useful diagnostic tool. The distribution of particle sizes can help determine the source of the particles and hence where to focus corrective actions.

When testing systems for particles, it may be important to include light impacts to the filter, piping, and components (e.g., moderate shocks with a rubber mallet or palm of the hand), releasing trapped particles into the gas stream, and detected by the instrument. This is important because during normal operation technicians or maintenance personnel may come into contact with piping systems and components and disturb them during the normal course of their work activities.

A good diagnostic tool for high particle counts from the line under test is to install a temporary POU filter at the inlet of the instrument to determine if the counts are real or an instrument anomaly.

31.7 QUALIFICATION/COMMISSIONING

At the completion of the certification process, approval of all related documentation, and confirmation that all purity and performance criteria has been achieved for a specific system, then that system can be considered qualified for use. If the system has been certified with the product gas then it can be turned over to the end user.

If the system will be used for a gas other than what was used in the certification process (i.e., He, H_2, O_2, and specialty gases), then it may be necessary to evacuate the system to remove the certification gas prior to introduction of the product gas. This decision depends on what impurities the end user considers as potentially damaging to the process.

For specialty gas systems with a gas control panel, the process tool is sometimes used to evacuate the line from the gas panel prior to the introduction of the product gas.

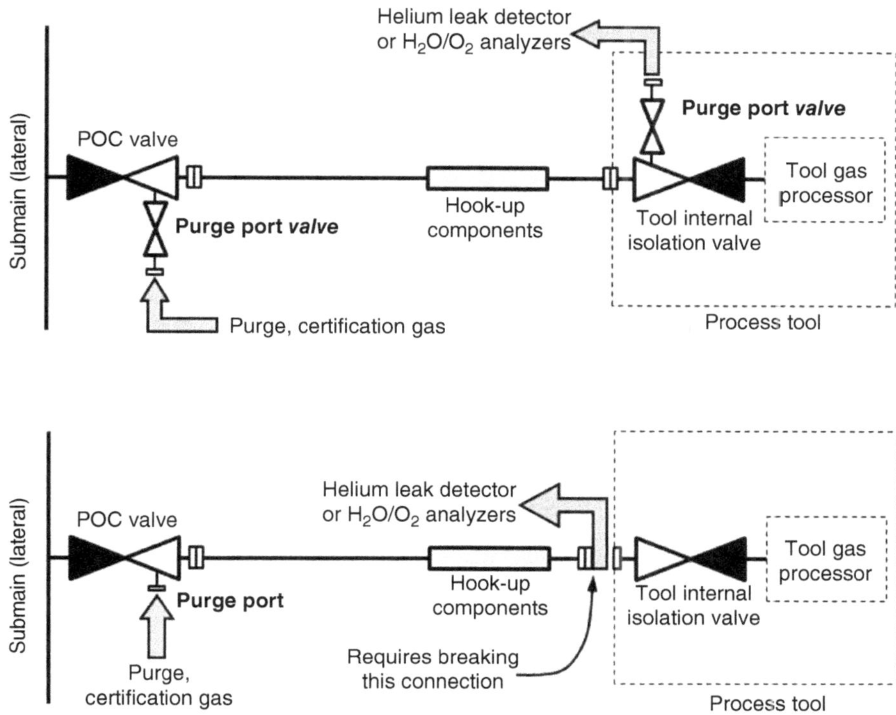

Figure 31.10 Tool-hookup with and without isolation valve purge port valves. (© *Air Liquide. Used with permission.*)

31.8 TOOL HOOKUP

This is the connection between the POC valve on the distribution submain or lateral and the tool inlet fitting. In between the two it is common to find a series of hookup components (sometimes called a "hookup stick") consisting of an isolation valve, sample valve, regulator, pressure gauge or transducer, and filter. Prefabricating, preconditioning, and pretesting these assemblies helps expedite the tool-hookup process.

There are a variety of concepts that can be used for tool-hookup lines, and the two main methods are the use of isolation valves with purge port valves versus isolation valves with only a purge port (Fig. 31.10). These can be found at the POC valve and/or the internals of the process tool.

Figure 31.10 (upper half) shows purge port valves at both the POC isolation valve and the isolation valve inside the process tool. This is an ideal configuration because the entire hookup line can be tested with minimal introduction of atmosphere and without breaking/making critical connections. This also reduces hookup certification time.

Figure 31.10 (lower half) is similar, but does not have purge port valves. This is often done to save cost, but it makes tool hookup more difficult and it also allows for the introduction of atmospheric impurities when making and breaking connections at the POC purge port and at the tool connection.

Since tool hookup occurs after the installation and certification of the product gas distribution systems, it is common for scheduling demands to impose considerable pressures on this phase of work. To mitigate mistakes or the introduction of impurities it is recommended that purge port valves be utilized as shown in Fig. 31.10 (upper half). Benefits of this concept will be realized throughout the lifetime of the facility by facilitating ongoing changes to tools and tool layouts.

31.9 OPERATION AND MAINTENANCE

Designing, installing, and turning over a high-purity gas system to a semiconductor manufacturer marks the end of one chapter (~2 to 3 years) and initializes another, more critical, chapter that will remain in play for the lifetime of the facility. If irregularities or errors occur during the course of the initial installation they can often be remedied with minimal impact to schedule and cost. If errors occur or contamination is introduced into an operating facility producing product, the consequences can be catastrophic resulting in millions of dollars of damaged wafers and millions of dollars of missed opportunity revenue. For these reasons, comprehensive and diligent control and management of the operation and maintenance of high-purity bulk gas systems is critically important.

31.10 REFERENCES

1. B. Jurcik and A. Norvilas, "Modeling Moisture in Transport of UHP Distribution Systems," *Solid State Technology*, 38 (11): 115, 1995.
2. C. Ronge, D. T. Murphy, and F. Shadman, "APIMS Technology: Sub ppb Measurement of Contaminants in Nitrogen with APIMS Trace+," *Proceedings Microcontamination Conference*, San Jose, CA, pp. 153–167, 1991.
3. J. McAndrew, M. Brandt, D. Li, and G. Kasper, "Establishing Moisture Test Methods for Process Gas Distribution Systems," *Microcontamination*, 9 (1): 33–37, 1991.
4. S. Chesters, H. C. Wang, and G. Kasper, "A Fractal Based Method for describing Surface Texture," *Solid State Technology*, 34 (73), Jan. 1991.
5. M. Mich, M. Miyoshi, K. Kawanda, and T. Ohmi, "Ultraclean Welding for High Grade Gas Handling Technology," edited by R. Novak, T. Ito, D. N. Schmidt, and D. Reedy *Contamination Control and Defect Reduction in Semiconductor Manufacturing II*, PV94-3, pp. 295–305, The Electrochemical Society, Pennington, NJ, 1994.

6. K. Grosser, "Ultra Clean Gas Delivery System," *Technical Proceeding Semicon/East*, pp. 7–15, Sep. 1989.

7. T. Jacksier, R. Tepe, and D. Vassallo, "Generation of Metal Carbonyl Standards for the Calibration of Spectroscopic Systems," US06153167, Nov. 2000.

31.11 FURTHER READING

I. Grant, M. J. Tinker, G. Mizuno, and C. Cluck, "Welding 304L Stainless Steel Tubing Having Variable Penetration Characteristics," Chulalongkorn University, Thailand, 1982.

Balazs NanoAnalysis, Electronics Specialty Gas Analysis, Application Note 0460.

Trace Analysis of Specialty and Electronic Gases, edited by W. Geiger and M. Raynor, Chap. 2, T. Jacksier, K. Tarutani, and M. Carre, "Sample Preparation for Metal Analysis in Gases by ICP-MS" Wiley, 2013.

T. Jacksier, R. Udischas, H.-C. Wang, and R. M. Barnes, "The Evaluation of Particle and Vapor Phase Contributions to Metallic Impurities in Electronic Grade Chlorine," *Analytical Chemistry*, 14: 2279, 1994.

NFPA 704, "Standard System for the Identification of the Hazards of Materials for Emergency Response," www.nfpa.org/704.

For example, International Code Council, International Mechanical Code, Chap. 5, Exhaust Systems, http://publicecodes.cyberregs.com/icod/imc/2012/icod_imc_2012_5_sec001.htm.

ANSI/CGA Code 13, "Storage and Handling of Silane and Silane Mixtures" Compressed Gas Association, Inc., 2015.

40 CFR Section 63, subpart BBBB, "National Emission Standards for Hazardous Air Pollutants for Semiconductor Manufacturing"

M. Dequesnes, M. U. Ghani, and R. Udischas, "Prediction of Pressure Drop in Gas Cylinders with Vapor Extraction at High and Low Flow Rates," Proceedings of the 2005 Annual AICHE Meeting, Cincinnati, OH, paper 71c.

H.C. Wang, R. Udischas, and B. Jurcik, "All Vapor Phase Delivery of Electronic Specialty Gases," Proceedings of SEMICON West 97 Gas Distribution Workshop, San Francisco, CA.

R. Udischas et al., "System and Method for Controlling Delivery of Liquefied Gases from a Bulk Source," US6363728, Apr. 2002.

G. Schmitt et al., "Evaluation of Air Liquide Gas Panel for Particles, Moisture and Live Gases," Microcontamination 93 Conference Proceedings, p. 453, 1993.

Chapter 32
Fundamentals of Ultrapure Water

BY

VYACHESLAY (SLAVA) LIBMAN *FTD Solutions LLC, Colorado Springs, Colorado*

32.1 INTRODUCTION

32.1.1 UPW System

Ultrapure water (also UPW or high-purity water) is water that is purified to exceptionally stringent quality specifications. Ultrapure water is a commonly used term in the semiconductor industry to emphasize the fact that the water is treated to the highest levels of purity for all contaminant types, including organic and inorganic compounds, dissolved and particulate matter, volatile and nonvolatile, reactive and inert, hydrophilic and hydrophobic, and dissolved gases. UPW, however, is very different from another commonly used term, deionized (DI) water.

Figure 32.1 illustrates a typical UPW system process flow, which has three main functional stages:

1. Pretreatment stage—to produce purified water, removing majority of the water impurities
2. Primary stage—completing deionization
3. Polishing stage—responsible for maintenance of recirculated UPW, as well as final particles, Total Organic Carbon (TOC), and often dissolved gases control

Recirculation of UPW is a key factor to ensure stability of the purity and quality of the water supplied. Traditionally the resistivity of UPW serves as an indication of the level of purity with a theoretical maximum of water resistivity of 18.18 MΩ·cm at 25°C.

32.1.2 UPW Quality Specifications

A number of organizations and groups develop and publish standards associated with the production of UPW. For microelectronics and power, they include Semiconductor Equipment and Materials International (SEMI) for microelectronics and photovoltaic, American Society for Testing and Materials International (ASTM International) for semiconductor and power, Electric Power Research Institute (EPRI) for power, American Society of Mechanical Engineers (ASME) for power, and International Association for the Properties of Water and Steam (IAPWS) for power. The most widely used requirements for UPW quality are documented by ASTM D5127 "Standard Guide for Ultra-Pure Water Used in the Electronics and Semiconductor Industries" [1] and SEMI F63 "Guide for ultrapure water used in semiconductor processing" [2]. ASTM D5127 provides sets of water quality specifications for each generation of semiconductor fabrications based on the node size of the respective generation. It refers to SEMI F063, as a guide responsible for the definition of the UPW quality parameters of the most advanced semiconductor technology generation. Table 32.1 provides recommended UPW quality, as published in the latest revision of SEMI F063 [2].

Bacteria, particles, organic, and inorganic sources of contamination can vary, depending on a number of factors including the feed water to the UPW, the piping materials to convey it, as well as the media and materials that make up the treatment process.

Bacteria. In past, bacteria was referred to as one of the most obstinate in this list to control [3]. Techniques that help in minimizing bacterial colony growth within UPW streams included occasional chemical or steam sanitization (which is still common in the pharmaceutical industry), ultrafiltration, ozonation, and optimization of hydraulic designs that ensure strict criteria for minimum flow [4] along with minimization of dead legs. In modern advanced UPW systems positive (higher than zero) bacteria counts are typically observed only in the newly constructed facilities. This issue is effectively addressed with intensive sanitization using ozone or hydrogen peroxide during system commissioning. Recently, bacteria control has been focused on minimizing nutrient sources for bacteria, maximizing purity of the installed system components, ensuring effective recirculation with minimum dead-leg, and providing efficient filtration at both treatment system and "point-of-use." With proper design of the polishing and distribution system positive bacteria counts are rarely detected throughout the life cycle of the UPW system.

Particles in UPW are the most critical and most difficult to control when considering advanced semiconductor industry. Particles can cause variety of different defects. Particles in UPW have always presented a major problem for semiconductor manufacture, as any particle landing on a silicon wafer can lead to a variety of serious defects in the semiconductor circuitry. When a pathway is short-circuited, the semiconductor device will not work properly; defects lead to manufacturing yield loss, one of the most closely watched parameters in the semiconductor industry. The most critical areas for particles control is front end of line (FEOL) semiconductor processing and EUV (extreme ultraviolet) lithography mask production [5]. In addition to FEOL and lithography newest generation of semiconductor manufacturing technology, involving 3-D device design, require tight particles control in BEOL (back-end of line) and device interconnects. This means that

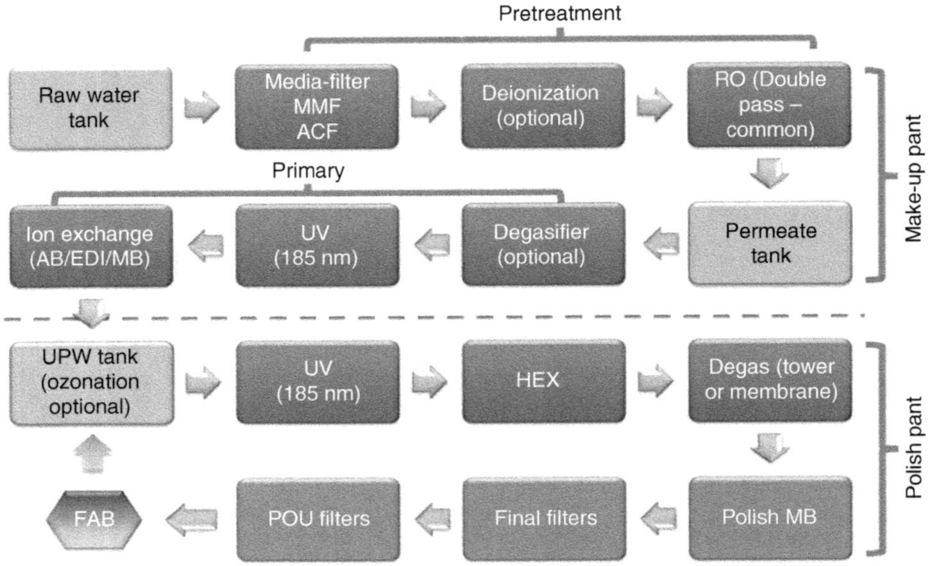

Figure 32.1 Typical UPW system configuration.

Table 32.1 Recommended Ultrapure Water Quality

Typical line width <0.065 μm			
Parameter	Performance	Parameter	Performance
Resistivity online @ 25°C (MΩ·cm)	>18.18	Residue after evaporation online (ppt)	<100
Temperature stability (K)	±1	Online particles > 0.05 μ size (#/L)	<200
Temperature gradient (K/10 min)	<0.1	Bacteria (CFU/L)	
TOC online (ppb)	<1	100 mL sample	<1
Dissolved oxygen online (ppb)	<10	Silica	
Dissolved nitrogen online (ppm)	8–18	Silica—total (ppb)	<0.5
Dissolved nitrogen stability (ppm)	± 2	Silica—dissolved (ppb as SiO_2)	<0.5
Ions and metals (ppt)			
Ammonium	<50	Chromium	<10
Bromide	<50	Copper	<1
Chloride	<50	Iron	<1
Fluoride	<50	Lead	<1
Nitrate	<50	Lithium	<1
Nitrite	<50	Magnesium	<1
Phosphate	<50	Manganese	<10
Sulfate	<50	Nickel	<10
Aluminum	<1	Potassium	<1
Antimony	<10	Sodium	<1
Arsenic	<10	Tin	<10
Barium	<1	Titanium	<10
Boron	<50	Vanadium	<10
Cadmium	<10	Zinc	<1
Calcium	<1		

particles control sensitivity is increasing as particles may affect larger number of the process steps.

The maximum allowed particle concentration in UPW has been updated based on the following considerations. The definition of the "killing" (critical) particle size is approximately half size of the manufacturing node linewidth. All particles are considered equally critical independently of their material of construction and their properties. Note: "killer" particle sizing and type criteria is expected to change as complexity of the device is increasing with 3D device configuration. There is empirically established correlation between the concentrations of the particles in UPW and the number of particles on the wafer, whereas maximum allowed number of the particles on the wafer is defined.

Sources of the particles include ion exchange resin leach-outs, pump materials, bacterial fragments, and other materials of the treatment and distribution system. Particles remain a high and exponentially growing risk as the node size shrinks. Online liquid particle measurement does not address killer particle size of the most advanced industry, which is approaching single-digit nanometer node size. At the same time, it is apparent that the killer size of the particles has approached limits of filtration capability of the most advanced final filters.

Total organic carbon (TOC) in ultrapure water can contribute to bacterial proliferation by providing nutrients, can substitute as a carbide for another chemical species in a sensitive thermal process, can leave unwanted residues on production wafers or sensitive immersion lithography lenses, or can cause refraction interference in the UPW during immersion lithography. TOC can originate from the feed water used to produce UPW, from the components used to convey the UPW (additives in the manufacturing piping products or extrusion aides and mold release agents), from subsequent manufacturing and cleaning operations of piping systems, or from contaminated system components.

Silica may occur in UPW in a form of reactive (dissolved) or colloidal silica. Dissolved silica in relatively high concentration is a potential cause for water spotting. Colloidal silica is considered to be particularly detrimental for the semiconductor manufacturing as it acts as particle, commonly present, and difficult to remove by filtration [5a]. When boron (borate—the weakest anion) is controlled below or near limits of boron detection the level of dissolved silica is usually bellow the limits of detection as well [5b].

Metallic and anionic contamination in UPW systems can be a cause of short- or long-term failure of electronic components in semiconductor chips and photovoltaic cells. Its sources are similar to those of TOC's. Depending on the level of purity needed, detection of these contaminants can range from simple conductivity (electrolytic) readings to sophisticated instrumentation such as ion chromatography (IC), atomic absorption spectroscopy (AA), and inductively coupled plasma mass spectrometry (ICP-MS).

32.1.3 Feed Water Sources

Water for semiconductor manufacturing can come from a variety of sources (typically delivered via municipality and commonly called *city water*). It is common that for redundancy reasons semiconductor facilities are fed from two independent water sources, often having different water quality characteristics. Most commonly the source is surface water or well water, in some cases seawater desalination. Driven by infrastructure limitations, cost, and environmental interest reasons, in some cases the reclaim water sources are used to feed UPW system. Understanding the water source and seasonal variability is a key driver to maintaining an ultrapure water quality for critical microelectronic manufacturing [30].

Key Parameters

Total organic carbon. UPW operating cost and quality can be affected by variation of organic compounds in the feed water. High molecular weight (HMW) compounds such as humic acids and polysaccharides are readily removed, but may cause membrane (i.e., RO or UF) fouling. Some organics decomposing may cause biofouling. However, low molecular weight (LMW) neutral compounds, such as fertilizer (urea) and pesticides (atrazine) commonly used in agriculture can have an adverse effect on final quality due to limited ability of conventional UPW system to remove LMW organics. Chlorination by a municipal supply will also increase trihalomethanes (THMs), which is also LMW neutral. Organic speciation can help determine treatment requirements. LC-OCD (liquid chromatography with organic carbon detector) [6] is most commonly used method of organic speciation in feed, UPW, and reclaim water in the semiconductor industry (see Fig. 32.2).

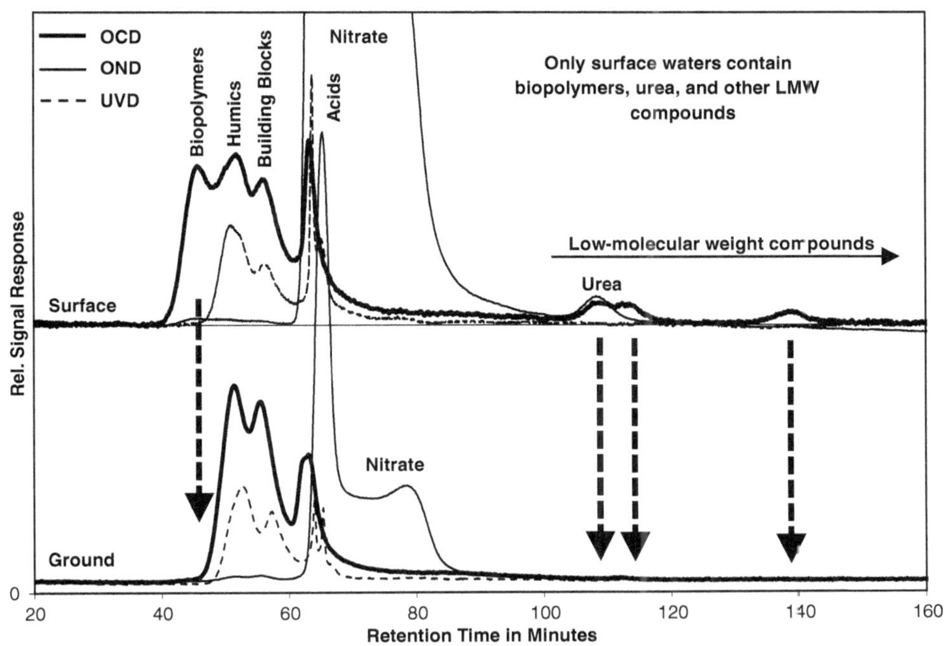

Figure 32.2 Typical feed water organic speciation (surface versus groundwater composition).

Particulates. These are typically measured as turbidity (NTU), silt density index (SDI), or as total suspended solids (TSS). Once understood, these can be effectively removed in the treatment system. When particulate loading is high, effective design of the pretreatment is required to ensure low operating cost.

Difficult ions. Ionic loading needs to be characterized to ensure cost effective and efficient treatment. Specifically, it is important to measure weakly dissociated ions such as boron, silica, and organic acids. Some underground waters may contain high concentration of silica that may limit RO recovery and foul membranes.

32.2 UPW PRODUCTION

32.2.1 Pretreatment

Pretreatment is responsible for removal of majority of the feed water impurities. Most of the maintenance and operation activities are focused on this portion of the treatment system.

The initial steps of pretreatment address removal of suspended solids and chlorine, as well as the temperature control, which are important for successful and reliable operation of the subsequent deionization processes. In addition, the degree of filtration upstream of these processes will be dictated by the level of suspended solids, turbidity, and organics present in the source water. The common types of filtration choices are depth filters (i.e., multimedia, sand, activated carbon), automatic back-washable mechanical filters, and ultrafiltration. The activated carbon may also be used to remove chlorine and other feed water oxidants (disinfection by-products), thus protecting reverse osmosis membrane or ion exchange resin. If activated carbon is not employed, then sodium bisulfite or sometimes sulfur dioxide is used to dechlorinate the feed water.

Typical pretreatment options are based on one of the three commonly used platforms:

1. Two pass reverse osmosis
2. Demineralization followed by reverse osmosis
3. HERO (high efficiency reverse osmosis) [7]

Choice of the platforms depends on incoming water quality, environmental requirements, and final UPW quality specification (Table 32.2).

The main focus of pretreatment is delivering cost-effective removal of dissolved and suspended solids. RO removes approximately 99 percent of TOC, leaving mainly low molecular weight neutral organics. Although RO is known to have imperfections in the membranes, a significant majority of particulates, including salts, particles of various sizes and types, bacteria, and viruses are removed. Use of ion exchange in front of the RO has become more popular in recent decade, as requirements for TOC and boron in final UPW have tightened and water conservation became imperative in the industry. These types of configurations provide lower total cost of ownership under such requirements than the double pass RO.

It should be noted that double pass RO still remains preferred platform for smaller systems, where lower operator involvement is preferred, where on-site chemical regeneration is not available, and where impact of lower water efficiency to the city infrastructure and environment is less of a concern. When double pass RO is used, pH adjustment to the levels of 8 to 9 allows for improved rejection of the anions of weak acids (borate, silicate, bicarbonate, and organic acids). This reduces the load to the primary mixed beds and improves their performance as well as the overall operating cost.

Another option of reducing operating cost if using decarbonator or vacuum degasifier between the first and the second RO passes. This allows CO_2 removal and reduces the cost needed for pH adjustment at the second RO pass. In some cases, this degasifier is also responsible for dissolved oxygen control.

32.2.2 Primary System

The deionization process following RO removes ions nearly completely. This deionization is typically based on the use of mixed ion exchange resin, achieving resistivity of greater than 18.1 MΩ·cm. Due to the high cost of regeneration of the primary mixed beds (MB), anion exchange beds or electro-deionization (EDI) are used upstream of the MBs to remove traces of boron, silica, carbonic acids, and organic acids. The efforts of using EDI to replace primary MBs did not find wide use in the industry, despite significant progress in the development of EDI technology. This is due to limited removal efficiency of EDI, compared to the MBs and due to its relatively high energy consumption.

A 185-nm TOC-destruct-UV is commonly used in the primary loop upstream to the MBs to reduce the level of TOC close to the final quality target level.

The commonly used operating practice is to produce primary product water at the level of ions and TOC as close as possible to the quality final UPW. This allows for reduction the ionic loads to the polish mixed beds, and as a result, improving UPW quality reliability and reducing operating cost (see more details in Sec. 32.2.3).

32.2.3 Polish Loop

UPW polish system design has been streamlined and relatively consistent in newer facilities [8]. Although the polish system involves less treatment steps, it is often more expensive than makeup due to higher capacity (consumption + recirculation), extensive use of high-purity materials, and expensive analytical instruments. The focus of the polish system is to provide the following:

- Oxygen and nitrogen control
- TOC polishing
- Particle removal
- Bacteria control
- Final quality monitoring
- Pressure and flow control in the distribution

Final UPW Quality Control in the Polish Loop

Polish loop configuration is generally more consistent in the industry. A typical UPW polish system begins at the UPW tank equipped with a nitrogen blanket. Some systems provide ozone addition in the tank or piping feeding the tank or leaving the tank. A 185-nm TOC-destruct-UV treatment provides additional TOC control and ozone decomposition, when applicable. Membrane or tower degasification provides oxygen and nitrogen control, as well as removal of volatile

Table 32.2 Advantages and Disadvantage of the Typical Pretreatment Platforms [8]

Configuration	Key features	Pros	Cons
Double pass RO	Chemical pretreatment, RO recovery 65–80%	Low chemical usage, no concentrated waste for disposal, less ops involvement	Lower rejection, high cost, organic/biofouling may cause ops problems
HERO	Complete hardness removal, CO_2 decarb, pH = 10.5–11. RO recovery 95–99.5%	High rejection of organics, boron, silica. Relatively low cost. High efficiency of water use. Less OPS involvement downstream RO.	Limited number of suppliers. High chemical consumption. Concentrated waste. Needs tight hardness control.
DI-RO	SAC-Decarb-SBA IX, RO recovery 95%	Boron and Silica removed in pretreatment. High efficiency of water use. Less OPS involvement downstream RO.	High chemical consumption. Concentrated waste.

organics. Polish mixed beds (typically non-regenerable) remove traces of contamination coming from the makeup plant or from the UPW return. They also remove by-products of ozone and UV oxidation. The last step of polish treatment is final filtration for removal of fine particles of 10 nm size and larger. Most of the advanced facilities use cross flow ultrafilters with or without cartridge prefilters.

Polish loop also includes final quality analytical equipment and sampling provisions used to qualify new system, online monitoring, as well as periodical (commonly monthly) sampling for bacteria, silica, metals, and anions typically tested by local or external laboratories. Advanced online metrology allows for tight control of dissolved oxygen, dissolved nitrogen, silica, boron, TOC, sodium, nonvolatile residue, and particles.

The main differences in the polish system designs can be described as follows:

1. *Ozonation.* Some systems use ozone injected upstream or inside of the UPW tanks. This provides additional option of UPW tank water disinfection as well as more effect TOC decomposition by 185 nm TOC-destruct-UV. However, some systems do not use ozonation, while maintaining successful operation.

2. *Materials used in the polish loop may slightly vary.* Some fabricators use polymer material lined pumps [9], while most have nonlined electropolished stainless steel pumps. Polish mixed may have different types of vessel lining materials, such as rubber, ETFE, and ECTFE.

3. A 185-nm TOC-destruct-UV may be designed for different output power, driving lower TOC levels, but potentially creating higher level of H_2O_2.

4. In some places degasification is done in the primary loop that provides some capital cost saving, but poses reliability risks.

5. Use of bacteria disinfection 254-nm UV became less popular in the recent decade, as tight TOC control and use of high-purity materials ensure absence of food for the microorganisms, while final filters designed for 10nm particles control guarantee bacteria removal. Since 254-nm UV do not provide residual disinfection, it became apparent that their use in UPW production cannot be justified.

32.3 UPW DISTRIBUTION SYSTEM

The distribution system is also a part of the polish loop. Its design is critical for the UPW quality at the point-of-use (POU) and may play major role in the cost of the overall system (Fig. 32.3).

One of the important factors determining UPW quality is the flowrate at the end of the pipes, where the flowrate is lowest and therefore the possibility of contamination is the highest (i.e., bacteria, pipe leach-outs, etc.). Initial UPW systems had piping materials of construction made of lower-purity materials (i.e., PVC) that required maintaining high minimum velocity criteria of 1 ft/s or even higher. As technology evolved, piping purity changed to high-purity materials (i.e., PVDF and PFA) that allowed reducing minimum flow velocity down to 0.5 fps. Since this criterion determined the significant portion of the system cost, additional investigations and industry learning allowed for further tightening the criteria [10]. A Reynolds number of 3000 to 4000 became popular, allowing a minimum flow criteria based on the turbulence factor. More sophisticated designs of UPW distribution and pressure control systems allowed for additional reduction in the recirculation to consumption ratio [10].

The right choice of lateral design, direct return versus reverse return (see Fig. 32.4) may influence pressure control and cost of the UPW system.

While choice of reverse return is generally preferred to ensure stable pressure control across the distribution system, it will add some piping cost. Effective engineering of the distribution system pay play significant role for both construction of the new and upgrade of the existing UPW systems. Such engineering includes hydraulic modeling, piping design optimization, and perfection of the pressure control scheme.

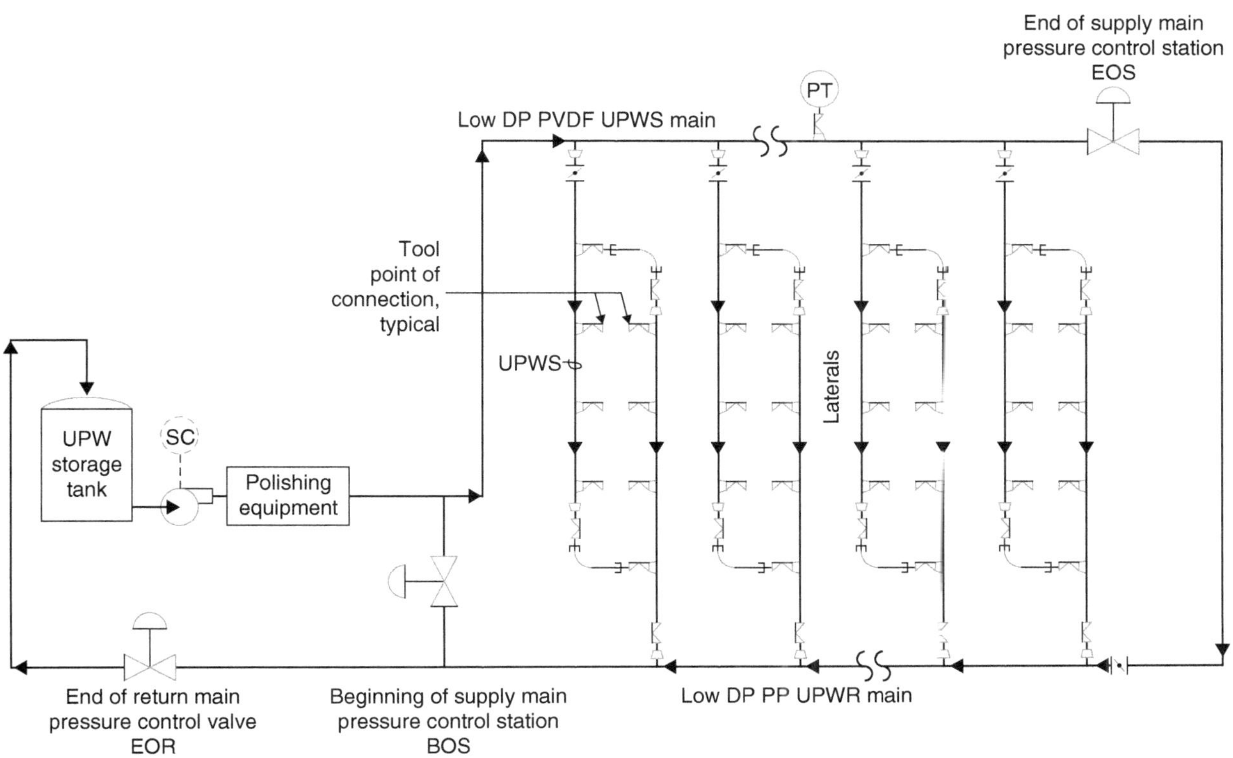

Figure 32.3 Schematic diagram of different methods of pressure control valves in the distribution system [10].

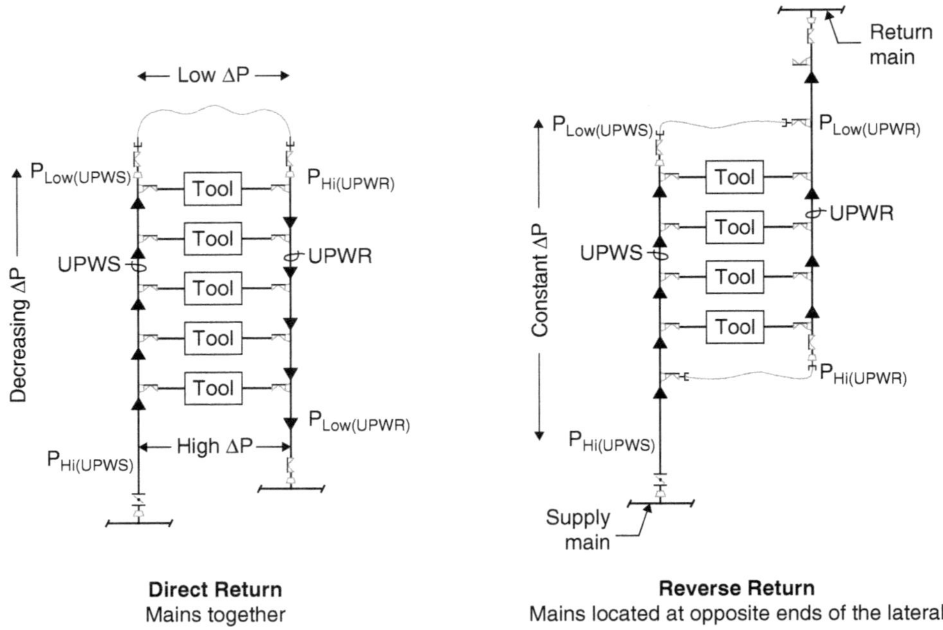

Figure 32.4 Typical "direct" and "reverse" return lateral configuration [10].

32.4 ANALYTICAL METHODS AND TECHNIQUES

32.4.1 Online Analytical Measurements

CONDUCTIVITY/RESISTIVITY

In pure water systems, electrolytic conductivity or resistivity measurement is the most common indicator of ionic contamination. The same basic measurement is read out in either conductivity units of microsiemens per centimeter (μS/cm), typical of the pharmaceutical and power industries or in resistivity units of megohm-centimeters (MΩ·cm) used in the microelectronics industries. These units are reciprocals of each other. Absolutely pure water has a conductivity of 0.05501 μS/cm and a resistivity of 18.18 MΩ·cm at 25°C, the most common reference temperature to which these measurements are compensated. An example of the sensitivity to contamination of these measurements is that 0.1 ppb of sodium chloride raises the conductivity of pure water to 0.05523 μS/cm and lowers the resistivity to 18.11 MΩ·cm [16, 17].

Ultrapure water is easily contaminated by traces of carbon dioxide from the atmosphere passing through tiny leaks or diffusing through thin wall polymer tubing when sample lines are used for measurement. Carbon dioxide forms conductive carbonic acid in water. For this reason, conductivity probes are most often permanently inserted directly into the main ultrapure water system piping to provide real-time continuous monitoring of contamination. These probes contain both conductivity and temperature sensors to enable accurate compensation for the very large temperature influence on the conductivity of pure waters. Conductivity probes have an operating life of many years in pure water systems. They require no maintenance except for periodic verification of measurement accuracy, typically annually.

SODIUM

Sodium is usually the first cation to break through a depleted cation exchanger. Sodium measurement can quickly detect this condition and is widely used as the indicator for cation exchange regeneration. The conductivity of cation exchange effluent is always quite high due to the presence of anions and hydrogen ion and therefore conductivity measurement is not useful for this purpose.

Online sodium measurement in ultrapure water most commonly uses a glass membrane sodium ion-selective electrode and a reference electrode in an analyzer measuring a small continuously flowing sidestream sample. The voltage measured between the electrodes is proportional to the logarithm of the sodium ion activity or concentration, according to the Nernst equation. Because of the logarithmic response, low concentrations in subparts per billion ranges can be measured routinely. To prevent interference from hydrogen ion, the sample pH is raised by the continuous addition of a pure amine before measurement. Calibration at low concentrations is often done with automated analyzers to save time and to eliminate variables of manual calibration [18].

DISSOLVED OXYGEN

Advanced microelectronics manufacturing processes require low single digit to 10 ppb dissolved oxygen (DO) concentrations in the ultrapure rinse water to prevent oxidation of wafer films and layers. DO in semiconductor water must be controlled to ppb levels to minimize corrosion.

Dissolved oxygen is measured by two basic technologies: electrochemical cell and optical fluorescence. Traditional electrochemical measurement uses a sensor with a gas-permeable membrane. Behind the membrane, electrodes immersed in an electrolyte develop an electric current directly proportional to the oxygen partial pressure of the sample. The signal is temperature compensated for the oxygen solubility in water, the electrochemical cell output and the diffusion rate of oxygen through the membrane.

Optical fluorescent DO sensors use a light source, a fluorophore, and an optical detector. The fluorophore is immersed in the sample. Light is directed at the fluorophore, which absorbs energy and then reemits light at a longer wavelength. The duration and intensity of the reemitted light is related to the dissolved oxygen partial pressure by the Stern-Volmer relationship. The signal is temperature compensated for the solubility of oxygen in water and the fluorophore characteristics to obtain the DO concentration value [19].

SILICA

Silica is a contaminant that is detrimental to microelectronics processing and must be maintained at sub-ppb levels. Silica is one of the first readily measurable species to be released by a spent anion exchange resin and is, therefore, used as the trigger for anion resin regeneration. At low levels, silica conductivity is low and therefore it is difficult to detect by conductivity.

Silica is measured on side stream samples with colorimetric analyzers. The measurement adds reagents including a molybdate compound and a reducing agent to produce a blue silicomolybdate complex color, which is detected optically and is related to concentration according to the Beer-Lambert law. Most silica analyzers operate on an automated semicontinuous basis, isolating a small volume of sample, adding reagents sequentially and allowing enough time for reactions to occur while minimizing consumption of reagents. The display and output signals are updated with each batch measurement result, typically at 10 to 20 min intervals [20].

Particles

The technique of choice to detect these single particles has been to shine a light beam (a laser) through a small volume of UPW and detect the light scattered by any particles (instruments based on this technique are called laser particle counters or LPCs). As semiconductor manufacturers pack more and more transistors into the same physical space, the circuitry linewidth has become narrow and narrower. As a result, LPC manufacturers have had to use more and more powerful lasers and very sophisticated scattered light detectors to keep pace. As linewidth approaches 10 nm (a human hair is approximately 100,000 nm in diameter) LPC technology is running out of steam, and new particle measurement techniques will be required. The smallest particles that can be measured with the most advanced LPC is 20 nm [21].

Nonvolatile Residue

Another type of contamination in UPW is dissolved inorganic material, primarily silica. Silica is one of the most abundant elements on the planet and is found in all water supplies. Any dissolved inorganic or non-volatile organic material has the potential to remain on the wafer as the UPW dries. Once again this can lead to a significant loss in yield. To detect trace amounts of dissolved inorganic material a measurement of nonvolatile residue is commonly used. This technique involves using a nebulizer to create droplets of UPW suspended in a steam of air. These droplets are dried at a high temperature to produce an aerosol of nonvolatile residue particles. A measurement device called a condensation particle counter then counts the residue particles to give a reading in parts per trillion (ppt) by weight [22].

TOC

Total organic carbon is most commonly measured by oxidizing the organics in the water to CO_2, measuring the increase in the CO_2 concentration after the oxidation or delta CO_2, and converting the measured delta CO_2 amount into "mass of carbon" per volume concentration units. The initial CO_2 in the water sample is defined as inorganic carbon (IC). The CO_2 produced from the oxidized organics and any initial CO_2 (IC) both together are defined as total carbon or TC. The TOC value is then equal to the difference between TC and IC [23].

Robust oxidation methods that supply sufficient oxygen include the following methods: ultraviolet light (UV) and persulfate, heated persulfate, combustion, and super critical oxidation. Some online TOC monitors use a membrane within the sample stream that can reject by-products of oxidized organic acids and halogenated organic compounds. These by-products increase conductivity, and as a result are then measured as "extra CO_2" downstream of the oxidation process which results in a overstated TOC level in the UPW. The membrane removes these by-products after the oxidation process but before the downstream measurement, which results in a more accurate TOC result. Some sites use this to their advantage by monitoring the difference in the TOC as measured by the two different monitors, which gives an indication of organic acids or halogenated organics present in the UPW.

Offline Lab Analysis

When testing the quality of UPW, consideration is given to where that quality is required and where it is to be measured. The point of distribution (POD) is the point in the system immediately after the last treatment step and before the distribution loop. It is the standard location for the majority of analytical tests. The point of connection (POC) is another commonly used point for measuring quality of UPW. It is located at the outlet of the submain or lateral take off valve used for UPW supply to the tool.

Grab sample UPW analyses are either complimentary to the online testing or alternative, depending on the availability of the instruments and the level of the UPW quality specifications. Grab sample analysis are typically performed for the following parameters: metals, anions, ammonium, silica (both dissolved and total), particles by SEM (scanning electron microscope), TOC (total organic compounds), and specific organic compounds.

Metal Analyses

Metal analyses are typically performed by ICP-MS (inductively coupled plasma mass spectrometry). The detection level depends on the specific type of the instrument used and the method of the sample preparation and handling. Current state-of-the-art methods allow reaching sub-ppt (parts per trillion) level (< 1 ppt) typically tested by ICPMS [24].

The Anion Analysis

The anion analysis for seven most common inorganic anions (sulfate, chloride, fluoride, phosphate, nitrite, nitrate, and bromide) is performed by ion chromatography (IC), reaching single digit ppt detection limits. IC is also used to analyze ammonia and other metal cations. However, ICPMS is the preferred method for metals due to lower detection limits and its ability to detect both dissolved and nondissolved metals in UPW. IC is also used for the detection of urea in UPW down to the 0.5 ppb level. Urea is one of the more common contaminants in UPW and probably the most difficult for treatment.

Silica Analysis

Silica analysis in UPW typically includes determination of reactive and total silica [25, 26]. Due to the complexity of silica chemistry, the form of silica measured is defined by the photometric (colorimetric) method as molybdate-reactive silica. Those forms of silica that are molybdate-reactive include dissolved simple silicates, monomeric silica and silicic acid, and an undetermined fraction of polymeric silica. Total silica determination in water employs high resolution ICPMS, GFAA (graphite furnace atomic absorption), and the photometric method combined with silica digestion. For many natural waters, a measurement of molybdate-reactive silica by this test method provides a close approximation of total silica, and, in practice, the colorimetric method is frequently substituted for other more time-consuming techniques. However, total silica analysis becomes more critical in UPW, where the presence of colloidal silica is expected due to likely silica polymerization in the ion exchange columns. Colloidal silica is considered more critical than dissolved in the electronic industry due to the bigger impact of nanoparticles in water on the semiconductor manufacturing process. Sub-ppb (parts per billion) levels of silica make it equally complex for both reactive and total silica analysis, making the choice of total silica test often preferred.

Although *particles* and *TOC* are usually measured using online methods, there is significant value in complimentary or alternative offline lab analysis. The value of the lab analysis has two aspects: cost and speciation. Smaller UPW facilities that cannot afford to purchase online instrumentation often choose offline testing. TOC can be measured in the grab sample at a concentration as low as 5 ppb, using the same technique employed for the online analysis (see online method description). This detection level covers the majority of needs of less critical electronic and all pharmaceutical applications. When speciation of the organics is required for troubleshooting or design purposes, liquid chromatography-organic carbon detection (LC-OCD) provides an effective analysis. This method allows for identification of biopolymers, humics, low molecular weight acids and neutrals, and more, while characterizing nearly 100 percent of the organic composition in UPW with sub-ppb level of TOC [27, 28].

Similar to TOC, SEM particle analysis represents a lower cost alternative to the expensive online measurements and, therefore, it is commonly a method of choice in less critical applications. SEM analysis can provide particle counting for particle size down to 50 nm, which generally is in-line with the capability of online instruments. The test involves installation of the SEM capture filter cartridge on the UPW sampling port for sampling on the membrane disk with the pore size equal or smaller than the target size of the UPW particles. The filter is then transferred to the SEM microscope where its surface is scanned for detection and identification of the particles. The main disadvantage of SEM analysis is long sampling time. Depending on the pore size and the pressure in the UPW system, the sampling time can be between 1 week and 1 month. However, typical robustness and stability of the particle filtration systems allow for successful applications of the SEM method. Application of energy dispersive x-ray spectroscopy (SEM-EDS) provides compositional analysis of the particles, making SEM also helpful for systems with online particle counters.

Bacteria

Bacteria analysis is typically conducted following ASTM method F1094 [29]. The test method covers sampling and analysis of high-purity water from water purification systems and water transmission systems by the direct sampling tap and filtration of the sample collected in the bag. This test method addresses both the sampling of water lines and the subsequent microbiological analysis of the sample by the culture technique. The microorganisms recovered from the water samples and counted on the filters include both aerobes and facultative anaerobes. The temperature of incubation is controlled at 28 ± 2°C, and the period of incubation is 48 h or 72 h, if time permits. Longer incubation times are typically recommended for most critical applications. However, 48 h is typically sufficient to detect water quality upsets.

32.5 UPW CHALLENGES OF THE SEMICONDUCTOR INDUSTRY

32.5.1 UPW Quality

The advanced semiconductor technology nodes of the future will require the most demanding UPW specifications.

Particles

The biggest challenge for UPW is particle control. This is due to the inability to detect "killer" particles and the concern that the final filters may be unable to remove those particles efficiently (Tables 32.3 and 32.4).

The size of "killer" particles is determined by half of the size of the node of semiconductor manufacturing. Table 32.3 indicates the size of the node and critical particle size as defined by ITRS (International Technology Roadmap for Semiconductor) [11]. Achieving effective 10-nm particles control at the level of 1000 particles/L (1 particle/mL) is particularly difficult when virgin ion exchange resin may produce 1E+6 particles/mL (see Table 32.4) and when best available and commonly used ultrafilters cannot completely remove 10-nm particles (Table 32.5).

Organics

The presence of dissolved organics in UPW has presented a significant challenge to semiconductor facilities over the past decade. Since TOC measurement techniques have improved, more organic compounds have become detectable in UPW. Urea is one of more common sources of TOC and is difficult to remove due to its low molecular weight and neutral (low charge) nature. Depending on conditions, RO typically removes between 20 and 50 percent of urea [10]. Urea may cause a T-topping effect on exposed chemically amplified resist [12]. State-of-the-art immersion lithography became a driver of a 1-ppb TOC specification in UPW. For most surface water sources, urea is always present [11]. TOC can be extremely difficult to control during seasonal urea excursions. Urea concentrations can rise as high as 100 ppb during seasonal excursions. A benchmarking study for organic speciation, involving 10 semiconductor grade UPW systems demonstrated that for all sites fed with surface water, urea concentration varied between 0.6 and 6 ppb [13]. Figure 32.5 represents typical LC-OCD results for UPW organic speciation. Using preconcentration, this technique allows for speciation of the organics at a level lower than 1 ppb as TOC. As

Table 32.3 Particles Target Level for Advanced Semiconductor Manufacturing [11]

Year of production	2015	2016	2017	2018	2019	2020
Dram ½ Pitch (nm) (contacted)	24	22	20	18	17	15
Critical particle size (nm) based on 50% of DRAM ½ pitch (nm) (contacted)	12	11	10	9	8.5	7.5
Number of particles (larger than critical size) (#/L)	1000	1000	1000	1000	1000	1000
Number of particles for EUV mask production (larger than critical size) (#/L)	100	100	100	100	100	100

Table 32.4 10 nm Particle Measured during Extensive Flush of New (Virgin) Ion Exchange Resin

High-purity virgin resin from major suppliers (anonymously listed)	Initial rinse particle concentration, 1E+6/mL (10 nm particles)	Particle concentration at 50 bed volumes of rinse, 1E+6/mL (10 nm particles)	Particle concentration at 500 bed volumes of rinse, 1E+6/mL (10 nm particles)
Analytical baseline		~1	
System baseline (no resin)	80	1	<1
Resin A	250	50	3
Resin B	60	2	<1
Resin C	135	40	6
Resin D	180	50	6
Resin E	180	115	15
Resin F	50	25	5

Table 32.5 Typical UPW Final Filter (Cross Flow Ultrafilters) Performance Testing per SEMI C079 Method

Particle size (nm)	Ultrafilter particle retention efficiency (%) measured at the target level of particles load (monolayer)		
	0.2 monolayer	0.4 monolayer	1.0 monolayer
33	99.7	99.5	99.7
26	99.6	99.6	99.5
22	99.3	99.5	99.4
19	99.3	99.1	99.2
16	98.9	98.7	99.0
14	98.9	97.7	98.6
12	98.2	97.1	96.3
9	94.2	87.1	81.4

shown in Fig. 32.5, the most common UPW organics are urea and low molecular weight organic acids. The acids can be by-products of 185-nm UV oxidation and may not be 100 percent removed by polish MBs. The ITRS benchmarking study indicated that long life time of the resin in the polish MB may result in biogrowth, producing biopolymers. In addition, new IX resin may also produce TMA and other organic compounds that may be difficult to treat, once again emphasizing the importance of selection of high-purity IX resin. As a result, reaching 1 ppb specified TOC level (see Table 32.1) is particularly challenges in those locations where seasonal agricultural runoffs may result in high level of urea in the feed water and where UPW system may add organic contamination (by-products or leach-outs).

Metals

Although the presence of metals in UPW is very low (at single digit parts per trillion or less) and modern UPW systems are very reliable, keeping metallic contamination under control, has been an important factor in UPW quality considerations and system design.

The key concerns and considerations include the following:

- High-purity piping materials used in UPW will leach out metals (calcium, iron, etc.), originating from the raw material and manufacturing processes. Similar contamination may be introduced via the construction environment and handling. Schedule pressures of fast-track projects may result in relatively high residual contamination in the range of 10 to 20 ppt to be present in water.
- High pH of SC1 cleaning may precipitate metals on the wafer that may be difficult to remove.
- Trace levels of sodium may result from IX regeneration (bigger concerns if regenerable resin is used in polishers).
- Stainless steel components downstream from IX beds may leach metals.
- Improperly chosen elastomers and other components may add metal contamination.
- Heat exchangers may have internal leaks.

Current analytical techniques can provide metal detection down to 0.2 ppt level (200 ppq). However, such tests can only be done in a grab sample mode and require very clean sampling methods. It is important that trace metals contamination is addressed at the stage of system design and project planning.

Hydrogen Peroxide

H_2O_2 as a by-product of 185-nm UV oxidation is an emerging concern in UPW technology [14]. Table 32.6 provides typical profiling of H_2O_2 in UPW system.

Although methods for catalytic decomposition of H_2O_2 in UPW exist [15], most of the UPW systems do not have those treatment steps currently installed. Adding catalytic beds to the existing system presents not only capital cost issue, but also space challenges. Typical level of H_2O_2 in UPW is 5 to 25 ppb, which is similar or higher than dissolved oxygen specification for most of the UPW systems. This present ongoing concern to the existing advanced semiconductor facilities.

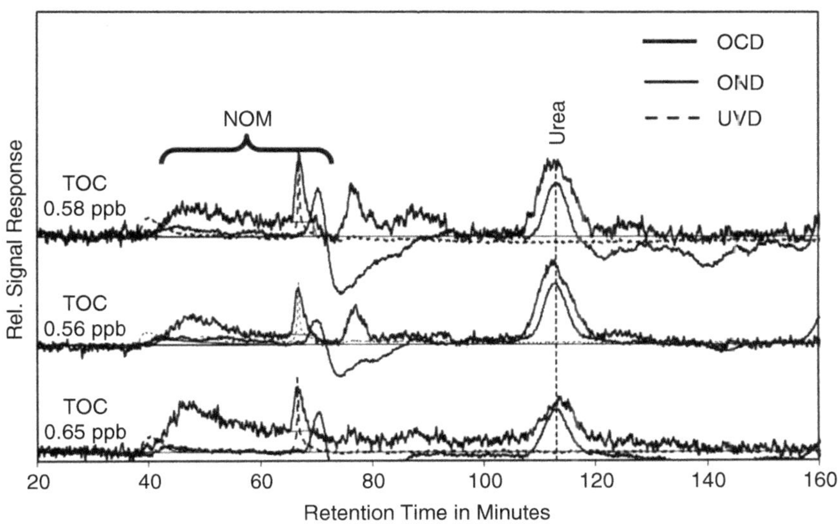

Figure 32.5 Typical organic speciation in UPW (in triplicates).

Table 32.6 H_2O_2 in UPW Treatment System

UPW system section	Unit operation	Hydrogen peroxide measured downstream the unit operation (ppb)
Pretreatment	Raw water tank	
	Activated carbon filter	
	1st pass reverse osmosis	
Primary	Permeate tank	
	2nd pass reverse osmosis	
	Electro deionization	
	Membrane degasifier	1
	185 nm UV	14–16
	Primary mixed bed	11–12
Polish	Ultrapure water tank	16–17
	185 nm UV	30
	Polish mixed bed	24
	Membrane degasifier	21–22
	Microfiltration	
	Ultrafiltration	20–22

32.5.2 Capacity Challenges

Despite significant efforts to minimize UPW and site water consumption in the semiconductor industry, UPW usage grew exponentially with the technology nodes. The diameter of the main UPW distribution piping coincidently matched the size of the wafer manufactured by the facility.

Table 32.7 illustrates curious observation of direct correlation of the largest diameter of the main UPW supply line with the size of the wafer in production. Increasing UPW consumption (lower line) drove increases in polish system capacity (flow rate of the recirculation—upper line), which determined the diameter of the largest UPW "main" line.

Significant growth in UPW consumption forced the industry to reconsider UPW system capacity sizing criteria. This means that any excess of the UPW polish system capacity needed to control cleanliness and pressure became extremely expensive. For example, with 200 mm wafer usage and smaller, the system capacity could be as high as 5 times the average consumption. Converting to 300 mm production resulted in expensive and sometimes not feasible configurations due to space constraints. Transition from flow velocity to Reynolds number criteria and more sophisticated distribution design provided partial solutions to the distribution capacity issue.

UPW consumption reduction efforts resulted in optimized tool and recipe design. However, increasing sensitivity to wafer contamination required increased UPW consumption, steadily growing from generation to generation of the semiconductor processes [31]. UPW consumption is expected to continue increasing, posing challenges to the local infrastructure. The only solution to this usage is going to be extensive water reclamation and reuse.

32.5.3 Water Management Challenges

One of the sources of water loss and capacity challenge is losses in UPW production. Some facilities lose 20 to 50 percent of raw water, depending on system configuration and incoming water quality. The focus on reduction in water consumption on the facility side is currently around increasing water efficiency of UPW system (i.e., higher RO recovery) and reclamation/recycling. This includes installation of IX processes in front of the RO or reclaiming wastewater from the UPW system. A key to optimizing water reclaim and recycling in modern semiconductor fabs is in proper management of waste streams at the manufacturing tools. Profiling and characterizing the waste streams by sampling and analyzing the tool drain discharge points can determine optimized drain segregation, which leads to higher water and chemical reuse/recycle with minimized treatment cost. In addition, online monitoring such as conductivity, pH, and TOC can be added at key points in the waste streams to collect data and divert based on threshold limits. This data can be used to divert reclaimed waste sources to from UPW makeup supply to lower risk areas such as cooling tower and scrubbers.

Most advanced existing and next generation factories are expected to consume 4 to 5 million gallons of water per day, most of which is high purity. It is important to note that such mega-fabs are typically added to

Table 32.7 Growth of Water Consumption with Increase of the Size of the Wafers

Wafer size for a given technology generation (in)	Typical UPW consumption (gpm)	Typical UPW distribution flow (gpm)	Largest main line pipe diameter in UPW distribution (in)
2	10	100	2
4	40	390	4
6	170	870	6
8	390	1540	8
12	1390	3470	12
18	3130	7810	18

already existing giant semiconductor manufacturing sites. It is expected to be extremely challenging to support such development from infrastructure, environmental, and purity points of view.

Additional non-UPW water consumption challenges driven by the evolution of semiconductor manufacturing are related to air pollution control and energy usage. Increasing requirements for air emissions require a growing number of highly efficient abatement systems, using water for absorption of the toxic gases and their decomposition bi-products. Smaller features of semiconductor wafers will require a new generation of lithography tools. Next generation of EUV (extreme ultraviolet lithography) will significantly increase fab energy consumption, and as a result, the cooling load. This will drive higher cooling tower evaporation. Increased evaporation will not only require higher water consumption, together with reclamation efforts, it will significantly concentrate salts and organics in the site effluent, posing environmental compliance issues. This will require additional technology solutions for wastewater treatment.

32.6 RECOMMENDATIONS FOR REACHING HIGH-QUALITY UPW

UPW technology has reached its maturity over several past decades of the UPW technology optimization. New treatment systems and combinations have been developed and optimized, while others removed or replaced. The industry consolidation happened for both semiconductor companies as well as for the supply chain. UPW technology suppliers are capable of delivering efficient and reliable UPW systems. At the same time there are specific areas that require considerations when building new or upgrading existing UPW system:

1. Material of construction in the polish and distribution loops: It is important to not only specify high-purity materials for these applications, but also ensure adequate testing, handling, and installation during the project execution. SEMI standard, such as SEMI F57, SEMI C90, and SEMI F061 provide relevant recommendations.
2. Critical components should be tested to ensure their best performance:
   a. SEMI C079 provides recommendations for testing UPW final filters as well as filters used at the POU applications.
   b. SEMI C093 provides recommendations for testing polish ion exchange resin to ensure minimum particles and other contamination leach-outs.
3. Schedule of the UPW system start-up should be reviewed with respect to the need date of the UPW usage. It common to expedite UPW distribution installation to begin with filling and flushing distribution piping as early as possible (using small mobile UPW system). This will provide sufficient time for reducing metal, TOC, and fluoride contamination in the distribution.
4. Qualification testing should be defined early in the project to provide effective logistics and adequate testing for the system commissioning.

32.7 ACKNOWLEDGMENTS

The author appreciates detailed peer review provided by Dan Wilcox, UPW technology veteran with many years of experience in water and wastewater treatment applications in semiconductor industry.

This chapter incorporates collective knowledge and development of the group of experts involved in the UPW technology roadmap development for semiconductor industry as well as in the development of the UPW system related standards and guidelines via SEMI (Semiconductor International) organization.

32.8 REFERENCES

1. ASTM D5127-13 Standard Guide for Ultra-Pure Water Used in the Electronics and Semiconductor Industries.

2. SEMI F63-0213 Guide for Ultrapure Water Used in Semiconductor Processing.

3. M. W. Mittlemann and G. C. Geesey, *Biofouling of Industrial Water Systems: A Problem Solving Approach,* Water Micro Associates, San Diego, California, 1987. https://www.abebooks.com/Biological-Fouling-Industrial-Water-Systems-Problem/17673081048/bd.

4. V. Libman, "Use of Reynolds Number as a Criteria for Design of High-Purity Water Systems," *Ultrapure Water,* Oct. 2006.

5. ITRS Annual Report 2013 Edition. International Technology Roadmap for Semiconductors. https://www.dropbox.com/sh/6xq737bg6pww9gq/AABi_Y8w8bDy-RDwxIrD6vrGa/2013Yield.pdf?dl=0.

5a. Don Grant, "The effect of particle composition on filter removal of sub-30nm particles from UPW." UPW Conference, Portland, OR, 2011.

5b. Godec R. Preventing the Release of Nano Materials from Depleting Ion Exchange Beds by Using an Online Boron Analyzer. UPW Conference, Portland, OR, 2011.

6. Stefan A. Huber, Andreas Balz, Michael Abert, and Wouter Pronk, "Characterisation of Aquatic Humic and Non-Humic Matter with Size-Exclusion Chromatography—Organic Carbon Detection—Organic Nitrogen Detection (LC-OCD-OND)," *Water Research,* 45: 879–885, 2011.

7. Saving Energy, Water, and Money with Efficient Water Treatment Technologies. Federal Energy Management Program. http://www.nrel.gov/docs/fy04osti/34721.pdf.

8. V. Libman, "Ultrapure Water (UPW)—Quality and Technology to Support Advanced Industries' Needs," Proceedings of 74th Annual International Water Conference, 2013.

9. F. Patrick, "Microelectronics—Driving Treatment System Innovation through Capital Expansion Projects," *Ultrapure Water Journal,* 32 2): 21–26, Mar./Apr. 2015.

10. V. Libman, D. Buesser, and B. Ekberg, "Next Generation of UPW Distribution System for the Next Generation of Semiconductor Factories," *Ultrapure Water Journal,* 29 (1): 23–30, 2012.

11. V. Libman, D. Wilcox, and B Zerfas, "UPW ITRS and SEMI: Synergy of Enabling Advanced Existing and Future Technologies," Ultrapure Water Conference, Portland, OR, Nov. 2015.

12. J. Rydzewski, "Identification of Critical Contaminants by Applying an Understanding of Different TOC Measuring Technologies," *Ultrapure Water Journal,* 2: 20–26, 2002.

13. V. Libman, "New Developments by UPW ITRS and SEMI Task Forces (2013)," UPW Conference, Portland, OR, Nov. 2013.

14. P. Rychen, J. Magnan, and M. Salamor, "Online Monitoring of Hydrogen Peroxide in Ultrapure Water," Ultrapure Water Conference, Portland, OR, Nov. 2015.

15. Y. Miyazaki, T. Fukui, H. Kobayashi, and K. Yamada, "Advanced Hydrogen Peroxide Removal Technology Using Nano-Sized Pt Catalyst Resin," UPW Conference, Portland, OR, Nov. 2013.

16. ASTM D1125 Standard Test Methods for Electrical Conductivity and Resistivity of Water.

17. ASTM D5391 Standard Test Method for Electrical Conductivity and Resistivity of a Flowing High Purity Water Sample.

18. ASTM D2791 Standard Test Method for On-line Determination of Sodium in Water.

19. ASTM D5462 Standard Test Method for On-Line Measurement of Low-Level Dissolved Oxygen in Water.

20. ASTM D7126 Standard Test Method for On-Line Colorimetric Measurement of Silica.

21. Rodier Nanoparticle Monitoring in Ultrapure Water at 20 nm and Below. UPW Conference, Phoenix, AZ, Nov. 2014.

22. D. ASTM D5544 Standard Method for On-Line Measurement of Residue after Evaporation of High Purity Water.

23. ASTM D5997—96 Standard Test Method for On-Line Monitoring of Total Carbon, Inorganic Carbon in Water by Ultraviolet, Persulfate Oxidation, and Membrane Conductivity Detection.

24. Albert Lee, Vincent Yang, Jones Hsu, Eva Wu, and Ronan Shih, "Ultratrace Measurement of Calcium in Ultrapure Water Using the Agilent 8800 Triple Quadrupole ICP-MS," Agilent Technologies, 2012. http://hpst.cz/sites/default/files/attachments/5991-1693en-ultratrace-measurement-calcium-ultrapure-water-using-agilent-8800-triple-quadrupole-icp.pdf.

25. ASTM D4517 Standard Test Method for Low-Level Total Silica in High-Purity Water by Flameless Atomic Absorption Spectroscopy.

26. ASTM D859 Standard Test Method for Silica in Water.

27. S. A. Huber, A. Balz, M. Abert, and W. Pronk, "Characterisation of Aquatic Humic and Non-Humic Matter with Size-Exclusion Chromatography—Organic Carbon Detection—Organic Nitrogen Detection (LC-OCD-OND)," *Water Research*, 45: 879–885, 2011.

28. Stefan Huber, Slava Libman, "Part 1: Overview of LC-OCD: Organic Speciation in Service of Critical Analytical Tasks of Semiconductor Industry," *Ultrapure Water Journal*, 31 (3): 10–16, May–Jun. 2014.

29. ASTM F1094 Standard Test Methods for Microbiological Monitoring of Water Used for Processing Electron and Microelectronic Devices by Direct Pressure Tap Sampling Valve and by the Presterilized Plastic Bag Method.

30. Avijit Dey and Gareth Thomas, *Electronics Grade Water Preparation*, Tall Oaks Pub, Inc., Littleton, CO, 2003. ISBN 0-927188-10-4.

31. V. Libman and A. Neuber, "Water Conservation Challenges Facing the Microelectronics Industry," *Ultrapure Water Journal*, 7: 36–43, 2008.

Chapter 33

Process Chemicals, Handling, and Abatement

BY

DANIEL FUCHS *Air Liquide Electronics US, Houston, Texas*

33.1 INTRODUCTION

The semiconductor industry has evolved with the advancement of technology and the exponential increases in transistor packing density and reduction of critical feature sizes. As the scale of feature sizes have decreased, the volume of chemicals has increased due to the additional process steps that are involved to manufacture features of increasing complexity. With the demand for increased volumes of chemicals supplied to the processing tools, methods for bulk chemical handling have been incorporated into the facility design. As the number of process steps has increased, so has the criticality of each step in the manufacturing process to prevent any excursion from altering device yield.

One of the most significant process advancements is the use and development of chemical mechanical planarization as a means to systematically homogenize the wafer's topography in preparation for additional processes as the geometries grow in the third dimension. Another example of enhancement in the semiconductor processing environment is highlighted in the modifications and enhancements to the RCA standard clean solutions.[1] Modifications that have been observed have been the inclusion of dilute hydrofluoric acid into piranha solutions, as well as ratio changes to the standard cleans formulations to meet the individualized process needs. In addition to the core set of wet process chemicals used in the manufacture of semiconductor devices, one of the more significant changes has been the increase in the use of proprietary formulations that assists the manufacturability of a particular fabrication step. While the majority of the formulations constituents fall into common chemical classes, proprietary chemicals are often minor constituent additives.

With these advancements wafer processing has experienced the advent of novel processing techniques and formulations befitting the manufacturability of semiconductor processes. Additional advancements that have enhanced the semiconductor manufacturing process include electronic grade chemical manufacturing with trace levels of metal impurities, stringent assay control of formulations for critical process steps (i.e., developers and buffered oxide etchants), and the traceability of chemical lots through the manufacturing process.

33.2 IMPORTANT CHEMICAL HAZARD TERMINOLOGY AND SYMBOLS

The chemicals used in semiconductor manufacturing vary in the degree of risk that they present to the operator as well as the facility under normal or emergency conditions. Some of the common hazards that can be found in a semiconductor processing location have been specifically identified and defined in NFPA 318, *Standard for the Protection of Semiconductor Fabrication Facilities*.[2] If not referenced specifically, the terms used in this chapter are defined by their normally accepted meanings as defined in *Merriam-Webster's Collegiate Dictionary,* 12th edition. Some of the potential chemical hazards that exist with process chemicals include:

Corrosive. A chemical that causes visible destruction of, or irreversible alterations in, living tissue by chemical action at the point of contact.[2]

Combustible liquid. A liquid that has a closed-cup flash point at or above 37.8°C (100°F).[2]

Flammable liquid. A liquid that has a closed-cup flash point that is below 37.8°C (100°F) at a maximum vapor pressure of 2068 mm Hg (absolute pressure of 40 psi) at 37.8°C (100°F).[2]

Flashpoint. The lowest temperature at which vapors above a volatile combustible substance ignite in air when exposed to flame.

Incompatible materials. Materials that, when making contact with each other in an upset condition, can react in a manner that generates heat, fumes, gases or by-products that could cause damage to life or property.[2]

Reactivity. Tending to change into something else when mixed with another substance.

Toxic. Containing or being poisonous material especially when capable of causing death or serious debilitation.

Water reactive. A material that explodes; violently reacts; produces flammable, toxic, or other hazardous gases; or evolves enough heat to cause self-ignition of nearby combustibles upon exposure to water or moisture.[2]

With the presence of a variety of process chemicals and waste streams, semiconductor manufacturing facilities must work in harmony with the jurisdictional authority for permitting, occupancy, and potential emergency planning.

33.2.1 Safety Data Sheets

Safety is of utmost importance in the handling of process chemicals, and Safety Data Sheets (SDS) are a critical tool in the transfer of information to all parties associated with chemical handling and disposal. In 2012, The United States Occupational Safety & Health Administration (OSHA) released an updated Hazard Communication Standard to align current processes with the United Nations Globally Harmonized System.[3] The changes in the classification standard update the basic informational content to a system that informs workers about the hazards that exist for chemicals under normal conditions of use versus a format centered on an emergency-based response. Safety data sheets for each chemical and specialty chemical formulation used in the semiconductor manufacturing process provide essential information in the following section-based format:

1. Identification
2. Hazard(s) identification
3. Composition information on ingredients
4. First-aid measures
5. Fire-fighting measures
6. Accidental release measures
7. Handling and storage
8. Exposure controls and personal protection
9. Physical and chemical properties
10. Stability and reactivity
11. Toxicological information
12. Ecological information
13. Disposal considerations
14. Transport information
15. Regulatory information
16. Other information

Each SDS is specific to the chemical supplier and the country of origin as the formulations and additives may vary based on the manufacturing location. The documented information provides important information to make the user aware of the levels of personal protective equipment that should be utilized when using and handling the material, identification of incompatible materials to prevent unwanted chemical reactions as well as the general methodology for handling the chemical.

Potential hazards of a specific chemical or formulation are summarized in a symbolic manner to easily identify the type and severity of a particular risk. Two of the most common pictograms used for the communication of hazards are the International Hazard Symbols and the National Fire Protection Association (NFPA) hazard diamond. The SDS may contain the NFPA hazard diamond, but if it is omitted the relevant information has been defined in various sections for health, flammability, instability, and special information sections to create the appropriate label.[4]

33.2.2 International Hazard Symbols

An SDS contains symbolic representation of the presence of a hazardous material as depicted by nine international hazard symbols. The nine symbols are (1) explosive, (2) flammable, (3) oxidizing, (4) corrosive, (5) acute toxicity, (6) hazardous to the environment, (7) health hazard/hazardous to the ozone layer, (8) serious health hazard, and (9) compressed gas. Examples of the symbols are shown on Great Britain's Health and Safety Executive webpage.[5] Often more than one symbol may be present to identify the various hazards of a chemical.

33.2.3 National Fire Protection Association Hazard Diamond

The NFPA has identified, defined and quantified potential hazards related to health, flammability, and reactivity and summarized the information in a segregated diamond symbol with specific hazards color coded and quantified, as noted in Table 33.1.

33.3 PROCESS CHEMICALS USED IN THE SEMICONDUCTOR MANUFACTURING PROCESS

A variety of chemicals are used to aid different process steps that have been described in detail in earlier chapters of this handbook. The fundamental sequence for manufacturing semiconductor devices is a multifaceted process based around the basic premises of cleaning, patterning, etching, material deposition via a variety of mechanisms or implantation, planarization, and repeating all or a subset of these fundamental steps in order to build the desired design. Common chemicals used in the semiconductor manufacturing process are shown in Table 33.2. To list every process chemical would be an enormous undertaking as new proprietary formulations continually enter the market. This chapter identifies typical process chemicals, as well as some of the significant constituents in the aforementioned formulations.

As the architecture of semiconductor design and layout has become more intricate, the number of process steps has increased; a significant number of process steps are attributed to cleaning applications. Cleaning is imperative for removing a variety of contaminants from the surface, which can be ascribed to organic films, ionic contamination, metals, and particles. A mantra within the semiconductor industry is prevention is the key to contamination control. Preventative cleaning methodology was adopted early in the manufacturing process (Fig. 33.1) when RCA detailed standard cleaning solutions.[1] As processes, designs, and the incorporated materials have evolved, so have the standard cleaning processes to meet the application requirements. Cleaning applications will vary based on the process step, and the cleaning procedures have to be compatible with the materials present at the patterned surface.

33.3.1 Aqueous Chemistries

Acids and bases are used extensively in cleaning and etching applications, and these chemicals constitute some of the largest volumes of process chemicals used in the facility. Acids and bases are the keystone of wet chemical processes and the various chemistries can be observed on-site in a variety of concentrations depending on the needs of the process tools. Some of the common process chemicals, and their corresponding weight percent (w/w), that may be encountered in a semiconductor facility are:

Table 33.1 General Summary of Chemical Hazards as Presented in the NFPA Hazard Diamond (Special hazards, e.g., alkaline, acidic, corrosive, oxidizer, and water reactive chemistry, will be noted on the hazard shield)

Rating number	Risk level	Health hazard (blue)	Flammability hazard—flashpoint (red)	Instability hazard (yellow)
4	Extreme	Can be lethal	Very flammable (below 73°F)	May explode under normal conditions
3	Serious	Can cause serious or permanent injury	Ignites under normal temperature conditions (between 73 and 100°F)	May detonate at high temperature or shock
2	Moderate	Can cause temporary or residual injury	Ignites with moderate heating (above 100°F, < 200°F)	Violent chemical change but does not detonate
1	Slight	Can cause significant irritation	Must be preheated before ignition (above 200°F)	Unstable if heated
0	Minimal	No hazard	Not flammable	Stable material

Table 33.2 Typical Semiconductor Process Chemicals

Chemical			
Description	Formula	Hazard classification	Applications
Ammonium hydroxide	NH_4OH	Corrosive, health hazard	Clean
Acetic acid, glacial	$C_2H_4O_2$	Corrosive, flammable	Etch
Cyclohexanone	$C_6H_{10}O$	Combustible, health hazard	Lithography
Ethyl lactate	$C_5H_{10}O_3$	Flammable	Lithography
Hexamethyldisilazane (HMDS)	$C_6H_{19}NSi_2$	Flammable, toxic, corrosive	Photoresist adhesion promoter
Hydrochloric acid	HCl	Corrosive, serious health hazard	Clean, etch
Hydrofluoric acid	HF	Corrosive, toxic, serious health hazard	Clean, etch
Hydrogen peroxide	H_2O_2	Oxidizer, corrosive, health hazard	Clean, slurry
Isopropyl alcohol (IPA)	C_3H_8O	Flammable, serious health hazard	Cleaning process
Nitric acid	HNO_3	Corrosive, oxidizer, health hazard	Etch
N-Methyl-2-pyrrolidone (NMP)	C_5H_9NO	Flammable, serious health hazard	Lithography
Phosphoric acid	H_3PO_4	Corrosive, health hazard	Etch
Potassium hydroxide	KOH	Corrosive, health hazard	Etch, slurry
Propylene glycol monomethyl ether acetate (PGMEA)	$C_6H_{12}O_3$ or $CH_3CH(OCOCH_3)CH_2OCH_3$	Flammable, serious health hazard	Lithography
Tetramethyl ammonium hydroxide (TMAH)	$(CH_3)_4NOH$	Corrosive, toxic, serious health Hazard	Developer
Sulfuric acid	H_2SO_4	Corrosive	Clean

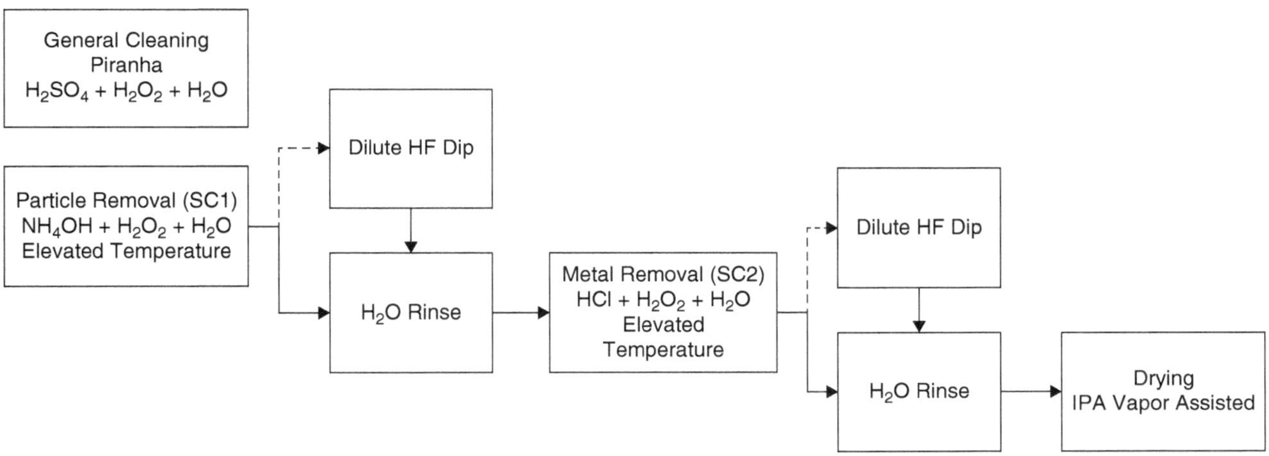

Figure 33.1 Critical front end of line cleaning processes based on the original RCA standard cleaning process. The standard clean process is shown as a solid path, and alternate methodologies are shown as a dashed line.

- *HF*. The typical concentration for this chemical is 49 percent, yet dilute HF solutions (50:1 and 100:1) are commonly used as SiO_2 etchants. It is also a constituent in different mixed acid etchant (MAE) formulations.
- *HCl*. The typical concentration of this process chemical is 37 percent; it is a constituent in SC_2 formulations for trace metal removal.
- *H_2SO_4*. A main component in piranha solutions for organic removal and cleaning applications, and it can be used to strip photoresist from oxide surfaces. Sulfuric acid is one of the higher volume chemistries used in wet process applications and the typical concentration supplied is 96 percent.
- *H_3PO_4*. Major component in aluminum etchants, also used as a nitride strip. The typical concentration of this chemical is 85 percent.
- *HNO_3*. Common constituent in MAEs, but it is often seen in its supplied concentration of approximately 70 percent.
- Acetic acid, glacial. Common constituent in mixed acid etchants.
- *NH_4OH*. Ammonium hydroxide, also known as ammonia water, is used for cleaning as an SC1 constituent and is generally supplied as a 29 percent solution.
- *TMAH*. Mainly used as a positive photoresist developer and the specific developer formulation may contain a surfactant. A typical concentration of 2.38 percent TMAH is used in photolithography applications (when the material is supplied in a ready-to-use concentration), otherwise 25 percent material is on-hand if the material is blended on-site; assay control of TMAH is critical.

- *KOH*. Can be seen in a variety of concentrations in a facility. Potassium hydroxide can be used as an etchant, but it is a common additive in CMP slurry formulations.
- *H_2O_2*. Typically, hydrogen peroxide is consumed at a very high rate, as it is used in a lot of different applications (e.g., standard cleans, CMP slurries, piranha). The typical concentration of this material is 30 percent, and it has one of the most stringent impurity thresholds.

In addition to the aforementioned process chemicals there are also specialty formulations used for a variety of process steps. Some of the most common formulations are

- *MAEs*. A variety of formulations to etch silicon surfaces. Often a mixture of HF, HNO_3, and acetic acid in a variety of ratios. Additional specialty MAEs can have minor constituent additives like fluoroboric acid.
- *Aluminum etchants*. Typically a mixture of phosphoric, nitric, and acetic acids mixed with water in a ratio of 16:1:1:2; the solution may or may not contain surfactants.
- *Buffered oxide etchants* (BOE). Composed of ammonium fluoride (NH_4F) and HF in a variety of different ratios, which is used to etch SiO_2. BOE formulations can contain surfactants.

Lastly, copper sulfate ($CuSO_4$) solutions cannot be omitted, as the electrochemical deposition of the metal is in integral part of damascene processes. Copper solutions are isolated from other chemicals as mobile copper ions can act as a poison to the design.

33.3.2 Solvents

Solvents are used in a variety of different processing steps ranging from vapor assisted drying (e.g., IPA), adhesion promoters for photoresists, photoresist thinners, edge bead removal after spin coating, as well as the final stripping of the photoresist after processing. Some of the most common solvents used in different lithography steps are shown in Table 33.3. There are a variety of proprietary photoresist strippers available in the market and the formulations can vary significantly. For example, the formulation may or may not contain hydroxylamines and different products are suitable for metallic and or low-k dielectric layers.

Similar to the variety of acid mixtures that may exist, there are numerous proprietary solvent mixtures that are used in the industry. For example, either dimethyl sulfoxide (DMSO) or n-methyl-2-pyrrolidone (NMP) is commonly used as a major constituent of a specialty formulated photoresist stripper. Additionally, PGMEA and ethyl lactate are commonly used as components in thinning and edge bead removal formulation; the former can be mixed with propylene glycol monomethyl ether (PGME).

33.3.3 Chemical Mechanical Planarization Slurries

Chemical mechanical planarization (CMP) slurries are a unique class of chemicals that are used in semiconductor processing, and the process has been discussed in Chaps. 15 and 35 of this handbook. In every other process, particles are detrimental to the process and can impact yield. In this case, particles are the mechanical portion of the planarization process to remove specific material from the wafer in preparation of the next process step. The chemical portion of the process can be aided by oxidizers and proprietary chemical formulations that have been designed for a particular process step. As feature sizes are decreasing there has been a trend in particle size distribution control; one of the advances has been the move from fumed to colloidal particles to tighten the particle size distribution. Additional advances in slurry formulations include tunable formulations, highly selective removal of specific materials, high removal rates, as well as the ability to dilute the material on-site.

Various types of slurries exist which are used in different stages of the manufacturing process suiting a particular application need. The most common types of CMP applications include bulk oxide removal, metal planarization (e.g., tungsten, copper, and aluminum), shallow trench isolation, through silicon via, nitride, and inter-layer dielectrics. The chemical composition of a CMP slurry is rather complex and can contain additives like anti-corrosion chemicals, surfactants, etc., in addition to the complex particles that constitute a percentage of solid material that is mechanically responsible for the planarization process.

In addition to the CMP slurries themselves, there are a variety of post-CMP cleaning solutions that are used to inhibit corrosion, remove particles, and residues. Some examples are citric acid solutions, ammonium hydroxide solutions, oxalic acid solutions, and benzotriazole-containing solutions.

33.4 GENERAL HANDLING OF PROCESS CHEMICALS AND SLURRIES

With a variety of different chemicals used in the manufacturing process, chemical properties are an important design consideration for the facility and its construction. A common example is housing and segregating chemicals that are deemed incompatible; therefore, it is necessary to segregate the chemicals based on their properties. One strategy that is commonly used is separate chemical rooms based on the chemical type. For example, acids, solvents, oxidizers, and bases are often stored and dispensed in separate chemical rooms; copper chemistries are isolated from other chemistries to prevent the possibility of cross-contamination. An example of room segregation is shown in Table 33.4.

Chemical segregation often goes further to isolate chemical types for waste disposal (fluorine-containing chemicals require additional waste treatment, per the jurisdictions guidelines). For example, fluorine-containing chemicals (HF, MAEs, and other materials) are segregated from the standard acid waste as there are additional process steps to mediate the excess fluoride in the waste stream.

Nonfluoride acid drains may also have additional features engineered into the waste disposal system so as to prevent and mitigate the potential for exothermic dilution reactions. Often, water dilution systems are engineered into the drain system to dilute acids upon diversion to drain to provide an excess source of water to dilute the chemistry. One common example is the use of sulfuric acid in various process steps. Mixing can, and does, occur in common drain lines, as other waste acids are mixed with sulfuric acid, there is additional water content (example 37 percent HCl contains 63 percent water) and there will be water added to the sulfuric acid resulting in an exothermic dilution of the acid. The excess heat can rise and cause piping and component damage in the

Table 33.3 Common Solvents Used in Different Lithography Steps

	Process			
Chemical	Adhesion promotion	Photoresist thinner	Edge bead remover	Stripper constituent
Cyclohexanone		X	X	
DMSO				X
Ethyl lactate		X	X	
HMDS	X			
NMP				X
PGMEA		X	X	

Table 33.4 Example of a Chemical Room Segregation Strategy for Semiconductor Process Chemicals

Room	Solvents	Acids	Fluorine-containing chemistries	Caustics	Oxidizers	Copper chemistries
Examples of chemistry	IPA	Sulfuric acid (H_2SO_4)	Hydrofluoric acid	Ammonia hydroxide	Hydrogen peroxide	
	PGMEA	Hydrochloric acid (HCl)	Dilute HF	Potassium hydroxide		
	HMDS	Phosphoric acid (H_3PO_4)	Some mixed acid etchants	TMAH		
		Nitric acid (HNO_3)				
		Some mixed acid etchants				

waste system if proper engineering considerations were not taken into account for this potential scenario.

Solvents and flammable materials require special handling considerations due to the inherent risks that may exist with the handling of the chemicals. Based on the material being handled, room designs have to take into account the potential of containing an electrically classified area. Segregated drains for solvents also provide a mechanism for the collection and removal of the chemicals to an off-site location for proper treatment and remediation.

Additional design considerations for any chemical room are compatible materials of construction of the facility's piping, drain, and exhaust systems as the distribution systems are an extension of the process chemicals themselves. Chemical distribution systems are commissioned for purity, and an excursion can impact the entire facility. Environmental control is a crucial consideration in the chemical handling process. For example, phosphoric acid and some BOEs are subject to freezing and crystallization. Depending on the location of the facility, special transportation and handling requirements are imperative. Another example is the out-gassing and loss of assay of ammonium hydroxide depending on the temperature.

Slurries for CMP are often housed in their own chemical rooms with the equipment necessary for the unique needs of CMP applications. The handling requirements of CMP slurries depend greatly on the specific type of the slurry being used as well as the process. While there are general guidelines on the best methods to handle and deliver a slurry to the point-of-use (e.g., humidification, prevention of dead legs, loop distribution methodology), there are a number of nuances to handling a specific slurry that may enhance process performance. Some of the common handling variables that are seen with CMP slurries are highlighted below:

- *Agitation.* CMP slurries should be agitated prior to use to resuspend the particles in solution. The type of agitation can vary between pump recirculation or mechanical agitation with a propeller. Different types of slurry should employ different strategies for homogenization of any settled material. In some cases, agitation is not recommended prior to use as these formulations may not contain any particles or have a surfactant additive that may be susceptible to foaming.
- *Blending.* There exists a variety of CMP slurries that require on-site blending to add additional elements to the formulation (e.g., H_2O_2 and ultrapure water). There are also other cases where the slurry may be supplied as a two-part formulation that requires on-site mixing. Slurry manufacturers often recommend a specific mixing order, constituent ratios and sequence of operations in order to maintain consistency in the slurry.
- *Chemical concentration monitoring.* With active components some CMP slurries may have a component that requires monitoring and dosing to maintain its concentration in the blend, an example is hydrogen peroxide.
- *Pre-aging.* There also exist formulations that require pre-aging of the material prior to use. These formulations are often mixed on-site and require an equilibration time prior to use in the chemical process.
- *Pot-life.* Additional slurries also may have a certain life time of viable use once blended, as there are often proprietary chemical additives that degrade over time.

As discussed in Chap. 35 of this handbook, both chemical and slurry distribution systems are designed to take into account the specific handling requirements of the chemistry to maintain fluid health and safety, and specific process needs of the customer to provide consistent and reliable material to the process with unparalleled uptime.

33.5 LOGISTICS

As semiconductor facilities and processes become more complex, the supply chain and logistics of the manufacturing process become more complex. Process chemicals enter the fabrication facility in a variety of different-sized containers and suitable materials of construction. Some of the common supply formats include drums, totes, and sometimes chemicals are provided by tanker; each method drives the design of the chemical distribution system. With an increase in the number of patterning and CMP process steps to fabricate more intricate geometries, chemical consumption requirements also increase. Logistically, the method in which process chemicals enter the facility is ever increasing in importance. For example, as the volume needs for particular chemicals has increased into the thousands to tens of thousands of liters of material per day, staging and handling requirements are increasing in importance and larger packaging formats are in being sought.

A common theme in semiconductor manufacturing is the best way to prevent contamination is to prevent it from entering the manufacturing process. In the logistical model where bulk chemical meets the facility requirement needs, analytical validation of the incoming material is imperative to not affect many downstream processes. Figure 33.2 provides an overview flowchart of a bulk hydrogen peroxide system. As bulk material arrives at the factory, it is imperative to validate the material prior to offloading and consumption as many different processes (e.g., cleaning and CMP processes) will rely on the purity of the incoming material to not alter the process in any manner.

33.6 ANALYTICAL VALIDATION

Analytical validation services are important to the manufacturing process as the information obtained can protect the manufacturing process from potential deviations in terms of metals, particulates or assay control for incoming material. Statistical control charts are extensively used to monitor process steps and any potential changes that may be encountered. Certificates of analysis from chemical suppliers are an important part of the supply chain as they provide information regarding control and specification limits. Ultimately, the information can be used to build a library based on key metrics and be transferred to process window changes based on potential lot-to-lot variability. One such metric is assay as there are numerous process steps in which etch rates or development times are correlated to concentration of solution.

In addition, laboratory results can be used to identify the type of contamination on the wafer so the source can be identified and remediated. System purity and the various types and potential sources of contamination are discussed in Chap. 35. Different techniques are used in the identification process: Spectrometry can be used to

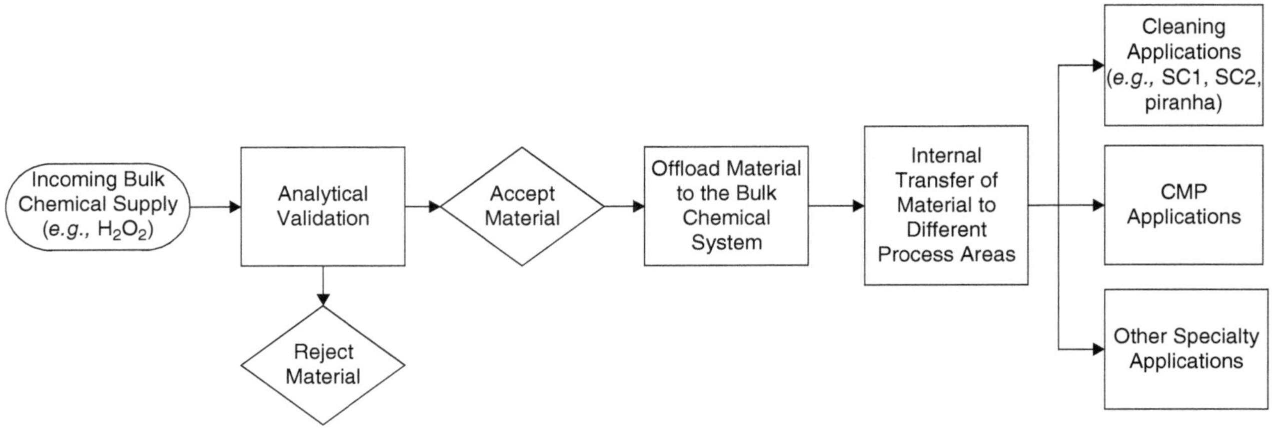

Figure 33.2 Flowchart highlighting chemical process flow and how the introduction of bulk chemicals into the manufacturing process can influence many different process steps.

differentiate between organic and metallic components; electron microscopy can be used for particulates. Some of the other techniques that are used include titration for assay control, Inductively Coupled Plasma Mass Spectroscopy (ICP-MS) to quantify trace metals, as well as additional analytical testing for anions (e.g., carbonates) in the materials of interest.

33.7 ABATEMENT

The handling of effluent streams is a key component in the handling of the process chemicals. As Table 33.1 highlighted some of the various hazards of some of the common process chemicals, engineering solutions are crucial to limit the exposure to the chemicals, or the products of the chemical reactions, to the operators or the general release into the environment. Chemical rooms and process tools are connected to scrubbed exhaust systems for neutralization of the chemicals via the facilities of various neutralization systems. Where scrubbed exhaust is not a suitable solution (e.g., flammable solvents), exhaust streams are connected to local abatement systems. Process chemicals that cannot be neutralized on-site often undergo collection into containers (tanks, totes, drums) that are then removed from site for disposal.

33.8 CONCLUSION

Wet chemical processes are an integral component in the manufacture of semiconductor devices, with the most prevalent process areas being cleaning, etching, lithography, and CMP applications. As processes have evolved, the mixtures and formulations have advanced to meet the unique process demands. While it is critical to protect the process, it is equally important to protect personnel, the facility, and the environment. Engineering controls and design considerations for handling and validating chemicals and slurries provide a mechanism for both a high-yield and safe facility.

33.9 ACKNOWLEDGMENTS

Thanks to Kristin Cavicchi, Kevin O'Dougherty, and Bryan Smith of Air Liquide Electronics U.S. LP.

33.10 REFERENCES

1. W. Kern and D. Puotinen, "Cleaning Solutions Based on Hydrogen Peroxide for Use in Silicon Semiconductor Technology," *RCA Review*, 31: 187–206, 1970.

2. NFPA 318: *Standard for the Protection of Semiconductor Fabrication Facilities*, 2015 edition.

3. Hazard Communication, Hazard Classification Guidance for Manufacturers, Importers and Employers. OSHA 3844-02-2016 (2016). Retrieved from https://www.osha.gov/Publications/OSHA3844.pdf.

4. OSHA Quick Card™ (2013). Retrieved from https://www.osha.gov/Publications/OSHA3678.pdf.

5. Great Britain's Health and Safety Executive (n.d.). Retrieved from http://www.hse.gov.uk/chemical-classification/labelling-packaging/hazard-symbols-hazard-pictograms.htm.

33.11 FURTHER READING

Chemical Hazards and Toxic Substances as listed by the United States Department of Labor and OSHA. Retrieved from https://www.osha.gov/SLTC/hazardoustoxicsubstances/

Misra, A., J. D. Hogan, and R. A. Chorush, *Handbook of Chemicals and Gases for the Semiconductor Industry*, John Wiley & Sons, Inc., Hoboken, New Jersey, 2002.

National Institute for Occupational Safety and Health (NIOSH) Pocket Guide to Chemical Hazards, as stated by the Center for Disease Control and Prevention (CDC). Retrieved from http://www.cdc.gov/niosh/npg/

NFPA 400, *Hazardous Material Code*, 2016 edition.

NFPA 704: Standard System for the Identification of the Hazards of Materials for Emergency Response, 2012 edition.

Chapter 34
Filtration

BY

BARRY GOTLINSKY *Nanometrex Solutions LLC, Santa Fe, New Mexico*

Filtration is a critical component for mitigation of suspended, and in specific instances, dissolved or homogeneous, contamination from semiconductor process liquids and gases. Removal of these contaminants, and control of risk from these contaminants, has a direct effect on yield and defect reduction. Virtually all semiconductor processes use filtration—in multiple locations ranging from bulk distribution to point-of-use in the process tool.

Filtration devices are often specific to the application, process, and the fluid being used, as design, as well as materials of construction, will affect suitability in varying process conditions and fluid types. In addition, the cleanliness of the filtration devices themselves is critical, as is the ability to withstand process conditions, which may involve elevated temperatures as well as aggressive fluids.

The engineering design of the filter will define various parameters such as fluid flow, size of the device, service lifetime, robustness, and uniformity of contaminant removal. The filtration medium, which may be composed of microporous films, melt blown fibers, drawn inorganic fibers, sintered powder, expanded polymeric materials, combinations of these, or other materials, defines the contaminant removal. Maximizing flow and contaminant removal, as well as uniformity of performance, are all performance goals, which many times require compromise or unique solutions.

The filtration device can also be a diagnostic tool to identify the contamination in the fluid, by virtue of its ability to remove, collect, and concentrate the contamination. This and the aforementioned aspects of filtration have been explored in this chapter.

34.1 CHEMICAL FILTRATION

34.1.1 Chemical Filtration: Filter Structure

Filters utilized for chemical filtration are typically composed of a polymeric membrane supported by woven or nonwoven materials and assembled into a pleated structure. The pleated structure is formed into a cylinder, around a polymeric core, with a polymeric cage to protect the filter during installation handling. The assembly typically has one blind (solid) end-cap, an open end-cap, which contains one or two o-ring grooves. The o-rings are utilized to provide integral seal to the housing, in the case of a replaceable filter element. In the case of a disposable capsule, the filter element is permanently sealed in the housing. The components are typically thermally welded, eliminating the use of adhesives. Typical filter construction examples can be found on websites of the filter cartridge manufacturers.

The membrane, which is essentially the working component, is made in sheet form. Depending on the polymer used, the membrane can be formed via solvent casting, expanded membrane process, or a combination of these. There are many membrane materials available, and the choice is highly dependent on the process application. This is defined by the fluid, process conditions such as temperature and flow, and cleanliness requirements.

Polymers used for filter membranes can include PTFE (polytetrafluorethylene), PVDF (polydivinylfluoride), nylon, HDPE (high-density polyethylene), PS (polysulfone), PES (polyethersulfone), PAS (polyarylsulfone), and various other polymers. In the case of particle removal from chemicals, the membrane's porous structure will define the particle removal. While the mechanism of particle removal is mainly sieving, the pore size of the membrane may not be directly related to the particle size removal. Other factors, including thickness of the membrane, tortuosity of flow through the membrane, pore structure and shape, and stacking of the pores, all have an effect on the particle capture. In addition, the polymer may interact with particles via surface charge (limited by high ionic strength of many chemicals) and surface adsorption.

Ideally the pore structure should be uniform, though in practice there is always a range of pore sizes in a membrane. While this is a controlling factor, ultimately the entire membrane package defines particle removal uniformity, including the factors mentioned in the preceding paragraph.

With removal ratings at or approaching single digit nanometer sizes, it is increasingly difficult to fabricate membranes with extremely fine pore structure, which allows for acceptable flow rate. Unique solutions have been devised including asymmetric membrane structures, which provide a continuously variable pore size across the membrane thickness. The degree of asymmetry can be controlled via the pore forming process, to optimize particle removal performance versus flux, or fluid flow. Typically the rated pore structure can be very thin, supported by a thicker region of the membrane where the more open structure improves flow and provides internal support within the membrane. Other techniques may include stacking of layers of membrane to optimize the particle removal performance as well as flow characteristics.

As the membrane is typically provided as a pleated pack in the completed filter, the membrane needs to be supported. In addition, a drainage layer is used to ensure proper flow across the membrane and prevent blockage of the membrane from fluid flow where the membrane pack pleated structure comes together in the final assembly.

The support and drainage materials are typically either woven or nonwoven polymeric materials, and can also be porous structures made from polymeric films. The support and drainage materials are very open materials, as they are providing structure, and not particle removal. Common materials include PTFE, PFA, HDPE, and PP (polypropylene).

The membrane, support, and drainage layers are in roll form prior to pleating. The rolls are unwound and introduced to a pleating machine. The design of the pleats can vary, with optimized pleat structure used for robustness, flow optimization, and other parameters.

The pleated pack is then cut to length, formed into a cylinder, and assembled with a core and end-caps. The core and end-caps can be PFA, HDPE, PP, or other polymers. The assembly is thermally welded.

The filter assembly is typically tested for integrity and cleanliness. Integrity testing is a nondestructive method of confirming no defects or damage, and it is based on gas diffusive flow through a wetted membrane. If the filter is intact, only diffusive flow will take place. If there is a defect or manufacturing damage, mass flow will be detected. Cleanliness may include metals and halide leachables, and particle cleanliness. This will depend on the cleanliness claims for the specific filters and the filter manufacturer's process controls.

Flow through the filter is from the external side of the cylinder (through the cage), across the membrane, and exiting through the core and out through the open end-cap. The open end-cap is sealed to a housing head, either via o-rings in the case of replaceable filter elements, or, in the case of disposable capsules, o-rings or thermally welded seals are used.

The housing is the vessel that contains the filter elements and the fluid being filtered. Upon startup, the upstream volume is vented to remove air from the vessel, which could cause air blockage.

Beyond the membrane material, there is the concern of wettability of the membrane. In the case of low surface tension fluids, such as many organic solvents, aqueous chemical mixtures containing surfactants, and aqueous mixtures containing miscible low surface tension solvents, wettability is not a concern, and either a hydrophilic or hydrophobic membrane can be used. Compatibility with the chemistry and process conditions is key.

However, in the case of aqueous fluids, and especially very high surface tension fluids such as, for example, HF, H_2O_2, or KOH, wettability is a real concern. In these cases, if the membrane is hydrophobic, the membrane must be prewetted with a low surface tension fluid followed by UPW rinse out of the wetting fluid. Isopropanol (IPA) is commonly used for prewetting. However, disposal of the IPA diluted in UPW, as well as the additional time and labor, makes filter change-out cumbersome and more costly.

To address this issue, many filter manufacturers have developed solutions eliminating the need to prewet. One of these solutions is to surface modify or treat the membrane surface to render the surface hydrophilic. In this way, the filter can be installed dry, and the high surface tension chemical introduced directly. This is an ideal solution provided both the membrane and surface modification (or treatment) is compatible and stable.

In cases of aggressive chemistries, a different solution is needed. One is to provide the filter prewetted and packaged in UPW. In this way, the installation only requires replacement of the UPW in the filter with the process aqueous chemistry. This solution does require the filter supplier to validate their process for shelf life so that bacterial growth is not an issue upon storage. In addition, the packaging must be robust so that either the poly bag or housing (in the case of a disposable capsule) remain integral without leaks.

While this solution eliminates the need for prewetting upon installation, a hydrophobic filter can still dewet upon use in high surface tension chemistry (e.g., HF, buffered oxide etch, or KOH), or hydrogen peroxide–containing mixtures, especially if flow is not continuous, as, for example, in a single wafer spray tool. To address this further issue, filter suppliers have developed non-dewet surface treatments, which can be accomplished even with inert polymers such as PTFE. In this case, the CWST (critical wetting surface tension) is increased over the base polymer's CWST to the level at which dewetting does not occur. This is a good solution, in that while it does not eliminate the UPW packaging, it does prevent dewetting in use while providing a membrane treatment, which is robust and compatible with more aggressive chemistries. As is always the case, the filter supplier must validate that the treatment is in fact robust, stable, and compatible.

Filters have also been developed which remove ionic contamination. While this is limited to very dilute aqueous fluids or certain organic solvents, these filters can combine particle removal as well as removal of cationic species. The technology of these filters can be composed of adding ion exchange moieties on to the filter membrane surface, or other techniques. The capacity of these ion exchange filters is calculated as milliequivalents (meq). Depending on the metal contamination of the fluid, the capacity may be shorter than the particle capture capacity of the filter. However, for ultrapure streams, these combination filters can provide prevention of excursions. As noted, these are not applicable beyond very dilute aqueous chemistries.

34.1.2 Chemical Filtration: Performance

The intention of filtration is to remove suspended solids from a liquid chemical stream. The filter retains the particles on the filter surface (except in the case of cross flow filtration—not typically used in chemicals), and these captured particles ideally are not released despite increased particle loading on the filter surface.

Filters are assigned particle retention ratings based on removal of particles under a specific set of conditions. These conditions are defined by the ability to measure particle concentrations at a given particle size. With the continued direction to ever-finer filters, this task becomes increasingly difficult due to particle measurement technology limitations.

When semiconductor geometries were in the micron size, utilization of filters with ratings of 100 to 200 nm were common. Determining particle performance of these filters involved either using an unfiltered chemical or a fluid spiked with particles such as poly-dispersed clay or mono-dispersed polystyrene latex (PSL) spheres. Light scattering particle counters were used to monitor the test filter influent and effluent particle concentrations. Due to limitations of the particle counter sensitivities, challenge particle concentrations typically exceeded those found in actual chemical streams in a semiconductor fab. Prior to particle challenges the cleanliness of the test filter was confirmed to establish a baseline particle concentration downstream of the test filter prior to particle challenge testing.

As semiconductor nodes increasingly introduced smaller geometry features, the filters became finer in removal ratings. To a filter rating of 50 nm, the spiked challenge method using PSL and light scattering optical particle counters (OPC) continued to be applicable, however, only in ultrapure water (UPW). The inability of particle counters to measure 50 nm particles in chemicals, as well as the stability of PSL in those chemicals, required that filter ratings be determined in UPW for chemical filters (Fig. 34.1).

As filters were introduced with ratings to 5 nm and finer, the conventional methods of determining particle removal performance were

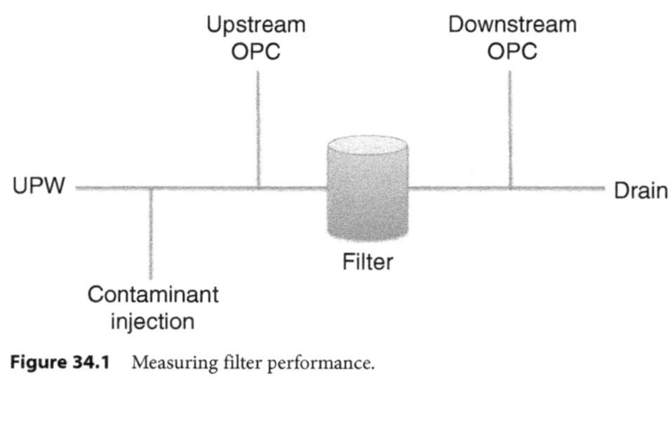

Figure 34.1 Measuring filter performance.

no longer applicable, in part due to the lower particle sensitivity limitations of laser light scattering particle counters, and the availability of suitable mono-dispersed PSL particles. Wafer inspection size capabilities were going beyond the ability of in-line fluid particle metrology at this point.

Filter manufacturers have methods to determine the pore distribution and pore sizes of membranes; however, these do not translate directly into particle removal ratings. While the mechanism of particle removal in liquid chemicals is mainly sieving, tortuosity path of the fluid flow through the membrane as well as sorptive surface effects also play a role in particle removal.

To achieve particle performance measurements, new methods have been developed to enable particle rating at single digit nanometer sizes. While these methods use particle challenges greater than the particle concentration in ultrapure chemical streams, these ratings have been shown to correlate with defectivity reductions on devices [1, 2].

One of these methods utilizes gold nanoparticles in UPW [3], which are commercially available as National Institute of Standards and Technology (NIST) traceable particles. The distribution of these gold nanoparticles is extremely mono-disperse, so a filter challenged with these particles provides a true measure of the ability to remove these particles. While there currently is no metrology to directly measure particle concentration in a fluid stream at single digit nanometer sizes, the gold nanoparticles can be measured for concentration via Inductively Coupled Plasma Mass Spectroscopy (ICP-MS). As gold is not a typical contaminant in UPW, interference is not an issue. Since the particle size of the gold nanoparticle is extremely uniform, and the shape is spherical, a simple calculation converts concentration of gold to particle concentration at the given size of the gold nanoparticles.

The gold nanoparticle method does have a major limitation, as these particles will readily adsorb to polymeric surfaces. Therefore, a ligand must be used to "disarm"—render the surface of the filter membrane nonadsorbing of the gold nanoparticle. Since each membrane may be of a different polymer, or in addition, have a unique surface treatment, for example, to render the surface non-dewetting, the choice of ligand must be determined individually for each filter type. This increases the complexity of the method.

An alternative method has been developed which utilized colloidal silica as the challenge particle [2]. As with other challenge materials, the concentration of particles used for the filter challenge is orders of magnitude greater than that found in ultrapure chemical streams. This method does not directly measure the particle concentration in UPW, and instead measures the particle concentration in an aerosol stream.

The instrument used to measure the colloidal silica particles in UPW initially passes a sampling of the UPW through an atomizer, which generates an aerosol stream. To prevent interference with the measurement, the UPW must be extremely clean with respect to suspended materials as well as nonvolatile dissolved contaminants. Any of these contaminants will be aerosolized along with the UPW-containing colloidal silica. The aerosol is dried and passed through a differential mobility analyzer (DMA), which separates the stream into discrete sizes. The separated aerosol is then passed through a condensation particle counter, which grows a liquid droplet on each particle in the aerosol stream, using each particle as a nucleation site, and then counts the enlarged droplets using laser light scattering method. The particle concentration derived from this method does not size particles, hence the requirement for no other contamination in the UPW challenge stream other than the colloidal silica.

The colloidal silica is commercially available as a poly-disperse distribution, and is not characterized by the suppliers. Therefore, the same metrology method must be used to characterize each sample of colloidal silica for suitability as a challenge material for the specific particle size rating to be determined for a filter.

New technology is being developed for particle measurement in fluid streams. These do have the promise of enabling more direct measurement of filter particle performance. One example of this is the acoustic counter [4].

34.1.3 Chemical Filtration: Usage Considerations

The choice of filter is highly dependent on the specific process. Factors including flow rate, materials of construction and compatibility, cleanliness, performance, and ease of use all come into play.

For high-temperature processes using aggressive chemistries such as SPM (sulfuric peroxide mixture) or silicon nitride etch (hot phosphoric acid), a disposable capsule would be the typical choice for these applications. Especially at high temperatures in the range of 150°C or higher, o-ring seals can be a concern with regards to safety and leakage. The disposable capsule eliminates the o-ring seal on the shell. Use of a disposable capsule also minimizes operator exposure to the aggressive chemistries during filter change-out, as the filter itself is not handled, remaining internal to the capsule.

Single wafer spray tools using small filters at the spray nozzle will also typically use small capsule filters, which can reduce holdup volume and ease installation.

Bulk chemical delivery systems will tend to use larger permanent housings with changeable filters.

In applications where there are additives in solutions, such as copper plating, the lack of additive stripping should be verified prior to using a particular filter type. This will be dependent on the filter membrane and the surface treatment or modification of the filter to render hydrophilic or non-dewetting.

Filter rating indicates particle removal performance; however, studies have shown [5] that using finer filtration will improve defects at a given size greater than the filter rating. So although it may appear intuitive to use a filter, which provides particle protection at the defect size, often there is justification to use filters with ratings finer than the defect size.

Service life of filters will be highly dependent on the contamination levels of the fluid being filtered. Differential pressure should be monitored to assess change-out prior to an increase in wafer defectively, and/or loss of process flow. Once a process has stabilized, and filter service life determined, PM (preventative maintenance) change-out schedules can be established. In extremely clean fluid streams, filters can be in service in excess of a year's time, while in POU applications where high levels of contamination are produced, the service life of the filter can be fairly short. In applications where soft, amorphous polymeric materials are captured by the filters, the filters should be changed-out at lower differential pressure to prevent extrusion of these soft materials through the filter medium.

34.2 ULTRAPURE WATER FILTRATION

The structure and ratings of microfilters for ultrapure water is essentially the same as that for chemical filters, with different materials of construction typically used. For UPW, these filters will have hydrophilic surfaces, either natively, such as nylon (type 6,6) or surface modified, such as hydrophilic PTFE, or PAS. The microfilters can be used for point of distribution (POD), point of use (POU), or on the main loop. Cleanliness is critical in terms of particles, resistivity rinse-up, total oxidizable carbon (TOC), metals, and anions. For POU applications, flow is also critical as space constraints often limit available space for filter installation.

As UPW systems typically are extremely clean, filtration ratings used for UPW have driven down to sub 20 nm filtration, with latest nodes requiring single digit nm filtration. This increases the need for advanced filter designs, which provide increased flow while maintaining the particle removal performance. Advances in filter ratings, as explained in the chemicals section of this chapter, are also critical to prove that the filters are capable of providing the particle protection needed by the UPW systems and POU UPW applications.

For large flow UPW applications, a different design of filter is often used, especially at POD. These filters are known as ultrafilters, and operate very differently from microfilters. Whereas microfilters operate by passing the fluid directly through the membrane, capturing the particles on the membrane surface, which is called "dead-end" mode, ultrafilters operate by passing the fluid across the surface of the membrane, with a small percentage of flow continuing across the membrane surface to a reject (or concentrate) port, and the majority of fluid through the membrane to a permeate (or product) port.

Known as cross-flow filtration, the majority of the particles does not remain on the membrane surface, but instead are swept off into the reject flow. This facilitates large flow volume and minimizes fouling of the membrane by particle capture. These membranes are typically rated by molecular weight cutoff, by measuring dextran retention, a method that does not directly correlate to particle size removal. However, studies have shown these filters to provide single digit nm particle capture [6].

Ultrafilters typically utilize a hollow fiber bundle, though could be of spiral wrapped design. The polymer used for the filtration membrane is typically hydrophobic, such as polysulfone, resulting in these filters being supplied prewet, with a preservative in solution. Upon installation, the preservative must be rinsed out with UPW prior to use.

Most UPW designs will use a combination of ultrafilters and microfilters, with the microfilters utilized closer to POU at lower flow and within the process tools themselves.

34.3 LITHOGRAPHY FILTRATION

While the structure of filters used for lithography applications is essentially the same as that used for chemical filtration, some distinct requirements are essential for lithography applications.

Many lithography fluids are solvent based, including photoresists, negative tone developers, and adhesion layers, though there are aqueous based fluids including positive tone developer, UPW immersion fluid, and UPW rinses.

The polymers used in filters must be compatible with lithography fluids, and also must not leach organics or residues into the solvent-based fluids. This is critical, as any material leaching from the filter that causes line edge roughness, t-topping, or other defects will greatly increase defect rates. As the rate at which these materials may leach from a filter may vary from filter to filter, or even as the filter is being used, defectivity as a result of this contamination will be difficult to predict. Most lithography filters use either nylon or HDPE membranes and HDPE support and hardware, as well as HDPE capsules.

Filters also must be clean with respect to metals, as metals introduced on to a wafer after dispense can diffuse into underlying layers during soft bake. Filters supplied for critical lithography applications are typically cleaned using robust processing by the filter manufacturer to minimize both metals as well as organic leachable contamination.

For dispense applications, another concern is microbubble formation [7]. Microbubbles increase defectivity by causing craters, haze, and other defects in the photoresist film. Even though solvent-based fluids have low surface tensions, microbubbles can still form, especially with increased viscosities of polymer solutions. Dispense capsule designs have incorporated venting mechanisms that eliminate air from collecting at filter membrane surfaces. These advanced designs also reduce startup times, greatly reducing fluid waste during priming and dispense qualification.

For dispense applications, it is critical to change-out dispense filters upon differential pressure increase. As polymeric contaminants, such as soft particles, gels, or slightly soluble materials collect on the filter surface, these contaminants can be forced to extrude through the filter membrane at high differential pressure. In addition, buildup of differential pressure can effect dispense rate in certain dispense pump designs.

Filters can also reduce defects that do not appear until post-etch, by adsorbing partially soluble contaminants in BARC (bottom antireflective coating) and photoresists that are somewhat unstable [8, 9]. This adsorption is dependent on the surface chemistry of the filter membrane, and polymers such as nylon have been proven to adsorb these contaminants and reduce defects such as line bridging [10, 11]. So in effect, the filter provides enhanced defectivity reduction beyond the rated particle removal.

Bulk lithography filtration at materials manufacturers utilizes essentially the same filters, in larger formats. As noted, the membranes typically are nylon or HDPE, with HDPE support and hardware. While particle removal is the primary goal of the filtration, adsorption of the partially soluble defects in certain fluid mixtures is also desirable.

Filtration for immersion UPW in an immersion DUV stepper/scanner is a special case as the UPW must be extremely pure and contaminant free, including TOC and metals that can leach from a filter, if not of the proper cleanliness levels. The use of all fluoropolymer filters in the POU immersion system providing UPW to the immersion stepper/scanner has been common practice to minimize contamination from the filter and provide efficient particle removal at single nm size levels.

As metals are a concern in lithography fluids, filters containing ion exchange functionality can be used as a polishing step to limit metals levels in these fluids. These filters, typically HDPE with ion exchange functionality added to the membrane surface, remove cations by conventional ion exchange methods, with a much higher surface area of the ion exchange functionality. The capacity is typically limited, rendering these technologies as a final polishing step of the fluids. These filters can also be used during dispense of wafer solvent precoat, to eliminate metals as a process variability.

With patterning continuing to drive down to ever-smaller feature sizes, and 193 nm ArF immersion DUV capable of 14 nm and smaller patterning, the filters used in lithography fluids currently are at 5 nm and finer particle removal capabilities. As finer and finer filters were used, there were concerns of stripping of polymers in lithography solutions. However, studies have shown that, even with the finer filtration utilized, no deleterious effects were found on the patterning process [12, 13]. Therefore, finer filtration is appropriate in reducing lithography defectivity. With the technical challenges of rating finer filters, the actual particle removal of these sub-10-nm filters is not readily measured, so porosity measurement methods are used by filter suppliers to determine relative pore size, and actual defectivity measurements after patterning are used to demonstrate defectivity improvements in the lithography process by going to the finer rated filters. Even though these filters are finer than current wafer inspection limits, the finer filters do have a positive effect on reducing defects from the patterning process.

34.4 CHEMICAL MECHANICAL POLISHING FILTRATION

Filtration for the CMP process is very different from other liquid filtration processes, as the end goal is not to remove all particle contamination, but rather only remove the particle contamination causing defects such as microscratches, and allowing all of the desired slurry particles to pass through the filter. Therefore, selectivity is most critical.

Filters for CMP applications have typically utilized melt blown polypropylene fiber filters, which have depth filtration. Melt blown filters are fabricated by extruding, at high temperature, a polymer through specific dies, with the resultant fibers being wound on a mandrel with a core. As the fibers cool, they fuse onto adjoining fibers, forming voids, which act as the filter pores through the depth of the medium. Because the fibers are fused, the porosity remains fixed, even under loading and increased differential pressure.

The capacity of these filters does need to be high as the mass of solids collected can be large. More recently, as slurry dispersions have become finer, membrane-based filters have also proven successful for CMP filtration.

Filters for CMP need to be selective to the point that the slurry dispersion passes through without effect on removal rates, and the filter retains agglomerates and other larger particle contaminants, minimizing defects such as microscratches.

34.5 GAS FILTRATION

34.5.1 Gas Filtration: Filter Structure

The mechanisms of gas filtration are very different from those of liquid filtration, because the fluid, in this case, is very low in viscosity. Whereas in liquids, the fluid will affect the motion and behavior of fine particles, in gases, the fluid stream has little effect on the motion of fine particles. The particles suspended in a gas are known as an aerosol. While the particles do flow with the gas in a pipe, their behavior within the pipe is random, at very small sizes. At very large sizes, the particles tend to settle due to mass and gravity, as there is not enough lift to keep those particles suspended, and under laminar flow conditions, there is little flow at the pipe walls. Moderately large particles are influenced by the gas flow and tend to follow the gas stream. Very fine particles will interact with gas molecules, resulting in brownian motion [14]. This brownian motion is a key factor in particle retention of filters.

Due to the unique particle behavior in gases, filters do not need to have pores as fine as the particles being removed. In fact, the smaller-sized particles are actually easier to remove by a filter, due to the statistical likelihood of a collision of the particle with the filter medium. Because these very fine particles have extremely low mass, Van der Waals force will retain the particle on the filter surface once captured. This mechanism is known as diffusional interception. In addition, once these very fine particles are captured, the filter can be more effective at particle removal due to dendritic capture of particles upon already captured particles.

At large sizes of particles, the tendency of the particle to be dragged by the gas flow and impact into the filter medium is known as inertial impaction. The particle has enough mass to maintain a trajectory straight to the filter medium even though the gas flow is essentially moving around the filter medium, as the gas passes through the filter.

There is a size region, which is most critical for particle capture by filters. This is known as the most penetrating particle size (MPPS). The MPPS varies by filter design and structure; however, most filters rated at 3 nm aerosol efficiency have a MPPS at approximately 100 nm [15]. When designing the filter structure, the most critical aspect in terms of aerosol efficiency is to maximize particle retention at the MPPS (Fig. 34.2).

Gas filters can utilize a membrane structure such as PTFE, as used in liquid filtration, to provide efficient particle capture. However, unlike in liquids, a 5-nm filter is not needed to provide less than 5 nm capture. A more open pore structure can be used, provided that capture at the larger MPPS is obtained. This reduces pressure loss at a given flow.

To further take advantage of the aerosol behavior in capturing particles, even more open structures can be used. A common solution is to use metallic fibers or powders in the design of filters. In the case of fibers, for example, 316L stainless steel, drawn fibers are used in a wet or dry lay down to form a sheet of filter medium. The sheet is then treated in a furnace to sinter the fibers, resulting in a fixed pore structure. The sheet is then fabricated into a filter design, which can be pleated, cylindrical, or other designs. Multiple layers of these sheets can used to increase the depth and tortuosity of the filter pack, increasing efficiency while allowing more open pore structures.

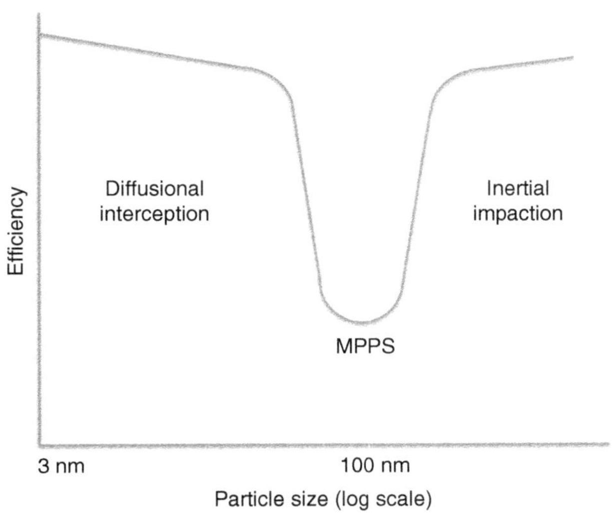

Figure 34.2 Aerosol regimes by size.

As with metallic fibers, metallic powders can be used by first forming into the appropriate shape, and then sintering in a furnace to yield fixed pore structures. For these metallic structures, stainless steel, nickel, and various alloys can be used, depending on the application in terms of compatibility of the metal.

Gas filters typically use 316L stainless steel housings. For bulk filtration, these are permanent housings into which cartridge filters are installed. Polymeric filters are commonly used for bulk filtration, due to the higher cost of metallic filters, and are sealed in the housings using o-rings on the filter cartridges, as used in liquid filtration.

Filtration of specialty gases in gas cabinets, as well as at POU and in process tools, generally use disposable assemblies where the filter is sealed in the shell. Here there is an advantage of metallic filters, as they can be welded in place into the stainless steel shell eliminating the o-ring seal.

The filter assemblies are supplied with metal gasket or compression fittings, with the former the more reliable seal in critical applications. The assemblies can also be supplied with tube ends, which are welded into place at the point of installation. In this case, the expectation is that the filter will last the life of the installation.

Filters used for specialty gases and at POU are leak tested upon installation. This is critical to ensure a hazardous gas will not leak into the work environment, and also to ensure that environmental contamination will not leak into the process gas stream. The helium leak test is used to confirm sealing upon installation. The test can be either inboard or outboard, using an appropriate sniffer to detect He.

34.5.2 Gas Filtration: Performance

As noted, gas filters are designed to remove particles at sizes including the MPPS. To determine filter efficiency, a polydisperse aerosol, which can be generated from a NaCl solution, aerosolized through a nebulizer and dried with a diffusional dryer to eliminate water droplets, is passed through a filter and the effluent of the filter is measured using a condensation particle counter (CPC), typically sensitive down to 3 nm in particle size. However, as the CPC nucleates a liquid droplet onto each particle and measures by laser light scattering, there is no size information in the particle concentration data.

In order to determine the size distribution of the generated aerosol, the aerosol can be charged and then separated into monodisperse "slices" via an electrostatic classifier as the charge for each particle will be proportional to the mass of the particle. Each monodisperse "slice" is then measured by the CPC, resulting in a size distribution of the aerosol. Ideally, the size distribution will range from 3 nm and maximize by particle number concentration at the expected MPPS for the filter [15].

The same monodisperse challenge sequence can be used to determine the MPPS for a filter by measuring the size at which the efficiency is minimized. At smaller and larger sizes than the MPPS, efficiency will increase, and the filter design is maximized at the MPPS for particle retention (Fig. 34.3).

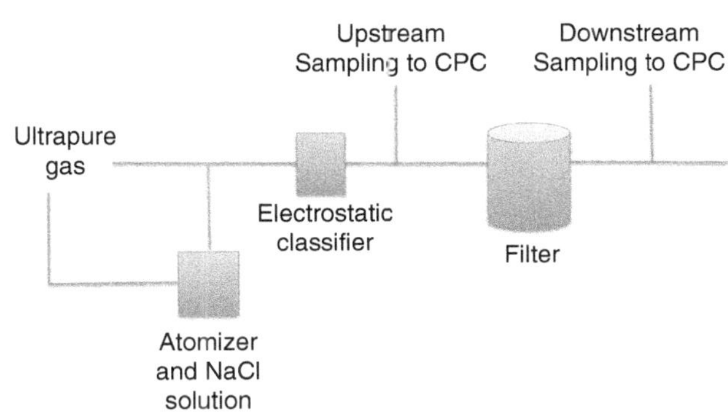

Figure 34.3 Aerosol efficiency measurement.

In many cases, the efficiency will also be affected by gas velocity through the filter, since at higher velocity; there is lower statistical probability of a particle collision with the filter medium, except in the case of large particles greater in size than the MPPS.

When rating a filter for gas filtration, there are a number of factors that must be stated, including

- Aerosol size distribution
- Efficiency at MPPS and particle size of MPPS
- Efficiency at various velocities within the operating flow range

While this is more complex than in liquid filtration, it does afford greater latitude in filter design, with the utilization of more open structures and advantageous use of tortuosity.

Filter cleanliness is also important in gas filtration. Filters need to be designed to not shed particles, and in many cases, not to outgas upon installation. Filters that have been preconditioned are available. In the preconditioned versions of filter assemblies, the filter assemblies have been purged and filled with an inert gas, so that startup is rapid, and contamination and extensive purge is not needed upon installation. This is important, especially at POU installations. However, in cases where the assembly will be welded in place or is part of a gas stick that will be purged in entirety, preconditioning may not be needed.

Materials compatibility for reactive or corrosive gases must be carefully considered. For example, sintered nickel filters and gaskets should not be used with CO (carbon monoxide) or carbonyl sulfate, as $Ni(CO)_4$ can form, which is volatile and could lead to leaks at the Ni gasket and resultant downstream contamination. Similarly, silver plated gaskets are not compatible with ozone systems as ozone is catalytically decomposed by silver. The Compressed Gas Association and SEMI guidelines or suppliers compatibility guides are good resources to consult on compatibility.

In addition to particle removal, the removal of gaseous contaminants from process gases is needed to prevent chamber contamination, unwanted side reactions, and yield effects. Filter suppliers often combine purification materials with particle filtration in the same assembly. The purification material is specific to the process gas and the contaminants removed and is typically a packed bed of purification material. Mechanisms of contaminant removal can include chemisorption and physisorption.

The purification filters are rated for removal efficiency, for a specific contaminant in a specific process gas, at a design flow. The capacities of the contaminants that are removed from a specific process gas, at a design flow rate are also stated. Peak flow rates may also be stated, as well as peak contaminant levels.

The performance of the purification material can be measured in a number of ways, depending on the process gas and the contaminants of interest. For example, a process gas can be spiked with a calibrated contaminant, and the contaminant concentrations measured by APIMS (atmospheric pressure ionization mass spectrometry), ion mobility spectrometry (e.g., for NH_3), CRDS (cavity ring-down spectroscopy), or other methods.

By combining purification material and particle filtration medium in the same assembly, both homogenous contaminants and aerosol contaminants are removed with a single filter assembly.

Compatibility of the purification material must be carefully considered to prevent unexpected exothermic reaction or downstream contamination. For example, use of an inert gas purifier which removes oxygen, in a pure oxygen stream, can result in a high temperature from the exothermic reaction and can lead to downstream contamination if volatile materials are formed from the high resultant temperature. The supplier's literature should be consulted for compatibility guidelines, as well as disposal instructions post-use.

34.6 USE OF FILTRATION AS A DEFECT ANALYSIS TOOL

Because filters capture contamination and concentrate this contamination with use, they provide a unique opportunity to concentrate and isolate contamination that may be leading to defects on devices.

There have been many instances where a factory has a defect excursion, and it is difficult to isolate the cause. Analyses of fluids sometimes do not indicate the potential contaminant due to low concentrations, or the fluid stream has diluted the contaminant. By evaluating a filter that has been in service during the defect increase, it is possible to determine possible contaminants. Materials suppliers investigating a particular lot of material, by evaluating the bulk filters used prior to bottling, have also used this method.

Filters that plug in service, or attain high differential pressure in service, causing inadequate flow in the process, can be used to assess the cause of the filter plugging. By evaluating the contamination plugging the filter, solutions to eliminate the contamination can be determined, or alternative filtration solutions developed.

The filter removed from service can be rinsed in the normal direction of flow with a suitable fluid to remove any hazards or, in the case of polymer solutions, solids that can interfere with the analysis upon drying. The filter is then dried under clean conditions, and the membrane carefully extracted from the filter cartridge. The membrane can be examined using a number of analytical techniques, including

- Visual inspection for staining, coloration, deposits, physical damage, or defects
- SEM/EDS (scanning electron microscopy/energy dispersive x-ray spectroscopy) for physical morphology and elemental composition of the contamination on the upstream filter surface
- FTIR (Fourier transform infrared spectroscopy) for organics identification (particularly if carbon is a major constituent found in EDS analysis, or polymeric or organic contamination is suspected)
- If possible, digestion followed by ICP-MS for metals analysis, or laser ablation ICP-MS
- Other specific methods, as warranted, such as Raman spectroscopy, XPS (x-ray photoelectron spectroscopy), ToF-SIMS (time of flight secondary ion mass spectrometry), or thermal desorption GC-MS (gas chromatography–mass spectrometry)

A suitable fluid can also be used to leach the filter membrane and the fluid analyzed by appropriate analytical techniques.

In the case of ion exchange membranes, the filter can be exchanged with a suitable acid, and the acid analyzed by ICP-MS to determine the concentration and identity of metals collected by the ion exchange membrane.

In gas processes, the filter can be analyzed for particulate contaminants, and in the case of gas purifiers, the remaining capacity can be determined, as an indirect method of measuring the total gaseous contamination the purifier has seen in service.

Any filter removed from service for analysis must be handled in the appropriate manner for materials exposed to particular liquids or gases, especially in the case of toxic, reactive, or corrosive fluids.

34.7 REFERENCES

1. T. Mizuno, A. Namiki, and S. Tsuzuki, "A Novel Filter Rating Using Less Than 30 nm Gold Nanoparticle and Protective Ligand," *IEEE Transactions on Semiconductor Manufacturing*, Vol. 22, No. 4, Nov. 2009.
2. G. Van Schooneveld and D. Grant, "Technique for Measuring the Particle Retention of Liquid Filters to 10 nm," AFS Spring Conference, Bloomington, MN, 2013.
3. T. Umeda, T. Mizuno, S. Tsuzuki, and T. Numaguchi, "New Filter Rating Method in Practice for Sub 30 nm Lithography Filter," *Proceedings of SPIE*, Vol. 7639, 2010.
4. S. Madanshetty, "Apparatus to Induce Acoustic Cavitation in a Liquid Insonification Medium," U.S. Patent 7253551, Aug. 7, 2007.
5. B. Gotlinsky, "Requirements for 10 nm Filters Intended for Sub 28 nm Technology Node Surface Cleaning Processes," *Proceedings of Sematech Surface Preparation and Cleaning Conference*, Mar., Austin, TX, 2011.
6. B. Gotlinsky and D. Blackford, "Using a Nonvolatile Residue Monitor to Investigate Ultrapure Water System Components and Optimize System Performance," *Microcontamination*, Vol. 12, No. 7, 1994.

7. T. Umeda, S. Tsuzuki, M. Boucher, H. Dinh, L. Ma, and R. Boten, "Development of Optimized Filter for TARC and Developer with the Goal of Having Small Pore Size and Minimizing Microbubble Reduction," *Proceedings of SPIE*, Vol. 6153, 2006.

8. B. Gotlinsky, M. Mesawich, and D. Hall, "The Effectiveness of Sub 50 nm Filtration on Reduced Defectivity in Advanced Lithography Applications," *Proceedings of Arch Chemicals Interface Microlithography Symposium*, San Diego, CA, 2003.

9. T. Umeda, S. Tsuzuki, T. Numaguchi, M. Sato, C. Yamamoto, and M. Sato, "Start Up Optimization for Point of Use Filter in Lithography Process," *ISSM International Symposium on Semiconductor Manufacturing*, Santa Clara, CA, 2007.

10. N. Brakensiek and M. Sevegney, "Effects of Dispense Equipment Sequence on Process Startup Defects," *Proceedings of SPIE*, Vol. 8682, 2013.

11. T. Umeda, F. Watanabe, S. Tsuzuki, and T. Numaguchi, "Filtration Condition Study for Enhanced Microbridge Reduction," *Proceedings of SPIE*, Vol. 7520, 2009.

12. B. Gotlinsky, J. Beach, and M. Mesawich, "Measuring the Effects of Sub 0.1 μm Filtration on 248 nm Photoresist Performance," *Proceedings of SPIE*, Vol. 3999, 1999.

13. M. Mesawich, M. Sevegeny, B. Gotlinsky, S. Reyes, P. Abbot, J. Marzani, and M. Rivera, "Microbridge and E-Test Opens Defectivity Reduction via Improved Filtration of Photolithography Fluids," *Proceedings of SPIE*, Vol. 7273, 2009.

14. K. Lee and B. Liu, "Theoretical Study of Aerosol Filtration by Fibrous Filters," *Aerosol Science and Technology*, Vol. 1, No. 2, 1982.

15. B. Gotlinsky, P. Connor, D. Capitanio, L. Johnson, and S. Tousi, "Testing of All Metal Filters for High Purity Semiconductor Process Gasses," *Proceedings of the Institute For Environmental Sciences*, 1991.

Chapter 35

Chemical and Slurry Handling Systems

BY

KRISTIN CAVICCHI *Air Liquide Electronics US, Houston, Texas*

35.1 INTRODUCTION

Process tools drive the semiconductor manufacturing process by managing a complex set of chemical reactions. A process tool's lifeblood consists of raw materials that include bulk inert gases, toxic and reactive gases, ultrapure water, process chemicals, and chemical mechanical polishing (CMP) slurries. The systems that generate, handle, purify, and/or measure each of these five classes of materials are typically designed and operated by a fab's facility management team and are referred to as "critical process systems." Each critical process system is comprised of standard products from specialized suppliers that are engineered and assembled into site-specific systems to meet the unique needs of the customer's physical facility and process requirements. As such, selecting and managing a critical process system requires an understanding of the relevant issues and choices that are common from fab to fab and an understanding of the facility or process-specific issues and options.

Automated chemical and slurry handling systems represent a specific field of expertise with tremendous implications to the safety, purity, and uptime of a fab's manufacturing processes. Chemical and slurry systems impact many process areas in a fab including CMP, lithography, wet etch, and clean.

Wet process chemicals are typically divided into five classes:

1. Acids—usually used for cleaning or etching; fluoride-containing chemicals may be broken out as a subset of the acid category
2. Bases—also used for cleaning; includes developers used as part of the lithography process
3. Oxidizers—used to create or augment reactions with other chemicals or on the wafer surface, such as facilitating photoresist removal or metal polishing processes
4. Solvents—used to aid drying or as strippers for other materials such as residue from back-end plasma etch processes
5. Slurries—suspended solids in a chemically active or buffered solution used in polishing processes to remove and/or planarize deposited materials

At a basic level, a chemical or slurry system consists of the following elements:

- A chemical source, typically a drum or tote
- A method to move the chemical
- A storage tank
- A filtration system
- A piping network, including valve boxes, to provide flow to multiple points of use
- A control system

In a 300- or 450-mm fab there are typically:

- 20 to 35 high-purity chemicals, 2 to 10 of which require blending or dilution
- 3 to 12 CMP slurries, most of which require blending or dilution
- 100 to 450 valve boxes
- 500 to 2000 points of use

35.2 IMPORTANT TERMS

In this chapter, the following terminology is used:

Point of use (POU)—a process tool or a portion of a process tool to which chemical or slurry is supplied

Equipment—typically one physical cabinet with a specific task such as blending, pumping, filtering, or a combination of tasks

Equipment controller—an electrical subsystem that manages the operational and safety-related tasks of a given piece of equipment

Distribution system—the network of tubing, valves, valve boxes, controllers, and wiring, typically located in the subfab, related to a chemical or slurry system

Distribution controller—an electrical computation device that manages the operational and safety-related tasks required to transport chemical from the equipment to the points of use; also known as a SCADA (supervisory control and data acquisition) system or FCS (factory control system)

Engine—that portion of the distribution system, often part of the equipment, which moves the fluid, for example, diaphragm pump, centrifugal pump, bellows pump, or pressure vessel

Daytank—a permanent on-site storage vessel that provides a buffer volume of chemical or slurry

35.3 HISTORY OF CHEMICAL AND SLURRY HANDLING SYSTEMS

While liquid chemicals have been part of the semiconductor manufacturing process from the beginning, automated chemical handling systems were developed during the late 1980s and early 1990s to lower costs and improve the safety, purity, and repeatability of wet processing.

35.3.1 Cost

The transition from manually handled small chemical packages to automated bulk systems provides direct cost benefits by reducing material and labor costs and by realizing the cost-of-ownership benefits in purchasing chemicals in bulk containers. Indirect cost benefits are derived from the increased safety and reproducibility benefits of automated chemical delivery. While these cost benefits are offset by the need to acquire and install delivery systems, automated chemical and slurry systems enable improved process results while saving cost and improving safety. All designs, capabilities, and attributes of chemical and slurry handling systems flow from this value proposition.

35.3.2 Safety

Initial mass-production wet processes relied on pouring chemicals by hand into baths in which wafers were immersed to perform a specific process step. This approach required significant manual handling effort and inherently exposed technicians and operators more directly to the aggressive and potentially dangerous chemicals. In addition to the exposure risk, the manual process was error-prone, creating the potential for adding chemicals in the wrong order or in inadvertent combinations. Increased consumption of chemicals also created strains on the manual pour process.

35.3.3 Purity

As geometries decreased, it became apparent that particle contamination, in particular, was a problem with direct-pour, manual processes. No matter how clean the source container, static materials are never as pure as materials that have been recently filtered. Automation offers the possibility of multiple filtration passes, but also introduces moving parts as sources of ionic and particle contamination. Designing for purity is still fundamental to the design of chemical management components, equipment, and systems.

35.3.4 Repeatability

As in all manufacturing processes, reproducible semiconductor manufacturing relies on the control and repeatability of key parameters. Controlling the process conditions permits the process engineer to focus on the inherent capabilities of each manufacturing step. Two such variables, purity and blend repeatability, are more repeatable when delivered through automated means.

35.4 CHEMICAL AND SLURRY HANDLING EQUIPMENT

35.4.1 High-Purity Chemical Systems

High-purity systems are used to move clear chemistries from the source container to the POUs. Clear chemistries include acids, bases, oxidizers, and solvents and exclude CMP slurries which contain suspended solids.

High-Purity Chemical System Conceptual Overview

High-purity chemical dispense systems feature a wide variety of designs. Customers and suppliers each make a unique series of value judgments and corresponding design choices that optimize various attributes of the chemical system. The key design drivers are uptime, purity, and cost. High uptime and low cost are competing goals that drive the high number of equipment variations from one device manufacturer to another device manufacturer. While the number of designs is large, each high-purity chemical system has the same core function: move chemical from point A to point B.

The most basic system will consist of a source container, typically a drum or tote, and a pump. The pump draws chemical from the drum and pumps it to the POUs. Given the importance of purity, almost all chemical dispense systems include filtration after the pump.

Two additional features are often added to a chemical system design to increase uptime: a daytank and redundancy. The daytank provides buffer volume when the source container is empty. A unit will typically transfer from the source container to the daytank and then dispense from the daytank to the POUs. Redundancy can be included on a component level, such as utilizing two pumps, or on a system level, such as providing two independent dispense engines.

Figure 35.1 provides a simple representation of a chemical dispense system that includes a daytank, filtration, and possibly engine redundancy if interconnects are included allowing both pumps to act as both the transfer and dispense pump.

Dispense Technologies

There are multiple methods available to dispense high-purity chemicals, each with strengths and weaknesses. Table 35.1 summarizes the most common techniques.

Engine Design and Redundancy Implementation

The primary mechanical chemical distribution module (CDM) design challenge is to manage the tradeoffs of cost and uptime. Uptime is related

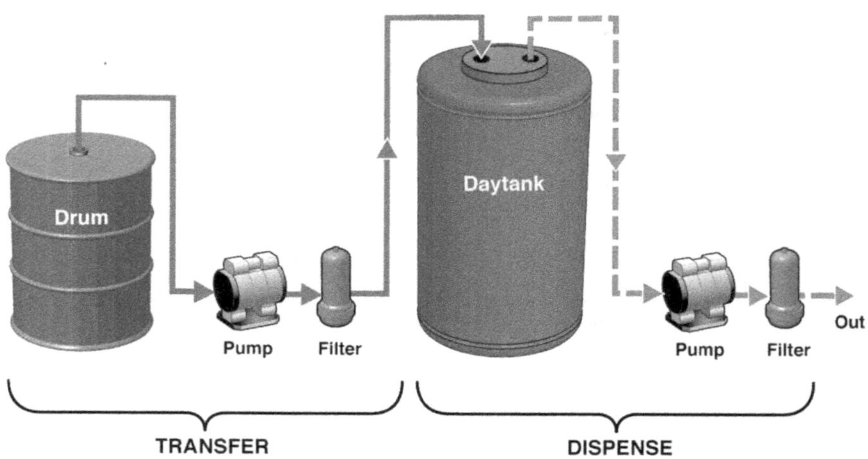

Figure 35.1 Representation of a high-purity chemical dispense system.

Table 35.1 Dispense Technology Characteristics

Dispense technology	Strengths	Weaknesses
Positive displacement pumps with pulsation dampener. Diaphragm pumps are the most common, but bellows pumps may also be used. Pumps are driven by compressed air.	Positive displacement pumps are reliable, self-priming and provided by many suppliers. They do not require significant footprint.	Moving parts under high pressure require maintenance, typically in the form of rebuilds, and contribute impurities to the fluid stream. In addition, pulse dampeners are required to smooth the pressure profile of the fluid.
Centrifugal pumps can be used to dispense chemical without the need for pulsation dampening.	Centrifugal pumps are capable of extremely high flow rates and are well-suited for real-time adjustments to control flow or pressure.	Centrifugal pumps are not self-priming, so additional system components are required. In addition, they are high-voltage components and have coated-magnetic impellers, which can cause catastrophic contamination events if breached.
Pressure dispense technology relies on pressurized vessels either filled by pumps or by vacuum. High-purity nitrogen is used to dispense the chemical from the pressure vessel(s). Pressure vessels can range from 10 to 1000s of liters.	Pressure dispense systems offer steady dispense pressure and large vessels can sustain high instantaneous flow rates.	Pressure vessels are expensive and require more space than pumps. There are limits associated with the fill-side of this technology. Vacuum fill systems have limited flow rates, and pump fill systems are exposed to the same weaknesses as pump systems. Some chemicals are susceptible to nitrogen absorption then dissolution at the POU.

to component reliability, and component unreliability is "managed" via redundancy. Subsystem redundancy is also used to reduce scheduled and unscheduled downtime. If the goal is to reduce unscheduled downtime, then redundancy must permit the unit to operate in the event of a failure long enough to permit replacement of the failed component when downtime can be scheduled. If downtime cannot be scheduled, the redundant design must permit replacement of the failed components while the system is operational. The difference relies on the quantities of various components and the use of valving for isolation and physical compartmentalization. Note the design differences shown in Fig. 35.2.

With the proper procedures, training, personal protective equipment (PPE), and isolation, design B (with isolation valves) could permit

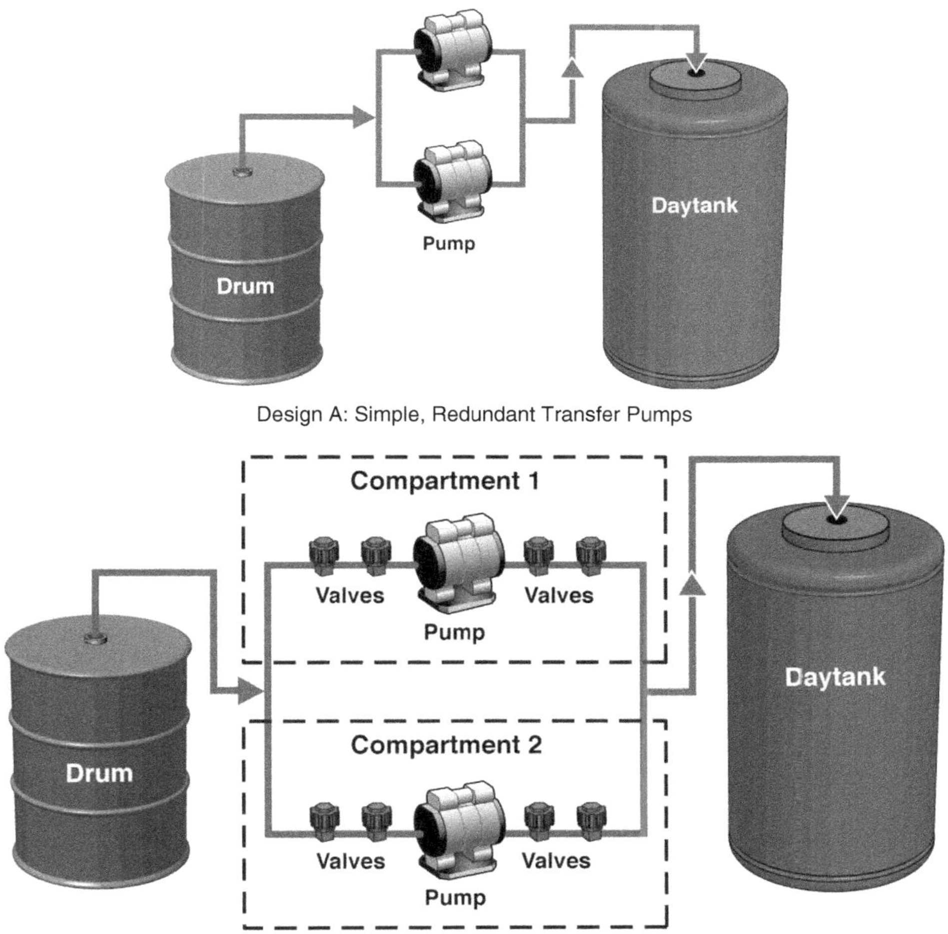

Figure 35.2 Redundant transfer pump designs.

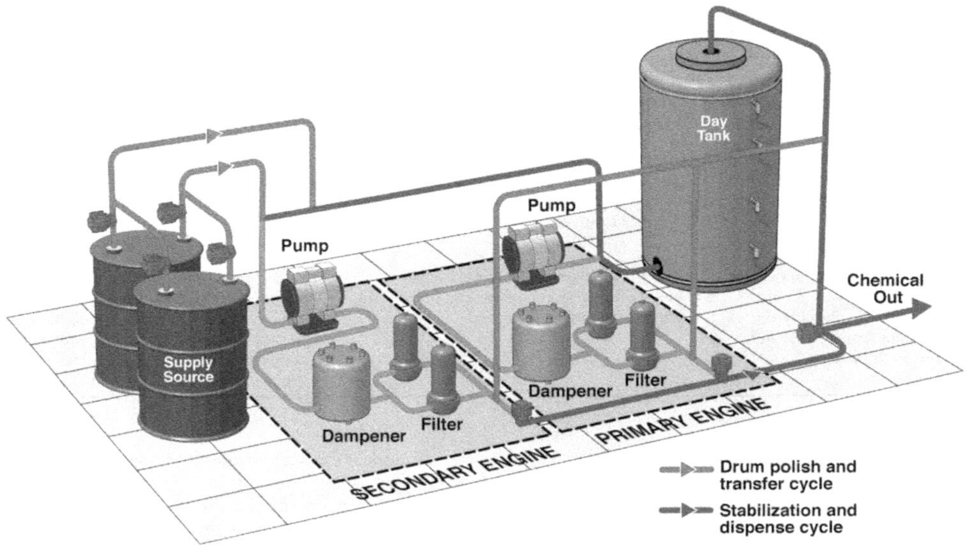

Figure 35.3 Chemical delivery system flow path that offers optimized redundancy. (*Used with permission of Air Liquide Electronics U.S. LP.*)

replacement of one pump while the other pump is operational. The design without isolation valves, or dual isolation valves, requires downtime for maintenance.

Referring to the transfer and dispense functions of a CDM and adding filtration to maintenance and uptime redundancy considerations, one can see that the desire for uptime to prevent scheduled or unscheduled downtime can quickly escalate the cost of a system. To prevent the need for four pumps in each system, one design choice known as optimized redundancy creates two engines, each consisting of a pumping method and a filter bank as shown in Fig. 35.3. Valving connects two flow paths for each pump. One path is from the supply drum to the daytank (the transfer function). The other path is from the daytank to the fab (the dispense function). This permits each engine to perform the transfer and dispense tasks. This design assumes that with a properly sized daytank, the system can operate on one engine for a prolonged period of time, permitting maintenance to the other.

Table 35.2 summarizes the range of redundancy solutions used in the industry today. This table assumes that the dispense methodology is a pump, but the same design choices can be applied to other dispense technologies. Choosing the optimal redundancy model is a highly subjective decision based on the particular uptime requirements of each fab. The desire to avoid even the slightest risk of chemical unavailability typically pushes most fabs to Level two, three, or four, but increased capability increases capital costs.

Table 35.2 Summary of Redundancy Solutions

	Design	Protection	Cost factor
Level zero	One pump	None	0.80×
Level one	Two pumps, one flow path, no isolation	One pump fails, the other is automatically brought online. Online maintenance is not typically possible.	1×
Level two	Optimized redundancy—two pumps, two flow paths, isolation	Same as one plus online maintenance	1.15×
Level three	Separate transfer and dispense flow paths, two pumps each	Redundant transfer pumps and redundant dispense pumps	1.80×
Level four	Two units, same capabilities	Two complete units	2.10×

System Purity

As opposed to CMP slurries, which contain abrasive particles suspended in a carrier fluid, high-purity chemicals are manufactured to minimize all forms of impurities, notably particles and ionic contamination.

Particle concentrations are measured by quantity at various sizes in a fixed volume of chemical, for example, x particles/milliliter greater than 0.1 μm. Filtration is the means for controlling particle concentration, and most systems reduce the particle level found in the source container; that is, a properly designed delivery system does not add particles to the chemical.

Ionic contamination is measured in parts per billion or parts per trillion by sophisticated instruments. Unwanted ions on the wafer can change the electrical properties of the device and result in a loss of die yield. Unlike ions in gas streams, ions in liquid chemical streams cannot typically be removed online by ion exchange methods, with a few exceptions. This fact shifts the ion control strategy from removal to prevention. Every component, tube, pump, and tank must be made to prevent ionic contamination from the start. Although Teflon® materials such as PFA and PTFE are inherently pure, they do not provide the robust mechanical properties of steel or aluminum, and thus create a design challenge between purity and physical characteristics of the machine such as uptime, cycle rating, and capacity.

The design challenge for a high-purity chemical system is to remove particles that exist in the source containers, while not creating sources of particle contribution, and to prevent the contribution of ionic contamination using clean but mechanically weak materials to design a machine with nearly 100 percent uptime and little or no required maintenance.

Filtration

Very little changed in the practice of chemical filtration from the mid-1990s until approximately 2012. Filters with a pore size of 0.05 μm were installed in almost all dispense systems. Despite rapidly shrinking critical dimensions and the availability of filters with tighter pores, industry standard practices remained unchanged.

Starting around 2012, end users began adopting several new practices. First, for critical chemicals, there was a shift toward smaller pore sizes. Filter pore size impacts dispense system design in that reducing pore sizes increases the pressure drop across a filter bank. In order to maintain the same levels of flow and pressure, dispense systems need to accommodate larger filter banks.

A second shift was the adoption of staged filtration. A staged filtration system utilizes filter banks with decreasing pore sizes in series. The premise is that the first filter bank will have more open pore sizes. Large particles that could rapidly clog a small-pore filter are removed in this stage. The second stage filters have a small pore size. Staged filtration may occur in two process steps, such as during transfer and then dispense. Alternatively, the dispense system may have two filter banks in series.

Finally, some end users began to explore the possibility of increasing the number of filter passes before chemical is delivered to the process tools. Most chemical dispense systems have recirculation circuits that can continuously filter the chemical. However, the number of times chemical passes through the filter bank depends on the tank size and rate of consumption. It is possible to institute a practice called *polishing*, which demands that chemical be filtered for a set amount of time or number of filter passes before it becomes available to the process tools. This can be implemented in a variety of ways, including drum polish, which requires that the drum be circulated through a filter bank for a set amount of time prior to being transferred to the daytank, or through the use of intermediate holding tanks designed to polish the chemical.

Regardless of the filter pore size or filtration design, a key to optimizing filter performance is stabilized filtration.[1] A stabilized system maintains a small flow rate across the filter bank even during periods when no chemical is being consumed. This stabilization flow rate prevents filter shock and the subsequent release of particles when the system receives a demand signal and ramps to a much higher flow rate.

Single-Wafer Process Tools

In the early days of automated chemical delivery, the typical process tool had a bath into which full lots of wafers were submerged. The bath would be changed after a set number of hours or wafer lots depending on the application. As such, chemical dispense systems would only pump chemical to the POUs during a bath change, and the pressure and flow rate at the POU was not critical.

Wet processing has gradually shifted from bath-based tools to single-wafer process tools, where a small volume of chemical is sprayed directly on the wafer. This trend has impacted chemical dispense system requirements in two ways: capacity and pressure control. Bath tools were typically staged such that not all baths would dump at the same time. As such, a chemical dispense system did not need to be sized to provide chemical to every POU at the same time. With single-wafer tools, the per-tool consumption rate is lower, but all tools may operate simultaneously. Dispense systems must be sized to meet higher demand.

As chemical is flowing directly onto the wafer, chemical dispense systems may be required to control the pressure in the distribution piping to provide repeatable performance at the POU. The two most common ways to control loop pressure are to control the back-pressure by installing a pressure control valve in the return line, or to continuously adjust the forward-pressure, generally through the use of a centrifugal pump with a variable speed control.

Peripheral Equipment for High-Purity Systems

In addition to the CDM, the typical chemical dispense system may also include one or more of the following peripheral subsystems and equipment:

- Drum, tote, and tank cabinets—enclosures that provide secondary containment for the supply container. Secondary containment is required if chemical from the container is circulated under pressure. The cabinet can also provide containment in the event of spill
- Bulk chemical management systems—tanker trucks (or lorries) fitted with special chemical tanks serve as a source of supply for high-consumption chemicals. More popular in Asia and Europe than in the United States, bulk management systems require a coupling system, storage tank, and transfer station. The coupling system is a CDM-like unit that houses specialized fittings that connect with a mated fitting on the truck. The coupling system cleans the fitting, permits sampling, and then pressurizes the tanker to transfer the chemical to the on-site storage tank
- Valve boxes—enclosed manifolds that permit supply from the main distribution line to individual POUs
- Blending equipment—specialized systems that dilute concentrated chemical or blend various components to meet highly controlled specifications; blenders serve as the chemical source for a dispense system

35.4.2 Chemical Blending Systems

Many chemicals used in the semiconductor industry are "ready to use," for example, the concentration of the chemical in the drum is suitable for wafer processing. Other chemicals must be diluted with ultrapure water or blended with additional chemicals to achieve the desired process result.

Reasons for Blending

A device manufacturer may choose to blend chemical on-site rather than purchase preblended supply containers. By far, the most common reason to do this is to reduce the cost of ownership of a process or chemical stream. Most chemical blends are, in fact, dilutions of a concentrated chemical with ultrapure water. Fabs may also pursue on-site blending if they require a proprietary mixture that is not commercially available, or if they wish to create a blend that has a short "pot-life," or becomes unstable over time.

Table 35.3 provides a summary of the most common on-site high-purity blending applications in the semiconductor industry. Prerequisites for on-site blending include the availability of concentrated chemical, the availability of suitable blending equipment, and the agreement of the fab process owners that the risks of blending on-site (vs premixed material from a chemical vendor) are acceptable for the process.

Some mixtures that have traditionally been blended on-site are moving from bulk blending at the facility level to POU blending in the process tools. This trend follows the transition from batch tools where one or more lots of wafers are exposed to chemical in a bath to single-wafer-processing where each wafer is exposed to fresh chemical, which then flows to the drain.

Cost of Ownership

On the surface, on-site blending appears to offer significant cost-saving opportunities compared to the purchase of pre-blended drums

Table 35.3 Common On-Site Blending Applications

Chemical	Typical target concentration	Application	Comments
Hydrofluoric Acid (HF)	Typically 10:1, 100:1, or 500:1 dilution of 49% HF	Front-end etch processes	HF is by far the most common chemical blended on-site
Tetramethyl ammonium hydroxide (TMAH)	2.38% (can be written as 0.263N)	Developer chemistry	Extremely tight blend specifications are required for on-site developer blending
Various post-CMP cleaning chemistries	Common examples include 2% ammonium hydroxide (NH_4OH) and <5% citric acid	Post-CMP cleaning chemistry	Citric acid blends tend to be multi-constituent blends rather than just DI water dilutions

of chemical. However, the additional costs that must be taken into consideration are the utility costs to run the blending equipment (DI water, CDA, Nitrogen, electricity, exhaust), the maintenance costs associated with adding another piece of equipment to the fab, and the maintenance costs specific to a unique piece of equipment (different spares, additional training, frequent laboratory testing of the blended chemical assay, etc.). In most cases, these costs are not prohibitive, but it demands an informed decision to justify them.

Critical Blending Parameters: Accuracy and Capacity

In addition to capacity, when considering an on-site blending process, the two most critical parameters are the accuracy and repeatability of the process.

Accuracy reflects how close the blend system's output is to the target concentration. Repeatability measures the blend system's ability to produce a blend of the same concentration over and over again. In most blending systems, the same concentration is used for a long time; thus, as a practical matter, most of the attention to the design of such blend systems is to ensure repeatability.

The accuracy and repeatability in Table 35.4 are listed in terms of relative error rather than absolute error. Absolute error has some unit of measurement, usually weight percent (e.g., 0.49 wt% HF ±0.005 wt%). Relative error is the absolute error as a percentage of the target value. It has no unit value. The equation for calculating relative error is shown below:

$$\text{Relative error} = \left(\frac{\text{absolute error}}{\text{target concentration}}\right) \times 100$$

As blended chemical generally flows into an online daytank, blending is a noncontinuous process. The capacity of a blend system is designed to exceed the estimated daily consumption of the facility, typically targeting 150 percent of the intended consumption or a duty cycle of 50 percent. Unlike a chemical dispense system, the buffer volume in the daytank and the high capacity of the blend system results in a significant amount of idle time. As a result, blend systems are typically not designed with redundancy. The buffer volume in the tank provides time for scheduled and unscheduled maintenance, and the high capacity allows the blend system to quickly refill the tank when it returns to service. In this way, the expense of redundancy can be avoided.

Batch versus In-Line Blending

Blending can be broken down into two general categories: batch and continuous processes. A batch process is one in which a discrete batch of some volume of blended chemical is produced. A batch process has a clear beginning and ending point. At the end, the batch is typically transferred to an external daytank, where it is held until it is needed by the fab.

In-line blending produces chemical blends in a continuous stream. Blended chemical may flow into a daytank or may be supplied directly to the wafer. In-line processes may or may not have large daytanks to provide a buffer volume of chemical.

Both batch and in-line blending have advantages and disadvantages. Batch blending can use a variety of blend technologies, or even combine multiple technologies in one system. As there are discrete batches, testing can be carried out on each batch to validate the concentration of the batch. However, batch blenders require a blend tank increasing the footprint of the system, and they can be capacity-limited as each batch requires time to both homogenize and transfer the chemical.

In-line blenders use flow-based technology. They can be very simple and have a small footprint for applications that do not require a high level of blend accuracy. In-line blenders can also be designed for very high capacity because no intermediate homogenization and transfer steps are required. In-line blenders get complicated when the application requires high accuracy. The equipment supplier must integrate metrology and develop techniques to accurately measure the concentration and adjust the blend in real time.

Table 35.4 Typical Accuracy and Repeatability Requirements for Common On-Site Blending Applications

Chemical/application	HF blending	TMAH developer blending	Post-CMP cleaning applications
Accuracy/repeatability	1%	0.1%	3–5%

Blend Technologies

While all blend systems conform to either a batch or in-line methodology, different blend technologies can be used to produce the blend. Table 35.5 provides a summary of the most common blend technologies.

Equipment suppliers must be capable of balancing the strengths and weaknesses of each technique against the requirements of the processes. Some equipment suppliers choose to focus on one technology and adapt it to the process requirements. Others use different techniques depending on the applications.

Blend Metrology

Metrology is the method by which blend accuracy is increased. Without metrology, the blend technologies listed above are generally limited to 2 to 5 percent relative error. While that may be acceptable for a very limited number of cleaning applications, most semiconductor processes require less than or equal to 1 percent error. The use of metrology to adjust or correct blend concentration is called feedback blending. Even if metrology is not required to improve the accuracy of a blend system, most end users require batch validation by an appropriate metrology device.

Feedback blending utilizes the concentration data from a metrology to adjust a batch to achieve the target concentration. One challenge associated with metrology selection is to identify a device capable of accurately measuring the concentration of a given blend. Blend specifications are approaching or exceeding the capabilities of many commonly used metrology devices, which means the variability in the concentration reported by the device is too large to control the process within the specification. Another challenge is to select devices that are easily and effectively integrated into the blend system. This includes using real-time measurements whenever possible, reducing the maintenance required for the metrology device, and selecting devices that are not susceptible to ambient changes.

Table 35.5 Strengths and Weaknesses of Various Blend Technologies

Blend technology	Strengths	Weaknesses
Volumetric Precise volumes measured in one or more vessels	Simple to understand and operate Robust design (few failure modes)	Limited accuracy and repeatability Limited blend ratio range Difficult to change blend recipes
Flow Flow measuring devices totalize flow or precisely control flow to control volumes of constituents	Simple to understand Can provide a high blend capacity Can be used as an in-line blend technology	Pressurized chemical source required Flow meters can be prone to drift and require frequent calibration
Weight Precise load cells measure weights of constituents in a batch process	Simple to understand Recipe changes are quick and easy Scalable for small or large capacity requirements	Load cells can be prone to drift Weight-based blending is always a batch process

The relationships among the chemicals that will be blended and the accuracy/repeatability requirements of the process impact the type of metrology that can be used. The equipment vendor must perform qualification tests, and end users must be comfortable with the type of metrology proposed. Typical metrology selections range from conductivity or pH probes to refractive index, titrators, spectrometers, and sonic velocity measurements.

35.4.3 Slurry Systems

In the mid to late 1990s the drive for increased metal interconnect layers and the commensurate desire for increased planarity of various deposited layers drove the wide development of chemical-mechanical planarization (CMP). As the name implies, semiconductor planarization relies on a combination of chemical and mechanical processes. CMP slurries typically consists of silica, alumina, or ceria particles suspended in a chemical mixture. Bulk blending and distribution methods are required to supply the process tools with slurry.

Slurry systems include many of the basic design features and capabilities found in high-purity chemical systems; however, the presence of abrasive particles introduces significant challenges caused by settling and other phenomena related to the presence of suspended particles in a complex chemical solution.

In addition to particle challenges, most slurry applications require blending, which led to the design of integrated blend and dispense systems. In the early days of CMP, almost all end users were using the same slurries. For equipment suppliers, it was relatively simple to configure the equipment as the small number of possible slurries was well understood. As the CMP process has matured, the variety of slurries being used and the relative complexity of blending them has increased. Many end users have customized products that need to be characterized in the field because limited information is available prior to starting up a new slurry dispense system.

Slurry Blend and Dispense Equipment Design Features

Slurry equipment designs are subject to the same challenges associated with high-purity chemical blend and dispense systems: balancing redundancy versus system cost, selecting components that are robust and do not introduce impurities into the fluid stream, blending accurately and monitoring with well-suited metrology devices, and meeting the evolving requirements of an industry where technology advances every year. In addition to these baseline design challenges, the presence of suspended particles in the fluids introduces additional design considerations required to maintain the health of the slurries.

Agitation In general, the solid particles in slurry tend to settle over time, requiring specific capabilities to control the homogenization of drums and tanks. The settling rate and re-dispersion requirements for each slurry are characterized by the slurry supplier or the equipment providers.

Drums can require little-to-significant stirring in order to homogenize the particles. In most cases, 10 to 20 min of initial stirring, followed by occasional stirring thereafter, is required. As another conflicting requirement, some slurries contain organics, which tend to cause foaming under excessive agitation that entrains gas. The stirring system must be configured and operated to prevent foaming at all drum levels, a particular challenge as the level of the chemical in the drum approaches the bottom of the container.

Tanks also require stirring for particle suspension as well as homogenization if the chemical requires blending. As an alternative to mechanical stirring, one can create agitation from pump circulation or the use of eductors, that is, static devices that cause turbulent flow.

Minimum Velocities If the velocity of the slurry in the distribution piping is not sufficient, the particles in the slurry may fall out of suspension (resulting in settling), and eventually clog the piping network. Minimum velocities are prescribed for each slurry with a range of one to several feet per second. The flow rate required to achieve the linear velocity is a function of the tube diameter.

Humidification As solutions of suspended particles, slurries pose challenges where there are gas-to-liquid interfaces, which occur in drums, daytanks, and pressure vessels. If the gas above the liquid level in a vessel is not moisture saturated, the liquid will tend to evaporate. When a slurry formulation evaporates, the remaining solids can create large particles. Large particles entering the dispense system can cause defects on the wafer. Humidification of the gas, typically nitrogen, satisfactorily solves the problem. Humidification systems may be high or low-pressure and rely on membrane cell, bubbler, or heating technologies.

Dead Legs The tubing in high-purity chemical systems is typically routed to promote the efficient use of space and easy maintenance access. In slurry systems, tubing layout requires careful consideration to avoid the creation of dead legs, that is, any section of tubing through which slurry does not typically flow. The most common dead legs occur in the downward legs of vertically oriented "tee" fittings. Dead legs can clog within days or weeks.

Filtration In contrast to high-purity chemical filtration, where there is a low concentration of small particles, slurries pose a distinctly different filtration challenge due to the high concentration of particles of various sizes, most of which are required by the process. Rather than seeking to remove all particles from the fluid stream, the primary objectives of slurry filtration are filter efficiency and life cycle. If a large slurry volume flows through a tight pore filter, it will clog quickly. If the pores are too large, the filter will only remove the largest particles, thus solving only part of the problem.

The most common approach to slurry filtration is to use a combination of main distribution filters and POU filters. The main loop filters are usually larger pore (3 to 20 μm), while the POU configuration is on the order of 1 to 3 μm. Slurry filtration system design must also consider practical issues such as ease of replacement, clog detection, and filter life extension.

POU Pressure Control Slurry is dispensed on the platen directly from the main slurry distribution loop that typically runs directly under the polisher. As there is no intermediate vessel between the source and use point of the slurry, the tool relies on the pressure from the slurry dispense system to control the flow rate of slurry delivered to the platen. While most tools have a pump or flow controller to help regulate flow, these devices are often sensitive to incoming pressure changes. Consider the following example. Assume "Tool One" is calling for chemical and the pressure in the main slurry dispense loop is 30 psi. If two more tools call for slurry, the pressure in the loop may drop to 28 psi as slurry is consumed. When the pressure drops, the slurry flow rate at Tool One drops, and the MRR or other key process factors may change.

The most effective way to control dispense pressure at the tool is to use a dynamically controlled pressure dispense system or a purpose-built flow controller at the tool that is not sensitive to pressure fluctuations. A dynamic pressure control system relies on a dispense engine that adjusts dispense pressure or loop back-pressure based on a signal from a mid-loop pressure sensor. As the loop pressure drops due to higher demand, the system compensates by increasing the dispense pressure or the back-pressure. As polishers have migrated from on-tool pumps to flow controllers, slurry dispense systems have needed to operate a higher pressures to meet the requirements of the on-tool flow controllers.

35.5 SYSTEM PURITY

System purity is critical as the levels of particles and ionic contamination in chemicals will have a direct effect on the process yields in the fab. Once ionic contamination is in the fluid stream, it cannot easily be removed. The only viable technique is preventing ionic contamination from entering the fluid stream. Process tools may or may not have filtration at the POU, so it is critical that the chemical dispense systems use the most effective particle removal technologies available.

35.5.1 Ionic Contamination

Not every element in the periodic table is of concern when measuring for ionic contamination. Only those likely to be present after the chemical manufacturing process or those that may be contributed by fab contamination events are of interest. Table 35.6 lists the 34 ions that are generally measured, their classification, and the likely sources of the ions.

Understanding contamination sources is critical to being able to troubleshoot contamination events. Due to the relatively high integrity of a modern chemical system, a failure event usually has one root cause.

Generally, ionic contamination levels are tested by subjecting a chemical sample to Inductively Coupled Plasma Mass Spectroscopy (ICP-MS). In ICP-MS, a plasma source converts the atomic contamination in the sample to ion form. A quadrupole filters the ions based on mass-to-charge ratio, and then a detector, coupled with software, identifies the type and concentration of elements present in the sample.

35.5.2 Particle Contamination

Particle levels are tested by providing a slipstream of chemical to an optical liquid particle counter after the chemical has passed through a filter. Most particle counters work by shining an illumination source of some wavelength through the fluid stream. The light is scattered, and the scattering is measured by a detector. Software calculates the number and size of particles in the fluid, based on the volume of chemical that is being sampled by the counter, and the scattering pattern.

Not all chemicals exhibit the same particle performance. Sulfuric acid and hydrofluoric acid may be dispensed by identical CDM's with the same filters, but monitoring will show more particles per milliliter in the H_2SO_4 than in the HF. Factors that impact particle performance include chemical viscosity, cleanliness of the manufacturing process, the presence of surfactants, the tendency to form microbubbles, and the index of refraction of the fluid relative of the index of refraction of the particle can impact the accuracy of the particle analyzer.

One of the most critical particle issues in the industry is the ability to measure small particles. Commercially available chemical particle counters in 2014 cannot measure particles smaller than 40 nm in size. According to the ITRS Roadmap,[2] the critical particle size dropped below 40 to 38 nm in 2005. Some equations can be used to extrapolate the number of particles at a small size based on the number counted at a larger size. When plotting particle cumulative concentration on the *y*-axis and particle size on the *x*-axis, the negative slope of the line is fairly predictable due to the normal particle distribution.

Calculated particle levels are not sufficient in the long term for several reasons. First, calculated particle levels are more accurate when the measured and predicted particle sizes are similar, meaning that measurable particle levels cannot be used to estimate the quantity of particles that are significantly smaller than the measured size. Second, as the critical particle size shrinks, so will the number of particles of measurable size (as predicted by the relationship between particle size and particle concentration). As the number of particles in the measurable channels approaches zero, the particle counting statistics suffer. As indicated by the 2013 update of the ITRS roadmap, new particle-measuring technologies are needed.

35.5.3 Sources of Contamination

All components in a chemical delivery system will contribute some level of ionic and particle impurity to the fluid stream. The way in which components contribute varies based on the type and function of a component.

The ionic levels a component contributes to the chemical is based on the cleanliness of the raw materials, the manufacturing process, and the surface area of the component that is exposed to chemical. The most common wetted materials in a chemical delivery system are PFA Teflon, PTFE Teflon, and electropolished 316L stainless steel. The primary concerns with Teflon components are the purity of the resins used to produce components, and the manufacturing processes. Typically, Teflon parts are used in the molds and machinery used to manufacture components for the semiconductor industry to reduce the possibility of cross-contamination. Most components are manufactured in a cleanroom environment, and the components are subjected to rigorous cleaning procedures before they are shipped. The primary issue with stainless steel components is to ensure that they are chemically compatible with the fluids they will be used to deliver. Some chemicals will corrode stainless steel or extract iron, which will result in increased ionic contamination levels. The surface area of a component is also a factor. A pump, for instance, has a relatively small surface area exposed to chemical as compared to a large-area component such as a daytank. While it is critical that all components be as clean as possible, cleanliness is especially important for components that contribute significant surface area to the system, such as tanks and tubing.

Table 35.6 Ion Contaminants Monitored by the Semiconductor Industry

Ions	Process concern	Likely contamination source
Gold (Ag), lithium (Li), potassium (K), sodium (Na)	Mobile ions	Ag: Rarely occurs Li: Catalyst in PFA manufacturing process K, Na: Environmental contamination (esp. people)
Nickel (Ni), copper (Cu), cobalt (Co)	Form silicides	Ni: Component of stainless steel Cu: Process cross-contamination Co: Cobalt electroless process; used as a dye
Aluminum (Al), barium (Ba), cadmium (Cd), calcium (Ca), chromium (Cr), iron (Fe), lead (Pb), manganese (Mn), magnesium (Mg), molybdenum (Mo), tin (Sn), titanium (Ti), vanadium (V), zinc (Zn)	Gate oxide killers	Al: Can be present in ammonium hydroxide Ba: Can occur in polymers exposed to chemical Cd: Occurs in alloys Ca, Mg: Environmental contamination (esp. people) Cr: Leaches out of stainless steel components Fe: Leaches out of stainless steel and polymers Pb: Can be leached from gaskets; PVC stabilizer Mn: Potassium permanganate added in distillation process Mo: Occurs in alloys Sn: Hydrogen peroxide stabilizer; common additive Ti: Occurs in fillers and alloys; polymer catalyst V: Maybe polymer catalyst (rarely occurs) Zn: Leaches out of polymers (drums, welds, etc.)
Antimony (Sb), arsenic (As), boron (B)	Dopants	Sb: Raw material for phosphoric acid manufacturing As: HF manufacturing B: DI water
Beryllium (Be), bismuth (Bi), gallium (Ga), germanium (Ge), niobium (Nb), silver (Ag), strontium (Sr), tantalum (Ta), thallium (Tl), zirconium (Zr)	Other Trace Metals	Rarely occur

Particle purity is also a function of the cleanliness of a component's raw material, but unlike ionic contamination, particle levels are affected by the behavior of the component. Components such as tanks, tubing, and filter housings are static. They have no moving parts, and generally are very clean from a particle shedding standpoint. Pumps and valves have moving parts, and the movement of diaphragms results in higher levels of particle shedding than with stationary components. When equipment suppliers evaluate components, they should study their shedding characteristics, especially for large surface area components and components with moving parts.

35.6 CONCLUSION

The key to successful chemical and CMP slurry management is to understand the behavior of the materials and design around them to provide perfect chemical or slurry to the tool 24 hours a day, 365 days a year. With safety and process protection as driving design parameters, the growing world of wet processing will continue to rely on critical process systems that handle, purify, blend, and control high-purity chemicals and CMP slurries.

35.7 ACKNOWLEDGMENTS

Thanks to Curt Anderson, Matt Fisher, Dan Fuchs, Kevin O'Dougherty, and Bryan Smith of Air Liquide Electronics U.S. LP as well as Dan Barsness for contributing to the content of this chapter. The illustrations were created by Bruce Tice of BTice.ink.

35.8 REFERENCES

1. D. C. Grant and W. R. Schmidt, "Particle Performance of a Central Chemical Delivery System," *Presented at the 7th Annual Microelectronics Symposium*, Bedford, MA, 1989.

2. International Technology Roadmap for Semiconductors, 2013 edition.

PART 6

OPERATIONS, EQUIPMENT, AND FACILITIES

Chapter 36
Yield Management

BY
DIETER RATHEI *DR YIELD software & solutions GmbH, Graz, Austria*

36.1 INTRODUCTION

36.1.1 What Is Yield Management?

Reaching a sufficiently high yield is a principal business objective for any semiconductor manufacturer. As yield may be influenced by each processing step as well as the interactions between these steps, it is necessary for the processing units like lithography, etching, and deposition, which are engaged in achieving high yields within their area of influence, to work in conjunction with a designated yield department. The yield department typically coordinates activities within the entire organization in order to achieve better yield results.

In this chapter, we discuss the fact that yield may be affected by both defects and process variability. Depending on the products being manufactured and the technologies being used, the focus of yield management may be different. In mature technologies or in factories shipping pure digital circuits, yield management will be focused mostly on defect monitoring and defect reduction. On the other hand, factories that run advanced process technologies or are specialized in analog products will place greater emphasis on process variability. The term "yield management" therefore also has different meanings in different organizations. In some companies (and in some older textbooks) it may be used synonymously with "defect management." We will discuss it here in a broader scope, including parametric variability.

36.1.2 The Role of the Yield Department

Regardless of whether defects or variability are responsible for yield issues, the yield department has to communicate with all the processing units to find and resolve the causes. The engineers in the yield department will analyze the data they have collected in their database to deduce the root cause of the problem. This may require additional data or knowledge from the process engineers. As the yield of the tested circuits may be affected by interactions between processing and chip design, as well as by the testing of the devices, the yield department may have to interact with both the chip designers and the test engineers. This may cause additional challenges, as in the case of the foundry business model, where the yield engineers are required to communicate with entities outside of their own company. Open communication may hereby be restricted by confidentiality issues and business concerns.

In some organizations, the yield department may also be the owner of the defect inspection tools in the manufacturing line, and therefore be responsible for performing the defect inspection steps. In other companies this is done by a separate defect group, and the yield department is only responsible for data analysis, communication, and coordination. Similarly, the failure analysis lab, which performs physical and chemical preparations of failed chips, may be part of the yield department or may be a separate group that has strong collaboration with the yield department.

36.2 FUNDAMENTALS OF YIELD MANAGEMENT

36.2.1 Definitions of Yield, Tests Performed

Yield in its broadest sense is always the ratio of good product that is delivered from a certain manufacturing step over the total amount of material that has been started, usually expressed as a percentage.

The "fab yield" for example, is defined as

$$Y_{\text{fab}} = \frac{\text{number of wafers out}}{\text{number of wafers started}} \times 100\%$$

Since wafers may break or may be scrapped due to obvious misprocessing, the fab yield may be less than 100 percent. However, the "fab yield" is rarely a topic of discussion in the yield department. The reason being that the root causes for broken wafers, such as inaccurately adjusted wafer handlers are typically evident and most undertakings to avoid these include only common sense actions. (An exception might be wafer breakage caused by process induced material stress.)

At the end of the wafer fabrication process, there is typically a parametric test (Fig. 36.1), where a few control structures (process control monitors or PCM) that are printed in the spaces between the circuits (the kerf) are tested [1]. This test is called the PCM test, kerf test, parametric test, or wafer acceptance test (WAT). It typically involves the measurement of transistor parameters like threshold voltages and leakage currents, as well as the resistance of chains of via contacts and the like. Some companies include wafers that fail this test in their definition of fab yield, and for foundries this test is the deciding factor as to whether a wafer is sold to the customer. In any case, the results from the parametric test are also an important input for the analysis of yield issues that become apparent at the next step, the functional test.

When people use the term "yield" in the context of semiconductors, they almost always refer to the functional yield that is the result of the functional wafer test, or sort test:

$$Y = \frac{\text{number of pass chips}}{\text{number of chips on the wafer}} \times 100\%$$

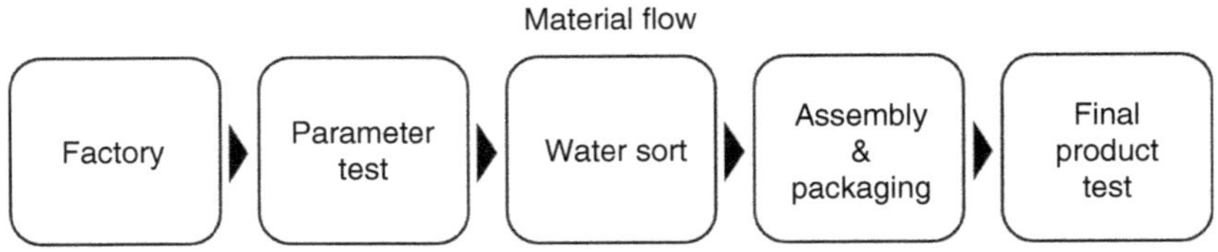

Figure 36.1 Material flow from fab to test.

At the functional wafer test, each chip on the wafer is contacted electrically by a prober and a test program is executed. This test program may perform between a few and several thousands of tests, which of course depend on the product being tested: The test program for a digital chip may check the logic of several distinctive functional blocks on the chip, with simple pass or fail results for each of these blocks, while testing memories may result in bit maps of every single memory cell. Most test programs will also include tests that measure electrical parameters like voltages, currents, and frequencies. The output of these tests are floating point numbers that have to be inside specified limits in order to judge the chip as a pass.

It is therefore important to note that the yield is affected by both the test program and the tester and prober setup. Issues like bad probe needle contacts can have a significant influence on the tested yield, and retesting of wafers is therefore not an unusual occurrence, especially for analog products.

The primary purpose of the wafer sort test is to eliminate defunct chips so that they are not packaged and assembled, as this is a rather expensive process. In the early times of chip manufacturing, failing chips were marked with a spot of ink. Today this information is delivered to the pick-and-place robot via a software file that is still nicknamed "ink map." After assembly, another functional test is performed, which is known as final product test or simply final test. Yields at this test are expected to be very high, as problematic chips should have been screened out during the earlier wafer sort test.

For the engineers in the yield department, however, the most important output of the sort test is the graphical representation of the test results, known as wafer maps. As each chip is allocated to a certain bin during the wafer sort test, the wafer test results can be displayed with colored maps, where each bin is represented by a distinctive color. The conventions for how the bins are named and how the chips are classified into certain bins, varies considerably between different organizations, but the vast majority use numbers as bins. In most companies pass chips are assigned the bin number 1, while higher bin numbers represent different failure mechanisms.

In addition, each tested parameter can be displayed on a parametric wafer map, where the color of each chip represents a value for the measured value (Fig. 36.2).

36.2.2 Systematic Yield, Random Yield

As the yield of a semiconductor wafer is limited by both random and systematic effects, it is common practice to split the overall yield into these two components, systematic yield, and random yield:

$$Y = Y_S \cdot Y_R$$

The random yield is considered the part that is limited by defects on the chips, and it can therefore be assumed that randomly distributed failing chips on the wafer are caused by defects. A defect is a localized, physical deviation from the desired structure of a chip that may affect the functioning of the chip. Defects can cause not only openings or shorts in the circuitry of the chip, but also extrusions (notches) or dents in the desired patterns on the chip. They can be caused by not only contamination, for example, particles from the environment surrounding the wafer or contamination from process chemicals, but also by undesired side effects of the manufacturing process itself. Finding the root cause of yield limiting defects is a core task that the yield and process engineers have to accomplish together.

A formula that expresses the random yield as a function of the electrical defect density D_0 on the wafer and the size of the chips on the wafer, is called a yield model:

$$Y_R = f(D_0, A)$$

By using such a yield model, it is possible to determine the systematic and random yield mathematically by using a method that is called windows clustering or cluster analysis [2]. In this method, we calculate how the yield would have been if the chip was twice as big for a wafer map with a given chip size. We assume that a defect that has killed a chip on our actual wafer would also have killed a chip twice the size, but there are only half as many chips on the virtual wafer, so the virtual yield decreases. This calculation can then be repeated with virtual chip sizes three times and four times the original size, and so on. We then get a set of equations with known values Y and A that can be used to calculate Y_S and D_0 and hence also Y_R:

$$Y_1 = Y_S \cdot f(D_0, A)$$
$$Y_2 = Y_S \cdot f(D_0, 2A)$$
$$Y_3 = Y_S \cdot f(D_0, 3A)$$

From these considerations we see a fundamental relationship that can also be inferred from the yield models in the next chapter: For any given technology and all conditions being equal, smaller chips will have higher yields than larger ones (Fig. 36.3).

36.2.3 Defects and Yield: Yield Models

If a defect causes an electrical failure like openings or shorts in the circuitry of the chip, we speak of an electrical defect, also called a fault. Yet not every defect causes a fault, as there may also be a chance that a

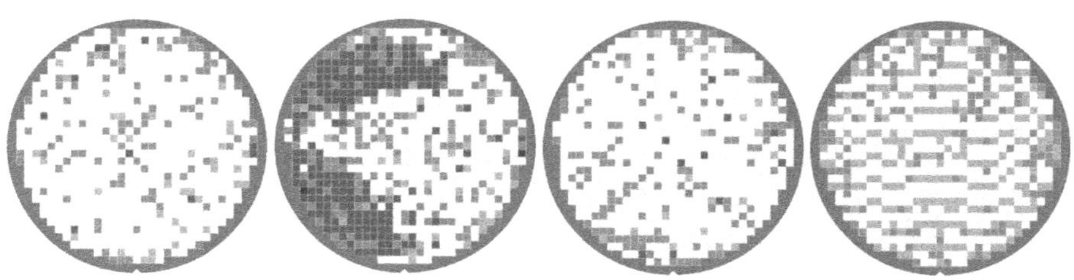

Figure 36.2 Some wafer maps [random defects, systematic issue, scratch (lower left), tester pattern]. (*Courtesy of DR YIELD.*)

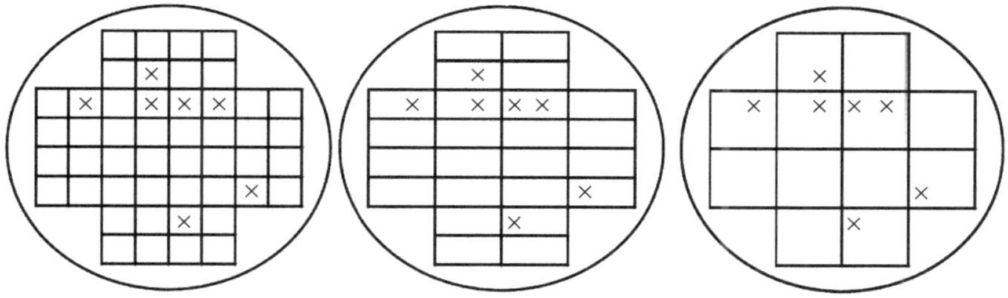

Figure 36.3 Cluster method illustrated: The same number of defects cause lower yield on larger devices. (*Courtesy of DR YIELD.*)

defect occurs in an electrically inactive area of the chip and hence has no impact.

The relationship between yield and defects is called a yield model. Most yield models express yield as a function of defect density D_0 and chip area A.

The simplest and yet most widely used yield model is the Poisson model:

$$Y_R = e^{-A \cdot D_0}$$

This can be derived as one of many special cases of the generalized yield integral from Murphy

$$Y_R = \int_0^\infty e^{-A \cdot D} f(D) dD$$

where $f(D)$ is the defect density function that describes the clustering of the defects on the wafer surface. If the defects are homogeneously spread over the wafer surface, that is, the defect density function is a singular, constant value then the integral can be simplified to the Poisson model. Many other distributions have been considered, analyzed, and used in the industry, most notable is the negative binomial distribution, which results in this model:

$$Y_R = \left(1 + \frac{A \cdot D_0}{\alpha}\right)^{-\alpha}$$

In this model α is a value that indicates how strongly the defects are clustered. For a discussion of the many yield models, the reader is referred to the literature [3–5].

36.2.4 Variability: Parametric Yield Loss

Any tested parameter has to lie within specified limits (lower spec limit and upper spec limit, LSL and USL) such that the product being manufactured fulfills its purpose. As there is always some variability in the manufacturing process, the results of the tests are generally distributed around certain target values, and one or both tails of the distribution may cut into the values outside of the specified limits. As the devices with values outside of the distribution are also failing devices, we speak of parametric yield loss in this case.

In many cases, the spatial distribution of a parameter shows a distinctive pattern, like a center-to-edge variation, or a notch-to-opposite notch gradient, hence the failing chips also appear in defined patterns, like rings or half-moon-shaped failure regions. Parametric failures are therefore typically part of the systematic yield loss. For the sake of referring to the problem before the root cause of these patterns is determined, they are often given a nickname by the yield engineers, for example, doughnut fails, croissant fails, and the like.

In the older technologies the predominant cause for yield loss was defects with parametric yield loss being a secondary issue. Parametric yield loss has, however, become a major concern for advanced technologies, as variability cannot always be scaled down at the same rate as the devices.

Causes for parametric yield loss are often marginal shifts of some processing chamber, like subtle changes in pressure or temperature during a process. Sometimes the signature of the problem on the wafer map already gives a good indication of which tool or chamber is involved (when there is a certain "foot print" of a tool, that is, a distinctive wafer map pattern caused, for example, by a unique arrangement of nozzles in a chamber), but more often finding the root cause requires a lot of investigative efforts by yield and process engineers (see Sec. 36.3.2).

36.2.5 Yield and Reliability

The probability that a randomly selected chip on a given wafer is free of detected electrical defects, is numerically equal to the yield. However, some defects do not immediately affect the functionality of the device, but because these structures are partly damaged by the defect they induce a quality risk. We can assume that a lower rate of detected defects may also indicate a lower rate of all defects and thus infer that higher yields relate to better reliability. This correlation has been confirmed in many studies [6].

36.3 METHODOLOGY: DEFECTS, DATA MINING, AND ENHANCEMENTS

36.3.1 Defect Control

Semiconductor factories use defect inspection equipment to scan wafers for defects several times optically or electron-optically during manufacturing. These dark field or bright field inspection tools give the locations of anomalies found on the wafer, but it is a nontrivial task to set up the defect inspection tools such that they actually discover the relevant defects. Sometimes an SEM review at the location reported by a defect inspection tool will not confirm that there is an actual defect at this location, and sometimes actual defects are not detected by the tool.

Hence, the sets of defects found by inspection tools, physical defects and electrically active defects are not identical. The defects that are used in the yield models are always the electrically active defects, and the value D_0 is therefore also called the electrical defect density.

Nevertheless, it is an important task to monitor and control these in-line defects, as this is an important control loop that allows one to react to a process problem. In many cases an undesired shift of a process condition, a high particle contamination, or other misprocessing of a wafer can be detected timeously by these in-line inspections and countermeasures can therefore be taken. In some cases, affected wafers can even be reworked in order to recover them.

As these inspections add to the total costs and cycle time in manufacturing it is important to design an effective sampling strategy. This will include considerations regarding at what steps in the process flow the inspections should take place, whether every lot is inspected or only one out of n lots, and exactly which wafers are inspected: a fixed sample, a random sample, or a full lot of inspection. The sampling strategy may very well define a dynamic sampling, for example, that a few wafers are always inspected at a certain step, but at a specified interval a full lot of inspection of all wafers takes place.

For yield analysis purposes, it is also desirable to have some wafers that are subject to all inspection layers in the manufacturing flow. This allows the correlation of electrical defects (faults) with defects on different layers and hence the calculation of "kill ratios" for the different layers. The kill ratio is the probability that a defect found in-line will have an electrical impact.

36.3.2 Data Mining

One of the most important tasks in yield analysis is the exploration of the available data. Most companies have a huge database where both test data and data from the factory are collected. This database, known as a yield management system (YMS), allows the engineers to access, visualize, and correlate the available data.

A review of a yield problem will in most cases start with a look at the wafer maps. The spatial signature of failing chips can give various insights into the cause of the failures. As mentioned earlier, sometimes a design feature of a processing tool may directly generate a certain "foot print" on the wafer. In other cases, the location of a failure may correlate with the occurrence of in-line detected defects at the same location. The fact as to whether a pattern always occurs at the same spot or whether it is found at certain radial symmetries might also provide further insight.

Some products also allow for more specific analyses: The failing cells of a memory chip, for example, can be displayed as a bit map, whereby the characteristics of a failure (single cell fails, bit-line fails, or word-line fails) are immediately visible. The bit map is also very useful for localization of the failing structures for detailed analysis in the failure analysis lab. On the other hand, failure localization on digital chips requires a far more challenging technique known as scan chain analysis. In this technique, special test patterns are applied on the chip in the hope that, together with design and layout information, the failing gates can be determined.

36.3.3 Tool and Chamber Signatures

Sometimes, even though a failure pattern does not appear localized on the wafers, a parameter may have elevated or lowered values on every second, third, or fourth wafer. In these cases, referred to as tool or chamber signatures, yield engineers will look at the tools used for wafer processing that have this exact number of chambers. Yield Management Systems (see Sec. 36.4), are typically fed with information from the Manufacturing Execution System regarding which lot was processed on which tool and which wafer was processed in which chamber. If one feeds a list of wafers that are affected by a certain problem into this system, it can respond with a list of processing tools (ideally, of course, it is only one tool) on which exactly all these wafers have been processed. This "tool commonality analysis" can be very helpful in identifying the root cause of a yield issue.

Some manufacturers go even further and instead of processing the wafers of each lot in the same order, the wafers are sorted into a different, characteristic order at several steps during the manufacturing process. This has the following advantage: In a manufacturing line there are many tools that experience continuous changes in conditions during processing of the wafers. If a continuous trend from wafer 1 to wafer 25 is detected at wafer test, it can be very difficult to identify the tool responsible for this trend. When wafers are placed into a characteristic order within the lot prior to critical process steps, any continuous trend during the process will translate into a characteristic pattern that can be identified by viewing the trend-by-wafer chart. The disadvantages of this technique are the additional costs and processing time necessary for reordering of the wafers.

36.3.4 Design of Experiments: Split Lots

The work of yield engineers is not only limited to maintaining a certain yield level, but also to continuously strive for improvements. Split lots are typically used to evaluate any desired changes to a process recipe, introduction of a new process sequence or qualification of a new chemical supply. Using split lots means that some of the wafers are processed using the new recipe, sequence, or chemistry, while the other wafers are processed according to the standard process, also referred to as process of record (POR).

To achieve a meaningful result at the end of this trial, it is desirable for process engineers and yield engineers to perform a statistical design of experiment. This involves a calculation of how many wafers are needed to prove that the new process is actually better or, in the case of cost saving measures, at least equal to the POR. Without a meaningful statistical design there is a risk that arbitrary variation is misinterpreted as a process improvement. It is therefore also important that other possible sources of variation are considered before a wafer split lot is created. For example, it would be unwise to try a new recipe on the wafers 4, 8, 12, 16, 20, and 24 when there are other process tools in the factory that have a four chamber architecture.

36.3.5 Process Change Review Boards

Any well-organized factory that is committed to maintaining the quality of its product will not simply allow any changes to the process or "optimization" of a recipe on a tool. Instead, it would appoint a process change review board (PCRB). This board, which is typically staffed with both senior engineers and managers, is tasked with approving each desired change of the POR. In most cases, even the trial runs have to be approved beforehand in order to avoid any risks like cross-contamination with newly introduced chemicals.

Once a new process recipe or chemistry has been evaluated, the results of the experiment are presented to and analyzed by the PCRB, who will ultimately approve or reject the submission. An important aspect of the qualification of a new process is that the data review involves real statistical analysis (like ANOVA) and not only a simple comparison of the wafer yield averages of the new versus the old process.

36.3.6 Excursion Prevention

A part of effective yield management also involves the implementation of systems to maintain yield and quality levels. The most commonly used system is statistical process control (SPC). This is mainly used in manufacturing where defect counts and in-line wafer measurements like critical dimensions are typically subject to SPC. Modern systems, however, also allow the application of SPC for electrical wafer test data. In SPC, control limits are calculated from past data statistics, and one or more control rules are then applied to any new measurements. The simplest rule is that an action, like a tool inspection, shall take place when a measured data point is outside the calculated control limits. Statistical process control aims at separating the normal, intrinsic variability that is inherent in any industrial manufacturing process from the so called special causes of variability, which are supposedly detected by means of violation of certain statistical limits. It is therefore an important aspect of SPC that the control limits are calculated statistically and are not set manually by engineering judgment [7].

SPC is sometimes considered part of a larger framework called advanced process control (APC). The tools used in APC include fault detection and classification (FDC), virtual metrology (VM), and run-to-run (R2R) control.

Fault detection and classification (FDC) uses statistical methods to monitor equipment parameters like temperature, pressure, and other sensor data during wafer processing. The objective of FDC is to eliminate undesired process conditions quicker than traditional methods.

Virtual metrology (VM) aims at exploiting sensor data from the equipment to calculate resulting wafer properties like deposited thickness layers.

Run-to-run (R2R) control refers to the continuous adjustment of process parameters based on measurements of previously processed lots or wafers. For more information on the wide-ranging topic of APC, please refer to Sec. 36.7.

36.3.7 Excursion Containment

Occasionally a wafer factory may encounter a major drop in yield for some or all products. As these are very costly events, all engineering resources are usually bundled in task forces to find the root cause of the issue as fast as possible. Finding the cause in a complex environment is very challenging and even once the problem has been found and fixed another major task nevertheless remains: There may still be many wafers in the work-in-progress that need to be assessed to see if they have been affected by the problem, and whether or not they are sellable products. In some cases additional tests need to be implemented swiftly to contain affected material.

36.4 SOFTWARE

36.4.1 MES, SPC, and APC Software

The central piece of software in a manufacturing line is the manufacturing execution system (MES). This software is used for tracking which lots are currently being processed on what manufacturing tool, dispatching lots to the next available equipment, and providing the correct process recipe for the equipment. The MES software thus has a huge database that not only tracks which lot is currently at what stage in the process flow and on what tool it is being processed, it also contains the process flows for each product and the recipes for each process step.

The MES software can also provide functionality for the SPC (statistical process control) or provide an interface to external SPC software, such that measurement values that are fed into the MES or SPC software can trigger a halt on a lot or a tool when the SPC rules demand it.

There are also many commercial tools available for APC, including FDC and R2R controllers. These systems typically also require integration with the MES software.

36.4.2 Yield Management Software

An important software tool for yield analysis is the yield management system (YMS). This is essentially a big database that is fed with all the measurement data from the factory and test, as well as the lot tracking information ("lot—equipment—history") from the MES software. The software clients that access the information in the YMS database can visualize the data in a variety of ways, including wafer maps, histograms, time trend charts, Pareto charts, correlation analyses, tool commonality analyses, and more.

There are many different YMS systems commercially available; however, some companies choose to develop their own YMS systems using their internal IT specialists. If a new YMS software needs to be introduced, a "make-or-buy" decision must therefore be made. There are some very reasonable arguments for buying an off-the-shelf software, like fast availability of the system, the typically vast set of available features, and overall lower costs—that is, if all or most of the features that have been developed by the YMS supplier are actually needed. On the other hand, if a chip manufacturer has numerous special requirements or special business processes that have little in common with other manufacturers, it may be better to develop a customized yield management system, such that these demands can be addressed in the software.

36.5 CONCLUSIONS AND FUTURE DEVELOPMENTS

In the past, the data generated from different types of equipment varied in format (and some tools did not even have an interface to access the data), we can assume that by way of more and more standardization, data integration in factories will become more elegant, effective, and extensive. A lot of effort has been required historically to simply combine data with different file formats and different coordinate systems on the wafers, into a data warehouse. With increasing standardization, we will see simpler data communication and faster interfaces, thus allowing access to even more data. For example, on the test floor there is a trend moving from the former separated testers and probers to an integrated "test cell" provided by a single vendor, which allows easier access to the data. Even more data can be collected from the front-end tools in the factory as many of them now allow access to tool-internal sensor data with a high sampling rate. This means, on the other hand, that for each relevant signal of a sensor, a lot of uneventful data from other sources are recorded.

The abundance of data available for analysis and control together with the expectation that the volume is still increasing has caused a current hype about data exploitation. The potential to gain additional efficiency by collecting more and more data certainly exists. However, the big challenge that remains is to gather useful information from this vast amount of data.

36.6 REFERENCES

1. P. Van Zant, *Microchip Fabrication*, 6th ed., McGraw-Hill, New York, 2014.

2. R. Ross and N. Atchison, "Yield Modeling," in: Y. Nishi and R. Doering (eds.), *Handbook of Semiconductor Manufacturing Technology*, CRC Press, Boca Raton, Florida, 2007.

3. A. V. Ferris-Prabhu, *Introduction to Semiconductor Device Yield Modeling*, Artech House, Boston, London, 1992.

4. D. Rathei, *Theoretical, Statistical and Empirical Review of Semiconductor Yield Modeling*, Graz University of Technology, Graz, Austria, 2003.

5. C. H. Stapper, "Fact and Fiction in Yield Modeling," *Microelectronics Journal*, 20 (1–2): 129–151, Elsevier Science Publishers Ltd, London, England, 1989.

6. F. Kuper, J. van der Pol, E. Ooms, T. Johnson, R. Wijburg, W. Koster, and D. Johnston, "Relation between Yield and Reliability of Integrated Circuits: Experimental Results and Application to Continuous Early Failure Rate Reduction Programs," *Proceedings IEEE International Reliability Physics Symposium* (IRPS), IEEE, Dallas, Texas, pp. 17–21, 1996.

7. D. C. Montgomery, *Introduction to Statistical Quality Control*, 7th ed., John Wiley & Sons, Hoboken, New Jersey, 2013.

36.7 FURTHER READING

A journal containing recent publications on yield is the *IEEE Transactions on Semiconductor Manufacturing*.

Current topics for yield engineering are also discussed at the annual Advanced Semiconductor Manufacturing Conference (ASMC, www.semi.org/asmc) and the International Symposium on Semiconductor Manufacturing (ISSM, www.semiconportal.com/issm).

Regarding advanced process control (APC), there are annual conferences in the United States, Europe, and Asia: www.apcconference.com, www.apcm-europe.eu, www.semiconportal.com/AECAPC. Furthermore, one can read more about R2R control, VM, and FDC in:

Joe Qin, S., G. Cherry, R. Good, J. Wang, and C. A. Harrison, "Semiconductor Manufacturing Process Control and Monitoring: A Fab-Wide Framework," *Journal of Process Control*, 16(3): 179–191, 2006.

Khan, A. A., J. Moyne, R. James, and D. M. Tilbury, "An Approach for Factory-Wide Control Utilizing Virtual Metrology," *IEEE Transactions on Semiconductor Manufacturing*, 20 (4): 364–375, Nov. 2007.

Moyne, J., E. del Castillo, and A. Hurwitz, *Run-to-Run Control in Semiconductor Manufacturing*, CRC Press, Boca Raton, Florida, 2000.

Chapter 37

CIM and Factory Automation

BY

CLINT HARIS *Entegris, Inc., Billerica, Massachusetts*

37.1 INTRODUCTION

37.1.1 What Is Factory Automation?

In order to stay competitive, IC manufacturers must continually improve the productivity of their factories. Increasing wafer size as well as building larger fabs with thousands of wafer starts per day are two methods that IC manufacturers use to minimize their manufacturing costs. To support the complex logistics requirements of these factories, companies utilize computer integrated manufacturing (CIM) and factory automation to ensure that the right material is delivered to the right place at the right time and that it is processed correctly. Software solutions utilized in semiconductor factories include the manufacturing execution system (MES), the manufacturing control system (MCS), and the advanced process control system (APC). These software systems are augmented with robotic solutions commonly known as automated material handling systems (AMHS) (Fig. 37.1).

An AMHS is composed of a set of components utilized to store and move material from one place to another in a factory without the need for human intervention. Stockers, a key component of AMHS, are placed in each bay of the factory and are utilized to store carriers as well as to act as a connection point between distinct transport systems. The material movement in an AMHS is accomplished by transport systems, which usually take the form of an overhead hoist transport (OHT), overhead shuttle (OHS), automated guided vehicle (AGV), rail guided vehicle (RGV), or conveyor system.

37.1.2 Drivers behind Factory Automation

There are several drivers behind the use of factory automation. These drivers include *shorter manufacturing cycle time, yield improvement, better tool utilization, personnel reduction, risk reduction,* and *ergonomic considerations.*

Manufacturing Cycle Time

By improving tool-to-tool delivery time and providing a method for automatically and predictably scheduling workflow, factory automation offers the potential to reduce manufacturing cycle time. Cycle time is an important metric for semiconductor fabs because faster time to market and faster delivery time of product to customers have a significant impact on the value of that product and enable the fab to achieve quicker yield improvement.

Yield Improvement

The amount of product that can be sold versus the amount of product that is started is known as the yield. Yield is one of the most critical attributes that semiconductor manufacturers strive to improve as it is directly related to their cost of manufacturing. In an effort to improve factory yield, fabs utilize factory automation to track thousands of different parameters across hundreds of thousands of wafers produced annually. The goal of tracking this data is to find ways to make process adjustments that can improve overall yield.

Tool Utilization

Ten to twenty percent of a semiconductor fab's output is often lost due to tools sitting idle waiting for material to be delivered. Manufacturers view automation as a means of eliminating variability in delivery times and improving tool utilization. Improved tool utilization through the use of automation can result in hundreds of millions of dollars of savings over the life of a factory.

Personnel Reduction

It is often not practical to provide the necessary number of employees or aisle space needed to support the manual movement of material between bays in fabs. For example, consider a 40,000 wafer start per month factory manufacturing a chip that requires 1500 process steps. For such a factory, one can calculate the number of *moves/hour* = 1500 process steps × 50,000 WSPM × (1/25 wafers per lot) × (12 months/year/365 days/year/24 h/day) × (safety factor = 2) = ~6500 moves/hour. If one assumes that an employee can move on average 15 moves per hour, ~400 employees would be required per shift just to move material. Minimizing the number of employees dedicated to menial tasks such as material movement is one of the drivers of AMHS.

Risk Reduction

In 2015, the value of a carrier of wafers can vary from $25,000 to more than $1,000,000. With thousands of carrier moves occurring hourly, the probability of damage with an operator-based transport solution is significant. The risk of human error causing a carrier to be dropped is another of the reasons fabs implement AMHS systems.

Ergonomic Considerations

A 200-mm wafer carrier weighs approximately 10 lb and a 300-mm wafer carrier weighs approximately 20 lb. While an individual can

Figure 37.1 Automated materials handling system.

easily lift these weights infrequently, experience in 200-mm fabs has shown that repetitive stress injuries result when individuals carry and frequently place 200 mm carriers to and from process tools. Therefore, with the conversion to 300 mm, most companies determined that it would be ergonomically unsuitable to have manual transport of carriers.

37.2 SEMICONDUCTOR FACTORY SOFTWARE

Unlike assembly lines in other industries, semiconductor facilities feature a complex, continuously changing, reentrant flow. A reentrant line utilizes the same tools multiple times during the manufacturing of a product. Additionally, a typical characteristic of semiconductor facilities is their continuous pushing of the limits of the capabilities of their process tools.

To overcome these challenges, semiconductor facilities utilize sophisticated software solutions to manage the flow and execution of production. At the highest level of abstraction, these software solutions are required to *control* factory processes to an appropriate level of *quality* (Fig. 37.2). The next two sections describe the typical elements of the factory *control* software and the factory *quality* software.

37.2.1 Factory Control Software

The primary factory control software systems are the manufacturing execution systems (MES), material control systems (MCS), and the scheduling and dispatch software. These software systems work in concert to manage the movement of wafers and ensure that each wafer is processed correctly.

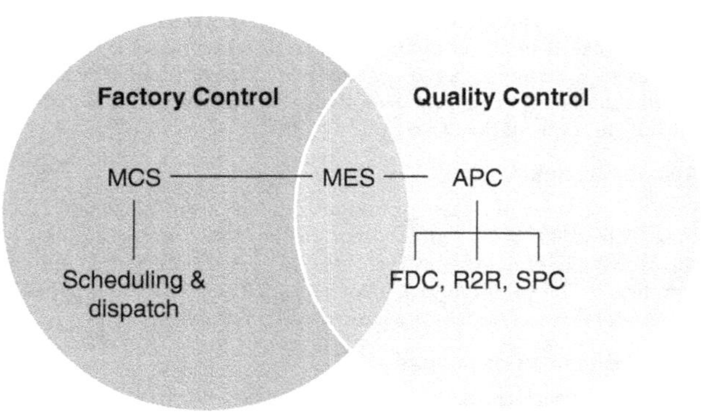

Figure 37.2 Semiconductor factory software systems.

Manufacturing Execution System

The MES manages a factory's process plan and tool recipes, interfaces with the MCS software to control material movement, and supports the continuous improvement of yield through advanced process control.

The MES is responsible for keeping a master database of all of the process plans as well as process recipes utilized in the fab. A process plan refers to the set of steps that a wafer receives during the manufacturing process. These steps are characterized by the manufacturing tool types (e.g., lithography tools, etch tools, implant, etc.) and the process plan stored in the MES contains the order in which the wafer is brought to each tool type.

Semiconductor process tools are capable of being optimized to run specific recipes. A tool recipe describes the control parameters utilized in a process step. Depending on the manufacturing tool, there are hundreds of different parameters that can be varied. Examples of control parameters include time, chamber temperature, temperature ramp speed, pressure, gas flow, wafer spin speed, etc.

Material Control Software

The MCS is a centralized real-time system that interfaces with the MES and controls the transport and storage of lots, reticles, and other materials in semiconductor fabs.

The MCS is utilized to translate the sequence of the MES process plan into move commands that can be understood by the AMHS.

Additionally, the MCS includes a database to identify and track material in the fab. Advanced tracking, batching, space partitioning, material starvation prevention, and contamination prevention algorithms are required in an MCS to ensure factory optimization. Along with 99.999 percent availability required for a mission critical software system, dynamic expansion and reconfiguration is also required in an MCS to enable the constant changes that occur in a fab.

Scheduling and Dispatch Software

The MCS is supplemented with scheduling and dispatch software that analyzes the status of all material in the factory and triggers the execution of where each lot should be delivered since often there are more than one process tool that can perform a particular process step.

The scheduling and dispatch system analyzes the state of the factory and triggers moves within the MCS to improve flow, reduce WIP, and increase on-time deliveries. The current status of the fab is continuously reviewed and the software determines when lots should be started, when lots will be completed and how to best balance customer ship requirements versus minimizing inventory levels. Rules are employed within the scheduling and dispatching software to optimize the performance of individual bays while also to considering more strategic overall aims (such as inventory levels or on-time delivery targets) when selecting a lot to process. As improvements are made to the overall dispatching logic, orders move more smoothly through production, resulting in reduced cycle time and cycle time variability as well as reduced operational variability across crews and shifts.

37.2.2 Quality Control Software

Semiconductor factories continuously strive to improve device yield and performance. The advanced process control (APC) system is the quality control software that performs this critical role in semiconductor factory. There are three primary components to an APC system: fault detection and control (FDC), run-to-run control (R2R), and statistical process control (SPC).

Advanced Process Control

An APC system monitors how process tools impact the physical and electrical characteristics of wafers and then adjusts the process recipes to improve device yield and performance over time. APC systems offer an enhancement over traditional SPC systems by enabling the fab to not only detect out-of-control behavior, but also proactively prevent this behavior from occurring.

The data for APC typically comes from information from the process tool or a metrology system that is located ex situ, in-line or in situ to the tool itself.

The shorter the time between the process and the metrology step, the faster that a change can be implemented and the higher the impact on cost savings. It is for this reason, that there is a push for more in situ metrology.

Fault Detection and Control

FDC systems watch for specific events that occur during the operation of manufacturing equipment that affect the quality of the wafer.

FDC systems track abnormal events in the manufacturing equipment and correlate them against wafer yield data to develop methods to prevent future yield excursions. FDC systems enable metrics to be created for preventative maintenance schedules and consumable lifetime. Additionally, FDC systems can be used for process qualification, tool, and chamber matching. Through these types of improvements, FDC systems lower factory cost by reducing scrap, rework, and unscheduled downtime.

Run-to-Run Control

Run-to-run control systems change the process recipes between runs to optimize yield and reduce variation. Run-to-run systems are used to continuously improve critical parameters such as device electrical performance, lithography CDs, CMP uniformity, and die yield.

Run-to-run systems use feed-forward techniques to adjust the recipe parameters of tool based on the results of a previous measurement of the wafer about to be processed. Additionally, they employ feedback control systems, which use a subsequent measurement step to change the recipe for a process tool.

Statistical Process Control

Statistical process control (SPC) is a monitoring system that measures the quality of a process and triggers an alarm when a deviation from statistical norms is detected. Also known as 6-sigma analysis, SPC is focused on quality control rather than on continuously changing and optimizing the processes themselves.

SPC systems monitor one or more variables and utilize control charts to detect abnormal variations in the process.

37.3 SEMICONDUCTOR AMHS

An AMHS is utilized to transport and store carriers within a semiconductor factory. There are a variety of carriers as well as AMHS components utilized in the semiconductor industry, which each has particular usage characteristics. The AMHS components can broadly be defined by two categories: the storage system and the transport system.

37.3.1 Wafer Carriers

As wafer sizes have evolved over the last few decades from 3 in or smaller wafers to state-of-the-art 300-mm wafers, the carriers that contain them have also changed. With the advent of 200-mm semiconductor production in the early 1990s, a significant emphasis was placed on designing carriers that could be automatically moved through a factory with an AMHS. This philosophy was extended with the design of 300-mm carriers, which were envisioned from the start to be optimized for AMHS.

200-mm Carriers

There are two cleanroom concept philosophies for 200-mm fabs that dramatically impact the style of carrier utilized: SMIF and open cassette.

SMIF A standard mechanical interface (SMIF) factory utilizes enclosed carriers that are opened by a standard loadport on each tool. In addition, each process tool in a SMIF factory contains a minienvironment that provides a clean environment for the wafers once they have been removed from the carriers. SMIF carriers are well suited to AMHS because they have a common automation interface as well as because the wafers are kept isolated from any potential contamination that might be generated in the movement of the carriers through the fab.

Open Cassette Open cassette factories utilize wafer carriers in which the wafers are exposed to the environment. As such, an open cassette factory must be kept cleaner than a SMIF factory. Open cassettes may be transported by themselves or they may be placed in a sealed storage box prior to transport. One of the key disadvantages to open cassettes and their storage boxes is that they were not designed with automation in mind. Therefore, custom interfaces and custom robots must be utilized to move open cassettes into and out of automated material handling systems.

300-mm Carriers

With the transition to 300 mm, a significant amount of effort was placed into standardizing the style of carrier utilized by fabs worldwide. Ultimately, a minienvironment friendly, automation ready, integrated carrier called a FOUP (front opening unified pod, Fig. 37.3 *left*) was selected as the industry standard. Similar to the 200-mm SMIF concept, the 300-mm FOUP seals the wafers in a controlled environment. Additionally, the same style of robotic handle used on SMIF carriers is utilized to enable FOUPs to be automatically transported through the factory. The difference between the FOUP and SMIF concept primarily exists in two areas: (1) the door is on the front of the FOUP versus the bottom of the SMIF carrier and (2) the FOUP contains an integrated wafer cassette, while a SMIF container has a removable wafer cassette.

37.3.2 Transport Systems

There are four primary styles of transport systems used in factories today: overhead hoist transport (OHT, Fig. 37.3 *middle*), overhead shuttle (OHS), automated guided vehicle (AGV, Fig. 37.3 *right*), rail guided vehicle (RGV),

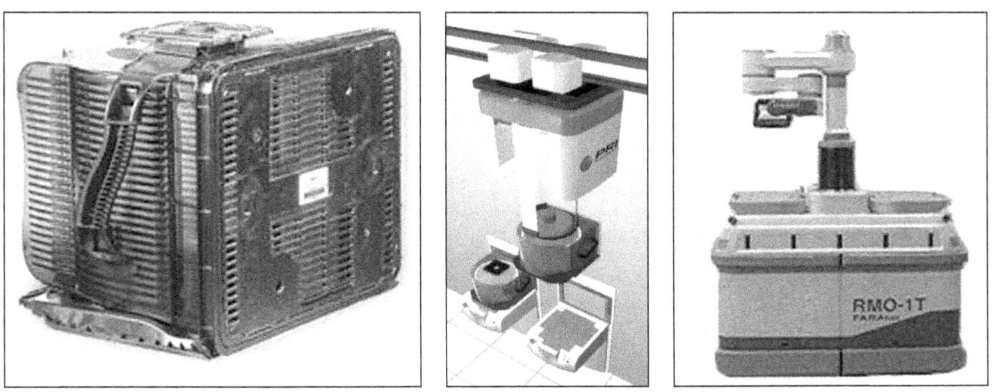

Figure 37.3 300-mm FOUP (*left*), overhead transport (*middle*), automated guided vehicle (*right*).

and conveyor systems. Vehicles that move carriers within a process bay are known as *intra*bay transport systems and those that move carriers between bays are known as *inter*bay transport systems.

Overhead Hoist Transport

An OHT is a vehicle that travels at ceiling height on a rail above the loadports of the tools and has the capability to raise and lower a carrier onto these loadports. OHT systems are primarily utilized in 300-mm factories; however, they can also be utilized for 200-mm SMIF fabs. One key advantage of OHT is the fact that it does not utilize fab footprint. Additionally, with appropriate sensors, an OHT can be used simultaneously with operators without worry of operator injury. Another key benefit of OHT systems is they are relatively flexible to support changes in factory layout. Last, as will be discussed later in this chapter, OHT is the only transport technology that is practical in a unified AMHS.

Overhead Shuttle

An OHS travels at ceiling height on a rail and transfers carriers from stocker to stocker. In other words, OHS is used as an interbay transport system. OHS systems are utilized in 200- and 300-mm factories. An OHS holds one or two carriers and is often used in conjunction with an OHT, RGV, or AGV to support intrabay moves. There are two styles of OHS, one utilizes an on-board robot to transfer carriers from a stocker to the OHS and the second utilizes a robot located inside of the stocker to load and unload carriers.

Rail Guided Vehicles

RGVs are ground-based vehicles that travel on a rail in the floor. RGVs can be used for interbay as well as intrabay delivery. RGVs are typically the fastest of the different styles of transport systems and therefore have often been utilized for the bays with the highest move rate requirements. One limitation of RGVs is the fact that they cannot coexist with fab personnel due to safety considerations. A second limitation arises from the inflexibility of RGVs to be moved to support layout changes within the fab. RGVs were utilized in 200 mm but are increasingly being replaced with OHT for 300-mm fabs.

Automated Guided Vehicles

AGVs are ground-based vehicles that can be programmed to travel anywhere in the fab. AGVs are used for both interbay as well as intrabay delivery. AGVs are slower than most other material handling systems; however, they are the most flexible support layout changes. With some limitations, AGVs may temporarily be utilized in bays in which fab personnel are working. RGVs were utilized in 200 mm but are not commonly found in 300-mm fabs.

Conveyors

Roller-based conveyor systems are placed at ceiling height and are utilized to transport both 200- and 300-mm carriers. While slower than many other material handling technologies, conveyors can support a high throughput of material due to the fact that the carrier movement is not limited by a limited number of vehicles. Conveyor technology is typically utilized for point-to-point applications such as the delivery of carriers between two separate fab buildings; however, fabs also utilize conveyors in conjunction with OHT in high-throughput areas of the fab.

37.3.3 Storage Systems

Stockers

Stockers are the traditional system utilized by fabs to store bulk amounts of carriers. Stockers consist of one or more input/output ports, a series of shelves in which carriers can sit and a robotic system that enables the carriers to be delivered to and from the shelves. Stockers come in a variety of shapes and sizes and their capacity can range from as low as 50 carriers to multifloor systems that can hold more than 1000 carriers.

Under Track Storage

Under track storage (UTS), also known as zero footprint storage (ZFS), is a storage option that can be used with OHT installations. UTS consists of a shelf located below the OHT track upon which carriers can be stored. UTS offers three significant benefits: (1) it is a zero-footprint storage solution, (2) it is highly reliable, and (3) it can improve the efficiency of the AMHS.

AMHS efficiency is improved with UTS because carriers for intrabay moves can be stored near the next process tool and thus the need to return the carrier to a stocker is eliminated. UTS, therefore, can be used to buffer surges in requested AMHS moves throughout the factory. This approach not only reduces the stocker utilization but can reduce congestion of the OHT vehicles because the UTS provides several interface points for the OHT instead of the two interfaces typically available at a stocker.

37.4 THE DESIGN OF AN AMHS

In building a semiconductor fab, several decisions must be made regarding the layout of the fab and how that layout will impact the AMHS. Given the high cost of semiconductor factories, multiple layout approaches are usually considered and complex simulation models are executed to determine which approach is most appropriate given the particular business constraints of the company.

37.4.1 Simulation

Following the initial approval of a fab, the first phase of activity traditionally centers around the fundamentals of how the fab will operate. In this phase, decisions such as how big the fab will be, exactly what mix of products will be run and what kind of layout will be used are made. In addition, operational philosophies are determined (e.g., is the fab going to try to minimize cycle time versus maximize wafers out, is the fab going to support research and development versus 100 percent production, etc.). One way in which automation systems play a key role in this phase is with the use of simulation tools to analyze and optimize fab operational philosophies.

37.4.2 Process Tool Layout Methodologies

Different fabs choose different tool layout methodologies and there is limited consensus on which approach is optimal. Nonetheless, there are several approaches which are commonly utilized for process tool layout.

Bay and Chase

In a bay and chase configuration, the process tools are mounted along walls in the aisles, which are typically half the width of the fab in length and extend off a long central aisle. This construction creates process bays, the cleanroom side of the wall, and chases, the tool side of the wall, hence the name. Operators access the tools from the bay side and maintenance technicians access the tools from the chase side. The chase side is commonly called the *gray area* due to its less clean status (Fig. 37.4).

Ballroom

The ballroom layout does not have the walls of the bay and chase layout. Instead, the process tools are all located within one large room, hence the name. This cleanroom is typically dirtier than one founding a bay and chase fab, which leads many ballroom configurations to require each tool to have a clean minienvironment in which the wafers are handled. When minienvironments are utilized, wafers are transported and stored in SMIF or FOUP carriers, which allow them to be transferred into the tool's minienvironment without being exposed to the cleanroom atmosphere.

There can be considerable variation in the process tool arrangements with the ballroom layout. Tools are generally arranged in rows but each row may not be the same length nor have the same regularity as in the bay and chase layout. It is possible to place the transport above the process tools in the ballroom but maintenance and move in requirements must be kept in mind. In this fab type, the stockers may be located in proximity to the tools they service, making them appear to be located randomly about the fab. At the other end of the spectrum a ballroom fab may have exactly the same layout as a bay and chase fab except there are no physical walls.

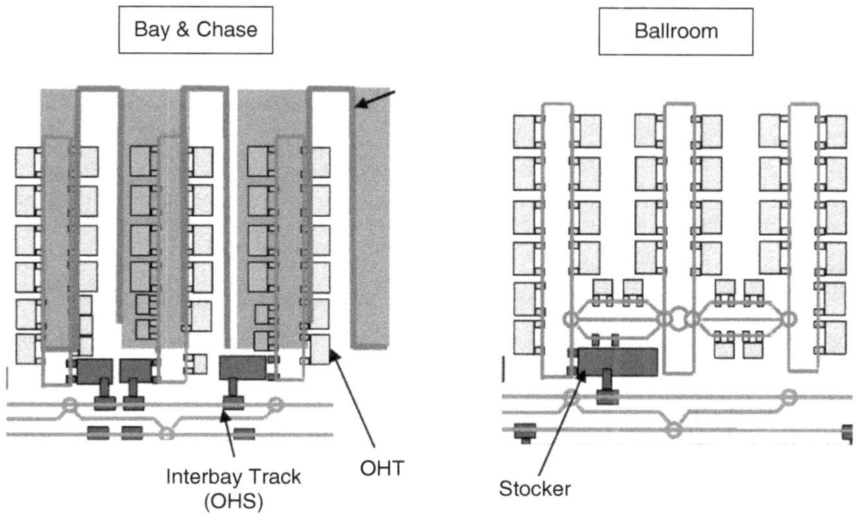

Figure 37.4 Bay and chase versus ballroom configurations.

Farm Layout

The term farm layout refers to one methodology of grouping process tools within the fab. In the farm layout, process tools of a particular type are all collocated within the same bay or adjacent bays of the bay and chase or in the same general area of the ballroom. Architect construction firms favor this layout because it simplifies the building of the fab and the facilities infrastructure required to support the process tools.

With few exceptions (some time sensitive steps are grouped, for example, cleaning before diffusion is frequently done in the diffusion bay) only one process step is performed on a wafer each time it visits a particular group of process tools. This layout puts a high load on the interbay portion of the material handling system since a wafer lot must be moved from one bay to another after each process step. The grouping of tools also results in very heavy traffic to and from the metrology area.

Hybrid Layout

The hybrid layout results when metrology tools are removed from the metrology bay of a farm layout and collocated with the appropriate process tools. The hybrid layout reduces the load on the interbay system (at the expense of the intrabay system), eliminates the heavy traffic metrology area, and has the potential to reduce the overall wafer manufacturing process time. Hybrid layouts require a larger number of metrology tools than farm layouts but that cost is offset by its somewhat smaller footprint requirement. Simulations show that fabs with a hybrid layout could potentially reduce manufacturing cycle time by 10 to 15 percent of fabs with farm layouts.

Modified Hybrid

The modified hybrid layout results when a variety of process tools along with metrology tools are grouped to allow a few (4 to 6) sequential steps to be performed in the same bay or area. For example, strip and clean tools can be located in the same area as implanters so that photoresist can be stripped and the wafers cleaned before they are sent to the next process step in diffusion or litho. In addition, metrology tools can also be included in the grouping of process tools.

The modified hybrid layout further reduces the load on the interbay system (at the expense of the intrabay system) and eliminates the heavy traffic metrology area. It requires a larger number of process tools and is more expensive to construct than the farm or hybrid layouts.

37.4.3 AMHS Layout Methodologies

Segregated Systems

A segregated system is an AMHS in which the interbay and intrabay transport systems are separated. In the segregated system, an OHS is typically used for interbay transport and OHT is used for intrabay transport although other combinations with PGV, AGV, and RGV are possible. For the segregated system each bay or set of linked bays has a dedicated intrabay system and each intrabay system operates independently from what is happening in other bays.

Material Flow in a Fab with Segregated AMHS Material flow for a segregated system is essentially the same in 300- and 200-mm fabs. A carrier is transferred from a process tool to a stocker dedicated to that process area. The carrier is then sent from that stocker to the stocker dedicated to the process area where the tool which will perform the next process step, is located. The carrier is pulled from the stocker when the process tool is ready. When the process step is complete the sequence repeats.

When more than one process step is performed within a bay or area, the carrier may move directly to the next process tool without returning to the stocker; however, the move sending the carrier to a tool cannot begin unless there is an empty load port on that tool. The higher the percentage of direct tool-to-tool moves, the lower the move rate the AMHS must support.

The benefit to use direct tool-to-tool moves is significant. Consider the simple case of lots moving to a tool with a 70 percent utilization; by definition the tool will have no material available 30 percent of the time, if it has two load ports one will be empty 51 percent of the time and if it has three load ports one will be empty 66 percent of the time. The use of tool groups where any one of several tools can perform the same process steps dramatically increases the probability of being able to make a move directly to the next tool. Simulations of various fabs have shown that, on average, 80 percent of all intrabay moves could be made directly tool to tool, avoiding the additional move of returning to the stocker.

A segregated AMHS has several detractors compared to an alternate approach known as a unified system. These include increased numbers of stockers, longer delivery time, and throughput limitations.

Unified System

In a unified system the transport for the intrabay and interbay systems is combined and a single system is used for all transport. This allows lots to be moved directly from tools in one bay to tools in another without passing through a stocker.

There are three types of unified AMHS. They are

1. ***Partially unified.*** This is a segregated AMHS with several groups of bays linked via OHT.
2. ***Fully unified.*** This is an AMHS using only OHT for both inter and intrabay moves. Lots can be moved to and from any tools in the entire fab without going through a stocker.
3. ***Combined unified.*** This is a full unified AMHS supplemented with a loop of OHS linking the stockers and used to reduce congestion on the OHT track.

Material Flow in a Unified System The integrated track of the unified system allows moves to be made directly between any two loadports in the fab. However, because a move to a tool cannot start unless a loadport on the destination tool is open, some completed lots are moved to a stocker or to an UTS before being moved to the next process step. That storage location is generally in close proximity to the bay or process area in which that tool is located; however, in theory any storage location can service any tool.

The exception to this material flow is for the case of hot lots. Some fabs hold a tool load port open when a hot lot is expected, allowing the carrier to be moved directly from one tool to another. Eliminating the need for transferring through stockers as is the approach in a segregated system dramatically reduces hot lot delivery time for unified systems.

Delivery times for normal lots are also improved in a unified system since at most they only pass through one stocker during a tool-to-tool move. Therefore, unified systems reduce the number of AMHS moves required for each "process step" move by one-third for normal lots and two-thirds for hot lots.

OHT Track Layouts in Unified Bays In fabs with unified AMHS, bay throughput is not limited to the throughput of the stockers as with a segregated AMHS. Carriers from any stocker can be delivered to any bay and carriers from several stockers can be sent to a single bay. The bay throughput is limited by the ability of the AMHS to move carriers through the bay as long as the capacity of the process tools is not exceeded.

Typical Applications for Unified Layouts Unified AMHS works well with hybrid or modified hybrid tool layouts. In a hybrid layout, metrology tools are placed in the same bay as the process tools they support. This increases the demand on intrabay material handling system. In a modified hybrid layout, metrology tools and other process tools, such as wet benches and strip, are distributed throughout the fab in the same bays or areas as the process tools they support. Therefore, a route might involve several sequential process steps in the same bay. This further increases the demand on the intrabay transport system and reduces the loading on the interbay movement system.

Benefits of Unified Transport There are two major benefits of unified transport:

1. *Shared storage, process, and metrology tools across bays.* Because unified transport allows bays to be linked, it is possible to reduce the number of stockers and process and metrology tools by sharing them across bays. In turn, this allows greater flexibility in the placement of stockers, process and metrology tools.
2. *Faster delivery of lots.* Unified transport is capable of delivering both production and hot lots faster than segregated transport. The main reason is that there are fewer hand-offs and only one stocker is required for a move between tools.

37.5 OPERATIONAL CONSIDERATIONS

With the multibillion dollar cost of 300-mm factories and the AMHS increasingly being recognized as a mission critical application, several key operational considerations must be made in the evaluation and selection of an AMHS. At a macro level, fabs must select an AMHS that meets their *performance requirements* while minimizing the overall *cost of ownership* of the system.

37.5.1 Performance Requirements

The International Technology Roadmap for Semiconductors (ITRS) highlights the roadmap for key performance criteria for AMHS (Table 37.1). Two important AMHS criteria often monitored are the *carrier delivery time and transport throughput*:

Carrier Delivery Time

The carrier delivery time is the average time it takes for a carrier to be delivered to a loadport after a request is made for the carrier to be moved to that loadport.

In general, factories are driving toward delivery times of 5 min or less. Delivery times can be somewhat reduced by adding additional vehicles and additional track; however, sophisticated software routing algorithms are required to prevent the increased amount of vehicles from running into traffic jams. Additionally, since delivery time is significantly impacted by the layout and process flow of each individual factory, it is critical that the layout is optimized to support fast delivery times in an automated factory.

Transport Throughput (moves per hour)

The transport throughput is the number of moves per hour between loadport pairs in the factory.

For an assumed 40,000 wafer start per month factory, more than 5000 moves per hour will be required by AMHS systems to support the ITRS. To increase the overall number of moves per hour, typically

Table 37.1 International Technology Roadmap for Semiconductors Key Metrics for AMHS

Year of Production	*2013*	*2014*	*2015*	*2016*	*2017*	*2018*	*2019*	*2020*
DRAM ½ Pitch (nm) (contacted)	28.3	26	23.8	21.9	20	18.4	16.9	15.5
Water Diameter (mm)	*300*	*300*	*300*	*300*	*300*	*450*	*450*	*450*
Transport E-MTTR (minutes) per SEMI E10	10	10	10	10	10	5	5	5
Storage E-MTTR (minutes) per SEMI E10	20	20	20	20	20	10	10	10
Transport MMBF (mean moves)	25k	35k	35k	35k	35k	50k	65k	65k
Storage MCBF (mean cycles)	100k	150k	150k	150k	150k	200k	300k	300k
Mean carrier removal time (second)	30	25	25	25	25	20	20	20
Mean carrier delivery time (second)	300	270	270	270	270	240	240	240
Downtime to extend system capacity when previously planned (minutes)	15	10	10	10	10	5	5	5
Time required to integrate process tools to AMHS (minutes per LP)	15	5	5	5	5	5	5	5

(Source: Courtesy of International Sematech.)

one must increase the number of vehicles in the fab. Once again, with increased vehicles it is absolutely necessary to have software routing algorithms that can dynamically adjust the travel path of vehicles to minimize traffic concerns.

37.5.2 Cost of Ownership

Cost of ownership for an AMHS is driven by three primary factors: *equipment cost, overall equipment effectiveness (OEE),* and *operations cost.*

Equipment Cost

The cost of an AMHS can vary greatly depending on the wafer size, factory size, and level of desired automation. 300-mm factories, which require both interbay and intrabay automation, often require \$25 to \$75 million worth of AMHS. Oftentimes, the installation is made in two to three phases with the primary phase consisting of the basic AMHS infrastructure in addition to enough vehicles and storage locations to support the early ramp of the fab, while the additional phases consist of adding incremental vehicles and storage locations.

Overall Equipment Effectiveness

The OEE of an AMHS is driven by the performance of the system as well as the availability of the system. As mentioned earlier in the chapter, the primary metrics for performance are average delivery time and transport throughput. The availability of an AMHS is measured by the reliability of the system (mean time between failure, MTBF) and the maintainability of the system (mean time to repair, MTTR). It is important to note that MTBF and MTTR metrics are most critical at a *system* level for an AMHS and not a component level. For example, if an individual vehicle fails and prevents the delivery of a particular carrier it is far less severe of a problem than if a vehicle fails and shuts down an entire bay or shuts down the entire fab.

Operations Cost

The operations cost of an AMHS is made up of the utilities and consumables needed to operate the system, the service and spare parts cost required for the system and the cost of taking up valuable factory footprint by the system. While AMHS do not typically utilize consumable parts, they do require a significant amount of electricity. In particular, due to their relatively inefficient power transfer method, inductively powered OHT systems can utilize in excess of \$1 million per year of electricity in a large 300-mm fab. Service and spare parts costs can also be significant due to the large number of moving parts and the need to have multiple technicians available to solve problems quickly due to the mission's critical nature of an AMHS. Last, the AMHS may require a large amount of factory footprint and thus the physical size of the AMHS may greatly impact the overall operations cost to the factory. Due to the high costs of operating a cleanroom, as well as the desire to fit as many process tools as possible into the fab, factory planners are increasingly utilizing a variety of zero footprint AMHS designs such as OHT and UTS to minimize the impact of AMHS footprint on operations cost.

37.6 FUTURE TRENDS

In the semiconductor industry CIM and factory automation were first used as a means to improve operator efficiency and to optimize factory output. In today's fabs while human intervention is still required to support error handling, complex decision making, as well as to manually correct equipment failures, the trend is for the CIM and factory automation systems to address each of these human-based operations automatically.

Error handling and complex decision making are frequently required in a fab due to the fact that the process flow is complex (many hundreds of steps), is reentrant (certain tools are used multiple times to process a wafer), is constantly changing (process recipes are frequently adjusted to improve yield and performance), and must be robust enough to support repeatable yields across multiple process tools. Additionally, customer demands often require the mix of products running in the fab to change frequently. In order to automatically support the decisions required in such a dynamic environment, increasingly sophisticated software and automated hardware systems are being deployed. These systems range from rule-based engines in which the AMHS reacts to a set of predetermined events to learning algorithms in which the CIM dynamically controls the process equipment and the AMHS to adapt their behavior over time to optimize overall system performance.

While automation designers continually strive to improve the reliability of their systems to minimize failures, increasingly fabs are looking for fault-tolerant systems that can avoid system-wide failure in the event of an individual component failing. For example, in the event that an AMHS vehicle fails in a bay, additional vehicles can be dispatched to service that bay and to prevent the overall system performance from degrading. As another example, a redundant MCS architecture is often utilized to ensure that if one instance of the MCS fails, a backup system is immediately available.

Last, fabs are pushing their automation designs to support higher throughputs, faster delivery times, larger number of wafers per month, smaller lot sizes, and shorter cycle times. To support these needs, the design and implementation of novel concepts such as single wafer transport AMHS, fab-wide conveyor systems, and linked process tool architectures that enable tool-to-tool wafer transport rather than carrier transport are all being considered for future factory designs.

37.7 FURTHER READING

Agrawal, G. K. and S. S. Heragu, "A Survey of Automated Material Handling Systems in 300-mm Semiconductor Fabs, Semiconductor Manufacturing," *IEEE Transaction*, Vol. 19, Issue 1, Feb. 2006.

Chapter 38
MES Fundamentals

BY

JULIE FRASER *Iyno Advisors, Inc., Cummaquid, Massachusetts*

38.1 THE ROLE AND PURPOSE OF MES

Manufacturing execution systems (MES) are multifunction software systems designed to manage production operations. This category of software was named in the 1980s and has evolved and grown since that time. Just as manufacturing operations and software technologies have changed, MES has evolved as well. As a result, the fundamentals of MES today overlap with, but certainly are not the same as the fundamentals of MES in the 1980s, or even in the early 2000s. (In fact, in batch process industries, the term "Manufacturing Operations Management" has taken hold to represent this broader and more modern style of production management software.)

In 1997, the Manufacturing Execution Systems Association (MESA) created its White Paper #6, *MES Explained: A High Level Vision for Executives**. This paper includes this (still valid) definition of MES:

> Manufacturing Execution Systems (MES) deliver information that enables the optimization of production activities from order launch to finished goods. Using current and accurate data, MES guides, initiates, responds to, and reports on plant activities as they occur. The resulting rapid response to changing conditions, coupled with a focus on reducing non value-added activities, drives effective plant operations and processes. MES improves the return on operational assets as well as on-time delivery, inventory turns, gross margin, and cash flow performance. MES provides mission-critical information about production activities across the enterprise and supply chain via bi-directional communications.

This paper also introduced the MES Functional Model shown at the bottom of this MESA International page: http://www.mesa.org/en/modelstrategicinitiatives/MESAModel.asp. Interestingly, MES solution providers did not offer all of these in a single MES at the time, and have built out over time so that they do. These 11 functions are still foundational to any MES. Often called the MESA-11 model, it includes these functions:

1. Product tracking and genealogy
2. Data collection and acquisition
3. Dispatching production units
4. Resource allocation and status
5. Process management
6. Quality management
7. Labor management
8. Performance management
9. Maintenance management
10. Operations/detailed scheduling
11. Document control

The MESA-11 model is on target, but does not reflect everything available in modern MES. In particular, today's MES often have many modules in each of these categories. All current MES also have strong tools to integrate information flows across multiple systems at one site as well as to support data transfer among multiple sites.

This model and definition also do not suggest everything a semiconductor company needs from MES to be effective at serving customers. While the categories in the MESA-11 model are correct, what is currently needed has many aspects not shown. Pressure is mounting as a wide range of semiconductor customers build out smart products and services for which semiconductors are mission critical. This is largely because of the new technologies both in semiconductor products and in IT systems—but is also due to the expansion of applications for semiconductors into nearly every other type of assembled product. For example, semiconductor failures in medical devices, avionics, and self-driving cars can be life-threatening, so quality control is vital.

Semiconductor companies adopted MES early, in the 1970s and 1980s. With high-value products too small for human inspection, enormous capital tied up in production equipment and facilities, and extremely complex production operations, automating the information flows was of critical importance. The business rationale was clear: lower costs, higher throughput and yield, and better control as well as accurate tracking of work in process (WIP).

MES became a standard in front-end wafer fabs. There is no question or specific business case needed. In a fabrication environment, MES is central to the factory's ability to operate. Given the mixed mode and complex reentrant and contingent routings materials take through processes, tracking the WIP would have been challenging at best without a system. The highly automated nature of fabs has also meant that MES coordinates with additional automation layers for engineering, equipment, and automation control.

Back-end operations likewise widely adopted MES for WIP tracking through complex splits and merges and to support engineering data. Yet MES is not as ubiquitous in these later stage operations. Some back-end operations still have a mix of paper and partial, legacy, or homegrown MES. Modern MES in probe, assembly, and test operations certainly

*© 1997, MESA International, White Paper #06: *MES Explained: A High-level Vision*, by Julie Fraser and significant contributions from the 30 participants who worked on this in the MESA community.

provides critical visibility and control. As automation gains traction in these operations, more sophisticated and comprehensive MES is required just as it is in the fabs.

38.2 EVOLUTION OF MES IN SEMICONDUCTORS

Semiconductor companies were among the first to see the benefits of MES and adopt it in their facilities. Inspired by books such as *Computer Integrated Manufacturing* by Joseph Harrington† and *Material Requirements Planning* by Joseph Orlicky‡, companies began to see what functions were needed to manage a semiconductor facility. Commercial MES comprehensive enough for semiconductor use was developed starting in the late 1970s. Jonathan Golovin founded Consilium in 1978 and Canadian company PROMIS came into being in a similar timeframe. Many other semiconductor companies developed their own MES; some also offered it for external sale, such as IBM and Texas Instruments.

Because a wide variety of systems were adopted over several decades, there is some contention about what MES is and does today. There are also many semiconductor companies with very old, unsupported commercial MES surrounded by custom code or bolt-on systems that, together, typically provide far less capability than modern semiconductor MES provides. Some companies also continue to use MES that was developed in-house. Some of these are more complete and modern than others.

Early days: WIP tracking was the major focus of MES originally. It is very difficult to understand exactly where each lot or batch of products is and their status without such a system. As a result, semiconductor fabrication facilities adopted this fabwide software to ensure good tracking and control. In Fig. 38.1, this original type of MES is shown on the left. The core capability was to define physical flows and product information and then allow WIP transactions to follow that defined flow. These systems could not easily detect events or create reports useful to others in the company, but were effective at tracking WIP.

Until recently and still widely used: Fabs typically use applications founded on a workflow engine that enabled them to develop a system to match the unique needs of the facility. These accommodate automation-enabled manufacturing management in fabs, but are very IT-labor intensive. Changes to accommodate new products or process improvements could take months to program, test, and commission into production. In assembly and test operations, most companies used extensible but complex tightly integrated operator-focused systems. These back-end MES systems typically require less programming, but offer many deep options that still require extensive IT time for configuration. In Fig. 38.1, these types of systems are shown in the middle, with fabs above and back-end below. All of these systems tend to help enforce routes, processes, and guide operators to reduce errors.

Now and future: MES redefined for the agile, automated fab or back-end factory as platform and applications. This is the current MES approach that is the focus of the remainder of this chapter. With a potent combination of current software technologies both as platform foundation and for user experience and continuous growth of out-of-the-box semiconductor-specific functionality, these systems are much better able to support today's business needs. These include deeper NPI support, a shift from operator guidance to increasingly automated execution with operator decision making. In this Internet of Things–enabled world, MES also needs to handle larger volumes of data coming from many "things" that are not directly in the control of the operation or MES.

One other trend over the past few years that puts extreme pressure on MES is the megafab, gigafab, and megafactory. These gigantic, high volume facilities by their nature bring data volume, process breadth, and special lot transfer challenges. Scalability for MES takes on a new dimension in these facilities. Each of these facilities is slightly different, but strives to use a single MES to ensure sound presimulated models and efficient operation.

Regardless of the scale of the facility, coherent, broad-scope MES is more critical than ever. Manufacturing IT teams are seeking to minimize software layers and providers in part to streamline the ability to support frequent NPI and ongoing process change. Between the Internet of Things (IoT), complex global manufacturing networks of sites and partners, and the need to manage higher mix and shorter runs, MES is both big data source and aggregator for production analysis. In Fig. 38.1, this is shown on the right.

38.3 MES SCOPE AND FUNCTIONS

There is a clear coordinating layer of software that encompasses at least one entire facility and focuses on efficient production. This is how we will scope MES. It is located as Level 3, Manufacturing Operations

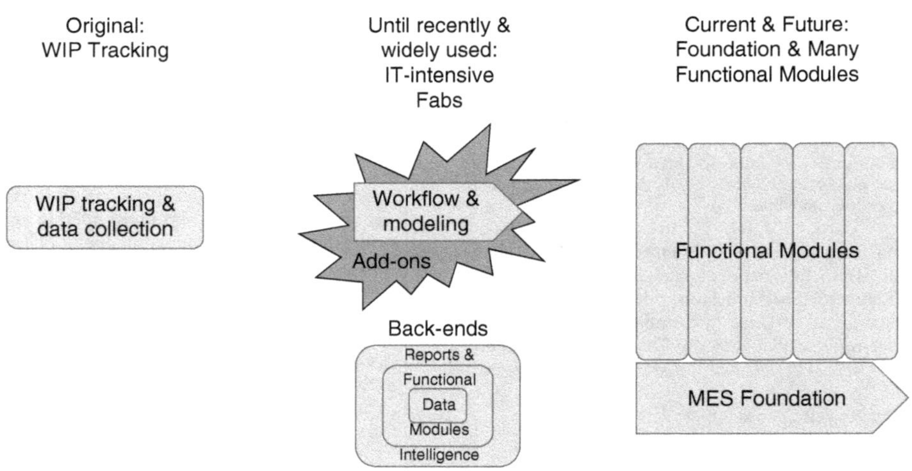

Figure 38.1 MES has not changed name or even the driving purpose over the past 30 years; yet the approach MES delivers and the capabilities these systems provide have changed dramatically.

† © 1974 Joseph Harrington, *Computer Integrated Manufacturing.*

‡ © 1975 Joseph Orlicky *Material Requirements Planning: The New Way of Life in Production and Inventory Management.*

Management (MOM), in the ANSI/ISA-95 Enterprise-Control System Integration Model, shown on this webpage: http://www.batchcontrol.com/s95/s95.shtml (and in slideshows from others). While some people argue that MOM is broader and better matches current MES product capabilities, in the semiconductor industry and some other discrete industries, it is typically referred to as MES.

This model applies not only across all types of semiconductor facilities, but across all manufacturing and production. So this will be a common language for the IT stack that companies can use with their customers and suppliers.

This automation and information systems "stack" specifically for a semiconductor fab includes an equipment engineering system layer parallel to the MES, including advanced process control, run to run, fault detection, and correction, etc., plus an equipment integration layer beneath the MES to connect into the equipment controllers. Some back-end facilities will not need all of these elements. For example, the MES may play the role of line or equipment control in back-end facilities. In fact, each company or facility may have a different scope of capabilities that it calls MES. Similarly, the solution providers have different capabilities included in their systems and overall company portfolios.

In many ways, MES is the top layer of automation in a semiconductor facility. It is the control system for the entire facility, end to end. Yet it is also increasingly an enterprise system, as companies deploy the same commercial software in all facilities. It is also an enterprise system in terms of delivering critical information to other enterprise applications and to partners both in the supply and customer bases.

Note also that the specific needs may vary by segment. Foundries that cater to specific semiconductor customers' needs with stringent service level agreements have different priorities and pressures, for example, than memory and simpler component makers whose products may be inventory items for a wide array of customers and distributors. Processor makers operate differently from either of those, striving to exactly meet overall market needs. Yet within those different focuses, the basic capabilities of the MES tend to be similar.

38.3.1 MES Core or Platform

At the core of nearly every commercial MES is a platform that delivers core system capabilities to an array of functional modules, which we'll discuss in the next subsection. Having a modern platform can make all the difference in a company's ability to be agile and fully leverage the MES functional modules.

The MES platform supplies the following capabilities to the overall system:

- *Modeling and workflow.* For MES to effectively support a plant and keep the processes flowing, it needs to have a model of the plant including the equipment and production process flows. Older MES are typically primarily a workflow engine that delivers this modeling and process support. In newer systems, the modeling and workflow are still central, but are not expected to be the mechanism by which functionality is developed. Rather, this modeling creates a coherent view of the facility while the workflow helps to express and enforce the standard operating procedures used in the wide array of functional modules.
- *Process change management.* As companies expand their global networks of facilities, the ability to change processes in a consistent, managed way across all sites in an enterprise becomes critical. Modern MES is able to deliver that capability more automatically than ever before.
- *Data management.* The MES core or platform also provides consistent and central data management for all of the functional modules to share. This data management capability creates a streamlined mechanism to gain an integrated end-to-end view of the information generated, stored, and used in one production facility—and increasingly across the end-to-end process of multiple internal and contractor facilities. The tricky part these days is that the data may come from autonomous IoT-enabled equipment or products, so the MES needs to have a way to manage this decentralized information effectively.
- *Role-based GUI.* Modern MES not only has a graphical user interface to make it more intuitive, but increasingly one that matches each user role. Further, some modern systems are ensuring that the GUI can be tailored for a specific set of users in a specific company or facility. Others are also ensuring responsive design to enable mobile users to have intuitive and full-featured access when they are either remote from the facility or performing work that has them moving throughout a facility.
- *Security.* Naturally, system security is of the utmost importance since the MES handles company-specific product and process information. This prized intellectual property (IP) must be protected with the most up-to-date security approaches at every level of the software stack. Particularly with so many partners often involved in production, the MES must build in assurances that all users are authorized and that data flowing to and from sensors and equipment as well as among value chain partners is protected. MES in semiconductor was early to move beyond the GUI and API level to ensure entire screens are only visible to authorized users and into object-level security to allow each user access only to the specific products and processes that pertain to their role.
- *Integration framework.* Being in the center of the IT stack also means that MES must have effective integration both up into enterprise systems and down into automation and equipment layers. What's new here is the level of automation including robots, as well as the need to connect into non-fixed data sources such as those from IoT both inside the facility and beyond. We will cover this further in Sec. 38.4, but here we will say that modern MES have a standard way of working with these other systems that can streamline both integration and validation of data flows.

Figure 38.2 is a conceptual view of modern MES capabilities. It shows that the MES platform provides core support capabilities to an array of functional modules. This diagram cannot show all of the functions and interdependencies; it is an overview that may break out differently for various commercial offerings.

38.3.2 MES Functions/Modules

Typically many MES functional modules sit on top of the MES platform. The modules can be relatively independent of each other, though some will be prerequisites for others. With this modular approach, MES software providers have been able to tailor their offerings for specific industries, customers, and facilities. Using a subset of modules delivers faster projects from conception to go-live, ensures that users are trained on a scope they can handle, and allows companies to migrate at their own pace away from former systems and processes for specific functions.

We cannot possibly list all of the functional modules that a company might need by the names they or their software provider would call them. However, this list provides an overview of typical functional areas deployed in the semiconductor industry through MES modules, grouped by discipline.

Manufacturing Management

This is the set of functionality used by operators and supervisors in the production facility directly and daily. Along with dispatching, these tend to be the functions that have been available in MES for decades. And yet, the realities of some of these areas have changed, and software functions must be capable of supporting these new realities. The descriptions below focus on the scope and capabilities needed for current and future semiconductor production success.

- *Data collection.* Today, much of the data gathering is automatic from equipment and the automated control systems that work with the equipment. The volume of data collected in semiconductor fabrication and test processes has always been very high, but the complexity of the devices and the processes involved has grown exponentially. So the MES data collection and management must be inherently scalable and available to manage new types and formats of data as change continues.
- *Tracking and visibility.* WIP tracking was the original focus of MES, and it continues to be essential. The complexity is only growing as hierarchies such as container-wafer-die in a fab or in the back-end, where container and wafer tracking in preassembly moves then into die-specific lead frames that go into different containers with different sizes (magazines, burn-in boards, test boards, etc.). Customers and operational improvement efforts now often demand genealogy information

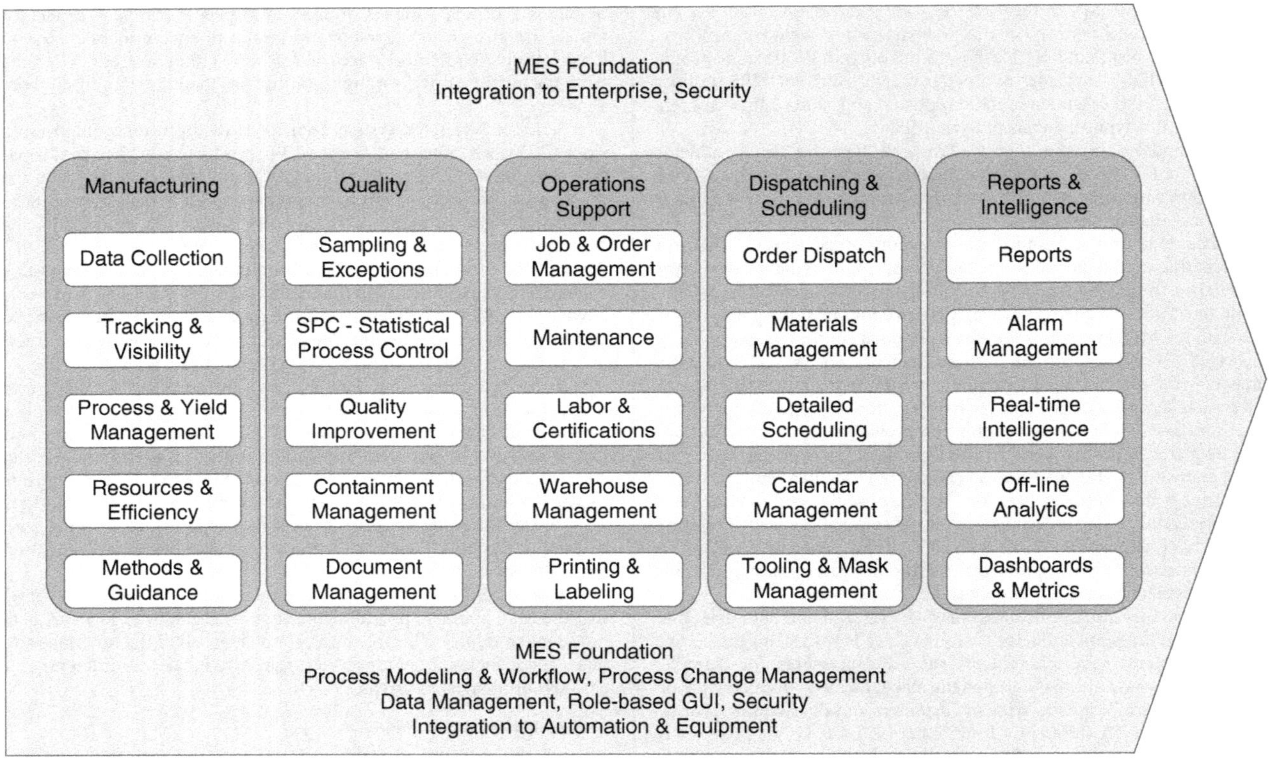

Figure 38.2 A conceptual view of the MES platform and MES functional areas, each of which might include several software modules that are optional.

to a very granular level. This adds urgency for MES software to provide tracking and visibility not only to the wafer or lot level, but to the source material (ingot), carrier, strip or single die or IC; the same is true for critical consumable materials and fixtures used in the process. This level of detail again requires the ability to handle far greater volumes of data. This also includes delivering real-time views that allow for in-process correction.

• *Process and yield management.* While the foundation layer of MES delivers the ability to model processes, the functional modules for process management deliver the in-process support. This encompasses tracking inputs from other modules such as SPC and event and nonconformance management. It also includes process enforcement and feeds actual data into intelligence systems for analysis.

• *Resources and efficiency.* As both back-ends and front-ends become more automated, the focus actually does not shift away from labor, machinery, materials, tooling, and other durables, but makes the real-time management of these resources more critical. The higher mix of products and need for more frequent changeovers also suggests that managing not only the actual resources but their reconfiguration to meet each set of needs is key to a speedy and profitable production operation.

• *Methods, instructions, and guidance.* As automation has taken hold, the number of operators and technicians in a facility—and the roles they play—has changed. With the barrage of new products and process improvements, improving methods is a constant effort and the people require real-time support.

Quality Management

Quality is one of the areas that new applications in industries such as automotive, medical, and avionics have impacted most strongly. These mission-critical applications involve human safety and thus the expectations for perfect quality are extremely high.

• *Sampling and nonconformance management.* Most companies sample their products during production, and record exceptions found either during sampling or at any point in the production and testing processes. The software needs to support a rigorous sampling process that is dynamic based on quality results. It also needs a way to record nonconformances or exceptions in some level of detail at any point in the process. Complete records of sampling and nonconformances are foundational both for quality assurance and for analysis to craft improvement mechanisms.

• *SPC.* Statistical process control is a foundational view of the many and complex processes in the journey a semiconductor makes from raw material to finished product. Having this statistical monitoring built into the MES rather than separate is critical for a comprehensive view of the process. It delivers visibility so the team can see when a process drifts out of the control limits.

• *Quality improvement.* This includes support for lean and six sigma activities as well as corrective and preventative actions (CAPA). These capabilities must connect tightly to the intelligence modules for effective analysis of trends, root causes, and effectiveness of the corrective and preventive improvement actions the company takes. As pressure mounts, the insights are ever more sophisticated and the improvement initiatives may focus on specific sites on a wafer and locations in a carrier or on a strip.

• *Containment management.* When a quality problem does occur, this set of functionality delivers a standardized best practice approach to minimize its effect. In addition to having this containment logic, it helps to analyze which other products might be involved, take them out of the normal production flow, and contain the problem. This can involve quality escapes in a recall situation or containing product that is still in the production process. To work comprehensively, this requires full backward traceability to understand the root cause of a faulty unit as well as forward traceability to identify any other unit that may have been processed in the same conditions. Given the complexity of the semiconductor supply chain these days, this function may need to include some outside-the-firewall capabilities to work effectively with suppliers.

- *Document management.* For quality to be effective, it is critical to have documentation of the processes to perform as well as of the actual results of those quality activities. Processes such as advanced process and quality planning (APQP) provide a plan during the product design that serves as a documentation framework during production. Many of the quality standards also focus on documenting standard operating procedures (SOPs) and creating enforcement by documenting signatures confirming that those SOPs were followed. Electronic SOPs are a critical element to ensure instant access to up-to-date standards.

Operations Support

- *Order management.* All activities in the plant are typically triggered by a work order that the MES creates based on a production order from the enterprise software system. This functionality must be able to keep up with the complexities of multiproduct facilities and more frequent changeovers to accommodate customer demand effectively.
- *Maintenance.* The ability to coordinate maintenance opportunities with production realities is critical. As the equipment monitors its cycles and performance, that data coming into the MES should be able to trigger maintenance activities. This functional area also helps handle the spare parts, tooling, and disposable materials needed to service the equipment. It also delivers technician instructions and certifications.
- *Labor and certifications.* Automating more of the process requires having the right employees with the right training at the right place. Managing labor to improve both its efficiency and the overall smooth running of the plant can be tricky when fewer employees are overseeing the operation. Ensuring up-to-date certifications as products, processes, and equipment change—and when new quality improvement experiments are in force—is, ironically, more important than ever. This is common in customer industries such as medical devices and aerospace; given how automated a wafer fab is, this may not be as critical there as only personnel with a certain level of training and skill typically are even in the facility.
- *Warehouse management.* Production typically involves some materials stored before they enter the production process, and often at various points during and after the process as well. Understanding where materials are at all times is more challenging as products are smaller and more diverse in each facility.
- *Printing and labeling.* Increasingly, companies must also have labels or printing that identify the materials clearly to humans or to automated identification devices. Labeling can be initiated for customer or regulatory compliance reasons not related to product function. MES must provide fast, efficient, and reliable printing and labeling.

Dispatching and Scheduling

- *Order dispatch.* Clearly, as orders come out of the order management function, they must be dispatched to the right area at the right time for each step of the process. The extremely capital intensive nature of semiconductor operations has focused dispatching on optimizing the utilization of the expensive equipment. Dispatching based on routings has always been part of MES, and with the many product variants, this is increasingly critical and complex. This function rests on the underlying process modeling and workflow capabilities. The rules for dispatching orders can be quite complex, and have historically been saved separately.
- *Materials management.* Perhaps the most obvious set of resources for manufacturing are the materials. Having a clear view of exactly what inventory of each specific material is on hand and how much is where at any given time is critical. This function links into the enterprise resource planning (ERP) or supply chain management materials management function, but takes it to a much more detailed level. As materials are consumed at a process step, this function gets very specific about how much and when, and calculates inventory remaining if any.
- *Detailed scheduling.* Beyond dispatching orders, scheduling can help to optimize the use of all of the resources in the facility—people, materials, containers, masks, reticles and tooling, and equipment. Each of these has some finite capacity, and this software introduces that capacity element to deliver a schedule that can be executed in the facility. By being part of the MES, the scheduler can stay up-to-date with events and changes as they occur and recalculate both when products will be through the process and also what resources can best be used to get them through the facility. Note that this capability is still not as widely used as other functions. This is an additional level of optimization; dispatching does keep key equipment running typically.
- *Calendar management.* MES provider allows the customer to set up a calendar with shifts, holidays, and related information either globally or for each facility. This is a foundation for scheduling as well as general operations and resource management.
- *Tooling and mask management.* Tools, reticles, and masks need to be tracked, maintained, cleaned, and organized, and MES can do that. Ensuring that all of these items are cared for efficiently and effectively is critical to operations running smoothly.

Intelligence and Reporting

Beyond actually managing the activities in the production operations, companies want to understand them more deeply. There are multiple levels at which this analyzed information is useful: during the production process in real time; for process and even product engineers in an offline mode, and for managers throughout the enterprise. Modern MES delivers reporting, analytics, and intelligence that support all of these audiences and their needs. This eliminates one of the most challenging aspects of older MES: that the data that goes in does not easily come back out.

- *Reports.* Naturally a wide variety of reports on the results of operations are available. Some are high-level summaries, others are far more in-depth on specific issues or steps in the process. Static reports are still widely used and every MES can deliver them—often with some degree of configurability.
- *Alarm management.* Related to exception management in the quality module, MES can deliver alarms when either a product or process measurement is not as expected. Comparing plan to actual seems straightforward but as more automation comes into play, the volume of data being monitored means that the alarm settings must be just right to capture true problems and allow normal fluctuation to pass without alarm. MES can also alarm when someone attempts to bypass the enforcement procedures.
- *Real-time intelligence.* Beyond simply alarms, MES today can bring together the many streams of information it is gathering to deliver deeper intelligence for operators and technicians. Once the company is monitoring the right data and bringing it into the MES, the choices and implications of each become far more obvious.
- *Offline analytics.* Process and manufacturing engineers in semiconductor gain a wealth of information when the company is using modern MES that older systems simply did not supply. In addition to seeing alarm and exception logs, they can run analysis against specific data sets of interest, all within the MES, but pulling on data from controls, equipment, IoT-enabled carriers, and many other sources. Test data is just a beginning, and while it is high volume, it is not necessarily as useful to engineering as it could be until it is combined with other data to see correlations. Increasingly, MES is becoming a manufacturing analytics platform. Even while true big data analytics including nonmanufacturing data is typically best handled in specialized platforms from the data within the facilities, MES analytics can provide great insight.
- *Dashboards and metrics.* Perhaps one of the most compelling functional areas developed over the past 10 years or so is the ability to see performance metrics and real-time dashboards. Having this function inside the MES delivers a rapid way to gauge the performance of a plant, a line, and a piece of equipment or an individual. Setting it up so that each person has a personalized dashboard view of just the few top

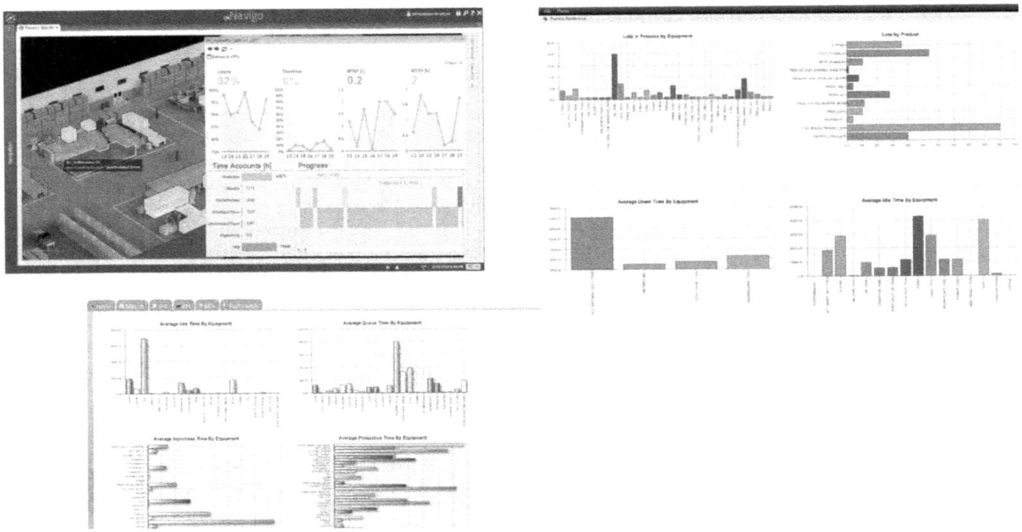

Figure 38.3 Modern MES provides dashboards and visual indicators of performance in a semiconductor facility. These are just a few of the screens available from commercial systems. (*Courtesy of Critical Manufacturing and Eyelit.*)

things in their span of control brings them the ability to prioritize what to do and evaluate the success of their activities. Figure 38.3 shows a few different examples of role-based dashboards.

38.4 MODERN MES CHARACTERISTICS AND FOUNDATIONS

Beyond the platform and modules of current semiconductor MES, there are also key ways these modern software products perform that are quite different from older systems. Some of these are crucial to effective and efficient support of today's highly distributed, global semiconductor operations. Others will become essential as product and production technologies continue to evolve.

38.4.1 Performance

As products become more complex, tracked to the device level, or applied in more demanding situations, volumes of data naturally rise exponentially. Ensuring that the MES' performance is adequate for the real-time guidance required is important. Product mix and volumes can also lead to a need for increased software performance. Some of the factors to consider for performance are

- Database structures and speed of data filing, access, and transfer mechanisms
- The ability of the MES offline analysis structure to correlate structured RDBMS data with design data and other types of data, such as that from wafer or strip maps, parametric test results, etc.
- Transaction speeds on normal hardware infrastructure as data volumes rise
- Ability to manage multiple sites and the transfers between them, particularly for operations the company is likely to maintain in-house

38.4.2 Scalability

One of the most critical factors for an MES' ability to provide ongoing support as a company and its product portfolio grow is the ability to handle more information and sustain system performance and speed. Many companies are also expanding through mergers and acquisitions, and having a common MES that can handle as much of the data as possible can be very beneficial. (This is true even if the MES does not run the newly acquired plants right away.) Scalability has to do with a system's architecture at every level. Ideal scalability allows the system's users to incrementally add more computers, faster computer processing units (CPUs), and software modules without system downtime or disruption and to model a process centrally and replicate it to other lines and sites.

38.4.3 Extensibility

Change is constant, and modern MES helps to accommodate that by allowing customers to tailor the system without writing custom code. While the extensive functionality and configuration of modern MES accommodates typically about 80 percent of a site's needs, the other 20 percent can be handled in this way. Highly trained Manufacturing IT staff is required to do this correctly, but it is within the system. Unlike traditional customization, this allows for company-specific or site-specific logic to be upgraded to new software versions with the rest of the system. This ability to extend the core product suite rather than creating separate add-on code has enormous benefits to the coherence, reliability and cost to sustain these modern MES systems. (As discussed in Sec. 38.3.1, another aspect of extensibility is the ability to configure the user interface for each role in the organization.)

38.4.4 Modularity

The fact that customers can select which modules they need for each site has many benefits. For one, it enables a single commercial system to be a good fit for a wider variety of facilities. As with other types of software, the benefits of standardizing on a single MES provider across the company can be compelling. There are now early examples where modularity and new functionality encompassed in those modules is allowing the same software to be deployed in both front-end fabrication and back-end sort, test and assembly operations. As time and quality pressure increases, that ability to tune into a single system to see the previous and subsequent stages' status and results will be increasingly valuable.

38.4.5 Logical Decentralization

Increasing automation means carriers, tooling, and equipment are now able to sense and have some intelligence about themselves with radio-frequency identification (RFID) and the industrial Internet of Things (IIoT). With this greater degree of autonomy and in-process decisions the process can be faster, more reliable and more responsive to unforeseen conditions. For the MES to play its central coordinating role, it must be able to handle this incoming data, regardless of where that intelligent device is at a given time. Traditional workflow models assumed that MES has central control, but this is no longer the case in automated facilities. So while the MES itself may still be a relatively centralized system, the logic to represent the process and workflow must accommodate a new degree of decentralization, monitoring, and reporting on activities it does not necessarily centrally control. Example applications with which the MES must collaborate in this way include the real-time dispatch software, and the material control system, both of which respond to changes in the workflow as they occur.

38.4.6 Modern Technology

In the past few years, a new world of computing has opened up, and MES must also accommodate these technologies such as IIoT, cloud, mobile, big data, social applications, and so on. While not all of these are equally important to most facilities, they all may play a role.

- IIoT: We have talked about IIoT extensively. MES must be able to manage the ever-growing data streams from IIoT-enabled products, carriers, tooling, and equipment. This is foundational, and not all systems handle it equally well.
- Cloud: Today cloud is primarily used for the functions that allow latency. It does not support operators as they do their jobs, nor any area where the connection to the automation systems and equipment are key. Yet for intelligence and analytics, for access to multiple facilities and so much more, cloud offers scalability and agility without sacrificing security. In many other industries, the focus of cloud software providers has actually created higher levels of security than an internal IT team and its staffing can possibly create, and semiconductor is likely to find some examples of that eventually as well.
- Mobile: It is clearly a sound approach for maintenance and supervisory personnel to use the MES. It is increasingly important for operators, also. As they work alongside robots and other automation, they may need to cover a large area and the cost of a tablet and secure Wi-Fi may be less than that of setting up fixed screens in all the possible work locations.
- Big data: Within the realm of production, data is an area that may or may not be fully covered in the MES today, but we expect it will be soon. To make the quality and speed improvements customers expect, semiconductor companies must better leverage many streams of data to automatically discover the patterns that drive high or low yield, fast or slow processing, and problems or successes in upstream or downstream operations. Separate big data systems will help to correlate data from across the company and the global network to the production data. We expect all of these to require special capabilities in data cleansing, normalization, correlation, and analysis.
- Social applications: There is a growing need to build out a knowledge base and allow various people in the global network to find others with expertise they need. It may not be not how the manufacturing IT teams will look at this. But the core capability of connecting with appropriate people to quickly resolve issues and to share best practices is critical. Particularly as some of the industry's most experienced professionals are nearing retirement age, this is an element to consider, even if few MES include this today.

38.4.7 Integration

MES has always been at the center of the IT stack in the semiconductor industry and the key interface between information technology (IT) and automation or operations technology (OT). The ISA-95 model referenced earlier in this chapter shows this relationship.

Automation: Equipment and Robotics

MES has always needed to connect into automation in fabs, and increasingly back-end facilities are also highly automated. As more robots and automated guided vehicles (AGVs) and advanced material handling systems (AMHS) are used, the landscape for MES to automation interoperability will continue to get more complex. So this ability to connect into the potentially wide variety of equipment and automated control systems is critical. Most equipment uses the widely deployed SEMI Equipment Communications Standard/Generic Equipment Model (SECS/GEM) standard developed in the 1980s and 1990s, but also the more recent SEMI Equipment Data Acquisition (EDA) suite of flexible, high-performance data collection standards (also known as "Interface A"). For direct application integration it may be useful to have open platform communications (OPC) and other types of application program interface (API) the MES developer makes available. Connections to automation and equipment must deliver fast and ultrareliable data exchange to ensure full control of the production process.

EES Applications

Equipment engineering has grown into a substantial software market. Equipment engineering systems (EESs) functionality, such as station/module control, fault detection and classification (FDC), process monitoring, run-to-run control (R2R), virtual metrology, predictive maintenance, recipe management, and analysis of yield and wafer map data ideally all integrates with the MES. Some of these such as R2R and FDC are also called advanced process control (APC). By ensuring that the EES and MES are exchanging information effectively, both the engineering and production disciplines are better assured of good outcomes. These more technical applications deliver yet another layer of software capability to connect the MES to equipment and the actual process. Some MES providers have their own EES as well—and others partner with separate companies but work closely together in an effort to make solutions that appear seamless to the semiconductor manufacturing customers.

ERP and Supply Chain

Naturally, MES must communicate with the order management and production planning capabilities in ERP or supply chain management (SCM) software. The demands coming in from these enterprise-level systems drive activities in the plant, and are the basis for orders, dispatching and scheduling in the MES. For order delivery date promising, materials management, and overall inventory balancing, it is critical that the MES send actual production information back to these systems as well. If the production planners and others using these systems gain visibility to not only the incoming materials and outgoing products, but also the timing and yield issues arising in the production processes, they can take a more realistic and proactive stance. Beyond the inventory and order management, the sales and customer collaboration teams that use enterprise systems will benefit from current production data being factored into the views from ERP.

PLM, EDA, and ALM

As customer collaboration demands rise and new product introduction times are squeezed, the need to integrate MES directly into the systems used by product research and development teams rises. In addition to traditional electronic design automation (EDA) systems to develop and test semiconductor designs, companies are increasingly using product lifecycle management (PLM) and application lifecycle management (ALM) systems to manage product and software data. The characteristics of actual products from a design as it goes into test runs are clearly critical input into that design team, as is information about the processes that created those product outcomes. In addition, strong interoperability between product design and manufacturing ensures that the production teams—as they become more fragmented and scattered—can always refer to the same R&D materials. Beyond the actual specs and design, the test parameters, the failure predictions, and designers' assumptions all may help production to make effective decisions that meet the design intent.

Partners' Systems

To accomplish device-level traceability across the global network, the MES ideally also gets information from suppliers to incorporate into the genealogy records. The brand owner must ensure they have a complete view of materials and previous steps. Particularly as special functions such as bumping and plating are outsourced, this picture can become quite complex. The ability to handle data in various formats and bring it into the context of the product, and also the production process that MES manages, can also be crucial to improvements over time.

38.4.8 Why These Characteristics Matter

One key driver that generates a need for this type of software is the increasing need for high mix in a single facility. The agility to change from one product to another on a weekly basis may not apply to every facility, but it will for more and more of them.

Some of these MES characteristics are also driven by the need for complete reliability, quality, and speed from R&D into scale-up and volume production. Particularly in situations where semiconductors are applied in mission-critical and human-safety related situations (self-guided vehicles, medical devices, avionics, etc.), reliability and quality are essential. For the many mobile applications, the product lifecycles are often so short that traditional design intent collaboration meetings would not be complete by the time the product must launch and start into a new version.

Some of these characteristics are simply how modern software works. This means that recent university graduates will only be versed in these systems with these characteristics and will be attracted and fulfilled by working with systems based on current technologies and concepts. Already we have seen that many semiconductor companies have struggled to recruit and retain younger manufacturing IT staff to replace the many who are retiring.

38.5 MES PROJECT CONSIDERATIONS

Semiconductor companies—and each facility—may be in different situations as they enter into the consideration of a new project for MES purchase and implementation. While the overall MES needed may not change much, the approach to the project and the core team needed to succeed is different. An MES selection and implementation project typically spans several years, though it can be quicker. Requirements gathering, review and fine-tuning as well as project planning to minimize or avoid downtime in the facility are critical to success of the project. Some of the key considerations for various common situations follow.

38.5.1 Greenfield

The price and risk of MES in this situation, in which the MES is just one element of the infrastructure of IT and OT that layer onto the physical plant, are comparatively low. The largest challenges in this situation tend to be that you do not know what you do not know. The plant and equipment are not yet in place, so some of the specifics about what the MES will need to model, interface to, and manage are unknown. Here, the characteristics of flexibility and quick model changes, interfaces to essentially any other system, and ability to quickly configure or extend the system be the foundational items to review. This situation may also involve a set of prospective users who have little to no experience with either MES or semiconductor production. So building an experienced core team with excellent communication and training skills, brought in to ensure a smooth and successful MES implementation will be essential.

38.5.2 Replacing Legacy MES

Most semiconductor facilities have an MES already, and they are producing products. The challenges with replacing older MES with a modern one include the risk of downtime, the need for master data in both systems, processing data for each lot across the long cycle, and historical information. While these pose real risks, replacing MES is no longer a guarantee of lost production time, as newer systems and approaches to the switch have been developed and successfully deployed in facilities around the world. However, some facilities simply do not have the remaining expected lifespan to make this type of project feasible. For most, it is a question of timing, and weighing the risk of inaction vs. the risk of switching out a system in a production facility. Some factors that suggest the need for a more modern system include

- The current system cannot scale to accommodate facilities' need for timely visibility and confident changes to products or processes.
- Transfers between facilities and processes within a single are causing undue issues.
- New requirements from customers or for new products cannot be met in the system without customization or within a few days.
- The system does not provide timely, in-context, ready-to-analyze data to engineers.
- Any aspect of the technology is out of support (hardware, operating system, application, etc.) or when spare parts are no longer available from the hardware supplier.
- When the skills required are not being taught in good computer science programs because they are so outdated (COBOL, FORTRAN, etc.).
- When making changes across multiple sites to ensure consistency and reliability takes many man-weeks.

The three main possible approaches for replacing a legacy MES with a modern one are (1) to simply switch everything over (big bang), (2) conduct a phased introduction of the new system for certain products or functions, and (3) running the old and new systems in parallel. There are examples of companies that have successfully deployed each of these in the past several years. In any of these cases, having a very solid program plan that factors in not only timing of procurement, design, setup, and testing, but also all of the tracking data needs for every product as well as the training needs for all personnel.

38.5.3 Replacing Paper

In some instances, companies have attempted to shift MES responsibilities into an ERP or other system that is not well suited to the tasks. The manufacturing team typically then implements an office application and paper-based system to ensure the process is tracked, managed, and the genealogy and WIP tracking is completed in some way. In these instances, the business case is clear and strong, as manual processes tend to generate additional errors and time lags in every process. However, the fact that there was not enough knowledge to discern the type of system needed to run production facilities suggests that some educational challenges are at hand. Companies must make a clear distinction between what the enterprise IT and manufacturing IT systems do, and also that the layers of software within the facility may also need some clarifying. The MES can be multi-plant or enterprise, but is very strong plantwide and designed for the timely and relatively detailed operation of a facility, which other systems are not.

38.5.4 Capabilities and Partners

Given the diversity of companies and facilities in the semiconductor industry, it is also clear that the project team may differ. Most of the megafactories will implement MES independently, not relying heavily on outside resources from either the software provider or consultants; some also have custom MES, so their team members are the world experts. Many semiconductor companies rely heavily on project expertise from the software provider and in some cases from consultants and systems integrators as well. Some companies have very few IT resources and may rely on an outside company—typically the software provider—to deliver MES in a turnkey project. The advantage of using a partner for the MES project is that they typically have seen many more recent situations and instances of MES implementations than the in-house team.

38.5.5 The Business Case

The basic elements of justifying the purchase are similar to MES justification in other industries, but fortunately semiconductor businesses rarely believe they can compete without MES. The MESA International Guidebook on *ROI and Justification for MES*[¶] is a valuable foundation for manufacturing and IT personnel to assemble a business case that finance professionals can accept.

[¶]© 2014, MESA International, *MESA Metrics Guidebook: ROI and Justification for MES*, by Darren Riley and 7 other authors who worked on this in the MESA community.

In all of these cases, it will be important to have a view of what is likely to be needed for a decade or more into the future. Some of the largest factors that have begun to change already are

- *Global value chain.* Beyond being a site-specific system, MES will increasingly need to manage multiple stages of production that are often performed at different sites, owned and partners'.
- *Internets of Things.* IoT-enabled carriers and equipment already in use in semiconductor manufacturing have already added many new requirements to MES, and new uses are likely. IoT also offers new applications and growth in volume to semiconductor companies, so companies need modern MES to scale up and also handle higher mix.
- *Device-level traceability.* Moving beyond lot tracking is already a reality and will expand to more products; this generates a tremendous amount of data as well as specific needs for the MES to be able to handle maps and very detailed genealogy data through many steps.
- *New product introduction.* Beyond the current handoff between design and manufacturing, companies will need the MES to integrate into the design systems effectively for fast ramp-up without compromise on quality.

The speed of change in the semiconductor market makes it essential to have manufacturing IT that can support the needs of the operation. Modern MES can provide that and will continue to grow and change as the market does.

38.6 FURTHER READING

A Reference Model for Computer Integrated Manufacturing (CIM) Purdue Research Foundation, 1989.

ANSI/ISA-95 standards and ISA-95 related publications: https://www.isa.org/store/products/?mstype=Publications&ms_product_type=Standards&mssearch=ISA-95.#/129a379272a3a4005515092db8d1a58b.

"Enterprise-Control System Integration Part 1: Models and Terminology," Research Triangle Park, North Carolina, USA: International Society of Automation. 2000. ISBN 1556177275.

"Enterprise-Control System Integration Part 3: Activity Models of Manufacturing Operations Management," Research Triangle Park, North Carolina, USA: International Society of Automation. 2005. ISBN 1556179553.

Harrington, J., *Computer Integrated Manufacturing*, Industrial Press, South Norwalk, Connecticut, 1974. ISBN 0831110961.

Kletti, J., *Manufacturing Execution System*, Springer, Berlin, 2010.

McClellan, M., *Applying Manufacturing Execution System*, St. Lucie Press, Boca Raton, Florida, 1997.

McClellan, M., *Introduction to Manufacturing Execution Systems*, MES Conference & Exposition, Jun. 4, 2001, Baltimore, Maryland (http://www.cosyninc.com/papers/3.pdf).

MESA International: http://www.mesa.org/en/index.asp.

Meyer, H., F. Fuchs, and K. Thiel, *Manufacturing Execution Systems*, McGraw-Hill, New York, 2009.

Orlicky, J., *Material Requirements Planning*, McGraw Hill, New York, 1975.

Vinhais, Joseph A., "Manufacturing Execution Systems: The One-Stop Information Source," *Quality Digest*. QCI International, Sep. 1998.

Scholten, Bianca, *MES guide for executives: why and how to select, implement, and maintain a manufacturing execution system,* Research Triangle Park, NC: International Society of Automation, 2009. ISBN 9781936007035.

"The Benefits of MES: A Report from the Field," Manufacturing Enterprise Solutions Association, MESA International, Chandler AZ.

Chapter 39

Advanced Process Control

BY

RAYMOND VAN ROIJEN *GLOBALFOUNDRIES, Hopewell Junction, New York*

39.1 INTRODUCTION

In the modern semiconductor manufacturing environment we routinely create wafers with millions or billions of individual components, mostly transistors, and we expect each of them to perform within rigorously defined specifications. The construction of these components depends on the execution of hundreds or thousands of process steps of various nature. We will be successful only if each of the steps on all these components is performed within very strict physical parameters. The tools of a modern fab are built to meet these demands, but by the nature of the industry, to make every new technology node exceed the previous one by a significant margin, we often require that tools operate at the limits of accuracy and reproducibility.

To enhance our capability to control product features, we want to rigorously quantify how tools are performing and leverage the knowledge we have about process and product (in the form of tool trace data and product characterization data) to adjust process parameters. Techniques to accomplish this have become indispensable in modern semiconductor manufacturing.

To keep tools operating within a narrow process window and respond to certain variations of incoming material, we can use techniques collectively known as advanced process control (APC). These techniques have in common that we use information available from the production process in real time to improve control of the tools, resulting in better yields on product. A wide variety of such methods can be found in literature. In the following we will discuss statistical process control (SPC), run-to-run (R2R) control, fault detection and classification (FDC), and virtual metrology (VM).

SPC uses statistical analysis of in-line product parameters to ensure that the product meets expectations.

R2R consists of adjusting process parameters on a tool based on some model of process control. The adjustment can be made for a lot being processed, but it can also be based on a single wafer.

FDC is based on a model of tool behavior derived from a study of a characterized process. Most modern tools provide an extensive set of trace data, that is, data of set conditions and measured data from the process over time. By comparing the actual data with an established model, we can verify the product is in specification.

VM uses the data recorded by tools during processing to predict product features, thus reducing or eliminating the need for product measurement.

39.2 STATISTICAL PROCESS CONTROL

To guarantee that products meet specifications, we measure a substantial number of critical parameters in line during manufacturing. For instance, the thickness of the gate oxide for the fabrication of MOSFET transistors is critical to functionality for multiple reasons [1]. Ideally every transistor has an oxide at the target value thickness (d_0), which will ensure it behaves according to the model used for circuit design. In the following, we assume wafers are processed in batches (of, e.g., 25 wafers). Generally, a number of instances of the oxide are measured, using an ellipsometric or x-ray photoelectron spectroscopy (XPS) tool, across a wafer and the process is repeated on multiple wafers in a batch. We can compile the data we acquire on multiple products to learn a number of things about our process. In its simplest form we use upper and lower limits d_{ul} and d_{ll} to control our process by sending an alarm or stopping a tool automatically when thickness d is outside the acceptable range. An example of a trend chart using such data is shown in Fig. 39.1.

In the example of Fig. 39.1 the process was stopped after the point exceeding the upper spec limit (first circled point). After determining the cause and fixing the issue, the process can be resumed.

However, while this will preclude recurrence of a failing process, it does not guarantee our products meet all standards. For instance, it is possible we manufacture many wafers with thickness d close to the upper or lower limit. Circuit design takes into account variability of parameters, but it assumes the thickness values (and most other parameters) to form a gaussian distribution around the target value, with a standard deviation σ. Therefore, a better control can be achieved by adding additional requirements to ensure that our process manufactures wafers that not only meet specification, but also has a distribution that meets expectation. In SPC it is common to use additional rules to achieve better control.

The so-called Western Electric rules are widely used for this purpose. A typical implementation uses limits at 3σ, the control limits mentioned above, but also shuts a process down for two consecutive points above 2σ (or below -2σ). These additional limits are sometimes referred to as upper and lower control limits (*ucl* and *lcl* in Fig. 39.1). The second circled set of points in the figure would stop the tool for violating these limits. If we find nine consecutive points above or below the target (third circle in Fig. 39.1) this would also cause an alarm. The numbers are chosen to have a low probability of sending a false alarm, that is, the chance that a well-controlled process fails the chart is very small.

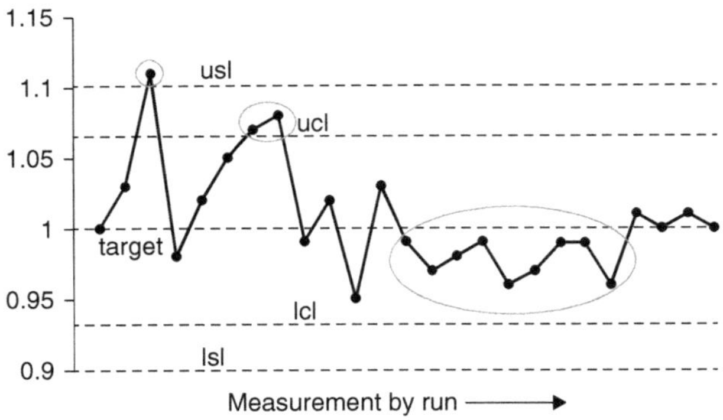

Figure 39.1 Trend chart of measurements with lower and upper spec limits and upper and lower control limits.

We characterize the distribution of a parameter by the numbers C_p and C_{pk}, where the process capability C_p is a measure of the variability

$$C_p = (d_{ul} - d_{ll})/6\sigma$$

where d_{ul} and d_{ll} are the upper and lower limits for the thickness, respectively. If d_m is the mean thickness then C_{pk} is the minimum of the difference between the mean and the upper limit and the mean and lower limit, divided by 3σ

$$C_{pk} = \min((d_{ul} - d_m)/3\sigma, (d_m - d_{ll})/3\sigma)$$

A high value of C_p means that our process is well controlled (it does not fluctuate excessively) and high C_{pk} signifies it is controlled and close to target. Requirements vary depending on the application and the maturity of a process, but a C_{pk} exceeding 1.33 is often considered to mean a process is in control.

During early technology development, we generally derive the dependence of the product properties on physical features. For instance, circuit limited yield (CLY) might depend on threshold voltage and on current of logic transistors, which in turn depend on gateoxide thickness d_{ox}. By maintaining a certain distribution of all relevant dimensions of the transistor, we can guarantee that yield will meet the standard.

Process control is not limited to measured features as discussed above. We are also watching for defects by using dedicated defect inspection tools. By using optical characterization, e-beam or other techniques, it is possible to detect defects, for example, particles of foreign material or other deviations on monitor or product wafers. As with the measurements, we can determine an acceptable level of such deviations and create an automatic alarm or a tool shutdown in case the level is exceeded.

39.2.1 Run-to-Run Control

By maintaining control charts for every operation which is critical and for which a measurement exists, we can ensure tools operate within the desired range, as long as the capability of the tools is sufficient and we only need to make incidental adjustments, for example, after regular maintenance or for malfunctioning tools. However, in some cases tool performance will drift slowly over time and the requirement on performance are such that control charts would cause alarms even for "normal" variation in performance. In such cases, rather than accepting frequent tool downs and interventions by engineers, we use R2R control of the process. R2R control is enabled by the ability of tools to be controlled by recipe download and comprehensive systems to monitor and affect the process [2, 3]. In R2R control the performance of the tool is continuously monitored by measuring the product after processing. For instance, for a deposition process the deposition rate is derived after every run. After calculating the deposition rate we can then adjust the next run to be on target.

In reality this exact method is less than ideal. Due to random fluctuation, caused by the process tool as well as the measurement, we would likely over- or undercorrect at every step. Since the distortion of every single datapoint is random, we can use averaging to get a better result. Exponential weighted moving average (EWMA) is generally found to be extremely useful for this purpose [4]. It weighs the most recent data highest, making the system responsive, but it uses older data points to avoid excessive response to a single data point. In a straightforward implementation

$$E = d_{n-1} - d_0$$
$$S = (E/x_{n-1})\lambda + S_{n-1}$$
$$x_n = d_0/S$$

where n is the sequential number of the run ($n - 1$ is the previous run), E is the deviation from target on the previous run, x is the adjustable process parameter, and λ is a weighing factor between 0 and 1. The adjustable parameter x is some process parameter with known (approximately) linear response, for example, process time, gas flow, power, etc. The constant S reflects the relationship between the process input parameter x and the response in output d.

The weighing factor λ should be chosen based on knowledge of the process. If we are correcting for a slowly drifting process, meaning the process changes significantly only after many runs, the factor can be quite small. On the other hand, if the process can exhibit large, abrupt run-to-run changes the factor should be larger. If we set λ as high as 0.5, a significant adjustment will be made based on the last readout.

This application can be a very powerful tool to improve process control and it can be readily expanded to account for many sources of variation. For example, we can take into account incoming variation caused by wafer condition or previous operations. Suppose that we target a total thickness d_0, which is the sum of some incoming thickness d_i and the process step of interest d_c

$$d_0 = d_i + d_c$$

If we know the incoming value d_i, we now have a dynamic target for the thickness

$$d_c = d_0 - d_i$$

which can be controlled in the same fashion as a static target.

As an example we use a trench in silicon etched by reactive ion etching (RIE), which is filled in a deposition step. Suppose that we are only interested in the resulting total thickness after RIE and deposition. In that case we can measure the depth of the trench and adjust the target for deposition based on the measured depth, as well as using R2R control of the deposition to improve the grown layer thickness. Finally, we could also be adjusting the RIE etch process time based on the measured trench depth after etching. The application of adaptive techniques like this has shown significant benefit in manufacturing [5–7].

A similar principle is followed by Kurihara et al. [8], improving gate CD control by including information about poly-Si thickness and shallow trench isolation (STI) step height in the R2R model.

39.2.2 Practical Considerations for Run-to-Run Control

Another useful expansion of the concept depends on our knowledge of tool and process behavior. In many cases the process parameter x will vary slightly from product to product because of pattern factor or other differences. This means we would have to have separate sets of historical data for each product to calculate the best predicted parameter x for each. However, if the variation is of similar sign and magnitude for each product, for example, if a change of S_a by 5 percent for product a also means S_b will change by 5 percent, or something very close to it, for product b, we can leverage this to apply feedback across products, using a product dependent factor F_a. We relate the parameter S_a to the reference by

$$S_a = F_a S$$

Now the equations above become

$$S = (E/(F_a\, x_{n-1}))\lambda + S_{n-1}$$

$$x = d_0/SF_a$$

The advantage of this approach is obvious: for every new run all of the most recent data are being used, which will result in the best estimate for the current run. Especially, if there is a diverse set of products, running at different intervals, the benefit will be substantial.

We need not limit the application of R2R control to physically measured properties of film thickness, and the like. We can leverage our knowledge of measured properties, compared to their target, to make adjustment to ion implant steps, which typically can be easily controlled and which have a well-understood impact on device properties. As an example, consider the case of feed forward control based on gate oxide thickness (d_{ox}) and gate length (L_g) data [9]. Since there is a well-known, inverse relation between these properties and MOSFET on current ($I_{d,sat}$), roughly

$$I_{d,sat} \sim c/(t_{ox}{*}\, L_g)$$

where c depends on the specifics of the technology, there is an opportunity to use feed forward to the device halo implant [1] to bring it closer to target.

For operations where uniformity across the wafer is a concern and the tool has the ability to vary cross wafer performance by recipe, we can use R2R control to optimize uniformity on each run [7].

One important issue potentially affecting many implementations of R2R control is measurement accuracy. Because measurements are used to direct the processing tool, we have substituted the accuracy and reproducibility of the processing tool with that of the measurement tool. Thus the accuracy of the measurement tool has become a main factor determining the quality of the process.

As we can see from the paragraphs above, the R2R control is a simple and very powerful way to improve control for tools and processes that exhibit some gradual drift over time. However, when R2R control is enabled, it is also possible that tools are misdirected because of some error in input data or other unforeseen situation. The challenge when setting up these controls is in accounting for any unusual circumstances that might cause the system to fail, such as faulty measurement data, missing historical data or an erroneous response to any type of tool error. A critical examination of potential failure conditions at the onset can prevent misprocessing. In most cases this task is more difficult and labor intensive than determining the coefficients of the feedback calculation.

In implementing these methods in a production environment it is possible to write a comprehensive code that encompasses all the elements discussed. However, it requires deep knowledge of tools, processes, and control systems to handle all elements in one single instance. Therefore, it is useful to code an application that handles all input and output (measurements, tool parameters) according to the methods outlined above, but to allow a selected set of input and system variables to be accessible to process engineers. Variables such as the target d_0 and the weighing factor λ might be stored in a so-called control file or generic cell controller [10]. Depending on the specifics of the process, there are a large number of variables that could be set and adjusted using the control file, such as limits of input parameters to respond to, a maximum to the change in one iteration of the process, absolute limits of the adjustable input parameters, etc. These limits are useful because we do not want to respond to erroneous measurement data, we do not want to make the process run outside the known process space and so forth [11].

When the R2R process control is made a permanent part of the control we should pay attention to the stream of data needed to ensure the system works. If measurement data expires on the system, we need to guarantee it does not do so when we might still need it. This is not likely to be an issue in successive runs of high volume manufacturing, which are likely to run the same process many times a day, but it could be a problem if a product dependent element in the calculation expires because it runs only infrequently. Depending on the specifics of the data system processing and measurement data might not be retrieved after a number of months or years. For instance, if we rely on some product dependent factor F_a being calculated from processing and measurement data and there is a long hiatus between the processing of a certain product, the feedback data for that product could expire. When designing the R2R control system, such limitations should be kept in mind.

39.3 FAULT DETECTION AND CLASSIFICATION

As noted above, R2R control typically relies on measurement of wafer features. The tools used in semiconductor manufacturing are capable of supplying a large number of signals relating to the processing itself and these signals can be leveraged to improve control. One established way to do this is known as fault detection and classification (FDC), where we create a model of each process step consisting of known critical parameters. Depending on the nature of the tool and the process the parameters can vary considerably: one can think of the length of time it takes to execute a part of a process recipe, the input power, the temperature, gas flow, valve position of a pressure regulating valve, etc. By studying the signals of an existing process the normal range of parameters can be established and summarized in a model. A simple but effective method is to calculate a mean value and allowed range for a parameter, or for instance an allowed range of ramp rates, depending on the process step in question. By comparing the actual values to the model it is possible to detect excursions. This can set off an alarm and inhibit a possibly malfunctioning tool. It also provides technicians and engineers with relevant information on the nature of the excursion, which is useful both for product disposition (Can the product be shipped? Is a recovery action needed?) and for determining the maintenance action needed to ensure proper functioning of the tool.

To see how FDC could be set up consider the illustration in Fig. 39.2. A wafer is heated in a tool to a target temperature for processing. The trace represents the wafer temperature, which starts with a transient reading, then rises and falls and finally rises again to the desired temperature. The target temperature is established by feedback on the tool itself and is as intended for all four wafers. In this example, all wafers reach the target temperature at the end of the cycle, but the time it takes do to so is not the same for all. Wafer 4, in particular, requires time t_2 to reach the target, which appears significantly longer than the time, at or close to t_1, for the other wafers.

Thus the trace of wafer 4 could be an indication something is wrong on the tool and it should be investigated before it is used again. However, the fact that some parameters for some wafers are different than others does not necessarily mean there is a malfunction, since they could fall within the normal variation. To answer this question we can study the

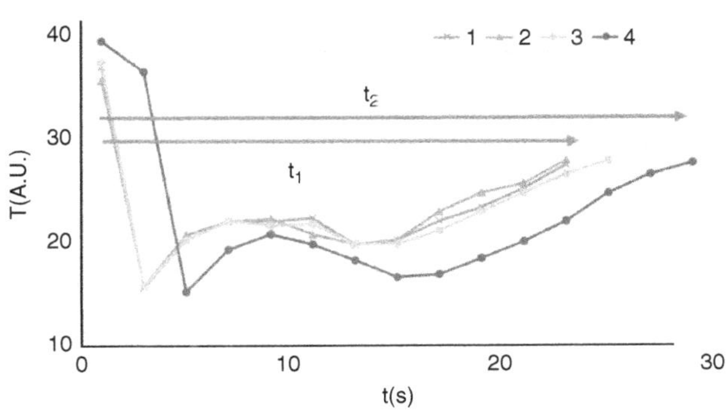

Figure 39.2 Trace data of wafer temperature in arbitrary units (AU) as a function of time, for four wafers. The time it takes to reach the desired temperature is at or close to t_1 for wafers 1 through 3, but the time for wafer 4 (t_2) appears significantly longer.

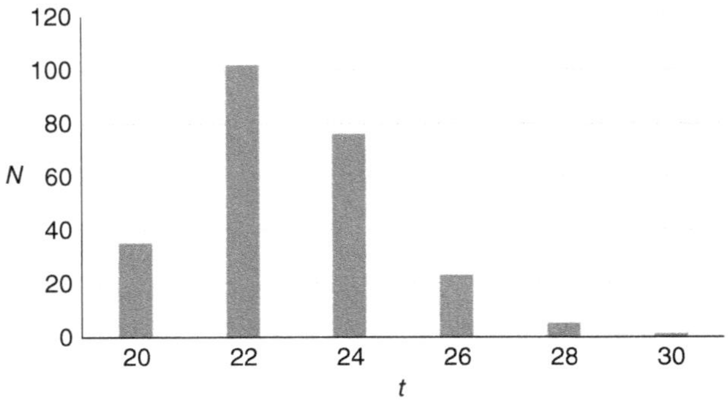

Figure 39.3 Distribution for a large number of wafers (N) of the time t to reach the target temperature.

distribution of time t_k for many wafers. If we plot the summarized data from Fig. 39.2 as a distribution of time, we might find something like the trend of Fig. 39.3. It reveals that time t_2 from Fig. 39.2 is not exceptional and should not be considered an excursion of the process.

It should also be noted that not all tool behavior lends itself to straightforward linear extrapolation. In some cases, a detailed understanding of the inner workings of a tool is required to take full advantage of the potential enhancement. See the rigorous treatment of the properties of a flywheel implanter and the nonlinear solution for a counterexample [12].

As we will argue below the sheer volume of data generated by modern toolsets provides a great opportunity for finding issues and preventing misprocessing, but the challenge lies in analyzing the data to find the indications of significant process deviation.

Though a method as illustrated above appears simple, the number of tools in a typical fab is large, and each operation on each tool in turn consists of a number of steps, such as pump down, pressure stabilization, plasma ignition, etc. At each step we can record many settings and sensor readings, for example, different gas flows, temperatures ramping up and down, plasma power, etc. Thus the number of variables to model, which is the multiplication of the tools by the different processes by all its substeps and all the parameters which are recorded, becomes very large and precludes an approach where each step is considered individually. Therefore, efforts to provide a practical and effective FDC focus on ways to find generally applicable statistical methods that allow extraction of the most significant signals and determining their limits.

On top of the challenging number of parameters that need to be monitored, we also have to allow for the variation of the conditions of different steps on each tool and the typical drift of parameters as tools go through their maintenance cycles. There is a substantial body of literature on this topic, with a focus on the issue of finding true excursions. It will be clear from the above that, given the large amount of data available and the desire not to overwhelm engineering teams with false positive readings, the aim is to find the optimal analytical technique to isolate those excursions that signify a deviation that threatens to impact product yield.

In [13] the authors describe a method to normalize data and account for correlations to identify true excursions, relying on principal component analysis (PCA) [14] to find the vectors best representing the variations.

Reference [15] also considers PCA for FDC, but notes that processes with nonlinear properties, such as some batch processes, pose a challenge for PCA. The authors combine PCA and k-nearest-neighbor rule to overcome this drawback, while simultaneously keeping the computational requirements in check.

Li et al. [16] review a number of available techniques and their strengths and weaknesses when analyzing process trace data. They propose a unified outlier detection method, which combines information theoretic and statistical techniques to meet the challenge of FDC.

39.4 VIRTUAL METROLOGY

With virtual metrology (VM), we take the extraction of relevant data from tool parameters one step further. Instead of measuring a product after processing, we estimate a critical feature by analyzing the known information about incoming wafers and the data collected during processing. An obvious advantage to this approach is that it eliminates the need for measurement, saving time, and money. We also get information on all wafers, instead of just the ones sampled for measurement.

To accurately model the physical features of the wafer after processing, we have to understand the impact of tool parameters on the product. By running a DOE (Design of Experiments) on, for example, an etch tool, we can find the sensitivity of the etch process to all available tool parameters. This information is incorporated in a VM module, which can now determine the most probable etch depth for each wafer and this information can be used for subsequent processing (feed forward) or to estimate the performance or yield.

Since the variety of processes and tools in semiconductor manufacturing is very large, VM typically relies on analysis of large datasets from specific operations to find the most relevant parameters for predicting the result of an operation. An example is described in [17]. After collecting the data of trench etching on a toolset for many runs and finding the most relevant parameters for predicting the trench depth, a model can predict the depth of the trench for every wafer in a lot, based on the measurement of just one.

As is pointed out in [18] VM provides us with data on every wafer in the lot, thus enabling us to apply R2R control at wafer instead of lot level, which will further reduce variability. The concept is refined in [19], which presents a study of VM and an illustration of the concept of the reliance index (RI) to apply the VM data weighted by a factor representative of the expected agreement with the actual result. As shown in Fig. 39.4, the RI helps to determine how much impact the VM should have on R2R control.

We should also account for tool drift and tool mismatch, that is, the gradual change of tool performance over time and the offset between tools performing the same operation. A good way to accomplish this is by keeping some measurement sampling of product, so that we can verify and correct the VM model on a regular basis. In [20] the authors present an overview of methods to ensure VM maintains an accurate model for metrology data.

While much of the discussion above focuses on correcting for tool drift and tool offset, it is also possible to apply these techniques for tool matching and predictive maintenance [21].

39.5 FUTURE TRENDS

The reader might have noticed that the methods described above share many features: they rely predominantly on the best available data of the production process, be it tool trace data or product measurement, and perform analysis and generate predictive output to improve manufacturing. All methods can, and often are used as stand-alone solutions to improve one process step, or a limited range of steps. It stands to reason that even better results can be achieved by analyzing and leveraging all data in aggregate. In [22] a detailed proposal is made to do just that by combining R2R control data and VM with scheduling capability (not discussed in this chapter; see, e.g., [23]) and yield management the productivity can be improved beyond what is achieved by piece-wise optimization of many smaller units. Moyne and Schulze refer to this as yield management enhanced advanced process control system (YMeAPC). Thus we can see the evolution of APC: going from signaling excursions to predicting behavior to optimizing the full production process.

In a general sense, much of what is discussed above relies on the extraction of some model of the manufacturing process and comparing that model to an instance of the present performance or generating a prediction of the characteristics of a tool or a product.

Extracting such models is one area where "big analytics" (or "big data") is trying to make its mark. Given the substantial amount of data generated in a fab, encompassing processing, logistics, measurement,

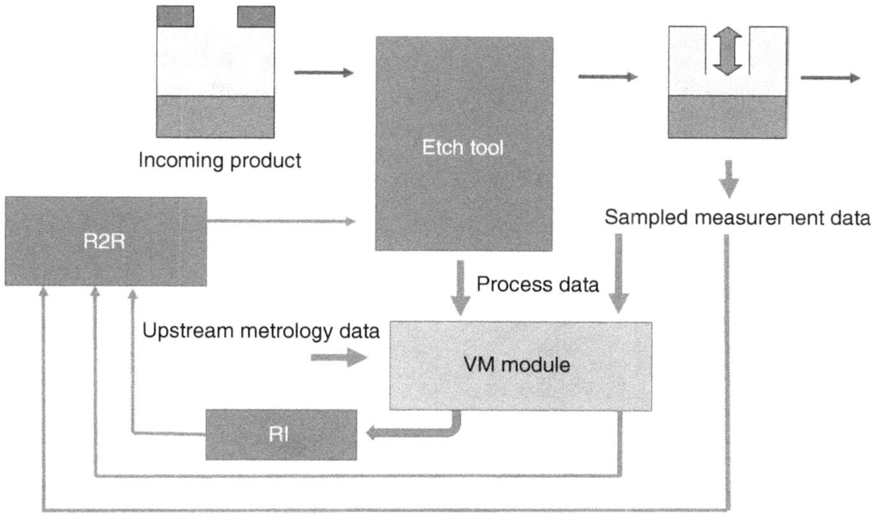

Figure 39.4 Schematic of feedback using VM with RI. The RI is used to determine the weight of the VM data in the R2R control.

test, and other data, it would seem that big analytics could harvest valuable insight beyond the "human" scale of much of what was detailed here. Indeed, big analytics is already used for logistics operations, since optimizing tool usage and product movement across the variety of tools in a fab is a task that is well suited for the capabilities of machine learning.

Reference [24] discusses the potential impact of big data and the advantages of emerging software platforms for analysis that have large scale as well as being affordable and fast.

With the increasing complexity of both the technology and the fab environment itself, it is likely that in future we will rely more and more on the assistance of smart datamining tools to find new connections and predictions to improve performance, yield, and throughput.

39.6 RESOURCES AND VENDORS

Statistical analysis and visualization:
Simca: http://umetrics.com/products/simca
SPSS: http://www.ibm.com/analytics/us/en/technology/spss/
JMP: http://www.jmp.com/
Open source solution *R:* https://www.r-project.org/
Applied Materials APC's services: http://www.appliedmaterials.com/
ASML APC's service: https://www.asml.com/
D R Yield: http://www.dryield.com/

39.7 REFERENCES

1. Yuan Taur and Tak H. Ning, *Fundamentals of Modern VLSI Devices*, Cambridge University Press, Cambridge, U.K., 1998.
2. L. F. Fuller, "IEEE/SEMI 1990 Advanced Semiconductor Manufacturing Conference and Workshop," pp. 108–111, 1990.
3. J. R. Moyne, R. Telfeyan, A. Hunvitz, and J. Taylor, "A Process-Independent Run-to-Run Controller and Its Application to Chemical-Mechanical Planarization," *Advanced Semiconductor Manufacturing Conference and Workshop, ASMC 95 Proceedings*. IEEE/SEMI, pp. 194–200, Cambridge, MA,1995.
4. S. Tseng and N. Hsu, "Sample-Size Determination for Achieving Asymptotic Stability of a Double EWMA Control Scheme," *IEEE Transactions on Semiconductor Manufacturing*, Vol. 18, Issue 1, pp. 104–111, Boston, MA, 2005.
5. R. van Roijen, C. Sinn, W. Afoh, E. Hwang, J. Scarano, S. Rangarajan, J. J. Brown, et al., "A Comprehensive Approach to Process Control," *23rd Annual SEMI Advanced Semiconductor Manufacturing Conference* (ASMC), pp. 278–283, Boston, MA, 2012.
6. Y. Sun, J. Reichelt, T. Bormann, and A. Gondorf, "A Multi-Step Wafer-Level Run-to-Run Controller with Sampled Measurements for Furnace Deposition and CMP Process Flows," *SEMI Advanced Semiconductor Manufacturing Conference* (ASMC), pp. 399–402, Saratoge Springs, NY, 2016.
7. R. van Roijen, M. Steigerwalt, J. D. Bell, E. Harley, A. Herbert, M. Fayaz, M. Brodfuehrer, et al., "Control of Epitaxial Growth of SiGe," *IEEE Transactions on Semiconductor Manufacturing*, Vol. 28, Nr. 4, pp. 480–485, 2015.
8. M. Kurihara, M. Izawa, J. Tanaka, K. Kawai, and N. Fujiwara, "Gate CD Control Considering Variation of Gate and STI Structure," *IEEE Transactions on Semiconductor Manufacturing*, Vol. 20, Nr. 3, pp. 232–238, 2007.
9. Z. Jian, L. S. Xiang, L. Ming, L. S. Yun, L. M. Chou, K. Suresh, T. M. Chin, et al., "An Integrated Advanced Process Control on RTP Gate Oxide Thickness and Feed-Forward Implant Compensation in Mass Production," *23rd Annual SEMI Advanced Semiconductor Manufacturing Conference* (ASMC), pp. 284–287, Saratoge Springs, NY, 2012.
10. J. Moyne, V. Solakhian, A. Yershov, M. Anderson, and D. Mockler-Hebert, "Development and Deployment of a Multi-Component Advanced Process Control System for an Epitaxy Tool," *IEEE/SEMI Advanced Semiconductor Manufacturing Conference* (ASMC), pp. 125–130, Boston, MA, 2002.
11. D. Boning, W. Moyne, T. Smith, J. Moyne, and Arnon Hurwitz, "Practical Issues in Run by Run Process Control," *1995 IEEUSEMI Advanced Semiconductor Manufacturing Conference and Workshop*, pp. 201–208, Cambridge, MA, 1995.
12. K. Hui and J. Mou, "Case of Small-Data Analysis for Ion Implanters in the Era of Big-Data FDC," *2013 24th Annual SEMI Advanced Semiconductor Manufacturing Conference* (ASMC), pp. 315–319, Saratoge Springs, NY, 2013.
13. J. Blue, A. Roussy, A. Thieullen, and J. Pinaton, "Efficient FDC Based on Hierarchical Tool Condition Monitoring Scheme," *2012 23rd Annual SEMI Advanced Semiconductor Manufacturing Conference* (ASMC), pp. 359–364, Saratoge Springs, NY, 2012.

14. R. A. Johnson and D. W. Wichern, *Applied Multivariate Statistical Analysis*, Prentice Hall, Upper Saddle River, New Jersey, Chap. 3, pp. 124–139, 2002.

15. Q. P. He and Jin Wang, "Large-Scale Semiconductor Process Fault Detection Using a Fast Pattern Recognition-Based Method," *IEEE Transactions on Semiconductor Manufacturing*, Vol. 23, Issue 2, pp. 194–200, 2010.

16. Zhiguo Li, R. J. Baseman, Y. Zhu, F. A. Tipu, N. Slonim, and L. Shpigelman, "A Unified Framework for Outlier Detection in Trace Data Analysis," *IEEE Transactions on Semiconductor Manufacturing*, Vol. 27, Issue 1, pp. 95–103, 2014.

17. G. Roeder, M. Schellenberger, L. Pfitzner, S. Winzer, and S. Jank, "Virtual Metrology for Prediction of Etch Depth in a Trench Etch Process," *2013 24th Annual SEMI Advanced Semiconductor Manufacturing Conference* (ASMC), pp. 326–331, Saratoge Springs, NY, 2013.

18. A. A. Khan, James R. Moyne, and D. M. Tilbury, "An Approach for Factory-Wide Control Utilizing Virtual Metrology," *IEEE Transactions on Semiconductor Manufacturing*, Vol. 20, Issue 4, pp. 364–375, 2007.

19. C. Kao, F. Cheng, and W. Wu, "Preliminary Study of Run-to-Run Control Utilizing Virtual Metrology with Reliance Index," *2011 IEEE Conference on Automation Science and Engineering* (CASE), pp. 256–261, Trieste, Italy, 2011.

20. J. Iskandar and J. Moyne, "Maintenance of Virtual Metrology Models," *2016 Annual SEMI Advanced Semiconductor Manufacturing Conference* (ASMC), pp. 393–398, Saratoge Springs, NY, 2016.

21. J. Moyne, M. Yedatore, J. Iskandar, P. Hawkins, and J. Scoville, "Chamber Matching Across Multiple Dimensions Utilizing Predictive Maintenance, Equipment Health Monitoring, Virtual Metrology and Run-to-Run Control," *2014 25th Annual SEMI Advanced Semiconductor Manufacturing Conference* (ASMC), pp. 86–91, Saratoge Springs, NY, 2014.

22. J. Moyne and B. Schulze, "Yield Management Enhanced Advanced Process Control System (YMeAPC)—Part I: Description and Case Study of Feedback for Optimized Multiprocess Control," *IEEE Transactions on Semiconductor Manufacturing*, Vol. 23, Issue 2, pp. 221–235, 2010.

23. J. Rothe, G. Gaxiola, L. Marshall, T. Asakawa, K. Yamagata, and M. Yamamoto, "Novel Approaches to Optimizing Carrier Logistics in Semiconductor Manufacturing," *IEEE Transactions on Semiconductor Manufacturing*, Vol. 28, Issue 4, pp. 494–501, 2015.

24. J. Moyne, J. Samantaray, and M. Armacost, "Next Generation Advanced Process Control: Leveraging Big Data and Prediction," *2016 Annual SEMI Advanced Semiconductor Manufacturing Conference* (ASMC), pp. 191–196, Saratoge Springs, NY, 2016.

Chapter 40

Airborne Molecular Contamination

BY

CHRIS MULLER *Purafil, Inc., Doraville, Georgia*

40.1 INTRODUCTION TO CHEMICAL CONTAMINATION AND DEFINITION OF AMC

Contamination control in the microelectronics industry still has a primary focus on the removal of airborne particulate matter. However, when the effects of airborne molecular contamination (AMC) came to the forefront with full-scale production of 0.25 μm (250 nm) semiconductor devices in the late 1990s, AMC control focused on applications such pregate oxidation storage environments, salicidation and contact formation processes, and DUV lithography environments. Wafers were still being exposed to the ambient cleanroom environment, where AMC sources both outside and inside the cleanroom presented challenges for their control.

As we moved from micrometers to nanometers the types and numbers of AMC that were targeted for control increased while the control levels continued to decrease. Although the wafer environment began to shrink down to front opening unified pods (FOUPs) and standard mechanical interface (SMIF) pods, and as wafer sizes approach 450 mm, the deleterious effects on devices with geometries that have moved past 20 nm require have become magnified and require tighter AMC control. All leading-edge manufacturers have now to varying degrees incorporated preventive airborne molecular contamination (AMC) measures into their operational, facility, and contamination control activities.

The term "airborne molecular contamination" (AMC) covers a wide range of chemical contaminants that can be present in cleanroom air. Outdoor air, manufacturing processes, fugitive emissions from process equipment, and chemical supply lines, cross-contamination between manufacturing areas, chemical storage areas, off-gassing from building and construction materials, accidental spills, and bioeffluents from cleanroom personnel can all contribute to the overall AMC load in the cleanroom. AMC can be detrimental to many processes and products and can also represent considerable health hazards to personnel.

AMC can be in the form of gases, vapors, or aerosols. Their chemical nature can be organic, inorganic, or mixed and can result in a large number of potential processing problems. Cases of uncontrolled boron and phosphorus doping, etch rate shifts, changes in the wafers' electrical properties, silicon carbide formation following pre-oxidation clean, stepper optics hazing, threshold voltage shifts, *high efficiency particulate air* (HEPA) or *ultralow penetration air* (ULPA) filter degradation, wafer metal and fab corrosion, and more have been reported. However, the effects of specific AMC on individual processing steps are still not adequately understood.

As semiconductor device geometry continues to decrease to 10 nm or smaller, the importance of chemical contamination has now become as important as particulate contamination. AMC can impact almost all aspects of submicron device fabrication, from overall fab operation to final device performance. New chemistries introduced to manufacturing processes have also been shown to cause unforeseen AMC-related effects. Hydrogen sulfide, for example, poses a significant threat to the metallization process with the transition from aluminum to copper. Copper and silicides are now common conductors for all types of integrated circuits and other electronic, optoelectronics, and nanotechnology devices. Many other new materials are being used including in photomasks for extreme ultraviolet (EUV) lithography, high-k and low-k dielectrics, metals, barrier layers, antireflective coatings, and atomic layer deposition. These materials can be new possible sources of AMCs, and effects on processes are not all understood yet. Monitoring and control methods, and understanding of reactive ion etch by-products and abatement are needed. Thus there is a shift toward equipment makers and fabs worldwide creating AMC-free manufacturing environments, including use of air filters, purifiers, FOUPs, pods, compacts, AMC-controlled factory interfaces, cluster tools, purged reticle storage and stocker areas, and purged lithography tools.

40.2 CLASSIFICATION OF AMC

The first consideration in the selection of an AMC control system should be an assessment for the type and quantities of AMC to be controlled. This can be done by direct gas monitoring or with semiquantitative techniques using passive or real-time monitors.[1,2] There can be dozens of contaminants present in a makeup or recirculation airstream, but there may only be relatively few that require close control. Contaminant classifications can help to group similar contaminant types, such as those with sources inside the fab (Table 40.1). This can help make the selection of the proper AMC control system more straightforward.

Table 40.1 Sources of Contamination

Process/equipment	Types of contamination*
Chemical mechanical polishing	Metals, organics
Resist strip	Metals, organics, oxides, residues
Wet etching and cleaning	Metals, organics, residues
Chemical vapor deposition	Metals, organics, halogens, hydrogen
Sputter deposition	Metals
Reactive ion etching	Metals, organics, halogens
Ion implantation	Metals
Oxidation and diffusion	Metals

*Particulate contamination is generated for all cases.

40.2.1 AMC Classifications

SEMI Standard F-21-1102[3] classifies AMC in cleanrooms by their chemical properties providing a way to characterize the environment by groups of materials that could have similar effects on an exposed wafer. The purpose of this standard is to classify cleanrooms with respect to their molecular (nonparticulate) contaminant levels. The classifications are defined as follows:

- *Acid (MA).* A corrosive material whose chemical reaction is that of an electron acceptor
- *Base (MB).* A corrosive material whose chemical reaction is that of an electron donor
- *Condensable (MC).* A chemical substance capable of condensation on a clean surface (excluding water)
- *Dopant (MD).* A chemical element that modifies the electrical properties of a semiconductive material

Another group of chemicals that has become more important for contamination control is refractory compounds. This category refers to volatile organic compounds (VOCs) or semivolatile organic compounds (SVOCs) containing, for example, sulfur (S), phosphorus (P), or silicon (Si) that can form surface molecular contamination on reticles, optics, inspection equipment, etc. Some of the more common refractory compounds are listed in Table 40.2.

Some chemical compounds of interest may be present in more than one category as shown in Table 40.3.

Given the unfavorable publicity that personnel exposure to potentially toxic and hazardous materials in semiconductor manufacturing has received, AMC control systems must also be designed to protect people as well as products. Thus, another way to characterize AMC would be based on the potential health effects to exposed personnel. In this light, we could generally classify AMC as below (in order of decreasing severity):

- *Toxic.* Substances that can cause damage to living tissue, impairment of the central nervous system, or in extreme cases, death

Table 40.2 Refractory Compounds Encountered in Semiconductor Fabs

Chemical name	Formula	CAS No.
Cyclopolydimethylsiloxane	$(\text{-Si(CH}_3)_2\text{O-})_n$	69430-24-6
Decamethylcyclopentasiloxane	$(\text{-Si(CH}_3)_2\text{-O-})_5$	541-02-6
Dodecamethylcyclohexasiloxane	$(\text{-Si(CH}_3)_2\text{-O-})_6$	540-97-6
Hexamethylcyclotrisiloxane	$(\text{-Si(CH}_3)_2\text{-O-})_3$	541-05-9
Hexamethyldisilazane	$(CH_3)_3SiNH\text{-}Si(CH_3)_3$	999-97-3
Hexamethyldisiloxane	$C_6H_{18}OSi_2$	107-46-0
Octamethylcyclotetrasiloxane	$(\text{-Si(CH}_3)_2\text{O-})_4$	556-67-2
Total phosphates	$(RO)_3P{=}O$	NA
Total siloxane and silicone compounds	Not applicable	NA
Tri(n-butyl) phosphate	$(C_4H_9O)_3P{=}O$	126-73-8
Tricresyl phosphate	$(CH_3C_6H_4O)_3P{=}O$	78-30-8
Triethyl phosphate	$(C_2H_5O)_3P{=}O$	45-40-0
Trimethylsilanol	$C_3H_{10}OSi$	1066-40-6
Tris(1-chloro-2-propyl) phosphate	$((CH_3)\text{-}(ClCH)CH_2\text{-}O\text{-})_3P{=}O$	13674-73-9
Tris(chloroethyl) phosphate (TCEP)	$(ClC_2H_4O)_3P{=}O$	306-52-5

- *Corrosive.* Substances that are likely to cause deterioration or damage to the interior of a building or its contents
- *Irritant.* Substances that can cause discomfort, and potentially permanent damage, to an exposed person
- *Odorous.* Substances that primarily affect the sense of smell

For personnel protection, various regulations and guidelines are in place pertaining to exposure to chemical contaminants. The guidelines for exposure to materials listed in Table 40.4 are those established for worker exposure in (typically) industrial environments.

Although none of these classifications are definitive, an environmental assessment can at least provide the AMC control system designer, a starting point with an indication of the types of contaminants that may be present and levels to which they must be controlled.

Even with a classification system in place, it does not always guarantee a quick and easy selection of the most effective and economical AMC control system. What it does do is help determine if absolute contaminant control is required or whether a system with a fractional efficiency could be employed.

40.2.2 AMC Effects

In much of the world today, the outdoor air contains levels of ozone, sulfur and nitrogen oxides, and VOCs high enough to cause problems

Table 40.3 Classifications of Molecular Contaminants and Examples

Acids	Bases	Condensables*	Dopants	Refractories
Hydrogen fluoride (HF)	Ammonia (NH_3)	Propylene glycol methyl ether acetate (PGMEA†)	Boron trifluoride (BF_3)	Trimethylsilanol (TMS)
Hydrogen chloride (HCl)	N-methyl pyrrolidone (NMP)	NMP	Boric acid (H_3BO_3)	Hexamethyldisiloxane
Sulfur dioxide (SO_2)	Triethylamine	Trimethylbenzene	Tris(chloroethyl) phosphate (TCEP)	Octamethylcyclotetrasiloxane (D4)
Hydrogen sulfide (H_2S)	Trimethylamine	Triethyl phosphate (TEP)	TEP	TEP
Acetic acid (CH_3COOH)	Ethanolamine	Butylated hydroxytoluene (BHT)		Tetrachloroethylene

*Condensables as defined by SEMI (boiling point > 150°C).
†PGMEA is at the limit of the SEMI definition. Some companies choose to use a different definition for condensables using benzene or toluene retention time from GC/MS chromatography as a lower limit, to include more species known to create problems (e.g., xylene, PGMEA).

Table 40.4 Chemical Substances Common to Semiconductor Manufacturing

	Contaminant types			
	Toxic	Corrosive	Irritants	Odorous
Representative compounds	Ammonia Arsine Boron trifluoride Chlorine Hydrochloric acid Hydrofluoric acid Methyl alcohol Nitric acid Methylene chloride Phosphine Phosphorus pentafluoride Silicon tetrafluoride 1,1,1-trichloroethane	Acetic acid Ammonia Chlorine Fluorine Hydrochloric acid Hydrofluoric acid Nitric acid Nitrogen dioxide Ozone Sulfur dioxide Sulfuric acid	Acetic acid Acetone Ammonia Chlorine Formaldehyde Isopropanol Methyl alcohol NMP Ozone PGME PGMEA Toluene Xylene	Acetic acid Ammonia Butyl acetate Chlorine Formaldehyde Hydrofluoric acid Hydrogen sulfide Isopropanol NMP PGME PGMEA Phosphine Xylene

in cleanrooms. There are elevated levels of sulfur and nitrogen oxides from automobile exhaust and combustion processes in urban areas. Atmospheric chlorine and boron can be found in coastal locations. Ammonia and amines may be present from agricultural activities. AMC are also present at trace levels in the cleanroom due to the off-gassing and/or release of process chemicals in and around the manufacturing facility (e.g., exhaust stacks), that can result in cross-contamination between process areas. Examples include amines from chemical mechanical planarization (CMP) processes are detrimental to lithography processes; acids from wet benches will impact metallization processes.

AMC can cause damage to raw materials, in-process and finished products, manufacturing, and mechanical equipment. Some of the effects that have been well documented are listed below:

- Deep ultraviolet (DUV) photoresists *T-topping*
- Uncontrolled boron or phosphorus doping
- Etch rate shifts
- SiC formation following pre-oxidation clean
- Gate dielectric failures
- Adhesion failure
- Threshold voltage shifts
- Resistivity shifts
- Nucleation irregularities
- Voids
- Wafer, mask, and stepper optics hazing
- Decreased optical transmission
- Optics hazing in inspection equipment
- High-contact resistance
- Corrosion on disk media and recording heads
- HEPA/ULPA filter degradation
- Ineffective wetting and cleaning
- Facility/equipment corrosion
- Reaction with corona ionizers forming dendrites, imbalances, long decay time, and nanoparticulates

40.3 AMC CONTROL CONSIDERATIONS

Although there will be differences in how AMC control will be applied, there are general guidelines that can be considered. Whenever possible fab locations should be selected that are not near point or mobile AMC and particulate sources, and not in highly polluted urban areas, near sewer plants, dumps/landfills, chemical plants, oil refining, heavy industry, etc.

Low emitting materials need to be used for critical environments and there are guidelines for characterizing outgassed organic compounds from cleanroom materials and components. Recommended practice IEST-RP-CC031.3[4] describes a test method appropriate for semi-quantitative determination and qualitative characterization of organic compounds outgassed from materials or components exposed to air or gases in cleanrooms or other controlled environments. The method is designed primarily to screen cleanroom materials and is relevant to industries that may experience adverse production yields as a result of gaseous organic contamination, mostly as VOCs and SVOCs. In the semiconductor industry, the deposition of outgassed compounds on hardware, products, and wafer surfaces is recognized as a source of processing problems and hardware failures.

40.3.1 Overview of AMC Control Systems

1. Makeup air systems will see primarily atmospheric contaminants if the outside air intake locations are chosen carefully. Zero downtime systems should be considered for makeup air handlers. A balance should be struck between the desired levels of AMC control versus pressure drop versus service life.

2. Recirculation air systems require that AMC control be chosen based on functional area requirements.

3. Exhaust air systems generate a significant number of complaints from neighboring facilities due to nuisance odors from exhaust abatement equipment. Continuous monitoring and regular maintenance is called for. Careful location of exhaust stacks to prevent re-entrainment, compliance with environmental regulations, and dispersion modeling, with due consideration to locations of air intakes and using stack heights and air velocities sufficient to disperse contaminants, is required for all production facilities.

4. Minienvironments with self-contained chemical contaminant control systems can significantly reduce AMC cross contamination and the occurrence of odors related to production processes. Added benefits include increased protection for the product and the reduction of the total exhaust requirements.

Source control for outdoor air contaminants is often not feasible or practical, therefore, ventilation control would be a likely option explored for general or specific AMC control. However, this may not prove viable because the increased use of dilution air is neither cost-effective nor energy efficient. Further, bringing in additional quantities of outside air could result in substituting one group of contaminants for another—those with sources outside the cleanroom for those that are internally generated. In areas with poor outdoor air quality, air cleaning must be employed.

A chemical filtration system as an integral part of a cleanroom HVAC system, can effectively reduce AMC to levels that are at or below the level of detection for the monitoring techniques employed. Properly applied, gas-phase air filtration also has the potential for energy savings.

Makeup air systems must typically be designed to control SO_x, NO_2, HNO_x, ozone, VOCs, and some site-specific contaminants such as hydrogen sulfide, organic sulfur compounds, chlorine, organophosphates, and ammonia. Chemical filtration equipment in recirculation systems must be designed to remove a wide array of acids, bases, hydrocarbons, and other VOCs. As a rule, organic compounds, although perhaps not the most damaging, are the most abundant types of AMC found in these facilities.

Table 40.5 Air Quality Survey Showing Levels of AMC in Different Areas of a Fab

Contaminant measured	Outside airstream	Recirculation airstream	Cleanroom air	Minienvironments
Acetic acid	3.0 mg/m^3	242 μg/m^3	8.8–44.2 μg/m^3	
Acetone	31–166 mg/m^3	3.9 mg/m^3	38.95 μg/m^3	
Ammonia				1.46–3.48 mg/m^3
Benzene	11.6–60.4 μg/m^3	3.13 μg/m^3	5.61 μg/m^3	
n-Butyl acetate	0.32 mg/m^3		2.63 μg/m^3	
Chlorine		1.2–10.9 μg/m^3	5.6–12.3 μg/m^3	
Decane		1.87 μg/m^3	2.26–6.11 μg/m^3	
Dichlorobenzene			0.60 μg/m^3	
Ethoxyethylacetate			1.67 μg/m^3	
Ethylbenzene	5.8–8.8 μg/m^3	0.97 μg/m^3	10.15 μg/m^3	
Ethyltoluene		0.99 μg/m^3	1.60–7.67 μg/m^3	
Formaldehyde	12.28–49.13 μg/m^3		38–65 μg/m^3	
Freon 113	4.4 mg/m^3			
Hydrogen sulfide				0.31–0.70 mg/m^3
Isopropanol	1.5 mg/m^3			
Isopropylbenzene			0.79 μg/m^3	
Methyl ethyl ketone	1.2 mg/m^3			
Methyl mercaptan				0.04–0.10 mg/m^3
Methylamine				0.15–0.38 mg/m^3
Nitrogen dioxide	0.291–1.457 mg/m^3			
NMP	0.65 mg/m^3	12–25 μg/m^3	30–40 μg/m^3	0.82 – 2.5 μg/m^3
Ozone	0.10–0.40 mg/m^3	20–80 μg/m^3	10–20 μg/m^3	
Styrene		1.19 μg/m^3	2.36 μg/m^3	
Sulfur dioxide	0.08–0.37 mg/m^3	15–200 μg/m^3	10–15 μg/m^3	
Tetrachloroethylene		0.99 μg/m^3	3.97 μg/m^3	
Toluene	33.3–157.6 μg/m^3	0.93 mg/m^3		
Trichloroethane		3.09 μg/m^3	40.98 μg/m^3	
Trichloroethylene			0.61 μg/m^3	
Trimethylbenzene		0.79–1.77 μg/m^3	1.80–9.91 μg/m^3	
Xylene	0.34 mg/m^3	7.3–153.9 μg/m^3	9.28–36.39 μg/m^3	
TVOC	195 μg/m^3	84–110 μg/m^3	129.67 μg/m^3	17.85 mg/m^3

NOTE: Data is reported in both milligrams per cubic meter (mg/m^3) and micrograms per cubic meter (μg/m^3).

An example of an air quality survey performed at one facility is shown in Table 40.5. It shows the types and amounts of AMC being introduced from the outside, the effect of dilution (if any) with ventilation air, as well as the contamination being introduced from inside the facility.

40.4 IMPLEMENTING AMC CONTROL

40.4.1 Three-Step Methodology for Optimum Control of AMC

AMC control has now become fully integrated into the cleanroom environmental management requirements of high-tech manufacturing facilities. The optimum control of AMC involves three steps:

1. *Assessment* of the air quality both outside and inside the facility to identify target contaminants as well as those that could affect the performance of the AMC control system
2. Selection and qualification of an AMC *control* system
3. Ongoing *monitoring* of both the controlled environment and the performance of the AMC control system[5]

Cleaning the outside air being used for ventilation and pressurization, removing tramp and fugitive emissions from recirculation airstreams, and cleaning process emission and exhaust streams are some of the many applications that require AMC control. Just as there are a wide variety of AMC control applications in the cleanroom, there seems to be an equal number of control options available to contamination control personnel.

With what may seem to be an overwhelming number of items to be considered, how is the contamination control engineer supposed to make the proper choice for a particular application? Following are some of the key issues to understand and the steps one needs to take to establish a successful AMC control program.

40.5 GAS-PHASE AIR FILTRATION PRINCIPLES

Gas-phase (chemical) air filtration, as applied in mechanical ventilation systems, uses two main mechanisms to remove airborne gaseous contaminants. One, which is a reversible physical process, is known as *adsorption,* and is typically used for removal of nonpolar organic contaminants and those with higher molecular weights (MW > 80). The other, which involves adsorption and irreversible chemical reaction(s), is termed *chemisorption*, and is used to control reactive chemicals such as acids and bases, as well as other compounds such as hydrides. Before chemical filtration media or equipment are discussed, it is appropriate to present a brief overview of gas-phase air filtration at this time.

40.5.1 Adsorption

Adsorption is a surface phenomenon and the removal capacity of an adsorbent is directly related to its total surface area. Therefore, it is important to develop as large an accessible surface area per unit volume as possible. Granular activated carbons (GACs) and activated aluminas are the most common materials that fulfill this requirement.

The rate at which adsorption occurs (and to a large degree the filtration removal efficiency) is dependent on the time required for the gases to diffuse from the external surface to the interior adsorptive sites. Therefore, this rate is inversely proportional to the size of the adsorbent. The removal capacity, however, depends on the total available surface area of the adsorbent. Since the internal surface area of the walls of the pores makes up practically all of the surface area of the adsorbent, the capacity is independent of the particle size.

Adsorbed water reduces the capacity of the media for the target gases due to a reduction in the number of available adsorptive sites. Therefore, adsorption occurs more readily at lower temperatures and humidity. The reverse is true for higher temperatures and humidity. Also, at higher temperatures, the rate of gas desorption increases. Care should be taken to avoid sudden temperature and humidity variations, such as when adsorption media are used in makeup (outdoor) handlers to avoid desorption of gaseous contaminants back into the supply airstream.

The adsorptive capacity of an adsorbent media is also a function of the contaminant concentration. The higher the contaminant concentration, the more that will be adsorbed, up to an adsorptive capacity of the media, usually for saturated organic compounds.

Because of the relatively weak forces holding the adsorbed gases on the adsorbent, adsorption is (essentially) totally reversible. Thus, the net rate of adsorption depends on the rate at which gas molecules reach the surface of the adsorbent, the percent of those making contact which are adsorbed, and the rate of desorption. However, many other factors can affect the removal of gaseous contaminants by physical adsorption. Among these are the type of adsorbent, the resistance to airflow, the adsorbent bed depth, the temperature, relative humidity, and velocity of the air through the bed, the characteristics of the contaminant(s) in the space around the adsorbent, and the removal efficiency required.

40.5.2 Chemisorption

Adsorbent materials do not adsorb all contaminant gases equally. For instance, higher molecular weight (>80) gases and nonpolar gases may be preferentially adsorbed over lower molecular weight gases and polar gases when present in the same airstream. Also, if these lower-weight/nonpolar gases were initially adsorbed, they may be desorbed by the introduction of a higher-weight/polar molecule. If the sorbent bed is of sufficient depth, these displaced molecules may again be adsorbed. However, at some point, they would be released back into the airstream. One way to improve the effectiveness of sorbents for these materials is by the use of various chemical additives (impregnants) that will react with these "less-adsorbable" gases. These impregnants react (essentially) instantaneously and irreversibly with these gases forming stable chemical compounds, which are bound to the media as organic or inorganic salts or released into the air as carbon dioxide, water vapor, or some material more readily adsorbed by other adsorbents. Therefore, it is very common to have a gas-phase air filtration system that uses a combination of unimpregnated and chemically impregnated adsorbent media.

In contrast to the reversible process of physical adsorption, chemical adsorption, or chemisorption is the result of chemical reactions on the surface of the adsorbent. Chemisorption is specific and depends on the chemical nature of both the adsorbent medium and the contaminants to be controlled. Some oxidation reactions have been shown to occur spontaneously on the surface of the adsorbent. However, a chemical impregnant is usually added to the adsorbent, which makes it more or less specific for a contaminant or group of contaminants, for example, activated carbon impregnated with potassium hydroxide (KOH) and/or potassium carbonate (K_2CO_3) for the removal of acid gases such as chlorine, hydrogen fluoride, and sulfur dioxide. Although some selectivity is apparent in physical adsorption, it can usually be traced to purely physical, rather than chemical, properties. In chemisorption, stronger molecular (valence) forces are involved. Many of the same factors that affect the removal of gases by physical adsorption also affect their removal by chemisorption.

The chemical reactions that occur during chemisorption are favored, to an extent, by higher temperatures and humidity. Higher temperatures increase the rates of reaction and the extra water enhances the ability of the adsorbed gases to contact the chemical impregnant. Gases that are adsorbed but not chemically reacted are subject to the same effects of temperature and humidity as for plain activated carbon, and as with plain carbon, desorption of these gases is also possible. High and variable humidity and the concurrent risk of condensation must also be controlled; therefore, it is preferred to install chemical filtration systems in the tempered airstream and not at the inlet of the makeup air handling units.

Chemisorption may be employed whenever physical adsorption by itself is inadequate or ineffective against a particular contaminant or group of contaminants. Desorption of target contaminants, once adsorbed and chemically reacted, does not occur.

40.6 DRY-SCRUBBING AIR FILTRATION MEDIA

40.6.1 Adsorbents/Chemisorbents

Practically all commonly available chemical filtration media today are manufactured from (granular) activated carbons, activated aluminas, and/or ion exchange resins/fibers; however, they do not remove all contaminant gases equally. Specific chemical additives, or impregnants, added during the manufacturing process impart special characteristics to the media and make them more or less specific for various chemical species. Many different chemicals are being used on activated carbon for general and specific applications. However, one of the more broad-spectrum chemical impregnants in common use, sodium permanganate ($NaMnO_4$), cannot be effectively used with activated carbon, and, therefore, is almost exclusively used as an impregnant on activated alumina (Table 40.6).

These impregnants react essentially instantaneously to form stable chemical compounds that are irreversibly bound to the media as inorganic or organic salts or released into the air as CO_2 and/or water vapor.

The optimum manufacturing technique for impregnated media is via a dry feed process in which the chemical-impregnant solution is added to the powdered adsorbent material(s) in such a way as to ensure a uniform distribution throughout in the finished product. The medium is then cured to have sufficient hardness to prevent particle attrition in handling and use, but still maintain a well-developed pore structure to allow gas molecules to be physically adsorbed and/or chemically reacted. These media are available in number of particle sizes and size distributions that can be well controlled in order to maintain product uniformity and predictability in use.

Sodium permanganate-impregnated alumina (SPIA) is often used in conjunction with plain or impregnated GAC to provide a very broad-spectrum gas-phase air filtration system. It is fairly common to have a chemical filtration system that uses a combination of unimpregnated and impregnated adsorbent media to provide removal of a wide range of AMC. However, this multimedia approach continues to be the exception rather than the rule, as many manufacturers prefer to focus their AMC control efforts on the control of a specific contaminant or contaminant type.

40.6.2 Adsorbent-Loaded Nonwovens

The optimum control of airborne contamination is achieved by the application of individual stages of both particulate and chemical filters in the air handling systems. Particulate control in semiconductor cleanrooms requires as many as five stages of filters up to and including HEPA/ULPA filters to achieve the required level of

Table 40.6 Common Types of Chemical Filter Media

Media composition	Target gases (types)
Granular activated carbon	Volatile and semivolatile organic compounds (VOCs/SVOCs), condensables, refractory compounds, organic acids, organic sulfur compounds (mercaptans)
Activated carbon w/wo activated alumina with potassium hydroxide (KOH) and/or potassium carbonate (K_2CO_3)	Chlorine, hydrogen chloride, hydrogen fluoride, sulfur dioxide, hydrogen sulfide, boron trifluoride
Activated carbon w/wo activated alumina with phosphoric acid (H_3PO_4)	Ammonia, NMP, amines
Activated alumina with sodium permanganate ($NaMnO_4$)	Sulfur dioxide, arsine, hydrogen sulfide, nitrogen oxides, low molecular weight organics
Activated alumina with sodium thiosulfate ($Na_2S_2O_3$)	Chlorine, chlorine dioxide
Cationic ion exchange resins	Ammonia, amines
Anionic ion exchange resins	Hydrogen chloride, hydrogen fluoride, sulfuric acid

cleanliness—especially for applications requiring ISO Class 4 or better. However, to achieve a comparable level of AMC control, most facilities provide for only 1 or 2 stages of chemical filters to treat outside (makeup) air and a single stage of chemical filters for recirculation air and in FFUs. Reasons for this apparent inequity in filter application include lack of space in the air handling systems, the cost of the chemical filters and increased energy costs associated with their use, but also due to a general lack of understanding of chemical filter technology and the requirements for the proper control of AMC.

One of the main compromises in using sorption (adsorption or chemisorption) for the control of chemical contaminants is between the objectives of efficient contaminant control and minimizing the resistance to airflow. In a gas-phase filtration system, the contaminant gases must first come in contact with the media before they can be adsorbed. By maximizing the contact efficiency of the system, one can virtually be assured that the maximum removal efficiency of the system is realized. However, it must be understood that maximizing the contact efficiency does not necessarily mean 100 percent contaminant removal efficiency. All it means is that the system has the opportunity to operate at its own particular maximum removal efficiency.

The smaller the media particles are, the quicker the contaminants from the airstream can reach the interior adsorption/chemical reaction sites and the smaller the concentration gradient that is needed to produce high removal efficiencies. Thus, the most efficient form for these media would be as a powder. However, one cannot economically blow air through a packed bed of powder.

Manufacturing technologies have been developed for a chemical filter medium that allows for the application of multiple plain and chemically active sorbents to a nonwoven bicomponent fiber matrix. The resulting product is a robust, pleatable roll good with many advantages over existing alternatives. The main benefits of this adsorbent-loaded nonwoven fiber (ALNF) product are the ability to use smaller medium particles and the maximization of the available surface area. These features combine to give higher initial and average contaminant removal efficiencies and lower pressure drops than competitive technologies.[6]

When comparing the performance of a 25-mm-deep (1-in) bed of a standard size (4 × 8 mesh, 3 to 4 mm, ~0.13 in) activated carbon and a smaller sized carbon (20 × 50 mesh, 0.3 to 0.8 mm, ~0.02 in) used in an adsorbent-loaded nonwoven product, the smaller medium had adsorbed almost 15 times more than the regular-sized medium. However, the pressure drop through the 20 × 50 mesh medium bed was 8½ times higher. Putting the smaller medium into the nonwoven fiber matrix and pleating it to a 25-mm depth reduced the pressure drop below that of the standard carbon and showed a removal capacity that was roughly equal.

Another important feature of this product is that it can also be produced with an integral particulate filter that practically eliminates retrofit costs. Particulate removal ratings as high as 90 to 95 percent at 0.3 μm (MERV 15-16, F7-F9)[7,8] are possible and allow for the use of existing filter hardware. The result is a unique and effective combination filter medium that can be pleated into essentially any standard filter size for the removal of both gaseous and particulate contaminants. It provides for flexible filter design, allowing easy application into new or existing HVAC systems. The main tradeoff when using this type of filter for removing AMC from outdoor air is a shorter service life due to the reduced amount of adsorbent media contained in the filter, typically one-third to one-tenth the amount contained in a comparable-sized granular media system. However, the applications for which this type of filter is typically recommended see lower concentrations of AMC and can therefore experience service lives of 1 to 2 years or more. They have proven particularly effective when used with fan filter units (FFUs). The design of the FFU should be evaluated for the addition of chemical filters and accommodations made to ensure even air flow across the filter and that pressure drop limits are not exceeded. Oftentimes an adapter (plenum box) is required between the top of the FFU housing and the chemical filters (Fig. 40.1). AMC can be removed from the cleanroom ambient air but even more important has been their use in wafer and reticle stocker cabinets, minienvironments, and process tools (Fig. 40.2).

40.6.3 Beaded Activated Carbon

A variation on this filter type takes beaded activated carbon or other adsorbent particles smaller than the 20 × 50 mesh particles described previously and puts them into a carbonized foam matrix. This produces a product that can be supplied as a flat sheet with thicknesses of up to 50 mm (2 in) that can be cut to specific sizes and shapes. Sheets with different adsorbent granules can be layered into a filter frame to provide performance against multiple contaminants. This media has a very low pressure drop and relatively high removal efficiency. Media loadings are comparable to the above; however, filter capacities are appreciably lower than comparable-sized bulk media filters. A major concern about these products is the media particle shedding that occurs. Cut edges especially are prone to shedding and most of these products are supplied with a nonwoven scrim that covers both sides of the media in a finished filter or totally encapsulates the sheet of media. Costs are significantly higher than the adsorbent-loaded nonwovens.

40.6.4 Ion Exchangers

Ion exchangers refer to any of a wide variety of synthetic polymers containing positively or negatively charged sites that can interact with or bind to an ion of opposite charge from a surrounding solution. Historically used in liquid filtration applications, ion exchange products are now the most common type of chemical filter used for ammonia control in semiconductor manufacturing. This is true for lithography tools, as a specialty filter for use in other specific AMC control applications, and in facility systems (e.g., reticle stockers). Ion exchange filters can be

Figure 40.1 FFUs with chemical filters (L, AMC filter adapter; R installed ALNF filters—one for acid gases and one for ammonia and bases). (*Photo courtesy of Purafil, Inc.*)

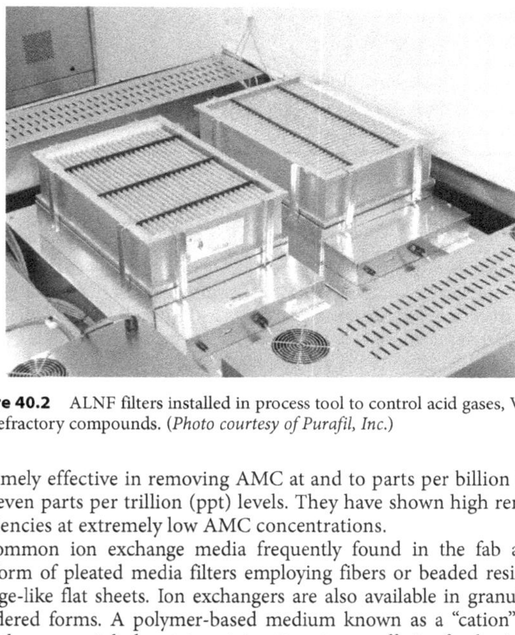

Figure 40.2 ALNF filters installed in process tool to control acid gases, VOCs, and refractory compounds. (*Photo courtesy of Purafil, Inc.*)

extremely effective in removing AMC at and to parts per billion (ppb) and even parts per trillion (ppt) levels. They have shown high removal efficiencies at extremely low AMC concentrations.

Common ion exchange media frequently found in the fab are in the form of pleated media filters employing fibers or beaded resins, or sponge-like flat sheets. Ion exchangers are also available in granular or powdered forms. A polymer-based medium known as a "cation" filter resin has a special chemistry giving it a strong affinity for basic contaminants (primarily ammonia and some amines). However, many gas-phase contaminants are not amines, but are organic or acidic in nature. As a result, there are many contaminants that would not be effectively removed by a cation ion-exchange media. It would require an "anion" exchange filter to eliminate acid gases such as HCl, HF, SO_2, HNO_3, H_3PO_4, acidic organic compounds, and some boron compounds. Neither type of ion exchanger would be very effective against organic compounds.

Ion exchange filters may be very good at a single targeted contaminant (such as ammonia), but they do not offer broad-spectrum AMC control, especially for acid removal. A better solution would be to employ carbon or alumina-based adsorbent filters that can offer a broader spectrum of control. Given the oftentimes unknown and variable nature of chemical acid contamination in the fab, broad-spectrum filters are better suited to address chemical contamination caused by external or internal sources. A single chemisorptive filter can accomplish both the physically and chemically adsorptive processes that are required to protect manufacturing processes and material from AMC.[9]

Ion exchange reactions are reversible reactions with equilibrium conditions being different for different ions. This reversibility is one the concerns about the wide use of ion exchangers in the fab. For example, the exposure of a cation filter resin to acid gases creates the potential for release of materials back into the airstream. This, along with being too specific and costing significantly more than other more common AMC filters, are the main reasons why ion exchangers are limited almost exclusively for use as ammonia/base filters in lithography applications. This includes wide use in both facility and process tool applications.

40.6.5 Extruded Carbon Composites

The newest chemical filter to gain acceptance for AMC control consists of a one-piece integral structure consisting of an extruded carbon composite structure (ECC, Fig. 40.3). It is composed of essentially 100 percent adsorbent materials that allow the entire structure to function as a chemical filter. Because of the large number of channels, the contact area between the adsorbent layer and the contaminated airstream is very large. Further, the channels are straight and parallel so that the flow is not obstructed and the pressure drop across the filter is extremely low. The size of the cells leads to turbulent flow and forces contaminated air into and through the gas-permeable cell walls of the structure. It also provides the residence time necessary to ensure optimum contact efficiency and the associated high initial and average removal efficiencies. As the air is forced through the block, removal of chemical contaminants takes place.

It is desirable to obtain high gas removal efficiencies from small bed volumes. Due to the high rate of contaminant diffusion to the interior of the cell wall and empty adsorption sites being continuously available for adsorption, this chemical filter provides better adsorption efficiencies and capacities than prior attempts with monolithic structures. This product has an extremely high adsorbent content and does not rely solely on activated carbon for contaminant removal.

Figure 40.3 Monolithic ECC chemical filter. (*Photo courtesy of Purafil, Inc.*)

Table 40.7 Gas Challenge Test Results for Hydrogen Sulfide (H_2S): Small-Scale Testing of 1 in (25 mm) Bulk Media versus a 2-in (50-mm) ECC against a Constant H_2S Challenge Concentration of 15 ppb at a Face Velocity of 500 fpm (2.5 m/s)

Sample	Volume	Weight	Density, g/cc (lb/ft³)	H_2S removed	H_2S capacity, g/cc	ΔP
Granular media	62.5 cc	42.1 g	0.6736 (42.05)	8.259 g	0.1305	187.5 Pa
Extruded carbon composite	82.5 cc	36.4 g	0.4412 (27.54)	7.200 g	0.0873	75.0 Pa

Table 40.8 Fab Makeup Air Monitoring Result Showing Generation of NO_2 Downstream of Chemical Filters. Control Limit is 0.5 ppbv for Each Anionic Species

	Anion monitoring data (ppbv)						
Sample information/sample area	F^-	Cl^-	Br^-	NO_2^-	NO_3^-	PO_4^{3-}	SO_4^{2-}
Makeup air unit (MAU) #1 air intake	8.01	1.96	0.12	2.33	1.85	0.00	0.97
Before MAU #1 chemical filter	0.72	0.91	0.02	**0.20**	0.77	0.00	0.11
After MAU #1 chemical filter	0.33	0.45	0.03	**3.04**	0.54	0.00	0.03
After MAU #1 HEPA filter	0.60	0.93	0.02	2.94	0.85	0.00	0.03
MAU room north	0.31	0.31	0.03	3.96	0.33	0.00	0.06
MAU #2 air intake	5.85	1.62	0.12	2.61	1.80	0.00	1.09
Before MAU #2 chemical filter	0.40	0.22	0.01	**0.53**	0.43	0.00	0.03
After MAU #2 chemical filter	0.19	0.09	0.00	**7.03**	0.19	0.00	0.02
MAU Room South	0.23	0.15	0.00	5.47	0.28	0.00	0.04

This delivery system exhibits the potential to provide better economics and performance of the AMC filter types most commonly used; these being 1 in (25 mm), 2 in (50 mm), and 12 in (300 mm) commercial pleated-type chemical filters and comparable economics and performance to 12 in (300) mm cleanroom grade pleated-type chemical filters and 1 in (25 mm) deep bulk-fill modules or canisters. Comparison test data is provided in Table 40.7.

40.6.6 Bonded Media Panels

Manufacturers have developed processes whereby activated carbon or other granular adsorbents are bonded and formed into monolithic panels using a polymeric binder or sintering processes. The panels are framed in an aluminized, stainless steel, or other specified (metal) frame with edge gasketing commonly provided. The panels can be made to specific sizes and shapes for retrofit or custom applications.

The main appeal of this filter type is that the media panels are self-supporting and there are no loose particles or visible dust coming from the media. However, despite manufacturers' claims about "zero dust" almost all panels as supplied are covered on both sides with a nonwoven polyester particulate filter media to prevent adsorbent granules from chipping or flaking off during shipping, handling, and installation. These panels are quite fragile which makes them prone to damage during shipping and installation.

Manufacturers report that this filter type retains a high level of open pore structure and no postactivation of the carbon is required after the bonding process. However, the reality is that most of these bonding processes often suffer from one or more serious drawbacks; the main one being the loss of a significant portion of adsorptive capacity due to the binder blocking a large percentage of the external and/or internal surface area of the media particle. Other problems include a less than optimum pore size distribution, lower removal efficiencies, reactions with chemical impregnants (when present), high pressure drops, and high manufacturing costs. Test results comparing loose granular media filled into a 1-in (25-mm) tray to commercially available bonded media panels have illustrated these deficiencies.

Note that strongly acidic chemical filters used to control ammonia can catalyze the hydrolysis of organic esters in air to make organic acids, such as butyl acetate + $H_2O \rightarrow$ butanol + acetic acid or ethyl lactate + $H_2O \rightarrow$ ethanol + lactic acid. If these organic esters are not removed before these filters, or the organic acids are not removed downstream of these filters, these reaction by-products could be introduced into the airstream downstream of the filters.

One item that has come under close scrutiny has been the observation that carbon-only or carbon-based filter media may not be acceptable for use in all applications due the potential for generation of NOx downstream of the filters. Over the last several years there have been many instances reported where air monitoring has shown that when exposed to NOx, catalytic surface oxidation reactions on the carbon (both plain carbon and impregnated carbons) cause the formation of nitrite ions (NO_2^-, nitrous acid), oftentimes exceeding the fab's control limits for cleanroom air (Table 40.8, Fig. 40.4). A number of "carbon-free" media have been developed that prevent this from occurring and their use is being written into current AMC control specifications.

When the use of multiple AMC control media are indicated, it is best practice to employ the individual media in separate filter stages. This allows for the individual media/filter types to be arranged such that the first stage would act as a "prefilter" to remove specific targeted AMC (e.g., activated carbon for VOCs) as well as other chemical species present in the airstream that are not specifically targeted but would affect the performance of the AMC control system. Subsequent stages of AMC filters would be incrementally more specific for different AMC classes (acid, base) or individual chemical compounds and serve the sample purpose analogous to the pre-, intermediate, final, and HEPA/ULPA filters for particulate control.

In applications where it is not possible to employ multiple stages of chemical filters due to physical or operational constraints, the media may be blended (if chemically compatible) into a single filter or filter module. The main tradeoff in using a blended media product is not necessarily in terms of the types and numbers of contaminants that can be addressed or of filter performance (initial and average removal efficiency), but rather terms in service life as compared to the use of single media filters.

40.7 CHEMICAL FILTRATION EQUIPMENT DESIGNS

40.7.1 Chemical Filters

Chemical filters used in AMC control applications are available in a wide variety of commercial designs, usually as packed-bed media filters where a dry, granular, or pelletized media is filled in the space between

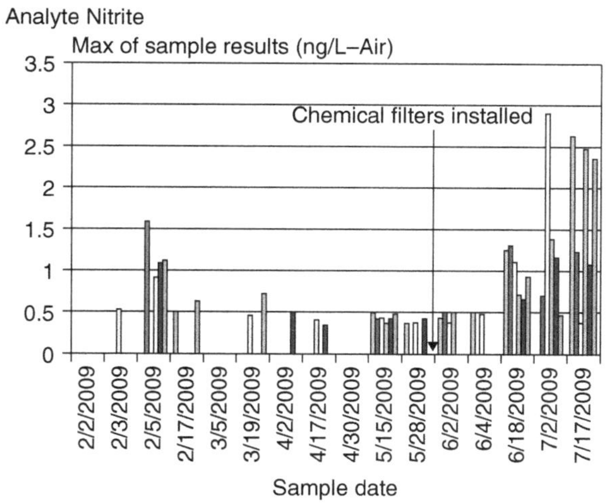

Figure 40.4 Levels of NO_2^- from different sampling locations before and after installation of carbon-based chemical filters.

perforated metal or plastic screens. Others incorporate pleated filters and media panels of the types described in Sec. 40.6. The filter types most commonly used in cleanroom HVAC systems include

- *Serpentine.* These contain thin-bed [0.375 to 0.500 in (9.4 to 13 mm)] convoluted filters and are used in medium-duty recirculation applications.
- *Thin-bed trays/panels.* These contain multiple filter trays or panels of thin depth [0.5 to 1.0 in (13 to 25 mm)] arranged in a "Z" configuration in a frame to achieve an extended surface area. Granular or pelletized media, adsorbent-loaded nonwovens, beaded activated carbon, and bonded carbon panels are all employed in this type of filter.
- *Intermediate bed-depth trays.* These are similar to thin-bed trays except bed depth is on the order of 0.875 to 2.000 in (23 to 50 mm). These are single filters designed for higher efficiency and longer service.
- *Intermediate bed depth V module.* Filters are in a "V" configuration, with a bed depth of 1.000 to 1.125 in (25 to 28 mm) and attain a low bed velocity by using extended surface area techniques.
- *Thick-bed trays.* These are single filters with a bed depth of 2 in (50 mm) or greater, oriented perpendicular to the airstream. When used with granular or pelletized media they have a high pressure drop and are used with low air velocities. Currently this type of filter is most often used with extruded carbon composites (ECCs) to provide a pressure drop lower than granular or pelletized media filters when used in makeup and recirculation air handling units. ECCs are also finding use in FFUs and minienvironments and as tool filters. They also offer advantages in retrofit applications where space is limited and AMC concentrations indicate larger amounts of media to meet removal efficiency and service life requirements.
- *Thick bed depth "V" module.* This unit has an average bed depth of 3 in (75 mm) and is similar to the intermediate depth "V" module. Used primarily in makeup air systems to remove high levels of AMC.
- *Radial flow canister.* This unit is a cylindrical filter with the granular or pelletized media contained in a radial design. These filters have an average bed depth of 0.5 to 2.0 in (13 to 50 mm) and are available in lengths of 12 to 24 in (300 to 600 mm). In some configurations they provide a favorable pressure drop versus media quantity ratio as compared to trays. A typical system consists of 16 cylinders on a 2 × 2-ft (600 × 600-mm) face area.
- *Pleated media filters.* These are single filters employing a pleatable chemical media in depths of 1 to 12 in (25 to 300 mm). These are designed for use with adsorbent-loaded nonwoven or pleatable ion exchange media. Advantages of pleated media filters include a high degree of cleanliness, low pressure drops, and dimensional flexibility.

The above filters can be installed in front, rear or side-access housings or frames, or other standardized equipment. They can also be installed in some self-contained air cleaner units.

40.7.2 Chemical Filtration Equipment

Manufacturers of chemical filtration systems offer many combinations of filters and equipment to handle distinct AMC control applications. Equipment is grouped into two main classifications, HVAC-integrable systems and self-contained systems.

HVAC-Integrable Systems

These are the most commonly applied chemical filter systems and are most often supplied in the form of front or rear-access systems designed to be easily integrated and adapted into a fab's existing air handling systems. These systems are effective for treating low-to-medium concentrations of AMC in both makeup and recirculation air handling systems.

These systems are usually installed as built-up banks of 2 × 2 ft (600 × 600 mm) or 1 × 2 ft (300 × 600 mm) frames which can be stacked in virtually any size configuration to meet the specific space requirements of most systems. Frames stacked on top or to the side of each other are fastened with rivets or bolts, or spot-welded. Frames are typically constructed of 16 to 20 gauge steel, aluminum, or stainless steel with or without special coatings are required for specific applications.

Canister filters are also employed on a front or rear-access frame system to allow for ease of access for the larger number of canisters in walk-in makeup or recirculation air handling units. Each canister typically includes a mounting assembly with three integral studs for mounting to matching cylindrical mounting flange (Fig. 40.5) ensuring a tight filter-to-frame seal.

The chemical filtration media are contained in ether refillable or disposable filter panels or modules. This modular approach offers a wide combination of media bed depths, media types, and particulate filtration. Frames can be used for either front (upstream) or rear (downstream) access applications and can be designed to hold virtually all types of chemical filters. Multiple stages of filters can be held in a single frame depending on the fastening device chosen. Each frame is typically supplied with a gasket around the inside flange return to provide a positive seal between the filter and the frame.

Self-Contained Equipment

The primary use of self-contained AMC control systems in semiconductor facilities is for the removal of hazardous production gases from process tool and gas cabinet exhausts, as well as providing "safe havens" for personnel in the event of a catastrophic release of a toxic material from process tools, gas cylinders, gas cabinets, storage tanks, and so forth.

Figure 40.5 Front/rear access frames with chemical filter. (*Photo courtesy of Purafil, Inc.*)

For waste gas treatment, these systems should be installed as close as possible to the tool outlet or gas cabinet. Exhaust streams from multiple tools or gas cabinets may be combined as long as the total combined gas challenge is not greater than the capacity of the system. Concentrations of hazardous gases are reduced well below threshold limit values (TLVs) when using the appropriate chemical filter media.

Emergency gas scrubbers (EGSs) have become necessary components of the health and safety plans for many facilities due to the use and storage of large quantities of toxic chemicals, including arsine (AsH_3), phosphine (PH_3), hydrogen fluoride (HF), chlorine (Cl_2), and boron trifluoride (BF_3). An accidental chemical release, even from a cylinder as small as 1 lb, can be fatal. As a result, manufacturers must take emergency precautions to protect personnel and the surrounding communities. EGSs are designed for the complete removal of accidentally released gases and with removal capacities for gas cylinders of 1 lb to 1 ton and greater in size.

EGSs can be designed with a single media bed for a single contaminant challenge or with multiple deep media beds to allow for the use of multiple chemical media, each targeting a specific contaminant or group of contaminants.

40.8 AMC MONITORING

Just as there are many ways to apply chemical filtration for the control of AMC, there are many different methods being used to measure AMC inside and outside the cleanroom environment and to evaluate the filter-system performance. Once an AMC control strategy has been decided upon and implemented, whether it involves the use of chemical filtration or not, one must be able to monitor the success or failure of meeting specified AMC control criteria. As a minimum any AMC monitoring program must provide the following:

- Type(s) of contaminants present and their relative levels
- Ambient air quality correlated to specific air purity classifications
- Chemical filtration system performance assessment
- Verification of the attainment of specified or standard AMC levels

The biggest problem today is not whether specified levels of chemical contaminants can be reached, but whether they can be accurately measured to ensure compliance with any standards or control criteria. There is a wide range of real-time (active) monitoring instrumentation available with many claiming detection limits in low ppb range and some claiming ppt levels.

There is an increased interest in portable systems as opposed to those that are installed facility-wide and accessed through the facility-monitoring system. Instrument manufacturers should be able to provide an understanding of the different technologies available, what to look for, and what would be most appropriate for a particular application. Table 40.9 provides an idea of the monitoring technologies available and the types of levels of different classes of chemical contaminants that can be measured. It is not meant to be inclusive, but rather to illustrate the point that one monitoring technique will most likely not provide information on all of the contaminants of concern.

These instruments and analytical techniques can provide real-time data on AMC levels but have a significantly higher cost when compared to other monitoring techniques. Because of this, a number of manufacturers now perform AMC monitoring using what may be referred to as "semiquantitative" monitoring.[2] These are analytical techniques that provide quantitative information on environmental air quality, but do not measure specific contaminants. They can involve passive or real-time monitoring using devices such as litmus paper, witness wafers, impingers, multisorbent tubes, surface contamination monitors, and reactivity monitors.

40.8.1 Impinger and Sorbent Tubes

Two of the more common methods used to monitor AMC in front-end-of-line (FEOL) processing include impingers and sorbent tubes. Impingers (or bubblers) are ultrahigh purity air samplers that can be used to monitor metallic, ionic (cations and anions), and amine contamination. Impinger sampling is typically used with ion specific electrodes, ion chromatography (IC), ion mobility spectroscopy (IMS), inductively coupled plasma mass spectrometry (ICP-MS), graphite furnace atomic absorption spectrometry (GFAAS), or other analysis techniques. Sampling with (multi-)sorbent tubes most often requires analysis using thermal desorption gas chromatography-mass spectroscopy (TD-GC-MS) or by gas chromatography-mass spectroscopy (GC-MS) alone to identify for organic contaminants. Detection limits will vary from element to element and species to species but are typically in the range of 0.5 to 0.005 ng/L of air. Longer sample times are required for lower detection limits.

40.8.2 Reactivity Monitoring

Of the semiquantitative monitoring techniques in wide use, the one that has seen the largest increase in use is reactivity (or corrosion) monitoring.[11] This is a low(er)-cost monitoring technique that has proven especially useful for establishing AMC baselines and identifying AMC episodes and their sources in back-end-of-line (BEOL) processing, where detection limits are not as critical as for FEOL semiconductor manufacturing. It can help reduce the number of reported health and safety incidents by allowing proactive investigation of potential problems. When incorporated into a facility's preventive maintenance program, it can reduce the number of AMC-related incidents that are reported with the ultimate goal being to eliminate production downtime due to AMC in the fab.

Employed in either real-time or passive form, reactivity monitoring is being used as a method of gauging AMC levels and to evaluate the effectiveness of AMC control strategies. Whether controlling AMC directly or indirectly, reactivity monitoring can provide an overall indication of the ambient air quality as well as AMC levels within the controlled spaces. Where chemical filtration is employed for AMC control, reactivity monitoring can provide information on filter performance in both quantitative and relative terms. One can directly quantify the reduction through a chemical filter system for a number of contaminant species with this technique. It can also provide evidence as to the

Table 40.9 Selection Matrix of Sampling Instruments and Analysis Methods in Relation to Expected AMC Concentrations[10]

<table>
<tr><th>Detection limits</th><th>(10^n g/m^3)</th><th>0</th><th>−1</th><th>−2</th><th>−3</th><th>−4</th><th>−5</th><th>−6</th><th>−7</th><th>−8</th><th>−9</th><th>−10</th><th>−11</th><th>−12</th></tr>
<tr><td rowspan="4">Chemical category</td><td>ACID</td><td colspan="4" rowspan="2">IC, UVS, RM</td><td colspan="2" rowspan="2">IMP, IC, UVS, CLS, IR, CPR, IMS, CRDS</td><td colspan="2" rowspan="2">IMP, DIFF, IC, UVS, IR, CLS, CPR</td><td>IMP, DIFF, IC, CLS</td><td>IMP, DIFF, IC, CZE, CLS</td><td colspan="2">IMP, CZE</td><td></td></tr>
<tr><td>BASE</td><td colspan="2">IMP, DIFF, IC, IMS, CLS</td><td colspan="3">IMP, IC, CZE</td></tr>
<tr><td>ORGANIC</td><td colspan="6">SOR, SB, GC-FID, GC-MS, IR</td><td colspan="3">SOR, WW, GC-FID, GC-MS</td><td colspan="4">TD-GC-MS</td></tr>
<tr><td>INORGANIC</td><td colspan="9">IMP, AA, AA-F, AA-GF, UVS, ICP-MS, CL, CLS, IR, ECS</td><td>IMP, WW, ICP-MS, CLS</td><td>ICP-MS</td><td colspan="2"></td></tr>
</table>

NOTE: The applicable analysis method for a given contaminant concentration is dependent on the sampling rate and duration. Key to abbreviations used in this table are listed at the end of this chapter.

types of contaminants present and a system's effectiveness against these contaminants.[12]

Reactivity monitoring (Fig. 40.6) is commonly used for evaluating long-term air quality trends inside and outside a facility and in performing environmental surveys as part of the development of an AMC control program. It can also be used to differentiate between classes of chemical contaminants and to provide estimates of the airborne AMC concentrations. The presence of low levels of chlorine and sulfur compounds and many other corrosive and problematic AMC, including sulfur dioxide, nitrogen dioxide, hydrogen sulfide, hydrogen fluoride, ammonia, and ozone (all of which have been cited as the cause of a number of AMC-related process effects) can be detected using this technique (Table 40.10).

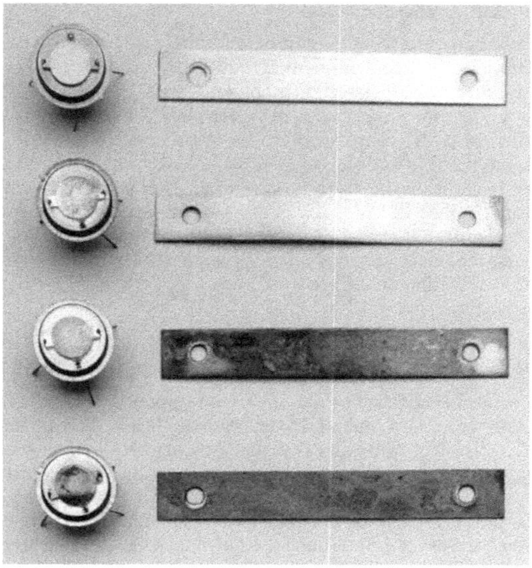

Figure 40.6 Picture of metal-plated quartz crystal microbalances (QCMs on left) used in environmental reactivity monitors (ERMs) and environmental reactivity coupons (ERCs on right) after exposure to corrosive gases. (*Photo courtesy of Purafil, Inc.*)

Table 40.10 Reactivity Monitoring Sensitivities for Various AMC Types

Chemical class	Chemical types	Detection limits (ppb)
Inorganic chlorine compounds	Cl_2, HCl	< 1
Halogen acids	F_2, HF, HBr, HI	< 1
Strong oxidants	O_3, ClO_2, HNO_3	< 2
Active sulfur compounds	H_2S, mercaptans, elemental sulfur	< 3
Sulfur oxides	SO_2, SO_3 (sulfuric acids)	< 10
Nitrogen oxides	NO, NO_2, N_2O_4	< 50
Ammonia and derivatives	NH_3, NMP, amines	200–500

An air quality classification scheme has been established for use with reactivity monitoring and it has gained wide acceptance throughout the semiconductor industry. Reactivity monitoring can provide information as to the types and relative concentrations of many contaminants problematic to manufacturing processes. It can provide the information required to determine if direct AMC control is indicated and, if so, what form it should take.

Whatever techniques and instrumentation are employed to detect and monitor AMC, there are a number of important characteristics to consider:

Sensitivity	A detection limit of 1 ppb/1 μg/m^3 is required for a majority of contaminants.
Accuracy	How close to the actual concentration?
Repeatability	Will you get the same answer every time for a given set of conditions?
Selectivity	Could there be interference from other compounds?
Identification	What specific contaminants are present?
Ease of use	Calibration required, prep time, complexity, and so forth.
Response time	Real-time (active) versus passive.
Size/weight	Lab-scale, portable, handheld.
Cost of ownership	Includes consumables, operator time, maintenance, and calibration.

40.9 AMC CONTROL APPLICATION AREAS

40.9.1 HVAC System Designs

Most air handlers employed in semiconductor cleanrooms are large custom units moving tremendous quantities of air used for both pressurization and recirculation. A typical cleanroom HVAC system might consist of makeup air handlers (MAUs) for pressurization air, recirculating air handling units (RAUs), or fan filter units (FFUs) to move the supply air to the cleanroom, ducts to route the air, plenums to distribute the air, and filtration systems to clean the air before it is introduced to the cleanroom.

Whichever type of air handling system is employed in a semiconductor fab, there are a number of locations in which chemical filtration can be applied. Ultimately, the nature of the contaminants, their source(s), and the cleanroom layout and design, all have to be considered for determining the best location for these filters and equipment.

Makeup (Outside) Air

One of the main applications of chemical filtration is to treat the air brought into the facility for ventilation and pressurization. This air oftentimes contains chemical contaminants of the types and quantities to be of concern to semiconductor manufacturing processes. Ozone, oxides of sulfur and nitrogen, and VOCs are the most common chemical contaminants encountered in outside air. Also, depending on the geographic location of the facility (e.g., urban environments, seaside, northern climates) and the activities being performed in the surrounding areas (e.g., industry, farming) significant sources of other contaminants such as boron, ammonia, chlorine, hydrogen sulfide, dust, fertilizer, and salt spray may also be present.

Makeup air handlers (MAUs) are a combination of air handlers and air treatment systems. These units contain the components necessary to prepare outside air for use in a cleanroom environment. Heating and cooling coils, humidification distribution systems, fans, filters, electrical systems, and controls can all be contained in these units. In high particulate areas, air (or water) washers can be used in makeup air handlers reduce the particulate load on filters.

Outside air is typically delivered at the rate of approximately 2 to 6 cfm/ft^2 for pressurization of the cleanroom floor space. In a "typical" 100,000-ft^2 ballroom-type semiconductor fab, this would require between 200K and 600K cfm of air. Because of these large amounts of outside air required and the corresponding large size of most MAUs, chemical filters are employed in built-up banks of front (or rear) access frames for ease of access and maintenance. Single banks of bulk media filters can be used, but it is common to see two or more banks employing different chemical filter media based on the types and levels of the contaminants expected.

Recirculation Air

Because of the tremendous quantities of air required to maintain a semiconductor cleanroom, most of the air within the space is recirculated to keep the amount of energy used by the HVAC to condition the outside air to a minimum. The MAUs deliver clean, pressurization air to the clean space either through ductwork or through a pressurized plenum. Perforated floor panels allow the air to flow into a large common plenum from which air is to be treated through recirculating air units (RAUs). These air handlers deliver air at a rate of 72 to 90 cfm/ft^2 of clean space. RAUs employing chemical filtration are thus used to treat or "polish" the air to and from the clean space as well as remove AMC from specific sources within the fab. Sources of AMC within the cleanroom include process chemicals and raw materials, process tools, spills, and so forth.

Because RAUs are many times comparable in size to the MAUs, the use of front (or rear) access frames is usually indicated. However, some fab designs employ the use of numerous smaller RAUs to compartmentalize and isolate airflows in order to prevent possible cross contamination. This smaller size could allow for the use of side-access housings integrated into the air handlers.

Fan Filter Units/Minienvironments

Fan filter units (FFUs) combine a fan and motor with an optional particulate prefilter and a HEPA/ ULPA filter and can be installed directly into a ceiling system. They can replace RAUs and can be used to achieve 100 percent coverage of the cleanroom, eliminate ductwork by drawing air from the interstitial area above the cleanroom, and reduce the outside air requirements by eliminating the pressure plenum. Chemical filters are often installed on top of FFUs to provide additional levels of AMC control to sensitive process areas.

Due to the relatively short lifetime of AMC filters (months to 2 years or more) compared to HEPA/ULPA filters (10 years or more), it is not practical or desirable to install these filters below FFUs. Replacement from the cleanroom side is complicated, disturbs normal fab operations, and could provide a source of particulate contamination. Therefore, almost all installations are done above the cleanroom ceiling.

An adapter frame may be required on top of the FFU to provide a mixing plenum above the fan if the FFU has not been specifically designed for the addition of AMC filters. Note that outgassing of any components, including filter motors, lubes, insulators, sound or vibration attenuating foams, foam sealants, gel seals, separators, lighting, etc., must be carefully controlled especially for all components downstream of AMC filters. Outgassing of cleanrooms materials is covered in IEST Recommended Practice 31.

Minienvironments are enclosures with engineered airflows that surround process tools. They are used to maintain cleanliness levels at the wafer environment that is independent of the external room environment. Steppers and reticle stockers are common examples of the use of minienvironments. Chemical filters can be integrated into the process tool supply air for final cleaning of the air with filters often installed on top of the minienvironment or into the exhaust duct to prevent fugitive emissions from being distributed throughout the facility.

40.9.2 AMC Application Areas

Table 40.11 provides a design summary for these applications and shows the common target contaminants along with the recommended chemical filtration media, filters, and equipment.

Figure 40.7 shows a schematic of a cleanroom with the most common locations for AMC control within the HVAC system and within the cleanroom itself. In addition to MAUs and RAUs, chemical filtration is applied to clean up the air to process tools. Tool exhaust air has been implicated as one of the major sources of fugitive chemical emission with the semiconductor cleanroom. Because of this growing numbers of process tool manufacturers provide gas-phase filtration as an integral part of their equipment design.

40.10 AMC CONTROL SPECIFICATIONS AND STANDARDS

AMC control specifications are often based upon individual chemical species or groups of chemical contaminants (acids, bases, chlorides, VOCs). This is not surprising as many of these were developed from an "industry-wide" perspective. Individual semiconductor manufacturers and/or equipment suppliers, due to intellectual property issues and competitive pressures, tend to use their own contamination control criteria that have been developed through their own experience, capabilities, and expectations. There is no general consensus on what may be considered acceptable levels of airborne chemical contamination, but the ranges have been narrowing over the last 20 years and this trend is expected to continue through successive device generations. Further, the actual wafer environment, where exposure to AMC could be expected, has shrunk from the general cleanroom ambient environment to SMIF Pods and FOUPs—each with their own set of AMC control recommendations.

40.10.1 SEMI Standards

SEMI Standard F-21-1102 titled "Classification of Airborne Molecular Contaminant Levels in Clean Environments"[3] classifies microelectronic clean environments with respect to their AMC levels. These standard classifications are used in the specification of semiconductor clean environments (including process tool environments) and of contamination control and measurement equipment performance. This standard determines environmental classifications by the maximum allowable gas-phase concentration of four specific contaminant categories: acids, bases, condensables, and dopants. The combination of a quantitative class for each of the four categories yields a classification describing the environment.

Table 40.11 AMC Control System Design Summary

Application areas	Outside air	Recirculation air	Filter fan unit (FFU), minienvironments, point-of-use filters	Emergency gas scrubbers, exhaust air
Target AMC	SO_X, NO_X, O_3, Cl_2, NH_3, VOCs	NH_3, NMP, amines, acids, alcohols, VOCs, AsH_3, BF_3	NH_3, NMP, amines, Cl_2, HCl, HF, VOCs	AsH_3, PH_3, BF_3, Cl_2, HCl, HF, ClF_3, VOCs
Media	GAC, impregnated carbons/ alumina, ion exchange, extruded carbon composites	GAC, impregnated carbon/alumina, ion exchange, extruded carbon composites, adsorbent-loaded nonwovens	Adsorbent-loaded nonwovens, ion exchange, beaded activated carbon	Specially impregnated carbons/alumina
Filters	Thick/thin bed trays/modules, canisters	Thin bed trays/ modules, canisters, pleated media filters, flat panel filters	Pleated media filters, flat panel trays	Bulk media
Equipment	Front/rear access cells, 2–3 passes	Front/rear access cells, 1–2 passes	Side access systems, integrated recirculation air filters, OEM filter housings	Deep-bed scrubbers, emergency gas scrubbers

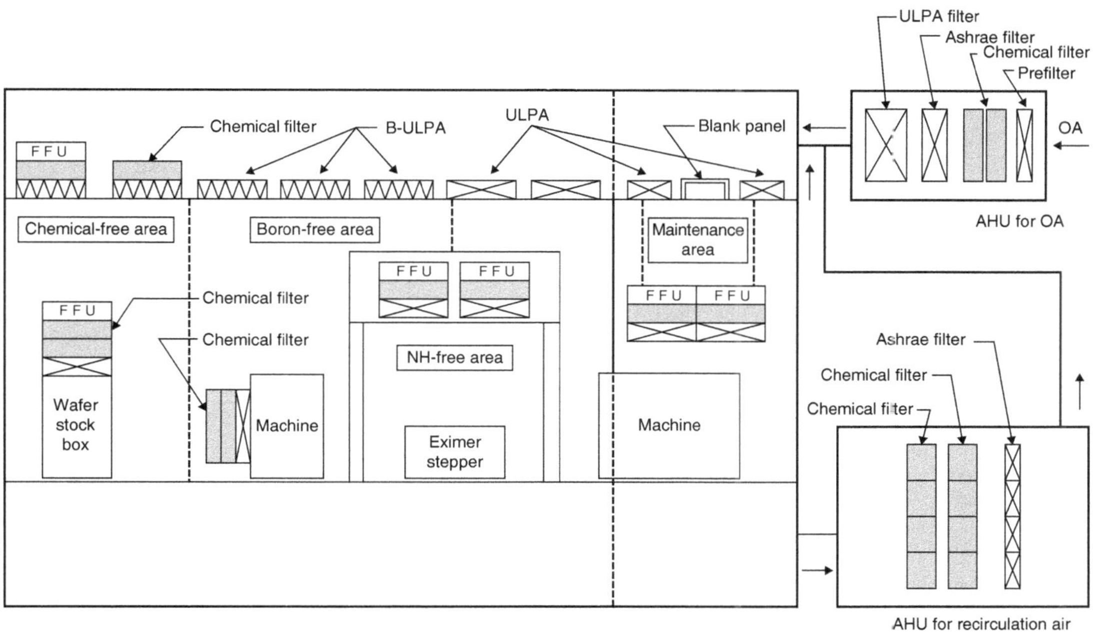

Figure 40.7 Possible locations for AMC control in a semiconductor manufacturing facility.

Standard F21 does provide a consistent means of communicating acceptable levels for groups of specific contaminants. In practice, however, it is rarely used in the specification of clean environments and contamination control and measurement equipment performance. This is primarily due to the lack of standard AMC classifications for semiconductor manufacturing environments and monitoring requirements in the standard. There is a need for better AMC monitoring instrumentation in the cleanroom to measure AMC at the ppt levels.

40.10.2 International Technology Roadmap for Semiconductors

To those charged with establishing and maintaining the appropriate controlled environments for leading-edge semiconductor manufacturing, AMC control can appear to be a moving target. Sulfur and nitrogen oxides, ozone, and organics from outside air, as well as acids, bases, dopants, and organics from sources inside the fab may all have to be considered in a successful AMC control program. But which contaminants should be targeted? What control levels should be considered? Fortunately there is a resource that facility and process engineers can use to learn about critical issues relative to AMC and its effects on semiconductor manufacturing.

According to the International Technology Roadmap for Semiconductors (ITRS)[13] the percentage of process steps affected by nonparticulate or molecular contamination is expected to increase. The impact of AMC on wafer processing can only be expected to become more deleterious as device dimensions decrease. Pregate oxidation, salicidation, contact formation, DUV photolithography, EUV masks, and atomic layer deposition (ALD) have been identified as particularly sensitive production steps.

Yield Enhancement

Yield enhancement[14] (YE) is represented by the functionality and reliability of integrated circuits produced on the wafer surfaces. YE for manufacturing of integrated devices addresses the improvement from research and development yield to mature yield. The YE chapter of the ITRS displays the current and future requirements for high yielding manufacturing of dynamic random-access memory (DRAM), microprocessors (MPUs), and flash memory. The YE chapter includes the focus topic of Wafer Environmental Contamination Control (WECC), where guidelines and technology requirements for AMC can be found.

Since the 2005 edition of the ITRS, the wafer environment includes the ambient space around the wafer at all times, whether the wafers are open to the cleanroom air or stored in PODs/FOUPs (front opening unified pods). AMC needs to be controlled in front-end and back-end of line operations in semiconductor fabs. This control may be achieved fab-wide or at certain critical processes, potentially also at different levels for different processes. A summary of yield enhancement WECC AMC interfaces by process areas as well as target levels of ambient acids, bases, condensables, dopants, and metals for specific process steps is shown in Table 40.12.

The YE Chapter of the ITRS and specifically the WECC technology requirements indicate target levels of ambient acids, bases, condensables, dopants, and metals for specific process steps. As the list of chemical contaminants to be controlled broadens one important WECC challenge is the accurate modeling of cleanrooms for AMC sources and distribution. This is because with the reduced air volumes many current cleanroom designs cannot dilute AMC as well as before—whether it comes from outside or inside the fab.

The ITRS takes advantage of expertise from around the world to help identify the technical challenges related to AMC and offers recommendations that can be used in establishing AMC control strategies and guidelines for advanced semiconductor device manufacturing. Requirements have been added for the control of total acids in addition to total bases in lithography applications. Subsequently, requirements have been added addressing specific acids—not just total acids. Additional changes for AMC control that have been proposed include total acids being reduced to < 20 pptv (5 pptv long term), total condensable organics being reduced to < 100 ppbv, and dopants being reduced to < 10 pptv.

Outgassing from materials of construction in the cleanroom, wafer processing equipment, postprocessed wafers, and wafer environmental enclosures as well as inadequate exhaust and fugitive emissions from chemicals used in wafer processing are the main sources of AMC. In some highly congested areas or regions with poor ambient air quality, makeup air is a significant source of AMC. Oxygen and water vapor as well as low concentration atmospheric contaminants [e.g., carbon dioxide (CO_2), ozone (O_3)] can also be considered as part of the AMC

Table 40.12 Yield Enhancement WECC AMC Interfaces by Process Areas (2016 Production Year)[15]

Airborne molecular contaminants in gas phase (pptV, V for volume)	
Lithography: point of entry (POE) to exposure tool	
Total inorganic acids	5,000
Total organic acids*[i]	2,000
Total bases	20,000
PGMEA, ethyl lactate	5,000
Volatile organics (w/GCMS retention times ≥ benzene, calibrated to hexadecane)† [ii]	26,000
Refractory compounds (organics containing, for example, S, P, Si)	100
Lithography: point of entry (POE) to track and inspection tools; temporary reticle pod storage	
Total inorganic acids	2,000
Total organic acids	2,000
Total bases	2,000
PGMEA, ethyl lactate	5,000
Condensable organics (definition to SEMI F21-1102, b.p. 150°C)	1,000
Refractory compounds (organics containing, for example, S, P, Si)	t.b.d.
Reticle storage (inside stocker, inside pod, inside exposure tool library, inside inspection tool)	
Total inorganic acids	< 200
Total organic acids	< 200
Total bases	< 200
Condensable organics (definition to SEMI F21-1102 b.p. 150°C)	< 100
Refractory compounds	t.b.d.
Gate/furnace area wafer environment (cleanroom FOUP ambient/tool ambient)	
Total metals (E^{+10} atoms/cm^2/week)	10
Dopants (E^{+10} atoms/cm^2/week; front end of line only)	10
Volatile organics (w/GCMS retention times ≥ benzene, calibrated to hexadecane)	20,000
Gate/furnace area wafer environment (FOUP inside)	
Total metals (E^{+10} atoms/cm^2/day)	0.5
Dopants (E^{+10} front end of line only)	0.5
Volatile organics (w/GCMS retention times ≥ benzene, calibrated to hexadecane)	2,000
SMC (surface molecular condensable) organics on wafers, ng/cm^2/day‡ [iii]	t.b.d.
Total SM (surface metals) on wafer, E^{+10} atoms/cm^2/day	0.5
Salicidation wafer environment (cleanroom FOUP ambient)	
Total inorganic acids	500
Total organic acids	5,000
Salicidation wafer environment (FOUP inside; wafer environment)	
Total inorganic acids	500
Total organic acids	5,000
Exposed copper wafer process environment (cleanroom ambient, tool inside)	
Total inorganic acids	500
Total bases	2,000
Total organic acids	500
Total other corrosive species § [iv]	1,000
H_2S	1,000
Total sulfur compounds	2,500
Exposed aluminum wafer process environment (cleanroom ambient, tool inside)	
Total inorganic acids	500
Total bases	t.b.d.
Total organic acids	t.b.d.
Total other corrosive species	1,000
H_2S	t.b.d.
Total sulfur compounds	t.b.d.

Table 40.12 Yield Enhancement WECC AMC Interfaces by Process Areas (2016 Production Year)[15] *(Continued)*

	Airborne molecular contaminants in gas phase (pptV, V for volume)
Exposed copper wafer environment (FOUP inside)	
Total inorganic acids	500
HCl	200
HF	5,000
HBr	t.b.d.
HNOx	t.b.d.
Total organic acids	100
Total bases	t.b.d.
Total other corrosive species	t.b.d.
H_2S	t.b.d.
Total sulfur compounds	5,000
Moisture (ppb)	t.b.d.
Exposed aluminum wafer environment (FOUP inside)	
Total inorganic acids	t.b.d.
HCl	100
HF	200
HBr	t.b.d.
HNOx	t.b.d.
Total organic acids	t.b.d.
Total other corrosive species	t.b.d.

NOTES: *Typical organic acids found in cleanroom environments that may be of concern include acetate, citrate, formate, glycolate, lactate, oxalate, and propionate. Others may also be of concern. These acids can be a significant load on acid removal filters.

†Ideally, continuous monitoring using online instrumentation would be preferred when practical since this can give both long-term averages and catch excursions. When online monitoring is not available, an average grab sample for at least 4 h, and not more than 24 h is recommended, to get an average, increase sensitivity of the analysis, and avoid short-term transient effects.

‡SMC organics: Single wafer shall be oxidized to make organic free, then wafer shall be exposed for 24 h and top side analyzed by TD-GC-MS with 400°C thermal desorption, and quantitation based on hexadecane external standard. TIC response factor per SEMI MF 1982-1103 (formerly ASTM 1982-99). Limits determined by above method are a guideline for many organics. Note higher limits can be used for process wafers oxidized or cleaned prior to subsequent process step. Processes such as gate oxide formation, or polysilicon deposition, may be more sensitive to organics, especially high boilers such as DOP. Silicon nitride nucleation may also be more sensitive than above for some processes. Please note dopants requirement is covered in earlier section. Contamination levels are time based, and samples should be exposed for a week's time for better sensitivity; ng/cm^2/week. Total contamination level on reticles that cause problems also vary with energy exposure. These guidelines subject to change with new data currently being generated.

§Other corrosive species include contaminants such as chlorine. Humidity is also of major concern, as it exacerbates corrosion. The humidity should be kept as low as possible in corrosive environments.

burden. Acid vapors in the air have been linked to corrosion on wafers, as well as with the release of boron from HEPA filters. The impact of amines on deep ultraviolet (DUV) photoresists is well-known examples of AMC affecting wafer processing. Hydrocarbon films of only a few monolayers may lead to loss of process control, especially for front-end processes. The impact of AMC on wafer processing can only be expected to become more deleterious with future device generations. A general list of AMC classes, representative chemical types, sources and effects is shown in Table 40.13.

ITRS 2.0

In 2012, it became evident that the ecosystem of the electronic industry had undergone a major change with the proliferation of the Internet, the worldwide spread of wireless mobile devices like smart phones and tablets and, most of all, the Internet had evolved from the Internet of Things (IoT) to the Internet of Everything. Sensors of all kinds continue to be added to the Internet and the range of remote operations is increasing by the day allowing remote operation of tools and remote medicine as an example. To fully address these changes the process of restructuring the ITRS was initiated in 2012 and completed in 2014. The 17 international technology working groups (ITWGs) were replaced in 2015 by seven focus teams in the ITRS 2.0. These are

1. System integration
2. Heterogeneous integration
3. Heterogeneous components
4. Outside system connectivity
5. More Moore
6. Beyond CMOS
7. Factory integration

Yield Enhancement was now covered as part of the Factory Integration Focus area and chapter, and more specifically the section on Wafer Environment Contamination Control section was maintained.

International Roadmap for Devices and Systems

In 2016, the traditional process of setting an industry roadmap for semiconductors has been taken on by the Institute of Electrical and Electronic Engineers (IEEE) and expanded to include all computing. The ITRS, first published in 1965, has been reorganized as the International Roadmap for Devices and Systems (IRDS) and this new effort aims to take a broad view of the needs of computing generally. It will address roadmap issues that include computer systems, architectures, and software as well as the chips and other components used in them.

The move comes at a time when the ITRS has lost some of its influence. The roadmap used to point a wide variety of chip makers toward a common set of technical milestones. In recent years, however, a consolidating set of chip makers have forged their own roadmaps, often naming nodes in ways more driven by marketing than engineering executives.

Advanced semiconductor industry has limited ability to monitor defects on the wafer. This requires that all critical utilities and components ensure their consist quality control at all levels from manufacturing through operations. This is critical for enabling existing and future advanced semiconductor manufacturing processes.

In the new IRDS AMC guidelines will be developed by AMC experts under a reorganization of the section on Wafer Environment Contamination Control (WECC+). Specific scopes will be defined in each area such as types of gases, chemicals, and components, critical processes, and priority focus areas and limitations. Interfaces with other groups

Table 40.13 AMC Sources and Effects[16]

AMC class	Contaminants	Sources	Effects
Molecular acids	Fluoride, chloride, bromide, sulfates, phosphates, nitrogen/oxygen compounds	Etch chambers, diffusions furnaces, CVD processes, wet benches using HCl, HF, BOE	Hazing of reticles, wafers, optics of exposure and metrology tools. Corrosion of Al and Cu metal lines. Inhibits CAR
Molecular bases	Ammonia, amines, amides, trimethylamine, triethylamine, cyclohexylamine, dimethylamine, methylamine, ethanolamine, morpholine	Ammonia sources: CVD, HMDS, CMP, slurries, wafer cleaning processes. TiN and Si_3N_4 films deposition Amines sources: photoresist strippers, polymers, epoxies, TMAH decomposition Amides sources: solvents such as NMP, dimethyl acetamide, polyimides	Neutralizes photoacids in resists. Reaction with acid vapors can cause hazing, particle formation, nitrides on wafer surface
Molecular condensables	Dibutyl phthalate, NMP, organophosphates, siloxanes, hexamethyldisiloxane, PGME, PGMEA	Outside air, process chemicals, outgassing from filters, sealants, adhesives, walls, floors, wafer shippers, FOUPs, pods, gaskets, sealing tape, bagging materials, flame retardants	Hazing of exposure tool optics and masks from silicones and HMDS byproducts. Delamination of PR and ARCs. Unwanted n-doping of wafers. Interference with thin film metrology
Molecular dopants	Boron, phosphorus, organophosphorus, arsenic, antimony	Outside air, degradation of HEPA and ULPA filters, exhaust from RIE, EPI, CVD processes, flame retardants	Unwanted n- and p-doping of wafers
Molecular metals	Organometallic compounds	Cross-contamination of wafers, plastic additives containing organotin and organobismuth compounds, corroding ductwork	Particulates in air and on wafers

will aid in the development of new AMC guidelines. Key deliverables will include

- Technology roadmap documents, including definition of the quality requirements, needs/risks, and potential solutions
- SEMI standards
- Conference presentations

The goals and objectives for WECC+ under the IDRS remains similar to those of the original ITRS in that risks will be identified and risk mitigation strategies proposed. However, with the IEEE's extensive resources, collaborative experimental studies can be initiated and conducted. Communication of the risks and their mitigation strategies to the industry will be improved and to facilitate solutions, SEMI standards will be updated or new standards developed. New technology trials (e.g., metrology) will be supported through benchmarking studies. All of this in an effort to continue refining the AMC requirements for advanced semiconductor manufacturing.

40.10.3 ISO Standard 14644-8[10]

ISO 14644-8 covers the classification of molecular contamination in cleanrooms and associated controlled environments in terms of airborne concentrations of the specific compound or chemical and provides a protocol to include test methods, analysis, and time weighted factors within the specification for classification. It considers only concentrations of airborne molecular contamination between 100 and 10^{-12} g/m^3 under normal cleanroom conditions of temperature, relative humidity, and pressure and is not relevant for application in those industries, processes, or production, where the presence of AMC is not considered a risk to the product or process.

This standard cannot be used to characterize the nature of airborne molecular contaminants nor does it provide a classification of surface molecular contamination. Most importantly, it will not set standard AMC levels or make recommendations for air cleanliness classes for specific operations. It does, however, provide guidance on monitoring methods.

40.11 SPECIFYING AN AMC CONTROL SYSTEM

As manufacturers have become much more sophisticated in their knowledge and understanding of AMC and its effects in the cleanroom. They have also developed a better general understanding of where AMC control should be applied and why. As their knowledge of AMC-related problems have increased, so too have their expectations for an AMC control system.

Some manufacturers' concerns about the proper selection of an AMC control system have become so acute that it is being reflected in their control specifications. One manufacturer may call for a minimum of 90 percent removal of target contaminants, while another will set AMC control limits at 1 ppb or less. Still another may require that the system must last a minimum of one year between filter changeouts. As strict as they may seem individually, there are manufacturers whose specifications require all three of the above criteria to be met.

Compounding the situation, some manufacturers insist on trying to find a "one filter fits all" solution for AMC control, demanding that a single filter should meet all of their control criteria for all contaminants of concern. However, chlorine requires one type of filter, ammonia another, and organic compounds still another. The type of filters/systems used for toxic gases would not be the same for odor control.

"Are you looking for high efficiency or long service life?" "Do you require absolute control of one contaminant or relative control of a group of contaminants?" These are only some of the questions that an AMC control system designer must consider, even though their customer may not understand the implications of not taking all of this into consideration.

40.11.1 Specification by Removal Efficiency

The removal efficiency of an AMC control system can be considered as the fraction of a single contaminant or group of contaminants that are removed either by physical or chemical means.

Many manufacturers are able to provide test data for their systems that show removal efficiency over time. However, this testing has been performed almost exclusively under accelerated conditions using high contaminant challenge concentrations that can be up to three to four orders of magnitude higher than what would be expected under actual use conditions.

Although realistic extrapolations of filter efficiency can be provided, many manufacturers are now requiring low-level gas challenge testing for efficiency ratings of the filters (Fig. 40.8). This is the best way to gauge filter performance; however, testing of this sort is more complicated, takes a long time to complete, and is much more expensive to perform.

One of the main problems in trying to provide an efficiency rating for a particular contaminant/filter combination is that any rating has to have a time component specified along with it. If a specification calls for 90 percent minimum removal efficiency, the end user is expecting that to hold true over the entire life of the filter.

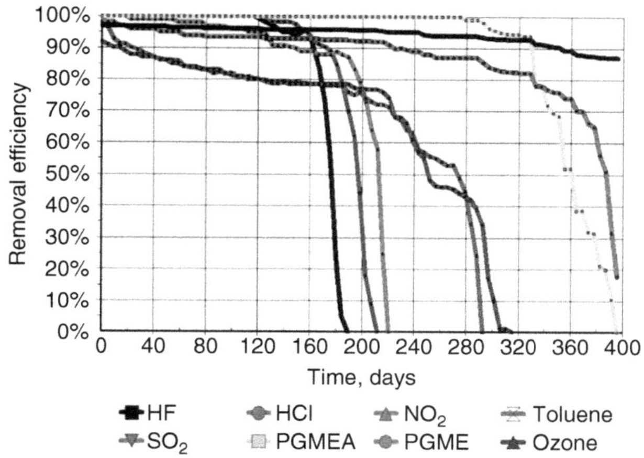

Figure 40.8 Removal efficiency of a single chemical filter against a 20 ppb challenge concentration for each of eight different chemical contaminants. The filter tested was a 24 × 24 × 12-in (600 × 600 × 300-mm) box-type filter employing an adsorbent-loaded nonwoven pleated media. (All gases were tested individually.)[17]

Most tests, however, are only run long enough to provide an "initial" efficiency, and the test is stopped after 1, 8, 24 h, and so on. All this does is provide information on how well the filter may work under a specific contaminant load, but it does not provide or predict an indication of performance beyond the stated test period.

An examination of the efficiency curves can provide some additional indication of future performance; however, it still cannot guarantee performance under actual-use conditions. Further, a filter that performs well against a single contaminant, for example, ammonia, can perform poorly when gas mixtures are considered.

40.11.2 Specification by Contaminant Limits

Some manufacturers have investigated specific AMC-related process problems so well that they have been able to set specific control levels for one or more contaminants. Common control specifications call for less than 1 ppb of ammonia in lithography bays or less than 1 ppb of chlorine or hydrogen fluoride (HF) in metallization processes.

Regardless of the ambient levels of ammonia, chlorine, or HF, the AMC control system is expected to keep the controlled environment at or less than 1 ppb over the service life of the system. Depending on the ambient levels, this could require a minimum working filter efficiency of 50, 80, 95 percent, and so on. If transient high-level episodes are probable, filter efficiencies greater than 99 percent may be required to maintain a 1 ppb level coming out of the AMC control system.

40.11.3 Specification of Service Life

Cost-of-ownership is always a major consideration when applying AMC control and can unfortunately be the primary factor when the final purchasing decision is made. Budget cycles, production schedules, capital expenditures, or simply a customer's preconceived notion of how long a filter should last, can be the basis for specifying filter life.

Testing to determine the working removal capacities of a certain AMC control product for a specific contaminant follows the same basic protocol as for efficiency testing. Using a known contaminant concentration at a specified airflow (typically the filter's maximum rated airflow), and running to a specific efficiency endpoint, one can calculate the amount of contaminant that has been removed.

The media's removal capacity is reported as a volumetric capacity (g/cc) or as a weight percent. This can then be used to estimate the media's consumption rates under a given set of conditions and provide an estimate of service life for a specific filter type. Capacities have to be determined for each contaminant of concern in order to provide a total consumption rate for a particular application. This is not practical; however, due to the uncertainty of the total contaminant load, the nature of the contaminants in question, and any associated safety concerns as well as the time and costs involved. Further, the effects other contaminants in the cleanroom may have on the AMC control system have to be accounted for.

Using observed removal capacities are useful in providing estimates of service life, and if a conservative approach is taken, these estimates can be valid tools for the determination of filter service life. Table 40.14 lists the results of testing of a number of different chemical filter media commonly used in adsorbent-loaded nonwoven fiber (ALNF) filters. The media tested are meant to provide an illustration of how different performance criterion might be measured and are not inclusive of all the media types that can be used for the gases tested.

40.11.4 Specification by Standardized Test Results

ISO Standard 10121-2.[19] This standard aims to provide an objective test method to estimate the performance of any full size gas phase air cleaning device (GPACD) for general filtration regardless of media or technique used in the device. In fact, the goal is to avoid relating the test data to internal parameters altogether. The standard encompasses the design of test equipment, analytical methodologies to measure adsorption, desorption, and avoid most common measurement mistakes. It also provides calculation methods and reporting guidelines for filter performance.

As discussed in Sec. 40.5.1, the filter removal capacity for a given removal efficiency depends on the contaminant concentration especially for physisorptive systems (VOC removal). It is therefore very important to compare the results for different chemical filters for identical inlet challenge concentrations. Additionally air flow, temperature, and relative humidity are all critical for determining filter performance. All these parameters are measured and reported when testing according to ISO 10121-2. This method provides an objective standardized benchmark of different AMC filter solutions.

Filter performance can be reported as removal efficiency (percent of challenge gas removed compared to the initial challenge gas concentration) versus time and/or by the amount of challenge gas removed (in grams) at a given removal efficiency. Both endpoints could be used as a way to set specifications for minimum filter performance when using this test method. Keep in mind that both parameters need to be specified in order to provide an accurate comparison of filter performance. For example, if specifying a chemical filter to remove ammonia: "the AMC filter shall maintain a minimum ammonia removal efficiency > 70 percent for 600 min," or "the AMC filter shall remove a minimum of 100 g of ammonia at a removal efficiency > 70 percent."

ASHRAE Standard 145.2.[20] Similar to ISO 10121-2, this provides a standard laboratory test method for assessing the performance of sorptive media chemical filters. The results of these tests can provide information to the facility and contamination control engineer useful for the design and selection of air-cleaning equipment and the design of air-cleaning systems for controlling indoor concentrations of AMC.

This standard describes a test procedure with quality control constraints to measure percent removal efficiency and removal capacity of chemical filters when challenged under steady-state conditions. The test is designed to simulate the capture performance of commercially available chemical filters under controlled, representative conditions. The filters to be tested using this standard are intended to remove gaseous contaminants that are present at low-to-moderate levels and nuisance odors, thereby protecting manufacturing processes and reducing corrosion. The test endpoint is chemical breakthrough that exceeds a minimum removal efficiency.

ISO 10121-2 and ASHRAE 145.2 provide tests that can be conducted at a fixed flow rate, temperature, relative humidity, and at an elevated challenge concentration relative to expected use levels. Such a test is shorter, reducing testing cost, but the test results are directly useful only for ranking different chemical filters. A substantial body of theory and data supports extrapolation of breakthrough times for many physically

Table 40.14 Breakthrough Capacity Test Results for Chemical Filter Media[18]
[Testing was performed for ALNF media under accelerated loading conditions using a 25 ppm challenge concentration except for formaldehyde (1.0 ppm) at a face velocity of 100 fpm (0.5 m/s)]

Contaminant gas	ALNF media tested*	Removal efficiency		Removal capacity (%)	Service life estimate‡ (h)
		Initial (%)	Avg. (%)†		
Acetic acid	GAC + PIA	94	64	15.6	1,641
Ammonia	AICA + PIA	99	50	10.9	10,197
Arsine	PIA	37	8	0.4	106
Chlorine	CICA + PIA	100	54	12.6	1,313
Ethylene	GAC + PIA	100	94	2.6	790
Formaldehyde	PIA	73	27	2.8	779
Hydrogen chloride	GAC + PIA	100	94	12.4	2,844
Hydrogen fluoride	CICA + PIA	100	97	7.5	3,050
Isopropyl alcohol	GAC + PIA	66	25	14.8	7,701
Methyl amine	AICA + GAC	98	34	1.6	1,212
Methyl ethyl ketone	GAC + PIA	78	36	13.8	4,202
n-Methyl pyrrolidone	AICA + GAC	99	85	32.4	3,036
Nitrogen dioxide	GAC + PIA	91	70	9.2	2,265
Ozone	GAC + PIA	91	67	4.5	1,108
PGME	GAC + PIA	80	49	7.9	1,417
PGMEA	GAC + PIA	88	55	14.8	1,619
Phosphine	PIA	60	14	0.3	542
Sulfur dioxide	CICA + PIA	100	95	13.4	4,197
Toluene	GAC + PIA	85	57	10.2	2,542

NOTES: *Where two media are listed, they were blended 50/50 vol/vol.
†Time-weighted average over the life of the filter.
‡Based on a 50 ppb challenge concentration.
KEY: PIA: activated alumina impregnated with sodium permanganate ($NaMnO_4$), AICA: acid-impregnated activated carbon + activated alumina, CICA: caustic-impregnated activated carbon + activated alumina, GAC: granular activated carbon.

adsorbed hydrocarbons across several orders of magnitude in concentration. Such extrapolation is not generally accepted for chemisorbed chemicals such as acid gases. Therefore, these test methods are best used only for comparison purpose or to specify AMC filtration devices, but the results obtained cannot be used to estimate actual filter service life.

Ultimately, AMC filter changeout schedules will often come from experience. Actual-use conditions often dictate that different media types in individual AMC control systems will need to be replaced at different intervals to maintain optimum performance. This is not the answer that manufacturers want to hear because they want to be able to set maintenance schedules and budgets. However, the reality is that one cannot predict, much less guarantee filter service life from testing on one or two target contaminants while not knowing the total contaminant load the system will actually see.

There is a trend toward increased real-time monitoring for contaminants of concern upstream and downstream of chemical filters to assess total filter lifetime with respect to AMC control specifications. Since the composition of AMC in the air can (and does) change day to day or due to leaks, spills, outdoor air quality, real-time monitoring is becoming more common and oftentimes required for the most critical process areas or tools. Real-time monitoring is now also available for FOUPs for acid or other residues.

40.12 FINAL CONSIDERATIONS

AMC control has become an essential design requirement for all new semiconductor manufacturing facilities, as well as for a large number of existing facilities. Subsequently, chemical filtration in many forms is being used in makeup and recirculation air handlers, fan filter units, minienvironments, and process tools.

Given the proper attention, one can successfully specify and implement an effective and economical AMC control solution for most applications. One cannot, however, go in with preconceived notions about how a particular system will perform under a given set of conditions, rather, one should determine what they ultimately want to achieve by the installation of such a system.

The following six points should always be considered in the design and specification of an AMC control system:

1. *Does AMC pose an immediate health threat to cleanroom personnel or is it primarily an odor control issue?* Personnel protection requires 100 percent control of the offending contaminants whereas odor control could be achieved with a system operating at an overall efficiency of 50 percent or less, depending on the odor thresholds for individual chemicals. Process protection requires very high performance, but that does not mean absolute control. Many specifications today call for removal efficiencies of at least 90 percent and this is realistic to expect, especially when multiple filter stages are employed in the facility air handling systems (makeup/outdoor air, recirculation air, process tool air).

2. *Accept filter performance testing for what it truly is.* That is, a comparison of different systems being offered for a particular application. Efficiency or capacity test data for a single contaminant cannot be used as an absolute predictor of system performance. It can, however, provide an indication of the relative performance differences between competitive systems.

3. *Request performance data for contaminant concentrations at or near expected use conditions if at all possible.* If the system performance and/or filter life estimates are based on accelerated test results, consider limiting gas challenge concentrations to no more than two orders of magnitude above what would be expected under actual use conditions.

4. *Filter service life should not be a part of a performance specification.* Filter life estimates should be required for all systems being considered, but again, these should be used as relative comparisons and not absolute values. There is no practical way to provide service life estimates without having performance data for the total contaminant load and not just a handful of target contaminants. Monitoring of filter performance using direct gas monitoring, reactivity monitoring, or by using grab samplers (impingers, adsorbent tubes, witness wafers, coupons, and the like) should be started along with system start-up and should continue as long as the AMC control system is in place.

5. *Use a multistage AMC control system.* Just as particulate control in cleanrooms requires several stages of filtration to achieve the desired cleanliness level in the protected space, the same should be considered for an AMC control system. A "prefilter" stage should be used to remove as much of the "junk" AMC as possible. This would help protect and preserve the "final filter's" ability to remove those contaminants of concern with good efficiency and capacity. If the types of contaminants targeted for control require different filters, serious consideration should be given to individual stages for each filter type. Blended media may be used, but at a possible tradeoff in average removal efficiency and overall service life.

6. *Look at local and/or optional controls for AMC.* For critical processes where wide variations in the types and levels of AMC are common, other approaches at a local level can be used to prevent air exposure, such as using nitrogen (N_2) or clean dry air (CDA) purges especially in lithography tools, cluster tools with vacuum to prevent air exposure, and purged stockers for wafers and reticles.

A properly designed, installed and maintained AMC control system can fairly easily achieve the removal efficiencies required for specific target contaminants. How long a system will take to meet specific performance criteria depends on the average and peak values of all the contaminants present, which must be considered in the final design of the AMC control system.

40.13 SUMMARY

AMC continues to grow in importance as a barrier to semiconductor manufacturers with the increased use of 300 mm wafers and copper interconnects. And as they continue toward the use of air gaps, low-k, high-k, porous low-k films, new materials and etchants, atomic layer deposition (ALD), 450 mm wafers and the 10 nm technology node AMC represents a wide range of chemical types that can result in a large number of potential processing problems. Many fabs are more actively addressing old and new types of AMC; however, the effects of specific contaminants on individual processing steps are still not very well understood. Direct correlations between process yield and individual contaminant concentrations, although one of the current "Holy Grails" of contamination control, are either rare or not published. When yield upsets do occur, they should be used as data points to set rational AMC control specifications for individual processes.

Just as the semiconductor industry has progressed in the last 10 years, so too has AMC control technology. This progress has presented itself in the form of

- Gas-phase filter media and chemical filters that can address essentially any AMC issue
- New delivery systems for the different types of chemical filter media available
- Filters and filtration systems that can be integrated into existing air handling equipment
- Monitoring instrumentation and devices that can provide real-time, accurate environmental assessments
- The technical competence to work with manufacturers to provide effective and economical solutions to AMC-related problems

AMC control technology has had to keep pace not only because the recommended control limits for the various chemical contaminants have dropped in some cases to parts-per-trillion (ppt) levels, but also because the types and numbers of contaminants specified to be controlled have increased. Early on the primary concern was for the control of ammonia and other bases in lithography bays. With copper processing came a requirement for the control of acids Now AMC control includes the "ABCs" of chemical contamination control: *A*cids, *B*ases, *C*ondensable organics, *D*opants, *E*lements (metal compounds, organometallics), *R*efractory compounds, *S*ulfur and silicon compounds (exact classes being somewhat arbitrary and overlapping). And the list continues to grow.[21]

With the rapid pace of technological advancements, semiconductor manufacturers have fully recognized the fact that sensitive electronic, optoelectronic, nanotech, and electrical components and equipment will be damaged if exposed to AMC. The IRDS has proposed that AMC will be the next technical challenge to overcome in order to maintain high reliability and yield for semiconductor devices. New requirements for AMC control are being added to each new edition of the Roadmap. "Next generation" semiconductor devices will require the strict control of AMC in order to ensure productivity, competitiveness, and profitability. Manufacturers are actively addressing AMC and its control, each according to the specific requirements of their own manufacturing processes and concerns. At the very least the incorporation of chemical filtration into makeup air handlers has become a requirement for these facilities. AMC control in the form of more filters designed for a specific application are being used with greater frequency for the most sensitive processes and equipment including lithography, diffusion furnaces, copper tools, and inspection equipment.

An assessment of outdoor air quality should be one of the first steps toward establishing any AMC control program. In review of air monitoring data from semiconductor fabs around the world, there are very few areas today where the outside air quality would meet specific air quality requirements with respect to AMC. Even ambient levels of AMC are high enough to be of concern.

As the industry's knowledge about the damaging effects of AMC on processes and materials continues to increase, essentially all facilities are implementing AMC control programs. Suppliers of specialized chemical filtration systems are being asked to assist in determining the types of AMC present, the specific control system(s) required, how best to determine if the systems are working and if so, then whether they are meeting specific design requirements.

As the necessity to control chemical contamination increases, more information and feedback from semiconductor manufacturers is going to be required for continuous improvement of AMC monitoring and control technologies. Intellectual property, new technology, and competitive considerations must be protected, but the suppliers of monitoring and control technologies must be able to address AMC issues with as much information as possible at their disposal. Only then can we be assured that we will keep pace with AMC requirements to enable new technologies with high yields and reliability.

40.14 REFERENCES

1. C. Muller, "Developments in Measurement and Control of Airborne Molecular Contaminants," *Proceedings of SEMICON Taiwan 2001*, Taipei, Taiwan, Sep. 17–19, 2001.
2. M. L. Kwan, C. Muller, and R. Thomas, "Semiquantitative Analysis Techniques for AMC Monitoring," *Proceedings of SEMICON Taiwan 2004*, Taipei, Taiwan, Sep. 13–15, 2004.
3. SEMI Standard F21-1102: 2002, "Classification of Airborne Molecular Contaminant Levels in Clean Environments," Semiconductor Equipment and Materials International, Mountain View, CA.
4. IEST-RP-CC031:2011, "Method for Characterizing Outgassed Organic Compounds From Cleanroom Materials and Components," Institute of Environmental Sciences and Technology, Arlington Heights, IL.
5. W. R. Jones, J. H. Knight, and C. O. Muller, "Practical Applications of Assessment, Control, and Monitoring of AMC in Semiconductor and Disk-Drive Manufacturing," *Proceedings of ESTECH 2002, 47th Annual Technical Meeting of the Institute of Environmental Science and Technology*, Anaheim, CA, Apr. 28–May 1, 2002.
6. M. C. Middlebrooks and C. Muller, "Application and Evaluation of a New Dry-Scrubbing Chemical Filtration Media," *Proceedings of the Air & Waste Management Association 94th Annual Meeting and Exhibition*, Orlando, FL, Jun. 24–28, 2001.

7. ANSI/ASHRAE Standard 52.2-2012, *Method of Testing General Ventilation Air-Cleaning Devices for Removal Efficiency by Particle Size*, American Society of Heating, Refrigerating and Air-Conditioning Engineers, Inc., Atlanta, GA.

8. EN 779:2012, "Particulate air filters for general ventilation. Determination of the filtration performance," European Committee for Standardization (CEN), Brussels, Belgium.

9. "Advantages of Carbon in Broad-Spectrum Chemical Filtration for Lithographic Processes," White Paper, Donaldson Company, 2001.

10. ISO 14644-8: 2013, "Cleanrooms and Associated Controlled Environments—Part 8: Classification of Air Cleanliness by Chemical Concentration (ACC)," International Organization for Standardization, Geneva, Switzerland.

11. C. O. Muller, "Reactivity Monitoring: An Alternative to Gas Monitoring for Semiconductor Cleanrooms?" In *Proceedings of the 45th Annual Technical Meeting of the Institute of Environmental Science and Technology*, Ontario, CA, May 2–7, 1999.

12. C. Muller, "Evaluating the Effectiveness of Airborne Molecular Contamination Control Strategies with Reactivity Monitoring," *Journal of the IEST*, Vol. 45, Annual Edition, 2002.

13. International Technology Roadmap for Semiconductors, 2013 Edition, http://www.itrs2.net/2013-itrs.html.

14. International Technology Roadmap for Semiconductors, 2013 Edition: *Yield Enhancement Chapter*, https://www.dropbox.com/sh/6xq737bg6pww9gq/AABi_Y8w8bDy-RDwxIrD6vrGa/2013Yield.pdf?dl=0.

15. International Technology Roadmap for Semiconductors, 2013 Edition: Table YE3: Technology Requirements for Wafer Environmental Contamination Control, https://www.dropbox.com/sh/qz9gg6uu4kl04vj/AADCoqXsn_31QOS2fQEFYt9Fa/Yield_2013Tables.xlsx?dl=0.

16. C. Muller and M. Tan, "From Microns to Nanometers: The ITRS and AMC Control," *Proceedings of ESTECH 2012*, Orlando, FL, Apr. 30–May 3, 2012.

17. G. O. Nelson, "Purafilter Gas Challenge Test Results: 2001–2003," Unpublished report presented to Purafil, Inc., Doraville, GA, 2003.

18. C. Muller, "Specifically AMC: Guidelines for Specification of AMC Control," *Cleanroom Technology Magazine*, Apr. 5, 2004.

19. ISO 10121-2:2013, *Test Methods for Assessing the Performance of Gas-Phase Air Cleaning Media and Devices for General Ventilation—Part 2: Gas-Phase Air Cleaning Devices (GPACD)*, International Organization for Standardization, Geneva, Switzerland.

20. ANSI/ASHRAE Standard 145.2-2011, "Laboratory Test Method for Assessing the Performance of Gas-Phase Air-Cleaning Systems: Air-Cleaning Devices," American Society of Heating, Refrigerating and Air-Conditioning Engineers, Inc., Atlanta, GA.

21. C. Muller, R. van Dijke, and A. Edeling, "Filtration for Advanced AMC Control for Leading-Edge Microelectronics Manufacturing," *Proceedings of Filtech 2013*, Oct. 22–24, 2013, Wiesbaden, Germany.

40.15 INFORMATION RESOURCES

Cleanroom Technology magazine: http://cleanroomtechnology.com/
Controlled Environments magazine: http://www.cemag.us/
IEEE Rebooting Computing: http://rebootingcomputing.ieee.org/
Institute of Environmental Science and Technology: http://www.iest.org/
International Confederation of Contamination Control Societies: http://www.icccs.net/
International SEMATECH Manufacturing Initiative: http://ismi.sematech.org/index.htm.
International Technology Roadmap for Semiconductors (ITRS 2.0) website: http://www.itrs2.net/
SEMATECH Standards and Guidelines: http://ismi.sematech.org/standards/index.htm.
Semiconductor Fabtech magazine: http://www.semiconductorfabtech.com/
Semiconductor Engineering: http://semiengineering.com/
SEMI Standards Publications: http://wps2a.semi.org/wps/portal/_pagr/116/_pa.116/121?docName=P001204
Solid State Technology: http://electroiq.com/semiconductors/

40.16 APPENDIX: ABBREVIATIONS FOR REFERENCED SAMPLING DEVICES AND ANALYSIS METHODS

AA = atomic absorption spectroscopy
AA-F = atomic absorption spectroscopy—flame
AA-GF = atomic absorption spectroscopy—graphite furnace
AES = atomic emission spectroscopy
API-MS = atmospheric pressure ionization—mass spectroscopy
CL = chemiluminescence
CLS = chemiluminescence monitoring system
CPR = colorimetric detection on chemically impregnated paper reel type analyzer
CRDS = cavity ring-down spectroscopy
CZE = capillary zone electrophoresis
DIFF = passive diffusive sampler
DSE-TXRF = droplet scanning extraction—total reflection x-ray fluorescence
ECS = sensors of electrochemical cell type
FTIR = Fourier transform infrared spectroscopy
GC-FID = gas chromatography—flame ionization detector
GC-MS = gas chromatography—mass spectroscopy
IC = ion chromatography
ICP-MS = inductively coupled plasma—mass spectroscopy
IMP = impinger (typically used with IC)
IMS = ion mobility spectrometry
IR = infra-red spectroscopy
MS = mass spectroscopy
RM = reactivity monitoring
SB = sample bag
SOR = sorbent tube sampling
TD-GC-MS = thermal desorption gas chromatography—mass spectroscopy
TOF-SIMS = time of flight—secondary ion mass spectrometry
TXRF = total reflection x-ray fluorescence spectroscopy
UVS = ultraviolet spectroscopy
VPD = vapor phase decomposition
VPD-TXRF = vapor phase decomposition—total reflection x-ray fluorescence
WW = witness wafer

Chapter 41

ESD Controls in Cleanroom Environments

BY

LARRY LEVIT *LBL Scientific, Alamo, California*

41.1 ELECTROSTATIC CHARGE IN SEMICONDUCTOR CLEANROOMS

Electrostatic charge grows to extremely high levels in the environment of a semiconductor fab. This effect is called triboelectrification is a surface phenomenon and is caused by contact and separation of two different materials. The presence of water absorbed into the surface of the material from the humidity in the air diminishes the effect.[1,2]

Many of the items in the cleanroom are constructed from plastics, which, in general, are extremely good insulators. Because it is necessary to employ these materials since they are inherent to the process. For example, quartz is required for reticles because of its optical properties, and Teflon is required because it is impervious to hydrofluoric acid. Also, the extreme cleanliness of the surfaces of objects in the fab denies static surface charge a conduction path to ground on an insulator.

These issues of increased static charge generation and decreased static charge dissipation result in significantly higher static charge levels in a semiconductor fab than in a conventional room. In fact, it is not uncommon for a fab with no safeguards against static charge to achieve levels of tens of kilovolts on a variety of critical surfaces.

41.2 PROBLEMS RESULTING FROM CHARGE IN CLEANROOMS

41.2.1 Contamination

The relationship between electrostatic charge and contamination control in high-technology cleanrooms is a strong one, representing a significant source of particle deposition on charged objects such as wafers. The rate of electrostatic attraction (ESA) of small particles to charged objects has been calculated for a controlled environment,[3] and particle deposition data recorded in cleanroom environments have been found to be consistent with the calculation.[4] The importance of ESA becomes greater as the particle size becomes smaller, and below 100 nm is the dominant effect in contamination levels as compared to sedimentation (gravity), aerodynamics (wind), and diffusion (Brownian motion).

Once ESA became known and accepted, it was believed that charges on a wafer caused contaminants to be drawn from the air and thus produce defects. Later it became clear that while this is true, another effect was equally important. Charge on adjacent objects in process tools also affect the trajectory of particles. This often causes them to be drawn off of the laminar flow in the process tool, thereby defeating the action of the laminar flow. This is the greatest weapon against microcontamination control and, without it, particles can approach the wafer and be drawn in by ESA.

One of several studies involved employing a high-voltage power supply to bias one wafer to a voltage of 2 kV and connecting another to ground electrically. In one of the studies, 200-mm wafers were exposed to the environment in an isolated part of an ISO class 3 cleanroom.[5] Human traffic through the area where the exposure occurred was eliminated by cordoning off the area. In order to ensure that a statistically valid sample was acquired, an exposure time of 6 weeks was employed. The wafers were scanned using a Tencor Surfscan with a particle size threshold of 0.2 μm. The Surfscan results are illustrated in Fig. 41.1. While the numbers of particles observed on the two wafers are not representative of real-world contamination levels for semiconductor processing, the ratio of the number of particles on the neutral wafer (3389) compared to the number of particles on the wafer at a voltage of 2000 V (22,764) is accurate. In this case, the ratio is 6.7:1. More recent studies have shown the same effect at 100 nm and at 50 nm.

The calculation of electrostatically driven contamination sums the components of the deposition velocity due to each deposition mechanism. Deposition velocity, loosely defined, is the rate at which particles move toward the surface of the object due to the forces on the particle. Particles deposit on wafers, flat panel displays, disk drive components, and equipment surfaces due to gravity, diffusion, and other forces. *High-efficiency particle attenuator* (HEPA) filtration attempts to keep particles out of the cleanroom, and laminar airflow is designed to minimize the sedimentation rate of contaminating particles by entraining them into the airflow. While a cleanroom is maintained at a high level of cleanliness, particles are still present due to people, equipment motion, and process.

The forces on the particle are aerodynamic (viscous drag), gravitational, diffusional, and electrostatic in nature. Of course, gravitational forces decrease as the size (mass) of the particle decreases. Similarly, the magnitude of the viscous drag on the particle is also smaller for smaller particle sizes. In contrast, the magnitude of the diffusional forces is greater for smaller particles because of the greater efficiency of momentum transfer to particles that are closer in mass to the gas atoms that strike them.

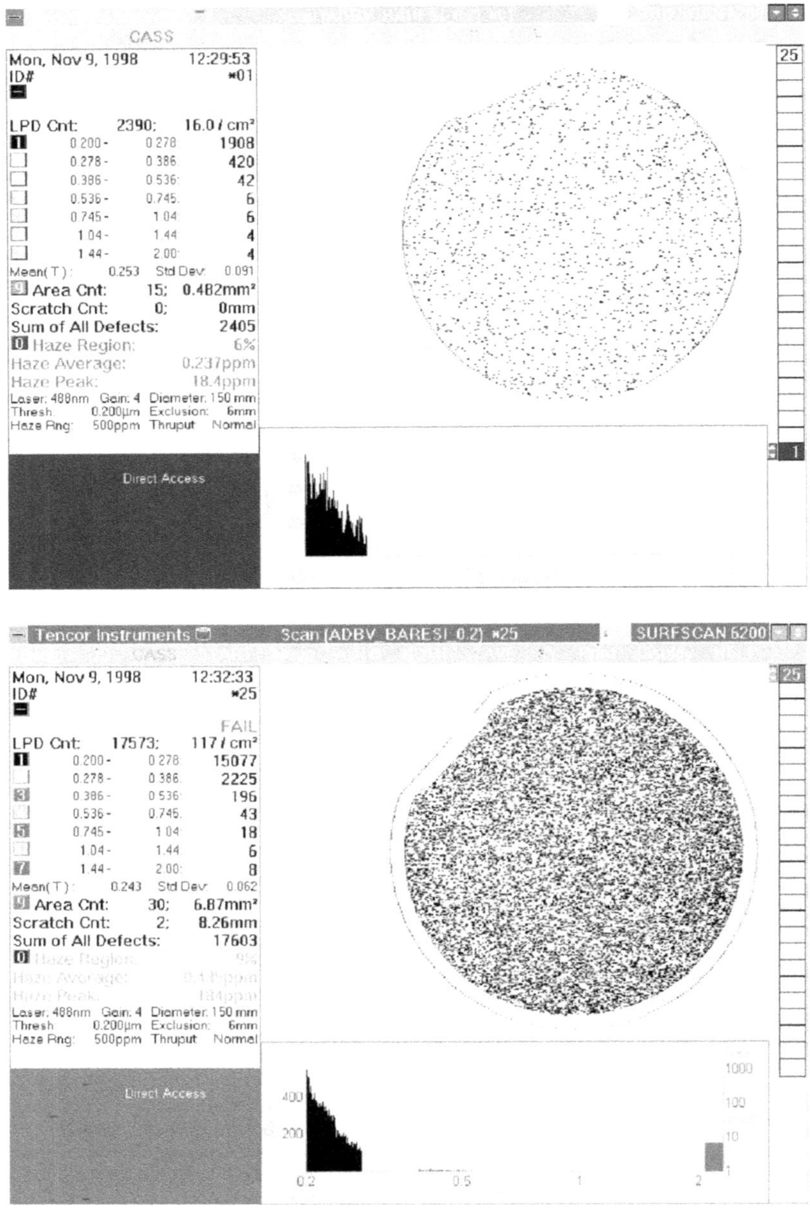

Figure 41.1 Surfscan results.

The calculation showed that the force of electrostatic attraction exceeds the other physical forces, for the conditions normally present in a cleanroom, surface voltages of several thousands of volts, and for particles in the micrometer to sub-micrometer range. The calculation assumed that the particles in the air were neutral on average but had a distribution of charges with a width that was monotonically related to the total number of electrons on the particle. Thus, larger particles have the possibility of more charge on them and can experience greater electrostatic forces but are also more massive. Thus, the deposition velocity due to electrostatic forces exhibits a maximum for a specific size.

Theoretically, increased particle deposition due to electrostatic forces will occur at any voltage but at field strengths of 500 V/in or more, the effects are significant and easy to measure. Field strengths of 10,000 V/in or more are not uncommon in cleanrooms.

Figure 41.2 illustrates some of the forces that act to attract particles to wafer surfaces.[5] For small particles (0.01 to 1.0 μm in size), electrostatic forces are the primary cause of increased particle deposition. What size of particle is to be expected in a modern high-technology cleanroom? Figure 41.3 gives a bit of insight into the expected size range. The figure shows that the nature of a HEPA filter is to stop virtually all particles that are incident upon it, but that the efficiency of the filter is lowest in the near sub-micrometer range (~0.1 to 0.2 μm). In this size range, and with the forces shown in Fig. 41.2, it is expected that the major contributor to contamination at 500 V/cm is electrostatic attraction.

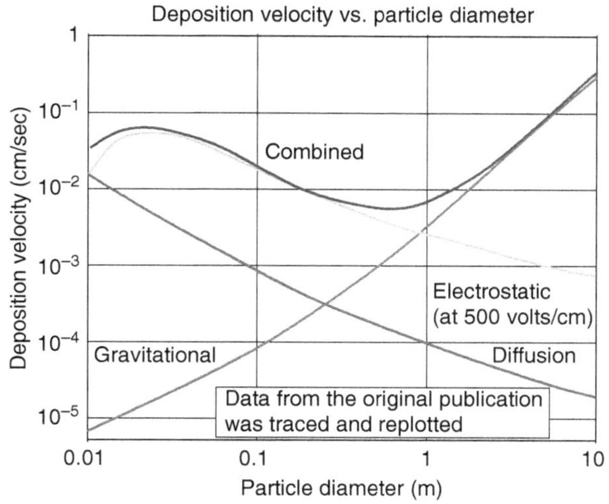

Figure 41.2 Forces acting on particles. (*Data from the original publication was traced and replotted.*)

How significant are the calculations and the measurements discussed previously? Is electrostatic attraction a significant factor in high-technology manufacturing? In a typical high-technology manufacturing cleanroom without an electrostatic charge control program, insulators in a room (e.g., reticles, wafers with oxide coating, and disk drive media) typically achieve voltage levels of 5 to 20 kV. Thus, based upon the data presented above, the majority of the contamination is electrostatically driven. Often this major factor goes unnoticed because the cleanliness levels are so good in the cleanroom. In subclass ISO class 3 cleanrooms, the number of *particle adders* (PWP) in a typical modern process can be as low as 0.1 to 1.0 (> 50 nm). Over the course of processing many layers of the product, this can represent 2 to 20 particles, which is still a very small number. If the average static charge level in the process is only 500 V, owing to static control delivered by the process tool manufacturer, in some parts of the process, the part of the contamination contributed by electrostatic attraction would likely be in the range of 0.25 to 10 particles. In a process with such low contamination levels, it is very difficult to devise a method for measuring the contribution due to one source. Nonetheless, the economic impact of removing this contamination source is profound.

41.2.2 Electrostatic Damage

Electrostatic attraction and its effect on contamination control is not the only negative effect of static charge. In addition, static charge is responsible for electrostatic damage, including the discharge from one object to another. Such a discharge can cause physical damage to a product that experiences a discharge owing to either electrical overstress (voltage punch-through) or damage caused by the energy deposit by a discharge (spark). This occurs frequently in back-end (BEOL) handling and testing owing to the fact that bonding pads have been created at this point, which makes it easy to apply voltages to the circuitry or drive current through circuit elements on the chip. Electrostatic discharge (ESD) is particularly important in photolithography. Damage to reticles is caused by either charge on the reticle or by induced discharges caused by electric fields from other charged objects in the proximate environment.[6]

The rapid uncontrolled transfer of static charge from one object to another is referred to as ESD. One aspect of such a discharge is that the source and destination of the electrical current are extremely localized (1 < μm). This means that the energy dissipated in the two objects that participate in the discharge are heated extremely locally. When such a discharge takes place between two electrical conductors, the speed of the transfer is typically in the nanosecond to sub-nanosecond regime. The nominal capacitance of objects in a fab (approximately tens of cm) is in the order of 10^{-11} to 10^{-10} F and metal-to-metal discharges have much less than 1 Ω of resistance. Thus, if they are conductors, they have RC time constants less than 1 ns. So even including inductive effects, metal-to-metal discharges are extremely fast. Discharges from charged individual die or packaged parts to a nearby ground are referred to as charged device model (CDM) and have durations of approximately 750 ps.

ESD damage can be caused by tribocharging of the victim object followed by a discharge to ground or to a large enough floating object that the capacitance to ground represents a low impedance ac path. This could be a single chip, which contacts a work surface which is not grounded. In this case, the object becomes charged by handling and then, when it is moved toward a socket or is approached by an end effector, the discharge occurs. The other effect that can lead to damage is called the field-induced model (FIM).[7] When a victim object (typically a die, wafer, or reticle) is subjected to an electric field, a potential difference is created between the victim and an adjacent ground. This damage mechanism is common in back-end processing. In this case, the insulating package becomes charged from handling and produces an electric field. This field drives a discharge from the die in the package to a ground pin on the test socket. See Fig. 41.4.

Measurements of the effect have been made by many authors by extending the time of exposure of charged wafers to the cleanroom atmosphere. Results of wafers charged to 500 to 2000 V have shown increased contamination levels by a factor if 3 to 7 as compared to neutral wafers.

In another example of FIM damage, a photomask is particularly sensitive to electric field. The image area on a photomask typically contains

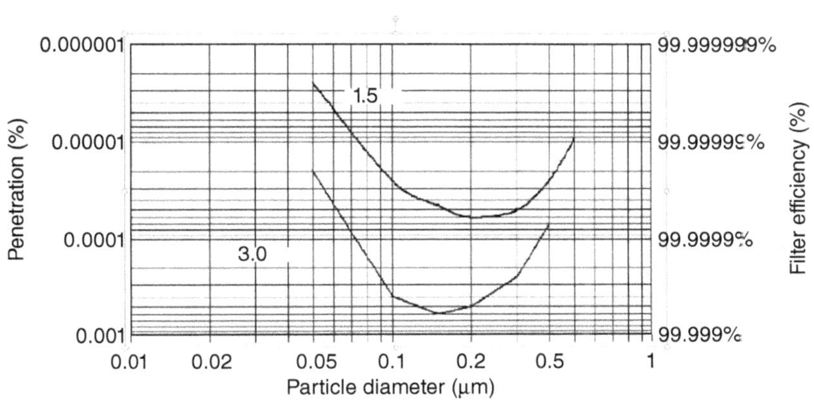

Figure 41.3 Size of particle to be expected in a modern high-technology cleanroom.

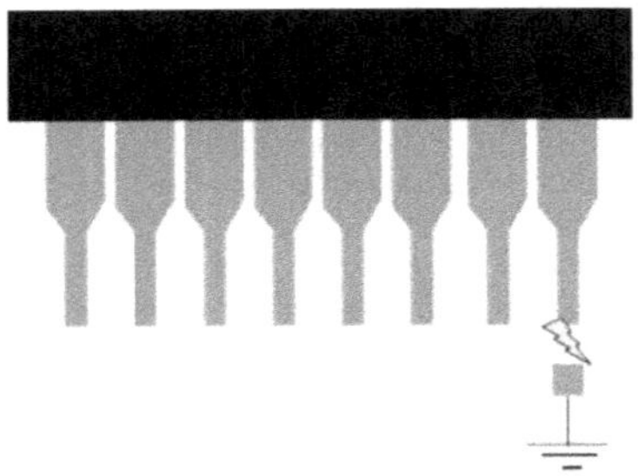

Figure 41.4 Field-induced damage driven by the field from a charged package.

many thousands of isolated conductive features, all supported on an insulating substrate and separated from their neighbors by very tiny gaps (measured in nanometers).

When an electric field traverses the image area, different features in the image have different voltages induced on them by the field. The potential gradient of the electric field becomes squeezed into the small gaps between the features, so this amplifies the ambient field strength. The amplification factor can be several thousand times in a typical photomask, so even a relatively weak electric field in its handling environment can be hazardous to a photomask.

Every time a photomask is exposed to an electric field the electrons within the conductive features move to cancel the field within the conductor. Because the features are conductors, the entire feature sits at a single voltage with all of the voltage gradient (field) being in the gaps between conductors. If the voltage difference between two isolated features rises high enough under the influence of the electric field, the insulating quality of the air above them can break down and electrons will jump from one feature to another in an electrostatic discharge (ESD). The energy dissipated in just a nanosecond can be sufficient to melt the conductive film, as can be seen in the micrograph of Fig. 41.5.

Such discharges are detrimental because they can cause the manufacturing process to suffer a sudden and precipitous yield loss. A single defect on a photomask from ESD results in mass production of the defect on the wafer. If the reticle contains two die images, there is an immediate 50% reduction in process yield. For this reason many semiconductor fabs adopt frequent inspections to check for any defects in the printed pattern. Despite the typically rapid detection of any printing fault, the financial cost of photomask ESD damage can easily amount to hundreds of thousands of dollars per event.

As microcircuits become ever smaller and more densely packed, the nature of field induction in the photomasks used to print circuit changes, as is shown in Fig. 41.6. This computer simulation indicates

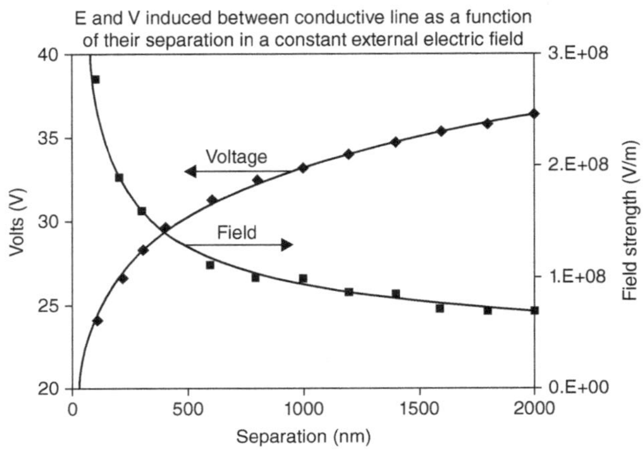

Figure 41.6 Two-dimensional finite element analysis of the induced potential difference and the field strength between two conductive lines on a reticle as a function of their separation, in a constant electric field. (*Courtesy of Gavin Rider.*)

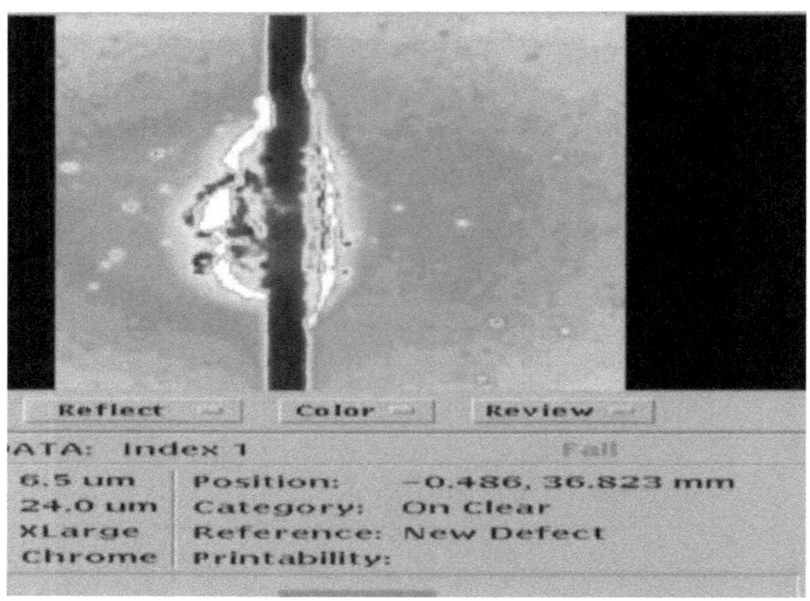

Figure 41.5 Another example of FIM damage.

how the electric field and the induced voltage between two conductive features held in a constant electric field changes as the separation between the features is reduced. The graph shows that by following Moore's law, it becomes increasingly difficult to induce a sufficiently high voltage between photomask features to cause ESD, but simultaneously the electric field between features increases in strength nonlinearly.

This evolution of field induction is very significant, because recently another physical process has been found to cause damage when photomasks are exposed to electric field.[7] The effect is called electric field-induced migration (EFM) and it takes place when the electric field strength is more than 100 times weaker than would be needed to generate ESD. This effect was first observed in reticle ESD studies by the author in early 2000, observations are being reported only now in production reticles.

Whereas ESD is a discrete and fairly easily detected event, EFM has the effect of continuously and cumulatively degrading the resolution of the image by perturbing the shapes and critical dimensions of certain parts of the pattern. The effect of EFM is to fuzz out the sharp edges of chrome line edges so the image projected on the surface of the wafer is no longer "black and white" but at the edges becomes "gray scale" and hence prints as poorly defined shapes. This process is driven by the magnitude of the electric field in the gaps between chrome features. But since the electric field that a photomask experiences in a production environment is constantly changing in both strength and direction, it is impossible to predict where in any particular photomask such damage will be induced. It is also impossible to know that the damage is happening without continuously monitoring every photomask in production, which is impractical.

Consequently, the damage from EFM builds up gradually until a point is reached when the wafers being printed fail routine inspection. The time evolution of degradation is likened to the building of stalactites and stalagmites in a cave. But wafers with marginally acceptable features will have been printed for a considerable time before any printed defects are detected, and this can mean that the production line is filled with partially completed devices that may fail at final test, or exhibit reliability issues under normal use. The financial implications of a yield loss due to the degradation of a photomask by EFM can be many times higher than a failure due to ESD.

The electrostatic countermeasures used in semiconductor production today are generally capable of preventing most photomask ESD, but the risk of degradation by EFM is always present and is constantly increasing as shown by Fig. 41.6. The only fail-safe way to address this risk is to shield photomasks from electric field using a Faraday cage, and when handling them outside the protective environment to handle them in a way that minimizes any field induction that may take place. Also, for locations in reticle production, cleaning and inspection tools as well as in the exposure tool, it is important to follow normal static charge control procedures for the environment of the stepper and the reticle manufacturing tools. This means the minimum use of insulators and the use of air ionization for the remaining insulators.

41.2.3 Electromagnetic Interference and Wafer Mishandling

There are two ways that wafers can be mishandled. One is through microprocessor errors caused by transient EMI due to ESD and the other is improper wafer moves caused by electrostatic attraction. Since both are electrostatic in origin, they are grouped together and mishandling.

The corruption of data or program words caused by transient EMI whose origin is electrostatic discharge somewhere in the proximate the environment. When a metal-to-metal discharge takes place, the frequency content of the discharge is extremely high (approximately multiple gigahertz).[8] For sub-nanosecond pulses (see Fig. 41.7), half of the energy in the discharge is radiated as an electromagnetic impulse. Such a pulse generates eddy currents that are broadcast into the fab. Because the bandwidth of the discharge extends into the microwave frequency range, it is extremely difficult to shield electronic equipment from this EMI. This impulse can induce voltage pulses in the traces of

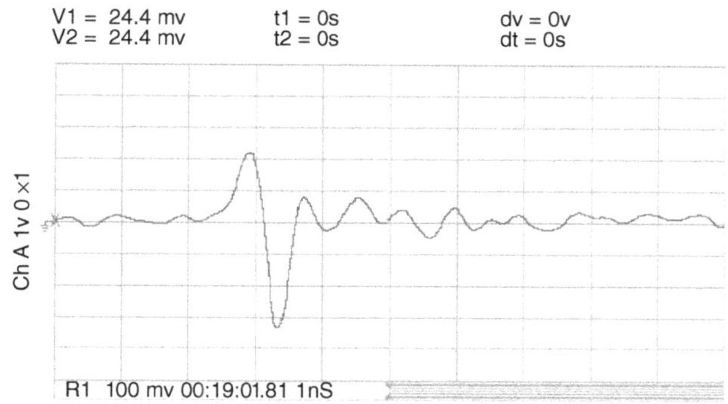

Figure 41.7 A sub-nanosecond discharge viewed using a scan converter with a 4.5-GHz bandwidth.

the microprocessor board controlling a robot in the process tool. This can result in moving the wafer to a ridiculous position such as against a minienvironment wall. A corrupted instruction will result in a tool stop with an error message. Most often this behavior is assumed to be a software bug. While this may be the case, the problem can also be the result of EMI transients generated by ESD.

In addition to ESD-induced EMI (electromagnetic interference), another problem in a fab is continuous EMI. This sort of EMI is usually broadcast by electronic circuitry in the fab. The amplitude of such EMI from a product and the immunity of a product to such EMI are regulated to much lower levels than those observed in association with ESD-induced EMI. Such EMI is normally regulated by the FCC in the United States and by CISPR (CE Mark) abroad.

The measurement protocol specified for this sort of continuous EMI is based upon a spectrum analyzer and is an averaging measurement. As such, the transient EMI that is generated by ESD does not register on a spectrum analyzer. A typical ESD event occurs at time scales of 10 s to 1 min and has a duration of much less than 1 μs. Thus, the duty cycle is 10^{-7} or less and the average radiated energy is negligible. For this reason, it is important to realize that a device that passes a conventional radiated power standard can still be a source of ESD-induced EMI. Similarly, if a radiated EMI study is done with a spectrum analyzer and antenna, there are no conclusions that can be drawn about the presence of ESD-induced EMI, as that test is blind to the effect.

It is of interest to estimate the peak current in such a discharge. If the object being discharged represents approximately 10^{-12} F and is charged to 5 kV

$$q = CV = 10^{-11}\ \text{F} \times 5 \times 10^{+3}\ \text{V} = 5 \times 10^{-8}\ \text{C}$$

Then, if the excursion is similar to the one above, 0.75 ns wide, the peak current is

$$i_{\text{peak}} = 2q/0.75\ \text{ns} = 67\ \text{A}$$

This is an extremely large current and an even larger di/dt and nearly impossible to shield against. This is an amazingly high current for a little spark and means that the magnitude of the radiated power is extremely high even though the total energy in the discharge is only a few hundred microjoules.

It is important to understand the mechanism for the interruption. While the discharges take place relatively often, typically one discharge (or ESD event) for each wafer passing through the tool, they rarely cause a problem even if they are large in amplitude. The ESD event must be accidentally timed to coincide exactly with a data strobe on the microprocessor board. When this coincidence occurs, the digital information (microprocessor instruction, address, or data word) becomes corrupted.

As a result, the processor either detects the error and stops or executes the bad data. In the case of the former behavior, the tool typically stops and displays an error message that may or may not be intelligible to the operator. This phenomenon is called lockup and usually requires that the tool be rebooted. The associated operations of the reboot process may require as much as 45 min to execute. During this time, the tool is unavailable. If the data are corrupted, the tool may make an incorrect decision based upon the information. This can result in strange behavior. For example, if the corrupted data word is a robot location coordinate, the tool may cause the robot to strike a wall and result in damage to the robot and the product.

Because the aberrant behavior requires the coincidence of two things, the rate of occurrence of the phenomenon is low. Also, because the events are stochastic in nature, their time distribution occurs with a Poisson distribution. It is not at all uncommon for the rate of such events to be on the order of one or two per week on average. Because of the nature of the Poisson distribution, these events have the appearance of occurring in clusters. There may be five in 1 week and then none for the next 2 weeks. This makes them extremely difficult to diagnose. To test the hypothesis that a given bad tool behavior is due to this effect, it is best not to look for the lockup effect but for the ESD-induced EMI instead, which occurs at a considerably higher rate and is therefore easier to diagnose. A discussion of instrumentation for this measurement is included in this section.

The other form of mishandling is the result of electrostatic attraction. The most dramatic example of this sort of mishandling is the result of attraction of a wafer to its nearest neighbor. When a new wafer enters a cassette or a boat it occasionally attracts the nearest neighbor wafer in a boat with such a strong attraction that the wafer is lifted off the end effector and the adjacent wafer is bent with the result that the two wafers bang against each other. This effect has been observed in most deposition vertical furnaces and occasionally in a wafer cassette. Often the driving factor is the presence of a ceramic end effector which becomes highly charged and affects the wafers that it is moving in.

In some cases, removal of a wafer from a cassette can draw the adjacent wafer along with it. This can result in a dropped wafer or an end effector striking the partially removed wafer and damaging it. To avoid this, it is necessary to use dissipative cassettes (or FOUPs) and to be sure that the nest provides a good ground for the cassette.

41.3 STATIC CHARGE GENERATION

Whenever two dissimilar surfaces placed in close contact are separated, one surface loses electrons and becomes positively charged, while the other surface gains electrons and becomes negatively charged. This is known as triboelectric charging. Any material, solid or liquid, may be charged triboelectrically by friction, contact and separation of materials, or fluid flow. The magnitude of the charge will be affected by the surface condition, area of contact, speed of separation or rubbing, and the humidity. Whether the material remains charged depends on its conductivity and the availability of a path for the charge to flow to ground. If charge is allowed to accumulate on a material it may attract and bond particles to its surface.

The principle of triboelectric charging is related to the work function of each material. This is defined as the energy required to remove an electron from the surface of the material. Since the work function is related to the electronic energy levels of the material, the work function of each chemical element or compound is unique. Thus, whenever two materials are placed in contact with each other, electrons are transferred from one material to the other. The amount of transfer is dependent upon the nature of the surfaces and the nature of the contact. Smooth surfaces with a significant amount of contact pressure and relative motion (sliding) result in a greater amount of charge transfer.

When the materials are separated, the capacitance between them decreases, so the potential difference (voltage) between them increases. Thus, the mechanism of triboelectric charging results from contact of dissimilar materials followed by their separation afterward.

The environment of a cleanroom lends itself to increased levels of tribocharging and decreased dissipation of the charge from the surface of insulators as compared with levels on insulators in a conventional room. First, any contamination that exists on the surface of an insulator provides a discharge path to ground for the charge. In a cleanroom, all objects are wiped down to eliminate this discharge mechanism.

Studies[10] have shown the relationship between humidity and tribocharging. In a location with 60 percent relative humidity, an equilibrium is achieved between water vapor and the surface of each material. At a lower humidity (~35 to 45 percent), the equilibrium condition dictates that less water vapor will be present in the surface of the material. This water vapor modulates the work function of the material, placing its value part way between the work function of the material and the work function of water. Thus, materials in a low humidity environment exhibit more triboelectric charging than those in a higher humidity environment. Since cleanrooms are maintained at low humidity, more charging is observed in the cleanroom that in the conventional room.

Static charge is also generated by induction. Static charge on an object can create or "induce" opposite polarity charges on the surface of another object by causing the positive and negative charges on the object to separate. As with charge generated triboelectrically, the induced charges can attract particles, and contact with ground can result in damaging ESD events and EMI.

An example of contamination resulting from induction is when a wafer is stored in a highly charged container (i.e., a FOUP), assumed negative for discussion purposes. If wafers without an oxide coating are held in an inner support structure that is conducting (as is the case with a FOUP) and connected to ground, the charge on the wafer-support structure will separate and the negative charge will be displaced from the wafers to the support structure and from there to ground. When the container is removed from the ground structure, the wafers will be charged positively and represent a contamination issue.

41.4 INSULATORS VERSUS CONDUCTORS

All materials can be characterized as either electrical conductors or as electrical insulators. A conductor is a material through which an electrical current can pass and an insulator is a material through which an electrical current cannot pass. There is a good deal of variation in the electrical properties of conductors (electrical resistivity) that allows electrical conductors to be further classified as conductors (good conductors) and as dissipative (poor conductors), but both are differentiated from insulators by their ability to sustain an electrical current. The parameter of electrical resistivity is a measure of the voltage that must be applied to a sample of the material to stand a unit current through a cubic sample of the material 1 m on a side. Surface resistivity is a similar parameter corresponding to currents passing along the surface of a material rather than through a bulk sample. It is defined in terms of the voltage required to pass a unit current through a square sample of the material. Resistivity is measured in units of Ω·m and the surface resistivity (relating to currents passing along the surface of a sample) is measured in Ω/sq as shown in Fig. 41.8.

41.4.1 Conductive Materials

Conductive materials are defined[11] as those having a surface resistivity less than 1 × 105 Ω/sq or a volume resistivity less than 1×10^4 Ω·cm. With a low electrical resistance, electrons flow easily across the surface

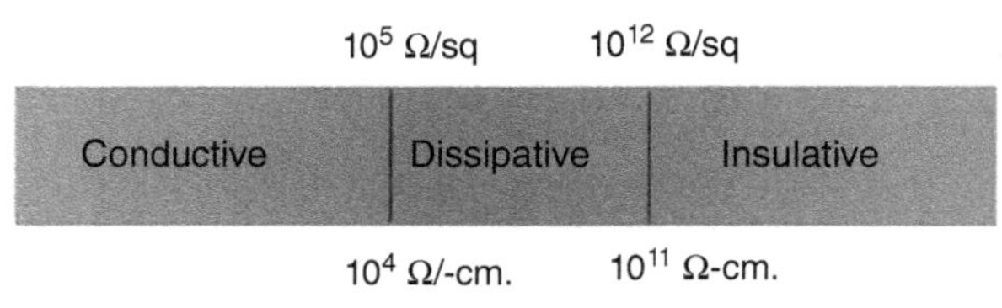

Figure 41.8 Surface resistivity.

or through the bulk of these materials. Charges travel to ground or to another conductive object that the material contacts. Metals like copper and aluminum have exceptionally low resistivities on the order of 10^{-8} Ω·cm, making them extremely good conductors of electricity.

41.4.2 Dissipative Materials

Dissipative materials have a surface resistivity equal to or greater than 1×10^5 Ω/sq but less than 1×10^{12} Ω/sq, or a volume resistivity equal to or greater than 1×10^4 Ω·cm but less than 1×10^{11} Ω·cm. For these materials, the charges flow to ground more slowly and in a somewhat more controlled manner than with conductive materials. Dissipative materials are used for ESD control, because they discharge so slowly that they often avoid ESD damage to microstructures that are manufactured in high-technology cleanrooms.

41.4.3 Insulative Materials

Insulative materials[12] are defined as those materials with a surface resistivity greater than 1×10^{12} Ω/sq or a volume resistivity greater than 1×10^{11} Ω·cm. It is clear that the ability of an insulator to dissipate surface charge is so poor that it can be ignored. For a typical 30 × 30 cm object sitting approximately 1 cm from a ground plane, the capacitance of the object is approximately 10^{-10} F so that the characteristic time to discharge will be given by the product of its resistance times its capacitance.[12,13] For a conductor, this corresponds to a time of much less than 1 ns. For a fairly good insulator (> 10^{13} Ω/sq), this corresponds to greater than 1000 s. For extremely good insulators like quartz, Teflon, or SiO_2, the material can be assumed to *never* discharge. Thus, in a cleanroom where high levels of charge are developed on such surfaces, they will build up over the period of days, achieving extremely high levels. Grounding one of these extremely good insulators accomplishes nothing, and the charge remains on the surface.

Grounding of conducting or dissipative materials is an important first step in cleanroom static charge control but of no help in dealing with insulators. Details of how to ground the materials, utilization of dissipative materials, selection of insulators that have similar work functions to the materials that they contact (called *tribo matching*), and techniques for dealing with static charge developed on insulators is a discipline called *electrostatic management.*

41.5 CLEANROOM ELECTROSTATIC MANAGEMENT

41.5.1 General Principles

A variety of methods have been developed to deal with static charge. The basic methods are grounding facility components (walls, floors, workbenches, and equipment), appropriate use of conductive and static-dissipative materials, and employing local or room ionization to control static charge on insulators. In addition to these basic methods, education of cleanroom personnel in the practice of static control and auditing for compliance with the static control program are essential for the program's success.

Modern cleanroom environments make extensive use of grounding with conductive and static dissipative materials in an attempt to control electrostatic charge. Grounding prevents the generation of static charge on materials that are connected to ground. If a conductive or static dissipative material does become charged, connecting it to ground will remove this stored charge. To be effective in controlling static charge, conductive and static dissipative materials must be provided with a reliable path for the static charge to flow to ground. The success of any grounding method depends on the integrity of the ground path. In critical applications, ground path monitoring and periodic verification may be important aspects of a static control program.

By creating the ground path, the static charge on objects is conducted to ground, rendering the object neutral and eliminating electric fields that had extended from the charged object. It is particularly important to ensure grounding of those surfaces within 300 mm (12 in) of static sensitive product. Care should also be taken with moving parts of equipment. A ground connection that exists when a robot handler is stationary may be lost when the robot is in motion. Flexible grounding connections are advisable rather than relying on conductive lubricants.

Static dissipative materials are used in the construction of cleanrooms and minienvironments to reduce the accumulation of static charge. The walls and floors of a cleanroom and panels of a minienvironment may be constructed of these materials. Static dissipative materials should be selected over conductive materials in order to slow down the charge removal process and prevent a damaging ESD event. The static-dissipative materials may be used for parts of equipment that come into contact with sensitive products. Dissipative reticle pods and dissipative work surfaces used for reticle inspection stations are two examples of locations, where dissipative materials are the best material choice.

In contrast with metal work surfaces or insulative minienvironment walls, the cost of the dissipative material is higher, but often the payback in terms of decreased ESD damage makes the dissipative materials a good investment. Also, in some cases, the properties of the dissipative material may be preferable. For example, metal reticle pods are impractical because of their weight and it is impossible to make a metal viewing port for a minienvironment.

Static dissipative packaging materials are also used to shield the product from static charge buildup as it is transported between processes in the cleanroom. This is a common practice in the *back end of the fabrication line* (BEOL). The slower discharge associated with dissipative packaging will have a lower peak current than the faster discharges.

Due to the deep submicron feature size of modern electronics, the time required to dissipate heat from the structure is extremely short[9] on the order of nanoseconds. Thus, a discharge from a dissipative object can be orders of magnitude large and not cause damage, whereas the same discharge from a conductor is often lethal to the product.

While it is important to employ static dissipative materials to minimize the peak current in a discharge, the use of dissipative materials does not ensure that the product will not charge. To be effective in any static control application, static dissipative materials require a secure path to ground. It will also be important to verify that the materials are cleanroom compatible. Without a path to ground, dissipative materials, while not an ESD hazard can still become charged and as such represent a contamination hazard. Particularly with construction materials used in cleanrooms, static dissipative properties may be sacrificed to achieve other features such as dimensional stability, low outgassing, or low cost.

Grounding methods and static dissipative materials are also used to ensure that charge does not accumulate or transfer from personnel to sensitive products or equipment. A variety of personnel grounding methods are employed using garments, booties, gloves, static-dissipative flooring, and wrist or heel straps. For successful static charge control, all of these methods must be monitored and verified periodically. It is important to note that dissipative booties do not provide protection unless the floor is dissipative or conductive and provides a path to ground for the person wearing them.

Unfortunately, cleanrooms must use many materials that are insulators, such as Teflon, various plastics, and quartz. Very often the insulating materials are an essential part of the product itself. It is not possible to remove the electrostatic charge on insulators by connecting them to ground. Most insulators are easily charged, retain their charge for long periods of time, and are often close to, or part of, the product. Teflon and quartz wafer carriers, oxide-coated wafers, epoxy *integrated circuit* (IC) packages, and glass hard disk media are examples of insulators in the cleanroom. The requirements of cleanrooms preclude the use of carbon particle or chemical additives to insulating materials to make them static dissipative. Antistatic sprays and solutions can create a contamination problem. Humidity control was proposed in the past as a static control method, but has been shown to be expensive and ineffective. Also, increased humidity plays havoc with photolithography chemistry. Neutralizing static charge on insulators (and isolated conductors) has become a necessity for achieving high product yields in cleanrooms. With few other methods available, some type of air ionization is often used. Using only the highly filtered cleanroom air, ionizers create clouds of both positive and negative air ions to neutralize static charges wherever they exist in the cleanroom environment.

41.5.2 Conductors and Insulators

The first principle of static charge control is to ground conductors whenever possible. A corollary to this rule is that conducting or static-dissipative material should be substituted for insulating material wherever possible. Reticles, for example, are made of quartz, which for optical reasons is required and cannot be replaced with a conductor. As a second example, there are cassettes and chemical containers made of Teflon, which is the material of choice due to its chemical immunity to most of the corrosive chemicals used in semiconductor processing.

In many cases, it is possible to replace conventional insulating plastics with dissipative material. While this plastic is somewhat more expensive, often the *return on investment* (ROI) for the dissipative material can be just a few days to a few weeks. This is particularly the case for dissipative plastic in load ports of process tools. In this case, the issue of contamination control is eliminated by the application of dissipative plastics. The ROI for dissipative plastic reticle boxes due to the prospect of reticle damage is even shorter than for dissipative inspection ports in a minienvironment.

While it is important to employ dissipative plastics when possible, it is also important to ground them. Because triboelectric charging is strictly related to the differences in the energy levels in dissimilar materials, conductors as well as insulators will charge. Without grounding, the dissipative plastics will still hold any static charge that is created in them.

41.5.3 Grounding

It is important to provide a path to ground for every noninsulator in the facility. These grounds come in a variety of formats, several of which will be described here. In each case, the objective is to provide a path to ground for any static charge that is generated in the facility.

PERSONNEL

Personnel are a constant source of static charge. The action of walking is inherently a static charge-generating mechanism. For that reason, it is a good ESD control practice to provide a path to ground for static charge. In manual applications such as reticle inspection and handling of dies, the conventional solution is the use of a wrist strap on the operator. This involves having the operator wear a dissipative strap that is connected by a wire to a ground point in the work area. Another approach is the use of ESD-safe chairs that ensure continuity of the operator to ground. This is done in concert with ESD-certified garments that have some number of dissipative threads sewn into the cloth. The wrist strap technique is not practical in a class 1 fab area as it requires exposed skin at the wrist to make the contact. Also, many operations in a class 1 fab are performed standing, so the use of ESD-safe chairs is also not applicable in class 1 applications. In these cases, the use of foot appliances may be used to ensure electrical contact between the operator and ground. Foot appliances are usually in the form of boots with dissipative soles or in the form of a heel strap that fits inside the operator's shoe and extends beneath the sole.

It is important to note that such appliances are of no use unless the floor is conductive or dissipative. A dissipative floor is typically made of a bulk material that is rolled into place and installed, or of a dissipative coating that is, in effect, painted onto the floor. Part of a good ESD control program involves periodically checking the continuity of the floor to ground.

TRANSPORTING AND STORAGE

Reticles, wafers and packaged parts are often transported through the facility. Packaged parts will fare best if they are carried in dissipative carriers by operators wearing proper footgear and walking on an ESD-safe floor. When carts (sometimes called personnel guided vehicles, or PGVs) are used, they should be electrically continuous to ground, either equipped with a drag chain or with ESD-safe wheels. The latter are made up of dissipative tires and have special bearings with conductive lubricant. Drag chains are simply light weight chains that hang down from a cart to provide electrical contact to the ESD-safe floor. Generally, it is good practice to use two chains to avoid discontinuity as the chain moves along the floor. This technique is highly effective in avoiding both EMI as the cart docks to a tool and damage to product as it is lifted from the cart by a grounded operator.

41.5.4 Air Ionization

When insulators cannot be avoided, the problem of electrostatic charge cannot be solved by simply grounding them. Insulators will not support an electrical current, so they cannot be handled by providing them continuity to ground in this conventional way. Air ionization is the tool for eliminating surface charge on an insulator. The action of air ionization is to make the air slightly conductive and provide a path to ground from all parts of the surface of the insulator. This technique is discussed in detail in the following section.

41.6 AIR IONIZATION FOR STATIC CHARGE CONTROL

Air is a good insulator, so any static charge that develops on a clean surface of a good insulator will remain on that surface indefinitely. Air ionizers are used to decrease the electrical resistivity of the air through the addition of ions that act as charge carriers. Air consists primarily of nitrogen, oxygen, and carbon dioxide, with water vapor and trace gases. Air ions are gas molecules in the air that have either lost or gained an electron. Air ionizers are used to introduce air ionization into the room. Ions naturally occur in the air, especially after a lightning storm, but in a cleanroom the HEPA or *ultralow penetrating air* (ULPA) removes the ions from the air. Because air ions are naturally occurring items, they have no negative effects on the health of operators.

Positive and negative air ions are present in normal outside air, where they are the result of radioactive decay of materials in the soil (e.g., uranium) and gases in the air (e.g., radon). But most air ions are stripped out of cleanroom air by high-efficiency air filtration. This makes the air in the cleanroom very insulative and encourages the generation of static charge. Air ionizers restore and increase the level of air ions in the filtered cleanroom air. When the ionized air comes in contact with a charged insulating surface, the charged surface attracts air ions of the opposite polarity. As a result, the static charge on the insulator is neutralized. Air ions of both polarities are required for neutralization, because both polarities of static charge are created in the cleanroom. This is shown in Fig. 41.9.

There are three methods that are commonly used to make air ions for charge control. These methods are corona, photoelectric, and radioisotope ionizers. While the ions created by each are identical, each method generally has its own application. The tradeoffs involve issues of cleanliness, ion distribution requirements, shielding issues, voltage balance (precision) level, and government regulations. Each type is discussed in the following sections.

41.6.1 Corona Ionization

The most common method used to create air ions in cleanrooms is *corona ionization*. In corona ionization, a very high electric field is created by applying high voltage to one or more sharp emitter points. This technique involves the use of high voltage (~5 to 20 kV). The voltage is applied to a set of sharp points. An intense electric field is established very near (~100 μm.) the points. This field accelerates free electrons to a sufficiently high energy to allow them to ionize molecules with which they collide. When the voltage on the point is positive, positive ions are repelled into the environment, and when the point is negative, negative ions are delivered. Corona ionizers are made with ac and with dc voltages. The basics of corona ionization for positive and negative polarities are illustrated in Fig. 41.10.

The corona process is different for positive and negative corona. As such, the process can only produce balanced (equal positive and negative ion populations) when adjusted either by the operator or by automatic circuitry. Manual adjustment will wander its voltage balance over time in response to barometric pressure, humidity or proximity of

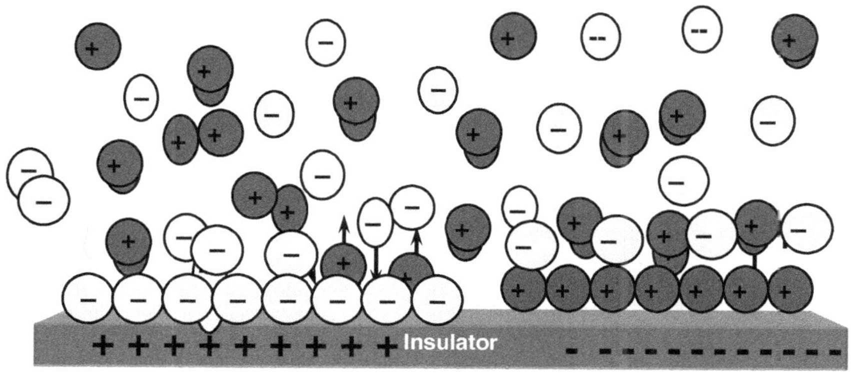

Figure 41.9 Neutralizing surface charges with bipolar air ionization.

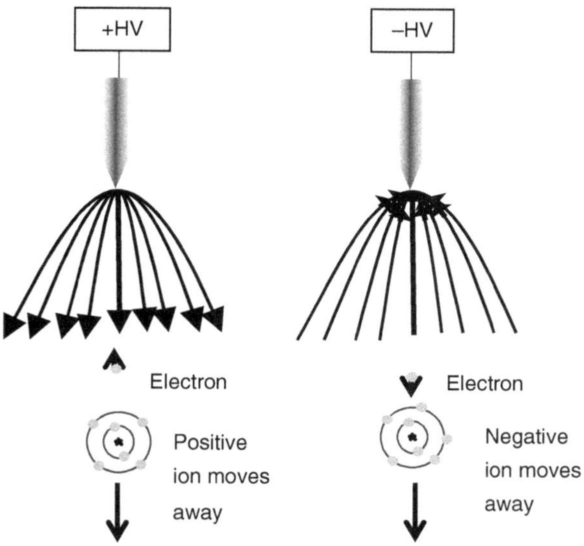

Figure 41.10 The corona ionizer.

objects like robots. The accuracy of most automatic adjustment circuits is usually within ±30 V, which is often good enough of the application. While environmental voltage wander and point wear both effect the balance of the ionizer, it is usually acceptable to readjust (calibrate) the ionizer quarterly.

Corona ionizers can be made using ac or dc voltages. Each has an advantage. Both will be described next.

Note that all corona ionizers interact with solvents in the air and precipitate material out of the air and onto the emitter points. This material must be cleaned off of the points for the ionizer to function. This cleaning may be needed as often as biweekly or as infrequently as semiannually. The wide variation is due to the concentration of airborne solvents in the work area.

One important fact that is critical in certain semiconductor tools is that virtually all corona ionizaters fail in nitrogen tools. Whereas in air, the negative ions are of oxygen and the positive ions are of nitrogen, in pure nitrogen, the action of the negative emitter point is to create free electrons. The very large voltages (and hence electric fields) in concert with the very light weight of the electrons as compared to the atomic weight of the nitrogen ions drives away the electrons and results in huge (kV) voltage balance errors. This becomes particularly important in nitrogen filled furnaces and in low temperature (liquid nitrogen) test handlers. For those applications, corona ionizers should not be used but be replaces with either alpha or photoelectric ionizers. See below.

The AC Ionizer

The ac ionizer uses a step-up transformer to create the high voltage bias (~7 kV rms) that is required to make ions. See Fig. 41.11. The transformer can be easily used to automatically create equal positive and negative voltage swings so that ac ionizers do not require any form of adjustment to work. Because of the simplicity of the ac ionizer, it is the least costly of most ionizer applications. Because the ac type ionizer produces the positive and negative ions in sequence from the same emitter point(s), these ions are separated in time by half of the period of the ac power line (i.e., 1/100 or 1/120 s). This means that the waves of positive and negative ions are rather close to each other, making loss through recombination a large factor. AC ionizers typically use fast airflow velocity to minimize recombination. This is not always desirable in a cleanroom environment. The airflow issue and the increased emitter point current, used to offset the recombination, limits the ac ionizer to ISO class 6 cleanrooms or above.

Experience shows that the actual offset voltage[10] of an ac ionizer as measured with a *charge plate monitor* (CPM) is typically less than 10 V. It has been shown,[11] however, that the CPM is not capable of recording voltage excursions faster than 1 Hz. In contrast, small objects like a single chip are capable of tracking the voltage excursions driven by

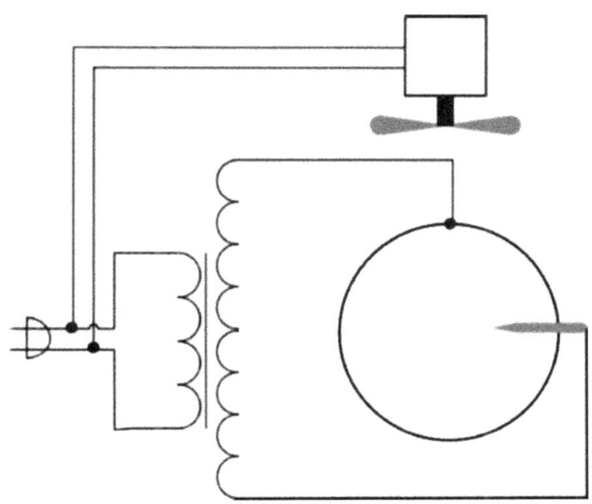

Figure 41.11 AC ionizer.

the time-varying electric fields from an ac ionizer. AC ionizers are not recommended for the handling of diced dies. AC ionizers do find application in many of the BEOL processing steps to protect wafers from ESD damage. This is in part due to their simplicity and in part due to their lower cost as compared with all other types of ionizers.

THE DC IONIZER

Employs a high-voltage power supply to create the necessary voltage. DC ionizers use separate emitter points for the positive and negative dc HV power supplies to create the ions. In order to provide equal numbers of positive and negative ions from separate sources, dc ionizers need some form of control to maintain this balance. Owing to the fact that the positive and negative emitters are well separated from each other, recombination is a lesser effect, and the dc ionizers can sometimes use lower airflow velocity to deliver the ions to the location where the ionization is required. Because of their lower recombination rates as compared with ac ionizers, dc ionizers are used in more contamination-critical applications. Their more efficient delivery allows them to be operated at lower currents and therefore to operate with cooler emitter tips. Thus, any contamination from the tip can be extremely well controlled. Many dc ionizers achieve ISO class 4 operation and are only limited by the turbulence and particulation issues of the fan and its motor.

Blower type ionizers (either ac or dc) with fans cannot be used in the front end of the line due to the prospect of contamination due to air turbulence. In the BEOL, however, fan type ionizers are common.

PULSED DC IONIZERS

The other notable use of dc ionizers in a ISO class 3 cleanroom is as a ceiling emitter. An array of ceiling emitters is mounted to the ceiling of the cleanroom, typically on 4- to 6-ft centers, and the laminar airflow from the HEPAs drives the ionization through the entire room. Normally the ionizers are operated in a mode called pulsed dc. This is very much like the steady-state dc mode described earlier, but the two polarities are turned on one at a time, with on times of 1 to 10 s. Pulsed dc provides even less recombination than steady-state dc. As a tradeoff, objects in the room swing through voltage levels of ±50 to ±200 V, making this method unacceptable for the most sensitive products such as magnetoresistive (MR) heads, but ideal for applications where such sensitivity is not an issue and where large levels of charge must be eliminated globally.

Pulsed dc technology is also available using an alpha particle–based ionizer (see below for details). This provides the benefits of pulsing (longer ion travel) without the wander issue or the cleaning requirements of corona ionizers.

41.6.2 Photoelectric Ionization

The photoelectric ionizer uses extreme UV to soft x-ray photons to create ions. When the energy of a photon exceeds the ionization potential of the medium it traverses, the photon can create an ion by collision with an atom. This is called the photoelectric effect.[12] The most common method of photoelectric ionization involves the use of very soft x-rays. These are x-rays in the range of 5 to 10 keV as contrasted with medical or dental x-rays, which are in the range of 80 to 175 keV. The lower energy makes them suitable for use as an ion source, because the air is rather opaque to photons in this energy range. A soft x-ray source has an approximately 100 percent efficiency for creating an ion within about 1.5 m for each photon it emits. While neither the licensing (in the United States by the FDA) nor the safety requirements are very difficult to manage, photoelectric ionization represents extra steps that are necessary for the use of this technology. This type of ionizer is not common in semiconductor processing due to the high cost of the product. It is, however, a very efficient method for generating and delivering ions in a minienvironment.

41.6.3 Alpha Ionization

The alpha ionizer method involves the use of ionizing radiation to make ions. While several forms of ionizing radiation sources are available, only α (alpha) sources are used for static control. The other forms of radiation source have a much longer range and thus require shielding in impractical amounts. Most commonly, Polonium 210 is employed. It produces α particles with a range of only 4 cm in air or 0.02 mm in aluminum[18] so that virtually none of the α particles travel far enough to represent a health hazard or create a shielding requirement. In fact, these alpha particles are stopped by the epidermis of a person so no alphas reach living tissue from exposure by proximity. The other aspect of ^{210}Po that makes it an ideal choice is that it decays to Pb^{206}, which is a naturally occurring stable isotope. Unlike ^{210}Po, many other sources decay into isotopes that may be active as well. Alpha ionizers can be used in high temperature (~100°C) and in vacuum applications. Alpha ionization produces exactly balanced ionization so it is useful for demanding applications like magneto resistive head manufacturing for the disk drive industry and for CDM class 0 semiconductor testing. Alpha particle ionizers require no cleaning and no balancing but must be replaced annually.

41.7 ELECTROSTATIC MEASUREMENT

41.7.1 Measuring Electric Field

The presence of electrostatic charge is indicated by an electric field. For that reason, identifying the presence of an electric field is an important diagnostic technique in dealing with static charge. An electrostatic field meter is the tool that is used for measuring the electric field at a surface. Presence of a field implies presence of electrostatic charge.

The field meter is calibrated in units of volts/length and typically quoted in kV/in for historical reasons. The field meter measures the field at a distance of 1 in from the object. To operate it accurately, it is necessary to ensure that it has a ground reference. This is normally supplied by a ground wire from the case to a grounded object nearby. Often, the instrument is used by a grounded operator with no gloves on or with dissipative gloves, and the ground can be made simply by contact of the operator's hand.

41.7.2 Measuring Ionizer Performance

Because the majority of ionizers use corona to make ions, a high voltage is required. With high voltage comes electric fields. These fields are responsible for making the ions, but they also drive the voltage on objects around them to a nonzero voltage. While the induced voltage is most often much lower than the voltages caused by static charge, it can be problematic nonetheless. For that reason, one of the parameters that must be measured to benchmark ionizer performance is the balance voltage. Loosely speaking, it is a measure of how accurately the ionizer's positive and negative voltages can be set equal. Balance voltage is measured using a charge plate monitor (CPM).[14] It is also used with pulsed dc ionizers to measure the variation of the balance voltage, referred to as the swing voltage.

The CPM is the instrument that is used to measure the performance of an ionizer. It includes a 6 × 6-sq in (150 × 150-sq mm) plate isolated from ground with a capacitance of 20 pF, as the voltage sensor. The balance voltage is the voltage to which the ionizer drives the plate. In addition, the CPM is used to measure the discharge time that is defined as the time to drive the plate from 1000 to 100 V.

41.7.3 Measuring ESD-Induced EMI

Because of the transient nature of ESD-induced EMI, it is impossible to detect it using a spectrum analyzer. The average energy broadcast by ESD events is much too low when the averaging time is less than a second or slower, as is the case for a spectrum analyzer. What is needed is an instrument that is some type of an event detector or a transient analyzer. The most common method for making this sort of measurement involves a digital sampling oscilloscope (DSO) and a wide bandwidth antenna (commonly a whip antenna on a ground plane with high-quality coax connecting it to the oscilloscope). This type of measurement has been reported elsewhere.[19,20] An advantage of this technique is that it can be used to determine the location of the ESD event and by measuring the pulse shape, it can differentiate ESD events from other sources of noise (e.g., ozone generators and gaseous discharge flash lamps). The signature of ESD- induced EMI is now a wave-shape, which is a sum of multiple, damped, and high-frequency sinusoids. The signal has a fast attack (~1 ns) and a slower decay (~50 ns). The signal is made up of a single very fast impulse and is often followed by several other cycles caused by reflections off various nearby surfaces in the fab. See Figs. 41.12 and 41.13. Note that Fig. 41.13 does not have either the bandwidth or the amplitude characteristic of ESD.

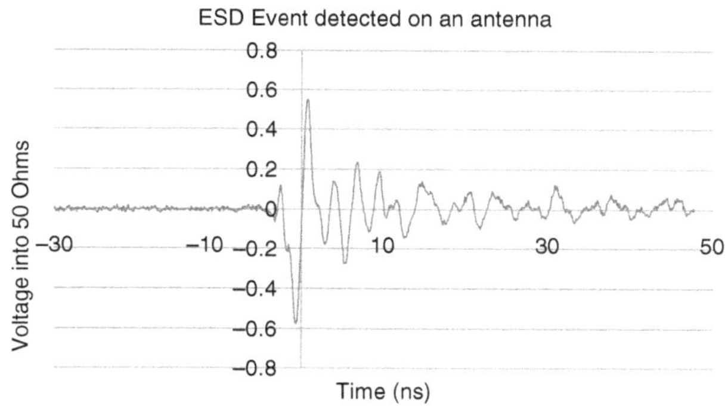

Figure 41.12 An ESD event.

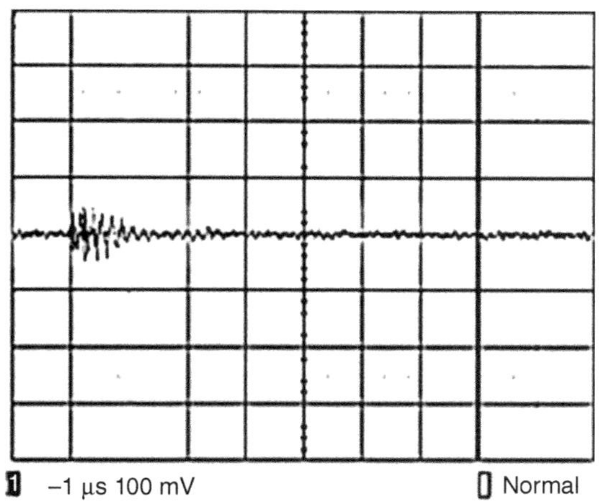

Figure 41.13 Not an ESD event.

Another technique for detecting the presence of EMI from ESD events is through the use of a device called an EMI locator. Such a device is a transient detector with a signal source based upon an integral antenna. These devices are tuned to a bandpass that is appropriate for ESD-induced EMI and are sensitive only to narrow signals (typically with durations <<1 µs). Such instruments are available from several companies.

The advantage of the DSO approach to measurements is that a DSO can be much more selective of pulse shape. In addition, it offers the possibility of determining the location of the event. Moreover, the DSO can be used with a probe to look for signals on conductors. Notably, these sorts of EMI transients can sometimes be found on the neutral line of the power to a process tool. The disadvantage of the DSO approach is that the technique is more costly to implement and much less portable than the ESD detector approach.

Whether an oscilloscope or an event detector is employed, it is important to use the instrument to make a diagnosis. The existence of a signal is useful information to have when a robot exhibits unwanted behavior. It is possible to identify the effect more completely by observing the environment while listening to the detector or watching the oscilloscope triggers. It is often possible to identify that the events are taking place at the same time as some other phenomenon. For example, the triggers might be occurring at the same time as a front opening unified pod (FOUP) passes by a certain location on the overhead track, or the events may coincide with a robot approaching a wafer in a given cassette. This suggests that there is a charged object involved and helps lead one toward a solution to unwanted robot behavior.

41.8 AIR IONIZER APPLICATIONS

41.8.1 Common Applications

Static charge can cause problems anywhere in the cleanroom environment. To control these problems ionizers are used in a variety of configurations. Applications can involve cleanrooms that operate in the ISO class 4 to ISO class 5 range and are typical of the BEOL. In contrast, front-end cleanrooms are most often characterized by high levels of automation and typically require cleanliness levels in the ISO class 3 or higher realms. These two scenarios require very different ionization solutions. Each of these situations is discussed below.

The most common method used in ISO class 3 ballroom-style cleanrooms is room air ionization. Such ionizers are mounted at or near the HEPA-covered ceiling and the laminar airflow is ionized as it passes the ionizer emitter points. The advantage of this type of system is that it brings static charge control to any surface that is exposed to the room airflow. Corona ionizers provide global coverage of the insulators moving through the room. These ionizers are either bars hung from the ceiling or ceiling emitters extending down from the ceiling. The distance that the emitter points stand off the ceiling is generally 30 to 100 cm. The purpose of this distance is to separate the sources of the ions from the grounded elements of the ceiling. A typical room system will provide global coverage with discharge times (90 percent reduction in charge levels) in the range of 20 to 120 s. High ceilings and low airflow velocities represent the upper level and 8-ft ceilings with 0.5 m/s (100 fpm) provide the fastest discharge times.

Many cleanrooms today employ minienvironments to protect the product from contamination. This configuration involves keeping silicon wafers in clean enclosures (standard mechanical interfaces or SMIF for wafers or reticles and FOUPs for 300-mm wafers) as they are transferred from one process tool to another. The individual tools are also fully enclosed to isolate contamination from the room. While this strategy blocks the contamination from contacting the product, it also blocks the ionization from contacting the products. Ions recombine as they pass through a HEPA filter, so room systems do not provide any electrostatic protection of objects within the process tools. To provide protection within the process tools, ion bars are mounted within the mini environments as illustrated in Fig. 41.14. The bar, most typically a pulsed dc corona bar, should be mounted below the HEPA filter within the minienvironment if possible. To minimize offset voltages, the bar should be located at least 30 cm from the product. If the direct HEPA location is not feasible (due to mounting particulars or due to a metal structure in the path of the airflow), a location should be chosen that is not immediately adjacent to a grounded metal structure, if possible. Care should be taken to select a location such that the airflow direction will carry the ions toward the object to be protected, as the ions are not efficiently drawn against the wind electrostatically for more than a few inches. For 300-mm applications, bars are located below the HEPA in the area of the factory interface module, as illustrated in Fig. 41.14. In general, the FOUPs shield the wafers from the laminar airflow, making it difficult to provide ionization. For this reason, and as the 300-mm wafers are larger than their 200-mm ancestors, two bars are used for 300-mm applications, and one bar is used for 200-mm applications.

The question often comes up as to whether room ionization is necessary in a minienvironment cleanroom. In fact, from a contamination standpoint, most of the electrostatically driven sedimentation occurs in the minienvironments and in the carriers. The electric fields from objects outside the carriers penetrate them and add to the electric fields inside. As the carrier moves through the room, or as charged objects

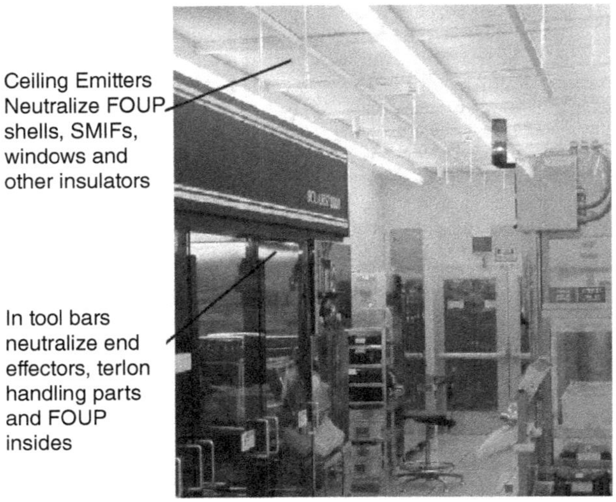

Figure 41.14 Ionizer applications in the ISO class 3 cleanroom.

in the room move past the carrier, the fields cause contamination within the carrier to mobilize, representing a secondary contamination issue. Also, charged objects within the room, but external to the minienvironments, can drive a discharge from one object to another. While this will not affect the contamination control of the product directly, it does generate EMI in the room and can easily cause irrational performance by the robotics. For these two reasons, many operations choose to install ionization both within the minienvironments and in the cleanroom itself.

41.8.2 Severe Discharge Applications

Severe discharge requirements in tools are the ideal applications for the photoelectric ionizer, and in particular in environments where it is possible to mount an ionizing bar at a distance of 30 cm or greater from the product, or in applications involving high levels of charge. Practically, this means either greater than 20,000 V or highly capacitive objects. Cleaning processes tend to produce high charge levels. Also, processes that place flat products (such as wafers or plates) on a flat surface and then raise them up a significant distance when the processing is complete are candidates for high charging.

In order to use the photoelectric ionizer, it is necessary to have an enclosure that serves as a complete shield against the photons. A thin shield is adequate (~2.5 mm of plastic or 750 μm of aluminum), but it is necessary to ensure that there are no hairline cracks in the structure such as doorways without an overlapping seal, as these can represent a way for the photons to escape. In addition to the shielding requirement, a physical knife switch is required that shuts down the ionizer when the door is open. If these requirements are followed, it is straightforward to get an FDA x-ray cabinet license for this type of ionizer. While this small amount of paperwork is not required for a corona ionizer, there are severe applications where this type of ionizer is the only solution that will provide protection for the process.

The other ionizer type that may be used in tight confines, or where static charge levels can be quite high, is the alpha ionizer. The alpha ionizer involves no high voltage and so has a balance voltage of zero. It can be used with the most static-sensitive products, such as MR heads. The alpha ionizer requires airflow to deliver the ionization to the item to be protected, so it needs either a blower (discussed below) or it needs to be placed in a laminar flow field. It can, however, be operated very close to the product with no significant effects. Since the range of Po^{210} alphas in air is 3.8 cm, placing the ionizer within a distance of 2 in of the source is perfectly acceptable. Again, at a short distance, the alpha ionizer is capable of dealing with extremely high charge levels. It requires minimal licensing, but, in practice, it does cause concern to manufacturing workers who do not understand it and require training by the ESD team on site.

41.9 CONCLUSIONS

Reduction of product defects to the levels acceptable in world-class cleanroom manufacturing requires continual attention to static charge control. Uncontrolled, static charge is responsible for increasing the number of particle-related defects, for damaging products and equipment with ESD, and for interfering with factory automation. As factory profitability increasingly depends on both high yields and high product throughput, solving static-related problems in cleanrooms has become a necessity.

Static charge control starts with the grounding of personnel, equipment, and any materials that come close to static-sensitive products. Insulating materials should be removed whenever possible and replaced with static dissipative equivalents. When insulators must be used to satisfy other process requirements, ionizers are used to neutralize static charges on the insulators. It should be remembered that insulators are very often a part of the product itself and cannot be eliminated.

Air ionization is one of the few methods available for controlling static charge in high-quality cleanroom environments. In some cases, it is the only method that can be used. Ionizers can reduce the number of contamination and ESD-related defects occurring in both minienvironments and in production equipment.

41.10 REFERENCES

1. S. Trigwell, C. Yurteri, and M. K. Mazumder, "Unipolar Tribocharging of Powder: Effects of Surface Contamination on the Work Function," *Proceedings of the Electrostatic Society of America Annual Meeting*, East Lansing, MI, Laplacian Press, San Jose, CA, pp. 27–35, Jun. 27–30, 2001.
2. L. Levit and W. Guan, "Measuring Tribocharging Efficiency in Varying Atmospheric Humidity and Nitrogen," *Proceeding of the Electrostatic Society of America Annual Meeting*, East Lansing, MI, Laplacian Press, San Jose, CA, pp. 43–50, Jun. 2001.
3. R. P. Donavan, *Particle Control for Semiconductor Manufacturing*, Marcel Dekker, New York, 1990.
4. J. J. Wu, R. J. Miller, D. W. Cooper, et al., "Deposition of Submicron Aerosol Particles during Integrated Circuit Manufacturing: Experiments," *Proceedings of the Ninth International Symposium on Contamination Control*, Los Angeles, ICCCS, pp. 27–32, 1988.
5. L. B. Levit, T. M. Hanley, and F. Curran, "In 300 mm Contamination Control, Watch out for Electrostatic Attraction," *Solid State Technology*, 43 (6): 209–214, Jun. 2000.
6. A. Steinman, "Static-Charge—The Invisible Contaminant," *Cleanroom Management Forum, Microcontamination*, 9 (1992): 46–512, Oct. 1992.
7. J. Montoya, L. Levit, and A. Englisch, "A Study of the Mechanisms for ESD Damage to Reticles," *IEEE Transactions on Electronics Packaging Manufacturing*, 24 (2): 78–85, Apr. 2001.
8. Gavin C. Rider, "Electrostatic Risk to Reticles in the Nanolithography Era," *Journal of Micro/Nanolithography, MEMS, and MOEMS*, 15 (2): 023501, 2016. doi:10.1117/jmmm.15.2023501
9. Semiconductor Equipment and Material International, SEMI, E78-0998, *Electrostatic Compatibility—Guide to Assess and Control Electrostatic Discharge (ESD) and Electrostatic Attraction (ESA) for Equipment*, SEMI, San Jose, CA, 1998.
10. L. B. Levit and W. Guan, "Measurement of the Magnitude of Triboelectrification in the Environment of the 157-nm Stepper," *Proceedings of the 21st Annual BACUS Symposium on Photomask Technology*, SPIE Publishing Services, pp. 307–312, 2001.
11. ESD Association, ESD ADV1.0-2003, *Glossary of Terms*. ESD Association, Rome, New York, Jun. 1, 1998.

12. D. Halliday, R. Resnick, and J. Walker, *Fundamentals of Physics Extended*, 6th ed., Wiley, New York, 2000.

13. Electronic Industries Alliance, EIA-541, *Packaging of Electronic Products for Shipment*, EIA, Arlington, VA.

14. L. B. Levit and A. Walash, "Measurement of the Effects of Ionizer Imbalance and Proximity to Ground in MR Head Handling," *Proceedings of EOS/ESD Symposium*, Vol. EOS-20, pp. 375–382, 1998.

15. Carl E. Newberg, "Analysis of the Electrical Field Effects of ac and dc Ionization Systems for MR Head Manufacturing," *Proceedings of EOS/ESD Symposium*, Vol. EOS-21, pp. 319–328, 1999.

16. Annalen der Physik, "On a Heuristic Point of View Concerning the Production and Transformation of Light," (öber einen die Erzeugung und Verwandlung des Lichtes Betreffenden Heuristischen Gesichtspunkt), Mar. 1905.

17. P. Trower, "High Energy Particle Data," *UCRL-2426*, Vol. II, p. 42 (1966 revision), 1966.

18. ESD Association, ANSI-EOS/ESD-S3.1-1991, *For the Protection of Electrostatic Discharge Susceptible Items-Ionization*, ESD Association, Rome, New York, 1994.

19. J. Bernier, G. Croft, and R. Lowther, "ESD Sources Pinpointed by Analysis of Radio Wave Emissions," *Proceedings of the 19th Annual EOS/ESD Society*, Santa Clara, CA, Paper 2A.2, Sep. 1997.

20. L. B. Levit, L. G. Henry, J. A. Montoya, F. A. Marcelli, and R. P. Lucero, "Investigating FOUPs as a Source of ESD-Induced Electromagnetic Interference," *Micro Magazine*, Apr. 2002.

Chapter 42
Vacuum Systems

BY
MICHAEL R. CZERNIAK *Edwards, Clevedon, United Kingdom*

42.1 INTRODUCTION

Vacuum pumps and point-of-use (PoU) gas abatement systems can be thought of as life-support systems for wafer processing tools, providing the reduced pressures required to satisfy the needs of the process recipe and subsequently treating the reagent and by-product gases, respectively. This chapter considers the challenges that pumps and abatement systems have to meet and the technological approaches that have been developed to satisfy these requirements, with commentary on their strengths and weaknesses, citing representative examples. The latest trends in combining vacuum pumps and PoU abatement systems into a single, fully integrated system are also covered.

42.2 VACUUM PUMPS

42.2.1 Pump Basics

Vacuum pumps are fundamental to most semiconductor wafer processing technologies (CVD, etching, etc.) by reducing the process chamber pressure to the correct level for the chemistry to proceed, typically in the range of below one to hundreds of mbar (1 millibar = 100 pascals ≈ 9.97×10^{-4} atmospheres). Reduced pressure is required to support the plasmas used in many etch and CVD recipes and helps improve uniformity over the entire wafer surface by increasing the gas velocity over the surface.

There has been a general trend of decreasing pressures and increasing flows required as wafer diameters have increased, with the result that pump rotational speeds have correspondingly increased[1] as well as eliminating the use of oil in the working volume of the pump mechanism as this would be a potential source of wafer contamination.

42.2.2 Pumping Speed and Boosters

The rate at which a vacuum pump can throughput gas varies according to the pressure of the inlet and is referred to its "speed" and this variation characteristic as its "speed curve." The pump outlet ("exhaust") is typically around atmospheric pressure for most primary pumps (i.e., a pump which can pump-down directly from atmospheric pressure; other pumps are referred to as secondary pumps, which require being "backed" by a primary pump; turbo pumps being a good example), and the low-pressure performance of a primary pump can be enhanced by adding a "booster pump" upstream of it. A booster pump is typically of the Roots type, which has pumping speed at low pressure and acts in much the same way as a "turbocharger" boosts the performance of an automobile engine. As a rule of thumb, a booster typically has approximately 10 times the peak pumping speed of the primary pump; it is used in combination with and reduces the lowest attainable ("ultimate") pressure by at least an order of magnitude.

42.2.3 Matching Pumps to Applications

The choice of the correct pump type and size is driven by the specific needs of the application. In the case of load locks and transfer chambers, these are usually considered as "clean" duty, that is, process gases are not usually present, and the challenge is often related to speed of pump-down from atmosphere and high pumping speeds at high pressure are often required, especially in flat panel displays (FPDs) due to the large chamber volumes dictated by the large glass sizes.[2] In practice, this generally requires a low volume ratio (i.e., of inlet to outlet displacement) and large capacity exhaust stages, which can result in high power consumption at ultimate vacuum. This may be offset to some extent by a pump with a high volume ratio and blow-off valves (i.e., bypassing the exhaust stage at high pressures). Etch and CVD processes, in contrast, involve significant flows of process gases at low pressure and often also contain many process by-products, so mechanical and chemical robustness of the pumping system is important. As these do not normally require high pumping speeds at high pressure, a high volume ratio can be utilized to reduce the size of the exhaust stages and therefore the power and running costs too. The low pressures associated with etch chemistries, typically around 70 microbar, are required to impart directionality to the etch species and require the additional use of turbo pumps. In any application, the magnitude of the required gas flow, usually expressed in standard liters per minute (slm) determines the necessary pumping speed of the pumping system.

In the case of powders and condensable by-products arising from the process chemistry (silicon dioxide, "silica," and ammonium chloride being notable examples), nitrogen purging of pumps can help dilute process gases, keep powders moving and minimize blockages. Care is needed not to overwhelm the pump's capacity (or that of any abatement devices downstream) or to cool condensable gases too much, and external electrical ("trace") heating of the exhaust pipework can minimize the risk of condensation. Many pumps, which are typically water cooled, can have their running temperature adjusted and optimized so as to minimize the risk of condensation. Connecting pipework is typically made from electro-polished seamless stainless steel tube, with a minimum number of O-ring joints (typically Viton, but also Kalrez or other fluorastomers where additional chemical resistance is required), use greater than or equal to 100 mm diameter for forelines into the pump, and greater than or equal to 40 mm for exhaust pipework.

The major pumping mechanisms used in semiconductor manufacture are now considered, including their suitability for different applications.

42.2.4 Turbomolecular Pumps

A turbomolecular pump (TMP) comprises two sets of angled blades, one fixed to the stationary body of the pump, the other attached to a high-speed rotating shaft (typically 20,000 to 90,000 rpm), so that momentum is transferred to gas molecules in a direction that guides them toward the exhaust, which is backed by a primary pump. Generally speaking, the greater the diameter, number of blades, and speed of shaft rotation, the higher the peak pumping speed. Most designs can be mounted in a variety of orientations.

Specific challenges posed by the semiconductor industry are as follows:

- Gas throughput is one of the most critical parameters in semiconductor production, especially on FPD and solar cell applications and when 450 mm wafers are introduced; the turbo pumps used are classed as "compound pumps" meaning that the pumps have a second molecular drag stage designed to maximize the throughput. If a Holweck stage is used, typically the longer the gas path in the Holweck the better the throughput of the pump.
- New processes with more corrosive gases. This has resulted in the development of magnetic "maglev" rotor bearings to replace traditional mechanical bearings, including electro-magnetic bearings which can autocorrect high-speed wobble and the use of special corrosion-resistant materials and coatings.
- Condensable gases, often as by-products of etch process chemistries, can form solids at the exhaust end of the pump; this can be minimized by electrically heating this part of the mechanism, thus keeping the process gas in a gas phase rather than condensing to a solid inside the turbo pump.
- The low molecular mass of light gases like hydrogen and helium mean that less momentum if transferred during collisions with the blades and they are more difficult to pump than heavier gases, often requiring a fully bladed version of maglev TMP and the use of larger-capacity backing pumps.

42.2.5 Cryo Pumps

This mechanism captures gas molecules by condensing them on a cryogenically cooled surface and therefore has a finite capacity of molecules that it can retain. It has the advantage of a very high effective pumping speed and therefore finds application in many large-volume process tools where achieving low vacuum quickly is important, such as ion implanters and coaters. A cryo pump typically comprises of a cryohead, which does the actual pumping, and a compressor that drives the refrigerant. The pumping speed depends of the gas species in question, and in particular it's boiling and freezing temperatures, and the temperature of the cryohead, which requires periodic regeneration by isolating it from the process chamber and allowing it to warm up to ambient temperatures, during which there is a rush of evaporating gas that can pose a challenge for downstream equipment like gas abatement systems if not carefully controlled. In practice, many applications that were traditionally cryo-pumped have switched to TMPs now that larger models are available, but remain the principal pumping mechanism for molecular beam epitaxy (MBE) via the cryoshield, as used for III-V R&D.

42.2.6 Dry Pumps

The first dry pumps were introduced during the 1980s[1] to overcome the issues posed by semiconductor processes for oil-lubricated rotary wet pumps, namely contamination of the oil by process by-products, which resulted in short pump life and high maintenance costs, and to minimize the risk of back-migration of oil mist back into the process chambers. Oil or grease is used to lubricate the rotating shafts but is located externally to the swept volume. This style of pump is suited to the semiconductor, FPD, solar, microelectromechanical system (MEMS), and LED industries. Gas sealing between the rotors and stators is achieved by maintaining very tight mechanical clearances. If particularly corrosive gases are present in the input gas, the use of surface coatings and/or special alloys may be required. Variable speed inverter drives allow different speeds of rotation to be used, optimizing the pump performance to the real-time application requirement and the opportunity to reduce power at times when the process tool is not processing wafers ("idle/sleep" modes) and even pressure control without the need for butterfly valves (which can become contaminated by process by-product). Rotational speeds are typically less than or equal to 6000 rpm, and nitrogen purging of the mechanism and/or exhaust helps the passage of process gases through the mechanism (especially light gases like hydrogen) and, by diluting them, through subsequent exhaust pipework. There are three primary mechanisms used, which are discussed below.

Roots Mechanism

This mechanism involves counterrotating shafts with multilobed rotors that mesh as they pass in the center of the swept volume in each stage of the pump, with the inlet stage (low vacuum) having the greatest length (and swept volume); this decreases in successive stages as the gas is compressed as it passes from stage to stage. The Roots mechanism is well suited to pumping large volumes of gas at low pressure and is therefore widely used for booster pumps and as the inlet stage of other primary pump mechanisms. This mechanism is characterized by generally low power consumption. Because the gas passes through each stage in the same direction (e.g., top to bottom), it needs to be directed to the opposite end of the next stage through passageways called interstages.

Roots-Claw Mechanism

In this type of pump, the Roots mechanism described above is combined with compressive claw stages to take full advantage of the benefits of each. A claw stage achieves compression during the rotation of the claws, not just between stages, by using the claws to act as rotating valves which pass over inlet and exhaust ports during the compression cycle, much like the valves in a combustion engine, allowing compression ratios in the order of 10:1. This is useful as the compression can be used to heat the gas, which can keep condensable gases in the volatile state; alternatively the heat can be removed by a water-cooling circuit if heat-sensitive gases are present. This valving also allows nitrogen dilution gas to be introduced between stages without back streaming and stalling the pump, permitting the concentration of volatile and corrosive gases to be controlled to reduce condensation and surface reactions. Roots rotors with a multiplicity of lobes can "chop" pulses of exhaust gas when used in the final pumping stage, thereby reducing exhaust pulsations without the need for a silencer as these tend to act as unintentional powder "traps." Because gas is transferred between opposing faces of each pumping stage (rather than top to bottom), there is no need for interstage passageways and their consequential risks of powder and condensation build-up. Figure 42.1 illustrates a typical Roots-claw mechanism.

Screw Mechanism

This is, in effect, a modern take on the ancient "archimedean screw" principle, but using two screws (one clockwise, the other anticlockwise) to achieve a continuous trapped volume that is either compressed at the exhaust end of a fixed-pitch screw, or progressively along the length of a variable-pitch screw design. A common feature to all screw mechanisms is generally good powder handling; this should be no surprise because archimedean screws are used in grain elevators for instance. In the case of the simplest fixed pitch screw they are generally characterized by a fairly flat speed curve and can be prone to stalling if the exhaust pressure rises due to blockages. Because all the compression happens at the final turn of the screw (i.e., the exhaust end), most of the mechanism remains relatively cold and is at risk from condensation of volatile materials, while the exhaust end of the mechanism may need significant cooling. A more complex variable-pitch screw is significantly more challenging to manufacture, but benefits in that the compression is spread out along

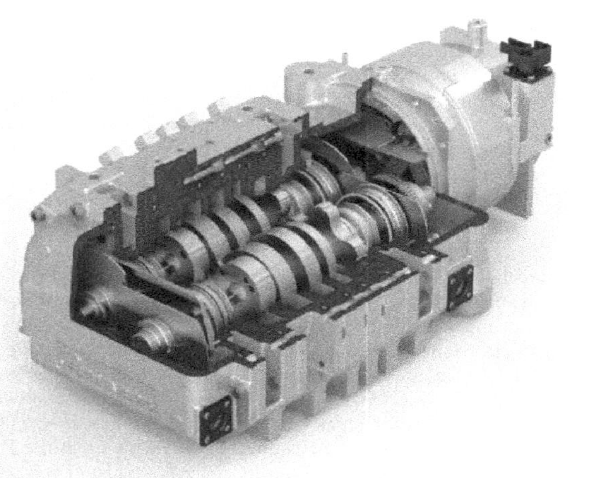

Figure 42.1 Roots-claw dry pump. (*Image Courtesy of Edwards.*)

the full length of the screw, producing a more even temperature distribution and reducing the risk of the condensation of solids. A further refinement can be to taper the diameter of the screw down toward the exhaust stage to give additional compression. An example of a variable-pitch screw is illustrated in Fig. 42.2.

42.3 POINT-OF-USE ABATEMENT

42.3.1 Abatement Basics

In the early days of the semiconductor industry, there was an adage that "the solution to pollution is dilution," that is, you could simply dilute any exhaust gas in a high air flow and the problem would "go away." Fortunately, this has long been recognized not to be true, and that such an approach simply results in dilute pollutants in the atmosphere. Indeed, many of today's precursor gases are so toxic or reactive that this approach would be dangerous in the extreme, and some greenhouse gases (GHGs) like CF_4 have a 100-year global warming potential (GWP) 7390 times that of CO_2, and an atmospheric half-life of 50,000 years.[3] CF_4 is an example of a perfluorinated compound (PFC). It has therefore long been the practice of the industry (including photovoltaic (PV), FPD, MEMS, and LED) to treat process exhaust gases before they are discharged into the atmosphere.

There are three generic classes of gas abatement (sometimes referred to as "scrubbers"), namely "central," "point of use" (PoU), and an intermediate "area abatement" class. A central system, usually in the form of a large wet scrubber (see Sec. 42.3.3) was one of the earliest approaches, and will still be found in most fabs as a safety backup, but tends to be "jack of all trades and master of none," that is, it is not optimized to any particular chemistry. Indeed, major drawbacks are the risk of mixing different, possibly incompatible chemistries, which could have unwanted cross-reactions, the considerable amounts of gas in the connecting pipes (unless PoU is also used), and the risk of shutting the whole fab down if there are problems with the central unit. These considerations flip in the case of PoU, where the unit can be optimized for the specific chemistries used on the tool with which it is associated (usually on a 1:1 basis), and these gases are dealt with locally, minimizing pipework concerns; this is particularly relevant with flammable or toxic gases. An additional benefit is the opportunity to implement utility savings whenever the associated process tool is not processing wafers ("idle/sleep" modes). Area abatement is an intermediate between the above scenarios (central and PoU) whereby compatible gases (i.e., which do not undergo adverse reactions) from a number of process tools are combined into a single PoU abatement system; silicon oxide etch particularly lends itself to this approach.

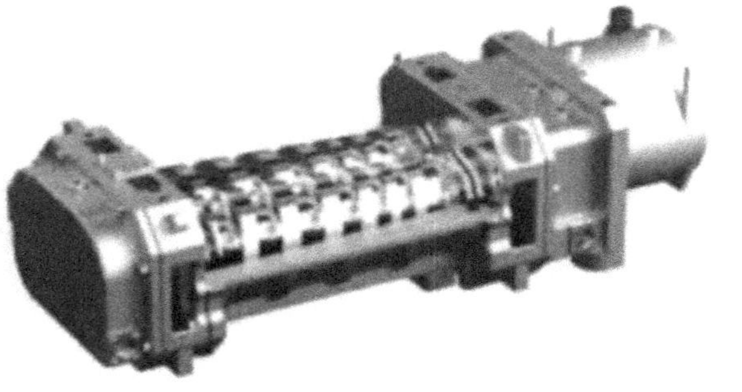

Figure 42.2 Variable-pitch screw dry pump. (*Image courtesy of Edwards.*)

42.3.2 Matching Abatement Technologies to Applications

There is no one-size-fits-all solution to exhaust gas abatement because the chemistries and size of gas flows vary so considerably and these are the principal axes for making a decision, along with the efficiency of gas removal required, usually referred to as destruction reaction efficiency (DRE) and expressed as a percentage. Other considerations might be meeting local emissions regulations, and the availability of utilities locally, for example, natural gas (methane, CH_4) is not universally available around the world and electricity supplies can be restricted in some territories. The principal technologies for PoU abatement are now considered, including how they measure up against these criteria and their suitability for specific process exhaust applications; simple dilution devices ("burn tubes," etc.) are not considered here as they generally do not react with process gases that are at concentrations below their flammable limits, which is frequently the case (most process exhausts are diluted with nitrogen gas from the vacuum pumps).

42.3.3 Wet Scrubbers

Wet scrubbers are classified as gas abatement devices, which contact the effluent gas stream with water to remove particles and/or soluble vapors and are available in a variety of sizes and configurations. This is usually achieved using a counterflow configuration with water flowing downward and gas upward. They are generally simple in design and were first deployed during the late Victorian era for removing ammonia from town gas produced from coal. Their disadvantages are that they only react with a limited range of gases (for instance, there is no reaction with PFC gases or hydrogen), and their reaction with certain gases can be explosive (e.g., trimethyl aluminium). Other vapors, such as ammonium chloride, tend to form sticky pastes when they come in contact with water which can lead to blockages and high maintenance, especially at the dry-wet interface at their inlets, and there is always the challenge and cost of safely disposing of the resulting contaminated water. Last, but certainly not least, water is increasingly becoming an expensive and scarce resource, so optimization of water usage is important. Nevertheless, a final water-washing stage is favored in combination with many other abatement technologies as a final "polishing" step, often to remove acidic gases or fine particles. Particle removal efficiencies can be improved by adding atomized water sprays or wet electrostatic precipitators. Adding chemical dosing can improve their efficiency, usually by making the water more acidic or caustic depending on the chemistry of the gas being treated, and a packed bed of balls or other packing media maximizes the contact surface area between gas and water and hence the scrubbing efficiency.

Wet scrubbers are widely used as central "house scrubbers" in semiconductor fabs and only use water, power, and nitrogen to operate. An example of a typical countercurrent flow (i.e., water flows down, gas flows up) PoU design is shown in Fig. 42.3.

There is a specific issue due to using plastic as a material of construction (often chosen due to its acid resistance) when used with an exhaust flow containing hydrogen. Plastics are usually electrically insulating, and the

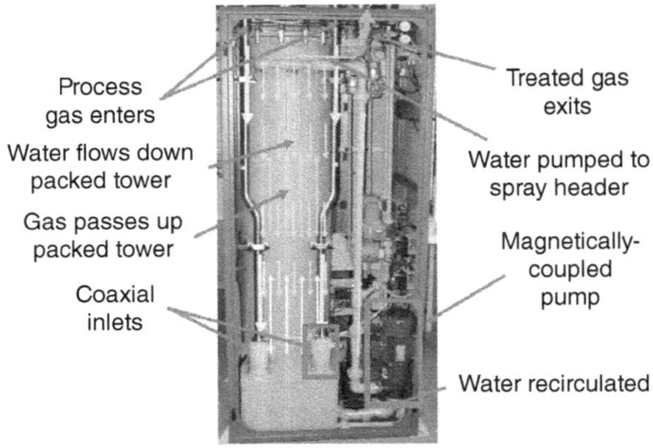

Figure 42.3 PoU wet scrubber. (*Image courtesy of Edwards.*)

passage of moving water over their surfaces can generate static electricity, that is, sparks. If there is also an oxidant present, perhaps from an air leak or from a chamber being roughed down from atmosphere, the conditions can arise where the flammability triangle is completed[4] (ignition source, fuel, and oxidant) and there is the risk of sudden unintentional reaction.

42.3.4 Cartridge Technologies

The basis of this type of abatement system is that a packed bed of solid reagent (i.e., with lots of interstitial spaces for gas to pass through and with high surface area) undergoes a surface reaction with a passing stream of exhaust gas that is to be treated, chemically "locking" the gases of interest in place in nonvolatile form for subsequent disposal when the cartridge is removed, making this technology very simple and rugged, and was one of the first improvements over wet scrubbers.

The simplest form simply involves an absorbent like charcoal, which contains many micropores that simply trap gas molecules, but they can often easily escape, so adding chemicals to the surface can induce reactions that chemically trap the gas of interest and enhance the efficiency (e.g., activated charcoal used in gas masks during the world wars). A single or multiple reagents can be used in the cartridge, depending on the gas challenge, and cartridges may be used at ambient temperature or electrically heated, depending on the chemistry. In general, the useful life of the cartridge will be determined by a number of factors, namely, exhaust gas flow (both in total, limited by the conductance of the cartridge itself, and the flow of target gas, which will consume the finite quantity of solid reagent and may also generate heat of the reaction is exothermic), whether the cartridge becomes contaminated by by-products, reducing the efficiency of gas reaction or, in extreme cases, blocking the passageways through the packed bed. Because of this, cartridges are generally most suited for treating process feed gases and etch process exhausts. Exceeding the stated maximum flows usually results in decreased efficiency due to reducing the residence time for reactions to occur. A special type of heated catalytic cartridge is capable of treating ammonia by reversing the Haber process and converting it to hydrogen, and nitrogen, for example, on GaN MOCVD (used in LED manufacture) exhausts, but this technology is challenging for PFC gases due to their chemical stability.

In general, as wafer diameters have increased over the years, process gas flows have also increased and cartridge technology has become less attractive, but they find particular application for emergency gas release containment, and their simplicity makes them well suited to R&D scenarios.[5]

42.3.5 Hot-Wet Abatement

These systems start to address some of the issues of cartridges in that these can readily be scaled to suit even large capacity requirements and are reasonably tolerant to high powder loading, making them generally suitable for CVD applications, but their poor DRE performance on PFC gases generally make them unsuitable for etch applications. They work by electrical heaters elevating the temperature of incoming gases above their flammable temperature in air, so that the process gas ignites in the presence of air injected into the system, usually forming solid powders as a by-product. A good example is the CVD precursor silane which, at elevated temperatures,[6] autoignites in air to form silicon dioxide powder ("silica") according to the reaction:

$$SiH_4 + O_2 \text{ (from air)} \rightarrow SiO_2\downarrow + 2H_2O$$

This technology is particularly suitable in areas of the world where fuel gas (methane, natural gas) or propane are not readily available, but as gas is a primary fuel, it is typically less expensive per kWh (kilowatt hour) than electricity. The hot reaction zone is typically followed by a wet scrubbing stage as described earlier. A second wet scrubber stage may be added at the inlet to the system to reduce the quantity of process by-product powder and/or act as a prehumidifier. If NF_3 is used as the fluorine source for cleaning the chamber, the gas arriving at the PoU abatement will be predominantly fluorine and SiF_4; the latter will react with the water to form fluorine and silica. The residual humidified fluorine then may react in the hot zone as follows:

$$2F_2 + 2H_2O \text{ (from moisture)} \rightarrow 4HF + O_2$$

Injected hydrogen or ammonia may be used as a substitute for moisture. The presence of corrosive hydrofluoric acid may require the use of special chemically resistant alloys in the heated section. Some designs add a PFC catalyst column after the final wet scrubber, but these are at risk from residual contamination chemically "poisoning" them. An example is shown in Fig. 42.4.

42.3.6 Plasma Abatement

Plasma abatement is a further refinement beyond hot-wet oxidizers by using an electrically generated plasma to generate much higher

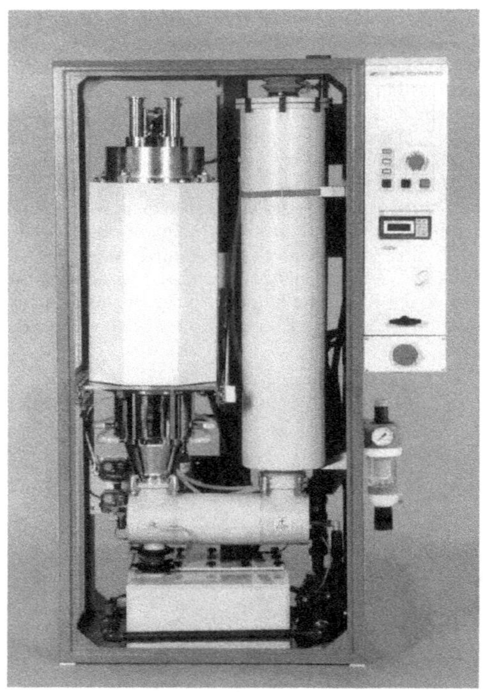

Figure 42.4 Hot-wet oxidizer system.

gas temperatures to induce even more chemistry, indeed sufficient to decompose chemically stable PFC gases like CF_4. A potential drawback is the risk of forming significant quantities of NOx; mono-nitrogen oxides formed by reactions between nitrogen from the process, added air and/or the abatement device itself, with oxygen from the process, or water vapor or air added in the abatement device. There are two categories of plasma devices depending on where they are physically located, namely, foreline devices located between the process tool and pumps, and exhaust abatement devices located in the more conventional post-pump exhaust pipe. These are so significantly different that they are discussed separately, with the various plasma techniques available considered for each scenario, which correspond to subatmospheric and atmospheric operation, respectively.

Foreline (Subatmospheric) Plasmas

There are three inherent advantages of foreline plasmas, namely, (1) that the process gas is at its most concentrated form as it has not yet been diluted by the nitrogen purge gas from the process pump, (2) it is easier to strike and maintain plasmas at reduced pressures, and (3) because they are typically mounted above the process vacuum pump, they can occupy zero additional floor-space (an advantage if retrofitting into a legacy facility). The net result is generally to reduce the average running power for a given flow of process exhaust gas. However, this is usually at a price; to provide chemical reagents for the decomposed gas species to react with (and prevent them simply re-forming the original gases), it is usual to introduce humidity (as discussed in the preceding section). There is a risk that water and/or dissolved by-products can find their way down the remaining foreline and into the vacuum pump, resulting in corrosion, blockages or, in a worst case, hydraulic lock which would typically stall the pump. Furthermore, hazardous air pollutants (HAPs) generated by the reactions, such as HF, still need to be treated prior to discharge to atmosphere, typically requiring the use of additional post-pump abatement, often cartridge or wet scrubber based; this adds extra cost and complexity. Typically one foreline plasma treats the exhaust from a single process chamber only, and are most commonly used on oxide etch applications to reduce PFC emissions. Two technologies dominate this category, namely. RF and microwave, which differ essentially in the way the plasmas are generated.

Exhaust (Atmospheric) Plasmas

This configuration is closer to that of other abatement technologies; being downstream of the process pumps, atmospheric plasmas are able to accept process exhausts from multiple chambers of a cluster process tool and include an integral wet scrubber after the plasma zone for HAPs removal. Air is usually added for combustion of CVD gases, and humidity for fluorine or PFC treatment. The technologies that have been deployed include RF, microwave, and DC plasma, with the latter being the most common. A high-temperature plasma flare is shown in Fig. 42.5.

Figure 42.5 Atmospheric DC plasma torch. *(Image courtesy of Edwards.)*

42.3.7 Combustion Abatement

This is probably the most common form of exhaust gas abatement in the semiconductor industry due to its wide applicability (etch and CVD processes, including PFCs), and the ability to scale the technology to deal with increasing gas flows such as those posed by the FPD industry. The fuel gas used is typically methane, although propane is a common alternative. Multiple inlet systems normal, and various backup configurations are available so that gas treatment (and hence wafer production) can continue while maintenance activities are carried out; this typically involves diverting the gas to be treated to another system. Combustion is usually followed by a HAPs wet scrubber stage, housed in a common extracted cabinet with common control system. There are two principle types of combustion abatement systems, which differ in the degree of separation between fuel gas and process gases to be abated, generically called *open flame* and *inward fired*, respectively.

Open-Flame Combustion Abatement

This is perhaps the simpler technology and involves one or more flames impinging on the process gas as it transits a combustion chamber into which suitable quantities of air are added to support combustion of not only the fuel gas but also the process gas. Maintaining the correct ("stoichiometric") mix of fuel and air is important to optimize fuel use (and cost), minimize NOx and CO formation, and achieve the desired DRE levels; this can be a challenge if gas composition/flows change rapidly. An example of a multi-inlet open-flame combustor is shown in Fig. 42.6.

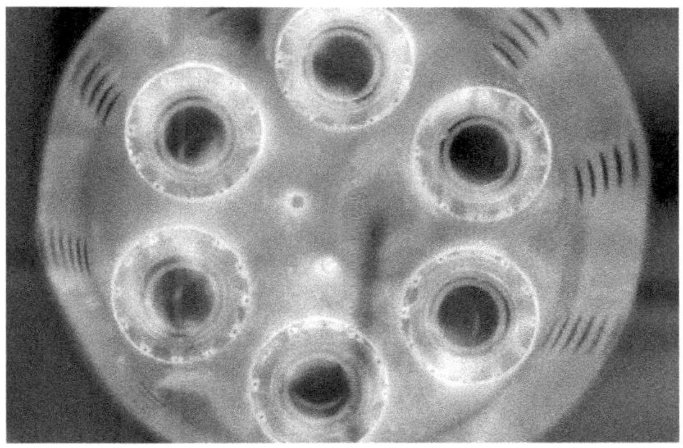

Figure 42.6 Open flame combustor—viewed from below; fuel is added concentrically around each of the six concentric nozzles and air is added through the large peripheral holes. *(Image courtesy of Edwards.)*

Inward-Fired Combustion Abatement

This is a subtly different approach and involves pinning an air-fuel flame to the inner surface of a cylindrical porous ceramic ("foraminous") liner, and passing the process exhaust gas through the isothermal high temperature zone in the middle. This keeps the fuel and process gas spatially separated and makes controlling the amount of residual oxygen in this zone straightforward, as well as NOx and CO levels. There are two additional specific benefits, namely, (1) the inward flow of fuel/air helps minimize the buildup of solid combustion by-products, and (2) the isothermal nature of the combustion zone helps eliminate cold gas paths that might reduce DRE. Figure 42.7 illustrates this technology.

Figure 42.7 Inward-firing combustor; cut-away illustrating fuel-air mixture passing through porous ceramic pad, creating a hot, humid, oxidizing isothermal zone into which process gases are injected from nozzles at the top.

42.3.8 Gas Recycling

Although this is not strictly speaking gas abatement, this technology involves harvesting process gas from the pump exhaust, separating the target gas of interest, purifying it to a particular specification, and then recycling it into the process, either directly (PoU) or indirectly (via a processing plant and supplying in cylinders). This sounds very appealing from an environmental standpoint, and was seriously investigated during the 1990s as regards PFC gases before suitable abatement routes were developed, but in reality there are three major obstacles: (1) obtaining sufficient purity (even if the recycled gas is 99.9999 percent pure, what exactly does the 0.0001 percent contamination consist of?), (2) the energy and effort to recycle the gas may exceed the financial and/or GHG (greenhouse gas) benefits of recycling (the issue that killed PFC recycling in the 1990s), and (3) a more human factor: "there's no way I'm putting gas from the exhaust back into my process tool!" However, there are growing pressures that may result in this being reconsidered, such as the increasing flows of hydrogen in particular being used by some processes and the global scarcity of certain gases (Xe and Ne being two examples).

Recycling usually involves filtration (physical and/or chemical), concentration (usually using cryogenic separation or pressure-swing absorption) and compression.

42.4 CONCLUSION AND FUTURE TRENDS

This chapter has looked at the development and use of vacuum pumps and exhaust gas abatement systems to support processes for manufacturing cutting-edge semiconductor devices, flat panel displays, and LEDs, citing numerous practical examples and applications.

A concept that has been developed and deployed over the last decade or so has been to close-couple vacuum pumps and gas abatement systems in a single, compact, extracted enclosure with common control and utilities systems, usually referred to as an "integrated system." The original motivation for doing so was safety, especially when using highly toxic gases like arsine and phosphine because any potential leaks would be contained, controlled and could be easily detected by monitoring of the extracted cabinet air. The use of integrated systems has expanded significantly since then, capitalizing on the many additional benefits of this approach, namely, (1) reduced number of utility hookups (faster and cheaper installs; less time to first silicon), (2) reduced footprint, particularly by stacking pumps above each other (allowing more equipment in a given fab space), (3) common control system (facilitates energy-saving idle modes), and (4) making connecting pipes as short as possible, reducing the risk of blockages and the power/cost of heating these lines. There are already many hundreds of such systems in use worldwide.

42.5 ACKNOWLEDGMENTS

The author would like to thank several of his colleagues for reviewing and contributing to this chapter: Neil Briault, Dr. Andrew Seeley, and David Turrell.

42.6 REFERENCES

1. S. Ormrod and N. Schofield, "Vacuum Pumps at the Heart of High Tech.," Compute Scotland, 2010, http://www.computescotland.com/vacuum-pumps-at-the-heart-of-high-tech-2808.php (accessed Oct. 15, 2015).

2. "How Do We Do It—Flat Panel Displays," http://www.appliedmaterials.com/company/about/how-we-do-it (accessed Oct. 15, 2015).

3. S. Solomon, D. Qin, M. Manning, Z. Chen, M. Marquis, K. B. Averyt, M. Tignor, and H. L. Miller (eds.), "IPCC, 2007: Climate Change 2007: The Physical Science Basis," Contribution of Working Group I to the Fourth Assessment Report of the Intergovernmental Panel on Climate Change, http://www.ipcc.ch/publications_and_data/publications_ipcc_fourth_assessment_report_wg1_report_the_physical_science_basis.htm (accessed Oct. 15, 2015).

4. Hydrogen Ignition, HySafe, http://www.hysafe.org/download/1042/BRHS_Chap3_hydrogen%20ignition%20version_0_9_0.pdf (accessed 15 October 2015).

5. D. Baker, P. Mawle, and R. Smith, "The Treatment of Organic Chlorides from Plasma Etch Processing," *Solid State Technology*, Mar. 1995, http://business.highbeam.com/412105/article-1G1-16793049/treatment-organic-chlorides-plasma-etch-processing (accessed Oct. 15, 2015).

6. V. Babushok, W. Tsang, D. Burgess, and M. Zachariah, "Numerical Study of Low- and High-Temperature Silane Combustion," 27th Symposium on Combustion, The Combustion Institute 1998, http://fire.nist.gov/bfrlpubs/fire99/PDF/f99076.pdf (accessed Oct. 15, 2015).

42.7 FURTHER READING

Biltroft, P., Chap. 7 in *Semiconductor Manufacturing Handbook*, 1st ed., McGraw-Hill, New York, 2005.

Harris, N. S., *Modern Vacuum Practice*, 3d ed., BOC Edwards, Crawley, 2004.

Shearer, M., *Semiconductor Industry: Wafer Fab Exhaust Management*, 1st ed., CRC Press, Informa, Boca Raton, FL, 2005.

Chapter 43
Control of RF Plasma Processing

BY
DAVID J. COUMOU *MKS Instruments, Inc., Rochester, New York*

43.1 INTRODUCTION

Many critical technologies rely on processes that utilize plasma-based material processing. Plasma processing is one of the cornerstone technologies for the semiconductor industry, and serves a vital role in the essential processing of material, broadly impacting other market sectors such as display technologies, energy efficient building materials, flexible electronics, bio-compatible materials and devices, and low cost photovoltaics. One of the major applications of plasmas in high-volume manufacturing is thin film growth and patterning. Electronic and optical devices require process efficient and economical manufacturing schemes for nanoscale pattern transfer down to less than 10 nm and deposition and surface activation technologies for self-organizing structures and atomic scale deposition. These semiconductor device manufacturing advances drive a higher level of integration whilst mandating repeatable target yields by minimizing plasma induced damage, optimizing process throughput, and gaining the technological advantage necessary to reach the next high-performance node with next-generation material. For decades, plasma processing has kept pace with these demands for the industry's adherence to Moore's law; however, the next technical hurdles to be addressed are becoming more fundamental.

Plasma-based material processing takes advantage of ionized gas generating unique chemistries and energetic particles to drive surface reactions under conditions not achievable in alternate regimes like thermal type systems. Plasma sources produce these conditions by introducing a prescribed gas mixture to electric fields that accelerate electrons. These electrons ionize and dissociate gas species through collisions, thereby creating reactive fluxes for surface reactions. In this chapter, we explore the underlying science of radio frequency (RF)–driven plasma systems. We ignore plasma chemistry, including vacuum, flow control and chemical kinetics, and focus on the physics of RF plasma sources so vital to semiconductor manufacturing. The capstone of this chapter is the advanced plasma processes, where we highlight many new and evolving technologies and schemes including pulsed RF plasmas as one such advanced technology. Prior to introducing these advancements, the fundamentals of various RF plasma sources are described in analytical form to convey how the physical system is harnessed for semiconductor manufacturing. We then proceed to describe plasma process control and diagnostic systems deployed to monitor plasma discharges, covering the breadth of purpose from rudimentary science to objectives related to high-volume manufacturing. After the meridian of this chapter with the compilation of advanced plasma processes, we conclude with a short discussion of dry etch and project future trends.

To meet the scope of this treatise, a survey from a vast body of work is presented. For this reason, an acknowledgment is warranted to a number of exceptional contributors to this field. The attempt at the provision of precise references and direction to additional information was a rigorous task for this development.

43.2 FUNDAMENTALS OF PLASMA GENERATION AND PROCESS CONTROL

In this section, the development of a physical model for low-temperature, high-density plasma reactors are described. The model for an inductively coupled plasma (ICP) source is first developed. This is then followed by a capacitive-coupled parallel plate. In each subsection, models, theories, and practices are explored.

43.2.1 Inductively Coupled Plasma Sources

The simplest form of an ICP source is shown in Fig. 43.1. The RF source produces a current to flow through the coil. The coil surrounds the cylindrical chamber formed by the tube. The tube is a dielectric material, traditionally in the form of a ceramic.

The genesis of ICP sources dates as far back as 1884 [1]. Due to the absence of an electrode and its contamination effect, this plasma source has proven to be attractive for applications requiring a clean and stable source. The inductively coupled discharge is either maintained by the axial electrostatic field or by an azimuthally electromagnetic field of the primary coil [2]. The electrostatic mode of operation appears at low RF power, while the electromagnetic mode of operation appears when the coil current is large enough to induce an azimuthal electric field that can maintain the ionization process. In this inductive mode, the gas discharge is sustained by induction from a time-varying magnetic field [3].

The physical model of an inductively coupled discharge can be described in terms of an electrical circuit. An air core transformer model is the traditional method used to describe an inductively coupled discharge [4–6]. Figure 43.2 depicts a transformer circuit to describe the mutual inductance of the coil assembly with plasma.

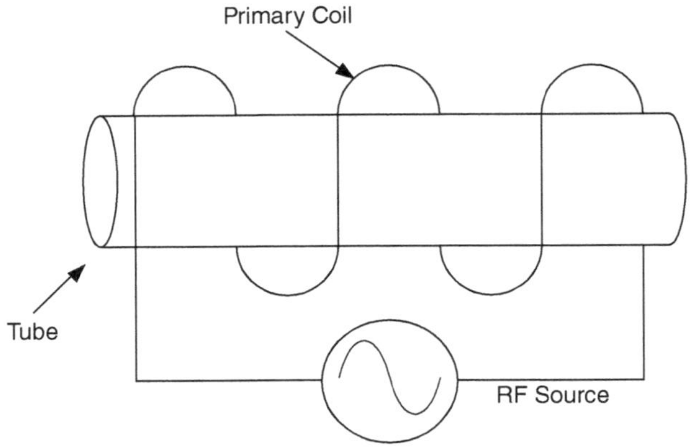

Figure 43.1 Physical model of an inductively coupled plasma (ICP) source. (*Courtesy of David J. Coumou.*)

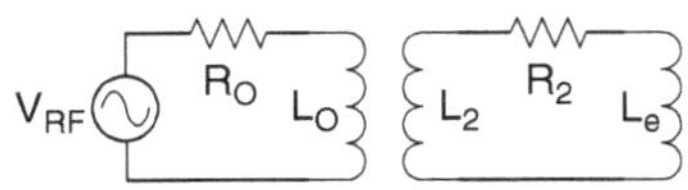

Figure 43.2 Electrical model of an inductively coupled plasma source. (*Courtesy of David J. Coumou.*)

The proceeding mathematical derivation for the electrical model of an inductively coupled discharge is transcribed from [3]. An air transformer consists of two windings. These windings are interlinked by a mutual magnetic field. The magnetic field produced by a current flowing in the primary coil interacts with the secondary coil. In the case of the inductively coupled discharge, the primary coil of the transformer is the coil that is wound about the tube, or in the planar coil case, positioned on a dielectric window in atmosphere. This coil is composed of N turns with inductance L_o and resistance R_o. The discharge is electrically conductive and considered a single turn secondary winding with inductance L_2. The discharge comprises an inductance L_e and a series resistance R_2 for the plasma. The two components of the discharge are the magnetic inductance and the electron inertia inductance. The magnetic inductance is due to the discharge current path and L_e results from the conductivity of the plasma [7].

In mathematical terms, the change in plasma resistance is given by

$$\rho = \frac{\omega^2 M^2 R_2}{R_2^2 + (\omega L_e + \omega L_2)^2} \tag{43.1}$$

where M represents the mutual inductance of the primary and secondary. The change of the plasma reactance is

$$\chi = \frac{\omega^2 M^2 (\omega L_e + \omega L_2)}{R_2^2 + (\omega L_e + \omega L_2)^2} \tag{43.2}$$

These components are used to define the secondary impedance as

$$Z_s = R_o + \rho + j\omega(L_o + \chi) \tag{43.3}$$

The RF voltage is the product of the RF current and Z_s. The power absorbed by the plasma is the product of the plasma resistance r and the square of the RF current. The reader is directed to [3] for a comprehensive development of the parameters described by the $\Re(\rho)$ and $\Im(\chi)$ components of the discharge impedance.

There are a number of advantages of using ICP sources: flexibility, high density, low transverse energy, low magnetic field, minimal device damage, and plasma confinement [8]. There are many chambers that use ICP sources for semiconductor manufacturing. An example of a plasma tool widely used in industry is the TCP 9600 tool and its product derivatives from Lam Research. This chamber design employs a planar inductor and is used for metal and polysilicon etch processing. Similarly, Applied Materials has the Producer and Centris tools that utilize a solenoid RF coil and are used for oxide etch processing. Readers who are interested in obtaining additional information about inductively coupled discharges are directed to Refs. [9–14] at the end of this chapter.

43.2.2 Electron Cyclotron Resonance Plasma Sources

An alternate form of ICP source is the electron cyclotron resonance (ECR) reactor. The principles of ECR plasma sources are explained in [15]. In summary, an ECR discharge is created from a magnetic field generated by a microwave radiation source. Electrons transverse in a circular motion in the direction of the magnetic field. The frequency of this circular motion is the cyclotron frequency. With the application of an electromagnetic field, energy is transferred to the electrons. Resonance occurs when the electron undergoes one circular orbit in one period of the applied electromagnetic field. Since the microwave frequency is typically 2.45 GHz, the hardware costs and radiation source availability are realizable for the production of commercial reactors and the necessary electromagnets and permanent magnets are reasonable in cost and size. Commercial uses of this technology in the production of semiconductor manufacturing are evident in both etch and deposition tools.

43.2.3 Capacitively Coupled Plasma Sources

Capacitively coupled plasma (CCP) sources are the most widely used for low-pressure material processing. Capacitively coupled plasma sources are used for reactive ion etch (RIE), plasma etch, physical vapor deposition (PVD), and plasma enhanced chemical vapor deposition (PECVD). Interestingly, the RIE chamber is a misnomer since the etching is a chemical process that is ameliorated by energetic ion bombardment of the substrate, rather than a removal process due to reactive ions alone [16]. The typical configuration for capacitively coupled plasma sources is a parallel plate arrangement of electrodes and is illustrated in Fig. 43.3. Not shown in this figure is the vacuum control system that is used to maintain a stable pressure in the chamber. A gas inlet is shown where the flow of chemistry into the chamber is controlled. By applying RF power, plasma is developed between the upper and the lower electrodes. For simplification, the upper electrode is powered in this model and the lower electrode is grounded. Some industrial applications for capacitively coupled discharges employ a second RF source at the lower electrode. A sheath is generated between the electrodes and the plasma. The sheath is a positively charged layer that confines electrons. The

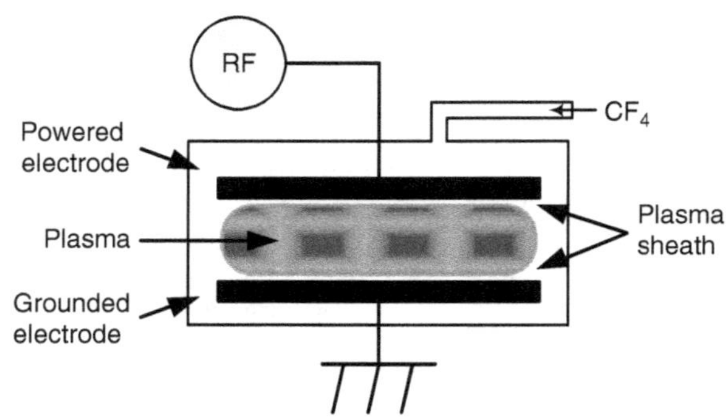

Figure 43.3 Physical model of a parallel plate plasma source. (*Courtesy of David J. Coumou.*)

electrons are confined because the electric field in the sheath is directed from the plasma [16].

The industry standardized the parallel plate configuration and refer to this chamber design as the Gaseous Electronics Conference RF reference cell [17]. It is commonly referred to as the GEC cell. Figure 43.4 illustrates the circuit model for this parallel plate chamber design. Reference [18] is utilized to describe the circuit model and [16] to correspond the circuit model to the plasma parameters. The shunt circuit elements (L_s, C_s, and R_s) will be addressed in the matching network discussion of this chapter. The circuit model includes the parasitic elements that are present in the powered and ground electrodes and in the chamber wall. Note, the powered electrode is sometimes referred to as the anode, and the grounded electrode is referred to as the cathode. C_{PE} is the parasitic capacitance for the powered electrode and is attributed to the insulator between the powered electrode and the ground shield. The transmission line that powers the electrode is represented by the series elements L_{PE} and R_{PE}. Typical chamber designs will have RF metrology installed in situ before the electrode. C_M represents the shunt capacitance that is introduced by the RF metrology. L_{PE} and R_{PE} may also address the effects of the RF metrology. Analogous to the powered electrode, elements C_{GE}, L_{GE}, and R_{GE} represent the parasitic elements of the grounded electrode. L_W is the self-inductance of the cavity between the chamber wall and the outer surface.

The inherent stray capacitance C_{chuck} between the lower and the upper electrodes and chamber side wall is not included in the circuit model in Fig. 43.4. The electrical properties of the chamber (L_W and C_{chuck}) and the transmission line that powers the electrode do not change with respect to the input RF power and plasma parameters. Specifically, C_{chuck} and the inductance of the rod L_{rod} can be measured with no plasma present. Obtaining a base pressure in the chamber and with an RF source set to a power that is insufficient to ignite plasma but drives the chamber with a voltage, the electrical characteristics of the chamber can be calculated from the measured voltage and current obtained from the RF metrology.

$$\frac{V_m}{I_m} = j\omega L_{\text{rod}} + (j\omega C_{\text{chuck}})^{-1} \tag{43.4}$$

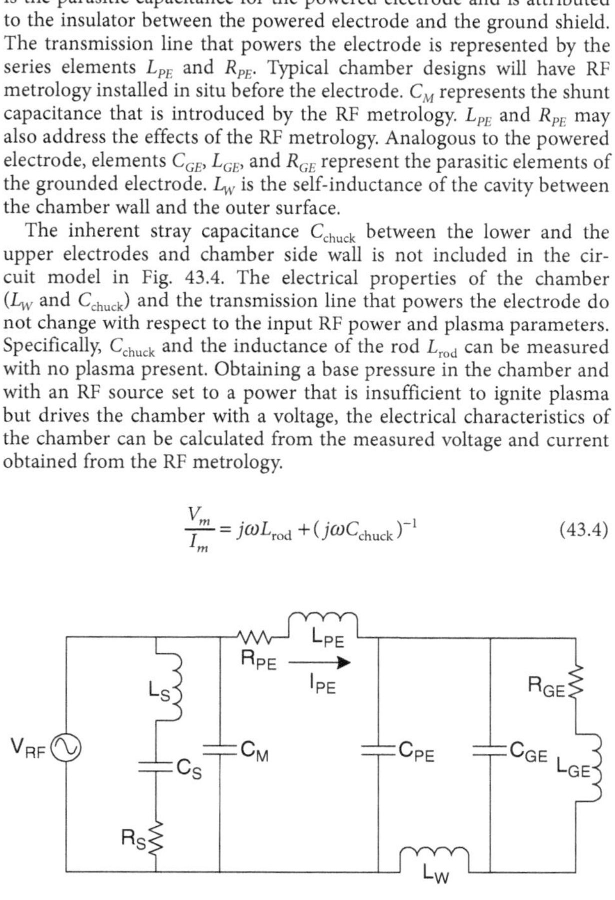

Figure 43.4 Electrical model of a capacitively coupled plasma source [18]. (*Courtesy of David J. Coumou.*)

V_m and I_m are the voltage and current measured by the RF metrology. V_e and I_e are the true voltage and current at the powered electrode. V_m, I_m, V_e, and I_e are all time-varying complex numbers. V_e and I_e are related to V_m and I_m by

$$V_e = V_m - j\omega I_m L_{\text{rod}} \tag{43.5}$$

$$I_e = I_m(1-\omega^2 L_{rod} C_{chuck}) - j\omega V_m C_{\text{chuck}} \tag{43.6}$$

The electrical properties of the chamber (L_W and C_{chuck}) and the transmission line have a negligible role in the analysis of plasma for parallel plate systems and are often omitted. C_s is used to define the equivalent capacitance of the plasma sheaths on the powered and grounded electrodes and R_s to define the equivalent resistance of the two sheathes. Since the plasma is typically operated at a frequency of 13.56 MHz, the dominant current-transport species in the plasma are the highly mobile electrons. This is represented as a resistive component R_b that is defined by

$$R_b = \frac{m\nu d_b}{A\eta e^2} \tag{43.7}$$

where A = electrode area
d_b = bulk thickness
e, m = electron charge and mass, respectively
η = bulk plasma density
ν = the electron-atom collision frequency

The bulk resistance is inversely proportional to the mobility and density of electrons, and proportional to the thickness of the bulk region. The sheath capacitance determines the effective sheath thickness S_o as

$$C_s = \frac{A\varepsilon_o}{2S_o} \tag{43.8}$$

where e_o is the vacuum dielectric constant and the gap distance between the two electrodes is given by $L = 2S_o - d_b$. The parallel sheath resistance represents the ion acceleration power (P_i) in the sheath in terms of the RF voltage drop (V_{sh}) across the sheath as

$$R_s = \frac{V_{\text{sh}}^2}{2P_i} = \frac{V_{\text{sh}}^2}{4I_i V_{\text{dc}}} \tag{43.9}$$

where I_i is the ion current to the electrode, and V_{dc} is the dc voltage across each sheath. Another approach to account for ion power loss is through an equivalent series resistance in terms of the discharge current, I, which yields

$$R_s = \frac{2P_i}{I^2} = \frac{4I_i V_{\text{dc}}}{I^2} = \frac{R_s}{1+(\omega C_s R_s)^2} \tag{43.10}$$

The plasma resistance and capacitance are obtained from

$$R = R_s + R_b \tag{43.11}$$

$$C = \frac{C_s(1+(\omega C_s R_s)^2)}{(\omega C_s R_s)^2} \tag{43.12}$$

Capacitively coupled plasmas have the disadvantage of high sheath voltages with low ion density and high ion bombarding energy. The ion bombarding energy cannot be controlled independent of the ion energy [16]. The addition of a magnetron to the capacitively coupled plasma system is an attempt to achieve these goals. The magnetically enhanced reactive ion etcher (MERIE) is described in the section that follows.

43.2.4 Magnetic Plasma Sources

Magnetrons are used to ameliorate plasma density, ion density, and uniformity for the production of semiconductor devices. They are found in sputtering systems that deposit metal (aluminum or copper) onto a target. The utilization and features of a magnetron are described in [19]. The efficacy of using a magnetron is related to the efficiency between power density and etch rate. Only the ion energy across the surface of the wafer is usable for etching. This results in a volume of unusable ion energy. Increasing the power density is generally not a viable solution because with the increase in power, there is a proportional increase in heat and loss, which induces device damage. A magnetron is used to generate a magnetic field in parallel with the wafer to confine the ion energy of the discharge. This magnetic field is orthogonal to the electric field generating the plasma. The electric and magnetic fields create an electron drift velocity component that traverses with a closed path radius in a direction orthogonal to both the electric and magnetic fields. The drifting electrons impinge on neutral molecules and cause additional dissociation to increase plasma density. The closed-loop path of the electron results in an electron storage ring that produces the high plasma density and limits electron mobility. The radius of the path r is proportional to the square root of the energy and the electron path is typically referred to as cycloidal.

There are several semiconductor manufacturing tools that use magnetron systems. Tokyo Electron Limited uses a dipole ring magnet assembly constructed with a plurality of magnetic segments [20]. The dipole ring has a nonmagnetic material between each magnetic segment. The magnet assembly uses a circular rotation on a track around the center axis about the outer periphery of the chamber. Tylan Corporation accomplished a similar effect that uses a linear motion in place of a circular rotation for the magnets [19]. Applied Materials obtained magnetrons results with a novel approach of using electromagnets spaced around the periphery of the chamber to generate and control the magnetic field [21]. Pulsing the magnets in a periodic fashion controls the magnitude and direction of the instantaneous magnetic field.

Material sputtering tools also benefit from magnetrons [22]. One example is a tri-magnetron assembly for a sputtering reactor [23]. Circular ring magnet assemblies are located on the top side of the chamber and the inner wall of the reactor vessel. These magnets rotate about the chamber axis. The third ring-shaped magnet assembly is located at the outer sidewall. The inner and outer magnet assemblies are parallel to the chamber axis and the roof magnet assembly is orthogonal to the chamber axis. The magnetic field components that are generated in this configuration yield several advantageous effects. Electron losses are reduced and the plasma density increased. Metal ions are extracted from the source material and their trajectories are guided toward the target for deposition. In a similar fashion, a magnetron uses a coaxial electromagnetic coil to create a magnetic field component from the target to the wafer [24]. This magnet assembly rotates around the chamber axis. The magnetron was improved [25] with the addition of auxiliary magnets that were added to the outer periphery of the chamber. The auxiliary magnets—permanent or electromagnets—proved to draw a portion of the unbalanced magnetic field toward the wafer and guided more of the ionized sputter particles.

The use of magnetrons is not excluded from ICP sources. Tokyo Electron Limited uses a stationary, axially symmetric permanent magnet assembly [26]. The magnetic field effect of this magnetron design concentrates the plasma, and by enhancing plasma confinement, improves uniformity.

43.2.5 Inductively Coupled Plasma versus Capacitively Coupled Plasma

By examining the analytical model for a CCP reactor in Sec. 43.2.3, the electron density and ion density are dependent on the sheath dynamics. The ion current, I_i, and V_{dc} are associated with the sheath resistance, R_s, defined in Eq. 43.9. The electron density is defined by the bulk plasma, R_b, in Eq. (43.7). Process performance increases with RF frequency corresponding with depreciable uniformity attributed to standing wave effects. The density of a capacitively coupled discharge is controlled by the amount of RF current that can pass through the sheath with the plasma sheath impedance dominating. The plasma density is proportional to the square root of the plasma power for electropositive gases [16]. The plasma efficiency decreases as the power and the sheath voltage increase. It is for this reason that an inductively coupled discharge has an advantage over capacitively coupled discharges. From Sec. 43.2.1, the analysis for an ICP source shows the reactance, c, is a function of the mutual inductance of the primary to the secondary loading by the plasma density, h_e [27]. An inductively coupled discharge tends to be more efficient because the plasma discharge can be generated and controlled in the proximity of its use. Notably excelling at diffusion; however as we will see later in this chapter, plasma coupling is related to r-2 effect as the distance from source to wafer and imposes a limit on the usable pressure range.

In summary, unlike a capacitively coupled plasma source, an ICP source provides almost independent control of ion flux and energy incident to the substrate. In an ICP scheme, the RF source governs the plasma density while the bias power controls the ion energy distribution function (IEDF). By source and bias power level control, etch rate via ion density and energy is achieved [28]. ICP sources are extensively used for gate and metal etch with a tendency of CCP sources utilized for dielectric etch. This traditional utilization is evolving toward a different balance. Semiconductor device manufacturing continues to achieve decreased feature sizes with a corresponding density increase along device area and volume. The consequence of thinning material layers is most evident in the fabrication of 3D devices; providing the opportunity for ICP-based tools to perform pattern etching for thin dielectric layers.

43.3 PROCESS CONTROL AND DIAGNOSTICS

Based on the derivations for inductively and capacitively coupled sources, it becomes evident that the control of the RF source is a principal control point of the plasma source. The significant physical component of the discharge is the delivery of RF power into the plasma. For the inductively coupled source, the power absorbed by the plasma was through the induced RF current. Conversely, the power absorbed by the capacitively coupled discharge was through the induced RF voltage. It is through these interactions that the RF source controls the stability of the plasma and the processing conditions. Typically, etch rate is a parameter that is controlled for a particular process. The block diagram in Fig. 43.5 illustrates a typical architecture for a plasma chamber. The RF source is contained in the RF generator. The RF generator's frequency must be compliant with Code of Federal Regulation 47, Part 18. The RF generator is typically designed for a 50-Ω load impedance. Since the impedance of the plasma and the transmission line to the chamber is not 50 Ω or matched to the impedance of the RF source, a matching network is located between the RF generator and the plasma chamber. It becomes imperative that RF power transfer through an impedance tuning network have a high RF efficiency so that the predominant dissipative loss in the impedance tuning circuit is the plasma. When RF power delivery is optimal and efficient, the RF delivery system ensures that the majority of RF power goes to electron and ion heating and thus into the process chemistry. This, in turn, ensures controllability and repeatability in the process and contributes to efficient, high yield device production.

The circuit topology of the matching network is generally a T, P, or L network. The purpose of the matching network is to maximize the power transfer from the RF source to the plasma by matching the impedance of the plasma to the impedance of the RF source.

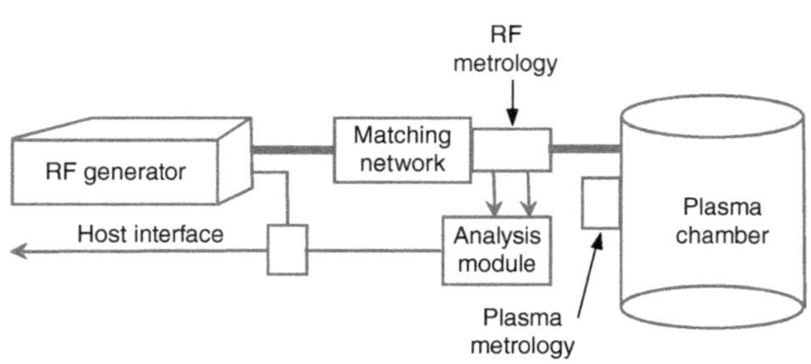

Figure 43.5 Typical plasma chamber block diagram. (*Courtesy of David J. Coumou.*)

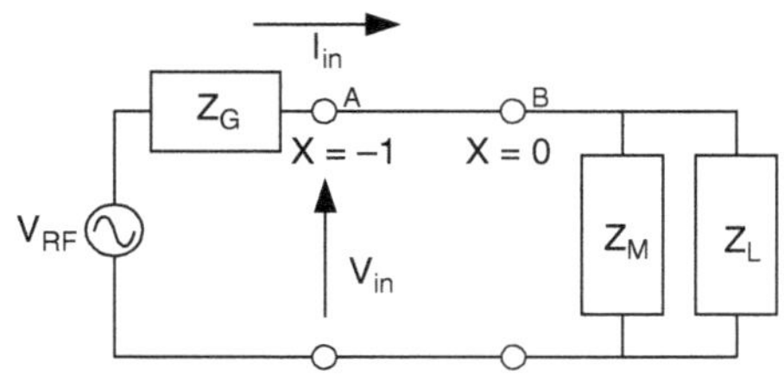

Figure 43.6 Plasma processing RF circuit model. (*Courtesy of David J. Coumou.*)

Figure 43.6 illustrates the electrical circuit of the RF source, the matching network, and the plasma. Z_G represents the source impedance, Z_M represents the matching network impedance at a distance l from the source, and Z_L represents the impedance of the plasma. Transmission line theory describes the voltage and current at a particular position in the transmission line as

$$V(-l) = Ve^{j\beta l}[1+\Gamma(-l)] \tag{43.13}$$

$$I(-l) = \frac{V}{Z_o}e^{j\beta l}[1-\Gamma(-l)] \tag{43.14}$$

where β is the phase constant and Γ is the reflection coefficient. If Z_{in} is defined as

$$Z_{\text{in}} = Z_M + Z_L \tag{43.15}$$

then the voltage on the transmission line can also be described as

$$V(-l) = Z_{\text{in}}I(-l) = Z_{\text{in}}\frac{V_{\text{RF}}}{Z_{\text{in}}+Z_g} \tag{43.16}$$

Substituting

$$Z_{\text{in}} = Z_o\frac{1+\Gamma(-l)}{1-\Gamma(-l)} \tag{43.17}$$

reduces the voltage on the transmission line to

$$V(-l) = V_{\text{RF}}\frac{1}{1+\dfrac{1+\Gamma(-l)}{1-\Gamma(-l)}} = \frac{V_{\text{RF}}}{2}(1+\Gamma(-l)) \tag{43.18}$$

Similarly, the current on the transmission line is

$$I(-l) = \frac{V_{\text{RF}}}{2Z_o}[1-\Gamma(-l)] \tag{43.19}$$

The time-averaged power at the input is then described by

$$P_{\text{av,in}} = \frac{1}{2}\Re(V_{\text{in}}I_{\text{in}}) = \frac{1}{2}\Re(Z_{\text{in}}I_{\text{in}}^2) \tag{43.20}$$

$$P_{\text{av,in}} = \frac{1}{2}\Re(V(-l)I(-l)) = \frac{1}{2Z_o}|V|^2(1-|\Gamma(-l)|^2) \tag{43.21}$$

The reflection coefficient is zero when the load (in this case Z_{in}) is equal to the characteristic impedance of the transmission line, Z_o. When this occurs, the time-averaged power at the input achieves a maximum

$$P_{\text{av,in}} = P_{\max} = \frac{|V|^2}{2Z_o} \tag{43.22}$$

The matching network usually has at least one tunable element since the impedance range of the plasma can vary based on power, pressure conditions, and chemistry of the discharge. The tunable element(s), usually a capacitor, of the matching network adjusts the impedance of the matching network to a load impedance that is within the operating specification of the RF generator. In mathematical terms, the matching network arrives to an impedance value that is a near equivalent to the complex conjugate of the plasma impedance. This cancels the phase component difference and requires a scaling of the impedance magnitude. When the matching network is tuned to the optimal load impedance of the RF generator, the transfer of power into the discharge is maximized.

Conventional tuning operation of the matching network is accomplished by circuitry internal to the matching network. The circuitry of the matching network has an RF detector and associated electronics to measure the magnitude and phase of the impedance or reflection coefficient. These measurements are used with an auto-tuning algorithm to command the tunable element(s) of the matching network. Traditionally, this is realized with analog and mixed signal circuitries [29]. This has some limitations due to the nonlinear equations that describe the transfer of power based on impedance. A more novel approach uses a digital controller with fuzzy logic to command the tunable elements [30]. Today, the industry is experiencing the advent of solid-state matching networks to replace the passive elements of the traditional matching network. A matching network using a pin-diode [31] configuration is one example of an early instantiation.

Another alternative to matching the impedance of the plasma to the RF source impedance is to use a fixed match with an RF generator using agile frequency [32]. Auto frequency tuning has the major benefits of speed, cost, and reliability by eliminating the movable tuning elements found in traditional matching networks. To use auto frequency tuning, it is necessary to incorporate a matching network with a fixed impedance between the generator and chamber. This matching network must be designed so that the process window of impedance can be tuned within a reasonable frequency range; typically ±10 percent for the mid-frequency range and ±5 percent for frequencies exceeding this range. The operation of auto frequency tuning generators is analogous to the operation of a tunable matching network. The tunable matching network adjusts the real and imaginary components of the impedance to a tune range within the operating specification of the RF generator. In the case of auto frequency tuning generators, the fixed match converts the real component of the impedance and the auto frequency tuning generator controls the trajectory of the imaginary, or reactive, component of the impedance. To ensure auto frequency tuning generators do not interfere with emergency safety and search and rescue radio transmissions, the following frequency bands are blocked: 490 to 510 kHz, 2170 to 2194 kHz, and 8354 to 8374 kHz.

Before proceeding to the next section and in the context of RF power spectrum and the analytical descriptions of RF plasma sources, we introduce, as a preamble to the material covered in the remaining chapter, the effect of RF on plasma processing. In Fig. 43.7, the effect of RF harmonic distortion is presented. The ion energy generated for a RF bias powered at 13.56 MHz is produced from a Monte-Carlo simulation with the amplitude and phase of the harmonic chosen as random variables. The effect of harmonic distortion relative to the 13.56 MHz RF power clearly shows an adverse effect on the percent variation of the ion energy from the mean. At approximately −45 dBc, the total harmonic distortion has shifted the ideal (no harmonic disturbance) mean ion energy by 1 percent. As a bridge between plasma fundamentals and putting into practice these RF plasma sources, process control is substantially critical to achieving desired semiconductor manufacturing results. This example conveys the criticality of process control of etch rate that is in principle a function of ion energy. The topics that follow within this chapter serve to relate the importance of process control, the advent of new techniques, and next generation methods.

43.3.1 RF Metrology for Plasma Chambers

There are numerous advantages of using Radio-frequency metrology for monitoring and controlling plasma processing chambers for the manufacturing of semiconductor devices [33]. RF sensors are important

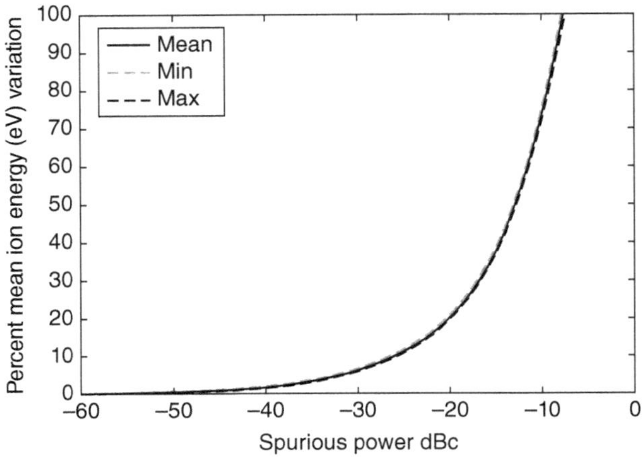

Figure 43.7 RF harmonic effects on plasma parameters critical to plasma processing. (*Courtesy of David J. Coumou.*)

because they are not intrusive, and they acquire and report data in real-time computation. This enables a fast response time to any changes that occur during plasma processing. RF monitors also give information about the discharge, which can be used to develop a physical understanding of the internal electrical characteristics of the plasma. At the very least, this information can be used to establish trends between the input settings and the electrical characteristics of the plasma source.

A number of sensors have been employed as RF metrology in plasma processing tools. These include voltage probes, current loops, diplexers and directional couplers [18, 34–39]. Directional couplers typically do not have adequate directivity for effective measurement in the typical operating impedance regimes of plasma chambers [40]. References [41] and [42] describe a robust voltage/current sensor design. This sensor can be configured in a coaxial line system with a fixed-characteristic impedance or a non-coaxial line system. When configured in the non-coaxial line system, the sensor is calibrated in a system that is a close replication of the final configuration. The construction of the probe is in the form of a coaxial line segment and designed with an aluminum-body outer conductor and a silver-plated copper inner conductor. The geometry of the inner conductor is square with chamfered corners, and the material of the dielectric spacer is boron nitride. The boron-nitride dielectric and the silver-plated inner conductor are designed specifically to permit the higher operating current at higher frequencies. The square shaft of the conductor improves coupling for increased signal-to-noise ratio. The square shaft also increases the surface area for optimal heat transfer from the inner conductor to the body of the probe. These design features enhanced the absolute and unit-to-unit repeatability accuracy. They also permit the use of the probe in high frequencies (>13.56 MHz), high-current applications without performance degradation at low frequencies, and power applications. The sensor comprises voltage and current pick-up assemblies. The voltage pick-up assembly converts the time-varying electric field created on the inner conductor to a small voltage signal representative of the line voltage(s). The current pick-up assembly converts the time-varying magnetic field created on the inner conductor to a small voltage signal representative of the line current(s). A version of this coaxial line section VI sensor is implemented in planar form with a shared detection surface for both the electric and magnetic field [43]. For lower frequency response, the planar sensor is ameliorated with a ferromagnetic material embedded into the printed circuit board sensor [44]. To process the RF information from these sensors, the voltage and current signals are connected to an associated analysis module.

A feature-rich signal processing architecture of an analysis module for an RF impedance/power sensor is described [40]. One of the significant advantages of this signal-processing architecture is its ability to autonomously track multiple RF sources. This is accomplished through a high-speed sampling and digital processing unit. The signal-processing architecture is analogous to a phase-lock-loop; however, in this case the implementation is in the form of analog and digital electronics.

Figure 43.8 illustrates before and after effects of an oxide etch. The illustration on the left in this figure shows the presence of oxide in a trench prior to initiating an etch. The circuit diagram of the capacitor-divider circuit is a simplification of the resulting change in the impedance measured by the voltage/current sensor and the accompanying analysis module. As the ion bombardment penetrates the trench and removes the oxide, the impedance of the plasma varies. Monitoring the harmonics generated from the plasma has proven to be an effective method of monitoring the etch process [39, 45, 46]. Real-time feedback control and monitoring of the impedance during the etch process has been demonstrated to yield an etch rate and endpoint [47]. Figure 43.9 yields an example of polysilicon etch when monitoring the sixth harmonic of 13.56 MHz. The T1 and T2 labels on the plot indicate the etch start and etch endpoint respectively from an optical emission spectroscopy sensor. The RF signals are normalized to illustrate that measuring multiple RF parameters can achieve a robust method of measuring the etch rate and endpoint. Analogous to controlling etch, a voltage/current sensor and the associated analysis module have proven to be an effective means of chamber excursion detection and clean optimization [38, 48].

Recall from Sec. 43.2.4, the utilization of a magnetic field to improve the density of the plasma, the magnetic field is rotated in the plane of the wafer. Figure 43.10 provides an impedance measurement from the voltage/current sensor and analysis module as evident from the characteristic change of the impedance phase. The period of the change in phase is the frequency of the rotation of the magnet system. Control of this type of measurement has been employed to control uniformity and density.

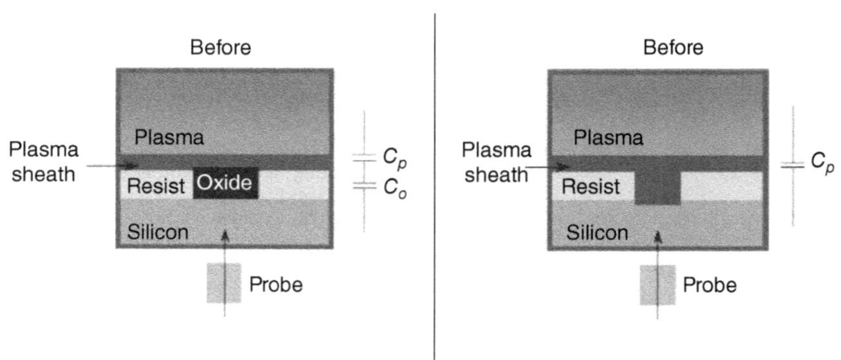

Figure 43.8 Oxide etch. (*Courtesy of David J. Coumou.*)

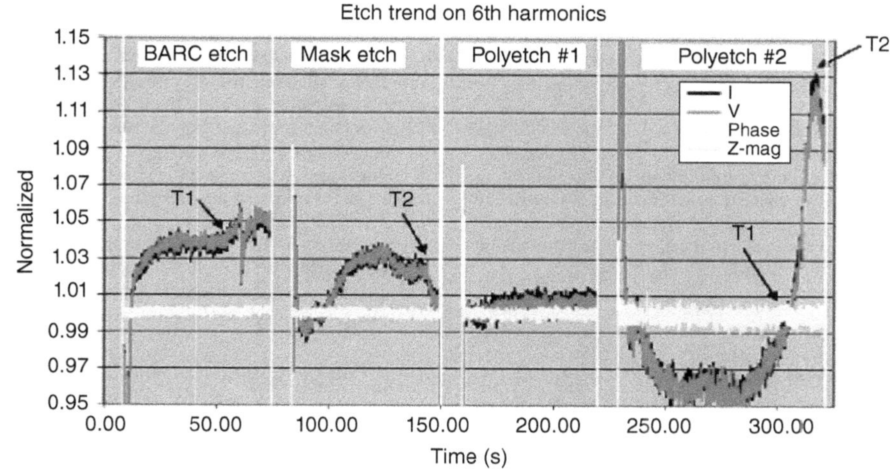

Figure 43.9 Endpoint detection using VI Probe sensor. (*Courtesy of David J. Coumou.*)

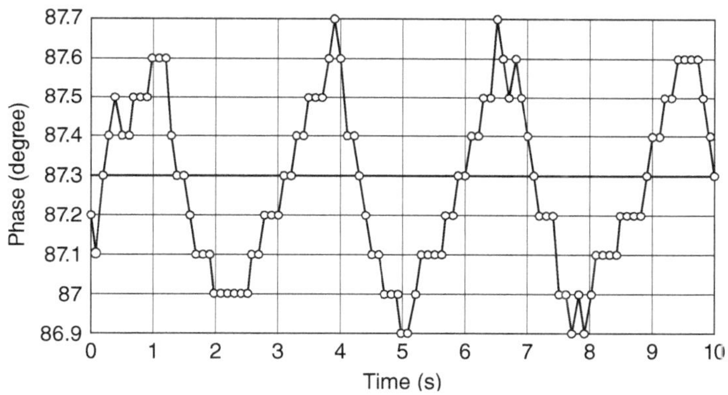

Figure 43.10 Phase impedance measurement for a MERIE plasma system. (*Courtesy of David J. Coumou.*)

These are just some applications that exemplify the inherent ability of an RF-impedance measuring system. In addition to these traditional applications, the high-speed sampling and digital processing unit has the ability to perform in more advanced applications. Some of these include the plasma stability and impedance measurements for RF-pulsing applications. These real-time measurements and analyses of process power and impedance contribute to advance process control (APC). This type of metrology is recommended as a necessary component of an APC system [49, 50]. The efficacy of using a multivariate analysis method of principal components on harmonic data from an RF sensor has been proven to be a resilient approach to APC [51].

43.3.2 Langmuir Probe

The Langmuir probe is a small-diameter cylindrical collector, similar to a piece of wire, and is inserted into the plasma. For a thorough review of this plasma sensor, the reader should consult [16] and [52]. When a negative potential (with respect to the plasma potential) is applied to the probe, the probe repels negative ions and electrons and attracts positive ions. This creates a cylindrical sheath around the probe surface. The total positive charge of the sheath is equal to the negative potential applied to the probe. The induced current into the probe is restricted to the rate at which the ions arrive at the sheath edge. Decreasing the negative potentials beyond this point, the probe current is predominantly ion current. When the probe is biased to a potential that is equal to the plasma potential, the probe current is induced from the mobile electrons in the plasma.

For probe bias voltages greater than the plasma potential, the probe current saturates at the electron saturation current.

Traditionally used by researchers, the Langmuir probe is a valuable instrument to understand plasma physics. One example utilizing the Langmuir probe to study ICP sources is given in [53]. Reliable-process endpoint is also achieved using the Langmuir probe [54]; however, it is undesirable to have this device installed on reactors used for high-volume semiconductor manufacturing. Surface contamination on the Langmuir probe adversely affect plasma parameter measurements [55, 56].

43.3.3 Optical Emission Spectroscopy

Optical emission spectroscopy (OES) utilizes an optical sensor to detect and measure the spectral emission from the plasma. Typically the sensors comprise a CCD image sensor and an optical filter. The OES system is the most widely used method for detecting oxide-etch endpoint in semiconductor manufacturing [57]. Patent applications and grants that employ this technology for etch endpoint and rate are numerous. Though they have proven to be a viable technology; the limitations of OES systems are well documented cited and justify alternative endpoint detection methods [58]. The case is made that current dual-damascene via etch requires exposed areas in the ranges of 2 to 0.5 percent, and with the demands of shrinking critical dimensions, the exposed area may decrease to 0.2 percent. Presently, the limitations encountered with OES are approximately 1 percent of the exposed oxide area.

43.3.4 Energy Analyzer

The influence of ion energy for plasma material interaction is well understood for deposition and etching plasma processing. To develop a plasma reactor or to devise a scheme to manipulate the distribution of the ion energies interacting with a surface, a method to measure the ion energy at the substrate is a necessary research instrument. A side view of an energy analyzer that would be used to measure the ion energies at a powered electrode is schematically shown in Fig. 43.11.

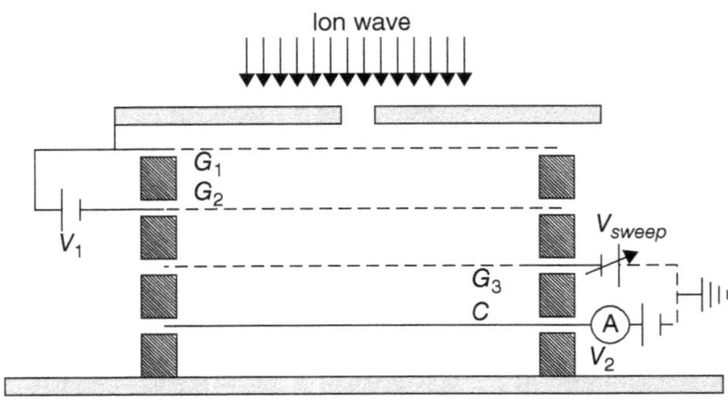

Figure 43.11 Side view of an energy analyzer sensor. (*Courtesy of David J. Coumou.*)

The energy analyzer is a sensor comprising a multiple grid layer to collect ions. Ions from a wave source enter through a hole above the grid. In the figure, a single hole is shown, but in practice, the sensor is configured with an array of holes. Electrically, the energy analyzer sensor floats at the same potential of the electrode that it is positioned on, and the ions crossing the sheath enter the sensor's opening(s) at right angles to the energy analyzer cavity. The grid is constructed with three biased grids, shown in the figure as G_1, G_2, and G_3. Each grid has holes with a diameter less than the array of holes the ions entered the cavity. To prevent the plasma from influencing the sensor measurement and to reduce the impairment of the sheath disturbance, the bias of G_1 is set to the same potential as the exterior sensor surface. The potential of G_2 resides at a DC voltage less than G_1 to repel plasma electrons from entering the cavity. G_3 is used to discriminate ions with different energies and has a sweeping potential, V_{sweep}, from the plasma's DC potential of G_1 to a few 100 V. Positive ions are repelled when the positive voltage between G_1 and G_3 exceeds the kinetic energy of the charged ion(s). When the bias of G_3 reaches the potential of G_1, the ion current is a maximum and is proportional to the ion flux at the exterior surface. As bias of G_3 sweeps more positively, the ion current is reduced as ions of specific energies are repelled until eventually the ion current goes to zero. Since the rate at which the ion current decreases is proportional to the IEDF, the derivative function of the measurement produces the IEDF. The ion energy is measured perpendicular to the ion entrance surface.

Across the electrode, multiple sensors may be arranged in a grid for spatial sampling of the ion distribution. The interested reader is referred to [59] for a commercially available retarding field energy analyzer that is widely adopted by the industry and academic research community.

43.3.5 Arc Detection and Arc Mitigation

Arc disturbances in an RF plasma are typically short transients with a duration of a few microseconds; they occur as discharges between the plasma and the electrode, the plasma and the chamber sidewall, or discharges within the plasma that are induced by the buildup of polymer structures. Arc events within a plasma chamber can occur during device manufacturing and such events may damage both the devices under construction and the plasma-based fabrication tools. As well, arc events can initiate undesirable chemistry in the process gas producing contaminants that impair device yields. Arcing events especially impinge semiconductor memory device manufacturing processes. A reliable means for detecting these occurrences and for intercepting the RF power delivery to mitigate the arc event is thus of critical importance to minimize arc-induced damage in these systems.

Signals and sensors employed for arc detection include measurements of the forward and reverse power from a directional coupler, measurements of the voltage and current from a VI sensor and the corresponding harmonic information. These methods employ signals, s, from the sensors for comparison against some threshold value, τ. When the signal exceeds the threshold, there is a probability that an arc event (Earc) occurred. A generalization of this detection can be expressed as

$$P(E_{\text{arc}} \mid s_1 > \tau_1, s_2 > \tau_2) \tag{43.23}$$

Later detection methods improved on this approach by applying the derivatives of the measured signals to determine the presence of an arc:

$$P(E_{\text{arc}} \mid s_1 > \tau_1, \dot{s}_2 > \tau_2) \tag{43.24}$$

These methods suffer from the fact that they are heuristic and nonquantitative. An alternate approach that has shown to be a highly robust detection with a subsequent mitigation is a method of measuring the energy of the arc disturbance [60, 61]. This solution uses tools from a communications equivalent paradigm to provide quantitative arc detection that measures the relative RF energy of the plasma arc transient.

A schematic of an RF power delivery system for an RF plasma source is shown in Fig. 43.5. As we discussed earlier in this section, in an impedance-tuned condition, the transmission line between the power supply and the match contains an RF power wave in the forward direction. In this operating condition, only the voltage and current traverses from the power supply to the matching network. With a nominal loss of the power through the matching network, the RF power is coupled from the match to the plasma chamber. In this portion of the RF circuit, a forward and reflected power wave traverses at the frequency of the RF generator. Due to the nonlinear characteristics of the plasma source, the RF spectrum contains an additional set of frequencies from the emission of the discharge. Though the information content is significantly richer with the harmonic and intermodulation distortion signals emitted from the plasma source, combined with the forward and reflected voltage and current waves, it turns out arc detection with a prematch sensor (say, in the RF power supply) provides analogous detection with respect to a postmatch sensor. Next, the detection method is presented.

A wideband correlation function collectively processes the signals that are sampled from the RF sensor, analogous to a digital communication system deploying a correlation receiver. The voltage and current signals from the RF sensor are digitally sampled and the power between these signals is derived from the cross correlation function. For arc detection, the signal processing is wideband to include the broader set of information signals contained in the RF spectrum. The cross correlation function is performed for nonoverlapping blocks of samples of the voltage and current signals with frequency f_{RF}, sampled at the rate f_s. Blocks contain k periods for $n = k\frac{f_s}{f_{\text{RF}}}$ total samples and duration $\Delta t = nf_s$. The power of the signals is obtained from the discrete correlation,

$$r_{vi}(m := 0) = p = \sum_{\forall n} (v_n - \text{E}[v_n])(i_n - \text{E}[i_n]) \tag{43.25}$$

with the mean of each block, E[·], subtracted from the sampled sequence. With the change in power that occurs during an arc event, the RF energy, $E = \Delta p \Delta t$, associated with the arc transient provides a quantitative measure.

The ability to detect arcs with substantial performance parity between prematch and postmatch enables a control protocol for arc suppression.

Coupling the arc detector, which detects the arc and measures the relative RF energy of the disturbance, with a countermeasure response from the RF power supply proportional to the measured arc energy affords an effective approach to suppressing the arc source, alleviating damage to manufacturing process yields. Figure 43.12*a* illustrates the effectiveness of the coordinated control of arc event detection and arc suppression. There are four signals captured in this scope plot. The signal indicated on the left by 1 is a signal from a Langmuir probe that measures the plasma's voltage potential change. The signal labeled 2 is the forward port signal from a directional coupler connected at the output of a 13.56-MHz RF power supply. The output of the coupler is connected to an impedance matching device with a VI probe sensor connected at its output and coupled to the reactor's powered electrode. The third and fourth signals in the scope plot are the response of the postmatch and prematch arc detection. The prematch arc detection algorithm is embedded in the controller of the RF power supply. The Langmuir probe, acting as a ground truth source for an arc transient, identifies the change in the plasma potential resulting from an arc. Using the RF energy disturbance algorithm, the arc detection response is determined by both detectors. With respect to the detection occurring at the RF power supply, the arc detector triggers the mitigation process by decreasing the power of the RF supply for time duration proportional to the measured energy. The momentary reduction in RF power and subsequent return to the RF set point is seen by the directional couplers forward port's response.

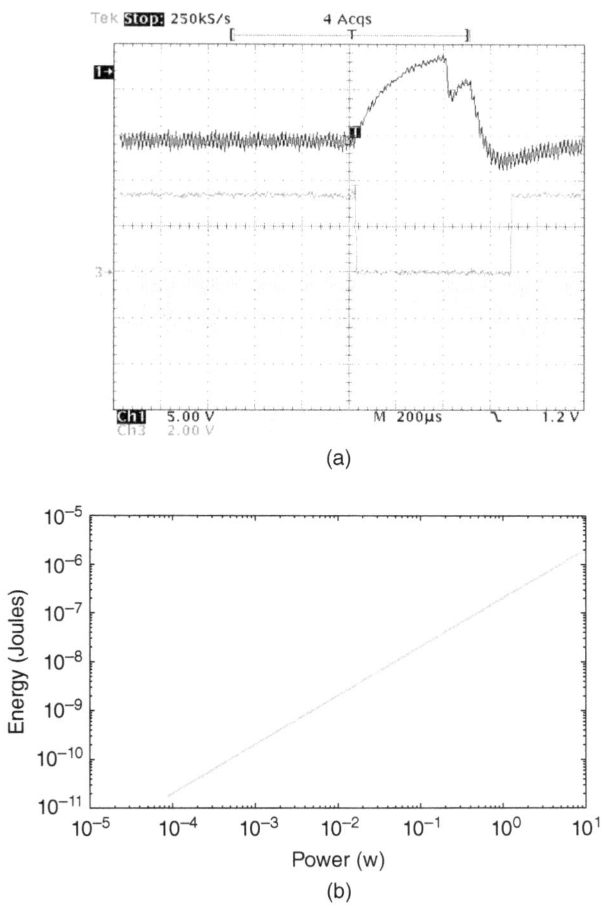

Figure 43.12 Arc detection and mitigation: (*a*) comparison of pre- and post-match arc detection with corresponding mitigation; (*b*) RF energy–based arc detection sensitivity. (*Courtesy of David J. Coumou.*)

In response to the detected arc, the RF power is suppressed for a period allowing the arc source potential to decrease. This can be seen by the Langmuir probe response. The sensitivity of the arc detector is shown in Fig. 43.12*b* with an approximate resolution of 86 μW.

43.4 ADVANCED PLASMA PROCESSING CONTROL

The RF configuration for modern plasma deposition and etch reactors vary by the number of RF power supplies and the corresponding driven frequency. It is well established for plasma assisted surface interactions the importance of the frequency drive and the phase relation of the RF with the sheath dynamics [62]. In terms of pattern etching, there are three primary factors to etching: (a) physical ion bombardment, E_P; (b) chemical etch by neutral radicals, E_C; and (c) ion-induced through ion bombardment, E_I [16]. The total etch rate is expressed as a function of $E_r = g\ (E_P,\ E_C,\ E_I)$ of these factors. It therefore becomes essential to control the ions to achieve the highest degree of plasma processing fidelity. As a generalization, the plasma sheath is parameterized by the power and frequency(s) of the RF power delivery system and their relative phase. Greater ion energies are produced as a function of lower frequency. While the ion energy increases with a lower RF frequency, the ion energy is more broadly distributed. As the frequency increases, the ion energy distribution narrows but results in lower ion energy production. It is desirable to achieve higher ion energy as a monoenergetic distribution function. In this section, we open with an overview of dual frequency systems to convey the density benefits of lower frequency excitation, where the IEDF is controlled, in conjunction with the higher frequency RF source, where plasma generation occurs. We next present RF pulsing as a means to control the IEDF in a time-varying manner by the plasma parameters. This is followed with schemes to tailor the IEDF function. To make these approaches feasible, we dedicate a subsection to discuss critical RF impedance tuning. The section concludes with electromagnetic approaches to solve uniformity challenges.

43.4.1 Dual Frequency Driven Plasma Sources

By convention, many plasma systems are excited by two RF power supplies with accompanying impedance tuning networks. The schematic of a dual frequency system CCP reactor is shown in Fig. 43.13. In this configuration, an RF power delivery system is coupled to each electrode. Different embodiments of this scheme might couple all RF power to a single powered electrode with a counter electrode grounded. The frequency selection of the coupled RF is chosen to provide independence between the plasma generation and the ion energy. The low-frequency controls the modulation of the ions. As the frequency scales, the ion effects diminishes and electron density increases.

With evaluating three different ion masses, ion energy distribution for a high-density plasma source commonly used for etching

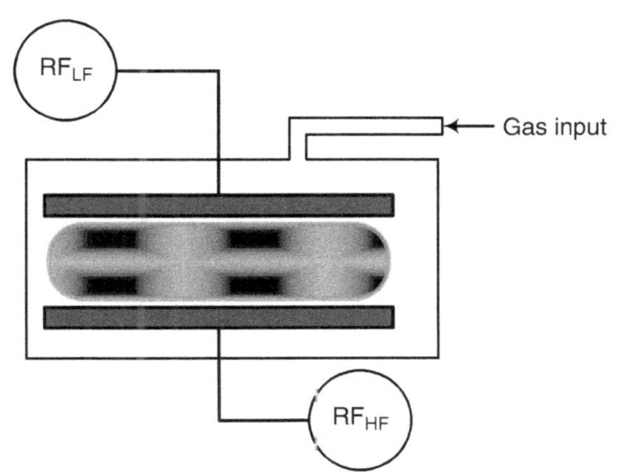

Figure 43.13 CCP reactor with dual-frequency excitation. (*Courtesy of David J. Coumou.*)

applications was explored [63]. The bias frequency was varied from 678 kHz to 60 MHz. For a low-frequency RF bias, the ions cross the sheath in a small fraction of the RF cycle and respond to the instantaneous sheath voltage. For higher frequencies, the sheath voltage oscillates during the period when the ions cross the sheath and respond to the average sheath voltage. The higher frequency resulted in a narrow ion energy distribution. The work concluded that ions with higher mass have narrower ion energy distribution than ions with lower mass and this relationship follows a square root of the mass of the ion. For a given ion mass, the ion energy distribution becomes narrower and increases the mean of the ion energy with the increase of the frequency of the RF bias. This finding was confirmed and generalized to find the ratio of the low frequency to the high frequency governs independent control of ion energy and flux, irrespective of pressure, geometry, and power [64]. These results can be summarized as the low frequency source drives the discharge voltage and couples to the ions. With a higher frequency with respect to the low frequency supply, the plasma source is dominated by the supply's current and excites more directly the electrons. The effect of multiple frequencies mixing in the formulation of the sheath introduces an intermodulation component that flattens the low-voltage portion of the time-varying sheath potential, thus producing a low-energy skew in the IEDF [65]. We will explore this effect in Sec. 43.4.3, when we describe tailoring IEDFs. Three frequency systems are also employed, which permit separate control of ion flux, ion energy, and surface bombardment energies [66].

With the ability to discriminate between ion energy and electron density with frequency selection, we will see in the proceeding sections the control of the IEDF by pulsing and tailored RF approaches.

43.4.2 Pulsing RF Plasma Sources

The relevance of RF pulsing in high-volume manufacturing of semiconductor devices has steadily increased. Much of the fundamental understanding of the physical properties of a pulsed source relative to a continuous wave plasma source dates back to 1995. Industry adoption remained stymied for over a decade and its more recent impact is prolonged. In this section, we present the fundamental physics of a pulsed discharge and correlate these benefits to semiconductor manufacturing. We conclude this section identifying the challenges posed to the RF power delivery system.

By pulsing RF, a variety of plasma processes occur at the first stages of RF turn-on and turn-off, as illustrated by the contrived example in Fig. 43.14. During the transient RF turn-on shown in Fig. 43.14*a*, the electron energy elevates above the steady state, electron (h_e) and ion (h_i) densities increase, and the thickness of the sheaths on the electrodes varies. During the on-cycle, the dynamics of the electron temperature (T_e) remains similar to the voltage potential (V_p) of the plasma [67, 68]. In this case, the mean of the IEDF is X. At the power turn-off depicted in Fig. 43.14*b*, electron energy decreases rapidly with corresponding reduction in electron and ion densities, and the plasma sheaths disintegrate with the decay of electron density. The mean of the IEDF drops to Y < X. In pulse operation, two control knobs are established for the process engineer: pulse frequency and duty cycle. These explicitly alter the plasma parameters (h_e, h_i, T_e, V_p) and the chemical interactions of dissociation rate and ion to neutral ratio. By adjusting the pulse frequency, the breadth of the IEDFs can be altered, thereby adjusting the disassociation rates and radical production. In Fig. 43.14, there is parity in the peak of the IEDFs for the on and off cycles. The ratio of the IEDF peaks is tailored by the duty cycle of the pulse rate. By time averaging, the IEDF can have a larger peak during the on cycle or the off cycle by adjusting the duty cycle.

The time dependence of the electron temperature and density ameliorates the selectivity and rate of etching to achieve greater process control. The benefits of this plasma control demonstrated a wide dynamic range of plasma chemistry control and reactive flux composition to the wafer surface by pulsing RF power delivery to an etch system [69]. The greatest process enhancements for pulsing are validated by surface interactions and device fabrication capability. To quantify these benefits, it is fundamental to put into perspective some limitations of continuous (CW) RF operation. In CW, the lowest energy impinging the surface at 15 to 30 eV can damage material in the 1 to 2 nm range. This is unacceptable for the silicon on insulator fabrication step and selective etch of ultrathin layers. For multiple dimension devices like FinFET, patterning of thin layers and future fragile material (e.g., graphene) will be a substantial challenge.

Plasma induced damage during etching arise from various plasma conditions: (1) surface due to highly energetic ions; (2) photon bombardment (UV and VUV); (3) plasma uniformity; and (4) differential charge at surface features. There are at least two notable manufacturing improvements achieved with RF pulse operation. The first of these improvements is in trench etching. A continuous plasma discharge has the proclivity for a charge buildup between the top and bottom of a trench. This results in negatively charged ions to be locally repelled from the trench opening resulting in undercutting at the bottom of the trench. A modulated discharge has an inherent pump like action to force negative ions and electrons into deep and narrow trenches. These neutralize the positive ions to improve etch quality. A second improvement obtained from modulating plasma is the reduction of "dust" particles. In literature, the polymerization of negative ions within the plasma is referred to as dust particles. The sheath of the plasma for continuously powered plasma retains negative ions, and after a period of time, these negative ions build up and

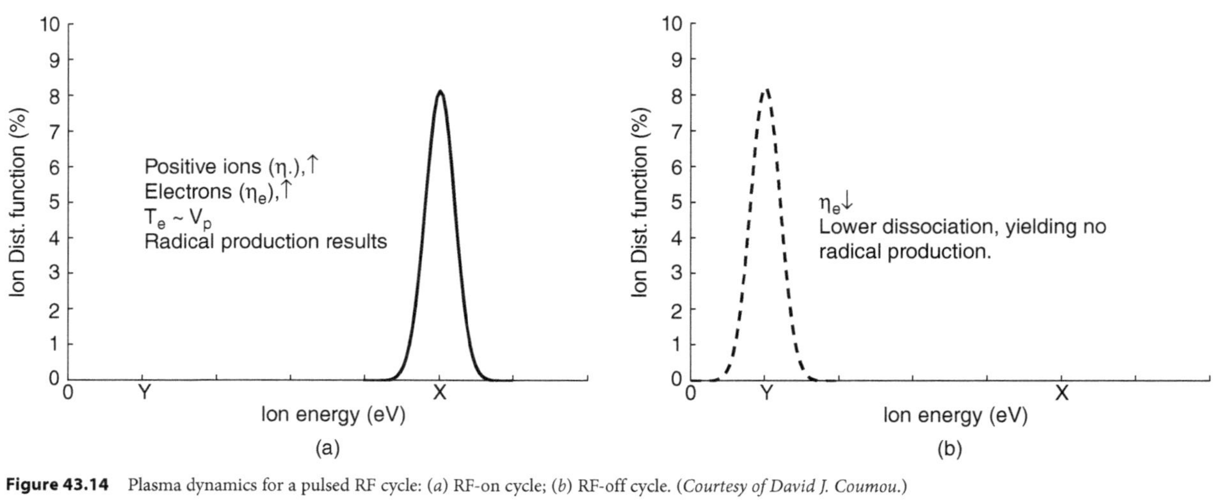

Figure 43.14 Plasma dynamics for a pulsed RF cycle: (*a*) RF-on cycle; (*b*) RF-off cycle. (*Courtesy of David J. Coumou.*)

polymerize to form contaminating particles. In modulated plasma, the sheath of the plasma periodically breaks up and reconstructs to release the negative ions before accumulation can occur. A further compendium of plasma processing benefits resulting from pulsed RF for plasma etching include: increased etch selectivity [70, 71]; improved vertical side-wall profile [70, 72]; reduction of trenching, notching, and charging damage [71]; increased etch uniformity [71]; reduction of aspect ratio dependent etch effects [71]; and reduction of heat flux to substrate [71, 73].

For deposition, the advantages of RF pulsing comprise: thin oxide-insulating layer with reduced ion-bombardment damage [74]; improved side-wall coverage [75]; and controlled deposition of conductive film [74, 76]. In particular, plasma assisted atomic layer deposition (PEALD) [77] has surpassed, in comparison, etch process capability and quality. For PEALD, RF pulsing occurs during the ALD cycle; not at the ALD rate. The benefit improves CW PEALD films that suffer non-uniform deposition properties. The sidewall deposition rate is different than the top or bottom of a trench. This is attributed to directional ion bombardment. RF pulsing addresses the problem by changing the ion directionality, enhancing the ion to radical ratio, and negative ion contribution. RF pulsing enhances PEALD with uniform dielectric film deposition at Angstrom level of material control.

The dynamics of RF pulsing affect the plasma impedance through the changes of the bulk plasma resistance and the sheath capacitance and resistance. At the pulse edges, the RF power decreases exponentially (not instantaneously) due to the energy stored in the matching network. This infers a significant impact on the RF power delivery system for reliable and repeatable pulse to pulse results for a substantial greater range of operation in contrast to continuous RF operation. Critical performance features for the RF power delivery system include (1) controlling the ramp of the pulse edges to manipulate T_e, V_p, and disassociation rate; and (2) synchronizing the source and bias RF power supplies in a multiple RF power delivery system. The requirement for pulse synchronization is to tailor a pulsing scheme to utilize negative ions and draw them into etched features. Perhaps the most challenge is RF power regulation as pulse frequency ranges from a few Hertz upward to tens of KHz.

43.4.3 Methods to Tailor the Ion Energy Distribution Function

IEDF plays an integral role in etching and deposition techniques on silicon wafers for chip manufacturing. The evolution commenced with the preponderance of dual frequency schemes that we described earlier in this section. High energy ions increase etches rates far superior than those of a linear combination of single-peak IEDF. In this section, we review methods, as an alternate to RF pulsing, to achieving a greater fidelity of IEDF control. We commence with the pioneering work of Wendt's research group showing the ability to form a monoenergetic IEDF and conclude with the more practical approach of the electrical asymmetry effect (EAE) [78].

Powering the bias electrode with an RF broadband spectrum of frequencies can tailor a wide range of IEDF shapes [79, 80]. The ability to control the IEDF with a narrow, single peak distribution, greater material interaction capability improves both pattern etching and material deposition. With this approach of IEDF generation, there are two limitations that stymie high-volume manufacturing adoption. In Sec. 43.3, we formulated the impedance matching circuits to couple RF power to the electrode of a plasma processing reactor. The circuits are based on single RF excitation signals and are not applicable to a broadband generated signal. For broad spectrum RF power delivery, these conventional matching networks are infeasible. The second aspect is formulating a broadband RF signal to generate a desired IEDF shape. The creation of the arbitrary waveform lacks the necessary intuition to create the specific IEDF target. A method that follows is based on an analytical expression to define the sheath dynamics and provides a model to correlated controlled variation of the RF power delivery system to generate a unique IEDF. Specifically, the width and skew of the IEDF are varied by adjusting inputs to RF power delivery system [81] and is based on the combination of multiple frequencies, a conventional pathway for controlling IEDF characteristics [82].

Driving one electrode at multiple harmonics provides even more than the opportunity to control the DC self-bias electrically by adjusting the phase between the driving frequencies. It enables tailoring the shape of the IEDF by controlling higher-order moments of the IEDF with customizing the sheath voltage waveform at the substrate. To tailor a specific IEDF, the equations that govern the sheath dynamics must be considered. The sheath thickness described as a function of time is $s(t) = s_1\,(1 - \sin(\omega t)) + s_2(1 - \sin(\omega nt + \phi))$, where $\omega = 2\pi f$ is the lower frequency of the dual frequency system and ϕ is the relative phase between the harmonic tones ($n > 1$). The amplitudes of the sheath oscillation are defined by $s_n = \dfrac{I_n}{e\eta_e n\omega A}\forall n$ with

where I_n = drive current associated with ω_n
η_e = electron density
A = electrode area
e = electron charge

Finally, the time dependent sheath voltage is described by $V_{\text{bias}}(t) = \dfrac{e\eta_e}{2\varepsilon_o}s^2(t)$. From this analytical model, the relative phase and current magnitude are controllable elements of the RF power delivery system. Power set points adjust the corresponding I_n, and through a digital phase lock loop [83], the frequencies of the dual RF power delivery system are harmonically derived, enabling phase locking. The sheath voltage is governed by the frequency, phase, and amplitude of the RF to produce unique IEDFs [81, 84] from the arbitrary waveform generation with the RF power delivery scheme.

In traditional dual-frequency CCP reactors, the distance between the excitation frequencies is substantial and is not phase-locked. In such plasma sources, the low frequency power controls the mean ion energy, the first statistical moment of the IEDF. The second-order moment of the distribution is proportional to the breadth (or width) of energies the distribution spans. With phase control between the harmonic related frequencies, the ability to control the third-order moment, skew, is achieved. Separate control of the mean ion energy and the ion flux is limited by the classical RF power supply scheme to narrow process windows, and higher order moments of the IEDF cannot be controlled.

Experimental results of the phase-locked, harmonic frequency drive scheme were demonstrated [31, 84]. To illustrate the control and tailoring of unique IEDFs, ion energy measurements were taken on a tri-frequency system using a digital phase and frequency control method [83] to generate the harmonic pair. The first powered electrode was biased with the harmonic phase-locked frequency drive at $f_1 = 13.56$ and $f_2 = 2f_1 = 27.12$ MHz. For plasma generation, the second electrode was powered by a 60 MHz RF power supply. On the bias electrode, ion energy measurements were taken from a retarding field energy analyzer (see Sec. 43.3.4). The empirical results for two current ratios are shown in Fig. 43.15. The ratio, x, normalizes the current for the low-frequency source, $I_1 = x$ and the current supplied by the second harmonic power supply as $I_2 = 1 - x$. By adjusting the phase relative to the output of each RF power supply by varying the component ϕ in the analytical model, the peak of the IEDF is shifted accordingly to demonstrate the skewness of the ion distribution. The figures show a narrow energetic distributions, and as x increases the distribution average energy increases. Simulated etch profiles using the EAE scheme is reported [85].

43.4.4 Impedance Tuning for RF Pulsed Sources

Since RF power principally provides the energy to accelerate electrons and ions to drive surface reactions [16], a central system to plasma processing schemes is the RF power delivery and associated components. Modern plasma deposition and etch reactor configurations are quite complex. Optimal and efficient RF power delivery faces a number of challenges in advanced thin-film manufacturing environments. These include, but are not limited to

1. Process conditions driven by a marginally stable plasma source

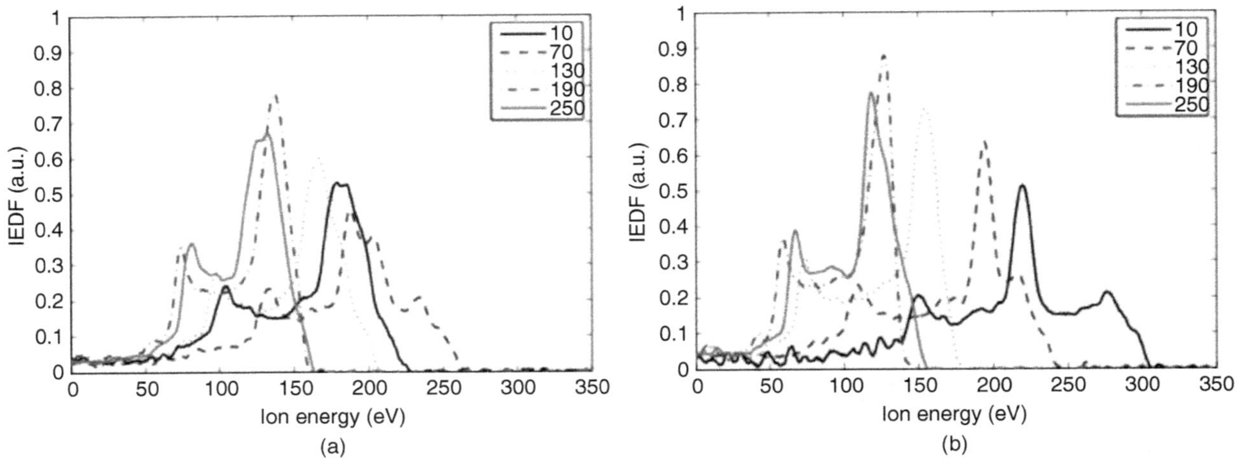

Figure 43.15 IEDF for phase-locked, harmonic drive: (*a*) $x = 0.25$; (b) $x = 0.33$. (*Courtesy of David J. Coumou.*)

2. Demand for deposition rates with higher precision to achieve thinning layers and controlled barriers
3. Impedance variations in pulsed RF generator operating modes
4. Greater dependency on the interaction of plasma sources driven by multiple RF power supplies

In the past two sections, we described methods to control RF plasma processing that are exemplary of the performance demands on the RF power delivery system. For tailoring the RF to achieve a target IEDF, the sheath dynamics interact with the RF power delivery system. For pulsing, the repetition of vacuum to plasma states in the reactor presents vexatious challenges to the RF power delivery system. This is best illustrated in Fig. 43.16 for a pulsed RF plasma source (argon). With no impedance correction scheme activated, the effect of pulsing (on/off operation) a discharge with a single 13.56 MHz RF power supply at 50 percent duty cycle for varying pulse rates can readily been seen as the peak percent of reflected power, a condition of the pulse rate. At low pulse rates, the percent of reflected power initially remains high and then monotonically decreases until about 100 Hz. As the pulse frequency continues to increase, the percent of reflected power monotonically increases before leveling off. This reflected power is considered a loss because it is not transferred to the plasma source performing the work function, and as a result, etching or deposition rates and quality are impinged.

In Sec. 43.3, we described impedance networks, which comprised of tunable matching networks and auto-frequency tuning RF power supplies, utilized to maximize the power transfer from the RF power supply to the plasma load. Traditionally, these impedance actuators

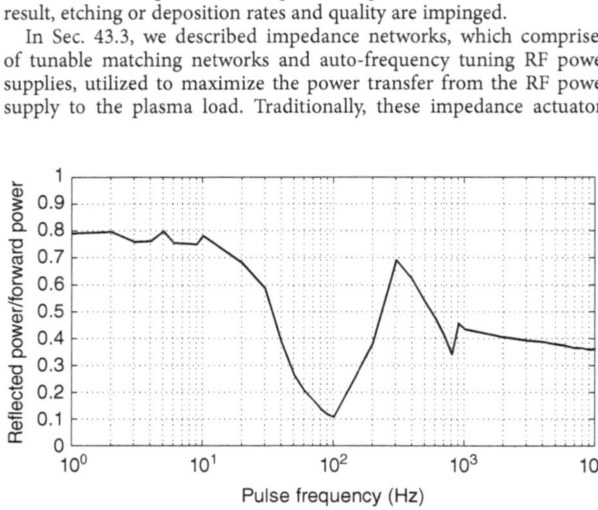

Figure 43.16 Percent of reflected power relative to the forward power set point of 800 W for various pulse rates. (*Courtesy of David J. Coumou.*)

are algorithmically driven with heuristic methods to find optimal tune condition to maximize this power transfer. These search-based control methods were suitable and widely accepted by industry, but recently yielded limitations commensurate with the ever increasing and continuously evolving process demands. An impedance controller, centralized in the RF power supply and coherent with the respective RF power regulation controller, is based on a convergence cost function to improve algorithmic aptitude, while removing systemic redundancy and purging complexity and cost. With brevity, this feedforward controller [86–88] is described next.

The feedforward controller resides in the RF power supply and iteratively computes, synchronous with the local control of RF power, actuators tuning commands until a criterion is satisfied. The criterion is based on a distortion quantity. The distortion impairing the RF power delivery is determined from a normalization procedure applied to the signals of an RF sensor contained in the power supply. The signals are treated as voltage and current signals, representative of the electric and magnetic fields coupled to the RF sensor. A ratio of these signals is computed for normalized impedance. Alternatively, an RF directional coupler can be used to measure the forward and reflected power waves; the associated ratio would represent the normalized reflection coefficient. In the case of the VI sensor, the complex quantities of voltage and current are digitally represented as $v(m) = |V|e^{j(\omega mT + \theta_v)}$ and $i(m) = |I|e^{j(\omega mT + \theta_v)}$, where T^{-1} is the frequency of interest. For normalization, the ratio between the product of these terms and the magnitude of the signals is computed. The numerator term is formed from the product of $v(m)$ and $i(m)^{-1}$ to yield $|V||I|^{-1}(\cos\theta + j\sin\theta)$, where θ is the phase difference between θ_v and θ_i. The numerator is normalized by the denominator term, $|V||I|^{-1}$, and reduces to $\cos\theta + j\sin\theta$. Power distortion is quantified by the sinusoidal functions $\cos\theta$ and $\sin\theta$, with the tuning actuator trajectories conveyed in the Smith Chart in Fig. 43.17. These distortion quantities are coupled to classical proportional-plus-integral-plus-derivative controllers to command the load and tune actuators. The load actuator is the variable shunt element in the matching network, whereas the tune actuator can be a combination of the variable series element in the matching network and the agile frequency of the RF power supply.

To demonstrate the feedforward controller deriving an impedance tune for significant impedance variation related to on/off pulsing, two plots are provided in Fig. 43.18. The experimental results were obtain from an ICP reactor configured for 3 mT pressure and 50 percent flow ratio of Ar and O_2 at a total flow rate of 125 sccm. The RF power applied to the bias was disabled, and the source coil assembly was pulsed at 300 W during the on cycle at 40 KHz and 50 percent duty cycle. The feedforward impedance controller was located in the 13.56 MHz power supply coupled to the coil antenna through an impedance matching network.

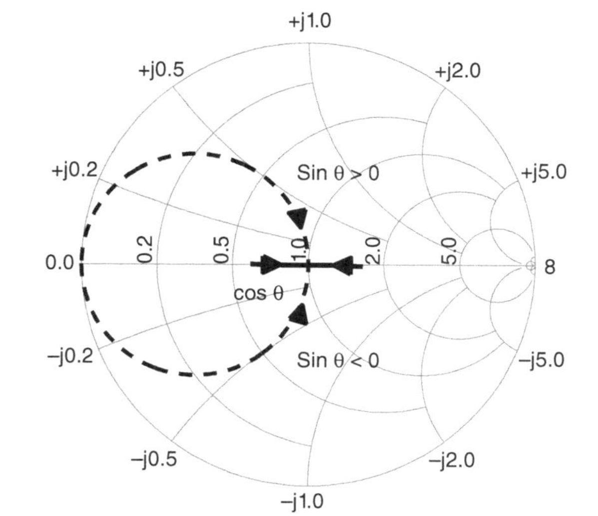

Figure 43.17 The impedance tuning actuator trajectories superimposed on a Smith Chart. (*Courtesy of David J. Coumou.*)

The impedance matching network has two variable capacitors, C_1 for the load element and C_2 for the tune element. The RF power supply has a high-speed digital connection to the matching network, updating command positions for C_1 and C_2 at a rate of 100 μS. The third impedance actuator is the frequency, f, of the RF power supply with a variation of ±5 percent. The plots were obtained from an oscilloscope coupled to a directional coupler positioned between the RF power supply and the matching network. Signal 1 shown in each plot is the forward power response. Signal 2 is the reverse power. The objective of the feedforward impedance controller is to minimize the reverse power to an optimal tune condition during the steady state portion of the pulse. For the results shown in Fig. 43.18*a*, the feedforward control was configured for C_1 and C_2 mode, adjusting only these elements during the pulse operation. The steady state reverse power is achieved in approximately 6 μS, or 50 percent of the on pulse time. With all three impedance actuators enabled (see Fig. 43.18*b*), the feedforward impedance control achieves a steady state condition in approximately 3 μS, or 25 percent of the on pulse time. The tri-actuator mode also reduces the reflected power to <1 W during the steady state, or by a factor 5 as compared to the C_1 and C_2 mode. The improvement related to the tri-actuator mode results from the actuation time of f, which is approximately 50 times faster than the update rate of the variable elements in the matching network and greater than the transition time from vacuum to plasma.

43.4.5 RF Mechanisms to Control Uniformity Effects

A processing challenge for ICP sources is spatial uniformity, a confluence of gas flow, pressure regulation, and RF field emissions from the top coil antenna and its coupling with the electrode bias. It is well known that these three entities can impair symmetrical flow and adversely impinge pattern etching across the wafer. In principle, the nonuniform electric field generates nonuniform density and impairs etch rate, thereby impinging a corresponding pattern on the wafer. Traditionally, this is referred to as the M-pattern effect. Industry has long countered this challenge by affecting the electric field with various approaches: (1) a permanent magnet assembly around the periphery of the substrate, (2) coil geometry arrangement, (3) RF source pulsing, (4) controlling the ratio of the coil currents, and (5) nonsimilar frequency assignments to the coils of the antenna structure. These techniques had various degrees of success, but have not been able to substantially impact the problem as a viable solution. In this section, we focus on a new approach by controlling the EM field emission from the coil antenna.

The electric field emission from a coil in the top coil antenna assembly toward the plasma source can be analytically described from a dipole emission as

$$E_{\phi} = \frac{I_x l \eta_o \beta^2 e^{-j\beta r}}{4\pi}((j\beta r)^{-3} + (j\beta r)^{-2} + (j\beta r)^{-1})\sin\phi \quad (43.26)$$

The electric field includes the electrostatic field, induction field, and radiation field terms. The main or primary energy absorption is Joule heating from the induction field, $(j\beta r)^{-2}$, which accelerates electrons. For a dual coil driven antenna, the amplitude and phase of the driven coil current, I_x, can be controlled respectively by the set point of the RF power supply and a digital frequency-phase locking circuit [83] coupling the frequency and corresponding phase, θ, output. For the inner coil, the RF current is configured to be controlled as $I_i(t) = |I_i| \sin(\omega t)$. For the outer coil, the RF current and phase is controlled as $I_o(t) = |I_o| \sin(\omega t + \theta)$. Previously, EM field simulations showed by amplitude and relative phase manipulation of the RF excitation for the source coil arrangement, the controllability of the constructive–nonconstructive interaction of the coil fields to positively affect plasma density and associated uniformity along the wafer area [89]. With constant current for $|I_i|$ and $|I_o|$, simulated EM field strength above the substrate is shown in Fig. 43.19*a*. By visual inspection, the uniformity of the electric field strength along the radius of the substrate improves as θ approaches 0 degree, and the inner to outer electric fields are nonconstructive as θ increases, resulting in a spatial bimodal emission.

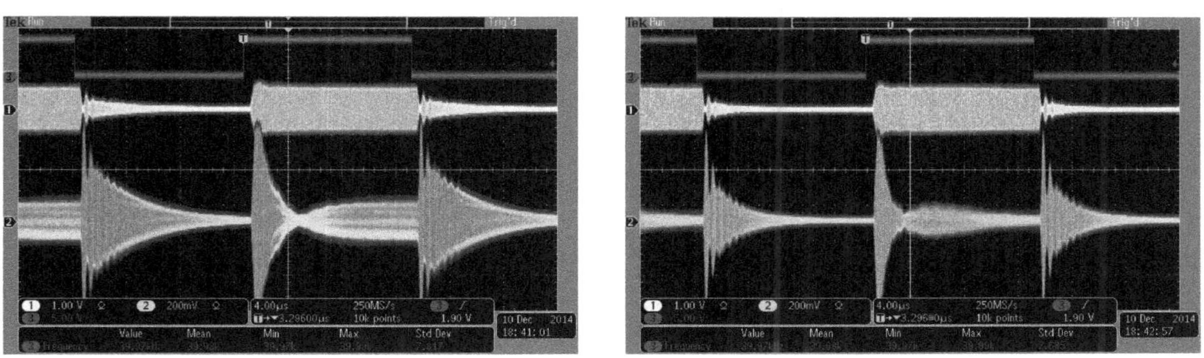

(a) (b)

Figure 43.18 The Feedforward impedance control for 40 KHz pulse frequency, 50 percent duty cycle: (a) C_1 and C_2 actuator control; (b) tri-actuator control (C_1, C_2, and f). (*Courtesy of David J. Coumou.*)

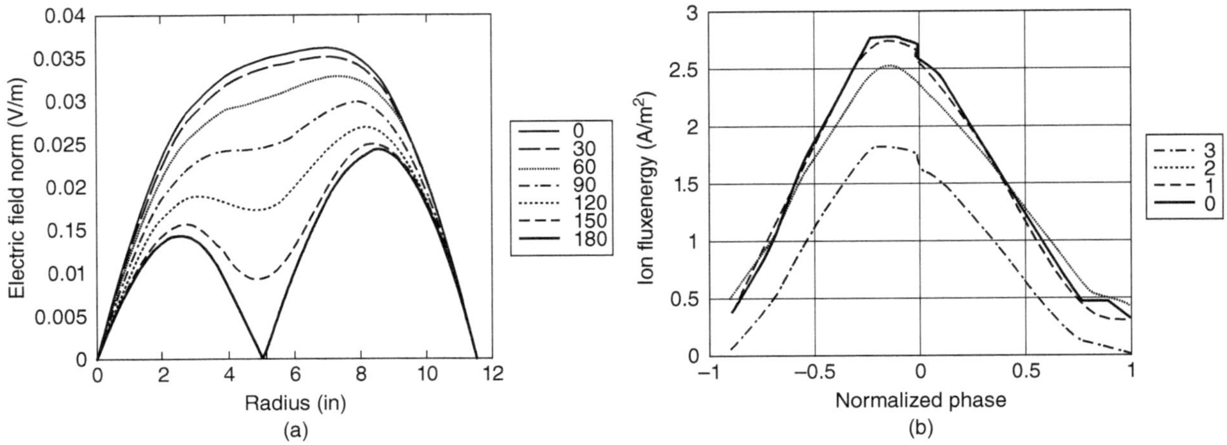

Figure 43.19 RF field driven plasma uniformity: (*a*) comparison of simulated inner to outer coil phase, θ, variation with constant coil current parity; (*b*) ion flux measurements along the substrate radius for varying θ constant coil current. (*Courtesy of David J. Coumou.*)

Figure 43.19*b*, ion flux measurements corresponding to the simulated field emissions antenna configuration are shown. These results were acquired using retarding field energy analyzer sensors at four positions along the radius. These measurements were taken for a constant RF power applied to the bias. In increasing order, the figure legend indicates the flux measurements relating to equidistant RFEA sensor positioning along the radius. The zeroth sensor is positioned at the center of the antenna coil assembly and the sensor labeled 3 in the legend is located at the substrate edge. It is important to note that in Fig. 43.19, the phase angles have different reference planes. For the E-field simulation, the phase angle is with respect to the relative phase of the inner and outer coil structures. For the ion flux measurements, the phase angle is controlled by the digital frequency-phase controller in the outer coil RF power supply. The ion flux measurements indicate a decreasing ion flux trend as the phase angle approaches an absolute maximum. This is attributed to the deconstruction effects of the electric field emissions from the inner and outer coil. As the phase angle approaches zero, the ion flux reaches a maximum for all sensors. The ion flux offset by the outer two sensors relative to the inner sensors results from lower current drive to the outer coil. This scheme provides a new pathway to control plasma generation and corresponding uniformity in an ICP source.

43.5 PROPERTIES OF DRY ETCH PROCESS

There are two types of methods of etching: chemical and physical. These combine to form four basic plasma processes: sputtering or physical, chemical, ion-energy-driven, and ion-enhanced inhibitor etching. Chemical etch is performed by free radicals that bombard the target with a near uniform angular distribution. This process is characterized by high etch rates and occurs in high-pressure conditions. Physical etch is performed by the sputtering of ions and is an anisotropic process. Anisotropy is defined as the ratio of the vertical etch to the horizontal etch. With this process only a marginal amount of sidewall removal occurs. This is the only etch process that can remove nonvolatile material from the target. As compared to chemical etch, the selectivity for physical etch is low. Selectivity is the ratio of the etch rate of the insulator to the etch rate of the semiconductor [16].

43.6 FUTURE TRENDS AND CONCLUSIONS

Radio frequency (RF) plasma generation and application is a critically important technology that is broadly applied throughout semiconductor device manufacturing. Processes that employ RF excitation include PVD, plasma etch and clean and plasma-enhanced chemical vapor deposition (PECVD). Looking toward the future, we first reflect on the past. In the 1980s, the semiconductor industry relied on single-frequency, capacitively coupled plasmas for manufacturing. The 1990s provided an array of technology choices with variants of ECR and ICP chambers. During this time all major etch companies were moving toward high-density plasma chambers. Presently, the CCP chamber appears to be the industry mainstay, with ICP sources increasing their share in etching.

The advances in etch capability of feature-size, selectivity, and critical dimension control are tremendous. The industry continues to be challenged with shrinking feature sizes. The present manufacturing capability of height-to-depth ratios ranges from 20:1 to as much as 60:1, but enhancements continue to improve with eventual adoption into high-volume processes. In this regard, etching lags deposition substantially. For instance, atomic deposition of nitride films is achieved with RF enabling technologies such as RF pulsing combined with the frequency tuning scheme presented in Sec. 43.4.4.

These technologies facilitated industry recognition of Lam Research's Vector tool for achieving process tool of record for PEALD. In comparison, mainstream adoption of RF pulsing for etch has been slower, possibly due to the many challenges commensurate with the benefits. With subtly, the advent of the industry acknowledging RF pulsing to achieve next-generation device fabrication capability, arbitrary waveforms for tailoring ion energies was comprehended. In Fig. 43.20, simulation results provide a comparison between a dual frequency scheme using the phase-locked, harmonic drive (reference Sec. 43.4.3) and single frequency driven bias at 0.4 and 3.2 MHz. As the bias frequency decreases, the ion energies increase as well as the width of associated distribution. The plot for 0.4 MHz achieves a higher set of ion energies, but the percent of the distribution is considerably broader in comparison to the density function for 3.2 MHz. The industry is preferential for a monoenergetic distribution given by the dual frequency plot of 13.56 and 27.12 MHz. With the right phase set point for the phase-locked, harmonic drive, the monoenergetic distribution is indicated by the solid line. With a preponderance of dual-frequency systems established for plasma processing systems, the industry embracing such a scheme is inevitable and will give way to other arbitrary waveform generating methods to tailor IEDF. In this manner, it is conceivable with greater acceptance of RF pulsing with customized pulse shapes, a pathway for advanced RF power delivery schemes will follow, lending greater economically driven performance for next-generation plasma processing to achieve future manufacturing capability.

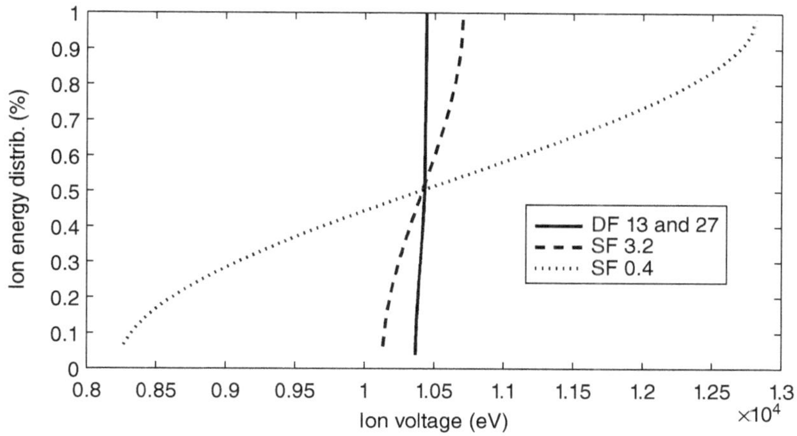

Figure 43.20 Cumulative density function of ion distribution for phase-locked, harmonic drive (DF) in comparison with single frequency (SF) 3.2 MHz and 0.4 MHz. (*Courtesy of David J. Coumou.*)

43.7 REFERENCES

1. W. Hittorf, *Wiedemann Ann Physics*, 21: 90–139, 1884.
2. K. A. MacKinnon, "On the Origin of the Electrodeless Discharge," *The London, Edinburgh, and Dublin Philosophical Magazine and Journal of Science*, 8: 605–616, 1929.
3. J. T. Gudmundsson and M. A. Lieberman, "Magnetic Induction and Plasma Impedance in a Cylindrical Inductive Discharge," *Plasma Sources Science Technology*, 6: 540–550, 1997.
4. J. J. Thomson, "On the Discharge of Electricity through Exhausted Tubes without Electrodes," *Philosophical Magazine*, 32: 321–336, 1891.
5. J. J. Thomson, "On the Discharge of Electricity through Exhausted Tubes without Electrodes," *Philosophical Magazine*, 32: 445–464, 1891.
6. J. Tykocinski-Tykociner, "Measurement of Current in Electrodeless Discharge by means of Frequency Variations," *Philosophical Magazine*, 13: 953–963, 1932.
7. R. B. Piejak, V. A. Godyak, and B. M. Alexandrovich, "A Simple Analysis of an Inductive RF Discharge," *Plasma Sources Science Technology*, 1: 179–186, 1992.
8. J. H. Keller, "Inductive Plasmas for Plasma Processing," *Plasma Sources Science Technology*, 5: 166–172, 1996.
9. I. M. El-Fayoumi and I. R. Jones, "The Electromagnetic Basis of the Transformer Model for an Inductively Coupled RF Plasma Source," *Plasma Sources Science Technology*, 7: 179–185, 1998.
10. P. Colpo, R. Ernst, and F. Rossi, "Determination of the Equivalent Circuit of Inductively Coupled Plasma Sources," *Journal of Applied Physics*, 85 (3): 1366–1371, 1999.
11. V. A. Godyak, R. B. Piejak, and B. M. Alexandrovich, "Electrical Characteristics and Electron Heating Mechanism of an Inductively Coupled Argon Discharge," *Plasma Sources Science Technology*, 3: 169–176, 1994.
12. M. M. Turner and M. A. Lieberman, "Hysteresis and the E to H Transition in Radiofrequency Inductive Discharge," *Plasma Sources Science Technology*, 8, 313–324, 1999.
13. R. B. Piejak, V. A. Godyak, and B. M. Alexandrovich, "The Electric Field and Current Density in a Low Pressure Inductive Discharge Measure with Different B-Dot Probes," *Journal of Applied Physics*, 81 (8): 3416–3421, 1997.
14. I. M. El-Fayoumi and I. R. Jones, "Measurement of the Induced Plasma Current in a Planar Coil, Low Frequency, RF Induction Plasma Source," *Plasma Sources Science Technology*, 6: 201–211, 1997.
15. W. M. Holber, *Handbook of Ion Beam Technology: Principles, Deposition, Film Modification and Synthesis*. Noyes, Park Ridge, New Jersey, p. 21, 1989.
16. M. A. Lieberman and A. J. Lichtenberg, *Principles of Plasma Discharges and Materials Processing*. Wiley, New York, 1994.
17. P. J. Hargis, Jr., K. E. Greenberg, P. A. Miller, J. B. Gerardo, J. Ft. Torczynski, M. E. Riley, G. A. Hebner et al., "The Gaseous Electronics Conference Radio Frequency Reference Cell: A Defined Parallel Plate Radio System for Experimental and Theoretical Studies of Plasma Processing," *The Review of Scientific Instruments*, 65 (1): 140–154, 1994.
18. M. A. Sobolewski, "Current and Voltage Measurements in the Gaseous Electronics Conference RF Reference Cell," *Journal of Research of the National Institute of Standards and Technology*, 100 (4): 341–351, 1995.
19. S. M. Bobbio, "A Review of Magnetron Etch Technology," *Proceedings of SPIE International Society of Optical Engineering*, 1185: 262–277, 1989.
20. J. Arami, "Plasma Process Device," U.S. Patent 6014943, 2000.
21. H. Shan, R. Lindley, C. B., Xue Yu Qian, R. Plavidal, B. Pu, Ji Ding, Z. Li, "Magnetically-Enhanced Plasma Chamber with Non-Uniform Magnetic Field," U.S. Patent 6113731, 2000.
22. W. D. Wang, P. Gopalraja, J. Fu, "Auxiliary Electromagnets in a Magnetron Sputter Reactor," U.S. Patent 6730196, 2004.
23. A. Subramani, U. Kelkar, J. Fu, P. Gopalraja, "Magnetron with a Rotating Center Magnet for a Vault Shaped Sputtering Target," U.S. Patent 6406599, 2002.
24. W. D. Wang, "Coaxial Electromagnet in a Magnetron Sputtering Reactor," U.S. Patent 6352629, 2002.
25. P. Ding, "Magnet Array in Conjunction with Rotating Magnetron for Plasma Sputtering," U.S. Patent 6610184, 2003.
26. J. Brcka, "Ring-Shaped High-Density Plasma Source and Method," U.S. Patent 6523493, 2003.
27. P. Chabert, N. S. Braithwaite, *Physics of Radio-Frequency Plasmas*. Cambridge University Press, New York, p. 231, 2011.
28. D. J. Hoffman and E. R. Gold, "Plasma Reactor Control by Translating Desired Values of M Plasma Parameters to Values of N Chamber Parameters," U.S. Patent 7901952, 2011.
29. A. R. Keane and S. E. Hauer, "Automatic Impedance Matching Apparatus and Method," U.S. Patent 5195045, 1993.
30. S. Harnett, "Fuzzy Logic Tuning of RF Matching Network," U.S. Patent 5842154, 1997.
31. R. W. Brounley, "Solid State Plasma Chamber Tuner," U.S. Patent 5473291, 1995.
32. J. Wilbur, "Ratiometric Autotuning Algorithm for RF Plasma Generator," U.S. Patent 6020794, 2000.

33. A. J. Miranda and C. J. Spanos, "Impedance Modeling of a Cl_2/He Plasma Discharge for a Very Large Scale Integrated Circuit Production Monitoring," *JVST A*, 14 (3): 1888–1893, 1996.

34. B. Andries, G. Ravel, and L. Peccoud, "Electrical Characterization of Radio-Frequency Parallel-Plate Capacitively Coupled Discharges," *JVST A*, 7 (4): 2774–2783, 1989.

35. V. A. Godyak and R. B. Piejak, "In Situ Simultaneous Radio Frequency Discharge Power Measurements," *JVST A*, 8 (5): 3833–3837, 1990.

36. N. St. J. Braithwaite, "Internal and External Electrical Diagnostics of RF Plasmas," *Plasma Sources Science Technology*, 6: 133–139, 1997.

37. H. Kawata, T. Kubo, M. Yasuda, K. Murata, "Power Measurements for Radio-Frequency Discharges with a Parallel-Plate-Type Reactor," *Journal of the Electrochemical Society*, 145 (5): 1701–1707, 1998.

38. K. L. Steffens and M. A. Sobolewski, "Planar Laser-Induced Fluorescence of CF_2 in O_2/CF_4 and O_2/C_2F_6 Chamber Cleaning Plasmas: Spatial Uniformity and Comparison to Electrical Measurements," *JVST A*, 17 (2): 517–527, 1999.

39. V. J. Law, A. J. Kenyon, N. F. Thornhill, V. Srigengan, I. Batty, "Remote-Coupled Sensing of Plasma Harmonics and Process End Point Detection," *Surface Engineering, Surface Instrumentation and Vacuum Technology*, 57: 351–364, 2000.

40. D. J. Coumou, "Advanced RF Metrology for Plasma Process Control," *Semiconductor International*, 2003.

41. K. S. Gerrish and D. F. Vona, Jr., "Baseband V-I Probe," U.S. Patent 5770922, 1998.

42. D. J. Coumou, "RF Power Probe Head with a Thermally Conductive Bushing," U.S. Patent 6559650, 2003.

43. T. Heckleman, D. J. Coumou, Y. K. Chawla, "Orthogonal Radio Frequency Voltage/Current Sensor with High Dynamic Range," U.S. Patent 8040141, 2011.

44. D. J. Coumou, et al., "On the Enhancements of Planar Based RF Sensor Technology," U.S. Patent 2014/0049250, 2014.

45. K. Ukai and K. Hanazawa, "End-Point Determination of Aluminum Reactive Ion Etching by Discharge Impedance Monitoring," *Journal Vacuum Science Technology*, 16 (2), 385–387, 1979.

46. M. Kanoh, M. Yamage, and H. Takada, "End-Point Detection of Reactive Ion Etching by Plasma Impedance Monitoring," *Japan Journal of Applied Physics*, 40 (3a): 1457–1462, 2001.

47. S. Bushman, T. F. Edgar, and I. Trachtenberg, "Radio Frequency Diagnostics for Plasma Etch Systems," *Journal of the Electrochemical Society*, 144 (2): 721–731, 1997.

48. E. Hanson, H. Benson-Woodward, and M. Bonner, "Optimising CVD through RF Metrology," *Plasma Monitoring*, 25–28, 1999.

49. D. W. Zhao and C. Spanos, "Towards a Complete Plasma Diagnostic System," *IEEE Conference Proceedings for the International Symposium on Semiconductor Manufacturing*, 137–140, 2001.

50. C. Schneider, L. Pfitzner, and H. Ryssel, "Integrated Metrology: An Enabler for Advanced Process Control (APC)," *Proceedings of SPIE*, 4406: 118–130, 2001.

51. A. T.-C. Koh, N. F. Thornhill, and V. J. Law, "Principal Component Analysis of Plasma Harmonics in Endpoint Detection of Photoresist Stripping," *IEEE Electronic Letters*, 35 (16): 1383–1385, 1999.

52. H. M. Mott-Smith and I. Langmuir, "The Theory of Collectors in Gaseous Discharges," *Physical Review*, 28: 727–763, 1926.

53. M. V. Malyshev, V. M. Donnelly, A. Kornbilt, N. A. Ciampa, J. I. Colonell, J. T. C. Lee, "Langmuir Probe Studies of a Transformer Coupled Plasma, Aluminum Etcher," *JVST A*, 17 (2): 480–492, 1999.

54. R. Murete de Castro, et al., "End-Point Detection of Polymer Etching Using Langmuir Probes," *IEEE Transactions Plasma Sciences*, 28 (3): 1043–1049, 2000.

55. T. L. Thomas and E. L. Battle, "Effects of Contamination on Langmuir Probe Measurements in Glow Discharge Plasmas," *Journal of Applied Physics*, 41 (8): 3428–3432.

56. E. P. Szuszczewicz and J. C. Holmes, "Surface Contamination of Active Electrodes in Plasmas: Distortion of Conventional Langmuir Probe Measurements," *Journal of Applied Physics*, 46 (12): 5134–5139, 1975.

57. P. Biolsi, D. Morvay, L. Drachnik, and S. Ellinger, "An Advanced Endpoint Detection Solution for <1% Open Areas," *Solid State Technology*, 39 (12): 59–67, 1996.

58. E. A. Hudson and F. C. Dassapa, "Sensitive End-Point Detection for Dielectric Etch," *Journal of the Electrochemical Society*, 148 (3): C236–C239, 2001.

59. D. Gahan, B. Dolinaj, and M. B. Hopkins, "Retarding Field Analyzer for Ion Energy Distribution Measurements at a Radio-Frequency Biased Electrode," *Review of Scientific Instruments*, 79 (3): 033502, 2008.

60. D. J. Coumou and E. Choueiry, "Reliable Arc Detection and Arc Mitigation in RF Plasma Systems," *Proceedings of APC/AEC Symposium XX*, October 2011.

61. D. J. Coumou, "Application of Wideband Sampling for Arc Detection with a Probabilistic Model for Quantitatively Measuring Arc Events," U.S. Patent, 8289029, 2012.

62. M. S. Barnes, J. C. Forster, and J. H. Keller, "Ion Kinetics in Low Pressure, Electropositive, RF Glow Discharge Sheaths," *IEEE Transactions on Plasma Science*, 19 (2): 240–244, 1991.

63. I. C. Abraham, et al., "Ion Energy Distributions versus Frequency and Ion Mass at the RF-Biased Electrode in an Inductively Driven Discharge," *JVST A*, 20 (5): 1759–1768, 2002.

64. P. C. Boyle, A. R. Ellingboe, and M. M. Turner, "Independent Control of Ion Current and Ion Impact Energy onto Electrodes in Dual Frequency Plasma Devices," *Journal of Physics D: Applied Physics*, 37: 697–701, 2004.

65. S. C. Shannon, D. J. Hoffman, J-G. Yang, A. Paterson, and J. Holland, "The Impact of Frequency Mixing on Sheath Properties: Ion Energy Distribution and Vdc/Vrf Interaction," *Journal of Applied Physics*, 97 (10): 103304, 2005.

66. M. T. Rahman, M. N. A. Dewan, M. R. H. Chowdhury, "Modeling of a Collisional Triple Frequency Capacitively Coupled Plasma Sheath," *Proceedings of 7th ICECE*, pp. 721–724, December 2012.

67. S. Banna, A. Agarwal, K. Tokashiki, H. Cho, S. Rauf, V. Todorow, K. Ramaswamy, et al., "Inductively-Coupled Pulsed Plasmas in the Presence of Synchronous Pulsed Substrate Bias for Robust, Reliable and Fine Conductor Etching," *IEEE Transactions on Plasma Science*, 37 (9): 1730–1746, 2009.

68. S. Banna, et al., "Pulsed High-Density Plasmas for Advanced Dry Etching Processes," *JVST A*, 30 (4): 1–29, 2012.

69. S. Samukaway and T. Mienoz, "Pulse-Time Modulated Plasma Discharge for Highly Selective, Highly Anisotropic and Charge-Free Etching," *Plasma Sources Science and Technology*, 5: 132–138, 1996.

70. W. L. Johnson and E. J. Strang, "Pulsed Plasma Processing Method and Apparatus," U.S. Patent, 7166233, 2007.

71. N. Hershkowitz, "Role of Plasma-Aided Manufacturing in Semiconductor Fabrication," *IEEE Transactions on Plasma Science*, 26 (6): 1610–1620, 1998.

72. S. E. Savas, "Apparatus and Method for Pulsed Plasma Processing of a Semiconductor Substrate," U.S. Patent, 6395641, 2002.

73. L. Dubost, et al., "Low Temp Pulsed Etching of Large Glass Substrates," *JVST A*, 21 (4): 892–894, 2003.

74. T. C. Chua, "Plasma Gate Oxidation Process Using Pulsed RF Source Power," U.S. Patent, 7214628, 2007.

75. J. Forster, Praburam Gopalraja, Liubo Hong, Bradley O. Stimson, "Pulsed-Mode RF Bias for Side-Wall Coverage Improvement," U.S. Patent 6673724, 2004.

76. K. G. Donohoe and G. S. Sandhu, "Method for Pulsed-Plasma Enhanced Vapor Deposition," U.S. Patent 5985375, 1999.

77. S. Swaminathan, D. M. Hausmann, Jon Henri, K. K. Kattige, A. J. McKerrow, V. Rangarajan, M. Sriram, P. Subramonium, S. Swaminathan, S. B. J. van, "Plasma Activated Conformal Dielectric Film Deposition," U.S. Patent 8637411, 2014.

78. B. G. Heil, U. Czarnetzki, R. P. Brinkmann, and T. Mussenbrock, "On the Possibility of Making a Geometrically Symmetric RF-CCP Discharge Electrically Asymmetric," *Journal of Physics D: Applied Physics*, 41: 165202, 2008.

79. S.-B. Wang and A. E. Wendt, "Ion Bombardment Energy and SiO_2/Si Fluorocarbon Plasma Etch Selectivity," *JVST A*, 19 (5): 2425–2432, 2001.

80. X. V. Qin, Y. -H. Ting, and A. E. Wendt, "Tailored Ion Energy Distributions at an RF-Biased Plasma Electrode," *Plasma Sources Science Technology*, 19 (6): 065014, 2010.

81. D. J. Coumou, D. H. Clark, T. Kummerer, M. Hopkins, D. Sullivan, and S. C. Shannon, "Ion Energy Distribution Skew Control Using Phase Locked Harmonic RF Bias Drive," *IEEE Transactions on Plasma Science*, 42 (7): 1880–1893, 2014.

82. E. Kawamura, V. Vahedi, M. A. Lieberman, and C. K. Birdsall, "Ion Energy Distributions in RF Sheaths; Review, Analysis and Simulation," *Plasma Sources Science Technology*, 8 (3): R45, 1999.

83. D. J. Coumou, "Phase and Frequency Control of a Radio Frequency Generator from an External Source," U.S. Patent 7602127, 2009.

84. E. Schüngel, E. Schüngel, S. Mohr, J. Schulze, U. Czarnetzki, and M. J. Kushner, "Ion Distribution Functions at the Electrodes of Capacitively Coupled High-Pressure Hydrogen Discharges," *Plasma Sources Science Technology*, 23: 015001, 2014.

85. Y. Zhang and M. J. Kushner, "Control of Ion Energy Distributions through the Phase Difference between Multiple Frequencies in Capacitively Coupled Plasmas," *Proceedings of the 67th Gaseous Electronic Conference*, 2014, Raleigh, NC.

86. D. J. Coumou, "Power Distortion-Based Servo Control Systems for Frequency Tuning RF Power Sources," U.S. Patent 8576013, 2013.

87. D. J. Coumou, "Coherent Feedforward Impedance Correction and Feedback Power Regulation in a Plasma Processing RF Power Delivery System," *Proceedings IEEE Control and Modeling for Power Electronics*, 1–4, 2013.

88. D. J. Coumou and L. J. Fisk, "Distortion Correction Based Feedforward Control Systems and Methods for Radio Frequency Power Sources," U.S. Patent 8781415, 2014.

89. D. J. Coumou, D. M. Brown, and S. C. Shannon, "Uniformity Control with Phase-Locked RF Source on a High Density Plasma System," *Proceedings 41th IEEE International Conference on Plasma Sciences*, May 2014, Washington, DC.

Chapter 44

IC Manufacturing Equipment Parts Cleaning Technology: Fundamentals and Applications

BY

ARDESHIR J. SIDHWA AND DAVE ZUCK *Quantum Global Technologies, LLC, Quakertown, Pennsylvania*

44.1 A HISTORIC PERSPECTIVE ON OUTSOURCED PART CLEANING

"Parts" are the tens of thousands of tool and chamber components, inside the thousands of process tools that become contaminated during the semiconductor manufacturing process. The process contamination results in yield loss and ultimately process failure, and the parts must all be removed and replaced by a cleaned part or replaced with a new part. Tool parts manufacturers include all major equipment original equipment manufacturers (OEMs), including Applied Materials, Lam Research, Tokyo Electron, ASML, ASM, and many others.

Before the 1990s many semiconductor manufacturers performed part cleaning activities "in-house" utilizing their own equipment and personnel. As the part cleaning process complexity increased (i.e., increasing numbers of substrate/deposition combinations), as part cleanliness requirements tightened and as fab space and facilities costs (e.g., deionized water, HVAC, waste treatment, etc.) became more expensive to build and operate, semiconductor manufacturers realized that in-house part cleaning was not a core competency. Fully loaded/fully burdened cost of ownership models that included labor, chemical, utilities, supplies, waste disposal, fab space, part performance, shortened part life/increased part purchases, etc., indicated that in-house part cleaning was not cost-effective compared to outsourced part cleaning. Semiconductor manufacturers realized that the technologies of tool part cleaning and the technology of manufacturing wafers were completely different sciences, requiring completely different skill sets.

As a result of the growing demand and opportunity for outsourced part cleaning (OPC) services, a new industry emerged. In the 1990s, among the first part clean companies established in this new market space were Shield Care and Dage (Scotland), Kachina (United States), KoMico (Korea), and CleanPart (Europe). By 2000, more than 100 part clean service providers were operating across all global semiconductor regions. Many of initial OPC providers were typically undercapitalized, capacity limited, utilized low technology methods, performed very little research and development, and were limited by regionalized supply chains. Most of the initial part clean providers did not fit the profile of other large established suppliers to the industry (e.g., tool OEMs, gas, chemical, and silicon suppliers). In fact, part cleaning had missed the "era of ultraclean" that began in the 1990s that forced all fab material suppliers to provide products with increasing levels of cleanliness supported by certificates of analysis.

Initial standards for recycled parts clean were largely based on visual inspection/cosmetic appearance. For example, in the 1990s and even to this day, tool owners say "The part looks great, let's run it in the tool!" As a result, the tool itself was and sometimes still is used for parts cleanliness validation. Test wafers and analytical methods such as KLA Tencor, total reflection x-ray fluorescence (TXRF), vapor-phase decomposition (VPD), and residual gas analysis (RGA) are performed on wafers and "in-line" on process equipment as indirect measurements of cleanliness of newly installed parts. Numerous conditioning wafers are still routinely used for chamber seasoning, running chambers until wafers meet the required level of particle and/or trace metal cleanliness. This practice not only puts the tools at risk but increases recovery time and decreases actual uptime, adding real cost in terms of time and expense.

By the early 2000s, major tool OEMs recognized that wafer yield and tool performance is dependent on controlling new part trace metal and particle cleanliness, and new part cleanliness specifications were established.

44.1.1 Introduction

Meeting defectivity requirements becomes increasingly challenging as technology nodes continue to scale down. Foreign particles and residues create killer defects at the component level and are expected to severely impact product yields, especially at sub-20-nm technology nodes. Additional complexity in the form of new processes and materials will further challenge component-level design and performance. Deepening technological requirements and the worldwide acceptance of sophisticated electronic devices have created an unprecedented demand for complex large-scale integrated circuits. Meeting these demands has required technological advances in materials and processing equipment. Significant increases in the number of individuals involved in integrated circuit design and an increased emphasis on effectively utilizing the computer and other highly sophisticated equipment aid the design and analysis of the manufacturing process.[1,2] The microscopic dimensional capabilities of current semiconductor manufacturing equipment facilitate the design of digital circuits which are very complex, yet extremely economical in space, power requirements, and cost, with outstanding processing speed. At the same time, however, these same dimensional capabilities make it possible for much smaller particles to contaminate and ruin an extended run of wafers before the contamination is discovered, causing major economic loss. It is therefore critical that there is a method to identify exactly what is causing those defects in the wafer manufacturing process. To avoid extensive economic loss, it is imperative to discover these defects in real time and as early as possible.

In this chapter, some of the major challenges and solutions toward contamination-free manufacturing (CFM) in a cost-effective manner are discussed. This chapter presents solutions and tips including coatings with superior adhesion properties and physical characteristics used to enhance chamber performance in PVD and etch process areas to address defectivity issues.

The character of the semiconductor industry is such that competition requires products that are designed, manufactured, and marketed in the most efficient manner possible. This has required improvements in fabrication technology that keep pace with the rapid improvements in the electronics industry.

44.2 PAST, CURRENT, AND FUTURE TECHNOLOGIES/APPLICATIONS

Parts cleaning requirements for sub-20-nm technology nodes have changed drastically. High-volume manufacturing semiconductor industries urge that with the challenges the industry faces from upcoming technologies, "things that didn't matter now do."[2] In the past, parts cleaning operations missed the ultraclean revolution as explained in Sec. 44.1. For geometries greater than 65 nanometers, parts that worked well were not necessarily "clean." At 32-nm and lower technology nodes, things that did not matter now do and this is particularly true with respect to parts cleaning.

Today, we are getting into an era of the so-called Internet of Things (IoT) or Internet of Everything (IoE). IoT and wearable technologies using MEMS microcontrollers, memory chip technologies, etc., are dominating today's market. By the year 2020, the wearable market is projected to become a $20 billion market.[3] The integration market with respect to wearables will also approach $100 billion by 2020.[3] The future goal is to make the wearable products seamlessly integrated with our human bodies. Other examples for current and future applications are autonomous transportation where cars will drive without human intervention, in-field crop inspection using autonomous drone technology and even having the equivalent of a medical center at your home, that is, you do not have to go to hospitals for an emergency—the hospital comes to you.

The common thread to supporting all current and future IoT and IoE applications is the microchip. A cleaning technology company's role in developing and manufacturing the microchip is indirect as the company "does not make microchips, but it does make the microchip manufacturing processes themselves run better."[3] This is accomplished by improving tool productivity through improving mean time between cleans (MTBC), defect reduction, and decreasing the cost of ownership for customers. A thorough understanding of manufacturing capabilities and process integration is essential for continued success in the parts cleaning business. Component parts are cleaned today at the same submicroscopic scale as wafers. Figure 44.1 shows the trace metal contamination and particle size limitations for different line geometries, especially for sub-28-nm technology nodes. Not all parts require the same detail but surface cleanliness is important to achieve higher wafer die yields and improvement in the MTBC performance, regardless of feature size.[2]

44.3 EQUIPMENT PARTS CLEANING TECHNOLOGY FUNDAMENTALS AND APPLICATIONS

Today's semiconductor industry is going through a paradigm shift, moving from 2D to 2.5D and 3D structures to meet the demands of Moore's law.[2]

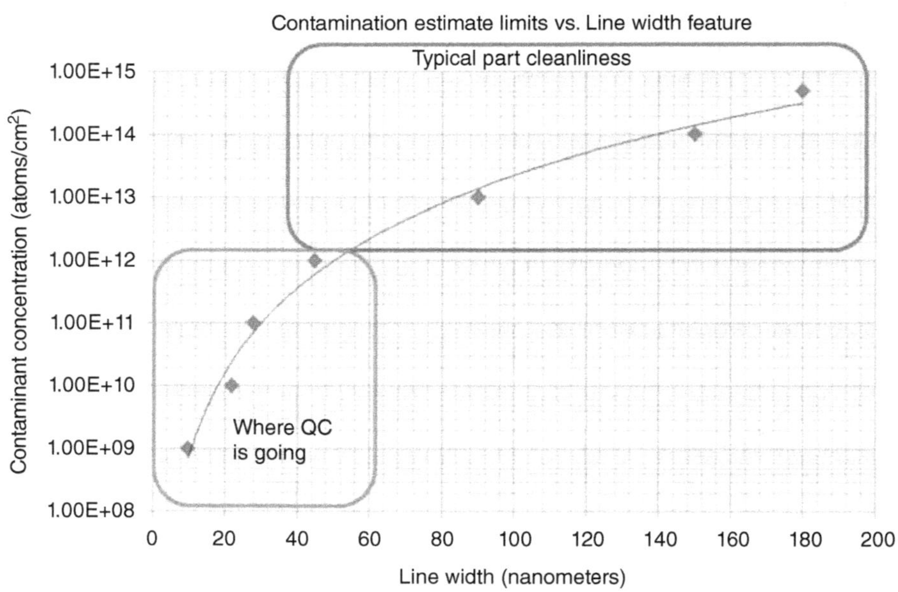

Figure 44.1 Trace metal contamination limits.

Moore's law mandates that the number of transistors per square inch on integrated circuits have to double approximately every two years since the invention of integrated circuits. Challenges presented by limitations in photolithography have caused the cost of transistors below 28 nm to increase, rather than decrease in order to measure up to Moore's law. Additional pattern and etch steps have been introduced to processes in order to produce smaller geometries. Semiconductor manufacturers have begun investigating other areas to decrease cost of ownership. One such critical area is in process chamber parts cleaning. The challenge today for semiconductor parts cleaning industry is identifying the relationship of part measurements to defects, and then providing improvements based on process control of these attributes. This involves huge investment, understanding of semiconductor applications and manufacturing processes. Table 44.1 shows the International Technology Roadmap for Semiconductors (ITRS) depicting production year of logic industry nodes, number of metallization layers, and wafer diameter. The ITRS roadmap also brings focus to ultrapure water and liquid process chemistries.[2] Unfortunately, the ITRS roadmap does not focus on particle sources and the sources or requirements of chemical contaminants, including trace metals and ionic species, for cleaned parts. Concentrating on contamination sources and minimization with respect to cleaned chamber components is very important.

44.3.1 Part Clean Processes and Technologies by Fab Module

Regardless of the technology node and/or device type being manufactured (logic, memory, analog, power, etc.), similar fab modules in all fabs have similar part cleaning/deposition removal needs for areas like PVD, CVD, etch, diffusion, EPI, lithography, and implant. Part cleaning processes and technologies required within each module vary significantly from each other and are dependent on module specific deposition and part substrates.

1. *PVD module parts (physical vapor deposition)*

   Part types. Stainless steel, aluminum, titanium shields, and parts.

   Typical deposition. Metals including Al, Cu, Ta, Ti, TiN, W, etc.

   Deposition removal/clean methods. Metal removal with inorganic chemistries including HF mixtures, hydrogen peroxide/ammonia mixtures, nitric acid mixtures, KOH, etc., metal passivation is performed with HF-containing passivation "bright dips," shield roughening utilizing various media bead blast methods, particle removal by ultrasonication, bake, and cleanroom bagging.

   Challenges. Removing deposition from aluminum and stainless steel shielding without over-etching metal substrates/reducing part life/changing part dimensions.

Table 44.1 Depicts a Roadmap Example by ITRS from 2013 through 2025[2,3]

Year of Production	2013	2014	2015	2016	2017	2018	2019	2020	2021	2022	2023	2024	2025
Flash 1/2 Pitch (nm) (un-contacted Poly) (f)	18.0	17.0	15.0	14.2	13.0	11.9	11.9	11.9	11.9	11.9	11.9	11.9	11.9
Logic Industry Nodes Range Labeling (nm)	"16/14"	"16/14"	"11/10"	"11/10"	"8/7"	"8/7"	"6/5"	"6/5"	"4/3"	"4/3"	"3/2.5"	"3/2.5"	"2/1.5"
Number of metal layers (includes ground planes and passive devices)	13	13	13	14	14	14	14	14	15	15	15	15	15
Wafer diameter (mm)	300	300	300	300/450	300/450	300/450	300/450	300/450	300/450	450	450	450	450
Quantum Clean Cleaning Capabilities	–	–	–	Phoenix, Hillsboro, Carrollton, Fremont, Scarborough, Taiwan, Korea, Singapore, and UK Facilities 450 mm Capable									
Metrology: Real Particle Detection Limit (nm) (MPU)	13	11	10	9	8	7	6	6	6	6	6	6	6
Wafer Environmental Contamination Control (Critical Particle size (nm))	20.0	17.9	15.9	14.2	12.6	11.3	10.0	8.9	8.0	7.1	6.3	5.6	5.0
Ultrapure Water													
Total organic carbon (ppb) [22] for immersion photolithography	<1	<1	<1	<1	<1	<1	<1	<1	<1	<1	<1	<1	<1
Critical Organics as C (ppb)	<1	<1	<1	<1	<1	<1	<1	<1	<1	<1	<1	<1	<1
Non-Critical Organics as C (ppb)	<3	<3	<3	<3	<3	<3	<3	<3	<3	<3	<3	<3	<3
Total silica (ppb) as SiO_2	<0.3	<0.3	<0.3	<0.3	<0.3	<0.3	<0.3	<0.3	<0.3	<0.3	<0.3	<0.3	<0.3
Number of particles > critical particles size (#/L)	1000	1000	1000	1000	1000	1000	1000	1000	1000	1000	1000	1000	1000
Liquid Chemicals													
All cleaning chemistries (aqueous and solvent): numbers of particles/ml >0.065 μm	0.87	0.63	0.44	0.31	0.22	0.16	0.11	0.08	0.06	0.04	0.03	0.02	0.01
All cleaning chemistries (aqueous and solvent): numbers of particles/ml >0.04 μm	4	2.7	1.9	1.3	0.9	0.68	0.47	0.33	0.24	0.17	0.12	0.08	0.06
All cleaning chemistries (aqueous and solvent): numbers of particles/ml >0.02 μm	30	22	15	11	7.5	5.4	3.8	2.6	1.3	1.3	0.94	0.66	0.47
All cleaning chemistries (aqueous and solvent): numbers of particles/ml > critical particles' size	30	30	30	30	30	30	30	30	30	30	30	30	30

2. *CVD module parts (chemical vapor deposition)*
   Part types. Al showerheads, ceramic chamber parts, anodized aluminum/yttria coated parts, bulk yttria parts.
   Typical deposition. Metal films (Ti, W, etc.), silicon compounds, dielectric compounds.
   Deposition removal/clean methods. Inorganic chemistries including HF mixtures for metal layers, peroxide/ammonia mixtures, and nitric acid mixtures, etc., for film removal, abrasive media methods for film removal, high temperature kiln process for ceramic parts, ultrasonication particle removal, bake, and cleanroom packaging.
   Challenges. Removing deposition from metal parts and from showerhead holes without etching substrate materials/enlarging holes, removal of deposition from yttria surfaces without compromise of yttrium surfaces.
3. *Etch module parts*
   Part types. Yttria-coated gas diffusion plates (GDP), bulk yttria showerheads, anodized aluminum/yttria-coated shields, metal shields, silicon and quartz rings, anodized Al chambers.
   Typical deposition. Silicon etch polymers (containing F, C, Si) metal etch polymers containing (e.g., Al, Ti, BCl_3, Cl), and AlF_3.
   Deposition removal/clean methods. Solvents (acetone, fluorinated solvents) and abrasive media for polymer removals, inorganic chemistries including HF mixtures, peroxide/ammonia mixtures, nitric acid mixtures, etc., for silicon and quartz clean, high temperature kiln process for ceramic parts, ultrasonic particle removal, bake, and cleanroom bagging.
   Challenges. Removing deposition from yttria surfaces without compromising yttria surfaces and anodized surfaces.
4. *Diffusion module parts*
   Part types/substrates. Fragile quartz tubes, quartz and SiC liners, quartz and SiC boats, quartz and SiC pedestals, quartz and SiC Baffle wafers.
   Typical deposition. Silicon and silicon compounds (e.g., polysilicon, silicon oxides, silicon nitrides).
   Deposition removal/clean methods. Inorganic chemistries include HF and nitric acids, bake, and cleanroom packaging.
   Challenges. Minimizing quartz devitrification and chemical damage of quartz during chemical clean, preventing breakage of fragile parts during handling/cleaning.
5. *ALD module parts (atomic layer deposition)*
   Part types. Aluminum showerheads, stainless steel, titanium, aluminum substrate shielding, ceramic shielding, Al and SST Valves.
   Deposition. By-products of ALD precursors, hafnium, Al_2O_3, etc.
   Deposition removal/clean methods. Inorganic chemistries including HF, nitric acid, and abrasive media blast methods for film removal, high temperature bake for ceramics, ultrasonic particle removal, bake, and cleanroom bagging.
6. *EPI module parts (epitaxial deposition)*
   Part types. Predominantly quartz parts.
   Typical deposition. Silicon film compounds doped with P and N compounds.
   Deposition removal/clean methods. Inorganic acids including HF, nitric, etc. Clean methods are ultrasonication, bake, and cleanroom bagging.
   Challenges. Epi processes are very sensitive to metallic contamination and cross contamination of N and P type dopants. Complete segregation of dopant types is required in the cleaning process.
7. *Lithography module parts*
   Part types/substrates. PTFE coater cups, SST photoresist lids, and dispense piping.
   Typical deposition. Various photoresist residues (positive, negative, BARC, etc.).
   Deposition removal/clean methods. Solvents including acetone, NMP (N-methylpyrrolidone), inorganic chemistries including KOH. DI water rinse, bake, and cleanroom bagging.
   Challenges. Photoresist materials are "messy" and part clean areas are difficult to maintain.
8. *Implant module parts*
   Part types and substrates. Assemblies and parts substrates that include stainless steel, ceramics, and carbon materials.
   Typical deposition. Hazardous residues containing arsenic, phosphorous, boron, and others.
   Deposition/film removal. Organic chemistries including hydrogen peroxide film removal, abrasive media film removal, high temperature kiln processes for ceramics, DI rinse, bake, and bagging.
   Challenges. Doping elements such as arsenic, phosphorus, and boron will cause significant process issues if cross contamination occurs with non-implant parts. Implant part cleaning processes must be completely isolated from other part cleaning processes. Additionally, implant compounds are extremely toxic and potentially carcinogenic. Extreme caution, PPE engineering controls must be utilized in implant clean processes. Residual deposition not removed can be sufficient to initiate hydride detection systems in fabs, and/or present a risk to fab personnel.
9. *Chemical hazards associated with part cleaning*
   The numerous inorganic chemistries utilized in part cleaning are extremely hazardous acids, bases, and oxidizers. Other materials such as isopropyl alcohol and acetone are flammable materials and pose significant fire risk. Extreme caution must be exercised using these materials. Extensive employee training and personal protective equipment (PPE) including acid gloves, aprons, face shields, and personal respirators must be utilized to prevent personnel exposure. Additionally, significant engineering controls must be utilized including ventilation systems, fume scrubbers, explosion proof electrical components in solvent areas, etc., to protect personnel, the facility, and the environment from the hazardous conditions and substances.

44.3.2 Source of Contamination

The part clean process faces the challenge of removing scores of different contaminants from scores of potential sources, many of which are actually introduced in the cleaning process itself.[1,2] The following is a partial list that details key contaminants, and their possible sources, which must be removed during the cleaning process:

1. Process deposition, process trace metals, and environmental contaminants
2. Cleaning process, methods, chemistries
3. Part substrate composition
4. People and environment (cleanroom air)
5. Methods and metrology of drying and mitigation of outgassing
6. Cleanroom wipes, gloves, and handling
7. Bagging material and packaging

Process Deposition, Process Trace Metals, and Environmental Contaminants (on Incoming Parts)

Parts pulled manually from process tools, handled and packed in noncleanroom environments, bagged in nonsealed bags shipped over the road in required cases arrive at cleaning facilities with extreme levels of environmental contaminants (Ca, Na, K). Cleaning processes must be created to recognize and remove these contaminants.[2,4,5]

The primary goal of parts cleaning is to remove the numerous deposition compounds detailed above. A common postclean contaminant is the residual film from the process itself. Residual process deposition occurs when the chemistries are not characterized and/or optimized to fully remove deposited film stacks from the part. Bright light inspection, ultraviolet light inspection, and magnification inspection is often required to discern the existence of residual process deposition. Residual process deposition leads to failure on installation, particle failures, adhesion issues/early chamber particle failure. Even parts that are completely visually "clean" following the clean process often exhibit high trace metal contamination that is originally introduced during chamber processing. Trace metal contamination that is not removed during the cleaning process is quite disruptive to some wafer processes and can result in yield loss, early part failure or poor mean time between failures (MTBF).

Cleaning Process, Methods, Chemistries

A majority of cleaning processes use a variety of inorganic acids, solvents, and/or abrasive media blast techniques. While essential for removing process deposition, all of these materials actually add significant levels of contamination to the part during the part clean process.

Inorganic Acids Typical part clean chemistries as noted above include hydrofluoric acid, nitric acid, ammonia, hydrogen peroxide, hydrochloric acid, sulfuric acid, and others. All chemistries utilized in cleaning are actually contaminants that must be fully removed. Additionally, the by-products of the chemical reactions between the deposition materials and the part substrates themselves create a multitude of new chemical compounds that also must be removed. The ionic charge associated with negative and positive ions can make them challenging to fully remove. Residual chemistries and excessive ionic species can create significant process issues including corrosive chamber conditions and accelerated etch and deposition rates. Table 44.2 shows an example of residual ionic compounds.

1. *Solvents.* Some solvents, such as acetone, have high vapor pressures and readily evaporate, while other solvents including perfluorocarbons and NMP have low vapor pressures and are difficult to remove from porous substrate surfaces such as yttria coatings.

2. *Bead blast media.* Perhaps no part cleaning steps add more metal and particulate contaminants to the parts than the deposition removal/substrate roughening media blast process. The common abrasive media utilized are industrial grade materials including aluminum oxide, silicon carbide, glass bead, and others. Environmental contaminant in the media includes different elements. Table 44.3 shows elemental impurities in aluminum oxide blast media.

Table 44.2 Residual Ionic Compounds Using IC Method

		DL	OCM Spec	Site 1		Batch 1		Batch 2	
Anions:									
Fluoride	(F^-)	3	300	63	64	580	630	1500	1200
Chloride	(Cl^-)	1.5	240	94	120	18	24	46	53
Nitrite	(No_2^-)	1.5	53	<1.5	<1.5	4.7	2.2	<3.5	4.3
Bromide	(Br^-)	0.5	10	<0.5	<0.5	<0.5	<0.5	<1.5	<1.5
Nitrate	(NO_3^-)	1.5	150	78	99	230	350	260	330
Sulfate	(SO_4^{2-})	1	100	<1	<1	<1	<1	<2.5	<2.5
Phosphate	(PO_4^{3-})	1	10	<1	<1	<1	<1	<2.5	<2.5
Cations:									
Lithium	(Li^+)	7.5	10	<7.5	<7.5	<7.5	<7.5	<20	<20
Sodium	(Na^+)	2.5	200	<2.5	<2.5	3.6	4.9	9.0	6.7
Ammonium	(NH_4^+)	3	100	<3	<3	3.8	6.8	36	41
Potassium	(K^+)	1.5	15	<1.5	<1.5	<1.5	<1.5	<4	<4
Magnesium	(Mg^{++})	2.5	440	76	80	44	44	94	110
Calcium	(Ca^{++})	1.5	200	<1.5	<1.5	2.1	5.8	45	38

Table 44.3 Elemental Impurities for Grit Blast Media Aluminum Oxide (99.6%)

Fe_2O_3	0.045% Max
TiO_2	0.008% Max
SiO_2	0.035% Max
CaO	0.040% Max
Na_2O	0.290% Max

In addition to the metal contamination present, the media itself is 100 percent pure particulate from sizes in the nanometer to millimeter range. Significant postblast processing must be utilized to remove the metal and particulate contamination introduced during the blasting process.

3. *Part substrate composition*

a. *Metal contaminants and part substrate.* Common part substrates themselves contain metal contaminants that can interfere with wafer processes in the sub-20-nm nodes. Examples of metal contaminants in common part substrates are shown below. Table 44.4 shows trace metal (TM) elemental impurities in 10^{10} atoms/cm^2 for a quartz substrate.

6061 aluminum is another substrate which has embedded intrinsic impurities by weight percent (wt%) within the substrate that can impact the cleaning process. Some of the impurities within the 6061 alloy material are listed in Table 44.5.

While the substrate contaminants cannot be removed by cleaning methods, surfaces can be passivated making the contaminants more inert to semiconductor processes.

b. *Particle control and part substrate.* Ultrasonic techniques must be developed that differentiate between "particles on the surface," and "particles lightly or tightly adhered to the surface," from "particles that originate from part substrate itself." While it is desirable to remove "surface particles" and "lightly or tightly adhered particles," it is not desirable to remove the substrate material. Depending on substrate "hardness" and surface morphology, different substrates shed substrate particles differently when subjected to various ultrasonic cleaning power and ultrasonic time. Figure 44.2 depicts an ultrasonic process for a quartz substrate part. Shedding substrate particles is not desirable as it leads to part pitting, removal of surface films and reduced part life. It is important to characterize each part by substrate composition to ensure there is a standard by which parts are considered thoroughly and completely cleaned of surface particles and lightly/tightly adhered surface particles. If not cleaned correctly, the part can shed defects inside the tool.

4. *People and environment (cleanroom air).* Environmental conditions and people are significant contamination sources that require control in parts cleaning processes. The total environment inside and surrounding the cleanroom definitely impacts final part cleanliness. Environmental influences include part management in the cleanroom, the use of proper HEPA filtration, and parts packaging procedures.[2,3,5] Figure 44.3 below shows measured number of defects resulting in the use of 0.3 μm HEPA filters against contamination particle size. Predictably, defect count decays exponentially with increasing particle size. Understanding of the defects below 100 nm is critical for cleaning at 10-nm technology nodes. The cleanroom environment plays an increasingly critical role in terms of part cleanliness, in the same way it is in wafer processing. Minienvironments with highly controlled conditions are used to control smaller sized particles.

5. *Methods and metrology of drying and mitigation of outgassing.* A poorly dried part will introduce challenges in vacuum chamber pump-down. Moisture and hydrocarbon residuals can significantly reduce the pump-down speed. These residuals can also adversely impact the chamber seasoning process, which can lead to film shedding and elevated defect counts. Figure 44.4 shows an example of the outgassing of a quartz part. Different part materials and texturing result in different outgas rates. In light of this, it is apparent that a single bake process condition will not work for all substrates. It is observed that textured parts outgas more than nontextured or smooth surfaced parts. This is due to the fact that textured parts have more surface area and topographical features. If the part is not dried correctly in an oven, excessive volatile material diffuses into the chamber under ambient and vacuum conditions extending pump-down time. Due to poor surface dryness, this part may also yield poor surface sticking coefficients which can also result in high defect counts and poor process control inside the chamber.

Table 44.4 Trace Metal from the Quartz Surface

Element	As	Ba	Bi	Ca	Cd	Co	Cr	Cu	Fe	Ga	Ge	K	Mg
TM (10^{10} atoms/cm^2)	80	40	30	150	50	100	100	100	100	80	80	150	250
Element	Mn	Mo	Na	Ni	Sn	Sr	Ta	Ti	Tl	V	W	Zn	Zr
TM (10^{10} atoms/cm^2)	100	60	250	100	50	70	30	100	30	120	30	90	60

Table 44.5 Aluminum 6061 Impurities

	Chemical Composition Limits								Others	
Weight %	Si	Fe	Cu	Mn	Mg	Cr	Zn	Ti	Each	Total
Minimum	0.4	–	0.15	–	0.8	0.04	–	–	–	–
Maximum	0.8	0.7	0.4	0.15	1.2	0.35	0.25	0.15	–0.05	0.15

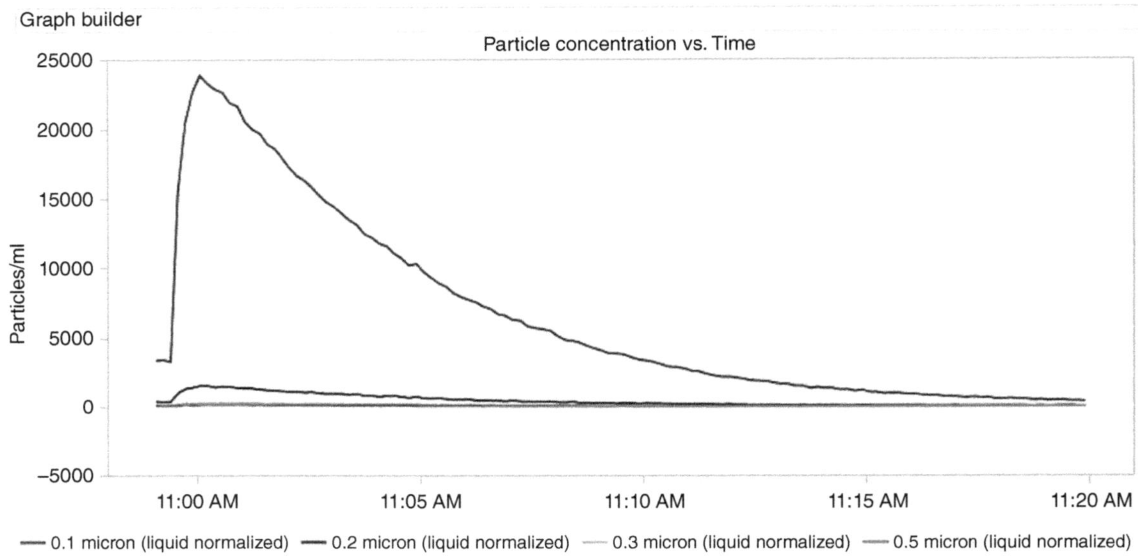

Figure 44.2 Depicts an ultrasonic process for a quartz substrate part.

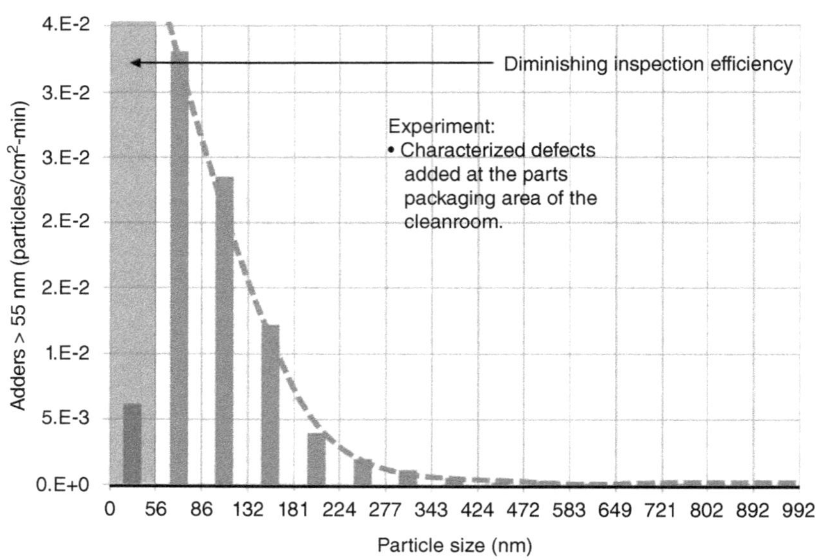

Figure 44.3 Particle size versus adders for a typical cleanroom environment with 0.3 μm HEPA filters.

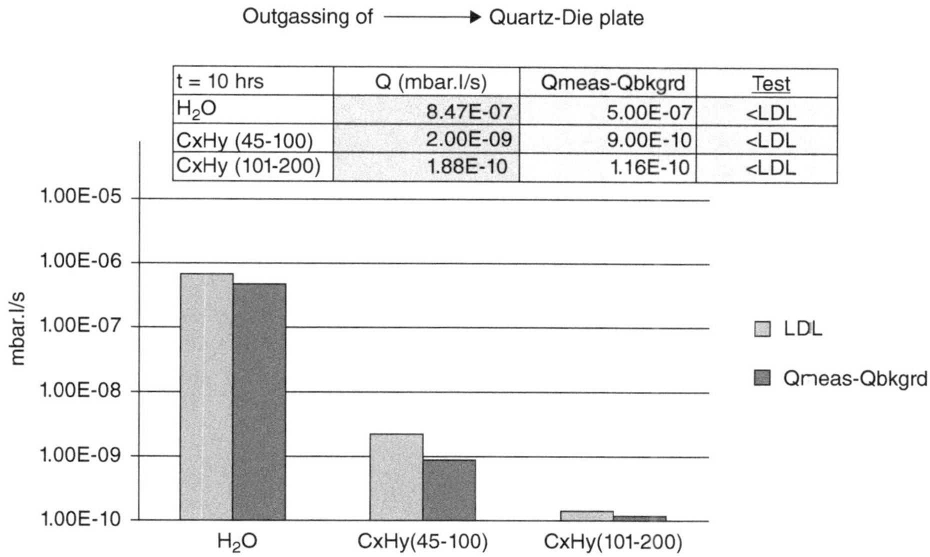

Figure 44.4 An example of the outgassing from a quartz part.

Figure 44.4 shows an optimum outgassing scenario, where the moisture, volatile, and nonvolatile hydrocarbons are below the lower detection limits (LDLs). It is projected that this part will pump down effectively inside the chamber and will perform well in terms of lower defects and process capabilities, as evidenced by low partial pressure of outgassed compounds.[6,7]

6. *Cleanroom wipes, gloves, and handling.* Cleanroom wipes, gloves, and part handling protocol all are crucial measures for controlling environmental contamination sources. Parts dried with ovens outgas significantly less than those dried with heat lamps.

Metallic and ionic contamination can impact manufacturing processes and final products. ICP-MS (inductively coupled plasma mass spectrometry) and IC (ion chromatographic) methods were utilized to detect trace metals in the wipers. Three different wipers from two different manufacturers were used for the metallic contaminants. Table 44.6 shows major metallic contaminants observed for different wipers. Note: Ti and Sb elements were detected as TiO_2 and Sb_2O_3, respectively.[2–4,5]

Soak and leach methods were used to measure particles from different set of wipers. Figure 44.5 shows particle count levels in four different cleanroom wipes. Stringer-type defects scan lead to chamber failures and high die yield loss.

Cleanroom gloves play a vital role in part handling but can also contribute considerable contamination. Figure 44.6 shows glove cleanliness measured through a specific defect transfer mechanism and it demonstrates that if the gloves are not clean they can transfer thousands of defects onto a good clean part. Part handling is also very critical, handling parts strategically and with care, just like wafers, can lead to better surface cleanliness and improvements in overall chamber performance.

7. *Bagging material and packaging.* Outgassing from poor part packaging can lead to severe tool pump down issues or defect failures. Poor packaging methods can contribute to defects and trapped moisture inside the bags which can impact part performance in the chamber.

Table 44.6 Metallic Contaminants for Different Wipers

Wiper	1	2	3
Wiper Weight (g)	7.88	5.43	7.35
Ti as TiO_2 (%)	0.28	0.31	0.25
Sb as Sb_2O_3 (%)	0.015	0.014	0.016
Other (%)	0.00042	0.0014	0.0015
Total (%)	0.29	0.32	0.27

44.4 PART SURFACE TREATMENT TECHNOLOGIES AND THEIR EFFECTS ON PROCESS PERFORMANCE

Part surface roughness and part surface composition have a direct impact on chamber particle performance in most chamber processes.

44.4.1 Twin Wire Arc Spray in PVD/Metal Sputter Applications

Metal sputtered onto tool shielding during metal deposition and metal sputtering processes tend to build film stress quickly, which causes the films to separate from the shielding, creating high chamber particles and process failure. In the late 1970s part clean developer Walt Warner at IBM discovered that metal shielding roughened with bead blast media had vastly superior metal adhesion properties versus smooth and spun metal shield surfaces. Better film adhesion led to significant improvement chamber life (MTBF) and increased device yield. The relatively low practical roughness limits achievable by mechanical roughening alone (Ra = 225–250 μin) led to the application of aluminum twin wire arc spray (TWAS) films on metal deposition shielding. The TWAS process involves feeding two high voltage/high amp aluminum wires together creating molten aluminum, and propelling the molten aluminum onto shielding surface with high pressure CDA and/or compressed gases. Figure 44.7 shows TWAS spraying process. Aluminum TWAS films achieved exceptionally high surface roughness (Ra = 800–1200 μin) significantly increasing film adhesion characteristics and increased chamber life.

44.4.2 TWAS Thermal Spray Process

Virtually all metal shielding in PVD and metal sputtering processes utilize a combination of mechanically roughened surfaces and aluminum sprayed TWAS surfaces, improving metal film adhesion and extending chamber life.

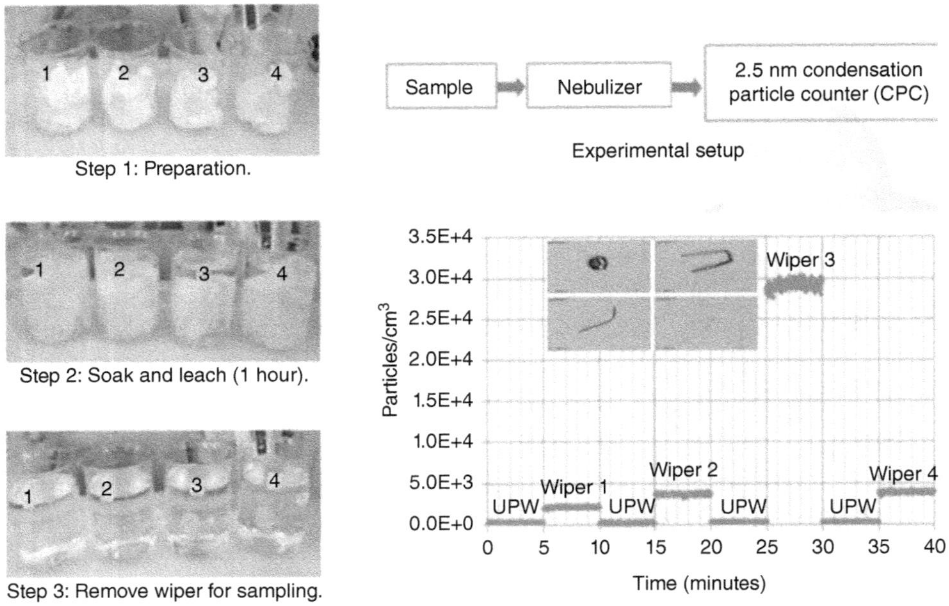

Figure 44.5 Analysis of different wipers.

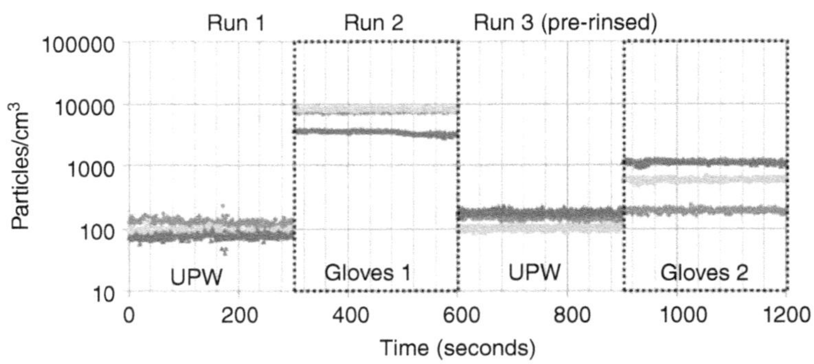

Figure 44.6 Gloves particles/residue shedding.

Figure 44.7 TWAS spraying process.

44.4.3 Thermal Spray Coating Applications in Etch Processes

In etch processes, part substrate materials are subjected to aggressive process chamber conditions and are actually etched by process plasma. The etching of part surfaces creates significant quantities of particles affecting yield and reducing part life. Figure 44.7 shows the etch rates of substrates and coatings under plasma etch conditions.

For larger geometry (> 100 nm) etch processes the part substrate material choices included anodized aluminum, bulk ceramic, and quartz. At higher geometries, the process tolerated the particles generated by part substrate etching. As line width decreased over the years, part substrate materials have evolved. For line widths less than 100 nm, the preferred surface material required was plasma yttria due to the materials low etching rate characteristics. At sub-20-nm geometries even plasma yttria began reaching the limits of suitability due to substrate etch rates. For sub-20-nm processes OEM are investigating and offering alternative proprietary plasma coatings.[8] Relative to yttria, alternative coatings have reduced porosity and increased hardness, which increases plasma compatibility and reduces surface etch rates. Some part clean companies have developed significant expertise in the plasma coating process as described in Sec. 44.5. By performing plasma coating services, part clean companies offer end users two significant advantages.

- Significant cost savings associated with refurbishment of end of life yttria coated parts. Fab owners can save hundreds of thousands of dollars with the reuse of parts formally discarded.
- Improved process results with the application of yttria alternative coatings. Process improvements including fewer particles and extended chamber MTBF.

44.5 THE PLASMA COATING PROCESS

Coatings and passivation can reduce the cost of ownership by extending the part life cycle after multiple clean process steps. Imagine if the recycled part is replaced with a new part after required chamber preventive maintenance step. Imagine the tremendous cost burden for replacing used parts with new parts after scheduled maintenance. Today, high-volume manufacturing semiconductor companies are saving millions of dollars every year by recoating and passivating part surfaces. The section below will discuss about coatings and passivation methods. Yttrium oxide–based coating mainly developed for etch chambers forms a protective layer on less noble substrates. A plasma spray tool is used to perform this coating.

The plasma spray process introduces yttrium oxide powder into a plasma plume, where it is fused and accelerated toward the target surface and deposited as a conformal coating on the substrate. Different coating methods are used to apply plasma coating to improve the MTBC and parts performance within the process chambers.

44.6 SUMMARY

Process chamber parts cleaning requirements have changed significantly since they were originally outsourced to specialized third party service providers some 25 years ago. These changes are driven through increasing numbers and combinations of materials and processes that allow improvements in device performance and footprint reduction. Trace metal, ionic, and particulate contamination levels that are near or even beyond current metrology capability are now found to be major impediments to chamber process performance and device yield, "Things that didn't matter before now do." The challenges moving forward include the development of increasingly complex selective stripping technologies, improved handling and packaging solutions, and continued improvements in real-time metrology that can detect even smaller contamination levels and allow predictable chamber performance to help drive the industry forward.

44.7 ACKNOWLEDGMENTS

John Deem is a senior member of technical staff at QuantumClean in Phoenix, Arizona. John has a chemical engineering degree from Arizona State University and over 15 years of experience in the semiconductor parts cleaning industry with emphasis on selective deposition removal and thermal spray coating development.

Osama Khalil is an R&D engineer at QuantumClean in Phoenix, Arizona. Osama earned his bachelor's and master's degrees in chemical engineering at Arizona State University in 2014 and 2015, respectively. Prior to coming to QuantumClean, he was a research fellow at the National Institute of Standards and Technology (NIST) in Gaithersburg, Maryland, and worked in ultrapure water engineering at Intel Corporation.

Special thanks. The authors would like to thank Scott Nicholas (CEO), Steve Dirugeris (CFO), and Michael Jenkins (VP, HR), our executive team members, and Kaveh Zarkar (director) for their insight and contributions to this publication.

44.8 REFERENCES

1. A. Sidhwa, D. Zuck, L. Bao, J. Fonshill, and M. Boomer, "Characterization of Cleaning Process via Implementation of Analytical Techniques for Aluminum Faceplate Used for SACVD and CVD Application Sematech," 2014 Surface Preparation Cleaning Conference (SPCC), Austin, TX, Apr. 22, 2014.
2. A. Sidhwa, D. Zuck, and J. Deem, "Importance of Precision Parts Cleaning for Sub-20-nm Technology Nodes," Semiconductor Technology Symposium (STS-Conf Semicon West), San Francisco, CA, Jul. 8, 2104.
3. A. Sidhwa, D. Zuck, J. Deem, O. Khalil, and S. Nicholas, "Implementation of Final Surface Finish™ (FSF™) by Using Cost-Effective Atomically Clean Surface (ACS™) Process for Sub-14-nm Technology Nodes," Semiconductor Technology Symposium (STS-Conf Semicon West), San Francisco, CA, Jul. 14, 2015.
4. A. Sidhwa, D. Zuck, and J. Deem, "Importance of Atomically Clean Surface™ (ACS™) Parts Used for Sub-20-nm Technology Nodes," TSIA sponsored by TSMC Tai-Chun, Taiwan, Sep. 12, 2014.
5. A. Sidhwa, D. Zuck, J. Deem, S. Deora, M. Samayoa, and A. Karumuri, "Advances towards Cost-Effective Contamination-Free Manufacturing (CFM)," Sematech 2015-Surface Preparation Cleaning Conference (SPCC), Saratoga Springs, NY, May 13, 2015.
6. A. Sidhwa, D. Zuck, T. French, F. Mitchell, and J. Lawson, "Surface Characterization of Extreme UV Lithography Tool Components and Parts by Using ACS™ Cleaning Process and High Performance Residual Gas Analyzer for Sub-20-nm Technology Nodes," 2014 EUVL Conference, Washington, D.C., Oct. 29, 2014.
7. A. Sidhwa, D. Zuck, T. French, F. Mitchell, J. Lawson, and C. Coulon, "Implementation of High Performance Residual Gas Analyzer for Sub-10-nm Technology Nodes for Extreme UV Lithography Tool Components and Parts," Precision Fair 2015, Veldhoven-Netherlands, Nov. 19, 2015.
8. O. Khalil, A. Sidhwa, J. Deem, and D. Zuck, "Finite Element Analysis of Silicon Nitride Parts Cleaning Process for in Silico Optimization," Applied Project, School for Engineering of Matter, Transport and Energy, ASU Arizona State University, IRA. A Fulton Schools of Engineering, Apr. 28, 2015.

Chapter 45

Equipment Design Challenges due to Increasing Hazards and Regulations

BY

MARK FESSLER *ASM America, Inc., Phoenix, Arizona*

45.1 INTRODUCTION: "THE PRODUCT COMPLIANCE PUZZLE"

Semiconductor equipment design engineers are faced with incredible design challenges to create the next generation process of record (PoR) toolsets for customers. These equipment design challenges are even more complex than you might think. *Why?* The answer is because semiconductor manufacturing equipment (i.e., for wafer processing) has many different types of hazards with the potential for serious harm to the workers, equipment, and environment. Harm, as defined in this chapter, implies human injury, process equipment damage, facility damage, or environmental impact.

The next generation tool designs are harboring even more aggressive hazards due to new reactive chemistries and increased sizes of 450-mm platforms. Each of the hazards listed below must be clearly identified by the equipment design teams and then they must either be removed, or have their effect reduced to a tolerable risk level. This is both an ethical and a regulatory requirement.

- Toxic/corrosive chemical hazards
- Flammable/pyrophoric chemical hazards
- Ionizing radiation (x-rays) hazards
- Nonionizing radiation (UV, IR, RF, etc.)
- High voltages/high current hazards
- Extreme hot/cold temperatures hazards
- Working at heights hazards
- Seismic zone protection hazards
- Laser hazards
- Magnetic field hazards
- Robot/automation hazards
- Mechanical crush hazards
- Loud noise hazards
- Ergonomic hazards

There are a multitude of different reasons (e.g., various equipment failures, human error scenarios, or external influences, etc.) why any of such hazards could lead to harm. Each of the causes should be clearly identified and managed. No one would disagree with this approach. Yet, even in the mature manufacturing environment of the semiconductor industry, we still continue to have hazardous conditions leading to harm within our semiconductor wafer fabrication plants (fabs). If we take an honest look at the data from reoccurring semiconductor EHS (environment, health, and safety) forums, EHS discussion groups, and from actual incident data from semiconductor fabs, we see that the harm due to human error happens much more frequently than the harm due to any equipment failure or external event. How should we prevent such harmful, hazardous events? Semiconductor equipment manufacturers, as well as our customers, must examine their choices. Should we improve the "human" or improve the "equipment design"?

45.1.1 Advanced Equipment Design Is the Focus

Obviously, the most effective answer, if time and resources were unlimited, is both. In the EHS world, the semiconductor industries' occupational safety professionals will tend to focus much more on the human error element by trying to improve the worker's situational awareness and implement behavior-based safety plans. In contrast, the emphasis of this chapter is specifically on advancing our equipment designs to mitigate the effects of the more commonly occurring (human-caused) hazard situations.

There are three suggested courses of action. First, do a better job of finding the hazardous conditions initiated by workers decisions, and engineer layers of protection to lessen the consequences of those decisions. Second, proactively improve our design engineering team's aptitude toward the numerous safety and environmental regulations. Implementing these documented compliance requirements can also help prevent known hazardous conditions. The third suggested course of action is to repeatedly ensure that any design change request is carefully analyzed to uncover inherent hazards resulting from the proposed design change (i.e., must ensure management of change (MoC) protocols are followed).

The good news is that the legal requirements (i.e., both country specific regulatory requirements and customer specific purchase contract requirements) for wafer process tools within the semiconductor industry are very well recognized. The industry specific SEMI S2 Guideline[1] provides the EHS design standard for semiconductor manufacturing equipment, while the European Machinery Directive[2] provides the essential health and safety requirements for the design and construction of machinery. Together they outline the minimum product safety design measures for global wafer processing equipment. Both of these important documents are considered "umbrella" design criterion. They both point to other product design standards to further specify particular design details, which can also be followed to help demonstrate presumption of conformance to the necessary design requirements. Figure 45.1 shows a simple visual of the two key product safety design "umbrellas" and how these can be applied to help protect against our semiconductor industry hazards.

A SEMI Standards Task Force undertook a major comparison effort[3] between these two documents. The work revealed that many of the Machinery Directive requirements are *not* sufficiently addressed within our own industry's suite of SEMI S2⁺ safety guidelines (e.g., S1, S2, S3, S6, S8, S10, S14, S22, etc.). Likewise, it is the author's belief, that SEMI S2⁺ documents offer very important *additional* design guidance that is certainly not included within the exact written requirements of the Machinery Directive. Thus, for any global equipment manufacturer, *both* of these umbrella documents (and the additional standards they reference) must be surveyed to ensure full product compliance. A distinctive requirement within both is that our design teams must consider realistically conceivable situations which can lead to potentially hazardous exposures: Reference their exact verbiage below:

- "Taking into account reasonably foreseeable misuse thereof"[4]
- "No reasonably foreseeable single-point failure condition or operational error"[5]

Yes, there will always be the rare (unforeseeable) hazardous events that might still occur. However, for any reasonably foreseeable hazardous exposures these must be prevented from causing harm! Equipment manufacturers are legally required to both understand, and resolve their designs' sensitivity to realistically conceivable events, which can lead directly to hazardous conditions. *Ok, this seems like a very a sensible approach.* If all of this is true, then why do these hazardous events keep occurring?

45.1.2 Today's Evaluation Methods: A False Sense of Security?

Our semiconductor industry frequently requires the use of an unbiased third party evaluator to perform a product safety audit to the design requirements specified within both SEMI S2 and the Machinery Directive. This helps to justify and provides some level of due diligence, by the equipment manufacturer to their customers, that their design is "safe" (e.g., low residual risk). In reality, however, our industries' tool designs are so very complex, and the safety audit reviews are so limited in time, that it is unrealistic to think that a third party auditor can come in during 1 week (or less) and learn and understand the many details to ensure that all unacceptable risks are adequately addressed. Remember that it usually takes our design engineering teams several years to develop these systems. At best, any third party auditor can only catch the "low hanging fruit" hazards, and other "readily observable" tool design deficiencies.

Semiconductor manufacturing equipment that is currently installed within the customer's fabs still contain hidden hazards. Note that these hazards exist, even after these process tools have been audited by a third party evaluator and found to be *compliant.* Continued accidents, equipment damage events, and near miss incidents within semiconductor manufacturing sites are clear evidence that the industry has not yet considered enough layers of protection in the equipment's design—especially due to reasonably foreseeable misuse during service and maintenance tasks.

The third party evaluators also acknowledge this huge comprehension challenge and regularly require the equipment manufacturer's engineering design teams to provide completed risk assessments (in advance of their on-site safety audits) to learn more about the equipment hazards/risks. Today, the third party auditors are not tasked with ensuring the risk assessment is all encompassing. Therefore, if there are any missing equipment hazards from the design team's risk assessment, it only exposes a deficiency from the design team's understanding, and not necessarily from the third party audit team. The auditors actually

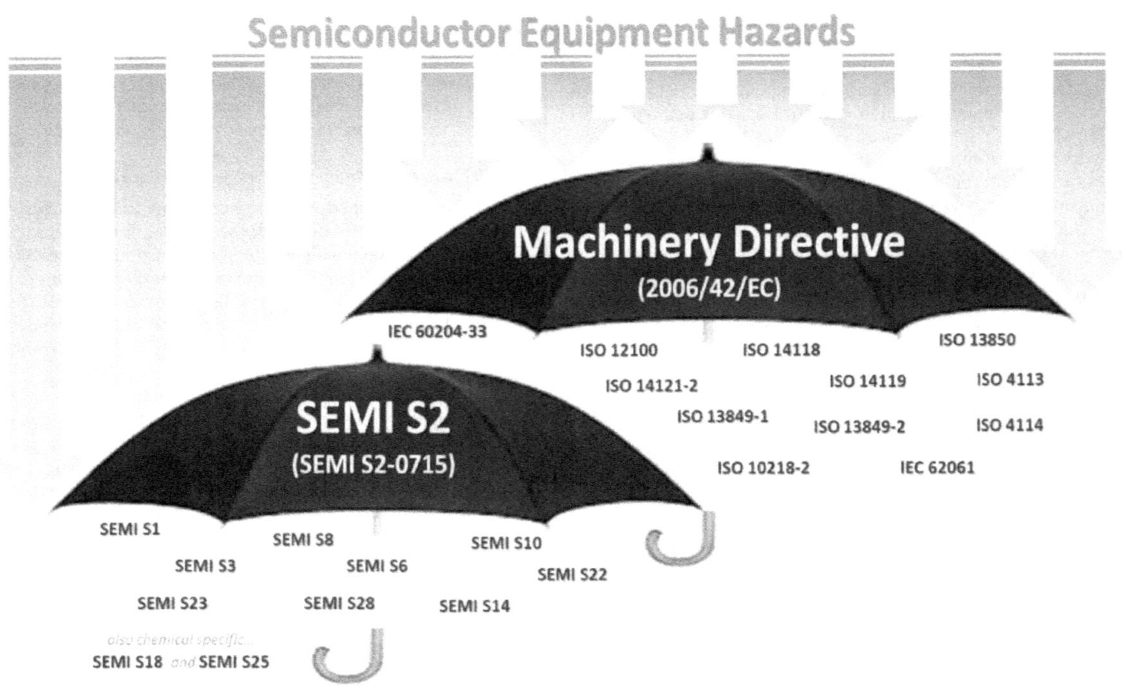

Figure 45.1 Design compliance "umbrellas of protection" and key related references.

get a big break (i.e., liability wise) as they can only audit such *hidden hazards* if they are told about them, or if they are fortunate enough to discover them during their limited evaluation time.

In fact, our industries' third party evaluators have very similar challenges to our design teams. To do an effective evaluation, they must first understand the complex equipment designs in order to find the foreseeable design deficiencies. There simply is not enough time during the current evaluation process to do this. The reality is that having third party certification does not provide the industry with a true validation that all of the potential hazards have been uncovered and mitigated to tolerable risk levels. The existing audit process is very narrow. We must acknowledge this. If we do not, then it gives our semiconductor industry a false sense of security. Based on reoccurring incidents on "fully compliant" tools, there is obviously more that could be done via the audit process to validate product compliance.

45.1.3 Engineering Department Skillsets: Prerequisites for Success

To make the current engineering design challenge even greater, there are many new safety and environmental design requirements. These have proliferated very fast, especially over the last 7 years, and it is difficult for our design teams to keep up with all of them. Admittedly, each new standard or new revision is well-purposed. They are generally updated to better address known safety and/or environmental hazards. Taken as a collective, however, they can be incredibly overwhelming. Today, a true knowledge gap of these regulatory design requirements exists, even within our world-class semiconductor engineering teams. Some of it is simply due to employee attrition, with proficient engineers moving on or retiring. Most of this gap, however, is due to the release of previously unstipulated requirements that now must be met, yet the engineers are still unaware. Regrettably, this information deficiency is growing wider every year as more regulations are released or updated.

Product safety engineering professionals can (of course) guide and instruct our design engineers to these new legal requirements, but this approach takes time for additional resource planning. Without a significant increase in the number of product safety engineering professionals for each engineering team, it is simply not feasible (or cost-effective) to check each and every engineering design decision. In reality, that level of QA/QC on the design engineering teams would bring the new tool development process to a screeching halt! This is simply not a workable solution moving forward.

The recommended solution, as outlined in this chapter, is to *coach up* the existing design engineers to the crucial requirements that their individual design specialties require. This also poses new challenges. Even at the world's very best engineering colleges/universities, the newly graduated design engineers (e.g., electrical, mechanical, chemical, software, etc.) are simply not well-trained on how best to comply with the international health and safety regulations within each country, or how to uncover inherent (hidden) design deficiencies within their own designs. Nevertheless, these new design team skills are undeniably essential to ensuring that semiconductor companies can provide safe (low residual risk) and regulatory compliant equipment to our customers globally.

Likewise, most engineering schools do not spend enough time to teach these future design engineers on critical business processes that are also needed to maintain their product's compliance. Effective engineering change control processes are historically incorporated to address technical feasibility and cost/benefits of a proposed design change. Rarely do they include all of the necessary hazard identification steps and related risk assessments (based on the proposed changes) to ensure we continue to maintain equipment regulatory compliance and keep our workers safe.

Such *prerequisite* skillsets within our engineering departments (i.e., to ensure product compliance) are of great importance, especially today. Our proud and successful semiconductor industry has done its best (through the last 25 years) in meeting the minimum design for safety/environmental requirements, but with the onslaught of the many new designs and continually changing regulatory requirements, it necessitates that our industry as a whole, now take an honest step backward and reconsider if there is a more effective, alternate approach to solving this difficult product compliance puzzle.

45.2 THE FUNDAMENTALS OF PRODUCT COMPLIANCE: "WHAT MUST BE DONE?"

This section will describe the fundamental (minimum) design criterion for wafer processing equipment within our global semiconductor industry. These requirements are driven by various legal responsibilities that each equipment manufacturer has to comply with (e.g., both country specific equipment regulations and customer purchase contracts). They are also anchored with the clear understanding of the hazardous situations that can exist throughout the entire life cycle of the equipment—manufacturing, installation, operation, maintenance, end of life decommissioning, including waste recycling/disposal. Each of the various hazard scenarios must then be prioritized based on the equipment manufacture review team's judgment of the risk. This process is not done just once. It is an iterative process that must be managed consistently if an organization wants to effectively mitigate their risks over the lifetime of their wafer process equipment.

45.2.1 Harm Resulting from "Reasonably Foreseeable Misuse"

As previously mentioned, recent discussions from reoccurring EHS forums, and our own fab experiences has shown us that incidents caused by humans can happen considerably more frequent than any caused by equipment failures (e.g., orders of magnitude more frequent)! We as humans make mistakes. No one will argue this fact. As equipment designers, however, we must always remember to be focusing on the types of *misuse* that we can anticipate happening. The foremost international standard on equipment risk assessment[6] defines reasonably foreseeable misuse as "the use of a machine in a way not intended by the designer, but which can result from readily predictable human behavior." This standard also provides some good examples of the different kinds of reasonably foreseeable misuse that must be taken into account when trying to uncover hazards within the equipment's design:

- Behavior resulting from lack of concentration or carelessness (i.e., human error)
- Behavior resulting from taking the line of least resistance in carrying out a task
- Behavior resulting from pressures to keep machinery running in all circumstances

When we read this list, we shake our heads, acknowledging that "yes ... those all makes sense." We all have personal examples we can recall where we have made similar choices. Therefore, it seems sensible that this is an area of our equipment designs that truly deserves more of our focused attention.

45.2.2 How Many Hazards Does Your Equipment Really Have?

This initial hazard identification step is critical. The author believes it is the most critical step along the long, winding road to product safety compliance. Remember that you cannot assess the risk if you do not know the hazard. There are already many good technical publications to guide our engineering teams on how to identify hazards intrinsic to their equipment designs.[7–11] Safety professionals can help to facilitate these team design reviews using many of these well-known methods.

The primary goal of these review sessions is to uncover hazardous situations that were not obvious to the engineers during the original design phase, but can still be introduced by either equipment/human failures, or through some external influences (e.g., power outage, earthquake, etc.). Design review teams should especially consider how a worker's "reasonably foreseeable misuse" can lead to any of the following damage categories:

Example Damage Categories	General Descriptions of Possible Harm or Financial Loss
Workers (employees/contractors)	Acute or chronic physical injury or medical illness
Individual wafer process equipment	Equipment damage, downtime, repair, or replacement cost
Individual facility/fab equipment	Equipment damage, downtime, repair, or replacement cost
Fab smoke/fire damage	Multiple equipment damaged, clean up time, or repair cost
Environmental damage	Contamination to the air, water, earth, etc.
Facility production time lost	Number of hours fab is down can be significant loss

This brings us to the so-called "hazard identification paradox,"[12] which states that no single discovery method will find all of them. It will take a *combination* of these different hazard identification techniques to improve the thoroughness of our design efforts. See Fig. 45.2, adapted from the original publication, to more accurately reflect the different techniques our industry should be considering.

If our engineering design teams skip these essential hazard identification steps (e.g., due to a lack of facilitator expertise, or reduced product development schedules) then they may *unintentionally* allow a serious hazardous situation to occur in the customer's fab. The initial harm caused by the design deficiency is not the only cost. We recognize the additional expense of recovering from a significant hazardous event. There is also the extra sunk cost of the design efforts to bring the equipment backup to full regulatory compliance (after an incident), especially if all tools in the field must be retrofitted.

Reasonably Foreseeable Equipment Failures

Equipment failures leading to harm were a big problem in the early days of the semiconductor industry. Since the 1990s, however, the entire industry has been highly focused on these incidents and has greatly improved our equipment's reliability performance. Today, the likelihood of a reasonably foreseeable equipment failure leading to harm in our high-volume production fabs is relatively low (regardless of whether the tool is up in production, or down for service/maintenance). Our designs seem to adequately consider the impacts of equipment failures, and include the appropriate layers of protection to prevent harm. This is not where most of our equipment design deficiencies are today.

Reasonably Foreseeable Misuse during Production Tasks

Additionally, most of the semiconductor industry is now extremely automated and they operate their high volume production lines with very minimal interaction by the operators. Similar to the frequency of equipment failures, reasonably foreseeable misuse incidents during production are also relatively low. This is not where most of our semiconductor equipment design deficiencies are today.

Reasonably Foreseeable Misuse during Maintenance/Service Tasks

This chapter promotes the concern that there is still a discernible gap related to the number of layers of protection provided by our equipment designers during maintenance and service tasks. The author strongly believes that this is where most of our equipment design deficiencies are today.

The *purest* of the product safety professionals (and even the global regulatory bodies) may argue that we should already be looking for such human-induced hazards during maintenance/service. The reality, I believe, is that our tool designs to mitigate such risks are simply not sufficient. *Why?* Many of the hazardous conditions during downtime only leverage administrative controls as the primary layer of protection against harm. Some well-accepted examples of administrative controls that are commonly applied during equipment downtime events are

- Isolation and deenergization of hazardous energies (e.g., lockout/tagout)
- Using written procedures and checklists for repetitive tasks
- Applying pretask planning (PTP) and job hazard analysis (JHA) for tasks that are undocumented
- Using warning signs and barricades around potentially hazardous areas
- Using of personal protective equipment (PPE) only as a last line of defense against exposures

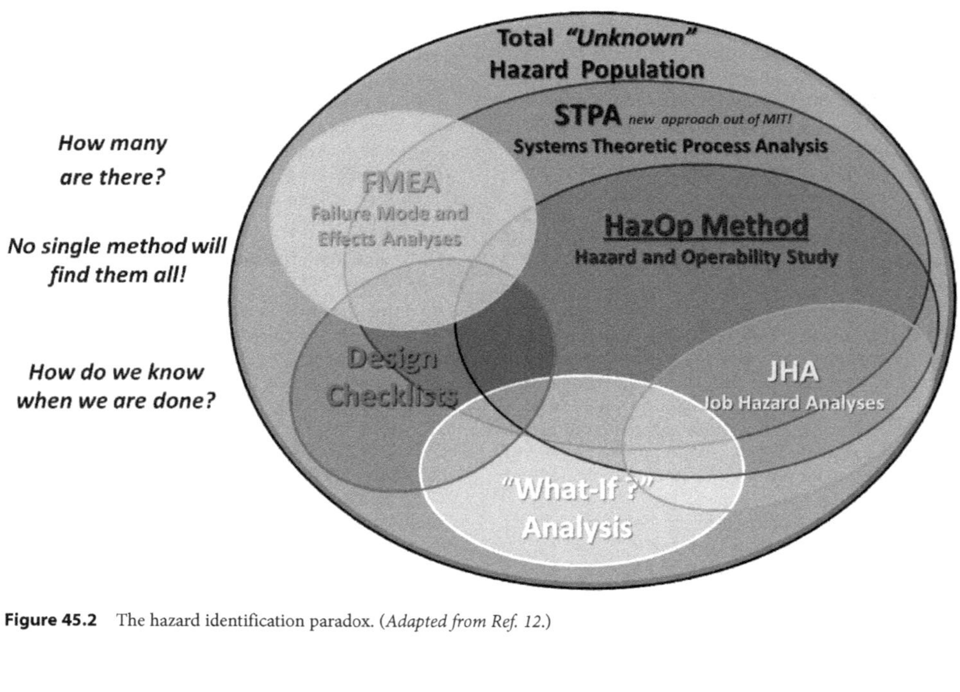

Figure 45.2 The hazard identification paradox. (*Adapted from Ref. 12.*)

From a legal point of view, these practices are typically deemed satisfactory by the various regulatory bodies all around the world. Furthermore, many other industrial sectors follow similar administrative controls for their downtime work on equipment. Regrettably, relying solely on administrative controls during downtime events is not a guaranteed layer of protection against harm. Recall that our industries legal design criteria[4,5] (mentioned in Sec. 45.1.1) mandate that our equipment designs prevent harm due to "reasonably foreseeable misuse." There is an obvious opportunity for improvement here. If we foresee the misuse of an administrative control, what should designers do to mitigate that risk? Considering additional engineered layers of protection (versus relying solely on the administrative controls) is the so-called "next layer of the onion" for semiconductor industry's advanced equipment designs that our equipment engineers should be focusing on.

45.2.3 Risk Assessment of Each Identified Hazard

Risk can be defined as the expected magnitude of losses from the hazard, expressed as a combination of 2 independent terms: the "severity of harm" and "likelihood of harm." These terms are first assessed, and then put into a risk matrix to estimate the overall risk of harm from each identified hazard. Equipment manufacturers are then mandated by our industry's umbrella design documents[13,14] to estimate the risk level of the hazardous situation (e.g., very high, high, medium, low, very low). The remaining details of the risk assessment process will not be explained here, as there are many good sources available that our semiconductor industry can reference.[15–17]

In order to achieve product safety compliance, equipment manufacturers must take corrective action(s) to mitigate the hazard's risk whenever the estimated risk level is deemed to be unacceptable by our industry standard[17] (i.e., medium, high, or very high). Very often, the review team estimates the current risk of a potential hazard, and then identifies a list of corrective actions. It is very important from a product compliance perspective that the estimated risk, *after* those corrective actions, is also estimated (e.g., the residual risk). Any hazard on the process equipment must be reduced to tolerable risk levels (e.g., low, or very low). This is the remaining risk that the company is willing to accept.

The severity of harm is usually assessed as different levels of personal injury. As an example, both the European Machinery Directive[18] and the international standard on equipment risk assessment,[19] consider the severity of harm *only* as physical injury or damage to health. In the semiconductor industry, however, it has been a long standing practice to include other elements of concern when estimating the severity of harm term. Our semiconductor industry's equipment risk guidelines[20,21] extend the scope of severity to also include damage to equipment, buildings, or the environment.

For each identified hazardous event, there is the potential to result in several different severities of harm. Each of these severities also has a corresponding likelihood of occurrence. A very good practical guidance reference[22] points out the importance of looking at a variety of severity/likelihood parings, stating that "the most severe can be improbable, and the most probable be inconsequential, so that using either could lead to an inappropriate estimation of risk." The review team will have to select the combination that gives the highest risk. It then goes on to give a great explanation of the numerous ranges that can occur. Thus, our engineering teams should estimate a range of credible, believable "severities of harm" and their corresponding "likelihoods of harm"; and then only use the specific combination that gives the team the overall highest level of "risk of harm."

Successfully completing risk assessments can be challenging, and even good product safety engineers can make critical mistakes when facilitating them. The author does not believe there is enough assistance within our SEMI risk guidelines on how to correctly estimate the severity and likelihood terms. The subjective nature of the practice leads to different interpretations, which then leads to inconsistencies between different risk assessments. There is opportunity for improving our SEMI risk guidelines to better educate our design engineers. SEMI's EHS Committee Task Forces are currently considering writing a new Related Information Sections with SEMI S10[20] and SEMI S14[21] to give such direction.

45.2.4 International Standards Provide "Lessons Learned"

The equipment designers within the semiconductor industry have gone through a progressive learning curve over the past 25 years. Current industry safety guidelines, standards, regulations, and directives all help to document what we have collectively learned. As previously mentioned, there are two governing equipment design documents[1,2] that must be understood and implemented to achieve product safety compliance. In the early years, these design and testing requirements were very basic. Each revision brought new insights into actual hazards that were occurring, and "we" (as an industry) agreed they should be corrected via new criterion being added, or clarified within the revision. Much more engineering design efforts were required to continue to meet the increase in product safety compliance requirements. Please refer to Fig. 45.3 (from a simple first-order approximation based on number of pages) that it's progressively more difficult to grasp and comply with these product safety regulations.

While these two umbrella design standards include essential design for health and safety criterion, they also point to *many* other guidelines and standards for specific design details—refer to Table 45.1. In total, these safety regulations listed have been a very positive resource to help to prevent known hazardous situations. Still, it is an enormous amount of additional material to digest in order to ensure we are designing correctly. The painful reality is that there are approximately 1400 pages of safety design criterion to be considered. This number grows even larger if the equipment designs fall under the scope of other safety directives/standards such as the high pressure codes/regulations.[23,24]

But wait … there's even more! Due to space limits, I have not even included the various product environmental regulations that are also legally required (now or in near future) to be met by our industries' global equipment manufacturers (e.g., EU RoHS Directive[25], EU REACH Regulation,[26] EU ErP Directive,[27] EU WEEE Directive,[28] EU F-Gas,[29] China RoHS,[30] etc.). These relatively new environmental regulations directly affect our industries equipment designs.

Wow! You now realize that product compliance for semiconductor equipment is very arduous. The real challenge is there are just too many requirements for just one person on the engineering team to understand and then act as the team's expert resource. There is just too much to know! What worked previously for our industry may not be the correct approach moving forward. Equipment manufacturers must share the responsibility across their different teams' design experts. Engineering management needs a better plan on how to educate their engineering teams when these design requirements are released or revised. Checks and balances must be incorporated into the existing design review and engineering change control processes to ensure we are asking the right questions in the design phase.

45.3 ENGINEERING DEPARTMENT SUGGESTIONS: "HOW CAN WE DO THIS BETTER?"

This section builds upon the fundamental equipment safety requirements outlined in Sec. 45.2 by introducing specific proposals for the further advancement of our semiconductor industry equipment designs. If we want to make our equipment succeed in solving this "product compliance puzzle," there are engineering department modifications (i.e., to key business processes) that should be considered.

45.3.1 What They Need to Teach in Schools: "A New Engineering Curriculum"

There is a U.S. government sponsored program called *Prevention through Design* (PtD),[31] with similar learning opportunities in Europe.[32] Still, an actual degree program for "product safety engineering" is currently lacking within the universities. Higher-learning institutions have many more degreed programs of study for "occupational safety" professionals. A very good reference[33] suggests a new skill development and curriculum model to help students uncover hazards and correctly risks them. Regrettably, most of these programs that do exist are currently targeting only the EHS students, and not the electrical, mechanical, or

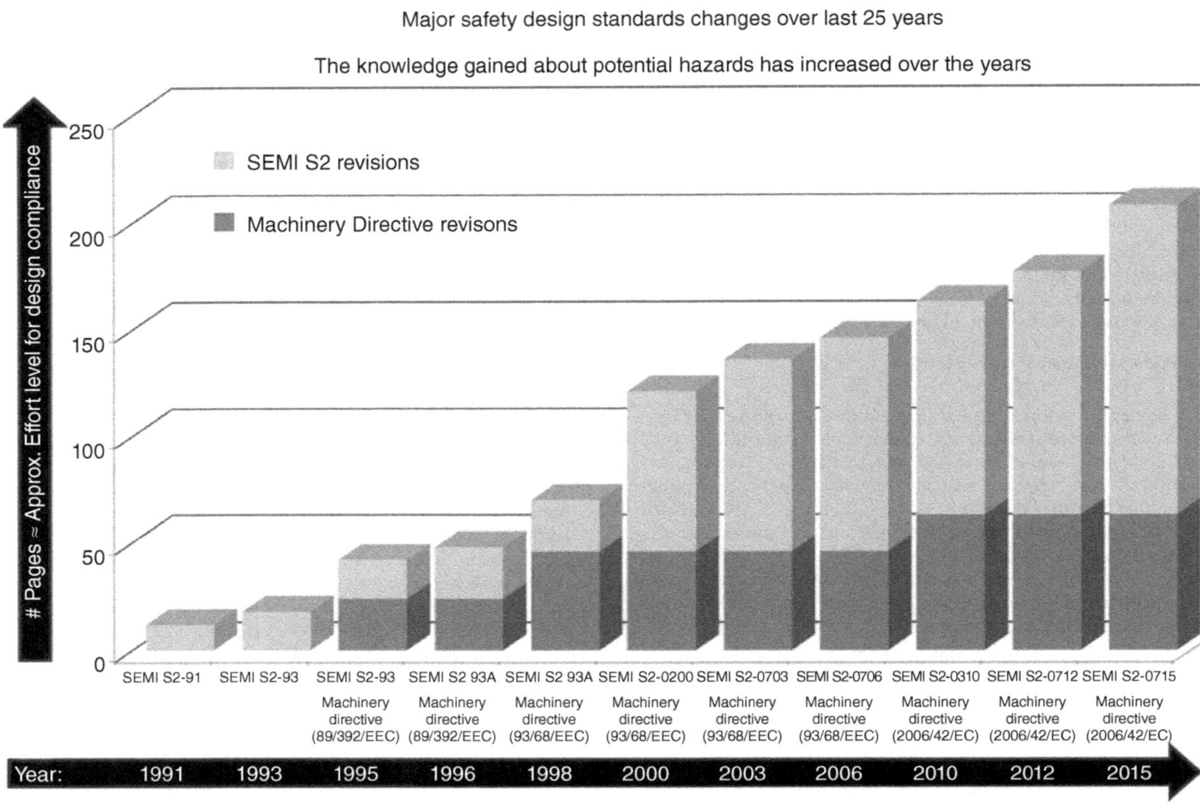

Figure 45.3 More design/training efforts are needed to comply.

software design students. The next step is getting similar curriculum to our future design engineers, as they do not have a direct resource within academia today.

Education solutions from other *hazardous* industries (e.g., nuclear, petrochemical, aerospace, etc.) are two-fold. The most direct approach is for companies to hire *external* private consulting firms[34–36] to provide and certify their equipment design engineers. While this approach can be useful to further the skill development of product safety engineers, it can also be quite expensive. Furthermore, these courses are not semiconductor industry specific. This external approach is also very unrealistic if want to educate entire design engineering teams. The alternative is to formally create "internal," industry specific education programs for our engineers. Regardless of the approach (i.e., external or internal), the primary emphasis should be on the following three *gaps* in our design engineer's aptitude:

Knowledge of Hazard Identification Methods

Some colleges/universities have recently started offering electives to explain these important methods, but it is not mandatory for all engineering graduates yet. Likewise, the private sector also has some relevant training available too, but it's still directed more toward the EHS professionals. We still have long way to go to help our engineering students learn the skills to *uncover* their hidden hazards.

Knowledge of Product Safety/ Environment Standards

Engineering curriculum at universities does not stipulate that their engineering students review the many international design for safety standards, or the various new environmental regulations within each country. Maybe there are a few forward thinking professors introducing some, but in general, this is an actual gap that institutions for higher learning should strongly consider closing. Thus, the only way our engineering teams can learn the product compliance standards (e.g., Machinery Directive, SEMI S2, RoHS, ErP, REACH, etc.) is to allow them the time and resources after they are hired. Note that SEMI has periodically offered SEMI Standards Technical Training Programs (SEMI STEPs),[37] but these would need to be greatly expanded to include training for all the listed design criterions from Table 45.1.

Knowledge of Management of Change Risks

Management of change (MoC)—this is probably one of the most important business processes within any engineering department. It is surprising that universities do not teach their design engineers on the critical importance of the formal change control processes, and how to do it efficiently and effectively. While you will see manufacturing/quality engineering students getting some guidance, it rarely makes its way into the other engineering disciplines. Even more rarely do they accurately include the documented requirements of their risk assessments within the management of change process.

These three areas above are all common stumbling blocks to ensuring any company's equipment design is compliant. No single company or industry has ever done this very well without a good deal of effort.

45.3.2 Better MoC Protocols Required

Any semiconductor manufacturing facility (fab or R&D lab) is subject to continual change. All of these changes have the potential to create new

Table 45.1 Key Safety Design Standards for Semiconductor Manufacturing Equipment (1477 pages)

		Key safety design standards for semiconductor manufacturing equipment	# pages
		2 Top Level "Umbrella" Design Standards—224 pages	
Directive (2006/42/EC)		European Union Machinery Directive	63
SEMI S2	(2017)	Environmental Health and Safety Guideline for Semiconductor Manufacturing Equipment	161
		Risk Assessment Standards—156 pages	
EN ISO 12100	(2010)	Safety of Machinery: General Principles of Design—Risk Assessment and Risk Reduction	77
ISO/TR 14121-2	(2012)	Safety of Machinery: Risk Assessment—Part 2: Practical Guide and Examples of Methods	38
SEMI S10	(2015)	Safety Guideline for Risk Assessment and Risk Evaluation Process	20
SEMI S14	(2016)	Safety Guidelines for Fire Risk Assessment and Mitigation for Semiconductor Manufacturing Equipment	21
		Electrical Design Standards—386 pages	
ISO 13850	(2015)	Safety of Machinery: Emergency Stop Function—Principles of Design	6
ISO 14118	(2000)	Safety of Machinery: Prevention of Unexpected Start-Up	13
ISO 14119	(2013)	Safety of Machinery: Interlocking Devices Associated with Guards	68
ISO 60204-33	(2011)	Safety of Machinery: Electrical Equipment of Machines Requirements for Semiconductor Fabrication	122
NFPA 79	(2015)	Electrical Standard for Industrial Machinery	98
SEMI S22	(2015)	Safety Guideline for the Electrical Design of Semiconductor Manufacturing Equipment	79
		Mechanical Design Standards—212 pages	
ISO 4413	(2010)	Hydraulic Fluid Power—General Rules and Safety Requirements for Systems and Their Components	46
ISO 4414	(2010)	Pneumatic Fluid Power—General Rules and Safety Requirements for Systems and Their Components	37
SEMI S6	(2007)	EHS Guideline for Exhaust Ventilation of Semiconductor Manufacturing Equipment	73
SEMI S8	(2015)	Safety Guidelines for Ergonomic Engineering of Semiconductor Manufacturing Equipment	56
		Robotic Design & Integration Standards—129 pages	
ISO 10218-1	(2011)	Robots and Robotic Devices—Safety Requirements for Industrial Robots—Part 1: Robots	43
ISO 10218-2	(2011)	Robots and Robotic Devices—Safety Requirements for Industrial Robots—Part 2: Robot Integration	72
SEMI S28	(2011)	Safety Guideline for Robots/Load Ports Intended for Use in Semiconductor Manufacturing Equipment	14
		Functional Safety Standards—254 pages	
ISO 13849-1	(2015)	Safety of Machinery: Safety Related Parts of Control Systems—Part 1 General Principles of Design	86
ISO 13849-2	(2012)	Safety of Machinery: Safety Related Parts of Control Systems—Part 2 Validation	79
IEC 62061	(2015)	Safety of Machinery: Functional Safety of Safety Related Electrical, Electronic and Programmable Electronic Control Systems	89
		Additional SEMI Design Standards—116 pages	
SEMI S1	(2015)	Safety Guideline for Equipment Safety Labels	21
SEMI S3	(2011)	Safety Guideline for Process Heating Systems	63
SEMI S23	(2016)	Guide for the Conservation of Energy, Utilities, and Materials Used by Semiconductor Equipment	32

workplace hazards, some of which are not immediately recognizable. Design and procedural changes can also make any of the existing/known hazards move to a higher risk level. Each request for change/modification should be carefully analyzed for inherent hazards that may result from that change. The MoC policy applies equally to the following five areas:

Modification to Any Processing Equipment's Design, or Operating Conditions

This primarily includes proposed design changes outside of the approved design and operation envelope (i.e., changes to temperature, pressure, flow rates, process chemical type, processing speeds/times, chemical distribution, electrical distribution, etc.). In addition, since our process equipment is also integrated, both upstream and downstream impacts to the fab should also be considered. This type of *shared* MoC (i.e., both customer and equipment manufacturer) is difficult. Integrated hazard analyses programs within fabs need immediate focus to address effects of a proposed equipment design change.

Modification to Any Facility Equipment's Design, or Operating Conditions

There are many different types of facility support equipment (e.g., chemical distribution equipment, electrical distribution equipment, chemical exhaust equipment, chemical abatement equipment, and waste removal equipment, etc.). As mentioned, these are also integrated together, so a more a holistic hazard identification approach is required to ensure fabs *continually* consider how any facility equipment's design change could impact the other equipment upstream and downstream of it.

Changes to Any Equipment Documentation

There is a very large volume of equipment documentation in our industry. This specifically includes any safety policies, manufacturing procedures, operating procedures, maintenance procedures, and associated work instructions. There are always opportunities to improve the instructions our workers must follow. An inadvertent change to clarify or fix one problem in the procedure may indirectly create an unforeseen hazardous situation. This is why the documentation must also be checked to ensure it is accurate and that it does not unintentionally introduce hazards due to a change in the instruction.

Changes in Legal/Regulatory Requirements

This chapter has already raised the issues about monitoring our equipment designs due to new safety or environmental regulatory requirements. What was once "good enough" last year, may not be satisfactory for full product compliance next year. This too must be monitored and managed.

Changes in the Knowledge of Health/Environmental Effects of Known Hazards

This last category of management of change applies more directly to industrial hygiene (IH) type of exposures (e.g., chemical inhalation, chemical skin contact, RF, IR, UV, noise, magnetic fields, etc.). These types of exposure limits changes are a common occurrence, and should also be monitored. This also may require going back and looking at previous equipment designs to see how they will hold up to the new exposure limit information.

45.3.3 More Design Focus during Maintenance/Service Tasks

One of the fundamental messages from this chapter is to have our entire industry seriously reconsider adding engineered layers of protection to protect against hazardous situations resulting from "reasonably foreseeable misuse"[4–6] during the downtime periods of service and maintenance tasks. If we can foresee it, then it can and will eventually happen. The major problem with the current approach (i.e., relying solely on administrative controls) is that they depend on humans, and humans will make mistakes.

The semiconductor industry has done a respectful job of incorporating equipment design features to mitigate risk of harm due to human mistakes during normal production uptime. Yet, most of these solutions during uptime are engineering controls, not administrative controls. This chapter's proposed resolution for the foreseeable downtime hazards is to simply design out them out from the beginning, instead of relying on the less effective administrative controls. Nowhere is this more evident than during the control of hazardous energies (CoHE). The complex term "CoHE" is more commonly referred to as "lockout/tagout" or "LOTO," but they are not the same. CoHE is a wide-ranging term describing the use of procedures, techniques, and specific equipment designs to prevent unintentional energizing or the release of stored hazardous energy, which can lead directly to harm. LOTO is simply one of the most common administrative methods in which to do this.

Common Examples of Hazardous Energies in Semiconductor Industry

- Distributed electrical (high voltages, high currents)
- Compressed gases (liquefied or pressurized)
- Electromagnetic radiation (x-ray, RF, IR, UV, lasers)
- Static magnetic fields (permanent magnets)
- Gravitational energy (e.g., suspended, hinged loads)
- Kinetic energy (moving robots, linear drives, gears)
- Stored electrical (capacitors, batteries)
- Pressurized liquids (hydraulic, pumped)
- Thermal/cryogenic energy (hot, cold temperatures)
- Chemical energy (heat of reaction, fire, explosion)
- Stored mechanical energy (springs, elastic seals), etc.

Inconvenient LOTO Locations Pose Design Concerns

Unfortunately, due to the equipment design and layout, certain LOTO procedures can be inconvenient, complicated, and slow. A common example of an inconvenient LOTO would be to LOTO a process chamber on wafer processing equipment with multiple process chambers. The worker(s) must first prepare the specified process module for shutdown. For most chemical systems this is not a trivial sequence of steps and may involve numerous pump/purges and/or flushing/draining of the chemical lines. The worker(s) must leave the main level of the fab and go downstairs (subfab). After leaving the cleanroom fab, they must de-gown their cleanroom suit, head covering, face mask, shoe coverings, and gloves (and other PPE as required). The worker(s) must then walk downstairs into the subfab and follow the documented LOTO protocols for isolating and deenergizing each of the multiple hazardous energies on the specific process chamber and any of its support modules, and then place individual padlock(s) on the appropriate energy isolating devices.

There are usually multiple locations of different types of hazardous energies (examples shown previously) which all need to be LOTO'd. Additionally, there are usually rows and rows of same/similar looking process equipment side by side, so the worker(s) must be careful to identify, then LOTO the correct process chamber to be worked at the fab level. They must then return upstairs and again follow the cleanroom protocols for re-gowning before entry into the fab. Next, the worker(s) must still complete the "verification" steps for each hazardous energy source at the fab level where the work is to be done. Once the maintenance or service task is completed, the worker(s) must exit the cleanroom, de-gown again, and go back downstairs to remove their personal locks within the subfab. They must then follow the cleanroom protocols for reentry into cleanroom, and then restart the equipment. If the situation is not corrected on the first attempt, the entire LOTO process in the fab/subfab has to be repeated until the situation is corrected. To help reduce incidents, our industry's process equipment designs should try to always ensure that there are no inconveniently located LOTO points. This should be a strategic focus of our engineering design teams to help prevent the reasonably foreseeable misuse (either intentionally or unintentionally) during required administrative LOTO, which may then lead to harm.

Semiconductor Industry Survey Highlights LOTO Differences

Based on a previous semiconductor industry survey[38] related to the lockout/tagout, the equipment suppliers, the end users (fabs), and the third party evaluators have each highlighted obvious differences of opinion related to the semiconductor industry's LOTO practices. There was a very wide discrepancy on the actual LOTO interpretations between different companies, and even differences within the same company, based on its different regions/countries where the equipment was installed.

This survey's results helped the SEMI EHS Committee to justify the formation of a new SEMI Task Force[39] to address these important LOTO interpretation inconsistencies from a global perspective. If we take an honest look at the incident data, we see that LOTO errors are still prevalent (even in our mature industry). If you disagree, just ask any safety professional at a semiconductor fab for their incident data due to improper LOTO! It is the hope that new design guidance within the normative text of SEMI S2 can be clarified, and other important design and administrative considerations added in a new Related Information section to prevent foreseeable misuse during maintenance/service tasks.

45.3.4 More Focus on Documenting Risk Assessments

Although there are already clear requirements for doing this, why is the risk assessment practice not being followed? Many EHS and engineering professionals within the semiconductor industry are familiar with the general concept of risk assessment, but most do not apply it formally on a regular basis.

Even if engineering teams are diligent enough to identify equipment hazards, it is less likely they will take the next step and estimate the hazard's risk of harm. This is expected because most EHS professionals are probably more at ease referring to the more *prescriptive* code requirements (i.e., a quote the law approach) instead of the subjective nature of estimating a hazard's risk level.

Another probable motive why risk levels are not frequently estimated may be the "fear of liability." Many EHS professionals, or their design engineering counterparts, feel too inexperienced to prepare them, or are too pressured to make a formal stand by documenting one. Ignoring the reality of the fact that a high risk hazard may actually be present within the equipment would be irresponsible. The entire risk assessment process allows for more effective communication up the management chain to the key decision makers. Without clear communication of our equipment risks, the upper management (who are unaware of the safety review details) may be misinformed and make the incorrect assumption that everything is ok, when it's not. The risk assessment practice takes time and effort to document correctly. Be careful. There are many ways to do a risk assessment "wrong." Better guidance is needed within our SEMI risk guidelines on the specific details of how to choose the severity of harm and likelihood of harm terms. This is an area of opportunity for future SEMI standard revisions. In addition, the overall risk assessment process should be taught in parallel with the new functional safety design criterion for safety interlocks. There are some important overlaps between the estimation of the hazard's risk level and the interrelated estimation of safety function's (safety interlock's) required performance level (PL_r) or safety integrity level (SIL) that should be considered.

45.3.5 An Advanced Engineering Design Program

Different engineering departments around the world all have similar problems, even if their process equipment has different technologies. This author proposes we re-evaluate our semiconductor equipment's product design for compliance programs based on the belief that we are stronger working together, than we are working independently (e.g., many global design teams are located in various geographic regions). Any new, advanced engineering design program should be based on the belief that the product compliance process is an ongoing journey.

Product Compliance Mission Statement

- Create a forum between the different engineering groups to share common problems.
- Remove roadblocks to product safety compliance.
- Remove roadblocks to product environmental compliance.
- Emphasize best known methods (BKMs) with focus on design review principles.
- Provide a forum for continued engineering training to address the changing global regulations.

Along this journey towards product compliance, there are certain milestones of progress that are laid out before us. Integrated and effective MoC is the ultimate goal at the end of our journey. Designing process equipment is not just done once, then you are finished! Checks and balances must be incorporated into the review processes to ensure we are asking the right questions with each design change. While any new program should continue to evolve, the author would like to share some of the key successes that advanced engineering design programs have had already, as well as some of the problem areas they have encountered along the way.

Engineers Want to Do the Best Job They Can!

This entire program is based on the common belief that our design engineers really want to do the best job they can do. In looking at the results from past product safety compliance audits, previous noncompliances frequently resulted from an honest lack of awareness of the design criterion. It is not that the design engineers knew they should do something and consciously avoided it—*No*! Usually, they were simply unaware of those design requirements beforehand. A product compliance program should help facilitate better collaboration and knowledge transfer between the design engineers, the engineering managers, and the product safety engineers to help avoid these issues.

First and foremost, there was an immediate need identified for the design engineers to receive *awareness* training to the many product compliance fundamentals already discussed in Sec. 45.2. As stated, even the best design engineers never learned these complex product compliance requirements when they went to engineering school, so if equipment suppliers want to be efficient and effective, we must create ongoing internal engineering training programs. We must prepare the current and the next generation of design engineers to be successful. This continued education effort is never really completed. It is forever ongoing and expanding.

Engineers want to solve real problems, and not just listen to just a lecture. Fortunately, these new training classes should not be your typical training PowerPoint presentations. The author has tried to spend dedicated time (in each course) working through some *real* life engineering design examples. Anybody can read a given safety standard or read the courses' PowerPoint slides. However, if you help an engineering group to solve a real problem during the class, then they can immediately see the positive benefit to this type of continuous learning. Good engineers also want to exceed expectations. We must expose them to these extra requirements that they are unaware of, or that have recently changed. The actual class itself is only the beginning. The learning process continues to take place after the class, through structured, reoccurring phone conferences, and followup e-mails. The strategy is to have these education classes updated annually. This enables refresher training for the engineers, to share the new/revised standards as they are released each year (training agenda available upon request).

Basic Engineering Training Examples This author's proposed program has defined a standard grouping of three *basic* engineering department training courses, which all the equipment design engineers should complete. To ensure committed participation, the training should be limited to a maximum of 10 per class. It was recognized that when providing training to larger class sizes, certain engineers becoming disinterested/disengaged.

- "A tour of safety design standards" (1 day class)
- "A tour of environmental regulations" (½ day class)
- "Hazards, risk, and management of change" (½ day class)

Specialized Training Examples

While the three standard courses above are intended for all design engineers, there are currently additional *special-interest* courses that were intentionally designed for design engineers with very specific job roles. The numbers of participants within these classes are much smaller than the *basic* training classes (usually less than five) and more focused examples are provided.

- "Advanced functional safety design" (4 day class)
- "Silane gas safety awareness" (½ day class)
- "High pressure design requirements" (½ day class)

Team Building Benefits

The advanced engineering design program's training classes themselves initiate the awareness of where engineers can go to look for certain regulatory requirements. The classes also provide opportunities for team building and can help create cross-department and cross-company alliances. For example, when a question or technical problem arises, an design engineer in Europe can now directly call an design engineer in United States, or Japan, that they met during the training. It helps bring good minds closer together to solve similar problems.

Common Engineering Program Roadblocks

The first roadblock the author encountered on this journey was the inherent difficulty of sharing ideas due to the different language barriers between Asia, United States, and Europe. Any global company will face similar challenges. Although original training material was first developed in English, effective knowledge transfer was hampered due to the differences in both spoken and written languages. The team members concluded that to maximize the engineer's aptitude, the language barriers must be removed. We recognized that the best way to train Japanese speaking engineers is with Japanese speaking trainers (and vice versa). As the product compliance teams expand globally to all the design teams worldwide, the necessary engineering training materials must be translated into the design engineer's respective languages (e.g., Japanese, Korean, Mandarin, German, English, etc.). It was also observed that the engineering directors and managers should select their strongest design engineers to participate in his program. Then these designated engineering coaches can go back to their own factories and provide training to their own design teams that they are comfortable with and already know very well.

Another key roadblock to making any program successful is the adequate allocation of resources. This authors' programs were no different. The complicated product compliance process discussed in this chapter reveals some vital facts about the expensive cost of doing insufficient engineering designs. Designing for both safety and environmental compliance in the development phase can be significantly less cost than reacting to field incidents and accidents, or import/export violations. Unfortunately, it is always hard to argue to allocate resources to prevent something that might happen in the future.

Our entire semiconductor industry shares these collective resource concerns. Face-to-face conversations with the decision makers are essential to fully scope the budget/resource problems. If we want to succeed in truly achieving product compliance moving forward, more effort and program planning will be required. Our semiconductor industry's leadership (i.e., from both the process equipment suppliers and the semiconductor fabrication companies) must make every effort to have an improved understanding about the consequences of living with the known risks of harm versus the unknown risks of harm and justifying the extra effort/cost (i.e., pay me now, or pay me *more* later!).

In order to make any advanced engineering design program successful, the final critical roadblock is how to validate the effectiveness of a companies' engineering change control process. Remember that in our semiconductor industry, product compliance is never truly finished until after the tool is deinstalled! Our equipment designs are *repeatedly* being upgraded to improve process performance. Once you have a fully compliant third-party evaluation report for a given model, then you are done, right? *No!* Product compliance is an iterative process, which must be managed over the life of the equipment to mitigate the foreseeable risks of harm, and to ensure the equipment remains in compliance. Yet, our industry does a fairly poor job of requiring the examination of the equipment's compliance after the initial audit report is published (i.e., the so-called *golden tool* design). What about the process equipment's level of safety or environmental compliance after engineering changes/upgrades have been made to it?

Unfortunately, the current third party audit protocols (i.e., SEMI S2 or Machinery Directive) spend minimal, to zero time assessing the company's engineering change control process to ensure that the original audited configuration will still remain compliant through its future design upgrades. *How should we evaluate this?* Should we be including this as a part of our current SEMI S2 or Machinery Directive evaluations, or consider doing this as totally separate MoC audit? If we do not verify the MoC, we may be further *magnifying* the semiconductor industry's confidence that we have effectively addressed all the reasonably foreseeable equipment hazards, when we may not have, yielding "a false sense of security"!

While it is easier to point out engineering problems, than to solve them, the engineering program roadblocks mentioned here should be considered, and then discussed within each of our own companies, and also at future semiconductor industry symposiums. These will not be easy conversations. It may take months, or even years to fully explain the many design challenges outlined here, before adequate engineering resources are to be allocated. Any advanced engineering design programs like it will require a constant vision toward the future to be successful.

45.4 CONCLUSION: "SOLVING THE PUZZLE"

This *Semiconductor Manufacturing Handbook* has reminded its readers of the monumental engineering design difficulties our industry faces as it attempts to invent the next, innovative process of records. In addition, this chapter has highlighted further product compliance obstacles. Equipment manufacturers are legally obligated to understand their equipment's potential hazards, then design to mitigate them to low residual risk levels. This is very problematic due to the high complexity of our advanced equipment and the continually changing regulatory requirements. It necessitates us to now take a step back, and reconsider if there is a more effective, alternate approach to moving forward. This section attempts to highlight the most critical of the design challenges already discussed. It will also challenge the semiconductor industry's status quo and recommend important changes. Such improvements can better prepare the design engineers and also to better direct the third party auditors who evaluate our wafer process equipment.

45.4.1 Functional Safety Engineering Design for Maintenance/Service Tasks

In recent years, a very strong new focus has been placed on the strict design and verification of our equipment's safety functions (i.e., historically referred to as safety interlocks). These safety function design requirements are also based on a hazard's risk assessment. As is evident from Table 45.1, these new functional safety standards for equipment[40–42] make up a large percentage of the total number of pages of published design criterion for semiconductor equipment compliance. They are not trivial standards to comprehend, implement, and verify.

Advanced equipment designs should be considering how functional safety circuits can mitigate the risk of harm, especially during the maintenance/service tasks which are more prone to mistakes. Complex and cumbersome LOTO's tasks (see the subsection "Inconvenient LOTO Locations Pose Design Concerns") can be replaced with robust functional safety control circuits to provide acceptable alternative "remote LOTO" methods for inconveniently located hazardous energy isolation devices. Following design guidance from other international standards,[43,44] the use of functional safety designs to prevent the unexpected energization of the equipment has been used with much success throughout the world. Remote LOTO designs can offer key advantages, especially when compared to the more inconvenient LOTO tasks. Remote LOTO designs can isolate and deenergize multiple hazardous energies simultaneously. Having them located conveniently where the task is to be performed encourages the proper application of LOTO. These designs can also have monitoring diagnostics to help the worker perform the critical "verification of isolation" and "verification deenergization" steps.

Unfortunately, U.S. LOTO regulations[45] only permit use of functional safety control circuits if they meet specific "minor service exemption" requirements. However, outside the U.S. employers are permitted to use functional safety designs to prevent the unexpected energization of the equipment. There has been precedence for the United States, allowing control circuits to control the hazardous energy,[46] but to fully sanction their use, requires the US regulations scope to be expanded. There are semiconductor industry task forces[47] and US national standards committees[48] currently in the process of updating their respective CoHE design requirements. This important work should be expedited in order to advance the safety performance of the equipment in our industry. This author feels strongly that the current OSHA regulations must be expanded to allow for such "remote LOTO" solutions, especially during maintenance/service tasks, where the LOTO steps are complex and prone to human error.

45.4.2 Advanced Engineering Design Programs Can Help Close the Knowledge Gaps

Today, there are just too many design requirements (recall Table 45.1) for a single product safety professional on an engineering team to monitor and review. The recommended solution, as outlined in this chapter, is to "coach up" the existing design engineers to the crucial requirements that their individual design specialties require. This also poses new challenges. Even at the world's very best engineering colleges/universities, the newly graduated design engineers (e.g., electrical, mechanical, chemical, software, etc.) are simply not well trained on how best to comply with the international health and safety regulations within each country, or how to uncover inherent (hidden) design deficiencies within their own designs.

Nevertheless, these new design team skills are undeniably essential to ensuring that semiconductor companies can provide safe (low residual risk) and regulatory compliant equipment to our customers globally. The suggested solution on how to solve the complex, product compliance puzzle was to create an overall "advanced engineering design" program. This will allow engineers to share common design problems, share BKMs, and will help remove the roadblocks to achieving full product compliance. This chapter has briefly described this author's image of this type of program. This type of advanced engineering design program is never really completed. Instead, it provides an ongoing platform for our design engineers, and their managers to utilize to help the engineering team maintain the product design's compliance over the entire life cycle of their equipment.

Recall that there are recognized roadblocks which must also be identified and overcome. The first challenge requires multilanguage content review and the corresponding training material translations in order to train our global design teams. Another key barrier is the adequate allocation of resources. Unfortunately, it is always hard to argue to provide resources to prevent something that might happen in the future. Still, exploiting the risk assessment process can provide more effective communication to the key decision makers about the consequences of living with *unknown* risks, which may in turn help our companies to justify these extra efforts/costs.

Last, we will still need to address the known deficiencies with our existing third party evaluation process of our equipment designs. We already recognize that it is very difficult for the third parties to understand all the details of our complex equipment designs. Likewise, it is almost impossible for them to validate the future effectiveness of a companies' engineering change control process. Thus, our industry as a whole must take an honest step back and seriously reconsider these advanced engineering design program roadblocks, to see if there is a more effective, alternate approach to solving this product compliance puzzle.

45.4.3 New Safety of Record Performance Metrics

As a final point, this chapter will close by proposing some potentially controversial ideas for the semiconductor industry to consider for the future in advanced equipment design. We have already a well-established means for comparing different competitor's equipment designs using the various process of record (PoR) metrics. Another way to compare different equipment designs is by their reliability performance metrics (e.g., MTTF, MWBF, %Uptime). Eventually, additional industry comparisons could evolve to include the process equipment's so-called *safety of record* (SoR). Would it not be nice if customers had the ability to compare different competitor's equipment safety performance? *Yes!* The original idea of comparing equipment SoRs was first proposed by this author in a semiconductor journal back in 2008 (available as a SESHA member, or upon request).[49]

The entire "safety of record" idea was born out of the previous work done back in the 1990s with SEMI's equipment reliability standard—SEMI E10.[50] This standard helped the semiconductor industry standardize on how reliability performance metrics were computed. It introduced the necessary statistical calculation knowledge for correctly analyzing interruption data on our process equipment. The semiconductor industry could potentially create a new SEMI "S" guideline to compute our equipment's safety performance [e.g., mean time between "harm" (MTBH), or mean wafers between "harm" (MWBH)] using the guidelines already outlined within SEMI E10. If this is indeed plausible, then safety performance comparisons could be made between different equipment models to determine "best in class." Safety performance could be based on the number of incidents of harm that are actually occurring on each model (e.g., number of injuries, first aides, near misses, equipment damage, etc.). This would give us an entirely new way to evaluate product safety compliance of our equipment!

After the minimum SEMI E10 reliability metrics were specified within customer purchase orders (POs) and warranty agreements, it rapidly drove semiconductor equipment manufacturers to improve their reliability programs. If these new SoR performance metrics could be included into customer POs/warranties, then it could also drive product safety programs to advance much more quickly than they are today. We could have data driven safety performance results to highlight which of the various fab's wafer process equipment has had poor product safety compliance execution. The data would reveal which designs performed well, and which equipment designs needed improvement. Fittingly, paraphrasing the astute words said over 100 years ago by William Thomson (a.k.a. Lord Kelvin): "If you can't measure it, You can't improve it!"[51,52]

The premise is that the equipment's advanced engineering design either keeps workers safe or it does not. The actual data from each tool set could demonstrate this. Successful implementation of advanced engineering design programs could be validated with their actual performance data. Once achieved, the proof of a companies' product compliance would no longer be based solely on receiving just a point in time third-party evaluation report (e.g., for just the "golden tool"). Nor would it be based on the theoretical evidence included within the companies' Technical File (TF). Instead, our industry's product safety compliance realizations would be justified based on the physical, real-life safety performance data of that equipment (i.e., MTBH or MWBH). This is essentially a way of fingerprinting our equipment's product safety designs. It could highlight true competitive advantages between different equipment suppliers who spend the extra time and resources to accomplish the very difficult tasks described within this chapter. This proposal should get both the EHS and engineering professionals thinking about what the next generation of product compliance should really look like.

Is this too difficult to imagine? Maybe it is. Maybe it is not. Time will tell.

45.5 REFERENCES

1. SEMI S2-0715, "Environmental Health and Safety Guideline for Semiconductor Manufacturing Equipment," Semiconductor Equipment and Materials International (SEMI), 124 pages, http://ams.semi.org/ebusiness/standards/SEMIStandardDetail.aspx?ProductID=1948&DownloadID=3592 (last accessed November 3, 2016).

2. Machinery Directive (2006/42/EC), Official Journal of the European Union, 63 pages, http://www.fmmi.at/uploads/media/MRL_neu_engl_02.pdf (last accessed November 3, 2016).

3. SEMI AUX031-1114, "S2 Mapping into the Machinery Directive (2006/42/EC) Essential Health and Safety Requirements," SEMI Semiconductor Equipment and Materials International, 84 pages, http://www.semi.org/en/sites/semi.org/files/docs/AUX031-00-1114.pdf (last accessed November 3, 2016).

4. Machinery Directive (2006/42/EC) Annex 1, *General Principles*, p. 35, Sec. 1.1.1 (i), p. 35, Sec. 1.1.2, p. 36, and Annex VII, p. 71, http://www.fmmi.at/uploads/media/MRL_neu_engl_02.pdf (last accessed November 3, 2016).

5. SEMI S2-0715, "Environmental Health and Safety Guideline for Semiconductor Manufacturing Equipment," Sec. 6.5, p. 10, Semiconductor Equipment and Materials International (SEMI), http://ams.semi.org/ebusiness/standards/SEMIStandardDetail.aspx?ProductID=1948&DownloadID=3592 (last accessed November 3, 2016).

6. ISO 12100 (2010), "Safety of Machinery—General Principles for Design—Risk Assessment and Risk Reduction," International Standards Organization, Sec. 3.24, p. 4 and Sec. 5.4 (c), p. 16, http://www.iso.org/iso/home/store/catalogue_tc/catalogue_detail.htm?csnumber=51528 (last accessed November 3, 2016).

7. Clifton A. Erikson, *Hazard Analysis Techniques for System Safety*, 2d ed., John Wiley & Sons, Inc., 2015; 640 pages, ISBN: 978-1-118-94038-9, http://www.wiley.com/WileyCDA/WileyTitle/productCd-1118940385.html (last accessed November 3, 2016).

8. *Guidelines for Hazard Evaluation Procedures*, 3d ed., American Institute of Chemical Engineers, Inc., 2008, John Wiley & Sons, Inc., 576 pages, ISBN: 978-0-471-97815-2, http://www.wiley.com/WileyCDA/WileyTitle/productCd-0471978159.html (last accessed November 3, 2016).

9. Brian Tyler, Frank Crawley, and Malcolm Preston, *HAZOP: Guide to Best Practice*, 2d ed. Institution of Chemical Engineers (IChemE), Rugby, 2008, ISBN: 9780852955253 (entire book as a reference), https://www.icheme.org/shop/books/safety/hazop%20guide%20to%20best%20practice.aspx (last accessed November 3, 2016).

10. D. H. Stamatis, *Failure Mode and Effect Analysis: FMEA from Theory to Execution 2003*, The American Society for Quality—ASQ Press, 488 pages, ISBN: 978-0-87389-598-9, http://asq.org/quality-press/display-item/?item=H1188 (last accessed November 3, 2016).

11. Nancy G. Leveson, *Engineering a Safer World—Systems Thinking Applied to Safety*, MIT Press, 560 pages, Jan. 2012, ISBN: 9780262016629, https://mitpress.mit.edu/books/engineering-safer-world (last accessed November 3, 2016).

12. Susan Cantrell and Pat Clemens, "Finding All Hazards," *Professional Safety—Journal of The American Society of Safety Engineers*, Vol. 54, Issue 11, p. 32, Nov. 2009. https://www.researchgate.net/publication/254508937_Finding_All_the_Hazards_How_Do_We_Know_We_Are_Done (last accessed November 3, 2016).

13. Machinery Directive (2006/42/EC), Official Journal of the European Union, Annex 1, General Principles 1, p. 35, and Sec. 1.1.2, p. 36, http://www.fmmi.at/uploads/media/MRL_neu_engl_02.pdf (last accessed November 3, 2016).

14. SEMI S2-0715, "Environmental Health and Safety Guideline for Semiconductor Manufacturing Equipment," Sec. 6.5 and Note 9, p. 10, Sec. 6.8, p. 11, Sec. 7.3, p. 12, Sec. 8.3.4.3, p. 13; Semiconductor Equipment and Materials International (SEMI), http://ams.semi.org/ebusiness/standards/SEMIStandardDetail.aspx?ProductID=1948&DownloadID=3592 (last accessed November 3, 2016).

15. SEMI S10-0815, "Safety Guideline for Risk Assessment and Risk Evaluation Process," Semiconductor Equipment and Materials international (SEMI), 20 pages, http://ams.semi.org/ebusiness/standards/SEMIStandardDetail.aspx?ProductID=1948&DownloadID=3590 (last accessed November 3, 2016).

16. SEMI S14-0309, "Safety Guideline for Fire Risk Assessment Mitigation for Semiconductor Manufacturing Equipment," Semiconductor Equipment and Materials international (SEMI), 21 pages, http://ams.semi.org/ebusiness/standards/SEMIStandardDetail.aspx?ProductID=211&DownloadID=1333 (last accessed November 3, 2016).

17. ISO 12100 (2010), "Safety of Machinery—General Principles for Design—Risk Assessment and Risk Reduction," International Standards Organization, 77 pages, http://www.iso.org/iso/home/store/catalogue_tc/catalogue_detail.htm?csnumber=51528 (last accessed November 3, 2016).

18. Machinery Directive (2006/42/EC), Official Journal of the European Union, Annex 1, Sec. 1.1.1 (a), p. 35, http://www.fmmi.at/uploads/media/MRL_neu_engl_02.pdf (last accessed November 3, 2016).

19. ISO 12100 (2010), "Safety of Machinery—General Principles for Design—Risk Assessment and Risk Reduction," International Standards Organization, Sec. 3.5, p. 2, http://www.iso.org/iso/home/store/catalogue_tc/catalogue_detail.htm?csnumber=51528 (last accessed November 3, 2016).

20. SEMI S10-1205, "Safety Guideline for Risk Assessment and Risk Evaluation Process," Semiconductor Equipment and Materials international (SEMI), Sec. 5.1.4, p. 2, App. 1, Table A1-1, p. 7, http://ams.semi.org/ebusiness/standards/SEMIStandardDetail.aspx?ProductID=1948&DownloadID=3590 (last accessed November 3, 2016).

21. SEMI S14-1205, Safety Guideline for Fire Risk Assessment Mitigation for Semiconductor Manufacturing Equipment," Semiconductor Equipment and Materials international (SEMI), Sec. 5.11, p. 3, 5.1.18, p. 4, App. 1, Table A1-1, http://ams.semi.org/ebusiness/standards/SEMIStandardDetail.aspx?ProductID=211&DownloadID=1333 (last accessed November 3, 2016).

22. ISO/TR 14121-2:2012(E), "Safety of Machinery—Risk Assessment—Part 2: Practical Guidance and Examples of Methods," Technical Report Second Edition, 2012-06-01, Sec. 5.4.2, p. 7, http://www.iso.org/iso/home/store/catalogue_tc/catalogue_detail.htm?csnumber=57180 (last accessed November 3, 2016).

23. Pressure Equipment Directive (2014/68/EU), "On the Harmonisation of the Laws of the Member States Relating to the Making Available on the Market of Pressure Equipment," Official Journal of the European Union, May 2014, 96 pages, http://eur-lex.europa.eu/legal-content/EN/TXT/PDF/?uri=CELEX:32014L0068&from=EN (last accessed November 3, 2016).

24. ASME B31.3-2014, "Process Piping," American Society of Mechanical Engineers, 536 pages, https://www.asme.org/products/codes-standards/b313-2014-process-piping-(1) (last accessed November 3, 2016).

25. RoHS2 Directive (2011/65/EU), "On the Restriction of Hazardous Substances in Electrical and Electronic Equipment," Jun. 8, 2011, Amended Annex II to Directive 2011/65/EU, Official Journal of the European Union, Mar. 31, 2015, 23 pages, http://eur-lex.europa.eu/legal-content/EN/TXT/PDF/?uri=CELEX:32011L0065&from=EN (last accessed November 3, 2016).

26. EU REACH Regulation (EC) No 1907/2006, "Registration, Evaluation, Authorisation and Restriction of Chemicals," European Union Regulation, Dec. 18, 2006, 525 pages, http://eur-lex.europa.eu/LexUriServ/LexUriServ.do?uri=CONSLEG:2006R1907:LATEST:EN:PDF (last accessed November 3, 2016).

27. EU (ErP) Directive (2009/125/EC), "Energy-related Products Directive," Official Journal of the European Union, Oct. 21, 2009, 26 pages, http://eur-lex.europa.eu/LexUriServ/LexUriServ.do?uri=OJ:L:2009:285:0010:0035:en:PDF.

28. EU WEEE Directive (2012/19/EU), "Waste Electrical and Electronic Equipment Directive," Official Journal of the European Union, Jul. 4, 2012, 34 pages, http://ec.europa.eu/environment/waste/weee/legis_en.htm (last accessed November 3, 2016).

29. EU F-Gas Regulation; Regulation (EU) No 517/2014, "On Fluorinated Greenhouse Gases," Official Journal of the European Union, Apr. 6, 2014, 36 pages, http://eur-lex.europa.eu/legal-content/EN/TXT/PDF/?uri=CELEX:32014R0517&from=EN (last accessed November 3, 2016).

30. China RoHS, "Management Methods for the Restriction of the Use of Hazardous Substances in Electrical and Electronic Products," http://zfs.miit.gov.cn/n11293472/n11294912/n11296182/14644981.html (MIIT Original Chinese Version—9 pages) (last accessed November 3, 2016).

31. "The State of the National Initiative on Prevention through Design—Progress Report 2014," Department of Health and Human Services, Centers for Disease Control and Prevention, National Institute for Occupational Safety and Health; NIOSH publication No. 2014-123, 40 pages, http://www.cdc.gov/niosh/docs/2014-123/pdfs/2014-123_v2.pdf (last accessed November 3, 2016).

32. "Priorities for Occupational Safety and Health Research in Europe 2013–2020," European Agency for Safety and Health at Work; Publications Office of the European Union, 2013, 107 pages, https://osha.europa.eu/en/tools-and-publications/publications/reports/priorities-for-occupational-safety-and-health-research-in-europe-2013-2020 (last accessed November 3, 2016).

33. David W. Wilbanks, "Prevention through Design—A Curriculum Model to Facilitate Hazard Analysis and Risk Assessment," *Professional Safety—Journal of The American Society of Safety Engineers,* Apr. 2015, pp. 46–51, https://www.onepetro.org/journal-paper/ASSE-15-04-46 (last accessed November 3, 2016).

34. CMSE®—Certified Machinery Safety Expert [Internet], https://www.pilz.com/en-US/services/trainings/cmse-certified-machinery-safety-expert (last accessed November 3, 2016).

35. Certified Functional Safety Engineer: Functional Safety of Machinery [Internet], http://www.tuvasi.com/en/trainings-and-workshops/tuev-rheinland-functional-safety-program/trainings-and-contents/functional-safety-of-machinery (last accessed November 3, 2016).

36. Functional Safety Process Hazard and Risk Analysis (PH & RA), [Internet], http://www.tuvasi.com/en/trainings-and-workshops/trainings/process-hazard-and-risk-analysis/182-prosalus (last accessed November 3, 2016).

37. SEMI Website [Internet], SEMI Standard Technical Education Programs (STEP), STEP and Other Standards Program Proceedings, http://dom.semi.org/web/wstandards.nsf/STEPProc (last accessed November 3, 2016).

38. Lockout/Tagout (LOTO) Survey, SEMI EHS Standards Committee, Sent: Tuesday, Mar. 26, 2013; results available, upon request, to any SEMI Standards Program Member.

39. SEMI EHS Committee—North America EHS Technical Committee, Chapter Meeting Summary and Minutes; SEMICON West 2013 Meetings, Jul. 17 , 2013, pp. 2, 14, 15, http://downloads.semi.org/standards/minutes.nsf/91eeb64567db378c88256dcf006a4252/2733c346620d255588257be800697c28/$FILE/EHS%20Minutes%20West%202013%20SF.pdf (last accessed November 3, 2016).

40. EN ISO 13849-1 (2008), "Safety of Machinery: Safety Related Parts of Control Systems—Part 1 General Principles of Design," European Committee for Standardization (CEN). British Standards Institution, 98 pages, http://www.standardsuk.com/products/BS-EN-ISO-13849-1-2008.php (last accessed November 3, 2016).

41. EN ISO 13849-2 (2012), "Safety of Machinery: Safety Related Parts of Control Systems—Part 2 Validation," European Committee for Standardization (CEN), British Standards Institution, British Standards Institution, 90 pages, http://www.standardsuk.com/products/BS-EN-ISO-13849-1-2008.php (last accessed November 3, 2016).

42. IEC 62061:2005+AMD1:2012+AMD2:2015 CSV, "Safety of Machinery: Functional Safety of Safety Related Electrical, Electronic and Programmable Electronic Control Systems," European Committee for Electrotechnical Standardization (CENELEC), British Standards Institution, 387 pages, https://webstore.iec.ch/publication/22797.

43. EN ISO 14118 (2000), "Safety of Machinery: Prevention of Unexpected Start-Up," International Standards Organization (ISO), 20 pages, http://www.iso.org/iso/iso_catalogue/catalogue_tc/catalogue_detail.htm?csnumber=22584 (last accessed November 3, 2016).

44. EN ISO 14119 (2013), "Safety of Machinery: Interlocking Devices Associated with Guards—Principles for Design and Selection," International Standards Organization (ISO), 68 pages, http://www.iso.org/iso/iso_catalogue/catalogue_tc/catalogue_detail.htm?csnumber=45291 (last accessed November 3, 2016).

45. Directive Number: CPL 02-00-147, "The Control of Hazardous Energy—Enforcement Policy and Inspection Procedures," U.S. Department of Labor Occupational Safety and Health Administration, Feb. 11, 2008, pp. 2-15, 2-16, and 3-25, 3-28, https://www.osha.gov/OshDoc/Directive_pdf/CPL_02-00-147.pdf (last accessed November 3, 2016).

46. OSHA Website [Internet], Selected Case Law: Docket Number: 91-2973, Occupational Safety and Health Review Commission and Administrative Law Judge Decisions, 04/26/1995—91-2973—General Motors Corporation, Delco Chassis Division, https://www.osha.gov/dts/osta/lototraining/caselaw/cl-gm-2973.html (last accessed November 3, 2016).

47. SEMI EHS Committee—Control of Hazardous Energy Task Force Report; North America EHS Technical Committee Chapter Meeting Summary and Minutes; SEMICON West 2013 Meetings, Jul. 17, 2015, pp. 34–35, http://downloads.semi.org/standards/minutes.nsf/91eeb64567db378c88256dcf006a4252/fe55c2f09a14ca3c88257eba00655783/$FILE/EHS%20Minutes%20West%202015%20SF.pdf (last accessed November 3, 2016).

48. ANSI Z244 Committee—ANSI Accredited Standards Committee—Control of Hazardous Energy—Lockout/Tagout and Alternative Methods. SECRETARIAT: American Society of Safety Engineers, http://www.asse.org/publications/standards/secretariats (last accessed November 3, 2016).

49. Mark J. Fessler, "Statistics: An Introduction to a New Way of EHS Analysis and Reporting," *SESHA Journal*, Spring 2008, p. 12, http://www.seshaonline.org/ejournal/articles/article.php3?id=52 (must be SESHA member to access article) (last accessed November 3, 2016).

50. SEMI E10-0814, "Specification for Definition and Measurement of Equipment Reliability, Availability, and Maintainability (RAM) and Utilization," Semiconductor Equipment and Materials international (SEMI), http://ams.semi.org/ebusiness/standards/SEMIStandardDetail.aspx?ProductID=1948&DownloadID=3446 (last accessed November 3, 2016).

51. William Thomson (often referred to simply as Lord Kelvin); Quotes: Lecture on "Electrical Units of Measurement" (May 3, 1883), published in *Popular Lectures*, Vol. I, p. 73, "I often say that when you can measure what you are speaking about, and express it in numbers, you know something about it; but when you cannot measure it, when you cannot express it in numbers, your knowledge is of a meagre and unsatisfactory kind; it may be the beginning of knowledge, but you have scarcely, in your thoughts, advanced to the stage of science, whatever the matter may be." https://en.wikiquote.org/wiki/William_Thomson (last accessed November 3, 2016).

52. Talk: William Thomson [Internet], "If you cannot measure it, you cannot improve it." https://en.wikiquote.org/wiki/Talk:William_Thomson (last accessed November 3, 2016).

Chapter 46

Cleanroom Design and Construction

BY

RICHARD V. PAVLOTSKY *cGMP Technologist, Woodland Hills, California*

46.1 INTRODUCTION

Manufacturing cleanroom is a complex organism whose purpose is to both support and protect manufacturing processes and personnel. Every cleanroom is composed of the walls, ceilings, floors, and air systems. For the purpose of this chapter, the cleanroom will be considered to be the classic cleanroom with central plant (HVAC, exhaust systems, utilities, etc.), and supporting process systems (ultra-high purity (UHP) gases, UHP water, process cooling, etc.).

When developing an overall manufacturing strategy, the cleanroom shall be considered as a process element. Just as each manufacturing process associated with high-tech manufacturing shall be studied and analyzed in depth in order to optimize its function for producing the finished product, the cleanroom shall be studied and analyzed for optimal support of the manufacturing process. The intent of this chapter is to provide an overview of the issues regarding functionality of the cleanroom as it relates to high-tech manufacturing.

46.2 CLEANROOM STANDARDS AND CLASSIFICATIONS

The U.S. FED-STD 209E provided guides on cleanliness, cleanroom design, construction, and certification that was canceled and superseded by ISO 14644 standards with the following parts:

Part 1 Classification of air cleanliness (Tables 46.1 and 46.2)
Part 2 Monitoring cleanroom performance
Part 3 Test methods
Part 4 Design, construction, and start-up
Part 5 Operations
Part 7 Separative devices
Part 8 Classification of air cleanliness by chemical concentration
Part 9 Classification of surface particle cleanliness
Part 10 Classification of surface cleanliness by chemical concentrations
Part 12 Specifications for monitoring air cleanliness by nanoscale particle concentration
Part 13 Cleaning of surfaces to achieve defined levels of cleanliness in terms of particle and chemical classifications
Part 14 Assessment of suitability for use of equipment by airborne particle concentration
Part 15 Assessment of suitability for use of equipment and materials by airborne chemical concentration

Definitions of Cleanroom

ISO Standard 14644 and British Standard 5295: "A room with control of particulate contamination, constructed and used in such a way as to minimize the introduction, generation and retention of particles inside the room and in which temperature, humidity and pressure shall be controlled as is necessary" (Table 46.3).

Certifications of the cleanroom during construction execution, including process equipment (tools) installation, typically occur at three stages: "as-built," "at-rest," and "operational."

1. "As-built" certification occurs after the installation of the terminal filters [high-efficiency particulate air (HEPA) or ultralow penetration air (ULPA)] with all ventilation systems fully operational and balanced (Table 46.4). Typically three tests are conducted during this certification: cleanliness classification, terminal filter integrity (air leakage from the actual filter due to damage or manufacturing issues and leakage at the grid/plenum filter seal), and air velocity/volume. This initial certification verifies that the cleanroom is in an acceptable condition for the process tools and provides valuable data that will be used to "fine tune" the ventilation systems.

2. "At-rest" certification occurs after the installation of the process equipment (tools), prior to their commissioning and qualifications, and without personnel present in the cleanroom. The purpose of this certification is to determine the effect of the process tools on the ventilation systems and to ensure that the installation of the process tools has not had a negative effect on the cleanliness of the cleanroom. Typically, cleanliness classification is the test performed during this certification (provided that it was preceded by an "at-rest" certification).

3. "Operational" certification occurs after the process equipment (tools) have been commissioning and with normal operating quantity of personnel in the cleanroom. The "operational" certification is the definitive measurement of the cleanroom functionality because it simulates actual operational conditions and tests for all critical functions. Typical tests include cleanliness classification, pressurization, airflow parallelism, temperature, and humidity uniformity. Other tests may be incorporated into the "operational" certification if desired by the client or if required to verify a function that has particular potential impact

Table 46.1 Cleanroom ISO Standard 14644-1:2015 (FED-STD 209E was superseded by ISO STD.)

	Maximum Concentrations (particles/m^3) for particles equal to and greater than considered sizes						EU GMP	
ISO Class	0.1 µm	0.2 µm	0.3 µm	0.5 µm	1.0 µm	5.0 µm	At rest	In Oper.
1	10							
2	100	24	10					
3	1,000	237	102	35				
4	10,000	2,370	1,020	352	83			
5	100,000	23,700	10,200	3,520	832		A, B	A
6	1,000,000	237,000	102,000	35,200	8,320	293		
7				352,000	83,200	2,930	C	B
8				3,520,000	832,000	23,900	D	C
9				35,200,000	8,320,000	293,000		
	EU Good Manufacturing Practice (GMP)	A (At rest/In operation)		3,520/3,520		20/20		
		B (At rest/In operation)		3,520/352,000		29/2,900		
		C (At rest/In operation)		352,000/3,520,000		2,900/29,000		
		D (At rest/In operation)		3,520,000/ND		29,000/ND		

ND: Not defined.

Table 46.2 Air Cleanliness Classification ISO Standards

	Measured particle size (µm) per cubic meter				
ISO	0.1 µm	0.2 µm	0.3 µm	0.5 µm	5 µm
1	10	2			
2	100	24	10	4	
3	1,000	237	102	35	
4	10,000	2,370	1,020	352	
5	100,000	23,700	10,200	3,520	29
6	1,000,000	237,000	102,000	35,200	293
7				352,000	2,930
8				3,520,000	29,300
9				35,200,000	293,000

Table 46.3 Typical Cleanroom Air Velocities (IEST-RP-CC012)

ISO	Airflow type	Airflow velocity, M/S	Air changes per hour M3/h
2	Unidirectional	0.3–0.5	
3	Unidirectional	0.3–0.5	
4	Unidirectional	0.3–0.5	
5	Unidirectional	0.2–0.5	
6	Non-U or Mixed		70 to 160
7	Non-U or Mixed		30 to 70
8	Non-U or Mixed		10 to 20

Table 46.4 Typical Cleanroom Filter Coverage

Cleanroom classification	Filter type	Airflow velocity Ft/Min	Filter coverage
1	FFU	60–100	80–100%
2	FFU	60–100	80–100%
3	FFU	60–90	60–100%
4	FFU	50–90	50–90%
5	FFU	40–80	35–70%
6	FFU	25–40	25–40%
7	FFU	10–15	15–20%
8	FFU	1–8	5–15%

on process. One of the key reasons for a "staged or phased" approach to certification is that this process allows for easier troubleshooting by isolating the facility from influence of process tools and personnel.

46.3 TYPES OF CLEANROOMS

Today's microelectronics facility cleanroom is a complex system whose purpose is to support and protect manufacturing processes and personnel. Cleanrooms generally constructed following basic conceptual configurations as

1. *Ballroom.* Ballrooms are typically open areas with ancillary functions at the perimeter (Figs. 46.1 and 46.2).
2. *Bay and chase.* Bay and chase cleanrooms are typically configured as clean "fingers" with supporting functions behind the chases walls in between (Figs. 46.3 and 46.4).
3. *Minienvironments.* A greater level of cleanliness in a localized area to effectively and economically support manufacturing functions may be achieved with Minienvironments (Fig. 46.5). Minienvironments offer significant savings in both the initial and life cycle cost of the cleanroom. Minienvironments are commonly constructed of a framing system (internal of the cleanroom structure) and equipped with fan-filter units (FFUs) and sidewall enclosure. The minienvironment isolates the equipment and process from the influence of surrounding conditions and particles. The FFUs take pre-conditioned and filtered air directly from the cleanroom, filter the air to a higher level of cleanliness, and distribute it appropriately to achieve a desired contamination control strategy.
4. *Combination of ballroom with installed minienvironments.* Most often, current design practice in cleanrooms utilizes the mix of these configurations (Fig. 46.5) in order to address the specific process requirements, economics, and space limitations. The objective of investing in a construction of a cleanroom is to achieve certifiable cleanliness in optimized space with most economical solution for a given manufacturing process type and known constrains.

46.4 AIRFLOW LAYOUTS AND PATTERNS

The airflow in cleanrooms may be vertical with supply air coming from the ceiling and return air being removed through the openings in the floor, vertical with supply air coming from the ceiling and return air being

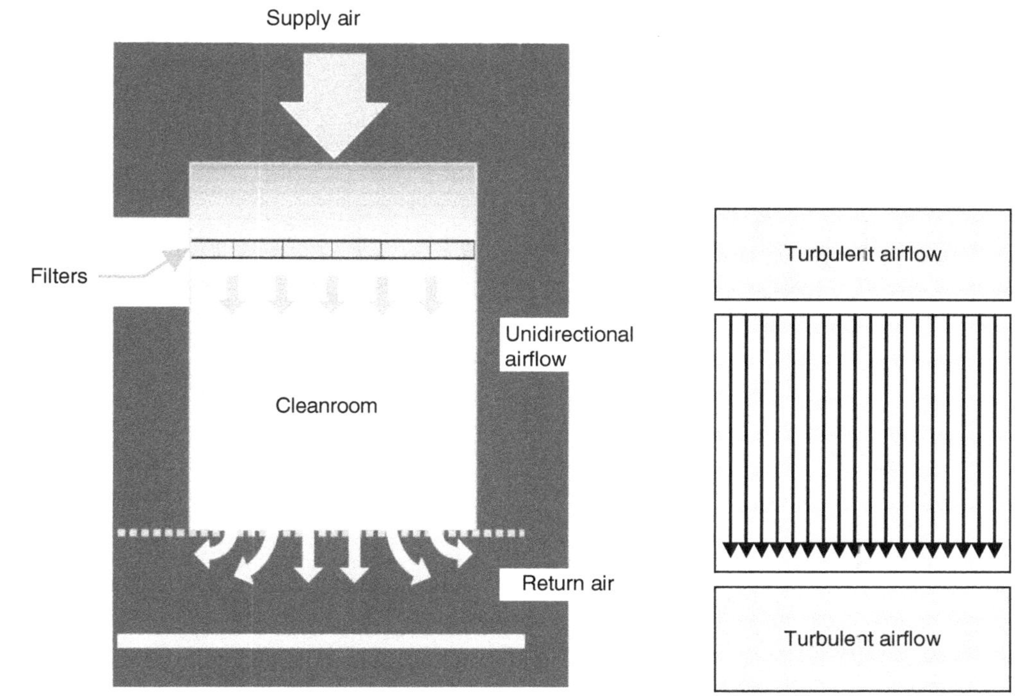

Figure 46.1 Ballroom concept unidirectional airflow.

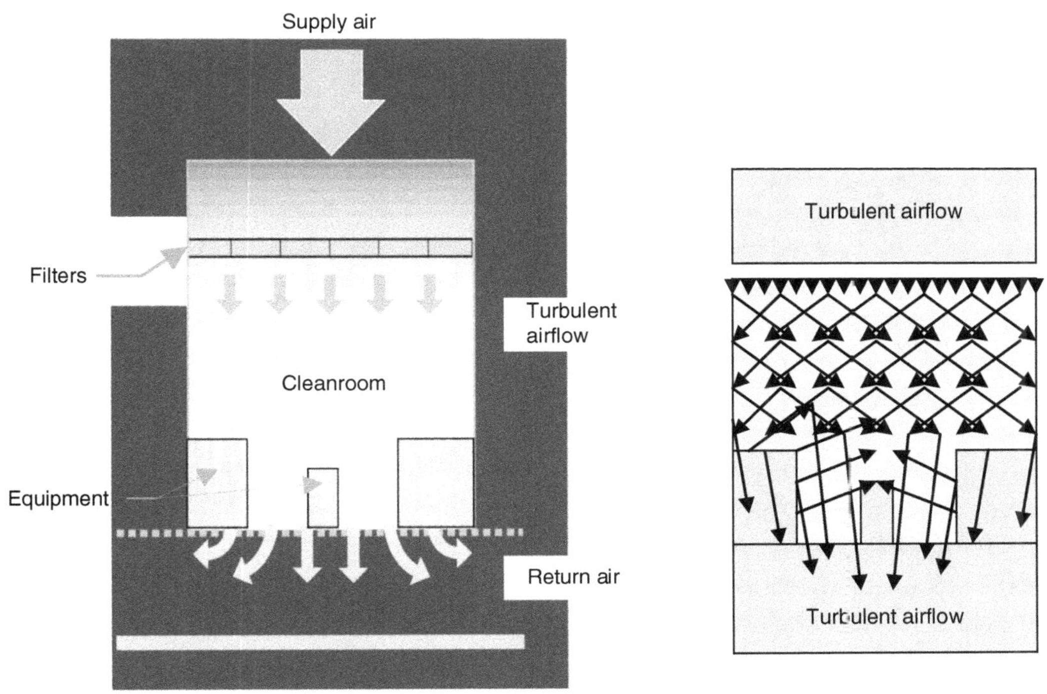

Figure 46.2 Ballroom concept turbulent airflow.

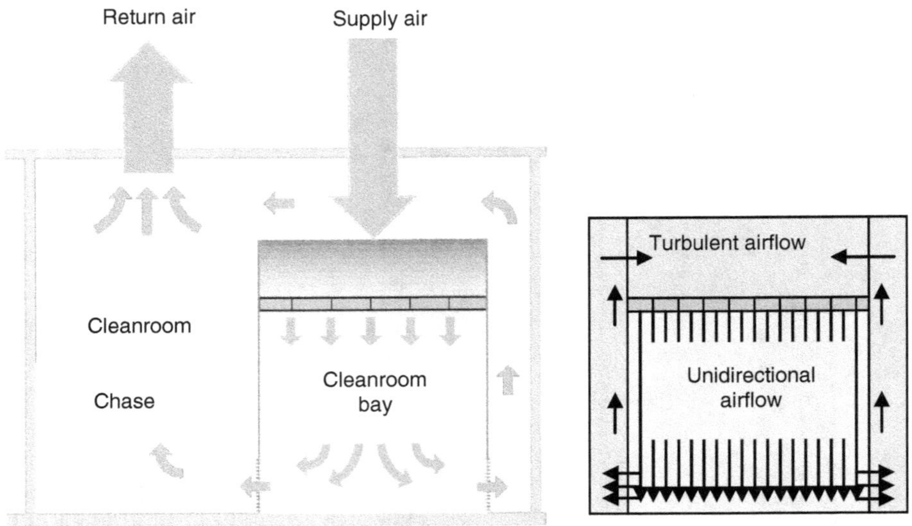

Figure 46.3 Bay and chase concept unidirectional airflow.

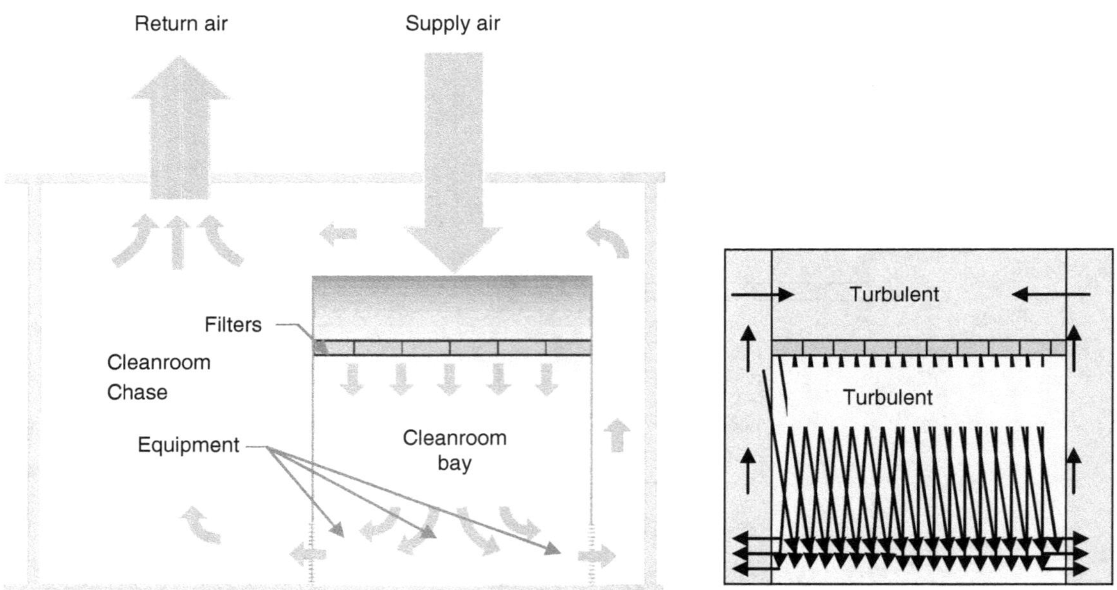

Figure 46.4 Bay and chase concept turbulent airflow.

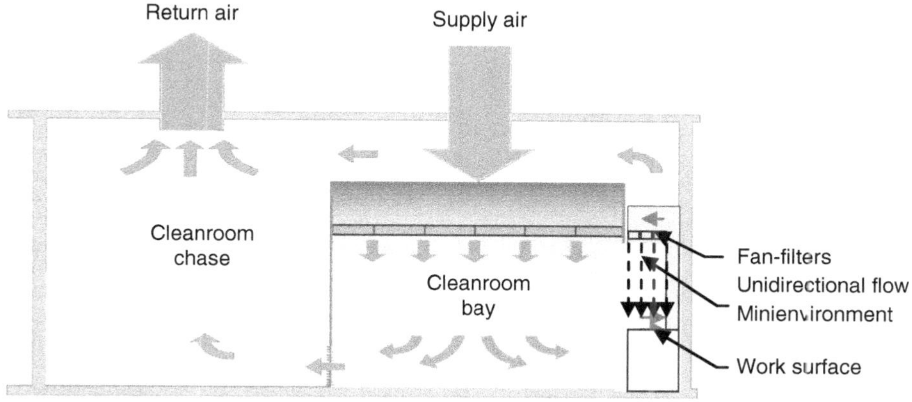

Figure 46.5 Bay and chase concept with minienvironment.

removed through the openings in the sidewalls, horizontal with supply air coming from the wall, and return air being removed through the openings in the opposite wall or a combination of high and low returns.

46.4.1 Unidirectional Airflow

When the air leaves the face of a clean HEPA filter, it travels at a uniform velocity, which allows the flow to be almost unidirectional (Table 46.3). The unidirectional airflow is limiting the dispersion of contaminants from internally generated sources. In the unidirectional airflow cleanroom the particles or molecules of contamination substance move along the airflow pass and exit boundaries of the clean process area. The unidirectional airflow cleanroom is easiest to model with computational fluid dynamic (CFD) software for particles travel and control (Fig. 46.6).

If unidirectional airflow is only necessary over small, sensitive areas of the cleanroom, unidirectional airflow may be handled with minienvironments. Source of contamination may be localized with glove boxes and filtration modules, etc. The majority of cleanrooms designed for total unidirectional flow can achieve unidirectional flow only at rest without workers, equipment, and room exhausts. The choice of vertical or horizontal airflow depends on room configuration

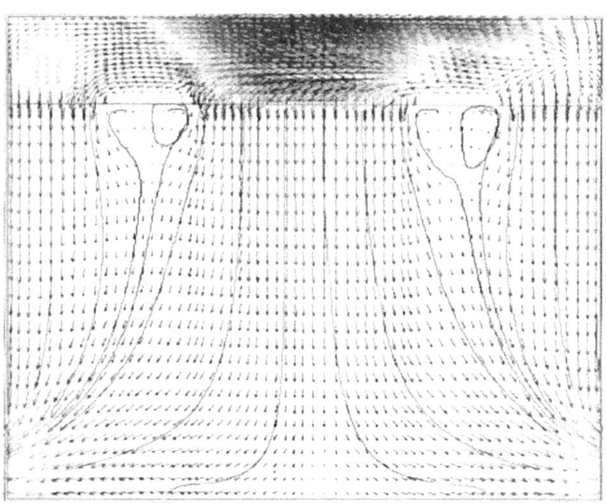

Figure 46.6 CFD model for unidirectional airflow with low side air removal.

and equipment layout. In many facilities, turbulent airflow with properly engineered exhaust and return air locations works fine for contamination removal. Raised floors were developed to help distribution of electrical wires, communication cables, utilities, piping, etc., between tools, equipment, and utility sources. Subsequently, for economy purposes, they were utilized for returning air from the bays to the chases in a bay and chase cleanroom configurations. In a pure unidirectional cleanroom the particles are expected to flow from working space toward the holes in the floor. Many successful class ISO 4 and class ISO 5 ballrooms operate with low wall returns, without raised floors, and the cleanroom airflow is designed to be turbulent.

Another option in product contamination control is to isolate an ultraclean area with a thin air jet moving at 500 to 800 fpm; it functions exactly the same way as the plastic curtain. Several modern cleanrooms were designed, built and operated with ultraclean areas cleaner than class ISO-4 inside the general turbulent airflow cleanroom class ISO-7.

46.4.2 Turbulent Airflow

When the air strikes a solid object, it changes direction. When this happens, the laminar flow may be disrupted, causing turbulent currents in the air. The main sources of air turbulence in the cleanroom are people and equipment. When a person walks in a unidirectional flow of air stream, the air gets pushed in front of the body and a partial vacuum is created behind. The surrounding unidirectional air is thus affected. This same effect is caused by moving equipment parts in the vicinity of an area of laminar unidirectional airflow. Stray particulates may accumulate in the turbulent air pockets and may eventually affect the products being manufactured.

For areas where extreme cleanliness is required, it is important to keep these areas as free as possible of people and other causes of air movement. The effects of obstructions to the unidirectional airflow are shown in Figs. 46.2 and 46.4.

It is important to realize that pure unidirectional airflow may exist only few feet from the air outlets (HEPA filters). In nonisothermal environment the airflow is affected by convection streams, ricocheted from solid objects, air streams, stray currents, etc.

Other elements of cleanroom performance are sweep, the rate of room air changes and percentage of particles removal per set time. These elements are dependent on many factors such as air temperature gradient, cleanroom shape and configuration, location of solid and moving objects on the floor, inlet and outlet location. Sweep is the most overlooked function in a cleanroom or in the design process equipment. Great emphasis is placed on particle counts with little attention to sweep. Sweep is an action of the airflow to capture contaminants from a surface and transport them away (return air) from the critical area of exposure. These captured contaminants are then removed from the airflow by the action of the filters. One of the most important functions of the cleanroom systems is to sweep internally generated contamination from the area of susceptibility. It is imperative to maintain proper velocity and pattern of airflow around the product and process.

Considering that turbulent airflow cleanroom is less expansive to construct and maintain and that the turbulent airflow cleanroom may have an advantage in room sweep and particles removal, the realistic approach may be to construct a turbulent flow ballroom with local unidirectional airflow minienvironments and barrier enclosures.

A typical manufacturing cleanroom is composed of cleanroom wall system, cleanroom ceiling and floor systems, and entrance/exit air locks with or without air showers. The required level of cleanliness is continuously maintained by the cleanroom air conditioning systems (supply, exhaust, and recirculation), cleanroom ionization/static electricity control systems, and cleanroom operational protocol.

46.5 AIR CHANGES

The air distribution devices in cleanrooms (usually ceiling filters) are designed to provide a uniform "shower" of cleaned, filtered air. The quantity of air supplied to the room brings to mind the analogy of a fire hydrant as opposed to an ordinary showerhead.

American Federal Standard 209 E (obsolete), Japanese JIS B9920, and ISO 14644-1 entitled "Cleanrooms and associated controlled environments—Part 1: Classification of air cleanliness by particle concentration" provide guidelines for cleanroom parameters, classification, and testing, but do not tell you how to get there economically. This is because each process and, therefore, cleanroom requirements are different and desired conditions may be reached utilizing different quantities of recirculating air.

There are many different designs to achieve the desired results. However, there is only one design that *fits* best. Finding the optimal and most economical solution for a project is a fundamental goal of the project concept for cleanroom air changes. As may be seen in Table 46.3, the air changes and quantity of recirculation vary significantly even for the given room class. The cleanroom class ISO 3–4, for example, may be achieved with 300 air changes or 540 air changes may not be enough.

The rate of room contamination and live particles generation shall be one of the major factors in cleanroom air quantity selection. The rate of particles removal from the room may be very important for one manufacturing process and may not have any effect on another. Other variables that may impact the quantity of recirculating air are room configuration, equipment location, equipment surface temperature, convective flux, type of the airflow (unidirectional over sensitive areas only or over the entire room), room operations and protocol, materials and chemicals used, etc. (Fig. 46.7). The room may need to be certified at rest only, certified at working conditions, validated for the process, or cGMP (current good manufacturing practice) validated. Federal standard 209(e), set by the U.S. General Services Administration (GSA), suggests that air in a class 100 (ISO 5) cleanroom shall be at 90 cfm/ft^2 or 90 fpm. However, it is possible to build a better than class 100 rooms with lower air movement. In fact, it has been done with as low as 45 cfm/ft^2.

46.6 ELEMENTS OF A CLEANROOM

Regardless of configuration, cleanrooms consist of common elements required to support and protect clean manufacturing process that is housed within the cleanroom. These elements include ceiling systems with HEPA filters, walls, floor systems, and the environmental conditioning systems. See Figs. 46.6 and 46.8.

46.7 CEILING SYSTEMS

Ceiling systems found in cleanrooms are the complex enclosure elements. The two most common systems are extruded aluminum grid with field-constructed plenum and modular prebuilt integrated plenum with grid. Both of these systems serve as the mounting for terminal filtration (HEPA or ULPA) or FFU and lighting. The ceiling system also

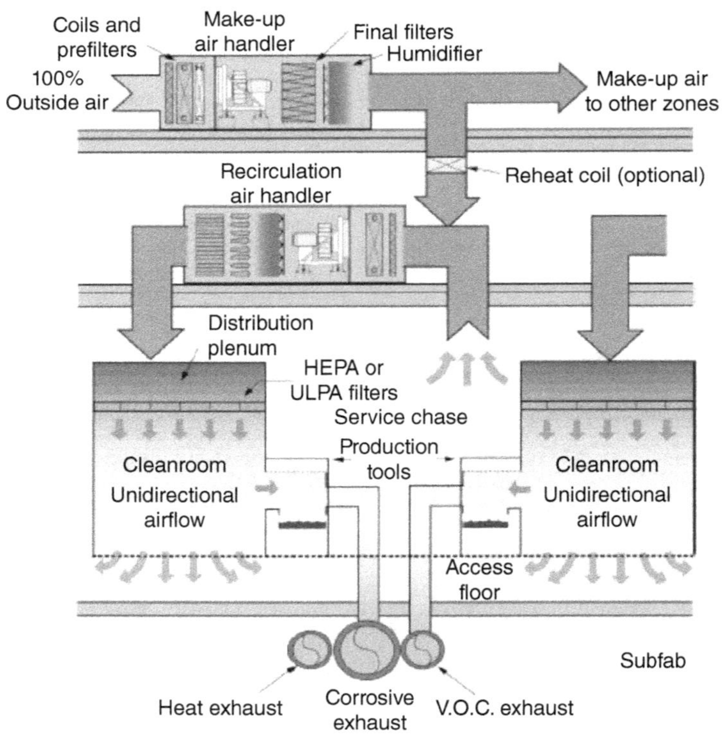

Figure 46.7 Cleanroom design.

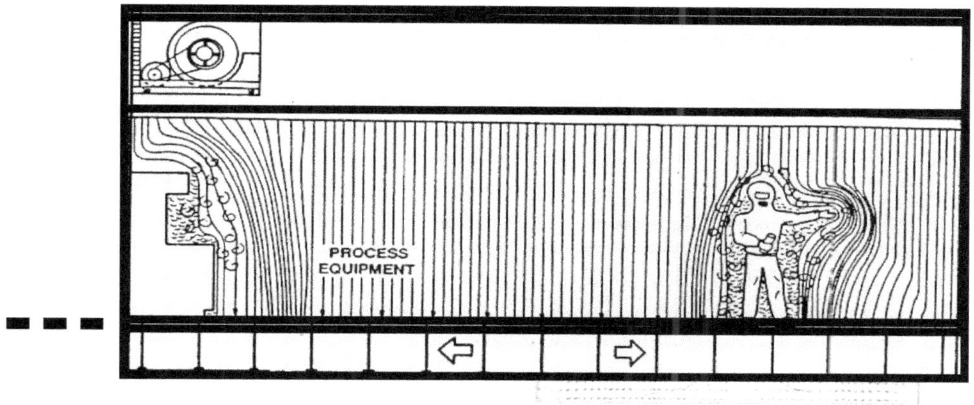

Figure 46.8 Unidirectional air as it meets obstacles.

may be supported from freestanding walls or columns or suspended from existing structure. The factors involved in the selection of the ceiling system are typically site specific depending on required free span, available vertical volume, access, and load carrying requirements.

46.8 WALL SYSTEMS

Wall systems for cleanrooms most commonly are of a premanufactured, modular design. They are constructed of extruded aluminum framing members and honeycombed aluminum panels. The finish is typically either anodized or epoxy coat. These systems are commonly used because they do not generate particles, have very low outgassing characteristics, and are easy to reconfigure. Such systems are well suited for the bulk heading of process tools and utilities connections. The drawbacks are initial costs and limited structural characteristics. Occasionally in less precise applications the "stick-built" wall systems are used. These wall systems are constructed of metal studs and metal faced drywall finished with epoxy coating. While such systems may be constructed at lower cost, they require a much more stringent construction protocol; they are less reliable to provide satisfactory results in a limited construction time. The stick-built wall systems offer good structural capabilities but are difficult to reconfigure without excessive particles generation and, therefore, have negative impact on the quality of operational cleanroom.

46.9 FLOOR SYSTEMS

Floor systems for the cleanrooms most commonly are perforated wafer concrete panels over the subfab, raised stand floor panels, or slab on grade with side air grilles. Typically these systems are provided with ESD (electrostatic discharge) or an ESD grade epoxy floor. Significant factor influencing the decision during selection of floor systems is airflow management. Where unidirectional airflow and parallelism are critical factors, the raised panel floor system or perforated wafer concrete panels over the lower floor (subfab) provide superior performance. The disadvantage of raised floor systems is relatively high cost and limited structural capability. When unidirectional airflow is not a major concern the slab on grade with sidewall air returns may perform effectively and at significantly lower cost.

46.10 ENVIRONMENTAL REQUIREMENTS

Environmental conditioning systems typically consist of recirculation and makeup air systems (temperature and humidity control of the cleanroom is accomplished via these systems), general exhaust, acids exhaust and scrubbing, solvents and VOC (Volatile organic compounds) exhaust and absorption, toxic gases exhaust, and destruction.

A proven element in cleanroom design is the use of AMC (Airborne Molecular Contamination) control systems. These systems incorporate cleaning of outside air used for ventilation and pressurization, removing VOC and fugitive emissions from the recirculation air system, and cleaning process emissions from the exhaust stream. AMC systems are typically classified by their chemical properties: acid, base, condensable, and dopants.

SEMI Standard F-21-1102 (see chapter 40 AMC) offers a cleanroom classification system, based on their molecular contaminant levels rather than particulate levels. AMC can address concerns regarding personnel exposure to toxic and hazardous materials. AMC systems typically consist of specific filtration and monitoring and can provide an effective removal rate of as high as 90 percent.

46.10.1 Exhaust Systems

Process exhaust systems typically are acid exhaust, solvents and VOC exhaust, toxic exhaust, heat exhaust, and general room exhaust. If ammonia is present in the exhaust air stream, some facilities prefer to run a separate ductwork and abatement for ammonia fumes. The most common abatement for acid exhausts is horizontal or vertical scrubber. Solvents and VOC exhausts require absorption, concentration, and removal of condensed solvents or on-site incineration. Toxic exhaust is usually abated by in-place or on-site high-temperature destruction. The abatement types should be carefully determined along with the quantities of required exhaust air streams. Good facility air management can help minimize quantities of exhaust air and, therefore, lower construction cost and energy waste during operation. It is important to maintain balance between potential needs for increase of exhaust quantity and expansion and economical value of installing such provisions initially or in the future. Far too often, oversized, expensive, lined stainless steel, or FRP exhaust systems have been built in consideration for future expansion in cleanrooms, while other utilities and mechanical systems, makeup air and cooling, for example, could not support such expansion. It is also common that acid exhaust systems have been built without consideration for future need, where velocity in the ductwork exceeds 4000 fpm with added airflow and new ductwork branches. Such systems were difficult or impossible to balance and expensive to operate.

46.10.2 Makeup Air Systems (See Figs. 46.6 and 46.8)

Typically the makeup air handling units (primary AHUs) provide the necessary makeup air for the recirculating air handling units (secondary AHUs). The makeup air AHUs consist of draw through centrifugal, vane axial or plug fans with filters, hot water coils for preheat and reheat, chilled water coils for cooling and dehumidification, and steam or adiabatic humidifiers. In addition, there are many available add-ons and variations, including static air mixers, steam preheat coils, ultrasonic humidifiers, brine, DX (direct expansion) or glycol subcooling coils for dehumidification, VOC absorption filters, sound attenuators, and VFD (variable frequency drives). The makeup AHUs typically discharge into a common header with ductwork laterals balanced to supply the required makeup air to the recirculation AHUs.

Air measuring stations shall be installed in the primary air and the secondary air supply main ducts to modulate the supply fans VFD drives or inlet vane dampers to maintain constant airflow. Magnehelic gauges shall monitor the loading of the HEPA filters, bag filters, and prefilters located at each air handling unit. In situations where ceiling fan-filters are provided, the makeup air shall be evenly distributed in the space above fan-filters. Air from the makeup AHUs enters cleanroom through recirculation AHUs.

46.10.3 Recirculation Air Systems (See Figs. 46.6 and 46.8)

Each recirculation air handling unit typically consists of an energy efficient centrifugal fan with filters and sensible (dry) cooling coil. Add-ons and variations are available, including reheat coils for zone temperature control, steam or ultrasonic humidifier for zone humidity control and constant volume control boxes. Proper selection of fan systems used for this application proved to be cost saving and energy conscientious. Multiple recirculation AHUs discharge into supply ducts feeding the cleanroom through ULPA or HEPA filters which typically cover 100 percent of the ceiling in the class ISO 4 and class ISO 3 areas. This vertical unidirectional flow passes down through the room, through the perforated raised floor tile into the return air space under the floor, and then up through vertical return air shafts, which are open to the ceiling, return air plenum. The air then reenters the recirculating air handling unit and the cycle repeats. In the case of ceiling fan-filters the sensible cooling may be provided by water or DX cooled fan-coil units located in the ceiling space above fan-filters.

46.10.4 Temperature Control

Temperature and humidity variations cause process equipment misalignment, impact the repeatability of the developed process, and eventually reduce the useful output of the product and increase the quantity of waste. It is understandable that the goal is the most stringent cleanroom temperature requirements, but the cost often dictates otherwise.

Often, in an attempt to lower the construction cost the engineer is asked to design a precise temperature control area within a large space where the temperature is allowed to swing ±4 to 6°F (i.e., the warehouse area with roll-up doors). Without hard walls and air locks it may be a very expensive alternative with virtually uncontrollable variations of humidity. Therefore, the quality of said cleanroom is substandard and cost remains relatively high. Common sense tells that cascading levels of cleanliness, temperature, humidity, and pressure are easier to achieve and maintain. The allowable tolerance should be carefully evaluated.

Mechanical equipment and control systems for cleanrooms with stringent temperature control requirements (66 to 72°F, ±0.1°F) may cost 20 to 50 percent more than a cleanroom with requirements (68 to 72°F) and set point (70°F, ±2.0°F).

A zone thermostat typically controls the design temperature in each cleanroom zone. It actuates the duct-mounted zone reheat or recool coil to satisfy the room sensible load conditions. In the case of fan-coil units the zone thermostat controls the temperature of air leaving the fan-coil unit coils in the zone.

46.10.5 Humidity Control

The design relative humidity for each cleanroom is controlled by a zone humidistat. If there is a high relative humidity in the room, the humidistat lowers the cooling coil discharge air temperature to provide more dehumidification. At the same time, the reheat coil provides heat to maintain the room temperature. If the relative humidity of any cleanroom falls below the design limit, the zone humidistat actuates the duct mounted zone humidifier. When precise humidity control is required it can typically be achieved with adiabatic humidification of makeup air in the air handler and by maintaining the cleanroom dew point. Local variation in humidity levels may be handled with ultrasonic humidifiers located in the ductwork plenum before the filtration terminals. The ultrasonic humidifiers for cleanrooms work well with UHP water quality. The design relative humidity for each cleanroom in that case is usually controlled by a zone humidistat. If the relative humidity falls below the design limit, the humidistat actuates the humidifier to increase the moisture content of the supply air. If there is a high relative humidity in the room, the humidistat lowers the makeup air handlers cooling coil discharge air dew point temperature to provide more dehumidification.

46.10.6 Air Pressure Differential

Cleanroom pressurization is necessary to protect the cleanroom against contamination from adjacent areas, to control the flow of unwanted contaminants, to prevent cross-contamination between areas, and to help maintain temperature and humidity at required levels. Typical air pressure differential between cleanroom and reference corridor and other areas of the facility is maintained at 0.25 to 0.005 in w.g. The higher number is usually more applicable to pharmaceutical facilities with cascading air pressures between areas to avoid cross-contamination. Such areas normally require series of cascading air locks between them with pharmaceutical doors that allow air to escape at a high velocity, therefore, creating a pressure differential. A well-designed microelectronics cleanroom normally operates at 0.02 to 0.005 in w.g. with a semihermetic air lock at the cleanroom entrance. The cloth change and gowning rooms often serve as said air locks. The presence of mechanical air showers at the entrance is more a question of facility culture and facility cleanroom protocol than a necessity. Many microelectronics and photolithography cleanrooms operate successfully with passive air pressure control and maintain only minimum air velocity of 50 to 100 fpm over entrance with the door fully open. Some biopharmaceutical facilities require an active differential pressure control and supplement air escaping through the door (at the time when the door is being opened) with makeup air automatically. Differential pressure monitors may be utilized for this task. Monitors may be mounted outside the cleanroom with a small LED inside the cleanroom, having an audible alarm and two indicators to show when pressure is normal or abnormal. The device detects negative pressure in biocontainment areas common in biopharmaceuticals as well as the positive pressure common to electronic facilities. It has a digital differential pressure display with resolution of 0.001 in W.C. and pressure/vacuum range of 0.5 in W.C. Differential pressure is indicated by illuminated LEDs and an audio alarm. Internal adjustable time delay prevents activation of the audible alarm by momentary door opening.

46.11 PROCESS CONTAMINATION CONTROL

46.11.1 Air Filtration

Cleanroom air typically is filtered with terminal HEPA filters that provide efficiency of 99.99 percent at 0.3 to 0.5 μm, whereas ULPA filters provide efficiency of 99.999 percent at 0.12 μm.

Other devices may also be required depending on cleanroom purpose, that is, HEPA filtration in recirculating and makeup air handlers, VOC absorption with charcoal type or similar synthetic filters, electrostatic filtration chambers, etc.

Fan-filters are no longer new to the cleanroom market, and properly applied shall provide an excellent and economical solution for many high-level cleanrooms especially in buildings with ceiling height limitations. Applying higher-grade terminal filters should be economically justified and weighted against first cost and lower pressure drop of more expensive filter media. It is not always obvious, but often true, that higher-grade filters with lower pressure drops are more economical in the long run, through the life of the cleanroom.

46.11.2 Prevention

Properly designed equipment and properly selected materials can minimize internally generated contamination. Uses of sliding surfaces and belts should be minimized or isolated from a process. Materials should be analyzed for shedding and outgassing. PFA (perfluoroalkoxy) is orders of magnitude cleaner than polypro while Teflon is an order of magnitude cleaner than PFA. Typically aluminum is far less contaminating than steel. The selection of anodized aluminum over stainless steel may remove serious contamination from the process.

46.11.3 Isolation

Properly designed through the wall installations, flexible or hard shielding of contamination sources. Utilization of a variety of system such as SMIF, vacuum processing, nitrogen purged processing, isolation carriers, transport tunnels, hanging shields, and robotics can dramatically reduce the contamination of the product or process. External isolation also includes isolation and consolidation of processes and the layout of equipment. For example, such processes as photolithography and ion implant shall be isolated from the rest of the manufacturing. The intent is to isolate the product from all internally induced contamination and cross contamination.

46.11.4 Sweep

Properly designed airflow and velocity shall ensure the effective sweep. If the air velocity is too low, there will not be a particles removal effect; if the air velocity is too high, an unexpected turbulence will result.

46.12 VIBRATION AND NOISE CONTROL

The size and weight of equipment affects vibration transfer and control. Concrete "waffle" slabs under the cleanroom floor worked sufficiently well for keeping equipment vibration from transfer to other areas of production or to metrology tools. Waffle slabs remain rigid even if holes are drilled in the floor for piping and conduits access to the cleanroom. Because the strength of the floor is in the grid system, rearrangement is feasible; additional holes can be punched in the floor without adversely affecting vibration considerations. All mechanical equipment should be isolated with springs, flexible connections, and isolated foundations to minimize the vibration effect. Quiet, energy efficient fans and motors allow to maintain the desired NC (Noise Criterion) level in the cleanroom. The vibration and sound consultant should be a part of the team to verify mechanical and architectural concept for vibration and noise control. It could be very expensive to fix it later. Many potentially costly measures may be eliminated or replaced with more economical solutions if these factors were considered from the beginning, from the start of the project with architects, mechanical engineers, client and consultants involved in the decision process.

46.13 MAGNETIC AND ELECTROMAGNETIC FLUX

Magnetic flux proposed to be the flow of the background oceanic particles of the galaxy. Galactic rotation is also proposed to be electromagnetic in nature. The word "flux" means flow, and we may think of magnetic field lines as lines of some type of fluid flow through an imaginary surface. The magnetic field magnitude is like a rate of flow and its direction is the direction of flow. The magnetic flux is like total volume of flow through the surface. Better yet, we may think of magnetic flux as the number of field lines passing through the surface. At a given speed, this force is greatest when the particle moves perpendicular to the magnetic field and zero when the motion is parallel to the magnetic field. At the most basic level, magnetic forces are exerted on moving charges by other moving charges, just as electrostatic forces are exerted by electric charges on other electric charges, whether or not they are moving. It appears from comparative studies of planets that the earth has a strong field because it rotates and because it has a molten metallic core. In theory, the field arises in internal electric currents that are induced by the earth's rotation and by circulation in its fluid core. Magnetic flux density has its own unit, the tesla (T). The typical value of the earth's field near its surface is about half of a gauss, which is about one 200,000th of a tesla. Occupational safety and health requirements limit magnetic field strength in areas open to the public to the 5 gauss limit. Magnetic shielding of the cleanroom can be very expensive. For example, 4- to 5-mm-thick shielding on the envelope of the cleanroom with magnetic (Fe-Si) steel, M15 type may lower the strength of magnetic flux from MRI equipment to 1.3 to 2.6 gauss. However, 14 mm of ordinary low carbon steel shielding may not be sufficient. Current semiconductors, metrology, and communication laboratories require the strength of magnetic flux in the area to be limited to 0.05 gauss or less.

46.14 ELECTROSTATIC CHARGE OF AIR AND SURFACES

"Static electricity" effect is present when surfaces in contact are separated. If the charge that arises from differences between the surfaces cannot escape to earth quickly enough then it is trapped and the charge will spread out over the surface of a material—it is "static." Retained electrostatic charge creates risks and causes problems in many areas of the industry. It can cause ignition of flammable gases and even shock personnel. It can make thin films and light fabrics cling, attract airborne dust and debris, damage electronic devices and upset the operation of precision equipment. The hazard that static presents in the case of flammable gases, vapors, and powders relates both to the capacitively stored energy in relation to minimum ignition energies and to the breakdown voltage of the minimum gap from which an ignition will propagate. Typically, the minimum ignition energies of common hydrocarbon gas/air mixtures are 0.2 mJ with a few kilovolts minimum breakdown voltages. With powders minimum ignition energies start at a few mJ. Shocks from electrostatic discharges become discernible around 1 mJ and are likely to be uncomfortable in the 10 to 100 mJ range. They will cause major muscular contraction above 1 J. Mechanical handling problems arise when electrostatic forces become comparable to gravitational or other constraining forces. This relates to the strength of local electric fields and hence on insulators to surface charge density. In general, electrostatic forces are weak, but dust will be attracted to surfaces at charge densities less than few mJ $\times$ 10^{-7} per m^{-2}. Electrostatic charge will be generated on people by such normal activities as walking across carpets, getting up from chairs, rubbing clothing against surfaces, etc. The levels of charge will be higher in low-humidity environments and where artificial fibers are extensively used. Body potentials up to 15 kV may be expected. Electrostatic discharges will occur from charged fabrics, a charged body, and any metal objects held in the hand, etc. These electrostatic discharges may involve high potentials and so may be able to jump several millimeters of air through gaps in equipment casings directly to internal circuitry. The discharges can involve currents up to several amps and involve frequency spectra extending up to several hundred megahertz—particularly where a metal conductor acts as the source of the discharge. Problems with the upset of microelectronic systems are also usually expressed in relation to a human body model discharge. System immunities of several kilovolts are likely to be needed (preferably over 15 kV for uncontrolled environments) because high potentials can readily be generated on personnel in normal working environments by movements across flooring, etc. To minimize risk and avoid potential problems it is necessary to ensure that static charge can dissipate more quickly than it is generated. For normal, manual handling and body motion activities this means the charge decay time needs to be ¼ s or less. A new concept, relevant to risk control, is that if static charge experiences a high capacitance on a material then only low surface voltages will be observed and potential problems and harm will be prevented.

46.15 LIFE SAFETY

Safety always shall be a major consideration in the design and operation of cleanrooms. Critical life safety systems and safety egress are required by codes and industry-accepted practices. Major underwriters such as Factory Mutual Global (FMG) publish guidelines for materials to be used in the construction of cleanrooms and generally will not insure a cleanroom that has not designed, constructed, and operated per the recommended practices. Typical safety systems for cleanrooms include toxic gas monitoring (TGM), leakage sensors in the containment piping, waste systems, enhanced (early detection), fire alarms, and oxygen level monitoring. These systems provide levels of alarming ranging from basic visual and auditory to automatic dial out to emergency responders and digital storage of 365/24/7 information about all supporting life safety systems alarms and operations.

46.16 COMPUTATIONAL FLUID DYNAMICS

One of the most powerful tools available for determining the optimal ventilation (recirculation, makeup air, and exhaust) for the cleanroom is computational fluid dynamic (CFD) modeling or airflow modeling (Table 46.5).

The very essence of the cleanroom contamination control is defined by ventilation. Electrical energy necessary to support HVAC equipment to support the cleanroom function represents a significant portion of the energy used by the cleanroom. With the development of high-powered personal computers, CFD modeling shall be used for even the smaller projects.

The CFD modeling allows for better anticipation of design inadequacies so that they can be remedied before the start of construction, better ability to quickly and efficiently reveal areas of opportunity for improved performance in operation of the cleanroom and ability to model various options for both planned and operating cleanrooms so that the most economical solutions can be pursued with a high degree of confidence.

Unfortunately, there are still too many cleanrooms being designed based on simplistic "rules of thumb." With the cost of manufacturing facilities increasing by "double digit" figures with each tool generation, it is projected that a new 300-mm facility will cost in the over $1800 per square foot.

"Time is of the essence" regarding the design and construction of a clean manufacturing facility.

Marketing analysis indicates that in a high-tech manufacturing, the first company to bring a new product (or an upgrade to an existing one) will maintain a 40 percent market share during the products life (providing that they maintain competitive pricing).

As a consequence of the above, enormous pressure is placed on the project team to deliver the cleanroom on or ahead of schedule; therefore, discovering problems at the late stage of the game during certification, validation and commissioning and hoping that the problems may be solved by simply rebalancing the air systems is an expansive misconception.

46.17 CLEANROOM DESIGN AND CONSTRUCTION

The design and construction of cleanrooms and contamination controlled facilities requires an experienced, dedicated, talented team capable of working in a collaborative manner in order to succeed.

Table 46.5 Cleanroom HVAC Requirements: Recommended to Meet and Exceed FED-STD 209E, Vertical Laminar Flow, and ISO 14644-1

ISO 14644-1 per cubic meter	Federal standard 209 E (SI)	Federal standard 209 E (USA) per cubic foot	Suggested minimum ceiling coverage	Suggested filter type	Suggested Minimum air velocity @ceiling level	Suggested minimum cleanroom air changes
ISO-3 Particle Counts ≥0.1μ - 1000 ≥0.5μ - 35	Class M1	Class 1 Particle Counts ≥0.1μ - 35 ≥0.5μ - 1	100% ceiling coverage	Ultra Low Penetrating Air (ULPA) filters providing efficiency of 99.99995% at 0.12 microns.	75 to 90 fpm (0.38 to 0.46 m/s)	500 to 640 ach
	Class M1.5	Class 1 Particle Counts ≥0.1μ - 35 ≥0.5μ - 1	100% ceiling coverage	Ultra Low Penetrating Air (ULPA) filters providing efficiency of 99.99995% at 0.12 microns.	75 to 90 fpm (0.38 to 0.46 m/s)	450 to 640 ach
	Class M2	Class 10 Particle Counts ≥0.1μ - 99.1 ≥0.5μ - 2.83	100% ceiling coverage	Ultra Low Penetrating Air (ULPA) filters providing efficiency of 99.99995% at 0.12 microns.	70 to 80 fpm (0.36 to 0.41 m/s)	420 to 600 ach
ISO-4 Particle Counts ≥0.1μ - 10,000 ≥0.5μ - 352	Class M2.5	Class 10 Particle Counts ≥0.1μ - 345 ≥0.5μ - 10	100% ceiling coverage	Ultra Low Penetrating Air (ULPA) filters providing efficiency of 99.99995% at 0.12 microns.	70 to 80 fpm (0.36 to 0.41 m/s)	420 to 600 ach
	Class M3	Class 100 Particle Counts ≥0.1μ - 991 ≥0.5μ - 28.3	80% ceiling coverage	High Efficiency Particulate Air (HEPA) filters providing efficiency of 99.999% at 0.12 microns.	50 to 70 fpm (0.26 to 0.36 m/s)	300 to 480 ach
ISO-5 Particle Counts ≥0.1μ - 100,000 ≥0.5μ - 3520	Class M3.5	Class 100 Particle Counts ≥0.1μ - 3450 ≥0.5μ - 100	75% ceiling coverage	High Efficiency Particulate Air (HEPA) filters providing efficiency of 99.999% at 0.12 microns.	50 to 70 fpm (0.26 to 0.36 m/s)	300 to 480 ach
ISO-6 Particle Counts ≥0.1μ - 1,000,000 ≥0.5μ - 35,200	Class M4.5	Class 1000 Particle Counts ≥0.1μ - 34,500 ≥0.5μ - 1000	40% ceiling coverage	High Efficiency Particulate Air (HEPA) filters providing efficiency of 99.999% at 0.12 microns.	30 to 50 fpm (0.15 to 0.25 m/s)	180 to 300 ach
ISO-7 Particle Counts ≥0.5μ - 352,000	Class M5.5	Class 10,000 Particle Counts ≥0.1μ - 345,000 ≥0.5μ - 10,000	30% ceiling coverage	High Efficiency Particulate Air (HEPA) filters providing efficiency of 99.99% at 0.3–0.5 microns.	20 to 30 fpm (0.10 to 0.15 m/s)	60 to 100 ach
ISO-8 Particle Counts ≥0.5μ - 3,520,000	Class M6.5	Class 100,000 Particle Counts ≥0.1μ - 3,450,000 ≥0.5μ - 100,000	15% ceiling coverage	High Efficiency Particulate Air (HEPA) filters providing efficiency of 99.99% at 0.3–0.5 microns.	15 to 20 fpm (0.08 to 0.10 m/s)	36 to 90 ach

Cleanrooms are often the most complex and costly element of the design for manufacturing facility and proper attention to all elements including, process flow, comfort, safety, and egress of people shall be given from the start. For example, clean materials and contaminated rejects shall not be stored in the same location, people safety egress and toxic chemicals delivery systems shall not cross each other, pressurized toxic liquid piping shall not be installed directly over the heads of facility workers or inside of the safety exits.

46.17.1 Ballroom with Minienvironments Technical Description (Example)

The Minienvironments

Facility shall have self-contained class ISO 3 minienvironments for each process tool, and a partial-HEPA-coverage vertical flow laminar clean ballroom that is capable of class ISO 5-Turbulent cleanliness (operating conditions) by use of turbulators on the output side of the HEPA filters to improve cleanroom air mixing and operating cleanliness.

The performance will be measured per ISO Standards, except for the performance conditions defined here.

Class 1 (ISO 3) Manufacturing Environment (Operating)

Definition "The Class 1 manufacturing environment is defined as less than 1 particle per cubic foot, of 0.03 microns and larger, measured at 3 inches above the top wafer location of all tool indexers."

Measurement The class 1 manufacturing environment is the airborne particle environment measured directly above the production wafer occupying the top slot of wafer indexers of the process tool. This location is within the microenvironment enclosure that is installed on the tool's atmospheric-side wafer handling and has tool covers permanently removed which would interfere with the laminar air field used to attain the class 1 Wafer Environment.

Measurement Conditions Measurements to demonstrate attainment of the class 1 manufacturing environment are made with the tool on, in a standby mode. The rest of the cleanroom shall be in normal manufacturing operations at the time of the measurement.

Class 100-(ISO 5) Turbulent Room Environment (Operating)

Definition "The Class 100 (ISO 5) Turbulent room environment is defined as less than 3520 particles per cubic meter, of 0.5 microns and larger, measured as defined for Class 100 (ISO 5) in Federal Standard 209E."

Measurement Conditions Measurements to demonstrate attainment of the Class 100 (ISO 5) Turbulent Room Environment with the cleanroom in normal manufacturing operations at the time of the measurement.

Class 10, 100 (ISO 4, 5) Turbulent and 10,000 Measurements

Excepting the class 100-Turbulent, these cleanliness classes are measured per Federal Standard 209 of the latest version.

The following is a description of the style and features needed to meet the requirements for the class 100-Turbulent clean ballroom.

Flexibility The class 100-Turbulent clean ballroom is a partial-HEPA-coverage ballroom with movable interior walls for yellow light separations and for protection from fumes.

The interior ceiling wall/structure system shall be highly flexible, suitable for moving on 1-ft steps. This allows the process area to be increased if ever needed in future. The movable cleanroom walls shall be a floor-standing structure. The interior wall system shall be nonprogressive.

Recirculating Air Path The air shall be fed to the ceiling HEPA filter boxes via a pressurized overhead ducting system. The return air from the clean ballroom shall passes through the perforated raised access floor, then through the open concrete waffle floor into the lower level, then into perimeter vertical-type recirculating-air fan/coil units, then into the overhead pressurized distribution ducting system.

Recirculated air from lithography cleanrooms shall not mix with the rest of the manufacturing facility cleanroom air, to provide protection from chemical fumes and smoke to the metrology, reticle, deep UV, and lithography equipment.

Cleanroom Ceiling Height Due to the need for overhead clearance for rail delivery systems and for servicing of vertical furnaces, the cleanroom shall have a minimum 12 ft clear height.

The lighting in the clean ballroom shall be flush lighting, not extending into the 12 ft clear height; however, the fire sprinkler heads may extend into the 12 ft clear height.

Tool Minienvironments The minienvironments are described in a separate section, after Class ISO %-Turbulent ballroom specification. The cleanroom ceiling-grid system shall easily accommodate the attachment and the hanging weight of the rail systems and tool mini enclosures, without additional modifications. The bottom face of the ceiling grid shall have an easy and structurally adequate connection method to hang rails and other equipment.

This system, including connections and handing equipment, shall be capable of meeting the local seismic zone requirements or shall meet seismic zone 4 if not otherwise specified.

Gowning Entry Near the employee entrance, a separate locker room will be used for outside clothes, storage of personal articles and shoes, and changing into clean building shoes.

All personnel entering the manufacturing building shall wear building shoes or shoe covers. Cleanroom entry shall be via a gowning room through an air-shower system. The gowning room shall act as an air lock, so the entrance and exit doors from gowning shall not be opened at the same time. The gowning room shall be full coverage unidirectional class ISO 2 area.

46.17.2 Airflow Systems Technical Description (Example)

Recirculating Air Units

The cleanroom system shall use perimeter vertical-type recirculating air fans with controlled chilled-water coils to remove sensible heat and to regulate in-coming room temperature. These fan units shall feed the pressurized air to the ceiling via a pressurized ducting system, then into the HEPA boxes. The HEPA filters shall be mounted in the ceiling grid with a cast-in-place type elastomer o-ring gasket HEPA filter box. These boxes shall be clamped in place to ensure that the flexible ducting connections to not tip the boxes and decompress one side of the seal. The fans used shall be low-velocity vertical fans. The fans shall be suitable to run 24 h per day, 7 days per week, for at least 1 year between scheduled maintenance. Belts may be allowable, if the maintenance schedule period is 1 year or greater. Prefilters shall be used during startup at the entrance to the fan housing. They shall be rated at 30 to 35 percent ASHRAE efficiency, and shall be of sufficient area to minimize pressure drop across the filter. After phase 1 tool installation is complete, the prefilters may be removed to save electrical running horsepower (electrical costs). The electric motors shall be wired with equally addressed electrical wiring, to minimize EMF (Electromagnetic field) generation. For cleanrooms with EMF limitations motors shall be fully enclosed types, with cast iron end caps to minimize EMF radiation. Alternatively, the EMF of the motors shall be measured and EMF calculations made to demonstrate that the EMF limitations of the areas are not exceeded. If electronic speed control equipment is used, it must not send electrical noise to the process tool by conduction or radiation, and there shall be no audible noise heard in the cleanrooms. Any cleanroom sound-attenuation filler material shall be a suitable inorganic mineral or glass fiber of a density sufficient to obtain the specified acoustic performance. Material shall be inert and moistureproof. Filler material shall be protected from particle erosion by airflow by the use of a sealed surface protection of 1-mil Mylar-type film (or equal). Recirculating air fan-module redundancy is not feasible due to the high number of fan units. As a result, the units must be extremely reliable and service access must be provided for easy replacement of the motor or fan-wheel or bearings, without shutdown of the rest of the cleanroom system. All servicing and repair required for the prefilters and fan/coil units shall be readily accessible without going into the cleanrooms. The HEPA filters, sprinkler heads, and lights shall be serviced from the cleanroom side.

Controls

The sensible cooling coil shall be provided as part of each fan unit, to remove heat for the control of the air temperature entering the room through the ceiling. The process module temperature is controlled by sensors located at 8 to 10 in below the HEPA ceiling. The temperature and humidity sensors should be mounted with only the probe exposed in the cleanroom. The probe must be located under active HEPA filter area, not structure. The requirement is to control the air temperature entering at the HEPA ceiling, and not at the process tools—where localized heat load changes (tools being turned on or off) might cause variations with the temperature control zone. A highly accurate wide-flow-range chilled water valve is required for sensible heat control, that is capable of accurately handling the flow for all tool-set heat load situations (50 to 100 percent tool sets) and for the initial tool set heat load situation. The start/stop fan control is normally provided through the building management system (BMS). This allows for sequential start-up to avoid overloading the electrical service after a power outage. Local manual override shall be also provided for maintenance and emergency shutoff needs. An air-pressure switch shall be incorporated in each fan unit output to detent a condition on non-normal airflow. Alarms shall be signaled automatically to the building management system. A liquid-level detector shall be incorporated to monitor internal leaks within the fan coil unit. Alarms shall be signaled automatically to the BMS.

Electrical

Each fan unit shall have a unit-mounted motor starter/disconnect which can be reached by the maintenance person at the fan service floor level. The motors shall be high-efficiency 3-phase types, and operate on the 3-phase 480 volts service, taken from a facility feeder, not a process tool feeder. The flush lighting in the cleanrooms shall be mounted in blank panel locations in the HEPA filter grid structure. It shall operate from a facility feeder, not a process tool feeder, to minimize conducted EMI to the process tools. Cleanroom and service areas shall have some lighting on the emergency generator service, to provide exit lighting. Also, battery powered lights shall be provided to ensure exit lighting in the case that the emergency generator fails to start.

Structural

The ceiling grid system shall include an easy attachment system that is capable of hanging rail transportation systems, without modification of the ceiling system. Also, there shall be a seismic resistance capability to meet the local requirements. If not otherwise specified, U.S. Seismic Zone 4 capability shall be met. This capability shall be present at all locations within the cleanroom, regardless of initial tool layouts, or the lack of support rails in the initial start-up. The design should avoid columns inside the cleanroom area by use of a long-span structural design. However, a design cost-reduction alternative, with a single row of columns, shall be also studied and the cost difference presented to the owner for decision. The air distribution plenum and the HEPA ceiling shall be supported from the overhead structure, not from the floor structure. The movable interior wall system shall be supported from the raised access floor, and shall sit on top of it. It shall be structurally and vibration independent from the overhead structure, so that vibration is not transmitted into the cleanroom floors. There is no soft wall joint required between the grid system and the movable interior cleanroom walls, which are sitting directly on the cleanroom raised floor. This has been found acceptable because normally there is not enough vibration power transmitted into the concrete facility floor by this route to take the concrete floor out of vibration limits. (This design point must be verified with the project vibration engineer, however.) The constructed system shall meet local seismic zone requirements.

Air-Balancing Dampers

It is desirable that a very low pressure drop recirculating system shall be built to minimize running horsepower. As part of this, the cleanroom system shall use a pressurized-duct design, with the individual HEPA boxes flow-matched at the factory to meet the flow rate specifications: flow adjustment dampers should not be required in the HEPA ceiling boxes. Also, the cleanroom floor shall not have slide dampers for aerodynamic adjustments.

Return Air Path

The cleanrooms air return shall be through the raised perforated floor. The return air shall be taken through the raised floor into the lower level, then to perimeter vertical-fans, and returned to overhead pressurized ducting system. The return-air path shall be constructed to be reasonably particle, pressure, and humidity tight. It shall meet the fire/smoke draft-stop requirement for the fume/smoke removal mode.

Differential-Pressurization Limitations

The cleanroom complex should have no significant pressure differentials between the connected similar cleanrooms, to avoid air currents stirring up dirt from the floors whenever doors are opened.

The following guide gives the maximum pressure differentials desired within the cleanroom complex:

1. No pressure differentials between ISO-3, ISO-4, and ISO-5 areas.
2. No pressure differentials between ISO-3, ISO-4, and ISO-5 areas and the perimeter clean corridors and the clean technical support rooms.
3. +0.05 to +0.3 in (water) from gowning rooms (air locks) to rest of building, outside cleanroom complex.
4. The differential pressures are referenced from areas ISO-3, ISO-4, ISO-5 of the cleanroom complex to the other area.
5. The + value means that the ISO-3, ISO-4, and ISO-5 room are at higher pressure.

The clean corridor connecting the process tool areas with the clean technical support rooms shall have a class ISO 5-Turbulent partial coverage ceiling.

46.17.3 Ceiling Grid Technical Description (Example)

Grid-to-Filter Sealing Method

The HEPA filter boxes are mounted in the ceiling grid with a cast-in-place type elastomer o-ring gasketed HEPA filter box. A gel-seal system shall also be allowed, if not at additional cost. The HEPA boxes shall be clamped in place to ensure that the flexible ducting connections to not tip the boxes and decompress one side of the seal.

Grid Construction

All metallic members shall be of a corrosion-resistant material or have a corrosion-resistant finish. An antistatic finish is not required, but is allowed if not at additional cost. The grid shall consist of 2 in wide construction. Grids wider than 2 in might be allowed but require prior user approval. Lighting shall be located in blank panel areas, not as part of the grid. The channel grid system shall be installed in a manner that it shall be possible to install or remove filters and other system components from the cleanroom side.

Structure

The ceiling grid system shall include an easy attachment system that is capable of hanging rail transportation systems, without modification of the ceiling system. There shall be a seismic resistance capability in the grid to meet the local requirements. This capability shall be present at all locations within the cleanroom, regardless of initial tool layouts, or the lack of rails in the initial startup. The entire unidirectional flow ceiling system and wall system shall meet the seismic requirements.

Fire Sprinkler Penetrations

Fire sprinklers shall penetrate in blank panel areas, not through the grids. The fire sprinkler penetrations shall be leak-tight and particle free. If allowed by local code, fire sprinkler pipes visible on the clean side of the cleanroom shall have a cosmetic cover (or other provision) to blend aesthetically with the HEPA ceiling system.

HEPA CEILING COVERAGE

The HEPA filters configuration of the ballroom shall be fixed, and is not intended to be changed with the tool configuration. This assumes that adequate air circulation has been provided for heat removal in each corral by the cleanroom designer.

UNIDIRECTIONAL FLOW CEILING HEIGHT

The unidirectional flow ceiling shall have an unobstructed clear height of 12 ft for automation monorails and for exceptionally tall process tools (such as vertical equipment service access).

UNIDIRECTIONAL FLOW CEILING GRID ALIGNMENT

The ceiling grid spacing shall be held to within ±0.25 in at all locations, matched to the 2-ft by 2-ft intervals of the already installed access floor face. The ceiling height shall be held to within ±0.25 in at all locations, matched to the already installed access floor face.

46.17.4 Cleanroom Lighting Technical Description (Example)

LIGHTING TYPE

Lighting fixtures in cleanrooms shall be sealed and fully compatible with the ISO-3, ISO-5 cleanliness requirement and the lamp ballasts fully compatible with the EMF limitation requirements.

Lighting fixtures shall be flush mounted, located in blank panel locations in the ceiling. (This is done to facilitate running monorails diagonally and to install microenclosures where needed.)

LIGHTING LEVELS

Lighting shall provide 90 to 85 ft-candles (initial) over the work surfaces at 30 in height along both side walls, when measured as white light, for both white and yellow light areas.

High-efficiency fluorescent lamps should be used. The preferred color is a warm white, rather than a cool white, if available in the high-efficiency lamps.

YELLOW LIGHTING

Yellow lighting shall be provided in all rooms and their related corridors and service corridors, where white light sensitive materials may be exposed. It might also be required in adjacent to white light sensitive rooms. The yellow light may be done by either gold tubes, or filter sleeves over white tubes. The gold tubes are preferable, if the lamps meet the user other performance criteria. The yellow sheet is normally sandwiched between window glass for door and window applications.

The yellow filters shall meet user performance criteria. The yellow filters used for tubes and windows shall meet the criteria of blocking all light from 250 to 450 nanometers, at energy levels as low as 20 mJ/sq. cm. The preferred filter will block all light from 250 to 650 nm.

LIGHTING LEVEL CONTROL

The lighting in the cleanrooms shall be capable of locally being reduced by 50 percent, or shut off in a local area. The user shall identify the size areas requiring this local light control during the cleanroom layout development.

All windows and doors at the perimeter of the yellow light rooms shall have yellow windows, so that no white light can stream into the yellow-light process areas, including their service sides.

46.17.5 Cleanroom Walls, Windows, Doors, and Flooring (Example)

PERIMETER WALLS CONSTRUCTION

The perimeter interior wall shall be permanent and shall be fire-rated. The cleanroom walls shall be constructed with interior cleanroom side face of painted aluminum. All edges shall be sealed. Holes cut through the wall shall also be carefully sealed to control particles and air leakage.

INTERIOR MOVABLE WALLS CONSTRUCTION

Movable interior walls shall be present in the cleanroom where required for light separation and fume/smoke separation purposes.

The interior movable walls shall be double-faced of non-fire-rated materials. These cleanroom walls shall be constructed with aluminum honeycomb panels, painted with antistatic epoxy paint. Alternative construction materials required the owner's prior approval; however, no flammable materials shall be used.

The wall system panels shall be field-cut capable for process support piping, duct penetrations, and process tool penetrations with a minimum of dust. The wall system must include the means of thoroughly sealing the cut edges against further emitting of particles. Interior cores of gypsum or particle board are not acceptable for these walls due to contamination when these walls are drilled and cut.

The wall system shall be noncombustible with a finish that conforms to at least class III, as defined by UBC. The cleanroom wall members shall be held to within ±0.25 in of true vertical. The cleanroom side of the wall shall be flush within ½ in with the edge of the ceiling grid and filter, to avoid turbulence at this interface, which can cause particle migration between areas.

REARRANGEMENT FLEXIBILITY

The interior cleanroom (non-load-bearing) walls and their support structure must allow for easy rearrangement, to allow for unplanned large process equipment rearrangements.

The interior cleanroom walls shall be nonprogressive type. Walls shall be easily movable and rearrangeable to accommodate new types of tools with significantly different footprints or other changes affecting the size of cleanroom.

WINDOWS

Windows in perimeter walls and doors shall be fire-rated glass.

Transparent panels in interior walls and doors shall be nonbreakable glass, plastic, or fire-rated glass. The wall panels and doors shall generally have windows in the top half. The windows shall be as wide as practical for the panel and door width, and extend from 3 ft 6 in above the access floor to 6 ft 6 in above the floor. Yellow filtered windows will be required in some of windowed panels, when they are used between areas sensitive to white light.

FINISH

The wall system finished surfaces must be nonparticle generating, and non-outgassing. Powder-type epoxy paint, factory cured, is preferred. Color shall be U.S. Standard 595A, color #27780 (oyster white), or user specified.

The complimentary color panels, mixed in patterns with the white panels, shall be used for reducing the stress effect on personnel from the starkness of the all white rooms, and in the yellow light room. It is possible to get complimentary colors for the wall panels that are available in standard epoxy paint colors.

DOORS

Doors used by normal manufacturing operations to pass into, out of, or between ISO-3, ISO-5 Class cleanrooms, shall be the automatic sliding type, constructed to be suitable for cleanroom use. Fiber-type weather stripping should be avoided, due to particle generation. Enclosed, exhausted, closure mechanisms should be used in clean locations.

Doors used for ISO-8 class areas and for maintenance may be swinging type, with their nonventing closure mechanisms located on the service chase side.

CAULKING PRACTICE

The material used for sealing cleanroom air ducts and walls, and for HEPA filter repair, is a process-critical material regarding the type of gases evolved during and after the specified cure time. These gases can affect the yield of the manufacturing operation.

Any caulk material to be used in the cleanroom and related ducting construction shall be tested at a lab for outgassing and the data reviewed by user's Q.C. to obtain written permission to use.

Example of the approved caulk is G.E. Silicone Caulk Type 162 (alcohol-cure type). The G.E. Type 162 caulk is more expensive, so it is imperative that the contractors understand that the cheaper types cannot be used. (The cheaper caulk types may evolve acetic-acid fumes during curing that will corrode the cleanroom and ducting surfaces.)

Clean Flooring

Flooring shall be perforated panels with chemical-resistant antistatic tile, rather than grating. The access flooring system shall be aluminum construction. No rustable or corrodible materials shall be used. Panels shall be set in place, and fastened for seismic resistance. The fastening method shall allow technicians to be able to easily remove and replace panels when required for tool installation or service. The floor system shall be suitable for site specific seismic requirements, including structurally allowing for fastening down of process tools to the flooring for seismic protection. Adjustable aluminum pedestals shall be used. Floor system shall meet all seismic requirements. The floor system shall be designed to support a live load of 350 psf. Where panels are cut for penetrations and at intersections with interior perimeter walls and interior columns, floor system shall still provide all support required to maintain panel floor load performance equal to that of uncut panels. Perforated panels shall have a minimum air-pressure drop configuration. Panels shall not have dampers.

Antistatic Flooring

Electrical conductivity flooring shall be provided throughout the ISO-3 to ISO-8 class areas. The floor tile surface shall be antistatic, with a surface resistivity of 106 to 5 × 10 Ω/unit area. The access floor system, including panels, shall be grounded during installation. The exceptions are assembly and component test areas that shall have ESD-type conductive-flooring (not antistatic) protection.

Floor Finish

All floor parts shall be finished to protect from corrosion and spilled chemicals. Perforated aluminum panels should be covered with antistatic-type sheet material. The panel system design should be carefully evaluated to ensure that edge-trim pieces are securely fastened and will remain intact with repeated panel removals during operations.

Floor Installation

Surface will be flat to ±0.25 in of the access floor plane, regardless of slope to the concrete floor underneath. The access floor will be installed to match ±¼ in to the geometry of the cast-concrete floor waffle structure throughout all clean bays and service chases. The concrete subfloor, beneath the access floor, should be painted with a solvent and corrosive-liquid resistant paint. This shall be done to protect the concrete floor from chemical spills damage.

Vibration-Sensitive Tool Mounting

Where highly vibration-sensitive process tools are installed, they shall be mounted directly to the concrete floor on vibration-stiff stands. These stands may be moved and secured to the floor along with moving the tools.

46.17.6 Final Adjustment and Certification Testing (Example)

The following procedures are used for the definitions and certification testing of the HEPA ceiling. Performance of the cleanroom filters and room environment shall fully meet the technical and performance requirements of this specification. The ISO 14644 standards should be used for definitions and testing methodology.

Pretest Conditions

The tests shall be performed during owner installation of the initial process tooling set. As a result, the cleanroom contractor's adjustments and testing activities must work around the owner's first priority process tool installation work. The cleanrooms and chases shall be first visually examined for any dirt left over and not cleaned. No debris, dust particles, machine parts, wires, stain, or paint marks, etc., should be visually seen at the part of particle count certifications. A thorough and in-depth visual cleanliness check shall be done, and a written report submitted—including the follow-up on the check list execution. All areas shall be checked along the clean air path, including all interstitial floors, wall structure from all sides, fans, ducts, under raised floor, and subfab. No penetrations, or holes, in the floor shall be open while testing or certifying. All doors shall be in their normal positions. The cleanroom contractor shall arrange with the owner that no construction work, tool hookup, or other personnel activity takes place locally while the contractor measuring the cleanliness levels, if that activity may affect the test results. It is expected to continue large tool installation and hookup in other parts of the class 100-Turbulent rooms, due to facility and tool set completion schedule requirements.

Balance of HVAC Fans

Purpose Adjustment of each cleanroom fan module, recirculating and fresh-air makeup fan unit's flow rate to within specifications, prior to doing all other performance testing. Setting of correct pressurization for that clean area to match local area exhaust flows, and attain correct differential pressurizations.

Percentage Tested All units shall be set and measured.

Performance Required The cleanroom shall be final air-balanced and set at the correct pressurization throughout.

All fans shall be operating together. Each fan shall be adjusted to produce the volumetric output equal to the design requirements for heat removal and cleanliness. Pressurization limits between cleanrooms shall be within the limits specified.

Measurement Procedure Flow rate measurement shall be made at the module fan inlet, using a duct velocity profile measured with a suitable calibrated instrument. This measurement shall most accurately determine the SCFM delivered into the fan. The total flow can be also determined by measurement of the SCFM output of each HEPA filter, using a collector method. Other methods, such as averaging velocity methods, are not accurate enough for this purpose, especially when turbulators are being used.

HEPA Filters Flow Settings

Purpose The purpose of this procedure is to ensure that the installed ceiling HEPA filter units each produce the correct unidirectional airflow rate.

Percentage Tested All HEPA filter units shall be adjusted and measured.

Performance Required This performance may be changed by the cleanroom contractor, during engineering design. The change then needs the approval of the owner.

1. *Output flow rate.* The HEPA filter units shall produce an SCFM flow of 90 ft/min of average room velocity.
2. *Output flow rate variation.* The total output flow of each HEPA filter unit shall be within ±10 percent.
3. *Measurement procedure.* The output flow measurements will be made using a calibrated collector volumetric unit. All tests will be made with all doors and wall hatches closed.
4. *Report.* These adjustment and measurement results data shall be delivered as part of the test report. These data shall include a description of the test setup and instruments used.

HEPA Ceiling Leak Testing

Test Purpose The purpose of this test is to ensure that the entire cleanroom ceiling, including HEPA filters, grid and seals, lights, sprinkler penetrations, room-to-ceiling edges, etc., of the HEPA ceilings are leak-free after installation.

Performance Required The leak-test scanning shall cover all HEPA filter faces, light fixtures, all seals and edges, all fixtures and their seals, and edge seals of all filters and light fixtures and blank panels. All leaks shall be detected, corrected, and retested to establish that the leak has been fixed.

Measurement Procedure After installation, passes will be made of the filter face, and the frame-to-gasket seal at the gap between the filter and the grid system. The testing challenge method and the scan and measurement method shall be demonstrated to user's Q.C. for approval.

Any test procedure, other than the one described here, requires prior approval of the owner. The test shall be done with a laser-type particle counter of 1 CFM flow rate, calibrated and sensitive to particles of 0.12 μm and larger. The probe head will be square, 1 in long in the direction of scan and 1.5 in wide. The probe head shall be an isokinetic type. The filter under test shall be subjected to a challenge of approximately 1×10^7 particles per cubic foot. The challenge media shall be as PLS (polystyrene latex spheres) of 0.12 μm size. All particle leak test challenges shall only be done on the upstream filter face. These data shall include a description of the test setup and instruments used. Any other test technique requires prior user approval. The scanning shall be accomplished by passing the strokes so that the entire area of the filter is sampled. A separate pass shall be then made around the entire periphery of the filter, along the adhesive bond between the filter pack and the rigid frame. The probe shall be held at 1 in from the filter face. An addition pass shall be made under the gap between the filter and the grid system. The scan shall be at a rate of less than 2 in/s. Any single particle reading requires that a 1-min count then be taken with the probe stationary under the location of concern. A reading of 2 to 3 counts, within 10 to 15 s, is a significant leak that requires repair. Methods using collector funnels for leak detection are not acceptable for this test. Other leak detection procedures require prior written user approval, from their designated quality assurance person. Leaks must be fully repaired and retested to prove that the leak no longer exists. Repairs may be done using G.E. Silicone Caulk Type RTV 162, or user approved equal. Excessively repaired filter units may be rejected at the discretion of the user's Q.C. either at time of shipment, or at unpacking in the field. A description of the test procedure, instruments, and calibration shall be part of the data delivered to user's Q.C. prior to the start of testing.

Clean Areas Certification Tests

Test Purpose The purpose of this test is to demonstrate that the cleanrooms run within specified stand-by cleanliness limits.

Performance Required The performance and technical requirements as set by the project and approved by the owner.

Procedure The measurements shall be made per ISO 14644 and IEST RP-006 procedures. The cleanroom certification measurements shall be made with all related facility systems running, with process tools off, and all personnel activities at rest.

The measurements shall include

1. *Particles count.* The cleanliness measurements shall be statistically measured/calculated as the statistical mean with a % confidence interval, using the method described in ISO 14644. In determining the statistical mean, a normal distribution may be assumed for the calculations.

The cleanliness measurement shall be made with a 0.1 or μm calibrated laser-type counter, of 1 CFM. Each particle measurement will be for a measured counter air volume of at least 10 ft^3.

The unidirectional ceiling shall not be challenged during this test. The recirculating fans, exhausts, and fresh air makeup system shall be operating normally. Particle measurements locations shall be per ISO standards. These locations shall be included in addition to the grid pattern measurements required per ISO standards.

2. *Temperature and humidity testing*
   - *Applicable.* All cleanrooms with temperature or humidity limitations. These performances shall be met in all the following conditions: (1) With the initial phase 1 tools set installed and operating; (2) with the phase 2 tools set installed and operating; (c) with the phase 1 + 2 tool set installed and operating.
   - *Purpose of test.* The purpose of this test is to demonstrate that the completed cleanroom operates within temperature and humidity limits of the individual cleanrooms.
   - *Percentage tested.* All cleanrooms with RH specifications limitations.
   - *Performance required.* Dew point, RH, and temperature performances shall be as specified.
   - *Procedure.* These temperature and humidity performance of the cleanrooms shall be measured directly below the HEPA filters, in the air entering the cleanroom.
   - *Temperature.* The intent is that the temperature should be controlled and measured at unidirectional air entry into the cleanroom, before the heat is added from the process tools.
   - The sensor location can be at any distance between 6 to 10 in below the HEPA filter face, providing it is in the normal unidirectional airflow from the HEPA filters. It should be more than 12 in from any structure, such as lighting or grid structure. The purpose is to measure/control the incoming air temperature to produce a stable air temperature envelope surrounding the process tool.
   - *Humidity.* The relative humidity should be measured at entry into the cleanroom, before the heat is added from the process tools.
   - The sensor location can be at any distance between 6 and 10 in below the HEPA filter face, providing it is in the normal unidirectional airflow from the HEPA filters. It should be more than 12 in from any structure, such as lighting or grid structure. The purpose is to measure/control the incoming air dew point temperature to produce a stable air humidity envelope surrounding the process tool. The room measurements shall be made over a continuous period of 5 days, 24 h per day. Data shall be recorded on a recorder of sufficient resolution to be able to read changes of less than 0.1°F or °C. The temperature measurement instrument shall have combined resolution and calibration to 0.1° accuracy (accumulated error), and be calibrated to a secondary NBS standard. The dew point measurement will be made over the "same continuous period of 5 days, 24 h per day." Data will be recorded on a recorder of sufficient resolution to be able to read changes of more than 0.3 percent RH. The dew-point measurement instrument shall have resolution and calibration to better than 0.3 percent RH (accumulated error), and be calibrated to a secondary NIST standard.

3. *Lighting level testing*
   - *Applicable.* Cleanrooms final acceptance test.
   - *Test purpose.* The purpose of this test is to establish that the cleanrooms provide adequate lighting levels in the cleanrooms.
   - *Percentage tested:* All lighted areas shall be measured.
   - *Performance required.* The performance requirements are that the entire cleanroom complex produce the required lighting levels listed in the design criteria and approved by the owner.
   - *Measurement procedure.* The lighting measurements shall be made at a height of 32 in from the floor, and 24 in from the side walls, along all areas where process equipment might be installed. The lighting measurements shall be also made in the operator corridor centerline at a height of 32 in.
   - The lighting measurements shall be made with a calibrated lightmeter, measuring in foot-candles or equal unit. All illumination measurements will be made in white light conditions, prior to installation of any yellow light. Alternatively an initial empirical calibration of the instrument shall be made between white light and the yellow light present in the facility under test. Then, lighting can also be measured under the installed yellow light.

4. Noise level testing
   - *Applicable.* Class rooms final acceptance test.
   - *Test purpose.* The purpose of this test is to ensure that the cleanrooms operate within noise limitations.
   - *Percentage tested.* All cleanrooms shall be measured.
   - *Performance required.* The cleanrooms shall not exceed the noise levels specified measured with all process tools off, the fresh air make-up system running, the facility exhausts fans running, and all cleanroom fans running at normal velocities.
   - *Procedure.* The noise measurements shall be made at a height of 32 in from the floor, and 24 in from the side walls, along all areas

where process equipment might be installed, and in the operator corridor centerline at a height of 32 in. Measurement shall be made at 10-ft intervals, along both sides and down the centerline of each cleanroom. The noise measurements shall be made with a calibrated acoustic-meter, measuring noise levels on an NC curve. A dB curve measurements shall not be acceptable.

5. Electromagnetic fields (EMF) testing
   - *Applicable.* All cleanrooms and scanning electron microscope (SEM) locations in quality assurance room.
   - *Test purpose.* The purpose of this test is to ensure that the completed cleanroom operates within electromagnetic fields (EMF) limitations.
   - *Percentage tested.* All cleanrooms and SEM locations in quality assurance room.
   - *Performance required.* The EMF limitations in the cleanroom locations, and in their associated service aisles, shall not be exceeded.
   - *Procedure.* The EMF measurements shall be made at a height of 32 in from the floor, and 24 in from the side walls, along all areas where process equipment might be installed in the cleanroom and in the associated service aisles. The EMF measurements shall be made with a calibrated instrument. EMF measurements shall be made with all cleanroom motors and lighting on, and all process tools off. The measurement procedure and calibrated instrument shall be approved by the owner prior to measurements.

6. *Vibration-level testing*
   - *Applicable.* All cleanrooms and die-sort rooms.
   - *Purpose of test.* The purpose of this test is to ensure that the cleanrooms and technical rooms operate within their vibration limits.
   - *Percentage tested.* All cleanrooms and die-sort rooms.
   - *Performance required.* Vibration performance limitations specified in basis of design.
   - *Procedure.* The vibration measurements shall be made with a calibrated instrument, with the sensor floor mounted directly to the concrete cleanroom floor. Measurements shall be made under the conditions of all cleanroom modules operating for the acceptance test. All cleanroom motors shall be running, and all process tools shall be off. Data shall be recorded of the floor vibrations levels both on the process-equipment slab and on the service aisle slab. Vibration measurement data are required on top of the perforated raised floor in areas where the floor has to be fastened down for access-floor standing vibration-sensitive process tools, but no specification has to be met.

46.17.7 Cleanroom Construction Protocol (Example)

Cleanroom construction shall be divided into six phases that include demolition, rough-ins, cleanroom perimeter, cleanroom enclosure, HEPA installation and certification, and post certification. From one phase moving to next, construction protocol control becomes more stringent. The following sections describe changes on controlling badge color, garments, housekeeping, cleaning of materials, equipment, tools, and protocol.

PHASE 0

Demolition if applicable (badge color: none)

Starts. With the demolition of the existing walls, ceiling, flooring, mechanical, and electrical services supporting the area

Ends. After completion of all demolition activities and the area has been "gross" cleaned

Activities. Removal of walls, ceiling, flooring, ductwork, piping, electrical conduit and wiring, etc., salvage materials as directed by construction manager

Garments. Normal work clothes and shoes

Housekeeping. Daily construction clean-up; removal of dirt and debris

Protocol. Normal construction procedures; use special care in handling materials scheduled for salvage and reuse

PHASE 1

Rough-ins (badge color: green)

Starts after the completion of the "gross" cleaning of the area

Ends at the completion of the installation of the exterior walls of the cleanroom area, when air can be supplied to the area for pressurization and at the completion of the "gross" cleaning of the area

Activities. Installation of supply and return ductwork, HVAC units, sprinkler piping, underfloor waste system piping, high-purity piping rough-in, electrical rough-in, data telecommunications wiring rough-in, replacement of raised floor tiles, installation of cleanroom walls

Garments. "Clean" work clothes and clean shoes

Housekeeping. Daily vacuuming of the clean zone with vacuums equipped with HEPA filters. Clean zone shall be mopped weekly (at minimum) with cleaning solution of high-purity water and cleaning agent designed to remove oil residues (isopropyl alcohol, Buckeye Blue or other approved product; product shall be approved by the construction manager)

Cleaning of material, equipment and tools. All material and equipment scheduled for incorporation into the cleanroom area below the raised floor and 10 ft above finished floor shall be free from contaminants when viewed under ambient light. All materials, equipment, and tools shall be free from dirt, debris and oil.

Prefabrication of components is required to the greatest degree possible.

Work activities not permitted in the cleanroom area and preparation zone (clean zone)
- Grinding
- Open flame cutting
- Plasma arc cutting
- Pipe threading
- Sanding, sandblasting, or other forms of air driven abrasive removal
- Sawing (excepting with hole saw; see below)

Work activities permitted with special requirements
- Drilling (immediate clean-up of debris. If drilling into concrete, "contain and capture" technique is required)
- Sawing with hole saw (immediate clean-up of debris)
- Brazing (requires the use of a HEPA filter equipped "smoke eater")
- Soldering (requires the use of a HEPA filter equipped "smoke eater")
- Welding (requires the use of a HEPA filter equipped "smoke eater")

Protocol
- All personnel shall enter and exit the clean zone through the designated entrance (as determined by the construction manager)
- There shall be no food, beverages, or tobacco products permitted in the clean zone
- All "gross" packaging (crating, cardboard, etc.) of materials and tools shall be removed prior to the material or equipment being moved to the preparation area
- No materials, equipment, or tools shall be stored in the cleanroom area or preparation area (clean zone)
- No aerosols or sprays are permitted in the cleanroom or preparation area (clean zone)
- All spills or other sources of contaminants are to be immediately cleaned up and cleaning materials properly disposed of
- All trash and debris shall be placed in proper containers for daily removal from the clean zone
- All materials, equipment, and tools shall enter the clean zone through the designated entrance. Materials, equipment, and tools shall be free from dirt, debris, and oil (construction clean). All materials and equipment scheduled for incorporation into the cleanroom areas below the raised floor and 10 ft above finished floor shall be free from contaminants when inspected under ambient light (level 1)
- Work activities shall be governed by the requirements indicated above

PHASE 2

Cleanroom perimeter (badge color: brown)

Starts at the completion of the "gross" cleaning of the cleanroom area at the end of phase 1 and air being supplied to the cleanroom area for pressurization

Ends at the completion of the installation of the high-purity piping rough-ins, electrical rough-ins, and other utility work in the chases, above the ceiling, and elsewhere within the boundary of the cleanroom area

Activities completion of the high-purity piping rough-ins, electrical rough-in, data telecommunications rough-in, ductwork drops to terminal HEPA filters

Garments

- "Clean" work clothes and shoes
- Disposable shoe covers (provided by construction manager)
- Disposable head covers (bouffant; provided by construction manager)
- Disposable shoe and head covers are to be changed whenever dirty or torn

Housekeeping

- All requirements of phase 1
- Daily mopping of cleanroom floor areas

Cleaning of materials, equipment, and tools. All materials, equipment, and tools scheduled for incorporation into the cleanroom area or for use in the installation shall be free of contamination when viewed under ambient light.

Work activities not permitted in the cleanroom area: as per phase 1

Work activities permitted in the cleanroom area with special requirements: as per phase 1

Protocol

- All requirements from phase 1.
- Air shall be supplied to the cleanroom area filtered with 95 percent ASHRAE filters to provide a minimum pressure differential of 0.03 in (positive) between cleanroom area and adjacent areas at all times. Ensure that air handlers and ductwork not being used for pressurization of the cleanroom area will not be contaminated by air being returned or exhausted from the cleanroom area.
- All materials, equipment, and tools entering the cleanroom area shall be free from contaminants when viewed under ambient light (level 1). Tools may remain in the cleanroom area if stored in cleaned tool boxes or other designated storage containers approved by the construction manager.

PHASE 3

Cleanroom enclosure (badge color: yellow)

Starts at the completion of all ductwork drops, piping, electrical, and data telecommunications rough-ins and at the completion of "gross" cleaning of the cleanroom area

Ends at the completion of the installation of the ceiling grid, light fixtures, ceiling blank panels, and the area has been "gross" cleaned

Activities testing of all piping systems, electrical distribution, and data telecommunication wiring, installation of wiring devices, installation of ceiling grid, sprinkler heads, light fixtures, ceiling blank panels, flooring

Garments

- "Clean" work clothes and shoes
- Disposable coveralls (provided by construction manager)
- Disposable shoe covers (provided by construction manager)
- Disposable head covers (provided by construction manager)
- Disposable face mask (beard cover if required) (provided by construction manager)
- Disposable latex gloves (provided by construction manager)
- Disposable garments may not be worn outside of the cleanroom area

Housekeeping. All requirements of phase 2

Cleaning of materials, equipment, and tools. All materials, equipment, and tools schedules for incorporation into the cleanroom area (cleanrooms and chases) shall be free from contaminants when viewed under high-intensity oblique white light (level 2)

Work activities not permitted in the cleanroom area

- As per phases 1 and 2
- Drilling (see below)
- Sawing (see below)
- Brazing (see below)
- Soldering (see below)
- Welding (see below)

Work activities permitted in the cleanroom areas with special requirements

- Drilling ("contain and capture" technique required)
- Sawing (hole saws only, "contain and capture" technique required)
- Brazing (chases only, with HEPA filter equipped "smoke eaters")
- Soldering (chases only, with HEPA filter equipped "smoke eaters")
- Welding (chases only)

Protocol

- All requirements of phase 2
- Personnel shall enter the cleanroom perimeter from the designated entrance. They shall remove "outer" garments and store as directed by the construction manager. They shall then proceed to the temporary gowning/staging area located adjacent to door #XX.
- In the temporary gowning area, personnel shall "gown up" as follows:
    - Head cover
    - Shoe covers
    - Coveralls
    - Face mask (beard cover if applicable)
    - Latex gloves
- After "gowning up" personnel shall then proceed to the cleanroom areas (cleanrooms and chases).
- All materials, equipment, and tools shall enter the cleanroom perimeter by the designated entrance. They shall enter the cleanroom area through the temporary gowning/staging area only.
- Tacky mats shall be installed at the entrance to the temporary gowning area from the preparation area and from the entrance from the temporary gowning area to the cleanroom area.
- All materials, equipment, and tools scheduled for incorporation into the cleanroom areas or for use in installation shall be free from contaminants when inspected under a high-intensity oblique white light and/or black light (level 2 cleanliness).
- Prior to the installation of any light fixtures, sprinkler heads, and blank ceiling panels in the ceiling grid, the grid shall be free of contaminant when inspected under high-intensity oblique white light.

PHASE 4

HEPA installation and certification (badge color: red)

Starts when all interior "finishes" of the cleanroom areas are complete and the cleanroom areas have been "gross" cleaned

Ends at the completion of the "as-built" certification of the cleanroom areas

Activities. Installation of the terminal HEPA filters, HEPA fan-filter units, air balancing, precision cleaning of the cleanroom areas, "as-built" certification of the cleanroom areas

Garments as required in phase 3

Housekeeping as required in phase 3

Cleaning of materials, equipment, and tools: as required in phase 3

Work activities not permitted in the cleanroom areas: as required in phase 3

Work activities permitted in cleanroom areas with special requirements. By permission and under the direction of the construction manager only

Protocol

- All requirements of phase 3.
- Terminal HEPA filters and fan-filter units shall remain in secured storage until immediately required for installation. They shall be handled and installed by personnel trained in their care and installation only. These personnel shall be approved by the construction manager.
- HVAC units serving the cleanroom areas shall be cleaned and have the final set of filters installed at the time of the installation of the terminal HEPA filters.
- HVAC systems serving the cleanroom areas shall be prebalanced (air) prior to the installation of the terminal HEPA filters.
- No other work is permitted in the cleanroom areas after the installation of the terminal HEPA filters has started.
- After completion of the installation of the terminal HEPA filters and fan-filter units, the cleanroom areas shall be final air balanced.
- After the completion of the final balancing, all surfaces of the cleanrooms shall have no more than 12 particles per square foot when inspected under black light or 72 particles per square foot when inspected under high-intensity oblique white light (level 3 cleanliness). This is the "precision cleaning" of the cleanroom areas and will be performed by and independent cleanroom cleaning contractor as directed by the construction manager.
- After completion of the "precision cleaning" of the cleanroom areas, the cleanrooms shall be "as-built" certified as per ISO 14644.

Phase 5

Postcertification (badge color: red with white stripes)

Starts after the completion of the "as-built" certification of the cleanroom areas.

Ends at the direction of the construction manager.

Activities by written permission of the construction manager only.

Garments as required by the owner's operating protocol and as directed by the construction manager.

Housekeeping by owner, personnel working in the cleanroom areas are required to comply with all requirements of the owner's operating protocol.

Cleaning of materials, equipment, and tools as required by the owner's operating protocol; materials equipment and tools shall no more than 12 particles per square foot when inspected under black light or 72 particles per square foot when inspected under high-intensity oblique white light.

Work activities not permitted in the cleanroom areas and permitted under special requirements by permission and under the direction of the construction manager.

Protocol: Personnel shall follow all requirements of the owner's operating protocol. Personnel shall enter the cleanroom area through the designated gowning rooms only.

Certification

In order to measure the effectiveness of the project team in meeting the functional requirements of the cleanroom, various tests are conducted to measure the performance of the critical systems against accepted reference standards. This testing process is commonly referred to as certification of the facility.

46.18 FURTHER READING

ASHRAE Handbook, Heating Ventilating and Air-Conditioning Applications, 2011, American Society of Heating, Refrigerating and Air-Conditioning Engineers, Inc., Atlanta, Ga.

Chubb, N., John Chubb Instrumentation papers for the IEEE-IAS meeting Oct. 1999 and for the ESA meeting at Niagara Falls, Jun. 2000.

International Organization for Standardization (ISO): ISO 14644 Series Cleanroom Standards.

"Design for Energy Efficient Low Operating Cost Cleanrooms" by Raj Jaisinghani of Technovation Systems, Inc.

ESD Association, Standard ANSI/ESD S20.20, "Development of an Electrostatic Control Program for Protection of Electrical and Electronic Parts, Assemblies and Equipment."

Gale, S.F., "FFUs: Setting a Course for Energy Efficiency," Cleanroom, Sep. 2004.

Jeng, M. S., T. Xu, and C. H. Lan, "Toward Green Systems for Cleanroom: Energy Efficient Fan-Filter Units," *Proceedings of Semiconductor Equipment and Materials International Conference (SEMICON) West 2004—SEMI Technical Symposium: Innovations in Semiconductor Manufacturing (STS: ISM)*, pp. 73–77, Jul. 2004. San Francisco, California, LBNL-55039.

International Standards Organizations, ISO–14644 Parts 1–8, "Cleanrooms and Associated Controlled Environments."

International Standards Organizations, ISO 14698, "Cleanrooms and Associated Controlled Environments-Biocontamination Control."

Institute of Environmental Science and Testing, IEST RP–CC006, "Testing Cleanrooms."

Institute of Environmental Science and Testing, IEST RP–CC012, "Considerations in Cleanroom Design and Construction."

International SEMATECH, "SEMATECH Guide for Documenting Process Tool Installation Time and Cost," SEMATECH Technology Transfer 98013447A-STD.

International SEMATECH, "SEMATECH Equipment Installation Sign-Off Procedures," SEMATECH Technology Transfer 98103579A-XFR.

Pavlotsky, Richard, Cleanroom design, Technical Features, CleanRooms.

Pavlotsky, Richard, Cost Savings Opportunities in Biotech Cleanroom Operations, CleanRooms East 2003, Boston, Massachusetts. Conference Proceedings.

Pavlotsky, Richard, The Whole Cleanroom Is Indeed the Sum of Its Parts. Technical Features, CleanRooms.

Pavlotsky, Richard, 15 Factors That Influence Cleanroom Design and Construction Cost CleanRooms.

Pavlotsky, Richard, Arriving at Approximate Cost for Pharmaceutical, R&D and Biopharmaceutical Facilities, CleanRooms.

Semiconductor Manufacturing Magazine: http://dom.semi.org/

Tschudi, William, and T. Xu, "Cleanroom Energy Benchmarking Results," sponsored by California Institute for Energy Efficiency, Pacific Gas and Electric Company, and Rumsey Engineers. Prepared by Supersymmetry.

Tschudi, Bill, Dale Sartor, and Tengfang Xu, "an Energy Efficiency Guide for Use in Cleanroom Programming," by Lawrence Berkeley National Laboratory, Northwest Energy Efficiency Alliance and California Energy Commission.

Xu, T., "Considerations for Efficient Airflow Design in Cleanrooms," *Journal of the IEST*, Vol. 47: 24–28, 2004. Institute of Environmental Sciences and Technology (IEST), Illinois; LBNL55970.

Xu, T., "Performance Evaluation of Cleanroom Environmental Systems," *Journal of the IEST,* Vol. 46: 66–73, Aug. 2003. Institute of Environmental Sciences and Technology (IEST), Illinois; LBNL-53282.

Xu, T., and M. S. Jeng, "Laboratory Evaluation of Fan-Filter Units' Aerodynamic and Energy Performance," *Journal of the IEST*, Vol. 47: 116–120, 2004. Institute of Environmental Sciences and Technology (IEST), Illinois; LBNL542507.

46.19 PROFESSIONAL ASSOCIATIONS

1. Institute of Environmental Sciences and Technology (IEST)
2. ASHRAE
3. semi.org
4. International Organization for Standardization (ISO)
5. Lean Construction Institute

Chapter 47

Vibration and Noise Design

BY

MICHAEL GENDREAU AND HAL AMICK *Colin Gordon Associates, Brisbane, California*

47.1 INTRODUCTION

Noise and vibration are environmental contaminants* that affect semiconductor processing and research.† Because the tools used in semiconductor manufacturing and analysis are sensitive to vibration and noise to varying degrees, excessive amounts of these contaminants can adversely impact yield, throughput, or the operating linewidth or resolution of the tools. In addition to the potential impact to tools, noise impacts personnel working in the cleanrooms and labs, and it is usually regulated in the external environment.

The vibration and noise performance of semiconductor and other advanced technology research and manufacturing buildings must be considered early in the conceptual design stage and throughout the design and construction of the project. This procedure, which applies not only to new buildings but also retrofit projects, service or process modifications, tool installations, etc., starts with site vibration and noise evaluations and reviews of manufacturer's tool specifications or selection of generic criteria. The structural design of the building, layout of the process and mechanical equipment, and the tool layout in the lab or cleanroom, are all critical to a successful vibration and noise design. When vibration and noise cannot be controlled solely by equipment layout, as is often the case, specification of appropriate vibration and noise control devices (isolators, silencers, etc.) will be necessary. Finally, appropriate care must be taken in the tool hook-up stage, so that the vibration and noise environment of the base building is not excessively elevated. This chapter discusses these and other considerations necessary to provide acceptable noise and vibration environments for semiconductor manufacturing and other advanced technology work (high-resolution metrology, nanotechnology research, LCD-TFT manufacturing, sensitive biotechnology work, etc.).

47.1.1 Vibration and Noise Sources

There are many sources of vibration and noise that may contribute to the operating production or research environment. Sources external to the building include general ambient conditions (due to many sources near and far); local traffic, rail, and aircraft; and adjacent industrial facilities. Internal sources include automation and product handling machinery; walking and other personnel activities; materials handling; electrical and mechanical equipment, piping, and ductwork; and the tools themselves.

Since the building owner will have limited control over off-site external vibration sources, and since there are limits to the mitigation that can be provided with the building itself, site selection and layout on the site with respect to external conditions is critical. With enough flexibility in the site selection and layout, adequate control of internal noise and vibration sources (and external noise sources) can usually be provided in the design and layout of the building and infrastructure.

47.1.2 Vibration and Noise Sensitive Areas and Equipment

The degree of vibration sensitivity of the tools may be gauged by comparison with more familiar vibration amplitudes. For example, the most demanding metrology or characterization equipment specifications may require environmental vibration amplitudes 10 times lower than typical microelectronics (photolithography) production floors, 100 times lower than typical laboratory (400x microscope) or semiconductor support function (CMP, implant, etch, wafer starts, etc.) floors, and 500 to 1000 times lower than just-perceptible vibration that might be experienced, for example, on a suspended office building floor under normal conditions. Most in-line semiconductor process equipment is not highly sensitive to noise, although a few tools have criteria within the range of typical operating cleanroom noise levels (NC-55 to NC-70‡). Metrology and nanotechnology characterization tools, such as Scanning Electron Microscope (SEM), focused ion beam (FIB), transmission electron microscope (TEM), etc., are typically more sensitive to acoustic noise, many requiring moderate noise levels (Noise Criterion or NC-30 to NC-50—noise levels more commonly found in an office environment than in a cleanroom). Some require very low noise levels (NC-15 to NC-30—which are similar to those required in recording studios, auditoria, or conference rooms). Special cases include physical science metrology suites and semiconductor research

*Vibration and noise are characterized as *energy contaminants* because they can impact fabrication or research as significantly as particulates or airborne contaminants, yet they exist only in the form of energy.

†Some aspects of research facilities—especially those for nanotechnology—are discussed here, because today's research becomes tomorrow's production. Many tools intended for research at the nanometer scale are now being adapted for use in production.

‡There are several standardized or commonly used noise criterion curves designed to rate human response to noise. Generic noise criteria for equipment have been developed, but as these are relatively recent and not yet standardized, it is common in the high-tech industry to reference one of these perception-based indices. See Refs. 1 and 2 for more information.

and development laboratories, which can require extremely stringent acoustical performance, often into the infrasonic (< 50 Hz) range.

The remainder of this chapter is divided into several subsections covering the following topics: measurement methodologies and criteria; vibration and noise sources; advanced technology building foundation and structural design; vibration and noise control in the mechanical, electrical, and process design; acoustical design; tool hook-up; purposes and timing of facility vibration surveys; maturation of the vibration and noise environment; and future trends.

47.2 MEASUREMENT METHODOLOGIES AND CRITERIA

Vibration and noise are fairly complex physical phenomena, which can be measured and represented numerically. Existing spaces may be assessed by comparing measured data with established criteria, or those criteria may be imposed on a design. There are established methodologies and standards for numerical representation of vibration and noise data, some of which are general, and others that are more applicable to either one or the other. In this section, we discuss the measurement approaches (generalities and then specialized), followed by an introduction to relevant vibration and noise criteria.

47.2.1 Measurement Methodologies

There are many possible ways to measure and represent vibration and noise. In any particular case, how one goes about it depends on the amount of detail one wishes to obtain. The general goal of measurement and signal processing is to adequately represent the nature of the environment, with appropriate complexity so that critical elements of the environment are not obscured. Both overly complex and overly simplified representations of the vibration or acoustical environment can obscure the results.

In general, vibration and noise can be described as continuous fluctuations in amplitude of an environmental variable over time. Vibration involves movement (generally of a volume or surface) and can be represented as acceleration, velocity, or displacement. Noise involves fluctuation in air pressure, and it is generally represented in units based on pressure. These are environmental variables; the representation of the vibration or noise associated with a particular source is typically and more usefully given in more inherent terms,[§] such as its sound power or the amount of force it applies to its support structure.

The fluctuations can occur at specific frequencies, or randomly over a wide range of frequencies, or a combination of the two. Thus, vibration or noise can be separated into several classes:

1. Tonal versus random[3]:
   a. Tonal vibration is periodic vibration at a single frequency. It is usually associated with periodic processes, such as rotating or reciprocating mechanical equipment, or magnetostriction in transformers. Likewise, tonal sound is at a single frequency, but there are a wide variety of causes, those given above and others, so that an environment such as an operating cleanroom can have many tonal components in the noise.
   b. Random (often known as "broadband") vibration and noise contains random energy at many frequencies. Sources of random vibration and noise include turbulent fluid flow, impacts such as due to footfall and door operations, etc.
2. Steady-state versus transient:
   a. Steady-state[¶] vibration or noise does not change significantly, on the average, over time. This is produced by continuous or relatively continuous sources, such as operating machinery, constant traffic, air circulation (HVAC) systems, etc.
   b. Transient vibration is time dependent. Sources include passing walkers and individual vehicles (trucks, trains, construction vehicles, etc.), and intermittent machinery such as robotics and automatic material handling systems (AMHS).

The vibration in semiconductor production floors, or on sites under consideration for the location of a factory, contains all of these elements. However, in the stiff floors used for production and research activities, the vibration environment is often best characterized as "steady-state random," that is, shaped by various random sources, and often including some steady-state tonal components associated with building mechanical equipment and tools.[4,5] Softer floors such as used for offices and production support functions (and stiff floors as well in more exceptional cases) will often in addition be punctuated by occasional transient vibration such as due to footfall.

The airborne noise in a cleanroom also contains a mix of random and tonal elements. The noise in a cleanroom without operating tools is mostly random, generated by air movement, and perhaps containing low-frequency tonal components associated with recirculation fan blade passage frequencies. The tools contribute a great deal of tonal and random noise to the cleanroom or lab environment.

A measurement characterizing the variation in amplitude over time—in representational form known as *time domain*—is necessary to record all of the information about the environment. However, given the complexity of the environment from the standpoint of frequency, time domain representations of the environment do not provide a simple readable representation of the significant components of the vibration or noise in buildings or the environment. There are exceptions to this, of course. If one is primarily interested in the response of a floor to a transient—especially if this excites a fundamental mode of the structure—it is readily visible in time domain data.

Perhaps the most useful aspect of time domain data is that it is completely reanalyzable and can be used to produce any of the other representations of data to be discussed presently. The reverse is not true, since basic time domain data cannot be reconstructed from other data representations, giving the time domain representation the distinction of having the greatest information depth.

Structures and tools are subject to resonance response, meaning that they are more sensitive to vibration and noise at certain frequencies and less sensitive at others. With many tools, the sensitivity to vibration or sound will be based upon whether or not internal resonances are excited. This suggests that it is important to have some knowledge of the frequency content of vibration or noise. When vibration or noise is represented as a function of frequency, it is said to be in the *frequency domain*. There are several ways to obtain data in the frequency domain.

Using Fast Fourier transform (FFT) methods, a snapshot of the time domain data can be transformed into a representation of amplitude as a function of frequency over the measurement period.[**] Definition of the parameters used in the FFT analysis is necessary in an accurate representation of the data, and in allowance of variations in representation, if needed. The following paragraphs note a few considerations in this regard:

- *Type of vibration environment.* Presence of the various types of vibration listed above may dictate measurement parameter settings:
  - *Stability of the environment during the sample period.* If the environment is steady state, representations of the *average* environment may be used, as long as the averaging time is long enough to represent a stable value. In the case in which environment is not steady state, representations of the *maximum* value obtained at each frequency may be used. If the environment contains significant transients, representations of the average value are usually not useful unless accompanied by time domain data for the sample period. In either case, maximum representations are useful if one wishes to know the maximum impact during the measurement period. It is especially important in this case to state the measurement time.
  - *Tonal versus broadband vibration.* The frequency resolution of the data may be influenced by the level of detail needed to identify

[§]By "inherent" we refer to quantities that are absolutely dependent on properties of the source, and are not dependent on or mediated by the environment.

[¶]Strictly speaking, continuous *random* vibration or noise is said to be "stationary," and the term "steady-state" is used for tonal content. However, we will use the latter term interchangeably.

[**]FFT analysis results in spectra with amplitudes spaced at uniform increments of frequency, which we call the spectrum's *resolution*.

tonal frequencies or resonances. Also, the type of frequency analysis used may be determined by the presence of tonal or broadband vibration. For example, while power spectral density (PSD) analysis is more general than representations of amplitude for arbitrary bandwidths in random vibration analysis, PSD is not defined for measurement of environments containing significant tonal vibration components.[4,5]

- *Measurement time.* Implications of the relevance of this are given in the bullet item above. The measurement time must also be long enough to adequately characterize the lowest frequency of the desired frequency range.
- *Frequency range.* This is the overall range of frequencies used in the analysis.
- *Bandwidth.* The measurement bandwidth affects the time for each individual sample[††] and the frequency resolution of the data. This also represents another potential limit in the mutability of the data, as relatively low resolution data can be constructed from higher resolution data, but not vice versa. Selection of appropriate bandwidth will be determined by the level of detail needed in frequency resolution. In FFT analysis, the bandwidth is constant over the entire spectrum and is some multiple times the frequency resolution. That multiple is dictated by the windowing function being used (see below).
- *Windowing functions.* Windowing functions, for example, *Hanning, rectangular, flat top,* etc., are used to mitigate sampling errors in the FFT process.[6] Selection of the appropriate windowing function is important in vibration analysis, for example, depending on whether better accuracy in frequency or better accuracy in amplitude is required.
- *Acceleration, velocity, displacement.* Various types of transducers measure one of these values, but the data can be integrated at the time of measurement or later into any of the other representations. Thus, there need be no preference for one format over another so long as significant errors are not introduced in the transformation or measurement, such as due to integration errors, inadequate dynamic range, etc.

Another form of spectral representation is the *proportional bandwidth* spectrum, in which the frequency bandwidth of each band is proportional to the center frequency of that band. The two most common forms are the octave band (in which each center frequency is double the previous frequency, or an *octave* higher) and the one-third octave band (in which there are three proportional bands per octave and the bandwidth is 23 percent of the band center frequency).

Proportional bandwidth spectra may be obtained either directly or indirectly. The former involves using a spectrum analyzer containing multiple parallel one-third octave band filters. Many of the bullet items above that apply to FFT spectra are also relevant to proportional bandwidth analysis, though the concept of windowing functions does not apply. The spectra may also be obtained indirectly by first carrying out an FFT measurement, and then transforming the constant bandwidth spectrum into a proportional bandwidth one. This process is called *synthesis* and is described in Refs. 4 and 5. This allows a single measurement to offer the level of detail required for diagnostic efforts as well as an associated spectrum in a simpler format for characterizing how a vibration might affect a tool.

Sound spectra are most often obtained directly, through the use of spectrum analyzers designed for that purpose. Octave band spectra are the most common, but one-third octave band and narrowband acoustic spectra are given in some tool specifications and are useful for diagnosis of an environment.

A third form of data representation used in acoustics is the single-value quantity, such as the A-weighted sound level, commonly denoted with units of dBA. There are several standard "weighting" curves—the most popular being A and C—which are used to emphasize content at some frequencies and deemphasize it at others. A-weighting is used to simulate how the human ear perceives sound at moderate amplitudes, deemphasizing content at low frequencies. The most important shortcomings of using A- and C-weighting in measurements pertaining to tools is that it is impossible to extract frequency characteristics from the data[1] and, based on comparisons of typical spectra,[2] the human biological weighting curve is not appropriate for machines, as discussed in more detail below.

As we have noted, building site and production or research environment vibration and noise can be classified as steady-state random much of the time. For this reason, and due to the frequency dependence of tool vibration sensitivity, basic analysis and representation of vibration and noise environments in advanced technology facilities is usually in the frequency domain, represented as an average (for steady-state vibration) or maximum value (for transient vibration), and using a constant narrow bandwidth, which may later be simplified in wider bandwidth formats for comparison with specific tool specifications or generic criteria. Use of PSD is not recommended when there are significant tonal components in the vibration environment.

47.2.2 Criteria—Equipment-Based Criteria versus Generic Criteria

There are several ways to approach criteria selection for a building design project. When the facility is being designed for specific analytical tools, and when the manufacturers can provide reliable criteria for these tools, engineering may be carried out accordingly. However, in many cases, specific requirements are often not identified or are not available at the time of initial design, or perhaps may not be known until after building commissioning, or it may be expected that they shall change in the future. In these cases, it is common practice to use in the design one or several generic criteria, which may conservatively represent entire classes of tools.

This practice lends itself to improved flexibility over the life of the facility, presuming this is useful. (One might argue that if a facility is unlikely to be renovated, then a tool-specific design is more cost-effective than a design based on generic requirements.)

47.2.3 Criteria—Generic Vibration Criteria

VC (Vibration Criteria)

The most popular generic vibration criteria are those developed in the early 1980s by Ungar and Gordon,[7] originally known as the "BBN Curves." They are now denoted VC-n, where n is a letter from A through G,[8] and in principle could be extended farther. The amplitude associated with each subsequent letter is one-half the amplitude of the previous letter. These criteria have been published in numerous places, including Recommended Practice RP-12 of the Institute of Environmental Sciences and Technology[9] and ASHRAE.[10]

The generic VC criteria were developed to represent the requirements of *families* of equipment or *process geometries*. They tend to form a lower bound for all equipment of a particular type working at a particular feature size. Thus, a facility designed generically for a process at 0.1 μm will be more likely to work for all tools working at that geometry than one designed specifically for one tool at that geometry, if that tool is not the most sensitive one of its kind.

The VC curves are shown in Fig. 47.1. The criteria are to be applied to data represented in rms (root mean square) velocity amplitude, as determined in one-third octave bands.[‡‡]

NIST—A Vibration Criterion

A special vibration criterion was developed for the design of the Advanced Measurement Laboratory (AML) at the National Institute of Standards and Technology (NIST) in Gaithersburg, MD.[§§] It has become popular for application in some of the more demanding spaces in nanotechnology research facilities such as those used for imaging and molecular manipulation. It is becoming more relevant in semiconductor research

[††]The measurement sample time is $\tau = 1/B = L/R_f$, where B is the frequency resolution in Hz (1/s), R_f is the frequency range, and L is the number of divisions of the range used in the analysis.

[‡‡]Extensive discussions of technical and historical perspectives of the VC curves may be found in Refs. 8, 5, 11, 12. The International Standards Organization (ISO) recommended criteria for workshops, offices, residences, and operating theaters, also shown in Fig. 47.1, are discussed in Ref. 13.

[§§]The evolution of this criterion is discussed in Ref. 8.

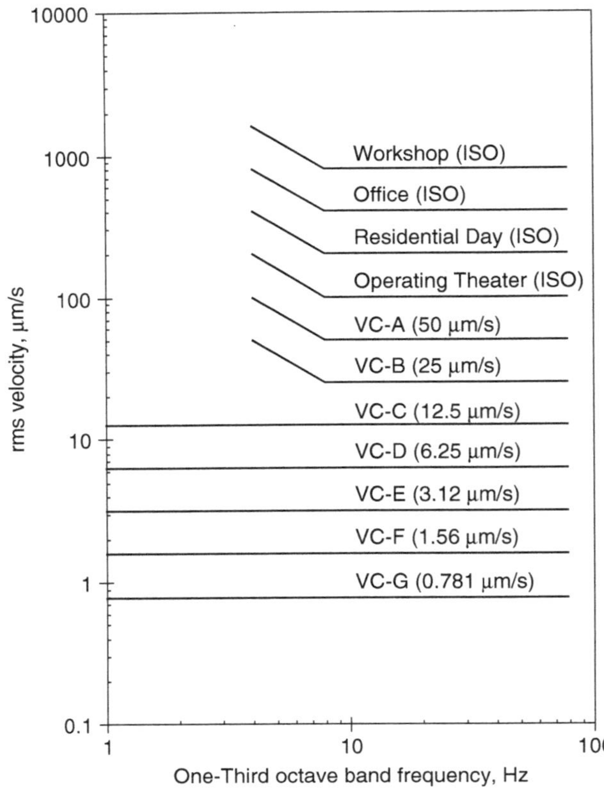

Figure 47.1 Graphical definition of VC criterion curves.[9]

as well, as production extends into nanometer scales, and as the industry begins to adopt some of the more demanding instruments. The NIST-A criterion may be defined as follows: 0.025 µm or 25 nm displacement for $1 \leq f \leq 20$ Hz, and 3.1 µm/s (or VC-E) velocity for $20 < f \leq 100$ Hz.

47.2.4 Criteria—Floor Dynamic Stiffness Requirements

Many photolithography and other tools, such as steppers, wafer probes, laser drills, etc., involve positioning mechanisms, which generate dynamic forces. Likewise, step-and-scan systems (or *scanners*), increase throughput and resolution (i.e., smaller linewidths) by carrying out the critical lithography operation "on the fly," while both reticle and stage are moving. In order to achieve this, a control system generates dynamic positioning forces to coordinate the positions of the reticle and stage.

In both cases, the dynamic forces must be resisted by the structure supporting the tool, and also by any supporting pedestal supporting the tool at access floor level. As a result, tool manufacturers specify "resistance" characteristics in addition to ambient vibration requirements. Those resistance characteristics are stated in terms of accelerance (the FFT of response acceleration divided by the FFT of the causative force) or mobility (responsive velocity divided by the force) or dynamic stiffness (causative force divided by response displacement). The dynamic measurements involved and the consequences for building designers were explored by Amick and Bayat.[14]

47.2.5 Other Forms of Vibration Criteria

There are other, less common forms of vibration criteria, including those involving time domain and others involving response spectra. It should be noted that time domain criteria are subject to the same caveats as time domain data representation: there is no consideration of frequency content. Vibrations at all frequencies are given equal importance and the importance of resonance response is disregarded. For further discussion, see Ref. 11.

47.2.6 Generic and Tool-Specific Acoustic Noise Criteria

Acoustic noise may be regulated with respect to impact to tools and to people. In the latter case, there are several established and commonly used families of rating curves available as octave band spectra, such as the NC (or NCB), RC, and NR curves.[10,15,16] These curves are based on empirical data derived from perceived differences in noise levels and are generally shaped in frequency in accordance with the human perception of balanced noise spectra. In addition, the single-value indices mentioned earlier (dBA, dBC, etc.) are sometimes used, but only in certain limited cases such as in compliance with occupational health standards or environmental noise regulations. Due to the lack of definition of relative frequency amplitudes, single-value indices should not be used in the design of regularly occupied interior acoustic spaces.

As tools are generally sensitive to noise as a function of frequency, similar to the way in which they are sensitive to vibration, tested tool noise sensitivity curves must be provided as a function of frequency. The use of a single-value acoustical criterion (such as dBA or dBC) is generally discouraged for tools, because it provides no information regarding the frequency distribution of a tool's sensitivity.[1] When the specific and tested sensitivity characteristics of the tools are not known, the generic perception-based criteria have often been used to represent relative tool sensitivity, which has the benefit of giving architects and engineers a first indication of the level of effort that will be required in the acoustical design. However, tools are not sensitive to noise in the same way humans are. Figure 47.2, representing proposed noise criterion curves for tools,[2] illustrates this by comparison of tool sensitivity with the NC curves for humans. The proposed tool curves were developed by taking the minimum values of many manufacturers' specification curves and sorting them as a function of the resolving power of the tools. Important variances in tool sensitivity from the NC curves include a lower high-frequency limit, flat mid-range response, and extended sensitivity below 63 Hz into the infrasonic range. The tool curves also only apply to the most sensitive metrology tools; for less-sensitive tools and for cleanroom tools, it is recommended that the NC or other relevant curves be used for human comfort.

Design to meet the infrasound requirements can be particularly challenging, requiring nontraditional design practice. There is a limit to the amount of laboratory test information available on the acoustical properties of materials (e.g., sound transmission loss, absorptive qualities, etc.), and noise control devices (e.g., silencers, absorptive treatments, wall designs, etc.), at frequencies less than 63 or 125 Hz.

47.3 VIBRATION AND NOISE SOURCES

47.3.1 Site Vibration Sources

The vibration environments of all potential building sites for advanced technology research are limited to varying degrees by groundborne vibration from a variety of potential sources: microseismic activity, transportation, other nearby facilities containing mechanical equipment, etc.,[17,18] examined the effects of locale on site vibrations, showing that proximity to developed areas also impacts the vibration environment.

A site vibration survey is generally recommended for any facility with requirements more stringent than VC-C, or any site near a significant vibration source. It is important that the survey be conducted by experienced personnel using instruments with appropriate sensitivity.[19] The survey can identify potential vibration-related problems at the site, and whether the site is compatible with the vibration requirements of the proposed facility.

Consideration of the site conditions is critical, because the future building owner may not have control over them (in contrast to sources within the building). In addition to existing conditions uncovered in

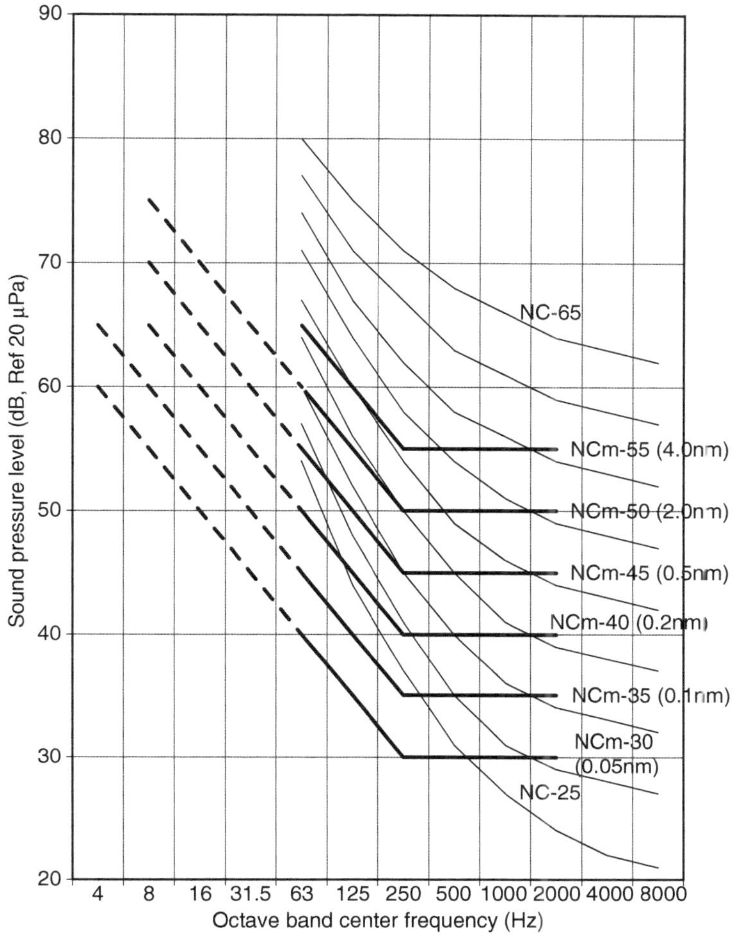

Figure 47.2 Tool criterion curves (NCm)[2] superimposed upon NC (noise criterion) curves for facilitating human perceptual comfort and communication.

the site survey, potential future impacts, such as construction, changes in transportation routes, and potential future changes in adjacent land zoning must be considered in the site selection process.

47.3.2 Vibration Sources within the Facility

There are many sources of vibration in an operating fab or research facility. Air handling devices and pumps are part of any technological building, as are the associated ducting and piping. These are under the purview of the facility designer, and are discussed in a later section of this chapter. However, there will also be interior sources that are not addressed by the building designer, because they are put in place after the facility is complete.

Gendreau and Amick[20] discuss the issue of *maturation*, which generally involves mechanical equipment (such as dry pumps) installed in connection with individual tools. Over the life of a facility, these sources can have a deleterious effect. In some cases, where these sources can be concentrated in one or several locations, it is possible to provide rooms that are acoustically and mechanically isolated from sensitive research areas, and which are flexible enough to allow an arbitrary set of equipment to be located therein. In other cases, such as when many tool support sources are widely distributed throughout a fab, it may or may not be considered economical to design the building to prevent their impact; in the latter case, it would be incumbent upon users and facilities staff to be vigilant and treat these sources appropriately with vibration isolation measures.

47.3.3 Noise Sources

Most of the noise that impacts sensitive tools and people has its source within the building, but occasionally external noise sources must be considered. Aircraft can be particularly troublesome to tools sensitive to low-frequency noise and to conversation in inadequately designed labs, conference rooms, and offices. Vehicular traffic may have to be considered in some cases. Impact from external sources can be treated in the building layout and shell design, but at certain sites, such as near an airport under a flight path, this can be prohibitively costly.

Internal noise sources include building mechanical and electrical equipment, and the activities of people. Noise from these sources can be transmitted via air- or duct-borne paths, or via structure-borne paths. The process and research tools can also be significant sources of noise, and the concept of maturation, or deleterious effects with the installation of tools and ageing of mechanical equipment, also applies from an acoustical standpoint. All of these sources can produce noise in the infrasonic (< 50 Hz) to the audio range (20 to 20,000 Hz) and above. Other sources of infrasound include door operation,[21] and vibration of flexible surfaces. Strategies for control of these noise sources are discussed in subsequent sections of this chapter.

47.4 FOUNDATION AND STRUCTURAL DESIGN

47.4.1 Soil and Foundation Design (with Respect to Environmental and Local Sources)

Especially with respect to very low-frequency vibration, there are limitations to the amount of improvement of the site ambient conditions that can be designed into the building foundations. In addition, due to the extreme complexity in characterizing site specific soils in the soil-structure dynamic interaction, the accuracy of predicting these effects is quite limited unless this can be improved with site-specific test data. Nevertheless, there is enough test data in the literature from which one can draw some general principles:

- The presence of a building on a site will suppress some of the vibrations observed in a vibration survey at an undeveloped site. This effect is frequency dependent and is difficult to quantify, though it appears that pile foundations may enhance attenuation in some frequency ranges.
- A solid slab will attenuate ground vibration about 25 to 50 percent or more at most frequencies above 10 to 50 Hz, depending on geotechnical conditions and the slab thickness. There is little difference between the three directions (vertical and two orthogonal horizontal axes).
- The use of a structural isolation break (SIB) in an on-grade slab will generally not improve performance significantly when the slab is the same thickness on both sides of the SIB, particularly if an "island" is created. It has been observed that vertical vibrations may decrease above a certain frequency associated with the slab thickness, but horizontal vibration amplitudes may *increase*. If an island is created that is significantly thicker than the surrounding slab, then some of the deleterious effects may be reduced.[22]
- Vibration resistance is maximized if a slab or foundation is underlain by a soil with a high dynamic subgrade modulus. This is achieved through the use of (1) dense virgin soil, (2) well-compacted engineered fill, or (3) well-densified granular fill.

47.4.2 Process Floors

The structural design of the floor that will support the vibration-sensitive equipment is of paramount importance in achieving a good quality of vibration performance. Low-vibration structures can be constructed of concrete, steel, or combinations of the two.

There are two generic floor concepts used in fab design:

- Slab-on-grade, where a concrete slab is placed directly on soil.
- Suspended floor, where the floor is suspended on columns or trusses.
  - The floor structure is usually of concrete¶¶ and may be of several generic types:
    - "Waffle" slab, which provides a flat surface on the top (or "topping" slab), and a pattern on the bottom that resembles a waffle, with some portions thicker than others. The thicker parts form a two-way framework of members that resist bending. There may be openings (or potential openings) for piping and other connections between the fab and subfab.
    - "Grillage," which has been described as a waffle without a topping slab. It is open on the top, and consists simply of beams running in two directions. This allows through-the-floor air flow.
    - "Flat" slab, which is simply a slab of constant thickness. Postconstruction penetrations pose challenges, as the reinforcement should not be cut during coring.
    - "Cheese" slab, which is a flat slab with cylindrical penetrations created with tubular forms at the time the concrete is poured.
    - Hollow-form slabs, which are concrete planks containing regularly spaced voids in various proprietary configurations. These can be either precast or constructed in situ.
  - Support may be provided in either of two ways:
    - Columns supporting the floor can form a direct path to the ground. This provides the stiffest floor, but limits the flexibility of the space beneath the floor (subfab).
    - The floor may be supported on a long-span steel truss, and there may or may not be columns between the floor and the truss.*** This provides a space beneath the floor that is free of columns, and is a popular approach for support of the upper process floors in the design concept known as the "stacked fab" and in LCD-TFT factories which require several process levels.

Suspended Floors

The design of suspended floors, and the types of vibration sources that control their performance, largely depends on the vibration requirements for the floor. Table 47.1 identifies typical structural design approaches to achieve various criteria.[24]

The design elements mentioned in Table 47.1 primarily address vertical vibration performance. For stiff suspended floors, input from the vibration consultant to control horizontal performance (using shear walls, moment framing, etc.) is also critical. For the less stringent floors, horizontal performance is usually dictated by the requirements of the structural engineer, although the vibration consultant will typically review this.

The vertical vibrations of fab floors may be excited by two types of loading. Vibrations may be generated by personnel walking (or other related activities), and they may be generated by the mechanical systems within the building (equipment, piping, and ductwork). In general, the vibration performance of floors with less stringent requirements (e.g., VC-A and VC-B) will be governed by walkers. The performance of "stiff" floors intended for the more stringent criteria (e.g., VC-D and VC-E) is more likely to be governed by mechanical systems. An exception is in the case of the massive AMHS systems used in LCD-TFT production, which generate transient forces high enough to significantly excite any of these floor types.

Table 47.1 Generic Criteria Assignable to Advanced Technology Spaces and the Structural Approaches to Achieving Them

Space type	Criterion	Structure
Research offices, computer modeling	ISO office	9–12 m spans, column or truss support, concrete slab on moderate-depth steel or concrete framing
Generic laboratory space, optical microscopes, epitaxy tools and ion implanters, CVD, CMP	VC-A/B	6.5–10 m spans, column or truss support, concrete slab on deep steel or concrete framing
Photolithography, nanofabrication	VC-D/E	Slab-on-grade; or concrete waffle or open grillage with 3.5–7.5 m column spacing
Metrology, surface characterization, SEM, SPM, AFM	VC-E or NIST-A	Slab-on-grade (NIST-A or VC-E); or concrete waffle or open grillage with 3.5–5 m column spacing (VC-E only)

¶¶Historically, the majority of these have tended to be "cast-in-place," but there is a growing trend toward using precast concrete elements, fabricated either at the site or elsewhere. There are several benefits to precast construction, one of which is generally improved quality control in the construction process. Additionally, there may be cost savings as well as a faster pace of construction, but it becomes more critical for the design team to develop details and construction sequences to ensure good connection and structural continuity between precast members. This is discussed at length in Ref. 23.

***When a concrete floor is supported on a long-span truss system, the "panel points," where the vertical and diagonal elements intersect the top and bottom horizontal elements (which are called the "chords") behave as "hard spots" providing support to the slab in the same manner as the columns beneath a conventional fab. The mechanics of these structures are discussed by Tang et al.[25]

Design of fab floors for walker-generated vibrations is usually approached using a derivation attributed to Ungar and White,[26] which is most readily accessible as Chap. 6 in a publication from the American Institute of Steel Construction (AISC),[27] even though it is equally applicable to concrete floors. That approach may be expressed as Eq. (47.1), in which V_w is the amplitude of vibrations generated by the person walking; k and f are the floor's stiffness and resonance frequency, respectively; C_w is a parameter that accounts for the characteristics of the footfall loading, including walker weight and speed; φ_w is the importance factor; and α and β are parameters that are adjusted for different floor types.

$$V_w = \frac{\varphi_w C_w}{k^\alpha f^\beta} \tag{47.1}$$

Design of floors for mechanically generated vibrations may be approached using a model developed by Gordon,[28] based upon observed statistical performance of actual fab floors.[†††] The vertical vibrations are represented by Eq. (47.2), in which V_m is the rms amplitude of vertical vibration (in one-third octave bands), k is the stiffness, C_M is an empirically derived coefficient based upon the statistical analysis of data from many fabs, φ_m is the importance factor, and γ is a parameter that is adjusted for different floor types. Generally the frequency associated with this vibration is the vertical fundamental frequency of the floor, usually between 10 and 50 Hz for stiff floors. The stiffness used in this model is the point stiffness at the middle of a typical structural bay (centered between four columns), which is usually the location with the lowest stiffness.

$$V_m = \frac{\varphi_m C_m}{k^\gamma} \tag{47.2}$$

Gordon's model predicts horizontal vibrations by Eq. (47.3), in which V_H is the rms amplitude of horizontal vibration (in one-third octave bands), k is the *global* stiffness of the structure supporting the process floor in the horizontal direction, and C_H is an empirically derived coefficient. The frequency associated with this amplitude is the horizontal (sway) frequency of the floor, usually between 2 and 8 Hz. The stiffness is calculated from the combined effect of columns, shear walls and any other structural components resisting lateral movement, and typically requires a computer model of the structure.

$$V_H = \frac{C_H}{\sqrt{k}} \tag{47.3}$$

Design of Floors Subject to Walker Vibration Impact

As noted above, relatively soft floors (VC-C and less stringent) tend to be more easily excited by the activities of personnel than by mechanical systems. The degree of impact of a walker is a function of the walker's speed (usually expressed as "pace rate" or paces per minute), gait, footwear type, and other factors.

Before the structural design based on walker impact can begin, the location of the source (person) and receiver (tool) within a bay or bays must be determined. Otherwise, impact is calculated in the "worst-case" condition, that is, walking at a relatively fast pace (say, 100 paces/min) immediately adjacent to the tool, assuming both to be located near the center of a bay. However, less conservative assumptions may be used in some instances to reduce project costs, if the relevant details are known. Some of these considerations are as follows:

- *Walker speed.* It is typical to assume a walker speed on the order of 100 paces/min for "high-speed" corridors. These are generally the straight, wide, general building traffic corridors outside the work area over which the tool user has no control. Quite often they have walls. From the standpoint of vibration control, it is preferable that high-speed corridors be separated from lab areas by at least one column line. For "ghost corridors," which typically parallel high-speed corridors but are inside the work area, it is common to assume a slower speed in the design, say, 85 paces/min, because the walker will exercise some care to avoid tools and other workers. For "low speed" corridors, generally obstructed areas in bays and between tools, one may assume even lower speeds, on the order of 70 to 85 paces/min. The affected users generally have more control over the traffic in the two latter cases.
- *Position of walkers and tools on the structure.* For layouts that are uniform with respect to the structural layout (column positions, etc.), one may wish to account for the relative stiffness of the bay at the position of the walker and/or tools, especially if either of these will always be located other than at mid-bay.

The prediction of walker-generated floor vibration amplitudes is an inexact science. Experimental studies have demonstrated that the variability of amplitudes generated by a "standard" walker (85 kg at 100 paces/min)—going to great lengths to maintain consistency—leads to a standard deviation of about 30 percent. When we consider a larger population at the same speed, the variation in footwear, gait, etc., leads to a standard deviation over 50 percent, even when there is a correction for variation in weight.

On-Grade Floors and Engineered Fill

The vibration performance of an on-grade floor depends on the site ambient vibration conditions, soil conditions, foundation type, and slab thickness. One key to good performance of an on-grade slab is the nature of the soil beneath it. For a given soil and compaction, the stiffness will vary approximately as $t^{1.5}$, where t is slab thickness. The ideal condition would be to have a quiet location, and support the slab directly upon bedrock or stiff virgin (undisturbed) soil. However, this is generally not possible. In many cases, it is necessary to fill in surface irregularities, which can be quite substantial in some cases. In other cases, it is necessary to have a granular drainage layer beneath the slab to prevent hydrostatic uplifting of the slab.

If imported fill is required, it should be examined by a soils engineer to ensure that it is of adequate quality and dimensional stability. If it can be characterized as "engineered fill," it should be compacted to 95 percent of maximum dry density as determined by the Modified Proctor Test, ASTM D-1557. If the fill material is granular for the purposes or drainage or leveling, then the above specifications are irrelevant. In that case, it is desirable to have the maximum possible densification of the material, to be specified by a soils engineer in a manner appropriate for the material being used.

Structural Isolation Breaks

In general, there is a trend away from the use of structural isolation breaks (SIBs) in slabs, whether on-grade or elevated. This shift in position has been based upon a growing collection of data suggesting—for frequencies less than about 40 Hz—that the positive contribution of an isolation break is minimal, and may actually be harmful.

In on-grade slabs, the only purpose a SIB serves is to attenuate high-frequency vibrations, not the low-frequency ones of primary concern. The low-frequency vibrations propagate through the soil. The reason for shifting away from the use of SIBs in this case is that they can actually make the horizontal vibrations worse if they are isolating too small a segment, and the isolated slab is of a thickness similar to that of the surrounding slab. All things considered, the solid slab is better than a jointed slab (or "island" in a larger slab) at the frequencies generally of concern for sensitive equipment and processes. However, at higher frequencies, particularly audible frequencies, they may be quite beneficial, and especially when the SIB can be located around the source of vibration (instead of the receiver—the sensitive tool), such as around a mechanical room.[‡‡‡]

[†††]The statistics associated with Gordon's model is discussed in Ref. 29.

[‡‡‡]Portions of the studies on which this position is based have been published. The effect of SIBs on suspended slabs is discussed in Sec. 7 in Ref. 30. The effect on slabs-on-grade is discussed in Ref. 22.

Support Floor Design for Source Mitigation

In the structural design, in addition to the design of the critical floors that will support vibration-sensitive equipment, it is important to consider the design of support and other floors that will carry mechanical and personnel loads. Typically, these floors are designed first by the project structural engineer to carry the design live load, and much of the time, this is adequate to serve the vibration design as well. However, many cases are documented wherein the vibration transmission from support areas was exacerbated by floor resonances. It is, therefore, recommended that these supplementary floor designs also be reviewed by a vibration consultant.

In particular, the following should be considered for support floors within or near the sensitive building:

- The fundamental resonance frequencies of the support floor systems (slab, columns, etc.) should be designed to avoid a matching with the fundamental rotational speed of supported mechanical equipment. For example, when designing a fan deck floor containing large fans rotating at 900 rpm, it would be best to avoid designs with a fundamental resonance near 15 Hz (equivalent to 900 rpm). Note that the same consideration is necessary for process floors that will support vibration generating tools, such as implanters, CMP tools, polishers, assembly tools, etc.
- Likewise, floors or structures supporting mechanical equipment should be stiff enough to provide adequate impedance to allow any vibration isolation systems supported on the floor to function properly.

The degree to which the support floor designs are critical depends on their proximity to vibration-sensitive areas, and whether they are separated by isolation breaks (to the degree that these provide a benefit, as discussed above).

47.5 VIBRATION AND NOISE CONTROL IN THE MECHANICAL/ELECTRICAL/PROCESS (MEP) DESIGN

No structural design will perform well with a poor mechanical design. Furthermore, it is not cost-effective to provide an excessively stout structural design with the intent that the mechanical design will be less relevant. The most cost-effective (and historically successful) method is to provide a balanced structural and mechanical design. With this in mind, in the structural design a specific quality in the mechanical design must be assumed. Critical aspects of the MEP design quality are discussed in this section.

47.5.1 MEP System Layout

The layout of the MEP systems is particularly important. From the standpoint of vibration impact, the best design adequately separates vibration-generating equipment from vibration-sensitive areas. As it is used here, *adequately separate* means that enough "natural" attenuation of the vibration forces is provided by radiation and materials damping (i.e., in the soil and building structures) in the horizontal and vertical path between source and receiver. This type of vibration reduction is separate from that provided by vibration isolation hardware, assuming that these are not required for adequately separated sources. In this design philosophy, isolation hardware, as discussed below, may be considered a compromise that effectively reduces the required separation distance, but at the cost of increased maintenance and amplification of vibration at the isolator resonance frequency, when this is relevant. Of course, long separation distances are not always practical due to the increased cost in building footprint and elevation, pipe lengths, thermal losses, etc., and therefore use of isolation hardware may become a more desirable compromise.

In addition, it may not be possible to isolate certain equipment (e.g., vibration-generating process tools), and separation distance may be the *only* practical means of isolation other than selecting source and receiver equipment for compatibility.

General Layout

The natural vibration attenuation (or amplification) provided in a building structure is a function of the specific structural design. Use of generic "rule-of-thumb" vibration transfer functions (e.g., "transfer mobility"[§§§]) are likely to cause either over- or underdesign of the MEP vibration control, which is in either case costly. It is possible, however, to calculate transfer mobilities within a building with a fairly high degree of accuracy using finite element analysis (FEA) methods, as long as the contribution from the soil-structure interaction is secondary. Calculations of paths involving the site soils are more accurate when supplemented with empirical propagation and soil stiffness data from the specific site in question. These considerations influence the MEP layout in a vibration-sensitive building, in addition to the specifics of the vibration sources, that is, forces generated (a function of the dynamic balance of the equipment and its internal support structures), number of sources, source locations, presence of external isolation hardware, etc.

In general, the structural analysis or experience of the vibration engineer will identify the amount of mechanical energy that can be applied at various locations within or nearby the vibration-sensitive structure. Uncertainties, for example, those due to specific dynamic balances or anomalous operating conditions, or that arise with the use of vibration isolators, may dictate the application of a reasonable amount of conservatism to ensure that impact to the sensitive floors is within acceptable amplitudes even under temporary unusual conditions.

For example, the structural design of a specific structure may reveal such things as: (1) the path attenuation from the subfab is less than from the fan deck; (2) a small amount of horizontal separation from the building provides significantly more attenuation than when sources are located directly below the process floor; or (3) vibration from piping attached the process floor support columns is better attenuated than vibration from piping attached directly to the underside of the process floor. These sample concepts, while true in most cases, are but examples of design considerations verified and informed by detailed structural analysis.

47.5.2 Recirculation Air Systems

Recirculation air systems could in some ways be considered the center of the cleanroom MEP system, as their presence and proximity define the degree of cleanliness achievable in the space. To prevent air (and hence energy) losses and reduce contamination, they must be immediately adjacent to the cleanroom. Yet this very concept is in opposition to vibration and noise control goals. Nevertheless, with good design practice, several types of recirculation air systems have been effectively employed in semiconductor and other advanced technology facility designs:

- *Packaged centrifugal or plug-type recirculation air handling (RAH) units.* This system usually requires a moderate number of units, typically ranging in flow rate from 4000 to 120,000 m^3/h per unit, most often installed above the cleanroom on a fan deck. Each unit is self-contained, containing the fan/motor assembly (sometimes multiple assemblies, such as with "fan wall" units), temperature and humidity control devices, and noise and vibration control hardware. Noise control features include special lining materials, "dump walls" (devices to block line-of-sight from the inlet or outlet to the fan wheel), silencers, optimized fan wheel size selection, and fan inlet flow straightening devices. Vibration control measures include isolation hardware under the fan/motor assembly within the cabinet, drive systems that allow improved dynamic balance (i.e., direct drive), and certain types of base designs. There are modular systems known as "fan coil units" that also fall into this category from the standpoint of vibration and noise.
- *Fan-filter units (FFUs).* FFUs are autonomous units usually consisting of a small plug-type or centrifugal fan (often variable in speed, which is controlled at the unit or remotely) typically operating in the 700 to 1700 m^3/h range, and a high-efficiency particulate air (HEPA) or ultralow penetration air (ULPA) filter at the discharge,

[§§§]"Transfer mobility" is the velocity spectrum that arises at one location due to a unit dynamic force applied at another location. See Ref. 30 for further explanation.

either 1.2 × 0.6 m or 1.2 × 1.2 m in area. Cleanroom ceiling grids are fitted out with FFUs in varying density depending primarily on the clean class required. FFUs do not provide temperature control; instead in cleanrooms using FFUs temperature control is provided by external coils or by makeup air units containing temperature controlling elements. In distinction from the other recirculation air handlers discussed, FFUs are normally not custom built. Often, hundreds of a manufacturer's standard units are installed in a ceiling space. The small cabinet size provides fewer options for internal noise and vibration control, although most manufacturers have taken steps in this regard.[¶¶¶] The use of an external silencer or vibration isolation is usually not feasible for FFU installations as their integral outlet filters make up the ceiling of the cleanroom. Therefore options for vibration and noise control are often limited to selection of the varying performance of standard units by various manufacturers, and by operating point selections.[31,32]

Based on this information and general industry experience, a few general conclusions may be given regarding recirculation air handler location and type selection:

- The location of the units is critical. The best location will depend on the specific structural design employed, but it is often found that location above the cleanroom level has its advantages, so long as the fan deck structural design avoids the fundamental operating frequency of the fans.
- From a vibration standpoint, in order of the greatest challenge to the least, industry experience shows that few large fans generates more vibration than many small fans (e.g., FFUs, fan-wall units) delivering the same capacity. This ranking not only accounts for the typical forces generated by the total number of units, but also typical mounting conditions and orientation, and other considerations. Exceptions are possible, of course.
- From a noise control standpoint, the three systems are more equal, except that there are typically fewer options for internal control of FFU noise.

47.5.3 Other Mechanical Equipment (Air Handlers, Compressors, Fluid Pumps, Vacuum Pumps, Cooling Towers, etc.) and General Considerations

The aforementioned conditions regarding layout, dynamic balance quality, drive type, and isolation also apply to other types of building mechanical equipment. Some general design principles may be mentioned in addition:

- Certain types of equipment in a particular class may be preferred. For example, a vertical in-line pump may generate less force at its base than an equivalent horizontal split case pump. Rotary screw compressors certainly generate less disruptive vibration than equivalent reciprocating compressors.
- The type of equipment used may influence the type of isolation required. To use the previous examples, the isolated horizontal split-case pump will certainly require an inertia base (as described below), but the equivalent in-line pump may not. The reciprocating compressor, if allowable at all in a particular case, may require an inertia base mass 5 to 10 times the supported equipment mass. (For well-balanced rotating equipment, an inertia base mass of at least 1 times the supported equipment mass is usually adequate.) The structural design of the mechanical areas will have to take these loads into account.
- The drive type (belt or gear drive versus direct drive) influences the practically achievable dynamic balance quality, and thus, the forces generated. In certain critical installations, only by the use of direct drive can the resultant forces be constrained to within manageable amplitudes.
- Variable frequency drives should be configured so that they do not allow operation of the equipment at certain critical frequencies, such as at or near resonance frequencies in the equipment and support structure or floor.
- It is not practical to design a facility such that one may locate any arbitrary mechanical load at any location within the building, even if it is provided with vibration isolation. Areas with compliant transmission paths to the process area (e.g., subfabs, the process floor itself, and other areas depending on the design) can only accept a limited amount of mechanical force before causing excessive vibration. The extent of this will depend on the structural design. Placement of equipment and support conditions must be considered early in the conceptual design of the facility.

47.5.4 Piping and Ductwork

Fluid flow in piping and ductwork generates vibration in proportion to the amount of turbulence in the flow. Turbulence is increased by obstructions in the flow (cross-sectional area changes, elbows, dampers, valves, etc.), but it is also a normal component in unobstructed flow, generally increasing in amplitude with increased flow rate and velocity.[33] For this reason, it is necessary to put restrictions on the flow velocity, size, support condition, and location (all mutually dependent variables) of piping and ductwork within vibration and noise sensitive facilities.

Flow turbulence is generally broadband in nature, producing energy at a wide range of frequencies. Figure 47.3 shows the change in vertical and horizontal vibration amplitude on a floor with a pumped water system powered on and off. This particular building had inadequately isolated piping systems attached to the underside of most of the floor bays, and thus easily excited the entire floor structure. Note the range of excitation frequencies, and that the structure is most strongly excited at its resonance frequencies (20 to 45 Hz vertical and 3 to 8 Hz horizontal).

Note in particular that the excitation reaches fairly low frequencies, and that these correspond with the horizontal modes of the floor. One may infer that the use of isolated piping supports with relatively high-frequency resonances (e.g., elastomeric isolators with typical resonances in the 7 to 20 Hz range) could be detrimental in this case. Thus, if possible, it is preferable to design the piping and ductwork systems so that isolation is not necessary. This is done by limiting flow velocities and pipe and duct diameters, and the use of rigid attachment to less critical elements of the structure such as the columns and subfab floor. This ideal design may not always be possible, in which case attachment using high-deflection (25 mm minimum) spring isolators may be practical, along with the requisite periodic inspection and maintenance.

There are no generic "rules of thumb" for piping support. Support in each case must be determined by the vibration specialist based on the pipe and duct layout, diameters, flow velocities, structural design of the floor, etc. The latter consideration may include determination of the capacity of the floor to resist mechanical excitation, which is a function of its stiffness with respect to the design criterion for the floor.

47.5.5 Central Utilities Buildings, Support Buildings, and Equipment Yards

Locating mechanical equipment in areas remote from the vibration-sensitive areas has the potential advantage of horizontal separation, thus requiring proportionally the installation of less mechanical isolation hardware. In addition to the attenuation factors through the fab and support or central utilities building (CUB) structures, ground transmission attenuation will play a role. Vibrations through the ground are attenuated by distance, especially at the higher frequencies associated with rotating machinery.[****]

[¶¶¶]From a vibration standpoint, the authors have yet to encounter a situation where the standard internal vibration isolation supplied with an FFU, usually a simple elastomeric isolator supporting the fan/motor assembly, was inadequate to control FFU vibration impact to vibration-sensitive floors located at lower levels, also see Ref. 31. However, if the FFUs serving a cleanroom must be suspended from another vibration-sensitive floor above, more extensive isolation may be required.

[****]Amick and Gendreau.[34] Although this presentation deals with construction-generated vibration, the same principles are applicable to mechanical vibrations from a CUB.

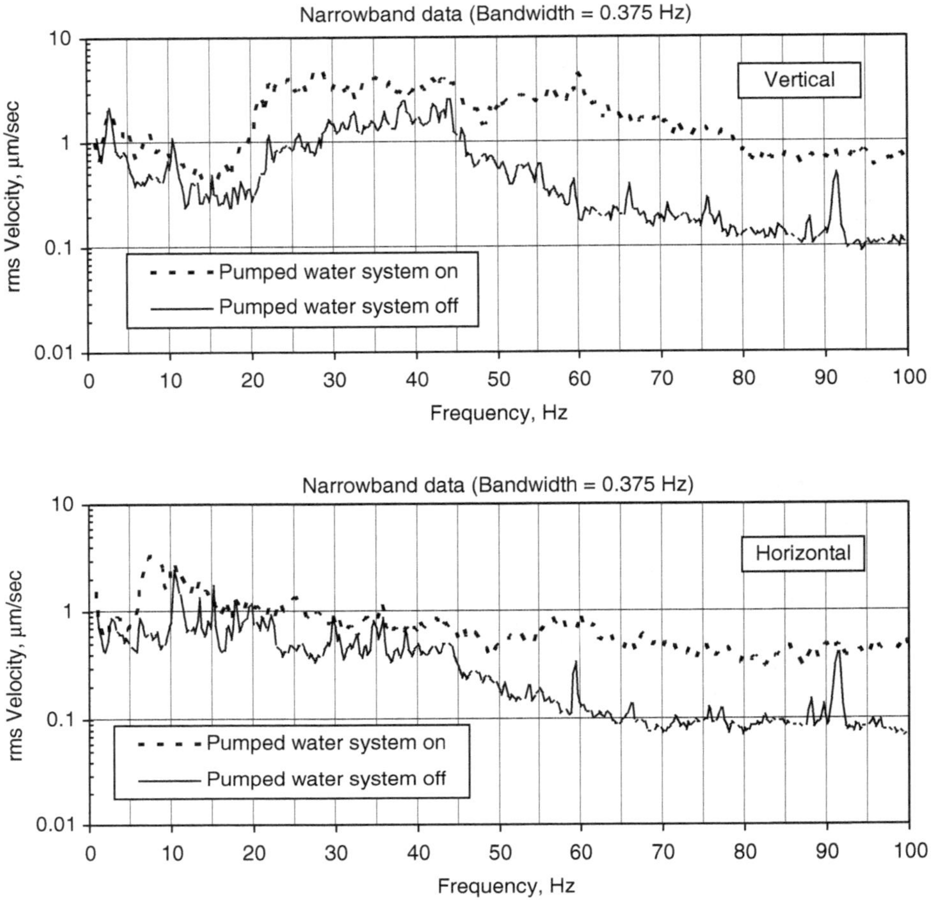

Figure 47.3 Vertical and horizontal vibration on stiff floor due to fluid flow in piping (average-plus-one-standard-deviation of 10 locations distributed over the floor).

From a noise standpoint, impact from separate utility areas to the sensitive areas is unlikely. However, the equipment located in these areas is typically extremely loud, and noise control considerations may be required with respect to occupational health and safety, as well as environmental noise regulations.

47.5.6 Electrical Equipment (Diesel Generators, CPS and UPS Systems, Transformers)

Certain items of electrical equipment are significant sources of vibration and noise, such as emergency generators, continuous or uninterrupted power systems (CPS and UPS), and transformers.

- Diesel emergency generators can usually be well separated from the sensitive areas by location in utility buildings. Nevertheless, as reciprocating machines, they generate significant unbalance forces, and the vibration impact must be considered. Furthermore, they are notorious sources of environmental noise. Most facilities use these to provide power to critical systems only and not process equipment. Others may use large banks of generators to allow continuous processing during power shortages. In the former case, the generators will be less of an impact if periodic testing can be scheduled to avoid sensitive operating times in a research facility, but this is rarely possible for a continuously operating production facility.
- CPS and UPS systems come in several varieties. The most innocuous systems use banks of passive batteries, which have no impact to the building from the standpoint of vibration and noise. Other uninterruptible power supply systems use energized elements that must be adequately isolated from the sensitive areas.
- Transformers generate noise and vibration at two times the main frequency plus harmonics due to magnetostriction of the core elements. This may be relatively less problematic for several reasons: (1) many tools are less sensitive to high-frequency vibration; (2) high-frequency vibration attenuates at a relatively high rate in the ground and in structures; and (3) relatively simple low-deflection elastomeric isolators are effective at reducing high-frequency forces. Nevertheless the vibration impact must be considered, as well as the radiated noise, which can be particularly intrusive in certain cases.

47.5.7 Machine Dynamic Balance Requirements

As already implied, the first order of vibration control is at the source. The amount of vibration generated by a rotating machine is a function of its static and dynamic balance quality, and alignment, if relevant (e.g., in horizontal split-case pumps). The dynamic forces due to imbalance are variable, controllable by initial balance or rebalance (and alignment) during the life of the equipment, and they typically worsen gradually during normal operation. Thus, periodic review and maintenance is required.

There are several standards commonly referenced in high-tech building design that regulate balance quality. International Standard ISO 3945[35,††††] lists several grades of balance classification for rigid and nonrigid (isolated) support. These are equated to vibration amplitudes measurable in various axes on the equipment bearing caps, which provides a convenient means of verification and monitoring.

Generally, the higher quality balance requirements are necessary in semiconductor and other advanced technology buildings, especially for the larger equipment located in the process building near the sensitive areas. This is especially the case for densely packed centrifugal or vaneaxial recirculation air fans serving cleanrooms. For example, depending on the layout and isolation quality, tolerances in the high ISO "good" range (0.5 to 0.7 mm/s rms) might be used for "critical" fans near the sensitive areas, but the lower end of the ISO "good" range (i.e., 1 to 2 mm/s rms) might be acceptable for other equipment found in smaller numbers such as major pumps, compressors, small fans, etc.

The balance quality achievable is a function of the type of equipment, among other things. For example, significantly higher balance qualities are achievable with direct-drive equipment than with belt drive equipment. In the former case, speed control, if necessary, is by the use of variable speed drive (an adjustment of line frequency going into the motor). Belt drive and geared equipment also have the disadvantage of producing tonal vibration at several frequencies, whereas in a well-balanced and aligned direct drive system the primary impact is at a single frequency related to the rotation speed. This issue is especially critical when the structural design attempts to account for the primary impact frequencies of the mechanical equipment by designing the structural resonances to avoid those frequencies.

Finally, it is necessary to establish a preventative maintenance or machinery health monitoring program to control increasing vibration due to bearing wear and other forms of disintegration, and more important, to prevent catastrophic failure.

47.5.8 Vibration Isolation Hardware

When equipment cannot be located at adequate distance from sensitive areas, vibration isolation hardware can reduce the transmission of vibration into the structure at selected frequencies. Transmitted vibration can also be radiated as noise from machine and architectural components, and vibration isolation can also reduce this effect. There are many types of isolation hardware and their short- and long-term effectiveness and maintenance requirements can vary dramatically by design. (They can also vary by accident if not properly selected, installed, and maintained.)

Isolation Hardware and Systems

In the simplest terms, vibration isolators work by superimposing the response of a single degree of freedom spring-mass-damper system upon the vibration or force spectrum of a source (machine, pipe, etc.). The idealized response of a simple spring-mass-damper system is typically shown in the literature[37] as a plot of transmissibility versus the ratio of the forcing frequency to the resonance frequency of the isolator. The curve shape will vary with the damping qualities of the isolator. When the curve is above the unity transmissibility line, amplification is indicated, and attenuation (reduction) is indicated when the curve drops below unity transmissibility. There are several important features of ideal isolators:

- There is a relatively low-frequency resonance at which the transmitted vibration is increased, especially when the forcing frequency closely corresponds to the isolator resonance frequency (frequency ratio = 1);
- Below resonance, there is either amplification (transmissibility > 1) or unity response.
- At frequencies greater than $\sqrt{2}$ times the frequency ratio, there is attenuation of the transmitted energy (transmissibility < 1).
- Also note that with higher damping, the amplification at resonance is reduced, but with a corresponding reduction of attenuation at high frequencies. Typical damping factors are $\zeta = 0.005$ for steel springs, $\zeta = 0.05$ for neoprene, and $\zeta = 0.33$ for a friction damped spring.

The fundamental resonance frequency of the system is a function of the *static deflection* of the isolator, defined as follows:

$$f_R = \frac{5}{\sqrt{\delta}} \quad (47.4)$$

where δ is the static deflection of the isolator in centimeters, defined as the difference between the unloaded and loaded height of the isolator. The static deflection of the isolator is a function of the isolator stiffness and the mass of the supported load, and is a common means of specifying isolator performance or efficiency.

Thus, isolators are useful when the primary vibration frequencies of the source—and the resonance frequencies of the critical building structures—are significantly higher than the isolator resonance frequency f_R.

The frequency response of an ideal isolation system is that of a single-degree-of-freedom (SDOF) mass on a spring element. The response of real isolation systems is more complex than the ideal (SDOF) response. In general, their performance is degraded from the ideal curve due to several factors. These factors include other resonances in the supported system and in the isolators themselves, such as *surge* (longitudinal) resonances in steel coil springs and resonances in the supported equipment base or platform. The latter case includes the dynamic response of the supported equipment frame or inertia base and compliance (resonance response) of the table structure of isolation tables. In addition, compliance of the building structure that supports the isolated system can degrade isolator performance. All of these effects tend to reduce high frequency isolator attenuation at certain frequencies or in ranges of frequencies.[38]

Isolators are commercially manufactured of various materials and for a wide range of operating and environmental conditions. The most common types are elastomerics (neoprene, rubber, fiberglass, etc.), steel springs, and pneumatic springs, described in more detail as follows:

- Steel springs are often selected for their durability and relatively simple response, in static deflections ranging typically from 10 to 100 mm ($f_R = 5.0$ to 1.6 Hz). Damping at resonance is relatively low unless separate damping elements are used. High-frequency response is degraded by surge frequencies, but sometimes a neoprene isolator is used in series with the steel spring to improve isolation at high frequencies.
- Elastomerics generally have a better damping response, with reduced attenuation at high frequency. The available static deflections are generally smaller for commercial systems, in the 1 to 10 mm range ($f_R = 16$ to 5 Hz). Hence, they tend to amplify at relatively high frequencies, and their isolation range begins at a higher frequency. They commonly are produced in pad or "puck" form, or in the higher-profile form operating in shear.
- Pneumatic isolators (sometimes known as *air springs*) are somewhat of a compromise between the other two types listed, typically consisting of a pressurized elastomeric vessel (analogous to a rubber tire) with a low resonance frequency ($f_R = 1$ to 2 Hz is common). These are available in commercial systems installed as units at the base of a vibration source or sensitive tool, or as elements contained in an *isolation table*. They usually require air pressure maintenance, which can add to the cost and complexity of the installation and maintenance of air springs. One drawback is that this type of isolation system can be degraded by pressure fluctuations in the room.[21]

There are other proprietary systems that may be considered, and surely others will be developed, that provide different isolation and cost benefits.

The vibration isolation systems described above are *passive isolators*. "Active" isolators employ elements that sense the incoming vibration

††††Also often referenced is American National Standards Institute ANSI S2.41,[36] which is in complete technical agreement with ISO 3945.

signal and circuitry that adds an inverted copy to it, thus partially canceling the vibration. The technology of these systems is rapidly advancing, but at the time of this writing they are generally only applicable within limited amplitude and frequency ranges, generally performing best at low frequencies. They are also somewhat limited in the dynamic force they can exert. They are often used in series with conventional passive isolators that work best at higher frequencies. Active and active-passive isolation systems are commercially available as base isolators and in isolation tables.

Equipment, piping and ductwork, and other vibration sources or sensitive equipment can either be supported from isolators attached at the base to the floor below, or suspended from above on isolation hangers. For the most part, the important technical features of these two types are similar. In addition to the isolator type and static deflection,[‡‡‡‡] other critical details that must be specified include material properties and corrosion resistance, attachment details, isolator locations, horizontal-to-vertical stiffness ratio, overload capacity, integration with seismic control devices, and other details that will increase the likelihood that the isolator will function as required after installation.

Some equipment is supplied with rigid framing allowing the equipment to be supported directly on springs. However, many systems require that a rigid base be situated between the equipment and the spring isolators. Base types include rigid steel frames and rails as well as relatively high mass *inertia bases*—usually composed of a steel frame filled with concrete.[§§§§] Inertia bases are specified to reduce vibration amplitudes on the isolated equipment in normal operation, and also to reduce the impact during start-up when the equipment speed momentarily corresponds with the isolator resonance. They also provide a very stiff platform for equipment with interconnected elements such as horizontal split-case pumps. Critical base details to specify include the base dimensions, total mass, and materials properties. The horizontal dimensions are typically defined by the equipment footprint plus any supplemental attachments to be supported on the base (pipe elbows, fan diffuser cones, etc.). The vertical dimension (depth) is typically a function of the horizontal dimensions, selected to provide adequate resistance to bending, and the required base mass.

It is important to control detrimental resonance effects by specification of the base details. The dynamic response of the supported system (steel frame, concrete inertia base, or even the equipment itself, as discussed below) changes when supported at discreet points—as when several isolators are used—rather than when it is continuously supported as it is without isolation. Resonances in the isolated support frame can lead to excessive vibration in the supported mechanical equipment, and can cause misalignment problems in interconnected systems (such as pumps attached to a motor by a shaft).

There are other types of isolation devices found in advanced technology facilities that must be mentioned:

- Flexible pipe, duct, and conduit connectors are used in these services between rigid and isolated components to allow flexibility in the isolated components, and to reduce transmission of vibration in the pipe and duct walls.[¶¶¶¶] Flexible connectors are sometimes used at service crossings of structural isolation joints. Not all flexible connectors are equal, and, in general, some types work better for microvibration control (e.g., twin-sphere elastomerics, multi-convolution stainless steel bellows) than others (e.g., stainless steel braided flexes). The latter type may be more suitable for expansion control, seismic flexibility, etc. When installed at isolated mechanical equipment, there are cases when flexible connectors may be substituted for the support of piping on isolators near the equipment (or vice versa).
- Thrust restraints are spring elements used primarily with fans to resist lateral thrust that would otherwise overcome the limits of base isolation and flexible connectors.
- Resilient penetration sleeves are used to maintain flexibility in isolated piping or ductwork when passing through a rigid wall, ceiling, or floor structure, or to prevent transmission of vibration into these structures.
- Resilient lateral guides are elastomeric or steel spring isolators used for isolated support of duct and pipe risers when these pass through floor/ceiling structure.
- Seismic control devices normally have the opposite performance goal of vibration isolators, which is to restrain equipment or piping under earthquake loading. However, certain of these devices are compatible with both the seismic restraint requirements and the isolation requirements and should be specified during the design stage along with the isolation devices. For example, certain types of base snubbers function without rigid contact except in a seismic event. Likewise, cable restraints for pipe and duct installations can be designed with adequate slack to allow the isolation supports to function. Seismic snubbers should be of a design that allows easy inspection for binding that would reduce isolation effectiveness during normal operation.

Isolator Installation and Common Faults

Unfortunately, many things can go wrong in the design, specification, installation, and maintenance of isolation systems. In a detailed inspection of any facility with a significant number of isolation devices employed, one is likely to find several of the following faults:

Design Faults

- Inadequate static deflection for equipment, based upon its rotational or low-frequency turbulence vibration
- The supporting building component (floor, ceiling, pipe rack, etc.) is excessively soft with respect to the spring stiffness
- Lack of consideration for long-term performance and reliability
- Improper isolator selection with respect to loading, causing either inadequate static deflection or overloading of isolator, either of which reduces or defeats isolator performance

Installation and Maintenance Faults In general, the isolated equipment should be able to move freely. If it does not, and if there were no design errors, one or more of the following problems may be indicated:

- Binding or shorting of isolation systems due to improperly adjusted isolators, presence of improperly adjusted or designed seismic snubbers, presence of shipping restraints or temporary supports, debris under isolation base, rigidly attached piping or conduits, rigidly attached pipe stanchions, binding between nested spring elements, binding at spring cages, etc.
- Spring misalignment or collapse
- Corrosion
- Poor load distribution on pad elements

It is important that the project vibration specifications define specific isolator types, operating parameters, and installation details in order to avoid these faults, which can result in higher vibration and noise levels.

47.6 ACOUSTICAL DESIGN

Noise can affect both personnel and tools within the facility, and if projected in excess into the environment, the attitudes of the neighboring communities. Control of HVAC noise and the careful selection of relevant architectural details can improve communications, speech privacy, and the general condition of work environments, and reduce long-term work-related hearing damage, fatigue, and general annoyance. As regards environmental noise, quite frequently facilities are required to show proof of compliance with state or local environmental noise regulations. It is best to consider noise control during the design stage, rather than to have to apply—as a retrofit—potentially more costly noise control measures later.

[‡‡‡‡]The isolator static deflection is usually specified as a minimum value, as higher deflections than required are usually not detrimental.

[§§§§]In the context of this discussion, inertia bases refer to massive bases supported or suspended from isolators. An inertia base in this case is to be distinguished from a *housekeeping pad*, which is a massive concrete pad attached to the ground or floor structure directly, and supporting isolated or unisolated equipment.

[¶¶¶¶]In certain cases flexible connectors can also reduce transmission of vibration in the fluid. For example, some types of elastomeric connectors can be used to reduce impeller pressure pulses in pumped water by absorbing some of the energy in the pulse by flexure of the connector walls. See Ref. 39.

47.6.1 Interior Noise Design

Cleanrooms

Cleanrooms pose particular challenges in noise control design. This is due to several unique characteristics of these environments from an acoustical standpoint. First, the air change requirements can be quite significant, calling for large volumes of air flow, and its attendant noise. Second, cleanrooms can be quite reverberant due to the acoustically reflective architectural materials from which they are typically constructed (concrete, cleanroom wall panels, stainless steel, etc.). Third, due to outgassing and contamination limitations, there are constraints on the use of traditional acoustical absorption materials. The situation is far from hopeless, however, and it is possible to model and construct the cleanroom to provide workable environments for people and most process tools.[40]

Cleanroom noise can be modeled using traditional acoustical hand calculations, or with computer-based ray tracing models. In either case, there are special considerations that must be taken. The models must include the unusual construction materials used in the room (honeycomb cleanroom wall panels, HEPA filters, etc.) to properly account for absorption and reverberation. Appropriate consideration of the general layout conditions, whether bay-chase or ballroom, must be taken. The HVAC sources to be considered include the inlets and outlets of the recirculation air systems, the makeup air system outlets, and any self-noise generated at HEPA filters (generally relevant only in cases where there is an attempt to achieve relatively low noise levels). Other noise sources associated with the base building systems include piping and valve noise, and coil pumps.

Using conventional cleanroom designs, it is generally very difficult to achieve HVAC noise levels below NC-55 to NC-60 in ISO classes 1 to 5 cleanrooms (IEST RP-12).[9] With appropriate considerations in the design, such as the use of low air velocities and adequate means of quieting fan noise, it is possible to achieve lower noise levels in cleanrooms rated at numerically greater ISO clean class values. In any case, the first means of noise control in cleanrooms is by fan specification and selection. As discussed in Sec. 47.5.2, means of noise control within or at the fan units depends on the types of units used. The use of external "packless" silencers or silencers with encapsulated fill and cleanroom compatible absorptive treatments is possible, but the efficiency of these means is reduced due to clean compatibility requirements, especially in the low-frequency regime (63 to 500 Hz) typically dominated by fan noise.

A potentially more difficult problem is controlling noise from the tools and tools support equipment (dry pumps, local chillers, etc.) located in the cleanroom or recirculation air path. These are added by the end user according to process or research requirements. Prepurchase noise requirements for these equipments, which might be used to maintain acoustic compatibility with the HVAC system design criterion, are unfortunately still uncommon. Thus it is not unusual to find that noise levels within the cleanroom are elevated significantly with the addition of the tools. In addition to the support equipment mentioned, primary sources associated with the tools include minienvironments, wafer handling equipment, servo actuators, and air leakage noise. Several ongoing pilot studies have shown that it is possible to contain this noise, especially with advanced planning. This includes designing appropriately ventilated pump and chiller rooms, requirement or provision of appropriate isolation and enclosure of mechanical sources, and the use of appropriate maximum noise specifications for individual equipment.

Clean Chases, Subfabs, and Support Areas

Cleanrooms often use support areas that also function as cleanroom air returns. Often the noise criterion can be reduced in these areas if they are not normally occupied and contain nonsensitive mechanical equipment. However, it is usually desirable to limit noise levels within chases that may contain a significant portion of the bulk-headed tools in a bay/chase arrangement.

Specialty Labs

The most stringent noise requirements in wafer fabrication and other advanced technology facilities are often found in the labs used for failure analysis, characterization, and other metrology. The types of tools used in these processes are often among the most sensitive to acoustic noise and infrasound.

Fortunately, or necessarily in cases where the acoustical requirements are very stringent, these labs often have no—or relatively less stringent—cleanliness requirements. If the temperature control requirements are stringent, this can add another challenge as specialty air delivery systems are required; however, these can be designed to be consistent with the acoustical requirements. Layout, as discussed in Sec. 47.7.1, is critical.

These labs will often require designs to control:

- Transmission of noise between labs and from corridors and mechanical areas outside the labs
- HVAC noise from fans, air terminal units, and diffusers, and considering both duct-borne and duct "breakout" noise[10]
- Reverberation within the room, by the use of appropriately specified and distributed absorptive ceiling and wall treatments
- Infrasound from HVAC systems, which can couple with the dimensional and other characteristics of the lab to cause excessive noise at the "Helmholtz" resonance frequency of the space, and other frequencies[41]

47.6.2 Noise in Mechanical Rooms and in the Outside Environment

The foregoing comments regarded the provision of "comfortable" environments for people and tools. In contrast, this section discusses environments that may be regulated in addition by outside authorities, such as occupational health organizations and local and national governments. Control of environmental noise and noise in mechanical rooms are more "traditional" acoustical design problems, so we shall only discuss aspects unique to advanced technology facilities.

The most significant noise sources in advanced technology facilities are located in the mechanical rooms and equipment yards, including exhaust fans, chillers, pumps, compressors, air dryers, boilers, cooling towers, and the like. When enclosed in the hard (concrete and steel) and reverberant environments typical of mechanical rooms and CUB buildings, they can produce significant hazards with respect to hearing damage. When these sources are exposed to the environment, as is necessary in the case of exhaust fans, cooling towers, makeup air and boiler combustion air intakes, they can produce formidable noise control problems, especially when the facility is located near urban or semiurban areas.[42]

Noise criteria in these cases are most often expressed as a maximum overall value in dBA. As an example, national or international hearing protection regulations might define a maximum noise exposure level of 80 to 90 dBA for an 8-h day in a mechanical room. Property line noise limits may range in absolute values from 40 to 70 dBA depending on the defined zoning use of the adjacent property, or they may be referenced to existing ambient conditions (i.e., before the addition of the new sources of noise). The actual regulations to which a facility is subject can vary significantly, of course, so these must be identified early in the design stage. The conditions under which these regulations apply, such as the measurement methodologies, meteorological conditions, etc., must also be identified.

In both cases, it must be emphasized that the criterion is not applied to individual sources, but to the totality of the noise from all relevant sources. In the case of semiconductor production facilities, there can be hundreds of individual sources that must be considered in the design, and the use of advanced computer modeling methods is necessary to adequately characterize the impact of all of these sources simultaneously, and to test various forms of noise control to provide an acoustically and cost-effective design. Thus, detailed analysis often reveals that the effective criterion for an individual source is significantly lower than the overall design criterion in a mechanical room or at a property line.

Traditional means of control of noise in these cases include layout, barriers and enclosures, silencers, application of absorptive materials, etc.

47.7 TOOL HOOK-UP

47.7.1 Tool Layout

The location of the functional areas within a cleanroom or laboratory within the building, and with respect to vibration sources outside the building, influences the vibration and noise environment achievable in the room. Judicious layout is the most cost-effective way (and sometimes the only way) to achieve a good vibration and noise environment. For example, the most sensitive labs should be located as far as possible from internal sources of vibration and noise such as mechanical and electrical equipment, elevators, major corridors, loading docks, etc. Position with respect to external sources—such as on-site and off-site roads and central utilities buildings—should also be considered.

Likewise, layout within the cleanroom bay or lab module is important for the most efficient control of noise and vibration. The designer should plan to separate the tools and tool support mechanical equipment in the lab or fab to reduce vibration and noise impact. Consider using a separate and acoustically sealed equipment chase to contain dry pumps, chillers, power conditioners, and other equipment that can disturb the tools and personnel with excess noise. The need for peripheral equipment chases will depend on whether these items of equipment are operated during times of critical tool use.

47.7.2 Passive Tool Pedestal Design

It is typically during the hook-up stage—and during any later tool installation—that rigid pedestals are designed to support tools at the elevation of the raised access flooring. These serve several purposes. From a structural standpoint, they are sometimes needed because the load bearing capacity of the access floor is inadequate (e.g., in the case of implanters). In other cases, due to the flexibility and relatively high vibration amplitudes found on the raised access floor, they are used as a separate rigid support to extend the good vibration environment of the structural floor to the elevation of the raised access floor.

Rigid pedestals are not intended to reduce the structural floor vibration environment, and care must be taken in their design so that they do not amplify the structural floor vibration within the critical frequency range. The pedestal may be custom-fabricated from conventional steel members such as structural tubes, or may be off-the-shelf prefabricated structures such as appropriately designed tripods. The pedestal must be independent from the access floor with a clear gap all around. The platform is normally designed by the structural engineer to carry all appropriate structural loads. It is then dynamically designed such that its fundamental resonance frequencies (in horizontal and vertical directions) are very high and removed from the supported tool's most sensitive frequency range, the modes of the fab structure to which it is attached, and the shaft speed of rotating equipment. In addition, the horizontal and vertical axes of the pedestal may have to meet foundation resistance (stiffness) requirements specified for certain tools. A detailed dynamic analysis using finite element modeling is generally necessary.

47.7.3 Active Tool Pedestal Design

Active isolation elements can also be used in pedestals and isolation tables. This technology was discussed briefly in the section "Isolation Hardware and Systems" earlier, and the same benefits and limitations apply when used in tool support pedestal design. Active isolation is not compatible with some types of tools.

47.7.4 Hook-up Equipment Vibration and Noise Control

Subfab systems associated with process level tools can generate substantial vibration unless their installation is handled carefully.[*****] Subfab systems include vacuum pumps (often stacked), small chillers and scrubbers, power conditioning equipment, and other mechanical and electrical tool support systems. While the power rating of these systems is typically relatively small (1 to 10 kW), they are often many in number and widely distributed.

Dry pumps have "dry" bearings and their installation must be such as to allow their easy removal for maintenance. For this reason the pumps are typically installed on carts with casters. The casters incorporate a locking foot to prevent accidental movement of the system when in service. Most manufacturers of these pumps provide on-board elastomeric vibration isolation mounts between the base of each pump and the frame of the cart. These isolators can work well in limiting the vibration transmitted to the subfab floor. However, isolation is not always provided, nor is it always as effective as one might wish. In these cases, soft neoprene-in-shear or steel spring mounts can greatly improve the isolation efficiency. An alternative to on-board isolation is to support the pump cart from the floor on springs, or from racks on hanger springs. This method of isolation can be more efficient than on-board isolation but it would, necessarily, complicate the issue of pump maintenance, etc. A very efficient means of isolating any arbitrary number of support mechanical systems is to provide an isolated floor in the equipment room. These typically consist of a concrete floor floating on spring or neoprene isolators.

It is equally important that piping and exhaust systems associated with the vacuum pumps include flexible elements to limit the transfer of vibratory forces between the pumps and the building structure and the tools served by the pumps. It is important also that gas lines and electrical connections be flexible.

[*****] For an example of the vibration impact associated with hook-up mechanical equipment, see Ref. 20.

47.8 PURPOSES AND TIMING OF FACILITY VIBRATION SURVEYS

It is quite common to carry out a vibration survey of a site prior to construction of a facility. This may be just the first of several vibration evaluations carried out over the life of a fab or research facility. There are a variety of logical milestones at which these surveys may be appropriate, as discussed by Amick et al.[43] These might include

- Fab construction process
  - Structural evaluation[†††††]
  - As-built (or final) evaluation[‡‡‡‡‡]
- Tool site evaluation
- Diagnostics
- Documentation of "current" state

47.9 MATURATION OF THE VIBRATION AND NOISE ENVIRONMENT

The term "maturation" refers to the change in the vibration environment over time (generally a worsening) due to a variety of reasons. Gendreau and Amick documented a number of aspects pertaining to this.[20] There are several key issues that contribute to the maturation of a facility after it passes beyond the "as-built" state:

- Hook-up
- Process changes
- Aging structure and mechanical installations and maintenance

47.10 FUTURE TRENDS AND SPECIAL CASES

47.10.1 Microelectronics Facilities

The vibration and noise sensitivity of microelectronics production equipment has remained fairly static for almost a decade, and it appears

[†††††] The intent of a structural evaluation is to validate specific structural parameters associated with the structural dynamics design, such as stiffness and resonance frequency of floors. (It is easier to measure footfall vibration, if desired, during this survey than during the as-built survey, since the footfall vibrations might be masked by mechanical vibrations in the completed and operational facility.) If deficiencies are found, they are more easily corrected at this stage of construction, rather than waiting until the project is complete.

[‡‡‡‡‡] The intent of a final evaluation is to document the performance of the vibration-critical portions of a fab at the time that the facility is "delivered" to the owner.

that industry pressure will continue to push tool makers to provide equipment that will work in current environments. If production technologies are forced to shift to new hardware concepts more in tune with those used in nanotechnology—particularly if they must be used in-line in a cleanroom environment, there may be some revision of vibration and noise requirements.

One trend that may affect tools is wafer size. Most current facilities are currently processing 300-mm wafers, and some facilities are now being designed to accommodate 450-mm wafers, with correspondingly larger tools, and the potential for generating even larger internal forces requiring resistance from the structure.

Another critical trend is associated with the requirement for new light sources capable of smaller linewidths. It will always be desirable to process linewidths at smaller and smaller dimensions. While this does not necessarily imply that photolithography scanners will require more stringent ambient vibration environments (indeed this appears to be the case over the last few generations due to engineering improvements), especially to be considered as they approach the EUV (extreme ultraviolet) light range, it is possible that the tools may, for example, impart greater dynamic forces to the process floor at times, requiring consideration of impact to other tools. These details should be reviewed and considered as the technology advances.

47.10.2 Flat Panel Display Facilities (LCD-TFT)

Flat panel display (FPD) facilities tend to get larger with each generation of production equipment, and there is no immediate sign this trend is going to change. The larger production equipment generally means larger materials handling systems, which generate larger forces, which require stiffer structures. At this time, design concepts appear to be keeping pace with the technology, though the larger forces associated with larger robots provide a significant challenge to designers.

47.10.3 Nanotechnology and Other Advanced Physics Facilities

It is difficult to predict where research at the nanometer scale is going, and what it will require with regard to vibration and noise environment. It is probable that equipment for imaging (SEM, TEM, STEM, AFM, etc.) will continue to become more sensitive to both, though the noise requirements of some of the most sensitive instruments are already at the edge of what is practical using today's technologies. One can only hope that tool vendors become better at internal vibration and noise isolation, as the more established vendors of photolithography tools have become.

After experience with the first generation of nanotechnology facilities, it has become apparent that great attention must be paid to investigation and mitigation design for multiple types of energy contaminants as well as the traditional airborne contaminants. Care must be taken when the requirements of different contaminants come into conflict, as may be the case between acoustics and EMI (Electromagnetic interference) or vibration and EMI. Some consensus has been reached with regard to the planning and design of these facilities, as presented in IEST Recommended Practice IEST RP-200.[44]

The acoustic noise requirements of some facilities are so demanding that the only way to meet the needs of a few cutting-edge systems (mostly dealing with microscopy) is to sequester them in their own sealed rooms, with no personnel present while they are being used. In addition, a new discipline has been developing with regard to infrasonic noise control in laboratories, to meet the requirements of the next generation of commercial tools, as well as acknowledged limitations on certain experiments due to low-frequency pressure fluctuations.

The situation is not as difficult with regard to vibration, though this discipline is also reaching its limits. Some technologies, including steppers, are fairly robust. Others, such as some of the E-beam photolithography tools and nanoprobes, impose stringent requirements that may limit the use of even moderately quiet sites. Current technology for active vibration control is not particularly adequate for reducing the vibration amplitudes of an already-quiet environment. Correcting this shortcoming may extend capabilities some distance into the future, but a great deal of technological development is still required.

47.11 ACKNOWLEDGMENTS

The authors wish to acknowledge the contributions of the "initiators" of this discipline, Colin Gordon and Eric Ungar, as well as our colleagues who have helped evolve the design and measurement approaches we use.

47.12 REFERENCES

1. M. L. Gendreau (1999, 2001), "(Specification of the) Effects of Acoustic Noise on Optical Equipment," *Proceedings of the International Society for Optical Engineering (SPIE)*, Vol. 3786, Denver, CO (July 1999) reprinted and updated in *Noise & Vibration Worldwide*, Vol. 32, No. 4, Apr. 2001.
2. M. L. Gendreau, "Generic Noise Criterion Curves for Sensitive Equipment," *Proceedings of Inter-Noise 2012*, New York, Aug. 19–22, 2012.
3. J. S. Bendat and A. G. Piersol, *Random Data: Analysis and Measurement Procedures*, 4th ed., Wiley-Interscience, New York, 2010.
4. H. Amick and S. Bui, "A Review of Several Methods for Processing Vibration Data," *Proceedings of the International Society for Optical Engineering (SPIE)*, Vol. 1619, Nov. 1991.
5. H. Amick, "On Generic Vibration Criteria for Advanced Technology Facilities: With a Tutorial on Vibration Data Representation," *Journal of the Institute of Environmental Sciences*, Vol. XL, No. 5, pp. 35–44, Sep./Oct. 1997.
6. K. G. McConnell, *Vibration Testing: Theory and Practice*, Wiley, New York, 1995.
7. E. E. Ungar and C. G. Gordon, "Vibration Challenges in Microelectronics Manufacturing," *Shock and Vibration Bulletin*, 53 (I): 51–58, May 1983.
8. H. Amick, M. L. Gendreau T. Busch, and C. G. Gordon, "Evolving Criteria for Research Facilities: Vibration," *Proceedings of SPIE Conference 5933: Buildings for Nanoscale Research and Beyond*, San Diego, CA, Jul. 31–Aug. 1, 2005.
9. Institute of Environmental Sciences and Technology IEST RP-12, "Considerations in Clean Room Design," *IEST-RP-CC012.3*, 2015.10.
10. American Society of Heating, Refrigerating and Air-Conditioning Engineers, Inc. ASHRAE, "Chapter 48: Noise and Vibration Control," *2015 ASHRAE Handbook: Heating, Ventilating, and Air-Conditioning Applications*, ASHRAE, 2015.
11. C. G. Gordon, "Generic Vibration Criteria for Vibration-Sensitive Equipment," *Proceedings of the International Society for Optical Engineering (SPIE)*, Vol. 3786, Denver, CO, Jul. 1999.
12. E. E. Ungar, D. H. Sturz, and H. Amick, "Vibration Control Design of High Technology Facilities," *Sound and Vibration*, Vol. 24, No. 7, pp. 20–27, Jul. 1990.
13. International Standards Organization ISO 2631, "Guide to the Evaluation of Human Exposure to Vibration and Shock in Buildings (1 to 80 Hz)," ISO 2631, 1981.
14. H. Amick and A. Bayat (2001, 2002), "Meeting the Vibration Challenges of Next-Generation Photolithography Tools," *Proceedings of ESTECH 2001*, 47th Annual Technical Meeting, IEST, Phoenix, Arizona, Apr. 2001; reprinted in *Sound & Vibration*, Vol. 36, No. 3, pp. 22–24, Mar. 2002.
15. W. E. Blazier, Jr., "RC Mark II: A Refined Procedure for Rating the Noise of Heating, Ventilating and Air-Conditioning (HVAC) Systems in Buildings," *Noise Control Engineering Journal*, 45 (6): 243–250, 1997.
16. L. L. Beranek, "Balanced Noise Criterion (NCB) Curves," *Journal of the Acoustical Society of America*, 86: 650–654, 1989.
17. M. L. Gendreau, H. Amick, and T. Xu, "The Effects of Ground Vibrations on Nanotechnology Research Facilities," *Proceedings of the 11th International Conference on Soil Dynamics and Earthquake Engineering (11th ICSDEE) and the Third International Conference on Earthquake Geotechnical Engineering (3rd ICEGE)*, Berkeley, CA, Jan. 2004.
18. C. G. Gordon, "A Study of Low-Frequency Ground Vibration in Widely Differing Geographic Areas," *Proceedings of Noise-Con 87*, State College, Pennsylvania, Jun. 1987.
19. H. Amick and C. G. Gordon, "Specifying and Interpreting a Site Vibration Evaluation," *Microcontamination*, Vol. 7, No. 10, pp. 42–52, Oct. 1989.

20. M. L. Gendreau and H. Amick, "'Maturation' of the Vibration Environment in Advanced Technology Facilities," *Proceedings of ESTECH 2004*, 50th Annual Technical Meeting, Institute of Environmental Sciences and Technology (IEST), Las Vegas, Nevada, Apr. 2004.

21. M. L. Gendreau, "The Effect of Varying Acoustic Pressure on Vibration Isolation Platforms Supported on Air Springs," *Proceedings of the 16th International Congress on Sound and Vibration (ICSV16)*, Kraców, Poland, Jul. 5–9, 2009.

22. H. Amick, T. Xu, and M. L. Gendreau, "The Role of Buildings and Slabs-on-Grade in the Suppression of Low-Amplitude Ambient Ground Vibrations," *Proceedings of the 11th International Conference on Soil Dynamics and Earthquake Engineering (11th ICSDEE) and the Third International Conference on Earthquake Geotechnical Engineering (3rd ICEGE)*, Berkeley, CA, Jan. 2004.

23. N. Tang, C. Wu, B. Xiong, M. L. Gendreau, H. Amick, Y.-L. Yin, and T.-L. Wu, "Dynamic Features of Precast Waffle Slabs in Cleanroom Design," *Proceedings of IMAC XXVI Conference and Exposition on Structural Dynamics*, Orlando, FL, Feb. 4–7, 2008.

24. H. Amick, M. L. Gendreau, and C. G. Gordon, "Facility Vibration Issues for Nanotechnology Research," *Proceedings of the Symposium on Nano Device Technology*, National Chiao-Tung University, Hsinchu, Taiwan, May 2002.

25. N. Tang, H. Amick, and M. L. Gendreau, "Long-Span Truss Structures for Low-Vibration Environments," *Proceedings of ASCE/SEI Structures 2009 Congress*, Austin TX, pp. 2829–2835, 30 Apr–2 May 2009.

26. E. E. Ungar and R. W. White, "Footfall-Induced Vibrations of Floors Supporting Sensitive Equipment," *Sound and Vibration*, pp. 10–13, Oct. 1979.

27. T. M. Murray, D. E. Allen, and E. E. Ungar, "Floor Vibrations Due to Human Activity," *Steel Design Guide Series 11*, American Institute of Steel Construction, p. 69, 1997.

28. C. G. Gordon, "The Design of Low-Vibration Buildings for Microelectronics and Other Occupancies," *Proceedings of the First International Conference on Vibration Control in Optics and Metrology, SPIE Proc.* Vol. 732, London, Feb. 1987.

29. H. Amick and A. Bayat, "Dynamics of Stiff Floors for Advanced Technology Facilities," *Proc. 12th Engineering Mechanics Conference*, American Society of Civil Engineers, pp. 318–321, May 1998.

30. H. Amick, M. L. Gendreau, and A. Bayat, "Dynamic Characteristics of Structures Extracted from In-Situ Testing," *Proceedings of the International Society for Optical Engineering (SPIE)*, Vol. 3786, Denver, CO, Jul. 1999.

31. C. G. Gordon and M. Q. Wu, "Noise and Vibration Characteristics of Cleanroom Fan-Filter Units," *Proceedings of ESTECH 1998*, Institute of Environmental Sciences and Technology (IEST), Phoenix, Arizona, Apr. 1998.

32. M. L. Gendreau, "Noise in Cleanrooms Served by Fan-Filter Units: Design Considerations," *Proceedings of the 11th International Congress on Sound and Vibration (ICSV11)*, St. Petersburg, Russia, Jul. 5–8, 2004.

33. R. D. Blevins, *Flow-Induced Vibration*, Van Nostrand Reinhold, New York, 1990.

34. H. Amick and M. L. Gendreau, "Construction Vibrations and Their Impact on Vibration-Sensitive Facilities," *Proceedings of the Sixth Construction Congress*, American Society of Civil Engineers (ASCE), Orlando, FL, pp. 758–767, Feb. 2000.

35. International Standards Organization ISO 3945, "Mechanical Vibration of Large Rotating Machines with Speed Range from 10 to 200 rev/s—Measurement and Evaluation of Vibration Severity In Situ," 1985.

36. American National Standards Institute ANSI S2.41, "Mechanical Vibration of Large Rotating Machines with Speed Range from 10 to 200 rev/s—Measurement and Evaluation of Vibration Severity In Situ," 1985.

37. M. R. Lindeburg, *Mechanical Engineering Reference Manual for the PE Exam*, 10th ed., Professional Publications, Inc., Belmont California, 1998.

38. E. E. Ungar, "Vibration Isolation," in L. L. Beranek and I. L. Ver, eds., *Noise and Vibration Control Engineering: Principles and Applications*, John Wiley & Sons, Inc., New York, 1992.

39. J. Paulauskis and W. E. Blazier, Jr., "Centrifugal Water Pumps," *Application of Manufacturers' Sound Data*, C. Ebbing and W. E. Blazier, Jr., American Society of Heating, Refrigerating and Air-Conditioning Engineers, Inc., 1998.

40. C. G. Gordon and A. M. Yazdanniyaz, "Noise Prediction and Control in Microelectronics Clean Rooms," *Proceedings of Inter-Noise 89*, Newport Beach, CA, Dec. 1989.

41. M. L. Gendreau, "Measurement Techniques Used to Verify the Cause and Nature of Low-Frequency Noise in Rooms," *Proceedings of the 16th International Congress on Sound and Vibration (ICSV16)*, Kraców, Poland, Jul. 5–9, 2009.

42. M. L. Gendreau and M. Q. Wu, "Environmental Noise Control for Semiconductor Manufacturing Facilities," *Proceedings of Inter-Noise 99, The 1999 Congress and Exposition on Noise Control Engineering*, Fort Lauderdale, FL, Dec. 1999.

43. H. Amick, M. Gendreau, and T. Xu, "On the Appropriate Timing for Facility Vibration Surveys," *Semiconductor Fabtech*, No. 25, Mar. 2005.

44. Institute of Environmental Sciences and Technology IEST RP-200, "Planning of Nanoscale Science and Technology Facilities: Guidelines for Design, Construction, and Start-up," IEST-RP-NANO200.1, 2012.

Index

www.ingramcontent.com/pod-product-compliance
Lightning Source LLC
LaVergne TN
LVHW082017200926
848481LV00010B/2873
9781265943424